The Contractors'Dictionary

The Contractors' Dictionary of Equipment, Tools, and Techniques

for Civil Engineering, Construction, Forestry, Open-Pit Mining, and Public Works

Compiled and Edited by

L. F. Webster

A Wiley-Interscience Publication

John Wiley & Sons, Inc.

New York • Chichester • Brisbane • Toronto • Singapore

Library of Congress Cataloging in Publication Data:

Webster, L. F. (Len F.), 1929-
      The contractors' dictionary of equipment, tools, and techniques : for civil engineering, construction, forestry, open-pit mining, and public works / compiled and edited by L.F. Webster.
        p. cm.
     ISBN 0-471-18115-3 (alk. paper)
     1. Engineering — Dictionaries. 2. Construction equipment — Dictionaries. 3. Forests and forestry — Dictionaries. 4. Strip mining — Dictionaries. 5. Surveying — Dictionaries. I. Title.
TA9.W43 1995
620'.003—dc20   94-24421

Printed in the United States of America

10 9 8 7 6 5 4 3 2

To my wife, who has driven with me so often in search of construction projects, and waited patiently as I disappeared into the mud, and to my children, who grew up to the sound of arcane building terms, and who endured the chore of appearing in countless construction photos 'to give a sense of scale...'

PREFACE

These definitions have been compiled to be of use to those working in a wide range of professions and trades involving architecture, engineering, surveying, building, heavy construction, forestry, surface and open-pit mining, and public works. Also included are many terms applicable to the tools and equipment employed in those specialities.

The definitions and descriptions have been kept purposely brief in the expectation that this book will be used as a source of first reference, rather than as a definitive text.

As the age of technology has flourished, so too has the range of devices and techniques that are employed to undertake the projects we consider essential to our way of life. *THE CONTRACTORS' DICTIONARY* includes terms in current use that span the centuries from Greek and Roman architecture to our present electronic age. While it is not a finite list — equipment and tools are continually being modified, materials developed, and techniques refined — it is nevertheless thought to be the only reference to cover the unique range of interests listed.

Throughout the dictionary, the symbols used as an abbreviation of a term have been included in their alphabetical position. How this has been done requires some explanation: There are as many rules about abbreviations as there are people who have tried to rationalize them — and they have all failed. Set a rule and common usage will usually defeat it.

In *THE CONTRACTORS' DICTIONARY*, yet another set of rules has been (mostly) followed. They are:

- An abbreviation (conc. for concrete) is given a period; a contraction (acct for account) is not. Exceptions are single-letter abbreviations such as j for junction, and the initials of organizations, EPA for Environmental Protection Agency, for instance.
- No dimension, be it abbreviated (ac for acre) or contracted (ft for foot) is given a period except the abbreviation for inch (in.).
- The abbreviation or contraction of a proper noun is capitalized (C for Celsius, Bhn

for Brinell hardness number). However, common usage (which can vary from trade to trade and among regions) calls for such as OD for outside diameter, DO for dissolved oxygen, and so forth

It should be noted that on drawings, it is considered good practice to include a list of the abbreviations and contractions used, and what they stand for. In text it is customary to spell out in full, the first time it is used, any term or name that will subsequently be abbreviated or contracted, followed immediately by the abbreviation or contraction placed in parentheses, for example 'shall be beveled (bev.) on all four sides.'

A word about spelling

The author was born in Britain (manœuver), has spent the majority of his life in Canada (manoeuvre), yet worked mostly with U.S.-produced technical literature (maneuver). This dictionary is written in American-English — with many misgivings about the use of *meter* for metric terms and definitions. (A meter is an instrument for measuring or recording; metre derives from the French *mètre*. The French invented the metric system of measurement, now known as SI, *Système international d'unités,* or International System of Units.) Canadian users of the dictionary, long familiar with the need to accommodate both British- and American-English (as well as their own hybrid form of Canadian-English) hopefully will have few occasions to wince.

ACKNOWLEDGMENTS

Many companies, organizations, and individuals have helped in the compilation of this dictionary. Their generosity in making available detailed information that in some cases has taken years to develop, an amount of which is copyright to those who developed it, is greatly appreciated. They include:

Allied Steel & Tractor Products
Alpine Press, Inc.
American Association of State Highway and Transportation Officials
American Concrete Institute
American Institute of Steel Construction
American Plywood Association
American Public Transit Association
American Road Builders Association
American Truck Historical Society
Asphalt Institute
Associated General Contractors of America
Association of Diesel Specialists
AT&T Bell Laboratories
Atlas Copco
Automotive Lift Institute
Bandag Incorporated
Barber-Greene, Division of Caterpillar Paving Products, Inc.
Berger Instruments, Division of High Voltage Engineering, Corp.
Blaw-Knox Construction Equipment Corporation
Brick Institute of America
Bridon American Corp.
British Columbia Assessment Authority
British Columbia Buildings Corporation
Canada Mortgage and Housing Corporation
Canadian Forestry Service
Canadian Particleboard Association
Canadian Plastics Institute
Caterpillar, Inc.
Caterpillar Paving Products, Inc.
Cedar Shake & Shingle Bureau
Ceramic Tile Institute

Concrete Equipment Company
Concrete Reinforcing Steel Institute
Construction Industry Manufacturers Association (CIMA)
Construction Safety Association of Ontario
Contractors Pump Bureau
Council of Forest Industries of British Columbia
Deep Foundation Institute
Dover Elevator Systems, Inc.
Dryden Oil Company
Eaton Corporation
Electrical Generating Systems Association
Energy, Mines and Resources Canada, Canada Centre for Mineral and Energy Technology
Expansion Joint Manufacturers Association
Fluid Sealing Association
Ford Heavy Truck, Inc.
Forest Engineering Research Institute of Canada
Forest Products Accident Prevention Association
Forestry Canada, Pacific and Yukon Region
General Electric Company
General Motors Corporation, Truck Division
Grove Worldwide
Hauk Manufacturing Company
Hercules Hydraulics
Imperial Eastman, Eastman Division of Pullman Canada Ltd.
Ingersoll-Rand Construction and Mining Group
Institute for Research in Construction (National Research Council Canada)
International Slurry Surfacing Association
International Society for Trenchless Technology
Jager Industries Inc.
JCB, Inc.
J I Case, A Tenneco Company
Korody-Colyer
Landoll Corporation
Leica Inc.
Link-Belt Construction Equipment Company
Metric Commission Canada
Mixer Systems, Inc.
Motor Vehicle Manufacturers Association of the United States, Inc.
National Asphalt Pavement Association
National Association of Surety Bond Producers
National Fire Sprinkler Association, Inc.
National Institute of Standards & Technology
National Lime Association
National Lubricating Grease Institute
National Paint & Coatings Association
National Particleboard Association
National Research Council of Canada
National Truck Equipment Association
North American Society for Trenchless Technology
Norton Company
Pacific Forestry Centre
Pall Industrial Hydraulics Company

Pilkington Glass Ltd.
Power Team Division, SPX Corporation
Research Energy of Ohio, Inc.
Resilient Floor Covering Institute
Rockwell International Corporation (Automotive Operations)
Rubber Association of Canada
Rubber Manufacturers Association
Scaffolding, Shoring & Forming Institute, Inc.
Sealant, Waterproofing & Restoration Institute
Simon-RO Corporation
Sokkia Corporation
Solid Waste Association of North America
Specialized Carriers & Rigging Association
Stanley Hydraulic Tools
Stihl Incorporated
Structural Board Association
Suspension Specialists Association
Talus Resources
Truck-Frame & Axle Repair Association
Truck Trailer Manufacturers Association
Turner, J&M, Inc.
United States Department of Agriculture, Forest Service
United States Department of the Interior, Bureau of Reclamation
United States National Archives and Records Administration
Workers' Compensation Board of British Columbia

The Contractors'Dictionary

a *Abbreviation for:* acceleration; acre; anode; are.

A *See* **Veneer grade.**

A *Abbreviation for:* ampere; Angstrom unit; astragal.

A-A Sanded plywood with A-grade face and back plies and D-grade inner plies, bonded with interior or exterior glue.

A-A EXTERIOR Exterior-type sanded plywood panel with A-grade face and back plies and C-grade inner plies, bonded with exterior glue.

ab *Abbreviation for:* aggregate base.

AB *Abbreviation for:* air blast; anchor bolt.

A-B Sanded plywood panel with A-grade face, B-grade back, and D-grade inner plies, bonded with interior or exterior glue.

ABACULUS Ceramic tile, one of a number of identical shapes forming a geometrical pattern.

ABACUS Flat slab, sometimes molded, the upper part of the capital of a column.

ABAMURUS Single masonry block forming a buttress.

ABANDONMENT 1. Relinquishing of the public interest in a right of way or activity thereon with no intention to reclaim or use again for highway (or transportation) purposes. *Also called* **Vacation; 2.** Cessation of work by a contractor and removal of all equipment and uninstalled materials from site.

ABATE Reduce in amount, degree, or intensity.

ABATEMENT 1. Waste of material to obtain a required size or shape; **2.** Diminishing or modifying a hazard to life or property created by defective trees; **3.** Reduction in environmental release of pollutants through improved waste management practices, including source reduction and recycling; **4.** Reduction of an expenditure.

ABATSON Baffle that deflects airborne sound waves downward.

ABATVENT 1. Tower roof having a slack rise; **2.** Device that breaks the force of, or prevents the entrance of wind to a duct, passage, chimney, etc.

ABATVOIX Reflector of airborne sound waves.

abbr. *Abbreviation for:* abbreviation.

abc *Abbreviation for:* aggregate base course.

ab ex. *Abbreviation for:* ab extra (from without).

A-B EXTERIOR Exterior-type sanded plywood panel with A-grade face, B-grade back and C-grade inner plies; bonded with exterior glue.

ABHESIVE Substance that resists adhesion, commonly applied to the mating surfaces of materials intended to temporarily protect the face of other materials to which an adhesive has already been applied.

ab init. *Abbreviation for:* ab initio (from the beginning).

ABLATION Formation of residual deposits by washing away of loose or soluble materials.

ABNEY LEVEL Hand held clinometer, used to measure slopes in percent.

ABORT Discontinue a turn (lift) in helicopter logging due to excessive load weight or mechanical difficulties.

ABOVE-CAB CARRIER Platform above the cab that can be in a fixed or hydraulically adjustable position for transporting vehicles.

ABOVE GRADE Any part of a structure or site feature that is above the adjacent finished ground level.

ABOVEGROUND BIOMASS Aboveground portion of a tree, excluding the root system.

abr. *Abbreviation for:* abrasion; abrasive.

abr *Abbreviation for:* alternate bars reversed.

ABRAM'S LAW Rule stating that, with given concrete materials and conditions of test, the ratio of the amount of water to the amount of cement in a mixture determines the strength of the concrete provided the mixture is of a workable consistency. *See also* **Water-cement ratio.**

ABRASION Frictional surface wear.

ABRASION-FINISHED SURFACE Surface obtained by sanding with discs coated with abrasive grit.

ABRASION RESISTANCE Ability of a surface to resist being worn away by rubbing or friction.

ABRASION RESISTANCE INDEX Measure of the abrasion resistance of a material relative to that of a standard material tested under the same specified conditions.

ABRASION TESTER Machine for determining the quantity of material worn away by friction under specified conditions.

ABRASIVE Substance used for wearing away or abrading (grinding, polishing, lapping) by friction, such as the natural material corundum, emery, diamond, sandstone, etc., and the manufactured or artificial materials aluminum oxide (Al_2O_3), carborundum, silicon carbide (SiC), and boron carbide (B_4C).

ABRASIVE PAPER Paper, or cloth, covered on one side with a grinding material glued fast to the surface. Used for smoothing and polishing. Materials used for this purpose include: crushed flint, garnet, emery, and corundum.

ABRASIVE WEAR *See* **Wear.**

abs. *Abbreviation for:* absolute.

ABS *Abbreviation for:* acrylonitrile-butadiene-styrene.

ABS *Abbreviation for:* antilock brake system.

absc. *Abbreviation for:* abscissa.

ABSCISSA Line or part of a line drawn horizontally on a graph by which a point is located with reference to a system of coordinates.

ABSENTEE OWNER Owner of real property who does not personally manage or reside at the location.

ABSOLUTE 1. Chemical substance free, for all practical purposes, of impurities; 2. Anything complete unto itself; 3. As exact as can be achieved.

ABSOLUTE AUCTION Sale at auction to the highest bidder, regardless of the amount offered.

ABSOLUTE COORDINATE Coordinate measured from established x, y, or z axes.

ABSOLUTE FILTRATION RATING Diameter of the largest spherical particle that will pass through a filter under specified test conditions. This is an indication of the largest opening in the filter element. It does not indicate the largest particle that will pass through the element, since particles of greater length than diameter may pass.

ABSOLUTE HUMIDITY *See* **Humidity.**

ABSOLUTE MANOMETER Instrument whose calibration can be determined from its measurable physical constants and which is the same for all vapors and gases.

ABSOLUTE MAXIMUM GRADE Steepest grade to which a road or highway is designed and built.

ABSOLUTE PRESSURE *See* **Pressure.**

ABSOLUTE SPECIFIC GRAVITY Ratio of the mass (referred to a vacuum) of a given volume of a solid or liquid at a stated temperature (in the U.S., 15.5°C (50°F)) to the mass (referred to a vacuum) of an equal volume of gas-free distilled water at a stated temperature.

ABSOLUTE VISCOSITY *See* **Viscosity.**

ABSOLUTE VOLUME In the case of solids, the displacement volume of the particles themselves, including their permeable and impermeable voids, but excluding any space between them; in the case of fluids, their volume.

ABSOLUTE ZERO Hypothetical temperature characterized by complete absence of heat, equivalent to approximately -273°C (-460°F).

absorb. *Abbreviation for:* absorbed; absorbing.

ABSORBED MOISTURE Moisture that has entered a solid by absorption and has physical properties not substantially different from ordinary water at the same temperature and pressure. *See also* **Absorption, Capillary water,** and **Gravitational water.**

ABSORBENCY Property of a substance to imbibe liquids.

ABSORBENT Material capable of taking in water, moisture, or other fluid.

ABSORBENT SEPARATOR *See* **Separator.**

ABSORBER Material or device for removing a specific fluid or gas from a substance to a designed degree, either chemically or mechanically.

ABSORBER PLATE In a solar energy system, that part of the collector system that transfers solar heat to a working medium.

ABSORBING DRAIN *See* **Well.**

absorp. *Abbreviation for:* absorption.

ABSORPTION 1. Weight of fluid taken in by a material when immersed. *See also* **Absorbed moisture;** 2. Capacity of a surface to reflect only a portion of the light that strikes it; shiny surfaces with a lighter hue or tone absorb less light; 3. Process in which a substance combines with or extracts one or more substances from a mixture of gases or liquids; 4. Property of a porous or fibrous substance to attract and hold liquids and gases.

ABSORPTION FIELD Effluent disposal system, particularly from septic tanks, consisting of a network of perforated pipes laid in coarse gravel in trenches that slope away from the tank outlet. The trenches are backfilled with progressively finer material, finishing with top soil.

ABSORPTION LOSS Water lost to the surrounding soil from unlined reservoirs, irrigation ditches, sloughs, etc.

ABSORPTION RATE Weight of fluid taken in by a material following immersion for a specific period in a liquid of known properties and temperature.

ABSORPTION TEST Test made to determine the ability of materials to absorb vapors or fluids under specified conditions.

ABSORPTIVE LINER Absorptive material used to line the inside face of formwork to improve the surface finish of concrete.

ABSORPTIVITY 1. Capacity of a material to absorb (light, heat, moisture, gases); 2. Fraction of a radiant energy absorbed by the surface that it strikes.

ABS PIPE *See* **Pipe.**

abstr. *Abbreviation for:* abstract.

ABSTRACT Brief background of all information and a history of transactions, typically of a plot from the date of the original deed.

ABSTRACT OF BIDS Summary of unit prices supplied by the owner, used as the basis for selecting the low bid on a tendered contract.

ABSTRACT OF TITLE Document showing the condensed history of the title to property, containing portions of all conveyances or other pertinent instruments relating to the estate or interest in the property, and all liens, charges, encumbrances, and releases.

abt *Abbreviation for:* about.

ABUT Meet or touch at one end; to be contiguous or adjoining. *Also called* **Butt.**

ABUTMENT 1. Solid wall placed to counteract the lateral thrust of a bridge deck, vault or arch; 2. In bridges, the end structure that supports the beams, girders, and deck of a bridge and sometimes retains the earthern bank, or supports the end of the approach pavement slab; 3. In prestressed concrete, the structure against which the tendons are stressed in producing pretensioned precast members or post-tensioned pavement; **4.** In dams, the side of the gorge or bank of the stream against which a dam abuts. *See also* **Arch.**

ABUTMENT CHEEKS Surfaces on each side of a mortise.

ABUTMENT PIECE Lowest structural member in framing that receives and distributes loads from an upright; a soleplate.

ABUTT *See* **Abut.**

ABUTTAL Parts in which land abuts on other land; boundaries.

ABUTTING JOINT Carpentry joint where the end grain of one piece forms an angle with the end grain of another piece.

ABUTTING TENONS Carpentry joint where two tenons meet in the center of a mortise.

ABYSSINIAN WELL Perforated tube driven into the ground through which groundwater is pumped.

ac. *Abbreviation for:* acute.

ac *Abbreviation for:* asphaltic concrete.

AC *Abbreviation for:* air conditioning; alternating (electrical) current; armored cable; asbestos cement; asphalt cement.

ACANTHUS Architectural ornament representing the leaf of this plant, especially on the capitals of Corinthian and Composite orders.

ACB *Abbreviation for:* asbestos-cement board.

acc. *Abbreviation for:* accordance; according.

accel. *Abbreviation for:* accelerate; acceleration.

ACCELERATED ABSORPTION TEST Test in which the end result is hastened by testing at conditions more severe than those anticipated in typical service.

ACCELERATED AGING Test for panels, applied finishes and other materials that attempts to duplicate long-term outdoor exposure with short-duration laboratory conditions.

ACCELERATED COMPLETION Requirement for a contractor to complete the work at an earlier date than was scheduled in the contract documents.

ACCELERATED CURING Curing concrete in an artificially controlled environment in which humidity and rate of temperature rise and fall are controlled, to speed up the rate of strength increase.

ACCELERATED CURING TANK Insulated water tank fitted with an immersion heater and thermostat to control the water temperature at 55°C (131°F) ± 2°C, plus a circulatory pump, used to store concrete cube molds.

ACCELERATED DELIVERY *See* **Accelerated design and construction.**

ACCELERATED DEPRECIATION 1. Depreciation calculated or taken at greater than the normal rate; **2.** Method where depreciation charges in earlier periods are greater than those in current or later periods. *See also* **Depreciation.**

ACCELERATED DESIGN AND CONSTRUCTION Technique whereby construction is started before completion of final detail drawings. *Also called* **Accelerated delivery, Fast-track construction, Fast tracking, Phased construction,** and **Phased design and construction.**

ACCELERATED LIFE TEST Method of designing to approximate in a short time the deteriorating effects obtained under normal service conditions.

ACCELERATED SERVICE TEST Service or bench test in which some service condition, such as speed, or temperature, or continuity of operation, is exaggerated in order to obtain a result in a shorter time.

ACCELERATED WEATHERING Testing of material samples to cycles of excess dryness, frost, heat, wetness, etc.

ACCELERATING ADMIXTURE *See* **Admixture.**

ACCELERATION 1. Increase in velocity; a measure of force (F = ma). It is the time rate of change of velocity, measured in g's; the acceleration of gravity; **2.** Increase of the speed of a moving vehicle by application of the surplus power from the engine; **3.** Period during which an elevator moves at an ever-increasing rate of speed, due either to gravity (down) or the force generated by the lift engine, usually referred to as the period between standstill and full speed; **4.** Quickening of the natural progress of a process such as hardening, setting, or strength development of concrete. *See also* **Accelerator.**

ACCELERATION CLAUSE Requirement that the balance of a loan becomes due immediately if contractual payments are not made or for breach of other stipulated conditions.

ACCELERATION RESISTANCE Inertial reaction opposing increase in speed of a vehicle. It is numerically equal to gross vehicle weight in pounds divided by 32.2 and multiplied by acceleration in feet per second per second. It may be expressed as pounds per thousand, or as a percent of gross vehicle weight.

ACCELERATION STRESS The additional stress imposed on a wire rope resulting from an increase in load velocity.

ACCELERATOR 1. Device that permits adjustment of the rpm rate of an engine or motor (*See also* **Decelerator**); **2.** Chemical that speeds up a reaction; **3.** Additive that hastens the rate of vulcanization of rubber compounds; **4.** Substance that, when added to concrete, mortar or grout, increases the rate of hydration of the hydraulic cement, shortens the time of setting, or increases the rate of hardening, strength development, or both; **5.** In heating systems, a centrifugal pump located in the return circuit by means of which it is possible to increase flow. *See also* **Acceleration.**

ACCELERATOR (QUICK-OPENING DEVICE) Device attached to the dry valve of an automatic sprinkler system to speed up the tripping of the dry pipe valve and to allow water to quickly be discharged from the sprinkler heads that have opened.

ACCENT LIGHTING Concentration of artificial light as part of interior design.

ACCENT TILE Ceramic wall or floor tile finished in a color and/or pattern and/or texture that makes it distinct from the tiles used for the main field of the work. *See also* **Field tile.**

ACCEPTABLE 1. Acceptable to the authority having jurisdiction; **2.** An installation or equipment is acceptable if it meets any of the following conditions:

(a) If it is accepted, or certified, or listed, or labeled, or otherwise determined to be safe by a qualified testing laboratory capable of determining the suitability of materials and equipment for installation;

(b) If no qualified testing laboratory accepts, certifies, lists, or determines the installation or equipment to be safe, if it is inspected or tested by a regulatory responsible authority and found to be in compliance with its requirements; or

(c) With respect to custom-made equipment or related installations that are designed, fabricated for, and intended for use by a particular customer, if it is determined to be safe for its intended use by its manufacturer on the basis of test data that the employer keeps and makes available for inspection.

ACCEPTANCE 1. Written indication that goods and/or materials supplied and/or installed are as specified; **2.** Consignee's receipt of a shipment that terminates a common carrier contract for transportation.

ACCEPTANCE INSPECTION Inspection by the owner's representative of work scheduled for full or partial payment.

ACCEPTANCE OF WORK Handing over of a completed construction project, excepting possibly any agreed-upon deficiencies or make-goods, with an obligation for payment less any holdback. *See also* **Interim acceptance, Final acceptance,** and **Partial acceptance.**

ACCEPTANCE PLAN Prescribed method of sampling, measuring and testing, together with criteria for the acceptability of a lot of material or construction.

ACCEPTANCE TEST Inspection that affirms if a material meets acceptance criteria.

ACCEPTED An installation or equipment item is 'accepted' if it has been inspected and found to be safe by a person qualified to make such an inspection, or by a qualified laboratory.

ACCEPTED BID Bid accepted by the owner or his representative and that forms the basis of a contract.

ACCEPTED ENGINEERING PRACTICES Practices that are compatible with standards of practice required by a registered professional engineer.

access. *Abbreviation for:* accessory.

ACCESS Way or means of approach.

ACCESS BALCONY Exterior passage giving access to a number of separate rooms or dwellings.

ACCESS CONNECTION Road (or private driveway) connecting to an arterial highway.

ACCESS DOOR Door giving access to machine rooms, machine and equipment spaces, or other normally closed areas.

ACCESS EYE *See* **Cleanout.**

ACCESS FLOORING Modular floor components that can be easily removed to give access to under-floor ducting.

ACCESS FOR THE PHYSICALLY CHALLENGED Special measures taken in design and construction to accommodate the special needs of the physically challenged.

ACCESSIBLE 1. Electrical and/or communications wiring capable of being removed or exposed without damaging the building structure or finish, or not permanently closed in by the structure or finish of the building. *See also* **Concealed,** and **Exposed; 2.** Equipment admitting close approach; not guarded by locked doors, elevation, or other effective means. *See also* **Readily accessible.**

ACCESSION Additions to property resulting from the annexation of fixtures or of alluvial deposits.

ACCESSORY Any supplemental device that cannot be classified as an attachment; a secondary part or assembly of parts that contributes to the overall function and usefulness of a machine.

ACCESSORY APARTMENT Separate living unit created within a single-family dwelling.

ACCESSORY SEAL Seal that is used for sealing accessory equipment. On various engines, 'accessory seal' pertains to a seal that is employed for sealing an accessory shaft in the gear box, such as a shaft for operating an oil pump, fuel pump, generator, starter, or de-oiler.

ACCESS RIGHT *See* **Right of access.**

ACCESS ROAD Paved or unpaved road, whose use requires authorization, either of public or private ownership, that provides a connection between a public road and a utility.

ACCESS SWITCH *See* **Switch.**

ACCESS TO EXIT Means of egress that leads directly from the floor area it serves, including any doorway leading directly from the floor area, to a public thoroughfare or an approved open space.

ACCIDENT Occurring by chance; a sudden, unexpected event identifiable as to time and place.

ACCIDENTAL AIR Air voids in concrete that are not purposely entrained and that are significantly larger and less useful than those of entrained air, 1.0 mm (0.04 in.) or larger in size. *Also called* **Entrapped air.**

ACCIDENTAL ERROR In surveying, a compensating error.

ACCIDENT INSURANCE Insurance indemnity for loss of time, loss of facility, medical expenses, and other defined perils and expenses due to accident.

ACCIDENT PREVENTION Planning and conscious acts taken to reduce the likelihood and opportunity for the occurrence of accidents.

ACCILIVITY Upward slope or steepness of a line or plane relative to horizontal.

accom. *Abbreviation for:* accommodate; accommodation.

ACCOMMODATION BRIDGE Temporary bridge.

accomp. *Abbreviation for:* accomplish.

AC CONTROL System where motion control is accomplished through the use of an AC motor. However, the control relays may be either AC or DC.

ACCORDION DOORS *See* **Door types.**

ACCOUNTABILITY Being answerable for the exercise of one's authority and the performance of one's duties.

ACCOUNT CODE STRUCTURE Assignment of alphanumeric identification to work components within individual cost packages.

ACCOUNT NUMBER Identification of a work package so that all costs incurred for its completion can be properly assigned and allocated.

ACCOUNT PAYABLE Sum owing to a creditor, usually resulting from the purchase of goods or services.

ACCOUNT RECEIVABLE Sum claimed against a debtor, usually resulting from the sale of goods or services.

ACCOUPLE Joining of one thing to another.

ACCOUPLEMENT 1. Timber tie or brace; **2.** Two columns or pilasters placed close together.

accred. *Abbreviation for:* accredit; accredited.

ACCREDIT Authorize; give credentials to; give authority to; certify.

ACCRETION 1. Increase in diameter or height of a tree; **2.** Addition to land from natural causes, such as alluvial deposits.

ACCRUAL METHOD Accounting procedure that recognizes revenues and expenses when they occur regardless of when cash is received or paid. *See also* **Cash method.**

ACCRUED ASSET Accumulating but not yet enforceable claim resulting from the provision of services or the expenditure of money (as during construction or development) that has been only partly performed, is not yet billable, and has not been paid for.

ACCRUED DEPRECIATION Difference between reproduction cost new, or replacement cost new, and the present worth of those improvements, both measured as of the date of appraisal.

ACCRUED EXPENSE Incurred expense for which there currently is no enforceable claim due to the project or contract not having been completed to a billable stage.

ACCRUED INTEREST Interest that has been earned but is not due or payable.

ACCRUED LIABILITY Accumulating but not yet enforceable claim by a consultant or contractor resulting from partially completed work which is not yet billable.

ACCRUED REVENUE Earned revenue resulting from partially completed work that cannot yet be billed.

acct *Abbreviation for:* account.

accum. *Abbreviation for:* accumulate.

ACCUMULATED DEPRECIATION Total by which an asset has been depreciated since being acquired or placed in service.

ACCUMULATING SHEAR Shearhead on a feller-buncher that is capable of accumulating and holding two or more cut stems.

ACCUMULATOR 1. Device that collects and stores energy; **2.** Reservoir in which hydraulic fluid is stored; **3.** Container in which fluid is stored under pressure; **4.** Device, containing nitrogen, that is used to absorb or possibly augment pressure peaks in a hydraulic system: hydraulic hammers have two accumulators (low pressure and high pressure). *See also* **Hydropneumatic accumulator.**

accur. *Abbreviation for:* accuracy.

ACCURACY Comparison of an indicated value to a known reference; the quality of accuracy is often expressed by stating the difference of two values as a percentage of the known reference value.

ACD *Abbreviation for:* automatic closing device.

ac/dc *Abbreviation for:* alternating-current or direct-current.

ace *Abbreviation for:* allowable-cut effect.

ACE LOCK Lock in which the tumbler pins are arranged in a circle.

acet. *Abbreviation for:* acetate; acetone; acetylene.

ACETATE One of a group of esters formed by combining alcohol with acetic acid, used as solvents.

ACETONE Colorless, volatile, flammable organic liquid, used principally as a paint remover and solvent.

ACETYLENE Welding gas composed of two parts of carbon plus two parts of hydrogen. When burned in an atmosphere of oxygen it produces a very high flame temperature.

ACETYLENE CYLINDER Storage container specially built to contain and ship acetylene.

ACETYLENE REGULATOR Automatic valve that reduces cylinder pressure to a constant torch pressure.

A-C EXTERIOR Exterior-type sanded plywood panel with A-grade face, C-grade back and inner plies, bonded with exterior glue.

acf/m *Abbreviation for:* actual cubic feet per minute.

ACID Compound having a pH of less than 7.0.

ACID- AND ALKALI-RESISTANT GROUT Grout formulated to resist prolonged contact with acids and alkalis.

ACID ATTACK *See* **Aggressive solution.**

ACID COPPER CHROMATE Waterborne wood preservative consisting of copper sulfate, sodium dichromate, and chromic acid.

ACID EMBOSSED GLASS Process whereby the surface of glass is obscured by treatment with hydrofluoric acid or its compounds.

ACID ETCHING 1. Imposition of a design on glass by the removal of one surface by the effect of acid attack on a controlled area, to a controlled extent; **2.** Removal of cement laitance from a horizontal surface of concrete by a 10% solution of hydrochloric acid in water; **3.** Treatment of a metal in an acid bath, usually phosphoric, to roughen the surface.

ACID GAS By-product of incomplete combustion having a pH value of less than 6.5.

ACIDIFICATION Forced or accelerated leaching by exposing a material to a higher concentration of solvents (e.g., acids) than would normally be encountered.

ACID NUMBER Measure of acidity. For a petroleum product, it is the weight of potassium hydroxide (KOH) required for neutralization.

ACIDITY See **pH**.

ACID PICKLE Dilute acid solution used to strip and clean metals.

ACID POLISHING Acid treatment of a glass surface in order to polish it.

ACID RAIN Form of air and water pollution caused by the release of acid gases from the combustion of fossil fuels in internal combustion engines, combustors, factories, mine processing plants, or coal and oil burning electrical generators.

ACID RESISTANT Having the ability to withstand the action of identified acids within specified limits of concentration and temperature.

ACID-RESISTANT BRICK See **Brick**.

ack. Abbreviation for: acknowledge.

ACKERMAN STEERING See **Steering**.

ACKNOWLEDGMENT Formally witnessed declaration that an instrument is the declared intent of the signator.

ACORN NUT See **Nut**.

acous. Abbreviation for: acoustical; acoustic.

ACOUSTIC Relating to sound and hearing.

ACOUSTIC ABSORPTIVITY See **Sound absorption coefficient**.

ACOUSTICAL BOARD Any type of special material, such as insulating boards, used in the control of sound or to prevent the passage of sound from one room to another.

ACOUSTICAL CORRECTION Measures taken to improve the acoustical performance of a completed room or auditorium.

ACOUSTICAL IMPEDANCE Mathematical expression for characterizing a material as to its energy transfer properties (the product of its unit density and its sound velocity pV).

ACOUSTICAL MATERIALS Sound absorbing or attenuating materials.

ACOUSTICAL REDUCTION FACTOR See **Sound reduction factor**.

ACOUSTICAL TILE Wall and ceiling tiles composed of materials, and of a design intended to attenuate sound waves.

ACOUSTICAL TRANSMISSION FACTOR Ratio of original sound volume to that which is transmitted through and beyond any obstruction to the sound waves, such as a wall, panel, or baffle.

ACOUSTIC CONSTRUCTION Construction techniques aimed at measurably reducing or attenuating sound transmission.

ACOUSTIC PILE TEST Site test to obtain information about the qualities of an installed reinforced concrete pile by passing sound waves through it.

ACOUSTIC PLASTER Gypsum plaster that, when applied to a wall or ceiling, has a high sound absorbency.

ACOUSTICS Study of sound; quality of sound.

ACOUSTIC TRACE Line on a vibration record that records a sound level.

acp Abbreviation for: asbestos-cement pipe.

acpt Abbreviation for: accept.

acq. Abbreviation for: acquire; acquisition.

ACQUISITION Process of obtaining a property or right of way. Also called **Taking**. See also **Conveyance, Dedication, Eminent domain, Expropriation, Negotiation, Option, Remainder,** and **Severality**.

ACRE Nonmetric unit of area, equal to 43,560 square feet or 4840 square yards. Symbol: ac. Multiply by 0.404 to obtain hectares, symbol: ha; by 404.856 to obtain square meters, symbol: m^2. See also the appendix: **Metric and nonmetric measurement**.

ACRE-FOOT Nonmetric unit of volume, equal to the amount of water, gravel, coal, or other minerals required to cover 1 acre to a depth of 1 ft, equal to 1613 yd^3. Symbol: ac/ft. Multiply by 1233.482 to obtain cubic meters, symbol: m^3. See also the appendix: **Metric and nonmetric measurement**.

ac/ft Abbreviation for: acre-foot.

ACROSS Technique of applying drywall board with the long edge perpendicular to the framing.

ACROTER Pedestal at the apex of a pediment that supports a statue or ornament.

AC ROTOR Rotating elements of an AC motor.

acrs Abbreviation for: across.

ACRYLIC Series of olefin acids; a colorless, pungent acid.

ACRYLIC ADHESIVE See **Adhesive**.

ACRYLIC EMULSIONS Clear, water-based repellents that form a film. Acrylic resins come from the polymerization of derivatives of acrylic acids, including esters of acrylic acid, methacrylic acid, acrylonitrile, and their copolymers. Acrylic resins vary from hard, brittle solids to fibrous elastomeric structures to viscous liquids. See also **Bonding admixture**.

ACRYLIC PAINT Water-based painting medium that dries alcohol-soluble; available as paints and as an ingredient in water-soluble finishes.

ACRYLIC PLASTIC SHEET Transparent sheet used for glazing.

ACRYLIC RESIN One of a group of thermoplastic resins formed by polymerizing the esters or amides of acrylic acid, used in concrete construction as a bonding agent or surface sealer, and as an ingredient of water-base (latex) paints and stains. Can be colorless or pigmented.

ACRYLONITRILE-BUTADIENE-STYRENE Plastic material, used in the manufacture of drainage pipe and fittings.

ACSR *Abbreviation for:* aluminum cable, steel reinforced.

act. *Abbreviation for:* action; active; actual; actuate.

actg *Abbreviation for:* acting; actuating.

ACTINIC GLASS Glass capable of excluding the actinic component of transmitted light.

ACTINIC SCREEN Materials that shut out the short-wave lengths of light, such as ultraviolet, and their chemical reactions.

ACTINOMYCETES Large group of mold-like microorganisms that give off an odor characteristic of rich earth and that are significant in the stabilization of solid waste by composting.

ACTION LEVEL Exposure level (concentration of the material in the air) at which certain OSHA regulations to protect employees take effect.

ACTIVATED ALUMINUM Form of aluminum oxide used as a drying agent.

ACTIVATED CARBON Porous carbon used to absorb odors.

ACTIVATED SLUDGE PROCESS Aerobic sewage treatment method in which air is introduced to raw sewage to promote mixing and bacterial digestion.

ACTIVATOR Compound ingredient used to increase the effectiveness of an accelerator.

ACTIVE EARTH PRESSURE Horizontal pressure of soil against a structure.

ACTIVE LAYER 1. Ground surface layer that moves seasonally owing to volume changes occasioned by temperature differentials; **2.** Area above permafrost that thaws each, or most years.

ACTIVE LIFE Period from the initial receipt of hazardous waste at a facility until the Regional Administrator receives certification of final closure.

ACTIVE PORTION Portion of a facility where treatment, storage, or disposal operations are being or have been conducted.

ACTIVE REPAIR TIME *See* **Machine time.**

ACTIVE SOLAR SYSTEM Solar system that uses mechanical devices such as collectors, thermal storage devices, transfer fluid, fans and pumps to collect, store, and distribute useful energy. *See also* **Hybrid solar system.**

ACTIVE SOLIDS Ingredients of a coating composition that are deposited following co-reaction or reaction with a substrate. Active solids are usually measured as a weight percent of the total.

ACTIVE STORAGE Volume of a storage facility available for use.

ACTIVITY Task performed over a period of time.

ACTIVITY DURATION Actual time or estimated time required to accomplish something.

ACTIVITY OF CLAY Ratio of plasticity index to percent by weight of the total sample of a clay that is smaller than 0.002 mm (0.00008 in.) in grain size.

ACTIVITY ORIENTED SYSTEMS Systems that present information in terms of activities (i.e., periods of time).

actvt. *Abbreviation for:* activate.

ACT OF BANKRUPTCY Act by a debtor that permits a creditor to initiate proceedings against him under bankruptcy legislation.

ACT OF GOD Circumstance assumed to be beyond human control.

ACTUAL *See* **Cost types.**

ACTUAL AGE (OF A PROPERTY) Actual number of years that have passed since a structure was built. *Also called* **Chronological age,** and **Historical age.**

ACTUAL CASH VALUE Actual market value of something insured at the time of its loss or damage.

ACTUAL DIMENSION Dimension as measured or calculated.

ACTUAL EVICTION *See* **Eviction.**

ACTUAL FIT Relationship existing between two mating parts, particularly the amount of clearance or interference that is present when they are assembled.

ACTUAL LOAD Weight of the load being hoisted and all additional equipment such as blocks, slings, sensors, etc. *Also called* **Working load.**

ACTUAL PRODUCTIVE TIME *See* **Machine time.**

ACTUAL SLOPE Slope to which an excavation face is excavated.

ACTUAL VALUE *See* **Market value.**

ACTUATOR Linear or rotary device that converts hydraulic or fluid energy into mechanical energy.

acty *Abbreviation for:* activity.

ACU *Abbreviation for:* air conditioning units.

ACULTURAL VIBRATION Vibration that is strange and unfamiliar to the observer.

ad *Abbreviation for:* advertise; advertising.

A-D Sanded plywood panel with A-grade face, D-grade back and inner plies, bonded with interior or exterior glue.

AD *Abbreviation for:* access door; air dried; area drain; as drawn; average deviation.

A/D *Abbreviation for:* advanced design.

ADAMANTINE DRILL Core drill, 100 to 760 mm (4 to 30 in.) in diameter, that uses chilled shot as a cutting agent.

ADAMANT PLASTER Quick-hardening gypsum plaster.

adapt. *Abbreviation for:* adaptability; adapter.

ADAPTER 1. Device that welds under a spring seat to increase the mounting height; **2.** End-ring shaped on the side toward the packing to conform to its shape, typically, convex V, concave V, etc.; **3.** Fitting that joins pipes of different materials or different sizes, or different threads; **4.** Structural fabrication or casting placed in a helmet to adjust to a different type or size of pile. *Also called* **Filler**, and **Insert**.

ADAPTER POT Adapter into which cylindrical injectors are inserted in order to be tested on a test stand.

add. *Abbreviation for:* addition.

ADDENDA Document issued before bid opening that clarifies, corrects, or changes the bidding documents or requirements of the contract documents.

ADDITION 1. Something added to what already exists; **2.** Material that is interground or blended in limited amounts into hydraulic cement during manufacture either as a 'process addition' to aid in manufacturing and handling the cement or as a 'functional addition' to modify the use properties of the finished product.

ADDITIONAL SERVICES Professional services, additional to those contracted for, that may be requested of architects or engineers by the owner.

ADDITION POLYMERIZATION Reaction that causes one molecule to hook on to another like molecule by addition to form a chain.

ADDITIVE 1. Material added to another to produce a modification of its properties. *See also* **Agent**; **2.** Usually refers to a system that allows controlled dispersement of a liquid additive; **3.** Chemical compound added to a base oil to alter its physical, chemical and performance characteristics.

ADDITIVE ALTERNATE *See* **Alternate bid.**

ADDITIVE CONSTANT In surveying, an amount added to the intercept on the staff and the multiplier that gives the distance between the telescope center and staff.

addl *Abbreviation for:* additional.

addns *Abbreviation for:* additions.

ADD-ON INTEREST Interest expressed in dollars per hundred per year, i.e., 10% add-on is $10/hundred/year.

addr. *Abbreviation for:* address.

ADDRESS Binary number identifying a memory location.

ADDRESSABLE SYSTEM SMOKE DETECTOR Smoke detection system in which individual detectors are programmed to signal their location when activated.

ADD TO SET Slight overhang given to the point of a saw tooth that widens the kerf, allowing the saw to run more easily.

ADF *Abbreviation for:* after deducting freight.

ad fin. Abbreviation for: ad finem (at the end/to the end).

ADFREEZING Adhesion of soil to a foundation unit resulting from the freezing of soil water. *Also called* a **Frost grip.**

adh. *Abbreviation for:* adhesive.

ADHEREND Item held to another by an adhesive.

ADHERENT Sticking, clinging action.

ADHESION 1. State in which two surfaces are held together by interfacial effects, which may consist of molecular forces, interlocking action, or both; **2.** Soil quality of sticking to metal parts of mechanical equipment: buckets, blades, tracks, etc.; **3.** Strength of bond between cured rubber surfaces or between a cured rubber surface and a nonrubber surface.

ADHESION FAILURE 1. Separation of two bonded surfaces at an interface by a force less than that specified in a test method; **2.** Separation of two adjoining surfaces owing to surface conditions.

ADHESION-TYPE CERAMIC VENEER Inner sections of a ceramic veneer held in place by adhesion of mortar to the unit and its backing. No mechanical anchors are required.

ADHESIVE Substance capable of holding materials together by surface attachment. Common adhesives used in construction include:

Acrylic: Two-part, fast-setting adhesive capable of adhering to an oily surface.

Anaerobic: One-part adhesive that cures without the presence of oxygen.

Animal glue: Made from the hide, bones, hooves and trimmings of animals. Animal glue is not waterproof. However, it develops a very high strength, up to 82,000 kPa (12,000 psi) in shear, when used to join wood faces.

Asphaltic: Asphalt-based mixtures that produce a range of thermoplastic adhesives, commonly used to bed floor and roofing materials and bond metals to rubber or glass. Their melting point can range from 10°C to 93°C (50°F to 200°F).

Casein glue: Made from the curd of milk; a phosphoprotein. It is highly resistant to humidity. Used mainly in carpentry and furniture making.

Cellulose: Air drying thermoplastic emulsion using water, ethyl acetate or acetone as a solvent. Used to bond wood, glass and similar dense materials with medium to good strength.

Ceramic: Rubber solvents and rubber plus resin-based emulsions that are used for bonding tile to a surface

Chlorinated rubber: Liquid adhesive offering medium to good water resistance, used to bond wood, metals, glass, and other dense materials.

Contact cement: Made from toluene and petroleum distillates, this transparent, nonstaining adhesive bonds instantly to itself, permanently and without clamping. In use, both surfaces to be bonded are coated with the cement, which is then allowed to dry before the two surfaces are brought together. Contact cement grows stronger and more heat resistant with age.

Cyanoacrylate: One-part, fast-setting clear adhesive.

Exterior: 100% waterproof adhesive bonding on all 'Exterior', 'Exposure 1' and most 'Interior'-type structural wood panels.

Foamed: One- or two-part adhesive that, when released from its container and exposed to the atmosphere, expands producing a substance whose apparent density has been significantly decreased by the presence of gaseous cells.

Glue: Form of adhesive manufactured from animal parts, used for relatively small parts and characterized by cool temperatures and quick setting.

Heat-activated: Dry adhesive film that becomes tacky or fluid following the application of heat and that bonds in that state through the application of pressure.

Heat-setting: Any adhesive that sets at temperatures above 100°C (212°F).

Hot-melt: Adhesive that is applied in a heated and molten state and that bonds as it cools and sets.

Interior: Moisture-resistant, but not waterproof adhesive.

Intermediate glue: Adhesive used to bond some interior-type plywood panels that has a moisture resistance midway between interior and exterior glues.

Intermediate-temperature-setting: Adhesive that sets at temperatures between 30°C and 100°C (87°F and 212°F).

Marine glue: Adhesive that is impervious to the effects of fresh and salt water.

Mastic: Term covering a wide range of special-purpose adhesives that set-up and dry when spread in thin layers, either exposed to air, or between materials. Mostly, the adhesives are used to bond materials of different composition and other physical characteristics. Typical applications are for gluing floor, wall and ceiling tiles, plywood panels, wood strapping to concrete walls, waterproof panels to gyproc or plywood, etc.

Moisture resistant: Adhesives that offer moderate resistance to wet and humid conditions, but that will likely fail on prolonged exposure to severe conditions.

Multiple-layer: Film-type adhesive having a different formulation on each side of a supporting film, designed to bond dissimilar materials.

Neoprene-rubber: Thermoplastic (with some thermosetting characteristics) adhesive containing water or volatile solvents, used to bond a wide range of dense materials such as wood, glass, metals, composite panels, etc.

Organic: Prepared organic material, ready to use with no further addition of liquid or powder.

Paste: Water-based, high-volume mucilage formed of organic or synthetic materials, commonly used for work such as hanging wallpaper.

Plastic cements: Cements used to join or bond plastic sheets, films and parts.

Plastic glues: Synthetic resin glues used on wood: epoxy resins for gluing light metals and other substances.

Plastic-resin glue: Made from urea-formaldehyde, this powder glue is mixed with water shortly before needed. It forms a very strong, waterproof bond, especially when used to join wood to wood, as in a joint, or when laminating plywoods. Has limited weather resistance.

Plywood: Interior- and exterior-type adhesives specially formulated for bonding together the veneers used in plywood construction.

Pressure-sensitive: Adhesive made so as to adhere to a surface at room temperature by briefly applied pressure.

Resin: Any of a number of adhesives based on resin formulations, including:

Cold-setting resin: Powdered resin-base glue that is mixed with water immediately prior to application and that sets and cures at room temperature.

Epoxy-resin glue: A two-part compound consisting of an epoxy and a catalyst which, when mixed together in the correct proportions (usually 1:1) react to form an extremely dense and solid material having superior adhesive properties when used in thin layers. The surfaces being bonded should be kept under pressure throughout the time that the epoxy-resin takes to cure (varying, according to the formulation, from seconds to hours).

Melamine resin: Cold-applied thermosetting, temperature-resisting and nonstaining adhesive, hot-pressed at 148°C (300°F) for cure, used to bond wood.

Phenolic resin: Thermosetting, water-resistant water-, alcohol-, or ketone-based adhesive that provides good to excellent bond for wood, medium to poor bonding for glass and metals. Resistant to a wide temperature range.

Polyvinyl-resin glue: A white, fast-setting, strong, and easy-to-use, water-soluble liquid glue used to create wood joints.

Resourcinol-resin: A two-part liquid glue that, when mixed together in the correct proportions, forms a strong, transparent film.

Synthetic: *See* **Synthetic resin,** below.

Urea: Water-, alcohol-, or alcohol-hydrocarbon-blend-based thermosetting adhesive.

Waterproof and boilproof resin: Thermosetting adhesive binder that, when fully cured, is not softened by heat or moisture.

Room-temperature setting: Adhesive that sets at temperatures between 20°C and 30°C (68°F and 86°F).

Rubber-based cement: A relatively weak glue that is moisture resistant.

Separate-application: Two-part adhesive where one part is applied to one component and the other to the second component to be joined.

Sodium-silicate: Water-based adhesive having good temperature resistance but poor water resistance, used to bond wood and metals. Dries at temperatures between 15°C and 93°C (60°F and 200°F).

Solvent (adhesive): Adhesive that uses a volatile liquid other than water as a vehicle.

Solvent-activated: Dry adhesive that becomes tacky on application of a solvent just prior to use.

Supported-film: Sheet adhesive incorporating a carrier that becomes part of the bond on application.

Synthetic resin: Adhesive formulated with a synthetic resin as its base.

Thermoplastic: Adhesives capable of being repeatedly softened when heated, but which maintain their bond when hard.

Thermosetting: Adhesives that are cured through the application of heat and that retain their bond when subsequently exposed to high temperatures.

Unsupported-film: Adhesive supplied in sheet or film form without an incorporated carrier.

Urea formaldehyde: Powder-type adhesive mixed with water or a catalyst just prior to use in making wood joints. Joint must be clamped to create effective bond.

White latex: Water-based adhesive used for installation of resilient floor and carpet.

Various descriptive adjectives are appended to the word adhesive to indicate certain characteristics. These include: Physical form (that is, liquid adhesive or tape adhesive); Chemical type (silicate adhesive, resin adhesive); Materials bonded (paper adhesive, metal-plastic adhesive, can label adhesive); and Conditions of use (hot-setting adhesive, for instance). *See also* **Mechanical adhesion,** and **Specific adhesion.**

ADHESIVE APPLICATION Technique of applying drywall board or other panels to studs using adhesives.

ADHESIVE BOND Adhering characteristics of adhesives.

ADHESIVE COATING Layer applied to any product surface to increase its adherence to an adjoining surface.

ADHESIVE NAIL-ON Technique of applying drywall board or other panels using an adhesive in combination with mechanical fasteners.

ADHESIVE RUPTURE Failure of the adhesive bond between otherwise separate components.

ADHESIVE SPREADER Tool having a working surface notched or serrated to a pattern recommended for the type of adhesive to be spread; leaves ridges or bands of adhesive on the surface to which the adhesive is applied.

ADHESIVE STRENGTH Shear and tensile strength of the bond created by an adhesive joining two objects.

ADHESIVE TAPE Flexible ribbon coated on one or both sides with an adhesive.

ADHESIVE WALL CLIP *See* **Fastener.**

ADHESIVE WEAR *See* **Wear.**

ADIABATIC Condition in which heat neither enters nor leaves a system.

ADIABATIC CURING Maintenance of adiabatic conditions in concrete or mortar during the curing period where heat is neither lost to or gained from the immediate surroundings.

ADIABATIC ENGINE Engine in which heat loss is minimized by retaining heat within the system through increased insulation of the combustion chamber.

ADIABATIC STORAGE CONDITION Rise in temperature due to chemical reaction lacking any possible heat loss or heat input.

ad inf. *Abbreviation for:* ad infinitum (to infinity).

ad init. *Abbreviation for: ad initium* (at the beginning; meanwhile).

ADIT Gallery driven more or less horizontally into a hillside for the purpose of mining. *Also called* a **Drift.**

adj. *Abbreviation for:* adjoining; adjust; adjustable.

ADJACENT Near to, but not necessarily in contact with.

ADJOINING Attached to; in contact with.

ADJOINING OWNER One who owns land, a building, or a portion of a building that abuts upon, or is within 3 m (10 ft) of a proposed building operation.

ADJUSTABLE ATTACHMENT Fastening device that allows for compensation in three planes, used to attach supports to structures.

ADJUSTABLE BIT *See* **Bit.**

ADJUSTABLE CLAMP Clamping device that can be altered in length or some other way to suit the job to be done.

ADJUSTABLE PARALLELOGRAM RIPPER *See* **Ripper.**

ADJUSTABLE RADIAL RIPPER *See* **Ripper.**

ADJUSTABLE-RATE MORTGAGE *See* **Mortgage.**

ADJUSTABLE RESISTOR Resistor with taps, sliding bands, or a wiper that, when moved, allows all or part of the resistor to be employed.

ADJUSTABLE, TIMBER, SINGLE-POST SHORE Individual timber used with a fabricated clamp to obtain adjustment. Not normally manufactured as a complete unit. *See also* **Fabricated single-post shore, Post shore,** and **Timber single-post shore.**

ADJUSTABLE TORQUE ARM Member used to retain axle alignment, and in some cases control axle torque. Normally one adjustable and one rigid torque arm are used per axle so that the axle may be aligned.

ADJUSTABLE WRENCH *See* **Wrench.**

ADJUSTED BASE COST Base construction cost adjusted for alternate components, plus or minus.

ADJUSTED TAX BASIS The net amount used to calculate depreciation and investment tax credit.

ADJUSTER 1. Specialist who settles a claim (as under an insurance policy); **2.** Elevator mechanic responsible for final inspection of new installations.

ADJUSTING NUT Threaded nut that, when turned, will vary some already established tolerance.

ADJUSTING RING Ring that adjusts release-bearing clearance in a clutch assembly to compensate for the wearing-away of disc linings.

ADJUSTING SCREW 1. Threaded screw that, when turned, will cause two connected faces to move further apart or closer together; **2.** A leveling device or jack composed of a threaded screw and an adjusting handle, typically used for vertical adjustment of shoring and formwork.

ADJUSTMENT 1. Method of eliminating inconsistencies in computed or measured quantities through application of predetermined corrections to compensate for errors. *See also* **Variance; 2.** Process of determining the cause and amount of loss under an insurance claim.

ad lib. Abbreviation for: ad libitum (at pleasure).

ad loc. Abbreviation for: ad locum (at the place; to the place).

admin. Abbreviation for: administrate; administrative.

Admin. Engr *Abbreviation for:* administrative engineer.

ADMINISTRATOR 1. Person having executive work or ability; **2.** Court-appointed person who administers the estate of a person who died intestate.

ADMIRALTY SHACKLE Heavy shackle at the tail tree that connects the skyline to the stub line (guyline extension).

ADMISSIBLE LOAD *See* **Safe bearing capacity.**

ADMIXTURE 1. Material added to another to achieve a specific effect; **2.** Materials added to mortar, concrete, and other cementitous mixtures or applied to the surface during the curing or setting process (as water repellent, colorant, accelerator, retarder, antifreeze, etc.). There are many types, including:

> **Accelerating admixture:** Admixture that causes an increase in the rate of hydration of hydraulic cement, and thus shortens the time of setting, or increases the rate of strength development, or both.

> **Air-entraining admixture:** Admixture that causes the development of a system of microscopic air bubbles in concrete, mortar, or cement paste during mixing.

> **High-rate water-reducing admixture:** Water-reducing admixture capable of producing large water reduction or great flowability without causing undue set retardation or entrainment of air in mortar or concrete.

> **Retarding admixture:** Admixture that causes a decrease in the rate of hydration of hydraulic cement, and lengthens the time of setting.

> **Water-reducing admixture:** Admixture that either increases the slump of freshly-mixed mortar or concrete without increasing water content or maintains the slump with a reduced amount of mortar, the effect being due to factors other than air entrainment.

ADOBE *See* **Construction types.**

ADOBE BRICK *See* **Brick.**

ADOBE SHOT *See* **Mud cap.**

adp. Abbreviation for: adaptable.

ADP Abbreviation for: apparatus dew point.

adpt. Abbreviation for: adapter.

ADS *Abbreviation for:* automatic door seal.

ADSORBED WATER Water held on the surfaces of material by electrochemical forces and having physical properties substantially different from those of absorbed water or chemically combined water at the same temperature and pressure. *See also* **Adsorption.**

ADSORBENT Material having the ability to contain another material within its structure without changing the character of the original material.

ADSORBER Material or device used to adsorb another material.

ADSORPTION Development (at the surface of either a liquid or solid) of a higher concentration of a substance than exists in the bulk of the medium; especially formation of one or more layers or molecules of gases, or dissolved substances, or of liquids at the surface of a solid (such as cement, cement paste, or aggregates), or of air-entraining agents at the air-water interfaces; also the process by which a substance is adsorbed. *See also* **Adsorbed water.**

ADT *Abbreviation for:* advanced design team.

adv. Abbreviation for: advance.

AD VALOREM According to value.

AD VALOREM **TAX** Tax or duty based on value and levied as a percent of that value, e.g. 10 mills per dollar (1.0%) of property value.

ADVANCE Payment made on account of, but before completion of, a contract, or before receipt of goods or services. *See also* **Deposit.**

ADVANCED CHARGE Charge, other than that for freight, advanced by one shipping line to another, or to the shipper, to be collected from the consignee.

ADVANCED REGENERATION Trees that have become established naturally under a mature forest canopy and which are capable of becoming the next crop after the mature crop is removed.

ADVANCE SPLICER Tapered steel unit for connecting pipe piles by driven friction.

ADVANCING A LINE Moving a fire hose toward a given area from the point where the hose-carrying equipment has stopped.

ADVANCING-SLOPE GROUTING Method of grouting by which the front of a mass of grout is caused to move horizontally through preplaced aggregate by use of a suitable grout injection sequence.

ADVANCING-SLOPE METHOD Method of placing concrete, as in tunnel linings, in which the face of the fresh concrete is not vertical and moves forward as concrete is placed.

ADVECTION Process of transport of a property solely by the velocity field in a fluid. For groundwater, advection refers to the natural state of motion.

ADVERSE POSSESSION Right of an occupant to acquire title in defiance of someone else's legal title following occupancy for a required statutory period in actual, open, notorious, exclusive use.

ADVERSE SLOPE Uphill incline for hauling or skidding logs or other loads.

advsy Abbreviation for: advisory.

advt *Abbreviation for:* advertisement.

ADZE Axe-like cutting tool used for dressing timbers, with the cutting edge at right angles to the handle.

A-E *Abbreviation for:* architect-engineer.

A-EM *Abbreviation for:* architect-engineer management.

AEOLIAN Carried or blown by the wind; deposited as loess.

AERATED CONCRETE Lightweight materials made from specially prepared cement, used for subfloors. The highly cellular nature of this material makes it particularly useful as a retardant to sound transmission. *See also* **Cellular concrete,** and **Foamed concrete.**

AERATION 1. Injection of air into a body of water; **2.** Introduction of air into soil in order to improve it as a growing medium for plants; **3.** Condition where air is present in hydraulic fluid, causing it to appear milky and components to operate erratically due to the compressibility of the trapped air; **4.** Entrainment of gas (air or combustion gas) in an engine coolant.

AERATION ZONE Depth of soil below ground to the top of groundwater level: divided into three layers, an upper soil belt, intermediate belt, and lower capillary fringe.

AERATOR Device that adds air to water.

AERIAL ATTACK Use of aircraft in fighting forest fires, mainly to drop fire retarding or extinguishing solutions on, or in the path of, the conflagration.

AERIAL LOGGING Yarding system employing an aerial lift of logs, such as by balloons or helicopters.

AERIAL PHOTOGRAMMETRY Interpreting information from aerial photographs.

AERIAL PHOTOGRAPHY Photos taken from the air at regular intervals and used in photo interpretation to provide information about land development, vegetation, and land forms.

AERIAL PLATFORM Mobile work platform that can be raised vertically. Available in a wide range of types and designs, may be hand or power propelled and actuated, either cable connected or with a self-contained power source. Power may be electric, gasoline-, propane-, or diesel-driven engines. The erecting system can be through a scissor action, sliding members, or telescoping cylinders. The work platform may rotate, and can be fitted with an extendable lip or may be moved horizontally over a limited range. The platform can also be equipped with electrical and/or hydraulic power to drive accessories.

AERIAL SURVEYING Surveying done through photo-interpretation of images taken from aircraft flying a grid pattern at a fixed altitude over the target area.

AERIAL TRAMWAY Personnel or material cars suspended from cables carried on towers. *Also called* a **Tramway.**

AERIAL TRIANGULATION Aerial photography that determines the triangulation necessary for the extension of horizontal and vertical survey control.

AEROBE Organism that uses oxygen to sustain life, used in water purification and, more particularly, sewage treatment.

AEROBIC Environment rich in free or uncombined oxygen.

AEROBIC DIGESTION Breakdown of organic components by microbial action in the presence of oxygen.

AEROBIC RESPIRATION Oxidation of organic compounds by oxygen.

AEROCRETE Low-strength, 6 894 kPa (1,000 psi) @ 28 days, concrete used for fill.

AERODYNAMIC INSTABILITY Oscillation in a structure caused by strong wind.

AERODYNAMICS Branch of physics that deals with the various forces associated with air or other gases in motion.

AEROLOGICAL INVESTIGATION Determination of surface data of an area of land based on aerial photographs.

AEROSOL Small, solid or liquid particles, from 0.01 to 100 microns in diameter, suspended in air.

AEROSTYLE Spacing of columns by four-to-five diameters.

AF *Abbreviation for:* armored front; asbestos felt; attenuation factor.

AF DIMENSION 1. Distance from the centerline of the rear axle to the end of the frame; **2.** Dimension between the center of the fifth wheel or center of gravity of the body and rear axle.

aff. *Abbreviation for:* affect.

AFF *Abbreviation for:* above finished floor.

AFFIDAVIT Written declaration sworn to and affirmed before a legally authorized witness.

AFFIDAVIT OF CLAIM Written declaration of relevant facts required when an insurance claim is made.

AFFINE STRAIN Homogeneous strain resulting when similar portions of a homogeneous body remain similar after deformation.

AFFINITY Force that causes the atoms of certain elements to combine and stay combined.

AFFORESTATION Establishment of forest crops by artificial methods, such as planting or sowing on land where trees have never before grown.

afft *Abbreviation for:* affidavit.

AFG-01 Performance specification developed by the American Plywood Association (APA) for glues recommended for use in the APA Glued Floor System; it requires that glues applied at the job site be sunlight resistant, strong under many moisture and temperature conditions, and able to fill gaps.

A-FRAME 1. Open, two-legged structure tapering from a wide base to a narrow, load-bearing top. *Also called* **Brace; 2.** Steel 'A'-shaped structure used to secure and support guylines on a snorkel (an extension to a loader used to reach distant logs for yarding). *See also* **Gantry; 3.** Building characterized by a pitched roof that extends to the foundation plate.

A-FRAME DERRICK Derrick having a boom hinged

from a cross member between the lower ends of two upright towers that meet at the top. The boom upper end is stayed to the junction of the side members, which in turn is guyed back.

A-FRAME GANTRY *See* **Gantry.**

A-FRAME JIB 1. Outermost attachment, connected to the fly section and supported by pendants; **2.** 'A'-shaped jib that provides a lifting machine with additional boom length and that offsets for up-and-over capability.

aft. *Abbreviation for:* after.

AFT BATTER Driving a pile at an angle from the vertical where the bottom of the pile is inclined toward the crane. *Also called* **Negative batter.**

AFTERBURNER Device used to burn or oxidize the combustible constituents remaining in effluent gases to destroy smoke and odors.

AFTERCOOLER 1. Device that cools a gas, usually after it has been compressed; **2.** Radiator-like device that removes heat from the intake air charge after it is compressed and leaves the turbocharger. Sometimes called an **Intercooler.** *See also* **Engine charge air cooler.**

AFTER-CURE Continuation of the process of vulcanization after the cure has been carried to the desired degree and the source of heat removed.

AFTERGLOW In fire resistance testing, the red glow persisting after extinction of the flame.

AG *Abbreviation for:* above grade; against the grain.

agcy *Abbreviation for:* agency.

AGE 1. Mean age of the trees comprising a forest, crop or stand. In forests, the mean age of dominant (and sometimes codominant) trees is taken. The plantation age is generally taken from the year the plantation was begun, without adding the age of the nursery stock; **2.** Of a tree, the time elapsed since the germination of the seed, or the budding of the sprout or cutting from which the tree developed.

AGE CLASS Any interval into which the age range of trees, forests, stands or forest types is divided for classification and use. Forest inventories commonly group trees into 20-year age class groups.

AGED RESIDUE System used to describe the condition of asphalt, after mixing with hot aggregate in a pugmill to represent the material actually placed on a pavement.

AGE HARDENING 1. Increase in consistency (hardening) of a lubricating grease with storage time. *See also* **Thixotrophy**; **2.** Capacity of some metals to harden over time.

AGENCY Relationship between principal and agent whereby the agent is employed to undertake certain acts on the principal's behalf when dealing with a third party.

AGENT 1. Person authorized to act for and/or represent another; **2.** General term for a material that may be used either as an addition or additive to another, typically to cement as an admixture in concrete, e.g., an air-entraining agent. *See also* **Additive.**

AGE/STRENGTH RELATIONSHIP Increase in strength of a material due to a maturing process and/or chemical reaction. *Also called* **Aging process.**

AGGLOMERATE 1. Coarse-grained pyroclastic rock; **2.** Gathering together into a floc of suspended colloidal particles.

aggr. *Abbreviation for:* aggregate.

AGGRADATION Building up of a portion of the earth's surface toward a uniformity of grade or slope.

AGGREGATE 1. Load-bearing component of a road structure, usually sand, gravel, crushed stone, slag and fines; **2.** Collection of granulated particles of different substances into a compound or agglomerated mass. There are many classifications, including:

Angular: Aggregate particles that possess well-defined edges formed at the intersection of roughly planar faces.

Coarse: Aggregate predominantly retained on the No. 4 (4.75 mm, 0.187 in.) sieve.

Crushed: Product resulting from the artificial crushing of gravel with a specified minimum percentage of fragments having one or more faces resulting from fracture.

Crushed gravel: Product produced by mechanically crushing large-size gravel with all resulting fragments having at leas one face resulting from fracture.

Crushed stone: Product resulting from the artificial crushing of rocks, boulders, or large cobblestones, substantially all faces of which possess well-defined edges and having resulted from the crushing operation.

Crusher-run: Aggregate that has been mechanically broken and that has not been subjected to subsequent screening.

Dense-graded: Aggregates graded to produce low void content and maximum weight when compacted.

Fine: Aggregate passing the 9.5 mm (0.375 in.) sieve and almost entirely passing the No. 4 (4.75 mm, 0.187 in.) sieve and predominantly retained on the No. 200 (0.074 mm, 0.0029 in.) sieve.

Gap-graded: Aggregate so graded that certain intermediate sizes are substantially absent.

Heavyweight: Aggregate of high density, such as barite, magnetite, hematite, limonite, ilmenite, iron, or steel, used in heavyweight concrete.

Lightweight: Aggregate of low density, such as (a) expanded or sintered clay, shale, slate, diatomaceous shale, perlite, vermiculite, or slag, (b) natural pumice, scoria, volcanic cinders, tuff, and diatomite, (c) sintered fly ash or industrial cinders, used in lightweight concrete.

Normal-weight: Aggregate that is neither heavyweight nor lightweight.

Open-graded: Aggregate in which the voids are relatively large when the aggregate is compacted.

Reactive: Aggregate containing substances capable of reacting chemically with the products of solution or hydration of portland cement in concrete or mortar under ordinary conditions of exposure, resulting in some cases in harmful expansion, cracking, or staining.

Refractory: Aggregate having refractory properties such that, when bound together into a conglomerate mass by a matrix, it forms a refractory body.

Single-sized: Aggregate in which a major portion of the particles are in a narrow size range.

Well-graded: Aggregate having a particle size distribution that produces maximum density, i.e., minimum void space.

AGGREGATE-ALKALI REACTION *See* **Alkali-aggregate reaction.**

AGGREGATE ATTRITION TEST *Also called* **Deval test.** *See* **Los Angeles abrasion test.**

AGGREGATE BASE COURSE Course used under highway surfacing to produce an even surface.

AGGREGATE BLENDING Process of intermixing two or more aggregates to produce a different set of properties, generally, but not exclusively, to improve grading.

AGGREGATE BRIDGING Unwanted irregular cavities on the surface of concrete with coarse particles at the top; occasioned by having too much small material within the coarse aggregate grade of a concrete mix.

AGGREGATE/CEMENT RATIO Ratio between the total dry mass of aggregate and the mass of cement in a concrete mix.

AGGREGATE-COATED PANEL Plywood panel coated with stone chips imbedded in a resin coating.

AGGREGATE CRUSHING VALUE Value that gives a relative measure of aggregate resistance to crushing when subjected to a gradually increasing compressive load.

AGGREGATE CRUSHING VALUE APPARATUS Apparatus designed to obtain the crushing value of a sample of aggregate, consisting of an open-ended heavy-duty cylinder of 150 mm (5.9 in.) internal diameter with a heavy-duty plunger and base plate, for aggregate sized between 10 mm (0.39 in.) and 14 mm (0.55 in.).

AGGREGATED SHIPMENT Shipment from different shippers to one consignee that is consolidated and treated as one consignment.

AGGREGATE FINISH Various sizes of stone, rock, pebbles, etc. used in the makeup of a panel to be placed on the facade of a building; not a smooth finish. The aggregate may be bonded together with concrete or mortar to give the desired or designed matrix.

AGGREGATE GUN Device used to apply aggregates to a surface. It consists of a hopper attached to a short round nozzle, which in turn is connected to a blower or compressed air source.

AGGREGATE INTERLOCK Effect of portions of aggregate particles from one side of a joint or crack in concrete protruding into recesses in the other side of the joint or crack so as to transfer load in shear and to maintain alignment.

AGGREGATE OPERATIONS LIABILITY Maximum amount an insurer will pay under a liability policy for claims for damages caused by the insured in the operation of a business.

AGGREGATE PROTECTIVE LIABILITY Maximum amount an insurer will pay under a liability policy for claims arising from acts of independent contractors.

AGGREGATE STRENGTH Strength derived by totaling the individual breaking strengths of the elements of a strand or rope. This strength does not recognize the reduction in strength resulting from the angularity of the elements in the rope, or other factors that may affect efficiency.

AGGREGATE TRANSPARENCY Mottled surface appearance of hardened concrete having dark-colored areas corresponding in shape and size to the coarse aggregate used in the mix.

AGGRESSIVE SOLUTION Any chemical that attacks the material on which it is spread: in concrete, a chemical that attacks the cement matrix, and to a lesser extent, the aggregate or hardened concrete. *Also called* **Acid attack.**

AGING 1. Process in which a material or complex changes over time: in some cases improving as it gets older, in others deteriorating; **2.** Analysis of accounts receivable in which the amounts are classified according to the length of time they have been outstanding, or due.

AGING PROCESS *See* **Age/strength relationship.**

AGITATING SPEED Critical speed at which the drum and blades of a concrete truck mixer should rotate for proper mixing.

AGITATING TRUCK Vehicle in which freshly mixed concrete can be conveyed from the site of mixing to the site of placement while being agitated. The truck body can either be stationary and contain an agitator or it can be a drum rotated continuously so as to agitate the contents.

AGITATION 1. Shaking or stirring so as to mix various constituents; **2.** Process of providing a gentle motion in mixed concrete being carried in a mixer truck, just sufficient to prevent segregation or loss of plasticity; **3.** Mixing and homogenization of slurries or finely ground powders by either mechanical means or injection or air. *See also* **Agitator.**

AGITATOR Device for maintaining plasticity and preventing segregation of mixed concrete by agitation. *See also* **Agitation.**

AGL *Abbreviation for:* above ground level.

A-GLASS Soda-lime-silica glass used for making drawn glass fibers. *See also* **Filament reinforcement.**

agn *Abbreviation for:* again.

agnt *Abbreviation for:* agent.

AGONIC LINE Line on a map along which the magnetic declination is zero.

agr. *Abbreviation for:* agricultural.

AGREEMENT Written agreement between two parties, i.e., between the owner and contractor pertaining to the work to be performed by the contractor, and the obligations for payment by the owner, and that is part of the contract documents.

AGREEMENT OF SALE Contract between a seller and buyer signed and dated by both parties setting out the terms and conditions of the arrangement. *Also called* a **Contract of purchase, Purchase agreement,** and **Sales agreement.**

AGRICULTURAL DRAIN Unsocketed, unglazed, earthenware, plastic, or concrete tiles, usually about 100 mm (4 in.) in diameter, laid end-to-end without closing the joints so as to drain the subsoil.

agrmt *Abbreviation for:* agreement.

AGRO-FORESTRY Practice of raising trees and agricultural products such as forage and/or livestock on the same area at the same time.

AH *Abbreviation for:* age-hardened; airfield heliport; Allen head.

ahd *Abbreviation for:* ahead.

AHU *Abbreviation for:* air handling unit.

AID ACCURACY Accuracy of an aid or system component before it is integrated into a system or installed accurately or applied to a machine.

AIR *See* **Standard air.**

AIR ATOMIZATION Use of air to break a liquid fuel into small particles prior to combustion. *See also* **Atomization, Impact atomization,** and **Injection atomization.**

AIR BARRIER Material used in the house envelope to retard the passage of air.

AIRBLAST Sound pressure wave from a blast traveling through the atmosphere.

AIRBLAST FOCUSING Concentration of sound energy in a small region at ground level due to refraction of the sound waves back to earth from the atmosphere.

AIR BLEED Pressurized air extracted from a gas turbine engine.

AIR BLEEDER Device for removal of air from a hydraulic fluid line.

AIR-BLOWN ASPHALT *See* **Asphalt.**

AIR-BLOWN MORTAR *See* **Shotcrete.**

AIR BLOW PIPE Air jet used in shotcrete gunning to remove rebound or other loose material from the work area.

AIR BOMB Chamber capable of holding compressed air heated to an elevated temperature.

AIR BOMB AGING Means of accelerating changes in the physical properties of rubber compounds by exposing them to the action of air at an elevated temperature and pressure.

AIR BOX Ventilation tube in a mine.

AIR BRAKE Vehicle brake system in which controlled compressed air operates chambers at the wheel positions to provide the force required to bring the brake shoes into contact with the brake drums or the discs in contact with the rotor.

AIR BREATHER Device permitting air movement between atmosphere and the component in which it is installed.

AIR BUBBLE Air trapped under a sealed surface (as with paint) or within a solid during the mixing and setting process (as with gypsum plaster).

AIR CHANGE Replacement of one complete volume of air (house, rented/leased area, plant, etc.) by natural or mechanical means, measured in air changes per hour.

AIR CHECK Surface marking or depression that occurs due to air becoming trapped between material being cured and the mold or press surface.

AIR CHORD Small diameter wire rope used as part of an elevator driving mechanism on door hangers, operators, or gates.

AIR CIRCULATION Movement of air from one area to another, either through convection or under pressure in a circulation system.

AIR CLASSIFICATION Process employing an air stream to separate materials by difference in density and aerodynamic properties.

AIR CLASSIFIER Mechanical device using air currents to separate solid components into 'light' or 'heavy' fractions.

AIR CLEANER Device that filters particles from air entering an engine or machine.

AIR CLEANER (DRY) Filter element made from material such as paper, fabric, or metal.

AIR CLEANER (WET) Filter in which air is cleaned through an oil-bath filter.

AIR COLLECTOR Solar collector that uses air as the heat transfer medium.

air cond. *Abbreviation for:* air conditioning.

air cond. eng. *Abbreviation for:* air conditioning engineer.

AIR CONDITIONING Process of heating or cooling, cleaning, humidifying or dehumidifying, and circulating air.

AIR CONTAMINANTS Airborne substances, solid, liquid, or gaseous, that are considered deleterious to life forms, or to the operation of machinery.

AIR CONTENT Volume of air voids in cement paste, mortar, or concrete, exclusive of pore space in aggregate particles, usually expressed as a percentage of total volume of the paste, mortar, or concrete.

AIR CONTENT TEST Test to give the precise air content of a sample of fresh air-entrained concrete: consists of an air-entrainment meter filled with fresh concrete in three equal layers, each being compacted with up to 25 strokes of a steel slump test tamping rod or by vibrating. Water is then introduced and its level noted. A pressure of 0.1 N/mm^2 is applied and the reduction in water level noted to give the volume reduction in the sample.

AIR CONTROL SUSPENSION SYSTEM Arrangement of valves and tubing connected to a trailer air system, designed to regulate ride, or mounting height, of an air suspension.

AIR-COOLED BLAST-FURNACE SLAG *See* **Blast-furnace slag.**

AIR-COOLED ENGINE Engine that is cooled by

means of air being forced about its heated parts.

AIR-COOLED FURNACE WALL *See* **Wall.**

AIR COOLER Device that lowers the temperature of air between its inlet and outlet stages.

AIR CURE Vulcanization without the application of heat. *Also called* **Self cure.** *See also* **Hot air cure,** and **Vulcanization.**

AIR CUSHION Blasting technique where a charge is suspended in a borehole, and the hole tightly stemmed so as to allow a time lapse between detonation and the ultimate failure of the rock.

AIR DAM Flexible air-deflecting panel, usually located below the radiator support on an on-highway vehicle.

AIR DEFICIENCY Lack of sufficient air in an air-fuel mixture, to supply the quantity of oxygen required to completely oxidize the fuel.

AIR DEFLECTOR Aerodynamically designed device attached to a truck, body or trailer, to reduce air resistance.

AIR DETRAINING COMPOUND Additive to a concrete mix to remove surplus air entrainment caused by excess workability agents.

AIR DIFFUSER Device that converts a stream of air into a diffuse and irregular flow to promote the mixing of primary air with secondary air, typically within a room or space.

AIR DOORS Doors that regulate air currents in mines.

AIR DRIED *See* **Seasoning.**

AIR-DRIED LUMBER Lumber that has been stacked in sheds so that air can circulate and facilitate drying.

AIR DRILL Rotary or reciprocating drill driven by compressed air. *See also* **Jackhammer.**

AIR DRY Applied substance (such as paint) that has dried to the stage where no further solvent will evaporate from the exposed surface.

AIR DRYING Process of partially drying of a sample to bring its moisture content near to equilibrium with atmosphere in the room in which further reduction, division, and characterization of the sample are to take place.

AIR DRY LOSS Moisture gain or loss from a sample that has been partially dried to bring its moisture content close to equilibrium with the atmosphere in the room in which further reduction and division of the sample is to take place.

AIR DUCT Pipe, tube, or passageway for conveying air, normally associated with heating, ventilating, and air conditioning.

AIR EMISSIONS Airborne solid particulates (such as unburned carbon), gaseous pollutants (such as oxides of nitrogen or sulfur), or odors emanating from any of a broad variety of sources.

AIR-ENTRAINED CONCRETE Concrete containing an air-entraining agent that causes small air bubbles to be trapped within the mix. The resulting material is more resistant to freeze-thaw cycles. Such concrete is usually designated by the letter 'A' after type, i.e., Type IA, Type IIA, etc.

AIR-ENTRAINED HYDRAULIC CEMENT Hydraulic cement containing an air-entraining additive in sufficient amounts to entrain air in mortar within specified limits.

AIR ENTRAINING Process of injecting air into material in a manner such that it forms minute bubbles of consistent size and dispersion. *See also* **Air entrainment.**

AIR-ENTRAINING ADMIXTURE *See* **Admixture.**

AIR-ENTRAINING AGENT Addition for hydraulic cement; admixture for concrete or mortar that causes entrained air to be incorporated in the concrete or mortar during mixing, usually to increase its workability and frost resistance.

AIR ENTRAINMENT Incorporation of air in the form of minute bubbles (generally smaller than 1 mm (0.04 in.)) during the mixing of either concrete or mortar. *See also* **Air entraining,** and **Entrained air.**

AIR ENTRAINMENT METER Apparatus that gives a precise indication of the air entrained in a sample of concrete mix.

AIR ESCAPE Mechanism that permits excess air to discharge from a water pipe. *Also called* **Air valve.**

AIR EXCAVATION EQUIPMENT Equipment designed to force compressed air through a wand so as to remove soil from around pipe, cable or other buried services by means of jetting.

AIR/FUEL RATIO Amount of fuel and air taken in by an engine during any intake stroke, expressed as a ratio of weight of air and weight of fuel.

AIR GAP Vertical distance from the top of the flood rim (highest point water can reach in a fixture) to the faucet or spout that supplies water to the fixture.

AIR GRATING Metal register, often ornate or decorated, over an air duct where it enters a space or room.

AIR-GROUND DETECTION Forest fire detection system combining fixed coverage of designated areas by ground detectors, augmented by aerial reconnaissance.

AIR GUN In the building trades, a gun for blowing materials onto the surface of a wall or ceiling, such as for insulation purposes.

AIR HAMMER 1. *See* **Pile hammer; 2.** Machine hammer driven by compressed air.

AIR HEATER 1. Heat exchanger through which air passes and is heated by a medium of a higher temperature, such as hot combustion gases oR steam; **2.** Device for warming the charge of intake air to a diesel's cylinders, especially during cold weather.

AIR HOSE Flexible hose connection between a tractor and trailer, used to convey compressed air to the trailer brake system.

AIR HOSE DUMMY COUPLER Fitting at the rear of a tractor, or to a converter dolly, to which the loose ends of air brake hoses are connected when not in use.

AIR INJECTION Inspection method using a high-pressure air probe to detect separations in the layers of a tire.

AIR INTAKE SILENCER Device to muffle the sound of incoming combustion air and objectionable noise originating in the intake manifold.

AIR JAM Pneumatic tool, operated by compressed air, for bucking rivets.

AIR JETS Streams of high-velocity air that issue from nozzles in a furnace enclosure to provide turbulence, combustion air, or a cooling effect.

AIR LEGS Air-activated pipe jacks, used to position light rock drills and to maintain pressure on the drills while in operation. *Also called* **Feed legs.**

AIR LEVELING VALVE Automatic control device used on air suspensions to regulate the flow of air in and out of the air springs to maintain the predetermined ride or mounting height.

AIR LIFT 1. Equipment whereby slurry or dry powder is lifted through pipes by means of compressed air; **2.** Pressure method of cleaning loose material out of an open-ended, cased pile, caisson, or cofferdam cell. *See also* **Blow out of pile**; **3.** Pump for raising fluids by forcing air into the bottom of an open discharge pipe submerged in a well or tank.

AIR-LIFT AXLE *See* **Axle.**

AIR-LINE PRESSURE REGULATOR Regulator that transforms a fluctuating air pressure supply to provide a constant lower pressure output.

AIR LOCK 1. Chamber that can be made airtight, separating two other rooms, chambers, or areas; **2.** Bubble of air, located in a pipe, that prohibits the movement of liquid.

AIR MAINTENANCE DEVICE Device that automatically maintains the required air pressure in a sprinkler system.

AIR METER Device for measuring the air content of concrete and mortar.

AIR MOTOR Device that converts compressed gas into mechanical force and motion, usually providing rotary mechanical motion.

AIR-OIL SYSTEM Hydraulic system employing air and oil under pressure to cause activation.

AIR-OIL TANK Storage vessel in an air-oil system that works as a reservoir from which oil is supplied under pressure.

AIR OVEN AGING Means of accelerating a change in the physical properties of rubber compounds by exposing them to the action of air at an elevated temperature at atmospheric pressure.

AIR-OVER-HYDRAULIC BRAKE *See* **Brake.**

AIR PERMEABILITY Property that allows the passage of air through a mass.

AIR-PERMEABILITY TEST Procedure for measuring the fineness of powered materials, such as portland cement.

AIR POCKET Space or void, created by trapped air, that accidentally occurs in concrete work or in a pipe line.

AIR POLLUTION Presence of unwanted material in the atmosphere, which includes any material present in sufficient concentrations for a sufficient time, and under circumstances to interfere significantly with the comfort, health, or welfare of persons, or with the full use and enjoyment of property.

AIR PORT Port on a diesel hammer that serves as air intake and exhaust port. *Also called* **Exhaust port.**

AIR PRESSURE WATER TANK Water storage tank, usually in the range of 100 to 200 gallons, in which air pressure from a mixer truck air system is used to pressurize the tank. The water is used to maintain the consistency of freshly mixed concrete and/or washdown the hopper and chutes after delivery of the payload.

AIR QUALITY CRITERIA Levels of pollution and lengths of exposure to pollution above which negative effects on human health and/or welfare may occur.

AIR QUALITY STANDARDS Level of pollutants prescribed by law that cannot be exceeded during a specified time in a defined area.

AIR RECEIVER Air storage tank of an air compressor.

AIR REGISTER 1. Grill mounted in the floor through which heated air is exhausted under pressure into a room or other space; **2.** Type of burner mounting that may admit secondary air to the combustion space.

AIR RESERVOIR Commonly identified as a wet or dry air tank (or two-stage dual compartment) that supplies air-operated vehicle equipment. The wet tank is supplied by an air compressor and accumulates the moisture condensation that accompanies the pressure-temperature changes of the compression-expansion cycles. The wet tank feeds the dry tank through a check valve, and the dry tank becomes the primary air source.

AIR RESISTANCE Measure of the retarding effect caused by air pressure produced by a vehicle in motion. This force is greatly affected by the speed of the vehicle and the speed and angle of the wind. Air resistance is negligible at speeds below 32 km/h (20 mph), but becomes a major resistance force at speeds over 72 km/hr (45 mph).

AIR RESTRICTOR INDICATOR Device applied in conjunction with a dry-type air cleaner to determine the maintenance interval of a filter cartridge.

AIR-RIDE SUSPENSION *See* **Suspension.**

AIR RIGHTS Property rights for the control or specific use of a designated airspace.

AIR RING Perforated manifold in the nozzle of wet-mix shotcrete equipment through which high-pressure air is introduced into the material flow.

AIR SEASONED Timber that is naturally dried and seasoned in the open air, protected from sun and rain.

AIR SEPARATOR Apparatus that pneumatically separates various size fractions of ground materials (fine particles are discharged as product; oversize is returned to the mill as tailing).

AIRSHAFT Unroofed area within the perimeter of a building into which windows and vents open to provide natural ventilation.

AIR SHIFTING Process that uses air pressure to engage different range combinations in a transmission's auxiliary section without a mechanical linkage to the driver.

AIR SHUTTER Adjustable shutter on a burner air

register by means of which the amount of air induced into a furnace through the register can be controlled.

AIR-SLACK Condition where a soft-body clay, after absorbing moisture and being exposed to the atmosphere, will spall a piece of clay and/or glaze.

AIR SLAKED Wetted by exposure to moisture in the atmosphere.

AIR SPACE 1. Cavity or space in walls, windows, or other enclosed parts of a building between various structural members; **2.** Landfill's remaining volume or capacity for waste disposal at a given time, not including volume occupied by cap, cover, and liner.

AIR SPRING *See* **Spring.**

AIR STARTING SYSTEM One that utilizes compressed air for engine or turbine starting.

AIR/STEAM HAMMER Impact pile driver powered by compressed air or steam; machine action may be single, double, differential, or compound.

AIR SUPPORTED STRUCTURE Structure supported by the positive air pressure maintained within its shell (usually consisting of a thin, flexible membrane).

AIR SURVEY Survey using aerial photography.

AIR TANKER Aircraft adapted to carry and discharge water and/or fire retardants.

AIRTIGHT Not capable of permitting the passage of air at designed pressures.

AIRTIGHTNESS Ability of the building envelope to resist infiltration and exfiltration of air.

AIRTIGHT WOOD STOVE Closed and sealed metal container, often decoratively finished, having one or more means of controlling air input and fume exhaust. Heat distribution is by radiation from the outer casing.

AIR-TO-AIR AFTERCOOLING Aftercooling system in which turbocharged intake air passes through a special heat exchanger mounted ahead of the radiator, to be cooled by outside air before reaching the intake manifold.

AIR-TO-AIR RESISTANCE Resistance to the flow of air, or the mixing of air, due to such factors as laminar flow, pressure differential, temperature differential, or other factors.

AIR TRACK Air-propelled, track-mounted rock drill.

AIR TUGGER Small winch run by compressed air.

AIR VALVE *See* **Air escape.**

AIR VESSEL Tank containing air under pressure.

AIR VOID Space in cement paste, mortar, or concrete filled with air (an entrapped air void is characteristically 1 mm (0.04 in.) or more in size and irregular in shape; an entrained air void is typically between 10 -mm and 1 mm (0.004 in.) in diameter and spherical, or nearly so).

AIR WASHER System for cleaning air, usually incorporating a water spray system.

AIR-WATER JET High-velocity jet of air and water mixed at the nozzle, used in the cleanup of surfaces of rock or concrete.

AIR WAVES Airborne vibrations or pulses, usually (but not always) accompanied by sound waves.

AIRWAY Space intended for the free movement of air, typically between roof insulation and roof sheathing.

AIR WELL Space within a building, enclosed by walls, partially or totally open to the outside air at the roof, and intended solely as a means of ventilation.

AISLE Part of a church, parallel to and divided by piers or columns or a screen wall from the nave, choir, or transept.

a.l. *Abbreviation for:* all length.

AL *Abbreviation for:* acceleration level.

alab. *Abbreviation for:* alabaster.

ALABASTER Massive, densely crystalline, softly textured form of practically pure gypsum.

ALARM BELL Bell that is rung automatically by a device set to detect an out-of-limit condition, or that can be manually activated to alert to a situation.

ALARM SIGNAL Signal indicating an emergency requiring immediate action, such as an alarm for fire from a manual box, a water flow alarm, an alarm from an automatic fire alarm, etc.

ALARM SWITCH *See* **Switch.**

ALARM VERIFICATION Automatic fire detection and alarm system that must report or confirm alarm conditions for a minimum period after being reset to be accepted as a valid alarm initiation signal.

ALBERTS LAY *See* **Lay types.**

ALBRONZE Alloy of aluminum and copper.

alc. *Abbreviation for:* alcohol.

ALCOHOL-BASE SOLVENTS Family of organic compounds used as solvents with shellacs and lacquers.

ALCOHOL EVAPORATOR Device that serves to keep a vehicle's air lines and valves free of ice during sub-freezing temperatures. Located between the compressor and wet tank, the unit permits controlled levels of vaporized alcohol to mix with the air stream, thus circulating into the tank and throughout the system. *Also called* an **Alcohol Injector.**

ALCOHOL INJECTOR *See* **Alcohol evaporator.**

ALCOVE Recess in a room, having the same floor level but not necessarily a continuous ceiling.

ALDEHYDE Specific chemical compound; a class of similar compounds.

ALERT-TO-STOP Section of an elevator control system that stops the car in response to a call.

ALGAL BLOOM Population explosion of algae in surface waters.

algn *Abbreviation for:* align.

algn. *Abbreviation for:* alignment.

ALIDADE Upper part of a surveying instrument. *Also called* a **Sight rule.**

ALIENATION Voluntary (by the owner) or involuntary (such as through condemnation or expropriation) transfer of title to and possession of real property.

ALIFORM Wing-like extensions, as to an abutment.

ALIGN Install and/or adjust to be in a straight line.

ALIGNED FIBERS Continuous short fibers aligned along the length of a cement paste or concrete member. *See also* **Filament reinforcement.**

ALIGNING CLIP Clip designed to fit over two similar sections with means to adjust one to align with the other, or to fit over a section and permit a second element to be aligned in or on the first.

ALIGNMENT Ground plan for the layout of a road, railway, canal, etc.

ALIGNMENT CHART FOR COLUMNS Nomograph for determining the effective length factor K for some types of steel columns.

ALIGNMENT WIRE *See* **Ground wire.**

A-LINE Dimensioned line in a tunnel beyond which rock projections are not permitted.

ALITE Name used to identify tricalcium silicate including small amounts of MgO, Al_2O_3, Fe_2O_3, and other oxides; a principal constituent of portland-cement clinker.

ALIVE (OR LIVE) Electrically connected to a source of potential difference, or electrically charged so as to have a potential significantly different from that of the earth in the vicinity. *Also called* **Current carrying,** and **Energized.**

alk. *Abbreviation for:* alkaline.

ALKALI Salts of alkali metals, principally sodium and potassium occurring in constituents of concrete and mortar. *See also* **Low-alkali cement.**

ALKALI-AGGREGATE REACTION Chemical reaction in either mortar or concrete between alkalis (sodium and potassium) from portland cement and other sources and certain constituents of some aggregates. Under certain conditions, deleterious expansion of concrete or mortar may result. *Also called* **Aggregate/alkali reaction.**

ALKALI-CARBONATE ROCK REACTION Reaction between the alkalis (sodium and potassium) in portland cement and certain carbonate rocks, particularly calcitic dolomite and dolomitic limestones, present in some aggregates. The products of the reaction may cause abnormal expansion and cracking of concrete in service.

ALKALI FLAT Flat area, usually the bottom of an undrained basin in an arid region, containing an excess of alkali.

ALKALINE STORAGE BATTERY *See* **Nickel-iron storage battery.**

ALKALI REACTIVITY (OF AGGREGATE) Susceptibility of aggregate to alkali-aggregate reaction.

ALKALI-SILICA REACTION Reaction between the alkalis (sodium and potassium) in portland cement and certain siliceous rocks or minerals, such as opaline chert, strained quartz, and acidic volcanic glass, present in some aggregates. The products of the reaction may cause abnormal expansion and cracking of concrete in service.

ALKYD Polyester resins made with some fatty acids as a modifier; a paint-thinner-soluble painting medium.

ALKYD PAINT Quick-drying paint with a polyester resin binder.

ALKYL ARYL SULFONATE Synthetic detergent from petroleum fractions.

ALL AGED Forest or stand containing trees of almost all age classes up to and including trees of harvestable age.

ALL-CONVENTIONAL DUMP TRAILER *See* **Trailer.**

ALLEN KEY Hexagonal-shaped bar with one long and one short leg at right angles. Available in various face sizes.

ALLEN WRENCH *See* **Wrench.**

ALLEY Narrow lane between houses or buildings giving access to their rear.

ALL HEART Lumber, free of sapwood.

ALLIGATOR CRACKS Interconnected cracks, as in an asphalt pavement or paint film, that form a pattern similar to an alligator's skin.

ALLIGATOR HOOD Truck design where access to the engine compartment is gained through an opening that hinges at the rear and raises at the front. *See also* **Butterfly hood.**

ALLIGATORING Failure of a paint film that resembles an alligator skin.

ALL-IN AGGREGATE Mixture of coarse and fine aggregate.

ALL LIVE TREES *See* **Tree classes.**

alloc. *Abbreviation for:* allocate; allocated.

ALLODIAL SYSTEM U.S. legal system that allocates to individuals full rights of property ownership.

allot *Abbreviation for:* allotment.

ALLOTMENT Action taken by administrative authority making funds available for obligations and expenditures for specified purposes and for certain periods. *See also* **Apportionment,** and **Appropriation.**

ALLOWABLE ANNUAL CUT Average volume of wood that may be harvested annually under sustained yield management: roughly equal to the amount of new growth produced by the forest each year including a proportion of the mature volume less deductions for losses due to fire, insects, and disease.

ALLOWABLE BEARING CAPACITY Maximum unit pressure to which a soil or other material should be subjected to guard against shear failure or excessive settlement. *See also* **Bearing capacity.**

ALLOWABLE BEARING PRESSURE Maximum allowable net loading intensity at the base of a foundation, taking account of the ultimate bearing capacity, amount and kind of settlement expected, and ability of a given structure to take up such settlement.

ALLOWABLE BODY-PAYLOAD Weight rating designated by a truck manufacturer for model types that are later equipped with some type of body (stripped chassis, chassis-cowl, or chassis-cab models, for example). This is the combined allowance for total weight of body and payload together.

ALLOWABLE CUT Volume of timber that may be harvested during a given period to maintain sustained production.

ALLOWABLE-CUT EFFECT Allocation of anticipated future forest timber yields to the present allowable cut. This is employed to increase current harvest levels (especially when constrained by even flow) by spreading anticipated future growth over all the years in rotation.

ALLOWABLE LOAD Ultimate load divided by the factor of safety. *See also* **Safe leg load.**

ALLOWABLE PAYLOAD Maximum load weight that may be carried by a truck without exceeding the manufacturer's designated maximum rating, or legal limit.

ALLOWABLE PILE LOAD Load permitted on any vertical pile, applied concentrically in the direction of its axis.

ALLOWABLE ROPE LOAD Nominal breaking strength of a rope, divided by a design factor.

ALLOWABLE SETTLEMENT Settlement for a specific structure that is limited by (a) allowable total settlement, and (b) allowable differential settlement, both for the structure as a whole and between parts of the structure.

ALLOWABLE STRESS Maximum permissible stress used in the design of members of a structure and based on a factor of safety against rupture or yielding of any type.

ALLOWANCE 1. When accepted by the owner, an allowance consists of the estimated cost of a specific category of materials or equipment that has yet to be finally specified or selected but that is to be furnished by the contractor; **2.** Specified difference in limiting sizes between mating parts; **3.** Rated tightness or looseness of mating parts; **4.** Sum granted as a reimbursement or repayment; **5.** Deduction from a gross weight or value.

ALLOY Metal to which trace elements are added to improve some of its desired qualities, i.e., the addition of trace elements such as copper, nickel, or chromium to steel, or the addition of manganese or copper to aluminum.

ALLOYING ELEMENTS Chemical elements added for improving the properties of finished products. Those present in quantities over 1% are:

In **alloy steels**: Nickel, chromium, manganese, molybdenum, vanadium, silicon, and copper.

In **super alloys (nickel base)**: Chromium, tungsten, molybdenum, cobalt, columbium, titanium, aluminum, and iron.

In **super alloys (cobalt base)**: Chromium, nickel, tungsten, molybdenum, columbium, titanium, and iron.

In **titanium alloys**: Aluminum, manganese, vanadium, chromium, tin, and zirconium.

ALLOY PIPE *See* **Pipe.**

ALLOY STEEL *See* **Steel.**

ALL-PURPOSE COMPOUND Drywall joint material that can be used as a bedding compound for tape, a finishing compound, laminating adhesive, or texturing compound.

ALL-PURPOSE TREAD Tread design suitable for on- or off-highway use at any wheel position.

ALL-RISK INSURANCE *See* **Insurance.**

ALL STRETCHER BOND *See* **Bond.**

ALL-WEATHER WOOD FOUNDATION Technique of using specially treated wood and plywood to replace poured concrete and concrete block foundations.

ALLUVIAL FAN Sloped, spreading deposit of boulders, gravel, and sand left by a stream where it emptied from a more confined channel.

ALLUVIAL SOIL Soil, sand, or gravel deposited by flowing water.

ALLUVIUM Sediments deposited by running or flowing water.

alm *Abbreviation for:* alarm.

aln *Abbreviation for:* align.

ALPHANUMERIC DISPLAY Type of digital readout that displays both letters and numbers, separately or mixed.

alt. *Abbreviation for:* alternate; alternating; alternator; altitude.

ALTERATION WORK Construction work where the state of an existing element is described together with written and graphic details of how the element is to be changed.

ALTERNATE BAY CONSTRUCTION Concrete for a large horizontal structure placed in alternate bays to control the development of cracks, allowing each bay to take up its initial shrinkage before the adjoining bays are placed.

ALTERNATE BID Amount stated in a bid to be added to or subtracted from the amount of the base bid under specified conditions, such as substitution of materials or change in scope of the project or method of construction. *Also called* **Additive alternate.**

ALTERNATE CAPACITY Capacity, other than basic capacity, of a piece of mechanical equipment such as a fork lift that adjusts for variations such as load center, fork height, and application of an attachment.

ALTERNATE LANE CONSTRUCTION Highway and large slab construction technique in which alternate lanes are constructed and allowed to harden followed by construction of the intermediate lanes.

ALTERNATE LAY *See* **Lay types.**

ALTERNATE LOAD CENTER Load center other than that used to describe the rated capacity.

ALTERNATE SOURCE OF POWER *See* **Emergency power.**

ALTERNATING CURRENT Electricity that reverses its direction of electron flow regularly.

ALTERNATING VOLTAGE Electrical voltage that goes through a series of different values, positive and negative, during a given period, that is continuously repeated.

ALTERNATOR Device for converting mechanical energy into alternating-current electrical energy.

ALTIMETER Aneroid barometer graduated to show elevation.

ALTITUDE (ANGLE) Vertical angle between the plane of the horizon and a line drawn to a fixed object. If above the horizon, the altitude is positive and is termed 'an angle of elevation'; if below it is negative and is termed 'an angle of depression.'

ALTITUDE LEVEL Level tube on the vertical circle of a theodolite, used to measure vertical angles.

ALTITUDE RATING Power recommended by a manufacturer for satisfactory operation at a given altitude.

altrn *Abbreviation for:* alteration.

alum *Abbreviation for:* aluminum.

alum. *Abbreviation for:* aluminized.

ALUMINA Aluminum oxide; mineral contained in clay used for brick making.

ALUMINA DIASPORE FIRECLAY *See* **Brick.**

ALUMINA FIBER Polycrystalline alumina filaments used as reinforcement for cement paste and concrete.

ALUMINA RATIO Ratio of the percentage by mass of alumina to that of ferric oxide in a sample of portland cement.

ALUMINATE CEMENT *See* **Calcium-aluminate cement.**

ALUMINATE CONCRETE Concrete made with calcium-aluminate cement, used primarily where high-early-strength and refractory or corrosion-resistant concrete is required.

ALUMINA WHITEWARE Any ceramic whiteware in which alumina is the essential crystalline phase.

ALUMINIZED Coating a metal by dipping it in a solution of molten aluminum.

ALUMINOUS CEMENT *See* **Calcium-aluminate cement.**

ALUMINUM Silvery, lightweight, easily worked metal that resists corrosion. *See also* **Alloying elements.**

ALUMINUM BRASS Brass to which aluminum has been added to improve resistance to corrosion.

ALUMINUM BRONZE Copper aluminum alloys, 4% to 11% aluminum, offering high tensile strength and resistance to corrosion, cast or cold worked.

ALUMINUM FOIL Thin, sheet aluminum used as an insulation.

ALUMINUM HYDRAULIC SHORING Pre-engineered shoring system comprised of aluminum hydraulic cylinders (cross braces) used in conjunction with vertical rails (uprights) or horizontal rails (walers), designed specifically to support the sidewalls of an excavation and prevent cave-ins.

ALUMINUM NAIL *See* **Nail.**

ALUMINUM OXIDE Abrasive used in sandpaper, sharpening stones, and grinding wheels. It is made by fusing bauxite clay in an electric furnace.

ALUMINUM PIPE Pipe fabricated of an aluminum alloy, restricted for use to service where no chemical reaction can occur, usually irrigation and potable water supply.

ALUMINUM STEARATE Complex salt or soap of aluminum and stearic acid used as a flatting agent, antisetting agent for pigments in paint and varnish, water repellents, and cement additives.

aly *Abbreviation for:* alloy.

aly st. *Abbreviation for:* alloy steel.

am. *Abbreviation for:* ammeter.

a.m. *Abbreviation for: ante meridiem* (before noon).

amb. *Abbreviation for:* ambient.

amdmt *Abbreviation for:* amendment.

AMBIENT AIR The surrounding air.

AMBIENT LIGHTING General lighting level, sufficient for all anticipated purposes, but which may be augmented by area or task lighting.

AMBIENT MOISTURE Amount of moisture permissible in the air surrounding a shipment of secondary material.

AMBIENT TEMPERATURE Surrounding air temperature.

AMBULATORY Covered or sheltered place for walking.

AMD *Abbreviation for:* air-moving device.

AMENDMENT 1. Written agreement signed by the owner and contractor after the date of the original agreement that specifically amends one or more provisions of the contract documents; **2.** Substance added to a soil to improve its physical properties such as texture, as opposed to fertilizer, that is added to improve chemical properties.

AMENITY AREA Area or areas within the boundaries of a project intended for recreation purposes and that may include landscaped site areas, patios, common areas, communal lounges, swimming pools, and areas for similar purposes.

AMERICAN BOND *See* **Bond.**

AMERICAN STANDARD WIRE GAUGE System of designating the diameter of wires of nonferrous metals by the use of numbers.

AMERICAN TABLE OF DISTANCES Table showing distances that explosives must be stored from other explosives, inhabited buildings, railroads, highways, and magazines, according to the amount of explosives stored. *Also called* **Quantity distance tables.**

AMERICAN WIRE GAUGE Gauge used for designating the diameter of solid copper wire. It is the same as the Brown & Sharp gauge.

AMMETER Gauge for measuring amperes in an electrical system.

AMMONAL Powdered nitroglycerine explosive.

AMMONIACAL COPPER ARSENITE Waterborne wood preservative consisting of copper hydroxide, arsenic trioxide, and acetic acid.

AMMONIUM CHLORIDE Chemical used with ammonium hydroxide and whiting to remove aluminum, bronze, and copper stains from the surface of concrete.

AMORPHOUS Lacking definite form.

amort. *Abbreviation for:* amortization.

AMORTIZATION 1. Process of writing off a charge, as for an item of equipment or machinery, by prorating its capital cost over time; **2.** Repayment of debt, principal and interest, in (usually) equal installments at regular periods over a given period; **3.** Procedure by which the capital cost of forestry-related projects, such as roads or bridges, is written off over a longer period as the timber volumes developed by the projects are harvested and extracted.

AMORTIZED COST Original cost of an asset less any portion that has been amortized or treated as an expense or loss.

AMOUNT OF MIXING Designation of extent of mixer action employed in combining the ingredients for either concrete or mortar. In the case of stationary mixers, the mixing time; in the case of truck mixers, the number of revolutions of the drum at mixing speed after the intermingling of the cement with water and aggregates. *See also* **Mixing time.**

AMOUNT SUBJECT Maximum value covered by an insurance policy that may be reasonably lost under an individual claim.

amp. *Abbreviation for:* ampere.

AMPACITY Current in amperes a conductor can carry continuously under the conditions of use without exceeding its temperature rating.

AMPERE One of the seven base units of the SI system of measurement: a unit of electrical current equal to that current which, if maintained in two straight parallel conductors of infinite length, or negligible circular cross-section, and placed 1 m apart in vacuum, would produce between these conductors a force equal to 0.2 -μN/m of length. Symbol: A. *See also the appendix:* **Metric and nonmetric measurement.**

AMPERE-HOUR Derived unit of electric charge over a specific length of time at 26.6°C (80°F). Symbol: A/hr. Multiply by 3.6 to obtain kilocoulombe, symbol: kC. *See also the appendix:* **Metric and nonmetric measurement.**

AMPERE PER METER A derived unit of magnetic field strength with a compound name of the SI system of measurement. Symbol: A/m. *See also the appendix:* **Metric and nonmetric measurement.**

AMPERE TURN Unit of magnetomotive force equal to one ampere flowing through one turn of a coil.

AMPHIBOLITE Coarse-grained, metamorphic rock containing more than 50% ferromagnesian minerals.

AMPHITHEATER Round or oval arena or building with an open space surrounded by rising rows of seats. (*See also* **Arena.**)

amp.hr *Abbreviation for:* ampere-hour.

ampl. *Abbreviation for:* amplifier; amplify; amplification.

AMPLIFICATION Process of increasing the strength of a signal.

AMPLIFICATION FACTOR Multiplier of the value of moment or deflection in the unbraced length of an axially loaded steel member to reflect the secondary values generated by the eccentricity of the applied axial load within the member.

AMPLIFIER Device for increasing the power or strength of electric waves or pulses.

AMPLITUDE Extreme range of a fluctuating quantity; in blasting, the height of a vibration or wave above the zero line on a vibration record.

AMS *Abbreviation for:* all machine screws.

amt *Abbreviation for:* amount.

AMYL ACETATE Slow acting lacquer solvent.

AN *Abbreviation for:* alteration notice.

ANAEROBIC Without oxygen; activity occurring in the absence of oxygen.

ANAEROBIC ADHESIVES *See* **Adhesives.**

ANAEROBIC BACTERIA Bacteria that live and are active in the absence of free oxygen; the working bacteria in an anaerobic waste disposal system such as a septic tank.

ANAEROBIC DIGESTION Breakdown of organic components by microbial components in the absence of oxygen.

anal. *Abbreviation for:* analog; analysis.

ANALOG Of or pertaining to the general class of devices whose output varies as a continuous function of its input.

ANALOG DISPLAY Display that provides a continuous indication relative to reference points.

ANALOGOUS ARTICLES Articles not detailed in a classification but having characteristics similar to those classified.

ANALOG PROCESSING Method of processing information using the representative value of a physical variable. (Digital processing, that uses direct numerical values, is many times faster and more reliable.)

ANALYSIS Investigation of a situation through detailed consideration of its essential elements. There are many approaches, including:

Benefit/cost: Evaluation of the profits, income, output, or other benefits against the cost of obtaining them.

Correlation: Measurement of the degree of relationship, if any, between variables.

Cost/benefit: *See* **Benefit/cost** above.

Cost/volume/profit: Study of the effects of changes in fixed costs, variable costs, sales quantities, sales prices, and/or sales mix.

Incremental: Evaluation of the changes, usually in revenue and expenses or cash flow, that would result from a course of action.

Input/output: Summary of the transactions between all economic units involved in a project, showing those consumed and the resulting product.

Network: Method of planning and scheduling a

project, commonly displayed in diagrammatic form, so as to identify the interrelated sequences that must be completed to finish a project.

Sensitivity: Measurement of the degree of change in one variable as a response to a measured change in another variable.

ANALYSIS SAMPLE Final subsample prepared from an air-dried laboratory sample but reduced by passing through a mill with a 0.5 mm (0.02 in.) or smaller final screen.

ANALYSIS SIEVED Particle size distribution, usually expressed as the weight percentage retained upon each of a series of standard sieves of decreasing size and the percentage passed by the sieve of the finest size. *See also* **Grading.**

anch. *Abbreviation for:* anchor.

ANCHOR 1. Piece of material, usually metal, used to attach building parts. *See also* **Wall anchor;** **2.** In prestressed concrete, to lock the stressed tendon in position so that it will retain its stressed condition; **3.** In precast concrete construction, to attach the precast units to the building frame; **4.** In slabs on grade or walls, to fasten to rock or adjacent structures to prevent movement of the slab or wall with respect to the foundation, adjacent structure, or rock. *See also* **Form anchor;** **5.** Supporting device for glass fixed to wall surfaces consisting of an L-shaped strip of metal; **6.** Any stump, tree, deadman, or rock used to secure a guyline.

ANCHORAGE 1. In post-tensioning, a device used to anchor the tendon to the concrete member; **2.** In pretensioning, a device used to maintain the elongation of a tendon during the time interval between stressing and release; **3.** In precast concrete construction, the device for attaching precast units to the building frame; **4.** In slab or wall construction, the device used to anchor the slab or wall to the foundation, rock, or adjacent structure; **5.** Hooks, bends, or embedment lengths that secure reinforcing bars in cast-in-place concrete. *Also called* **Anchorage device.**

ANCHORAGE BOND STRESS Bar forces divided by the product of the bar perimeter or perimeters and the embedment length. *See also* **Anchorage distance.** *Also called* **Development bond stress.**

ANCHORAGE DEFORMATION OR SEATING Loss of elongation or stress in the tendons of prestressed concrete due to the deformation or seating of the anchorage when the prestressing force is transferred from the jack to the anchorage. Also known as **Anchorage loss.**

ANCHORAGE DEVICE *See* **Anchorage.**

ANCHORAGE DISTANCE Distance behind a wall at which an anchor block must be placed so as to ensure that it will not slip relative to the tie wall and that it anchors effectively. *See also* **Anchorage bond stress.**

ANCHORAGE LOSS *See* **Anchorage deformation or seating.**

ANCHORAGE ZONE 1. In post-tensioning, the region adjacent to the anchorage subjected to secondary stresses resulting from the distribution of the prestressing force; **2.** In pretensioning, the region in which the transfer bond stresses are developed.

ANCHOR BAR Part of a tow sling or hitch that is placed under or against a vehicle to be towed, fastened to the vehicle with two chains.

ANCHOR BEARING PLATE Steel bearing plate through which tapered holes are drilled to take prestressed concrete tendon anchor wedges.

ANCHOR BLOCK 1. Block of wood built into masonry walls to which partitions and fixtures may be secured; **2.** Massive concrete block buried in the ground to provide an anchorage for tie rods, guy wires, or suspension bridge cables.

ANCHOR BOLT 1. Bolt that fastens columns, girders and other structural steel members, or wooden sill plates to concrete or masonry; **2.** Metal bolt or stud, headed or threaded, either cast in place, grouted in place, or drilled into finished concrete, used to hold various structural members or embedments in the concrete, and to resist shear, tension, and vibration loadings from various sources such as wind, machine vibration, etc. *Also called* **Foundation bolt,** and **Hold-down bolt.**

ANCHOR CABLE Line used to tie down a yarder to prevent tipping on a heavy pull.

ANCHORED WALL Wall that is fixed at its base but supported against overturning by tie rods or struts near its top.

ANCHOR LOG Log buried in the ground used to secure a guyline. *Also called* a **Deadman.**

ANCHOR NUT *See* **Nut.**

ANCHOR PILE *See* **Pile.**

ANCHOR PIN Steel rod, 38 mm to 50 mm (1.5 in. to 2 in.) in diameter, having an eye or threaded end, grouted or wedged into a rock hole, used as a guyline anchor.

ANCHOR PLATE Metal piece, often perforated, through the center of which hangers may be inserted.

ANCHOR POINT Advantageous location from which construction of a fire line can begin.

ANCHOR TENANT Principal tenant in a commercial or retail complex.

ANCHOR TREE Tree used as a guyline anchor.

'AND' DEVICE Control device that has its output in the logical '1' state if all the control signals assume the logical '1' state.

ANDIRON Metal frame used to hold logs above the hearth of a fireplace to increase air circulation necessary for combustion.

ANEMOMETER Instrument for measuring fluid velocity.

ANEROID Without liquid; device on a diesel engine that compensates for the thinner, less dense air found at higher altitudes, and that leans out the amount of fuel delivered when the air is less dense.

ANFO Ammonium nitrate-fuel oil mixture, used as a blasting agent.

ang. *Abbreviation for:* angle.

ANGLE Difference in direction between two lines that meet at a single point.

ANGLE BAR Vertical bars at the angles of windows.

ANGLE BEAD Small wood or metal molding placed at an external angle formed by plastered surfaces in order to preserve the corner from accidental fracture. *Also called* **Angle staff.**

ANGLE BRACE Tie fixed across the inside of an angle in a framework in order to stiffen the framework of a structure. *See also* **Pipe brace.**

ANGLE BRACKET Type of support that has two faces, usually at right angles to each other, used to increase the strength of a joint or juncture.

ANGLE BRICK *See* **Brick.**

ANGLE BUTTRESS *See* **Buttress.**

ANGLE CLOSER Cut brick used to close the bond at a corner.

ANGLE DIVIDER 1. Measuring tool consisting of three blades, two of which are hinged from the top of a central shaft, the other being at right angles and centred on the bottom of the shaft; **2.** Tool used by a tilesetter to determine the degree of an angle to cut.

ANGLEDOZER Tractor dozer whose blade can be pivoted about a vertical center pin so as to cast its load to one side or the other.

ANGLE FLOAT Finishing tool having a surface bent to form a right angle, used to finish reentrant angles in plaster, mortar, stucco, concrete, etc.

ANGLE FIFTH *See* **Fifth wheel.**

ANGLE GAUGE Hand tool used to set off and measure angles on work.

ANGLE INDICATOR Mechanical or electrical crane accessory that measures the angle of the boom above horizontal.

ANGLE IRON Structural steel shape in the form of a 90° angle, frequently used to support masonry over a span such as a door or window opening.

ANGLE IRON GUIDE *See* **Side channel.**

ANGLE JIB Multiple-section boom-tip extension supported by pendants and assembled from steel angle iron.

ANGLE JOINT Jointing of members not in the same straight line or in the same plane.

ANGLE OF BEND Inclusive angle formed by a curved connection, measured by radial lines from the beginning and end of the bend to the center.

ANGLE OF CONTACT Portion of a sheave contacted by a rope, measured in degrees of contact. *Also called* **Arc of contact, Angle of contact,** and **Angle of wrap.**

ANGLE OF INTERNAL FRICTION *See* **Angle of natural repose.**

ANGLE OF LAY Angle developed at the intersection of a structural element and a line parallel to its axis.

ANGLE OF NATURAL REPOSE Angle to the horizontal at which a material will no longer be affected by gravity and slide downward of its own accord. *Also called* **Angle of internal friction.**

ANGLE OF NIP In a roll crusher, the angle between

tangents to the roll surfaces at the widest point at which they will grip a stone. *Also called* **Nip.**

ANGLE OF OPERATION Maximum deviation from horizontal at which an engine operates in a given application.

ANGLE OF THREAD Angle included between the sides of a thread measured in a plane through the centerline.

ANGLE OF TURN *See* **Intersection elements.**

ANGLE OF WALL FRICTION Angle between the direction of the resultant active earth pressure on the back of a retaining wall and the normal to the back of the wall. The angle is considered positive when the resultant reaction is oriented such that its tangential component acts in an upward direction.

ANGLE OF WRAP *See* **Angle of contact.**

ANGLE POST Corner post in post-and-beam or half-timbered construction.

ANGLE RAFTER *See* **Rafter.**

ANGLE RIB Curved piece forming a mitre for a curved ceiling.

ANGLES *See* **Error of closure.**

ANGLE SPRING CLUTCH *See* **Clutch.**

ANGLE STAFF *See* **Angle bead.**

ANGLE STOP VALVE *See* **Valve.**

ANGLE TIE Roof timber tying wall plates together and supporting one end of a dragon beam.

ANGLE TILE Plain roof tile shaped to a right angle, used at the ridge or hip.

ANGLE TROWEL Trowel used to finish an outside or inside corner.

ANGLE VALVE *See* **Valve.**

ANGSTROM Obsolete unit of length, equal to 0.1 nanometer, that should not be used with the SI system of measurement. Symbol: Å. *See also the appendix:* **Metric and nonmetric measurement.**

ANGULAR AGGREGATE *See* **Aggregate.**

ANGULARITY Angular difference between mating or adjacent parts, such as between the connecting rod and cylinder wall, or between the connecting rod and crankshaft throw of a reciprocating engine.

ANGULARITY NUMBER Number between 0 and 12 that gives the average roundness of particles in a sample of aggregate.

ANGULAR MISALIGNMENT Minor angle formed between the intersecting axes of two pipes.

ANGULAR PARTICLE Angular shape of some coarse aggregate particles.

ANGULAR ROTATION Bending of an expansion joint along its centerline.

ANGULATED ROPING System of stage suspension in which the upper rope sheaves or suspension points are closer to the plane of the building face than the corresponding attachment points on the stage, thus causing the unit to press against the face of the building during its vertical travel.

anhyd. *Abbreviation for:* anhydrous.

ANHYDRITE A mineral, anhydrous calcium sulfate; gypsum from which the water of crystallization has been removed, usually by heating above 160°C (325°F). Natural anhydrite is less reactive than that obtained by calcination of gypsum.

ANHYDROUS Devoid of water.

ANHYDROUS LIME Quicklime.

ANILINE POINT Lowest temperature at which a liquid is completely miscible with an equal volume of freshly distilled aniline.

ANIMAL GLUE *See* **Adhesive.**

ANION Negative ion.

ANIONIC EMULSION Emulsified asphalt in which the asphalt globules are electronegatively charged.

ANISOTROPIC SOIL Soil mass having different properties in different directions at any given point, referring primarily to stress/strain or permeability characteristics.

anl *Abbreviation for:* anneal.

ann. *Abbreviation for:* annual.

ANNEAL Heating metal or glass to a specific temperature and then cooling it slowly in oil so as to reduce its brittleness and increase its ductility.

ANNEALED WIRE 1. Soft, pliable wire used as tire wire; **2.** Wire applied to wire rope in preparation for cutting or socketing. *Also called* **Binding wire** and **Seizing wire.**

ANNEX Building added to, or used in relation to a previously existing structure or complex.

ANNEXATION Legal process by which an incorporated municipality expands its boundaries.

annod. *Abbreviation for:* annodized.

ANNUAL 1. Event that occurs at approximately, or exactly the same time every year; **2.** Contract or agreement valid only for one year; **3.** Plant with a life span of one year's duration.

ANNUAL ALLOWABLE HARVEST Quantity of timber scheduled to be removed from a particular management unit in one year.

ANNUAL GROWTH Average annual increase in the biomass of growing-stock trees of a specified area.

ANNUAL GROWTH RING Growth layer put on by a tree in a single growth year, including springwood and summerwood. *Also called* **Growth ring.**

ANNUAL MAXIMUM (ELECTRICAL) DEMAND *See* **Demand (electric).**

ANNUAL SYSTEM MAXIMUM (ELECTRICAL) DEMAND *See* **Demand (electric).**

ANNUITY Series of equal payments, made at equal intervals of time.

ANNULAR Ring-shaped.

ANNULATED COLUMN Cluster of small-diameter columns serving as one structural element.

ANNULET Semicircular molding circumscribing a column.

ANNULUS 1. Circular space or groove; the space between two concentric circles; **2.** Space between a drill string, pump column, or casing and the wall of a borehole or outer casing.

ANNUNCIATOR Device, usually electronic or electromagnetic, but also mechanical, that indicates the origin of a signal or message.

ANODE Positive terminal of an electrical circuit.

ANODIZING Process of coating the surface of aluminum or titanium and their alloys by an anodic oxidation process to form both a decorative and protective coating.

AN PRILLS Small spheres or pellets of ammonium nitrate (as opposed to flaked, granular, or powdered).

ans. *Abbreviation for:* answer.

ant. *Abbreviation for:* antechamber; antenna.

ANTECHAMBER Room from which another room is reached.

anti. *Abbreviation for:* antilogarithm.

ANTI-ACTINIC GLASS Glass that absorbs heat.

ANTICLINE Upfold of layered rocks in the form of an arch.

ANTICORROSIVE COATING Coating, usually a paint, containing pigments to inhibit and delay corrosion of metal.

ANTICREEP DEVICE Means of compensating for any loss of fluid pressure in the operating system of a hydraulic elevator that maintains the car within 25 mm (1 in.) of the landing.

ANTI-EXTRUSION RING Rigid or semirigid ring employed at one or both ends of a packing set, primarily to prevent extrusion into clearances. *Also called* a **Bull ring.**

ANTIFIX Vertical ornament at the eaves of a tiled roof, or a cresting at the ridge.

ANTIFOAM ADDITIVE Chemical added to oils to help control air release and reduce the foaming that results from aeration in service.

ANTIFREEZE Chemical mixed with the liquid coolant of an engine that reduces its freezing point.

ANTIFREEZE SPRINKLER SYSTEM System, usually 20 heads or less, connected to a wet pipe sprinkler system to protect small unheated areas of heated occupancies.

ANTIFRICTION BEARING *See* **Bearing.**

ANTIHAMMER DEVICE Air chamber, such as a closed length of pipe or coil, designed to absorb the shock caused by a rapidly closed valve.

ANTIHEAVE MEASURE Precaution taken to prevent soil heave.

ANTILOCK BRAKE SYSTEM Brake system that electronically monitors wheel speed and helps the driver control the vehicle by avoiding wheel slip on low-friction surfaces and in emergency situations by applying and re-applying the brakes.

ANTILOCK VALVE Part of an electromechanical antilock brake system that receives signals from the logic controller and releases air from, or admits air to the brake chambers on a given axle to adjust the brake actuation force.

ANTIOXIDANT Additive, usually incorporated in relatively small proportions to retard oxidation, of lubricants for instance.

ANTI-OZONANT Compound ingredient used to retard deterioration caused by ozone.

ANTIPICK LATCH Spring latch having a parallel bar that prevents it from responding to external pressure from other than the proper key.

ANTIQUING Chemical process used to darken copper and brass, or to artificially age a painted surface.

ANTIROTATION DEVICE Device such as a key used to prevent rotation of one component relative to an adjacent component. *See also* **Shear pin** and **Torque pin.**

ANTISAG BAR Vertical tie connecting the peak of a truss to the horizontal tie beam.

ANTISIPHON TRAP Trap in a drainage system designed to preserve a water seal by preventing siphonage.

ANTISIPHON VALVE *See* **Valve.**

ANTISLIP DIFFERENTIAL *See* **Differential.**

ANTISTATIC *See* **Static conductive.**

ANTISTATIC FLOOR Concrete floor in which the steel reinforcement has been suitably grounded.

ANTI TWO-BLOCK SYSTEM Device that alerts a crane operator of an impending two block situation through an audio-visual warning system. *See also* **Kruger H system,** and **Two-block warning system.**

ANTIWEAR ADDITIVE Chemical added to oils to reduce wear, generally by reacting with metal surfaces in contact to provide a protective layer.

anv. *Abbreviation for:* anvil.

ANVIL 1. Fixed steel block that provides support and resistance for the cutting blade of a single-action tree shear; **2.** Fixed jaw in a swaging device that supports the back of the tooth while the die or roller presses against the front of the tooth; **3.** Part of a power-operated hammer that receives the blow of the ram and transmits it to the pile being driven. *See also* **McDermid plate; 4.** Steel plate on top of cushion material in a drive cap that the impact block strikes. *Also called an* **Anvil block, Penny, Striker plate,** and **Top plate.**

ANVIL BLOCK 1. Movable piece of steel in a paving breaker, between the air piston stem and the steel (bit). *Also called* **Anvil; 2.** *See* **Driving cap.**

anx *Abbreviation for:* annex.

AP *Abbreviation for:* access panel.

APA GLUED FLOOR SYSTEM Floor system developed by the American Plywood Association (APA) in which a single layer of APA-Rated Sturd-I-Floor panels (or subflooring in the case of double-layer construction) is glue-nailed to wood joists. The bond is so strong that floor and joists behave like an integral unit, greatly increasing floor stiffness and greatly reducing floor squeaks and nail popping. Only construction adhesives conforming to APA specification AFG-01 are recommended. *See also* **T-Beam.**

APA PERFORMANCE-RATED PANEL Panel product developed by the American Plywood Association (APA).

APA RATED SHEATHING American Plywood Association (APA) Performance Rated Panel designed and manufactured specifically for residential and other light frame wall sheathing, roof sheathing, and subflooring applications.

APA RATED SIDING Grade designation covering American Plywood Association (APA) proprietary siding products.

APA RATED STURD-I-FLOOR American Plywood Association (APA) Performance Rated Panel designed and manufactured specifically for residential and other light-frame single-floor (combined subfloor-underlayment) application for use under carpet.

APA RATED STURD-I-WALL Construction system in which American Plywood Association (APA) Rated Siding panels are attached directly to studs.

APARTMENT Room or suite of rooms used as living quarters; a dwelling unit of a multi-family house.

APARTMENT BUILDING Type of multiple dwelling comprising three or more dwelling units with shared entrances and other essential facilities and services and with shared exit facilities above the first story.

apch *Abbreviation for:* approach.

apchg *Abbreviation for:* approaching.

APERTURE Any opening; in construction, an opening left for a door, window, ventilation, etc.

APEX Pointed top of a pyramid or cone.

APEX STONE Top stone in a gable end.

API Scale adopted by the American Petroleum Institute (API) to indicate the specific gravity of a liquid; a scale constructed in such a way that gravity readings are higher for less dense liquids. Therefore the API gravity for a liquid rises as its temperature rises.

APOPHYGE Slight curve at the top and bottom of a column where the shaft joins the capital and fillet.

app. *Abbreviation for:* appended; appendix; appoint.

appar. *Abbreviation for:* apparatus.

APPARATUS Collective term for equipment specially designed to fight fires.

APPARENT DIRT CAPACITY In fluid filter evaluation, the amount of dirt that can be added to the filter test system before the terminal differential pressure is reached.

APPARENT POWER Product of current and voltage, expressed in Kilovolt Amps (kVA).

APPARENT SPECIFIC GRAVITY *See* **Specific gravity.**

APPARENT VISCOSITY Ratio of shear stress to rate of shear of a non-Newtonian fluid, calculated from Poiseuille's equation and measured in poises. The apparent viscosity of most greases varies with changing rates of shear and temperature, and must therefore be reported as the value at a given shear rate and temperature.

app. *Abbreviation for:* approved; applicable; application; applicator; applied.

APPEARANCE Those characteristics of a lubricating grease that are observable by visual inspection only. This general term includes characteristics described under **Bloom, Bulk appearance, Color,** and **Texture.**

APPLIANCE 1. Utilization equipment, generally other than industrial, normally built in standardized sizes or types, that is installed or connected as a unit to perform one or more functions; **2.** Device to convert fuel into energy, including all required components, controls, wiring, and piping.

APPLICATION FOR PAYMENT Written request by a contractor for partial payment under the terms of the contract; may be for work done, materials on site, etc.

APPLICATION LIFE Period of time during which a sealant, after being mixed with a catalyst or exposed to the atmosphere, remains suitable for application. *Also called* **Working life.**

APPLICATOR Fire hose pipe or nozzle for applying foam or water fog to fires. *Also called* **Foam applicator,** and **Fog applicator.**

APPLIED COST Cost that has been assigned to a product or activity.

APPLIED MOLDING Molding attached to a wood-faced wall so as to give the effect of paneling.

APPLIED TORQUE Torque transmitted to a fastener assembly as indicated by a torquing tool.

APPORTIONMENT 1. Administrative assignment of funds based on a prescribed formula, by a governmental unit to another governmental unit for specific purposes and for certain periods, or between buyer and seller through the proration of property expenses such as insurance and taxes. *See also* **Allotment,** and **Appropriation; 2.** When more than one insurance policy is in force, determination of the amount to be contributed by each to satisfy a claim.

appr. *Abbreviation for:* apprentice; apprenticeship; approve.

APPRAISAL Estimation of the value of real property.

APPRAISAL APPROACH Methods for estimating the value of real property. There are several methods, including:

    Capitalization: Estimation of value according to capitalization of productivity and income.

    Cost: Appraisal based on the depreciated new replacement cost of improvements, plus the current market value of the site.

    Income: Appraisal based on a property's anticipated future income.

    Market comparison: Appraisal based on analysis of the recent sales prices of similar properties.

    Summation: Adding together the parts of a property, each part being considered separately, e.g., value of the land considered as vacant, plus the cost of reproduction of structures, less depreciation.

APPRAISAL INCREASE Credit resulting from an increase in the recorded value of a fixed asset arising from a reappraisal.

APPRAISAL REPORT Report presenting a value estimate of real property. It should contain the data on which the appraisal is made together with any analysis of the data, plus conclusions leading to the estimate.

APPRAISED PRICE Price of a particular timber sale based on the estimate of the timber's actual market value; the minimum accepted price of a sale.

APPRAISED VALUE Market value: the amount a qualified assessor calculates a property will fetch on the open market.

APPRAISER One who is experienced and/or qualified in the procedures necessary to complete an appraisal of real property.

APPRECIATION Increase in the value of property and/or goods.

APPRENTICE One who enters upon an agreement to serve an employer for a stated period for the purpose of receiving instruction and learning a trade.

APPRENTICESHIP Period of indenture during which an apprentice in what are commonly known as the skilled crafts and trades, those require a wide and diverse range of skills and knowledge, is given instruction and experience, both on and off the job, in practical and theoretical aspects of the work.

APPROACH Combination of land forms and/or structures that lead towards an entrance, monument, etc.

APPROACH ANGLE 1. Ramp clearance angle for the front of a vehicle, measured from the forward edge of the front tire contact patch to the lowest part of the vehicle forward of the tire; **2.** Angle made between the plane of a platform and the ground plane on a carrier body.

APPROACH NOSE End of a traffic island, or area between diverging roadways, that faces approaching traffic passing to one or both sides.

APPROACH PIT Narrow pit adjacent to that part of a structure that is to be underpinned.

approp. *Abbreviation for:* appropriate; appropriation.

APPROPRIATION 1. Act of a legislative body that makes funds available for expenditures with specific limitations as to amount, purpose, and period. *See also* **Allotment,** and **Apportionment; 2.** Allocation of funds to a project. *Also called* **Feasibility budget.** *See also* **Project budget; 3.** Setting aside private land for public use.

APPROVED Sanctioned, endorsed, accredited, certified, or accepted as satisfactory by a duly constituted and nationally recognized authority or agency.

APPROVED EQUAL Material, equipment, or method approved by the architect or engineer as being an acceptable substitute for its specified equivalent. *Also called* **Or equal.**

APPROVED RIM WIDTH Rim width approved for use with a given tire size.

approx. *Abbreviation for:* approximate; approximately.

approxn *Abbreviation for:* approximation.

appt *Abbreviation for:* appointment.

APPURTENANCE 1. Subordinate but necessary accessory; in plumbing, for instance, the fittings, valves, traps, etc., that are necessary to complete a house drain; **2.** Something external and separate to a structure that adds to its real value and intrinsic enjoyment.

APRON 1. Flat member of the inside trim of a window, placed against the wall immediately under the sill to cover the rough edge of the wall finish. *See also* **Wall; 2.** Extension of the concrete floor of a garage or other structure beyond the face of the building; **3.** Short ramp with a slight pitch; **4.** Front gate of a mechanical scraper body that can be raised and lowered; **5.** Rearmost crossmember on the body sub-frame of a vehicle; extends down from floor level and overlaps the ends of the lengthwise beams. Also known as a **Rear bolster.** *See also* **Brace,** and **Wall.**

APRON CONVEYER *See* **Conveyer.**

APRON FLASHING Flashing that makes a waterproof joint between a sloped surface and a vertical surface, as at the front or downslope face of a chimney stack where it penetrates the external face of a sloped roof.

APRON MOLDING Raised ornamentation on the lock rail or apron rail of a door.

APRON PIECE Timber fixed in the walls of a staircase to carry the landing joists and support the carriage pieces.

APRON WALL *See* **Wall.**

aprt *Abbreviation for:* airport.

APS *Abbreviation for:* accessory power supply; auxiliary power supply.

apt *Abbreviation for:* apartment.

APTERAL Building with columns at the ends, but not along the sides.

APU *Abbreviation for:* auxiliary power unit.

AQUASTAT Automatic switching device consisting of a metal- or liquid-filled heat-sensitive element designed to detect a temperature drop or rise of boiler water.

AQUEDUCT Artificial channel to convey water; usually elevated for most of its length.

AQUEOUS Water-based solution or suspension.

AQUIFER Underground formation of sands, gravel, or fractured porous rock, that is saturated with water, and that supplies water for wells and springs.

AQUIFER TEST Procedure in which measured quantities of water are removed or added to a well and the resulting changes in head are measured during and/or after the period of discharge or recharge.

AQUITARD Layer of soil or rock that is relatively impermeable and restricts the flow of water from one aquifer to another.

AR *Abbreviation for:* as required; as rolled.

ARABESQUE Surface decoration, usually based on geometrical patterns and including classical vases, sphinxes, flowing lines, tendrils, etc.

arb. *Abbreviation for:* arbor.

ARBITRATION Settlement of a dispute between parties to an agreement or contract using the experience and knowledge of an independent expert or panel. *See also* **Binding arbitration.**

ARBOR 1. Lattice structure providing shade, usually covered in part with vines; **2.** Short shaft or spindle on which another rotating part is mounted. *Also called* **Spindle.**

ARC Flow of electricity through a gaseous space or air gap. *See also* **Geometry of circular curve.**

ARCADE 1. Range of arches carried on piers or columns, either free-standing or attached to a wall; **2.** Passage, roofed over, often arched; **3.** Any covered walk, especially with shops on either side.

arc cos. *Abbreviation for:* inverse cosine.

arc cosec. *Abbreviation for:* inverse cosecant.

arc cot. *Abbreviation for:* inverse cotangent.

ARC CUTTING Making a kerf in metal using the energy of an electric arc.

ARC FURNACE *See* **Furnace.**

ARC OF CONTACT Portion of the circumference of a grinding wheel touching the work being ground. *See also* **Angle of contact.** *See also* **Angle of contact.**

arch. *Abbreviation for:* architect; architectural; archive.

ARCH 1. Supporting device towed behind or mounted on a skidder, used to lift one end of a log or logs to reduce skidding resistance and/or transfer the weight of the load to the skidding vehicle; **2.** Curved supporting structure. One of the earliest means of supporting complex deck and roof structures, the arch has evolved over the centuries from a simple semicircle to a wide range of complex forms. From the earliest times, various parts of the arch were identified and their structural properties defined. These include the following:

> **Abutment:** Solid wall that counteracts the lateral thrust of an arch.
>
> **Axis:** Median line of the arch ring.
>
> **Center:** Point(s) from which the shape of the arch is struck.
>
> **Crown:** Highest or central part.
>
> **Extrados:** Upper or outer curve of the arch.
>
> **Haunch:** Side between the crown and piers or wall below the springing line.
>
> **Impost:** Member in the wall on which the end of an arch rests; the upper face of the impost is the springing line.
>
> **Intrados:** Inner curve or underside of the arch; the soffit.
>
> **Keystone:** Stone at the summit locking the whole together.
>
> **Rib:** Load-bearing main member of a ribbed arch.
>
> **Ring:** Load-bearing part of an arch.
>
> **Span:** Clear distance between piers or supports.
>
> **Springing line:** Level at which an arch 'springs' from its supports.
>
> **Voussoirs:** Wedge-shaped stones or blocks com-

prising the shape of the arch between the springing block and the keystone.

The various forms into which the arch has evolved have been generally grouped into the following classifications:

Back: A concealed arch carrying the backing of a wall where the exterior facing is carried by a lintel.

Basket: So-named because it resembles the handle of a basket, this arch is formed by a segment of a large circle that is continued left and right by two segments of much smaller circles.

Bell: Semicircular arch supported on quarter-round corbels.

Drop: Form of lancet arch in which the centers of the two halves are located within the span.

Elliptical: Arch described from three or more centers.

Equilateral: Produced by two curves, each with a radius equal to the span and meeting at a point. *Also called* **Pointed arch.**

Flat: An arch whose top and bottom are flat, or practically so. It involves the same principles of stress and strain as a segmental arch, but has voussoirs (wedge-shaped pieces) extended up and down to reach the horizontal lines.

Floor: *See* **Jack** below.

Four-centered: An equilateral arch of four arcs — the two outer and lower ones springing from centers on the springing line; the two inner and upper arcs from centers below the springing line.

Gauged: Arch built from gauged bricks, i.e., soft bricks sawn to shape and rubbed smooth on a stone or another brick. The bricks are laid with a fine joint, often of lime putty.

Gothic: High and narrow arch, often with a point at its apex instead of a keystone.

Groin: Rib form accenting the intersection of two vaulted surfaces and dividing a vault into bays.

Horseshoe: Can be either pointed or round horseshoe.

Interrupted: Pediment in the form of an arch with the central portion cut away.

Inverted: Arch having its keystone at the lowest point, used to concentrate a load over a foundation.

Jack: Arch having horizontal, or nearly horizontal upper and lower surfaces.

Keel: Inflected or ogee arch.

Lancet: Equilateral arch where the radii is much larger than the span.

Major: Arch that spans greater than 1.8 m (6 ft) and supports equivalent to uniform loads greater than 1327 kg/m (1,000 lb/ft). Typically has a span ratio greater than 0.15.

Minor: Arch with a maximum span of 1.8 m (6 ft) and loads not exceeding 1327 kg/m (1,000 lb/ft). Has rise to span ratio less than or equal to 0.15.

Ogee: A pointed arch usually comprised of four arcs: two centered inside the arch, two centered outside. This produces a compound curve made up of convex and concave parts.

Pendentive: Arch that cuts off the corners of a square building so that the superstructure may become an octagon or dome.

Ramp: Arch with one abutment higher than the other.

Rear: Arch on the inside of a wall spanning a door or window opening.

Relieving: Often of rough construction, placed in a wall above a lintel, flat arch, or true arch to relieve it of the accumulating weight and to discharge stress to the piers or abutments. *Also called* **Discharging arch,** and **Safety arch.**

Roman: Semicircular arch.

Rough: Relieving arch, usually of brick, located either in a location where the quality of final finish is unimportant, or in a wall that will sequently be covered with another material: stucco or roughcast, for instance.

Segmental: Type of masonry construction where the curve of the arch through an arc, or segment, or circle, is always less than a semicircle.

Semicircular: Arch where the intrados forms a half circle.

Soldier: Flat arch formed of bricks standing on end, supported by a metal bar or angle iron.

Stilted: Where the springing line is raised by vertical piers above the impost level.

Trimmer: Usually a low-rise arch of brick, used for supporting a fireplace hearth.

Trussed: Arch composed of wood or steel sections arranged and joined so as to conduct the imposed load to its supporting piers or impost.

Tudor: A pointed arch whose shanks start with a curve near to a quarter circle and continue to its apex in a straight line.

ARCH BAND Visible part of a rib in vaulting.

ARCH BAR Structural support for a flat arch, between springing points along the soffit.

ARCH BRICK *See* **Brick.**

ARCH BUTTRESS *See* **Buttress, Flying.**

Arch. Engr. *Abbreviation for:* architect and engineer.

archeol *Abbreviation for:* archeological.

ARCHIMEDES SCREW Large, slow-pitch spiral-feed screw, sometimes contained within a tube, set at an incline, that, when turned, lifts and carries water and other material to a higher elevation.

ARCHING Skidding logs or trees using a mounted or trailing arch.

ARCHITECT A qualified professional who designs and oversees the construction of buildings and structures; one who is licensed to practice architecture; the owner's representative. An architect may also contract with an owner to perform, in addition to architectural responsibilities, engineering and contract administration services as necessary for the project. The term architect includes its officers, directors, agents and employees, all of the architects consultants, and any other person for whom it may be liable.

ARCHITECT-ENGINEER Architect, engineer, architectural firm, engineering firm, issuing project drawings and specifications, or administering the work under contract specifications and drawings, or both, with full responsibility for the design and integrity of the structure. *Also called* **Engineer-architect.**

ARCHITECTURAL CONCRETE Concrete that will be permanently exposed to view and that therefore requires special care in selection of the concrete materials, forming, placing, and finishing to obtain the desired architectural appearance. *Also called* **Exposed concrete,** and **Fair-faced concrete.**

ARCHITECTURAL DESIGN Overall design, often preliminary or in sketch form, prepared by an architect, showing the general scope of a project and usually including three-dimensional sketches.

ARCHITECTURAL DRAWING Drawing that shows the general design and form of buildings by means of elevations, plans, and sections; shows the various materials, their dimensions and details of fixing; shows fixtures and finishes.

ARCHITECTURAL MODEL Three-dimensional scale model of a project, often including the site and an indication of its setting. *See also* **Architectural rendering.**

ARCHITECTURAL PROGRAM Written statement included in the documentation of the conceptual design. On the basis of the written function program, that reflects the client's or user's needs, the architect prepares a further written program including: (a) a statement of the net areas of all rooms and additional areas normally included in the 'gross area' of the building, (b) basic standards for and performance requirements of the various systems such as mechanical, electrical, plumbing, refrigeration and others as may be required, and (c) a statement of area requirements to accommodate the plans to any long-term requirements stated in the functional program. *See also* **Conceptual design, Functional program,** and **Requirements.**

ARCHITECTURAL RENDERING Depiction of a project as completed, usually showing the finished site and an indication of its setting. *See also* **Architectural model.**

ARCHITECTURAL SPECIFICATION *See* **Specification of works.**

ARCHITRAVE 1. Lowest of the three main parts of an entablature. *See also* **Entablature; 2.** Main beam resting on the abacus or the capital of a column; **3.** Molded frame surrounding a door or window; **4.** Various parts surrounding a doorway or window; **5.** Molding round the exterior of an arch.

ARCHITRAVE CORNICE An entablature without a frieze.

ARCHIVOLT 1. Soffit with molded edges; **2.** Molding on the face of an arch, parallel to the intrados.

ARCH POUR Process in which the upper half of a tunnel lining is poured separately from the lower half. *See also* **Full-circle pour,** and **Invert pour.**

ARCHWAY Passageway under an arch.

ARC QUENCHER Device that eliminates or reduces the arc that may be formed when electrical current-carrying contacts are opened or closed.

arc sec. *Abbreviation for:* inverse secant.

arc sin. *Abbreviation for:* inverse sine.

ARC SPECTROGRAPHY Spectrographic identification of elements in a sample of material heated to volatilization in an electric arc or spark.

arc tan. *Abbreviation for:* inverse tangent.

ARCUATED Building that is structurally dependent on the use of arches.

arc/w *Abbreviation for:* arc weld.

ARC WELDING Electrical welding process in which intense heat is obtained by the arcing between a welding rod and the metal to be melted. The molten metal from the tip of the electrode is then deposited in the joint and, together with the molten metal of the edges, solidifies to form a sound and uniform connection.

ARE Obsolete metric unit of area, equal to 100 m², that should not be used with the SI system of measurement. Symbol: a. *See also the appendix:* **Metric and nonmetric measurement.**

AREA 1. Square measurement; **2.** Amount of surface enclosed by defined boundaries, expressed as a square measurement. *See also* **Area of building, Floor area, Gross area, Net area,** and **Net room area; 3.** Uncovered space, such as an open court; **4.** Sunken space around the basement of a building, providing access and natural lighting and ventilation. *See also* **Wall.**

AREA DIFFERENTIAL SYSTEM Valve operated by means of opposing pistons of different displacement.

AREA DIVIDER Raised ridge on a built-up flat roof system, running with the slope, used where it is necessary to divide the overall area to allow for thermal expansion and contraction.

AREA DRAIN Drain set in a floor that cannot be otherwise drained.

AREA ESTIMATING METHOD Method of estimating costs by multiplying the adjusted gross floor area by a unit cost.

AREA IGNITION Simultaneous or quick successive firing of all parts of a forest area to be burned in order to produce a fast, hot spread and clean burn. *See also* **Backfire, Controlled burning,** and **Prescribed burning.**

AREA OF BUILDING Maximum projected horizontal area of a building at or above grade within the outside perimeter of the exterior walls or within the outside perimeter of exterior walls and the center line of fire or party walls. *See also* **Floor area, Gross area, Net area,** and **Net room area.**

AREA OF CONTACT Total area of the grinding surface of a grinding wheel in contact with the work being ground.

AREA OF STEEL Cross-sectional area of bars required for a given concrete section. *See also* **Effective area of reinforcement.**

AREA REGULATION Method of controlling the annual or periodic acreage harvested from a forest, despite fluctuations in fiber-yield volumes.

AREA SALVAGE Timber sales in which the regulating agency for the forest area sells dead timber within a given area. Usually covers more than one operating season and requires the operator to return annually to remove any dead timber present.

AREA SEPARATION WALL Solid or hollow wall comprising materials of specific fire-resistance, and constructed in such a manner as to produce a fire-rated partition to prevent the spread of fire between adjoining occupancies.

AREA TECHNIQUE See **Sanitary landfilling.**

AREA WALL 1. Masonry surrounding or partly surrounding an area; **2.** Retaining wall around basement windows below grade. See also **Wall.**

AREAWAY Open subsurface space adjacent to the foundation walls.

ARENA 1. Central open space of an amphitheatre in which performances take place; **2.** Any building, part of a building, or open space in which contests or displays take place. See also **Amphitheater.**

ARENACEOUS Composed primarily of sand; sandy.

ARGILLACEOUS Composed primarily of clay or shale; clayey.

ARID CHAMBER Test facility simulating high temperature, low humidity, and solar radiated heat and light.

arith. Abbreviation for: arithmetic.

ARITHMETIC LOGIC UNIT (ALU) Logic circuit that performs arithmetic and logical functions on data and words.

ARKANSAS OILSTONE Natural stone quarried in the Ozark Mountains of Arkansas and capable of producing a fine cutting edge.

arm. Abbreviation for: armature; armored.

ARM Section of an excavator attachment between the boom and the bucket.

ARMATURE Part that revolves in an electric motor or dynamo.

ARMATURE GROWLER Test device used to test the armature of an engine starter motor.

ARM CYLINDER Hydraulic cylinder that controls movement of the arm section of the attachment on an excavator/loader.

ARM HOIST See **Truck hoist.**

ARMOR COAT Asphalt paving involving two thin layers of different-size aggregate sprayed with asphalt.

ARMORED CABLE Insulated wire having an additional flexible metallic protective sheathing, often referred to as **BX cable.** See also **Cable.**

ARMORED FRONT Steel plate covering the bolts or set screws holding a cylinder to its lock that are normally exposed when the door is ajar.

ARMORED HOSE Hose with a protective covering, applied as a braid or helix, to protect it from physical abuse.

ARMORY Room or building in which armaments are manufactured, repaired, or stored.

arm.-pl. Abbreviation for: armored plate.

ARM RESTRAINER Device that restrains pivotal movement of a swinging arm.

ARM'S LENGTH Transaction among individuals, each of whom acts in their own best interest.

arnd Abbreviation for: around.

arr. Abbreviation for: arrange; array; arrester; arrival.

ARRAY Series of related elements or components placed in relationship to each other.

ARRESTER Something that impedes or stops; as sparks from leaving the top of a chimney.

ARRIS Sharp edge formed by the angular contact of two plane or curved surfaces.

ARRISSING TOOL Tool similar to a float, but having a form suitable for rounding an edge of freshly placed concrete, mortar, plaster, etc.

ARRISWAYS Tiles or slates laid diagonally, usually in bands as a decorative element.

ARROW Representation of an activity with an indication adjacent to the arrow (above or below) of its duration, used in both CPM and PERT systems. See also **Critical Path Method (CPM), and Program Evaluation and Review Technique (PERT).**

ARROW DIAGRAMMING Technique used in CPM or PERT to display the network sequence and timing of activities.

ARS Abbreviation for: asbestos roof shingles.

ARSON Intentional and malicious burning of property.

art. Abbreviation for: article; arterial.

ARTERIAL HIGHWAY See **Highway.**

ARTESIAN AQUIFER Aquifer confined above and below by material of lesser permeability and in which the water level stands above the top of the water body it draws upon.

ARTESIAN WATER Subsurface water under sufficient head of pressure to raise the water in wells above the groundwater table.

ARTESIAN WELL Drilled well that penetrates a strata containing water under pressure.

ARTICLE 1. Manufactured item that (a) is formed to a specific shape or design during manufacture, (b) has end-use function(s) dependent in whole or in part upon its shape and design during end use, and (c) does not release, or otherwise result in exposure to a hazardous chemical or promote a hazardous condition under normal conditions of use; **2.** Subdivision of a document; **3.** Primary subdivision of a specification. May be further divided into paragraphs, subparagraphs, and clauses.

ARTICULATED 1. Jointed, capable of bending through a hinged joint; **2.** Truck designed to pull a trailer; **3.** Type of steering mechanism.

ARTICULATED DROPCHUTE Device consisting

of a succession of tapered metal cylinders so designed that the lower end of each cylinder fits into the upper end of the one below, used to confine or direct the flow of a falling stream of fresh concrete. *See also* **Dropchute**, and **Flexible dropchute.**

ARTICULATED DUMP TRUCK On- and off-highway dump truck having the load-carrying dump body and its associated frame, suspension, and drive wheels connected to the operator's compartment, engine compartment, front suspension, and steering wheels through an articulated joint that gives a limited range of vertical and horizontal movement.

ARTICULATED FLOOR Concrete floor slab reinforced continuously at mid depth with sufficient cross-sectional steel mesh fabric to resist tensile stresses at flexible joints provided at intervals.

ARTICULATED JIB Tower crane jib having a pivot point in its central area. *Also called* a **Pivoted luffing jib.**

ARTICULATED JOINT Joint with movement limited by restraints.

ARTICULATED STEERING *See* **Steering.**

ARTICULATING-BOOM PLATFORM Aerial work platform that is raised by a boom having two or more hinged sections.

ARTICULATING HOIST *See* **Truck-mounted crane.**

ARTICULATION 1. Jointed; **2.** Machine in two sections connected by a vertical hinge or kingpin.

ARTIFICIAL ABRASIVE An abrasive that is manufactured; not a natural mineral.

ARTIFICIAL CONTAMINANT Contaminant of known composition and particle size distribution that is introduced into a fluid system or fluid system components for test purposes. *See also* **Built-in contaminant, Contaminated,** and **General contaminant.**

ARTIFICIAL FUEL Man-made fuel, including all manufactured and by-product fuels. Examples are water gas, blast furnace gas, gasahol, coke, etc.

ARTIFICIAL HARBOR Protected area created by building one or more breakwaters or by placing artificial barriers around a body of water.

ARTIFICIAL HORIZON True horizontal, as indicated by reflection, using a container of mercury when measuring altitudes with a sextant.

ARTIFICIAL LAKE Man-made basin (other than a swimming pool), 23 m² (250 ft²) or greater, permanently containing water to a minimum depth of 60 mm (2 ft).

ARTIFICIAL OBSTRUCTION Any barrier other than a natural obstruction that impedes or prevents passage or flow.

ARTIFICIAL RECHARGE Recharge of an aquifer at a rate greater than that due to natural circumstances, as a result of human activity, for instance.

ARTIFICIAL REGENERATION Establishing a new forest by planting seedlings or by direct seeding (as opposed to natural regeneration).

ARTIFICIAL STONE Special concrete unit, some-times artificially colored, intended to resemble natural stone, made by mixing chippings and dust of natural stone with portland cement and water. This mixture is placed in molds and cured before use.

ARTISAN Skilled craftsman or tradesman.

AS *Abbreviation for:* aftersight; air suspension; asbestos sheathing; as-sheared; automatic sprinkler.

asb. *Abbreviation for:* asbestos; aggregate subbase.

asb.-c. *Abbreviation for:* asbestos-covered.

ASBESTOS Fibrous mineral used as a heat and fire barrier.

ASBESTOS BLANKET Small asbestos-filled foil blanket wrapped around pipes being welded to reduce heat loss.

ASBESTOS CEMENT Fire-resisting weatherproof building material made from portland cement and asbestos fibers in forms such as plain and corrugated sheets, shingles, pipes, etc.

ASBESTOS-CEMENT BOARD Fire-resistant sheet made from asbestos cement.

ASBESTOS-CEMENT PIPE *See* **Pipe.**

ASBESTOS-CEMENT PRODUCTS Products manufactured in rigid material composed essentially of asbestos fiber and portland cement.

ASBESTOS-DIATOMITE Mixture of asbestos fibers, diatomaceous earth, and portland cement used to make insulating, fire-resistant building products.

ASBESTOS FELT Sheet of matted asbestos fibers, the basis for saturated asbestos felt and asbestos asphalt-coated roofing products.

ASBESTOS FIBER Naturally occurring fibrous rock used in the production of asbestos-cement goods.

ASBESTOS-FREE PRODUCTS Range of products that perform the same functions as those that have historically contained asbestos fibers but which are manufactured with alternative materials.

ASBESTOS PLASTER Nonflammable pipe insulation containing asbestos fibers, diatomaceous earth, and gypsum plaster.

ASBESTOS SHINGLES Type of shingle made for fireproofing.

asb.-p. *Abbreviation for:* asbestos-packed.

asb.-pr. *Abbreviation for:* asbestos-protected.

AS-BUILT DRAWING Scale drawing that shows a project as it was built and that incorporates details of any additions, deletions, or variations from the drawings used at the time of accepting a bid to construct the project and/or as approved by the client.

A-SCALE Sound level measurement scale that discriminates against low frequencies and approximates the human ear.

ASD *Abbreviation for:* advise shipping date.

AS DRAWN Prestressing wire supplied from the mill with a natural curvature.

ASEISMIC REGION Region not liable to earth tremors.

A SELECT *See* **Lumber grades.**

asg. *Abbreviation for:* assign.

asgd *Abbreviation for:* assigned.

asgmt *Abbreviation for:* assignment.

ash. *Abbreviation for:* ashlar.

ASH Constituents of a fuel burned in an engine or furnace that can wear or clog the system. *See also* **Solid waste.**

ASH CONTENT Measure of the residue which remains after combustion under specified conditions.

ASH DUMP Flat, metal door with a flange, built flush with the inner hearth of a fireplace, used for dumping ashes into the ash pit.

ASHLAR Rectangular units of burned clay or shale, or hewn blocks of masonry with an even facing; masonry constructed of this (as opposed to rubble or unhewn stone). Completed ashlar work is generally grouped into the following:

> **Coursed:** Ashlar set to form continuous horizontal joints.
>
> **Irregular coursed rubble:** Rubble walls built to courses of various depth.
>
> **Random ashlar:** Ashlar set with stones of varying length and height so that neither vertical nor horizontal joints are continuous.
>
> **Stacked ashlar:** Ashlar set to form continuous vertical joints.

ASHLAR FACING Outer skin of a faced or veneered wall comprising solid, rectangular masonry with sawed, dressed, or squared beds, and mortar joints.

ASHLAR LINE Main line of the surface of a wall of the superstructure.

ASHLAR MASONRY Masonry composed of rectangular units of burned clay or shale, or stone. Generally larger in size than brick and properly bonded, it has sawed, dressed or squared beds and joints laid in mortar. *See also* **Ashlar, Random.**

ASH PIT Area directly below the inner hearth of a fireplace or furnace where ashes from the fire are collected via the ash dump.

ASH SLUICE Trench or channel in which water flushes residue from an ash pit to a storage or disposal point.

A-SIDE Left side of an elevator car or hoistway when facing the hoistway from a landing.

AS IS Without guarantee as to condition.

ASKAREL Generic term for a group of nonflammable synthetic chlorinated hydrocarbons used as electrical insulating media.

asmb. *Abbreviation for:* assemble; assembly.

ASME-TYPE LP GAS CYLINDER Fuel container for liquefied petroleum gas made and inspected under the American Society of Mechanical Engineers (ASME) Boiler and Pressure Vessel Code, Section V-iii, for Unfired Pressure Vessels. Where used for storage of an industrial truck's fuel supply, it may be permanently or removably affixed to the vehicle.

A-SPACER Reinforcing bar support that is shaped like an 'A.'

ASPECT Compass direction to which a slope faces. *Also called* **Exposure.**

ASPECT RATIO 1. In any rectangular configuration, the ratio of the lengths of the sides. **2.** Ratio of tire height to tire width.

ASPEN Primary waferboard wood source: a fast-growing hardwood tree that primarily grows in Canada and the northern U.S. states.

ASPENITE Proprietary name for a waferboard product.

ASPERITIES Minute imperfections on a seal face or surface of the mating ring that are the result of normal surface-finishing processes.

asph. *Abbreviation for:* asphalt.

ASPHALT 1. Dark brown to black cementitious material in which the predominating constituents are bitumens that occur in nature or that are obtained in petroleum processing; **2.** The binder commonly used in modern asphalt paving mixtures. It is produced by distillation of petroleum crude oil and also is found in natural deposits. There are many types, including:

> **Air-blown:** Asphalt treated by blowing air through it at elevated temperature to give it characteristics desired for certain special uses such as roofing, pipe coating, undersealing portland cement concrete pavements, membrane envelopes, and hydraulic applications.
>
> **Catalytically blown:** Air-blown asphalt produced by using a catalyst during the blowing process.
>
> **Cutback:** Asphalt cement that has been liquefied by blending with petroleum solvents (*Also called* diluents). Upon exposure to atmospheric conditions the diluents evaporate, leaving the asphalt cement to perform its function of cementing and waterproofing. It is designated as follows:
>
>> **Medium-curing (MC) asphalt** Cutback asphalt composed of asphalt cement and kerosene-type diluent of medium volatility.
>>
>> **Rapid-curing (RC) asphalt** Cutback asphalt composed of asphalt cement and a naptha or gasoline-type diluent of high volatility.
>>
>> **Road-oil** A heavy petroleum oil, usually one of slow-curing (SC) grades.
>>
>> **Slow-curing (SC) asphalt** Cutback asphalt composed of asphalt cement and oils of low volatility.
>
> **Emulsified:** Emulsion of asphalt cement and water that contains a small amount of an emulsifying agent, a heterogeneous system containing two normally immiscible phases (asphalt and water) in which the water forms the continuous phase of the emulsion, and minute globules of asphalt form the discontinuous phase. Emulsified asphalts may be of either the anionic, electronegatively charged asphalt globules, or cationic, electropositively charged asphalt globule types, depending on the emulsifying agent.
>
> **Hard:** Solid asphalt with a normal penetration of less than 10.

Mineral-filled: Asphalt containing finely divided mineral asphalt passing a No. 200 sieve (0.074 mm, 0.0029 in.).

Natural (native): Asphalt formed naturally, and generally found in lakes at or near the surface of the earth. It is mined with special equipment and can be used with little or no further processing.

Oxidized: *See* **Air-blown,** above.

Powdered: Hard, solid asphalt having a normal penetration grade of less than 10. For use, it is pulverized to a very fine state with 100% passing a #100 mesh (0.149 mm, 0.0059 in.) sieve.

Rock: Porous rock such as sandstone or limestone that has become impregnated with natural asphalt through geologic process.

Sand: Asphalt mix incorporating aggregate that will pass a 13 mm (0.5 in.) sieve.

Saturating: Type of asphalt used to saturate felt used in roofing materials.

Semisolid: Asphalt that is intermediate in consistency between emulsified and cutback asphalt products and solid or hard asphalt; that is, normally has a penetration between 10 and 300.

Solid or hard: Asphalt having a normal penetration of less than 10.

ASPHALT BASE COURSE Foundation course or pavement layer consisting of mineral aggregate bound together with asphalt material on which successive course(s) are placed. *See also* **Base course.**

ASPHALT BINDER COURSE Intermediate course between base and asphalt courses, usually coarse-graded aggregate asphaltic concrete, but seldom containing mineral matter passing a # 200 sieve.

ASPHALT BLOCK PAVEMENT pavement in which the surface course is constructed of asphalt/aggregate blocks. These blocks are laid in regular courses, as in the case of brick pavements.

ASPHALT BLOCKS Asphalt concrete molded under high pressure. The type of aggregate mixture composition, amount and type of asphalt, and the size and thickness of blocks are varied to suit usage requirements.

ASPHALT CEMENT 1. Cement prepared by refining petroleum until it is free of water and all foreign material, except the mineral matter naturally contained in the asphalt. It should contain less than 1% ash; **2.** Fluxed or unfluxed asphalt specially prepared as to quality and consistency for direct use in the manufacture of bituminous pavements, and having a penetration at 25°C (77°F) of between 5 and 300, under a load of 100 kg (220 lb) applied for 5 seconds.

ASPHALT COLOR COAT Asphalt surface treatment of mineral aggregate selected to produce a desired color.

ASPHALT CONCRETE High-quality, thoroughly controlled hot mixture of asphalt cement and well-graded, high-quality aggregate, thoroughly compacted into a uniform dense mass.

ASPHALT CUTTER Attachment that fits to a backhoe or excavator stick, or to the floor of a bucket, or to a toolcarrier, consisting of a sharp metal wheel that can be pressed into an asphalt surface and that cuts as it is drawn along. Typically will cut to depths up to 125 mm (5 in.).

ASPHALT CUTTINGS An expression referring to material produced by cold planing/milling operations.

ASPHALT DIP Coating and lining for steel culverts.

ASPHALT EMULSION Emulsion where asphalt is the dispersed liquid or internal phase and water is the dispersing liquid or external phase, commonly called oil-in-water emulsion.

ASPHALT EMULSION SLURRY SEAL Mixture of a slow-setting emulsified asphalt, fine aggregate, and mineral filler, with water added to produce slurry consistency.

ASPHALT EMULSION STABILIZATION Mixing of an asphalt emulsion with aggregates to improve cohesion of the particles and provide waterproofing.

ASPHALT EXPANSION JOINT Felt composition strip used to separate materials subject to extensive thermal expansion, such as concrete road slabs, sidewalks, etc.

ASPHALT FILLER Type of asphalt used to fill cracks and joints in pavement and structures. *Also called* **Asphalt joint filler.**

ASPHALT FOG SEAL Light application of a slow-setting asphalt emulsion diluted with water and without mineral aggregate cover.

ASPHALTIC ADHESIVE *See* **Adhesive.**

ASPHALTIC CONCRETE High type asphalt mix used for construction of roadways for high traffic volumes and axle loads, generally plant mixes produced from dense-graded aggregates with close tolerance and from 4% to 9% asphalt cement.

ASPHALTIC MASTIC Mixture of asphalt and mineral materials, used especially for roofing and for dampproofing.

ASPHALT INTERMEDIATE COURSE Asphalt pavement course between a base course and an asphalt surface course. *Also called* a **Binder course.**

ASPHALT JOINT FILLER *See* **Asphalt filler.**

ASPHALT JOINT SEALER Asphalt product used for sealing cracks and joints in pavements and other structures. *See also* **Preformed asphalt joint sealers.**

ASPHALT LEVELING COURSE Course (asphalt aggregate mixture) of variable thickness, used to eliminate irregularities in the contour of an existing surface prior to superimposed treatment or construction.

ASPHALT MACADAM Type of pavement construction using a coarse, open-graded aggregate that is usually produced by crushing and screening stone, slag, or gravel. Such aggregate is called macadam aggregate. Asphalt may be incorporated in macadam construction either by penetration or by mixing.

ASPHALT MASTIC Mixture of asphalt and fine material in such proportions that it may be poured hot or cold into place and compacted by troweling to a smooth surface.

ASPHALT MIXING PLANT Plant used to proportion and mix asphalt paving materials, generally at a central site.

ASPHALT OVERLAY One or more courses of asphalt construction on an existing pavement. The overlay may include a leveling course to correct the contour of old pavement, followed by uniform course or courses to provide needed thickness.

ASPHALT PAINT Asphaltic product in liquid form, sometimes containing small amounts of other materials such as lampblack and mineral pigments.

ASPHALT PAVEMENT Pavements consisting of a surface course of mineral aggregate coated and cemented together with asphalt cement on supporting courses such as asphalt bases; crushed stone, slag, or gravel; on portland cement concrete, brick, or block pavement.

ASPHALT PAVEMENT RECLAIMING To bring into a condition for renovation or other use by recovering existing asphaltic pavement.

ASPHALT PAVEMENT REMOVAL Removal of all of an existing asphalt mat.

ASPHALT PAVEMENT STRUCTURE Pavement structure with all its courses of asphalt-aggregate mixtures, or a combination of asphalt courses and untreated aggregate courses placed above the subgrade or improved subgrade.

ASPHALT PAVING MACHINE Self-propelled machine that receives material (asphalt, aggregate, or concrete) from a dump truck, windrow elevator, or other device and distributes or spreads it to a specified width and thickness on the area being paved. *Also called* **Lay-down machine**, and **Spreader**.

ASPHALT PAVING PLANT Plant, usually mobile, designed to heat and mix crushed aggregate with heated asphalt to a specified mix and consistency in preparation for its transport to the area being paved.

ASPHALT PENETRATION Measure of the hardness or consistency of asphalt expressed as the distance that a standard needle will penetrate a sample under known conditions of temperature, loading, and time.

ASPHALT PLANK Premolded mixtures of asphalt, fiber, and mineral filler, sometimes reinforced with steel or fiberglass mesh. It is usually made in 0.9 to 1.4 m (3 to 8 ft) lengths and 150 to 300 mm (6 to 12 in.) widths. Asphalt planks may also contain mineral grits that maintain a sandpaper-like texture throughout their life.

ASPHALT PRIME COAT Application of low-viscosity liquid asphalt to an absorbent surface, used to prepare an untreated base for an asphalt surface.

ASPHALT PRIMER Application of a low-viscosity cutback asphalt product to an absorbent surface. It is used to prepare an untreated base for an asphalt surface. The prime penetrates into the base and plugs the voids, hardens the top, and helps bind it to the overlying asphalt course.

ASPHALT ROOFING Roofing and waterproofing material composed of saturated asbestos or rag felt cemented together with asphalt or tar pitch.

ASPHALT-SAND Mixture of sand and asphalt cement and cutback or emulsified asphalt. It may be prepared with or without special control of aggregate grading and may or may not contain mineral filler. Either mixed-in-place or plant-mix construction may be employed. Asphalt-sand is used in construction of both base and surface courses.

ASPHALT SEAL COAT Thin asphalt surface treatment used to waterproof and improve the texture of an asphalt wearing surface. Depending on the purpose, seal coats may or may not be covered with an aggregate. The main types of seal coats are aggregate seals, fog seals, emulsion slurry seals, and sand seals.

ASPHALT SHINGLE *See* **Shingle**.

ASPHALT SOIL STABILIZATION Treatment of naturally occurring, nonplastic or moderately plastic soil with cutback or emulsified asphalt at ambient temperatures. Aeration and compaction of the asphalt-soil mixture produces water-resistant base and subbase courses of improved load-bearing qualities.

ASPHALT STABILIZATION Mixing of asphalt with sand or soil to improve granular cohesion, waterproofing, or other factor.

ASPHALT SURFACE COURSE Top course of an asphalt pavement. *Also called* an **Asphalt wearing course**.

ASPHALT SURFACE TREATMENT Several types of asphalt or asphalt-aggregate applications, usually less than 25 mm (1 in.) thick, to a road surface. The types range from a light application of emulsified or cutback asphalt to single or multiple surface layers made up of alternating applications of asphalt and aggregate.

ASPHALT TACK COAT Very light application of asphalt, usually asphalt emulsion diluted with water. It is used to ensure a bond between the surface being paved and the overlying course.

ASPHALT WEARING COURSE *See* **Asphalt surface course**.

ASPIRATE To draw by suction. *See also* **Engine, Naturally aspirated**.

ASPIRATOR MIXER Gas-air proportioning device that uses the venturi principle to cause the flow of combustion air to induce the proper amount of gas into the air stream, used with low-pressure and zero-pressure gas. *Also called* a **Suction-type mixer**.

ASR *Abbreviation for*: automatic sprinkler riser.

assem. *Abbreviation for*: assemble; assembly.

ASSEMBLAGE Combination of two or more parcels of land.

ASSEMBLY Group of materials or parts, designed to act as a coordinated whole.

ASSEMBLY DRAWING Drawing that shows the composite parts of a unit and how, and in what order they should be combined and fastened.

ASSEMBLY LANGUAGE Language used in computer programming that uses mnemonics for the binary words of machine language.

ASSEMBLY OCCUPANCY Occupancy or use of a building by a gathering of persons for civic, political, travel, religious, social, educational, recreational, or like purpose, or for the consumption of food or drink.

ASSEMBLY ROD External threaded rod or bolt used in machine assembly.

ASSEMBLY SEQUENCE Designated sequence in which parts or components must be assembled and/or fastened.

ASSEMBLY TIME Time necessary to assemble a group of materials and/or parts designed to form a complete unit.

ASSEMBLY TORQUE Designed torque applied at final assembly.

ASSESSED VALUE Value of property, established for the purpose of levying a tax.

ASSESSMENT Valuation and listing of property for the purpose of levying a tax upon it.

ASSESSMENT RATIO Ratio of assessed value of property to its established or estimated market value.

ASSESSMENT ROLL Listing of all taxable property within a property taxing jurisdiction.

ASSESSOR One who determines the value of property for tax purposes.

ASSET Something of value that is owned.

ASSIGN Transfer property or rights to another.

ASSIGNED RISK Any risk that an underwriter refuses to insure, but which it is required by statute to cover, and that is then assigned to a pool of insurers to divide coverage liability.

ASSIGNEE Person to whom a contract or agreement is assigned.

ASSIGNMENT Method by which a right, or contract, is transferred from one person to another.

ASSIGNMENT OF CONTRACT Document specifying that payments due on a contract be paid to other than the contractor or subcontractor.

ASSIGNER Person who assigns or transfers something to another.

ASSIZE Cylinder-shaped block of masonry forming part of a column; a layer of stone in a building.

assn *Abbreviation for:* association.

assoc. *Abbreviation for:* associate; associated.

Assoc. Std. *Abbreviation for:* associated standards.

AS-SPUN Prestressing strand supplied without heat treatment.

asst *Abbreviation for:* assist; assistant.

asst. *Abbreviation for:* assorted.

Asst. Dir. *Abbreviation for:* assistant director.

ASSUMABLE MORTGAGE *See* **Mortgage.**

ASSUMED LIABILITY Liability flowing from agreement between people, as against liability arising from common or statutory law.

ASSUMPTION OF MORTGAGE Obligation by the purchaser of property to be liable for the outstanding terms of an existing mortgage.

assur. *Abbreviation for:* assurance.

ASSURANCE LEVEL Minimum percentage of pressure containing devices of a verified design that will sustain 10 million applications of its rated fatigue pressure.

assy *Abbreviation for:* assembly.

A-STAND 1. On a geared machine, the structure that supports the outboard bearing and end of the drive sheave shaft; **2.** On a gearless machine, the structure that supports either end of the rotating armature shaft assembly and its bearings.

ASTEL Plank used for overhead partition in tunneling.

ASTRAGAL Small molding round the top or bottom of columns, often decorated with a bead-and-reel enrichment.

asym. *Abbreviation for:* asymmetrical.

ASYNCHRONOUS COUNTER Solid-state binary counter involving the use of flip-flops.

ASYNCHRONOUS MACHINE AC current-carrying machine in which the rotor does not turn at synchronous speed.

at. *Abbreviation for:* airtight,

AT *Abbreviation for:* air temperature; asphalt tile.

ATB *Abbreviation for:* asphalt tile base.

AT GRADE That part of a structure or site feature located at the same elevation as the adjacent ground level.

AT-GRADE INTERSECTION *See* **Intersection types.**

ATHEY WAGON Large, bottom-dump trailer.

ATHLETIC SURFACING Surfacing manufactured of synthetic materials, used to cover surfaces intended for athletic events.

atm. *Abbreviation for:* atmosphere; atmospheric.

ATMOMETER Device that measures the rate of evaporation of water to the atmosphere.

ATMOSPHERE Refers to a standard weight of air. A standard atmosphere for purposes of direct comparison is considered to be sea level atmosphere of 1 kg/645 mm^2 (14.7 lb/in.^{2}) at 15°C (59°F), or measured as 749 mm (29.95 in.) of mercury on a barometer.

ATMOSPHERE FOR TESTING Air at ambient humidity and temperature.

ATMOSPHERIC CORRECTION Correction applied to a measured distance to allow for the effects of air temperature and atmospheric pressure on the speed of light.

ATMOSPHERIC INVERSION Condition in which the air temperature increases with height, preventing the normal tendency of heat to rise from the ground.

ATMOSPHERIC PRESSURE *See* **Pressure.**

ATMOSPHERIC-PRESSURE STEAM CURING Steam curing of concrete products or cement at atmospheric pressure, usually at maximum ambient temperatures between 40°C and 95°C (100°F and 200°F).

See also **Autoclave curing.** *Also called* **High-tem-perature steam curing,** and **High-pressure steam curing.**

ATMOSPHERIC (RATIO) REGULATOR Dia-phragm-type regulator that maintains a gas pressure at atmosphere or 'zero' pressure. *Also called* a **Zero governor.**

ATMOSPHERIC STORAGE TANK Storage tank designed to operate at pressures from atmospheric to 3.5 kPa (0.51 psi) gauge.

atm. press. *Abbreviation for:* atmospheric pressure.

ATOM Smallest unit of an element that retains the properties of that element.

ATOM ARC WELDING Welding with heat gener-ated by hydrogen atoms created by an electric arc recombining to form hydrogen molecules.

ATOMIZATION Disintegration of solids or fluids into minute particles; of a fuel into many small parti-cles prior to injection into a combustion chamber. *See also* **Air atomization, Impact atomization,** and **In-jection atomization.**

ATOMIZE Convert to a fine spray.

ATOMIZING AIR That part of the air supplied through a burner (usually about 10%) that is used to break the oil stream into tiny droplets. The atomizing air is also used for combustion after it has broken up the oil stream.

a/trans. *Abbreviation for:* automatic transmission.

ATRIUM Originally, in Roman domestic architecture, an inner court open to the sky and surrounded by the roof. Later, a covered portico, especially before a church. More recently, a multistory vestibule or entrance, usu-ally with at least one glass wall and a major portion of the roof also glazed.

att. *Abbreviation for:* attached; attend.

ATTACHED Permanently joined; cemented; fastened; (wall) common to another structure.

ATTACHED COLUMN Column attached to a wall.

ATTACHED DWELLING One of three of more dwellings separated from its neighbors by a fire wall, and containing its own entrance. It may be further divided vertically to contain up to two separated living spaces, each having its own entrance. *Also called* **Row housing.** *See also* **Semidetached dwelling.**

ATTACHED PIER Pier completely attached to a wall or other support.

ATTACHMENT 1. Device specifically designed to perform a function or purpose and for a particular vehicle, machine, or item of equipment, that does not form part of the basic vehicle, machine, or equipment; **2.** Legal seizure of property, typically to force payment of a debt.

ATTACHMENT PLUG Device that, by insertion in a receptacle, establishes connection between the electri-cal conductors of the attached flexible cord and the conductors connected permanently to the receptacle.

ATTACK LINE Line of hose used to fight a fire.

ATTENUATE Make slender or thin; lessen in force, power or severity; dilute. *See also* **Sound attenuation.**

ATTERBERG LIMITS Arbitrary water contents (shrinkage limit, plastic limit, liquid limit) determined by a standard test that define the boundaries between the different states of consistency of plastic soils. *Also called* **Consistency limits.** *See also* **Plasticity index,** and **Plastici limit.**

ATTERBERG TEST Method for determining the plasticity of soils.

ATTEST To witness by observation and signature.

ATTIC Space within a pitched or sloped roof; room within that space.

ATTIC BASE Base to a Doric or Ionic column con-sisting of torus, scotia, and lower torus with fillets between.

ATTIC FAN *See* **Fan.**

ATTIC TRUSS *See* **Truss.**

ATTIC VENTILATORS Louvers and other vents, passive or power-operated, intended to vent to air-space within roofs.

attn *Abbreviation for:* attention.

ATTO Prefix representing 10^{-18}. Symbol: a. Used in the SI system of measurement. *See also the appendix:* **Metric and nonmetric measurement.**

ATTORNMENT Formal agreement by a tenant of a new landlord.

ATTRITION Removal of material from a surface by friction wear.

AUCTION Sale of property and/or goods to the high-est bidder.

aud. *Abbreviation for:* audible; auditorium.

AUDIBLE ALARM Horn, siren, bell, or buzzer that is used to attract the attention of a machine operator when a fault occurs.

AUDIBLE SIGNAGE Mechanical annunciation of information or directions. *See also* **Signage.**

AUDIBLE SOUND Sound frequencies between ap-proximately 15 and 20,000 Hz with sufficient sound pressure to be detectable.

AUDIOGRAPH Printout from a machine that deter-mines the rate at which sound levels diminish over distance; when used in an enclosed area this is an indication of the sound absorption quality of the mate-rials used in construction.

AUDIOMETRIC ROOM Room constructed and fin-ished of materials that, in combination, will absorb over 99% of sound generated within the room, isolated from the remainder of the structure by springs to elimi-nate the transfer of vibrations.

AUDIOVISUAL WARNING SYSTEM Alarm de-vice that signals the operator of equipment of out-of-tolerance conditions or potentially hazardous situa-tions. These include such circumstances as low engine oil pressure, high engine coolant temperature, high hydraulic and/or transmission oil temperature, two-block condition, potential overload, etc.

AUDIT Examination to determine the reliability of a record or claim, or to evaluate compliance with rules or policies or with conditions of an agreement.

AUDITORIUM That part of a public hall dedicated to an audience.

AUDIT TRAIL Route by which data can be tracked, forward or backward, through a processing cycle.

aug. *Abbreviation for:* augment.

AUGER 1. Rotating drill having a screw thread to carry cuttings away from the surface being drilled. *See also* **Bit**; **2.** Hand-operated soil sampler; **3.** One of two devices, mounted on the rear bulkhead of a tractor (and in front of a screed). The augers are powered by a drive common to the conveyor on the respective side of the machine and work in concert with the conveyor to distribute paving materials laterally in front of the screed. *See also* **Scraper**.

AUGER BORING Technique for conduit installation using a nonsteerable rotating cutting head attached to auger flights within a casing.

AUGER FLIGHT Section of a continuous-flight auger, quick-coupled to other sections to form a continuous auger.

AUGERED PILE *See* **Pile**.

AUGER TBM Type of microtunneling machine in which the excavated soil is transferred to the drive shaft through auger flights passing through the product pipe.

AUGER-TYPE SCRAPER *See* **Scraper**.

AUSTENTITE Steel alloy that is heat treated in a specific way to make it particularly tough.

auth. *Abbreviation for:* authority; authorization; authorized.

AUTHORITY Sum of the right and powers assigned to a position or vested in a person.

AUTHORIZED APPROPRIATION *See* **Cost types**.

AUTHORIZED PERSON Person approved or assigned by the employer to perform a specific type of duty or duties or to be at a specific location or locations at the jobsite. *See also* **Competent person**, and **Designated person**.

auto. *Abbreviation for:* automatic.

AUTOCLAVE Pressure vessel in which an environment of steam at high pressure may be produced, used in curing of concrete products and in the testing of hydraulic cement, and for vulcanizing rubber products.

AUTOCLAVE CURING Curing of hydraulic cement-bound products in an autoclave at maximum ambient temperatures generally between 170°C to 215°C (340°F to 420°F).*See also* **Atmospheric pressure steam curing.** *Also called* **High-pressure steam curing.**

AUTOCLAVE CYCLE 1. Time interval between the start of the temperature-rise period and the end of the blowdown period; **2.** Schedule of time and temperature-pressure conditions or periods that make up the cycle.

AUTOGENOUS HEALING Natural process of filling and sealing cracks in concrete or in mortar when kept damp.

AUTOGENOUS LENGTH CHANGE Length change caused by autogenous volume change.

AUTOGENOUS VOLUME CHANGE Change in volume produced by continuous hydration of cement, exclusive of effects of applied load and change in either thermal condition or moisture content.

AUTOGRAPHIC DILATOMETER Instrument that records instantaneous and continuous variations in dimensions and some other controlled variable, such as temperature or time.

AUTOIGNITION TEMPERATURE Temperature at which a material ignites in contact with a hot surface, in the absence of any other source of ignition.

AUTOMATIC Self-acting, operating by its own mechanism when actuated by some impersonal influence.

AUTOMATIC ADVANCE Device that senses engine speed and, in a diesel engine, automatically varies the pump-to-engine timing or, in a gasoline engine, advances the timing of spark ignition relative to piston top-dead-center.

AUTOMATIC BATCHER *See* **Batcher**.

AUTOMATIC BRAKE *See* **Brake**.

AUTOMATIC BUCKET CONTROL Device that permits an operator to stop the bucket at a preset digging angle to shorten the cycle time and free the operator to concentrate on maneuvering.

AUTOMATIC CIRCUIT RECLOSER Self-controlled device for automatically interrupting and reclosing an alternating current circuit with a predetermined sequence of opening and reclosing followed by resetting, hold closed, or lockout operation.

AUTOMATIC CLUTCH *See* **Clutch**.

AUTOMATIC CONTROL Arrangement by which a system reacts to change or an unbalance in one or more variables and compensates by adjusting the other variables to restore the system to the desired balance.

AUTOMATIC COUNT Particle count obtained by an electromechanical or electronic device, as opposed to visual microscopic counting techniques.

AUTOMATIC COUPLER Trailer connection that may be connected without manual actuation.

AUTOMATIC COVER Clause in some types of property insurance giving immediate coverage for a limited period on property newly acquired by the policy holder.

AUTOMATIC ELEVATOR Type of elevator that does not require an attendant.

AUTOMATIC FEEDER CONTROL Device that controls the level of mix in the screw chamber ahead of an asphalt paver screed, generally by use of a paddle riding on the material that is attached to a potentiometer.

AUTOMATIC GRADE CONTROL Electronic device that controls or maintains a given screed elevation as dictated by a designated reference.

AUTOMATIC GROUTER Foam-rubber-faced, pressurized steel form that forces grout between and around stone or brick.

AUTOMATIC OPERATION Mode of operation of a system wherein it performs its specific function without operator adjustments after the initial setup.

AUTOMATIC PILE HAMMER *See* **Pile hammer**.

AUTOMATIC REINSTATEMENT Insurance policy provision calling for the original coverage to apply following settlement of a claim, as opposed to the coverage being reduced by the amount of loss paid.

AUTOMATIC SEAL Seal that is activated by the pressure of the fluid that it seals.

AUTOMATIC SLOPE CONTROL Electronic device that controls or maintains a given slope by sensing changes (or errors) from level (zero) grade.

AUTOMATIC SPRINKLER Nozzle, attached to a water supply system, fitted with an orifice with a flow control device and a deflector that will distribute water over a given area at a prescribed rate and in a designed manner. There are several types, including:

> **Frangible-bulb:** Fire protection sprinkler head incorporating a bulb filled with a liquid and a bubble of air. When the liquid is heated it absorbs the air causing the bulb to break, allowing water to flow from the wet-type distribution system through the sprinkler head.

> **Frangible-pellet:** Fire protection sprinkler head incorporating a pellet of solder that melts at a given temperature, releasing the control device of the sprinkler and allowing water from the distribution system to flow through the sprinkler head.

> **Solder-link:** Sprinkler head connected to a water supply system by a link constructed of an alloy of tin, lead, cadmium, and bismuth in proportions that will melt at a determined temperature. When connected to a wet-type fire-protection sprinkler system, such a link will melt if the ambient temperature at its location rises to the preset limit, allowing water to pass from the distribution system and through the sprinkler head.

AUTOMATIC SPRINKLER SYSTEM An integrated fire protection system of underground and overhead piping designed in accordance with fire protection engineering standards and including a water supply, such as a gravity tank, fire pump, reservoir, or pressure tank and/or connection by underground piping to a municipal supply. The system is usually activated by heat from a fire and discharges water over the fire area. There are several types of system:

> **Combination system:** One that combines two or more defined systems.

> **Deluge system:** Used where simultaneous operation of all sprinklers is required for complete discharge over a total area.

> **Dry system:** Used in unheated areas subject to freezing.

> **Preaction system:** Used where accidental discharge or where water leakage in the absence of fire can cause damage to contents, usually of high or critical value.

> **Wet system:** Used only in heated areas not subject to freezing.

AUTOMATIC TRACTION CONTROL (ATC) Optional, traction-control device that is available on both 4- and 6-channel ABS-equipped vehicles.

AUTOMATIC TRANSFER DEVICE Mechanism that automatically moves a load to and from the platform of a material lift or dumbwaiter.

AUTOMATIC TRANSMISSION *See* **Transmission.**

AUTOMATIC WELDING Welding where the arc is mechanically moved while controls govern the speed and/or direction of its travel.

AUTOMATION Use of mechanical and/or electronic processes to complete actions or processes otherwise carried out or controlled by humans.

auto. recl. *Abbreviation for:* automatic reclosing.

AUTO TRANSFORMER Iron core transformer that has a single taped winding serving as both the primary and secondary.

aux. *Abbreviation for:* auxiliary.

AUXILIARY Anything additional or supplementary: a helper, standby generator, etc.

AUXILIARY BOOM NOSE Boom nose used in conjunction with an auxiliary hoist to provide the operator with the versatility of tandem line operations. The auxiliary boom nose is connected to the main boom nose by bolts. It can be removed to reduce boom weight, increasing the machine's lifting capacities. *Also called* a **Rooster sheave.**

AUXILIARY CREEPER TRANSMISSION Additional transmission used in a mobile crane carrier to provide a low gear ratio for slow, careful movement around the jobsite, plus higher ratios for normal use.

AUXILIARY ENERGY SUBSYSTEM In a solar energy system, equipment using energy other than solar both to supplement a solar system and to ensure adequate energy backup.

AUXILIARY EQUIPMENT 1. Equipment, additional to the basic chassis, that is required for a piece of equipment or vehicle to perform its design function(s); **2.** Accessory equipment necessary for the operation of a process train.

AUXILIARY FUEL PUMP Pump separate from the prime mover, usually used where the main fuel storage is some distance from the engine fuel pump.

AUXILIARY FUEL TANK Fuel tank installed in addition to a standard-equipment tank.

AUXILIARY HEAT Heat that is supplemental to that provided by a conventional system.

AUXILIARY HOIST Second lifting device that increases the versatility of the prime lifting machine (crane, etc.) by allowing two separate lift lines to operate independently. Traditionally the auxiliary hoist has powered a single part of the line and freed the main hoist for multiple reeving for increased lifting capacity.

AUXILIARY LANE *See* **Lane.**

AUXILIARY LIFTING SHEAVE Unit that connects to the boom head machinery and is used for reeving the winch rope for a second (auxiliary) winch drum.

AUXILIARY LOAD Dynamic load, other than the basic design load, that a structure must safely withstand.

AUXILIARY-POWER ELEVATOR Elevator having a source of mechanical power, such as shafting, in common with other machinery.

AUXILIARY RAFTER Second principal rafter in large queen-post trusses.

AUXILIARY REAR SPRING *See* **Spring.**

AUXILIARY REINFORCEMENT Any reinforcement in addition to that participating in the prestressing function.

AUXILIARY ROPE-FASTENING DEVICE Device attached to an elevator car or counterweight, or to the overhead dead-end rope-hitch support, designed to function automatically to support the car or counterweight in case the regular wire-rope fastening fails at the point of connection.

AUXILIARY SECTION Back section of a range-type transmission where range shifting occurs (most likely air-operated). This section houses the auxiliary drive gear, auxiliary mainshaft assembly, auxiliary countershaft, and synchronizer assembly.

AUXILIARY SPRING *See* **Spring.**

AUXILIARY TOWING LIGHTS Stop, tail, or turn-signal lights attached to the trailing end of a towed vehicle, operated as part of the towing vehicle lighting system.

avail. *Abbreviation for:* available.

AVAILABILITY FACTOR Measure of the time that a machine is available to be used against the time it is unavailable due to servicing or mechanical failure: actual working time divided by available working time, in hours.

AVAILABLE CARBON Carbon not combined chemically with oxygen in any way.

AVAILABLE FAULT CURRENT Maximum short-circuit current that can flow in an unprotected circuit.

AVAILABLE HEAT Gross quantity of heat released within a combustion chamber minus both the dry flue gas loss and the moisture loss. It represents the quantity of heat remaining for useful purposes.

AVAILABLE HYDROGEN Hydrogen not chemically combined with oxygen in any way.

AVAILABLE OXYGEN Quantity of atmospheric oxygen dissolved in a given amount of water.

AVALANCHE PROTECTOR Guard plates that prevent loose material from sliding into contact with the wheels or tracks of a digging machine.

avdp. *Abbreviation for:* avoirdupois.

ave *Abbreviation for:* avenue.

AVERAGE BOND STRESS Force in a bar divided by the product of the perimeter and the development length of the bar.

AVERAGE COST FLOW *See* **Cost flow.**

AVERAGE COVER Sum of the product of the cross-sectional area of each reinforcing bar provided to resist tension due to ultimate loads and the distance from the nearest exposed face, divided by the total area of the bars.

AVERAGE DAILY TRAFFIC *See* **Volume.**

AVERAGE DIMENSION Average of corresponding dimensions of random sample units.

AVERAGE (ELECTRIC) DEMAND *See* **Demand (electric).**

AVERAGE END AREA Method of calculating the volume of earthwork between two cross sections where the volume equals the average cross-sectional areas multiplied by the distance between the two cross sections.

AVERAGE FROST PENETRATION Numerical average based on historical data of the depth of frost penetration for a given area or locality.

AVERAGE HAUL Average distance material is hauled from the cut to the fill (sometimes calculated from a mass diagram). *See also* **Free haul, Haul, Overhaul,** and **Station yards.**

AVERAGE OVERALL TRAVEL SPEED *See* **Speed.**

AVERAGE RELATIVE HUMIDITY Average of the maximum and minimum relative humidities.

AVERAGE RUNNING TIME *See* **Speed.**

AVERAGE SEA LEVEL Mean average elevation of sea water at a designated location, used as a reference for other points measured above or below seal level.

AVERAGE SPOT SPEED *See* **Speed.**

AVERAGE YARDING DISTANCE Total yarding distance for all turns divided by total number of turns for a particular setting.

AVERAGING GRADE REFERENCE Device that is attached to and towed by an asphalt paver that uses randomly spaced, articulated springs or wheels or a long ski to average the elevation of the grade being paved, used as a reference for automatic screed controls.

AVERAGING SKI An integral part of the automation system used to average an asphalt reclamation planing pass to a consistent profile. *Also called* **Leveling ski.**

avg. *Abbreviation for:* average.

AVIATION SNIPS *See* **Tin snips.**

av. l. *Abbreviation for:* average length.

AVOIDED COST Tipping fees that are avoided by reducing the volume of solid waste destined for a landfill or combustor, using waste reduction, recycling, and/or composting.

AVULSION Natural transfer of land from one parcel to another, as when a stream or river changes its channel.

av. w. *Abbreviation for:* average width.

av. wt. *Abbreviation for:* average weight.

AW *Abbreviation for:* Axle width.

A/W *Abbreviation for:* all weather; actual weight.

AWARD Give, allocate, designate; the decision to accept a proposal or offer.

AWARDING AGENCY Entity for whom contract work is done.

AWARD LETTER Letter sent in response to an offer, accepting that offer.

AWARD OF CONTRACT Formal process, in writing, that accepts a tender or bid and gives the contractor/supplier authority to proceed.

awd *Abbreviation for:* award.

AWG *Abbreviation for:* American wire gauge.

AWL Small, sharp-pointed hand tool, used for making starting holes for nails or screws.

awn. *Abbreviation for:* awning.

AWNING Roof-like cover for a door or window, often of canvas on a metal frame.

AWNING WINDOW Frame containing one or more sashes, each of which is installed in a vertical plane and is hinged to permit the bottom of the sash to open outward. *See also* **Sash.**

AWS *Abbreviation for:* all wood screws.

AWWF *Abbreviation for:* All-Weather Wood Foundation.

ax. *Abbreviation for:* axis.

AXE Hand tool, part of a faller's safety equipment that can be used for both chopping and pounding (driving in wedges, for instance), as well as to plumb the lean of a tree. A heavy, slightly curved steel wedge some 100 to 200 mm (4 to 8 in.) long, attached at its widest point to a haft ranging in length from 500 to 900 mm (20 to 36 in.) long.

AXED WORK Stone dressed with an axe.

AXIAL LOADING Force applied along the length of a member.

AXIAL MOTION Motion occurring parallel to the centerline of the object; in the case of an expansion joint, parallel to the centerline of the bellows and can be either extension or compression.

AXIAL-PISTON HYDRAULIC PUMP *See* **Pump.**

AXIAL SEAL Shaft seal in which the packing member approaches the surface to be sealed in an axial direction. The primary seal function is at right angles to the axis of rotation.

AXIS 1. Real or imaginary line on which an object supposedly (as in a pictorial projection) or actually (as with a shaft) rotates; **2.** Center line of a tunnel or shaft. *See also* **Arch.**

AXIS LINE Point in a building from which an elevator is located.

AXIS OF ROTATION 1. Vertical line representing the axis around which an object rotates; **2.** Vertical line about which a crane or derrick rotates. *Also called* **Swing axis.**

AXIS OF SYMMETRY Line about which something geometrically balanced is developed, and on which the center of gravity is located.

AXIS OF WELD Imaginary line along the center of gravity, and perpendicular to a crosssection of the weld metal.

AXLE 1. Bar or rod supporting wheels; **2.** Assembly consisting of a housing, carrier, and wheel ends that joins a vehicle's wheels, transmits power to the wheel, helps support the vehicle's load, and serves as a mounting platform for the brakes. There are many configura-

tions, including:

Air lift: Single, air-actuated vertically pivoted axle that can be lowered to increase the number of load-bearing axles and thus a vehicle's load-carrying capacity.

Bogie: Two or more axles mounted to a frame so as to distribute the load between axles and permit vertical oscillation of the axles.

Dead: Load-carrying, nondriving axle.

Double-reduction: Tandem planetary gear sets that provide two-stage drive-shaft speed and higher levels of torque.

Drive: Any axle that is turned by engine power and therefore able to move the vehicle.

Drop: Axle in which the centerline of the spindles is higher than the centerline of the beam portion.

Dual-drive tandem: Both axles have drive mechanisms that are connected to the engine power unit.

Flip: Additional trailer axle that can be swung vertically through 180° when not needed to distribute a load to sit inverted on the trailer bed.

Front: The front axle of a vehicle has two primary purposes: support the front of the truck and load, and provide good maneuverability through steering control. It may also provide full- or part-time driving power.

Full-floating: Rear axle design that provides support for the payload and vehicle weight by the axle housings. The wheels are driven by splined shafts that 'float' within the axle housing.

Lazy: *See* **Tag,** below.

Live: Axle that provides the means of support for the wheels at each end and which connects the wheels with members that rotate with them.

Multi: Axles spaced along the length of a trailer chassis so as to uniformly distribute the load being carried.

Nonpowered: *See* **Dead axle,** above.

Planetary: Axle with specific gearing to provide increased torque for improved gradability on uneven ground.

Powered: Axle that supports a portion of the vehicle weight and transmits a driving force to the wheels.

Pusher tandem: Where only the rearmost axle is a driving type and the forward unit is free-rolling (load-carrying only), commonly called **Dead** (above).

Rear: The rear axle has two primary purposes: to support the rear of the truck and load, and to transmit driving motion to the rear wheels (if it is a driving axle).

Rear double-reduction: Axle having a primary reduction through a hypoid or split bevel and ring gear and a secondary reduction through a set of herringbone or helical gears.

Semi-floating: Rear axle design that provides for support of the payload and vehicle weight by the axle shaft, through the wheel bearing to the axle housing.

Set-back: Vehicle configuration where the front steering axle is purposely located further toward the rear. This permits heavy loads to be applied to the front axle and increases maneuverability.

Set-forward: Vehicle configuration where the front steering axle is set as close to the front of the vehicle as the design and wheel and tire size will permit.

Single-drive: Rear axle assembly with one single-drive axle unit.

Single-reduction: Type of rear-drive axle featuring one driving pinion and ring-gear set to turn the axle shafts.

Spread-tandem: Two-axle assembly in which the axles are spaced to allow maximum axle loads under existing regulations: the distance generally being more than 1400 mm (55 in.).

Steering: Powered or nonpowered axle through which directional control of a vehicle is applied.

Swing: Drive axle arrangement where the differential is mounted rigidly on the vehicle frame and the axle shafts are allowed to 'swing' as the vehicle moves up and down while running.

Tag: Nonpowered rear axle on a tractor-trailer combination. *Also called* **Lazy**, above, and **Trailer**, below.

Tandem-drive: Two axles mounted as a group with a means of distributing the imposed load.

Three speed: Rear axle design in which three-speed final-drive ratios are obtained as follows: two-speed axles in tandem are operated in either low, intermediate, or high range. In low range, the range selector of both axles is shifted to low range. In intermediate range, the range selector of the forward axle is shifted to high, while the selector of the rear axle is shifted to low. The power divider unit of the forward axle splits the difference of the ranges, thus effecting the intermediate range. In high range, the range selector of both axles is shifted to high.

Three-speed tandem: Two regular-type matched two-speed rear axles on a single suspension with an inter-axle differential.

Trailer: Nondriving rear axle behind a driving axle or axles. *Also called* **Tag axle**, above.

Trailing tandem (tag): Where the forward unit of the tandem is a driving type and the rear unit is free rolling.

Triaxle: Three axle assembly with a means of distributing an imposed load.

Trunnion: Stub axle pivoted at or near its mid-point about a horizontal axis transverse to its own centerline.

Two-speed: Rear axle design that uses the same type gearing as double-reduction axles and a shifter collar and actuating mechanism that allows the operator to lock out or engage the secondary reduction. This facilitates the shift

from fast ratios to slow ratios.

AXLE CAMBER Designed convexity of a trailer axle that compensates for road crowning.

AXLE HOUSING Covering to the internal mechanisms of a driving axle that also supports part of the load.

AXLE LOAD Portion of the gross weight of a vehicle transmitted to a structure or a roadway through the wheels supporting a given axle.

AXLE LOCKOUT Locking of the rear axle in a position parallel to the surrounding ground. Lockout is achieved through an automatic lockout valve that is activated when the machine (crane, etc.) is on rubber and the boom is more than 6° from the centerline of the machine's front lifting quadrant. This feature increases the crane's stability on rubber for over-the-side lifts.

AXLE MOUNTING CENTERS Distance between the right and left axle mounting pads required to bolt an axle to the vehicle.

AXLE OSCILLATION Rocking motion of an axle when passing over rough terrain. In some designs this motion is regulated through hydraulic cylinders that are mounted to the axle. This produces a relatively stable superstructure (typically a crane lifting platform) while maintaining four-wheel (or more) contact with the ground.

AXLE PULLEY *See* **Sash pulley**

AXLE RATING *See* **Gross axle weight rating.**

AXLE RATIO Number of revolutions made by the drive pinion (input) for each revolution of the ring gear (output). Specifically, the number of revolutions of the driving gear compared to the number of revolutions of the driven gear.

AXLE RING GEAR Large bevel-driven gear on a driving axle's gearset.

AXLE SEAT Suspension component used to support and locate the spring on an axle. Formerly called a **Spring chair.**

AXLE SHAFT Component that transmits power from the carrier assembly to the wheel end.

AXLE SHIFT MECHANISM Control mechanism used to handle movement of a shift fork or sliding clutch in an inter-axle differential or a driver-controlled, full-locking main differential.

AXLE SPINDLE Machined journal or shaft at the end of an axle on which wheel bearings and seals are mounted and about which the wheel rotates.

AXLE STEEL Steel from carbon-steel axles from railroad cars.

AXLE-STEEL REINFORCING BARS Deformed reinforcing bars rolled from carbon-steel axles from railroad cars.

AXLE-TO-FRAME Distance from the center of the rear axles(s) to the end of the frame. *See also* **Chassis dimensions.**

AXLETREE Bar connecting two opposite wheels.

AXLE WEIGHT Weight carried by an axle and transmitted to the ground.

AXLE WIDTH Distance between the front wheels, measured from the centerline of the front tires.

AXONOMETRIC PROJECTION Three-dimensional view of a building or other object produced by geometrical drawing where the plan is set up to a convenient angle and the verticals are projected to scale. The result is that all dimensions on a horizontal plane, and all verticals, can be scaled, but diagonals and curves on a vertical plane are distorted.

ayd *Abbreviation for:* average yarding distance.

az *Abbreviation for:* azimuth.

AZEOTROPE Mixture whose vapor and liquid phases have identical compositions at a given temperature.

AZIMUTH Bearing or direction of a horizontal line measured clockwise from north (grid, magnetic, astronomic, etc.) and expressed in degrees. *See also* **Error of closure.**

AZIMUTHAL PROJECTION Map on which all points are shown on their true bearings from one central point.

AZIMUTH SENSOR *See* **Azimuth transducer.**

AZIMUTH TRANSDUCER Device that measures the azimuth, or rotational position of a superstructure of a crane or item of mobile equipment with respect to its base. *Also called* an **Azimuth sensor.**

AZIMUTH TRAVERSE Series of lines of sight related to each other by measured angles only.

B

B *See* **Veneer grade.**

B *Abbreviation for:* beam; button.

Ba *Abbreviation for:* barium.

BA *Abbreviation for:* bright annealed.

BABBITTED FASTENING Method of providing wire rope attachments in which the ends of the wire rope strands are bent back and are held in place in a tapered socket by means of poured molten babbitt metal.

BABBITT METAL Soft lead alloy, used in some plain-metal bearings and in wire rope end finishing.

BACHELOR DWELLING UNIT Unit with not more than one bedroom providing sleeping, eating, food preparation and sanitary facilities for one or two adults. Other essential facilities and services may be shared with other dwelling units. *Also called* **Efficiency apartment.**

BACK 1. Roof or top of an underground opening; **2.** Ore located between a given level and the surface, or between two levels; **3.** Top surface of a horizontal or inclined timber; **4.** Convex or upper part of a saw tooth; **5.** Surface opposite the face side of a piece of timber; **6.** Surface opposite the face side of a drywall panel.

BACKACTER *See* **Drag shovel.**

BACK ARCH *See* **Arch.**

BACKBAND Molding applied to the side of a door or window casing.

BACK BEAD Choker on the butt rigging furthest from the yarder. Usually the responsibility of the newest crew member.

BACK BEDDING *See* **Back putty.**

BACK BLADING Pushing soil in a windrow back into the position from which it came, usually with a grader.

BACK BLOCKING Strips of drywall paneling glued to the main paneling between studs and joists in horizontal hanging to reinforce the joint.

BACKBOARD Temporary board on the outside of a scaffold.

BACK BREAK Rock broken beyond the limits of the last row of holes.

BACK BUTTER Application of mortar on the back of a wall or floor tile to supplement the adhesive spread on the setting bed.

BACK CLEARANCE Angle between the face and the back of a saw tooth in the direction of the circumference or cutting circle.

BACK CLIP Metal clip that attaches to the back of drywall paneling, fitting into slots in (steel) framing.

BACK COATING Asphalt coating applied to the back of shingles or roll roofing.

BACK CONNECTED Where connections are made to normally unexposed surfaces of components.

BACK CORNER Location where the tailblock on the haulback side turns the haulback around the corner.

BACKCUT 1. Cut made into finish flooring or molding so that the piece will lie flat. **2.** Final falling cut. The backcut progresses until the tree starts to fall in its intended direction.

BACK DRAIN Perforated drain pipe, usually of 100 mm (4 in.) internal diameter, placed longitudinally behind the bottom of a retaining wall and surrounded by a stone filter, used to reduce water pressure acting against the rear of the wall.

BACK DROP *See* **Drop manhole.**

BACK END 1. Furthest point away from a landing or yarder in a setting; usually refers to the tailblocks. **2.** The furthest active logging area away from camp; **3.** End of a road system; **4.** Far end of a logging claim.

BACK-END MATERIALS RECOVERY Engineered system that provides for collection of discrete reusable materials from mixed wastes that have been burned or treated.

BACK-END SYSTEM Combination of system components that changes the chemical properties of a waste or converts its components into energy or compost. *See also* **Front-end process.**

BACKER Nondecorative laminate used on the back of composite panel constructions to protect the substrate from changes in humidity and to balance the panel construction.

BACKFILL 1. Material used to replace earth, etc., previously excavated, commonly into a trench or pier excavation, around and against a basement foundation. *Also called* **Soil fill.** *See also* **Fill,** and **Structural fill;** **2.** Loose earth or sand used to fill the space between the cylinder and the wall of a jack hole for a hydraulic excavation.

BACKFILL BLADE Small dozing blade fitted to a machine having a prime function other than dozing (trenching, excavating, etc.) so as to permit an operator to replace material excavated from a trench or hole.

BACKFILL CONCRETE Nonstructural concrete used to correct over-excavation, fill excavated pockets in rock, or prepare a surface to receive structural concrete.

BACKFILL DENSITY Percent compaction required or expected for backfill.

BACKFILL GROUT *See* **Low-pressure grout.**

BACK FILLING 1. Rough masonry built behind a facing or between two faces; **2.** Filling work over the extrados of an arch; **3.** Brickwork in spaces between structural timbers. *Also called* **Brick nogging,** and **Nogging.**

BACKFIRE 1. Fire started upwind of an existing brush fire, to burn against and cut off its spreading. *See also* **Area ignition, Controlled burning,** and **Prescribed burning; 2.** Explosion in the intake or exhaust passages of an engine.

BACK FLAP Type of hinge with large wings, used for screwing on the face of both door and frame.

BACKFLOW Flow of water or other liquids, mixtures, or substances into the distributing pipes of a potable supply of water from any source other than that intended, produced by the differential pressure existing between two systems, either or both of which are at pressure greater than atmospheric.

BACKFLOW PREVENTER Device used to protect potable water supplies against potential contamination due to cross-connections with nonpotable water supplies.

BACK FOLLOWER Piston, or portion of a piston, to which a cup is bolted.

BACKFORM *See* **Top form.**

BACKFURROW First cut of a plow, from which the slice is laid on undisturbed soil.

BACKGROUND CONTAMINATION Total of the extraneous particles that are introduced in the process of obtaining, storing, moving, transferring, and analyzing a fluid sample.

BACK GUTTER Gutter formed behind a chimney stack at its up-slope intersection with a pitched roof. *Also called* **Chimney gutter.**

BACK GUY Line behind the spar tree, opposite the main line, that takes most of the pull in yarding logs.

BACKHAND WELDING Welding in the opposite direction as the gas flame is pointing.

BACK HAUL 1. Line that pulls the bucket of a drag scraper backward from the dump point to the excavation; **2.** Load carried by a hauler or truck on the return journey from point of discharge or delivery of a different load.

BACK HEARTH *See* **Hearth.**

BACK-HITCH GANTRY *See* **Gantry.**

BACKHOE Bucket, boom, and stick assembly that digs by pulling the bucket toward itself. Used as an attachment to a self-propelled machine such as a tractor-dozer; as a companion tool on a loader-backhoe; as an attachment on a tool-carrier; as an attachment to mount on the bed of a pickup truck, etc. *Also called* **Hoe,** and **Pullshovel.**

BACKHOE/LOADER Multipurpose machine equipped at its front end with a bucket for loading, and at its rear with a backhoe. The bucket can often be interchanged with a range of types and sizes, and with other attachments such as snowplows, blades, forks, etc. The backhoe is usually a permanent fixture and also can be equipped with a range of attachments and buckets.

BACKHOE TAMPING Processing step, often used in direct-dump transfer systems, in which a conventional backhoe is used to compact waste contained in an open-top transfer trailer.

BACKING 1. Lengths of 50 x 100 mm (2 x 4 in.) fastened between framing studs to act as support for objects that might later be fixed to the finished interior face of the wall, such as cupboards; **2.** Some material placed on the root side of a weld to aid control of penetration; **3.** In tire work: the removable protective material used on the application side of tread rubber and repair materials to preserve cleanliness and tackiness; **4.** Layer of liner material on the underside of a sheeted product for mechanical reinforcement; **5.** Soft rubber layer between a tube and/or cover and the tire carcass to provide adhesion; **6.** Fire burning against the wind.

BACKING BOARD 1. Final board in a converted log, to which the carriage dogs were attached; **2.** Drywall panel used as the first or base layer in a multilayer system; **3.** Base layer of drywall in ceilings to which acoustic tile is adhered or attached.

BACKING DEALS Vertical planks behind the curbs in a mine shaft.

BACKING HIP RAFTER Beveled corner of a hip rafter that lines up with adjacent roof surfaces.

BACKING MATERIAL Disposable material applied to a self-adhesive surface to prevent it sticking to itself or to other surfaces prior to use or application.

BACKING RING Metal strip used when making a butt-welded joint on pipe to prevent melted metal from entering the pipe.

BACKING TIER Tier of rough brickwork that backs up the face tier of an exterior wall.

BACKING UP Laying the inside portion of a brick wall after the facing has been built header high.

BACK IRON Steel plate that supports and reinforces the cutting iron of a smoothing plane.

BACKLASH Clearance between gear teeth meshing with the teeth of another gear that permits some motion of one gear before power is transmitted to the other.

BACK LINE 1. That part of the haulback between the home tree and the corner block; **2.** Boundary line marked by blazed or painted trees indicating the cutting area.

BACK LINTEL Steel or concrete lintel over an opening that cannot be seen on the front face of the wall.

BACK LOG Used in a box crib, the back log holds the fill end of the tieback logs together and is laid parallel to the face log.

BACKLOG 1. Events, actions, or materials that are overdue for completion or supply; **2.** Forest land area where silviculture treatments such as site preparation and planting are overdue.

BACK-MOUNTED TILE Wall and floor ceramic tile packaged in sheet form mounted on material that forms part of the application.

BACK OBSERVATION *See* **Backsight.**

BACK ORDER Order for goods, or part of an order that cannot be filled from inventory and that must await later delivery.

BACK-OUT *See* **Nail pop.**

BACKPACK Personal equipment used to fight a brush or forest fire, consisting of a 19 L (5 gal) pump-tank extinguisher carried on a person's back and supported by straps, and having a hand-operated pump and spray nozzle.

BACK PLASTERING Plaster applied to one face of a lath system following application and subsequent hardening of plaster applied to the opposite face. *See also* **Parge.**

BACKPLATE Small plate fixed to the inside of a door through which the cylinder connecting screws and bar of a lock are passed. *See also* **Subplate.**

BACK PRESSURE *See* **Pressure.**

BACK PRIMING Application of a coat of primer to the back of a panel.

BACK PROP Raking strut used to transfer the mass of timber to the ground in deep trenches; usually placed below every second or third frame.

BACK PUTTY Mastic material placed in rabbets before installing glass, to provide a bed for the glass. *Also called* **Back bedding.**

BACKREST Wood bench opposite the tensioner, used to support a circular saw blade when checking tension.

BACK RING Split or multi-segment ring in a circumferential seal assembly.

BACK RIPPER Short spikes pivoted on the back of the blade of an angledozer, bulldozer, or other machine and mounted to face toward the rear. They bite into the surface when the machine moves backwards and float on the surface when the machine moves forward. *See also* **Ripper.**

BACKSAW *See* **Saw.**

BACKSET Horizontal distance from the front face of a door lock to the centerline of the keyhole or door knob.

BACKSIGHT In surveying, the first sight taken on a datum (which determines the placing and/or elevation of the instrument) plus any sight taken toward the last station passed. *Also called* **Back observation.**

BACKSIPHONAGE Backflow of water from a plumbing fixture or vessel or other sources into a water supply pipe due to a negative pressure in the pipe.

BACKSIPHON PREVENTER Plumbing fixture designed to prevent the backflow of water into a potable supply.

BACKSPAR Tree rigged at the back end to provide lift for a skyline or yarding lines. *Also called* **Tail spar.**

BACKSTAMP Approved agency mark on the back of a plywood panel.

BACKSTAY *See* **Brace.**

BACK-STEP WELDING Welding small sections of a joint in a direction opposite to the progression of the weld as a whole.

BACKSTOP 1. Sturdy structure to hinge a log loading boom to, and to protect the equipment from damage; **2.** Operator guarding package mounted on the boom of a grapple yarder; **3.** Device used to limit the angle of a boom, jib, or mast at their highest recommended angle.

BACKUP 1. Second unit of equipment installed or held ready in case of failure of the prime unit; **2.** Any

object or structure the operator of a backhoe may use to push against, moving material into either the backhoe or loader bucket; **3.** That part of a masonry wall behind the exterior facing.

BACKUP MATERIAL Material placed in a joint cavity behind a sealant to control the joint depth of the selant without inhibiting joint closure. Often made of polyethylene or polyurethane foam.

BACKUP RING 1. Ring that bridges a clearance to minimize seal extrusion. *Also called* **Anti-extrusion ring, Bull ring,** and **Junket ring; 2.** Steel ring placed on the inside of a pipe to back-up a butt or bevel weld. *Also called* a **Chill ring.**

BACKUP (TURN) (Logging) turn to which a helicopter can be directed in the event of abort or difficulties with the first turn.

BACK VENT Branch vent connected to the main vent stack of a plumbing system and extending to a location near a fixture trap to prevent the trap from siphoning.

BACK WALL Wall facing an observer who is standing at the entrance to a room, shower, or tub shower.

BACKWARD PASS Calculation of the latest finish and latest start dates within the **Critical Path Method (CPM).**

BACKWARD STABILITY Resistance to overturning of a crane or lifting device in a rearward direction.

BACKWATER Water held back from its natural flow by a natural or artificial obstruction.

BACKWATER VALVE *See* **Valve.**

BACTERIAL CORROSION Destruction of a material by chemical processes brought about by the activity of certain bacteria that produce substances such as hydrogen sulfide, ammonia, and sulfuric acid.

BAD AIR Contaminated air, or gas, or air containing insufficient oxygen that constitutes a hazard.

BAD DEBT Account or receivable considered uncollectible.

BADGER 1. Tool used to remove excess mortar from the inside joints of drain tile as it is laid; **2.** Large wood rabbet plane used in cabinetry.

BA DIMENSION Distance from the foremost part of the front bumper of a vehicle to the center line of the front axle. *See also* **Chassis dimensions.**

baf. *Abbreviation for:* baffle.

BAFFLE 1. Vane set at an angle to the main direction of the flow of vapor or water to impede the flow or influence its direction; **2.** Vane designed into a hydraulic tank to reduce the turbulence of hydraulic oil reentering the reservoir. By reducing turbulence, contaminants and air have time to separate from the hydraulic fluid.

BAFFLE CHAMBER Chamber following a furnace combustion chamber, in which baffles change the direction of and/or reduce the velocity of the combustion gases in order to promote the settling of fly ash or coarse particulate matter.

BAFFLE PIERS Obstructions set in the path of fast-flowing water to break up the stream and to dissipate energy.

BAFFLE SCREW High-hardness screw used to divert fuel under high-pressure away from the soft main housing of a pump.

bag. *Abbreviation for:* baggage.

BAG DAM Low-rise, temporary dam constructed of bags filled with a mixture of soil or sand and cement.

BAGHOUSE Asphalt mixing plant component that captures fine material contained in dryer and drum mixer exhaust gases as they pass through bags made of special filter material. The captured dust is generally returned to the mix. *See also* **Fabric filter collector.**

BAG OF CEMENT Quantity of portland cement, 39.7 kg (87.5 lb) in Canada, 42.6 kg (94 lb) in the U.S. For other kinds of cement, a quantity indicated on the bag.

BAG PLUG Inflatable device used to block and seal the outflow of a drain when testing.

BAIL 1. U-shaped member of a bucket; **2.** U-shaped portion of a socket or other fitting used on wire rope; **3.** Frame equipped with sheaves and connected to the gantry, used in conjunction with the boom hoist drum and bridle to alter a crane's boom angle.

BAILER Hollow cylinder used to remove rock chips and water from holes made while churn drilling.

BAILING BUCKET Bucket-like tool for removing water from the hole during drilling or in preparation for concrete placement. *See also* **Bucket.**

BAKED FINISH Paint, varnish, or enamel compounded to withstand baking at temperatures above 65.5°C (150°C) to develop the desired finish.

BAKELITE An early type of plastic, used in the production of electrical devices, door and cupboard furniture, etc.

bal. *Abbreviation for:* balance.

BALANCE 1. Device used to counteract the weight of a window sash for ease of operation. *See also* **Dynamic balance,** and **Static balance; 2.** Excess of debits over credits, or credits over debits in an account.

BALANCE BAR Large, heavy beam extending along the upper edge of a lock gate and extending back a considerable distance beyond the hinge, used to open and close the gate when the water level is the same on each side of the gate.

BALANCE BOX Counterbalance weight opposite the jib and load of a crane.

BALANCE BRIDGE Bridge that can be tilted vertically about a balance point. *Also called* a **Bascule bridge.**

BALANCED BID Bid where all the details, individual prices, and total are appropriate for the work to be done. *See also* **Bid,** and **Unbalanced bid.**

BALANCED CIRCUIT Three-wire electrical circuit having the same load on each side of the neutral wire.

BALANCED CONSTRUCTION Composite panel construction that will not warp when subjected to uniformly distributed moisture changes.

BALANCED CUT Cut of a tile at the perimeter of an area that will not take a full tile. The cuts on opposite sides of such an area should be of the same size; also, the same size cuts on each side of a miter.

BALANCED CUT AND FILL Earthworks in which the volume of material to be excavated equals the volume of material needed as fill.

BALANCE DIAMETER Diameter of a face seal at which the resulting force is considered to be acting.

BALANCED LOAD 1. Load capacity at simultaneous crushing of concrete and yielding of tensile steel; **2.** An electrical load connected so that equal currents are taken from each side of a three-wire system; **3.** Condition when the load on the car side of a dumbwaiter or elevator car sheaves equals the weight of the counterweight.

BALANCED PISTON Piston ring, the geometric design of which includes face, bore, and/or relief grooves or cuts that are employed to minimize pressure forces and thus to reduce friction forces.

BALANCED REINFORCEMENT Amount and distribution of reinforcement in a flexural member such that in working-stress design the allowable tensile stress in the steel and allowable compressive stress in the concrete are attained simultaneously, or such that in strength design, the tensile reinforcement reaches its specified yield strength simultaneously, with the concrete in compression reaching its assumed ultimate strain of 0.003.

BALANCED TENSION State when the tensioning forces on individual groups of tendons in a prestressed concrete member are balanced about the centroid member.

BALANCED TRAFFIC Condition where the number of up calls in an elevator group automatic supervisory system equals the number of down calls.

BALANCED WINDERS Stair winders not radiating to a common center to give increased step area at the narrow end.

BALANCE OF TRADE Difference between the dollar, or money value of a country's exports and its imports of merchandise or goods and services.

BALANCE POINT 1. Point about which an object will pivot; **2.** Point where all excavated material has been used as fill and all additional fill must be imported or additional excavation exported.

BALANCE-POINT SPEED In sleeve metering pumps, the point where full fuel delivery and rated speed are correct.

BALANCE SHAFT Spinning shaft designed to cancel out vibration forces caused by rotation of the major moving parts of an engine.

BALANCE WEIGHTS One of several weights attached to the underside of an elevator platform, or to the car frame, to balance it within the guide rails so as to distribute, as evenly as possible, the pressure of the individual guide shoes on the guide rail surfaces.

BALANCING Testing for balance; adding or subtracting weight to put a piece into either static or dynamic balance.

BALANCING A SURVEY Eliminating any error of closure by apportioning corrections throughout the traverse.

BALANCING SUBGRADE Trimming subgrade until there are several areas that are still too high or too low, but that, when trimmed, will average out to the finished grade tolerance required.

balc. *Abbreviation for:* balcony.

BALCONET False balcony consisting of a doorway and railing but lacking usable floor space.

BALCONY Accessible platform, projecting from or recessed into a wall and equipped with a protective balustrade or railing.

BALD HEADED Logs outhauled without chokers; refers to butt rigging.

BALER Machine that compacts materials, such as solid waste, into high density cubes.

BALES CATCH Type of ball catch.

BALK Squared, roughly hewn timber over 225 mm (9 in.) square. *Also called* **Baulk.**

ball. *Abbreviation for:* ballast; balloon.

BALLAST 1. Material used to provide increased weight; **2.** Liquid pumped into a tire for added traction; **3.** Rock or gravel hauled in on a subgrade for roadbuilding.

BALLAST TAMPER Machine for compacting ballast under railroad ties.

BALLAST WEIGHT Material added to increase the gross weight of a machine.

BALL BEARING Bearing in which a shaft bears on small spherical metal balls that turn freely as the shaft revolves.

BALL CLAY Secondary clay, commonly characterized by the presence of organic matter, high plasticity, high dry strength, long vitrification range, and a light color when fired.

BALL COCK Valve or faucet controlled by a change in water level and generally consisting of a device floating near the surface of the water that operates the valve.

BALLED AND BURLAPPED Plant material that has been dug with a solid ball of earth around the roots, that is then wrapped and tied in burlap for shipment.

BALL-HOOTING Sliding or rolling logs down a hillside by means of log jacks or peaves.

BALLISTIC SEPARATOR *See* **Separator.**

BALL JOINT Ball and socket connection that allows a limited hinge movement in any direction.

BALL MILL Horizontal, cylindrical, rotating mill charged with large grinding media. *See also* **Rod mill.**

BALL MILLING Method of grinding and material mixture, with or without liquid, in a rotating cylinder or conical mill partially filled with grinding media such as balls or pebbles.

BALLOON Ball capping a pier or post as an ornament.

BALLOON CONSTRUCTION *See* **Frame construction.**

BALLOON MORTGAGE *See* **Mortgage.**

BALLOON PAYMENT Final payment on a loan where the amount is greater than the preceding installments.

BALL-PEEN HAMMER *See* **Hammer.**

BALL TEST Test to determine the consistency of freshly mixed concrete by measuring the depth of penetration of a cylindrical metal weight with a hemispherical bottom. *See also* **Kelly ball.**

BALL VALVE *See* **Valve.**

BALONEY CABLE Heavy insulated electric drag cable, used to supply power to mobile cableway towers, cranes, shovels, and tunneling equipment.

BALSA Lightest known wood. Little used in construction.

BALUSTER Short pillar or post helping to support a rail or coping.

BALUSTRADE 1. Protective barrier approximately 900 mm to 1200 mm (35 in. to 47 in.) high at the edge of openings in floors or at the side of stairs, landings, balconies, mezzanines, galleries, raised walkways, or other locations; may be solid or have openings in it; **2.** Side of an escalator above the steps and including the skirt panels, interior panels, decks and handrails.

BALUSTRADE BRACKET One of several structural steel members that support an escalator. *Also called* **Deck bracket.**

BALUSTRADE LIGHTING Lighted panels running the full length of the balustrade of an escalator, mounted parallel to and immediately above the skirt.

BANBURY MIXER Specific type of internal mixer used to incorporate fillers and other ingredients into rubber or plastic.

band *Abbreviation for:* banded.

BAND 1. Group of reinforcing bars distributed in a slab or wall or footing. *See also* **Strips**; **2.** Flat member having little projection; **3.** Moldings encircling the shaft of a column; **4.** Any continuous ornamentation along a wall or around a building; **5.** Metal ring that is welded, shrunk, or cast on the outer surface of a hose nipple; **6.** Thin strip of metal used as a boltless clamp; **7.** Less than full width lengths of drywall panel used in horizontal applications where required.

'BAND AID' JOINT Sealant joint composed of a bond-breaker tape over the joint movement area with an overlay of sealant lapping either side of the tape sufficiently to bond well. Used where extreme movement occurs and conventional joint design is not possible, i.e., metal joints, deep V joints.

BANDALET Small band encircling a column shaft.

b&b *Abbreviation for:* bell and bell.

B&B *Abbreviation for:* grade B and better.

BAND BRAKE *See* **Brake.**

b&ec *Abbreviation for:* beaded on edge and center.

BAND CHAIN Surveyor's graduated steel tape, usually 30 m (100 ft) or more in length.

BAND CLAMP *See* **Clamp.**

BANDED Masonry built in bands or stripes of different textures or colors in alternating courses.

BANDER Worker who bolts the metal bands on corrugated metal pipe.

b&f *Abbreviation for:* bell and flange.

B&F *Abbreviation for:* buck and frame.

BANDING 1. High-tensile steel strapping applied to timber piles to prevent splitting while driving, or applied to the above-ground portion to join them so as to form a single unit when in position; **2.** Inlay covering the edge of a veneer or the end of a cores.

BAND IRON Thin metal strap used as a form tie, hanger, etc.

B&O *Abbreviation for:* back-out punch.

B & O Pin mounted on a wooden handle with a head for hitting, used as a hammer, or to knock out rivets or bolts.

BAND PLY The first inner cord ply of a tire.

b&s *Abbreviation for:* bell and spigot.

B&S *Abbreviation for:* beams and stringers.

BAND SAW *See* **Saw.**

BAND SHELL Parabolic-shaped sound amplifying and reflecting enclosure above and to the rear of a bandstand.

BANISTER Handrail of a staircase.

BANJO 1. Tool used for bucking up rivets; **2.** Banjo-shaped fuel line end that accepts a hollow line bolt.

BANJO TAPER Tool that dispenses drywall tape and compound simultaneously. *Also called* a **Bazooka.**

BANK 1. Steep rise or slope; **2.** In excavation, a mass of soil rising above digging or trucking level; **3.** Soil to be excavated from its natural position; **4.** Logs cut or skidded above the required daily production and held in reserve; **5.** Group of two or more elevators that operate as a unit; **6.** Grouping of common devices mounted in a common frame or in a common panel.

BANK CUBIC METER/CUBIC YARD One cubic meter (yard) of material as it lies naturally. *See also* **Loose cubic meter/cubic yard.**

BANKER 1. Worker who hauls ties or other timber from forest to storage place on the bank of a river from where they are floated; **2.** Small platform on which concrete is mixed by hand.

BANK GRAVEL Natural mixture of fines, sand, gravel, and cobbles.

BANKIE *See* **Marine borer.**

BANK MEASURE Measurement of material in its original place in the ground.

BANK PLUG Stake driven into ground to be graded, left to show some 600 mm (24 in.) above existing grade. String lines connecting nails driven into a series of plugs indicate desired finished grade or a fixed measure above desired finished grade.

BANK PROTECTION Measures taken to prevent or reduce erosion of an earthen sloping face.

BANK-RUN GRAVEL Material 150 mm (6 in.) maximum to less than 6 mm (0.25 in.) minimum, excavated from a naturally occuring deposit (finer material being screened out).

BANKRUPTCY Inability to pay due debt.

BANK SEAT Reinforced concrete mass that supports one end of a bridge at the top of an embankment.

BANK SLOPING Working a bank, usually with a grader or blade-equipped dozer, to form a uniform slope to the required profile.

BANK STORAGE Water absorbed by the banks of a waterway or stream that is returned to the stream as the water level falls.

BANK YARD Cubic yard of soil or rock measured in situ, before excavation.

BANQUETTE 1. Footbridge over a road; **2.** Sidewalk about 450 mm (18 in.) wide; **3.** A berm.

bar. *Abbreviation for:* barrier; barometer.

BAR 1. Relatively long, thin piece of metal; **2.** Rigid and straight section of metal with a cross section small in relation to its length; **3.** Element, normally composed of steel, with a nominally uniform cross-sectional area, used to reinforce concrete; **4.** Part that carries the cutting chain on a chain saw. *Also called* the **Blade; 5.** Non-SI unit of pressure, equal to 100 kilopascals, permitted for use with the SI system of measurement for a limited time. Symbol: bar. *See also the appendix:* **Metric and nonmetric measurement.**

BARBED Deformation of a metal shaft, typically a nail, consisting of a repetitive pattern of indentations and/or ridges.

BAR BENDER Tradesman who cuts and bends steel reinforcement; machine for bending steel reinforcement.

BAR-BENDING BENCH Heavy-duty bench on which reinforcing bar can be bent and shaped.

BAR-BENDING SCHEDULE Schedule detailing:

(a) Positioning of reinforcing bars in the structure;

(b) Number of identical structural members;

(c) Bar mark;

(d) Diameter, or size of the bar;

(e) Number of bars of one type in each structural member;

(f) Total number of bars of each type;

(g) Total straight length of the bar;

(h) Shape and bending dimensions of the bar;

(i) Special remarks.

BARBER CHAIR Tree that splits when being felled.

BAR CHAIR Support for reinforcing bars during concrete placing.

BAR CHART Graphical representation of frequencies or magnitudes with vertical or horizontal bands drawn in proportion of the frequencies or magnitudes involved.

BAR CLAMP *See* **Clamp.**

BARDON HOOK 1. Hook used with wire rope slings for gripping trees or logs to be skidded; **2.** Type of choker hook.

BARE CONDUCTOR *See* **Conductor,** and **Electrical conductor.**

BARE DUCK Duck surface of a hose wherein the exposed duck surface is free of any rubber coating.

BARE-FACED TENON Tenon shouldered on only one side.

BAR-END CHECK Check on the ends of reinforcing bars to determine whether they fit the devices intended for connecting the bars. *See also* **Mechanical connection.**

BARE ROOT Plants that are planted with their roots bare and unprotected by such means as burlap.

BARE SITE Designated area of land, cleared of any existing structures, debris, and unwanted vegetation.

BARE SOIL Exposed soil, free of vegetation.

BAR FABRICATOR Company that is capable of preparing placing drawings, bar lists, and storing, shearing, bending, bundling, tagging, loading and delivering reinforcing bars.

BARGAIN PURCHASE OPTION *See* **Option.**

BARGAIN RENEWAL OPTION *See* **Option.**

BARGE Overhanging edge of a roof up the slope of a gable.

BARGEBOARD Board placed against the incline of a gable, supported on soldiers projecting beyond the face of the wall, that hide the ends of the horizontal beams and other roof members. *Also called* **Facer board,** and **Verge board.**

BAR HANGER Supporting bracket for a sink hung on a wall.

BARITE Mineral, barium sulfate ($BaSO^4$), used in either pure or impure form as concrete aggregate, primarily for the construction of high-density radiation-shielding.

BARIUM SULFATE TEST Quick chemical color test to determine approximately the pH value of soil.

BAR JOIST Trussed joist consisting of steel top and bottom members with heavy wire or rod lacing.

BARK Outer layer of a tree, comprising the inner bark or thin, inner living part and the outer bark, or cork-like layer composed of dry, dead bark.

BARKER Machine used to remove bark from pulpwood.

BARK GAUGE Instrument for measuring the bark thickness of a live tree in making growth studies.

BARKING DRUMS Large drums in which small softwood logs are revolved against knives to remove the bark preparatory to pulp manufacture.

BARKING IRON Tool with a narrow-shaped, curved blade used in removing bark by hand. *Also called* a **Spud.**

BARK POCKET Opening between annual growth rings of a tree that contains bark and which appears as dark streaks on radial surfaces and as rounded areas on tangential surfaces.

BARK RESIDUE 1. Bark removed from a log; **2.** Portions of wood and foreign matter such as sand, grit, or stones that may be imbedded in the bark.

BAR LIST Bill of materials, where all quantities, sizes, lengths, and bending dimensions of reinforcing steel are shown.

BAR LOCK Type of door interlock used with manually operated elevator doors.

BAR MARK Identification given reinforcing steel bars showing, in abbreviated or notational form, its cross-sectional shape, type of steel, diameter or equivalent size, and bent shape.

BAR MAT Assembly of steel reinforcement composed of two or more layers of bars placed at angles to each other and secured together either by welding or tying. *Also called* **Mat.**

BARMINUTOR Combined bar screen and comminutor, used to trap and shred solids in the primary influent to sewage treatment plants.

BARN Agricultural building for the housing of livestock and storage of feedstuff, equipment, etc.

BARN-DOOR GATE Type of tailgate that is hinged along its outer vertical edges to allow it to fold laterally back against the sides of the body, usually split in the middle and made in two parts but can be of one-piece construction and hinged on one end only.

BARN-DOOR HANGER Set of pulleys over a door and attached to it by a steel strap carrying the weight of the door; the pulleys run on a rail hung from the lintel.

BAR NUMBER A number (approximately the diameter of a reinforcing bar in eighths of an inch) used to designate the bar size, e.g., a #5 bar is approximately 15 mm (0.625 in.) in diameter. Bar numbers are rolled onto the bar for easy identification.

BAR OFF Unloading bars from a truck bed by levering individual bundles over the side with pinch bars.

BAROMETER Instrument for measuring atmospheric pressure, usually in centimeters, millimeters, or inches of mercury column.

BAROMETRIC DAMPER *See* **Damper.**

BAROMETRIC PRESSURE Atmospheric pressure at a specific place according to the current reading of a barometer.

BAROQUE Architectural style characterized by ornate curves, carved ornament, and elaborate scrolls.

BAR PLACER Worker who handles and places reinforcing steel and bar supports. *See also* **Iron worker, Metal lather,** and **Rebar contractor.**

barr. *Abbreviation for:* barrier.

BARRACK 1. Temporary building to house workers; **2.** Building to house soldiers.

BARRAGE Dam to divert a flow of water or to increase its depth.

BARRANCA Steep bluff.

BARREL 1. Cylindrical container with sides that bulge outward and having flat ends. *See also* **Drum; 2.** Water passage in a culvert; **3.** Weight measure for portland cement corresponding to four bags (159 kg (350 lb) net in Canada, 170.5 kg (376 lb) net in the U.S.); **4.** Non-metric unit of volume used in the petroleum industry equal to 42 U.S. gallons, 159 liters, or 135 kg; **5.** Cylindrical component of a hydraulic cylinder.

BARREL HITCH Knot used to lift a barrel or keg.

BARREL KEY Key with a round stem, hollow at the bottom, that fits over a pin in a lock keyway to ensure proper alignment. *Also called* a **Hollow-post key**, and **Pipe key**.

BARREL LIGHT Roof window of curved shape.

BARREL NUT *See* **Nut**.

BARREL NIPPLE Short length of pipe, outside threaded on each end, bare in the middle.

BARREL POST *See* **Drill pin**.

BARREL ROOF Roof with semicylindrical ceiling. *See also* **Barrel-vault roof**.

BARREL SWIVEL Swiveling device in butt rigging.

BARREL VAULT *See* **Vault**.

BARREL-VAULT ROOF Thin concrete roof in the form of a part of a cylinder. *See also* **Barrel roof**.

BARETTES 1. System of piling involving the excavation of large rectangular panels under bentonite; similar technique used in the construction of diaphragm walls; 2. Large, hollow oblong piles used to carry heavy loads to great depths.

BARRICADE Obstruction to deter or prevent the passage of persons or vehicles.

BARRIER Obstruction to the spread of fire.

BARRIER FILTER Filter that removes particles from a fluid stream with a porous structure in the flow path.

BARRIER-FREE HARDWARE Range of hardware designed not to obstruct openings or passageways used by the physically challenged.

BARRIER PLANTING Strategic use of trees, shrubs, and other vegetation to form a physical, visual, acoustic, or other type of screen.

BARRIER RAIL Metal or concrete continuous section used to maintain separation between high-speed traffic lanes going in opposite directions, or as a perimeter between a traffic lane and the highway edge.

BARRING AND WEDGING Removal, by prying with bars and the insertion of wedges, of the final portion of unsuitable rock from the foundation surface of a dam.

BARRON TOOL Long-handled combination hoe and rake used in building fire lines in light fuels. *See also* **McLeod tool**, and **Pulaski tool**.

BARROW PIT *See* **Borrow pit**.

BAR SCHEDULE List of reinforcement showing the shape, number size, and dimensions of every different element required for a structure or portion of a structure.

BAR SCREEN Frame containing a series of similar-size bars or rods placed parallel with each other.

BAR SPACING Distance between parallel reinforcing bars, measured center-to-center of the bars perpendicular to their longitudinal axes.

BAR SUPPORT Hardware used to support or hold reinforcing bars in proper position to prevent displacement before and during concreting. *See also* **Bat, Chair**, and **Slab bolster**.

bas. *Abbreviation for:* basin.

BASAL Usually considered the bottom 2 m (6 ft) of the stem of a tree. *See also* **Butt**.

BASAL AREA Cross-sectional area of a tree, measured at breast height. Used as a method of measuring the volume of timber in a given stand.

BASAL AREA FACTOR Number of units of basal area per hectare (or per acre) represented by each tree.

BASAL AREA PER HECTARE (ACRE) *See* **Land classification (forest management)**.

BASALT Extrusive igneous rock.

BASALT WARE Black unglazed vitreous ceramic ware having the appearance of basalt rock.

BASCULE 1. Span, pivoted to rotate vertically about an axis that may be fixed or movable; 2. Apparatus in which one end is counterbalanced by the other.

BASCULE BRIDGE Bridge having one or more sections balanced to pivot vertically. *See also* **Balance bridge**.

BASE 1. Concrete subfloor slab or 'working mat,' either previously placed and hardened or freshly placed, on which floor topping is placed in a later operation; 2. Underlying stratum on which a concrete slab, such as pavement, is placed. *See also* **Baseboard, Mud slab**, and **Subbase**; 3. One or more rows of tile installed above the floor level; 4. Wood, plastic, or metal trim applied at the internal junction of floor and wall to protect and cover the joint. *See also* **Baseboard**; 5. Colored or white paste or heavy liquid added to paints to tint them and to which is added a vehicle to make a finished paint. 6. *See also* **Molding**.

BASE AND MERIDIAN Theoretical survey lines used to locate and describe the location of land. The baseline lies east-west; the meridian north-south.

BASE ANGLE Metal angle fastened to the perimeter of a foundation, used to support and close wall panels.

BASE BEAD *See* **Base screed**.

BASE BID Amount for which the bidder offers to do the work or supply the goods, not including work for which alternate bids are also submitted.

BASE BLOCK Squared block that terminates a molded baseboard at an opening.

BASEBOARD Board placed around a room against the wall next to the floor to properly finish the junction. *Also called* **Base**, and **Skirting**.

BASEBOARD HEATER Radiator shaped like a decorative baseboard having openings at top and bottom through which hot air circulates.

BASE CAP *See* **Molding**.

BASE COAT 1. First coat of paint applied to a surface after the primer coat. *Also called* **Ground coat**. *See also* **Full coat, Glaze coat**, and **Mist coat**; 2. Plaster coat or coats applied prior to application of the finish coat.

BASE COURSE 1. Bottom or bed course of brick or masonry block; 2. Layer of specified select material of planned thickness constructed on the subgrade or subbase of a pavement to serve one or more functions, such as distributing loads, providing drainage, or minimizing frost action. It may be composed of crushed stone, crushed slag, crushed or uncrushed gravel and sand, or a combination of these materials. It may also be bound with asphalt. *See also* **Asphalt base course,** and **Pavement structure.**

BASE FLOOD Flood having a 1% chance of being equaled or exceeded in a given year.

BASE FLOOD ELEVATION Maximum elevation that a base flood would reach.

BASE LAYER First layer of drywall panel in a multilayer application.

BASE LOAD Minimum load over a given period of time.

BASELINE 1. Reference line to which all other measurements are subordinate; **2.** Centerline ofconstruction; **3.** Main traverse or surveying line from which lines and grades for work are plotted

BASE LOG First log in a cribbing face (face log) resting directly on the ground or on the mud sills.

BASEMENT That part of a building, the floor of which is wholly or partially below ground level.

BASEMENT SOIL *See* **Subgrade.**

BASEMENT MACHINE Installation where the elevator machine is placed at the bottom of the hoistway, or adjacent to it, rather than directly over it.

BASEMENT ONE-TO-ONE SINGLE WRAP Elevator hoist rope arrangement with one end of each hoist rope fastened to the counterweight. The ropes then continue up and over the idler sheaves in the overhead and down the hoistway and around the machine driver. The ropes then continue up the hoistway, over the idler sheaves in the overhead, down, and fasten to the car.

BASEMENT WALL Foundation wall enclosing usable space below a structure.

BASE METAL Metal to be welded, cut, or brazed.

BASE MOLDING Molding used to trim the lower edge of interior baseboard. *See also* **Molding.**

BASE-MOUNTED FUEL TANK Fuel tank that is incorporated into the system subbase of a generating system.

BASE MOUNTING Structure forming the lowest component of a crane or derrick, transmitting loads to the ground.

BASE NUMBER Amount of acid required to measure all or part of an oil's basicity, usually expressed in weight of potassium hydroxide (KOH) equivalent.

BASE OIL *See* **Base stock.**

BASE PLATE 1. Plate of metal or other material placed under pavement joints and the adjacent slab ends to prevent the infiltration of soil and moisture from the sides or bottom of the joint opening; **2.** Steel plate used to distribute vertical loads, as for bridge beams, building columns, scaffolding legs, or machinery.

BASE SCREED Preformed metal screed with perforated or expanded flanges to provide a ground for plaster and to separate areas of dissimilar materials. *Also called* **Base bead.**

BASE SECTION 1. Lowermost element of a telescopic boom; it does not telescope but contains the boom and foot-pin mountings and the boom-hoist cylinder upper-end mountings; **2.** Lowermost section of a lattice boom, tower, or jib.

BASE SHEET Asphalt-saturated-and-coated composition roofing, used in built-up roofing.

BASE SHOE *See* **Molding.**

BASE STATION Fixed central radio unit to which mobile units are ancillary.

BASE STOCK Refined petroleum or synthetic fluid into which additives are blended to produce the finished product. Also called **Base oil.**

BASE STOCK COST FLOW *See* **Cost flow.**

BASE THICKNESS Axial dimension of a base.

BASE TRIM Molding or extruded plastic section used to decorate the meeting of a wall and floor and to mask that junction.

BASIC BOOM Minimum length of boom that can be mounted and operated on a mobile crane, usually comprising a boom base and tip section.

BASIC CAPACITY (OF A FORK TRUCK) The capacity at 600 mm (24-in.) load center for loads up to 9071 kg (20,000 lb), and 914 mm (36 in.) for loads over 9071 kg (20,000 lb) at a specified fork height.

BASIC CREEP Creep that occurs without migration of moisture to or from the concrete. *See also* **Drying creep.**

BASIC JIB Jib attachment made up of only the base and top section of the jib.

BASIC MODULE Unit of dimension used to coordinate the size of building components and elements.

BASIC SAFETY CIRCUIT Section of an elevator control wiring that includes a number of mechanical switch contacts and relay contacts in series. Usually includes the final limits, emergency stop button, governor contacts, and a safety-operated switch. Failure of any or all of such functions will cause the elevator to stop. *Also called* **Safety circuit.**

BASIN 1. Ground depression with sloping sides; **2.** Receptacle for water; in plumbing, a lavatory.

BASKET 1. Wire assembly to support and space dowel bars and expansion joints in concrete slabs on ground. *See also* **Load transfer assembly; 2.** Recessed strainer fitted into the drain opening of a sink; **3.** The conical portion of a socket into which a broomed-rope-end is inserted and then secured.

BASKET ARCH *See* **Arch.**

BASKET CRIB Protective enclosure of interlocking timbers.

BASKET WEAVE Pattern resembling the weave used in basket making.

BASKET WORK Materials (masonry, wood block, wood inlay, etc.) placed in a pattern resembling a basket weave, i.e., checkerboard.

basmt *Abbreviation for:* basement.

BAS-RELIEF Sculpture with slight projection from

the background.

BASSANITE Calcium sulfate hemihydrate. *See also* **Hemihydrate,** and **Plaster of paris.**

BASTARD Anything not standard, uniform, typical, or true.

BASTARD FILE Medium file, neither coarse nor fine.

BASTARD GRANITE Laminated or foliated rock similar to granite, used in wall construction.

BASTARD MASONRY Facing stones of a rubble wall, dressed and laid like ashlar, but not fully dressed.

BASTARD POINTING *See* **Pointing.**

BASTARD-SAWN 1. Lumber that is plain-sawn, flat-sawn, or slash-sawn; **2.** Hardwood lumber in which the annual rings make angles of 30° to 60° with the face of the piece.

BASTION Projection from a wall having two flanks.

bas. wt *Abbreviation for:* base weight.

bat. *Abbreviation for:* battery; batten.

BAT (also spelled BATT) 1. Piece of a brick with one end whole, the other end broken off; **2.** Unit length and width of mineral insulation; **3.** Semi-rigid section of mineral wool anchored to paper that is fixed between framed members. *Also called* **Friction-fit batt. 4.** *See also* **Bar support,** and **Slab bolster.**

BATCH 1. Measured quantity of material or materials; **2.** Quantity of either concrete or mortar mixed at one time.

BATCH BOX Container of known volume used for measuring constituents of a batch of either concrete or mortar in proper proportions.

BATCH CYCLE TIME Time required for a batch plant to proportion, mix, and discharge a single batch of mix. Cycle time multiplied by batch size theoretically equals plant capacity.

BATCHED WATER Mixing water added by a batcher to a cementitious mixture either before or during the initial stages of mixing. *Also called* **Batch water,** and **Mixing water.**

BATCHER Device for measuring ingredients for a batch of concrete, may be of the following types:

Automatic batcher: Batcher equipped with gates or valves that, when actuated by a single starter switch, will open automatically at the start of the weighing operation of each material and close automatically when the designated weight of each material has been reached, interlocked in such a manner that (a) the charging mechanism cannot be opened until the scale has returned to zero, (b) the charging mechanism cannot be opened if the discharge mechanism is open, (c) the discharge mechanism cannot be opened if the charging mechanism is open, (d) the discharge mechanism cannot be opened until the designated weight has been reached within the allowable tolerance, and (e) if different kinds of aggregates or different kinds of cements are weighed cumulatively in a single batcher, interlocked sequential controls are provided.

Manual batcher: Batcher equipped with gates or valves that are operated manually, with or without supplementary power (pneumatic, hydraulic, or electrical), the accuracy of the weighing operation being dependent on the operator's observation of the scale.

Semiautomatic batcher: Batcher equipped with gates or valves that are separately opened manually to allow the material to be weighed but that are closed automatically when the designated weight of each material has been reached.

BATCH-FED INCINERATOR *See* **Incinerator.**

BATCH HOUSE Place where batch materials are received, handled, weighed, and mixed.

BATCHING 1. Measuring the ingredients for a mix of materials to a consistent formula; **2.** Measuring the ingredients for a batch of concrete or mortar by weight or volume and introducing them into the mixer.

BATCH LOADER *See* **Refuse truck.**

BATCH MIXER Machine that mixes batches of either concrete or mortar.

BATCH PLANT Installation for batching or for batching and mixing materials such as concrete or asphalt. *Also called* **Mixing plant.**

BATCH-TYPE FURNACE Furnace that is shut down periodically to remove old charge and add a new charge, as opposed to a continuous-type furnace.

BATCH WATER *See* **Batched water.**

BATCH WATER HEATER In a domestic solar heating system, a water storage tank in an insulated container fitted with a glazed cover, used to heat and store water.

BATCH WEIGHT Weights of any one of the various materials (cement, water, the several sizes of aggregate, and admixtures if used) that comprise a batch of concrete.

BATHTUB 1. Tub for human bathing; **2.** Concrete container on a dumpcrete truck.

batt. *Abbreviation for:* batten.

BATT *See* **Bat.**

BATTEN 1. Narrow strip of wood, often used to cover joints, especially as decorative vertical members over plywood or wide boards; **2.** Strip of wood used for attaching roofing tiles, shakes, shingles, etc., to a roof or wall, also internally for attaching panelling and for fixing generally.

BATTEN BOARD Similar to laminated wood (plywood) but with the core formed of strips of any width up to 75 mm (3 in.) running at right angles to the grain of the outer veneers.

BATTEN PLATE Steel plate element used to joint two parallel components of a built-up steel column, girder, or strut, rigidly connected to the parallel components and designed to transmit shear between them.

BATTEN STRIP Piece of wood placed over a joint between two panels.

BATTER Inclined (from ground level upwards), receding slope of the face of a wall that is wider at its base than at its crown.

BATTER BOARD Right-angle corner post and board, built on the job to elevate, hold and maintain sight lines or string delineating the actual wall lines of a building, used during the period of excavation and construction of the foundation.

BATTER BRACE Inclined compressive member set at the end of a truss to add strength and support.

BATTER LEVEL Clinometer used to measure the slope of earth cuts and fills.

BATTER PILE *See* **Pile.**

BATTER STICK Tapering stick used in conjunction with a plumb rule or level for building battered surfaces.

BATTERY 1. In blasting, a loose expression for a **Blasting machine**; **2.** Component consisting of two or more cells electrically connected for producing electric energy. (Common usage permits this designation to be applied also to a single cell used independently. In the following definitions, unless otherwise specified, the term battery will be used in this dual sense.) Industrial batteries are identified by the letters E, EE, EO, and EX and are defined as:

  E: Battery assembled as a unit with a cover, for use in Type E or ES trucks, that do not have a covered battery compartment.

  EE: Battery assembled as a unit with a cover that can be locked, for use in Type EE trucks that do not have an enclosed battery compartment.

  EO: Battery assembled as a unit without a cover, for use in Type E trucks with a covered battery compartment, in TYPE ES trucks with an enclosed battery compartment, or in TYPE EE trucks with an enclosed battery compartment with locking means.

  EX: Battery assembled as a unit with a cover that can be locked, for use in EX trucks that do not have a locked battery compartment.

BATTERY BANK Several storage batteries connected together in series or parallel to provide the desired voltage or amperage in storage capacity.

BATTERY CELL CURRENT CAPABILITY Battery cell's ability to provide a given magnitude of current for a specified time to a given cell and voltage measured with the load applied.

BATTERY CELL FAILURE Condition existing when a cell will provide less than 80% of its original rated current as specified in the manufacturer's cell performance data sheet.

BATTERY CHARGER Equipment that is capable of restoring the charge in a storage battery.

BATTERY COMPARTMENT Chamber in which a storage battery is housed and mounted.

BATTERY-ELECTRIC TRUCK Electric truck in which the power source is a storage battery.

BATTERY-FURNACE WALL *See* **Wall.**

BATTERY, LEAD ACID One that uses lead as a negative electrode, lead dioxide as a positive electrode, and sulfuric acid as an electrolyte; nominal voltage 2 volts per cell.

BATTERY LIFTING SLING Tool used in the shop for handling a battery. Lifting slings are generally made for two-point suspension; however, some are still made for the old four-point lift.

BATTERY, NICKEL CADMIUM Battery that uses trivalent nickel oxide as a positive electrode and cadmium for a negative electrode with an alkaline electrolyte; nominal voltage 1.2 volts per cell.

BATTERY RATING Rating based on the ability of a fully-charged storage battery to meet a sustained demand of defined load. A 20-hour rate, for instance, indicates the lighting capability of a battery. The fully charged battery is brought to a temperature of 26.6°C (80°F) and is discharged at a rate equal to 1/20th of the published 20-hour capacity in ampere-hours, until the voltage falls to 1.75 volts per cell. Cold ratings of -17°C (0°F) indicate the cranking ability of a fully charged battery at low temperatures expressed: (a) by the number of minutes required for the battery to reach a terminal voltage equivalent to 1.0 volt per cell when discharged at a rate of 300 amperes with an initial electrolyte temperature of -17°C (0°F); or (b), by the terminal voltage of a fully charged battery taken 30 seconds after the start of a discharge at 300 amperes with an initial electrolyte temperature of -17°C (0°F). The latter rating is applied to batteries intended for diesel service.

BATTERY WARMER Heater used in extremely cold climates to insure that the battery electrolyte solution does not freeze.

BAULK *See* **Balk.**

BAUMÉ SCALE One of several useful measurements of specific gravity of a liquid. *See* **Specific gravity.**

BAUXITE Rock composed principally of hydrous aluminum oxides; the principal ore of aluminum, and a raw material for the manufacture of calcium-aluminate cement.

BAY 1. Division of a wall between columns, buttresses, or other structural elements; **2.** Small, well-defined area of concrete laid at one time in the course of placing large areas such as floors, pavements, or runways; **3.** Portion of a roof between two adjacent trusses; **4.** Curved shoreline enclosed between headlands or promontories.

BAY AND PROMONTORY Bay, shaped by a ridge line, and a promontory that is the dominant upland feature that shapes the bay. *Also called* a **Headland.**

BAY WINDOW *See* **Window.**

BAZOOKA *See* **Banjo taper.**

b/b°—*See* **Coarse aggregate factor.**

BB *Abbreviation for:* ball bearing; beam bolster; best steel.

B-B Sanded plywood panel with B-grade face and back and D-grade inner plies; bonded with interior or exterior glue. A utility panel for interior or protected applications.

bb/b *Abbreviation for:* body bound bolts.

BBC DIMENSION Distance from the foremost point on the front bumper of a vehicle to the back of the cab. *See also* **Chassis dimensions.**

bbd *Abbreviation for:* blackboard.

B-B EXTERIOR Exterior-type sanded plywood panel with B-grade face and back and C-grade inner plies, bonded with exterior glue. A utility panel with solid, paintable surfaces on both sides.

bbl *Abbreviation for:* barrel.

B-B PLYFORM Concrete form grades of plywood paneling with a high reuse factor. Sanded both sides and mill-oiled unless otherwise specified. Also manufactured as HDO for very smooth concrete finish.

BBU *Abbreviation for:* beam bolster upper.

bc *Abbreviation for:* begin curve; back of curb.

BC *Abbreviation for:* bar chair; bottom chord; building code.

BC DIMENSION Distance between back of cab and installed body.

B-C EXTERIOR Exterior-type plywood panel with sanded B-grade face, C-grade back, and C-grade inner plies, bonded with exterior glue.

B-D Plywood panel with sanded B-grade face, D-grade back, and D-grade inner plies, bonded with interior or exterior glue.

bd *Abbreviation for:* bead; board; bond.

BD *Abbreviation for:* bus duct.

BDC *Abbreviation for:* bottom dead center.

bd ft *Abbreviation for:* board foot.

bdl. *Abbreviation for:* bundle.

b. dn *Abbreviation for:* bend down.

bdr *Abbreviation for:* border.

BDT *Abbreviation for:* bone-dry ton.

BDU *Abbreviation for:* bone-dry unit.

bdy *Abbreviation for:* boundary.

bdzr *Abbreviation for:* bulldozer.

be. *Abbreviation for:* bell end.

B/E *Abbreviation for:* bill of exchange.

BEACH 1. Gently sloping, sand and pebble edge to a body of water; **2.** Face of a hydraulically filled earth dam.

BEACH PIPE Hydraulic fill pipe used to carry a slurry of earth and water to construct an earth dam.

BEACON Light, usually flashing, that marks a geographical location or an obstruction to navigation.

BEAD 1. Semicircular molding; **2.** Appearance of a finished weld; **3.** *See* **Tire construction**; **4.** Strip of sealant, caulk, or adhesive; **5.** Metal or plastic sections intended to reinforce and finish corners, angles, edges, and ends of drywall panels; **6.** Type of choker.

BEAD-AND-BUTT Door panel, flush on one face with the frame and recessed on the other, without molding.

BEAD-AND-REEL Semicircular molding decorated along its length with bead forms, circular or longer, arranged in patterns.

BEADING Undesirable ridge at finished drywall joints. *Also called* **Ridging**.

BEAD MOLDING Molding like a series of beads.

BEAD SEAT Flat part of the rim where the tire bead rests.

BEAD-TO-BEAD MEASUREMENT Distance on a tire from the heel of one bead over the crown to the heel of the other bead.

BEAD TOE Part of the bead that faces the inside of the tire.

BEAK 1. Conical end of an anvil. *Also called* a **Horn**; **2.** *See* **Drip**.

BEAM 1. Structural member supported at two or more points, but not throughout its full length, transversely supporting a load, subjected to axial load and flexure but primarily to flexure. *See also* **Girder, Girt, Joist, Ledger, Purlin, Spandrel beam,** and **Stringer; 2.** Graduated horizontal bar of a weighing scale on which the balancing poises ride.

BEAM AND COLUMN Structural system consisting of a series of rafters supported by columns.

BEAM-AND-JACK OUTRIGGER *See* **Outrigger**.

BEAM-AND-SLAB CONSTRUCTION Reinforced concrete floor system in which a solid slab is supported by reinforced concrete beams or girders, that may be an integral part of the slab.

BEAM-AND-SLAB FLOOR Reinforced concrete floor slab supported by beams.

BEAM BLOCK Hollow concrete masonry unit in which short webs or a channel allow placement of reinforcement, which is then embedded in grout or concrete to form a continuous beam.

BEAM BOLSTER Continuous wire bar used to support the reinforcing bars in the bottom of reinforced concrete beams.

BEAM BOLSTER UPPER Welded wire support for the upper layer of bottom reinforcing bars in reinforced concrete beams, and the top bars in bridge deck slabs.

BEAM BOTTOM Soffit or bottom form for a beam.

BEAM CLAMP Formed steel plates connected by bolts that can be attached to the flange of a beam and used to suspend elevator hoisting equipment.

BEAM-COLUMN Structural steel member whose primary function is to carry loads both transverse and parallel to its longitudinal axis.

BEAM COMPASS Drafting instrument used to draw large-radius circles and arcs.

BEAMED CEILING Ceiling construction where the supporting beams are exposed to view.

BEAM FILLING Brick nogging or masonry between floor and ceiling joists at their supports, used as a fire stop.

BEAM FORM Retainer or mold so erected as to give the necessary shape, support, and finish to a concrete beam.

BEAM FORM-CLAMP Any of various types of fastening units used to hold the sides of beam forms.

BEAM GIRDER Two or more pieces of lumber fastened together by cover plates, bolts, or welds to form a single structural member.

BEAM HANGER Wire, strap, or other hardware device that supports formwork from structural members. *Also called* **Beam saddle.**

BEAM HAUNCH Poured concrete section that extends beyond a beam to support the landing door sill of an elevator hoistway.

BEAM PAD Pad attached to a beam to protect a traveling cable from rubbing against it.

BEAM POCKET 1. Opening in a vertical member in which a beam is to rest; **2.** Notch formed at or near the top of a foundation wall to receive and support the end of a beam or girder; **3.** Opening in a column or girder where forms for intersecting beams will be framed.

BEAM SADDLE *See* **Beam hanger.**

BEAM SCHEDULE 1. List giving number, size, and placement of structural steel beams used in a structure; **2.** Table giving the quantity, size, and mark number of reinforced concrete beams; the quantity, size, length, and mark numbers of bars and stirrups (including stirrup spacing), and where specified, the stirrup support bars and beam bolsters.

BEAM SIDE Vertical or sloping side of a beam.

BEAM TEST Method of measuring the flexural strength (modulus of rupture) of concrete by testing a standard unreinforced beam.

bear. *Abbreviation for:* bearing.

BEARER 1. Any small member that is used primarily to support another member or structure; **2.** Horizontal scaffold member upon which the platform rests and which may be supported by ledgers or runners; the transverse member that joins scaffold uprights, posts, poles, and similar members.

BEARER BRACKET *See* **Roofing bracket.**

BEARING 1. That portion of a beam or truss that rests upon a support; that part of any structural member of a building that rests upon its supports; **2.** Part in which a shaft or pivot revolves. There are many types, including:

> **Antifriction:** Bearing consisting of inner and outer rings separated by balls or rollers held in position by a cage.
>
> **Needle:** Type of antifriction bearing that uses small-diameter rollers between wide faces.
>
> **Pilot:** Small bearing that keeps the end of a shaft in line.
>
> **Solid:** One-piece bushing or sleeve.
>
> **Throwout:** Bearing that permits the throwout collar of a clutch mechanism to slide along the clutch shaft without rotating with it.
>
> **Thrust:** Bearing designed to withstand an axial load.

BEARING AREA *See* **Circumferential bearing area,** and **Projected bearing area.**

BEARING BLOCKS Small blocks of stone built into a wall to support the ends of beams.

BEARING CAPACITY 1. Applied load per unit area of surface that a soil can support. *See also* **Allowable bearing capacity; 2.** Maximum load a pile can sustain by soil resistance; **3.** Maximum load that a material can support before failing.

BEARING FACE Load-carrying flat surfaces of a structural component.

BEARING LOAD Compressive load transmitted to a structure by a fastener.

BEARING OF LINE Direction of a line in relation to a known datum.

BEARING OVERLAY Plating used on the surface of a plain bearing.

BEARING PARTITION Interior partition supporting any vertical load in addition to its own weight. Requires its own footing or foundation support.

BEARING PILE *See* **Pile.**

BEARING PLATE Plate placed under a loaded member to distribute the load so its weight will not exceed the bearing strength of the supporting member.

BEARING PRELOAD Initial load placed in the bearings to keep rollers and cups in contact for better load distribution.

BEARING PRESSURE Load on a bearing surface divided by its area.

BEARING STRATUM Soil or rock stratum on which a concrete footing or mat bears or which carries the load transferred to it by a concrete pile, caisson, or similar deep foundation unit.

BEARING STRENGTH 1. Maximum bearing load at failure divided by the effective bearing area; **2.** Amount of weight that a soil or subgrade can safely support.

BEARING STRESS Compressive stress applied to a bearing face or other surface in bearing.

BEARING SURFACE Sectional area of a structural member's surface that supports a load.

BEARING TREE Tree marked to identify the nearby location of a survey corner. *Also called* a **Witness tree.**

BEARING WALL *See* **Wall.**

BEARING/WEAR RINGS Soft metal or plastic rings placed in grooves on a piston or in a cylinder head to prevent contact between hard metal surfaces.

BEARING WIDTH Width of the member providing support to a truss or similar load-transferring device.

BEATER Ironworker's sledge hammer.

BEATING BLOCK Wooden block used to embed tiles in a flat plane.

BEATING IN Setting tiles in a mastic bed by beating on them using a wooden block that cushions each blow.

BEAVER TAIL Rearmost, hinged section of a lowbed trailer that slopes between the bed and the ground to facilitate loading of vehicles.

BEAVER-TAILING Burying the whole bar of a chain saw while cutting.

BECKET End attachment to facilitate wire rope installation.

BECKET LOOP Loop of small rope or strand fastened to the end of a larger wire rope. Its function is to facilitate wire rope installation.

BECQUEREL One of 17 derived units with special names of the SI system of measurement: a unit of radioactivity identified as one disintegration per second (1 Bq=1 s⁻¹). Symbol: Bq. *See also the appendix:* **Metric and nonmetric measurement.**

bed. *Abbreviation for:* bedding.

BED 1. Plane of stratification more than 12 mm (1/2 in.) thick in sedimentary rock; **2.** Horizontal layers of mortar, cement, sand, earth, etc., upon which blocks of stone, bricks, paving slabs, etc., are laid; **3.** Recess formed by the mold to hold a plaster casting; **4.** Base for machinery; **5.** Bottom of the body of equipment designed for hauling; **6.** Intended position in which a tree will be felled; **7.** To level and buffer the ground along the line on which a tree is to be felled to minimize shattering of the timber.

BED AND HAUNCHING Weak concrete, not less than 150 mm (6 in.) thick, partly surrounding a pipe laid in a trench; the concrete reaches the sides of the trench and extends up to the horizontal diameter of the pipe.

BED COAT First coat of jointing compound applied over drywall tape, bead, and fastener heads.

BEDDING 1. Supports or prepared ground on which buried pipe is laid; **2.** Raised mound on which seedlings are planted (technique common to the southeastern U.S.).

BEDDING PLANES Rock formation formed by layering of rock as it was deposited, as in igneous flows or in separated sedimentary deposits.

BEDDING PUTTY Putty placed in the rebate of an opening into which glass is bedded.

BED DOWEL Dowel placed in the center of a bed or a stone.

BED JOINT 1. Horizontal layer of mortar on which a masonry unit is laid; **2.** Horizontal joint, or one perpendicular to the line of pressure (as in the radiating joints in an arch); **3.** Joint between two horizontal courses of brick; **4.** Caulking joint that has something embedded in it.

BED LOAD Volume of sand and gravel or debris moved along the bed of a stream by the flow of water.

BED LINER Abrasion-resistant, rubber, plastic, or composition liner for the bed of a pickup or dump truck. *Also called* **Truck-bed liner.**

BED MOLDING Any molding used to cover the joint at the intersection of a wall and projecting cornice.

BED OF A SLATE Opposite side to the face.

BED PIECE Skid under a pile of lumber.

BED PLATE Metal platform on which a machine is placed.

BEDROCK Continuous solid rock that everywhere underlies the regolith (noncemented rock fragments and deposited materials).

BEDROOM Room designed and equipped for sleeping.

BEDROOM COMMUNITY Predominantly residential community, usually in the suburbs.

BEEHIVE Shape resulting when a strand of a steel cable under tension breaks and is pushed back.

BEEHIVE BURNER *See* **Burner.**

BEETLE HAMMER *See* **Hammer.**

BEL Nonmetric unit of the relation between amounts of signal power and differences in sound-sensation levels: bels equal the common logarithm of the ratio of the two powers.

BELFAST TRUSS Bowstring girder constructed of relatively small timbers, used for spans up to 15 m (50 ft).

BELFRY Upper room or story in a tower in which bells are hung.

BELL 1. Enlargement of the lower end of a pier or pile excavation to increase its bearing area. *Also called* **Underream; 2.** Expanded termination at one end of a pipe section into which the regular diameter end of the next pipe or pipe fitting fits; **3.** Component that slides on a choker cable between the two knobs.

BELL AND SPIGOT Joint system used on large-diameter cast-iron, plastic, or clay pipe, one end of each length being swelled to form a bell into which the narrow end (spigot) of another length will fit, the resulting joint (in cast iron work) being caulked with oakum and lead. *Also called* a **Bell joint.**

BELL ARCH *See* **Arch.**

BELL-BOTTOM PIER HOLE Type of shaft or footing excavation, the bottom of which is made larger than the cross section above to form a belled shape.

BELL CRANK 1. Lever whose two arms form an angle at the fulcrum; **2.** Triangular plate hinged at one corner.

BELLED ANCHOR Having a butt or bottom-end shaped like a bell, often used on concrete piers or caissons.

BELLED CAISSON Caisson with an enlarged base.

BELLED EXCAVATION Part of a shaft or footing excavation, usually near the bottom, that is bell shaped.

BELLEVILLE SPRING Dished washer, made from strong and elastic steel, that deforms across its face to provide a springing action.

BELL HEATER Tank heater for heavy oil.

BELL HOUSING *See* **Clutch housing.**

BELLING BUCKET Drilling bucket tool with expanding cutters that can enlarge the bottom of a drilled hole to form a bell or underream. *Also called* **Underream bucket.** *See also* **Bucket.**

BELL JOINT *See* **Bell and spigot.**

BELLMOUTH Flared end of a tube.

BELLMOUTH OVERFLOW Progressively larger overflow structure from a reservoir.

BELLOWS 1. Diaphragm-like device that, upon deflection, produces a partial vacuum on one side of the diaphragm and atmospheric pressure upon the other; **2.** Flexible element of an expansion joint, consisting of one or more convolutions, formed from thin metal.

BELLOWS SEAL Metal bellows used in a shaft seal, in place of packing for valves, and in place of gaskets in long pipelines, to allow for expansion and contraction of the line.

BELLOWS-TYPE THERMOSTAT Thermostat regulated by a bellows device.

BELL TILE Clay pipe sections with one end enlarged to loosely join with the next pipe section.

BELLY Sag in a line.

BELLY DUMP Material hauler, prime mover or trailer, that discharges its load from the bottom.

BELLY HOOK Helicopter's load hook directly attached to its frame.

BELLY PAN Steel pan-shaped sheet mounted on the underside of heavy earth-moving equipment to protect the upper structure, engine, and hydraulic components.

BELOW GRADE Any part of a structure or site feature that is below the adjacent finished ground level.

BELT 1. Flexible endless band that passes around two or more pulleys; for transmitting motion or power, or for conveying material; **2.** *See* **Tire construction.**

BELT CONVEYOR *See* **Conveyor.**

BELT COURSE Narrow, continuous horizontal course of masonry, sometimes slightly projected from the face of the wall. *Also called* a **Sill Course,** and **String Course.**

BELT-DRIVEN MACHINE *See* **Driving machine.**

BELTED CONSTRUCTION Tire construction with several belts of steel or fabric that encircle the tire.

BELT ELEVATOR Elevator where the prime mover is connected to the machine by one or more belts, all parts of which act together, and where reversal of the elevator is accomplished by reversal of the prime mover.

BELT HIGHWAY *See* **Highway.**

BELT LOADER Traveling or movable belt conveyor with a cutting edge or disk.

BELT MARK Blemish on the face side of drywall panels caused during manufacture.

BELT-PALLET-TYPE MOVING WALKWAY *See* **Moving walkway.**

BELT SANDER *See* **Sander.**

BELTSTONES or -COURSES Horizontal bands or zones of stones encircling a building or extending through a wall.

BELT TIGHTENER Pulley system connected to a conveyor belt to adjust the slack.

BELT-TYPE MOVING WALKWAY *See* **Moving walkway.**

BELVEDERE Raised turret from which to view scenery.

BENCH 1. Working level or step in a cut that is formed in several layers; **2.** Horizontal ledge in a quarry face along which holes are drilled vertically. *Also called* **Pretensioning bed; 3.** Low scaffolding used by the hanging crew when applying drywall panels to ceilings; **4.** Strong worktable, often shaped or equipped with fittings necessary for the trade or task.

BENCH CLAMP *See* **Clamp.**

BENCH DOG Movable stop at one end of a work-bench, used to restrain material being worked on.

BENCHING 1. Process of excavating whereby terraces or ledges are worked in a stepped shape; **2.** Concrete around a half-round drainage channel in a manhole, sloping up from the tile edge to the manhole wall; **3.** Berm above a ditch.

BENCH MARK 1. Definite mark of known elevation, put on some permanent object, to be used as a reference and serve as a datum. *See also* **Monument; 2.** Marks of known separation applied to a specimen, used to measure strain (elongation of a specimen).

BENCH STAND Off-hand grinding machine, with either one or two wheels mounted on a horizontal spindle, attached to a bench.

BENCH TERRACE More-or-less level step in a slope of cut.

BENCH TEST Modified service test in which the service conditions are approximated in the laboratory. *Also called* **Simulated service test.**

BEND 1. Pane of glass that has been bent to fit an opening; when bent in two directions it is called a **Double bend; 2.** Curvature of a hose from a straight line; **3.** Short piece of curved pipe; **4.** Intertwining two ropes to make one continuous rope.

BENDERGAIN Allowance incorporated when measuring the length of sheet metal, conduit, or other material that must be of accurate length when bent.

BENDING Manipulation of flat glass in a kiln to form curved shapes or bends. *Also called* **Flexure.**

BENDING FORCE Amount of stress required to induce bending around a specified radius and hence, a measure of the stiffness.

BENDING MOMENT Bending effect at any section of a structural element, equal to the algebraic sum of the moments of the vertical and horizontal forces, with respect to the centroidal axis of a member, acting on a freebody of the member. *Also called* **Moment.**

BENDING-MOMENT DIAGRAM Graphical representation of the variation of bending moment along the length of the member for a given stationary system of loads.

BENDING SCHEDULE List of reinforcement showing the shapes and dimensions of every bar and the number of bars required.

BENDING STRENGTH Ability of a sheet to bend but not crease.

BENDING STRESS Force causing a deflection in shape or position of any member of a structure..

BEND RADIUS Radius of a bent section of hose measured to the innermost surface of the curved portion.

BENDS *See* **Caisson disease.**

BEND TEST Test used to determine the ductility of a metallic component that is subject to bending.

BENEFICATION Improvement of the chemical or physical properties of a raw material or intermediate product by the removal or modification of undesirable components or impurities.

BENEFICATION SYSTEM In recycling, the mechanical process of removing contaminants and cleaning scrap; the treatment of a material to improve its form or properties.

BENEFICIARY Person who receives, or is due to receive, benefits resulting from defined activities.

BENEFIT Anything contributing to an advantage or improvement. *See also* **General benefit**, and **Special benefit.**

BENEFIT/COST ANALYSIS *See* **Analysis.**

BENEFIT-COST RATIO Ratio obtained by dividing the anticipated benefits of a project by its anticipated costs. Either gross or net benefits may be used as the numerator.

BENT 1. Self-supporting frame with one or more columns, usually at right angles to the length of the structure it supports; **2.** Two posts and a roof timber, used in tunnel timbering; **3.** Roof truss complete with its supporting columns; **4.** Line of piling built as a structural unit, transverse to the length of a bridge and supporting the load of the superstructure. *Also called* a **Pile bent.**

BENT BAR Reinforcing bar bent to a prescribed shape. *See also* **Hook, Hooked bar, Stirrup,** and **Tie.**

BENT CAP Concrete beam or block extending across and encasing the heads of columns, comprising the top of a bent for a bridge span. *See also* **Pile cap.**

BENTGRASS *See* **Grass.**

BENTONITE Clay composed principally of minerals of the montmorillonoid group, characterized by high adsorption and very large volume change with wetting or drying; the basis of driller's mud, used to prevent collapse of holes augered into cohesionless soils.

BENTONITE/CEMENT PELLET Bentonite clay and cement, mixed 1:3 by volume, used to backfill boreholes with grout without need for grout mixers or pumps.

BENTONITE SLURRY TRENCH Trench excavated through common material and filled with bentonite slurry to maintain the dug sides.

BENT-UP BAR Used in a reinforced concrete beam where a main tensile bar in the bottom of the beam is bent up, at an angle, to the top of the beam, where it resists compressive forces.

BENZENE Coal-tar distillate.

BENZINE Solvent produced from petroleum.

BENZOL Light, coal-tar distillate used as a solvent.

BEQUEATH Give under the terms of a will.

BERLINER Type of terrazzo topping using small and large pieces of marble paving, usually with a standard terrazzo matrix between the pieces. *Also called* **Palladiana.**

BERM 1. Shoulder of a paved road or ditch; **2.** Artificial ridge of earth; **3.** Earth embankment, often combined with fencing or planting to create a visual and/or sound barrier.

BERNOULLI'S THEORY If no work is done on or by a flowing, frictionless liquid, its energy, due to pressure and velocity, remains constant at all points along the streamline.

BERTH Docking place for a ship.

BERTHING IMPACT Live forces imposed on piers during vessel docking.

BEST BID/TENDER Bid to complete work or supply services/materials that most closely meets the needs of the owner. This may not necessarily be the lowest price submitted.

BETA X (β_x) Filtration ratio obtained during a multipass test. Under specified test conditins, it is the number of particles in an influent fluid greater than C (in μm) per unit volume, divided by the number of particles in the effluent fluid in the same size range and unit volume.

BETA (β) RATIO In fluid filtration, the ratio of a number of particles greater than a given size (z μm) in a given volume of influent fluid to the number of particles greater than the same size (x μm) in the same volume of effluent fluid.

BETTERMENT Improvements, adjustments, or additions to a highway that more than restores it to its former good condition and that results in better traffic serviceability without major changes in its original construction.

betw. *Abbreviation for:* between.

bev. *Abbreviation for:* bevel; beveled.

bev. bd *Abbreviation for:* beveled board.

BEVEL The angle that one surface or line makes with another, when they are not at right angles. *See also* **Sliding T bevel.**

BEVELED EDGE 1. Shallow-angle factory-formed shape at the long edges of drywall panels that, when butted against the corresponding edge of an adjacent panel, form the v-groove in which the joint will be formed. *Also called* **Chisel edge. 2.** *See also* **Wood chisel.**

BEVEL GEAR *See* **Gear.**

BEVELING Edge finishing of flat glass glass to a bevel angle.

BEVEL OF COMPOUND-BEAD Bead of compound applied with a finished surface that slants away from the glass or panel so as to shed water.

BEVEL SIDING Board used for exterior wall covering that is thicker along one edge. When placed on the wall, the thicker edge overlaps the thinner edge of the siding below to shed water. The face width of bevel siding is from 90 mm to 285 mm (3-1/2 in. to 11-1/4 in.).

BEVEL SQUARE Type of try-square with a blade that can be set to any angle.

BEVEL WELD Butt weld where the pieces to be joined have one or more edges beveled, because the material is too thick for the electrode to penetrate for a required weld.

bev. sid. *Abbreviation for:* beveled siding.

BEZEL Sharpened edge of a cutting tool.

bf *Abbreviation for:* both faces; buff.

bf. *Abbreviation for:* boiler feed.

BF *Abbreviation for:* back feed; bottom face.

BFCT *Abbreviation for:* boiler feed compound tank.

BFP *Abbreviation for:* boiler feed pump.

bfr *Abbreviation for:* buffer.

bfr. *Abbreviation for:* before.

BFW *Abbreviation for:* boiler feed water.

BG *Abbreviation for:* below grade; below ground.

BH *Abbreviation for:* Brinell hardness; boiler house.

bhd *Abbreviation for:* bulkhead.

Bhn *Abbreviation for:* Brinell hardness number.

bhp *Abbreviation for:* brake horsepower.

BIAS Systematic error that is consistently negative or consistently positive. The mean of errors resulting from a series of observations that does not tend toward zero.

BIAS ANGLE Smaller included angle between the warp threads of a cloth and a diagonal line cutting across the warp threads.

BIAS CUT Cut of a textile material made diagonally at an angle less than 90° to the longitudinal axis.

BIAS-PLY TIRES Vehicle tires in which the main plies or cords run diagonally across the tread; the cords are generally made of fabric, as opposed to steel.

BIAS SEAM Seam at which bias-cut fabrics are joined together.

BIAT Timber bearer giving support to guard rails, decking, walkways, etc.

BIAXIAL BENDING Simultaneous bending of a member about two perpendicular axes.

BIB Tap or faucet threaded for connection of a hose. *Also called* a **Hose bib,** and **Sill cock.**

BICENTRIC CYLINDER LOCK Lock having two pin-tumbler cylinders, either of which, fitted with the correct key, will open the lock, permitting master keying without reducing the security of the locking mechanism.

BICK IRON Light anvil used for sheet metalwork.

BICYCLE Carriage or trolley used on a skyline.

BID Submitted tender or bidding price for the purchase of something, completion of specified work and/or supply of specified materials. *See also* **Balanced bid,** and **Unbalanced bid.**

BID ABSTRACT List of bidders to a project with details of bid items per item. *Also called* **Bid tabulation.**

BID BOND *See* **Bond.**

BID BOND INSURANCE *See* **Insurance.**

BID CALL Published invitation to submit bids for specified work or materials at a designated place and time.

BID DATE Date, and time at which bids must be filed. *Also called* **Bid time.**

BIDDER One who is qualified, and who has taken out a bid package, in order to prepare a price for the supply of goods and/or materials or to complete a job. *Also called* **Tenderer.**

BIDDER'S LIST List of those who have taken documents in order to prepare a bid. *Also called* **Tenderer's list.**

BIDDING DOCUMENTS Range of documents, including an invitation to bid, instructions to bidders, bid form, proposed contract documents, and drawings.

BIDDING PERIOD Elapsed time between the final day for taking bid documents and the prescribed time to submit a completed bid.

BIDDING REQUIREMENTS Specific details establishing procedures and conditions for the submission of bids, contained in the bidding documents.

BID DOCUMENTS Documents, including working drawings, specifications, contract details, etc., that are available to all contractors bidding the same job. *Also called* **Tender documents.**

BIDET Plumbing fixture designed to facilitate washing the perineal area of the body.

BID FORM Form supplied to a bidder that must be completed and that becomes part of the bidding documents. *Also called* **Tender form.**

BID GUARANTEE Security against completion of the contract and securing of any required bonds, required to be submitted with the bid documents.

BID OPENING Process of opening sealed bids and evaluating the submissions according to the terms indicated in the bid notice. *Also called* **Opening of tenders.**

BID PACKAGE Documents and other submissions that make up a bid to complete work and that may include offers from more than one company for parts of the work, and/or details of materials or construction techniques that vary from those in the bidding document. *Also called* **Tender package.**

BID PRICE Sum stated in a bid for which the bidder offers to complete the work.

BID QUANTITIES Quantities in a bid schedule for unit-price items; extended by unit prices to arrive at a total bid price for each unit-price bid item.

BID SECURITY Deposit of a bid bond, bank draft, cash, cashier's check, certified check, or other specified instrument required to be submitted with a bid to serve as a guarantee that the bidder, if awarded the contract, will complete the contract in accordance with the bidding and contract documents.

BID TABULATION *See* **Bid abstract.**

BID/TENDER BOND *See* **Bond.**

BID/TENDER NOTICE Advertisement calling for bids.

BID TIME *See* **Bid date.**

BIENNIAL PLANT Plant with a life span of two seasons.

BIFOLD DOORS *See* **Door types.**

BIFURCATION 1. Division of one penstock into two or more smaller units, each of which is separately connected to a turbine scroll case; **2.** Phenomenon whereby a perfectly straight steel member, under compression, may assume a deflected position or may remain undeflected, or a steel beam under flexure may either deflect and twist out of plane or remain in its in-plane deflected position.

BIG-END BORE Bore of the connecting rod of an engine into which the crank shaft fits

BIGHT 1. Any part of a line between its two ends; **2.** The hazardous zone contained within lines, either slack or under tension; **3.** The area made hazardous when slack cable is tensioned. **4.** An unintentional bend or deviation in a line caused by trees, stumps, or other obstacles preventing the line from running straight.

BIG-STICK LOADER 1. Steel frame located either midway on the bed of a bobtail pulpwood truck or directly behind the cab; **2.** Short, rotatable horizontal boom attached to a center post mounted on a pulpwood truck.

BILATERAL TOLERANCE Tolerance specified in two directions from a basic size.

BILLED WEIGHT Weight shown in a freight bill.

BILLET 1. Piece of semifinished steel, nearly square in section, formed by hot rolling an ingot or bloom, from which reinforcing bars (among many other products) are rolled; **2.** Heavy steel plate supporting a column.

BILLET STEEL Steel, either reduced directly from ingots or continuously cast, made from properly identified heats of open-hearth, basic oxygen, or electric-furnace steel, or lots of acid-Bessemer steel, and conforming to specified limits on chemical composition.

BILLET-STEEL REINFORCING BARS Reinforcing bars rolled from steel billets, in contrast to rail or axle steel.

BILLING DEMAND *See* **Demand (electric).**

BILL OF LADING Itemized list of goods contained in a shipment.

BILL OF MATERIALS List detailing the materials and products to be used in a project, usually forming part of the bidding document. *Also called* **Bill of quantities.**

BILL OF QUANTITIES *See* **Bill of materials.**

BILL OF SALE Contract for the sale of goods.

BILTMORE STICK Stick graduated in such a way that the diameter of a standing tree may be estimated when the stick is held across the main axis of the tree at the distance from the eye for which the stick is graduated (usually 60 cm (24 in.)). *See also* **Cruiser's stick.**

BIMETALLIC ELEMENT Product formed of two metals have different coefficients of thermal expansion, used as a temperature sensor, typically in a thermostat.

BINARY Counting system having two as its base.

BINARY CIRCUIT Circuit having a dual-state switching operation of on or off; high or low, etc. *Also called* **Digital circuit.**

BINARY COUNTER Counter circuit that tallies the number of pulses entering a circuit.

BINARY EXPLOSIVE Two-component explosive in which the components are not considered explosive until mixed together.

BINARY NUMBER SYSTEM Base-2 system of notation that employs 0 and 1 only.

bind. *Abbreviation for:* binding.

BIND Compression created in a falling or bucking cut due to uneven terrain or contact pressure from other trees or logs.

BINDER 1. Brick that extends only a part of the distance across the thickness of a wall; **2.** That part of a medium that facilitates bonding; **3.** Cementing material, either hydrated cement or products of cement or lime and a reactive siliceous material (the kinds of cement and curing conditions govern the general kind of binder formed); **4.** Material such as asphalt, resin, and others forming the matrix of concretes, mortars, and sanded grouts; **5.** Fines that hold gravel together when it is dry; **6.** Material used to bind aggregate particles together, preventing the entrance of moisture, acting as a cushioning agent, and in some cases, waterproofing an entire road surface; **7.** Deposit check that makes a contract valid; **8.** Wire rope placed around the load on a log truck or rail car, secured by a cinch to prevent spillage of logs. *Also called* a **Wrapper; 9.** Material used to join other materials (e.g. wafers) together; **10.** Written agreement providing immediate insurance coverage that is binding until a formal policy can be completed; it has an effective date and specified time limit, designates the insurer, amount, and type of coverage, and specific perils insured against.

BINDER COURSE Layer of asphalt mix placed between the base and the surface courses. *See also* **Asphalt binder course.**

BINDER SOIL Material consisting primarily of fine soil particles (fine sand, silt, true clays and colloids) and having good binding properties. *Also called* **Clay binder.**

BINDING ARBITRATION Settlement of a dispute between parties to an agreement or contract using the experience and knowledge of an independent expert or panel who, after hearing submissions by the various parties, will decide upon the form of settlement, which will be binding upon all parties to the dispute. *See also* **Arbitration.**

BINDING-IN WIRE Wire used to anchor a hose to a nipple, usually applied during the construction of the hose. *Also called* **Nipple wire.** *See also* **Annealed wire.**

BINDING MEDIUM Liquid that carries pigment particles and holds them together.

BINDING WIRE Soft wire used for tying together reinforcing bars where they cross each other. *See also* **Annealed wire.**

BIN WALL Series of connected metal, concrete, or brick bins, usually filled with earth or gravel and forming a retaining wall, abutment, etc.

bio. *Abbreviation for:* biological.

BIOAERATION Technique and associated equipment by which oxygen is introduced to a biological waste by a device that promotes mixing and development of aerobic bacteria within the mass.

BIOCHEMICAL OXYGEN DEMAND (BOD) Measure of the amount of oxygen used by microorganisms to break down organic waste materials in water.

BIOCONVERSION Conversion of organic waste via biological decomposition to produce usable gas, liquid fuels, or compost products.

BIODEGRADABLE Substance susceptible to deterioration by bacterial attack.

BIOGASIFICATION Resource recovery process for the extraction of methane resulting from anaerobic decomposition of organic matter.

BIOGEOCLIMATIC CLASSIFICATION Delineation of biotic regions or zones on the basis of vegetation, soils, topography, and climate.

BIOLOGICAL MATURITY In stand management, the age at which trees or stands have peaked in growth rate and are determined to be merchantable. *See also* **Rotation age.**

BIOLOGICAL SHIELDING Shielding provided to attenuate or absorb nuclear radiation, such as neutron, proton, alpha, and beta particles, and gamma radiation. The shielding is provided mainly by the density of the concrete, except that in the case of neutrons, the attenuation is achieved by compounds of some of the lighter elements, e.g., hydrogen and boron. *See also* **Boron-loaded concrete,** and **Shielding concrete.**

BIOLOGICAL TREATMENT Reduction of waterborne organic wastes by bacteria.

BIOMASS Total woody material in a forest. Refers to both merchantable material and material left following a conventional logging operation. In the broad sense, all of the organic material in a given area; in the narrow sense, burnable vegetation is used for fuel in a combustion system.

BIOMASS HARVESTING Harvesting of all material including limbs, tops, unmerchantable stems and stumps, usually for wood energy.

BIOSPHERE Portion of the earth and its atmosphere capable of sustaining life.

BIOTA Animal and plant life of a region.

BIPARTING DOORS *See* **Door types.**

BIPOLAR POWER SUPPLY Power supply having two poles, positive and negative.

BIRDCAGE Colloquialism descriptive of the appearance of a wire rope forced into compression: the outer strands form a cage and, at times, displace the core. The condition is caused by the sudden release of tension and resultant rebound of rope from an overloaded condition. These strands and wires will not return to their original positions.

BIRD SCREEN Wire mesh used behind louvers to prevent birds from entering a building.

BIRD'S EYE Distinctive grain pattern, visible in hardwood veneers, that resembles a bird's eye.

BIRD'S-EYE VIEW Perspective view in which the station point is vertically above the object.

BIRD'S MOUTH *See* **Plate cut.**

BIRMINGHAM WIRE GAUGE Series of numbers that rate the thickness of sheet steel and wire diameters; the smaller the number the thicker the metal.

BISCUIT Unglazed tile.

BISCUIT CHIPS Glazed-over chips on the edge or corner of the body of a tile.

BISCUIT CRACKS Fractures in the body of a tile visible both on the face and the back.

BISECT Divide or cut into two equal parts.

BISQUE Refined mixture of clay, water, and additives that has been shaped and dried ready for firing in a kiln to produce ceramic.

BISTIBLE Of or pertaining to the general class of fluidic devices that maintain either of two position operating states in the presence or absence of the setting input.

BIT 1. That part of a key that is cut to fit a lock; *See also* **Working tool; 2.** Part of a drill or excavating machine that cuts, usually of the following types:

> **Adjustable:** Wood bit with a mechanism that permits adjustment of the cutting diameter.

> **Auger:** Bit fashioned with a pointed, conical, fine-pitch starting thread followed by slack-pitch flights to evacuate the augered material.

> **Carbide tipped:** Bit having cutting faces fabricated of tungsten carbide steel welded to the shank.

> **Chopping:** Bit that cuts by being raised and dropped while also rotating.

> **Circle cutter:** Drill bit (used to cut a pilot hole) with an attached horizontal arm having a cutting blade at the end. The cutting blade can be moved along the arm to create holes up to some 200 mm (8 in.) in diameter.

> **Combination drill and countersink:** Bit that drills and countersinks in one operation. Commonly available in sets covering screw sizes #6, #7, #8, #9, #10, and #12.

> **Coring:** Bit that grinds the outside ring of a hole, leaving an inner core intact for sampling.

> **Countersink:** Flared bit that opens the mouth of a drilled hole to the shape of a truncated cone, permitting the head of a screw to be driven flush with or slightly below the surface.

> **Diamond:** Rotary bit with industrial diamonds or diamond grit embedded into its cutting face(s).

> **Disposable:** Bit that is discarded when worn.

> **Drag:** Diamond or fishtail bit that cuts by rotation of fixed cutting edges or points.

> **Fishtail:** Rotary bit having cutting edges or knives.

> **Fluted:** Straight-shank bit with a flute milled along its length. For wood boring or reaming.

> **Forstner:** For boring wood. This complex bit is equipped with a starter spur that permits a counterhole to be bored from the opposite direction once the location of the center has been identified from the appearance of the spur. This design of bit also permits the boring of a larger hole where a smaller hole already exists, without the necessity of plugging the smaller hole. A Forstner bit will also bore a clean hole on the end grain of wood.

> **Glass:** Slow-speed bit with a cutting edge (often spear-shaped) designed for cutting or grinding through glass.

High-speed: Twist bits for use on wood and metal that can be operated at high speed without overheating or damage to the cutting edge.

Masonry: A drill bit fitted with tungsten-carbide faces to the cutting edges.

Multiuse: Bit that can be sharpened for reuse when worn.

Plug: Diamond bit that grinds out the full diameter of a hole.

Plug and dowel cutters: Produce plugs and dowel lengths from common stock.

Plug-center: Bit used to widen holes. The center part is a plug inserted in the hole already drilled.

Roller: Bit that contains cutting elements that are rotated inside as it turns.

Screwdriver: A wide range of special-purpose bits for driving and removing slot-, Phillips-, and Robertson-head screws.

Spade: Wood bit comprising a sharpened, fluted point central to two sharpened blades.

Twist: For wood and metal boring.

BITCH Steel or iron clip used to secure heavy excavation timbers that cross each other; similar to a **Dog anchor** but with one end at right angles to the other.

BIT DEPTH GAUGE Adjustable device that fits on a drill or onto the bit held by a drill and that permits the depth to which the bit will penetrate to be predetermined.

BITE 1. Depth of cut per tooth per unit of measure; **2.** Distance a log or cant advances into the saw between successive teeth; **3.** Initial attack of a solvent on the surface to which an adhesive or other solvent-based substance has been applied; **4.** See **Nip.**

BIT EXTENSION Attachment, one end of which fits into the chuck of a drill and at the other accepts a drill bit.

BIT GAUGE Attachment that fits over a drill bit and that can be clamped to predetermine the depth that can be drilled.

BIT KEY LOCK Lock operated by a key having a wing bit.

BITT Vertical timber on a dock (or ship's deck) to which cables, ropes, lines, etc. are made fast.

BITTING Shapes on a key that set the tumblers of a lock.

bitum. *Abbreviation for:* bituminous.

BITUMEN Mixture of hydrocarbons of natural or pyrogenous origin, or a combination of both; frequently accompanied by nonmetallic derivatives that may be gaseous, liquid, semisolid, or solid; and that are completely soluble in carbon disulfide.

BITUMEN CONTENT Percent of binder used in a bituminous mix.

BITUMINOUS Term used to designate materials that are derived from petroleum, coal tar, etc.

BITUMINOUS BASE COURSE Foundation for surface and binder courses composed of aggregate bound together with bituminous material.

BITUMINOUS BINDER COURSE Intermediate course between a base course and a surface course.

BITUMINOUS CEMENT Black, solid, semisolid or liquid substance at natural air temperatures that is appreciably soluble only in carbon disulfide or some volatile liquid hydrocarbon, being composed of mixed indeterminate hydrocarbons mined from natural deposits, produced as a residue in the distillation of petroleum, or obtained by the destructive distillation of coal or wood.

BITUMINOUS COATING One of several types of application of a bituminous material, including:

Bituminous prime coat: Low viscosity asphalt sprayed on an untreated base in preparation for construction of an asphalt pavement.

Contrast seal: Seal coat designed primarily to provide color or texture contrast with an adjacent surface.

Fog seal: Thin application of bituminous material without cover aggregate.

Prime coat: An application of a low viscosity liquid bituminous material to coat and bind mineral particles preparatory to placing a base or surface course.

Seal coat: Thin treatment consisting of bituminous material usually limited to sand-seal, chip seal, slurry seal, contrast and fog seal.

Slurry seal: Seal coat consisting of a semifluid mixture of asphaltic emulsion and fine aggregate.

Tack coat: Application of bituminous material to an existing surface to a provide bond with a superimposed course.

BITUMINOUS CONCRETE Designed combination of dense graded mineral aggregate filler and bituminous cement mixed in a central plant, laid and compacted while hot.

BITUMINOUS LEVELING COURSE Mix of sand and asphalt, used to crown an old pavement or a rough base before surfacing.

BITUMINOUS MACADAM Coarse stone, gravel, or slag fragments, coated with bitumen either in place or in a mixing plant.

BITUMINOUS MAT Layer of bituminous-cemented aggregate, from 19 to 32 mm (0.75 to 1.25 in.), or more, produced by repeated surface treatments and aggregate covering, using viscous bitumens.

BITUMINOUS MATERIAL DISTRIBUTOR BODY Insulated tank-truck body with means for distributing hot bituminous material under pressure. It is usually equipped with a means for heating the contents.

BITUMINOUS MIX-IN-PLACE Mixture of aggregates and liquid bituminous material, mixed directly on the ground using blade graders, harrows, rotary tillers, or a traveling plant machine.

BITUMINOUS PAINT Natural asphaltic or coal-tar bitumens dissolved in organic solvents, sometimes containing softening agents, pigments, synthetic resins, or inorganic fillers.

BITUMINOUS PAVEMENT Layers of aggregate over 19 mm (0.75 in.) thick, coated and cemented together with bituminous material.

BITUMINOUS PLANT MIX Aggregates and bituminous materials mixed in a central plant before being hauled to the point of application.

BITUMINOUS PRIME COAT *See* **Bituminous coating.**

BITUMINOUS STABILIZATION Technique of mixing soils, gravels, and sand with liquid bitumen.

bk *Abbreviation for:* back; book.

bkcs. *Abbreviation for:* bookcase.

bkd *Abbreviation for:* baked.

bkdn *Abbreviation for:* breakdown.

bkgd *Abbreviation for:* background.

bkr *Abbreviation for:* breaker.

bk rk *Abbreviation for:* book rack.

bks *Abbreviation for:* barracks.

bkt *Abbreviation for:* bucket; bracket.

bl *Abbreviation for:* body length.

bl. *Abbreviation for:* back blade; bleed.

BL *Abbreviation for:* base line; bottom layer; building line.

B/L *Abbreviation for:* bill of lading.

BLACK BASE Asphalt-stabilized bases, ranging from coarse stone to sand mixes, and from road mix to high-type plant mixes.

BLACK BOLT Machine bolt coated with black iron oxide.

BLACK CORE Most fireclays and brick clays contain carbonaceous material; if a brick shaped from such clays is fired too rapidly, this carbonaceous matter will not be burned out before vitrification begins. The presence of carbon and the consequent reduced state of any iron compounds in the center of the fired brick result in a black core or 'black heart.'

BLACK IRON PIPE *See* **Pipe.**

BLACK LABEL *See* **Shake,** and **Shingle.**

BLACK POWDER Mixture of carbon, sodium, or potassium nitrate, and sulfur, used for blasting. *Also called* **Powder.**

BLACKTOP Generally, asphalt pavement.

BLACKWALL HITCH Loop of rope running back on itself around the shank of a lifting hook; a knot that will not slip under tension.

BLADE 1. Equipment part or accessory that digs and pushes broken rock or earth but that does not carry it; typically fitted to a tractor dozer, trencher, or excavator; **2.** That part of the chain saw upon which the cutting chain travels. *See also* **Bar.**

BLADE MIXING Mixing of materials, on site, with the blade of a motor grader.

BLAINE APPARATUS Air-permeability apparatus for measuring the surface area of a finely ground cement, raw material, or other product.

BLAINE FINENESS Fineness of powdered materials, such as cement and pozzolans, expressed as surface area (usually in square meters per kilogram) as determined by the Blaine apparatus. *See also* **Specific surface.**

BLAINE TEST Method for determining the fineness of cement or other fine material on the basis of the permeability to air of a sample prepared under specific conditions.

BLANKET 1. Layer of soil and broken rock left or placed over a blast to confine or direct the throw of fragments; **2.** Inorganic insulation in widths to fit snugly between studs and joists, supplied in long rolls.

BLANKET BOND *See* **Bond.**

BLANKET COVERAGE Insurance protection afforded a class of property or person which may fluctuate in quantity, quality, or location.

BLANKET CRIME POLICY Policy that insures against dishonesty, losses of money and securities on, and off the premises, etc.

BLANKET FIDELITY BOND *See* **Bond.**

BLANKET MORTGAGE *See* **Mortgage.**

BLANKET PURCHASE ORDER Standing order for the purchase of goods and/or services.

BLANK FLUE Vertical duct built into a chimney but either not leading from a fireplace or not venting to the atmosphere, built for aesthetic reasons, to conserve materials, to reduce weight, etc.

BLANK KEY Key that has not been cut or shaped to fit a lock.

BLANK WALL Wall without openings.

BLANK WINDOW (DOOR) Window or door opening that has been walled up.

BLAST Operation of breaking rock by means of explosives. *Also called* **Shot.**

BLAST AREA Area in which explosives loading and blasting operations are being conducted.

BLASTER *See* **Shot firer.**

BLAST-FURNACE GAS Gas of low Btu content recovered from a blast furnace as a by-product and used as a fuel.

BLAST-FURNACE SLAG (*See also* **Cementitious materials**). Nonmetallic product, consisting essentially of silicate and aluminosilicates of calcium and other bases, that is developed in a molten condition simultaneously with iron in a blast furnace. There are several distinct types:

> **Air-cooled blast-furnace slag:** Material resulting from solidification of molten blast-furnace slag under atmospheric conditions. Subsequent cooling may be accelerated by application of water to the solidified surface.

> **Expanded blast-furnace slag:** Lightweight, cellular material obtained by controlled processing of molten blast-furnace slag with water, or water and other agents, such as steam or compressed air or both.

Granulated blast-furnace slag: Glassy, granular material formed when molten blast-furnace slag is rapidly chilled, as by immersion in water.

BLAST GATE Shutoff air valve, not used for flow control.

BLASTHOLE Hole, 100 mm (4 in.) or more in diameter, drilled in rock or other material for the placement of explosives. *Also called* **Borehole.**

BLASTHOLE DRILL *See* **Drill.**

BLASTING AGENT Any material or mixture, consisting of a fuel and oxidizer, intended for blasting, not otherwise classified as an explosive, and in which none of the ingredients are classified as an explosive, provided that the finished product, as mixed and packaged for use or shipment, cannot be detonated by means of a No. 8 test blasting cap when confined.

BLASTING CAP Copper shell containing a charge of detonating agent that is ignited from the spark of a fuse or by an electrical current.

BLASTING FUSE Fuse having a predictable rate of burn, used for igniting a blasting charge. *Also called* **Fuse.**

BLASTING GELATIN Jelly-like high explosive made by dissolving nitroglycerin into nitrocotton.

BLASTING MACHINE Hand- or battery-operated generator used to supply electrical current to blasting circuits. *Also called* a **Battery.**

BLASTING MAT Woven wire or cable mat, or shape made from an assembly of tires, used to cover a shot to hold down flying material. *Also called* **Mat.**

B/L att. *Abbreviation for:* bill of lading attached.

BLAZE To mark a tree with an axe for layout or trail identification.

bld *Abbreviation for:* build.

bldg *Abbreviation for:* building.

bldr *Abbreviation for:* builder.

BLEACH Removing discolorations or stains from wood, or making wood lighter in color using oxalic acid or other bleaching agent.

BLEB Small blister or bubble.

BLEED 1. Removal of unwanted air or fluid from passages, pipes, etc. *See also* **Purge**; 2. Giving up of color when in contact with water or a solvent. *Also called* **Migration.**

BLEEDER Intentional leak, usually used to reduce pressure in an impulse line.

BLEEDER TILE Drainage pipe connecting an external foundation drain to an internal drain, passing through or under the foundation of the building.

BLEEDING 1. Surface exudation; 2. Exudation of resin, gum, creosote, or other substance from lumber; 3. Autogenous flow of mixing water within, or its emergence from, newly placed concrete or mortar caused by the settlement of the solid materials within the mass. *Also called* **Water gain**; 4. Formation of a film of asphalt on a pavement surface due to upward movement of the asphalt in the mix. *Also called* **Flushing**; 5. Separation of liquid lubricant from a lubricating grease for any cause. *See also* **Bulk appearance**; 6. Discoloration at finished drywall joints; 7. Property of some paints to transfer their color into subsequently applied coats; diffusion of coloring matter through a coating from a substrate.

BLEEDING CAPACITY Ratio of the volume of water released by bleeding to the volume of paste or mortar.

BLEEDING CHANNEL Small vertical open channel caused by excessive bleeding from placed concrete.

BLEEDING CREOSOTE Creosote, as on a pile surface, that resembles fresh paint that has been poured rather than brushed. On hot days, creosote will 'bleed' out of a pile.

BLEEDING RATE Rate at which water is released from a paste or mortar by bleeding.

BLEED OFF To pass by or circumvent.

BLEED WATER Excess water in a concrete mixture that surfaces after placing.

BLEMISH 1. Any superficial defect that causes visible variation to a surface; 2. Visible variation from a consistently smooth and uniformly colored surface of hardened concrete. *See also* **Bug hole, Efflorescence, Honeycomb, Laitance, Lift joint, Popout, Rock pocket,** and **Sand streak.**

BLEND 1. To mix or make homogeneous; 2. Mixe of different genetic raw materials to form a water-repellent product.

BLENDED CEMENT Hydraulic cement consisting essentially of an intimate and uniform blend of granulated blast-furnace slag and hydrated lime, or an intimate and uniform blend of portland cement and granulated blast-furnace slag, or portland cement and pozzolan, produced by intergrinding portland cement clinker with the other materials, or by blending portland cement with the other materials, or by a combination of intergrinding and blending.

BLENDER Machine comprising a rotating drum in which materials are thoroughly mixed, or blended, or in which components are coated with a substance.

BLENDING 1. Application of a binder and additives to particles in the manufacture of particleboard; 2. Process of mixing fluid lubricant components for the purpose of obtaining desired physical properties. (This is different to compounding, which is regarded as the mixing or otherwise combining of lubricant components with other components for the purpose of securing chemical and/or physical properties not usually obtainable by blending fluid lubricant components alone.) *See also* **Compounding.**

bli. *Abbreviation for:* blind.

BLIND 1. Process of covering the subbase or base of a road with fine material to fill interstices; 2. Attachment used to obscure or prevent light from entering a window. *See also* **Shutters.**

BLIND ARCH Relieving arch concealed behind a wall facing.

BLIND AREA Area or zone that cannot be seen from an observation point.

BLIND BOND *See* **Bond.**

BLINDBOX Casing for an external window blind.

BLIND FASTENER Self-forming device designed to fasten or joint together two or more pieces of a structure from one side of the workpiece.

BLIND HEADER Concealed brick header in the interior of a wall, not showing on the faces. *See also* **Header.**

BLINDING 1. Application of a layer of weak concrete or other suitable material to reduce surface voids, or to provide a clean, dry working surface; **2.** Filling or plugging of the openings in a screen or sieve by the material being separated; **3.** Compaction of soil immediately over a tile drain so as to reduce the tendency of fines to wash into the tile.

BLIND JOINT Joint with no parts visible.

BLIND MORTISE Mortise that does not pass through the lumber and that encloses a stub tenon.

BLIND NAILING Nailing in such a way that the nailheads are not visible on the face of the work; usually at the tongue of tongue-and-groove or matched boards.

BLIND NUT *See* **Nut.**

BLIND RIVET Fastener for joining light metal that can be inserted and formed from the face side of the work. *Also called* a **Pop rivet.**

BLIND STOP Rectangular molding, 20 mm x 35 mm (3/4 in. x 1-3/8 in.) in section, or greater, used in the construction of a window frame. Serves as a stop for storm window and screen or combination windows, and to resist air leakage.

BLIND WALL OR FOUNDATION A hidden structural element that is designed to perform in a way, or for a purpose different to similar elements in the same structure.

B-LINE Pay line for excavation in a tunnel.

bl.ir. *Abbreviation for:* black iron.

BLISTER 1. Raised area on the surface, or a separation between layers, usually creating a void or air-filled space; **2.** Relatively large gas-filled cavity in glass.

BLISTERING 1. Irregular raising of a thin layer at the surface of a material; **2.** Spalling of placed mortar or concrete during or soon after completion of the finishing operation, or in the case of pipe, after spinning; **3.** Bulging of a finish plaster coat as it separates and draws away from the base coat; **4.** Formation of bubbles or blisters on a painted surface while the paint is still elastic; **5.** Development during firing of enclosed or broken microscopic vesicles in a body, or in a glaze or other coating.

blk *Abbreviation for:* black; blank; block.

blkt *Abbreviation for:* blanket.

BLOATED Swollen, as in certain lightweight aggregates, as a result of processing.

BLOCK 1. Masonry unit larger than the size of a brick in any dimension; **2.** *See* **Construction types**; **3.** Pulley device through which rope, cable, or chain is run to obtain mechanical advantage; **4.** Term applied to a wire rope sheave (pulley) enclosed in side plates and fitted with some attachment such as a hook or shackle. *See also* **Crown block, Sling block, Snatch block,** and **Traveling block**; **5.** Log cut to a designated length, usually 1.2 m to 2.4 m (4 ft to 8 ft), for veneer production. *Also called* a **Bolt**; **6.** Short, thick piece of wood (100 mm (2 in.) or thicker), used to transfer the load between two structural members; usually used between the small stringer end and the sill or cap to elevate the stringer.

BLOCK AND CROSS BOND *See* **Bond.**

BLOCK AND TACKLE Comprising chain hoists, windlasses, and winches, or a combination of these, used to give mechanical advantage for lifting or pulling.

BLOCK ANGLE Square of tile specially made for changing direction of the trim.

BLOCK BEAM Flexural member composed of individual blocks that are joined together by prestressing. *See also* **Segmental member.**

BLOCK BRIDGING *See* **Solid bridging.**

BLOCK CUTTING PATTERN Felling pattern started along the timber's front face next to a roadside. A felling machine works back and forth along the face. When sufficient timber has been felled to allow skidding, the cutting machine begins a second pass along another side of the block.

BLOCK END *See* **End reinforcement.**

BLOCK HEATER Heater device located in the cylinder block water jacket of a water-cooled engine to warm engine coolant.

BLOCKHOLE Hole drilled into a boulder to allow placement of a small charge to break the boulder.

BLOCKHOLING Breaking of boulders by firing a charge of explosives that has been loaded in a drill hole.

BLOCKING 1. Method of bonding two adjoining or intersecting walls, not built at the same time, by means of offsets whose vertical dimensions are less than 200 mm (8 in.); **2.** Timber infilling in a stud-frame wall, behind the drywall or other facing, to which to subsequently attach fittings such as kitchen cupboards, counters, etc.; **3.** Piece of wood placed behind a joint between two panels to make the structure more rigid; **4.** Timber block used as a distance piece between a waling and the temporary lining of an excavation to permit insertion and erection of reinforcement in retaining walls; **5.** Supports used to prevent shipments from shifting during transportation.

BLOCKING BEAM Steel beam placed on a machine beam to elevate the machine room equipment.

BLOCKING COURSE Course of stone placed on top of a cornice crowning a wall.

BLOCKING DIODE Device that allows an electric current to flow in one direction only.

BLOCKING HAMMER *See* **Hammer.**

BLOCK MATING SURFACE Surface of an engine block that mates to a corresponding surface on the cylinder head.

BLOCKOUT Space within a concrete structure under construction in which fresh concrete is not to be placed.

BLOCK-OUT CONCRETE Concrete placed after a main structure is poured, to complete ancillary or auxiliary elements.

BLOCK PAVING Vehicle or pedestrian wearing surface constructed of blocks of granite, concrete, or other material.

BLOCK PENETRATION Penetration at 25°C (77°F) of a sample of lubricating grease that is sufficiently hard to hold its shape, determined on a freshly prepared face of a cube cut from a block of the grease. *See also* **Penetration.**

BLOCK PLAN Plan of a building showing the outlines of existing and proposed buildings.

BLOCK PLANE *See* **Plane.**

BLOCK PURCHASE Use of one or more blocks for mechanical advantage.

BLOCKWORK Construction formed from precast concrete blocks.

BLOOM 1. Visible exudation or efflorescence on a surface; **2.** Loss of luster of a coating, usually caused by condensation of moisture on the film before drying or curing is complete. *See also* **Appearance**; **3.** Discoloration or change in appearance of the surface of a rubber product caused by the migration of a liquid or solid to the surface; **4.** Surface color (usually blue or green) of a lubricating oil or grease when viewed by reflected daylight at an angle of about 45° from the surface. Bloom is associated with the absorption of ultraviolet light in the oil and may not be visible if the sample is viewed by artificial light.

BLOT Green mark or stain on the face of a tile.

BLOTTER Disk of compressible material, usually of blotting paper stock, used between a wheel and flanges when mounting.

BLOW 1. Localized delamination caused by steam pressure buildup during the hot pressing process of particleboard manufacture. Steam may result from excessive moisture, excessive glue spread, or high press temperatures; **2.** Area of facing or backing paper separated from the core of drywall during manufacture. **3.** In air distribution, the distance an air stream travels from an outlet to the point where motion along its axis has reduced to a velocity of 15 m/min (50 ft/min); **4.** Distance an air stream travels from a unit heater without perceptible temperature change or loss of velocity. **5.** *See also* **Boil.**

BLOW-BY Gases and unburned combustion by-products that escape past the piston rings, into an engine crankcase.

BLOW COUNT 1. Number of blows of a pile driving hammer required to insert a pile into the ground a specified distance; **2.** Blows on a soil sampler in a standard penetration test. *See also* **Driller's stroke**, and **N Value.**

BLOWDOWN 1. Stand of trees blown down by wind. *Also called* **Wind throw**; **2.** Removal of water from a system or vessel by the injection of air under pressure.

BLOWDOWN PERIOD Time taken to reduce pressure in an autoclave from maximum to atmospheric.

BLOWER 1. Mechanical device that moves air under pressure; **2.** In diesel engines, refers typically to the Rootes type of low-pressure air pump used to force air into a two-stroke diesel engine; **3.** Slang term for a turbocharger; **4.** Tank vehicle accessory (that may be either tractor-mounted or trailer-mounted) used in the transfer by air of dry bulk products at pressures normally below 34.47 kPa (5 psi).

BLOWER DRIVE Fan shaft on a turbocharger.

BLOWER FAN Fan positioned in an engine cooling system such that air passes through the fan before entering the radiator.

BLOWHOLE 1. Hole in a casting or weld caused by gas entrapment during cooling. **2.** *See* **Bug hole**; **3.** Unwanted, individual cavity, usually less than 10 mm (0.4 in.) in diameter, on the surface of concrete, usually caused by an air void being trapped against the face of formwork.

BLOWING 1. Upward movement of soil material in the base of a cofferdam or excavation due to groundwater pressure; normally associated with insufficient toe penetration of the sheeting. *See also* **Blowout, Boiling**, and **Piping**; **2.** Plastering defect in which a conical piece may be blown out of a finished surface owing to an imperfectly slaked particle of quicklime in the mixture. *Also called* **Pitting.**

BLOWLAMP Torch burning gas, kerosene, or other fuel under pressure that produces a high temperature, jet-shaped flame.

BLOWN INSULATION Low-density, loose insulation material that is mechanically installed (blown).

BLOWN JOINT Plumbing joint made using a soft metal, such as lead.

BLOWN OR OXIDIZED ASPHALT *See* **Asphalt.**

BLOWOFF 1. Vent to atmosphere of a fluid or vapor held under pressure; **2.** Outlet for the discharge of sediment or water from a sewer.

BLOWOUT 1. Sudden release of air from a heading being driven under air pressure. *See also* **Blowing**; **2.** Basin excavated in shifting sand or other easily eroded surface material by wind action.

BLOWOUT OF PILE Method of removing the soil from the interior of open-ended piles. *See also* **Air lift.**

BLOWPIPE Alternate name for an oxycetylene torch.

BLOW SAND Wind-borne sand.

BLOW UP 1. Localized buckling or shattering of rigid pavement, caused by excessive longitudinal stress; **2.** Enlarged portion of a photograph, drawing, map, etc.; **3.** Sudden increase in forest fire intensity or rate of spread. *Also called* **Flareup.**

blp. *Abbreviation for:* blue-line print.

blr *Abbreviation for:* boiler.

blstg *Abbreviation for:* blasting.

blstg pwd. *Abbreviation for:* blasting powder.

blt *Abbreviation for:* borrowed light; built.

BLUE BLUING Dye applied to gear teeth to check points of contact between them and meshing gears.

BLUE BUTT Log that is bigger than average.

BLUED Metal, heat treated to result in an oxidized,

bluish surface.

BLUE (WATER) GAS Artificial fuel made by passing steam over incandescent carbon (usually coke). This forms a mixture of hydrogen and carbon monoxide.

BLUEGRASS *See* **Grass.**

BLUE LABEL *See* **Shake,** and **Shingle.**

BLUEPRINT 1. Generic term for a construction drawing; **2.** Reproduction of a drawing in which the background is blue and the lines of the drawing are shown in white. *See also* **Whiteprint.**

BLUE STAIN A discoloration of lumber due to a fungus growth in unseasoned wood. Although it mars the appearance, the stain has no serious affect on the lumber's strength. *See also* **Brown stain,** and **Chemical brown stain.**

BLUE STAIN FUNGUS Most common form of fungal stain occurring in sapwood. Conifers are most susceptible but it may also occur in light-colored heartwood of perishable timbers. Commonly develops in dead trees, logs, lumber, and other wood products until the wood is dry. Reduces the grade of wood, but does not significantly reduce the strength. Some blue-stain lumber is highly valued for speciality products.

BLUE TOP Grade stake, the top of which is painted blue, used to indicate finish grade level.

BLUFF Abrupt rise of rock on the terrain.

BLUNGING Wet process of blending, or suspending ceramic material in liquid by agitation.

BLUSHING 1. Hazing or whitening of a paint film, caused by absorption and retention of moisture in the drying phase; **2.** Patches of whiteness that appear on a cured resin.

blvd *Abbreviation for:* boulevard.

blw *Abbreviation for:* blow.

blwg *Abbreviation for:* blowing.

blwr *Abbreviation for:* blower.

blws *Abbreviation for:* bellows.

bm *Abbreviation for:* board measure; beam; bell mouth; brass mounted.

BM *Abbreviation for:* bench mark.

B/M *Abbreviation for:* bill of materials.

bmep *Abbreviation for:* brake mean effective pressure.

BMER NUMBER Building Material Evaluation Number: a sign of a product meeting specified CMHC or CCMC required standards.

bmkr *Abbreviation for:* boilermaker.

bmpr *Abbreviation for:* bumper.

bnch *Abbreviation for:* bench.

bnch bd *Abbreviation for:* bench board.

bnd *Abbreviation for:* bend.

bnh *Abbreviation for:* burnish.

bnr *Abbreviation for:* burner.

bnt *Abbreviation for:* bent.

bnth *Abbreviation for:* beneath.

BO *Abbreviation for:* blackout; blowoff; blowout.

BOARD Finished lumber that is less than 50 mm (2 in.) thick, and from 100 mm to 300 mm (4 in. to 12 in.) wide. *See also* **Lumber.**

BOARD AND BATTEN Type of external siding composed of wide boards nailed vertically to the sheathing so there is an average 13 mm (1/2 in.) space between them, with narrow 50 mm to 75 mm (2 in. to 3 in.) battens then nailed over the spaces.

BOARD BUTT JOINT Shotcrete construction joint formed by sloping a gunned surface to a 25 mm (1 in.) board laid flat.

BOARD FOOT A unit of measurement for wood. It is the equivalent of a board that is 300 mm (1 ft) square and 25 mm (1 in.) thick.

BOARDING Boards butted together, or closely laid and fastened to a framework.

BOARD INSULATION *See* **Rigid insulation.**

BOARD KNIFE Hand tool used to score and trim drywall.

BOARD-MARKED FINISH Architectural surface finish to concrete, produced by using rough-sawn lumber for construction of the formwork.

BOARD MEASURE System of measurement for lumber, the unit of measure being one board foot, that is represented by a piece of lumber 300 mm (1 ft) square and approximately 25 mm (1 in.) thick.

BOARD RULE Scaled measuring stick, used for computing board feet.

BOARD SAW *See* **Saw.**

BOAST To surface stone with hand tools.

BOASTED WORK Masonry finished with a boaster.

BOASTER Mason's broad-faced chisel, used for facing stone and cutting bricks.

BOAT PATCH *See* **Repairs.**

BOATSWAIN'S CHAIR Seat attached to a suspended rope, designed to accommodate one user in a sitting position.

BOATSWAIN'S STAND Suspended stand designed to accommodate one user in a standing position.

BOB *See* **Plumb line.**

BOBBIN CABLE HANGER Insulated hanger, used to support traveling cables.

BOBTAIL 1. Tractor operating without a trailer; **2.** Two-axle truck.

BOBTAIL TRUSS *See* **Truss.**

boc *Abbreviation for:* back or center; back outlet central; blowout coil.

BOCA NATIONAL BUILDING CODE *See* **Building code.**

BOD *Abbreviation for:* biochemical oxygen demand.

BODIED LINSEED OIL Linseed oil thickened by

heating or chemicals.

BODILY INJURY LIABILITY INSURANCE Protection against loss caused by bodily injury, sickness, or death sustained within a defined area by other than employees.

BODY 1. Structure mounted on the chassis of a vehicle designed for cargo or passengers. This does not normally include the cab unless the cab is an integral part of the body, as in a school bus; **2.** Load-carrying part of a hauler or scraper. *See also* **Quarry body** and **Rock body**; **3.** Structural portion of a ceramic article or the material or mixture from which the article is made; **4.** Fluid thickness of a medium. *Also called* **Consistency** or **Viscosity.**

BODY BRICK *See* **Brick.**

BODY HINGE Attaching hinge that connects a dump body to the hinge pin to form a pivot axis about which the body will rotate during dumping operations.

BODY LENGTH Distance from the foremost point of a vehicle body to its rearmost point.

BODY PLIES *See* **Tire construction.**

BODYING IN Staining, filling, and first priming prior to finish painting.

BODY SHELL Primary exterior portion of a dump body consisting of sheet metal sides and floor.

BODY UNDERSTRUCTURE (SUBFRAME) Crossmembers and longitudinal members under the floor of a dump body.

BODY WAVES Seismic waves that travel through the mass or body of a rock material.

BODY WEIGHT Unmounted weight of a (machine or vehicle) body with applicable options.

BODY WIRE Round or flat wire helix embedded in a hose wall to increase strength and to resist collapse.

boe *Abbreviation for:* back outlet eccentric.

b of b *Abbreviation for:* back of board.

BOG Soft, spongy ground, usually wet and composed of vegetable matter.

BOGEY *See* **Undercarriage.**

BOGIE Assembly of two or more axles arranged to allow both vertical wheel displacement and an equalization of loading on the wheels. *See also* **Tandem.**

BOGIE AXLE *See* **Axle.**

BOHEMIAN BOOM Flat boom constructed on log loaders to heel logs.

BOIL Vertical flow of water and soil into the bottom of an excavation.

BOIL BOARD Board nailed to formwork to prevent concrete from 'boiling out' while being placed.

BOILED LINSEED OIL Linseed oil incorporating lead, manganese, or cobalt salts in quantities sufficient to make the oil harden rapidly when spread in thin coats. *See also* **Linseed oil.**

BOILER 1. Approved vessel in which water can be heated or steam generated; **2.** Appliance intended to supply hot water or steam for space heating, processing, or power purposes.

BOILER BLOW-OFF Outlet on a boiler to allow for the entering or discharging of water or sediment in the boiler.

BOILER HEATING SURFACE Total of the surface area of all heat-transfer appliances connected to the hot water or steam distribution system served by a boiler.

BOILER HORSEPOWER *See* **Horsepower.**

BOILER RATING Heating capacity of a boiler.

BOILING *See* **Blowing.**

BOILING POINT Temperature at which a specified liquid boils; 100°C (212°F) for water at sea level.

BOIL POINT *See* **Open-bubble point.**

BOLE Tree stem that has grown to a substantial thickness, capable of yielding sawtimber, veneer logs, logs, or large poles..

BOLECTION MOLDING Molding used to cover the joint between two members with different surface levels.

BOLLARD 1. Wooden, concrete, or metal post used to prevent vehicles from entering a pedestrian area; **2.** Post on a dock to take a ship's mooring lines.

bols. *Abbreviation for:* bolster.

BOLSTER 1. Chisel used for cutting brick. *See also* **Set**; **2.** Transverse members commonly used to define the horizontal support for the diagonal braces of a vertical mast. *See also* **Bunk**; **3.** Bottom member of a sling for a hydraulic elevator; **4.** Wooden block with a hole, used to support sheet metal work being punched; **5.** Short structural steel member placed horizontally on top of a column to support other structural members.

BOLSTER ASSEMBLY Bottom horizontal member of a hydraulic elevator car sling to which the platen plate attaches.

BOLT 1. Steel rod machined with a thread for all or part of its length up to an enlarged head, that may be of a range of shapes for specific uses; **2.** Short section of a tree. *Also called* **Block**, and **Veneer bolt.**

BOLT BAG Bag in which an ironworker carries bolts.

BOLT CIRCLE Circle scribed around the arbor hole of a circular saw that has one or more holes laid out to receive bolts.

BOLT CUTTER Tool used by iron workers to cut bars on the job. Smaller (wire) mesh cutters are used to cut welded wire fabric or bundle wires.

BOLTED FRAME ASSEMBLY Assembly with bolts, washers, and nuts instead of rivets or welding.

BOLT HOLE CIRCLE Circle on a flange face around which the center of the bolt holes are distributed.

BOLTING UP The phase of steel erecting that consists of completely bolting all connections of a structural frame.

BOLT-ON-HOUSING OUTRIGGER *See* **Outrigger.**

BOLT SLEEVE Tube surrounding a bolt in a concrete wall to prevent concrete from adhering to the bolt, and acting as a spreader for the formwork.

BOLT SOCKET Sockets in concrete members, used

for lifting or securing members and/or accessories.

BOM *Abbreviation for:* bill of materials.

bona fide In good faith, without fraud.

BOND 1. Adhesion and grip of concrete, mortar, or grout to reinforcement, masonry units, or to other surfaces against which it is placed, including friction due to shrinkage and longitudinal shear in the concrete engaged by bar deformations; **2.** Adhesion of cement paste to aggregate; **3.** Adherence between plaster coats or between plaster and a substrata produced by adhesive or cohesive properties of plaster or supplemental units in masonry; **4.** Arrangement of units in masonry and paving; **5.** Junction of weld metal and base metal; **6.** Holding or gripping force between reinforcing steel and concrete; **7.** Electrical connection from one conductive element to another for the purpose of minimizing potential differences or providing suitable conductivity for fault current or for mitigation of leakage current and electrolytic action; **8.** Material in a grinding wheel that holds the abrasive grains together; **9.** Promissory note. An instrument of prequalification, representing that the principal has been examined by the surety and found to be qualified to complete the obligation or undertaking in question. There are several types, including:

Bid/tender: Sum of money required to be put in escrow at the time of preparing and submitting a bid/tender for work.

Blanket: Bond covering a group of persons, articles, or properties.

Blanket fidelity: Bond that covers an employer's losses due to dishonest acts of employees.

Completion: Form of surety that contains the promise of a third party that a project will be completed or, in the case of the construction company defaulting, the cost of completion will be covered.

Contract: Approved form of security, executed by the contractor and his surety or sureties, guaranteeing complete execution of a contract and all supplemental agreements and the payment of all legal debts pertaining to the construction of the project.

Contract performance: Security furnished to the contracting agency to guarantee completion of the work in accordance with the contract.

Contract surety: Three-party instrument by which one party (surety) guarantees or promises a second party (obligee) the successful performance by a third party (principal).

Fidelity: Surety bond that reimburses an employer for loss sustained through dishonest acts of his employees.

Guarantee: Assurance that the contractor will either complete the work as specified or pay all obligations, or both.

Labor and material payment: Surety to provide for the payment of labor done and materials supplied to a project.

Lease: Bond guaranteeing that a lessee will erect a structure as described in the lease.

Liability: Protection for the assured against court-directed damages.

License: Surety against loss or damages resulting from operations permitted and licensed by law.

Lien: Assurance given by a contractor protecting an owner against liens against his property resulting from work done there; surety that mechanic's or material liens will be filed against a bond and not against the project itself.

Maintenance: Bond in favor of an owner guaranteeing for a specified period, rectification by the contractor of defects in workmanship of materials.

No lien: Bond that denies the right to file a lien against the protected contract or resulting works.

Performance: Bond required by a client to insure that a contractor fulfills contractual obligations.

Release of lien: Mechanism whereby a surety bond is issued to a lienholder to permit the clearance of title to a project.

Release of retained percentage: Owner's protection against loss resulting from premature release of the retained percentage of the contract sum following completion of the work.

Subcontract: Performance and payment bond given by a subcontractor to the general contractor.

Subdivision: Surety provided to a public authority by a subdivider guaranteeing construction of all required public facilities and improvements.

Supply: Bond furnished by a manufacturer and/or supplier/distributor guaranteeing that materials or equipment will be delivered as contracted for.

Surety: Promise to be liable for the debt, default, or failure of another.

Termite: Bond protecting against damage to a structure by termites.

Union wage: Bond provided by a contractor guaranteeing a union that he will pay scale wages and remit any benefit funds withheld.

10. In brickwork, the arrangement of individual bricks so as to distribute the mortar joints and to join vertical planes of brickwork into a cohesive whole. Some common bond patterns include:

All-rowlock: *See* **Rat-trap,** below.

All stretcher: Bond showing only stretchers on the face, each stretcher divided evenly over the stretcher below it.

American: Bond consisting of from five to seven stretcher courses between header courses. *Also called* **Common American.**

Blind: Bond used to tie the front course to the wall in pressed brickwork where it is not desirable that any headers should be *Seen* in the face work.

Block: Masonry units laid so that vertical joints are centered on the blocks of the preceding course.

Block and Cross: A combination of the two bonds the face of the wall is in cross bond and the backing is block bond.

Chain: The building into masonry of iron bars, chain or heavy timbers.

Common American: *See* **American,** above.

Common: Several courses of stretchers followed by one course of either Flemish or full headers.

Cross: Bond in which the joints of the second stretcher course come in the middle of the first; a course composed of headers and stretchers intervening.

Diagonal: A form of raking bond where the bricks are laid in an oblique direction in the center of a thick wall, or in paving.

Double-Flemish *See* **Flemish double,** below.

Dutch: The arrangement of bricks forming a modification of English bond made by introducing a header as the second brick in every alternating stretcher course, with a three-quarter brick beginning the other stretcher course.

English: The exposed face pattern consists of courses of stretchers (long, 228 mm (9 in.) face), alternating with courses of headers (short, 114 mm (4.5 in.) face), with the vertical joints between stretchers centered on the headers. A 50 mm (2 in.) closer is placed next to the corner header.

English garden wall: A wall, 228 mm (9 in.) or more thick, where the exposed face pattern consists of three courses of headers followed by one course of stretchers, all vertical joints staggered. *See also* **Garden-wall,** below.

Flemish: The exposed face pattern consists of courses where headers and stretchers alternate, with the header centered on the stretcher below.

Flemish cross: Any bond having alternate courses of Flemish headers and stretcher courses, the Flemish headers being plumb over each other and the alternate stretcher courses being crossed over each other.

Flemish double: Exposing Flemish bond on both sides of a wall.

Flemish double cross: A bond with odd numbered courses made up of stretchers divided evenly over each other. The even numbered courses are made up of Flemish headers in various locations with reference to the plumb of each other.

Flemish garden: Bricks laid so that each course has a header to every three or four stretchers.

Garden wall: A name given to any bond particularly adapted to walls two tiers thick; a bond consisting of one header to three stretchers in every course. *See also* **English garden wall,** above.

Header: Bond showing only headers on the face, each header divided evenly on the header under it.

Herringbone: Bricks laid in an angular or zigzag fashion.

Longitudinal: Occasional courses of bricks laid as stretchers in a wall otherwise laid to a bond such as header bond.

Monk: Flemish bond modified to show on the face two stretchers and a header repeating in each course.

Perpend: Signifies that a header extends through the whole thickness of the wall.

Raking: Brick laid in an angular or zigzag fashion.

Rat-trap: Wall built with two courses of stretchers standing on edge, alternating with one course of headers standing on edge. *Also called* **All-rowlock.**

Running: Same as stretcher bond.

Stack: Bond pattern in which the head joints form a continuous vertical line.

Stretcher: Wall built to show all stretchers or long faces.

bond. *Abbreviation for:* bonderize; bonding.

BOND AREA Nominal area of interface between two elements across which adhesion develops or may develop, as between concrete and reinforcing steel.

BOND BEAM A horizontal reinforced course or beam designed to strengthen a masonry wall and reduce the possibility of cracking.

BOND BREAKER 1. Material used to prevent adhesion of newly placed concrete and the substrate. *See also* **Form oil,** and **Release agent; 2.** Material used in joints for the purpose of preventing sealant adhesion to the rear joint surface. This allows the sealant to have maximum extension and compression.

BOND COAT Material used between the back of a tile and the prepared surface.

BOND COURSE Course consisting of units that overlap more than one wythe of masonry.

BOND DURABILITY Ability of a glue line to hold materials together under adverse conditions.

BONDED HOLLOW-WALL MASONRY Cavity wall, built of masonry units, in which the inner and outer walls are tied together by bonders.

BONDED MEMBER Prestressed concrete member in which the tendons are bonded to the concrete either directly or through grouting.

BONDED POSTTENSIONING Posttensioning construction in which the annular spaces around the tendons are grouted after stressing, thereby bonding the tendon to the concrete section.

BONDED TENDON Prestressing tendon that is bonded to the concrete either directly or through grouting.

BONDER Masonry unit that ties two or more wythes (leaves) of a wall together by overlapping. *See also* **Header.**

BONDERIZE Chemical treatment of iron and steel by boiling in a solution of manganese dihydrogen phosphate and subsequently applying a coating of paraffin oil to produce a dark, corrosion-resistant protective coating.

BOND FIRECLAY *See* **Plastic fireclay.**

BONDING 1. Adhesion of one film to another, or to a substrate; **2.** Reliable connection to assure electrical conductivity; **3.** Three-way contractual relationship between a principal, surety, and obligee, related to financial stability.

BONDING ADMIXTURE Substance applied to a suitable substrate to create a bond between it and a succeeding layer, as between a subsurface and a terrazzo topping or a succeeding plaster application. *See also* **Acrylic emulsions.**

BONDING AGENT Independent agent representing bonding or surety companies.

BONDING COMPANY Private entity that executes surety bonds, payable to an owner, securing performance of a contract, or securing payment for labor, materials, or other obligations.

BONDING COMPOUND Substance capable of adhering to similar, or dissimilar substances, causing them to act in unison.

BONDING CONDUCTOR Cable or wire that grounds cable sheaths or metal frames of electrical equipment.

BONDING JUMPER Reliable conductor to assure the required electrical conductivity between metal parts required to be electrically connected.

BONDING LAYER Layer of mortar, usually 3 mm to 12 mm (1/8 in. to 1/2 in.) thick, that is spread on a moist and prepared, hardened concrete surface prior to placing fresh concrete.

BOND LENGTH *See* **Development length.**

BOND OBLIGATION Obligation relative to labor, material, equipment and other specified services provided directly to a project.

BOND PLASTER Specially formulated gypsum plaster designed as a first coat application over monolithic concrete.

BOND PREVENTION Measures taken to prevent adhesion of concrete or mortar to surfaces against which it is placed.

BOND STONE Stone projecting laterally into a backup wall, used to tie the wall together.

BOND STRENGTH 1. Resistance to separation of mortar or concrete from reinforcing and other materials with which it is in contact; **2.** Collective expression for all forces such as adhesion, friction due to shrinkage, and longitudinal shear in concrete engaged by the bar deformations that resist separation; **3.** Ability of an adhesive assembly to resist a measured load applied in tension, compression, flexure, impact, or shear.

BOND STRESS 1. Force of adhesion per unit area of contact between two bonded surfaces such as concrete and reinforcing steel or any other material such as foundation rock; **2.** Shear stress at the surface of a reinforcing bar, preventing relative movement between the bar and the surrounding concrete.

BONE DRY As dried in an oven at 105°C (221°F) for 24 hours.

BONE-DRY TON Wood pulp or residue that weighs 2,000 pounds at zero percent moisture content. *Also called* **Oven-dry ton.**

BONE-DRY UNIT Wood residue that weighs 2,400 pounds at zero percent moisture content.

BONES Rocks from an aggregate base that have been worked to the surface and separated from finer material.

BONEYARD Storage place for old, used, or worn-out equipment or machinery.

BONGOSSI African hardwood used as a hammer cushion.

BONING ROD T-shaped instrument for obtaining levels and falls.

BONNET 1. Protective cap or covering. *See also* **Pile driving cap**; **2.** Upper portion of a gate valve into which the disk rises when the valve is opened; **3.** Wire mesh, usually in the shape of a sphere, used to cover the top of vent pipes; **4.** Roof over a bay window.

BONNET PACKING Pliable material around a valve stem that prevents water from leaking.

BONNET TILE *See* **Roof tile.**

BONUS 1. Two logs in one choker; **2.** Anything above or beyond that expected or contracted for.

BONUS-AND-PENALTY CLAUSE Provision in a construction contract for payment of a bonus for early completion of the project, and for a financial penalty for failure to complete on time.

BONUS FOR EARLY COMPLETION Additional money to be paid a contractor for completion of the works ahead of the date stipulated for completion by the owner at the time the contract was awarded.

BOOK INVENTORY Inventory shown in the records and presumed to be on hand. *See also* **Physical inventory.**

BOOK OF SPECIFICATIONS *See* **Specification of work.**

BOOK TILE Clay tile used for roof and wall covering.

BOOK VALUE Capital amount at which property or equipment is shown in the accounts. Normally it is the original cost less depreciation plus additions (not including maintenance or repairs).

BOOM 1. Spar or beam projecting from the mast of a derrick, supporting or guiding the weights to be lifted; **2.** Member hinged to the top of the rotating superstructure of a crane, supported and manipulated by hydraulically activated cylinders that allow the boom to be lowered, raised, extended, and retracted; **3.** Attachment comprising a cantilever member, with a load-carrying hook at the free end; **4.** Assembly of the base, middle, and tip sections used as the telescoping extension for a three-section hydraulic attachment; **5.** Assembly of the base, inner-mid, outer-mid, and manual sections used as the telescoping extension for a four-section manual hydraulic attachment; **6.** Assembly of the base, inner power, outer power, and tip sections used as the telescoping extension for a four-section, full-power hydraulic attachment; **7.** Lattice structure consisting of multiple sections, pinned or bolted together to a specific length, and their support system; **8.** Main lifting structure of the hoe assembly of a backhoe-loader; **9.** Beam, supported by cables and hinged to the deck front of a revolving shovel. *See also* **Jack boom, Lattice boom,** and **Live boom**; **10.** Steel projection on mobile equipment to attach a snorkel or lead blocks so as to gain height or distance; **11.** Raft of logs or a loose bag of logs in the water; **12.** Logs connected together to form a pocket to confine logs into a raft.

BOOM ALIGNMENT SYSTEM Adjustable

aluminum/brass-alloy pads mounted on the sides of boom sections to allow easy boom alignment, permitting this procedure to be employed in routine machine maintenance.

BOOM ANGLE Angle of the longitudinal axis of the boom above or below horizontal.

BOOM ANGLE/LENGTH INDICATING SYSTEM Crane-mounted device that measures and displays the angle and length of a variable-length, telescoping boom.

BOOM ANGLE LIMIT ALARM Visual and/or audible indication that one of the boom limits (upper or lower, minimum or maximum of angle) has been reached.

BOOM ANGLE SENSOR *See* **Boom angle transducer.**

BOOM ANGLE TRANSDUCER Device that measures boom angle. It may be mechanical, electrical, or hydraulic. *Also called* **Boom angle sensor.**

BOOM BASE SECTION Lowermost section of a telescopic boom. It does not telescope, but contains the boom pivot pin mountings and the boom lift cylinder upper end mountings as well as the telescoping sections of the boom and their cylinders or other extension devices.

BOOM CENTERLINE Imaginary line parallel to the longitudinal axis of the undeformed boom structure.

BOOM CHORD Main corner structural member of a boom.

BOOM CRANE *See* **Crane,** and **Crane arm.**

BOOM CYLINDER Cylinder(s) that control movement of the boom section of an attachment.

BOOMER Lever used to boom down or cinch up a load, particularly one carried on a flatbed truck.

BOOM EXTENSIONS Additional lengths of boom that can be telescoped from the base section and that are supported by tapered metallic pins that connect the extensions to the main boom. Boom extensions may be easily differentiated from jibs by their lack of supporting pendant lines, and limited offset capabilities.

BOOM FOOT Base of a boom where it attaches to the upper revolving frame.

BOOM FOOT PINS Connection where the lower or base section of a crane boom is hinged to the revolving superstructure.

BOOM HARNESS Arrangement of blocks and sheaves on the boom point of a crane to which the topping lift cable is reeved for raising and lowering the boom.

BOOM HEAD 1. Portion of the boom housing the upper load sheaves; **2.** Part that connects the feed beam to the drill boom, allowing orientation of the feed beam for drilling.

BOOM HOIST 1. Hydraulic method of raising and lowering a boom to different boom angles; **2.** Rope drum and its drive, or other mechanism, for controlling the angle of a boom.

BOOM HOIST CYLINDER Hydraulic ram controlling the derricking motion of the boom.

BOOM HOIST LINE Wire rope that operates the boom hoist system of derricks, cranes, draglines, shovels, etc.

BOOMING IN (AND OUT) *See* **Luffing.**

BOOM INSERTS Removable sections of a lattice boom.

BOOM JACK Boom that supports sheaves carrying lines to a working boom.

BOOM LATTICE Structural truss members at angles to and supporting the boom chords of a boom.

BOOM LENGTH Distance from the center of the boom foot pin to the center of the boom peak load hoist sheave pin, parallel to the longitudinal centerline of the boom.

BOOM LENGTH INDICATING SYSTEM Crane-mounted device that measures and displays the length of a variable-length, telescoping boom.

BOOM LENGTH SENSOR *See* **Boom length transducer.**

BOOM LENGTH TRANSDUCER Device that measures the length of a variable-length, telescoping boom. *Also called* **Boom length sensor.**

BOOM MAST *See* **Mast.**

BOOM LIGHTS Lights mounted on a grapple yarder boom for night yarding.

BOOM NOSE Extreme end of the fly section where the boom sheaves are located in the boom nose to allow for multiple reeving.

BOOM OVERHANG Distance between the outermost tip of the boom and the outermost portion of the machine's carrier when the crane is in a travel configuration.

BOOM PENDANTS Nonoperating rope or strand with end termination to support the boom.

BOOMPLACER Type of mobile concrete pumping and placing machine.

BOOM POINT Boom head section containing the boom point sheaves for the main hoist line and hook block.

BOOM POINT SHEAVE Principal sheave at the end of the main boom over which the main hoist line is reeved.

BOOM POSITIONING TIME Amount of time required to locate the drill centralizer over the next hole to be drilled. In angle drilling, it also includes the time to set the proper guide-angle position.

BOOM SCRAPER Boom equipped with a chain of buckets that swing through an arc to handle aggregates from a stockpile behind a batch mixer.

BOOM SECTION 1. Base, middle, inner, mid, outer-mid, manual or tip segments that are used as part of a telescoping attachment; **2.** Individual lattice structures that are pinned together to form a boom attachment. Crane lattice booms are usually in two basic sections: top and base. Such booms may be lengthened by insertion of one or more additional extension sections.

BOOM STOP Device used on crane-type or long booms to limit the maximum boom angle, and to prevent the boom from being pulled, or falling over backwards.

BOOM SUSPENSION System of lines and fittings, either fixed or adjustable for length, that supports the boom and controls its angle.

BOOM TIP SECTION Uppermost section of a sectional lattice boom, including the upper load sheaves. *Also called* **Head section**, and **Tapered tip**.

BOOSTER 1. Auxiliary device that increases force or pressure; **2.** Chemical compound, not containing an initiating device, used for intensifying an explosive reaction.

BOOSTER CABLES Large-capacity, insulated wires fitted with hand clamps at each end, used to connect an outside battery to a machine for a temporary supply of starting current.

BOOSTER CYCLE Period during which additional hydraulic pressure is exerted to push the last charge of solid waste into a transfer trailer or a container attached to a stationary compactor.

BOOSTER FAN Fan installed in heating ducts to increase air movement.

BOOSTER HEATER Heater used to raise the temperature of oil from that required for pumping to that required for atomization. The booster heater is usually located close to the burner where atomization takes place.

BOOSTER PUMP Pump that operates in the discharge line of another pump, either to increase pressure or to restore pressure lost between the prime pump and booster pump for any cause.

BOOSTER RELAY VALVE Relay-type valve used to accelerate the application and release of pressurized air to towed vehicles or relay emergency valve.

BOOT 1. Projection from the side of a concrete beam or floor slab, usually to support brickwork. **2.** Lath set behind the hub by the grade setter when there are obstructions blocking the line of site to the hub. The grade setter draws a horizontal line on the lath 300 mm (12 in.) or more above the hub and shoots grade from this line.

BOOTLEG 1. Situation in which a blast fails to cause total failure of the rock because of insufficient explosives for the amount of burden, or because of incomplete detonation of the explosives; **2.** That portion of a blasthole that remains relatively intact after having been charged with explosives and fired.

BOOT LINTEL L-shaped reinforced concrete lintel supporting a skin of brickwork, used where it is not desired to show the full depth of the lintel on the exterior face.

BOOTSTRAP TEST Test loading a pile by jacking against the reaction provided by a beam spanning between two anchor piles driven either side of the test pile.

BOOT TRUCK Oil truck with a spray rack, used for spraying asphalt oil. *See also* **Distributor truck**.

bor. *Abbreviation for:* borough.

BORAX Flux used for hard soldering and brazing.

BORDEAUX CONNECTION Steel-wire rope thimble fitted with a link, used to join wire rope to short-link crane chain.

BORDER 1. Piece around the edge of anything; **2.** Earth berm built to contain irrigation water in a field.

BORE Internal diameter of a cylinder, hose, or pipe.

BORED PILE Pile formed by boring a hole into the ground and filling it with concrete. *See also* **Pile, Drilled**.

BORED-TYPE BLOCK Engine block containing no cylinder sleeves, with pistons and rings inserted directly into the cylinder block bores.

BOREHOLE 1. Interior of a well; **2.** *See* **Blasthole**.

BOREHOLE PRESSURE Peak effective pressure caused by expanding gases that acts behind the detonation head on the cylindrical surface area of a borehole during an explosion; approximately equal to one half of the detonation pressure.

BORER HOLES Holes made by wood-boring insects.

BORESCOPE Fiber-optic instrument used for viewing hidden cavities: the cavity is illuminated by light passing down the optical fibers with the image transmitted to the head of the instrument through parallel fibers and displayed via a lens for viewing or recording, photographically or electronically.

BORING 1. Hole in the earth produced by various methods; **2.** Method of exploring subsurface conditions by drilling or otherwise advancing a cased or uncased hole into the earth; **3.** Rotary drilling; **4.** Starting a cut in the center of a log using the tip of the chain-saw blade. *Also called* **Plunge cut**.

BORING SYSTEM Combination of equipment designed to bore a hole horizontally through soil, working from an excavation, maintaining a predetermined direction and angle. *See also* **Directional boring system**.

BORON CARBIDE Hardest material (B_4C) made commercially by man.

BORON FRITS Clear, colorless, synthetic glass produced by fusion and quenching, containing boron. *See also* **Boron-loaded concrete**.

BORON-LOADED CONCRETE High-density concrete including a boron-containing admixture or aggregate, such as the mineral colemanite, boron frits, or boron metal alloys, to act as a neutron attenuator. *See also* **Biological shielding**, **Colemenite**, and **Shielding concrete**.

BORROW Suitable material from off-site sources, used primarily for fill.

BORROWED LIGHT Glazed part of an internal partition that derives its natural light from external windows in the wall or roof.

BORROW PIT Excavation from which material is taken for use in another location. *Also called* **Barrow pit**.

BORTS Industrial diamond chips on the face of a rock-boring bit.

BOSS 1. Ornamental knob or projection. **2.** Surface failure of a metal part where parts of the surface are raised above the previously flat or continuous surface; **3.** Enlarged part of a shaft; **4.** Cone for opening pipes; **5.** Keystone to a dome; **6.** To work a malleable metal by beating with a tool, causing it to 'flow.'

BOSSAGE Rough-faced stone left projecting from the face of a wall to be sculptured later.

BOSSING MALLET Wooden mallet with an egg-shaped head, used to shape sheet metal.

BOSSING STICK Shaping tool used to shape and dress malleable metal sheet, such as sheet lead.

bot. *Abbreviation for:* bottom.

botmg *Abbreviation for:* bottoming.

BOTTLE-NOSE DRIP Rounded edge of the drip of a metal roof.

BOTTOM ASH Nonairborne combustion residue from burning fuel in a boiler.

BOTTOM BIND One of five basic conditions characterized by the lay of a tree when it is lying over some solid object or when supported on one end with a substantial overhang on the other, causing the top side to be in tension and the bottom side in compression. *See also* **Drop, End pressure, Side bind,** and **Top bind.**

BOTTOM BRACE *See* **Brace.**

BOTTOM CHORD *See* **Chord.**

BOTTOM CRACK INDUCER Compressive-fiber, plastic, or timber fillet placed at the bottom of a concrete road slab joint to reduce the cross-sectional depth and help induce a crack.

BOTTOM DEAD CENTER Point in piston travel where the piston is at the lowest point in the cylinder bore.

BOTTOM DUMP *See* **Hopper body.**

BOTTOM-DUMP SCRAPER *See* **Scraper.**

BOTTOM-DUMP TRUCK *See* **Truck.**

BOTTOM HEADING Excavation of the bottom half of a tunnel after the top half has been excavated for the full tunnel length.

BOTTOM HEAVE Upward movement of soil in the base of a large excavation.

BOTTOMING Final stage of an excavation, usually hand worked to insure a smooth finish to the correct level.

BOTTOM PLATE Lower horizontal member of wood-frame construction that is nailed to the bottom of wall studs and to the floor framing members.

BOTTOM PRIMING Method in which the primer is placed at the bottom of a column of explosives.

BOTTOM TERMINAL LANDING Lowest landing served by an elevator or material lift.

BOTTOM U-BOLT PLATE Plate that is located on the bottom side of a spring or axle and that is held in place when U-bolts are used to clamp the spring and axle together.

BOUGE To knock out dents in raised sheet metal by working over a stake.

BOULDER Rock, usually rounded by weathering and abrasion, greater than 200 mm (8 in.) in size.

BOULDER CLAY Stiff, hard, usually unstratified clay that contains boulders.

BOULEVARD Wide street, usually landscaped.

BOULEVARD STRIP 1. Portion of a street right-of-way that lies between the curb and the property line; **2.** Grassed or planted area between curb and sidewalks. *See also* **Median.**

BOUNCE CHAMBER PRESSURE Pressure in the air compressed by the upward moving piston of a closed-end diesel hammer. The pressure is read by a gauge connected to the upper chamber by a hose.

BOUNDARY LAYER Layer of air immediately adjacent to the surface of a body in motion, typically a moving vehicle.

BOUNDARY LUBRICATION *See* **Lubrication.**

BOUNDARY PIEZOMETER Disk piezometer used to measure pore water pressure at the interface between a structure and the adjacent soil.

BOUNDARY PRESSURE CELL Hydraulic, pneumatic, or electrically operated pressure cell having at least one flat surface.

BOUNDARY SURVEY Measured and plotted survey of the perimeter of a plot showing distances, bearings, and angles.

BOUNDARY WALL Wall or permanent fence marking the surveyed boundary of a property.

BOURDON TUBE Metallic tube of elliptical cross-section, shaped in an arc or spiral with one end attached to an indicating, recording, or controlling device, used in industry to measure pressures.

BOW 1. Distortion in a board or structural wood panel that deviates from flatness lengthwise, but not across the faces. *See also* **Cup; 2.** Tail part of a key that is held between the fingers.

BOWING Deviation from a straight line, measured at the point of greatest distance from the straight line.

BOW IRON Device on a sidewalk elevator that unlocks and operates the sidewalk doors.

BOWL 1. Body or bucket of a scraper; **2.** Moldboard or blade of a dozer; **3.** Exterior shell of an expansion ring-type coupling; **3.** A case that is closed at one end and mates with a filter head. *Also called* **Shell.**

BOWLINE Rope knot, widely used in construction, that never slips or jams.

BOW SAW *See* **Saw.**

BOW-STRING TRUSS Truss with convex upper chords, mostly used in bridge construction.

BOW TROWEL Plasterer's finishing trowel with a slight curve, used for finishing drywall joints.

BOW WAVE Moving ripple that occurs in an asphalt mat ahead of a breakdown roller, especially on tender mixes.

BOW WINDOW *See* **Window.**

BOX 1. Load-carrying element of a truck or trailer; **2.** Accessible metal case in which electrical conductors meet, are joined, or terminate; **3.** Transmission of a vehicle.

box. *Abbreviation for:* boxed; boxing.

BOX BEAM Hollow beam formed like a long box.

BOX BRACE Reinforcing, usually of sheet metal, formed into square or rectangular crosssections, used to stabilize various elements of a dump body or hoist.

BOX CAISSON Rectangular, open-ended, reinforced-concrete box, built on shore and floated to site and sunk, in which work can be done while the bottom of the structure is below water level.

BOX CASING Inside lining to cabinetwork.

BOX COLUMN Built-up hollow column of square or rectangular section.

BOX CORNICE False, hollow, enclosed cornice that hides the soffits of rafters.

BOX CULVERT Tunnel-like reinforced concrete structure consisting of single or multiple openings, usually square or rectangular in cross section.

BOX CUTTER Hand tool for cutting holes for electrical boxes in drywall.

BOX DRAIN Small, rectangular, brick or concrete drain.

BOXED GUTTER *See* **Box gutter.**

BOXED HEART Term used when the pith in a tree falls entirely within the four faces of a piece of wood, anywhere in its length.

BOX FORM Lumber box built in a wall to form an opening, as for duct work.

BOX FRAME Hollow framing of a double-hung sash window constructed to receive the weights balancing the sliding sashes.

BOX GIRDER 1. Structural section having a hollow rectangular section; **2.** Bridge span having a top and bottom slab with two or more walls forming one or more rectangular voids.

BOX GUTTER Rain gutter built into the roof, consisting of a horizontal wooden trough lined with galvanized sheet metal or copper. *Also called* a **Boxed gutter, Concealed gutter,** and **Parallel gutter.**

BOXING Method of mixing liquids (such as paints) for uniformity by pouring them back and forth from one container to another.

BOXING UP Closing in, nailing sheathing to studs and rafters, etc.

BOX LEAD *See* **Lead.**

BOX NAIL *See* **Nail.**

BOX OUT To form an opening or picket in concrete by a box-like form.

BOX PILE *See* **Pile.**

BOX SHEAR TEST *See* **Shear test.**

BOX SILL Header nailed on the ends of joists and resting on a wall plate.

BOX STAIRS Staircase built between parallel walls and supported by them.

BOX SYSTEM Structural steel system that lacks a complete vertical load-carrying space frame and in which lateral forces are resisted by braced frames or shear walls.

BOX THREAD Female side of a tapered thread.

BOX WRENCH *See* **Wrench.**

BOYLE'S LAW Statement that the absolute pressure of a fixed mass of gas varies inversely as the volume, provided the temperature remains constant.

bp *Abbreviation for:* blueprint; boiler pressure; boiling point; brass plug.

BP *Abbreviation for:* barometric pressure; base pay; base plate; bearing pile; bolted plate; brass pin.

bprf *Abbreviation for:* bulletproof.

br. *Abbreviation for:* brace; bracing; branch; brush.

BR *Abbreviation for:* bedroom.

BRACE 1. Inclined piece of framing or other lumber applied to a wall or floor to stiffen the structure. Often used on walls as temporary support; **2.** Structural member used to provide lateral support for another member, generally for the purpose of assuring stability or of resisting lateral loads; **3.** Tie that holds one scaffold member in a fixed position with respect to another member. *Also called* **Backstay; 4.** Telescoping structural member used to attach the bottom of the leads to the crane base and to position or batter the leads in or out, left or right. Can be mechanically, hydraulically, or pneumatically operated; it is used with either fixed leads or semifixed leads. *Also called* **A-frame, Apron, Bottom brace. Kicker, Spider, Spotter, Spreader bar, Stinger,** and **Strut; 5.** Hand-operated boring device consisting of a crank fitted with jaws that can grip a bit.

BRACED BEAM Steel frame in which resistance to lateral loads or frame instability is primarily provided by a diagonal, a K-brace, or other auxiliary system of bracing.

BRACED COLUMN Column tied to a structure or supported in one or more planes by a wall or other bracing so as to resist all lateral forces in that plane.

BRACED FRAME Structural wood or steel frame in which frame instability and resistance to lateral loads is provided by one or more systems of diagonal bracing.

BRACED FRAMING Heavy timber framing in which the frame is formed and stiffened by the use of posts, girts, and braces.

BRACED WALL Wall that is tied at right angles to a structure, or otherwise provided with stiffening or bracing.

BRACE PILE Batter pile connected to a structure in a manner to resist lateral forces.

BRACE ROD 1. Steel rod used in roofs and walls to transfer tensile stresses; **2.** Supports for the outer corners of an elevator platform, each of which tie to upper portions of the stile.

BRACES AND PLATES Wide range of metal shapes used to reinforce or join members, including:

> **Flat corner brace:** L-shaped piece drilled on both legs for mounting on surfaces at right angles in the same plane.
>
> **Inside corner brace:** L-shaped piece drilled on both legs for mounting on an inside corner.

Mending plate: Flat metal plate predrilled to receive screws or bolts.

T-plate: T-shaped flat metal plate, predrilled on both legs.

BRACING 1. System of structural members, usually diagonal to the principal members, positioned to stiffen a structure; **2.** Support attached to a portion of a structure so that it will not distort or twist; **3.** System of a horizontal and/or inclined structural members fastened to the piles of a bent, group, or row to increase stability by resisting or distributing lateral forces to the structure; used extensively in trenches and sheet-pile cofferdams as compression stuts. *Also called* **Shoring. 4.** Strengthening or support for a weakened or defective tree.

BRACKET 1. Supporting piece of stone to carry a projecting weight, often highly carved; **2.** Support, generally projecting from or fastened to a wall.

BRACKETED STAIR Stair with decorative scrollwork to its exposed end.

BRACKET FORM SCAFFOLD Wood or metal bracket attached to a wall form, upon which scaffold planks rest, for the purpose of providing an elevated working unit for those engaged in erecting the wall or placing concrete.

BRAD *See* **Nail.**

BRADAWL Straight awl with a chisel edge, used to make small holes in which to start brads.

BRAID Hollow or solid structure of round or polygonal cross-section, produced by interlocking yarn strands or filaments that are disposed in direction, oblique to the axis of the braid. *Also called* **Wire braid.**

BRAID ANGLE Angle developed at the intersection of a braid strand and a line parallel to the axis of a hose.

BRAIDED CHANNEL Large flood channel of a fast-flowing stream made up of many small stream beds crossing each other. These stream channels are associated with low stream banks. They are also associated with frequent runoffs and/or ice flows. *See also* **Channel.**

BRAIDED HOSE Hose in which the reinforcing material has been applied as interlaced spiral strands.

BRAIDED NYLON ROPE High-strength, abrasion- and rot-resistant rope fabricated of nylon fibers.

BRAIDED PLY Layer of braided reinforcement.

BRAIDED SLING Flexible sling, the body of which is made up of two or more wire ropes braided together.

BRAIDED WIRE Electrical conductor composed of a number of thin wires twisted or braided together.

BRAIDER Machine that interweaves strands of yarn or wire to make a hose carcass.

BRAIDER DECK Base plate upon which the bobbin carriers of a braider machine travel.

BRAID-OVER-BRAID Braid made by more than one pass through a multiple-carrier braider.

BRAID SMASH Defect in a braided reinforcement caused by one or more of the ends of reinforcing material breaking during the braiding operations.

BRAKE 1. Machine for bending, flanging, folding, and forming sheet metal; **2.** Device for slowing, stopping, or holding an object. There are several types, including:

Air-over-hydraulic: Air brake subsystem that uses compressed air to transmit force from the driver's control to a hydraulic brake system.

Automatic: Drum brake system that is applied automatically any time the drum control lever is in neutral.

Band: Circular external contracting-type brake having a strap lined with heat- and wear-resistant friction material.

Caliper disc: Self-adjusting brake system with a caliper with brake pads on both sides of a rotating disc. When the brakes are applied, the pads contact the disc.

Cam: Wheel brake in which the brake shoes and attached linings are forced against the drum by the rotation of a cam.

Centrifugal: Device on hoist drums that throws the brake shoes outward and into contact with a fixed brake drum if the hoist drum rotates beyond a specified speed.

Contracting band: Brake that surrounds the brake drum and steering clutches on some track machines. The brake contracts against the drum to stop the machine.

Disc: Brake assembly that utilizes friction between fixed and rotating discs, or between discs and shoes.

Drag: Brake on a revolving shovel that stops and holds the drag drum.

Drum: Rotating cylinder with a machined inner or outer surface upon which a brake band or shoe presses.

Electric: System that uses an electric signal or electric current to actuate the service brake.

Emergency: Independently actuated secondary brake system used for parking, or in an emergency, when the service brake is inoperative.

Enclosed hydraulic hoist: An internal hoist brake that will engage only when an elevator controls are in the down mode. To release the brake it is necessary for hydraulic pressure to build to the point where the spring will release the brake and allow the machine to hoist down.

Engine: Braking device that uses engine compression pressure as a retarding medium.

Expander tube: Brake consisting of a nonrotating assembly that contains a series of brake blocks mounted to and forced outward by an expanding tube. When the tube is inflated, the blocks contact the brake drum.

Expanding shoe: Brake that uses two separate moving shoes. The shoes are pushed outward by one or two hydraulic cylinders, and released by a spring.

Friction: Brake that operates by the friction created between two surfaces rotating or sliding on each other.

Full air: Where compressed air is used to provide

the force necessary to expand the brake shoes by cam or wedge against the brake drums with air pressure supplied direct to chambers at the wheel position.

Full disc: Brake system in which a series of steel plates are splined to the housing. When the brake is actuated, the discs and plates are compressed, creating friction and braking action. Some are dry-type, others run in oil, and in some external oil cooling reduces heat buildup.

Jaw: *See* **Tooth**, below.

Multiple-disc swing: Series of clutches and discs together in a brake pack that ensures smoother braking and long, trouble-free brake service.

Overrun: Brake on a towed vehicle set to operate as soon as the towing vehicle slows and the towed vehicle tends to push into it.

Power: Booster device that helps to provide greater application of pressure to the vehicle brakes with reduced application effort by the operator.

Self-energizing: Brake action that is applied partly by friction between its lining and the drum.

Service: Principal mechanism for retarding and stopping a vehicle.

Steering: Brake that slows or stops the drive mechanism on one side or the other of a tractor.

Swing: Friction device to hold a revolving superstructure in any desired position relative to the chassis.

Swing park: Self-contained brake used for holding the upper in any position during normal, stationary crane operation.

Tooth: Brake mechanism used to hold a shaft from rotation by means of a tooth or teeth engaging with fixed sockets. *Also called* a **Jaw brake.**

Vacuum assist: Standard-type hydraulic brake with a pressure-assist cylinder having a vacuum chamber that, when atmospheric pressure is allowed to one side of the piston or diaphragm, drives a plunger in the hydraulic system increasing the effective pedal pressure.

Wedge: Wheel brake in which the brake shoes and attached linings are spread apart from the action of a wedge being forced between their opposing ends, pushing them against the drum.

BRAKE BOOSTER Mechanical means for increasing pressure to the wheel cylinders of a vehicle braking system while decreasing the brake pedal effort.

BRAKE CHAMBER Chamber that converts air pressure to mechanical force for applying vehicle brakes; when pressure is exhausted, a return-spring retracts the brake shoes, releasing the brakes.

BRAKE COIL Coil that operates a brake, used to prevent an elevator from moving when the car is at rest.

BRAKE CONTROL VALVES Valves actuated by the ABS unit's electronic control unit to ensure that each wheel is optimally braked. On a tractor, it is referred to as a modulator valve, and on a trailer as an ABS relay valve.

BRAKE DRUM 1. Machined drum, part of a wheel assembly, against which the brake shoes ride when

hydraulic pressure is transmitted through the braking system from the brake pedal; **2.** Smooth-surfaced sheave mounted on an elevator hoist machine drive shaft, contacted by the brake shoes when the brake magnet is de-energized.

BRAKE DUST SHIELD Nonrotating cover attached to the brake spider that restricts entry of foreign matter into a brake drum.

BRAKE FADE Loss of vehicle braking ability. Fade can be caused by mechanical deficiencies, such as improperly fitted shoes, water or oil on the linings, or by excessive heat.

BRAKE HOP *See* **Wheel hop.**

BRAKE HORSEPOWER *See* **Horsepower.**

BRAKE HOSE Flexible conduit that conveys brake-actuating air or fluid.

BRAKE HOSE COUPLER *See* **Gladhand.**

BRAKE LIMITING VALVE Valve in a vehicle's brake system that limits the pressure passed on to a given axle's brakes for controlling braking on wet and/or slippery road surfaces.

BRAKE LINE Conduit that transmits a brake actuating signal, air or fluid.

BRAKE LINING Composite material having a high coefficient of friction fastened to the brake shoes.

BRAKE MAGNET Solenoid that, when energized, causes the brake shoes to move away from an elevating device brake drum.

BRAKE MEAN EFFECTIVE PRESSURE Theoretical value calculated from known (measured) horsepower values: the average pressure that needs to be exerted during the engine's power stroke to produce a power output equal to the brake horsepower. A value that is often used to compare one engine with another.

BRAKE PACK Internal brake on a skidder or machine transmission.

BRAKE POWER Power available at an engine's output member(s) for doing useful work.

BRAKE POWER UNIT Device in a brake system that, on receipt of an actuating signal from the operator, provides the energy required to actuate the brakes, either directly or through an auxiliary device.

BRAKE SHOES Parts of a brake system, faced with friction material, that can be brought into contact with the brake drum to slow and arrest movement.

BRAKE-SPECIFIC FUEL CONSUMPTION Rate of fuel consumed by an engine, divided by the flywheel power output.

BRAKE SPIDER Backing plate to which brake shoes are anchored.

BRAKING TORQUE Torque produced on the axle housing during braking, and transferred to the frame by the rear axle's suspension or torque arms.

BRANCH Soil-or-waste pipe that (a) is in one story, (b) is connected at its upstream end to the junction of two or more soil-or-waste pipes, or to a soil-or-waste stack, and (c) is connected at its downstream end to another branch, a soil-or-waste stack, or a building drain.

BRANCH CIRCUIT 1. Circuit conductors between the final overcurrent device protecting the circuit and the outlet(s); **2.** Circuit leading from the main oil circulating loop to a burner or group of burners.

BRANCH FITTING Special pipe shape that permits the junction of one or more branch pipes to a main, typically a Tee, Y, Tee-Y, double-Y, or V. *Also called* **Pipe branch.**

BRANCH INTERVAL Vertical distance between the connection of pipes to a main drain, waste-and-vent stack.

BRANCH LINE Piping into which fire protection sprinklers are fitted and fed by a cross main.

BRANCH VENT Vent that connects with one or more vents from fixtures and that leads to a vent stack.

BRANCHWOOD Wood portions of a tree, excluding the stem and roots.

BRAND Mark or symbol identifying or describing a product and/or manufacturer. *Also called* **Printed brand.** *See also* **Stamp.**

BRANDING AXE *See* **Stamp hammer.**

BRASS 1. Alloy of copper and zinc; **2.** Inscription plates set into the face of tombs and memorials.

BRASS PIPE *See* **Pipe.**

BRASS SCRIBER Jointed brass rods used to mark tinplate.

BRASS TONGS Tongs made of brass or having brass working ends, used to dip nonferrous metals in acid pickle.

braz. *Abbreviation for:* brazier, brazing.

BRAZING Similar to welding: a metal rod (bronze filler rod) is used with a lower melting point than that of the metals being joined (base metals).

brdg. *Abbreviation for:* bridge.

brdsy. *Abbreviation for:* birds-eye.

BREACH OF CONTRACT Actions by one who has signed a contract in violation of the terms and/or conditions set out in the document.

BREAK 1. To twist open or disconnect; **2.** Opening a (reinforcing steel) spiral to the round shape and forcing it toward the opposite direction so that it will remain circular for placing; **3.** Fragmentation of solid rock resulting from the detonation of an explosive mixture in drill holes at the heading of a tunnel; **4.** A short period of rest; **5.** Discontinuity; **6.** Abrupt change in direction of a wall; **7.** In a tire, a crack extending into or through the fabric (an impact break is usually in the shape of an X or star and can be seen from inside the tire. A flex circumferential break runs parallel to the beads.)

BREAKAWAY CHAIN *See* **Chain.**

BREAKAWAY CURRENT Electrical current drawn by the engine starter motor to break the motor away from a stalled condition.

BREAKAWAY TORQUE Torque required to overcome static friction when 100% of a locking feature is engaged and a fastener is not seated.

BREAKAWAY VALVE Safety valve designed to protect the air supply of a tractor or leading trailer and to automatically apply the brakes of the trailing unit in the event of accidental separation.

BREAKDOWN 1. Initial compaction of an asphalt mat behind the paver that is intended to achieve maximum density in the shortest time frame; **2.** List of materials to be used, and often their cost, and hours necessary to complete a task.

BREAKER Electrical device that opens a circuit under conditions of overload or shorting.

BREAKER PLY Open mesh fabric used to anchor a hose tube or cover to its carcass and to spread impact. *See also* **Leno breaker.**

BREAKERS *See* **Tire construction.**

BREAKER STRIP/PLY Band or strip of rubber-coated, bias-cut tire cord that encircles a tire between the top steel or fabric ply and the tread. *Also called* **Impact ply,** and **Shock Ply.**

BREAKER TOOL Any large steel chisel, mole, etc., that fits into the chuck of a pneumatic, hydraulic, or electric tool for the purpose of breaking up or dressing materials.

BREAK GROUND To commence a project; start an excavation.

BREAK-IN FAILURE Foundation failure where the soil beneath it is uniformly compressed, and bulges uniformly around the outside of the foundation.

BREAKING Initial separation of water from an asphalt emulsion, that can be detected by a marked color change from brown to black, and often by the release of fairly clear straw-brown water. This results in the deposition of the base asphalt on an aggregate or paved surface.

BREAKING CONTACT Contact on a relay, contactor, or switch that is open when the device is energized; closed when the device is de-energized.

BREAKING JOINT *See* **Broken joint.**

BREAKING STRENGTH The ultimate load at which a tensile failure occurs in the sample being tested (synonymous with actual strength). *See also* **Minimum acceptance strength,** and **Nominal strength.**

BREAKING TAPE Technique for measuring the horizontal distance down a slope whereby a convenient horizontal measurement is transferred to ground level using a plumb bob.

BREAK IRON Iron attached to the top of a plane bit to curl and break the shavings.

BREAK JOINT Laying masonry or placing sheet material or other coverings in a manner that the vertical or short joint of one course or layer is not in line with that of the preceding course or layer.

BREAK LINES On a drawing, lines used in pairs to indicate that information is not included in the area between them.

BREAKLOOSE TORQUE Loosening torque required to initially move a fastener from the assembled torque conditioning.

BREAK-OFF TEST Test that measures the strength

of concrete near to the surface, based on measurements of the flexural strength of concrete in an annular crosssection parallel to, and at a specific distance from, the concrete surface.

BREAKOUT Force to inaugurate sliding.

BREAKOUT FORCE Maximum vertical upward force created by the curling action of a bucket attached to an excavator, loader, or backhoe, measured 100 mm (4 in.) behind the tip of the bucket cutting edge.

BREAKOVER Fire edge that crosses a control line or natural barrier intended to confine a forest fire.

BREAKUP Period of time in the spring when melting snow creates soft soil conditions and high water in streams. Logging must usually be curtailed during this time. *See also* **Freeze-up.**

BREAKWATER Structure designed and placed so as to break the force of waves, tide, or flood.

BREAST 1. Projecting portion of a wall, as with a chimney; **2.** Masonry forming the back of a recess and the parapet under a window sill; **3.** Moldboard of a plow or dozer.

BREAST BOARD Temporary barrier used to prevent a digging face from caving or flowing into a tunnel or excavation.

BREAST DERRICK Type of derrick, similar to an A-frame. The mast consists of two side members spaced father apart at their base than at their head, and linked top and bottom by rigid spacers. The mast is maintained at its forward angle by guys; the load is raised and lowered by lines passing through a sheave or block attached to the top crosspiece.

BREAST DRILL Hand tool used for drilling holes in which a bevel gear transmits hand power to the bit, and where down-pressure is applied by the operator leaning his breast on a pad on top of the drill.

BREAST TIMBER Leaning brace from the floor of an excavation to a wall support.

BREAST WALL One built to prevent the falling of a vertical face cut into natural soil.

BREATHABLE COATING Any coating that cures to form a film that is permeable to water vapor. *Also called* **Permeable coating.**

BREATHER CAP Cap mounted on top of hydraulic tanks to provide proper ventilation for the system. The component's dual function is to allow air entrained in the hydraulic oil to escape and to maintain atmospheric pressure inside the hydraulic tank.

BREATHER PLUG Plug or cap to an enclosed space that may contain permanent vents or a valve to permit adjustment of pressure between the space and atmosphere.

BRECCIA Rock composed of angular fragments of older rock cemented together.

BREECHING Flue pipe or chamber for receiving flue gases from one or more flue connections and for discharging these gases through a single flue connection.

BREEZE Usually clinker, also finely divided material from coke production.

BREEZE BLOCK Lightweight, precast concrete building block having breeze as an aggregate.

BREEZEWAY Covered passage, open at each end, that passes through a dwelling or between two structures.

brg *Abbreviation for:* bearing.

BRICK 1. Block of bonded abrasive, used for such purposes as rubbing down castings, scouring castings, general machine shop use, scouring chilled iron rolls, polishing marble, and work of like nature; **2.** *See* **Construction types; 3.** Clay, kneaded, molded, or extruded and baked (usually in a kiln: thus, fired). Conventionally 230 x 115 x 75 mm (9 x 4-1/2 x 3 in.) nominal, but of many other proportions. Usually with a depression (frog, or keyway) in one 230 x 115 mm (9 x 4-1/2 in.) face or with holes piercing that face to hold mortar. Brick types include the following:

Acid-resistant: Brick suitable for use in contact with chemicals, usually in conjunction with acid-resistant mortars.

Adobe: Large, roughly-molded, sun-dried clay brick.

Alumina-diaspore fireclay: Brick consisting mainly of disapore or nodule clay and having an alumina content of 50%, 60%, or 70% (±2.5%).

Angle: Any brick shaped to an oblique angle to fit a salient corner.

Arch: 1. Wedge-shaped brick for special use in an arch; **2.** Extremely hard-burned brick from the arch of a kiln; **3.** Brick with a curved face suitable for wells and other circular work.

Body: The best bricks in the kiln, those that are baked hardest with the least distortion.

Building: Brick for general building purposes, not especially treated for texture or color. *Also called* **Common brick.**

Bullnose: Brick with one vertical edge rounded.

Calcium-silicate: Concrete product made principally from sand and lime that is hardened by autoclave curing.

Capping: Specially shaped bricks used to finish the exposed top of a brick wall.

Chuff: *See* **Salmon,** below.

Clinker: Very hard-burned brick whose shape is distorted or bloated due to nearly complete vitrification.

Common: *See* **Building brick,** above.

Compass: Curved or tapering brick, used in curved work. *Also called* **Feather edged brick.**

Double bullnose: Brick, tile or other building component with the same convex radius on two opposite or adjacent sides.

Dry-press: Brick formed in molds under high pressure from relatively dry clay (5% to 7% moisture content).

Economy: Brick whose nominal dimensions are 100 x 100 x 200 mm (4 in. x 4 in. x 8 in.)

Engineered: Brick whose nominal dimensions are 100 x by 81 x by 200 mm (4 in. x 3.2 in. x 8 in.).

Engineering: Brick manufactured to very close tolerances and uniformity of material, burned to the point of partial vitrification.

Facing: Brick made especially for facing purposes, often treated to produce surface texture. They are made of selected clays, or treated, to produce desired color.

Feather edged: *See* **Compass brick,** above.

Fire: Brick made of refractory ceramic material that will resist high temperatures.

Flare header: Brick the header of which is a darker color than the field of the wall.

Floor: Smooth, dense brick, highly resistant to abrasion, used as finished floor surfaces.

Gauged: (a) Brick that has been ground or otherwise produced to accurate dimensions; **(b)** A tapered arch brick.

Half bat: Brick cut in two along its length showing a full-length and full-depth face.

High-duty fireclay: Fireclay brick that has a pyrometric cone equivalent not lower than cone 31-23, or that does not deform more than 1.5% at 1350°C (2,460°F) in the standard local test.

Hollow: Masonry unit of clay or shale whose net cross-sectional area in any plane parallel to the bearing surface is not less than 60% of its gross cross-sectional area measured in the same plane.

Insulating: Firebrick having a low thermal conductivity and a bulk density less than 550 kg m³ (70 lb ft³), suitable for lining industrial furnaces.

Intermediate-duty fireclay: Fireclay brick that has a pyrometric cone equivalent above cone 29 or that does not deform more than 3% at 1350°C (2,460°F) in the standard local test.

Jumbo: Generic term indicating a brick larger in size than the standard.

King closer: Brick cut diagonally to have one 50 mm (2 in.) end and one full-width end.

Nonmodular: Bricks that do not fit into the 100 mm (4 in.) modular grid system of measurement.

Norman: Brick whose nominal dimensions are 100 x 64 x 300 mm (4 in. x 2.5 in. x 12 in.).

Pallet: Brick that is rebated or grooved to hold a pallet.

Paving: Vitrified brick especially suitable for use in pavements where resistance to abrasion is important.

Perforated: Brick with vertical perforations through the frog.

Place: *See* **Salmon brick,** below.

Pressed: Those that are pressed in a mold by mechanical power before they are burned or baked.

Roman: Brick whose nominal dimensions are 100 x 50 x 300 mm (4 in. x 2 in. x 12 in.).

Rubber: Soft-textured dense brick that can be rubbed (against another brick or on an abrasive surface) to form a shape, such as a wedge, for use in an arch.

Salmon: Generic term for an under-burned brick that is more porous, sightly larger, and lighter colored than hard-burned brick; usually pinkish-orange in color. *Also called* **Chuff** and **Place brick.**

Sand-struck: *See* **Soft-mud brick,** below.

SCR: Brick measuring 50 x 150 x 300 mm (2 x 6 x 12 in.) that lays up three courses to 200 mm (8 in.) including joints. It is used to form 150 mm (6 in.) thick walls.

Sewer: Low absorption, abrasive-resistant brick intended for use in drainage structures.

Soft-mud: Brick produced by molding relatively wet clay (20% to 30% moisture), often hand produced. When the insides of the molds are sanded to prevent the clay from sticking, the product is termed a **Sand-struck brick;** when the molds are wetted to prevent sticking, the product is a **Water-struck brick.**

Splay: Brick beveled at 45° at one end.

Standard: Common brick measuring 57 x 95 x 203 mm (2.25 x 3.75 x 8 in.); permissible variables are ± 1.5 mm (0.0625 in.) in depth, 3.1 mm (0.125 in.) in width, and 6.3 mm (0.25 in.) in length.

Stiff-mud: Brick produced by extruding a stiff but plastic clay (12% to 15% moisture) through a die.

Super-duty fireclay: Fireclay brick that has a pyrometric cone equivalent above 33 on the fired product, shrinks less than 1% at 1598°C (2,910°F), and does not incur more than 4% loss in the panel spalling test preheated to 1648°C (3,000°F).

Textured: Brick with a rough finish.

Water-struck: *See* **Soft-mud brick,** above.

Wire-cut: Extruded brick cut to length by a wire prior to drying and firing.

BRICK-AND-BRICK Method of laying brick so that units touch each other with only enough mortar to fill surface irregularities.

BRICK ASHLAR Wall of ashlar facing backed with bricks.

BRICK AXE Bricklayer's hammer.

BRICK BUGGY Cart used to transport masonry units to the mason; may be powered or hand pushed.

BRICK CEMENT Waterproofed masonry cement.

BRICK CONCRETE Concrete made with broken brick as a major portion of the aggregate.

BRICK CORE Nonload-bearing brick infilling between the top of a lintel and the soffit of a relieving arch.

BRICKEARTH Any natural material suitable for making bricks.

BRICK FACING *See* **Construction types.**

BRICK GRADE Designation for durability of the unit, expressed as:

SW: Severe weathering.

MW: Moderate weathering.

NW: Negligible weathering.

BRICKLAYER'S SQUARE SCAFFOLD Supported scaffold composed of framed squares that support a platform.

BRICK MASON Bricklayer.

BRICK NOGGING *See* **Back filling.**

BRICK-ON-EDGE Brick laid with its long face uppermost, either as a decorative string or to form a coping.

BRICK SEAT Ledge on a wall or footing to support a course of masonry.

BRICK SET Bricklayer's cutting chisel.

BRICK TROWEL Most popular size of trowel measuring 127 mm (5 in.) wide and 280 mm (11 in.) long.

BRICK TYPE Designation for facing bricks that controls tolerance, chippage and distortion. Expressed as:

FBS, FBX and **FBA:** For solid brick;

HBS, HBX, HBA and **HBB:** For hollow brick.

BRICK VENEER *See* **Construction types.**

BRICK-VENEER TILE Ceramic tile finished to resemble a clay brick.

BRICKWORK Use of bricks to form a shape.

BRIDGE 1. Structure spanning between two elevations above a lower elevation; **2.** Wire that is heated by an electric current in an electric blasting cap so as to ignite a charge; **3.** Shunt connection between cap wires in a blasting circuit; **4.** Section of a double-drop lowbed or drop-frame trailer between the rear drop and the rear end. **5.** Straightedge used as a starting line for the laying of tile; **6.** In a drainage structure, a span of 6 m (20 ft) or more, having a designed clearance over the highest expected floods.

BRIDGE ABUTMENTS Portion of a bridge construction designed to withstand a thrusting force.

BRIDGE APPROACH Road segment at each end of a superstructure (bridge) guiding and carrying traffic on and off the bridge.

BRIDGEBOARD String of a stair consisting of a notched board for supporting the treads and risers of a wooden stairway.

BRIDGE CABLE Structural rope or strand; the all-metallic wire rope or strand used as the catenary and suspenders on a suspension bridge.

BRIDGE CAP Upper surface of a bridge pier on which bearings or rollers are seated.

BRIDGE CLEARANCE Elevation difference between the lowest part of the superstructure and the water surface directly under a bridge. The required clearance is stated at design flood.

BRIDGE CRANE *See* **Crane.**

BRIDGE DECK Load-bearing structure that is supported on a bridge superstructure and serves as the roadway or other traveled surface.

BRIDGE FORMULA Method of computing legally allowable gross vehicle weight, based on allowable weight per foot of overall wheelbase length.

BRIDGE-FURNACE WALL *See* **Wall.**

BRIDGE HEIGHT Elevation difference between the deck surface and the deepest point in the channel under the bridge.

BRIDGE LOAN Mortgage financing between the termination of one loan and the beginning of another.

BRIDGE LOCATION 1. Site of a bridge; **2.** Process of selecting and surveying a bridge site.

BRIDGE LOCATOR Person who searches for a bridge site and selects and surveys it.

BRIDGE MAINTENANCE Process of inspecting and repairing a bridge to reduce deterioration and insure safety.

BRIDGEMEN *See* **Pile crew.**

BRIDGE OPENING Distance between two abutments. In the case of single spans it is sometimes used as a synonym for span.

BRIDGE PIER Intermediate support to a bridge deck.

BRIDGE REAMER Reamer used to increase the size of rivet holes or make two or more holes line up by making them bigger.

BRIDGE SEAT Portion of a pier or abutment on which a bridge superstructure is mounted.

BRIDGE SOCKET Wire rope or strand end termination made of forged or cast steel that is designed with baskets (having adjustable bolts) for securing rope ends. There are two styles:

Closed type: Has a U-bolt with or without a bearing block in the U of the bolt.

Open type: Has two eye bolts and a pin.

BRIDGE WALL Partition wall between chambers of a furnace system over which pass the products of combustion.

BRIDGE WALL COVER Refractory blocks spanning the space between bridge walls.

BRIDGING 1. Arrangement of small wooden or metal pieces between timbers, such as joists, to stiffen and hold them in place. *Also called* **Cross nogging,** and **Joist bridging; 2.** Method of bracing partition studding and floor joists by the use of short strips of wood. *See also* **Cross bridging; 3.** Situation where the continuity of a column of explosives in a borehole is broken, either by improper placement, as in the case of slurries or poured blasting agents, or where some foreign matter has plugged the hole; **4.** Condition of a filter element loading in which contaminant spans the space between adjacent sections of a filter element, thus blocking a portion of the useful filtration area; **5.** Ability of some caulks and adhesives to deform and to continue to fill or cover an opening or joint subject to movement; **6.** Diagonal bracing installed between parallel structural steel members to resist twisting.

BRIDGING FLOOR Floor carried on common joists only.

BRIDGING JOIST Common joist.

BRIDGING PIECE Bearer between or across common joists that carries a partition.

BRIDLE 1. Frame equipped with sheaves and connected to a boom by stationary ropes called pendants. The bridle is used in conjunction with the boom hoist drum and bail to alter the crane's boom angle; **2.** In riveting, a light wire or cable with one eye attached to the handle of the gun and the other eye fastened to the snap wire; **3.** Method of choking a log from opposite sides by using two chokers; **4.** Method of securing a guyline to two stumps with a block and strap. *See also* **Floating harness.**

BRIDLE CABLE Anchor cable that is at right angles to the line of pull.

BRIDLE HITCH Connection between a bridle cable and another cable or a sheave block.

BRIDLE IRON Metal strap attached to one timber and supporting another meeting it at an angle (usually 90°). *Also called* a **Stirrup,** and **Stirrup iron.**

BRIDLE SLING Sling made up of multiple wire-rope legs with the top ends gathered in a fitting that passes over a lifting hook.

BRIGHT Metal finish resulting from cleaning or from plating.

BRIGHT GLAZE *See* **Glaze.**

BRIGHTNESS Degree of brilliance; luminous aspect of a color, as distinct to its hue.

BRIGHT ROPE Wire rope fabricated from wires that are not coated.

BRIGHT STOCK Heavy lubricant base stock derived from further refining of the residues left after vacuum distillation of certain types of crude oil.

bril. *Abbreviation for:* brilliant; brilliance.

BRILLIANCE Clearness of a varnish, lacquer, or paint; reflective quality of a surface.

BRINELL HARDNESS NUMBER Measurement of surface hardness determined by pressing a carbide ball of standard diameter at a standard load into the surface of a material. Dividing the applied force (in kg) by the area of the indentation (in mm²) provides the brinell hardness value.

BRINELL HARDNESS TESTER Machine used for testing the indentation hardness of metals, except very hard materials like tool steels.

BRINELLING Surface failure characterized by the indentation of the surface so it is no longer smooth enough to be serviceable.

BRING-UP TIME Time required to raise a cold furnace, and its charge, if any, to operating temperature.

BRIQUETTE or BRIQUET 1. Masonry unit smaller than a standard brick in one or more dimensions; **2.** Molded specimen of mortar with enlarged extremities and reduced center having a cross section of definite area, used for measurement of tensile strength.

BRISANCE Ability of an explosive to break or shatter material by shock.

BRISE-SOLEIL Sun break, usually consisting of thin vertical or horizontal members that permit the passage of air.

brit. *Abbreviation for:* brittle.

BRITISH IMPERIAL GALLON *See* **Imperial gal-**lon.

BRITISH THERMAL UNIT Nonmetric unit of energy, equal to the quantity of energy required to raise 1 lb of water through 1°F. Symbol: Btu. Multiply by 1055.06 to obtain joules, symbol: J. *See also the appendix:* **Metric and nonmetric measurement.**

BRITISH THERMAL UNIT PER GALLON (IMP) Nonmetric unit of heat. Symbol: Btu/gal. Multiply by 238.08 to obtain kilojoules per cubic meter, symbol: kJ/m³. *See also the appendix:* **Metric and nonmetric measurement.**

BRITISH THERMAL UNIT PER GALLON (US) Nonmetric unit of heat. Symbol: Btu/gal (US). Multiply by 278.717 to obtain kilojoules per cubic meter, symbol: kJ/m³. *See also the appendix:* **Metric and nonmetric measurement.**

BRITISH THERMAL UNIT PER HOUR Nonmetric unit of heat. Symbol: Btu/hr. Multiply by 0.292 072 to obtain watts, symbol: W. *See also the appendix:* **Metric and nonmetric measurement.**

BRITISH THERMAL UNIT PER POUND Nonmetric unit of heat. Symbol: Btu/lb. Multiply by 2.326 to obtain kilojoule per kilogram, symbol: kJ/kg. *See also the appendix:* **Metric and nonmetric measurement.**

BRITISH THERMAL UNIT PER POUND °F Nonmetric unit of heat. Symbol: Btu/lb°F. Multiply by 4.1868 to obtain kilojoules per kilogram °C, symbol: kJ/(kg°C). *See also the appendix:* **Metric and nonmetric measurement.**

BRITISH THERMAL UNIT PER SQUARE FOOT PER HOUR Nonmetric unit of work. Symbol: Btu/(ft²/hr). Multiply by 3.154 60 to obtain watts per square meter, symbol: W/m². *See also the appendix:* **Metric and nonmetric measurement.**

BRITISH THERMAL UNIT PER SQUARE FOOT PER HOUR PER DEGREE FAHRENHEIT Nonmetric unit of work. Symbol: Btu/(ft²/hr)°F. Multiply by 5.678 29 to obtain watts per square meter per degree Centigrade, symbol: W/(m²°C). *See also the appendix:* **Metric and nonmetric measurement.**

BRITISH THERMAL UNIT PER (2000-lb) TON Nonmetric unit of heat. Symbol: Btu/ton. Multiply by 1.163 to obtain kilojoules per tonne, symbol: kJ/t. *See also the appendix:* **Metric and nonmetric measurement.**

BRITTLE *See* **Texture.**

BRITTLE FACTOR Abrupt cleavage in steel with little or no prior ductile deformation.

BRITTLENESS Quality of a material that causes it to develop cracks with little bending.

brk *Abbreviation for:* break; brick.

brkt *Abbreviation for:* bracket.

brlp *Abbreviation for:* burlap.

bro. *Abbreviation for:* broach.

BROACH 1. Enlarge or ream a hole, usually in stone; **2.** Rough dressing to stone; **3.** Type of stone-mason's chisel; **4.** Shearing cut produced by a tool with one or more cutting edges; produces machined shapes not easily obtained by turning or other work methods.

BROACHING Rock excavation method involving holes drilled along the proposed break line with the rock remaining between the holes being shattered with a broach. The rock is then removed by wedging.

BROAD-BASE TERRACE Long ridge of earth, 4.5 to 9 m (15 to 30 ft) wide and 254 to 760 mm (10 to 30 in.) high with sloping sides, a rounded crown, and broad shallow channel along the upper side, designed to control erosion by diverting surface water along the contour at low velocity.

BROADCAST To toss granular material, such as sand, over a horizontal surface so that a thin, uniform layer is obtained.

BROADCAST BURNING Controlled forest burn, where the fire is intentionally ignited and allowed to proceed over a designated area within well-defined boundaries, for the reduction of fuel hazard or for site preparation. *Also called* **Slash burning.**

BROAD KNIFE Wide, flexible finishing knife, used for applying drywall joint finishing compound.

BROB Iron fastening having its head bent at right angles to the shaft, used for connecting timbers.

BROKEN JOINTS 1. Manner of laying masonry units so as to avoid vertical joints in adjacent courses from lining up. *Also called* **Breaking joint; 2.** Distribution of joints in lumber sheathing, flooring, lath, and panels so no two adjacent end joints are directly in line. *Also called* **Staggered joint,** and **Step joint.**

BROKEN-JOINT TILE Single-lap tile.

BROKEN PEDIMENT *See* **Pediment.**

BROKEN RANGE Masonry construction in which the continuity of the courses is broken at intervals.

BROKEN-RANGE ASHLAR Uncoursed rubble.

BROKEN STEP-CHAIN SWITCH *See* **Switch.**

BROKEN WHITE Off-white color, usually a cream tone.

BROKER Middleman or limited agent, often acting for both parties to a transaction.

BRONZE Alloy of about 88% copper and 8% to 10% tin, plus zinc.

BRONZE INSERT Removable component of a mobile connection that prevents steel-to-steel contact.

BRONZING Formation on a paint film of a metallic-appearing haze.

BROOM FINISH Surface texture obtained by stroking a broom over freshly placed concrete or freshly applied mortar or plaster. *See also* **Brushed surface.**

BROOMING 1. Separation of fibers at the butt or tip of a timber pile, generally caused by excessive or improper driving; **2.** Roughening of an undercoat of cement paste or plaster, often by brushing with a straw broom.

BROTHERS Two-leg or four-leg chain or rope sling.

BROW *See* **Landing.**

BROW LOG Log placed parallel to a road or track to protect equipment during loading and unloading or at log dumps. It takes the bump of logs accidentally dropped or swung out of control.

BROWN COAT Second coat in three-coat plaster application.

BROWN OUT To complete application of base coat plaster.

BROWN OXIDE Brown mineral pigment having an iron oxide content between 28% and 25%. *See also* **Limonite.**

BROWN STAIN Rich brown-to-deep-chocolate-brown discoloration of the sapwood of some pines caused by a fungus. *See also* **Blue stain,** and **Chemical brown stain.**

BROWNSTONE Sandstone with a reddish-brown color.

BROWSE Buds, shoots, and leaves of woody plants that can be eaten by livestock or wild animals.

brs *Abbreviation for:* brass.

Br. Std *Abbreviation for:* British Standard.

brt *Abbreviation for:* bright.

brt. *Abbreviation for:* brightness.

BRUSH 1. Trees and shrubs less than 100 mm (4 in.) in diameter; **2.** Component of an electric motor, generator, or other rotating electrical device that carries electrical current from the moving to the static parts.

BRUSHABILITY Ability of a compound, such as paint, to be brushed.

BRUSH COATING Waterproof coating of one or more layers of asphalt, pitch, or waterproofing compound, applied to the exterior face of below-grade concrete foundation walls. May also be spray applied.

BRUSHED SURFACE 1. Sandy texture obtained by brushing the surface of freshly placed or slightly hardened concrete with a stiff brush for architectural effect or, in pavements, to increase skid resistance. *See also* **Broom finish; 2.** Plywood siding surface treatment; brushed or relief-grain surfaces that accent the natural grain pattern to create striking textures.

BRUSHING 1. Silviculture treatment to remove brush and weed species that compete with seedlings for sunlight, water, and soil nutrients. *See also* **Conifer release; 2.** Act of applying paint using a brush.

BRUSH LINE One of a series of fine, parallel surface blemishes to glass that resemble brush marks on a painted surface. *Also called* **End line.**

BRUSH OUT To clear an area of limbs, saplings, and debris. *Also called* **Swamp out.**

BRUSH RAKE Lightly constructed rake blade with a high top; an attachment for a crawler-tractor, used in mechanical site preparation to penetrate and mix soil and tear roots.

brz. *Abbreviation for:* bronze.

brzg *Abbreviation for:* brazing.

bs. *Abbreviation for:* backset, base; beam spacer, both sides.

BS *Abbreviation for:* British Standard.

B/S *Abbreviation for:* band saw; bill of sale

B SELECT *See* **Lumber grades.**

bsfc *Abbreviation for:* brake-specific fuel consumption.

bshg *Abbreviation for:* bushing.

B-SIDE Right side of an elevator car or hoistway when facing the hoistway at a landing.

bskt *Abbreviation for:* basket.

bsmith *Abbreviation for:* blacksmith.

bsmt *Abbreviation for:* basement.

bsp *Abbreviation for:* black steel pipe.

BSR *Abbreviation for:* building space requirements.

bstd *Abbreviation for:* bastard.

bstr *Abbreviation for:* booster.

BSU *Abbreviation for:* beam spacer upper.

bt *Abbreviation for:* bent.

BT *Abbreviation for:* ball tip;bolt tolerance; bus tie.

BTB *Abbreviation for:* back-to-back; bituminous-treated base.

bthrm *Abbreviation for:* bathroom.

bt hse *Abbreviation for:* boat house.

btl. *Abbreviation for:* bottle.

btr *Abbreviation for:* better.

B-TRAIN *See* **Trailer.**

Btu *Abbreviation for:* British thermal unit

Btu/hr *Abbreviation for:* British thermal units per hour.

btwn *Abbreviation for:* between.

BU *Abbreviation for:* back up.

BUBBLE 1. Air bubble in a level gauge; **2.** Void caused by trapped air in the core of drywall panels.

BUBBLE POINT Pressure at which the first steady stream of gas bubbles is emitted from a wetted filter element under specified test conditions.

BUBBLE TUBE Slightly arched, sealed glass tube partially filled with a fluid; used to determine true horizontal that, when the tube is properly adjusted, is indicated by the position of the bubble within the tube. *Also called* **Level tube.**

BUCK 1. *See* **Harvest functions; 2.** Framing around an opening in a wall.

BUCKER 1. *See* **Harvesting machines (single function); 2.** Ironworker who 'bucks' up or backs up a hot rivet, using a dolly bar, so that the driver can form a head on it.

BUCKET Fitting or attachment to an item of mechanical equipment, such as an excavator, backhoe, or dragline, that digs, lifts, loads, and carries material. *See also* **Bailing bucket, Belling bucket,** and **Slat bucket.** There are many types of bucket attachment, including:

  Claw: Bucket with positive clamping jaws for garbage pickup, snow removal, debris cleanup

or any loose bundled loads.

  Extreme service: Designed for tougher trenching applications such as fragmented rock, frozen ground, caliche, etc. It is equipped with pockets on the rear to accept optional ripper shanks.

  Four-in-one: *See* **Multipurpose,** below.

  General-purpose: Designed for general excavation digging dirt and mass excavation.

  Grading: Bucket with a long, flat floor and straight lip, used for finish work in housing developments, concrete pours, landscaping, and light dozing.

  Heavy-duty: Designed for rougher conditions dense clay and light rock.

  High dump: Gives extended dump height of light materials.

  Light material: For excavating, loading, and easy-digging in light material. Can also be used as a finishing/cleanup bucket. Can be equipped with a bottom cutting edge.

  Loose material: Designed for snow, woodchips, hay, coal, etc. Can be fitted with an independently controlled top clamp for such material as brush, silage, or compost.

  Mass excavation: Designed for volume truck loading with a shorter tip radii and greater bite width.

  Multipurpose: Loader or backhoe clam bucket whose floor is hinged to the top of the bucket back, and is under separate control. It enables the operator to perform multiple tasks with one bucket, such as dig, load, doze, grade, backfill, grab and, with flip-over pallet forks, lift. *Also called* **Six-in-one bucket.**

  Rock-ripping: Specially designed for extreme digging and rock conditions, with a staggered tooth design that allows the center tooth to enter the ground first at a 45° angle, allowing it to use all the machine force to rip. The two teeth on either side enter the ground next. The outer teeth, which project outward, enter the ground last and slice the trench wall, leaving it clean and straight.

  Severe-duty: Designed for highly abrasive applications: shot rock and demolition.

  Side-dump: Able to dump forward or to the left; particularly useful in close quarters or to reduce turning time.

  Slat: Openwork bucket made of bars instead of solid plates, used for digging sticky soil.

  Trenching: A narrow but deep bucket whose bite width is usually dictated by the pipe diameter.

BUCKET AUGER Cylindrical rotary drilling tool with a hinged bottom containing a soil cutting blade. Spoil enters the bucket which is then lifted out of the hole and swung aside where the contents are dumped by releasing the latch on the hinged bottom. *Also called* **Bucket loader,** and **Drilling bucket.**

BUCKET CAPACITY (*See also* **Bucket rating.**) Actual volume of material that can be carried in an equipment bucket. It is rated in two ways:

  Heaped: Volume in the bucket under the strike-off

plane plus the volume of the heaped material above the strike-off plane, having an angle of repose of 1:1, without any consideration for any material supported or carried by the spillgate or bucket teeth.

Struck: The volume actually enclosed inside the outline of the sideplates and rear and front bucket enclosures without any consideration for any material supported or carried by the spillgate or bucket teeth.

BUCKET CAR Rail car that holds up to four concrete buckets on a cableway.

BUCKET CYLINDER Cylinder that controls movement of an attachment bucket. *See also* **Curl and crowd force.**

BUCKET DOCK Landing platform for concrete buckets from a cableway.

BUCKET ELEVATOR Endless chain with self-discharging buckets or skips attached, used to move granular material from one elevation to another.

BUCKET LADDER EXCAVATOR *See* **Trencher.**

BUCKET LOADER Usually a chain-bucket loader, but also a tractor-loader. *See also* **Loader.**

BUCKET PUMP Large-capacity reservoir-and-pump for lubricants.

BUCKET RATING (*See also* **Bucket Capacity.**) Volume of material that a bucket is designed to hold. There are two ratings:

Heaped: Bucket struck capacity plus the additional material needed to create a 2:1 angle of repose (1:1 for hydraulic excavators) with a struck line parallel to the ground.

Struck: Volume of material in a bucket after a load is leveled by a straight edge resting on the cutting edge and the back of the bucket (not including see-through rock guards).

BUCKET SHEAVE *See* **Sheave.**

BUCKING *See* **Slashing.**

BUCKING UP To hold a hot rivet in place with a bucking-up tool while the head is being formed on the other side by the driver.

BUCKLE Bend under strain.

BUCKLED PLY Deformation in a ply that distorts its normal plane. *Also called* **Wrinkled ply.**

BUCKLE GUY Guyline attached at the midpoint of a yarding spar to prevent bending.

BUCKLING 1. Failure by lateral or torsional instability of a structural member, occurring with stresses below the yield of ultimate values; **2.** Pulling away of a panel edge from its support structure.

BUCKLING LOAD Load at which a perfectly straight steel member under compression assumes a deflected position.

BUCKS 1. Metal frame forming the entrance to an elevator hoistway; **2.** Framing around an opening in a wall.

BUCKSHOT Tough earthern material that, when dry, shatters when struck into irregularly shaped particles about the size of buckshot, 3 mm (1/8 in.), or more.

BUCKSKIN LOG Log that has no bark.

BUCKSTAY Vertical member, usually in cross-connected pairs.

BUCKUP Aid to erecting hollow metal frames, consisting of two telescoping tubes with clamps at either end.

bud. *Abbreviation for:* budget.

BUDGET Plan or schedule adjusting the management of finances or activities for a given period. Budgets can be designed to achieve a range of objectives, including:

Capital: Budget for proposed additions to capital assets and their financing.

Cash: Budget of cash receipts, payments, and periodic balances.

Fixed: Budget prepared for a single level of activity.

Flexible: Budget prepared for a range of levels of activity.

Project: Budget for authorized appropriation of funds, generally based on an estimate of the scope of work in the construction phase of a project, and including the cost of the feasibility phase.

BUDGETARY CONTROL Process of planning, executing, and evaluating a project or program through the use of a budget.

BUDGET ESTIMATES Estimates of anticipated costs prepared in order to establish a budget.

BUDGET MORTGAGE *See* **Mortgage.**

BUDGET VARIANCE *See* **Variance.**

BUFF Polish the surface of metal using a fabric wheel and buffing compound.

BUFF CONTOUR Shape of a buffed tire.

BUFFED SURFACE Specially prepared surface of a tire casing that provides proper adhesion between it and new rubber.

BUFFER 1. Previously shot material, not removed, lying against a face to be shot; **2.** Movable metal plate used in tunnels to limit scattering of blasted rock; **3.** Device designed to stop a descending elevator car or its counterweight from traveling beyond its normal limit by storing, absorbing, or dissipating the kinetic energy of the moving objects; **4.** Circuit or component that isolates one electrical circuit from another; **5.** Machine used to rasp old tread from a tire; **6.** Screen of vegetation, a berm or other physical barrier erected between the public and a disposal, transfer, or recycling function.

BUFFER BLASTING Secondary fragmentation produced by crushing, achieved by setting off a row of holes nearest the face first and the other parallel rows behind it in succession, or where the last row of boreholes has a reduced burden, spacing, and explosives load. *See also* **Rotational firing.**

BUFFER CHANNEL Channel iron placed on the pit floor of an electric elevator to support the buffers and guide rails.

BUFFER POINT Steel point at the end of a feed beam that is pressed against the rock face to stabilize the unit while drilling.

BUFFER ROW Row of explosives with reduced spacing and explosives load; in buffer blasting, the buffer row is the last row adjacent to the planned excavation limit.

BUFFER SCREW Screw used to eliminate engine roll or surge at idling speed.

BUFFER SOLUTION Solution containing two or more properties that, when combined, resist marked changes in pH after moderate quantities of either strong acid or base have been added.

BUFFER STRIP Strip of land (often including undisturbed vegetation) where disturbance is not allowed or is closely monitored to preserve or enhance aesthetic or other qualities along or adjacent to roads, trails, watercourses and recreation sites. *Also called* **Green strip, Leave strip,** and **Streamside management zone.**

BUFFER SWITCH *See* **Switch.**

BUFFER ZONE 1. Transitional area between two areas of significantly different land use; **2.** Chamber or cavity that is adjacent to a mechanical seal and that is filled with buffer fluid.

BUFFING Grinding a surface to obtain dimensional conformance or surface uniformity.

BUFFING COMPOUND Mixture of mild abrasive and polish, used with a buffing wheel to polish metal.

BUFFING RADIUS Distance from the center or pivot point of a buffing rasp assembly and the arc it describes.

BUFFING TEMPLATE Shaped guide used to determine the contour of a buffed tire.

BUFFING WHEEL Fabric disks sewn together to form a wheel, used at high speed on the arbor of a bench grinder or similar machine to polish metal.

BUFF LINE Dividing line in a tire cross section between the buffed surface of the original tire and the new retread rubber.

BUG Signaling device, carried on a belt, used to transmit signals to a yarder.

BUGGED FINISH A smooth finish produced by grinding with powder abrasives.

BUGGY Manual or powered vehicle used to transport fresh concrete.

BUG HOLE Small regular or irregular cavity, usually not exceeding 12 mm (0.5 in.) in diameter, resulting from entrapment of air bubbles in the surface of formed concrete during placement and consolidation. *See also* **Blemish, Blowhole,** and **Sack rub.**

BUILD Application of coats of paint to a required film thickness.

BUILDER 1. Contractor skilled and experienced in building construction; **2.** Machine used to apply tread rubber to a tire casing.

BUILDER'S LEVEL *See* **Level.**

BUILDER'S RISK INSURANCE Type of property insurance covering buildings under construction against fire and lightning.

BUILDER'S TAPE Steel or plastic tape, usually 15 or 30 m, or 50 or 100 ft in length, housed in a circular case, with a hook or other device on the lead end for fastening to one end of the span to be measured.

BUILDER'S WARRANTY Guarantee on the quality of construction offered by a developer or building contractor.

BUILDING 1. Complete structure, finished in all details for its intended purpose; **2.** Act of putting together materials in order to create a structure.

BUILDING AREA Greatest horizontal projected area of a building at or above grade within the outside perimeter of the exterior walls or within the outside perimeter of the exterior walls and the center line of fire walls.

BUILDING BLOCKS Hollow or solid, modular blocks of conventional masonry materials, larger than bricks.

BUILDING BRICK *See* **Brick.**

BUILDING BYLAW *See* **Building code.**

BUILDING CODE Regulations governing building design and construction, established by regulatory authorities. Building codes are promulgated and administered by various levels of Government, and originate also from a number of quasi-governmental and private organizations. They include:

> **Building bylaw:** Building regulation enacted by a local authority for application within its jurisdiction.
>
> **City building code:** Regulations enacted by a municipal authority governing the design and construction of all types of buildings and structures. In Canada, these regulations supplement standards established in the National Building code.
>
> **County and township building code:** In the U.S., some counties and townships have enacted ordinances (regulations) governing construction work done outside incorporated cities.
>
> **Housing code (Canada)** *See* **National building code,** below.
>
> **Housing code (U.S.)** National set of standards that all housing must meet on final construction. May be augmented by state or local requirements. *Also called* **Minimum standards bylaw.**
>
> **Minimum standards bylaw** *See* **Housing code (U.S.),** above, and **National building code,** below.
>
> **Model building code:** In the U.S., three organizations have sponsored model building codes. These are:
>
> **BOCA National Building Code** Sponsored by the Building Officials and Code Administrators International.
>
> **Southern Standard Building Code** Sponsored by the Southern Building Code International Congress.
>
> **Uniform Building Code** Sponsored by the International Conference of Building Officials.

National Building Code: In Canada, a publication of the National Research Council of Canada that sets forth the minimum national standards for construction of buildings. These standards may be altered by provincial or local authorities. The National Building Code forms the basis for all building regulations in Canada. In the U.S., the BOCA (Building Officials and Code Administrators International) National Building Code forms the basis for all building regulations.. Buildings erected on federal property are exempt from the requirements of local codes. *Also called* **Housing code,** and **Minimum standards bylaw.**

Provincial building codes: Building regulations enacted by the provincial authorities in Canada that may supplement the National Building Code, and local building codes. Typically they include an electrical code, plumbing and gas installation codes.

State building codes: In the U.S., most states have passed ordinances that apply to buildings of a particular type of construction or use; these are not uniform across the country.

Township building code: Construction regulations that supplement national or model building codes, applicable to specific jurisdictions.

BUILDING COMMITTEE Persons appointed or assembled to develop and manage a project.

BUILDING CONSTRUCTION Subject, and technique of building.

BUILDING CONTRACTOR Person knowledgeable of, and skilled in building construction and who contracts to build.

BUILDING COOPERATIVE Constituted organization in which members pool their resources to build homes that they will own and occupy.

BUILDING DEPARTMENT 1. Department of local government responsible for the regulation of construction within its jurisdiction; **2.** Department of local government responsible for municipal works, including maintenance; **3.** Department within an organization responsible for the maintenance of physical property.

BUILDING DESIGN Design of works and structures, having particular concern for the materials to be used and the construction techniques.

BUILDING DRAIN That part of a plumbing drainage system that receives discharge from soil, waste, and other stacks inside a building. *Also called* **House drain.**

BUILDING DRAINAGE SYSTEM *See* **Drainage system.**

BUILDING ELEMENT Major functional part of a structure: foundation, floor, wall, roof, services, etc.

BUILDING ENVELOPE 1. Outer skin of a building that limits thermal transfer; **2.** Outer structure of a building.

BUILDING EQUIPMENT 1. Mechanical devices used in construction; **2.** Services, plant equipment, fixtures, and furniture used in a completed building.

BUILDING FACE That part of an exterior wall of a building that faces in one direction and is located between ground level and the ceiling of its top story, or, where a building is divided into fire compartments, the exterior wall of such a compartment that faces one direction.

BUILDING FACE ROLLER Staging guide roller designed to contact a portion of the outer face or wall structure of a building.

BUILDING HEIGHT 1. Total number of stories in a building; **2.** The vertical distance from grade to the highest point of the roof surface if a flat roof, to the deck of a mansard roof, or to the mean height between eaves and ridge for a pitched roof.

BUILDING IN Incorporation of premade or prefabricated equipment and fittings as a structure is completed.

BUILDING INSPECTOR Person charged with administration and enforcement of the applicable building codes. *Also called* **Building official.**

BUILDING LAYOUT 1. Simplified line drawing showing the principal areas of a building; **2.** Setting out on the ground the exterior walls of a building.

BUILDING LINE 1. Line set back from the centerline of a road, in front of which no construction may take place; **2.** Main line used to lay out a building, usually the front wall of a structure.

BUILDING LOAN AGREEMENT Contract in which a lender advances money to an owner with provision for payments at defined stages of construction.

BUILDING LOT Parcel of land available for development within local zoning regulations.

BUILDING MAIN Water supply piping beginning at the source of supply and ending at the first branch inside the building.

BUILDING MAISONETTE Form of horizontal multiple housing in which one dwelling shares three party walls with adjacent dwellings, one wall of which may be an internal corridor. Access to the dwelling is at grade, either to the exterior or to the corridor, or both.

BUILDING MATERIALS Any of the full range of raw and manufactured materials used for building.

BUILDING OFFICIAL *See* **Building inspector.**

BUILDING OPERATIONS All or any of the tasks and functions necessary to complete a structure in the manner shown on the contract drawings using the materials listed and described in the contract specification.

BUILDING OWNER Person, or his agent, who owns a site, and who authorizes work to be done and guarantees payment for the work up to an agreed sum.

BUILDING PAPER A thick, water resistant (frequently bitumen coated) breather-type sheathing paper used under siding and between double floors.

BUILDING PERMIT Permit issued for a fee by local authorities that authorize various stages of construction.

BUILDING PROGRAM Schedule of stages and/or operations for the completion of a building project.

BUILDING REGULATIONS Regulatory requirements, adopted by a local authority, governing the design and construction of physical structures.

BUILDING RESTRICTIONS Provisions in a building code or other applicable ordinance or regulation that affect the siting, orientation, size, appearance, construction, or other aspect of a building.

BUILDING SEWER Pipe that is connected to a building drain 900 mm (36 in.) outside of a wall of a building to conduct sewage, clear water waste, or storm water to a public sewage disposal system.

BUILDING SITE Parcel of land suitable for building, or on which a building is or may be built, including all surrounding land specifically allocated to the building. *Also called* **Plot.**

BUILDING STORM DRAIN Drainage system used to receive rainwater, surface water, and groundwater, connected to the building sewer outside the building.

BUILDING STORM SEWER Piping that connects to the end of a building storm drain to receive and convey the contents to a public storm sewer or combined sewer.

BUILDING SUBDRAIN Portion of a building drainage system that can drain by gravity into the building sewer.

BUILDING TRADE Any of a number of skilled and semi-skilled crafts for which training and/or certification is necessary.

BUILDING TRAP Trap that is installed in a building drain or building sewer to prevent circulation of air between a drainage system and a public sewer.

BUILD-TO-SUIT Offer by a landowner to construct on the property a building specified by a potential tenant.

BUILDUP 1. Portion of the tread of a tire covering the shoulder and blending into the sidewall; **2.** Term referring to an item that has been formed from several smaller components that have been fastened together; **3.** Gunning of shotcrete in successive layers to form a thicker mass; **4.** Accumulation of residual hardened concrete in a mixer; **5.** Amount of a weld face extended above the surface of joined metals; **6.** Cumulative effects of drying of brush and forest areas that increases the danger, and the potential for acceleration of a fire with time.

BUILT-IN Collective term for furniture that is fitted, or built in as part of a structure.

BUILT-IN CONTAMINANT Initial residual contamination in a component, fluid, or system. Typically: burrs, chips, flash dirt, dust, fiber, sand, moisture, pipe dope, weld spatter, paints and solvents, flushing solutions, incompatible fluids, and operating fluid impurities. *See also* **Artificial contaminant, Contaminant,** and **Generated contaminated.**

BUILT-IN DIRT Material passed into the effluent stream composed of foreign materials incorporated into a filter medium.

BUILT-IN GARAGE Garage contained within the structure of a house, with direct access to the dwelling.

BUILT-IN-PLACE FORMS Concrete formwork assembled *in situ.*

BUILT-UP Structural member made up of two or more parts fastened together so as to act as a single unit.

BUILT-UP BEAM Beam formed by bolting or nailing two or more planks together.

BUILT-UP COLUMN Column composed of more than one piece.

BUILT-UP MEMBER Member made of structural metal elements that are welded, bolted, or riveted together.

BUILT-UP ROOF Roof composed of several (usually three to five) layers of asphalt felt laminated with coal tar, pitch, or asphalt. The upper, exposed surface is finished with a wearing surface of pea gravel, crushed stone, or crushed slag embedded in bitumen. *See also* **Glass-fiber felt, Organic felt, Roofing felt,** and **Tarred felt.**

BUILT-UP TIMBER Timber made by fastening several pieces together and forming one of larger dimension.

BULB EDGE Extreme lateral edge of sheet glass, formed as it is drawn.

BULB OF PRESSURE Area around and below a friction pile or group of piles where pressure on the soil is increased by driving a pile and by the load imposed through it.

BULGE Doughnut-like swelling below a nail head.

BULK APPEARANCE Visual appearance of grease when the undisturbed surface is viewed in an opaque container. *See also* **Appearance,** and **Texture.** Bulk appearance should be described in the following terms:

> **Bleeding:** Showing free oil on the surface of the lubricating grease (or in the cracks of a cracked grease).
>
> **Cracked:** Showing surface cracks of appreciable magnitude. In describing such a lubricating grease, the number and size of the cracks should be included.
>
> **Grainy:** A surface composed of small granules or lumps of constituent thickener particles.
>
> **Rough:** Surface composed of many small irregularities.
>
> **Smooth:** Surface relatively free of irregularities.

BULK CEMENT Cement that is transported and delivered in bulk (usually in specially constructed vehicles) instead of in bags.

BULK DENSITY Mass of material (including solid particles and any contained water) per unit volume including voids. *See also* **Specific gravity.**

BULKER 1. Vehicle constructed to haul bulk cement; **2.** Truck-mounted tank having a side metering device for proportioning bulk cement into a windrow.

BULK FREIGHT Freight not in packages or containers.

BULKHEAD 1. Partition or wall securely dividing one part of a structure from another; **2.** Box-like structure that rises above a roof or floor to cover a stairway or an elevator shaft; **3.** Partition blocking fresh concrete from a section of the form or closing the end of a form; **4.** Wall or partition erected to resist ground or water pressure; **5.** Sloping door or doors affording entrance to a cellar from outside a building; **6.** Heavy barrier between cab and load to protect the driver of a log truck.

BULKHEAD FORM 1. Form separating the concrete in a dam into a series of individual blocks; **2.** Forms placed in a vertical plane in the top and sides of a tunnel pour to form a construction joint.

BULKHEAD POUR Concrete placed in a short section of tunnel between two bulkhead forms or between a previous bulkhead pour and a bulkhead form.

BULKHEAD WALL Retaining wall for a marginal wharf.

BULKING 1. Increase in the bulk volume of a quantity of sand in a moist condition over the volume of the same quantity dry or completely inundated; **2.** Increase in the volume of materials due to manipulation, such as excavating, loose spreading, or addition of moisture.

BULKING AGENT Material used to add volume to another material to make it more porous to air flow.

BULKING CURVE Graph of change in volume of a quantity of sand due to a change in moisture content.

BULKING FACTOR Ratio of the volume of moist sand to the volume of the same sand when dry.

BULK LOADING Loading of unbagged cement in containers, specially designed trucks, railroad cars, or ships.

BULK MODULUS 1. Ratio of the change in average stress to the change in unit volume. *See also* **Modulus of compression**; **2.** Measure of resistance to compressibility of a fluid. The reciprocal of the compressibility of this fluid.

BULK REDUCTION Reduction in volume of compacted concrete to approximately two-thirds of the total volume of the dry unmixed material.

BULK SPECIFIC GRAVITY *See* **Bulk density**, and **Specific gravity.**

BULK SPREADER Machine for carrying cement or other material and spreading it on a prepared surface for soil stabilization.

BULK STRENGTH Strength of a cartridge of dynamite in relation to the same sized cartridge of straight nitroglycerine dynamite.

BULKY WASTE *See* **Solid waste.**

bull. *Abbreviation for:* bulletin.

bull. bd *Abbreviation for:* bulletin board.

BULL BLOCK High-lead block used on wooden spars for the mainline.

BULL BUCKER Person who runs a felling and bucking crew.

BULL CHOKER Large, heavy choker used for heavy pulls or lifts.

BULL CLAM Bulldozer fitted with a curved bowl, hinged to the top of the front of the blade.

BULLDOG CLIP Type of floor clip.

BULLDOG GRIP U-bolt threaded at each end.

BULLDOZER Tractor fitted with a front pushing blade.

BULLET-PROOF GLASS Generic term for a lamination of thin sheets of glass cemented together under heat and pressure. Its resistance to penetration is conditional upon the number of layers.

BULL FLOAT Tool comprising a large, flat, rectangular piece of wood, aluminum, or magnesium, usually 200 mm (8 in.) wide and 1060 to 1725 mm (42 to 60 in.) long, and a handle 1.2 to 4.9 m (4 to 16 ft) in length, used to smooth unformed surfaces of freshly placed concrete.

BULL GANG 1. Work crew, usually on the day shift, used to lay rail track, install pipe, clean up the tunnel, and do other routine chores behind the face area; **2.** Workers who erect and rig the home spar. *Also called* **Rig-up crew.**

BULL GEAR Toothed driving wheel that is the largest or strongest in a mechanism.

BULLHEAD Plumbing tee in which the branch is longer than the run.

BULL HEADER Brick laid on its edge showing only the end of its face in the wall.

BULL HOOK Heavy hook on the butt rigging to which chokers are attached. *Also called* **Butt hook.**

BULLNOSE 1. Convex rounding of a member, such as the front edge of a stair, brick, or tile; **2.** *See* **Brick.**

BULLNOSE BLOCK Concrete or clay masonry unit with one or more rounded exterior corners.

BULLNOSE CORNER Type of bullnose finish with a convex radius on two adjacent edges or faces. *See also* **Brick.**

BULLNOSE PLANE *See* **Plane.**

BULL PIN Tapered piece of steel, used to line up holes in steel members in connections.

BULL POINT Pointed, steel hand drill, used with a striking hammer to break rock.

BULL POLE Steel pipe mounted to project laterally from the base of a derrick mast, used to swing the derrick manually.

BULL RING D-shaped tie-down ring mounted at or near floor level or at intervals along the sides of flatbed trucks and trailers. *See also* **Anti-extrusion ring**, and **Backup ring.**

BULL'S LIVER Inorganic silt of slight plasticity; quakes like jelly from vibration.

BULL STICK 1. Pipe or other form of long handle attached at right angles to the base of a derrick mast to provide leverage to rotate the mast and boom, sometimes with rope falls; **2.** Steel bar for punching holes under stumps for the placement of explosives; **3.** Tool used in splicing wire.

BULL STRETCHER Brick laid on edge so as to show the broad side of the brick on the face of the wall.

BULL WHEEL 1. Large driving wheel or sprocket; **2.** Horizontally-mounted circular frame mounted to the base of a derrick mast to receive and guide the ropes used for swinging.

BULWARK Side of a ship above the upper deck.

BUMBOAT Work boat, typically a small boat equipped

with fenders for pushing, bollards for pulling, and a hoist for handling pipes or lines and anchors of a dredge.

BUMMER Small truck with two small wheels and a short pole, used in skidding logs. *Also called* **Dolly.**

BUMPER 1. Device, other than an oil or spring buffer, designed to stop a descending elevator car or counterweight from traveling beyond its normal limit; **2.** Structure attached to the rearmost portion of a truck body to provide rear end protection. *Also called* **Platform floor extension.**

BUMPER OUTRIGGER Outrigger located on the front of a crane carrier that provides additional stability for 'on outrigger' capacities when deployed. *Also called* **Fifth outrigger.**

BUMPER SPIKE Toothed stop at the front of a chain saw body, used to hold the saw against wood.

BUMPER STEP Light-truck rear bumper with a flat top surface to provide a step for entry into the truck body.

BUMPER UP Worker who assists a riveting crew by backing the holder up with a second hammer.

BUNCH *See* **Harvest functions.**

BUNCHING A 'wave' of tread rubber that forms in front of the tread contact point of a tire with the road. Occurs when a tire is under load and moving.

BUNDLE 1. Unit or stack of wood panels held together for shipment with bands. The stack size varies throughout each industry with the average stack running about 760 to 830 mm (30 to 33 inches) high; **2.** Two pieces of drywall panel packaged face-to-face.

BUNDLE BUCKING Cutting bundles or truck loads of tree-length wood or long wood into short lengths.

BUNDLED BARS Group of not more than four parallel reinforcing bars in contact with each other, usually tied together.

BUNDLE OF BARS Bundle of steel reinforcing bars consisting of one size, length, or mark (bent) of bar with the following exceptions: (a) very small quantities may be bundled together for convenience; and (b) groups of varying bar lengths or marks (bent) that will be placed adjacent may be bundled together. Maximum weight of bundles is dependent on regional practices and site conditions. *See also* **Lift.**

BUNDLE STRAP Wire rope clamped around log loads (bundles) to contain the logs for transport by water.

BUNDLING TAPE Tape used to secure the ends of drywall panels to form a bundle.

BUNGALOW Single-story house. Corruption of the Hindustani word bangla, belonging to Bengal. Coined during the British administration of India to describe a lightly built or temporary house erected mainly for administration staff and families of the military.

BUNK Steel movable members on logging trucks and trailers, rail cars or at dryland sort yards for supporting a load of logs during sorting or transport.

BUNKER 1. Protective embankment, mostly below ground; **2.** Device that protects and feeds the belt feeder of crushing and screening plants; **3.** Storage for ore and stone.

BUNK LOG Any log resting on the bunk.

BUNSEN-TYPE BURNER Gas burner consisting of a straight tube with a gas orifice at one end. Primary air enters through adjustable openings around the gas orifice and is entrained by the gas jet. The gas-air mixture burns with a short intense flame as it emerges from the tube.

BUOYANCY Tendency to float or rise in water or air.

BUOYANT FOUNDATION Reinforced concrete raft foundation designed so that its total mass, including all superload, is approximately equal to the mass of soil or water displaced.

b up *Abbreviation for:* bend up.

bur. *Abbreviation for:* bureau.

BURDEN Generally considered to be the distance from an explosive charge to the nearest free or open face at the time the hole detonates. Technically there may be an apparent burden and a true burden, the latter being measured always in the direction in which displacement of broken rock will occur following firing of an explosive charge.

BURGLARPROOF Doors and locks designed to be impregnable against forced intrusion.

BURGLAR RESISTANT Doors and locks capable of resisting force or unauthorized intrusion for a limited time.

BURIED CABLE (HEATING SYSTEM) *See* **Heating system.**

BURIN Tool used to engrave metal.

BURL Hard, woody outgrowth on a tree, more or less rounded in form, usually resulting from the entwined growth of a cluster of adventitious buds; in wood, a localized severe distortion of the grain, generally rounded in outline, usually resulting from overgrowth of dead branch stubs, varying from 12 mm to 75 or 100 mm (1/2 in. to 3 or 4 in.) in diameter.

BURLAP Coarse fabric of jute, hemp, or less commonly, flax, for use as a water-retaining covering in curing concrete surfaces. *Also called* **Hessian.**

BURN 1. Cut with a torch; **2.** Pulverize with extremely heavy explosive charges; **3.** Burnt-over area of timber slash.

BURN CUT Narrow section of rock pulverized by exploding heavy charges in parallel holes. *See also* **Cut.**

BURN-CUT DRILL Heavy drill used for the large-diameter holes in the center of a tunnel face.

BURNER Device that positions a flame in the desired location by delivering fuel (and sometimes air) to that location. Some burners also atomize the fuel, and some mix the fuel and the air. There are several basic types, including:

   Conical: A vertical, hollow, cone-shaped combustion chamber that has an exhaust vent at its apex and a door in its side at its base through which waste materials, usually wood, are charged; air is drawn through louvers to the burning solid waste inside the cone. *Also called* **Beehive burner,** and **Teepee burner.**

   Primary: Burner that dries out and ignites a fuel

in the primary combustion chamber.

Refuse: Device for either central or on-site volume reduction of solid-waste by burning.

Secondary: Burner installed in the secondary combustion chamber to maintain a minimum temperature to complete the combustion of incompletely burned gases.

BURNER BLOCK Refractory block with one or more orifices through which fuel is admitted to a furnace.

BURNER REFRACTORY Refractory block with a conical or cylindrical hole through its center. The block is mounted in such a manner that the flame fires through this hole. The block helps to maintain ignition, and reduces the probability of flash-back or blow-off.

BURNING (THE WORK) In grinding, a change in the work-piece being ground caused by the heat of grinding, usually accompanied by a surface discoloration.

BURNING AREA Horizontal projection of an air-admitting grate, hearth, or both, upon which fuel is deposited for combustion.

BURNING CONDITIONS Environmental conditions that affect fire in a given fuel association.

BURNING INDEX Number related to the effort needed to contain a forest fire of a particular fuel type within a rating area.

BURNING OFF 1. Burning of excess vapors or gasses under controlled conditions; **2.** Removal of old paint with a torch and scraping.

BURNING OUT Setting fire inside a control line to consume fuel between the edge of a forest fire and the control line. *Also called* **Clean burning,** and **Firing out.**

BURNING RATE Quantity of fuel per unit of time that is charged to a furnace, or the amount of heat released during combustion.

BURNING REINFORCEMENT Cutting reinforcing bars with an oxyacetylene torch.

BURNISH 1. Make or become shiny by rubbing; **2.** Pressing or polishing the edges of masking tape to prevent paint or glaze seepage.

BURNISHING TOOL Fine file or tool coated with a mild abrasive, used to clean contact points.

BURNT CLAY Clay fired in a kiln to a high temperature to make brick, tile, earthenware, etc.

BURNT LIME Quicklime; calcium oxide.

BURR Rough ridge or edge left on metal that has been worked.

BURRED Rough surface on metal caused by dirt, insufficient lubrication, or extreme surface loading.

BURRED WHEEL Steel or iron wheel with rough slivers or projections of metal around the rim edges.

BURRING 1. Act of removing burrs from metal; **2.** Cracks on a pile top caused by hammer blows during driving. *Also called* **Mushing; 3.** Of pulpstone, passing over the face of a pulpstone with a special tool to develop a pattern to provide a freer cutting surface.

BURRING REAMER Plumbing tool, used to remove the burr left inside a pipe by a pipe cutter.

BURST Rupture caused by internal pressure; the inside-out differential pressure that causes outward structural failure.

BURST PRESSURE *See* **Pressure.**

BUS Conductor or group of conductors that serve as a common connection for two or more circuits.

bus *Abbreviation for:* business.

bush. *Abbreviation for:* bushing

BUSHELING 1. Method of payment for performing piecework other than volume of timber cut, such as car-loading; **2.** Cutting wood on a per-cord or per-thousand basis.

BUSH HAMMER Tool or bit with a specially grooved head, usually in the form of rows of pyramidal points, used to lightly chip a surface in a uniform pattern.

BUSH-HAMMER FINISH Finish obtained by means of a bush hammer.

BUSH-HAMMERING Method of obtaining an even, rough texture on concrete after it has set, or on stone.

BUSHING 1. Tapped fitting used to reduce the size of an end opening of a fitting or valve. *See also* **Split bushing; 2.** Insulating structure including a through conductor, or providing a passageway for such a conductor, with provision for mounting on a barrier, conducting or otherwise, for the purpose of insulating the conductor from the barrier and conducting current from one side of the barrier to the other; **3.** Material, usually lead or babbitt, that sometimes serves as a lining for the hole in a grinding wheel; **4.** Metal cylinder between a shaft and its support. *See also* **Split bushing.**

BUSHING FITTING *See* **Fitting.**

BUSINESS AND PERSONAL SERVICES OCCUPANCY Occupancy or use of a building or any part for the transaction of business or the rendering or receiving of professional or personal services.

BUSINESS INTERRUPTION INSURANCE *See* **Insurance.**

BUSINESS PLAN Plan identifying markets, customers, expenditures, and finances required to carry out the identified 'business' based on projected revenues and costs over a specific period of time.

BUSINESS TAX Local or municipal tax, based on property assessment, levied upon the occupant of real property who conducts a business from that property.

BUSTER Air tool, used to 'bust' heads off rivets.

BUS WIRE *See* **Electric blasting circuitry.**

but. *Abbreviation for:* button; buttress.

BUTADIENE Petroleum derivative capable of forming polymers such as artificial rubber.

BUTANE Petroleum hydrocarbon compound with a boiling point of approximately 0°C (32°F). It is often combined with propane and sometimes referred to as liquid petroleum gas. *See also* **Commercial butane.**

BUTT 1. Literally to join end-to-end, but more commonly meaning to meet or touch with one end (either end-to-end or end-to-side or -face). *Also called* **Abut;**

2. Lower 2 m (6 ft 6 in.) of the stem of a tree. *See also* **Basal; 3.** *See* **Wood chisel.**

BUTT CUT Sloping cut made in existing pavement by a cold planer/milling machine. These cuts can be made parallel or perpendicular to a roadway. *Also called* **Conform, Header cut,** and **Taper.** *See also* **Butt log.**

BUTTE Steep-sided formation with a nearly flat top (usually igneous intrusions that have been exposed to erosion).

BUTTERED JOINT A very thin mortar joint made by scraping a small quantity of mortar with the trowel on all joining edges of the brick and laying it without the usual mortar bed.

BUTTERFLY DAMPER *See* **Damper.**

BUTTERFLY HOOD Where vehicle engine access is gained by folding up the sides of the hood. *See also* **Alligator hood.**

BUTTERFLY ROOF Two shed roofs connected at their lower edges.

BUTTERFLY VALVE Damper or throttle valve in a pipe, consisting of a disk that rotates about its diameter as an axis.

BUTTERING 1. Placing mortar on a masonry unit with a trowel; **2.** Spreading of a bond coat (followed by a mortar coat, a thin-setting bed mortar, or an organic adhesive) to the backs of ceramic tile just before the tile is placed; **3.** Process by which the interior of a concrete mixer, transportation unit, or other item coming in contact with fresh concrete is provided with a mortar coating so that the fresh concrete coming in contact with it will not be depleted of mortar.

BUTTERING TROWEL Trowel with a blade that is 114 mm (4.5 in.) wide and 178 mm (7 in.) long.

BUTTERY *See* **Texture.**

BUTT GAUGE Marking device, used to mark the position of hinges.

BUTT-HINGE Door hinge screwed to two faces that butt together.

BUTT HOOK *See* **Bull hook.**

BUTTING UP Two logs or timbers placed next to each other end-to-end.

BUTT JOINT 1. Any joint made by fastening two members together without overlapping. *Also called* **End joint; 2.** Junction where the ends of two timbers, pipes, or other materials meet in a square-cut joint; **3.** Joint having opposing faces that may move toward or away from each other.

BUTT LOG First log cut above the stump. *Also called* **Butt cut.**

BUTT OF A PILE 1. Larger head end of a tapered pile; **2.** The upper end of a pile as driven.

BUTT OFF 1. To cut off a piece of a log because of a defect; **2.** To square the end of a log.

BUTTON 1. Small bar, turning on a central fastener, used to fasten doors; **2.** Part of a weld that is torn out in destructive testing of a weld.

BUTTON-BACKED TILE Ceramic tile having raised dots or squares on the back. The protrusions serve to separate the tiles in the kiln during manufacture, ensuring a more uniform distribution of heat.

BUTTON-BOTTOM PILE *See* **Pile.**

BUTTON-HEADED SCREW *See* **Half-round screw.**

BUTT PLATE End plate of a structural member.

BUTTRESS 1. Ridge of wood that develops in the angle between a lateral root and the butt of a tree, that may extend up to a considerable height; **2.** Support built against or projecting from a wall to give extra strength and to conduct stresses in a predetermined direction. Usually of the following forms:

> **Angle:** Two buttresses meeting at an angle of 90° at the angle of a building.
>
> **Clasping:** One that encases the angle of a building.
>
> **Diagonal:** One that meets the angle of a building at an enclosed angle of 135°.
>
> **Flying:** Type of masonry structure in which a detached buttress or pier at a distance from the wall is connected to the wall by an arch or portion of an arch.
>
> **Hanging:** Buttress not rising from the ground, but supported on a corbel.
>
> **Setback:** A buttress set slightly back from the angle of a building.

BUTTRESS DAM Dam supported on its downstream side by a series of buttresses.

BUTTRESS DRAIN Stone- or rock-filled trench drain constructed in steps down the surface of a soil slope, used to assist in removing water from the soil slope and to provide weighty buttresses capable of resisting any local tendency to slip.

BUTTRESS FILL Compacted fill placed so as to prevent movement or sliding of an earthwork or foundation.

BUTTRESS SCREW THREAD Screw thread designed to carry a heavy axial load in one direction only.

BUTT RIGGING System of swivels, chain-like links, shackles, and bull hooks that connect the haulback and mainline and to which chokers are fastened.

BUTT ROT Decay or rot characteristically confined to the base or lower bole of a tree.

BUTT SEAM Joint of two panels that do not interlock.

BUTT STRAP Metal plate covering a butt joint and connecting the two members by bolting, welding, or riveting.

BUTT SWELL Unusually large flare of the tree trunk above the stump.

BUTT VENEER Veneer marked by distinctive curly figuring, caused by roots coming at all angles to the trunk.

BUTT WELD Weld where two pieces are butted together and fused.

BUTT-WELDED SPLICE Reinforcing bar splice made by welding the butted ends.

BUTT-WELD STEEL PIPE *See* **Pipe.**

BUTYL ACETATE Slow-working lacquer solvent.

BUTYL RUBBER Synthetic compound used for tire manufacture made from 98% to 99% isobutylene and 2% to 1% isoprene. Noted for excellent weather and chemical resistance and vibration absorption.

BUTYL STEARATE Colorless, oleaginous, and practically odorless material used as an admixture for concrete to provide dampproofing.

BUY-BACK AGREEMENT Contractual provision in which a seller agrees to repurchase a property at a stated price upon occurrence of a stated event within a specific period.

BUYER'S MARKET Condition where the supply of a product or service exceeds the demand. *See also* **Seller's market.**

BUYER'S RISK Probability of accepting poor or unsuitable material or construction as a result of using a particular acceptance plan. *See also* **Seller's risk.**

BUY-OUT INSURANCE *See* **Insurance.**

buz. *Abbreviation for:* buzzer.

BUZZ-OUT Removal of damaged material prior to the repair of a tire.

BUZZER Electric signaling device.

bv *Abbreviation for:* butterfly valve.

BV *Abbreviation for:* back view.

bw *Abbreviation for:* both ways.

b/w *Abbreviation for:* back window

BW *Abbreviation for:* butt weld.

BW DIMENSION Distance between the dual rear wheels from the outside of the outer wheels.

Bwg *Abbreviation for:* Birmingham wire gauge.

bx *Abbreviation for:* box.

BX CABLE *See* **Armored cable.**

bxd *Abbreviation for:* boxed.

bxg *Abbreviation for:* boxing.

BYATT Horizontal timber supporting decking and walkways in trench excavations.

byd *Abbreviation for:* beyond.

BYE CHANNEL Spillway leading water around a reservoir when it is full.

BYLAW Local law, made by an organized community for application within its boundaries.

byp. *Abbreviation for:* bypass.

byp. vlv. *Abbreviation for:* bypass valve.

BYPASS 1. Secondary passage of fluid, air or electrical flow, in addition to the main flow path; **2.** On an engine's cooling system, a small coolant passage that routes coolant from one side of the water pump to the other side of the pump, external to the pumping system. This equalizes pressure within the system and prevents the suction of the pump from collapsing the coolant hoses when the thermostat has not yet opened; **3.** On a lubrication system, a spring-loaded valve that opens to allow lubricant to flow freely through the system if the filters should become so clogged as to prevent flow, or when lubricant is so cold it cannot yet pass through the filter elements; **4.** On the injection system, the passages through which fuel can pass and return to the fuel storage tank, on a continuous basis, without being injected into the engine; **5.** Arterial highway that permits traffic to avoid part or all of an urban area.

BYPASS CHANNEL Man-made channel or conduit designed to carry excessive flows from a stream.

BYPASS FILTER 1. Filter that removes contaminants from a fluid by bypassing a percentage of the flow through its filter medium; **2.** Filter incorporating a ball valve which, when closed, isolates the filter from the circuit into which it is coupled, permitting media exchange or other service to be completed without closing down the circuit; **3.** Component that allows the proper hydraulic flow to continue when the system's filter becomes sufficiently clogged to inhibit the normal flow of filtered hydraulic oil.

BYPASS HIGHWAY *See* **Highway.**

BYPASS INDICATOR *See* **Indicator.**

BYPASS OIL FILTER *See* **Partial flow filter.**

BYPASS SLIDING DOOR Two or more doors hung on parallel tracks that stack when all units are opened, and that form a solid, interlocking closure when all units are closed.

BYPASS SWITCH *See* **Switch.**

BYPASS VALVE *See* **Valve.**

BY-PRODUCT COKE OVEN GAS Gas given off during the process of making coke from coal. It consists chiefly of hydrogen and methane.

C

C 1. Thermal conductance. C is similar to k, but applies to the actual thickness of a material; it is a measure of the rate of heat flow for the thickness of a material for a given area at one degree F difference between the inner and outer surfaces; **2.** *See* **Veneer grade.**

c *Abbreviation for:* candle; cathode; channel; clear; cycle.

C *Abbreviation for:* hundred; carbon; Celsius; Centigrade.

°C *Abbreviation for:* degree Celsius.

ca *Abbreviation for:* cab-to-axle.

ca. *Abbreviation for:* cable; calcium; centarc.

CA *Abbreviation for:* certificate of approval.

CAB 1. Decorative room in which people ride in a passenger elevator; **2.** Part of a vehicle that encloses the driver and vehicle operating controls. There are several types:

Cab-beside-engine: Truck or tractor with the driver's compartment and controls located beside the engine.

Cab-over-engine: Where the cab is located high enough to be entirely over the engine. This puts the driver at the extreme front of the cab and allows the shortest possible cab for maximum payload capacity. The design provides good engine accessibility as the cab can be tilted forward to expose the engine and engine-related components.

Conventional cab: Where the engine, steering gear, hood, and front fenders are all located ahead of the firewall (the separating panel between engine compartment and cab interior).

Low forward entry: Forward control truck chassis that has a tilt cab. *Also called* **Low cab forward.**

Short conventional: When compared to the normal conventional, the short conventional has a shortened hood and front fenders.

Tilt: Vehicle designed with its engine beneath a cab having provision for being tilted forward on a pivot near the front bumper to provide easy access.

cab. *Abbreviation for:* cabinet.

CABANA Change room; specifically, as available to an outdoor swimming pool.

CAB-BESIDE-ENGINE *See* **Cab.**

CAB CONTROL LEVERS Handles to control the action of power-take-off and hydraulic valves, usually located in the truck cab convenient to the driver's hand.

CAB EXTENDERS Aerodynamic-improvement devices for on-highway tractor applications; used in conjunction with a full-roof fairing system.

CAB FORWARD *See* **Forward control.**

CAB GUARD *See* **Canopy.**

CABINET Built-in, or free-standing item of furniture with shelves, drawers (sometimes both), and enclosed with doors.

CABINET MAKER Worker skilled in the manufacture of fine furniture.

CABINET PROJECTION *See* **Orthographic projection.**

CABINET SCRAPER Hand tool made from a piece of flat steel, used for final smoothing of surfaces prior to sandpapering.

CABINETWORK 1. Finely-finished woodwork, whether as furniture or as part of the building; **2.** Interior woodwork, the product of a joiner or woodworking plant.

CABIN HOOK Hook-and-eye fastener.

CABLE 1. Group of tendons used in prestressed concrete; **2.** Term loosely applied to wire rope, wire strand, and electrical conductors. *See also* **Armored cable; 3.** Electrical conductor.

CABLE CHAIN Chain that operates over a smooth sheave (as opposed to a roller chain, that operates over a sprocket).

CABLE CLAMP *See* **Cable clip.**

CABLE CLIP U-bolt used to clamp parts of a cable together. *Also called* **Cable clamps** or **Crosbie clip.**

CABLE CONNECTOR Length of insulated cable terminating at each end in a casting or a lug and used to connect one cell of a storage battery to another. *See also* **Connector,** and **Terminal connector.**

CABLE CONTROL UNIT High-speed tractor winch having one to three drums, each under separate control, used to operate nonhydraulic dozer and towed equipment.

CABLE CUTTER Hydraulic, mechanical (guillotine), or powered abrasive wheel tool for cutting wire rope.

CABLE DRILL *See* **Drill, churn.**

CABLE EXCAVATOR Cable-operated excavation machine that works between a mast and an anchor or between a head tower and a foot tower.

CABLE FILLER Material used in multiple-conductor cable that occupies the spaces formed by the assembly of the insulated conductors.

CABLE FOLLOWER Device that maintains the proper lay of wire rope on a hoist drum, preventing the tangling of wire rope and increasing its life.

CABLE HOOK Round hook with a wide beveled face. *See also* **Grab hook, Grab, Pintle hook, Round hook, Safety hook,** and **Swivel hook.**

CABLE JACKET Outside cover of an electrical cable.

CABLE-LAID WIRE ROPE 1. Type of wire rope consisting of several wire ropes laid into a single wire rope; **2.** Wire rope made up of six wire ropes wrapped around a fiber or wire rope core.

CABLE-LAID-WIRE-ROPE ENDLESS SLING 1. Cable-laid-wire-rope sling made endless by joining the ends with one or more mechanical fittings; **2.** Cable-laid-wire-rope sling made from one length of rope wrapped six times around a core with the ends hand-tucked inside the wraps; **3.** Wire-rope sling made from cable-laid rope with eyes fabricated by pressing or swaging one or more metal sleeves over the rope junction.

CABLE LOGGING Yarding system employing winches, blocks, and cables.

CABLE MOLDING Molding that imitates a twisted rope.

CABLE PROFILE Profile of the stressed tendons in a concrete element.

CABLE PULLOUT UNLOADING METHOD Procedure in which a landfill tractor assists in the unloading of a transfer trailer by pulling a cable network especially placed within the transfer van for this purpose.

CABLE SAG Track-cable sag at center span of a cableway.

CABLE SHEATH Protective covering applied to cables.

CABLE SKIDDER *See* **Harvesting machines (single function).**

CABLE SOCK Double-helical cylindrical wire cage fitted over the end of a prestressing tendon permitting it to be drawn through the preformed duct in a concrete member.

CABLE SPREADER Device used to separate traveling cables for hanging.

CABLE-SYNCHRONIZED BOOM Fully synchronized boom actuated by one control lever, one telescopic cylinder, and the employment of a wire rope system for extension and retraction of the fly section.

CABLE TOOL DRILLING Vertical earth or rock drilling technique that employs a full string of components including: drill bit, drill stem, drilling jars, swivel socket, and a cable that is used to raise and lower the assembly, imparting a crushing action.

CABLE TOOL DRILLING LINE The wire rope used to operate the cutting tools in the cable tool drilling method (i.e., rope drilling).

CABLE WRAP Amount of the drive sheave actually in contact with the cable.

CABLE YARDING Taking logs from the stump area to a landing using an overhead system of winch-driven cables to which logs are attached with chokers.

CABLE ZONE Limiting zone along the length of a prestressed concrete beam within which a cable must be positioned.

CABLING Infilling of the flutes of a column by shapes resembling cylinders.

CAB-OVER-ENGINE *See* **Cab.**

CAB PROTECTOR *See* **Canopy.**

CAB SHIELD *See* **Canopy.**

CAD *Acronym for:* computer-aided design; computer-aided drafting.

CADASTRAL SURVEY Survey establishing the lengths and directions of boundary lines and the area they enclose; a survey establishing such lines on the ground.

CADD *Acronym for:* computer-aided design and drafting.

CA DIMENSION Distance from the back of the cab to the centerline of the rear axle or bogie.

CADMIUM PLATING Type of zinc plating used to rust-proof metals.

CAFETERIA Self-service restaurant.

CAGE 1. Rigid assembly of reinforcement ready for placing in position; **2.** Circular frame that limits the sideways motion of balls or rollers in a bearing; **3.** Enclosed platform used to transport workers and light materials between storys of a building; **4.** *See also* **Lead, offshore.**

CAGE LEAD *See* **Lead.**

CAGE SCREEN Filtration screen in the form of a drum, hinged so that it may be raised or lowered for cleaning.

CAGING Wood or metal framing around pipes, columns, beams, and other irregular shapes to provide support for drywall.

CAIRN Mound of stones used as a marker or monument.

CAISSON Structural chamber used to keep soil and water from entering into a deep excavation or construction area.

CAISSON BOX Caisson closed at the bottom but open at the top.

CAISSON DISEASE Decompression sickness manifest by cramping, induced by too-rapid decease of environmental air pressure after a stay in a compressed atmosphere, as in a caisson. *Also called* **Bends.**

CAISSON FOUNDATION Foundation system where holes are drilled into bearing strata, through earth or below water, and then filled with concrete.

CAISSON HAMMER *See* **Pile hammer.**

CAISSON PILE *See* **Pile.**

CAISSON WEIGHTS Steel ingots or concrete blocks, used to help sink a caisson or as pile test weights. *See also* **Kentledge,** and **Load-test weights.**

ca.j *Abbreviation for:* caulked joint.

CAKE Layer of clay or clayey soil built up on the wall of a boring filled with slurry (drilling mud, bentonite, etc.), having the effect of forming an impermeable lining to prevent (or diminish) loss of water from the hole. *Also called* **Filter cake.**

cal. *Abbreviation for:* calibrate; calibrated; calorie.

CALAMINE DOOR *See* **Door types.**

calc. *Abbreviation for:* calculate.

CALCAREOUS Containing calcium carbonate, or less generally, containing the element calcium.

CALCIMINE White or colored lime wash used to wash ceiling and other interior plaster.

CALCINE To alter the composition or physical state by heating below the temperature of fusion.

CALCINED GYPSUM Gypsum partially dehydrated by heat.

CALCITE Mineral having the composition calcium carbonate and a specific crystal structure; the principal constituent of limestone, chalk, and marble; used as a major constituent in the manufacture of portland cement.

CALCIUM Silver-white metallic element of the alkaline-earth group, occurring only in combination with other elements.

CALCIUM-ALUMINATE CEMENT Product obtained by pulverizing clinker, consisting essentially of hydraulic calcium aluminates that results from fusing or sintering a suitably proportioned mixture of aluminous and calcareous materials. *Also called* **Aluminate cement, Aluminous cement,** and **High-alumina cement.**

CALCIUM CARBONATE Naturally occuring calcite, used for making lime and portland cement.

CALCIUM CHLORIDE Crystalline solid; in various grades, used as a drying agent, as an accelerator of concrete, a deicing chemical, and for other purposes. *See also* **Admixture, accelerating.**

CALCIUM CHLORIDE STABILIZATION Incorporation of calcium chloride to aggregate mixtures prepared for road base and wearing surfaces, and for shoulders, so as to control the moisture content during construction and throughout the useful life of the construction.

CALCIUM HYDROXIDE *See* **Hydrated lime.**

CALCIUM OXIDE Quicklime.

CALCIUM-SILICATE BRICK *See* **Brick.**

CALCIUM-SILICATE HYDRATE Any of the various reaction products of calcium silicate and water, often produced by autoclave curing. *See also* **Dicalcium silicate,** and **Tricalcium silicate.**

CALCIUM STEARATE Product of the reaction of lime and stearic acid, used as an integral water repellent in concrete.

CALENDAR DAY Any day shown on a calendar, beginning and ending at midnight. *See also* **Day,** and **Working day.**

CALENDER Machine equipped with three or more heavy, internally heated or cooled rolls, revolving in opposite directions, that is used for continuously sheeting or plying up rubber compound, or frictioning or coating fabric with rubber compound.

CALENDER CUT Cut or break in sheet roofing material caused by a fold in the sheet before it passes through the calender.

CALF DOZER Small bulldozer.

CALF'S TONGUE MOLDING Molding with a series of pointed, tongue-like projections in line in relief against a plane or curved surface.

CALIBER Bore or internal diameter of a cylinder or pipe.

CALIBRATE To check and adjust the graduations of a measuring instrument.

CALIBRATED SAND METHOD Weight of a cubic measure of soil, in place.

CALIBRATION Precise adjustment against, or to an established measure or standard.

CALIBRATOR Person or thing that calibrates.

CALICHE Gravel, sand, and desert debris cemented by calcium carbonate or other salts.

CALIFORNIA BEARING RATIO Ratio of the force per unit area required to penetrate a soil mass with a 193.5 mm² (3 in.²) circular piston at the rate of 1.27 mm (0.05 in.) per min to the force required for corresponding penetration of a standard material. The ratio is usually determined at 0.1 in. penetration.

CALIFORNIA SWITCH Portable platform that rides on a track in a tunnel, used for passing cars and trains.

CALIGNUM Liquid plastic product, used as a wood filler.

CALIPER 1. Instrument comprising a pair of curved legs fastened together at one end, used to measure the internal or external diameter or thickness of an object or part; **2.** Diameter of the trunk of a tree, normally measured at a height of 300 mm (12 in.) above grade, **3.** An optical instrument that incorporates two parallel lines of sight separated by a variable baseline; **4.** Wire loosly held within a flexible armored sleeve, frequently used to remotely actuate a device.

CALIPER DISC BRAKE *See* **Brake.**

CALK Fish-tailed bar built into masonry.

CALL FOR BIDS Request to submit a price for the completion of specified work.

CALL FOR PROPOSALS Invitation to qualified bidders to complete a project in which the bidder may be requested to include proposals for structural systems, construction details, financing, project timing, etc. *Also called* **Invitation for proposal call,** and **Proposal call.**

CALL FOR TENDERS Invitation to qualified contractors and/or suppliers to meet the requirements of a proposal.

CALLICHE *See* **Material density.**

CALLING THE PILE Indicating that a pile has reached specified penetration and/or blow count; that the pile is 'home.'

CALL OPTION *See* **Option.**

CALLOUT Note on a drawing with a leader to the feature.

CALLUS Tissue that develops around a wound on the stem or a root of a tree.

CALORIE Obsolete unit of energy, equal to 4.19 joules, that should not be used with the SI system of measurement. Symbol: cal. *See also the appendix:* **Metric and nonmetric measurement.**

CALORIFIC VALUE Heat, measured in calories or Btu, released by combustion of a unit quantity of fuel.

CALORIFIER Sealed tank in which water is heated, usually by a heat exchanger in the form of a submerged coil carrying steam or hot water.

CALORIMETER Instrument for measuring heat exchange during a chemical reaction, typically as the quantities of heat liberated by the combustion of a fuel or hydration of a cement.

CALOROMETRIC VALUE Indication of the mass of organic compounds present in fine aggregate.

CALYX Core barrel without hard-metal cutting teeth, with which rock is cut or ground up by chilled steel shot that roll and are ground up under the steel edge of the rotating barrel. *Also called* a **Shot barrel.**

CAM 1. Irregular surface on a rotating device, designed to convert rotary motion into linear motion in an engine, to work the intake and exhaust valves or to actuate the plunger(s) on an injector system; or a sliding piece with an irregular surface, used to impart exactly timed motion to adjoining parts. *See also* **Camshaft; 2.** Portion of a lock that is turned by the key or when the bolt knob is turned.

CAM (BASE CIRCLE) Zero reference axis of any camshaft or the zero lift portion of a cam lobe.

CAMBER 1. Slight convexity above the horizontal plane; in a beam, truss, or deck, to allow for self weight plus imposed load; to a road or travelled surface, the amount of rise between the crown and one perimeter; **2.** Lean from a vertical line of the front wheels of a motorized vehicle (outward lean on the top of a wheel is positive camber).

CAMBER ARCH An arch with a flat horizontal extrados, and a cambered intrados with a rise of about 1.0%.

CAMBER BEAM Beam cambered on its upper face.

CAMBER ROD Tensioning rod below a trussed beam.

CAMBER SLIP Curved wooden member temporarily used in the centers of a flat brick arch to give the soffit a slight midspan camber above the springing line.

CAMBIUM One-cell-thick layer of tissue, between the bark and wood of a tree, that repeatedly subdivides to form new wood and bark cells. *See also* **Phloem.**

CAM BRAKE *See* **Brake.**

CAM BUSHINGS Bearing surface used to support a camshaft.

CAM DURATION Amount of time (expressed in crankshaft degrees) during which a cam holds a valve open.

CAME Shaped metal strip used in stained glass work.

CAMEL-BACK BODY Truck body with the floor curving downward at the rear.

CAMEL-BACK RIDGE Paired knolls of near equal size that occur along a ridge line.

CAM FOLLOWER Tube-shaped device that rides on the camshaft lobes in a reciprocating engine and provides a place to install push rods.

CAM LIFT Amount of valve movement caused by a cam lobe at its point of maximum movement.

CAMOUFLET Blasthole made larger by chambering.

CAM OVERLAP Period of engine rotation (expressed in crankshaft degrees) during which the camshaft is holding both the intake and exhaust valves open.

CAMPANILE Bell tower separate from the main building.

CAMSHAFT Shaft of which one or more cams are an essential part. *See also* **Cam.**

CAMSHAFT JOURNAL Round machined part on a camshaft that supports it in the engine block.

CAMSHAFT RUN-OUT Variance from true of a camshaft, indicated when a camshaft is rotated while supported on V-blocks and measured by a dial indicator.

can. *Abbreviation for:* canal.

CANAL Channel built to carry water and, in some cases, vessels.

CANAL TRIMMER Equipment that performs the final shaping of the bottom and sides of an earthen canal in preparation for placing a waterproof membrane.

CANARY Iron rod, 2 m (6 ft 6 in.) long with a hook on one end and a handle on the other, used to pull cable or chain under bundles of pulpwood logs for binding or hookup.

canc. *Abbreviation for:* cancel.

CANCELLATION CLAUSE Provision in a contract, typically a lease, which confers upon one or more or all of the signators the right to terminate their obligations under defined conditions and/or circumstances.

cand. *Abbreviation for:* candelabrum.

C&B *Abbreviation for:* cost and budget.

CANDELA One of the seven base units of the SI system of measurement: it is the luminous intensity, in the perpendicular direction, of a surface of 1/600 000 m² of a black body (full radiator) at the temperature of freezing platinum under a pressure of 101.325 kPa. Symbol: cd. Derived units are illumination lux (symbol: lx), and luminous flux lumen (symbol lm). *See also the appendix:* **Metric and nonmetric measurement.**

CANDELA PER SQUARE METER A derived unit of luminance with a compound name of the SI system of measurement. Symbol: cd/m². *See also the appendix:* **Metric and nonmetric measurement.**

CANDLEPOWER Measure of luminous intensity, based on the lighting capacity of a standard candle.

C&P *Abbreviation for:* cord and pulley.

CANDY SIDE 1. Well-equipped, smoothly operated, and efficient logging show; **2.** Site with big logs or flat ground; therefore one that is easy to work.

CANISTER STAKE Cylindrical anvil with a flat end, used in sheet metal work.

CANKER Area of dead tissue in a woody stem, caused by fungi, bacteria, or dwarf misletoe, marked by sloughing of the tissue that leaves an open wound surrounded by a zone of callus.

CANKER FACE Area of dead tissue in a canker or a tree stem.

CAN LEAD *See* **Lead.**

CANNIBALIZE Remove part of a machine or installation for use elsewhere. *Also called* **Rob.**

CANOPY 1. Roof-like structure over an opening in an exterior wall; 2. Sheet metal projection to the forward section of a dump body and extending over the truck cab to provide protection from falling debris and to enable payloads to be piled high towards the front of the vehicle for extra capacity and better weight distribution. *Also called* **Cab guard, Cab protector,** and **Cab shield;** 3. Forest cover of branches and foliage formed by tree crowns; 4. Portion of an elevator cab, located above and supported by the walls, containing the ceiling and enclosing the top of the cab.

CANOPY GUARD *See* **Overhead guard.**

CANSIM (Canadian Socio-Economic Information Management System) Fully computerized data bank of Statistics Canada's current and historical information from its most widely used series of statistics.

CANT 1. Log that is squared on two or more sides. *See also* **Flitch;** 2. Set at a slope or slant from the horizontal or vertical; 3. Molding made up of plain surfaces and angles.

CANT BAY Three-sided window having splayed side lights.

CANT BRICK Brick with a return face at other than 90°.

CANT DOG Short, stubby peavey.

CANTED WALL Wall joining another at an angle.

CANT HOOK Stout wooden lever used in rolling logs. Differs from a peavey in that it has no spike in the end of the stock.

cantil. *Abbreviation for:* cantilever.

CANTILEVER 1. Self-supporting projection without external bracing; 2. Beam longer than its breadth and more than twice its depth projecting from a wall or pier; 3. Structural member that projects beyond the line of its support.

CANTILEVER ABUTMENT Cantilever wall supporting the end of a bridge above it and backfilled material behind it.

CANTILEVER ARM Cantilever bridge overhanging from the support into the central span, carrying one end of the suspended span.

CANTILEVER BRIDGE Bridge comprising two or more cantilevers whose projecting ends meet but do not support each other.

CANTILEVER FOOTING Combined footing with individual column footings joined by a beam and the footing under the boundary or external column made slightly larger.

CANTILEVER FORMWORK Climbing forms.

CANTILEVER OUTRIGGER *See* **Outrigger.**

CANTILEVER TRUCK Self-loading, counterbalanced or noncounterbalanced industrial truck equipped with cantilever load engaging means, such as forks.

CANTILEVER T-WALL Reinforced cantilever retaining wall in the shape of an inverted tee.

CANTILEVER WALL 1. Sheet pile wall stabilized by the length of penetration on the free side; 2. Retaining wall stabilized by the mass of material retained.

CANTING TABLE Saw table that can be tilted about the arbor (as against tilting the arbor within the table).

CANTLEDGE Applied load: to give stability to a crane, provide a reaction to a jack, push down a plate in a plate bearing test, etc. *See also* **Kentledge.**

CANT STRIP 1. Projecting molding near the bottom of a wall to direct rain away from the foundation; 2. Wedge or triangular-shaped piece of lumber installed in the deck of a flat roof around the perimeter or at the junction of the roof and an adjoining wall. *See also* **Chamfer strip.**

canv. *Abbreviation for:* canvas.

CANYON TECHNIQUE *See* **Sanitary landfill.**

CAP 1. Detonator initiated by a burning fuse or electric current; 2. Smooth, plane surface of suitable material bonded to the bearing surfaces of test specimens to insure uniform distribution of a load during strength testing; 3. Pipe plug with female threads; 4. Roof or top piece in a three-piece timber set used for tunnel support; 5. Fitted or threaded piece to protect the top of a pile while being driven; 6. Block, strip or other covering, plain or molded, forming the top of a wall, pier, newel post, partition, or column; 7. Upper half of the top plate in wood-frame walls and partitions; 8. Cone of sheet iron on steel, with a hole in the apex through which a chain passes, fitted over the end of a log to prevent catching on stumps in skidding; 9. Layer of clay or other highly impermeable material installed over the top of a closed landfill to prevent entry of rainwater and minimize production of a leachate.

cap. *Abbreviation for:* capacity.

CAP *Acronym for:* clean air package.

CAPACITANCE Property of any system or device to store electrical potential energy.

CAPACITOR Device capable of storing electrical energy consisting of two conducting surfaces separated by an insulating material.

CAPACITOR-DISCHARGE IGNITION CONTROL Built-in engine speed governor designed to help prevent engine failure from over-revving.

CAPACITY 1. Maximum or rated volume or load-carrying limit; 2. Maximum load in pounds, or the maximum load in pounds at a given load center, that an industrial truck can safely transport and/or stack to a specified height; 3. Measure of the rated volume of a particular mixer or agitator; 4. Maximum number of vehicles that have a reasonable expectation of passing over a given section of lane or a roadway in one direction or in both directions for a two-lane or a three-lane highway, during a given time period under prevailing roadway and traffic conditions. *See also* **Design capacity; 5.** Maximum amount of insurance that a carrier will write on a single risk; **6.** Maximum ability of a resource or organization to provide goods or services. There are several ways to measure this, including:

Ideal, or theoretical: Maximum level of production, making no allowance for interruptions.

Idle: Portion of capacity that is not used to its maximum, due to lack of demand, errors in planning, etc.

Normal: Maximum level of production under normal circumstances.

Practical capacity: Maximum level at which a plant or organization realistically operates most efficiently.

CAPACITY CHART 1. Plate or plates on a crane that give rated lifting capacities for the unit under different load conditions and setups; **2.** Chart on an excavator that gives lifting capacities for different load heights and radii.

CAPACITY CURVE 1. Graphic, two-dimensional presentation of the data representing some aspect of performance; **2.** Graph showing the volume contained in a storage vessel at any given water level.

CAPACITY FACTOR Ratio of the average load on a machine or equipment for the period of time considered, to the capacity rating of the machine or equipment.

CAPACITY INSULATION Ability of masonry to store heat as a result of its mass, density, and specific heat.

CAPACITY REDUCER Device that permits an adjustment of the output capacity of an air compressor without alteration to any other operating condition.

CAPACITY REDUCTION FACTOR *See* **Phi factor.**

CAP AND BASE CONSTRUCTION Type of tire tread construction in which the cap or antiskid compound differs from the base or subtread compound.

CAP AND PLUMB POST Used with deteriorated pile bents, usually timber; existing piles are cut off at good material, a cap placed across them, and support built up to the structure.

CAP BLOCK *See* **Drive cap, and McDermid plate.**

CAP CABLE Short cable (tendons) introduced to a concrete member to prestress the zone of negative bending only.

CAP DELAY Electric blasting cap that explodes at a set interval after passage of an electric current.

CAPILLARITY Movement of a liquid in the interstices of a porous material due to surface tension.

CAPILLARY FLOW Flow of moisture through a capillary pore system.

CAPILLARY FRINGE Zone immediately above the water table where some or all of the interstices are filled with water at a pressure less than atmosphere.

CAPILLARY GROOVE Space between two surfaces, large enough to prevent capillary movement of water; a water break.

CAPILLARY MOVEMENT Movement of groundwater as a result of capillary attraction.

CAPILLARY PRESSURE Internal pressure that helps earth to stand in an excavation being drained from outside the excavated area. *Also called* **Seepage force.**

CAPILLARY SPACE Void resembling a microscopic channel small enough to draw liquid through it by the molecular attraction of the water adsorbed on its inner surface (capillarity).

CAPILLARY STRESS Pore water pressure less than atmospheric value produced by the surface tension of pore water acting on the meniscus formed in void spaces between soil particles.

CAPILLARY TUBE Tube with a very small bore that conducts a small quantity of liquid, under pressure, to a readout. Typical system components are: a capillary tube and dial gauge for an oil pressure readout.

CAPILLARY WATER Subsurface water held above the water table by capillary attraction. *See also* **Absorbed water, and Gravitational water.**

CAPITAL 1. Head of a column, pier, pilaster, etc., widened for decorative purposes or to distribute loads; **2.** Plant, equipment, and related facilities used to produce a flow of goods and services; **3.** Owner's interest in the assets of a business or project.

CAPITAL BUDGET *See* **Budget.**

CAPITAL EXPENDITURE Improvement that is not a repair and that will have a life of more than one year.

CAPITAL GAIN Profit on the sale or disposition of a capital asset.

CAPITALIZED VALUE Value based on present worth plus anticipated future benefits.

CAPITAL LEASE *See* **Lease.**

CAPITAL LOSS Loss on sale or disposition of a capital asset.

CAP MATERIAL *See* **Cushion block.**

CAPPED END Hose end covered to protect its internal elements.

CAPPING 1. Finishing or crowning piece on built-up work; **2.** Anything that closes, seals, stops, or crowns.

CAPPING BEAM 1. Beam along the top of a sheet pile wall that aids in the distribution of pressure; **2.** Beam along the top of a row of single piles to tie them longitudinally.

CAPPING BRICK *See* **Brick.**

CAPPING PIECE Horizontal timber over the ends of two walings butted together that takes the thrust of a strut, transferring it to the walings.

CAP ROCK Harder and more impervious rock sometimes found above more weathered and softer rock.

CAP SHEETS Sheets of heavy roofing felt coated on both sides with asphalt and on one side with mineral granules, mica, or other organic materials, used granule-side-up as a capping layer in a built-up roofing construction.

CAPSTAN *See* **Winch.**

CAPSTONE Stone that caps, seals, and covers the top

of a load-bearing and face wall of an elevation. A crowning stone of a structure; a coping.

CAPSULE ANCHOR Steel bolt securely bonded into concrete over its embedment length by the interaction of a polyester resin and hardening agent.

CAPTIVE NUT *See* **Nut.**

CAPTURE Water that wholly or in part replaces that withdrawn from an aquifer.

CAPWAP *Acronym for:* case pile wave analysis program.

CAR ANNUNCIATOR Device that proclaims, audibly and/or visually, the landing at which an elevator car is standing at when at rest.

carb. *Abbreviation for:* carburetor; carburize.

CARBIDE *See* **Tungsten carbide.**

CARBIDE-TIPPED BIT *See* **Bit.**

CARBODY Section of crawler crane base mounting that carries the superstructure and to which the crawler side frames are attached.

CARBOLOY (Trademark) Hard compound of particles of tungsten carbide bonded in a matrix of cobalt, used for dies, cutting tools, and wearing surfaces.

CARBON-ARC WELDING Electric arc welding with one electrode a carbon rod, the other the piece being welded, which is grounded.

CARBONATION Reaction between carbon dioxide and a hydroxide or oxide to form carbonate, especially in cement paste, mortar, or concrete.

CARBONATION SHRINKAGE Shrinkage resulting from carbonation.

CARBON BLACK Finely divided amorphous carbon characterized by a high oil absorption and low specific gravity, produced by burning natural gas in a supply of air insufficient for complete combustion.

CARBON DIOXIDE One of the products of combustion.

CARBON FIBER Fibers of carbon used as reinforcement for cement paste and concrete.

CARBONITE Explosive mixture composed of nitrobenzene, potassium nitrite, sulfur, and kieselguhr.

CARBONIZATION Reduction of hydrocarbons resulting in the formation of carbonaceous residue.

CARBON MONOXIDE Colorless, odorless and poisonous gas; a by-product of the burning of carbon or carbon-based fuels, as in an internal combustion engine.

CARBON RESIDUE 1. Free carbon that remains when an oil is evaporated; **2.** Measure of the coke-forming tendency of oils at high temperatures.

CARBON STEEL *See* **Steel.**

CARBORUNDUM GRIT Hard fine grit troweled into the surface of a concrete floor to produce a nonslip surface.

CARBURATED BLUE (WATER) GAS Water gas with gaseous hydrocarbons added, originally for the purpose of increasing the flame's luminosity, but now added to increase the Btu content of the water gas.

CARBURIZE Increase in the surface hardness and abrasion resistance of iron or steel parts by heating the metal to a high temperature in an environment rich in carbon.

CARCASS 1. Completed, self-supporting principal structure of a building; **2.** Fabric, cord, and/or metal reinforcing section of a hose, as distinguished from the hose tube or cover.

CAR COUNTERWEIGHT Weights roped directly to an elevator car that, on drum-type installations, will equal the car weight.

CARDONIC LEAD *See* **Lead.**

CAR DOOR/GATE CLOSER Device that automatically closes a hand-opened elevator car door.

CARE, CUSTODY, AND CONTROL Exclusion, standard in liability insurance, where the liability does not apply to property in the care or custody of the insured or over which he has physical control.

CAR ENCLOSURE Top and walls of an elevator car, resting on and attached to the platform.

CAR FRAME Supporting frame to which an elevator car platform, upper and lower sets of guide shoes, car safety and hoisting ropes or hoisting-rope sheaves, or the plunger of a direct-plunger elevator are attached. *See also* **Overslung car frame, Subpost car frame,** and **Underslung car frame.**

CARGO Freight carried by a vehicle.

CARGO CONTROL *See* **Tiedown assemblies.**

CARGO WEIGHT RATING Value specified by the manufacturer as the cargo-carrying capacity of a machine or vehicle, exclusive of the weight of the occupants.

CAROLITHIC COLUMN Column with a foliated shaft.

carp. *Abbreviation for:* carpenter.

CAR PASSER Portable tracks used for switching cars in a tunnel.

CARPENTER One who builds and repairs wooden articles or the wooden parts of buildings.

CARPENTER ANT Large ant that lives in hollowed-out spaces in wood, but which does not consume large quantities of cellulose.

CARPENTER'S BRACKET SCAFFOLD Supported scaffold consisting of a platform supported by brackets attached to building or structural walls.

CARPENTER'S HAMMER *See* **Hammer.**

CARPENTRY WORK Work done by tradesmen in cutting, framing, and joining lumber in construction.

CARPET COURSE *See* **Hot-mix seal coat.**

CARPET STRIP *See* **Molding.**

CAR PLATFORM Structure that forms the floor of an elevator car and that directly supports the load.

CARPORT Shelter for a car, adjacent to or attached to a dwelling, roofed but not completely enclosed.

carr *Abbreviation for:* carrier.

carr. *Abbreviation for:* carriage; carried.

CARREL Enclosed space, usually with a desk and shelving, for individual study in a library.

CARRIAGE 1. Sliding or rolling base or supporting structure; **2.** Support structure for forks or attachments, generally roller mounted, traveling vertically within the mast of a cantilever truck; **3.** Wheeled device that rides on a skyline, used for hauling logs; **4.** Frame and wheel sheaves that are pulled along the track cable of a cableway by the conveying line; **5.** Support for the steps of a stairway.

CARRIAGE BACKREST EXTENSION *See* **Load backrest extension.**

CARRIAGE BOLT 1. Bolt intended for wood-to-wood use, but which also can be used for fastening wood to metal; **2.** Bolt with a round head, machine-threaded along only a portion of its shaft.

CARRIAGE CABLE 1. Cable for traveling a carriage on a cable; **2.** Traction cable.

CARRIAGES An inclusive term for the supports, steps and risers for a flight of stairs.

CARRIAGEWAY Traveled portion of a road.

CARRIER 1. Rotating or sliding mounting or case; **2.** portion of a crane located below the turntable bearing; **3.** Wheeled chassis that is the base mounting for mobile truck and rough terrain cranes; **4.** Removable drive-axle assembly consisting of the carrier housing, ring gear, pinion, differential, and related parts.

CARRIER CAB Housing that covers the driver's station on the carrier of a truck crane.

CARRIER CAP Part that holds the bearings and adjusting rings of an axle assembly in correct location; supports the main differential, located on the housing. Comes as a set with the axle housing.

CARRIER FRAME Main structure of the carrier section of a crane.

CARRIER OR CRAWLER MOUNTING Structure forming the lowest element of a mobile crane, transmitting loads to the ground.

CARRIER ROLLER Rollers or track mechanism that are not power driven but which are used to guide the track along the top of the side frame.

CARRYING CAPACITY Electrical load that a cable or fuse can carry without overload.

CARRYING CHANNEL Principal support in a suspended ceiling system.

CARRYING CHARGE Cost associated with the ownership (as distinct from the acquisition) of property.

CARRYING CONTAINER *See* **Waste container.**

CARRYOUT COLLECTION *See* **Collection.**

CARRY POSITION For bucket-equipped mobile mechanical equipment, the vertical distance from the point of contact with the ground of the nearest wheel or track to the center line of the bucket hinge pin, with the angle of approach at 15°.

cartog. *Abbreviation for:* cartographer; cartographic.

CARTOGRAPHER One who prepares charts or maps.

CARTOON Full-size drawing, worked up from a preliminary sketch, from which details are transferred to a working surface.

CARTRIDGE 1. Wrapped stick of dynamite or other explosive; **2.** *See* **Element.**

CARTRIDGE ELEMENT Porous device that performs the actual process of filtration.

CARYATID Supporting column in the shape of a draped female figure.

CAS *Acronym for:* clean air system.

CASE 1. Framework of a structure; **2.** Box in which a lock-operating mechanism is contained; **3.** Surface of steel. *See also* **Case hardening.**

CASED Covered or enclosed with other materials, usually of higher quality.

CASED PILE Pile consisting of a permanent casing or shell that is subsequently filled with concrete.

CASE DRAIN LINE Line connecting fluid from a component housing to the reservoir.

CASE HARDENING 1. Adding chemicals to the surface of a metal to make its surface (case) tougher. Carburizing is one method of case hardening. However, nitriding and other methods also are case hardening methods; **2.** Stressed condition in a board or timber characterized by compression in the outer layers accompanied by tension in the center of the core, the result of too severe drying conditions.

CASEIN Component of emulsion paint.

CASEIN GLUE *See* **Adhesive.**

CASEMENT Hinged part of a window, attached to the upright side of the window-frame.

CASEMENT WINDOW *See* **Window.**

CASE METHOD Method that electronically determines static soil resistance that uses a field analyzer, which processes measurements of force and acceleration from the pile top. *Also called* **Pile dynamics.**

CASE PILE WAVE ANALYSIS PROGRAM Computer wave equation program that determines the magnitude and distribution of dynamic and static resistance along a pile from field measured strain and acceleration (force and velocity).

CASE WARD Obstruction integral to the case of a warded lock.

CASH ALLOWANCE Sum included in a contract to cover the cost of prescribed items not fully detailed, the amount subject to change orders based on actual expenditures.

CASH BUDGET *See* **Budget.**

CASH DISCOUNT *See* **Discount.**

CASH FLOW Difference between cash receipts and cash expenditures over a given period.

CASH METHOD Accounting procedure that recognizes revenues and expenses only when cash is received or paid. *See also* **Accrual method.**

CASING 1. Finished wood that is installed around a post or beam; **2.** *See* **Molding;** **3.** Open-end steel pipe installed by drilling, driving, or vibrating; **4.** Support to the wall of a hole; **5.** Pipe dropped or driven into an augered hole to prevent caving temporarily; **6.** Tire structure, less tread and sidewall rubber.

CASING BEAD *See* **Molding.**

CASING BLOWS Blows, usually of a 136 kg (300 lb) hammer falling 450 mm (18 in.) on to a soil sampler casing while making a soil boring.

CASING FACTOR Load carried by the casing only of a tire, not including the load carried by the air pressure.

CASING NAIL *See* **Nail.**

CASING OFF Technique of pile installation consisting of driving an open-ended tube of larger internal diameter than the finished pile, excavating the material from within the tube, then installing the pile through the tube.

CASING REEL Winch on a churn drill used for handling casing and miscellaneous chores. *See also* **Reel.**

CASING SPIDER Frame and wedge set that supports the top of a casing string while new sections are added.

CAST To form in a mold.

cast. *Abbreviation for:* casting.

CASTABLE REFRACTORY Packaged, dry mixture of hydraulic cement, generally calcium-aluminate cement, and specially selected and proportioned refractory aggregates which, when mixed with water, will produce refractory concrete or mortar.

CASTELLATED Having turrets or battlements.

CASTER 1. Angle that a kingpin makes with a vertical line when viewed from the side; **2.** Placing of an axle to locate the center of weight either ahead or behind the ground contact point of a tire to provide easier steering; **3.** Swiveled wheel for mounting on movable objects. May contain a braking system for industrial applications.

CASTER FRAME Wheel-guiding frame with a swivel connection to the machine or vehicle that rests on it. *Also called* a **Fork head.**

CASTING 1. Fabrication of masonry panels involving the combining of masonry units and grout in a form; **2.** Metallic form produced by pouring molten metal into a shaped container or cavity.

CASTING PLASTER Fast-setting gypsum plaster used to anchor marble to walls, set spots, or mix temporary 'hot mud.'

CASTING YARD Area where precast concrete members are fabricated and allowed to cure before use.

CAST-IN-PLACE Mortar or concrete that is deposited in the place where it is required to harden as part of the structure.

CAST-IN-PLACE CONCRETE *See In situ* concrete.

CAST-IN-PLACE PILE *See* **Pile.**

CAST IRON Iron containing more than 1.7% carbon and poured into molds while molten. *See also* **Gray cast iron, Iron, Malleable cast iron,** and **White cast iron.**

CAST IRON PIPE *See* **Pipe.**

CASTLE NUT *See* **Nut.**

CASTOR OIL Natural vegetable vehicle used in paints where it is important to retain color purity, in the manufacture of alkyd resin paints, and as a plasticizer.

CAST SPOKE Type of wheel consisting of a one-piece hub-and-spoke casting to which a brake drum is bolted; demountable rims are used with this type of wheel.

CAST STEEL Steel that has not been forged or rolled after casting.

CAST STONE Concrete or mortar cast into blocks or small slabs in special molds so as to resemble natural building stone.

CAST-WELD ASSEMBLY Cast parts fixed in an assembly by welding.

CASUALTY INSURANCE Insurance against losses caused by injuries to persons or their property.

cat. *Abbreviation for:* catalog; category.

CATALYST Substance that initiates a chemical reaction and enables it to proceed under milder conditions than otherwise required and that does not, itself, alter or enter into the reaction.

CATALYTICALLY BLOWN ASPHALT *See* **Asphalt.**

CATALYTIC COMBUSTION SYSTEM Process in which a substance is introduced into an exhaust gas stream to oxidize vaporized hydrocarbons or odorous contaminants; the substance itself remains intact.

CATAMARAN Wooden float with sides, for carrying wood piles to a floating driver.

CATASTROPHE Sudden and severe calamity: for insurance purposes, an event that causes a loss of extraordinary size.

CATCH Light-duty latch.

CATCH BASIN Receptacle connected with a sewer or drain tile into which water from a roof, floor, etc., will drain.

CATCHER Worker who catches hot rivets thrown by the heater.

CATCH FEEDER Subsidiary waterway in a catchment area.

CATCHING CAN Cone-shaped can to catch rivets.

CATCHMENT AREA Watershed or drainage area.

CATCH PIT Accessible receptacle in a drainage system in which grit and other debris is deposited.

CATCH POINT *See* **Hinge point.**

CATENARY Curve made by a flexible line hung between two points at the same elevation.

CATENARY SCAFFOLD Suspension scaffold consisting of a platform fastened to two essentially horizontal and parallel wire ropes, that are secured to structural members.

CATENATE Link together.

CAT EYE Blemish on or near the surface of drawn glass, usually with a leading and trailing tail, one or both of which are open to the surface.

CAT FACE 1. Scarred tree trunk with no bark on it, caused by internal rot or damage; **2.** Blemish or rough depression in a finish plaster coat caused by variations in the base coat thickness.

CATHEAD 1. Notched wedge placed between two formwork members meeting at an oblique angle; **2.** Spindle on a hoist; **3.** Large, round retention nut used on she bolts; **4.** *See* **Winch**; **5.** Light frame and sheave at the top of a material tower through which the lifting cable is operated; **6.** Rotating power-driven spool or winch head. *See also* **Driller's stroke.**

CATHEDRAL Large, imposing church.

CATHEDRAL CEILING 1. Ceiling formed by the finished underside of the roof. *Also called* an **Open roof; 2.** Ceiling higher than that of adjacent rooms, often sloped upward from two parallel walls to a ridge.

CATHEDRAL GLASS Obscure sheet glass patterned to resemble the indentations made by hammer blows.

CATHODE Electrical term for the negative terminal.

CATHODIC PROTECTION Form of corrosion protection wherein one metal is caused to corrode in preference to another.

CAT HOOK Normally, the hook connected by a shackle to the eye of the mainline.

CATIONIC EMULSION Emulsified asphalt in which the asphalt globules are electro-positively charged. There are several formulations, including:

  CMS: Cationic medium setting.

  CRS: Cationic rapid setting.

  CSS: Cationic slow setting.

CAT LADDER Board with cleats fastened to it to give firm foothold when the board is laid between two surfaces at different elevations. *Also called* a **Duckboard,** and **Roof ladder.**

CAT LINE Fire line developed by a bulldozer or scraper.

CAT RUN Member, 50 x 100 mm (2 x 4 in.), across the bottom chord of roof trusses, connecting them to each other.

CAT'S EYE Glass or plastic prism set in the road surface or on posts alongsie the carriageway designed to reflect headlight illumination back to the driver of a vehicle to advise or warn of a position or condition: the centerline of the road or the perimeter of a curve, for instance.

CATSKINNER Operator of a crawler-tractor.

CAT'S-PAW 1. Simple, nonslipping knot used on fiber or wire rope, where the line is run through an eye and looped back on itself to make a quick connection; **2.** Hand tool consisting of two curved prongs at the end of a stout metal shaft, used for removing nails. *See also* **Nail claw.**

CATTLE PASS Large-diameter culvert under a road or highway, used to pass livestock from one side to the other.

catw. *Abbreviation for:* catwalk.

CATWALK Narrow elevated walkway.

CAUDIL DRIVE POINT PILE *See* **Pile:**

CAUL Flat metal plate on which wood particles are formed into mats, conveyed, and pressed during particleboard manufacture.

CAULK 1. Waterproof sealant used to fill joints or seams; available as putties, ropes, or compounds extruded from cartridges; **2.** To fill a joint with mastic, usually under pressure.

CAULKING 1. Process of filling in cracks or cavities and expansion joints using an elastic material, usually from a caulking gun; **2.** Material used for joint sealing where minor or no elastomeric properties are required.

CAULKING COMPOUND Soft, plastic material consisting of pigment and a vehicle, used for sealing joints in buildings and other structures where normal structural movement may occur. Such compounds retain their plasticity for extended periods after application by gun or knife, or in the form of preformed extruded shapes.

CAULKING GUN Hand- or power-operated tool that holds prepackaged cartridges of caulk.

CAULKING RECESS Space between the bell of one piece of pipe and the spigot of the pipe length to which it is joined that is filled with caulking to seal the joint.

CAULKING TOOL Tool used for driving a caulking compound into seams and crevices to make joints air- and watertight.

CAULKS Short spikes driven or screwed into the soles of boots to give footing while walking on wood or logs.

CAULLESS SYSTEM Particleboard manufacturing process in which particle mats are formed and conveyed on moving flexible plastic sheets and then pressed directly between the press platens without the use of caul plates.

CAUSEWAY Raised way or road, usually across wet or marshy ground, or through shallow water.

caust. *Abbreviation for:* caustic.

CAUSTIC Typically, a liquid that is the opposite of an acid in terms of pH measurement. Caustic liquids are often used to etch materials that are acid resistant. Caustics have a lesser effect on ferrous materials, and are often used in cleaning processes for engine castings, forgings, and formed parts.

CAUSTIC SODA Chemical that removes old paint from masonry (also known as sodium hydroxide, or lye).

cav. *Abbreviation for:* cavity.

caveat emptor Let the buyer beware; the buyer must examine the goods or property and buy at his/her own risk.

CAVE-IN Sudden collapse of a trench or excavation wall.

CAVETTO *See* **Molding.**

CAVING SOIL Soil that tends to fall into an uncased or unshored hole or excavation.

CAVITATION 1. Starvation of pumps created by a blockage or restriction that prevents the oil supply from reaching the pumps. The extreme low-pressure vacuum developed can cause pieces of metal to be pulled from parts of the pump; **2.** Formation of a void due to reduced pressure in lubricating grease dispensing systems.

CAVITATION DAMAGE Pitting of concrete caused by implosion, i.e., the collapse of vapor bubbles in flowing water that form in areas of low pressure and collapse as they enter areas of higher pressure. *See also* **Erosion.**

CAVITY Female portion of a two-piece matching mold.

CAVITY BLOCK Hollow precast concrete masonry unit that, when laid, forms a cavity wall.

CAVITY FLASHING Damp course crossing the gap of a cavity wall.

CAVITY WALL *See* **Wall.**

cb *Abbreviation for:* cab-to-body.

CB *Abbreviation for:* cast brass; catch basin; cement box; circuit breaker; covered base.

cbal. *Abbreviation for:* counterbalance.

cblmn *Abbreviation for:* cableman.

cbn *Abbreviation for:* cabin.

cbore *Abbreviation for:* counterbore.

CBR *Abbreviation for:* California bearing ratio.

C/B RATIO Ratio of the weight of water absorbed by a masonry unit during immersion in cold water to the weight absorbed during immersion in boiling water. The result is an indication of the probable resistance of a brick to freezing and thawing. *Also called* **Saturation coefficient.**

cc *Abbreviation for:* cubic centimeter.

CC *Abbreviation for:* cast copper; color code.

C-CLAMP *See* **Clamp.**

CCN *Abbreviation for:* contract change notification.

C-C PLUGGED EXTERIOR Exterior-type touch-sanded plywood panel with C-plugged-grade face, C-grade back and inner plies, bonded with exterior glue. Commonly used for extreme moisture conditions.

cc/STROKES The measurement in cubic centimeters of fuel delivered for a specified number of fuel injector strokes.

cct *Abbreviation for:* circuit.

CCTV *Abbreviation for:* closed-circuit television.

CCW. *Abbreviation for:* counterclockwise.

cd *Abbreviation for:* cord.

cd. *Abbreviation for:* candela; code.

CD *Abbreviation for:* center distance; clothes dryer; cold drawn.

CDD *Abbreviation for:* critical degree of deformation.

CDD AND D INSURANCE Comprehensive dishonesty, disappearance, and destruction insurance coverage against employee dishonesty, loss of money and securities, and theft of materials and equipment.

C-D PLUGGED Touch-sanded plywood panel with C-plugged-grade face, D-grade back and inner plies, bonded with interior or exterior glue. Used for built-

ins, cable reels, and walkways.

cdrill *Abbreviation for:* counterdrill.

CDS *Abbreviation for:* cold-drawn steel.

ce *Abbreviation for:* cab to end-of-frame.

CE *Abbreviation for:* circular error; control equipment.

CE DIMENSION Distance from the back of the cab to the end of the frame.

ceil. *Abbreviation for:* ceiling.

CEILING Construction covering the underside of a floor, roof, or other horizontal interior surface.

CEILING BEAMS Beams supported by the lower chords, spanning between trusses and supporting the ceiling construction.

CEILING JOISTS Horizontal wood members, usually 50 mm (2 in.) nominal thickness, used for support.

CEILING MORTAR Extra-rich wall mortar.

CEILING OUTLET Electrical outlet for a ceiling lighting fixture.

CEL *Abbreviation for:* center east line.

CELITE Calcium aluminoferrite constituent of portland cement.

CELL 1. Hollow space in a building tile or block; **2.** Galvanic cell for the generation of electric energy in which the cell, after being discharged, may be restored to a fully charged condition by an electric current flowing in a direction opposite to the flow of current when the cell discharges; **3.** General term for the microscopic units of wood structure, including wood fibers, vessel segments, etc.

CELLAR That portion of a building between two floor levels that is partly or wholly underground and that has more than one-half of its height, from finished floor to finished ceiling, below grade.

CELL CONNECTOR An electric conductor used for carrying current between adjacent storage cells. *See also* **Connector.**

CELL HEIGHT Vertical distance between the top and bottom of compacted solid waste enclosed by natural soil or cover material in a sanitary landfill.

CELL-TYPE INCINERATOR *See* **Incinerator.**

CELLULAR ABUTMENT Hollow cellular mass, or reinforced concrete abutment, used when very low, even-bearing pressures are required and/or where pining can be avoided.

CELLULAR BLOCK Masonry block having uniformly distributed pores throughout its mass.

CELLULAR COFFERDAM Cofferdam enclosed by a wall consisting of a series of filled cells. *See also* **Cofferdam,** and **Crib.**

CELLULAR CONCRETE Lightweight product consisting of portland cement, cement silica, cement-pozzolan, lime-pozzolan, or lime-silica pastes, or pastes containing blends of these ingredients, and having a homogeneous void or cell structure, attained with gas-forming chemicals or foaming agents (for cellular concretes containing binder ingredients other than, or in addition to, portland cement, autoclave curing is usually employed). *See also* **Aerated concrete, Foamed concrete,** and **Preformed foam.**

CELLULAR CONSTRUCTION Concrete elements in which part of the interior volume is made up of voids.

CELLULAR RAFT Concrete raft in which the the intersecting beams form a number of cells.

CELLULAR STEEL DECK Structural steel floor system comprising two layers of metal sheet shaped to form cells and welded together.

CELLULOSE 1. Carbohydrate that is the primary constituent of wood and that forms the framework of wood cells; **2.** First modern plastic; a mixture of solid camphor and nitrocellulose under heat and pressure.

CELLULOSE ADHESIVE *See* **Adhesive.**

CELLULOSE DEBRIS Fine wood debris that is often evidence of the presence of wood borers and termites.

CELLULOSE SHEET Composite floor panel made from cork dust, sawdust, wood flour, and pigments, with gelatinized nitrocellulose on a backing of jute.

CELLULOSIC Natural high-polymeric carbohydrate found in the fibrous matter of woody plants.

CELSIUS Modern name for the centigrade scale on which water boils at 100° and water freezes at 0°. *See also* **Centigrade.**

cem. *Abbreviation for:* cement; cementery.

cem. ab. *Abbreviation for:* cement-asbestos board.

cem. & cu. *Abbreviation for:* cement and cure.

CEMENT 1. Mixture of alumina, silica, lime, iron oxide, and magnesia that is burned in a kiln, then pulverized to a fine powder. It is used for making concrete or mortar. *See also* **Hydraulic cement; 2.** Unvulcanized raw or compounded rubber in a suitable solvent, used as an adhesive or sealant.

CEMENT-AGGREGATE RATIO Ratio of cement to total aggregate, either by mass or volume.

CEMENTATION Binding together of the particles of a sedimentary rock by such mineral cements as calcite, silica, and iron oxide.

CEMENTATION PROCESS Process of injecting cement grout under pressure into certain types of ground (e.g., gravel, fractured rock) to solidify it.

CEMENT BACILLUS *See* **Ettringite.**

CEMENT-BODY TILES Tiles with the body made from a mixture of sand and portland cement. The surface may be finished with portland cement, spheroids of marble, or other materials.

CEMENT-BOUND MACADAM Road consisting of a base of broken stone, crushed slag or gravel, and either a grout or mortar filler, formed by rolling the base to a compacted mass having an even surface, and then rolling in the cementitious filler.

CEMENT-COATED NAILS Nails coated with cement to increase their holding power.

CEMENT CONTENT Quantity of cement contained in a unit volume of concrete or mortar, preferably expressed as weight. *Also called* **Cement factor.**

CEMENTED END Hose end sealed with the application of a liquid coating.

CEMENTED SOIL *See* **Soil types.**

CEMENT FACTOR *See* **Cement content.**

CEMENT GEL Colloidal material that makes up the major portion of the porous mass of which mature hydrated cement paste is composed. *See also* **Gel.**

CEMENT GROUT Mortar or cement mixed with water and sand to the consistency of thick cream; used for bedding bearing plates, setting anchor bolts, and filling and smoothing foundation cracks.

CEMENT GUN Machine for the pneumatic placement of mortar or small-aggregate concrete. In the 'dry gun,' water from a separate hose meets the dry material at the nozzle of the gun. With the 'wet gun,' the delivery hose conveys the premixed mortar or concrete. *See also* **Shotcrete.**

CEMENTITIOUS Having cementing properties.

CEMENTITIOUS MATERIALS Cements and pozzolans used in concrete and masonry construction. *See also* **Blast-furnace slag, Hydraulic cement, Masonry cement,** and **Mortar.**

CEMENT KILN A kiln in which ground and proportioned raw mix is dried, calcined, and burned into clinker at a temperature of 1420°C to 1650°C (2,600°F to 3,000°F).

CEMENT LATEX Flexible, jointless floor-surface compound consisting mainly of portland cement, rubber latex, and color pigment together with some decorative granular material.

CEMENT-LIME CONCRETE Concrete to which a small amount of lime has been added to increase its workability.

CEMENT-MODIFIED SOIL *See* **Soil stabilization.**

CEMENT MORTAR Mortar in which the cementitious material is primarily portland cement.

CEMENT PAINT Paint consisting generally of white portland cement and water, pigments, hydrated lime, water repellents, or hygroscopic salts.

CEMENT PASTE Constituent of concrete consisting of cement and water. *See also* **Neat cement paste.**

CEMENT PLASTER *See* **Plaster** and **Stucco.**

CEMENT ROCK Natural impure limestone that contains the ingredients for production of portland cement in approximately the required proportions.

CEMENT SCREED Layer of cement mortar laid on a concrete slab.

CEMENT SILO Tall, usually cylindrical, airtight tower for the storage of cement powder, equipped with a means for dispensing measured quantities at its lower end.

CEMENT SLURRY Liquid cement-water mix, used as a wash or for injection.

CEMENT STABILIZATION *See* **Soil cement,** and **Soil-stabilization.**

CEMENT WASH Solution of cement in water, used

for painting exposed concrete surfaces.

cem. fin. *Abbreviation for:* cement finish.

cem. fl. *Abbreviation for:* cement floor.

cem. mort. *Abbreviation for:* cement mortar.

cem. pl. *Abbreviation for:* cement plaster.

cen. *Abbreviation for:* center; central.

CENTER 1. The axial member of a strand about which the wires are laid; **2.** Inner ply or plies of a plywood panel whose grain runs parallel with that of the face and back plies.

CENTER BEARING 1. On a mixer truck, usually a rubber-encased, self-centering bearing used to provide support to the mixer driveshaft(s) at intermediate points along its routing; **2.** Self-centering, rubber-cased bearing used to support a long driveshaft at an intermediate point in the driveline; suspended by a bracket mounted to a crossmember.

CENTER CONTROL TRUCK Powered industrial truck in which the operator's position is located near the longitudinal center of the truck.

CENTER FIRING Technique of broadcast burning, used to create a strong updraft, in which fires are set in the center of an area followed by additional fires being set progressively nearer the outer control lines as indraft builds to draw them into the center.

CENTER-FRAME PLATFORM *See* **Trailer.**

CENTER GAP *See* **Core gap.**

CENTER-HOLE LAPPING Cleaning or lapping of center holes with a bonded abrasive wheel cemented onto a steel spindle.

CENTERING Framework used in arch and similar construction that is later removed. *See also* **Formwork.**

CENTERING RING Extension of a gasket for the purpose of locating it centrally on a flange.

CENTERLESS GRINDING Grinding the outside or inside diameter of a round piece not mounted on the center.

CENTERLINE 1. Line on a drawing about which the elements are balanced; **2.** Midway between the guardrails of a bridge; the final survey line of a bridge or road indicating the middle of the final structure; **3.** Inked line or indentation in the center of the tread rubber of a tire that aids in positioning.

CENTERLINE MARKING *See* **Pavement marking.**

CENTERLINE OF ROTATION *See* **Axis of rotation.**

CENTER MATCHED Tongue-and-groove lumber with the tongue and groove at the center of the piece rather than offset (as in standard matched). *See also* **Standard matched.**

CENTER OF GRAVITY Point in a body about which the weights of the various parts balance; point within a machine around which its weight is evenly distributed. The three measurements necessary to determine the center of gravity of a vehicle are:

> **Horizontal:** Measured fore-and-aft from a refer-

ence plane.

> **Lateral:** Measured from the center line of the vehicle to the side.

> **Vertical:** Measured up and down from a reference plane.

CENTER OF MASS Cross-sectional line of a cut or fill that divides the volume into halves.

CENTER OF MOMENTS 1. Point at which a body tends to rotate; **2.** Point, arbitrarily selected, for determining the resultant moment of a series of forces.

CENTER OF PRESSURE Point, or an area, subjected to hydraulic or pneumatic pressure over which the whole force due to pressure is taken to act.

CENTER-OPENING DOOR *See* **Door types.**

CENTER PIN Fixed vertical shaft in revolving machinery around which the machinery deck rotates.

CENTER PLATE Plate that is connected to the flywheel or clutch cover (depending on clutch design) and that separates the driven disks, increasing clutch capacity by providing additional surface friction area between the driven disks.

CENTER POINT DESIGN Where the vertical centerline of a king pin and the vertical centerline of a wheel intersect at the loaded tire at ground level.

CENTER PUNCH *See* **Punch.**

CENTERS 1. Distance measured between two or more adjacent blastholes, without reference to hole locations as to row; **2.** Conical steel pin(s) of a machine upon which the work is centered and rotated during grinding, turning, polishing, etc.

CENTER STAKES Stakes indicating the centerline of construction.

CENTER-TO-CENTER Spacing of elements or components, as in studs for frame construction being on 400 mm (16-in.) centers. *See also* **On center.**

CENTI Prefix representing 10^{-2}. Symbol: c. Used in the SI system of measurement. *See also the appendix:* **Metric and nonmetric measurement.**

CENTIGRADE Scale of temperature that features 0° and 100° as the freezing and boiling point of water, respectively. *See also* **Celsius.**

CENTIGRAM Unit of mass, equal to 1/100 gram. Symbol: cg. Used in the SI system of measurement. *See also the appendix:* **Metric and nonmetric measurement.**

CENTILITER Unit of volume, equal to 1/100 liter (0.6102 in.³) Symbol: cL. Used in the SI system of measurement. *See also the appendix:* **Metric and nonmetric measurement.**

CENTIMETER Unit of length, equal to 1/100 meter (0.3937 in.). Symbol: cm. Used in the SI system of measurement. *See also the appendix:* **Metric and nonmetric measurement.**

CENTIPOISE Unit of absolute (dynamic) viscosity.

CENTISTOKE Unit of kinematic viscosity.

CENTRAL AIR CONDITIONER Air conditioning system that, from a single, central location, serves an entire structure, or a complex of rooms and spaces.

CENTRAL BUSINESS DISTRICT Nonspecific area containing a community's principal retail, commercial, governmental, and service functions. Commonly called 'Downtown.'

CENTRAL HEATING Heating system in which a number of rooms or spaces are heated from a central source.

CENTRAL INCINERATOR *See* **Incinerator.**

CENTRALITY Part that is centered relative to another part.

CENTRALIZER Device that lines up a drill steel or string between the mast and the hole.

CENTRAL LUBRICATION System whereby a number of remotely located points may be lubricated from one location, either manually as required, or automatically at preset intervals.

CENTRAL-MIXED CONCRETE Concrete that is completely mixed in a stationary mixer from which it is transported to point of use.

CENTRAL MIXER Stationary concrete mixer from which the freshly mixed concrete is transported.

CENTRAL VACUUM SYSTEM Vacuum cleaning plant permanently installed in a building, connected to all floors through a network of pipes. *Also called* a **Vacuum cleaning plant.**

CENTRIC LOAD An axial load.

CENTRIFUGAL AIR CLEANER Precleaning system in an engine that removes large dust particles.

CENTRIFUGAL ATOMIZING OIL BURNER Burner in which oil is thrown by centrifugal force from a rotating cup into an air stream, causing the oil to break into a spray.

CENTRIFUGAL BLOWER Supercharger or turbocharger device that uses a fan to compress air. (NOTE: on older, more constant speed engines, centrifugal blowers were often gear driven by the engine. All modern engines use exhaust-driven centrifugal blowers called **Turbocharger.**)

CENTRIFUGAL BRAKE *See* **Brake.**

CENTRIFUGAL CLUTCH *See* **Clutch.**

CENTRIFUGAL COMPRESSOR Air compressor employing a rotary, as against a reciprocal, action.

CENTRIFUGAL FAN *See* **Fan.**

CENTRIFUGAL FORCE 1. Outward force exerted by a body moving through a curve; **2.** Force generated by eccentric weight(s) rotating at a specific frequency inside the drum of a vibratory compactor.

CENTRIFUGAL GOVERNOR Engine speed-governing device that depends upon the action of centrifugal flyweights to monitor and control rpm.

CENTRIFUGALLY CAST CONCRETE *See* **Centrifugal process,** and **Spun concrete.**

CENTRIFUGAL PROCESS Process for producing concrete products, such as pipe, that uses an outer form that is rotated about a horizontal axis and into which concrete is fed by a conveyor. *Also called* **Spinning process.** *See also* **Centrifugally cast concrete, Dry-cast process, Packerhead process, Spun concrete,**

Tamp process, and **Wet-cast process.**

CENTRIFUGAL PUMP *See* **Pump.**

CENTRIFUGAL PUMP CAPACITY Factor that varies directly as the speed or the diameter of the impeller.

CENTRIFUGAL PUMP HEAD Factor that varies as the square of the speed or as the square of the impeller diameter.

CENTRIFUGAL PUMP POWER Factor that varies as the cube of the speed or the cube of the impeller diameter.

CENTRIFUGAL SEPARATOR *See* **Separator.**

CENTRIFUGE In model pile testing, a device to approximate gravity-induced horizontal stresses to compensate for scale.

CENTRIFUGE KEROSENE EQUIVALENT Aggregate test used in the Hveem mix design method to estimate the required percentage of asphalt for a particular mix.

CENTRIFUGE MOISTURE EQUIVALENT Water content retained by a presaturated soil that has then been subjected to a force equal to 1,000 times gravity for one hour.

CENTRIFUGE VOLUME Volume of a liquid or solid, or both, separated from a volume of liquid exposed to centrifugal force.

CENTRIPETAL FORCE 1. Inward force exerted on a body to keep it moving in a curved line; **2.** Force exerted on a moving vehicle rounding a curve as a result of superelevation of the carriageway.

CENTROID Center of mass.

CENTROLINEAR PERSPECTIVE *See* **Linear perspective.**

CEP *Abbreviation for:* circular error probable.

cer. *Abbreviation for:* ceramic.

CERAMIC ADHESIVE *See* **Adhesive.**

CERAMIC BOND Development of fired strength as a result of thermochemical reactions between materials exposed to temperatures approaching the fusion point of the mixture.

CERAMIC COLOR GLAZE An opaque colored glaze of satin or gloss finish, obtained by spraying a clay body with a compound of metallic oxides, chemicals, and clays that is then burned to a high temperature, fusing the glaze to the body, making them inseparable.

CERAMIC FRICTION MATERIAL Metallic-based (ceramic and bronze) friction material found on disk drives.

CERAMIC MOSAIC TILE Unglazed tile formed by either the dust-glazed or plastic method, usually 6.4 to 9.5 mm (0.25 to 0.365 in.) thick, and having a facial area of less than 3872 mm^2 (6 in.^{2}) and that is usually mounted on sheets approximately 0.3 x 0.6 m (2 x 1 ft). Ceramic tile may be of either porcelain or natural clay composition and may be either plain or with an abrasive mixture throughout.

CERAMIC PROCESS Production of articles or coatings from essentially inorganic, nonmetallic materials, made permanent by heat at temperatures sufficient to cause sintering, solid-state reactions, bonding, or whole or partial conversion to the glassy state.

CERAMICS Science and art of clay working and various related industries. The use of vitrified bonds brings abrasive wheel manufacture under this classification.

CERAMIC TILE Vitreous clay tile used for surface finish, normally used where excessive exposure to moisture could occur.

CERAMIC VENEER Architectural terra-cotta, characterized by large face dimensions and thin sections.

cert. *Abbreviation for:* certificate; certified; certify.

CERTIFICATE FOR PAYMENT Certification by an authorized professional that work to a specific stage has been completed and that specific expenditures have been made, qualifying a contractor for an interim or final payment under the contract for the works.

CERTIFICATE OF COMPLIANCE Certification by an authorized authority that a completed project is in compliance with the appropriate regulations and requirements.

CERTIFICATE OF INSURANCE Document issued by an insurance company that verifies coverage.

CERTIFICATE OF OCCUPANCY Document issued by a competent authority certifying that a building complies with relevant public standards and permitting occupancy for its designated use.

CERTIFICATE OF TITLE Written legal opinion declaring that a title to land is vested as stated, following a review and examination of the abstracts or chains of title.

CERTIFICATION OF APPRAISER That part of an appraiser's report containing any limiting conditions, conclusions, and a signed certification as to the correctness of the conclusions.

CERTIORARI Review in a competent court of the rulings of an inferior court, tribunal board, or officer exercising judicial functions.

CESSPOOL Chamber below grade for collecting and holding disposal from house drains.

CETANE INDEX Prediction of the cetane number of diesel fuel based on physical property measurements.

CETANE NUMBER Comparison measurement of a fuel's volatility, the comparison being with the characteristics of cetane, one of the many fractions of crude petroleum. The higher the cetane value, the easier the fuel is to ignite. High cetane values are important in winter or cold climates. The cetane number is the volatility standard for diesel fuels, in the same way in which octane is the volatility standard (in this case, for antiknock qualities) for gasoline engine fuels.

cf. *Abbreviation for:* confirm.

CF *Abbreviation for:* centrifugal force; cold finish; cost and freight.

CFBK *Abbreviation for:* cut for bit key.

CFC *Abbreviation for:* cut for cylinder.

cfd *Abbreviation for:* confirmed.

cfi *Abbreviation for:* continuous forest inventory.

cfl *Abbreviation for:* continuous forest land.

cfm *Abbreviation for:* cubic feet per minute.

CFP *Abbreviation for:* concrete-filled pipe.

C-FRAME C- or U-shaped frame that connects the blade to a dozer.

cfs *Abbreviation for:* cubic feet per second.

CFT *Abbreviation for:* cut for turn knob.

CG *Abbreviation for:* center of gravity; coarse grain.

CGA DIMENSION Distance from the center of gravity of a truck body and payload to the center of the rear axle (midpoint between the axles for a tandem).

cgs *Abbreviation for:* centimeter/gram/second.

ch. *Abbreviation for:* chain.

CH *Abbreviation for:* case harden; coat hook.

CHAFER DUCK Duck of approximately square-woven construction made with single or ply yarn warp and filling.

CHAFER FABRIC Layer of fabric covering the bead of a tire to eliminate friction and wear between the bead and the rim.

CHAFFING FATIGUE Fatigue initiated in a surface damaged by rubbing against another body.

CHAIN 1. Flexible series of joined links, usually of metal. There are many types, including:

Breakaway: Safety chain that holds a vehicle and towed unit together if the regular fastening opens or fails.

Leaf: Silent chain designed for low-speed, heavy-duty work.

Logging: Chain made up of links of round bar pieces curved and welded to interlock, with a grab hook at one end and a round hook at the other.

Measuring: Continuous measuring tool consisting of a given number of links of known length, or a continuous ribbon of metal, fabric, metallized fabric, or plastic, marked in units of length from end to end.

Roller: Sprocket-driven chain made up of links connected by hinge-pins and sleeves.

Silent: Roller-type chain in which the sprockets are engaged by projections on the link side bars.

Stud-type: Roller chain in which the inner links are connected solidly by nonrotating bushings.

2. Tow line or drive belt made of interlocked links; **3.** Unit of length that measures 66 ft; **4.** 66-ft-, 100-ft-, 50-m-, or 75-m-long surveyor's steel measuring tape (formerly a chain comprised of 66 or 100 1-ft-long hinged chain links, *Also called* a **Gunter's chain**). *See also* **Engineer's chain**, and **Surveyor's chain.**

CHAINAGE Length measured by chain or steel tape, usually during a survey.

CHAIN ASSEMBLIES Chain with all hardware and coupling devices.

CHAINBLOCK Lifting tackle comprising an endless chain fed over the smaller of two wheels of different diameter but turning on the same shaft. *Also called* a **Chainfall, Chainhoist,** and **Differential pulley block.**

CHAIN BOND *See* **Bond.**

CHAIN BOOK Field record of measurements taken by a surveyor.

CHAIN BOX Box recessed in the floor of, or mounted under a trailer and designed for the storage of tire chains, load binders, and other accessories.

CHAIN BRAKE Manually- or inertia-activated device to stop the rotation of the chain of a chainsaw in a kickback situation.

CHAIN-BUCKET DREDGER Ladder-bucket dredging machine.

CHAIN BUCKET LOADER Mobile loader that employs a series of buckets on a chain to excavate and load material. *Also called* a **Bucket loader.**

CHAIN CATCHER Device that helps reduce the risk of operator contact by a chain when it breaks or comes off the bar of a chainsaw.

CHAIN-DRIVE MACHINE *See* **Driving machine.**

CHAINFALL *See* **Chainblock.**

CHAIN GRATE *See* **Grate.**

CHAIN GUARD Removable cover or scabbard that covers the bar and chain of a chainsaw when the saw is not in use.

CHAINHOIST *See* **Chainblock.**

CHAIN HOOK *See* **Binder.**

CHAINING 1. Survey term that originally meant measuring, using a Gunter's 66 ft or 100 ft chain but that is now used to denote measuring with either a chain or a tape; 2. Method of skidding pulpwood on short, steep slopes by wrapping a chain around several bunches of wood and dragging them crosswise down the grade.

CHAIN LINK FENCE Woven fence, normally made of steel wire, attached to posts and rails.

CHAINMAN Member of a survey party who carries the chain to the point to be measured.

CHAIN OF LOCKS Series of interconnected hydraulic locks on a waterway.

CHAIN PIPE VISE Portable vise that uses a clamped length of chain to hold and restrain a length of pipe being worked on. *Also called* **Chain tongs.**

CHAIN SAW *See* **Saw.**

CHAIN SHEAVE Sheave having a groove in the form of the links of the chain that runs over it.

CHAIN SLING Length of chain, supported at its ends, used to lift heavy objects.

CHAIN SPROCKET Toothed wheel that drives the chain of a chainsaw.

CHAIN SURVEY Survey technique that relies on the taking and plotting of lengths only; no angles are read, but are deduced from offsetting measurements.

CHAIN TONGS *See* **Chain pipe vise.**

CHAIR *See* **Bar support, Bat,** and **Slab bolster.**

CHAIR BRACKET Formed metal bracket used to attach guide rails to a structure. *Also called* a **Hat bracket.**

CHAIR-RAIL *See* **Molding.**

CHALET Small, often wooden house, used for recreation or as guest accommodation.

CHALK Soft limestone composed chiefly of the calcareous remains of marine organisms.

CHALKING 1. A common effect of weathering on paint in which the surface oils are deteriorated leaving loose color particles or powder. *Also called* **Frosting;** 2. Formation of a loose powder resulting from the disintegration of the surface of concrete or of an applied coating such as cement paint; 3. Formation of a powdery surface condition due to disintegration of surface binder or elastomer by weathering or other destructive environments. *Also called* **Frosting.**

CHALK LINE 1. Mark to indicate a straight line and for temporary reference; 2. String line covered in chalk powder that is stretched between reference points slightly above the surface to be marked. When the line is 'snapped' by raising the center vertically against the applied tension and letting go, the chalk powder is deposited on the surface below in a straight line. *Also called* **Snapline.**

CHALK TEST Method of locating cracks in a metallic vessel or member by first applying a penetrating liquid, wiping it off and then applying chalk or whiting. Penetrant seeping out of cracks into the whiting or chalk causes a distinct color change.

cham. *Abbreviation for:* chamber; chamfer.

CHAMBER 1. Enclosed space; 2. Private room, one for personal use.

CHAMBERING Process of enlarging a portion of a blasthole, usually the bottom, by firing a series of small explosive charges. *Also called* **Springing.**

CHAMFER Bevel symmetrically applied to an edge, usually at an angle of 45°; called a hollow or concave chamfer when the surface is concave.

CHAMFERED *See* **Rustication.**

CHAMFER STOP Ending of a chamfer, usually decorated.

CHAMFER STRIP Either a triangular or curved insert, placed in an inside form corner to produce either a rounded or flat chamfer or to form a dummy joint. *See also* **Cant strip, Fillet,** and **Skewback.**

chan. *Abbreviation for:* channel.

CHANCE 1. Logging unit, such as a timber sale or a specific drainage area; 2. Any unit of operation in the woods that has particular reference to its topographical characteristics.

CHANDELIER Ornate, branching light fixture suspended from a ceiling or roof.

CHANGED CONDITIONS Job conditions that differ substantially from those represented in the plans and specifications and/or the contract documents.

CHANGE HOUSE Building or room in which workers working in tunnels or in constantly adverse conditions change into protective clothing, dry and store such clothing and special equipment.

CHANGE IN THE WORK Modification to the work as described in the contract documents.

CHANGE KEY Key cut to operate only one lock in a series, all of which may be operated by a master key.

CHANGE OF STATE Change from one phase to another: solid to liquid, liquid to gas, etc.

CHANGE ORDER Form issued by the architect (or owner's agent) to a construction contractor, after execution of the agreement, authorizing a change in the work or directive to accelerate, and stating (a) the change in the work including any changes in the method or manner of performance; (b) adjustment of the contractor's profit or fee and the contract sum; (c) adjustment of the guaranteed maximum price, if any; and (d) adjustment of the contract time. *Also called* **Notice of change.**

CHANGE POINT Place in leveling where a backsight is taken to establish the level of the instrument, and from which foresights are taken to establish ground levels.

CHANGE ROADS To move rigging and running lines in order to yard logs progressively from the next unlogged area in the felled and bucked timber.

CHANNEL 1. Long groove or furrow; **2.** Tube-like passage for liquids. **3.** Three-sided, U-shaped member in a sash or frame to receive glass or panel inserts. *See also* **Hoist frame side rail,** and **Lift member; 4.** Bed of a stream or waterway. *See also* **Braided channel, Flood channel,** and **Meandering channel; 5.** Formed or rolled metal shape in the shape of a U.

CHANNEL BEAM Structural steel member, U-shaped in cross section.

CHANNEL GROOVE Siding texture consisting of shallow grooves cut into plywood panel faces during manufacture.

CHANNELING 1. Voids in the shoulder area of a tire between the tread and the buffed surface; **2.** In lubricating greases, the (usually desirable) tendency to form a channel by working down of lubricating grease in a bearing, leaving shoulders of unworked grease that serve as seal and reservoir; **3.** In liquid lubricants and flow-type lubricating greases, the tendency at low temperatures for these materials to form a plastic structure sufficiently strong to resist flow under gravitational forces only. (Similar to, but not identical with, the pour point of liquid lubricants, it is measured on empirical tests.); **4.** *See* **Hoist frame side rail,** and **Lift members.**

CHANNEL IRON Steel section having a web with two flanges extending in the same direction.

CHANNELIZED INTERSECTION *See* **Intersection types.**

CHANNEL PACKING Gasket of channel cross section.

CHANNEL PIPE Fired clay pipe, semicircular or three-quarter-round in crosssection, used principally in manholes gathering branch drains.

CHANNEL TERRACE Contour ridge built of soil moved from its uphill slope, used to divert surface water.

CHANNEL TILE Under-tile for Spanish and Italian tiling.

char. *Abbreviation for:* character; characteristic.

CHARACTERISTIC CONCRETE STRENGTH Strength below which no more than 5% of concrete tests should fall.

CHARACTERISTIC DEAD LOAD Mass of a structure.

CHARGE 1. Conversion of electrical energy into chemical energy within a battery; **2.** Quantity of fuel fed to a furnace.

CHARGED AND DRY Condition of a storage battery when it is assembled with charged and dried plates, dry separators, and with no electrolyte.

CHARGED AND WET Condition of a storage battery when it is filled with electrolyte and is fully charged.

CHARGE MARK *See* **Lead stop.**

CHARGE PRESSURE *See* **Pressure.**

CHARGE RATE Electrical current expressed in amperes at which a battery is charged.

CHARGES Debts, obligations. *See also* **Direct charges,** and **Indirect charges.**

CHARGE WEIGHT Amount of explosive charge, in pounds or kilograms.

CHARGING Putting materials into a mixer, furnace, or other container where they will be further processed.

CHARGING CHUTE Overhead passage through which waste materials and solid fuels are introduced into a furnace.

CHARGING GATE Horizontal movable cover that closes the opening of a top-charging furnace.

CHARGING PLUG Half of a terminal device on the ends of the positive and negative terminal connectors of a storage battery. The connection is readily made to the load or charging circuit, each of which is equipped with a charging receptacle. *See also* **Plug.**

CHARGING PORT Port in the hydraulic head of an injection pump through which fuel passes to fill the pumping chamber.

CHARGING RECEPTACLE Half of a terminal device on the ends of the positive and negative terminal connectors of a storage battery. *See also* **Receptacle.**

CHARLES' LAW The volume of a fixed mass of gas varies directly with absolute temperature, provided the pressure remains constant.

CHARPY TEST Impact test employing a notched test-piece supported at both ends and broken by a blow on the face immediately behind the notch.

CHARRING Surface scorching or burning.

CHART OF ACCOUNTS List of account numbers and designations.

CHASE 1. Continuous recess built into a wall to receive pipes, ducts, etc.; **2.** To unhook chokers at the landing.

CHASE HOLE Opening in rough flooring, used to pass materials between levels.

CHASE MORTISING Mortise cut in lumber already in position, the bottom of which is curved allowing the tenon to be inserted from the side.

CHASER 1. Tool for cleaning out and removing burs from threads; **2.** Worker who unhooks chokers at the landing. *Also called* the **Landing worker.**

CHASE WALL Internal partition enclosing mechanical systems.

CHASE WEDGE Wooden wedge used in bossing lead.

CHASING Grooving or indenting the face of a material.

CHASSIS 1. General term that represents: (a) the entire vehicle including cab and sheet metal, less body; (b) the entire vehicle less cab and sheet metal; or (c) frame and axles without cab or drivetrain; **2.** Assembly that supports the upperstructure of a crane, consisting of an undercarriage frame, swing bearing, and the travel and steering mechanism.

CHASSIS CAB An incomplete vehicle consisting of a chassis upon which is mounted a complete occupant compartment. The vehicle is capable of performing work by the addition of a body/load-carrying structure.

CHASSIS DIMENSIONS Designations commonly used to describe a truck and its components, comprising:

  AF: Axle to end of frame.

  BA: Bumper to axle.

  BBC: Bumper to back of cab.

  BL: Body length.

  CA: Cab to axle.

  CB: Cab to body.

  CE: Cab to end of frame.

  FH: Frame height.

  LA: Load to axle.

  OAL: Overall length.

  WB: Wheelbase.

CHASSIS WEIGHT Actual weight of a fully equipped vehicle without body and driver, but including oil, coolant and maximum fuel. *See also* **Tare weight.**

CHATEAU Country house.

CHATS Gangue material that is found intimately mixed with lead-zinc ores and that is closely akin to chert.

CHAT SAWED A textured stone finish, obtained by using chat sand in the gang sawing process.

CHATTEL Personal property.

CHATTEL MORTGAGE *See* **Mortgage.**

CHATTER Wavy condition across the width of a particleboard panel caused by sanding. The markings are parallel to one another between 6 mm and 12 mm (0.25 in. and 0.5 in.) apart and perpendicular to the sander grit markings. Sometimes they can be felt, but always can be seen.

CHATTER MARK Surface imperfection on the work being ground, usually caused by vibration between the wheel and the work.

CHC *Abbreviation for:* continuous high chairs.

CHCU *Abbreviation for:* continuous high chairs upper.

chd *Abbreviation for:* chord.

CHEATER BAR Extension to the handle of a tool to gain leverage.

CHECK 1. Lengthwise separation of wood, usually extending across the rings of annual growth, commonly resulting from stresses set up in the wood during seasoning; **2.** Design made up of alternating squares or rectangles of contrasting color, texture, or material; **3.** Device or structure that regulates or controls; **4.** Paint failure characterized by small surface cracks.

CHECK CHAINS Heavy-duty chain and spring assembly designed to prevent dump body kick-up at the end of the hoist stroke and to assist in returning the body to a level position. *Also called* **Restraining devices, Snubber chains,** and **Stop chains.**

CHECK DAM Dam that divides a drainage course into two or more sections with reduced slopes.

CHECKER One who verifies or confirms the work of another.

CHECKER-BLOCK PAVING Surface of concrete paving blocks with open sections in which grass can be grown, in order to create an overall checkerboard pattern.

CHECKER-WORK Decoration formed of alternating squares or rectangles of contrasting materials.

CHECK FILLET Curb, formed of asphalt, on a pavement, used to control the flow of surface water.

CHECKING 1. Development of shallow cracks at closely spaced but irregular intervals on the surface of plaster, cement paste, mortar, or concrete. *See also* **Craze cracks,** and **Crazing; 2.** Minute cracks in rubber caused by aging and oxidation; **3.** Act of cracking in paint; **4.** *See also* **Face checking.**

CHECK RAIL Meeting rail in a sliding or double-hung window sash that meets another check rail when in the closed position and that is of sufficient thickness to overlap the other.

CHECK RAIL WINDOW Frame containing at least a pair of sashes that are engaged when closed. The sashes are installed in a vertical plane and designed to be moved either vertically or horizontally.

CHECKS 1. Numerous, very small surface cracks in metal or other material created during processing; **2.** Cracks or splits in wood, usually across the rings of annual growth.

CHECK SCALER Worker who rescales logs in order to detect errors in the initial scaling.

CHECK VALVE *See* **Valve.**

CHEEK WEIGHT Overhauling weight attached to the side plate of a lower load block. *See also* **Overhauling weight.**

CHEEKS Facing matched members, part of a larger construction.

CHEESINESS Condition of paint that is soft and incompletely dry.

chem. *Abbreviation for:* chemical; chemically.

CHEMICAL Any element, chemical compound, or mixture of elements and/or compounds.

CHEMICAL BOND Bond between materials that is the result of cohesion and adhesion developed by chemical reaction.

CHEMICAL BROWN STAIN Chemical discoloration of wood that sometimes occurs during air- or kiln-drying of several species, apparently caused by the concentration and modification of extractives. *See also* **Blue stain,** and **Brown stain.**

CHEMICAL CONTAMINANT Foreign material in a fluid which is either in solution or in a gas or liquid bulk phase.

CHEMICAL CURE In regards to tires: vulcanization activated by chemical agents without the application of heat.

CHEMICAL GAUGING Regulating the quantity of water flow through determination of a chemical solution introduced upstream at a known rate and concentration.

CHEMICAL GROUT Low-viscosity grout composed of chemical compounds in liquid form or in a liquid solution that can be injected into soil or fissured rock to serve as a water cutoff or to improve load-carrying capacity.

CHEMICALLY CURED SEALANT Sealant that cures by chemical reactions usually involving the formation of cross-linked polymers.

CHEMICALLY PRESTRESSED CONCRETE Concrete made with expansive cement and reinforcing under conditions such that the expansion of the cement induces tensile stress in the reinforcing so as to produce prestressed concrete. *Also called* **Self-stressing concrete.**

CHEMICALLY PRESTRESSING CEMENT Type of expansive cement containing a higher percentage of expansive component than a shrinkage-compensating cement that, when used in concretes with adequate internal or external restraint, will expand sufficiently, due to chemical reactions within the matrix, to develop the stress necessary for prestressing the concrete. *See also* **Expansive cement.**

CHEMICAL NAME Scientific designation of a chemical in accordance with the nomenclature system developed by the International Union of Pure and Applied Chemistry or the Chemical Abstracts Service.

CHEMICAL OXYGEN DEMAND Measure of the oxygen equivalent of the organic matter in a sample of sewage, liquid waste, leachate, or polluted water that is susceptible to oxidation by a strong chemical oxidant.

CHEMICAL PRECIPITATION Settling out of a solid fraction from a solution, caused by the addition of a chemical, hastened by floccing.

CHEMICAL PULPING Process in which wood fibers are separated by removing the lignin and certain other wood components through the use of chemicals.

CHEMICAL RESISTANCE Resistance of a plastic substance to chemical acids and alkalines.

CHEMICAL SHRINKAGE Change in the volume of a material due to chemical action or reaction, typically in a concrete mix due to hydration.

CHEMICAL THINNING Any forest thinning in which the unwanted trees are killed by chemical poisoning; band or frill girdling may be done at the same time.

CHEMISE Nonretaining wall that protects a soil slope against weathering.

CHERNACK LOOM Four-shuttle circular loom for the production of seamless hose reinforcement.

CHERRY PICKER 1. Small, wheeled, self-propelled personnel crane, most commonly a rough-terrain type; **2.** Log loader that loads roadside logs left by the roadbuilders.

CHERT Very fine-grained siliceous rock in a variety of colors, characterized by hardness and conchoidal fracture in dense varieties; the fracture becomes splintery and the hardness decreases in porous varieties.

CHEVRON DRAIN *See* **Herringbone drain.**

chf. des. eng. *Abbreviation for:* chief design engineer.

chg. *Abbreviation for:* change; charge.

chgbl. *Abbreviation for:* changeable.

CHICAGO BOOM Swinging boom attached to a column; a form of derrick.

CHICAGO CAISSON Shaft, 900 mm (3 ft) or larger in diameter, with vertical timber sheathing, sunk in increments as circular bracing and additional sheathing are installed.

CHICKEN LADDER *See* **Crawling board.**

CHICOT Wholly or partly dead tree, top, or large branch that may break off and fall without warning, constituting a major hazard to loggers working in the woods.

CHILLER Air conditioning unit consisting of a compressor, condenser, and evaporator unit only.

CHILL RING *See* **Backup ring.**

chim. *Abbreviation for:* chimney.

CHIMNEY That part of a building that contains flues for drawing off smoke and fumes from heating appliances. *See also* **Factory-built chimney.**

CHIMNEY BAR Bar above a fireplace opening that supports the front of the chimney breast.

CHIMNEY BLOCK Hollow precast-concrete masonry units used in conjunction with vitrous clay flue liners.

CHIMNEY BREAST Structure projecting from the face of a wall and containing the flue or stack.

CHIMNEY COWL Vented, metal termination to a flue that either revolves according to wind velocity, or turns to face downwind, used to increase the updraft in the flue.

CHIMNEY EFFECT Tendency of warmed air or a gas to rise through a vented vertical shaft or duct.

CHIMNEY FLASHING Any kind of metal or composition material placed around a chimney where it penetrates through a roof, to cover the joint and prevent water from entering.

CHIMNEY FLUE Passage housed in a chimney through which products of combustion are carried from a fuel-burning appliance to the exterior. *See also* **Chimney lining.**

CHIMNEY GUTTER *See* **Back gutter,** and **Chimney saddle.**

CHIMNEY LINER Conduit containing a chimney flue, used as a lining of a masonry or concrete chimney.

CHIMNEY LINING Fire-clay or terra-cotta shapes made for use inside of a chimney. *See also* **Flue lining.**

CHIMNEY PAD Footing for a chimney.

CHIMNEY PIECE Decorative treatment about and above a fireplace.

CHIMNEY POT Short, exposed extension of a flue above the top of a chimney.

CHIMNEY SADDLE Peaked flashing between a chimney and the roof to shed moisture around the chimney. *See also* **Cricket.** *Also called* **Back gutter,** and **Chimney gutter.**

CHIMNEY SHAFT Chimney with one flue.

CHIMNEY STACK Structure containing several flues.

CHIMNEY THROAT Narrow horizontal opening between a fireplace and its flue that constricts the flow of gasses, regulating the draft.

CHINESE FINGERS Split wrap-around cable grip, used to support electrical wires.

CHINK 1. Small fissure or crack; **2.** To fill cracks or small fissures in or between materials.

CHIP *See* **Harvest functions.**

CHIP BLASTING Blasting of shallow-depth ledge rock.

CHIP CONTROL FILTER Filter intended to prevent only large particles from entering a component immediately downstream. *Also called* **Grit control filter,** and **Last-chance filter.**

CHIP CRACKS Crazing of a paint surface with the resulting material curling up away from the surface in concave-shaped fragments.

CHIP LOAD Quantity of wood removed by an individual saw tooth as it cuts on each pass.

CHIP OUT Condition along the top or bottom face edges of particleboard where the fines or flakes are removed, or torn out of the surface.

CHIPPER *See* **Harvesting machines (single function).**

CHIPPER DECK Infeed of a chipping machine, including the chain that feeds the material to be chipped.

CHIPPER DISCHARGE Direction chips leave the chipper housing. May be horizontal, overhead, or from the bottom.

CHIPPER INFEED Series of rollers at the front portion of the chipping machine where the material to be chipped enters.

CHIPPER KNIFE Replaceable piece of steel with sharpened edges; attaches to a rotating drum.

CHIPPING 1. Treatment of a hardened concrete surface by chiseling; **2.** Loosening of shallow-depth rock by blasting or pneumatic hammers.

CHIPPING HAMMER *See* **Hammer.**

CHIPS 100% fractured stone, usually passing a 12 mm (0.5 in.) square-mesh sieve but retained on a #8 sieve.

CHIP SEAL Fine crushed rock spread on asphaltic oil and rolled.

CHIP SEPARATOR Attachment to whole-tree chippers that separates acceptable chips from unacceptable bark, limbs, and foliage.

CHIPSPREADER Self-contained hopper designed to be mounted on a truck chassis or in a dump body, complete with conveyor belts, spread gates, and/or a broadcasting mechanism by which gravel chips or other granular material can be evenly and uniformly spread over a road surface by a truck moving at a specified rate of travel.

CHIP UNIT Chip volume equal to one cord of pulpwood.

CHISEL *See* **Cold chisel,** and **Wood chisel.**

CHISEL EDGE *See* **Beveled edge.**

chk *Abbreviation for:* check.

chkd *Abbreviation for:* checked.

chkr *Abbreviation for:* checker; checkered.

chlor. *Abbreviation for:* chlorinate; chlorination; chlorine.

CHLORINATED POLYETHYLENE MEMBRANE Flexible plastic sheet used as a cover sheet to a setting mortar bed or substrate to serve as a waterproofing, curing, or isolating membrane.

CHLORINATED POLYVINYL CHLORIDE (CPVC) PIPE *See* **Pipe.**

CHLORINATED RUBBER Vehicle used in the manufacture of solvent-thinned, pigmented paints.

CHLORINATED RUBBER ADHESIVE *See* **Adhesive.**

CHLORINATION Application of chlorine to potable water supplies as a disinfectant.

CHLORINE Nonmetallic chemical element of the halogen (salt-producing) family; one of several chemicals used as an oxidant in water sterilization and purification.

chnl *Abbreviation for:* channel.

CHOCK 1. Block or wedge used under and against an object to restrain it from rolling, sliding, or swinging; **2.** Timber member used as a separator between piles or timbers.

CHOKE 1. Adjustable restriction in the throat of a carburetor that controls the flow of air so as to provide a richer mixture for starting a cold gasoline engine; **2.** To pass a line or choker around a log or other object and pull it tight.

CHOKED Condition in which a log is attached to a skidding unit by means of a wire rope or chain choker.

CHOKER 1. Sling hitch that is self-tightening; **2.**

Short length of chain or cable with a noose, used to haul logs, pull out stumps, etc.; **3.** Road shoulder designed to remain higher than the subgrade level; **4.** Term sometimes used to describe a sling.

CHOKER HITCH Method of attaching a sling to a load by forming a noose in the sling that tightens around the load.

CHOKER HOOK 1. Fastener on the end of a choker that forms the noose; **2.** Hook that can slide along a chain. *Also called* a **Round hook.**

CHOKER LINE Short piece of line that closes a grapple.

CHOKER ROPE Short wire rope sling that forms a slip noose around an object that is to be moved or lifted.

CHOKER SETTER Rigging crew member who sets chokers under the direction of a rigging slinger.

CHOP Movable outer face of the jaw of a carpenter's bench vise.

CHOPPED FIBER Continuous fiber cut into short lengths and used as a multidirectional reinforcement in cement paste and concrete.

CHOPPING BIT *See* **Bit.**

CHOP SAW *See* **Saw.**

CHORD 1. Principal horizontal member in a rigid framework; **2.** Outer members of a truss that define the envelope or shape. There are two distinctions:

> **Top chord:** An inclined or horizontal member that establishes the upper edge of a truss and that is subjected to compressive and bending stress.

> **Bottom chord:** The horizontal (and inclined, e.g., scissor trusses) member defining the lower edge of a truss, carrying ceiling loads where applicable. This member is subject to tensile and bending stresses.

3. Main corner structural member of a boom; **4.** Straight line joining any two points on an arc, curve, or circumference. *See also* **Geometry of circular curves.**

CHORD MEMBERS Upper and lower flange members of a truss, called the upper and lower chords, respectively.

CHORD MODULUS *See* **Modulus of elasticity.**

chpbd *Abbreviation for:* chipboard.

chr. *Abbreviation for:* chrome.

chrg. *Abbreviation for:* charge; charged; charging.

CHROMACITY Light, characterized by dominant wavelength and color purity.

CHROMATE Salt of chromic acid.

CHROMATING Protective coating to metal of lead or zinc chromate.

CHROME BORE Any cylinder bore that is plated with chromium as a method of increasing the wear life of the cylinder.

CHROMETRY Land measurement.

CHROMIUM Grayish-white metal resistant to corrosion; used to plate metal. *See also* **Alloying elements.**

CHRONOLOGICAL AGE *See* **Actual age.**

CHRONOTHERM Thermostat and clock combination that can be set to activate a furnace or airconditioner at predetermined times and/or temperatures.

chs. *Abbreviation for:* chase.

CHUCK 1. Part of a drill that holds the bit, or rotates the steel; **2.** Device for holding grinding wheels of special shape or the work piece being ground.

CHUCK KEY L- or T-shaped steel tool having a toothed conical base, used to crank a chuck so as to open or close its gripping fingers.

CHUCK LEAD *See* **Lead.**

CHUFF BRICK *See* **Brick.**

CHUNKING Separation of the tread of a tire from its casing, in particles ranging from very small to several square inches in area.

CHUNK OUT Remove log chunks, branches, and debris from a landing or work area.

CHUNK SAMPLE METHOD Determining weight per cubic measure of soil in place.

CHUNK UP Clean up and pile debris after logging an area.

CHURN Vessel used for making rubber cement, in which rubber compounds are stirred into solvents.

CHURN DRILL *See* **Drill.**

CHUTE 1. Sloping trough or tube for conveying free-flowing materials from a higher point to a lower point; **2.** Outfeed portion of a chipper. *Also called* a **Discharge spout,** and **Flow trough.**

CHUTE-FED INCINERATOR *See* **Incinerator.**

CHUTE GATE Opening in the tailgate of a dump truck body to allow a limited flow of material during dumping, normally fitted with a sliding door, control handle, and lip for directing the flow. Chutes may also be located on the body side or in the floor.

CI *Abbreviation for:* cast iron; channel iron; cloth inserted.

CI cnd. bx *Abbreviation for:* cast-iron conduit box.

cif *Abbreviation for:* cost, insurance, and freight.

cin. blk *Abbreviation for:* cinder block.

CINCH ANCHOR Steel bolt with sleeves, used to place bolts in existing concrete.

CINDER BLOCK *See* **Concrete masonry unit.**

CIP *Abbreviation for:* cast-iron pipe.

cir *Abbreviation for:* circular.

cir. *Abbreviation for:* circle.

circ. *Abbreviation for:* circuit; circulate; circulation; circumference; circumstance.

CIRCLE Rotary table that supports the blade of a grader and regulates its angle.

CIRCLE CUTTER Adjustable scribe, used for cutting circular holes in drywall. *See also* **Bit.**

CIRCLE REVERSE Mechanism that changes the angle of a grader blade.

CIRCLIPS Snap rings or keepers.

CIRCUIT 1. Complete path of an electrical current; **2.** Path of a detonating electric circuit or detonating fuse; **3.** Arrangement of interconnected component parts. *See also* **Regenerative circuit, Sequence circuit,** and **Servo circuit.**

CIRCUIT BREAKER Automatic mechanical device that serves the same purpose as a fuse, i.e. to prevent overheating in a circuit through overloading.

CIRCUIT VENT Vent pipe that is connected at its lower end to a branch and at its upper end to a vent stack, or is terminated in open air.

CIRCULAR FREQUENCY Method of expressing the frequency of vibration by relating one vibratory cycle to a full circle, or π radians. Circular frequency is stated in angular measure per unit of time, usually radians/second, and 1 Hz = π rad/s.

CIRCULAR MIL Unit used in measuring the cross-sectional area of wires; equal to the area of a wire having a diameter of one mil, or one thousandth of an inch (0.001 in.).

CIRCULAR PLANE Compass plane.

CIRCULAR PLUG *See* **Repairs.**

CIRCULAR SAW *See* **Saw.**

CIRCULAR STAIR *See* **Spiral stair.**

CIRCULAR-TYPE CELLULAR COFFERDAM Structure constructed of interlocking steel-sheet piling consisting of circular cells joined with connecting arcs. The arcs are installed after the cells are completed; the cells are then filled with granular soils. *See also* **Diaphragm-type cellular cofferdam.**

CIRCULAR WOVEN JACKET Textile reinforcing member produced on a circular loom for such types of hose as fire hoses.

CIRCULATING HOT WATER SYSTEM Hot water supply system consisting of a storage tank connected to the potable supply and equipped with a means to heat the water (immersion electrical element, gas-fired furnace, etc.), plus a pump to continuously circulate heated water through the hot-water distribution system, drawing its supply from the top of the tank and returning it to the bottom of the tank, plus a cold water connection to the bottom of the tank for makeup water to replace hot water drawn from the system. *See also* **Hot water supply system,** and **Noncirculating hot water system.**

CIRCULATING LOOP Main loop in which oil is circulated from oil storage tanks to the branch circuit, then back to the storage tank.

CIRCULATING WOOD STOVE Double-walled fireplace in which air is circulated, by convection or with the aid of circulating fans, between the two walls, gathering heat from the firebox and venting it to the room through grilles.

CIRCULATION Flow of air, fluid, or gas throughout a system or circuit. *See also* **Reverse circulation.**

CIRCULATION CONNECTION Port, normally in a stuffing box or gland, through which a fluid is introduced.

CIRCUMFERENTIAL BEARING AREA Area of the entire facing or cylindrical wearing surface of a bearing; equal to πD x Length. *Also called* **Bearing area,** and **Projected bearing area.**

CIRCUMFERENTIAL CRACK Crack in a tire running parallel to the beads (usually occurring in a tread groove).

CIRCUMFERENTIAL SEAL Seal composed of a continuous ring, or of one or more split or segmented rings.

cisp *Abbreviation for:* cast-iron soil pipe.

CISTERN Reservoir or tank, often underground, for storing water.

CITY AND REGIONAL PLANNING Land and community planning that embraces areas larger than a single community, usually on a cooperative basis between adjacent communities and/or jurisdictions.

CITY BUILDING CODE *See* **Building codes.**

CITY GROWTH *See* **Urban growth.**

CITY PLANNER *See* **Urban planner.**

civ. *Abbreviation for:* civic; civil.

civ. eng. *Abbreviation for:* civil engineer.

CIVIC CENTER Part of a community where the principal public buildings and facitilies are located.

CIVIL AUTHORITY CLAUSE Provision in fire insurance that covers the policyholder against loss caused by civil authorities while attempting to prevent the spread of fire.

CIVIL ENGINEER Professional engineer specializing in public works and heavy engineering works.

CIVIL ENGINEERING CONTRACTOR Contractor who specializes in structures and plant.

cke *Abbreviation for:* centrifuge kerosene equivalent.

ckt *Abbreviation for:* circuit.

cL *Abbreviation for:* carload; clad; cladding; class; classification; clearance; close; clutch.

CL *Abbreviation for:* centerline.

CLADDING 1. External covering or skin applied to a structure; **2.** Thick layer of weld applied on a surface to improve resistance to corrosion and abrasion.

CLADDING GLASS Glass specially manufactured for curtain walling.

CLAIM 1. Demand or assertion by one of the parties to the agreement seeking an increase, decrease, or other adjustment of the terms of the contract documents; **2.** Demand for payment under an insurance policy.

CLAIMANT Individual or entity contracted to furnish labor, materials, or equipment to a contractor or subcontractor for use within a contract.

CLAIM CHUTE Trough-like structure to which concrete is conveyed by the discharge apron of a mixer truck.

CLAM BUNK Payload bed of a forwarder, equipped with top-opening hydraulic jaws.

CLAM-BUNK SKIDDER *See* **Harvesting machines**

(single function).

CLAMP 1. Knob or lever on a surveying instrument that tightens the axis so that the tangent screw can turn the instrument; **2.** Attachment to an industrial truck that uses a clamping action to engage the load from opposite sides. *See also* **Clamp arm**, and **Coupler**; **3.** In hose, a metal fitting or band used around the outside of a hose end to bind the hose to a coupling, fitting, or nipple. *See also* **Ferrule**, and **Hose clamp**; **4.** Range of tools and devices designed to hold two or more pieces or parts in a predetermined position. They include:

Band: Canvas, cloth, or metal band or strip drawn tight through a buckle.

Bar: Long metal bar with a stationary stop at one end and a sliding, adjustable stop at the other.

Bench: Including vises, bench hooks, bench stops, dogs, and holdfasts.

C-clamp: Metal clamp shaped like a letter 'C' with a bolt threaded through one extremity and directed at the other.

Corner: Pair of clamps mounted on a jig at an angle of 45° to facilitate gluing miter joints.

Hold-down: Clamp that can be attached to a workbench and that has a vertical adjustment screw.

Parallel: Two blocks of wood joined in opposite directions by two bolts passing through threaded holes.

Pinch dog: U-shaped metal piece with two sharp, pointed legs, used to hold wood items together while they are being worked on.

Pipe: Paired devices, one of which is internally threaded to attach to the end of a pipe, the other of which will slide over the pipe, locking itself in place where desired. Generally made for 12.7 and 19 mm (0.5 and 0.75 in.) pipe.

Screw: Matched hardwood jaws connected by large screw spindles.

Spring: Flat metal pieces joined by a spring-equipped hinge.

Three-way: Similar to a C-clamp but with an additional screw in the center of the throat. Used to apply right-angle pressure.

Vise: Tool fitted to a workbench and consisting of large flat jaws that can be drawn together by a worm. A vise with steel jaws is known as a machinist's vise; one made of or faced with wood as a woodworker's vise.

Web: Hard nylon or webbing belt that can be looped around an object, tightened, and held under tension by a buckle-like device.

CLAMP ARM Auxiliary member that bolts, or is welded to a clamp arm support. Various types are available for clamps to suit them for specific loads such as bag, drum, carton, paper, roll, tire, etc.

CLAMPING FORCE Amount of pressure exerted by the spring or springs to engage the clutch.

CLAMPING SCREW Screw used in a clamp to exert pressure.

CLAMSHELL 1. Bucket with two jaws that are held open by the opening line, and that clamp together to load under their own weight when lifted by the closing line; **2.** Shovel equipped with a clamshell bucket.

CLAPBOARD Overlapping, wedge-shaped, horizontal boards. The upper edge is thinner and fits into a rabbet struck on the inside of the thicker edge of the following board.

CLAPPER BRIDGE Bridge constructed of large slabs of stone, some forming rough piers, with longer slabs making the roadway.

clar. *Abbreviation for:* clarification; clarify.

CLARIFIED SEWAGE Sewage from which all or part of the suspended matter has been removed.

CLARIFIER Mechanism that removes, by gravity (sometimes with chemical additives), the particles suspended in a liquid.

CLASPING BUTTRESS *See* **Buttress**.

class. *Abbreviation for:* classification; classifier; classify.

CLASS (OF CARGO-CARRYING VEHICLE) System that classifies a commercial cargo-carrying vehicle by its maximum gross vehicle weight rating (for straight trucks) or maximum gross combination weight (for combined vehicles). The classes are:

Class 1: 0–2721 kg (0–6,000 lb).

Class 2: 2722–4535 kg (6,001–10,000 lb).

Class 3: 4536–6350 kg (10,001–4,000 lb).

Class 4: 6351–7257 kg (14,001–16,000 lb)

Class 5: 7258–8845 kg (16,001–19,500 lb).

Class 6: 8846–11 793 kg (19,501–26,000 lb).

Class 7: 11 794–14 968 kg (26,000–33,000 lb).

Class 8: 14 969 kg (33,001 lb) and above.

CLASS (OF CONCRETE) Arbitrary characterization of concrete of various qualities and usage, usually by compressive strength.

CLASS (OF FIRE)

Class A: Fire involving combustible materials such as wood, cloth, and paper.

Class B: Fire involving a flammable liquid, fat or grease.

Class C: Fire involving energized electrical equipment.

Class D: Fire involving a combustible metal.

CLASS (OF WILDFIRE) Size of a wildfire:

Class A: Fire of 0.1 ha (0.25 ac) or less.

Class B: Fire greater than 0.1 ha (0.25 ac) but less than 4.0 ha (10 ac).

Class C: Fire greater than 4.0 ha (10 ac) but less than 40 ha (100 ac).

Class D: Fire greater than 40 ha (100 ac) but less than 120 ha (300 ac).

Class E: Fire greater than 120 ha (300 ac) but less than 400 ha (1,000 ac).

Class F: Fire greater than 400 ha (1,000 ac) but less than 2000 ha (5,000 ac).

Class G: Fire of 2000 ha (5,000 ac) or more.

CLASSIFICATION 1. Grouping of subjects or materials according to a description, size, etc.; **2.** Group, or family designation based on wire rope construction with common strengths and weights listed under the broad designation.

CLASSIFICATION OF TRUCKS BY GROUND CONTACT Truck classification system based on the number of wheels and the number of driving wheels of the vehicle. If a truck is designated as a 4x2, it has four wheels and two driving wheels; a 4x4 has four wheels and four driving wheels; 6x4 six wheels and four driving wheels. (Wheels are considered a unit whether they have single or dual tires.)

CLASSIFIED EXCAVATION *See* **Excavation.**

CLASSIFIED PRODUCT Product listed and bearing the label of an approved laboratory.

CLASSIFIER RAKE Machine for separating coarse and fine particles of granular materials temporarily suspended in water; the coarse particles settle to the bottom of a vessel and are scraped up an incline by a set of blades, while the fine particles remain in suspension to be carried over the edge of the classifier.

CLASS RATE Insurance premium rate applicable to a specific class of risk.

CLAUSE Section of a contract dealing with a particular subject, condition, or requirement.

CLAW Hand-held, claw-headed bar used to bend small-diameter reinforcing bars round a mandrel on a bar-bending bench.

CLAW BUCKET *See* **Bucket.**

CLAW HAMMER *See* **Hammer.**

CLAW HATCHET Hand ax with a claw as part of the splitting blade, opposite a striking face, used in shingling and lathing.

CLAY *See* **Soil types.**

CLAY BINDER *See* **Binder soil.**

CLAY CONTENT Percentage of clay by dry weight of a heterogeneous material.

CLAY LOAM *See* **Soil types.**

CLAY MASONRY UNIT Solid or hollow masonry unit formed of clay in its plastic state, and burned or fired in a kiln.

CLAY MORTAR MIX Finely ground clay used as a plasticizer for masonry mortars.

CLAY SIZE FRACTION Portion of soil that is finer than 0.002 mm (0.00008 in.); not a positive measure of the material's plasticity or its clay characteristics.

clch *Abbreviation for:* clinch.

cld *Abbreviation for:* cooled.

cldg *Abbreviation for:* cladding.

CLEAN 1. Free of foreign matter; **2.** Sand or gravel lacking a binder; **3.** New or properly cleaned filter element.

CLEANABLE Filter element which, when loaded, can be restored by a suitabale process to an acceptable percentage of its original dirt capacity.

CLEANABILITY Ability of a cleanable filter element to withstand repeated field cleanings and retain adequate dirt capacity and service life.

CLEAN AGGREGATE Sand or gravel, free from clay or silt.

CLEAN BILL OF LADING Bill of lading signed by the carrier for receipt of goods in good condition (no damage, loss, etc. apparent), that does not bear such notation as 'Shipper's load and count.'

CLEAN BURNING *See* **Burning out.**

CLEAN FILL Uncontaminated, inert solid material used to bring a site to grade.

CLEANING EYE Access eye, opening, or cleanout.

CLEANLINESS LEVEL 1. Analogue of contamination level; **2.** Measure of the level of dirt or other insoluble material contamination in fresh or used oil. commonly expressed as the number of particles of different size ranges in a fixed volume of fluid.

CLEANOUT 1. Opening in the first course of a cavity masonry wall for the removal of mortar droppings; **2.** Opening under a fireplace for removal of ashes; **3.** Any opening or orifice left so as to permit the removal of debris from the space behind; **4.** Plumbing unit with a removable plate or plug affording access into the pipe for rodding, usually provided at pipe bends. *Also called* **Access eye,** and **Rodding eye.**

CLEANOUT DOOR Metal door placed below a flue at the chimney bottom, opening to the cleanout with access to the ash pit.

CLEAN ROOM Room in which the controlled atmosphere is filtered to remove 99.99% of all dust and contaminants and in which a positive atmospheric pressure is maintained to prevent the infiltration of unfiltered air.

CLEANUP Treatment of horizontal construction joints to remove all surface material and contamination down to a condition of cleanness corresponding to that of a freshly broken surface of concrete.

CLEAR 1. Log without knots; first quality log; **2.** Paint containing no colored pigments.

CLEARANCE 1. Gap or space between adjoining or mating surfaces. *See also* **Concrete cover;** **2.** Space between moving and stationary objects.

CLEARANCE FIT Fit that results when a smaller object is inserted in a larger mating hole.

CLEARANCE HOLE Hole drilled slightly larger than the bolt it is to accommodate. *Also called* **Clearing hole.**

CLEARANCE LIGHTS Lighting to indicate the overall width of a machine or vehicle.

CLEARANCE POCKET Space of controlled volume giving the effect of greater or less cylinder clearance and thus changing compressor capacity.

CLEARANCE RADIUS One half of the diameter of a circle at the furthest projection of the boom as a machine maneuvers through its tightest turning mode.

CLEARANCE SEAL Seal that limits the leakage between a rotating or reciprocating shaft and a stationary housing by means of a controlled annular clearance between the two.

CLEAR AND GRUB Removal of all vegetation, trees,

and rubble, or anything that will interfere with construction inside the limits of the project.

CLEAR APERTURE Unobstructed diameter of the objective lens of a survey instrument.

CLEAR CERAMIC GLAZE Similar to a ceramic color glaze except that it is translucent or slightly tinted with a gloss finish.

CLEAR COATING Invisible glossy film or penetrant applied to substrates to protect, repel, or resist water and the hydration of minerals.

CLEARCUT When all trees and saplings of a logging area are felled, bucked, and removed.

CLEAR GLAZE *See* **Glaze.**

CLEARING 1. Area from which the natural brush and vegetation have been removed; 2. Cutting down and removing trees and brush.

CLEARING HOLE *See* **Clearance hole.**

CLEARING THE SITE *See* **Site clearing.**

CLEAR LENGTH Portion of the tree between the ground and the point where the lowest limbs join the trunk.

CLEAR LUMBER *See* **Lumber grades.**

CLEAR SPAN Distance between opposite faces of supports.

CLEAR TITLE Title to something, clear of encumbrances or disputes.

CLEAT 1. Strip of wood or metal, fastened across a door or other object to give it additional strength; 2. Strip of wood nailed to a wall, usually for the purpose of supporting some object or article fastened to it; 3. Small board connecting formwork members or used as a brace; 4. Ladder crosspiece placed on edge to assist in ascending or descending a ladder.

CLEAVAGE Tendency of rock to split along definite, parallel, and closely spaced planes.

CLEAVAGE FAILURE Failure of cured concrete characterized by a clean fracture.

CLEAVAGE MEMBRANE Membrane providing separation and slip between a mortar setting bed and the backing or base surface.

CLEAVAGE PLANE Line along which rock will break easily when dug or blasted.

CLEFT STONE Stone that is split rather than sawn from a large block, producing a naturally sheared face.

CLERESTORY An outside wall of a room or building carried above an adjoining roof and pierced with windows.

CLERESTORY WINDOW Window that occurs in the wall of a clerestory.

CLEVELAND OPEN CUP Test used to determine the flash point of powdered asphalts, asphalt cements, and road oils.

CLEVIS 1. Shackle; 2. Split end of a rod, drilled for insertion of a pin through the two sections.

clg *Abbreviation for:* ceiling.

c/lgth *Abbreviation for:* cut to length.

CLIENT Person or entity enaging another to perform a duty or hold a responsibility; one to whom those engaged to carry out some act are responsible.

CLIMATOLOGY Study of regional climatic variations and changes; planning and siting of structures with regard to local climate.

CLIMAX FOREST Forest community that represents the final stage of natural forest succession for its locality, i.e., for its environment. Often identified as those forests that can reproduce indefinitely, i.e., in their own shade.

CLIMAX SPECIES Plant species that will remain essentially unchanged in terms of species composition for as long as the site remains undisturbed.

CLIMB CUT Where a circular saw blade rotates in the same direction as the material is fed during the cutting process. *Also called* **Climb sawing,** and **Power cut.**

CLIMBING EQUIPMENT Irons with sharp spurs, strapped to the legs at the ankle and below the knee, and a heavy, leather safety belt with a wire-cored manila rope. Used by riggers and linemen to climb trees and wooden poles. *See also* **Climbing rope,** and **Climbing spurs.**

CLIMBING FORM Form that is raised vertically for succeeding lifts of concrete in a given structure.

CLIMBING FRAME Frame used with climbing tower cranes to conduct climbing and/or working reactions to the structure supporting the crane.

CLIMBING LADDERS 1. Separate ladder attached to a scaffolding structure, or a ladder built into a scaffolding frame, used to reach the working platform; 2. Pair of members suspended from the climbing frame, used as jacking supports when some climbing cranes climb.

CLIMBING LANE *See* **Passing lane.**

CLIMBING ROPE Manila rope, usually with a steel cord, attached to the belt of a high rigger and looped around a permanent support to provide a brace. *See also* **Climbing equipment,** and **Climbing spurs.**

CLIMBING SPURS Irons with sharp side spurs, strapped to the legs at ankle and below the knee, used by riggers to climb trees for topping and rigging, and by linemen to climb utility poles. *See also* **Climbing equipment,** and **Climbing rope.**

CLIMB SAWING *See* **Climb cut.**

CLINCH 1. Process of securing a driven nail by bending down the point; 2. To fasten firmly by bending down the ends of protruding fasteners.

CLINCH NAIL Made from open hearth or Bessemer steel wire, this type of nail is only used in places where it is desired to turn over the ends of the nails to form a clinch, as in the case of battens or cleats.

CLINK 1. Pointed steel bar about 300 mm (12 in.) long, used to break up pavement; 2. Seam between adjacent bays of flexible metal roofing.

CLINKER Partly fused products of a kiln; other vitrified or burnt material.

CLINKER BLOCK Relatively inexpensive lightweight precast concrete building block incorporating clinker as an aggregate.

CLINKER BRICK *See* **Brick.**

CLINOMETER Instrument for measuring the inclination of a line from horizontal.

CLIP 1. Fitting for clamping two parts of wire rope to each other; **2.** Wire or sheet metal device used to attach various types of lath to supports or to secure adjacent lath sheets; **3.** Portion of a brick cut to length; **4.** Small fastening device.

CLIP ANGLE Connector made from steel used to fasten the post or frames to the sill in post or frame bents.

CLIP COURSE Course of brick resting on a clip joint.

CLIP JOINT Joint of abnormal thickness (maximum 12 mm (0.5 in.)) to bring courses up to a required height.

CLIPPED HEADER A bat placed to look like a header for the purpose of establishing a pattern. *Also called* a **False Header.** *See also* **Header.**

CLIP WRENCH *See* **Wrench.**

clk *Abbreviation for:* caulk.

cln *Abbreviation for:* clean.

clnd *Abbreviation for:* cleaned.

clo. *Abbreviation for:* closed; closet; closure.

CLOD BUSTER Drag towed by an excavator to break up lumpy soil.

CLOGGING INDICATOR Indicator that is activated when a predetermined pressure differential across a filter is reached.

CLOISTER Covered walk, often around a quadrangle, with a wall on the outer side and a colonnade or windows on the inner side.

CLOISTER VAULT *See* **Vault.**

CLOSE BOARDED Wall or roof covered with boards butted together.

CLOSE-COUPLED Roof comprising common rafters joined at the plate, level with tie beams.

CLOSE-CUT HIP Hip, or valley, in which the shingles, tiles, or slates are mitered to meet on the hip or valley line.

CLOSED AREA Forest area in which specified activities, such as burning or entry, are temporarily disallowed due to acute fire hazard.

CLOSED BIDDING *See* **Selective bidding.**

CLOSED BURNER Sealed-in burner that, in most cases, supplies all the air for combustion through the burner itself.

CLOSED CANOPY Forest stand where the crowns of the main level of trees forming the canopy are touching and intermingled so that light cannot reach the forest floor.

CLOSED-CELL FOAMS Products not allowing air to pass through.

CLOSED-CENTER CIRCUIT Condition where a fluid only flows through the main control valves when a control valve spool is actuated. This can be done in two ways: (a) using a pressure-compensating pump, or (b) using a fixed-displacement pump, an unloading valve, and an accumulator.

CLOSED-CENTER VALVE *See* **Valve.**

CLOSED-CIRCUIT GROUTING Injection of grout into a hole intersecting fissures or voids that are to be filled at such volume and pressure that grout input to the hole is greater than the grout take of the surrounding formation, excess grout being returned to the pumping plant for recirculation.

CLOSED-CIRCUIT OIL SYSTEM System in which oil may be pumped completely through the circulating loop and back into the oil storage tank.

CLOSED CONTAINER Container so sealed with a lid or other seal that neither liquid nor vapor will escape at ordinary temperatures.

CLOSED-CYCLE GAS TURBINE ENGINE Closed-cycle engine that has a working fluid independent of the atmosphere.

CLOSED-END MORTGAGE *See* **Mortgage.**

CLOSED-END NUT Type of plumbing cap nut.

CLOSED IMPELLER Blower impeller consisting of a series of rotating blades or vanes, similar to the old-fashioned paddle wheel. In the case of an open impeller, the impeller blades rotate between the stationary walls of the blower housing. These walls tend to channel the air so that most of it flows out through the tips of the blades, but some air slips out sideways from between the blades and short-circuits back to the impeller inlet. A closed impeller has cover plate disks attached to the sides of the blades, and thus short-circuiting is minimized.

CLOSED LIST OF BIDDERS *See* **Invited bidders.**

CLOSED SANDWICH-TYPE PANEL Sandwich panel in which all edges are closed except for weep holes and vents.

CLOSED SOCKET Wire rope end termination consisting of a basket and ball made integral.

CLOSED SPECIFICATIONS Specifications stipulating the use of specific products and/or processes.

CLOSED SPOOL VALVE Valve that locks the hydraulic oil in an actuator when that particular device's control lever is placed in neutral. Closed spool valves are commonly employed in hydraulic cylinder circuits to maintain the position of a component when that component is not in the process of being actuated.

CLOSED STACK Unvented plumbing system.

CLOSED STAIR Stair walled on each side and closed by a door at one or both ends.

CLOSED SURVEY Survey traverse that starts and ends at the same point, whose position has been fixed by other surveys.

CLOSED SYSTEM 1. Building system having interchangeability only of its own subsystems and components; **2.** Heating or cooling piping system shut off from the atmosphere except for a vented expansion tank.

CLOSED-TOP VAN *See* **Truck.**

CLOSED TRAVERSE Survey that opens and closes on the same station.

CLOSED TYPE BRIDGE SOCKET *See* **Bridge socket.**

CLOSED VALLEY Roof valley in which the courses of shingles completely cover the valley lining, forming a secret gutter.

CLOSE FIT Part machined to a defined tolerance giving a precise clearance between mating parts.

CLOSE-GRAINED WOOD *See* **Grain.**

CLOSE JOINTED Rock containing joints that are near together; typically, columnar basalt.

CLOSE NIPPLE Pipe nipple so short that its two sets of threads meet in the middle. *See also* **Nipple.**

CLOSER 1. Last masonry unit laid in a course. Can be a whole unit or a portion of a unit. *See also* **Brick; 2.** Sheet pile cut or made to close a cofferdam when a standard pile will not fit the gap; **3.** Machine that lays strands around a core to form rope.

CLOSE-RATIO TRANSMISSION *See* **Transmission.**

CLOSE STRING Finish to the outer edge of a flight of stairs having a built-up curb string on which the balusters are set, and against which the treads and risers stop.

CLOSE STRING STAIR Stairway construction where the outer exposed string has a straight edge parallel to its lower edge, positioned so the outer ends of the steps are entirely covered.

CLOSET 1. Small room or cupboard; **2.** Room containing a toilet and basin.

CLOSET BEND Elbow drainage fitting connecting a water closet to a branch drain.

CLOSET BOLT Bolt used to attach a water closet to the closet flange.

CLOSE TIMBERING Planks placed adjacent to each other against the ground.

CLOSET LINING Tongue-and-groove boards used for lining clothes closets, often of red or aromatic cedar.

CLOSET SPINDLE Door knob spindle having a turn knob fastened to one end and provision for attaching a knob to the other end.

CLOSET SPUD Connector between the base of a ball-cock assembly in a water closet tank and the water supply pipe.

CLOSE UTILIZATION Maximum stump height of 300 mm (12 in.); minimum top dib (diameter inside bark) of 100 mm (4 in.). *See also* **Utilization standards.**

CLOSING COSTS Expenses that are normally incurred by sellers and buyers to complete the transfer of ownership of real property and that are in addition to the agreed selling price of the property.

CLOSING DATE Date (and usually time) when invitations to bid on a contract close. *Also called* **Final date.**

CLOSING DAY Day specified in an agreement for the sale/purchase of real property on which the formalities of transfer of title are concluded.

CLOSING LINE Wire rope that performs two functions: (a) closes a clamshell or orange-peel bucket, and (b) operates as a hoisting rope. *See also* **Crowd, Digging line, Drag,** and **Dragline.**

clo. stp *Abbreviation for:* closure strip.

CLOSURE 1. Device or assembly for closing an opening through a fire separation, such as a door, shutter, wired glass, or glass block, and including all components such as hardware, closing devices, frames, and anchors; **2.** Quarter or three-quarter brick to close when required; the end of a course, as distinguished from a half-brick.

CLOSURE FITTING *See* **Fitting.**

CLOSURE STRIP Flexible strip, one half of which is formed to the contour of ribbed panels, used to close openings created by the junction of metal panels and flashings.

CLOTH IMPRESSION *See* **Fabric impression.**

CLOTH INSERT Term applied to low-strength, small-diameter hose reinforced with a ply or plies of lightweight fabric.

CLOTH-INSERTED TUBING Small-diameter hose reinforced with a ply or plies of lightweight fabric.

CLOUDING Loss of luster in a painted surface, usually caused by a porous undercoat.

CLOUD (ON TITLE) Outstanding claim that adversely affects the marketability of title to real property.

CLOUD POINT 1. Temperature at which waxes in diesel fuel or a lubricant appear; **2.** Point at which fuel will clog the filters in the system.

CLOUDY Incomplete covering of a painted surface.

CLOUT NAIL *See* **Nail.**

CLOVE HITCH Knot used to hoist material, with an extra half-hitch to prevent slipping.

CLOVERLEAF *See* **Interchange types.**

clp *Abbreviation for:* clamp; clip.

clr *Abbreviation for:* clear.

clrg *Abbreviation for:* clearing.

clsd *Abbreviation for:* closed.

clstr *Abbreviation for:* cluster.

clt *Abbreviation for:* cleat.

CLUB HAMMER *See* **Hammer.**

CLUSTER Lighting fixture having fittings for two or more lamps.

CLUSTERED COLUMN Group of shafts having their bases and capitals joined to form one support.

CLUSTER GEAR *See* **Gear.**

CLUSTER HOUSING Dwelling units grouped in relative proximity, leaving open spaces as common areas.

CLUTCH 1. Device, mounted on an elevator car door,

that pulls the hoistway doors as the car door moves open or closed; **2.** Friction, electromagnetic, hydraulic, or pneumatic device that connects and disconnects two shafts revolving in line with each other. There are many types, including:

Angle spring: Conventional clutch that uses a set of six coil springs set an an angle to the input shaft. The springs apply pressure to the clutch disks, forcing them to engage the flywheel.

Automatic: Clutch whose engagement is controlled by centrifugal force, vacuum, or other power without attention from the operator.

Centrifugal: Clutch that is kept in engagement only by centrifugal force so that it automatically disconnects the power train below a fixed speed of revolutions.

Denture: *See* **Jaw clutch,** below.

Diaphragm-spring: Metal plate that acts as a diaphragm-style spring to engage or disengage a clutch.

Disk: Coupling that can be engaged to transmit power through one or more disks squeezed between a back-plate and a movable pressure plate.

Eddy current: Magnetic torque transmitting component in which the slip is varied by the voltage impressed in its electromagnetic coils, and no mechanical friction contact takes place between the driving and driven member.

Engine: Friction clutch in an engine flywheel.

Fluid: Hydraulic coupling that does not increase torque.

Flywheel: Friction clutch in an engine flywheel, oil-type or dry-type.

Freewheeling: *See* **Overrunning clutch,** below.

Friction: Device that uses friction disks for the transfer or transmission of engine power to the operating functions of a machine.

Jaw: Toothed hub and sliding toothed collar that can be engaged to transmit power between two shafts having the same axis of revolution. *Also called* **Denture clutch,** and **Positive clutch.**

Lockup: Clutch that can be engaged to provide a non-slip mechanical drive through a fluid coupling.

Oil: *See* **Wet clutch,** below.

Overrunning: Coupling that transmits rotation only in one direction, and disconnects when the torque is reversed. *Also called* **Freewheeling unit.**

Positive: *See* **Jaw clutch,** above.

Reversing: Forward-and-reverse transmission that is shifted by a pair of friction clutches.

Safety: Clutch that will slip under loads that might damage the machine. *See also* **Slip clutch.**

Slip: Friction clutch that protects a mechanism by slipping when the applied load exceeds a pre-set limit. *Also called* **Safety clutch.**

Wet: Clutch that operates in an oil bath. *Also called* **Oil clutch.**

CLUTCH BRAKE 1. Device that slows a jackshaft when the clutch is released, permitting a more rapid gear shift; **2.** Device on the input shaft of nonsynchronized transmissions that stops gears from rotating so that first or low gear may be engaged at a standstill.

CLUTCH COLLAR Major component in a locking main differential, moves and engages its large inner splines with the outer splines on the differential case, locking the axle shaft to the differential case.

CLUTCH HOUSING Transmission component that surrounds and protects the clutch and connects the transmission case to a truck's engine. *Also called* a **Bell housing.**

CLUTCH-SHIFTED TRANSMISSION *See* **Transmission.**

clv. *Abbreviation for:* clevis.

cm *Abbreviation for:* centimeter.

CM *Abbreviation for:* construction manager.

C/M *Abbreviation for:* center matched; concrete masonry.

cmai *Abbreviation for:* culmination of mean annual increment.

CMP *Abbreviation for:* corrugated metal pipe.

CMPA *Abbreviation for:* corrugated metal pipe arch.

cmpnd *Abbreviation for:* compound.

CMS *Abbreviation for:* cationic medium setting.

CN *Abbreviation for:* change notice.

cnc. *Abbreviation for:* concave.

cncl *Abbreviation for:* cancel; conceal.

cncld *Abbreviation for:* concealed.

cncr. *Abbreviation for:* concurrent.

cncv. *Abbreviation for:* concave.

cnd. *Abbreviation for:* conduit.

cnd.bx *Abbreviation for:* conduit box.

CNE *Abbreviation for:* center northeast.

CNL *Abbreviation for:* center north line.

cnsl. *Abbreviation for:* console.

cntgcy *Abbreviation for:* contingency.

cntn *Abbreviation for:* contain.

cntr *Abbreviation for:* container; counter.

cntr/bal. *Abbreviation for:* counterbalance.

cntrl *Abbreviation for:* control; controlled; controller.

cnvr *Abbreviation for:* conveyor.

cnvx *Abbreviation for:* convex.

CNW *Abbreviation for:* center northwest.

c/o *Abbreviation for:* care of.

Co. *Abbreviation for:* company.

CO *Abbreviation for:* certificate of occupancy; change order; clean out; cutout.

COACH SCREW *See* **Lag bolt/screw.**

COAGULANT Substance that, when added to water, causes suspended solids to bulk together or floc.

COALESCENCE Capacity of materials to grow together, fusing and binding.

COALESCING SEPARATOR *See* **Separator.**

COAL-TAR-BASE SEALANT Emulsion that forms an impermeable barrier between pavement and rain or snow, resisting penetration or displacement by water.

COAL-TAR EMULSION Coal-tar pitch emulsified with water.

COAL-TAR ENAMEL Bituminous liquid used to coat metal vessels prior to burying them.

COAL-TAR PAINT Cold-applied bituminous coating consisting of coal-tar pitch dissolved with solvents and fillers.

coam. *Abbreviation for:* coaming.

COAMING Raised frame, as around a roof light or the hatchway in a deck, to keep out water.

COARSE AGGREGATE *See* **Aggregate.**

COARSE-AGGREGATE FACTOR Ratio, expressed as a decimal, of the amount (mass or solid volume) of coarse aggregate in a unit volume of well-proportioned concrete to the amount of dry-rodded coarse aggregate compacted into the same volume. *Also called* b/b°.

COARSE FUEL *See* **Heavy fuel.**

COARSE GRADED AGGREGATE Sample of aggregate containing a high proportion of large particles.

COARSE-GRAINED SOIL Soil in which the larger graine sizes, such as sand and gravel, predominate.

COARSE-GRAINED WOOD *See* **Grain.**

COARSE GRAVEL Rock fragments ranging in size from 20 to 60 mm (0.75 to 2.5 in.).

COARSER RESIDUE Wood processing and plant residue that is suitable for chipping; for example, slabs, edgings, and veneer cores.

COARSE SAND Rock fragments ranging in size from 0.6 to 2 mm (0.024 to 0.78 in.).

COARSE STUFF Plaster made with either hydrated lime or lime putty, used for the first and second coats of multicoat work.

COAT Film or layer, as of paint or plaster, applied in a single operation.

coat. *Abbreviation for:* coated; coating.

COATED ABRASIVE Paper or cloth having abrasive grains bonded into the surface.

COATED BAR Bar on which a coating has been applied, usually to increase resistance to corrosion.

COATED MACADAM Road material consisting of aggregate containing very little fines and coated with tar.

COATED WIRE NAILS Nails coated with various resin gums to increase withdrawal resistance.

COATING 1. Material applied to a surface by brushing, dipping, mopping, spraying, trowelling, etc., to preserve, protect, decorate, seal, or smooth the substrate; **2.** Foreign or deleterious substances found adhering to another.

COATING ASPHALT Layer of asphalt applied to roofing material into which granulated surfacing material is embedded during manufacture.

COAXIAL Having a common axis; mounted concentrically.

COAXIAL COIL SPRINGS Springs in the driven clutch disk that help to neutralize engine torsion vibrations.

COBALT A rare mineral used as an alloying substance to make ferrous materials harder and more heat-resistant. *See also* **Alloying elements.**

COBBLE 1. In geology, a rock fragment between 63 and 254 mm (2.5 and 10 in.) in diameter; **2.** Coarse aggregate for concrete in the nominal size range 76 to 152 mm (3 to 6 in.); **3.** Small and roughly squared or egg-shaped stone.

COBBLESTONE Rock fragment, usually rounded or semirounded, with an average dimension between 76 and 300 mm (3 and 12 in.).

COBI MANDREL Segmented steel mandrel designed to be expanded inside a thin, helical pile shell to hold it while driving.

COBI PILE Thin steel shell used for cast-in-place concrete piles, that is installed with a mandrel, and that makes no structural contribution to the completed pile.

COBWEBBING Production during spraying of fine strands instead of normal atomized particles.

COCK Control valve.

COCKING Tipping sideways.

COCKLE Crease-like wrinkle in a surface intended to be flat. *Also called* **Wrinkle.**

CO-COMPOSTING Composting of municipal solid waste and wastewater treatment plant sludge.

COCONUT OIL Natural vegetable oil, used in the manufacture of alkyd resin paint.

COD *Abbreviation for:* cash on delivery.

CODE Any systematic collection or set of rules pertaining to one main subject and drafted for the purpose of securing uniformity of standards of workmanship or for maintaining proper standards of procedure.

CODE OF PRACTICE Collection of rules pertaining to a specific subject setting out procedures, lines of authority, powers such as purchasing authority, etc.

CODISPOSAL Disposal of municipal solid waste and ash in a municipal waste landfill.

CODOMINANT In forest stands with a closed canopy, those trees whose crowns form the general level of the canopy and that receive full light from above, but comparatively little from the sides; in young stands, those trees with above average height growth.

COE Cab-over-engine vehicle configuration.

coef. *Abbreviation for:* coefficient.

COEFFICIENT Factor that contributes to produce a result.

COEFFICIENT OF ACOUSTICS Property of a material that attenuates sound, i.e. reduces echoes.

COEFFICIENT OF COMPRESSIBILITY Change in the ratio of voids per unit increased from pressure.

COEFFICIENT OF DISCHARGE Factor used in figuring flow through an orifice. It takes into account the facts that a fluid flowing through an orifice will contract to a cross-sectional area that is even smaller than that of the orifice, and that there is some dissipation of energy due to turbulence.

COEFFICIENT OF EARTH PRESSURE Ratio of the active earth pressure, normal to a plane surface, to the corresponding pressure in a fluid of the same density.

COEFFICIENT OF EARTH PRESSURE AT REST Ratio of the earth pressure, normal to a plane surface, to the corresponding pressure in a fluid of the same density, when there is no movement.

COEFFICIENT OF EXPANSION Rate at which a material expands with a rise in temperature.

COEFFICIENT OF FRICTION Ratio of the forces required to move an object resting on a horizontal plane to the weight of the object. *See also* **Coefficient of kinetic friction.**

COEFFICIENT OF HEAT TRANSMISSION Constant that represents the ability of a certain material to transmit heat.

COEFFICIENT OF KINETIC FRICTION Ratio of the tangential force sustaining motion at constant velocity in a sliding system to the load perpendicular to the motion. *See also* **Coefficient of friction.**

COEFFICIENT OF LINEAR EXPANSION Fractional change in length of a material for a unit change in temperature.

COEFFICIENT OF PASSIVE EARTH PRESSURE Ratio of the passive earth pressure, normal to a plane surface, to the corresponding pressure in a fluid of the same density.

COEFFICIENT OF PERFORMANCE Measure of heat pump efficiency; the ratio of useful heat energy transferred, to the amount of energy put into the system.

COEFFICIENT OF PERMEABILITY Average velocity of water through the total area of soil under a hydraulic gradient of 1.

COEFFICIENT OF PERMEABILITY TO WATER Rate of discharge of water under lamina flow conditions through a unit cross-sectional area of a porous medium under a unit hydraulic gradient and standard temperature conditions, usually 20°C (68°F).

COEFFICIENT OF RESTITUTION Nondimensional coefficient of cushion materials that accounts for the energy-absorbing characteristics of the material. Symbol: e. The value of 'e' lies between 1 and 0.

COEFFICIENT OF SKID RESISTANCE Ratio of skid resistance to wheel load under stated conditions.

COEFFICIENT OF STORAGE Volume of water that an aquifer receives per unit surface area per unit change in head.

COEFFICIENT OF SUBGRADE FRICTION Coefficient of friction between a slab and its subgrade.

COEFFICIENT OF SUBGRADE REACTION Ratio of load per unit area on soil to the corresponding deformation. *Also called* **Modulus of subgrade reaction, and Subgrade modulus.**

COEFFICIENT OF THERMAL EXPANSION Change in linear dimension per unit length, or change in volume, per unit volume per degree of temperature change.

COEFFICIENT OF TRACTION Ratio of the maximum kilograms (pounds) pull (before the drivers slip) to the weight on the drivers.

COEFFICIENT OF UNIFORMITY Ratio between the number of particles having a diameter that is larger than 60% by mass of the particles in a soil, or aggregate sample of that diameter, to the effective size which is larger by 10% by mass of the particles.

COEFFICIENT OF VARIATION The standard deviation expressed as a percentage of the average. *See also* **Standard deviation.**

COEFFICIENT OF VELOCITY Ratio of measured velocity to the theoretical discharge velocity.

COFFER Ornamental recessed panel in a ceiling or soffit.

COFFERDAM Structure that enables construction work to be carried out below water level; a caisson. A cofferdam does not have to be entirely watertight. It may be cheaper or more practical to permit some flow into the working area with the excess water then being removed with pumps. *See also* **Cellular cofferdam, and Crib.**

COFFERED CEILING Ceiling made up of panels in more than one plane.

COFFERING Decoration of a ceiling or soffit consisting of recessed squares or other shapes.

COFFING HOIST Portable chain hoist.

COG 1. One of a series of teeth on the rim of a wheel; **2.** Wheel with a series of teeth on its rim, externally or internally; **3.** Notch on a beam that fits into a corresponding groove on another beam, making a joint; **4.** Nip formed on a roofing tile.

COGENERATION 1. Generation of heat and/or electricity using more than one type of fuel; **2.** Development of more than one type of usable energy from a fuel, i.e., electricity and heat from waste wood fiber.

COGGED JOINT Joint formed by one member being notched so as to fit into a corresponding groove in another.

COHESIOMETER Device used for determining tensile strength, measuring the force required of a cantilever beam to bend or break a specimen.

COHESION 1. Ability of a material to maintain its strength when unconfined; **2.** Bonding between particles of certain fine-grained soils that enhances shear strength and that is independent of confining pressure.

COHESIONLESS SOIL Soil or sand that, when unconfined, has little or no strength when air-dried and that has little or no cohesion when submerged.

COHESIVE SOIL *See* **Soil types.**

COIL Circular bundle or package of wire rope that is not affixed to a reel.

COINCIDENT (ELECTRIC) DEMAND *See* **Demand (electric).**

COINSURANCE *See* **Insurance.**

COKE OVER GAS Gas saved for use as a fuel when coke is made from coal in by-product ovens. It consists chiefly of hydrogen and methane.

COKING Diesel engine plugging from the formation of solid carbon due to incomplete burning of diesel fuel in the combustion chamber(s).

col. *Abbreviation for:* collar; collumn.

COLCRETE PROCESS Method of placing cement grout under pressure, consisting of cement and water combined in a tank to form a colloidal water/cement grout that then passes into a second tank containing sand. The whole grout is then delivered under pressure through a hose.

COLD APPLIED Capable of being applied without heating, as constrated to hot applied. Cold-applied products are supplied in a liquid state; hot-applied products are furnished as solids that must be heated to liquify them. *See also* **Hot applied.**

COLD CATHODE Instant-starting-type of fluorescent lamp that uses cylindrical electrodes. *See also* **Hot cathode.**

COLD CHISEL Hand chisel formed from tempered tool steel, capable of chipping and cutting mild steel, masonry, etc. There are many types including:

> **Diamond point:** Similar to a flat but ending in a flat-faced square section (used for sharp corners, etc.).
> **Flat:** Short, relatively-thick, hexagonal or octagonal steel bar ground to a two-faced taper ending with a steep-angled cutting edge.
> **Plain:** Chisel built to withstand rugged use in stonecutting, 31 to 70 mm (1.25 to 2.75 in.) wide.
> **Point:** Chisel used to remove small stone knobs that project off a stone face; the point chisel is drawn to a point on one end.
> **Round nose:** Similar to a flat but ending in a curved face (used for cutting grooves).
> **Tooth:** Steel chisel in which the blade forms a series of teeth, used for dressing stone.

COLD CRANKING ABILITY Represents a battery's sustained constant current in amps that produces a minimum terminal voltage under load of 1.2 volts per cell at 30 seconds.

COLD CRANKING SIMULATOR Viscometer used to predict the ability of an engine lubricant to allow cranking during cold starts.

COLD DECK Pile of yarded logs left for later transportation.

COLD-DRAWN WIRE REINFORCEMENT Steel wire made from rods that have been hot rolled from billets, cold-drawn through a die; for concrete reinforcement, of small diameter such as in gauges not less than 2 mm (0.080 in.) nor greater than 15 mm (0.625 in.).

COLD FACE Surface of a refractory section not exposed to the source of heat.

COLD FEED Introduction of compounded rubber into processing equipment without milling.

COLD FEED SYSTEM Equipment used to proportion and feed aggregate to an asphalt mixing plant prior to drying.

COLD FLEX *See* **Low-temperature flexing.**

COLD FLEXIBILITY Relative ease of bending following exposure to specified low temperature conditioning.

COLD FLOW 1. Continuous deformation under stress. *See also* **Deformation,** and **Drift;** 2. Change of physical dimensions of a sealant after polymerization or original set has taken place.

COLD-FORMED MEMBERS Structural members formed from steel without the application of heat.

COLD JOINT 1. Joint or discontinuity resulting from a delay in concrete placement of sufficient time to preclude union of the material in two successive lifts; 2. Condition where, due to a delay in tile setting, the bed has lost its plasticity.

COLD-JOINT LINES Visible lines on the surface of formed concrete indicating the presence of joints where one layer of concrete had hardened before subsequent concrete was placed.

COLD-LAID ASPHALT CONSTRUCTION Plant mixes that may be spread and compacted at atmospheric temperature.

COLD-MIX ASPHALT Asphalt mix, produced either on site or at a central plant, that is mixed and placed at ambient temperatures. Cutbacks and emulsions are examples.

COLD PATCH Mixture of aggregate and liquid bitumen that can be stored indefinitely and used cold to repair small potholes.

COLD PLANER Self-propelled construction machine (either rubber-tired or crawler mounted) that is specifically designed to cut a pavement to a predetermined depth, grade, or slope, and that reduces the pavement material in size in the process, using a rotating drum equipped with special cutting tools. *Also called* **Grinder,** and **Milling machine.**

COLD PLANING/MILLING Process in which a rotating drum, equipped with special working tools, cuts the pavement to a predetermined depth and reduces the cuttings to a minimum size in the process.

COLD-POURED SEALING COMPOUND Multipart compound mixed and poured into a horizontal concrete joint at ambient temperature.

COLD-PROCESS ROOFING Specially formulated asphalt emulsions used in built-up roofing in place of hot emulsions.

COLD RECYCLING Reuse of reclaimed asphalt pavement (RAP) by recycling without heating, generally on-site, for construction of bases or low-volume roadways.

COLD ROLLING Bending steel plates to form light structural sections, without application of heat.

COLD SETT *See* **Sett.**

COLD STRENGTH Compressive or flexural strength of refractory concrete determined prior to drying or firing.

COLD-TRAIL To build a fire line right along the edge of a wildfire when it is burning slowly or very little, taking advantage of favorable burning conditions.

COLD-WATER PAINT Paint in which the binder or vehicle portion is composed of latex, casein, glue, or some similar material dissolved and dispersed in water.

COLD-WATER TEST Soundness test in which a standard-size pat of cement paste is stored in cold water for 28 days. If after this period the cement pat shows no sign of cracking, disintegration, or warping, the cement is considered sound without risk of lime expansion.

COLD-WEATHER CONCRETING Precautions taken before mixing or placing concrete at temperatures below 4.5°C (40°F).

COLD WELDING Use of high pressure and no outside heat to force metal parts to fuse.

COLD-WORKED STEEL REINFORCEMENT Steel bars or wires that have been rolled, twisted, or drawn at normal ambient temperatures.

COLD WORKING Bending or otherwise deforming metal at a temperature lower than its recrystallization temperature.

COLEMANITE Mineral hydrated calcium borate. *See also* **Boron-loaded concrete.**

coll. *Abbreviation for:* collateral; collect; collector; college.

collab. *Abbreviation for:* collaboration.

COLLAPSE 1. Failure of a structure resulting in its complete or partial disintegration; **2.** Flattening of single cells or rows of cells of heartwood during the drying or pressure treatment of wood, characterized by caved-in or corrugated appearance; **3.** The outside-in differential pressure that cuses structural or filter medium failure of a filter element.

COLLAPSED SLUMP Slump test in which the cone of concrete collapses on being released from the mould.

COLLAPSE INSURANCE Indemnity against loss or damage caused by collapse of a building or structure.

COLLAPSE PRESSURE Outside-in differential pressure that causes structural failure.

COLLAPSIBLE ANCHOR *See* **Wall anchor.**

COLLAPSIBLE MANDREL *See* **Mandrel.**

COLLAPSIBLE PANS Telescopic centering.

COLLAPSIBLE SOIL Soil in which the voids ratio rapidly decreases on the introduction of water.

COLLAR 1. Tie beam positioned high up the slope of a roof; **2.** Flange mounted on a saw arbor to support the blade, on one or both sides; **3.** Mouth of an opening of a borehole or shaft; **4.** Act of starting a borehole; **5.** Sliding ring mounted on a shaft so that it does not revolve with it; **6.** Unloaded portion of a blasthole

extending from the surface down to the top of the explosives column.

COLLAR BEAM *See* **Collar tie.**

COLLAR-BEAM ROOF Roof frame composed of two common rafters supported by wall plates and meeting at the ridge, tied together by a horizontal beam connecting points more than halfway between the plate and the ridge.

COLLAR BRACE *See* **Collar tie.**

COLLARING Starting a drill hole; the phase from first applying the bit against a surface until the hole is deep enough to hold the bit from slipping out of it.

COLLAR JOINT Vertical, longitudinal (often mortared) joint between wythes of masonry.

COLLARLESS SAW Saw that floats on the arbor, being keyed directly to the arbor itself.

COLLAR TIE Horizontal member used to provide intermediate support for opposite roof rafters, usually located in the middle third of the rafters. *Also called* **Collar Beam,** and **Collar brace.**

COLLATERAL Property pledged as security for a debt.

COLLATERAL LOAD All building dead loads other than the structural frame: plumbing and electrical systems, finishings, etc.

COLLATERAL SECURITY Security given in addition to the principal security and only to be resorted to after the principal security has been realized upon.

COLLECTING SYSTEM Drains and sewers between point(s) of influent and discharge to a main sewer.

COLLECTION Act of removing accumulated containerized solid waste from the generating source. There are several methods, including:

> **Alley:** Picking up of solid waste from containers placed adjacent to an alley.

> **Carryout:** Crew collection of solid waste from on on-premise storage area using a carrying container, carry-cloth, or a mechanical method.

> **Contract:** Collection of solid waste carried out in accordance with a written agreement in which the rights and duties of the contractual parties are set forth.

> **Curb:** Collection of solid waste from containers placed adjacent to a thoroughfare.

> **Franchise:** Collection made by a private firm that is given exclusive right to collect for a fee paid by customers in a specific territory or from specific types of customers.

> **Municipal:** Collection of solid waste by public employees and equipment under the supervision and direction of municipal authorities.

> **Private:** Collection of solid waste by individuals or companies from residential, commercial, or industrial premises; the arrangements for the service are made directly between the owner or occupier of the premises and the collector.

> **Setout/setback:** Removal of full and the return of empty containers between the one premise storage and the curb by a collection crew.

COLLECTION CENTER Site designed to accept secondary materials from homes, businesses, institutions, and industrial sites.

COLLECTION FREQUENCY Number of times refuse collection is provided in a given period of time.

COLLECTION LINE Plumbing drain.

COLLECTION METHOD Means by which refuse is collected from its point of origin. There are several methods, including:

> **Daily route:** Each collection crew is assigned a weekly route that is divided into daily routes.

> **Definite working day:** Collection proceeds along a route for a length of time adopted for a working day. The next day, collection begins where the crew stopped the day before. This procedure continues until the whole route is covered, whereupon the crew returns to the beginning of the route.

> **Group task:** Responsibility for collecting on assigned routes is shared by more than one crew. Any crew that finishes a particular route works on another until all are completed.

> **Inter-route relief:** Regular crews help collect on other routes when they finish their own.

> **Large route:** Each crew is assigned a weekly route. The crew works each day without a fixed stopping point or work time, but it completes the route within the working week.

> **Reservoir route:** Several crews are used to pick up on a centrally located route after having collected on peripheral routes.

> **Single load:** Areas or routes are laid out that normally provide a full load of solid waste. Each crew usually has at least two such routes for a day's work. The crew quits for the day when the assigned number of routes is completed.

> **Swing crew:** One or more reserve work crews go anywhere help is needed.

> **Variable-size crew:** A variable number of collectors is provided for individual crews, depending on the amount and condition of work on particular routes.

COLLECTION STOP Stop made by a vehicle and crew to collect solid waste from one or more service sites.

COLLECTION SYSTEM Diverse system that collects and transmits all forms of sewage and wastewater, plus in some instances storm water, to a discharge point or treatment facility.

COLLECTIVE SYSTEM Elevator having the ability to 'store' both hall and car calls, sort the calls, and stop for them as they are reached by the car traveling in either direction.

COLLECTOR Part of a solar system that intercepts the sun's rays and converts them directly to a form of transportable energy.

COLLECTOR ANGLE Angle at which a solar collector must be tilted relative to the horizontal to maximize collection of solar radiation.

COLLECTOR BOX Transition piece between a rain-water gutter and downspout.

COLLECTOR EFFICIENCY Ratio of solar energy arriving at a collector to that collected by the device.

COLLECTOR RING Assembly of slip rings for transferring electrical energy or an electric signal from a stationary to a rotating member.

COLLECTOR SUBSYSTEM Assembly necessary for absorbing solar radiation, converting it into useful thermal energy, and transferring the thermal energy to a heat transfer fluid.

COLLECTOR TILT Angle of a solar collector assembly or the roof supporting it to the horizontal.

COLLET See **Flange.**

COLLIMATED Filaments rendered parallel.

COLLIMATION Adjustment of the line of sight in a surveying telescope in relation to the cross hairs; the process of aligning the optical and mechanical axes or surfaces of an instrument.

COLLIMATION ADJUSTMENT Action of bringing the line of collimation of a surveying telescope into near coincidence with the axis of collimation. *See also* **Error of collimation.**

COLLIMATION ERROR Error in a survey attributable to misalignment of the telescope.

COLLIMATION LINE Line of sight that passes through the cross hairs of the reticule of a survey instrument.

COLLIMATION METHOD Survey method in which the height of the instrument is determined first by taking a sight on a point of known elevation, calculating all subsequent elevations relative to the instrument, and closing the survey by again sighting on a point of known elevation. *Also called* **Height of instrument.**

COLLOID Suspension of finely divided particles in a continuous medium producing a substance that is in a state of division, preventing passage through a semipermeable membrane.

COLLOIDAL CONCRETE Concrete in which the aggregate is bound by colloidal grout.

COLLOIDAL DISPERSION Finely divided mixture that is not chemically combined.

COLLOIDAL GROUT Grout in which a substantial proportion of the solid particles have the size range of a colloid.

COLLOIDAL MIXER Mixer designed to produce colloidal grout.

COLLOIDAL PAINT Paint made with pure pigments and without fillers. The pigments are not ground, but are reduced to extremely fine particles and colloidally suspended in the vehicle.

COLLOIDAL PARTICLE Electrically charged particle dispersed in a second continuous medium.

COLLUVIAL SOIL Soil deposited mainly through the action of landslides.

colog. *Abbreviation for:* cologarithm.

COLONIAL RING-AROUND FIREPLACE Fireplace with only the facing of bricks stacked up at the jambs, with an exposed soldier course over the top.

COLONNADE Series of regularly spaced columns.

COLONNETTE Small column, usually nonload bearing and applied as decoration.

COLOR Aspect of the appearance of an object dependent upon the spectral composition of the incident light, the spectral reflectance or transmittance of the object, and the spectral response of the observer. *See also* **Appearance, Hue, Lightness,** and **Saturation.**

COLOR CODING Identification of similar parts of a system according to an applied color.

COLORED CEMENT Mixture of cement and pigments to produce a colored product.

COLORED GROUT Commercially prepared grout consisting of carefully graded aggregate, portland cement, water dispersing agents, plasticizers, and color-fast pigments.

COLORIMETRIC VALUE Indication of the amount of organic impurities present in fine aggregate.

COLORING PIGMENT Substance capable of imparting color to another substance.

COLOR (OF LUBRICATING GREASE) Shade and intensity shown when lubricating grease is viewed under conditions such as to eliminate bloom. *See also* **Bloom.**

COLOR (OF LUBRICATING OIL) Shade shown when viewed under transmitted light only.

COLOR RETENTION Permanence of a color under a specific set of conditions.

COLOR SYSTEMS 1. Codification of colors according to a range of constituent pigments, giving the amount of each so that any given color, hue, or tint can be made from a fixed base; **2.** Design of a coordinated and/or contrasting series of colors as part of a scheme for final decoration.

COLPROVIA Patented asphalt mix formula utilizing powdered asphalt.

COLUMBIUM *See* **Alloying elements.**

COLUMN Upright member, circular or rectangular in section, often slightly tapering, with a height at least three times its least lateral dimension, that receives pressure in the direction of its longitudinal axis.

COLUMN ANCHORS Metal straps used so that a column base is clear of grade level where moisture may collect.

COLUMN CAPITAL Enlargement of the head of a column below a slab, intended to increase shearing resistance.

COLUMN CLAMPS Steel straps fastened around formwork for a column to prevent spreading.

COLUMN CURVE Curve expressing the relationship between axial column strength and slenderness ratio.

COLUMN FOOTING Additional reinforcement of a concrete slab, possibly with an increase in the depth of the slab, so as to accommodate the additional loads transferred by a column.

COLUMN FORM Formwork used for a reinforced concrete column.

COLUMN HORSE Wood or metal supports, used in groups of two or more, to hold the main steel reinforcement in a convenient position for placing ties while prefabricating column, beam, or pile cages.

COLUMNIATION Arrangement of columns in relationship to each other and to other building elements.

COLUMN LOAD Continuous charge of explosives in a blasthole, with no stemming between charges.

COLUMN SCHEDULE 1. List giving number, size, and placement of structural steel columns used in a structure; **2.** Table giving the mark number and size for the column, number of reinforcing steel pieces and size of verticals, ties or spirals, and any bar mark numbers required.

COLUMN SECTIONS Rolled-steel sections used as columns in a structural-steel frame.

COLUMN SIDE One of the vertical panel components of a column form.

COLUMN STRIP Portion of a flat slab over columns and consisting of the two adjacent quarter panels on each side of the column center line. *See also* **Edge strip,** and **Middle strip.**

COLUMN TIES Reinforcing bars bent into square, rectangular, circular, or U shapes for the purpose of holding column vertical bars laterally secure for the placement of concrete.

COLUMN VERTICALS Upright or vertical reinforcing bars in a concrete column.

com. *Abbreviation for:* common; commercial.

COMB 1. Ridge of a roof; **2.** Hand tool used to dress the face of stone; **3.** Tool used to simulate wood grain when applying stain or paint; **4.** Tool for scratching wet plaster to form a key for subsequent coats.

comb. *Abbreviation for:* combination; combine; combined; combustible; combustion.

COMB CHISEL FINISH Architectural rough-textured surface finish for concrete, produced by crushing or spalling off the cement skin with a comb chisel.

COMBED JOINT Right-angle joint consisting of a series of tenons engaging in mating mortises, typically used in cabinetry to form drawers, etc.

COMBINATION Joining of two or more parts, materials, elements, etc., each of which has a separate use, and each of which performs differently when combined.

COMBINATION (AUTOMATIC SPRINKLER) SYSTEM *See* **Automatic sprinkler system.**

COMBINATION BURNER Burner capable of burning either gas or oil.

COMBINATION DOOR *See* **Door types.**

COMBINATION DRILL AND COUNTERSINK *See* **Bit.**

COMBINATION FIXTURE Fixture designed to be both a kitchen sink and a wash basin.

COMBINATION LINTEL Reinforced or prestressed concrete lintel, shaped for a load-bearing inner leaf of a wall with a formed steel tray to carry an outer facing.

COMBINATION LOCK Lock operated by a dial that can be set to a series of numbers and/or letters, as against being operated by a key.

COMBINATION MEDIUM Filter medium composed of two or more types, grades, or arrangements of filter media to provide properties that are not available in a single filter medium.

COMBINATION PLANE *See* **Plane.**

COMBINATION PLIERS Pincer-like hand tool that can perform more than one function: gripping and bending or cutting, for instance.

COMBINATION POLICY Insurance policy combining the contracts of two or more insurers where each provides a different kind of coverage.

COMBINATION RISER Main vertical water supply pipe of an automatic sprinkler system, supplying both sprinklers and hose outlets.

COMBINATION SCREED Type of asphalt paver screed that uses both a tamper bar and vibration to achieve compaction of the mix.

COMBINATION SQUARE Hand tool combining an inside try-square, outside try-square, miter square, plumb and level, depth gauge, marking gauge, straightedge, bevel protractor, and square head.

COMBINATION STORM AND SCREEN *See* **Window.**

COMBINATION SUBFLOOR Panel material that provides structural rigidity and load panel capacity to the structure, and also acts as a surface for finished floorings.

COMBINATION VEHICLE Truck or tractor coupled to one or more trailers (including semitrailers).

COMBINATION WRENCH *See* **Wrench.**

COMBINED-AGGREGATE GRADING Particle size distribution of a mixture of fine and coarse aggregate.

COMBINED CARBON Carbon that is chemically combined with oxygen.

COMBINED DRAIN Drainage piping that carries both sanitary sewage and storm water. *See also* **Drain.**

COMBINED HYDROGEN Hydrogen that is chemically combined with oxygen.

COMBINED FOOTING Structural unit or assembly of units supporting more than one column.

COMBINED MECHANISM Mechanism determined by a plastic analysis procedure, that combines elementary beam, panel, and joint mechanisms.

COMBINED SEWER Sewer intended to carry both sanitary sewage and stormwater, or industrial wastes and stormwater.

COMBINED SHEAR REINFORCEMENT Combination of bent-up bars with links to resist shear stresses.

COMBINED STRESSES Stress more complex than simple tension, compression, or shear.

COMBUSTIBLE 1. Flammable, easily ignited; 2. Material that fails to conform to ULC-S114-1975, *Standard Method of Test for Determination of Non-Combustibility in Building Materials;* 3. Material that can be burned.

COMBUSTIBLE CONSTRUCTION Type of construction that does not meet the requirements for non-combustible construction.

COMBUSTIBLE DUST Dust particles ignitable and liable to explode.

COMBUSTIBLE FIBER Finely divided combustible vegetable or animal fiber, or thin sheet or flake of such material, which in a loose, unbaled condition presents a flash fire hazard.

COMBUSTIBLE LIQUID Liquid having a flashpoint at or above 37.8°C (100°F) but below 93.3°C (200°F).

COMBUSTIBLE WASTE *See* **Solid waste.**

COMBUSTION 1. Burning of fuel in the presence of oxygen or other gas; 2. Chemical change, especially oxidation, accompanied by the production of light and heat.

COMBUSTION AIR Air that enters an engine and is mixed with a fuel for the combustion process.

COMBUSTION CHAMBER Cavity between the piston and cylinder head when the piston is at top-dead-center (TDC).

COMBUSTION DELAY Time between impact and combustion in an impact atomization diesel pile hammer.

COMBUSTION GASES Mixture of gases and vapors produced by combustion.

COMBUSTION PRODUCT Matter resulting from combustion, such as flue gases, water vapor, and ash.

COMBUSTION VENTING Necessary fresh air fed to an oil- or gas-burning appliance.

COME-ALONG 1. Ratchet tool with chain and hook, used for pulling; 2. Hoe-like tool with a blade about 100 mm (4 in.) high and 500 mm (20 in.) wide and curved from top to bottom, used for spreading concrete.

COMFORT LINE Distance above floor level at which a constant temperature is maintained by a heating/cooling system.

COMFORT STATION Public toilet.

COMFORT ZONE Temperature range over which a majority of people feel comfortable.

comm. *Abbreviation for:* commission; committee; communications; commutator.

COMMENCEMENT OF WORK Date when work on a project begins on site and the contractor assumes responsibility.

COMMERCIAL BUTANE Mixture of easily liquefiable hydrocarbon gases (consisting principally of butane) that is sold as 'butane.' It is one of the components of raw natural gas, but is also derived from petroleum refining processes. *See also* **Butane.**

COMMERCIAL FOREST LAND *See* **Land use classes (forest management).**

COMMERCIALLY SMOOTH Degree of smoothness of an article that is acceptable in accordance with industry practice.

COMMERCIAL MATCHING Of paint, matching

colors within a defined tolerance that will not usually be visible to the naked eye.

COMMERCIAL PLANT Asphalt mixing plant capable of producing a variety of mixes with minimal changeover time. Usually refers to large stationary batch-type plants located in metropolitan areas.

COMMERCIAL POWER Electrical energy provided on a user-pay basis over an independent distribution network. *Also called* **Grid power.**

COMMERCIAL PROPERTY Property designed for use by owners and tenants whose occupancy enables them to engage in business.

COMMERCIAL THINNING Silviculture treatment that thins out an overstocked stand by removing trees that are large enough to be sold as products, such as poles or fence posts (as opposed to juvenile spacing). It is carried out to improve the health and growth rate of the remaining crop trees.

COMMERCIAL TIRE Tire designed for trucks and for industrial use.

COMMERCIAL WASTE *See* **Solid waste.**

COMMINUTOR Device that reduces the size of solid material, typically installed at the influent of a waste treatment facility.

COMMISSION Money paid to an agent as compensation for finding a buyer and completing a sale.

COMMISSIONING Procedures aimed at taking a completed but inanimate plant or project to the point where it can perform according to design and as specified.

COMMITMENT 1. Pledge or promise. *See also* **Cost types; 2.** Risks accepted by an insurer.

COMMITMENT DOCUMENT Contract, change order, or purchase order issued for the supply of goods and services. The document legally binds the issuer and commits them to a financial obligation.

COMMITTED COST Value of commitment documents issued for the scope of the work within a cost class at a reporting date. A cost class may be partially or completely committed at a particular point in time.

COMMODE 1. Portable washstand; **2.** Piece of furniture containing a chamber pot.

COMMODE STEP Horizontally curved riser, usually at the foot of a stairway.

COMMON Grade of lumber containing numerous defects, rendering it unsuitable for high-class finish.

COMMON AMERICAN BOND *See* **Bond.**

COMMON AREA Area available for use by all owners or tenants.

COMMON BOARD Lumber less than 50 mm (2 in.) thick.

COMMON BOND *See* **Bond.**

COMMON BRICK *See* **Brick.**

COMMON BRICKWORK Wall built out of ordinary and cheaper classes of brick, where appearance is not an important consideration.

COMMON CARRIER Trucking firm that hauls for hire.

COMMON COST Cost of raw materials, facilities, or services employed in two or more projects.

COMMON DOVETAIL Dovetail joint with both members showing end grain.

COMMON DUCT 1. Duct containing two or more building services; **2.** Duct common to two or more elevator cars.

COMMON ELEMENT Any part or facility (hall, entrance area, swimming pool, parking area) of a (condominium, apartment, etc.) development that is owned in common by the unit owners, or that is available to all owners or tenants. *Also called* **Common property.**

COMMON EXCAVATION *See* **Excavation.**

COMMON GROUND 1. Land in the public domain and available to everyone for specific uses; **2.** Strip of wood, often rough, that serves as a base for plaster, cabinetry, sheeting, etc.

COMMON NAIL *See* **Nail.**

COMMON JOIST Board set on edge to span between walls.

COMMON NAME Designation or identification such as code name, code number, trade name, brand name, or generic name.

COMMON PARTITION Nonload-bearing wall, finished on both faces, dividing areas used for different purposes or by different tenants.

COMMON PROPERTY *See* **Common element.**

COMMON RAFTER *See* **Rafter.**

COMMON SEWER Sewer to which subdrains from various properties are connected.

COMMON TRUSS *See* **Truss.**

COMMON VENT Vertical vent pipe serving two fixture drains installed at the same level in a vertical stack.

COMMON WALL *See* **Wall.**

COMMON WIRE Feed or return wire that connects circuits in parallel.

COMMUNAL AMENITY AREA Area within the boundary of a project, used by the residents on a shared basis for their enjoyment and recreation. *Also called* **Communal space.**

COMMUNAL SPACE *See* **Communal amenity area.**

COMMUNICATION LINE Conductor and its supporting or containing structures that is used for public and private signal or communication service, and that operates at a potential not exceeding 400 volts to ground or 750 volts between any two points of the circuit, and the transmitted power of which does not exceed 150 watts.

COMMUNITY Area within a city, primarily residential, that uses such facilities as schools, shopping, and recreation as part of the common resource. Such an area has ill-defined and continually changing boundaries and is more an expression of those who live within it than of lines drawn on a map or plan.

COMMUNITY CENTER Area within a community

where public, social, and recreational facilities are located.

COMMUNITY PLANNER *See* **Urban planner.**

COMMUNITY PLANNING *See* **Urban planning.**

COMMUTATOR Part of an armature on which the brushes ride; a ring of adjacent copper bars, insulated from each other and to which the wires of the armature of a direct current generator are attached.

COMMUTATOR SEGMENTS Copper segments of the commutator connected to the conductors within an armature.

comp. *Abbreviation for:* compass; component; compose; composite; composition; compound; compress; computation.

COMPACT Reduce in bulk; densify.

COMPACT COARSE SAND *See* **Soil types.**

COMPACTED Materials are considered compacted if their relative compaction exceeds 90%.

COMPACTED-CONCRETE PILE *See* **Pile.**

COMPACTED CUBIC YARD Measurement of material after it has been placed and compacted as fill, equivalent to 0.765 m³.

COMPACTED EARTH Area of bare soil, made dense by artificial means or by pedestrian and vehicular traffic.

COMPACTED FILL Fill that is placed and compacted in layers under controlled conditions to achieve a uniform and dense soil mass capable of supporting a defined load. *See also* **Fill.**

COMPACTED VOLUME Volume of a material, such as soil, after it has been compacted a known percent.

COMPACT FINE SAND *See* **Soil types.**

COMPACTING FACTOR Ratio obtained by dividing the observed mass that fills a container of standard size and shape when allowed to fall into it under standard conditions of test, by the mass of fully compacted material that fills the same container.

COMPACTION 1. Increase in density from loose material to compacted material; **2.** Compression of soil or other granular material into a dense state through the application of weight or mechanical manipulation. *See also* **Consolidation.**

COMPACTION PIT TRANSFER SYSTEM Transfer system in which solid waste is compacted in a storage pit by a crawler tractor before being pushed into an open-top transfer trailer.

COMPACTION RATIO Initial volume of solid waste divided by the final volume attained after compaction.

COMPACTIVE EFFORT Energy transferred into a material that presses the particles together, expels air and moisture from the mass, and fills the voids to make the material more dense.

COMPACTOR 1. Machine designed and used specifically to compact materials. It densifies material through the application of static force, or dynamic force combined with static force; **2.** Self-propelled or towed vehicle used, to densify materials through the application of static force, or centrifugal force combined with static force; **3.** Equipment used to compact waste materials. There are several types, including:

> **Mobile:** Vehicle with an enclosed body containing mechanical devices that convey and compress solid waste in the main compartment of the body.

> **Sanitary landfill:** Vehicle equipped with a blade and wheels with load concentrators to provide compaction and a crushing effect.

> **Stationary:** Machine that reduces the volume of solid waste by forcing and compressing it into a container.

COMPACTOR ATTACHMENT Attachment designed for backhoe/loaders and excavators that provides compaction and/or driving capability; may be attached directly to the stick or attached to the bucket.

COMPACTOR COLLECTION TRUCK Enclosed vehicle provided with special mechanical devices for loading the refuse into the main compartment of the body and for compressing and distributing the refuse within the body.

COMPACTOR/FINISHER Machine, part of a fixed-form paving train, consisting of rotary strike-off paddles, vibrating compaction beam, and oscillating finishing beam.

COMPACT SECTION Steel section capable of developing a fully plastic stress distribution and possessing a rotation capacity of approximately 3 before the onset of local buckling.

COMPARATOR Device used to make comparisons.

COMPARTMENT 1. Forest management subdivision of a block of land, usually of continuous land ownership; **2.** Subdivision of an enclosed space.

COMPASS 1. Instrument that shows the magnetic meridian and fixes the magnetic azimuth of a line of sight through a sighting implement; a graduated horizontal circle and a pivoting magnetic needle; **2.** Drafting instrument, used for drawing circles.

COMPASS BRICK *See* **Brick.**

COMPASS PLANE *See* **Plane.**

COMPASS ROOF Roof having an exterior surface that forms a continuous arc.

COMPASS SAW *See* **Saw.**

COMPASS WINDOW Bay oriel window.

COMPATIBILITY Ability of materials to be in contact indefinitely without any adverse effect to either. Compatibility does not imply adhesion.

compd *Abbreviation for:* compounded; compacted.

compen. *Abbreviation for:* compensate; compensated; compensating.

COMPENSABLE INTEREST Property right, that if acquired for highway purposes, would entitle the owner to receive just compensation.

COMPENSATED FOUNDATION Foundation slab constructed at such a depth below existing ground level that the mass of the structure and foundation is balanced by the mass of the soil excavated.

COMPENSATING CHAIN Welded-link chain used

for hoist rope weight compensation.

COMPENSATING DIAPHRAGM Fitting to the telescope of a survey instrument that alters the space between stadia hairs when a sloping sight is made.

COMPENSATING DRIVE Free-wheeling unit in the front propeller shaft of a four-wheel-drive transmission that allows the front wheels to go farther than the rear on curves.

COMPENSATING ERROR Combination of two or more errors, usually in calculation or measurement, that combine to cancel each other. *See also* **Systematic error.**

COMPENSATING ROPE Wire rope installed to obtain hoist rope weight compensation.

COMPENSATING-ROPE SHEAVE Sheave mounted in an elevator pit that maintains tension on and guides the compensating ropes.

COMPENSATING-ROPE SHEAVE SWITCH *See* **Switch.**

COMPENSATION 1. Payment made as an adjustment for a benefit denied. In the case of land expropriation, it is in the form of payment based upon the market value of the land taken at the time of the expropriation; **2.** Use of wire ropes or chains suspended from the underside of an elevator car to counterbalance the transfer of weight of the hoist ropes from one side of the hoisting machine to the other as the car moves in the hoistway.

COMPENSATOR Device used to restore a level line of sight to an instrument that is slightly out of level. The compensator uses the force of gravity to pull a set of prisms or mirrors into the correct position to cause the line of sight to be level. There are two common types of compensators: wire hung and ball bearing.

COMPETENT PERSON 1. One who has been proven through examination or demonstrated experience and ability to have a broad knowledge of a subject or skill; **2.** One who is capable of identifying existing and predictable hazards in the surroundings or working conditions that are unsanitary, hazardous, or dangerous to employees, and who has authorization to take prompt corrective measures to eliminate them. *See also* **Authorized person** and **Designated person.**

COMPETITION 1. Difference between two or more proposals for the same contract; **2.** Struggle among trees and other vegetation, generally for limited nutrients, light, and water present on the site.

COMPETITIVE BID CONTRACT Contract awarded as a result of competitive bidding.

COMPETITIVE BIDDING/TENDERING Process in which several independent, qualified contractors or suppliers are invited to bid on a contract.

compil. *Abbreviation for:* compilation.

compl. *Abbreviation for:* complete; completed.

COMPLETED-CONTRACT METHOD OF ACCOUNTING *See* **Method of accounting.**

COMPLETED OPERATIONS INSURANCE Indemnity against accidents that occur following completion of a project that has been turned over the to owner or abandoned.

COMPLETED VALUE INSURANCE Policy written at the start of a project in an amount, usually derived from the contract sum, less specified exclusions, and adjusted to the final insurable cost on completion.

COMPLETED VEHICLE Vehicle that requires no further manufacturing operations to perform its intended function, other than the addition of readily attachable components such as mirrors or tires and rim assemblies, or minor finishing operations such as painting.

COMPLETE FUSION Fusion that has occurred over the entire surface of a base metal being welded.

COMPLETE JOINT PENETRATION When weld metal completely fills the grooves and fuses with the base metal through its entire thickness.

COMPLETELY KNOCKED DOWN Shipment of specific product parts to a remote location from which a complete product or component can be assembled. *See also* **Partially knocked down.**

COMPLETE TREE Every component of a tree from leaves or needles to root hairs.

COMPLETE TREE HARVESTING Harvesting of a complete tree, including the roots.

COMPLETION BOND *See* **Bond.**

COMPLETION DATE Date upon which the work described in the contract documents must be completed.

COMPLETION OF PROJECT *See* **Project completion.**

COMPLEX SOAP Soap wherein the soap crystal or fiber is formed, usually by co-crystalization of two or more compounds:

(a) Normal soap such as metallic stearate or oleate;

(b) Complexing agent such as metallic salts or short-chain organic acids such as acetic acid or lactic acid, or the inorganic salts such as carbonate or chloride, which brings about a change in grease characteristics, usually recognized by an increase in dropping point.

COMPONENT 1. Manufactured product that serves a specific function; **2.** For plywood applications, a glued and/or nailed structural assembly of plywood and lumber, such as a box beam or stressed-skin panel; **3.** Prefabricated building sections in panelized construction.

COMPONENT METHOD Estimating technique based on the cost of completed building components, rather than unit costs.

COMPOSITE 1. Designation of a classic order of architecture that, in the capital to a slender fluted column, combines the scroll-like ornaments of the Ionic capital with the acanthus leaves of the Corinthian order; **2.** Combinations of materials differing in composition or form on a macroscale. The constituents retain their identities in the composite. Normally the components can be physically identified and there is an interface between them.

COMPOSITE BEAM Steel beam structurally connected to a concrete slab so that the beam and slab respond to loads as a unit. *See also* **Concrete-encased**

beam.

COMPOSITE COLUMN 1. Concrete compression member reinforced longitudinally with structural steel shapes, pipe, or tubing with or without longitudinal reinforcing bars; **2.** Steel column fabricated from rolled or built-up steel shapes and encased in structural concrete, or fabricated from steel pipe or tubing and filled with structural concrete.

COMPOSITE CONCRETE FLEXURAL MEMBERS Concrete flexural members consisting of concrete elements constructed in separate placements but so interconnected that the elements respond to loads as a unit.

COMPOSITE CONSTRUCTION *See* **Construction types.**

COMPOSITE JOINT *See* **Pile joint.**

COMPOSITE LINER Landfill liner that includes both a synthetic liner and compacted clay components.

COMPOSITE MASONRY Masonry construction in which at least one wythe has structural characteristics different from the others and where the wythes are bonded in a manner to act as a single structural whole.

COMPOSITE ORDER OF ARCHITECTURE Column, base, shaft, and capital combining features of the Ionic and Corinthian designs. *See also* **Order.**

COMPOSITE PANEL Veneer-faced panel with a reconstituted wood core.

COMPOSITE PILE *See* **Pile.**

COMPOSITE RATING Single liability insurance premium covering the five standard divisions of construction coverage: payroll, number of elevators, amount of sublet work, gross receipts, and contract cost.

COMPOSITE SAMPLE Sample obtained by blending two or more individual samples of a material.

COMPOSITE SEAL Seal composed of two or more materials of differing flexibility or hardness.

COMPOSITE-TYPE PAVEMENT *See* **Pavement structure combination.**

COMPOSITE WALL *See* **Wall.**

COMPOSITION BOARD Building panels manufactured from various forms of wood fiber that is bonded and subjected to heat and pressure.

COMPOSITION ROOFING Roll-roofing material, available in various weights, consisting of asbestos felt saturated with asphalt and finished with colored mineral granules.

COMPOSITION SHINGLES Shingles fabricated from composition roofing material.

COMPOSITION SIDING Siding panels made from a wide range of materials, most of which include a prefinished external weathering surface, usually patterned and colored.

COMPOSITION TILE Hard tile surfacing unit made from a mixture of chemicals. The finished surface can be the mixture of chemicals or can be marble chips to create a terrazzo finish. The unit is made hard by the set of the chemicals and the product is not fired as in the manner of a ceramic tile.

COMPOUND Mixture of materials that are combined to give desired properties when used in the manufacture of a product. *See also* **Mixture.**

COMPOUND BATTER PILE *See* **Pile.**

COMPOUNDING Chemical process that combines rubber components to achieve tire qualities such as wear, traction, and density. *See also* **Blending.**

COMPOUND INGREDIENT Material added to another to form a mix.

COMPOUNDING SHEAVE *See* **Sheave.**

COMPOUND MATERIAL Substance composed of two or more separately identifiable materials.

COMPOUND PIER Pier comprising several shafts.

COMPOUND ROPING Arrangement of elevator hoist ropes in which one end of each hoist rope passes from the dead-end hitch in the overhead, down and under a car sheave(s), up and over the drive sheave(s), down around a counterweight sheave, and up to another dead-end hitch in the overhead. *Also called* **Two:one roping.**

COMPOUND SUPERCHARGING Use of two or more superchargers or turbochargers to pressurize the intake air for an engine.

COMPOUND TRANSMISSION *See* **Transmission.**

COMPOUND WALL Wall constructed in two or more skins of different materials.

compr *Abbreviation for:* compressor.

compr. *Abbreviation for:* compress; compression; compressive.

COMPREHENSIVE GENERAL LIABILITY Insurance policy that automatically includes all forms of general (as against specific) liability.

COMPREHENSIVE MATERIAL DAMAGE INSURANCE Indemnity against loss or damage from such causes as fire, flood, theft, vandalism, windstorm, etc.

COMPRESSED AIR Air compressed to a pressure greater than one atmosphere. *See also* **Free air,** and **Standard air.**

COMPRESSED-AIR ATOMIZING OIL BURNER Burner in which compressed air is used to break the oil stream into a spray of tiny droplets.

COMPRESSED-AIR CAISSON Caisson with a working chamber in which the air is maintained above atmospheric pressure.

COMPRESSED FIBER TUBE Compressed paper or fiber tubes used to form straight, or nearly straight, holes in concrete.

COMPRESSED GAS Any contained mixture or material with either an absolute pressure exceeding 275 kPa (40 psi) at 21°C (69.8°F) or an absolute pressure exceeding 717 kPa (104 psi) at 54°C (130°F), or both.

COMPRESSIBILITY 1. Ratio of the percentage change in the volume of a material or fluid to a percentage change in pressure; **2.** Soil property that allows it to deform under load.

COMPRESSIBILITY EFFECT Change of density of a gas under conditions of compressible flow.

COMPRESSIBLE FLOW Flow of high-pressure gas or air that undergoes a pressure drop sufficient to result in significant reduction of its density.

COMPRESSION 1. Force that tends to densify a granular material and shorten a structural member; **2.** Compacting effect of the weight of steel-wheel rollers, measured at the bottom of the roll in kg per mm (lbs per linear inch) of roll width.

COMPRESSIONAL WAVE Seismic wave whose motion is compression-dilation, or push-pull, generated by rock's resistance to compression.

COMPRESSION BAR Steel reinforcing used to resist compresion forces.

COMPRESSION BRAKE Braking device that uses engine compression to retard vehicle speed.

COMPRESSION FAILURE Deformation of wood fibers by excessive compression along the grain in direct end compression or in bending.

COMPRESSION FITTING *See* **Fitting.**

COMPRESSION FLANGE Widened portion of an I, T, or similar cross-sectional beam that is shortened or compressed by bending under normal loads.

COMPRESSION FORCE *See* **Force.**

COMPRESSION GRIP Metal sleeve swaged or extruded on to a prestressed concrete strand.

COMPRESSION IGNITION Ignition of fuel due to the heat of compression; the basic operating principle of any diesel engine.

COMPRESSION JOINT Joint in which sealant or sealing faces are always subjected to a compression stress due to a closing tendency of the joint faces. *Also called* **Manipulative joint.**

COMPRESSION LEAD *See* **Lead.**

COMPRESSION MACHINE Load-applying machine, usually of four-column construction with a fixed upper head holding a ball-seated platen and a ram holding a lower platen.

COMPRESSION MEMBER Any member in which the primary stress is longitudinal compression.

COMPRESSION PILE *See* **Pile.**

COMPRESSION PRESSURE Force exerted by the compressed charge in the combustion chamber space of an engine when the piston is at the top of the compression stroke.

COMPRESSION RATIO Difference in volume within a cylinder between BDC (bottom dead center) and TDC (top dead center); the squeeze factor within a cylinder.

COMPRESSION REINFORCEMENT Reinforcement designed to carry compressive stresses.

COMPRESSION RING 1. Any piston ring used primarily to contain the pressures of compression and combustion; **2.** Two top rings, and occasionally more, of a set of piston rings.

COMPRESSION ROLL Drive wheel of a steel-wheel roller. *Also called* **Drive roll.**

COMPRESSION SEAL Compartmentalized or cellular sealant that, under compression between joint faces, provides a seal.

COMPRESSION SET Change occurring in a pliable or mastic substance which, when deformed, prevents full recovery.

COMPRESSION SPRING *See* **Spring.**

COMPRESSION STRENGTH Strength of a material to resist a compressive force, applied parallel or perpendicular to the grain, where applicable.

COMPRESSION TEST Test made on a specimen to determine its compressive strength; unless otherwise specified, compression tests of mortars are made on 50 mm (2 in.) cubes and of concrete on cylinders 150 mm (6 in.) in diameter and 300 mm (12 in.) high.

COMPRESSION VALVE *See* **Valve.**

COMPRESSION WEB MEMBER Truss member that is subject to compression stress.

COMPRESSION WOOD Wood formed on the lower side of branches and inclined trunks of softwood trees, identified by relatively wide annual rings, usually eccentric. *See also* **Reaction wood.**

COMPRESSIVE STRENGTH Measured maximum resistance of a material to axial compressive loading, expressed as force per unit cross-sectional area, or the specified resistance used in the design calculations.

COMPRESSIVE-STRENGTH AVERAGE Average compressive strength of a given class or strength level of a material; of concrete, defined as the average compressive strength required to statistically meet a designated specific strength.

COMPRESSIVE STRESS Stress that resists a force attempting to crush a body.

COMPRESSIVE STROKE Distance traveled by a diesel hammer ram compressing cylinder gas: measured from the bottom of the air intake port to the point of impact.

COMPRESSOMETER Device used to determine the strain and deformation characteristics of a standard cylindrical sample of hardened concrete.

COMPRESSOR Machine for densifying air or gas from an initial intake pressure to a higher storage or discharge pressure.

COMPRESSOR (BRAKE) Engine-driven compressor that supplies air for a brake system. Using one or more pistons, the unit pulls in atmospheric air, compresses it, and sends it into the wet tank.

COMPRESSOR (FREON) Engine that compresses and pumps refrigerant through an air-conditioning system.

compt *Abbreviation for:* compartment.

compt. *Abbreviation for:* compute.

comptr *Abbreviation for:* comptroller.

COMPUTER-AIDED DESIGN Use of computer-generated solutions to assist in the determination of design problems.

COMPUTER-AIDED DESIGN AND DRAFTING Use of a computer to generate solutions to assist in the determination of design problems and to drive a drafting machine.

COMPUTER WAVE EQUATION PROGRAM
Computer program used to solve the wave equation as applied to pile driving, and specifically to attempt to predict the optimum hammer size and pile length to acquire a desired ultimate bearing value of a pile. *See also* **Diesel 1 program.**

conc *Abbreviation for:* concentric.

conc. *Abbreviation for:* concentration; concrete.

CONCAVE Sloping inward and downward in a continuous curve.

CONCAVE WELD FACE Weld having the center of its face below the weld edges.

conc. blk *Abbreviation for:* concrete block.

conc. clg *Abbreviation for:* concrete ceiling.

CONCEALED Rendered inaccessible by the structure or finish of the building. *See also* **Accessible,** and **Exposed.**

CONCEALED GUTTER *See* **Box gutter.**

CONCEALED HEATING Space heating device that is not readily visible after installation.

CONCEALED NAILING Nailing so that the head of the nail is concealed by the next unit to be installed. *See also* **Secret nailing,** and **Toe nailing.**

CONCENTRATED LOADS *See* **Loads.**

CONCENTRATING COLLECTOR *See* **Solar collector.**

CONCENTRATION YARD Pulpwood yard providing facilities for unloading trucks, storage, and loading for shipment.

CONCENTRIC Circles with a common center.

CONCENTRICITY Being concentric; circles positioned so as to have a common center.

CONCENTRICITY INDICATOR Dial indicator and fixture used to check concentricity.

CONCENTRIC TENDONS Tendons following a line coincident with the gravity axis of the prestressed concrete member.

CONCEPTUAL DESIGN Stage in the life of a project that culminates in the preparation of a document containing a functional program, an architectural or space program, a concept estimate and a set of design standards. *See also* **Architectural program, Functional program,** and **Requirements.**

conc. fl. *Abbreviation for:* concrete floor.

concl. *Abbreviation for:* conclusion.

concn *Abbreviation for:* concentration.

CONCORDANT TENDON Tendon, in a statistically indeterminate structure, that is coincident with the pressure line produced by the tendon.

CONCORDANT TENDON PROFILE Profile of a stressed tendon that causes a continuous structure, such as a continuous beam, to deflect in exactly the same way as the structure in an unloaded condition.

CONCRETE Composite material that consists essentially of a binding medium within which are embedded particles or fragments of aggregate, usually a combination of fine aggregate and coarse aggregate; in portland-cement concrete, the binder is a mixture of portland cement and water.

CONCRETE BLOCK Basic modular building material, usually 200 x 200 x 400 mm (8 x 8 x 16 in.) and divisions of these numbers, made from cement, fine aggregates, and sand. Used for structural walls and foundations.

CONCRETE BLOCK BAR SUPPORTS Precast concrete blocks, with or without tie wires, used to support bars above the subgrade or to space bars off vertical forms and above horizontal forms.

CONCRETE BLOCK PAVING Paving consisting of small, individual, high-strength precast concrete units, laid to a continuous bond and locked in place between edge restraints.

CONCRETE BREAKER Compressed-air tool specially designed to break up concrete.

CONCRETE BRICK Solid concrete masonry unit, usually 100 x 100 x 300 mm (4 x 4 x 12 in.).

CONCRETE BUCKET Bucket for handling freshly mixed concrete fitted with a bail or bridle, to be attached to the rope of a crane and hoisted to the dumping location.

CONCRETE CONTAINMENT STRUCTURE Composite concrete and steel assembly that is designed as an integral part of a pressure retaining barrier that, in an emergency, prevents the release of radioactive or hazardous effluents from nuclear power plant equipment so enclosed.

CONCRETE COVER Distance from the face of the concrete to the reinforcing steel. *Also called* **Clearance, Concrete protection,** and **Fireproofing.**

CONCRETE CURING MAT Burlap mat laid on fresh concrete and kept wet so as to improve the curing process.

CONCRETE-ENCASED BEAM Steel beam totally encased in concrete cast integrally with the slab. *See also* **Composite beam.**

CONCRETE FINISHING MACHINE 1. Machine mounted on flanged wheels that rides on the forms or on specially set tracks, used to finish surfaces such as pavements; **2.** A portable power-driven machine for floating and finishing of floors and other slabs.

CONCRETE FLATWORK General term applicable to concrete floors and slabs that require finishing operations.

CONCRETE FOOTING Widened section at the base or bottom of a foundation wall, pier, or column.

CONCRETE FORM 1. Box-like assembly of wood or metal panels into which concrete is placed. *Also called* **Formwork; 2.** Special concrete facing unit, or precast concrete pipe, used as permanent formwork.

CONCRETE MASONRY UNIT Hollow or solid unit made of portland cement and suitable aggregates. (Units often take their name from the type of aggregate used in their manufacture, i.e., **Cinder block, Lightweight concrete block, Slag block,** etc.)

CONCRETE MIXER Rotating metal drum pivoted at an angle, into which the ingredients for making concrete are placed and which contains flights to promote mixing. When the rotational direction of the drum is reversed, the flights lift and eject the prepared concrete.

CONCRETE NAIL Hardened metal spike having a stout shaft that can be hammered into brick or concrete.

CONCRETE PAINT 1. Mixture of cement powder and water, applied as a thin paint to finished concrete, to give it a uniform color, and to protect joints against weathering; **2.** Specially formulated paint used to treat finished concrete (particularly floors) to make it dust-free and washable.

CONCRETE PAVEMENT RECLAIMING To bring into a condition for renovation or other use by recovering existing concrete pavement.

CONCRETE PAVER 1. Concrete mixer, usually mounted on crawler tracks, that mixes and places concrete pavement on the subgrade; **2.** Precast concrete paving brick.

CONCRETE PAVING Surface of cast-in-place portland cement concrete, normally installed on a base of crushed stone or gravel.

CONCRETE PENETROMETER Hand-held instrument used to determine the setting time of the mortar fraction of a concrete mix that has a slump value greater than zero.

CONCRETE PILE Pile made of concrete, either pre-stressed, reinforced, or unreinforced; precast or cast *in situ.*

CONCRETE PIPE *See* **Pipe.**

CONCRETE PROTECTION *See* **Concrete cover.**

CONCRETE PUMP *See* **Pump.**

CONCRETE PUMPABILITY Ability of a concrete mix transmit sufficient pressure to overcome the resistance created by friction and head in the pump, joints, bends, and pipes through which it is conducted.

CONCRETE REACTOR VESSEL Composite concrete and steel assembly that functions as a component of the principal pressure-containing barrier for a nuclear fuel's primary heat extraction fluid in a nuclear power generating plant.

CONCRETE SPREADER Machine, usually carried on side forms or on rails, designed to spread concrete from heaps already dumped in front of it, or to receive and spread concrete in a uniform layer.

CONCRETE TILE *See* **Roof tile.**

CONCRETE VIBRATING MACHINE Machine that consolidates a layer of freshly mixed concrete by vibration.

CONCRETOR Worker skilled in the mixing, placing, and finishing of concrete.

CONCUSSION Shock waves or air waves caused by an explosion or heavy blow upon a hard object or surface.

cond *Abbreviation for:* conducted.

cond. *Abbreviation for:* condenser; condition; conditional; conduct; conductivity; conductor.

CONDEMNATION Process by which property is ac-

quired for public purposes through legal proceedings under power of eminent domain. *See also* **Inverse condemnation.**

CONDENSATION Transformation of the vapor content of the air into water on cold surfaces.

CONDENSATION GROOVE Channel that collects condensation forming on the interior face of large glass surfaces and conducts it to a condensation gutter.

CONDENSATION GUTTER Channel or orifice through which the moisture of condensation collected in a condensation groove is discharged to the exterior.

CONDENSER 1. Machine that compresses air; **2.** Apparatus for storing or intensifying an electric charge; **3.** Equipment that reduces vapor to a liquid.

CONDENSER-DISCHARGE Blasting machine that uses batteries to energize a series of condensers, whose stored energy is released into a blasting circuit.

CONDITION Provision in a contract detailing obligations of the parties.

CONDITIONAL SALES CONTRACT Contract for the sale of real property in which delivery is made to the buyer but title remains vested in the seller until all conditions of the contract have been fulfilled.

CONDITIONED SPACE Space in a building that is either heated or cooled by mechanical means.

CONDITIONING 1. Exposure of a specimen under specified conditions (e.g., temperature, humidity) for a specified period of time before testing; **2.** Stressing a filter to simulate system operating conditions.

CONDITIONS OF BID Conditions contained in the instructions to bidders detailing, among other things, how the bids are to be prepared and submitted, time and place of filing, etc.

CONDITIONS OF CONTRACT Fundamental terms that collectively describe the rights, obligations, procedures, and recourses of contracting parties.

CONDOMINIUM System of individual property ownership of units in a development complex combined with joint ownership of common areas of the structure and land. *Also called* **Strata title.**

CONDOMINIUM UNIT Individual dwelling unit within a condominium development to which an individual has ownership (as distinct to joint ownership with other unit owners to common areas and facilities).

CONDUCTION Transfer or travel of electricity, heat, light, or vibration through a body by molecular action.

CONDUCTIVE Rubber having qualities of conducting or transmitting heat or electricity. (Most generally applied to rubber products capable of conducting static electricity.)

CONDUCTIVE MORTAR Tile mortar to which specific electrical conductivity is imparted through the use of conductive additives.

CONDUCTIVE TILE Tile made from special body compositions that result in specific properties of electrical conductivity while retaining other normal physical properties of ceramic tile.

CONDUCTIVITY Ability of a material to carry electricity, heat, light, or vibration.

CONDUCTOR 1. Any connector, such as a pipe, for carrying rainwater from the roof of a building to the ground. *See also* **Downspout**; **2.** Substance capable of transmitting electricity, heat, light, or vibration. *See also* **Bare conductor**, and **Electrical conductor**; **3.** First casing of a cased borehole.

CONDUCTOR PIPE Round, square, or rectangular pipe used to lead water from a roof to the sewer.

CONDUCTOR SHIELDING An envelope that encloses the conductor of a cable and provides an equipotential surface in contact with the cable insulation.

CONDUIT Tube, usually steel, plastic, or aluminum, through which electrical conductors are run. *See also* **Flexible metal conduit.**

CONDUIT BOX Metal box located between the ends of lengths of conduit (up to four can be accommodated) where electrical conductors are spliced or attached to an electrical fitting, such as an outlet.

CONDUIT BUSHING Short threaded sleeve fastened to the end of the conduit inside the conduit box. The inside of the sleeve is rounded out on one end to prevent damage to electrical conductors.

CONDUIT COUPLING Short metal tube, threaded on the inside, used to fasten two lengths of conduit end-to-end.

CONDUIT ELBOW Short piece of tubing bent to an angle, usually of 45 or 90 degrees.

CONDUIT FITTINGS All of the auxiliary items and fittings used or needed to assemble a conduit system.

CONDUIT SYSTEM An array of tubes, connections, and fittings in which an electrical distribution system can be housed.

CONE Packing, convex-conical on one face and concave-concical on the other, operating against any inner reciprocating or rotating member.

CONE BOLT Form of tie rod for wall forms with cones at each end inside the forms so that a bolt can act as a spreader as well as a tie.

CONE CRUSHER Rock crusher that operates by use of two concentric cones, one of which gyrates.

CONE FOUNDATION Column foundation where the underlying soil is first shaped into a cone and then covered by a uniform thickness of reinforced concrete.

CONE OF DEPRESSION Area of soil around an underground suction point depresed below normal elevation and dried out.

CONE PENETRATION TEST Exploratory test for soil or unsound rock.

CONE PENETROMETER Instrument used to determine the consistency of masonry mortar in the field.

CONE WHEEL Small wheel shaped like a bullet nose, used for portable grinding.

conf. *Abbreviation for:* conference; conform; conformance; conforming.

config. *Abbreviation for:* configuration; configure.

CONFIGURATION Arrangement of parts or elements within an overall layout or design.

CONFINED AQUIFER Aquifer in which groundwater is confined under pressure that is significantly greater than atmospheric pressure.

CONFINED CONCRETE Concrete containing closely-spaced, special transverse reinforcement that is provided to restrain the concrete in directions perpendicular to the applied stress.

CONFINED GROUNDWATER *See* **Groundwater.**

CONFINED REGION Region with transverse reinforcement within beam-column joints.

CONFLAGRATION Raging fire; in the forest, often with a fast-moving front.

CONFORM *See* **Butt cut.**

CONGÉ *See* **Molding.**

CONGLOMERATE Rounded boulders and pebbles cemented together with a clayey material in naturally occurring deposits.

congr. *Abbreviation for:* congruent.

CONICAL BURNER *See* **Burner.**

CONICAL POINT *See* **Pile point.**

CONICAL ROLL Roll joint in metal roofing formed over a triangular section of wood.

CONICAL ROLLER Device that attaches the upper to the carrier and allows the upper to rotate on the carrier.

CONIC PROJECTION *See* **Linear perspective.**

CONIFER Resinous tree with cone-like fruits and needle-like or scaly leaves; generally evergreen with a few deciduous exceptions.

CONIFER RELEASE To 'release' established coniferous trees from a situation in which they have been suppressed, by thinning out undesirable trees and shrubs that have overtopped them. *See also* **Brushing.**

CONING AND QUARTERING Manual method of reducing the size of an aggregate sample without altering its characteristics. Results are dependent upon the skill of the person performing the work.

conj. *Abbreviation for:* conjunction.

CONK Hard, spore-bearing structure of a wood-destroying fungus that projects beyond the bark of a tree.

conn. *Abbreviation for:* connect; connection; connector.

conn. diag. *Abbreviation for:* connection diagram.

CONNECTING BAR Bar attached to the rear of the cylinder of a plug lock.

CONNECTING ROD Mechanical connection between the piston and the crank in an internal combustion engine, used to convert reciprocal motion to rotary motion, or vice versa.

CONNECTING WIRE 1. Any wire used in a blasting circuit to extend the length of a leg wire or leading wire; **2.** *See* **Electric blasting circuitry.**

CONNECTION Combination of joints used to transmit forces between two or more members, characterized by the type and amount of force transferred (moment, shear, end reaction). *Also called* **Splice.**

CONNECTION BOX Box attached to or recessed into the nose of a trailer and containing supply (emergency) and control (service) air connections, electrical receptacles, and electrical circuit protectors.

CONNECTOR 1. Item for quick connection composed either of male and female halves or two halves that differ only in their method of mounting; **2.** Two halves of a battery connector, that are known as the charging plug and charging receptacle. *See also* **Cable connector, Cell connector,** and **Terminal connector; 3.** Device used in blasting to initiate a delay in a detonation cord circuit, connecting one hole in the circuit with another, or one row of holes to other rows of holes. **4.** Sealed, corrosion-resistant links between the electronic control unit, wheel speed sensors, and modulator or relay valves of an anti-lock braking system (ABS).

CONNECTOR FITTING *See* **Fitting.**

CONNECTOR PLATE Galvanized steel plate with teeth punched out on one side, that is hydraulically pressed into either side of a wood truss joint to fasten chord and web members together.

consec. *Abbreviation for:* consecutive.

CONSENT OF SURETY Written consent of the surety on a performance bond and/or labor and material payment bond to contract changes such as change orders, reduction of retainage, etc.

CONSEQUENTIAL DAMAGES Loss of value of a parcel, no portion of which is acquired, resulting from highway improvement.

CONSEQUENTIAL LOSS Loss not directly caused, but that may arise from prior damage.

conser. *Abbreviation for:* conserve; conservation.

CONSERVATION Protection, improvement, and wise use of natural resources according to principles that will assure utilization of the resources to obtain the highest economic and/or social benefits.

CONSIDERABLE NEGATIVE ALLOWANCE *See* **Fit classification.**

CONSIDERATION Anything of value given to encourage closing of a contractual agreement.

CONSIDÔRE HINGE Hinge formed in a large concrete arch during its construction that allows the arch to drop slightly under its own action when the formwork is removed.

CONSIGNEE Person to whom something is shipped.

CONSIGNMENT 1. Shipment; **2.** Shipment of goods made under an agreement whereby the consignee undertakes to sell the goods as an agent on behalf of the consignor, who retains title to the goods until they are disposed of.

CONSIGNOR Person by whom articles are shipped.

CONSISTENCY 1. Relative mobility, or ability of freshly mixed concrete or mortar to flow. *See also* **Body; 2.** Degree to which a plastic material such as a lubricating grease resists deformation under the application of force. It is therefore a characteristic of plasticity, as viscosity is a characteristic of fluidity.

CONSISTENCY FACTOR Measure of grout fluidity, roughly analogous to viscosity, that describes the ease with which grout may be pumped into pores or fissures.

CONSISTENCY INDEX Ratio of the difference between the liquid limit and natural water content to the difference between the liquid limit and the plastic limit.

CONSISTENCY LIMITS *See* **Atterberg limits.**

CONSISTOMETER Apparatus for measuring the consistency of cement pastes, mortars, grouts, or concretes.

consol. *Abbreviation for:* colsole; consolidate; consolidated.

CONSOLE 1. Ornamental bracket, usually of greater height than projection; **2.** Any frame containing an array of control devices (lighting, heating, musical, electrical, etc.).

CONSOLIDATION 1. Process of inducing a closer arrangement of the solid particles in freshly mixed concrete or mortar during placement by the reduction of voids, usually by vibration, centrifuging, rodding, tamping, or some combination of these actions; **2.** Treatment of a stone surface with a liquid solution, that is commonly brush or spray applied and that employs both organic and inorganic chemicals. *See also* **Compaction.**

CONSOLIDATION SETTLEMENT Settlement of loaded soil that occurs over a period of years due to dissipation of excess pore pressure as water, or air and water are expelled from the voids in the soil.

CONSOLIDATION TEST Method of determining the coefficient of permeability of soils.

con. spec. *Abbreviation for:* construction specification.

const *Abbreviation for:* constant; construct.

const. *Abbreviation for:* construction.

CONSTANT Factor or quality that does not vary.

CONSTANT DOLLAR Represents the value of expenditure or production measured in terms of some fixed base period's prices. (Changes in constant dollar expenditure or production can only be brought about by changes in the physical quantities of goods purchased or produced.)

CONSTANT-HEAD PERMEATOR Apparatus and method of determining the permeability of soils.

CONSTANT MESH Gears that are always in mesh with each other, one rotating freely on its shaft unless locked to it by a shiftable collar or dog.

CONSTANT MESH TRANSMISSION *See* **Transmission.**

CONSTANT PRESSURE OPERATION *See* **Operation.**

CONSTANT RATE OF PENETRATION Load test accomplished by applying a load to a pile in such a manner that the pile top (head) experiences a constant rate of displacement and where, at short time intervals, the force and displacement are recorded, from which a force displacement curve may be plotted; a definition of failure may be developed from this. Typical displacement rates range from 0.010 to 0.020 inches per minute. *Also called* **CRP.**

CONSTANT RATE OF PENETRATION TEST Test in which a pile is made to penetrate the soil from its installed position at a constant speed with the force applied at the top of the pile being continuously measured.

CONSTANT RATE OF UPLIFT TEST Test in which a pile is extracted from its installed position in soil at a constant speed while the force applied to the top of the pile to maintain the rate of uplift is continuously measured.

CONSTANT RATE SPRINGS Spring assembly (leaf-type, coil, torsional bar, or air) that has a constant rate of deflection.

constr. *Abbreviation for:* constructor.

constr. eng. *Abbreviation for:* construction engineer.

constr. mgmt *Abbreviation for:* construction management.

CONSTRUCTION 1. Result of work on site, done according to a prepared plan or design; **2.** Geometric design description of a wire rope's cross section. This includes the number of strands, the number of wires per strand, and the pattern of wire arrangement in each strand.

CONSTRUCTIONAL STRETCH Stretch that occurs when a rope is loaded due to the helically laid wires and strands creating a constricting action that compresses the core and generally brings all of the rope's elements into close contact.

CONSTRUCTION AND DEMOLITION WASTES Waste building materials and rubble resulting from construction, remodeling, repair, and demolition operations on houses, commercial buildings, pavements, and other structures.

CONSTRUCTION CONTRACT Written agreement between an owner and a contractor, referencing all contract documents, calling for construction of the described and illustrated project.

CONSTRUCTION COST Any of the cost types (appropriation, commitment, expenditure, or estimate to completion) associated with the scope of the work.

CONSTRUCTION CONTRACTOR Corporation or individual who has entered into a contract to perform construction work.

CONSTRUCTION DEFECT Anything about a construction project that is at variance to the contract documents.

CONSTRUCTION DOCUMENTS *See* **Contract document.** *See also* **Working drawings.**

CONSTRUCTION ESTIMATOR Person who estimates the costs of materials and labor necessary to complete a project.

CONSTRUCTION INSPECTION *See* **Site inspection.**

CONSTRUCTION JOINT *See* **Joint.**

CONSTRUCTION LOAD Load to which a permanent or temporary structure is subjected during construction.

CONSTRUCTION MANAGEMENT Project delivery system directed by a construction manager, a member of the owner-architect/engineer-construction manager team, whose responsibility is to clarify time and cost consequences of design decisions and their construction practicability and to manage the bidding, award, and construction phases of the project.

CONSTRUCTION MANAGEMENT FIRM Corporation that performs the duties of the construction manager and his staff. Especially on large construction projects, such firms take over the role of a general contractor by providing day-to-day supervision and direction of activities by general and subtrade construction contractors.

CONSTRUCTION MANAGER Title of a position on the project team of a person responsible to the project manager for directing the construction of a project within authority and responsibility limits established by the project manager. On smaller projects these duties may be the responsibility of the project manager.

CONSTRUCTION PHASE That part of a project life during which working drawings are prepared, contracts are tendered and awarded, and the construction work is carried out. *Also called* **Construction process.** *See also* **Feasibility phase.**

CONSTRUCTION PROCESS *See* **Construction phase.**

CONSTRUCTION PROGRAM Plan that sets out the sequence in which construction operations will be carried out, materials delivered to site, strength of labor force at any point in time, expenditures, cash flow, etc.

CONSTRUCTION PROGRESS REPORT *See* **Progress report.**

CONSTRUCTION SCHEDULE Sequence of events that will *See* the completion of a construction program. *Also called* **Schedule of work,** and **Work schedule.**

CONSTRUCTION SHEATHING Panel product composed of wood that is essentially dependent upon certain mechanical and physical properties for successful performance as subflooring, roof, or wall sheathing in specified light frame construction.

CONSTRUCTION SITE *See* **Job site.**

CONSTRUCTION STAGE PLANNING Process of determining the principal stages necessary to complete a construction project, sequencing them, and ensuring that there is adequate provision for materials and labor and that there is no conflict in operations between one stage and another.

CONSTRUCTION STAKEOUT *See* **Stakeout.**

CONSTRUCTION START Point in a project when new construction, alterations, renovations, or demolition begin, following such work as site surveying.

CONSTRUCTION STATUS REPORT *See* **Progress report.**

CONSTRUCTION SURVEYING *See* **Surveying.**

CONSTRUCTION TIME OVERRUN *See* **Time overrun,** and **Overrun.**

CONSTRUCTION TYPES Broad classes construction, including:

Adobe: Type of construction in which the exterior

walls are built of blocks made of soil mixed with straw and hardened in the sun.

Block: Type of construction in which the exterior walls and bearing walls are made from concrete block or structural clay tile.

Brick: Type of construction in which the exterior walls and bearing walls are made of brick or a combination of brick and other unit masonry.

Brick facing: *See* **Brick veneer,** below.

Brick veneer: The outside facing of brickwork used to cover a wall built of other materials that serves as a wall covering only and carries no structural loads. *Also called* **Brick facing.**

Composite: Type of construction using members produced by combining different materials.

Double-wall: Framing technique used to increase the space for insulation by introducing a second wall that is generally nonload-bearing.

Drywall: Construction technique where the interior wall finish is applied dry, generally in the form of sheet materials or panels, most commonly as precast gypsum-plaster sheets.

Factory-built: Construction system where the complete house or a portion of it are assembled in a plant, then transported to the site for final erection.

Fire-resistive: Floors, walls, roof, etc., constructed of slow-burning or noncombustible materials recognized by building codes or local regulations to withstand collapse by fire for a stated period of time.

Formed-steel: Construction using sheet or strip steel formed into structural panels, decks, studs, joists, and other fabricated shapes.

Light-gauge steel: Construction comprising sheet or strip steel less than 4.6 mm (0.18 in.) thick, formed into structural panels, decks, studs, joists, and other structural shapes.

Monolithic concrete: Process in which the concrete for walls, floor, beams, etc. is poured in one continuous operation.

Post-and-beam: Construction system in which post and beam framing members are the principal load-bearing units.

Prefabricated: The manufacture of whole buildings or components at a central location for transportation to site.

Reinforced concrete: Type of construction in which the principal structural members such as floors, columns, and beams are made of concrete placed around isolated steel bars or steel meshwork in such a manner that the two materials act together in resisting force.

Steel frame: Type of construction in which the structural parts are of steel or dependent on a steel frame for support.

Wood frame: Type of construction in which the structural parts are of wood or dependent upon a wood frame for support. In codes, if brick or other noncombustible material is applied to exterior walls, the classification for this type of construction is usually unchanged.

CONSTRUCTION WARRANTY *See* **Warranty.**

CONSTRUCTION WORK Work that contributes to a physical structure (as distinct to decoration, that contributes to the appearance of a structure).

CONSTRUCTIVE EVICTION *See* **Eviction.**

CONSTRUCTIVE NOTICE Information imputed by law to a person who could have discovered the fact from public records.

CONSTRUCTIVE TOTAL LOSS Where the cost of repair of damaged property exceeds its actual cash value, resulting in it being a total loss, even though repairable.

CONSTRUCTOR Person who contracts with an owner or his agent to undertake a project.

consult *Abbreviation for:* consultant.

CONSULTANT One who offers expert and specialized knowledge and opinion.

consult. engr *Abbreviation for:* consulting engineer.

CONSULTING ARCHITECT Architect who offers expert opinion but who has subordinate responsibility to the prime consultant or contractor.

CONSULTING ENGINEER Professional engineer who offers expert opinion but who has subordinate responsibility to the prime consultant or contractor.

CONSUMER PRICE INDEX Measure of the rate of change in a typical basket of consumer goods and services, including housing, food, and health care. It can be broken down into food and nonfood items, goods and services, and other components.

CONSUMER WASTE Materials that have been used and discarded by the buyer, or consumer, as opposed to waste discarded in-house during the manufacturing process.

cont *Abbreviation for:* contact.

cont. *Abbreviation for:* container; contents; continue; continued; continuous; control; controller.

CONTACT Part of a contactor relay or switch that completes or opens a circuit.

CONTACT AERATOR Biological waste treatment process in which air, usually in the form of bubbles, is introduced under pressure into the liquid waste in a specially shaped tank.

CONTACT BED In wastewater treatment, a bed of broken rock over which the waste flow is distributed so as to come in intimate contact with aerobic bacteria.

CONTACT BREAKER Electric overload switch.

CONTACT CEILING Ceiling that is secured in direct contact with the construction above without use of furring.

CONTACT CEMENT *See* **Adhesive.**

CONTACT CORROSION Oxide coating on contact surfaces that inhibits conductance.

CONTACTING-RING SEAL Type of circumferential seal employing a ring that is spring-loaded radially inward against a shaft.

CONTACTOR Electromechanical device for repeatedly establishing and interrupting an electrical power circuit.

CONTACT PITTING Erosion of contact material caused by arcing at the moment of opening or closing of current-carrying contacts.

CONTACT PRESSURE Pressure acting at, and perpendicular to the contact area between a footing and the soil, produced by the weight of the footing and all forces acting on it. *Also called* **Soil pressure.**

CONTACT SPLICE Means of connecting reinforcing bars in which the bars are lapped and in direct contact. *See also* **Lap splice.**

CONTAINER Nonspecific term for a receptacle capable of closure.

contam. *Abbreviation for:* contaminant; contamination.

CONTAMINANT Material that, by reason of its action upon, within, or to a person, is likely to cause physical harm. *See also* **Artificial contaminant, Built-in contaminant,** and **Generated contaminant.**

CONTAMINANT CAPACITY *See* **Dirt capacity.**

CONTAMINATION LEVEL Quantitative term specifying the degree of contamination.

CONTEMPLATED CHANGE NOTICE (CCN) Form issued to a construction contractor as a means for obtaining his quotation on the price change associated with a contemplated change in the work included in his contract.

CONTENTS RATE Fire insurance on the contents of a building, but not the building itself.

CONTINGENCY Component of the authorized appropriation of estimated final cost types for the scope of work associated with a cost class. The contingency is an estimated allowance for the cost of unknowns or changes. The anticipated award price of a cost class contains contingencies for escalation and estimating error. It is added to the estimated award price for further contingencies for design changes and contractor claims. NOTE: The reserve for scope changes is not a contingency in this same sense; rather it is an allowance that is transferred into specific cost classes when the scope of work in the class is amended. The authorization appropriation for that class is correspondingly amended following the transfer. *Also called* **Contingency allowance.**

CONTINGENCY ALLOWANCE *See* **Contingency.**

CONTINGENCY FUND Cash or investments set aside or reserved for unforeseen expenditures.

CONTINGENT AGREEMENT Agreement that relies upon some future condition or circumstance in order to remain valid, i.e., an agreement to complete work being contingent on the owner successfully securing financing.

CONTINGENT RENTAL Rental based on a factor other than the passage of time, e.g., percentage of sales, amount of usage, prime interest rate, price indices, etc.

CONTINUING COOPERATIVE Program in which individuals join to develop and operate housing to be owned collectively and occupied by the members of the cooperative.

CONTINUITY 1. Complete, continuous and effective connection throughout an electrical circuit; 2. Maintenance of standards or quality throughout a whole.

CONTINUITY EQUATION Statement that the mass rate of fluid flow into a fixed space is equal to the mass flow rate out. Hence, the mass flow rate of fluid past all cross sections of a conduit is equal.

CONTINUITY STEEL Steel reinforcement in *in situ* in concrete placed between, over, or around precast concrete units to form a homogeneous concrete structure.

CONTINUOUS BEAM Beam that rests on three or more supporting members.

CONTINUOUS-FEED INCINERATOR *See* **Incinerator.**

CONTINUOUS-FLIGHT AUGER String of helical augers and a cutting head, used to bore a hole in earth into which a pile section may be set, concrete cast in place, or a tieback grouted. *Also called* **Vertical earth-boring machine.**

CONTINUOUS FOOTING Combined footing of prismatic or truncated shape supporting two or more columns in a row.

CONTINUOUS FOREST INVENTORY Timber sampling system that provides for periodic remeasurement of specific stands or plots of individual trees to show status and periodic change over time for the forest as a whole and major subdivisions therein.

CONTINUOUS FOUNDATION Combination footing and foundation wall where the footing transmits the foundation loads to the soil, and the foundation wall extends a minimum distance above finished ground line around the entire perimeter of the building.

CONTINUOUS FURNACE Furnace operated on an uninterrupted cycle, in which the charge is being constantly added to, moved through, and removed from the furnace, as opposed to a batch-type furnace.

CONTINUOUS GIRDER Structural steel member supported at more than three points and extending over the supports (as distinct from a series of independent girders or beams).

CONTINUOUS GRADING Particle size distribution in which all intermediate size fractions are present.

CONTINUOUS HEADER Timber beam comprised of two 50 x 150 mm (2 x 6 in.) members turned on edge running around an entire house with its upper surface at the level of the plate. This header is strong enough to act as a lintel over all normal wall openings (up to approximately 2.4 m (8 ft) in width), eliminating some cutting and fitting of stud lengths and separate headers over openings.

CONTINUOUS HIGH CHAIR Welded wire bar support consisting of a top longitudinal supporting wire with evenly spaced legs welded thereto, used to support reinforcing bars near the top of the slab. *See also* **Individual high chair,** and **Support bar.** *Also called* **High chair.**

CONTINUOUS HORSEPOWER *See* **Horsepower.**

CONTINUOUS LINK One in a series of reinforcement links, continuous along their length and helical in form, used as column transverse reinforcement.

CONTINUOUSLY REINFORCED PAVEMENT Pavement with continuous longitudinal reinforcement and no intermediate transverse expansion or contraction joints.

CONTINUOUS MIXER Mixer into which the ingredients of a mixture are fed without stopping, and from which the mixed product is discharged in a continuous stream

CONTINUOUS-MIX PLANT Asphalt plant that continuously proportions and mixes aggregate and asphalt utilizing interlocked feeding systems and a separate mixing pugmill.

CONTINUOUS SAMPLING Sampling without interruptions throughout an operation or for a predetermined time.

CONTINUOUS SLAB Slab that extends as a unit over three or more supports in a given direction.

CONTINUOUS VENT Upward continuation of a drain to produce a vent.

CONTINUOUS WASTE Two or more fixtures connected to the same trap.

CONTINUOUS WASTE AND VENT Open vertical continuation to atmosphere of a waste collection stack connected to a drain.

CONTINUOUS WELD Weld made in one operation.

CONTOUR 1. Line that defines or bounds something; **2.** Line on a topographical map that connects all points of identical elevation.

CONTOUR FELLING Timber felled parallel to the ground contour line.

CONTOUR GAUGE Series of slender metal or plastic rods about 100 mm (4 in.) long, held parallel in, and free to slide through a narrow metal frame about 150 mm (6 in.) wide. Used to create a template of an irregular surface such as a molding.

CONTOUR INTERVAL Vertical distance represented by two consecutive contour lines. This interval is normally constant throughout a contour map.

CONTOUR LINE 1. Imaginary line on the surface of the ground, every part of which is at the same elevation; **2.** Line on a drawing that identifies points of common elevation.

CONTOUR MAP Topographic map that portrays relief by means of contour lines.

CONTOUR SCALING Crust forming across the surface of sandstones and limestones that follows the contour of the surface rather than the bedding planes of the stones. The result of direct pollution, the pores of the stone are blocked by formations of recrystalized calcium sulfates.

contr *Abbreviation for:* contract; contraction; contractor.

CONTRACT Agreement in law; a commitment document. The term contract is used to refer to the document itself. Contracts are often classified and described by the terms of payment they contain: stipulated price contract, lump sum contract, unit price contract, cost-plus contract, etc. *See also* **Contract item,** and **Contract time.**

CONTRACT ADMINISTRATION Responsibilities of the owner's designated agent, architect, and engineer during the construction phase of a project.

CONTRACT AMENDMENT *See* **Contract change.**

CONTRACT BOND *See* **Bond.**

CONTRACT CARRIER Trucking firm that has a hauling contract with a company or companies.

CONTRACT CHANGE Alteration to a completed and signed contract that must be agreed to and acknowledged in writing by all parties represented in the original contract. *Also called* **Contract amendment.**

CONTRACT COLLECTION *See* **Collection.**

CONTRACT DATE Date that a legally binding contract between two or more parties is signed by all parties, not withstanding any other date contained within the text of the contract. *Also called* **Date of agreement.**

CONTRACT DOCUMENT(S) 1. Document forming part of a contract; **2.** Documents that define the responsibilities of the parties involved in bidding, purchasing, supplying, and installing and/or erecting components and materials necessary for completion of a project. *Also called* **Construction documents.**

CONTRACT DOCUMENT(S) STAGE Stage in a project where planning has developed to the point where documents (working drawings, specifications, etc.) have been completed and bids for completion of the work can be called.

CONTRACT HAULER Independent truck owner or a driver working for the contractor who hauls to or from a site.

CONTRACTING BAND BRAKE *See* **Brake.**

CONTRACTING OFFICER Person designated by the owner as his official representative with specific responsibilities and authority with regard to the project.

CONTRACTING OUT Subcontracting to another for work, material, or supplies.

CONTRACTION 1. *See* **Deformation; 2.** The shrinkage or squeezing of a material from thermal contraction.

CONTRACTION JOINT *See* **Joint.**

CONTRACTION-JOINT GROOVE FORMER Equipment, forming part of a concrete paving train, consisting of a vibrating guillotine blade suspended from a mechanically propelled chassis.

CONTRACTION-JOINT GROUTING Injection of grout into contraction joints.

CONTRACT ITEM Specific unit of work for which a price is provided in a contract. *See also* **Contract.**

CONTRACT LIMIT LINE Line establishing the legal limit of the area inside which construction work is to be carried out

CONTRACT LOGGING Operator doing all or part of the logging for a company.

CONTRACT OF PURCHASE *See* **Agreement of sale.**

CONTRACTOR One who agrees to supply materials and/or perform certain types of work, to an agreed schedule, and to defined standards, for a specified sum of money. *See also* **Subcontractor**, and **Superintendent**.

CONTRACTOR DEFAULT Failure by a contractor to perform or otherwise comply with the terms of the construction contract.

CONTRACTOR'S AFFIDAVIT Notarized statement by a contractor relating to payment of debts and claims, release of liens, etc. that require specific procedures for the protection of the owner.

CONTRACTOR'S EQUIPMENT FLOATER Insurance against loss or damage to a contractor's tools and equipment that are customarily used away from the contractor's premises.

CONTRACTOR'S LIABILITY INSURANCE Indemnity of a contractor against specified claims that may arise as a result of his, or his subcontractor's work.

CONTRACTOR'S OPTION Provision of a construction contract giving the contractor freedom to select certain materials or use certain procedures of his own option.

CONTRACT OVERRUN (UNDERRUN) Difference between the contract price, including all approved extras and variations to the original specifications, and the final cost of the works.

CONTRACT PAYMENT BOND Security furnished to the contracting agency to guarantee payment of the prescribed debts of the contractor covered by the bond.

CONTRACT PERFORMANCE BOND *See* **Bond.**

CONTRACT PERIOD Number of days between the date of signing of the contract and the date stipulated for completion of the work or, if no completion date is specified, or has been legally changed since commencement of the works, the date that the project is handed over to the owner or his representative.

CONTRACT PRICE Sum of money for which a contractor agrees to complete the work described in the contract documents.

CONTRACT RENT Designated payment for the use of property and facilities as defined in a lease.

CONTRACT SURETY BOND *See* **Bond.**

CONTRACT TIME Number of working days or calendar days allowed for completion of a contract. If a calendar date of completion is shown in the proposal in lieu of a number of working or calendar days, the contract shall be completed by that date. *Also called* **Period for completion**, and **Time for completion**. *See also* **Contract.**

CONTRACTUAL LIABILITY Responsibilities and liabilities assumed by the signatory to a contract.

CONTRAFLEXURE Shape assumed by a structural frame member, opposite to the direction of the inflected stress when stress is placed on the member.

CONTRAST SEAL *See* **Bituminous coating.**

contrib. *Abbreviation for:* contribute; contribution.

CONTRIBUTION Payment or obligation to pay all or part of an insurance claim.

contr. mgr *Abbreviation for:* contract manager.

CONTROL 1. System or systems governing starting, stopping, direction, acceleration, speed, and retardation. *See also* **Servo control**; **2.** Product of known characteristics that is included in a series of tests to provide a basis for evaluation of other products; **3.** In fighting forest fires, the overall program to control a fire and suppress fire losses (a forest fire is under control when it no longer threatens additional destruction).

CONTROL ALGORITHM Formula programmed into an Electronic Control Unit to control individual brake actuation under conditions of slip or impending slip.

CONTROL BLASTING Various techniques used to limit the amount of backbreak developed during the blasting phase of an excavation cycle by reducing the level of ground shock vibrations.

CONTROL CENTER Location to which communication and telemetry lines and/or signals are directed and from which information and/or control signals can be sent.

CONTROL CONSOLE Panel, easily visible from the operator's station, containing dials and gauges that display the principal operating conditions of mechanical equipment and machines.

CONTROL FACTOR Ratio of the minimum compressive strength to the average compressive strength.

CONTROL FLUME *See* **Flume.**

CONTROL JOINT 1. Groove cut or tooled into the top of a slab, usually to about one-fifth of its thickness, to predetermine the location of natural cracking caused by shrinkage as the concrete cures; **2.** In concrete block masonry, a continuous vertical joint without mortar, filled with a pliable or mastic caulking, to allow for thermal expansion and contraction of a wall of unusual length; **3.** Space or gap intentionally left between adjacent surfaces of the same material to relieve stress and allow for movement.

CONTROL-JOINT BLOCK Concrete masonry units that facilitate construction of vertical shear-control joints.

CONTROLLED-AIR INCINERATOR *See* **Incinerator.**

CONTROLLED BURNING Use of fire to destroy logging debris, reduce buildup of dead and fallen timber that may pose wildfire hazard, control tree diseases, and clear land. Other functions include clearing a buffer strip in the path of a wildfire. *See also* **Area ignition, Backfire,** and **Prescribed burning.**

CONTROLLED-DESIGN CONCRETE Concrete for which working stresses are based on the ultimate strength determined from preliminary ultimate-strength tests to establish the mix design.

CONTROLLED FILL Suitable inert fill material placed on natural ground in thin layers under controlled compaction after any existing weak and compressible soil has been removed.

CONTROLLED FREEFALL Feature available on Gearmatic hoists. Freefall is controlled through the use of a hydraulically controlled, spring-applied hoist brake. As hydraulic flow increases, hydraulic pressure forces the spring to apply the brake that slows down the speed of the freefall hoist.

CONTROLLED-GAP SEAL Seal designed to maintain a constant clearance with a shaft.

CONTROLLED LOW-STRENGTH MATERIALS Materials that result in a compressive strength of 8272 kPa (1,200 psi) or less. *Also called* **Low-strength materials.**

CONTROLLED PROCESS One in which the mean and variation of a series of tests of the product remain stable, with the variation being due to chance only.

CONTROLLED PRODUCT Product which, because of flammability, toxicity or similar hazard-related reasons requires special handling, labelling and a Material Safety Data Sheet.

CONTROLLED STRUCTURE Process of manufacturing grinding wheels whereby the relationship between the abrasive and bond is definitely controlled.

CONTROLLER 1. Device or group of devices that serves to govern, in some predetermined manner, the electric power delivered to the apparatus to which it is delivered; **2.** Device that detects a change in a process variable, and then automatically uses an external source of power to amplify the detected signal and to energize a mechanism that will correct the deviation in the process variable until it returns to a preset value. *See also* **Regulator.**

CONTROL LEVER Device for imparting motion into control linkage.

CONTROL LEVER LOCKOUT Option available on Krueger systems that monitors the functions available on each specific Krueger system fitted to a crane. As a machine enters a potentially dangerous situation, the control lever lockout will electronically sense the impending condition and restrict all operations that could cause further danger to the operator. Crane movements that will improve the situation, such as hoisting down, booming up and telescoping in, will remain functional and permit the operator to rectify the situation.

CONTROL LINE Construction of natural barriers to a forest fire.

CONTROL OF ACCESS Condition where the right of owners or occupants of abutting land or other persons to access, light, or air in connection with a highway is fully or partially controlled by public authority. *See also* **Full control of access,** and **Partial control of access.**

CONTROL POINT Point having a precise enough location to act as a datum for other surveys.

CONTROL STATION(S) Command position on either side of a tractor that allows the operator to steer the machine, control most of the paver (roller, or other machine) functions, and follow the available steering reference. On a paver, the control/seat location is near the rear of the tractor so the operator can observe the material supply in front of the screed as well as the hopper and then steer and control the functions as necessary.

CONTROL STOP Device installed in supply piping to regulate or shut off the flow of water to a flush valve.

CONTROL SUBSYSTEM Assembly of devices used to regulate the processes of collecting, transporting,

storing, and using solar energy.

CONTROL SURVEY Survey that establishes arbitrary horizontal and/or vertical stations used as reference points.

CONTROL VALVE *See* **Valve.**

CONTROL ZONE Section of a furnace within which temperature is controlled by the throttling action of a single valve.

contr. spec. engr *Abbreviation for:* contract specification engineer.

conv. *Abbreviation for:* convection; convert; convertible; convex; conveyor.

CONVECTION Transportation of heat by movement due to the ascension of air or liquid when heated and its descension when cooled.

CONVECTIVE COLUMN Rising warm air above a continuing heat source.

CONVECTOR 1. Heat-transfer surface designed to transfer its heat to surrounding air, largely by convection currents; **2.** Heating device in which the air enters through an opening near the floor, is heated as it passes through the heating element, and enters the room through an upper opening.

CONVECTOR RADIATOR Heating system in which the heat from a prime source (steam, hot water, electrical coil, etc.) is transferred to the atmosphere through attached fins.

CONVENIENCE OUTLET Electrical outlet into which may be plugged portable equipment such as lamps or electrically operated equipment.

CONVENTIONAL CAB *See* **Cab.**

CONVENTIONAL CUT Where the saw rotates in a direction counter or opposite to the direction of feed. *See also* **Counter sawing.**

CONVENTIONAL DESIGN Design procedure using moments of stresses determined by widely accepted methods.

CONVENTIONAL DIFFERENTIAL *See* **Differential.**

CONVENTIONAL FACE One of two types of face (undercut) commonly used to fall a tree. The face is taken from the butt of the tree facing the direction the tree is intended to fall and consists of a horizontal cut penetrating approximately one-third of the diameter with a second cut angling down from above to meet the first at the maximum point of penetration. *Also called* **Horizontal face cut,** and **Undercut.** *See also* **Humboldt face.**

CONVENTIONAL FOREST PRODUCTS All commercial roundwood products except fuelwood. Includes boards, dimension lumber, pulp, and paper products.

CONVENTIONAL MORTGAGE *See* **Mortgage.**

CONVERSE-LABARRE FORMULA Formula to give the efficiency of a group of piles relative to their overall bearing capacity.

CONVERSION 1. Transformation of a forest from one forest type to another, favoring a particular species or group of species through practices such as cutting, planting, or weeding; **2.** Process of sawing or otherwise changing the shape of timber; **3.** Transformation of timber into any kind of product; **4.** Process of changing the use of a property, either from one type of use to another (residential to commercial, for instance) or from one type of occupancy to another (storage to living, for example); **5.** Transformation of wastes into other forms, such as steam, gas, or oil.

CONVERSION BURNER Fuel burner installed in a heating unit originally designed to use a different fuel: gas to oil, oil to gas, etc.

CONVERSION COST Cost of changing an existing resource from one form or function to another, e.g., the cost of direct labor and facility overhead incurred in transforming raw materials into a finished product.

CONVERSION FACTOR Number enabling the units of one system to be transformed into the units of another system.

CONVERSION PROCESS Process of chemical change in materials so that the identity of the original material is lost.

CONVERTED SEMI-TRAILER See **Trailer.**

CONVERTER DOLLY See **Dolly,** and **Trailer.**

CONVERTER LOCKUP Locking together the driving and driven elements in the torque converter of an automatic transmission above a given speed, to eliminate any loss of transmitted engine power from slippage in the converter.

CONVERTIBILITY Facility designed into a product or system that permits it to be installed or used in more than one manner.

CONVERTIBLE See **Trailer.**

CONVERTIBLE ADJUSTABLE-RATE MORTGAGE See **Mortgage.**

CONVERTIBLE DIESEL PILE HAMMER See **Pile Hammer.**

CONVEX BEAD Glazing compound bead having a convex exposed surface.

CONVEX WELD Weld with the face above the weld edges.

CONVEYANCE 1. Legal document stating an exchange in interest in real property from one person or entity to another; **2.** Any unit for transporting explosives or blasting agents, including but not limited to trucks, trailers, rail cars, barges, and vessels. See also **Acquisition, Dedication, Eminent domain, Expropriation, Negotiation, Option, Remainder,** and **Severality.**

CONVEYING HOSE See **Delivery hose.**

CONVEYING LINE Endless line on a cableway, used to transport buckets from the loading to the discharge location, and back. Also called **Travel line.**

CONVEYOR Device for moving materials, usually a continuous belt, articulated system of buckets, confined screw, or a pipe through which material is moved by air or water. There are several types, including:

Apron: One or more endless chains fitted with interlocking or overlapping plates that transport materials on their upper faces.

Belt: Endless, pulley-driven, rubber-covered fabric belt supported on rollers, that conveys material placed on its upper surface.

Decline: Conveyor that transports and discharges to a lower elevation.

Feeder: Pushing device or short belt that supplies material to a crusher or belt conveyor.

Screw: Revolving auger flights that move bulk materials along a tube or trough.

CONVEYOR SWING ANGLE Measurement in degrees a conveyor will swing left and/or right of a machine centerline.

convl Abbreviation for: conventional.

CONVOLUTE Twist, coil, roll, or wind materials.

convt Abbreviation for: convenient.

COOLANT Fluid, used to transport heat from one point to another; liquid that is used to cool an engine and that is contained in the water jacket.

COOLANT HEATER Device used to heat an engine coolant at cold ambient temperatures.

COOLANT RECOVERY TANK Special reservoir used to collect coolant overflow from heat expansion or surging in the system. Also called a **Surge tank.**

COOLER Heat exchanger that removes heat from a fluid.

COOLING COIL Coiled pipe or tubing that acts as a heat exchanger to cool a contained liquid or cool an atmosphere by contained refrigerant.

COOLING ELEMENT Heat transfer surface containing a refrigerant.

COOLING (HEATING) AIR CONDITIONER Air handling and treating unit that will ventilate, circulate, clean, heat, or cool, and possibly humidify or dehumidify air within a room, space, or structure.

COOLING MEDIUM Substance used, without change in state, to lower temperatures.

COOLING POND Enclosed body of water used to cool an inflow through surface evaporation, convection, and radiation.

COOLING RANGE Difference between a fluid entering a device and its temperature leaving that device.

COOLING SPRAY Water spray directed into hot gas or air for cooling.

COOLING STRESSES Stresses resulting from uneven distribution of heat during cooling.

COOLING SYSTEM Group of interrelated components to effect the transfer of heat.

COOLING SYSTEM CAPACITY Amount of coolant designated by a manufacturer to completely fill a cooling system.

COOLING TOWER Device through which heated water is passed so as to reduce its temperature, in an open or closed system, by convection or conduction.

COOLING UNIT Refrigeration coil that can be installed in the ductwork of a forced-air heating system for summer cooling.

COOLING WATER Water used to condense a refrigerant.

COOL WHITE *See* **Fluorescent lamp.**

coop. *Abbreviation for:* cooperative; cooperation.

COOPERATIVE Corporation formed for the benefit of its members, e.g., the owners of a multiple-unit residential building.

COOPERATIVE HOUSING Housing financed, built, operated, or managed by a voluntary, nongovernment association of people. *See also* **Housing cooperative.**

coord. *Abbreviation for:* coordinate; coordinating; coordination.

COORDINATE SYSTEM System used to locate a point on a plane with reference to a point called the origin, through which two axes are established that represent north-south and east-west lines.

COORDINATE Measured distance from fixed lines, used to establish the relative location of a point.

cop. *Abbreviation for:* copper; coping.

COPAL Resin used in oil-based varnishes.

COPE 1. To cut, dress, or notch one member so that it will fit against or over another; **2.** To join two molded strips at an angle without a miter; **3.** To finish the top of a wall with materials, and in a manner to shed water.

COPED JOINT Fitting woodwork to an irregular surface. In moldings, cutting the end of one piece to fit the molded face of another at an interior angle.

COPING 1. Capping or overhanging ledge, usually with a sloping upper surface, designed to throw off water; **2.** Process of cutting parts of the web or flange of a steel member for fitting purposes; **3.** Sawing stone with a grinding wheel.

COPING SAW *See* **Saw.**

COPING STONE/SLAB Specially shaped brick, stone, or concrete slab, used to provide a weather protection to the top of a wall.

COPOLYMER Polymer consisting of molecules containing large numbers of units of two or more chemically different types in irregular sequence.

COPOLYMER LATEX Manufactured paint vehicle.

COPPER *See* **Alloying elements.**

COPPER AND BRASS NAILS Used in buildings for attaching similar metals to wood.

COPPER NAPHTHENATE Wood preservative.

COPPER PIPE STRAPS Straps fabricated of copper, used to secure copper pipe to the structure of a building.

COPPER WATER PIPE *See* **Pipe.**

COPPICE In silviculture, a tree cutting method in which renewal of a newly cutover area depends primarily on vegetative reproduction, like sprouting.

COPPICE REGENERATION Ability of certain hardwood species to regenerate by producing many new shoots from a cut stump.

COQUINA Type of limestone formed of sea shells in loose or weakly cemented condition, found along present or former shorelines.

cor. *Abbreviation for:* corner.

corb. *Abbreviation for:* corbel; corbeled.

CORBEL Projecting block of stone, brick, or timber. It can support a beam or lintel, or additional corbels. *See also* **Oversailing courses.**

CORBEL OUT To build one or more courses of brick or stone from the face of a wall, each course projecting further from the face than the preceding course on which it rests.

CORBEL TABLE Projecting cornice or parapet, supported by a series of corbels a short distance apart that carry a molding, above which is a plain piece of projecting wall forming a parapet, and covered by a coping.

CORD 1. Non-SI unit of measure of stacked wood that measures 1.22 x 1.22 x 2.44 m (4 x 4 x 8 ft) or 3.625 m^3 (128 ft^3) of wood, bark, and empty space within the stack; **2.** Any timber product delivered to a receiving facility in short-length form, 2.44 m (8 ft) or less in length, and intended for use as a raw material in the manufacture of pulp and pulp products; a cord is approximately 2359 kg (5,400 lb) for soft hardwood, 2540 kg (5,600 lb) for mixed hardwood, and 2631 kg (5,800 lb) for hard hardwood. Provisions do not apply to pulpwood damaged by insects or other causes, or to timber sold in bulk on the stump; **3.** One of a group of strands forming tire plies.

CORD BREAKER Openly-spaced cord fabric to spread impact or to improve cover adhesion, or both.

CORDON String course.

CORDS PER MAN-HOUR Quotient derived by dividing the total number of cords of wood that a crew produces by the number of man-hours required for the production.

CORDUROY ROAD Road constructed of logs laid crossways to the direction of travel; may comprise more than one layer.

CORDWOOD Wood cut in lengths of 1.22 m (4 ft) or less, to be used as fuel. Also applies to other products measured in cords.

CORE 1. Center of a plywood or other laminated panel whose grain runs perpendicular to that of the outer plies. Plywood cores are either sawn lumber (lumber-core plywood) or crisscrossing layers of veneer (veneer-core plywood). *See also* **Cross band; 2.** In composite panels, a layer of reconstituted wood; **3.** Gypsum contained between the face and back papers of a drywall panel; **4.** Filling of undressed stone used in the interior of a masonry wall; **5.** Axial member of a wire rope about which the strands are laid; **6.** Soil material enclosed within a tubular pile after driving; **7.** Mandrel used for driving casings for cast-in-place piles; **8.** Structural shape used to internally reinforce a drilled-in caisson; **9.** Cylindrical sample of hardened concrete or rock obtained by means of a core drill; **10.** Molded open space in a concrete masonry unit or precast concrete unit; **11.** Tubular-fin structure that acts as a heat exchanger; **12.** Remaining wood after a veneer peeling operation is completed; **13.** Plug of a cylinder lock.

CORE AREA Cross-sectional area of a column bounded by the links.

CORE BARREL Cylindrical rock-drilling tool that

cuts an annular space around a central cylindrical core of rock, that can then be removed to allow the hole to be deepened.

COREBOARD Thick, drywall panel; may have square, rounded, or tongue-and-groove edges.

CORE BORING *See* **Dry sampling.**

CORE BOX Box divided into narrow strips, used to hold cores from a borehole in the order in which they were extracted.

CORE CATCHER Spring device that keeps sand samples from falling out of a soil sampler.

CORE CUTTER Attachment at the foot of a core barrel that grips the core and breaks it at its root when the core is withdrawn for examination.

CORED BEAM 1. Beam whose cross section is partially hollow; **2.** Beam from which cored samples of concrete have been taken.

CORED HOLE Void cast in concrete by leaving a removable core in place during placement and removing it shortly after, during the set-up phase.

CORED PILE Pile inserted into an excavated hole.

CORE DRILL *See* **Drill.**

CORE DRILLING Exploratory drilling that involves cutting cylinders of rock or soil and bringing them to the surface for inspection and analysis.

CORE (FURNACE) WALL *See* **Wall.**

CORE GAP Open veneer joint extending through or partially through a plywood panel.

CORE PLUG *See* **Expansion plug.**

CORE SEPARATION Delamination of the core of a composition (multi-layered) board panel, normally at the center line, caused by steam pops or poor glue distribution and cure, or of a drywall panel.

CORE STRENGTH Compressive strength of a concrete core cut from an existing concrete structure, prepared and tested in accordance with a recognized procedure.

CORE TEST Compression test on a concrete sample cut from hardened concrete by means of a core drill.

CORE WALL Wall of impervious material built inside an earthern dam to reduce percolation.

CORING Act of obtaining cores from soil, rock, or concrete structures for examination and testing.

CORING BIT *See* **Bit.**

CORING TEST Soils test made by augering a core sampler into the ground and withdrawing an undisturbed plug for analysis.

CORINTHIAN Designating the most elaborate of the three orders (Doric, Ionic, Corinthian) of classical Greek architecture, distinguished by a slender, fluted column and a bell-shaped capital decorated with a design of acanthus leaves. *See also* **Order.**

CORK Bark of the cork oak tree.

CORKBOARD Granules of cork bark compressed and baked into sheets and blocks; used for decoration and sound deadening.

CORKING Method of connecting one beam to another, with the first beam bearing across the second.

CORK-SETTING ASPHALT Type of asphalt adhesive used to fasten cork sheet to a surface.

CORK TILE Surfacing product (floors, walls, ceilings) manufactured from the granulated bark of the cork oak tree, bonded together under heat and pressure.

CORKSCREW STAIR Spiral stair.

corn. *Abbreviation for:* cornice.

CORNER 1. Shape formed by the side and back lines of a setting; **2.** To cut through the sapwood on both sides in felling trees to prevent splitting; **3.** Left or right side face of the holding wood left after the backcut and undercut are completed; **4.** Corner of the falling 'face.' *See also* **Corner nipping.**

CORNER BATT Three corner points from which a concrete form is started.

CORNER BEAD 1. Preformed metal or plastic strip placed on internal and external corners of drywall prior to mudding, as reinforcing. **2.** Wood or plastic molding finished three-quarter round or angular, used on an internal corner as a finish.

CORNER BLOCK 1. Masonry unit with a flat end for construction of the end or corner of a wall; **2.** When two or more tailblocks are used, the corner blocks are the outermost blocks that change the direction of the haulback. *Also called* **Tailblock.**

CORNER BOARD Built-up wood member against which the ends of horizontal siding are butted, installed vertically on an external corner of a framed structure.

CORNER BRACE Diagonal structural member used in a corner to make a frame more rigid.

CORNER CLAMP *See* **Clamp.**

CORNER GUARD Metal, wood, rubber, plastic, or other material shaped to an exterior corner in an interior wall to serve as a guard against damage from such moving traffic as hospital beds and carts, internal delivery carts, etc.

CORNER LOT Building lot abutting upon two or more streets at their intersection.

CORNER MOLDING *See* **Molding.**

CORNER NIPPING Felling technique of partially cutting the extreme outside holding wood corners to prevent root pull, slabbing and variance from the desired falling direction. *See also* **Corner.**

CORNER POST 1. Several studs spiked together to form a corner in a frame building; **2.** Method of mounting rails in opposite corners of a hoistway, usually to accommodate doors in adjacent hoistway walls.

CORNER REINFORCEMENT 1. Metal reinforcement for plaster at reentrant corners to provide continuity between two intersecting planes; **2.** Concrete reinforcement used at wall intersections or near corners of square or rectangular openings in walls, slabs, or beams.

CORNER SIDE YARD Side yard that adjoins a public thoroughfare.

CORNERSTONE Stone in the base of a building, specially prepared to be prominent and usually containing carving giving information about the structure.

CORNER TOOL Finishing knife that permits the simultaneous application of jointing compound and the finishing of 90° internal drywall angles.

CORNER WEAR Tendency of a grinding wheel to wear on a corner so that it does not grind sharp corners without fillets.

CORNICE 1. Projecting molding along the top of a building, wall, arch, etc.; **2.** Ornamental molding round the wall of a room, just below the ceiling.

CORNICE RETURN Type of cornice trim where the sloping line of a gable roof meets the vertical line of the wall of the building. The fascia is started across the face of the wall below the gable end and is then returned upon itself about 600 mm (2 ft) from the corner.

CORNICE TRIM Exterior finish on a building where the sloping roof meets the vertical wall.

CORONA Brow of a cornice, projecting over the bed moldings to throw off water.

corp. *Abbreviation for:* corporation.

corr. *Abbreviation for:* corrected; correction; correspond; correspondence; corresponding; corridor; corrode; corrosive; corrugate.

CORRAL Piled enclosure used to position a caisson as it is being sunk.

CORRECTED POWER Observed power adjusted to standard atmospheric conditions.

CORRECTIVE ACTION Action or work done to remedy a fault.

correl. *Abbreviation for:* correlation.

CORRELATION *See* **Analysis.**

corr. engr *Abbreviation for:* corrosion engineer.

CORRIDOR 1. Interior passage giving access to rooms or spaces; **2.** Cleared strip for a skyline or guyline; **3.** Strip of land between two terminals within which traffic, topography, environment, and other characteristics are evaluated for transportation purposes.

CORRIDOR PATTERN *See* **Ribbon development.**

CORRIDOR SKIDDING Logging procedure using cable yarders in which narrow clearcuts are made through a stand. Cables are strung in these clearcut corridors to transport logs from the woods to the landing. Between corridors only a portion of the trees in the stand are removed, and these harvested trees are skidded to the corridor.

CORROSION 1. Destruction of metal by chemical, electrochemical, or electrolytic reaction with its environment; **2.** Chemical or galvanic decomposition of the wires in a rope through the action of moisture, acids, alkalines, or other destructive agents.

CORROSION EMBRITTLEMENT Loss of ductility or workability of a metal due to corrosion.

CORROSION FATIGUE Effect on metal of repeated stress in a corrosive atmosphere characterized by shortened life of the part.

CORROSION INHIBITOR 1. Chemical compound, either liquid or powder, that effectively decreases corrosion of steel reinforcement before being embedded in concrete, or in hardened concrete if introduced, usually in very small quantities, as an admixture; **2.** Property of packing that actively inhibits corrosion of adjacent metal surfaces.

CORROSION-RESISTANT STEEL In wire rope manufacture, chrome-nickel steel alloys designed for increased resistance to corrosion.

CORROSIVE SUBSTANCE Solid, liquid, or gas which when contacting living tissue damages the tissue, or when contacting other materials and certain chemicals, causes fire or accelerated deterioration of the material or chemical.

CORROSIVE WEAR *See* **Wear.**

CORRUGATED 1. Having a cross-section or profile comprising a regular series of repeated geometric shapes, most commonly semi-circular; **2.** In wire rope terminology, term used to describe the grooves of a sheave or drum after these have been worn down to a point where they show an impression of a wire rope.

CORRUGATED COVER Longitudinally ribbed or grooved exterior hose covering.

CORRUGATED GLASS Glass rolled to give a corrugated profile.

CORRUGATED HOSE Hose with a carcass fluted radially or helically to enhance its flexibility or reduce its weight.

CORRUGATED IRON Sheet steel formed with parallel corrugations to increase stiffness; used as a roof and wall covering and for other building purposes.

CORRUGATED KEY Key having longitudinal corrugations along its shank to correspond to the profile of a keyway.

CORRUGATED PLEATS Series of folds in a filter medium, usually of uniform height and spacing.

CORRUGATED ROOFING Roofing sheets of a variety of materials, having a section profile of continuous corrugations in a range of pitches. *See also* **Sheet-metal roofing.**

CORRUGATIONS Parallel grooves and ridges, developed in a wide range of profiles, that impart semirigidity to a thin, flat sheet.

CORUNDUM Natural abrasive of the aluminum oxide type, of higher purity than emery.

cos *Abbreviation for:* cosign.

cos⁻¹ *Abbreviation for:* inverse cosign.

cosec *Abbreviation for:* cosecant.

cosec⁻¹ *Abbreviation for:* inverse cosecant.

COST ACCOUNTING Accounting method concerned with the classification, recording, analysis, reporting, and interpretation of expenditures identifiable with the production and distribution of goods and services.

COST APPROACH Determination of the value of a property, including land and buildings, made first by estimating the value of the land as if it were vacant, adding to it the estimated cost to reproduce the buildings as new, then deducting an amount for depreciation of the existing property. *See also* **Depreciated cost method.**

COST/BENEFIT ANALYSIS *See* **Analysis.**

COST BREAKDOWN Detailed listing or analysis of the constituent costs that contribute to a single cost item.

COST CENTER Subdivision of an organization with which costs can be identified for the purposes of managerial control.

COST CHECK Periodic estimate review to give an up-to-date analysis of cost in relation to the progress of design or construction.

COST CLASS Any of the subdivisions of the total scope of the work in a project to which costs are assigned.

COST CODE System that allocates codes to defined classes and types of work and materials; an aid to estimating and cost control.

COST CONTROL Accounting procedure designed to produce a range of reports giving information about total expenditure, total income, cash flow, accounts receivable, etc.

COST ESTIMATE Approximation of cost within a given percentage of anticipated actual cost.

COST FLOW Method of allocating costs between inventory and cost of sales. There are several methods, including:

> **Average:** Method where the cost of an item is determined by applying a weighted average of the cost of all similar items at a point of time or over a period.
>
> **Base stock:** Method where the cost of items sold or consumed during a period is determined by assuming that a predetermined minimum quantity is carried in inventory permanently and at a fixed price.
>
> **FIFO:** First in, first out, where the cost of items sold or consumed during a period is computed as though they were sold or consumed in order of their acquisition.
>
> **LIFO:** Last in, first out, where the cost of items sold or consumed during a period is deemed to be at the cost of the most recent acquisitions.
>
> **NIFO:** Next in, first out, where the cost of items sold or consumed during a period is deemed to be at the cost of the next acquisition.
>
> **Specific identification:** Where the actual cost of each item is determined separately.

COST OF CAPITAL Investment required to create and maintain productive capital.

COST OF REPRODUCTION, NEW Cost of exact duplication of a property with the same, or closely similar materials, as of a certain date.

COST OF WORK Costs incurred in the completion of work described and illustrated in the contract documents.

COST OVERRUN Cost beyond that budgeted for or contracted for.

COST PER KM/MILE Total cost (including such factors as depreciation, insurance, downtime, repairs and service, replacement parts, etc.) involved in running a vehicle or equipment item for one kilometer or mile, averaged from the costs incurred over as large a number of kilometers or miles as possible.

COST PER TON PER MINUTE Unit used in cost comparisons between transfer and direct-haul operations.

COST PLAN Sets out the total cost limit of a budget, sub-divided into meaningful sections, each with its own cost and outline specification stated. It provides the frame of reference required as the first principle of a valid cost control system.

COST-PLUS Form of contract for construction work wherein the construction contractor is reimbursed for the costs he incurs in performing the work, plus a lump sum or proportional fee, hence cost-plus. This type of contract is favored where the scope of the work is indeterminate or highly uncertain and the kinds of labor, material, and equipment needed also are uncertain. Under such an arrangement it is necessary to maintain complete records of all time and materials expended by the contractor on the work.

COST-RATIO METHOD OF ACCOUNTING *See* **Method of accounting.**

COST REVIEW Planned, systematic reassessments of the estimated final cost of the scope of work in a cost class. Each design review is accompanied by a reforecast of cost. *See also* **Forecast to complete.**

COST TYPES Associated with the scope of the work in a cost class are four different cost types, having to do with whether a cost is authorized, contractually committed, expended, or estimated. These types are:

> **(a)** authorized appropriation;
>
> **(b)** commitment;
>
> **(c)** expenditure, or actual; and
>
> **(d)** forecast to complete, or estimate to complete, or uncommitted.

Types (a) and (d) contain contingencies. The sum of type (b) and type (d) is the estimated final cost. Before any commitments are made within a cost class, the type (d) cost contains a component called the anticipated award price, that in turn contains an allowance for escalation. Type (c) costs are further subdivided into payments and retentions (holdbacks).

cot *Abbreviation for:* cotangent.

cot⁻¹ *Abbreviation for:* inverse cotangent.

cot. *Abbreviation for:* cotter; cotton.

COTENANCY Any of several forms of joint or multiple ownership.

COTTER PIN Split pin whose ends can be flared after it is inserted in a hole or slot to fasten two or more pieces together.

COTTON MATS Cotton-filled quilts fabricated for use as a water-retaining covering in curing concrete surfaces.

cot. web. *Abbreviation for:* cotton webbing.

COULOMB One of 17 derived units with special names of the SI system of measurement: a unit of electric charge transported in 1 s by a current of 1 A. (1 C=1 A•s). Symbol: C. *See also the appendix:* **Metric and nonmetric measurement.**

COUNT In fabric, the number of warp ends, or the number of filling picks, or both, in a square inch of fabric.

COUNTER 1. In the opposite or contrary direction or manner; **2.** Member of a truss system that acts only for a particular partial loading, and that has zero stress when the truss is completely loaded.

COUNTERBALANCE 1. Dead weight that offsets a static or live load; **2.** Force offsetting another.

COUNTERBALANCE VALVE *See* **Valve.**

COUNTERBALANCED-FRONT SIDELOADER TRUCK Self-loading high-lift industrial truck (equipped with a fixed or tiltable elevating mechanism) capable of transporting and tiering a load in both the counterbalanced forward position and any location up to and including 90° from the longitudinal centerline of the truck, while possessing the capability of traversing the load laterally.

COUNTERBALANCED TRUCK Industrial truck equipped with a load-engaging means wherein all the load during normal transporting is external to the polygon formed by the wheel contacts.

COUNTERBALANCE PRESSURE-CONTROL VALVE *See* **Valve.**

COUNTERBATTENS 1. Battens fastened to the back of boards to stiffen them; **2.** Battens nailed over a boarded and felted roof, positioned parallel to and over the rafters, to which tiling or slating battens are nailed.

COUNTERBLAST Explosive charge ignited to limit and contain the effects of a larger charge, initiated almost simultaneously with the main explosion.

COUNTERBORE Lip cut in the top of an engine block in which a cylinder sleeve is fitted.

COUNTERBORING To enlarge a hole through part of its length by boring.

COUNTERBRACE Web member of a truss that is designed to resist either tension or compression.

COUNTERCEILING False or suspended ceiling.

COUNTERFIRE Fire set between a main forest fire and a backfire to hasten spread of the backfire.

COUNTERFLASHING 1. Flashing used at the junction of a vertical surface and a sloped or horizontal surface to prevent moisture entry; **2.** Flashing applied above another flashing to shed water over the top of the under flashing and to allow differential movement.

COUNTERFLOOR Lower of a two-layer floor system; a subfloor.

COUNTERFLOW Liquid flow in the opposite direction to the main flow, as in the return flow from a heat exchanger.

COUNTERFORT Strengthening wall in the middle of an abutment.

COUNTER GAUGE Mortise gauge having two scribes.

COUNTERJIB Horizontal member of a tower crane

on which the counterweight and, usually, the hoisting machinery are mounted. *Also called* **Counterweight jib.**

COUNTERKNIVES Piece of steel in a portable chipping machine that breaks a chip into desired lengths. Found behind, and similar in appearance, to a chipper knife.

COUNTERPROPOSAL Response made to an original proposal giving alternatives.

COUNTERSAWING *See* **Conventional cut.**

COUNTERSHAFT Shaft that receives power from a parallel mainshaft and transmits it to working parts or to a different part of the mainshaft.

COUNTERSHAFT TRANSMISSION *See* **Transmission.**

COUNTERSINK 1. Metal punch used to drive nail heads below the surface without marking the surround area; **2.** Recess drilled in the top of a hole so that the head of a screw or bolt will set flush with the surface; **3.** To drive the head of a nail below the surface with a punch so that the resulting hole can be filled and finished. *See also* **Bit.**

COUNTER UNIT Floor mounted cabinet having a top that forms a working surface.

COUNTERWEIGHT 1. Nonworking load used to supplement the weight of a machine in order to provide improved stability; on equipment used for lifting, the counterweight is commonly attached to the rear of the revolving superstructure opposite the boom nose, or on the counterjib of a climbing crane; on the crankshaft of an engine, a weight that counteracts the inertia forces developed by the pistons and the connecting rods. *See also* **Hydraulically extendible counterweight,** and **Power removable counterweight; 2.** Dead weight on traction elevators that counterbalances the weight of an elevator car, plus nearly half of the capacity load.

COUNTERWEIGHT CLEARANCE Distance from the counterweight to any stationary object.

COUNTERWEIGHT FILLER Any of a number of pieces of metal stacked and bolted together with headers to form a stacked counterweight, or assembled in a frame to form a counterweight.

COUNTERWEIGHT HEADER In a stacked counterweight, the weight section larger than a filler that extends around the counterweight guide rails to guide the counterweight.

COUNTERWEIGHT JIB *See* **Counterjib.**

COUNTERWEIGHT ROD Rod with threaded ends that passes through the filler weights and holds them in place with washers, plates, and nuts.

COUNTERWEIGHT SHEAVE Sheave(s) mounted on a counterweight frame when two:one multiple roping is used.

COUNTING CALIBRATION FACTOR Ratio of the effective filtration area on a test pad membrane to the area counted.

COUNTY BUILDING CODES *See* **Building codes.**

COUPLED LENGTH *See* **Hose assembly.**

COUPLED ROOF Pitched roof with common rafters

and no tie beams.

COUPLER 1. Device for connecting reinforcing bars or prestressing tendons end-to-end; **2.** Device for locking together the component parts of a tubular scaffold (also known as a **Clamp**); **3.** Device by which a trailer is attached to a tractor via a fifth-wheel plate. *See also* **Clamp, End-bearing sleeve,** and **Mechanical connection.**

COUPLER HEIGHT 1. Vertical distance from the floor to the effective horizontal centerline of a coupler; **2.** Distance from the ground to the bottom of an unladen trailer upper coupler plate.

COUPLER PLATE Flat plate on the underside of a trailer upper coupler through which the kingpin protrudes and which rests directly on a tractor fifth wheel.

COUPLING 1. Connecting or joining two or more distinct parts; **2.** Device attached to the end of a hose to facilitate connection to a suitable fitting and to insure a passageway; **3.** Transfer of energy from an explosive reaction into the surrounding rock, considered perfect when there are no losses due to absorption or cushioning; **4.** Pipe fitting containing female threads on both ends; **5.** Device that joins a drill steel to another drill steel or to a striker bar.

COUPLING AGENT That part of a sizing or finish that is designed to provide a bonding link between the reinforcement and the laminating resin.

COUPLING PIN Insert device used to connect lifts or tiers of formwork scaffolding vertically.

COUPLING RATIO Square root of the ratio of the volume of the borehole (excluding the volume of the collar) divided by the volume of explosive material.

COUPLING SLEEVE Device fitting over the ends of two reinforcing bars for the eventual purpose of providing transfer of either axial compression or axial tension or both from one bar to the other. *See also* **End bearing sheave,** and **Mechanical connection.**

COURSE 1. One of the continuous horizontal layers of units, bonded with mortar in masonry; **2.** In concrete construction, a horizontal layer of concrete, usually one of several making up a lift.

COURSE BED Stone, brick, or other building material upon which other material is to be laid.

COURSED ASHLAR *See* **Ashlar.**

COURSED RUBBLE *See* **Rubble masonry.**

COURT Open space, unoccupied from the ground or intermediate floor to the sky, contiguous with the building and on the same lot; it is intended primarily for the provision of light and air, but may serve for entrance to the building. It is entirely enclosed by walls or enclosed on three sides having one side partially or totally open to a street, yard, or abutting property.

cov. *Abbreviation for:* cover; covered.

COVE 1. Molding with a concave face. *See also* **Molding**; **2.** Sloped or arched junction of a wall and ceiling.

COVE BASE Specially made glazed tile with a projecting curved lip at the bottom, made for the finished floor to fit against.

COVE BRACKETING Bracketing at the junction of

the walls and ceiling that forms a circle or ellipse. *Also called* **Pendentive bracketing.**

COVED CEILING Ceiling that is formed at the edges to give a hollow curve from the wall to the ceiling instead of a sharp angle of intersection.

COVE LIGHTING Artificial lighting where the lighting source is hidden from sight by a cove or other projection with the light being reflected from above.

COVENANT Requirement written into deeds and other instruments relating to land ownership that speaks to performance or nonperformance of certain acts.

COVENANT RUNNING WITH THE LAND Covenant restricting or limiting property rights to land that cannot be extinguished.

COVER 1. In reinforced concrete, the least distance between the surface of the reinforcement and the outer surface of the concrete; **2.** Outer component, usually intended to protect the carcass of a product.

COVERAGE 1. Measure of the amount of material required to cover a given surface; **2.** Extent of protection provided by insurance.

COVER BLOCK Small precast mortar or plastic block used inside formwork to ensure proper cover to reinforcement.

COVERED CONDUCTOR *See* **Electrical conductor.**

COVERED ELECTRODE Metal rod used in arc welding that has been covered with materials to aid the process.

COVER FILLET Molded strip used to cover the joint between two components, usually sheet materials.

COVERING POWER Measure of the capacity of a fluid or plastic substance to cover a surface.

COVER MATERIAL Soil used to cover compacted solid waste in a sanitary land fill.

COVER PLATE Closure that caps an otherwise open space, box, receptacle, etc.

COVER RING Ring mounted on the outside diameter of a primary seal ring to cover the gap in the latter.

covers. *Abbreviation for:* conversed sine.

COVER SEAM Spiral or longitudinal joint formed by the lapping of hose cover stock.

COVER TYPE Category of forest defined primarily by its vegetative composition and/or locality factors.

COVER WEAR Loss of material during use due to abrasion, cutting, or gouging.

COVE TRIM Wood, tile, metal, or plastic used to mask a right-angled junction.

covg. *Abbreviation for:* coverage.

COWL 1. Hood-shaped covering of a chimney or ventilating shaft, used to reduce down-draft; **2.** That part of a vehicle cab directly below the windshield between the firewall and instrument panel or door.

COYOTE BLASTING Practice of drilling blastholes (tunnels) horizontally into a rock face at the foot of the shot, used where it is impractical to drill vertically.

cp. *Abbreviation for:* candlepower; charging plug; concrete pipe.

CPA *Abbreviation for:* cost planning and appraisal.

cpbl. *Abbreviation for:* capable.

cpd *Abbreviation for:* compound; coped.

CPE MEMBRANE *See* **Chlorinated polyethylene membrane.**

CPF *Abbreviation for:* claw plate, female.

CPFF *Abbreviation for:* cost, plus fixed fee.

CPFL *Abbreviation for:* clamping plate, flanged.

CPIF *Abbreviation for:* cost, plus incentive fee.

cpl. *Abbreviation for:* complete; couple.

cplg *Abbreviation for:* coupling.

cplr *Abbreviation for:* coupler.

C-PLUGGED *See* **Veneer grade.**

CPM *Abbreviation for:* claw plate, male; Critical Path Method; cycles per minute.

CPP *Abbreviation for:* clamping plate, plain.

cpunch *Abbreviation for:* counterpunch.

CPVC (CHLORINATED POLYVINYL CHLORIDE) PIPE *See* **Pipe.**

CQ *Abbreviation for:* commercial quality.

cr *Abbreviation for:* charging receptacle.

CR *Abbreviation for:* cold-rolled; complete round; cooling rate; corrosion resistant; crossroads.

CRA *Abbreviation for:* cold-rolled and annealed.

CRAB 1. Formed metal component designed to hold an injector or injector assembly into its bore in the head of a diesel engine; **2.** A hand winch.

CRACK 1. Complete or incomplete separation produced by breaking or fracturing; **2.** To open a valve slightly and then close it immediately. *See also* **Reflection crack.**

CRACK CONTROL REINFORCEMENT Reinforcement in concrete construction designed to prevent opening of cracks, often effective in limiting them to uniformly distributed small cracks.

CRACKED *See* **Bulk appearance.**

CRACKED SECTION Section designed or analyzed on the assumption that the material has no resistance to tensile stress.

CRACKING Sharp break or fissure in the surface, generally caused by strain and environmental conditions.

CRACKING AND SEATING Cracking of portland cement concrete into small pieces, rolling the pieces to seat them on the base or subgrade, and overlaying with asphalt concrete to reduce reflective cracking.

CRACKING LOAD Load that causes tensile stress in a member to exceed the tensile stress of the concrete.

CRACKING OF HYDROCARBONS Splitting of hydrocarbon molecules into both heavier and lighter molecules. This results from the application of pressure, a catalyst, heat in the absence of air, or various

combinations of the three.

CRACKING PRESSURE Pressure at which a pressure-operated valve begins to pass fluid.

CRADLE 1. A carriage; **2.**Support bracket, that may be hinged to the load it carries; **3.** Large metal brackets or a wood framework made to hold small chunks, poles, or pieces of pulpwood being bundled for transport; **4.** Frame in which individual bolts are towed; **5.** Steel frame, usually made of angle iron, in which a battery is assembled for use in an industrial truck having its own integral battery compartment; **6.** Side of a log carved out to fit another log placed across it.

CRAFT Occupational skill involving the manual completion of specialized work.

CRAMMING 1. Clamping parts of wood or other materials permanently, or temporarily so as to work on them; **2.** Plugging a pipe before making a repair.

CRAMP Metal bar with one or both ends turned at right angles to the body of the bar in order to enter holes in the faces of adjacent stones to hold them in place.

CRANE Hoisting device. There are several types, including:

> **Boom:** Crane equipped with a boom and cables that will pick up a load, move it to another position, and set it down again. May be static or mobile.

> **Bridge:** Lifting unit that can maneuver horizontally in two directions.

> **Climbing:** Tower crane having the ability to elevate itself by its own structure, or to increase its height using the structure that surrounds it.

> **Monorail:** Lifting unit, suspended from a single rail, that can only move in one horizontal direction.

> **Tower:** T-shaped structure where the horizontal member is balanced across, and rotates about the vertical tower and mounts the hoisting engine at one end, plus counterweights, and the hoisting rope sheave(s) at the other.

CRANE ARM *See* **Boom.**

CRANE CONFIGURATION Physical arrangement of a crane as prepared for a particular operation, in conformance with the manufacturer's operating manual and load rating chart.

CRANE POST Upright mast of a derrick..

CRANE RAILS Tracks supporting and guiding the wheels of a bridge crane.

CRANE RATING MANUAL Compilation of the necessary information needed to plan lifts with a crane. It includes instructions such as the allowable lifting capacity charts, working range diagrams, working area diagram, etc.

CRANE TOWER Tower on a derrick that carries the mast and crane machinery.

CRANK 1. Part connected at right angles to a shaft so as to transmit motion; **2.** To start an engine by operation of a crank; **3.** Short straight offset, or double bend, in a reinforcing bar.

CRANK BRACE Carpenter's hand brace, used with a

bit to drill holes.

CRANKCASE Metal casing enclosing the crankshaft of a reciprocating engine and acting as a reservoir for the engine ubricating oil.

CRANKCASE DILUTION Dilution of an engine lubricating oil due to unburned fuel which finds its way past the piston rings, down the cylinder walls and into the crankcase.

CRANKED BEAM Beam having a double bend, all in the same plane.

CRANK PIN Cylindrical bar attaching a connecting rod to a crank.

CRANKSHAFT Shaft turning a crank, or turned by a crank; in a reciprocating engine, to transmit the movement of the piston(s).

CRANKSHAFT COUNTERBALANCE One or more weights attached to a crankshaft throw so as to eliminate forces tending to produce vibration.

CRANKSHAFT-DRIVEN TRANSMISSION See **Power takeoff.**

CRANKSHAFT JOURNAL Round part of the crankshaft onto which the connecting rod is fastened.

CRANK WHEEL Expression used to designate wheels for grinding crankshafts.

CRASH CUSHION Traffic barrier used to safely shield fixed objects or other hazards from approximately head-on impacts by errant vehicles.

CRATER Depression in the face of a weld, usually at the point where an arc weld was terminated.

CRATERING Depression in a coating film, usually caused by air or solvent trapped in the coating forming bubbles that break after the film has set sufficiently to prevent leveling.

CRAWLER Machine mounted on movable tracks.

CRAWLER FRAME Part of the base mounting of crawler cranes, attached to the carbody and supporting the crawler treads, track rollers, drive, and idler sprockets.

CRAWLER TRACK One of a pair of roller-chain tracks used to support and propel a machine.

CRAWLING 1. Propelling a track-equipped machine or vehicle by causing the movable tracks to rotate; **2.** Parting and contraction of the glaze on the surface of ceramic ware during drying or firing, resulting in unglazed areas bordered by coalesced glaze; **3.** Wet film defect that results in a paint film pulling away from certain areas, or not wetting some areas, leaving them uncoated.

CRAWLING BOARD Supported scaffold consisting of a plank with cleats spaced and secured at equal intervals, for use on sloped surfaces such as roofs. *Also called* a **Chicken ladder.**

CRAWL SPACE Shallow space below the main floor of a house, sometimes as part of an enclosed basement.

CRAWLWAY Space high enough to crawl through, usually giving access to building services.

CRAZE CRACK Fine random crack or fissure in a surface of plaster, cement paste, mortar, or concrete.

See also **Checking.**

CRAZING Development of craze cracks existing in a surface. *See also* **Checking, Craze crack,** and **Map cracking.** *Also called* **Pattern crack.**

CRAZY PAVING See **Flagstone.**

CREAMING Logging operation where only the best trees in a stand are cut.

CREAM PAPER Sized and calendered paper used as the face paper of drywall panels. *Also called* **Ivory paper.**

CREASING HAMMER See **Hammer.**

CREASING IRON Anvil with grooves across it, used in sheet metal work.

CREEP 1. Movement of aggregate particles under roller pressure during compaction; **2.** Unique movement of a wire rope with respect to a drum surface or sheave surface resulting from the asymmetrical load between one side of the sheave (drum) and the other. It is not dissimilar to the action of a caterpillar moving over a flat surface. It should be distinguished from slip, which is yet another type of relative movement between the rope and the sheave or drum surface; **3.** See **Deformation, inelastic; 4.** Down slope, almost imperceptible movement of the regolith of the earth's surface; **5.** Permanent deformation caused by stress or heat, or both. *See also* **Drift.** *Also called* **Plastic loss; 6.** Action of a belt alternately losing speed on the driving pulley and gaining speed on the driven pulley; **7.** Sporadic or continuous motion of a shaft driven only by drag in a fluid coupling or other disconnecting device; **8.** Very slow travel or motion of a machine or part. *See also* **Plastic flow.**

CREEP DEFLECTION Deflection of a beam due to creep.

CREEPER 1. Traveling derrick used in bridge construction; **2.** Brick next to an arch that is cut to fit the curvature of the extrados.

CREEPER LANE Reduced-speed lanes of a multilane highway, usually on steep up and down grades, often reserved for trucks hauling heavy loads.

CREEP FACTOR Ratio of creep strain to elastic strain.

CREEPING Effect of the finish coat of paint to form globules, caused by incompatible materials or as a result of the undercoat having too high a gloss.

CREEPING FIRE Forest fire burning with a low flame and moving slowly.

CRENELATED Having a profile formed of regularly spaced depressions, notches or other shapes; a battlement.

CREOSOTE Oily liquid distilled from wood tar, used as a preservative on wood.

CREOSOTED PILE Timber pile impregnated with a specified amount of coal-tar creosote by an approved process.

CREOSOTING Use of creosote, often under pressure, to preserve lumber.

CRESCENT Road, housing, or other structures laid out to follow the curve of an arc.

CRESCENT PUMP *See* **Pump.**

CRESCENT WRENCH *See* **Wrench, clip.**

CREST 1. Toe of the face created by a previous shot in blasting; **2.** Maximum amplitude of a wave in the upward direction above zero; **3.** Point of highest elevation on a hill. *See also* **Military crest; 4.** Outerfold of a pleat.

CREVICE SEAL Seal at the bottom of a cylinder liner preventing engine coolant from leaking into the crankcase.

CRF *Abbreviation for:* capital recovery factor.

crg. *Abbreviation for:* carriage.

CRH *Abbreviation for:* cold-rolled, hard.

CRHH *Abbreviation for:* cold-rolled, half-hard.

CRIB 1. Crate-like framing used to support a structure. *See also* **Cellular cofferdam,** and **Cofferdam; 2.** Any of various frameworks of logs or timbers, used in construction work; **3.** Wooden lining on the inside of a shaft; **4.** Openwork of horizontally cross-piled, squared timbers, or beams, used as a retaining wall.

crib. *Abbreviation for:* cribbing.

CRIBBING High strength stacking material (i.e., wood, steel matting, etc.) under outrigger pads to form a support for a level lifting foundation in uneven terrain, or to distribute the weight of a crane over a larger surface area.

CRIBBING FOUNDATION AREA Area occupied by the base log, mud sills, or foundation logs.

CRIB DAM Barricade formed of bays or cells that are filled with impervious materials.

CRIB HEIGHT Elevation difference between the top of the cap or sill and the bottom of the base log at the center line.

CRIB WALL Retaining wall constructed of rectangular interlocking precast concrete or timber members, forming a cellular structure, laid on top of each other and filled with soil or broken rock.

CRIBWORK Construction of crib-like structures.

CRICKET 1. Small, false roof or the elevation of part of a roof behind an obstacle, as a watershed built behind a chimney or other roof projection; **2.** Watertight flashing over such a construction.

CRIMP 1. To offset the end of an angle or metal strip so it can overlap another piece; **2.** To indent the ends of pipe, tubing, flanges, etc.; **3.** In fabric, the sinusoidal curvature impressed in the warp or filling during weaving; **4.** Difference in distance between two points on a yarn as it lies in a fabric, and the same two points when the yarn has been removed from the fabric and straightened under tension.

CRIMPED WIRE 1. Wire deformed into a curve that approximates a sine curve as a means of increasing the capacity of the wire to bond to concrete; **2.** Welded wire fabric crimped to provide an integral chair. *Also called* **Deformed wire.** *See also* **Indented wire.**

CRIMPING TOOL Hand tool that applies a predetermined amount of pressure to parts held within its jaws, typically to apply metal beads or clinch light metal parts together.

CRIPPLE Any part of a frame that is cut less than full size, typically as a cripple studding under a window opening.

crit. *Abbreviation for:* criteria; critical.

CRITERIA Performance level, standard, or other arbitrary description.

CRITICAL ACTIVITIES OR WORK ITEMS Activities or work items that, if not completed by the indicated time for them, will correspondingly increase the total project duration by the extent of the delay in completion of that activity or work item.

CRITICAL FIBER LENGTH Double the length of fiber embedment that would cause fiber failure in a pull-out test.

CRITICAL FIBER VOLUME Minimum volume of fibers that, after matrix cracking, will carry the load that the composite sustained before cracking.

CRITICAL FLOW Rate at which a fluid flows through an orifice when the stream velocity at the orifice is equal to the velocity of sound in the fluid. Under such conditions, the rate of flow may be increased by an increase in upstream pressure, but it will not be affected by a decrease in downstream pressure.

CRITICAL LOAD Load at which bifurcation occurs, as determined by a theoretical stability analysis.

CRITICAL PATH Route through the network logic sequence on which all activities must be completed within their expected timings or the schedule for the overall project will change (i.e., activities or work items with zero float).

CRITICAL PATH METHOD Activity-oriented time control system used in situations where activities and their duration can be well defined on a project, usually in one location and controlled by one organization. *Also called* **CPM.** *See also* **Arrow,** and **Program evaluation and Review technique (PERT).**

CRITICAL POINT State point at which liquid and vapor have identical properties.

CRITICAL PRESSURE *See* **Pressure.**

CRITICAL RADIUS 1. Minimum radius to which a reinforcing bar can be curved when used in a circular structure, such as a chimney; **2.** Minimum radius to which a reinforcing bar should be bent when forming a bend, hook, or link.

CRITICAL SATURATION Condition describing the degree of filling by freezable water of a pore space in cement paste or aggregate that affects the response to freezing, usually taken to be 91.7% because of the 9% increase in volume of water undergoing the change to the state of ice.

CRITICAL SILENCER Exhaust silencer that is applied in sensitive noise-control areas.

CRITICAL SPEED Engine or spindle speed (rpm) at which destructive harmonic vibration occurs.

crk *Abbreviation for:* cork; crack; crank.

crk *Abbreviation for:* cracked.

crk bd *Abbreviation for:* cork board.

cr. moly. *Abbreviation for:* chrome molybdenum.

CROCUS 1. Gneiss or similar rock in contact with granite in a quarry; **2.** Fine polishing powder, used in buffing metal.

CROCUS CLOTH Cloth used to polish metal.

CROOK 1. Abrupt bend in a tree or log; **2.** Distortion of a board or log in which there is a deviation, edge-wise, from a straight line from end to end.

CROP TREE Tree in a young stand selected to be retained until final harvest.

CROSBIE CLIP See **Cable clip.**

CROSS Pipe fitting with four female openings at right angles to one another.

crossb. Abbreviation for: cross band.

CROSS BAND 1. Layers of veneer at right angles to the face plies. See also **Core;** **2.** To place layers of wood with their grain at right angles, to minimize warping.

CROSS BAR See **Lift bar.**

CROSSBEARER See **Mudsill.**

CROSS BOND See **Bond.**

CROSS BRACE 1. Horizontal member of a shoring system installed perpendicular to the sides of the excavation, the ends of which bear against either uprights or wales; **2.** One of two diagonal scaffold members joined at their center to form an 'X,' used to brace frames and/or uprights.

CROSS BRACING Crossing members usually designed to act only in tension, often used in scaffolding systems. See also **Swaybrace,** and **X-brace.**

CROSS BRIDGING Arrangement of small wooden pieces, 25 x 75 mm, 50 x 50 mm, or 50 x 100 mm (1 x 3 in., 2 x 2 in., or 2 x 4 in.), between joists at intervals through a span, to stiffen the structure. See also **Bridging.**

CROSS CONNECTION 1. Any link between a potable water supply and contaminated water; **2.** Connecting roadway between two nearby and generally parallel roadways.

CROSSCUT Cutting across the grain.

CROSSCUT SAW See **Saw.**

CROSS-DITCH Shallow channel laid diagonally across the surface of a road so as to lead water off the road and prevent soil erosion. Also called a **Water bar.**

CROSS-FALL Lesser fall on a surface set to a fall in two directions.

CROSS FEED Surface grinding; the distance of horizontal feed of the wheel across the table. See also **Down feed,** and **Feed.**

CROSS FURRING Furring used to support lath in suspended ceilings.

CROSS-GRAINED WOOD See **Grain.**

CROSS HAIR Optical effect of a line that divides a telescope's field of view into halves or quarters.

CROSS HATCHING Parallel lines placed close together and at opposite angles. See also **Hatching.**

CROSSHAUL Loading logs by rolling them with a cable.

CROSSHEAD 1. Device that allows valves to be operated in pairs. See also **Valve bridge;** **2.** Connection between the piston and connecting rod in a large reciprocating engine; **3.** Top portion of a sling.

CROSSHEAD EXTRUDER Extruder so constructed that the axis of the emerging extruded product is at right angles to the axis of the extruder screw. Commonly used for applying the cover to braided or spiraled hose.

CROSSHEAD GUIDE PIN Pin in a cylinder head that guides the crosshead.

CROSSING THE LEAD Falling a tree at an angle across the established lead or falling pattern.

CROSS JOINT 1. Joint between the two ends of adjacent bricks. See also **Head joint;** **2.** Joints at the end of individual formboards between subpurlins.

CROSS-LAMINATION Practice in plywood and other laminar construction of orienting the grain of alternating layers of veneer at 90°. In oriented-strand board (OSB), alternating layers of wood strands are also oriented at 90°. Cross-lamination provides strength and stiffness across both the width and length of the panel and helps reduce shrinkage and swelling.

CROSS LAY See **Lay types.**

CROSS-LINKING Setting up of chemical links between molecular chains. When extensive, as in thermosetting resins, cross-linking makes one infusible super-molecule of all the chains.

CROSS MAIN Automatic sprinkler system water supply pipe that is connected to and feeds the branch lines, and which in turn is fed by a feed main or riser.

CROSSMEMBER Vehicle frame structural member that keeps the side rails in parallel alignment and free of buckling or excessive twisting.

CROSS NOGGING See **Bridging.**

CROSSOVER Connection between two piping runs in the same piping system, or the connection of two different piping systems that contain potable water.

CROSSOVER POINTS Points of rope contact where one layer of rope on a drum crosses over the previous layer.

CROSS-POLING Short lengths of poling boards placed horizontally across a gap between runners or sheeting in an excavation and tucked in behind them, used where runners or sheeting cannot be driven continuously and vertically.

CROSS SECTION 1. Section of a body perpendicular to a given axis of the body; **2.** Drawing showing such a section; **3.** Horizontal grid network set up on the ground to determine the contours, quantities of materials, etc., by means of elevations of the grid points; **4.** Vertical surface ground section measured at right angles to a center line; **5.** Maximum width of a tire.

CROSS-SECTION LEVELING Readings taken at right angles to and on both sides of the line of survey, usually at regular intervals and at constant distances from the line, used to produce a representation of the ground's surface on each side of the centerline of work.

CROSS SPRINGER Transverse ribs of a vault.

CROSS SUPPORT Lateral line used to provide intermediate support for a multispan skyline. *Also called* **Jack line,** and **Support line.**

CROSS-TEE Light-gauge metal member resembling an upside-down tee, used to support the abutting ends of formboards in insulating concrete roof construction.

CROSSTIE Sawn timber (part of a bridge deck) placed and fastened transversely over the stringers to distribute the load, providing a flat surface to carry the planking.

CROSS TONGUE Loose tongue, usually of plywood, glued into grooves cut in butting members to stiffen an angle joint.

CROSS VAULT *See* **Vault.**

CROSS VENTILATION Act of causing fresh air to circulate through open doors, windows, or gratings at opposite sides of a room or space.

CROSSWALK Marked lane for passage of pedestrians, bicycles, etc., across a road.

CROSS-WALL CONSTRUCTION Box-like form of concrete construction where the loads are taken on cross, or through walls.

CROSS WRAP Overlapping layer or layers of narrow tensioned wrapper fabric spiraled circumferentially over the outside of a hose to obtain external pressure during vulcanization. *See also* **Herringbone wrap,** and **Wrapped cure.**

CROTCH GRIP Method of chainsaw starting in which the saw is firmly held between the thighs with the handle 'locked' behind one knee.

CROTCH LINE Loading method that uses two lengths of rope suspended from the end of a loading line and terminating in the end hooks.

CROTCH VENEER Veneer cut from the crotch of a tree, forming an unusual grain effect.

CROWD Mechanical thrust by an earth- or rock-moving/drilling machine; of the bucket of an excavator or power shovel when forced into a bank; or the downward thrust of a kelly bar or auger caused by mechanical reactions against the weight of the boring rig. *See also* **Closing line, Digging line, Drag line,** and **Stick.**

CROWDING FORCE Force that pushes the bucket into the bank. It is a function of power available, hydraulic power, and linkage design.

CROWDING THE LINE In masonry construction, laying bricks in such a way as to prevent the line or string used to guide a course from being clear of the face of the brickwork.

CROWD ROPE Wire rope used to drive or force a power shovel bucket into the material that is to be handled.

CROWN 1. Apex of an arch ring. *See also* **Arch**; 2. Upper side of a bowed joist or timber; 3. Upward curve at the middle of a road to allow for water runoff; 4. Curved roof of a tunnel; 5. Portion of the tire between one edge of the tread and the other; 6. Branch- and foliage-bearing portion of a tree; 7. Point in a trap where the direction of flow changes from upward to downward.

CROWNABLE EXTENSION Screed extension capable of being bent in the middle to establish a crown point at some off-center position. Used when specifications call for a shoulder slope that is different from the main road.

CROWN ADJUSTMENT Device that allows each side of an asphalt paver screed to be adjusted independently to achieve the required crown in the pavement surface.

CROWN BAR Beam installed above the roof sets in a tunnel.

CROWN BLOCK Sheave set suspended at the top of a derrick. *See also* **Block.**

CROWN CLASS All trees in a stand whose tops or crowns occupy a similar position in the canopy or crown cover.

CROWN CONTROL Method of 'bending' a screed plate at the center point to shape the finished asphalt mat either higher or lower than the ends. A positive crown, sometimes called a rooftop crown, is used at midpoint of a road to control water drainage. A negative crown leaves a finished mat that is low at the center point and is often used in alley construction to cause water to drain away from the adjacent buildings.

CROWN COVER Ground area covered by a crown, as delimited by the vertical projection of its outermost perimeter.

CROWN DENSITY Thickness, both spatially (depth) and in closeness of the growth (compactness) of an individual crown as measured by its shade density. Collectively, crown density should properly be termed canopy density, as distinct from canopy cover.

CROWNED BOARDS Condition where the center of the width of a sanded board is thicker than the two long edges.

CROWNED DRAINAGE Where the centerline of a structure, typically a street or highway, is raised above its longitudinal perimeter so as to cast water to the edges. *See also* **Valley drainage.**

CROWNED JOINT Finished profile of a drywall joint, the center of which is slightly higher than the adjacent surface and that is feathered out to meet that surface.

CROWN FIRE Fire burning in the tops of trees. *See also* **Crown out.**

CROWN GLASS Glass made in large circular disks by blowing and spinning.

CROWN HEIGHT Vertical distance of a standing tree from ground level to the base of the crown, measured to the lowest live branch whorl or to the lowest live branch (excluding epicormics), or to a point halfway between the two.

CROWN LENGTH Vertical distance of a standing tree from the tip of the leader to the base of the crown, measured to the lowest live branch whorl or to the lowest live branch (excluding epicormics), or to a point halfway between the two.

CROWN LENGTH RATIO Of a standing tree, the ratio of crown length to tree height.

CROWN MOLDING *See* **Molding.**

CROWN OUT Forest fire burning principally as a surface fire that intermittently ignites the crowns of trees and shrubs as it advances. *See also* **Crown fire.**

CROWN POST Short post near the center of a hammerbeam roof.

CROWN RADIUS Measurement of tire tread curvature between the shoulders. Expressed as a percentage, it indicates the 'flatness' of the tire tread area.

CROWN THINNING Removing superfluous live growth in a tree crown to admit light, reduce weight, and lesson wind resistance.

CROWN VENT Vent connected to a trap at the crown.

CROWN WEIR Point in the curve of a trap directly below the crown; the point at which the water level will normally remain when the fixture is not discharging through the trap.

CROWN WIDTH Distance from shoulder to shoulder on the buffed contour of a tire.

CROW'S FOOT Lath set by a grade setter with markings to indicate the final grade at a defined point.

CROW'S-FOOT WRENCH *See* **Wrench.**

CROW'S STEPS Projections in the form of a series of steps on the sloping sides of a gable.

crp *Abbreviation for:* crimp.

CRP *Abbreviation for:* constant rate of penetration.

crs. *Abbreviation for:* course; cross.

CRS *Abbreviation for:* cationic rapid setting; cold-rolled steel.

CRT *Abbreviation for:* cathode ray tube.

CRUCIBLE FURNACE *See* **Furnace.**

CRUDE *See* **Crude oil.**

CRUDE OIL Unrefined oil in its natural state as it comes from the ground. *Also called* **Crude,** or **Petroleum.**

CRUDE SEWAGE Untreated domestic waterborne waste.

CRUISE Survey of forest land that includes the location, volume, species, size, and quality of timber stands; estimates obtained from such a survey. *See also* **Estimate.**

CRUISER One who conducts surveys of timber land. *Also called* an **Estimator.**

CRUISER'S STICK Graduated stick to measure diameter and heights of trees and occasionally to scale logs. *See also* **Biltmore stick.**

CRUMBER Dozer blade that follows the ladder or wheel of a ditching machine to clean and shape the bottom of the excavation.

CRUMBING SHOE Metal arm-like attachment on wheel trenchers to keep loose earth at the trench bottom pulled back into the digging buckets.

CRUMB RUBBER Ground or shredded rubber.

CRUMBS Ragged chunks of drywall on cut ends or cutouts.

CRUMMY Vehicle used to transport a crew to and from the woods.

CRUSH DRESSING Process of using steel rolls to form or dress grinding wheels to a wide variety of shapes.

CRUSHED GRAVEL *See* **Aggregate.**

CRUSHED STONE *See* **Aggregate.**

CRUSHED ZONE Region immediately surrounding a blasthole where the compressive ground stress due to an explosion has exceeded the dynamic compressive strength of the rock.

CRUSHER Machine that reduces rock and other materials to a smaller size. There are several types, including:

> **Gyratory:** Crusher with a central conical member having an eccentric motion in a circular chamber tapering from a wide top opening.

> **Hammermill:** Rock crusher employing hammers or flails on a rapidly rotating axle.

> **Jaw:** Fixed and movable jaw combination, widely spaced at the top and close at the bottom, the movable jaw of which continuously moves toward and away from the fixed jaw.

> **Primary:** Jaw-type crusher that reduces very large rocks to a size that can be processed by a secondary and any subsequent crushers.

> **Roll:** Crusher having two large spring-loaded rolls that counter-rotate, the product size being determined by the space between the rolls.

> **Secondary:** Crusher that receives broken material from a primary crusher and further reduces its size.

CRUSHER-RUN AGGREGATE *See* **Aggregate.**

CRUSHER SCREENINGS Crushed stone or gravel with a gradation from 0 to 6 mm (0 to 0.25 in.).

CRUSHING FAILURE Failure of hardened concrete due to crushing and always accompanied by the formation of debris.

CRUSHING STRENGTH Load at which material fails due to compressive failure, divided by its cross-sectional area.

CRUSHING TEST Test employing a compressive force during which a material is caused to fail.

CRUSH PLATE Expendable strip of wood attached to the edge of a form or intersection of fitted forms to protect the form from damage during prying, pulling, or other stripping operations. *See also* **Wrecking strip.**

CRUST Topmost zone of the Earth. *Also called* the **Regolith.**

crv. *Abbreviation for:* curve; curved; curving.

CRYOGENICS Study of material at very low temperature.

CRYPT Underground burial chamber, often below a church, sometimes with an ornate commemorative structure above ground.

cryst. *Abbreviation for:* crystal.

CRYSTAL Solid symmetrical particle, bounded by plane surfaces.

CRYSTALLINE Made up of crystals.

CRYSTALLINE GLAZE *See* **Glaze.**

CRYSTALLIZATION (POLYMER) Arrangement of previously disordered polymer segments of repeating patterns into geometric symmetry (that results in a reversible hardening of a rubber compound).

CRYSTALLIZE To convert into crystal.

cS *Abbreviation for:* kinematic centistokes.

CS *Abbreviation for:* carbon steel; cast steel; caulking seam; circular spike; commercial standard; control switch; crushed stone.

c/s *Abbreviation for:* cycles per second.

CSB *Abbreviation for:* concrete splash block.

csc *Abbreviation for:* cosecant.

csc⁻¹ *Abbreviation for:* inverse cosecant.

C-SCALE Sound measurement scale that has only slight discrimination at low frequencies.

CSE *Abbreviation for:* center southeast.

C SELECT *See* **Lumber grades.**

csg *Abbreviation for:* casing.

c/shft *Abbreviation for:* crankshaft.

csk *Abbreviation for:* countersink.

csk-o *Abbreviation for:* countersink other side.

CSL *Abbreviation for:* center south line.

csmith *Abbreviation for:* coppersmith.

csmt *Abbreviation for:* casement.

CSS *Abbreviation for:* cationic slow setting.

cstg *Abbreviation for:* casting.

cstr *Abbreviation for:* canister.

C STUD Formed metal stud. *See also* **Stud.**

CSW *Abbreviation for:* center southwest.

c/sz. *Abbreviation for:* cut to size.

ct *Abbreviation for:* cab-to-tandem; coat; court.

ct. *Abbreviation for:* coats.

CT *Abbreviation for:* ceramic tile; coal tar; column tie; cork tile.

ctb *Abbreviation for:* cement-treated base.

CTC *Abbreviation for:* center-to-center.

ctd *Abbreviation for:* coated.

CTE *Abbreviation for:* center-to-end.

ctf. *Abbreviation for:* certificate.

ctg *Abbreviation for:* crating.

ctg. *Abbreviation for:* cartridge; cartridge; cotangent.

ctge *Abbreviation for:* cartage.

ctn *Abbreviation for:* carton.

c to c *Abbreviation for:* center-to-center.

ctr *Abbreviation for:* center; cutter.

ctrg *Abbreviation for:* centering.

ctsk *Abbreviation for:* countersunk.

ctwt *Abbreviation for:* counterweight.

cu *Abbreviation for:* cubic.

cub. *Abbreviation for:* cubical.

CUBE 1. Product of the length multiplied by the breadth multiplied by the width; **2.** Inside dimensions of a truck body or trailer (so-called 'high-cube' equipment is designed to offer the maximum interior load space for its exterior length and width); **3.** Pallet of concrete blocks or a banded stack of bricks.

CUBE STRENGTH Load per unit at which a standard cube fails when tested in a specified manner.

CUBE TEST Test to find the compressive strength of a representative sample of a material.

CUBICAL AGGREGATE Angular aggregate having a cubic appearance.

CUBICAL EXPANSION Volume increase of hose when subjected to internal pressure.

CUBICAL PIECE (OF AGGREGATE) One in which length, breadth, and thickness are approximately equal.

CUBIC CENTIMETER Unit of volume equal to 1 mL. Symbol: cm^3. Used with the SI system of measurement. Multiply by 0.061 02 to obtain cubic inches, symbol in.^{3}; by 0.035 195 to obtain fluid ounces, symbol: fl oz; by 1.0 to obtain milliliter, symbol: ml; by 0.001 to obtain liters, symbol: L. *See also the appendix:* **Metric and nonmetric measurement.**

CUBIC CONTENT In construction, the cubic measure contained within the walls of a room or combination of rooms, used as a basis for estimating materials, costs, etc.

CUBIC FOOT Nonmetric unit of volume. Symbol: ft^3. Multiply by 0.028 316 85 to obtain cubic meters, symbol: m^3; by 28.316 85 to obtain liters, symbol: L. *See also the appendix:* **Metric and nonmetric measurement.**

CUBIC FOOT PER HOUR Nonmetric unit of flow rate. Symbol: ft^3/h. Multiply by 28.316.85 to obtain liters per hour, symbol: L/h; by 0.007 865 79 to obtain liters per second, symbol: L/s. *See also the appendix:* **Metric and nonmetric measurement.**

CUBIC FOOT PER MINUTE Nonmetric unit of flow rate. Symbol: ft^3/min. Multiply by 0.000 471 947 4 to obtain cubic meters per second, symbol: m^3/s; by 0.471 947 4 to obtain liters per second, symbol: L/s. *See also the appendix:* **Metric and nonmetric measurement.**

CUBIC INCH Nonmetric unit of volume. Symbol: in.^{3}. Multiply by 16.387 064 to obtain cubic centimeters, symbol: cm^3; by 16.387 064 to obtain cubic millimeters, symbol: mm^3; by 0.016 387 064 to obtain liters, symbol: L. *See also the appendix:* **Metric and nonmetric measurement.**

CUBICLE Small, enclosed space within a room, usually lacking a door or windows and with walls that do not reach the ceiling.

CUBIC METER A derived unit of volume with a compound name of the SI system of measurement.

Symbol: m³. Multiply by 0.2759 to obtain cords (of stacked wood). Symbol: cd; by 1.3080 to obtain cubic yards, symbol: yd³; by 35.3147 to obtain cubic feet, symbol: ft³; by 219.97 to obtain (imperial) gallons, symbol: (imp)gal; by 264.17 to obtain (U.S.) gallons, symbol: (U.S.)gal. *See also the appendix:* **Metric and nonmetric measurement.**

CUBIC METER PER MOLE A derived unit of molar volume with a compound name of the SI system of measurement. Symbol: m³/mol. *See also the appendix:* **Metric and nonmetric measurement.**

CUBIC METER PER SECOND A derived unit of volume flow rate with a compound name of the SI system of measurement. Symbol: m³/s. *See also the appendix:* **Metric and nonmetric measurement.**

CUBIC SCALE Estimate of the cubic-measure volume of wood fiber in a tree, log, or other wood product.

CUBIC YARD Non-SI unit of volumetric measurement equal to a volume measuring 3 ft by 3 ft by 3 ft, equivalent to 0.765 m³. Multiply by 0.764 555 to obtain cubic meters, symbol: m³. *See also the appendix:* **Metric and nonmetric measurement.**

CUBIC YARD BANK MEASUREMENT Unit of excavation in a cut or natural bed.

CUBIC YARD COMPACTED MEASUREMENT Unit of excavation in a fill or embankment, after compaction.

CUBIC YARD LOOSE MEASUREMENT Unit of excavation in a machine, stockpile, or in uncompacted fill or embankment.

CUBIC YARD STRUCK MEASUREMENT Unit of capacity of the bucket, body, bowl, or dipper of a machine, excluding the teeth, measured by striking off the ends and sides of the container by a straight edge.

cu cm *Abbreviation for:* cubic centimeter.

cu ft *Abbreviation for:* cubic foot.

cu in. *Abbreviation for:* cubic inch.

CUL-DE-SAC *See* **Dead end,** and **Street.**

CULL Tree or log, or part of a log, that is unmerchantable because of defects.

CULLET Broken glass.

CULLING Sorting materials (especially brick and lumber) for like size, color, quality, etc.

CULLS 1. Any material laid aside from the process of culling, usually for reason of defect in quality of material or manufactiring quality standards; **2.** Rot, usually in the core of a tree.

CULMINATION OF THE MEAN ANNUAL INCREMENT Point in the growth cycle of a tree or stand at which the mean annual increment for height, diameter, basal area, or volume is at its maximum. At this point MAI (mean annual increment) equals PAI (periodic annual increment).

cult. *Abbreviation for:* cultural; culture.

CULTURAL VIBRATION Vibration that is commonplace and familiar to the observer.

CULVERT Pipe or channel to carry water under a roadway or other obstruction. It is either a short bridge less than 6 m (20 ft) in length, or a structure with enough fill over it so that little of its strength is needed to support the traffic load. *See also* **Drainage ditch.**

cu m *Abbreviation for:* cubic meter.

cum. *Abbreviation for:* cumulative.

CUMULATIVE BATCHING Measuring more than one ingredient of a batch in the same container by bringing the batcher scale into balance at successive total weights as each ingredient is accumulated in the container.

CUMULOSE SOILS Accumulated organic matter.

CUNEIFORM PILE Tapered or step-tapered pile.

CUNIT Unit of volume consisting of 100 ft³.

CUP Crosswise distortion of a structural wood panel from its flat plane. *See also* **Bow.**

cup. *Abbreviation for:* cupboard.

CUPBOARD Shallow storage area fitted with shelves, either part of the structure or freestanding.

CUPHEAD Type of fastener head, the center of which has a bowl-shaped depression.

CUPOLA Small, dome-shaped structure built on a roof.

CUP PACKING Cup-shaped sliding piston packing.

CUPPED TAPER Condition where the outer edge of the beveled edge of a drywall panel is in the same plane as the surface.

CUPPING Distortion of a board where there is a flatwise deviation from a straight line across the width of the board.

CUP WHEEL Grinding wheel shaped like a cup or bowl.

cur. *Abbreviation for:* current.

CURB 1. Edging to a pavement or other road surface. *See also* **Lowered curb** and **Rolled curb; 2.** Roof in which the slope is broken on two or more sides; so called because a horizontal curb is built at the plane where the slope changes; **3.** Raised concrete strip poured on the bottom, and at each side of a tunnel invert, outside of the minimum concrete line, used as a reference and support for invert and arch pours.

CURB AND GUTTER MACHINE Slipforming machine designed to continuously form a combined curb and gutter profile, or either, as necessary.

CURB BOX Cylindrical casting placed in the ground over a municipal stop that permits insertion of a special key to turn the stop. *Also called* a **Service box.**

CURB COCK Control valve installed in a house service between the municipal stop and the structure. *Also called* a **Curb stop.**

CURB COLLECTION *See* **Collection.**

CURB FORM Retainer or mold used in conjunction with a curb tool to give the necessary shape and finish to a concrete curb.

CURB GUARD Protrusion of rubber circling a tire to protect the cord body from scraping against curbs and other obstructions.

CURB LOADING ZONE Roadway space adjacent to a curb and reserved for exclusive use of vehicles during loading or unloading of passengers or property.

CURB RADIUS Track of a vehicle negotiating a turn in its tightest steering mode, measured from the center of the circle to the outside of the outermost tire.

CURB REVEAL Result of a cutting pass alongside a curb that lowers the surface of the existing pavement in relation to the curb.

CURB SHOE Device bolted to the blade of a grader when grading curbs to help the blade match the profile of the curb bottom.

CURBSIDE Right or passenger side of a vehicle when viewed from the rear (opposite of **Roadside**).

CURBSIDE COLLECTION Programs where recyclable materials are collected at the curb, often from special containers color-coded as to type of content, to be brought to various processing facilities.

CURB STOP *See* **Curb cock.**

CURB-TO-CURB TURNING DIAMETER Smallest circle within which a vehicle will clear a curb 150 mm (5.9 in.) high while executing its sharpest practicable turn.

CURB TOOL Tool used to give the desired finish and shape to the exposed surfaces of a concrete curb.

CURB WEIGHT Empty weight (no payload) of a fully equipped truck, including oil, water, and fuel. Also known as **Chassis weight,** and **Tare weight.**

CURE 1. Change of physical properties by chemical reaction; usually accomplished by heat and/or catalysts, with or without pressure. *Also called* **Set**; **2.** In asphalt, the entire process of breaking and set until the final mixture of emulsion and aggregate has lost all moisture due to evaporation or dehydration.

CURE TEMPERATURE Temperature range in which chemical reaction will occur that changes physical properties.

CURE TIME 1. Time required to produce vulcanization at a given temperature; **2.** Period between application of a curing agent and the time when the material reaches its design physical properties.

CURING 1. Maintenance of a satisfactory moisture content and temperature in concrete during its early stages so that desired properties may develop. *See also* **Electrical curing**; **2.** Process of heating or treating a rubber or plastic compound to convert it from a thermoplastic or fluid material into a solid, relatively heat-insensitive state. When heat is employed, the process is called vulcanization.

CURING AGENT Material that promotes setting, hardening, maturing, or some other process; a catalyst.

CURING BLANKET Built-up covering placed over freshly finished concrete.

CURING CHAMBER Chamber constructed of normal building materials in which to steam cure concrete.

CURING COMPOUND Liquid that can be applied as a coating to the surface of newly placed concrete to retard the loss of water or, in the case of pigmented compounds, also to reflect heat so as to provide an opportunity for the concrete to develop its properties in a favorable temperature and moisture environment.

CURING-COMPOUND SPRAYER Equipment, part of a concrete paving train, used to spray the surface of the concrete with a curing compound.

CURING RIM Rim used to support a tire and keep the curing tube in place while curing occurs.

CURING TUBE Special, heavy-duty tube placed within a tire during curing.

CURL 1. Rotational movement of a bucket on hinges to dig or retain a load. *Also called* **Roll-back,** and **Tuck-in**; **2.** Grain pattern produced in wood when sawn at the junction of a branch and the stem of a tree; **3.** Distortion of an originally essential linear or planar member into a curved shape, such as the warp of a slab due to creep or to differences in temperature or moisture content in the zones adjacent to its opposite faces.

CURL AND CROWD FORCE Bucket penetration into a material is achieved by the bucket curling force and stick crowd force. Rated digging forces are the digging forces than can be exerted at the outermost cutting point as produced by:

> **Bucket cylinder:** The digging force generated by the bucket cylinder(s) and tangent to the arc of radius with the bucket positioned to obtain maximum output moment from the bucket cylinders and connecting linkages.

> **Stick cylinder:** Digging force generated by the stick cylinder(s), tangent to the arc of radius, with the stick in a position to obtain maximum output moment from the arm cylinder.

CURLING Distortion of an original essentially linear or planar member into a curved shape.

CURLY-GRAINED WOOD *See* **Grain.**

CURRENT 1. Force within water or air that causes it to move in a specific direction; **2.** Flow of electricity.

CURRENT CARRYING *See* **Alive.**

CURRENT-CARRYING PART Conducting part intended to be connected in an electric circuit to a source of voltage.

CURRENT LIMIT Control function that prevents an electric current from exceeding its prescribed limits.

CURRENT METER Instrument used to measure the rate of flow in a stream or the current in an open body of water.

CURSTABLE Ornamented masonry course producing a string.

curt. *Abbreviation for:* curtail; curtain.

CURTAILMENT Point in the length of a reinforcing bar at which it is no longer required to resist tensile stress and must be consequently extended by the minimum length required to satisfy anchorage length, or bent up to assist in resisting shear.

CURTAILMENT DIAGRAM Diagram, similar to a bending moment diagram, used to determine the curtailment of reinforcing bars in a concrete beam.

CURTAIL STEP Lowest step in a flight, with a curved end.

CURTAIN Single layer of vertical and horizontal reinforcing bars in a concrete wall. *See also* **Double curtain.**

CURTAIN DRAIN *See* **Intercepting drain.**

CURTAIN (FURNACE) WALL *See* **Wall.**

CURTAIN GROUTING Injection of grout into a subsurface formation to create a water barrier.

CURTAINING Sagging in a paint film or finish coat, resembling the curved bottom edges of drapery swags.

CURTAIN REINFORCEMENT Mat of orthogonal reinforcing steel in a member such as a wall, known as a double curtain (of reinforcement) when there is a mat at each face.

CURTAIN WALL *See* **Wall.**

CURVATURE Rotation per unit length due to bending.

CURVATURE AND REFRACTION Corrections to surveying measurements that account for the curvature of the earth and the bending of light rays passing through the atmosphere.

CURVATURE FRICTION Friction resulting from bends or curves in the specified prestressing cable profile.

CURVE ALLOWANCE Size correction made when marking out and cutting a bend in sheet metal due to measuring around the neutral axis.

CURVED-CUT *See* **File.**

CURVED PANEL Stressed-skin or sandwich panel curved to various degrees; used in roof construction.

CURVED RIPPER SHANK *See* **Ripper shank.**

CURVILINEAR GABLE Roof gable in the form of geometric curves.

cush. *Abbreviation for:* cushion

CUSHION 1. Device that provides controlled resistance to motion by use of a gradually reduced flow area. *Also called* **Cylinder cushion; 2.** Layer of sand and gravel equal in depth to the width of a foundation, placed beneath a foundation to cushion low-bearing-capacity soil below. *Also called* **Pile cushion.**

CUSHION BLASTING Technique of firing a single row of holes along a neat excavation line to shear the web between closely drilled holes, fired after production shooting has been accomplished. *Also called* **Post-splitting, Slabbing,** and **Trim blasting.**

CUSHION BLOCK Material inserted between a pile hammer and the pile driving cap of a pile to minimize damage at the point of impact. *Also called* **Cap material.** *See also* **Dolly.**

CUSHION-EDGED TILE Tile on which the facial edges have a distinct curvature that results in a slightly recessed joint.

CUSHION FLOOR Vinyl or resilient sheet-type floor with a heavy, soft back that is soft to walk on and that bridges small imperfections in the subfloor.

CUSHION GUM Tacky rubber compound used for adhesion, under-tread repair of a tire, and buildup.

CUSHION SOLID TIRE Solid tire usually having a high rounded profile in cross section.

cust. *Abbreviation for:* custodian; custody; custom.

CUT 1. Depth to which material is to be excavated; **2.** Volume of excavation for a given cut; **3.** To lower an existing grade. *See also* **Gross cut,** and **Net cut; 4.** One season's output of logs; **5.** Drilling pattern at a tunnel face that provides relief for an explosive charge. *See also* **Burn cut; 6.** Portion of a land surface or an area from which earth or rock has been or will be excavated; the distance between an original ground surface and an excavated surface; **7** Section of highway located below natural ground elevation, thereby requiring excavation of earthen material; **8.** Indentation made in a key blank that corresponds to the tumbler setting.

CUT-AND-COVER Means of developing a sanitary land fill or refuse dump whereby a cut is made, alternate compacted layers of refuse and excavated material from a stockpile are placed in the area of the cut that, when developed to the design profile, is capped with additional fill.

CUT-AND-FILL Construction or development process involving excavation of cuts and using the material for adjacent fills. In a balanced cut-and-fill, the volume of excavated material equals the volume of material required for fill, with allowance for swell or shrink from cut to fill.

CUTAWAY DIAGRAM *See* **Diagram.**

CUTBACK Petroleum solvent used to liquefy an asphalt for construction operations. The resulting material is called a **Cutback asphalt.**

CUTBACK ASPHALT *See* **Asphalt.**

CUT BANK Excavated bank from a ditch line to the top of the undisturbed slope of a road.

CUT BLOCK Specific forest area with defined boundaries authorized for harvest.

CUT BRICK Brick cut to shape with a bolster or with a mason's trowel.

CUT END End of a drywall panel showing the gypsum plaster core.

CUT FLOORING NAIL *See* **Nail.**

CUT-FULL LUMBER Lumber manufactured larger than the nominal dimension to allow for shrinkage.

CUT IN To insert or fit something after the work is finished.

CUT-IN Device that connects to electrical circuits.

CUT LOOSE 1. Release a load; **2.** Unhook rigging or chokers.

CUT NAIL *See* **Nail.**

CUTOFF 1. Condition where a portion of a column of explosives has failed to detonate because of bridging, because of a shifting of the rock formation, or due to an improper delay system; **2.** Prescribed elevation at which the top of a driven pile is cut or left; **3.** Action of sawing the backcut right into the undercut, keeping no holding wood, when felling a tree.

CUTOFF DEPTH Depth reached by cofferdam walls or sheet piling below an excavation.

CUTOFF ELEVATION Elevation of the top of a pile as shown on the contract drawing.

CUTOFF SHOE Device used to reduce the paving width of an asphalt paver screed in small increments.

CUTOFF TRUSS *See* **Truss.**

CUTOFF WALL Structure used to stop or divert a flow of water.

CUTOFF WHEEL Thin wheel, usually made with an organic bond, for cutting off.

CUTOUT Assembly of a fuse support with either a fuseholder, fuse carrier, or disconnecting blade.

CUTOUT BOX Enclosure designed for surface mounting and having swinging doors or covers secured directly to and telescoping with the walls of the box proper.

CUTOVER Land that has previously been logged.

CUT PERIOD Interval between major harvesting operations in the same forest stand.

CUT RESISTANT Having that characteristic of withstanding the cutting action of sharp objects.

CUT SECTION Any section of a pavement where material is removed or excavated to lower the elevation.

CUT STONE Finished, dimensioned stone ready to set in place.

CUTTER 1. Tool mounted on a handle with a head for hitting and a beveled end for cutting rivets; **2.** Worker who falls, bucks, and/or limbs timber; **3.** *See also* **Working tool.**

CUTTER DRUM Rotating cylinder to which working tools are attached and which performs a planing/milling action.

CUTTER DRUM DRIVE System through which power is transmitted to the cutter drum.

CUTTERHEAD Set of revolving blades for fragmenting hard material, fitted to the head of the suction line of a hydraulic suction dredge.

CUTTERS Part of a grinding wheel dresser that comes in contact with the wheel and does the cutting.

CUTTING 1. Lowering a grade; **2.** Excavating; **3.** Process of felling trees; **4.** Area on which trees have been, are being, or are to be cut.

CUTTING BAR Grooved bar on a chain saw that carries the chain.

CUTTING CIRCLE Circle described by the outer rim or extremity of the teeth of a circular saw.

CUTTING FLAME Flame used to cut metal by the rapid oxidation process at a high temperature produced by a gas flame accompanied by a jet action that blows the oxides away from the cut.

CUTTING FLUID Fluid used in metal cutting to improve finish and tool life and/or dimensional accuracy.

CUTTING IN Making a clean, sharp edge to an area being painted, usually at a change of color.

CUTTING IRON Sharpened steel blade of a carpenter's plane.

CUTTING LIST List of the numbers of pieces of wood, steel, reinforcing bar, etc., giving the length, size, and other pertinent data.

CUTTING OIL Cooling oil sprayed against a tool when cutting metal; may be combined with other substances to improve cutting efficiency.

CUTTING-OUT PIECE Short piece of timber that may be cut out to facilitate the striking of timbering.

CUTTING PATTERN Pattern to which trees are felled during a cutting operation.

CUTTING PERMIT Document that contains the authority to harvest trees on a woodlot licence.

CUTTING PLAN Overall operating plan for felling the timber on a given strip or cutting block.

CUTTING PLIERS Pliers equipped with nipping faces for cutting thin wire.

CUTTING RATE Amount of material removed by a grinding wheel per unit of time.

CUTTINGS Particles of soil or rock excavated and brought to the surface by a rotary drill. *See also* **Spoil.**

CUTTING SCREED Sharp edged tool used to trim shotcrete to a finished outline. *See also* **Rod.**

CUTTING SHOE Additional metal placed as an inside or outside cast steel ring or welded plate at the bottom of an open-end pile or caisson to strengthen the tip.

CUTTING SURFACE Surface or face of the wheel against which material is ground.

CUTTING TIP Part of an oxygen cutting torch from which the gases are released.

CUTTING TOOL *See* **Working tool.**

CUTTING TORCH Device that controls and directs the ignited gases and oxygen needed for cutting and removing metal.

CUTTING UNIT Area of timber designated for harvest.

CUT-UP Tree or log left standing or suspended with the falling or bucking cuts not completed.

CUTWATER Wedge-shaped upstream face of a pier or a bridge designed to break the flow of water.

cu yd *Abbreviation for:* cubic yard.

cv. *Abbreviation for:* cove.

CV *Abbreviation for:* check valve.

CV1S *Abbreviation for:* center vee one side

CV2S *Abbreviation for:* center vee two sides.

cvd *Abbreviation for:* covered.

cw *Abbreviation for:* cold water.

CW *Abbreviation for:* clockwise; curtain wall; cut washer.

CWL *Abbreviation for:* center west line.

CWP *Abbreviation for:* circulating water pump.

cwt *Abbreviation for:* counterweight; hundredweight.

CWT *Abbreviation for:* cold water temperature.

cy. *Abbreviation for:* cycle.

CY *Abbreviation for:* calendar year.

CYANOACRYLATE ADHESIVE See **Adhesive.**

cyc. *Abbreviation for:* cyclone.

CYCLE Complete set of individual operations that a machine performs repetitively.

CYCLE TIME 1. Time for a machine to complete one cycle, i.e., load, haul, dump, and return; **2.** *See* **Traffic signal.**

CYCLICAL LOAD TEST Test to estimate the distribution of the axial load along the length of a beam or pile.

CYCLOPEAN See **Rustication.**

CYCLOPEAN CONCRETE Mass concrete in which large stones, each of 45 kg (100 lb) or more, are placed and embedded as the concrete is placed. *See also* **Plum,** and **Rubble concrete.**

cyl. *Abbreviation for:* cylinder; cylindrical.

CYLINDER 1. Bored hole in an engine block in which pistons and rings are inserted; **2.** Storage vessel; container for holding the gas used in welding; **3.** Open caisson or monolith of cylindrical form; **4.** Lock housing containing a tumbler mechanism and a keyway; **5.** Device (*Also called* **Linear actuator**) that converts fluid energy into linear mechanical energy. It consists of a piston and piston rod within a steel cylinder. In hydraulic systems, the piston is fitted with end seals and the cylinder is equipped with a port or ports to allow the entrance and exit of fluid. There are several types, including:

Double-acting: Cylinder with ports at both ends in which fluid pressure can be applied to the movable piston rod in either direction to enable both extension and retraction of the rod under power.

Double-rod: Cylinder with a single piston and a piston rod extending from each end.

Dual-stroke: Cylinder combination that provides two working strokes.

Nonrotating: Cylinder in which relative rotation of the cylinder housing and the piston and piston rod, plunger, or ram, is not recommended.

Plunger: Cylinder in which the piston has the same cross-sectional area as the piston rod. *Also called* **Ram cylinder.**

Ram: *See* **Plunger cylinder,** above.

Single-acting: Cylinder with a port at one end only and in which the fluid pressure can be applied to the piston in only one direction to extend the rod. The cylinder retracts by gravity, spring or cable.

Single-rod: Cylinder with a piston rod extending from one end.

Slave: Small cylinder whose piston is moved by the piston rod of a larger cylinder.

Tandem cylinder: Two or more cylinders with interconnected rod and piston assemblies.

Telescoping cylinder: Cylinder with nested multiple tubular rod segments that provide a long working stroke in a short retracted envelope.

CYLINDER ASSEMBLY Casing, plunger, and one or more guide bearings and associated packing, plus hydraulic or pneumatic inlet and outlet valves.

CYLINDER BLOCK Cast metal mass that includes the cylinder and water jackets (or cooling fins, in the case of air-cooled engines).

CYLINDER CASING Housing in which a plunger operates, actuated by hydraulic or pneumatic pressure.

CYLINDER CUSHION See **Cushion.**

CYLINDER GUIDE BEARING Bearing that provides for alignment of the plunger in a cylinder assembly.

CYLINDER HEAD 1. Removable section of an engine covering the upper end of the cylinder bores and containing the combustion chambers and, in the case of overhead valve engines, the valve assemblies; **2.** End of a hydraulic cylinder through which the piston rod extends. *Also called* **Header.**

CYLINDER LIFT 1. Point at which the cylinder of a double-acting air or diesel hammer begins to lift, or 'float'; **2.** Point at which the force of fluid in the hammer overcomes the weight of the hammer casing or cylinder. *Also called* **Hammer uplift.**

CYLINDER LINER Circular metal sleeve that is pressed into a cylinder bore to provide a sliding surface for the piston. A liner that contacts the coolant is called a **Wet liner,** and one that does not is called a **Dry liner.**

CYLINDER LOCK Lock in which the cylinder can only rotate when a set of interior tumblers are correctly positioned by a matching key.

CYLINDER RING Collar or washer used in a lock under the head of a cylinder.

CYLINDER SCREW Set-screw in the front of a cylinder lock that prevents the cylinder from being turned after installation.

CYLINDER SLEEVE Wear liner or sleeve between the piston and cylinder wall or water jacket.

CYLINDER STRENGTH Compressive strength of a concrete cylinder with a diameter-to-length ratio of 1:2 made, cured, and tested in accordance with standard procedures.

CYLINDER TEST Compressive strength test, similar to a concrete cube test, using a 150 mm (5.9 in.) diameter concrete cylinder, 300 mm (11.8 in.) high.

CYLINDER WHEEL Grinding wheel of similar characteristics to a straight wheel but with a large hole size in proportion to its diameter, and usually several inches in height.

CYLINDRICAL GRINDING Grinding the outside surface of a cylindrical part mounted on centers.

CYLINDRICAL IMPELLER Round impeller used to pump fluids.

CYLINDRICAL PROJECTION See **Linear perspective.**

CYMA RECTA See **Molding.**

CYMA REVERSA See **Molding.**

D

D *See* **Veneer grade.**

d *Abbreviation for:* penny; displacement; density; differential.

D *Abbreviation for:* down; diameter; dimensional.

DA *Abbreviation for:* double acting.

DAB Dressing the face of stone with a pointed tool.

DABBER Soft, round-tipped brush, used for applying spirit varnish.

DADO 1. Lower few feet of an interior wall, or the interior face of an external wall, that is colored or finished differently to the remainder; 2. Rectangular groove across the width of a board or plank; 3. Joint in which one board fits into a rectangular groove cut across the width of the surface of a second.

DADO JOINT Recessed groove on the face of a board that receives the end of a perpendicular board.

DAILY ACTIVITY LEVEL Estimate of the activity of a man-made forest fire, rated as none, low, normal, high, and extreme.

DAILY CAPACITY Amount of waste capable of being processed or landfilled each day at a facility.

DAILY COVER Soils placed on top of landfilled waste at the end of each day, or at the completion of a landfill cell, as control against rodents and vectors.

DAILY ROUTE *See* **Collection method.**

DAIS Raised platform, commonly to raise a speaker above an audience.

DALE *See* **Glen.**

DAM 1. Barrier constructed to hold back water; 2. Placement of products to hold something back.

dam. *Abbreviation for:* damage.

DAMAGES Sums that can be recovered from those who cause loss by another; from a contractor by an owner for failure to complete a project by the date specified in the contract documents, for instance.

DAMMING Use of a pliable insulation to restrict or channel the flow of melted babbitt when forming babbitted bearings or during the necking of a wire rope.

DAMP Either moderate absorption or a moderate covering of moisture. Implies less wetness than that connoted by 'wet' and slightly wetter than 'moist.' *See also* **Moist,** and **Wet.**

DAMP COURSE Literally, a 'damp prevention course': a layer of impervious material placed in a position and in a manner that will prevent the passage of moisture from the ground into the structure of a building. *Also called* **Dampproof course.**

DAMPER 1. Metal device installed over the firebox in a fireplace to control the draft from the firebox into the chimney flue; 2. Device for cancelling or reducing vibrations in a journal or crankshaft; 3. Manualy or automatically controlled valve or plate in a breeching, duct, or stack, that is used to regulate a draft or the rate of flow of air or other gases. There are several types, including:

Barometric: A hinged or pivoted plate that automatically regulates the amount of air entering the system, thereby maintaining a constant draft ahead of the damper.

Butterfly: A plate or blade installed in the system that rotates on an axis for regulating the flow of air or gases.

Guillotine: An adjustable gate, used to regulate the flow of gases, installed vertically in a duct.

Sliding: A plate, normally installed perpendicular to the flow of air or gas in a conduit, and arranged to slide across it.

DAMPING 1. Moderation or attenuation of the frequencies of sound or vibration; or reducing the frequency of oscillations; 2. To slow the action of a meter pointer when returning to zero from near full scale.

DAMPING MOTOR Small DC motor that permits the voltage to a generator shunt field to be increased and decreased gradually to provide smoother acceleration and deceleration.

DAMPING PARAMETER Measure of the time required as a function of the maximum pressure excursion of the power supply output to attain essentially steady-state operation after an abrupt disturbance. Specifically, it is the transient recovery time divided by the maximum energy excursion.

DAMP LOCATION Partially protected locations under canopies, marquees, roofed open porches, and like locations, and interior locations subject to moderate degrees of moisture, such as some basements.

DAMPPROOF COURSE *See* **Damp course.**

DAMPPROOFING 1. One or more coatings of a compound that is impervious to water; 2. Treatment of concrete or mortar to retard the passage or absorption of water, or water vapor, either by application of a suitable coating to exposed surfaces, or by use of a suitable admixture or treated cement, or by use of preformed films such as polyethylene sheets under slabs on grade. *See also* **Vapor barrier** and **Waterproofing.**

DAMPPROOF MEMBRANE Layer or sheet of impervious material within a floor, wall, ceiling, or roof to prevent passage of moisture.

DAMP WOOD TERMITES *Zootermopsis:* a large termite that feeds primarily on dead and decaying wood, particularly in moist environments.

d&a *Abbreviation for:* disassemble and assemble.

d&c *Abbreviation for:* disconnect and connect.

D&H *Abbreviation for:* dressed and headed.

D&M *Abbreviation for:* dressed and matched.

D&SM *Abbreviation for:* dressed and standard matched.

d&w *Abbreviation for:* discharged and wet.

DANGER TREE Tree that, because of position, deterioration, or physical damage, endangers persons.

DANGLING LEADS Cable ends of the terminal, free from the edge of a battery, being connected to an unmounted connector or charging plug.

DAP Incision or notch cut in timber, into which the head of a pile or other timber is fitted.

DARBY Hand-manipulated straightedge, usually 0.9 to 2.4 m (3 to 8 ft) long, used in the early stage leveling operations of concrete or plaster, preceding supplemental floating and finishing.

DARCY'S FORMULA Formula used to determine the pressure drop due to flow friction through a conduit.

DART VALVE Drain for a well bailer that opens automatically when rested on the ground.

DASH-BOND COAT Thick slurry of portland cement, sand, and water flicked on surfaces with a paddle or brush to provide a base for subsequent portland cement plaster coats; sometimes used as a final finish on plaster.

DASHPOT Mechanical piston device used to moderate the movement of an arm, lever, or rod through the compression of air or oil.

DATA Facts, particularly those expressed numerically.

DATE CODE Any combination of numbers, letters, symbols or other methods used by a manufacturer to identify the time of manufacture of a product.

DATE OF AGREEMENT *See* **Contract date.**

DATE OF COMMENCEMENT OF WORK Date that work is authorized by the owner's agent to start on site.

DATE OF SUBSTANTIAL COMPLETION Date certified by the owner's agent or consultant when the project may be occupied for the purpose for which it was intended, excepting specific tasks and/or details as noted at that date as requiring to be completed.

DATUM 1. Known starting point; **2.** Any level surface taken as a plane of reference from which to measure elevations.

DATUM LINE In surveying, the base line from which all lines or levels are taken.

DAUB Clay or mud.

DAUBING 1. Technique of throwing a rough coating of plaster onto the surface of a wall to simulate a rough stone finish; **2.** Dressing the surface of a stone with a special hammer that produces a pattern of small holes.

DAY Non-SI unit of time, equal to 24 hours, 1440 minutes, and 86 400 seconds, permitted for use with the SI system of measurement. Symbol: d. *See also the appendix*: **Metric and nonmetric measurement.** *See also* **Calendar day** and **Work day.**

DAY LABOR Workers paid on a daily basis for work performed.

DAYLIGHT FLUORESCENT *See* **Fluorescent lamp.**

DAYLIGHTING Falling a wider strip of timber along a grade, generally to cherry-pick more logs, but sometimes to let the grade dry up better. *See also* **Widening.**

DAY RATE Method of payment for work done based by the day or hour rather than by the piece.

DAYS OF GRACE Days allowed by law for payment of a debt after its due date.

DAY TANK Small fuel tank, usually adjacent to or in close proximity to an engine-driven fuel pump, that stores a ready fuel supply near the engine.

DAY WORK ACCOUNT Terms of a contract that specify the method of payment for works not included in the scope of the contract, but which the construction contractor is obliged to perform at the request or direction of the owner or his agent. Generally, such day-work-account work is paid for on unit-price or cost-plus terms.

dB *Abbreviation for:* decibel.

db *Abbreviation for:* disc brake.

DB Clg *Abbreviation for:* double-beaded ceiling.

DBFS *Abbreviation for:* dull black finish slate.

DBG *Abbreviation for:* distance between guides.

dbh *Abbreviation for:* diameter breast height.

dbl. *Abbreviation for:* double.

dbp *Abbreviation for:* drawbar pull.

dbrn *Abbreviation for:* decibels above reference noise.

DC *Abbreviation for:* decimal classification; direct current.

DC&M *Abbreviation for:* dressed and center matched.

DC GENERATOR Electric generator that transforms mechanical energy into direct current electric energy.

DCI *Abbreviation for:* ductile cast iron.

DCP *Abbreviation for:* design change proposal.

D-CRACKS Progressive formation on a concrete surface of a series of fine cracks at rather close intervals, often of random patterns, but in slabs on grade paralleling edges, joints, and cracks, and usually curving across slab corners. *Also called* **D-line cracks.**

dd *Abbreviation for:* degree-day.

ddlk *Abbreviation for:* deadlock.

DEAD AXLE 1. Fixed shaft acting as a hinge pin; **2.** Fixed shaft or beam on which a wheel revolves. *See also* **Axle.**

DEAD-BEAT GOVERNOR *See* **Governor.**

DEADBLOW HAMMER *See* **Hammer.**

DEADBOLT Door lock, knob-operated inside, key-operated outside. Also called **Deadlatch.** *See also* **Night latch.**

DEAD CENTER 1. Point, referenced on any rotating

object, centered upon the axis of rotation; **2.** Point, referenced off the crankshaft of an engine, where any specific piston is either at the bottom of its travel or the top of its travel.

DEAD END 1. In the stressing of a tendon from one end only, the end opposite that to which the load is applied; **2.** Point of fastening of one rope end in a running rope system, the other (live) end being fastened at the rope drum; **3.** Short street or passageway open at one end only. *Also called* **Cul-de-sac.** *See also* **Street; 4.** Branch of a drainage piping system that ends in a closed fitting.

DEAD-END ANCHORAGE Anchorage at the end of a tendon that is opposite the jacking end.

DEAD-END HITCH Termination point and fastening of wire rope or cord.

DEAD-END INSTALLATION Installation of a device at the dead, or stationary, end of a hoisting line.

DEAD END OF A LINE End of a pipeline that does not lead back to an oil storage tank, so that the oil in the end of the line cannot be recirculated.

DEAD END SYSTEM Fluid system that does not contain a return line to a storage tank; therefore the fluid cannot be pumped around in a closed circuit.

DEADENING 1. The use of insulating or sound-deadening or absorbing materials to reduce the passage of sounds through walls and floors; **2.** Area on which timber has been killed by fire, flooding, insects, or disease.

DEAD FRONT Without live electrical parts exposed to a person on the operating side of the equipment.

DEAD FUELS Naturally occurring forest fuels in which the moisture content is governed almost exclusively by relative humidity and precipitation.

DEADHEADING Traveling without load, except from the dumping area to the loading point within a work cycle.

DEADLATCH *See* **Deadbolt.**

DEAD LEG Hot water connection in which the water is stationary except when it is being drawn off, and thus will cool when motionless.

DEAD LEVEL Absolutely level.

DEAD LIGHT Window that is fixed in place and that will not open.

DEAD-LINE In drilling, the end of the rotary drilling line fastened to the anchor or dead-line clamp.

DEAD LOAD 1. Tare weight or unladen weight of a vehicle or piece of mobile equipment; **2.** Steady, constant, and unmoving load. *See also* **Live load,** and **Loads; 3.** Weight of a superstructure such as floors, roofs, and walls.

DEAD LOCK Lock worked only by a key.

DEADMAN 1. Anchor for a guyline, usually a beam, block, or other heavy item buried in the ground, to which the line is attached. *See also* **Anchor log; 2.** Pile, cluster of piles, or buried timber or a wall driven to withstand a horizontal force as through a tie rod fastened to a retaining wall; **3.** Anchor used to keep pipework from separating.

DEAD PLATE GRATE *See* **Grate.**

DEAD REEL Storage reel. *See also* **Reel.**

DEAD SHORE Vertical shore used for temporary support of the upper parts of a wall, the lower parts of which are required to be removed in the process of underpinning.

DEAD WEIGHT RELIEF VALVE *See* **Valve.**

DEADWOOD Timber from dead standing trees.

DEAD ZONE 1. Space where air remains static, uninfluenced by air currents; **2.** Area immediately above and below the floor level in which leveling of an elevator is not effective.

DEAERATING TANK Tank capable of removing entrained air and/or combustion gas from the circulating coolant of an engine.

DEAERATOR Heater that removes oxygen, carbon dioxide, and ammonia from boiler feedwater to reduce its potential for corrosion.

DEAFENING *See* **Pugging.**

DEAL *See* **Flitch.**

DEBARK *See* **Harvest functions.**

DEBARKER *See* **Harvesting machines (single function).**

DEBARKING Removing the outer protective layer (bark) from trees or parts of trees.

DEBONDING Procedures whereby specific tendons in pretensioned construction are prevented from becoming bonded to the concrete for a predetermined distance from the ends of the flexural members.

DEBRIS Rubbish or waste.

DEBRIS JAM Congested debris obstructing the free movement of water in a stream.

DEBRIS REMOVAL Insurance policy clause requiring the insurer to remove debris resulting from damage to the insured property.

dec. *Abbreviation for:* decimal; declination; decorate; decoration; decorative.

DECA Prefix representing 10. Symbol: da. Used in the SI system of measurement. *See also the appendix:* **Metric and nonmetric measurement.**

DECADENT Decaying logs or timber where 30% or more of the wood is unusable because of defect.

DECAL Pattern or lettering mounted on specially-prepared paper that can be permanently transferred to another surface. *Also called* **Decalomania.**

DECALITER Unit of volume, equal to ten liters. Symbol: daL. Used in the SI system of measurement. *See also the appendix:* **Metric and nonmetric measurement.**

DECALOMANIA *See* **Decal.**

DECAMETER Unit of length, equal to ten meters. Symbol: dam. Used in the SI system of measurement. *See also the appendix:* **Metric and nonmetric measurement.**

DECANTATION TEST Test to determine the actual proportion of loam, silt, or other material in a sample of sand. *Also called* **Silt test.**

DECANTING 1. Method for decompressing under emergency circumstances in which workers are brought to atmospheric pressure with a very high gas tension in the tissues and then immediately recompressed in a second and separate chamber or lock; **2.** Pouring off a liquid from a container without disturbing any sediment.

DECAY 1. Falling pressure; **2.** Decomposition of wood and other organics by fungi.

DECAY RATE Ratio of pressure decay to time.

DECELERATION 1. Continuous slowing from a higher speed to a slower speed; **2.** Period during which an elevator moves at an ever decreasing rate of speed, usually referring to that period from full speed to leveling speed.

DECELERATION LANE Lane provided on a multilane highway to permit vehicles to slow for an exit.

DECELERATION RESISTANCE Inertial reaction opposing decrease in speed of a vehicle. It is numerically equal to gross vehicle weight in pounds divided by 32.2 and multiplied by deceleration in feet per second per second. It may be expressed as pounds, or as pounds per thousand, or as a percent of gross vehicle weight.

DECELERATION STRESS The additional stress that is imposed on a wire rope as a result of a decrease in the load velocity.

DECELERATOR Device that permits adjustment of the rpm rate of an engine or motor. *See also* **Accelerator.**

DECENTER Lower or remove centering or shoring from an arch.

DECI Prefix representing 10^{-1}. Symbol: d. Used in the SI system of measurement. *See also the appendix:* **Metric and nonmetric measurement.**

DECIBEL Non-SI measure of sound level. One-tenth of a bel, the number of decibels denoting the ratio of the two amounts of power and being ten times the logarithm to the base of 10 of this ratio: symbol dB. A unit of measure of noise level in which the faintest sound we can hear, called the threshold of hearing, is 0 dB, and the loudest sound the human ear can tolerate, called the threshold of pain, is 140 dB.

DECIBEL REDUCTION Lessening of the actual intensity of sound.

DECIDUOUS Woody plants that lose their leaves each year.

DECIDUOUS CONCRETE Condition where corroding reinforcement has pushed off lumps of concrete.

DECILITER Unit of volume, equal to one tenth of a liter. Symbol: dL. Used in the SI system of measurement. *See also the appendix:* **Metric and nonmetric measurement.**

DECIMETER Unit of length, equal to one tenth of a meter, or 10 centimeters, or 100 millimeters. Symbol: dm. Used in the SI system of measurement. *See also the appendix:* **Metric and nonmetric measurement.**

DECISION TREE Graphical representation of the relationship between decisions and chance events.

DECK 1. Exposed flat portion of a roof or floor; **2.** Traveled floor of a bridge; **3.** Formwork upon which concrete for a slab is placed; **4.** In blasting, a smaller charge or portion of a blasthole loaded with explosives that is separated from the main charge by stemming or an air cushion; **5.** To store logs; a pile of yarded logs; **6.** Load-carrying area of a platform, lowbed, or chassis-type trailer.

DECK BOARD Capping to the balustrade of an escalator.

DECK BRACKET *See* **Balustrade bracket.**

DECK CHARGE Blasthole charge comprising cartridges interspaced with stemming.

DECKING 1. Prefabricated units forming the horizontal structure of a floor or roof; **2.** Sheathing material for a deck or slab form; **3.** Shaped metal forms installed over structural steel members to facilitate attachment or installation of other materials; **4.** Technique of separating explosive charges in a blasthole using inert material that prevents passing of concussion, each charge having a primer.

DECK LOAD Explosive charges spaced well apart in a borehole and fired by separate primers or by detonating cord.

DECK-ON-HIP Hipped roof terminating in a flat roof.

DECK PLATE Metal plate with a raised pattern, intended to provide stable footing in exposed and/or hazardous locations.

DECK SCREEN *See* **Screen.**

DECLINATION Bearing of magnetic north as fixed by the positive pole of a magnetic needle not exposed to any artificial influence.

DECLINE CONVEYOR *See* **Conveyor.**

decn *Abbreviation for:* decision.

DECODER Device that converts a digital signal into an analog output.

decomp. *Abbreviation for:* decompose; decomposition; decompress; decompression.

DECOMPOSITION 1. Bacterial breakdown of organic material; **2.** Chemical alteration of organic material or minerals.

DECOMPRESSION 1. Controlled release of pressurized air or gas; **2.** Procedure of gradually lowering high air pressures in which men have been working.

DECOMPRESSION PRESSURE CONTROL VALVE *See* **Valve.**

DECOMPRESSION VALVE Device that lowers compression in the cylinder(s) for easier engine starting.

decon. *Abbreviation for:* decontaminate; decontamination.

DECONTAMINATION Process of removing unwanted material or substances; the reduction of contamination to an acceptable level.

DECONTAMINATION AREA Enclosed area adjacent and connected to a regulated area and consisting of an equipment room, shower area, and clean room that is used for the decontamination of workers, materials, and equipment.

DECORATED Adorned, embellished, or made more attractive by means of color and/or surface detail.

DECORATION Embellishment of a surface area through the addition of patterns, designs, colors, or other enriching media. In tile work this can take the following forms:

    Inglaze: Ceramic decoration applied on the surface of an unfired glaze and matured with the glaze.

    Overglaze: Ceramic or metallic decoration applied to and fired on a previously glazed surface.

    Underglaze: Ceramic decoration applied directly on the surface of ceramic ware and subsequently covered with a transparent glaze.

DECORATIVE TILE Tile with a ceramic decoration on the surface.

DECORATOR PLANK Fabricated planks of both the extension and fixed length types, used for supporting one user and limited material, usually used with ladder-jack scaffolds, trestles, extension trestles, platforms, or stepladders.

DECOUPLED CHARGE Charge that has a smaller diameter than the blasthole in which it is loaded; the coupling ratio is less than one.

DECOUPLING Separating components or elements so as to prevent, or retard the transmission of structurally borne sound, heat, or loads.

decr. *Abbreviation for:* decrease.

DECREE Order issued by an authority; a judicial decision.

DECREMENATION Removal of a load in steps at completion of test loading. *See* **Incremental loading.**

ded. *Abbreviation for:* dedicated; deduct; deduction.

DEDICATED POWER SUPPLY Electrical power supply reserved for a specific purpose.

DEDICATION Gift of land by its owner for public use. *See also* **Acquisition, Conveyance, Eminent domain, Expropriation, Negotiation, Option, Remainder,** and **Severality.**

DEDUCTIBLE CLAUSE Insurance clause detailing the amount or percentage to be deducted from any loss.

DEDUCTION Amount deducted from an agreed contract sum by change order.

DEE Steel item used to connect a wire rope or shackle to a knob-type line terminal.

DEED Written instrument conveying real property or interest therein, usually under seal. *See also* **Quitclaim deed,** and **Warranty deed.**

DEED OF TRUST Instrument whereby real property is given as security for a debt and in which three parties are involved: the lender, the borrower, and a trustee.

DEED RESTRICTION Limitation incorporated into a deed of title that restricts the use to which the land can be put.

DEENERGIZED Free from any electrical connection to a source of potential difference and from electrical charges; not having a potential difference from that of earth.

DEEP FOUNDATION 1. Foundation formed at a depth greater than 3 m (10 ft); **2.** Foundation unit that provides support for a building by transferring loads either by end-bearing to a soil or rock at considerable depth below the building, or by adhesion or friction, or both, in the soil or rock in which it is placed.

DEEP LIFT Placing newly mixed concrete in formwork, usually through trunking, in a deep layer to minimize the number of horizontal joints.

DEEP SEAL TRAP Trap with a seal of 100 mm (4 in.) or more.

DEEP SUMP Extra depth engine oil sump that allows the equipment to operate on slopes up to 30°.

DEEP-WELL PUMP Centrifugal pump used to raise water more than 7.62 m (25 ft).

def. *Abbreviation for:* defect; defective; definite; definition.

DEFAULT Failure to meet an obligation; failure to make a sum of money available when contractually required, as with a mortgage payment.

DEFECT 1. Imperfection that, by its size, shape, location, or makeup, reduces the useful service of a part; flaw or blemish; **2.** Work, materials, equipment, or system that is unsatisfactory, defective, deficient or does not comply with the contract documents; does not meet the requirements of any reference standard, test, or approval; or has been damaged before the architect's recommendation of final payment; **3.** In a tree, a fault or point of weakness caused by nonpathological or pathological agents; **4.** In lumber, an irregularity occurring in or on wood that will tend to impair its strength, durability, or utility.

DEFECT IN TITLE Recorded instrument that would prevent a grantor from giving clear title.

DEFECTIVE WORK Work (actually performed on site or incorporated into a supplied product) that does not meet the quality specified in the contract documents.

DEFERRED MAINTENANCE Physical deterioration that a prospective purchaser would anticipate having to correct following purchase of a property.

DEFICIENCY Item or work missing or incomplete as described in the specifications and contract documents.

DEFICIENCY JUDGMENT Judgment given when the security for a loan does not entirely satisfy the debt upon its default and sale upon foreclosure.

DEFICIENCY OF AIR Supply of air that is inadequate for complete combustion of a fuel.

DEFINITE WORKING DAY *See* **Collection method.**

defl. *Abbreviation for:* deflect; deflection; deflector.

DEFLAGRATION Explosive reaction that consists of a burning action at a high rate of speed along with which occur gaseous formations and pressure expansions.

DEFLECTED PILE Pile that has deflected away from its intended direction during driving.

DEFLECTED TENDONS Tendons that have a path that is curved or bent with respect to the gravity of the axis of the concrete member. *Also called* **Draped tendons,** and **Harped tendons.**

DEFLECTING SADDLE Strong metal saddle used for deflecting the direction of a prestressing tendon.

DEFLECTION 1. Movement of a point on a structure, structural element, or pavement, usually measured as a linear displacement transverse to a reference line or axis; **2.** Bending of a beam or any part of a structure under an applied load; **3.** Difference in radius between a loaded and unloaded tire; **4.** Sag in the ground profile; **5.** Amount of sag in a line measured at mid-span, expressed as a percentage of the horizontal span length. *See also* **Geometry of circular curves.**

DEFLECTION RATE Measure of spring movement in relation to the load applied.

DEFLECTOR SHEAVE *See* **Sheave.**

DEFOLIATORS Insects that destroy foliage, or chemicals that cause plants to drop their leaves.

DEFORMATION Change in dimension or shape. There are several conditions, including:

Contraction: Decrease in either length or volume

Creep: Time-dependent deformation due to sustained load.

Expansion: Increase in scope, length, or volume.

Inelastic: Deformation not proportional to the applied stress.

Length change: Increase or decrease in length.

Shrinkage: Percent decrease of a material's volume when compacted, or from the escape of any volatile substance, or by a chemical or physical change in the material.

Time-dependent: Deformation resulting from effects such as autogenous volume change, thermal contraction or expansion, creep, shrinkage, and swelling, each of which is a function of time.

Volume change: Either an increase or decrease in volume due to any cause.

DEFORMED BAR Reinforcing bar with irregular surfacing for producing a better bond with grout than can be obtained with a smooth bar. *Also called* **High-bond bar.**

DEFORMED HOT-ROLLED HIGH-YIELD BAR High-yield round-steel reinforcing bar, the surface of which is deformed with indentations during hot rolling.

DEFORMED HOT-ROLLED MILD STEEL BAR Mild steel round reinforcing bar, the surface of which is deformed with indentations during hot rolling.

DEFORMED PLATE Flat piece of metal, thicker than 6 mm (0.25 in.), having horizontal deformations or corrugations, used in construction to form a vertical joint and provide a mechanical interlock between adjacent sections.

DEFORMED REINFORCEMENT Metal bars, wire, or fabric with a manufactured pattern of surface ridges that provide a locking anchorage with surrounding concrete.

DEFORMED TIE BAR *See* **Tie bar.**

DEFORMED WIRE *See* **Crimped wire.**

DEFROST Mechanical or chemical removal of ice or frost from a surface.

deg. *Abbreviation for:* degree.

degC *Abbreviation for:* degree Celsius.

degF *Abbreviation for:* degree Fahrenheit.

DEGRADATION 1. To lower in quality, value, price, etc., or to place in a lower classification; **2.** Lowering of any portion of the earth's surface by erosion.

DEGRADE Any defect that lowers the grade or quality of a log.

DEGREASED CONDITION Condition in which prestressing wire must be delivered to ensure maximum bond with the surrounding concrete.

DEGREASING Removing oil and grease from a surface using either solvents or vapors.

DEGREE ANGLE Unit of plane angle equal to the circumference of a circle divided by 360. Symbol: °. Multiply by 0.017 453 to obtain radians, symbol: rad. *See also the appendix*: **Metric and nonmetric measurement.**

DEGREE CELSIUS Non-SI unit of measure for temperature where the freezing point of water is 0 and the boiling point 100. Symbol: °C. Permitted for use with the SI system of measurement. Add 273.15 to obtain kelvin, symbol: K; multiply by (°C x 9/5) + 32 to obtain degrees Fahrenheit, symbol: °F. *See also the appendix*: **Metric and nonmetric measurement.**

DEGREE DAY Daily measure of the difference between the average outside temperature and 18°C (64.4°F). The seasonal sum of degree days below 18°C (64.4°F) is used in calculating heating requirements.

DEGREE FAHRENHEIT Nonmetric unit of temperature in which 32 is the freezing point of water and 212 the boiling point. Symbol: °F. Multiply by (°F - 32) x 5/9 to obtain degrees Celsius, symbol: °C. *See also the appendix*: **Metric and nonmetric measurement.**

DEGREE-HOUR Measure of strength gain of concrete as a function of the product of temperature multiplied by time for a specific interval. *See also* **Maturity factor.**

DEGREE OF ARC Non-SI unit of angle (180° = π·rad), equal to 60 minutes of arc, and 3600 seconds of arc, permitted for use with the SI system of measurement. Symbol: °. *See also the appendix*: **Metric and nonmetric measurement.**

DEGREE OF COMPACTION Measure of the density of a soil sample, estimated by a standard formula.

DEGREE OF CONSOLIDATION Settlement after time, divided by the final settlement.

DEGREE OF CURVE Number of degrees at the center of a circle, subtended by a chord of 30.48 m (100 ft) at its rim. *See also* **Geometry of circular curves.**

DEGREE OF DENSITY Measure of compaction.

DEGREE OF SATURATION Ratio of the weight of water vapor associated with a weight of dry air to the weight of water vapor associated with a similar weight of dry air saturated at the same temperature. *See also* **Saturation.**

DEGREE OF SLOPE *See* **Expression of slope.**

DEHOTTAY PROCESS Refinement of the ground freezing process for shaft sinking or foundations in which liquid carbon dioxide is pumped through pipes installed in the ground.

DEHUMIDIFIER 1. Adsoption or absorption equipment for removing moisture from the atmosphere; **2.** Air cooler or washer for lowering the moisture in air pumped through it.

DEHUMIDIFY Reduction, by any process, of the quantity of water vapor or moisture content in the air of a room.

dehyd. *Abbreviation for:* dehydrator.

DEHYDRATION Removal of chemically bound, adsorbed or absorbed water from a material.

DEICER Chemical, such as sodium or calcium chloride, used to melt ice or snow on slabs and pavements.

del. *Abbreviation for:* delineate; delineation; delaminate; delete; deletion.

DELAMINATION 1. Separation along a plane parallel to a surface, as in the separation of a coating from a substrate or the layers of coating from each other; **2.** In the case of a concrete slab, a horizontal splitting, cracking, or separation of a slab in a plane roughly parallel to, and generally near the upper surface; **3.** Separation of the face layer of a composite panel from the core, or a laminate from a substrate.

DELAVAL TEST *See* **Los Angeles abrasion test.**

DELAY 1. Blasting cap that does not fire instantly but has a predetermined built-in lag or delay. *See also* **Short period delay; 2.** Time lost while traffic is impeded by some element over which the driver has no control. *See also* **Fixed delay,** and **Operational delay.**

DELAY BLASTING Blasting that uses delays or delay caps.

DELAYED MIXING Process in which fuel and air leave a burner nozzle unmixed and therefore mix relatively slowly, largely through diffusion.

DELAY ELEMENT Portion of a blasting cap that causes a delay between the instant of impressment of electrical energy on the cap and the time of detonation of the base charge of the cap.

DELAY TIME *See* **Machine time,** and **Scheduled operating time.**

DELETE Omit.

DELIMB *See* **Harvest functions.**

DELIMBER *See* **Harvesting machines (single function).**

DELIMBER-BUCKER *See* **Harvesting machines (multifunction).**

DELIMBER-BUNCHER *See* **Harvesting machines (multifunction).**

DELIMBER-SLASHER *See* **Harvesting machines (multifunction).**

DELIMBER-SLASHER-BUNCHER *See* **Harvest-** ing machines (multifunction).

DELIMBING Removing branches from a tree.

DELIMBING GATE Metal grid used with a skidder for removing limbs.

DELIQUESCE Dissolve and become liquid by absorbing airborne moisture.

DELIVERED AIR/FUEL RATIO Mass of delivered air divided by the mass of delivered fuel.

DELIVERY 1. Volume of fluid discharged by a pump in a given time, usually expressed in liters per minute (L/min) or gallons per minute (gpm); **2.** Transfer of ownership and possession.

DELIVERY BOX Enclosure for the control and measurement of water.

DELIVERY HOSE Hose through which shotcrete, grout, or pumped concrete or mortar passes. *Also called* **Conveying hose,** and **Material hose.**

DELIVERY LEAD TIME Amount of time ahead of when a supplied material or product is needed to be used on a job that it must be delivered to the site.

DELTA 1. Body of sediment deposited where a fast-flowing body of water empties into standing or the slow-current water of a lake, bay, or ocean. **2.** *See* **Geometry of circular curves.**

DELTA T 1. Difference between any two temperature readings; **2.** In solar heating, the difference between outside ambient temperature and either inlet or outlet temperature of the fluid passing through the solar collector.

DELUGE SYSTEM *See* **Automatic sprinkler system.**

DE-LUGGER Machine used to cut the lugs from tires prior to buffing.

DELUXE COOL-WHITE FLUORESCENT *See* **Fluorescent lamp.**

DELUXE WARM-WHITE FLUORESCENT *See* **Fluorescent lamp.**

dem. *Abbreviation for:* demand; demonstrate; demurrage.

DEMAND (ELECTRIC) Rate at which electric energy is delivered to or by a system, part of a system, or a piece of equipment. It is calculated in various ways, including:

Annual maximum: The greatest of all demands of the load under consideration that occurred during a prescribed demand interval in a calendar year.

Annual system maximum: The greatest demand on an electric system during a prescribed demand interval in a calendar year.

Average: The demand on, or the power output, of an electric system or any of its parts over any interval of time, as determined by dividing the total number of kilowatt hours by the number of units of time in the interval.

Billing: The demand upon which billing to a customer is based, as specified in a rate schedule or contract. It may be based on a contract year, a contract minimum, or a previous maximum and, therefore, does not necessarily coincide with the actual measured demand of the billing period.

Coincident: The sum of two or more demands that occur in the same demand interval.

Instantaneous peak: The maximum demand at the instant of greatest load, usually determined from the readings of indicating or graphic meters.

Integrated: The demand averaged over a specified period, usually determined by an integrating demand meter or by the integration of a load curve. It is the average of the continuously varying instantaneous demands during a specified demand interval.

Maximum: The greatest of all of the demands of the load under consideration that has occurred during a specified time period.

Noncoincident: The sum of two or more individual demands which do not occur in the same period interval. Meaningful only when considering demands within a limited period of time, such as a day, week, month, a heating or cooling season, and usually for not more than one year.

DEMAND CHARGE Specified charge to be billed on the basis of the billing demand, under an applicable rate schedule or contract.

DEMAND HORSEPOWER *See* **Horsepower.**

DEMAND INTERVAL Period of time during which the electric energy flow is averaged in determining demand, such as 60 minute, 30 minute, 15 minute, or instantaneous.

DEMAND LOAN Loan repayable at the will of the lender and to which days of grace are not applicable.

DEMARCATION 1. Setting and marking the limits; 2. In masonry, a fixed line.

DEMISE To convey real property for a defined period.

DEMISED PREMISES Property subject to lease.

DEMISING CLAUSE Clause in a lease whereby the owner leases and the tenant assumes the property.

DEMISING PARTITION Partition between two rental areas.

DEMISING WALL *See* **Wall.**

DEMOGRAPHIC STUDY *See* **Population study.**

DEMOLDING Removal of molds from concrete test specimens or precast products. *See also* **Strip.**

DEMOLISH To pull down and destroy.

DEMOLISHED MATERIALS Miscellaneous materials that result from a demolition process, some of which may be suitable for recycling and/or reuse.

DEMOLITION Breaking and removal of buildings and structures.

DEMOLITION PERMIT Permit to demolish a taxable structure, issued by the taxing authority.

DEMOUNTABLE BOX Steel case with cover in which a battery is assembled for use on an industrial truck that does not have its own integral battery compartment.

DEMOUNTABLE RIM Type of rim that can be detached from the spokes of a wheel; used with cast-spoke wheels.

DEMOUNTABLE RIM CHORDING Tendency of a demountable rim to assume the shape of the five- or six-spoke wheel on which it is mounted.

DEMOUNTABLE RIM OFFSET Lateral distance from a demountable rim surface that contacts the spacer band to the rim centerline.

DEMOUNTING PARTITION Wall system designed to be assembled and disassembled with the minimum damage to or loss of components.

DEMULSIBILITY Measure of the ability of an oil to separate from water.

DEMULSIBILITY TEST Test used to predict the relative rate at which asphalt globules in rapid-setting emulsified asphalts will break when spread in thin films on soil or aggregate.

DEMURRAGE 1. Charge made for the delayed completion of a contract; 2. Charge by a gas supplier to the gas user as rent of a gas cylinder (the user is commonly allowed free use for a number of days with a daily charge applying from then on); 3. Charge made by a carrier for a delay in loading or unloading a cargo beyond the time specified.

DENATURED ALCOHOL Alcohol used as a solvent for shellac-based media and dried latex and acrylic film. *Also called* **Denatured solvent,** and **Solvent alcohol.**

DENATURED SOLVENT *See* **Denatured alcohol.**

DENDROLOGY Study and identification of trees.

DENIER Yarn sizing system for continuous-filament synthetic fibers. The denier of filament yarn is the weight in grams of a length of 9000 m of that yarn.

denom. *Abbreviation for:* denominate; denomination; denominator.

dens. *Abbreviation for:* densify; density.

DENSE CONCRETE Concrete containing a minimum of voids.

DENSE GRADED Asphalt mix in which the aggregate gradation has been carefully controlled to produce a dense, tight mat with maximum mechanical strength.

DENSE-GRADED AGGREGATE *See* **Aggregate.**

DENSE-TYPE SPRINKLER SYSTEM *See* **Fire sprinkler system.**

DENSITY 1. Ratio of the mass of a substance to its volume, commonly expressed as kg/m^3 (lb/yd^3). *See also* **Material density;** 2. Number of similar things within a given area or volume; 3. In urban planning, the number of people or the number of residences per unit area; 4. Mass of an explosive per unit volume, expressed in gm/cc. 5. Number of vehicles per mile on a traveled way at a given instant. 6. Ratio of the combined weight of soil cover and the underlying solid waste to the combined volume of the solid waste and the soil cover. 7. *See also* **Stand density.**

DENSITY CHISEL Hardened-up cold chisel for removing highly compacted material.

DENSITY CONTROL 1. Control of the density of a manufactured material; **2.** For concrete in field construction, control of specified values as determined by standard tests.

DENSITY (DRY) Mass per unit volume of a dry substance at a stated temperature. *See also* **Specific gravity.**

DENSITY ZONING Laws or regulations restricting land use intensity.

'DENTAL' WORK Work with hand tools in a drilled hole to break up and remove boulders or other rock intruding into the hole.

DENTIL Series of ornamental square blocks, often incorporated into a complex molding.

DENTURE CLUTCH *See* **Clutch.**

DENUDATION Wearing down and disintegration of rock masses by rain, frost, wind, running water, and other surficial agencies.

dep. *Abbreviation for:* departure; depend; dependent; deposit; depository.

DEPARTURE Difference in easting from a survey instrument to the point being measured.

DEPENDENCY Relationship of a minor building to a major one in a single composition.

DEPLETION ALLOWANCE Deduction from taxable income derived from a wasting asset.

DEPOSIT 1. Anything laid down, such as rocks, minerals and ores; **2.** Process of laying down; **3.** Percentage of a bid sum or of a contract sum, or a fixed monetary sum required as a pledge at time of tendering or bidding for a contract; **4.** Percentage of a total price paid in advance of receipt of the goods to secure acquisition of the commodity. *Also called* **Advance.**

DEPOT 1. Bus or railroad station; **2.** Storage or collection center.

depr. *Abbreviation for:* depreciate; depreciated; depreciation.

DEPRECIATED COST METHOD *See* **Cost approach.**

DEPRECIATION 1. Loss in value or of useful life, for any cause; **2.** In the case of machinery and equipment, loss resulting from wear, obsolescence, inadequacy, or any other cause. *See also* **Accelerated depreciation.**

DEPRECIATION ACCRUAL RATE *See* **Depreciation rate.**

DEPRECIATION FACTOR Percentage of the adjusted tax basis that is depreciated over the depreciation period.

DEPRECIATION RATE Annual percentage by which it is estimated that a loss in value of machinery, equipment, or property occurs. *Also called* **Depreciation accrual rate.**

DEPRESSANT Substance or device that lessens undesirable properties.

DEPRESSED MEDIAN Median that is lower in elevation than the traveled ways for traffic in opposite directions.

DEPRESSED SEWER Section of a sewer that flows below the grade line and therefore runs full and at greater than atmospheric pressure.

DEPRESSION Defect in a finished composite panel that appears as a concave area on the surface.

dept *Abbreviation for:* department.

DEPTH FILTER Filter medium that primarily retains contaminants within torque passages.

DEPTH FILTRATION Filtration that primarily retains contaminant within tortuous passages.

DEPTH GAUGE Instrument for determining the depth of holes or recesses.

DEPTH KEY Key that enables a locksmith to cut blanks made for a special lock according to a code.

DEPTH OF CUT Measurement of a cut from the pavement surface to the bottom of the cut.

DEPTH OF FIXITY Distance from the ground surface to the depth at which a pile is held firmly by the soil. *Also called* **Point of fixity.**

DEPTH OF FUSION Depth to which base metal is melted during welding.

DEPTH OF THREAD Depth between the crest and the root of a thread, measured at right angles to the centerline.

DERAILMENT SWITCH Device activated by the counterweight of a hoist to indicate that the counterweight has left its guides.

deriv. *Abbreviation for:* derivation; derivative.

DERIVATIVE CODE Numerical sequence that relates the tumbler arrangement of a lock to the depth of the key cuts necessary to make it operate.

DERRICK 1. Hoisting device, usually static, used for lifting or moving heavy weights; **2.** Structure consisting of an upright framework with a hinged arm that can be raised and lowered and that can be swung around to different positions for handling loads. *See also* **A-frame derrick.**

DERRICKING Operation of changing boom angle in a vertical plane. *See also* **Luffing.**

DERRICK TOWER GANTRY Steel staging making three towers, one of which is the crane tower for the mast and jib of the derrick, the remaining towers form the legs and their cantledge; the towers are tied together at their tops by the derrick legs.

des. *Abbreviation for:* design; designer.

DESALINATION Process to convert brackish or salt (sea) water to fresh water.

DESANDING PLANT Equipment used to filter or desand contaminated bentonite slurry prior to its reuse.

descr. *Abbreviation for:* describe; description.

DESCRIPTION Written depiction, of the location and dimensions of property, for instance.

des. dftsmn *Abbreviation for:* design draftsman.

des. engr *Abbreviation for:* design engineer.

desgnr *Abbreviation for:* designer.

des. gp ldr *Abbreviation for:* design group leader.

DESICCANT Substance capable of absorbing or adsorbing water or water vapor.

DESICCANT COOLING Method of lowering temperature by using materials such as activated carbon to absorb moisture from the atmosphere.

DESICCATE Lower or remove the moisture content of a material.

DESICCATION Process of shrinkage or consolidation of fine-grained soil, produced by an increase of effective stresses in the grain skeleton accompanying the development of capillary stresses in the pore water.

DESICCATOR TEST Quality control test performed on particleboard to monitor formaldehyde emissions from the panel product.

desig. *Abbreviation for:* designation.

DESIGN Creation that embodies ideas, aims, and objectives; pictorial depiction from which to work.

DESIGN ALTERNATIVES At each of the design stages (concept, preliminary, and working drawings), identification by the designer of several technical solutions that all satisfy the functional requirements and the standards that constrain his choices. The alternative designs may be compared in terms of cost effectiveness, often using an analysis of trade-offs to arrive at an optimum. Ideally, the most economic combination of design alternatives is the one that is finally chosen and implemented.

DESIGNATED PERSON *See* **Authorized person,** and **Competent person.**

DESIGNATED REPRESENTATIVE Individual or organization to whom written authorization to act on another's behalf has been given in writing.

DESIGN BEARING PRESSURE As applied to foundations, the pressure applied by a foundation unit to a soil or rock, that is not greater than the allowable bearing pressure.

DESIGN-BUILD Construction technique whereby the contractor designs the structures as well as builds them.

DESIGN CAPACITY 1. Maximum volume that something is designed to contain, bear, process, etc.; **2.** Maximum number of vehicles that can pass over a lane or a roadway during one hour without operating conditions falling below a preselected design level.

DESIGN CHANGE Change in the overall design or a component of the design after the design has been approved by the owner, or following completion of the contract documents.

DESIGN CONTINGENCY Estimated allowance for the cost of changes to the design to make the scope work.

DESIGN DEVELOPMENT STAGE Stage in the design process following conceptual, sketch, or preliminary designs and prior to final working drawings.

DESIGN DOCUMENTS Documents prepared by the designer (plans, design details, and job specifications).

DESIGNED MIX Concrete mix where the grade of concrete required, its minimum cement, content and any special properties required are specified. The concrete supplier is then free to chose the mix proportions to satisfy the specification.

DESIGN ENGINEER One who conceptualizes the overall design and details its parts.

DESIGN ENGINEERING FIRM Professional organization responsible for the design, plans, and specifications to fulfill the scope of work to be performed to successfully complete the design of a project. The firm may also monitor and observe the construction of the project.

DESIGNER Person responsible for the design.

DESIGN FACTOR Ratio of the nominal strength of a product or component to the total working load.

DESIGN FLUORESCENT *See* **Fluorescent lamp.**

DESIGN HEATING LOAD Total heat loss from a building under the most severe potential conditions of weather and use.

DESIGN HOURLY VOLUME *See* **Volume.**

DESIGN LIFE Period of time during which a product or assembly will meet the minimum design standards and requirements.

DESIGN LOAD 1. Factored load; **2.** Load that a pile is intended to carry without excessive movement and with an acceptable factor of safety against plunging failure. *See also* **Maximum intended load.**

DESIGN LOAD CAPABILITY Level of loading that a structural member is designed to sustain, with appropriate safety factors, against collapse, deflection or local damage.

DESIGN SPEED *See* **Speed.**

DESIGN STAGE Stage in the development of drawings that will become part of the contract documents following conceptual and sketch design and prior to working drawings.

DESIGN STRENGTH 1. Nominal strength of a member multiplied by a strength reduction (Phi) factor. *See also* **Nominal strength,** and **Phi factor; 2.** Resistance (force, moment, stress, as appropriate) provided by an element or connection; the product of nominal strength and the resistance factor; **3.** Strength of a material, as used in calculations, so that allowable stress is not exceeded under the applicable loading conditions.

DESIGN TEAM Group whose members contribute their individual talents, specialities, and skills toward the resolution of a design problem.

DESIGN VALUE NUMBER Means of rating or classifying materials as to their properties, stability, strength, etc.

DESIGN VOLUME *See* **Volume.**

DESIGN WEIGHT Maximum weight to which a component may be loaded, without the danger of failure and/or premature wear taking place.

DESIGN WORKING PRESSURE Maximum working pressure for which a system, or part of a system, is designed.

DESOLDERING WICK Flux-coated copper braid used to absorb melted solder.

des. spec. *Abbreviation for:* design specialist.

dest. *Abbreviation for:* destroy; destruction.

DESTRESSING Reverse process of stressing a pre-stressed concrete tendon, necessary for the dismantling of temporary work.

det. *Abbreviation for:* detach; detail; detailed.

DET *Abbreviation for:* double end trimmed.

DETACHABLE BIT Hardened-steel rod, threaded at one end, screwed to the end of a rock-drill steel.

DETACHABLE CONTAINER *See* **Waste container.**

DETACHABLE CONTAINER SYSTEM Partially mechanized self-service refuse removal procedure with specially constructed containers and vehicles. It is mechanized in that special equipment is used to empty the containers and haul refuse to the disposal site. It is self-service when the customer deposits the refuse in the container.

DETACHED Standing alone; not sharing a wall with any other building.

DETACHED HOUSE Dwelling standing completely within the boundaries of the lot on which it is con-structed.

DETAIL DRAWING Large-scale drawing detailing small parts of a larger whole.

DETAILER Draftsman who prepares drawings of de-tails as part of a set of drawings or for shop and site use.

DETAILING Preparation of working drawings.

DETECTOR 1. Mechanical or electronic device that senses and signals an event or presence; **2.** *See* **Traffic signal.**

DETECTOR CHECK VALVE *See* **Valve.**

DETENT Part that stops or releases motion by me-chanical or other means.

DETERGENT 1. Chemical used to clean and degrease surfaces; **2.** Additive used in engine oils to remove and hold in suspension foreign matter that finds its way into the lubrication system.

DETERGENT ADDITIVE Additive used to prevent deposits from forming on surfaces (may also help to remove previously formed deposits).

DETERIORATION Physical manifestation of failure of a material (e.g., cracking, delaminating, flaking, pitting, scaling, spalling, staining, etc.). *See also* **Dis-integration,** and **Weathering.**

detm. *Abbreviation for:* determine.

detmn *Abbreviation for:* determination.

DETONATING CORD Plastic-covered core of high-velocity explosive used to detonate charges or explo-sives in boreholes and underwater, e.g., Primacord.

DETONATION 1. Explosive reaction consisting of the propagation of a shock-wave through an explosive, accompanied by a chemical reaction that furnishes energy to sustain the shock-wave propagation in a stable manner, followed by gaseous formation and pressure expansion; **2.** Explosion in the combustion chamber of an engine when a mixture of vaporized fuel and air is ignited. When this occurs ahead of the flame front, raising the pressure and temperature of such gases, a self-ignition point is reached resulting in a knock or 'pinging' sound.

DETONATION PRESSURE Pressure exerted by gases as they are first produced at the detonation head.

DETONATOR Device attached to explosives, used to initiate an explosion.

DETRIMENTAL SETTLEMENT Settlement or cracking of a structure due to the stress deformation of the underlying soil.

DETRITUS Loose material produced by the disinte-gration of rocks or of the skeletal remains of organ-isms.

DETRITUS SLIDE Tendency for detritus to move downhill.

DETRITUS TANK Detention chamber where the flow of liquid waste is slowed, allowing sediment and other settleable solids to sink to the bottom where they can be removed by mechanical equipment.

dev. *Abbreviation for:* develop; developing; develop-ment; deviate; deviation; device.

DEVAL MACHINE Machine for carrying out an at-trition test on a sample of aggregate.

DEVELOPED LENGTH Length of connected pipe and fittings, measured along the centerline.

DEVELOPED SURFACE Curved surface graphically shown as flattened out in two dimensions.

DEVELOPER One who undertakes to own, manage, and pay for the entire design, procurement, and con-struction of a building project by agreement, for satis-factory return on investment. The developer may al-ready own the land, or will procure the land as part of his undertaking. He may also finance the works both during and after construction. The contract with the developer is generally for the provision of an end product that meets agreed standards of space, quality, and function for a stipulated price, leaving the devel-oper to choose the specific materials and configuration that satisfy the standards.

DEVELOPMENT 1. Work that sees land put to a higher and better use. *See also* **Land development; 2.** Outline on a flat surface in two dimensions of a shape that will eventually become three-dimensional; **3.** Re-pair of damage to a formation caused by drilling to increase the porosity and permeability of the materials surrounding a well intake; **4.** Work done to a dug or drilled well to improve the flow in the aquifer in the immediate vicinity.

DEVELOPMENT BOND STRESS *See* **Anchorage bond stress.**

DEVELOPMENT COST 1. Expenditures made in converting research findings, plans, and specifications into an actual project; **2.** Expenditures made to bring a mineral property or other natural resource into produc-tion.

DEVELOPMENT LENGTH Embedment length re-quired to develop the design strength of reinforcement at a critical section (formerly called **Bond length**).

DEVELOPMENT OBJECTIVES Short-term (often 5-year) planning objectives for a specific Management Area.

DEVELOPMENT PLAN Specific plan outlining har-vesting, road construction, protection and silviculture activities over the short-term (often 5 years) in accordance with the approved Forest Management Plan.

DEVIATION Variation from something intended or planned; difference in the designated plan location or elevation. *See also* **Coefficient of variation,** and **Standard deviation.**

DEVICE 1. Something made for a specific purpose; **2.** Unit of an electrical system that is intended to carry but not utilize electric energy.

DEVIL Firegrate used to heat asphalting tools.

DEVIL'S FLOAT Wooden float with two nails protruding from the toe, used to roughen the surface of a brown plaster coat.

DEVISE Real estate willed to another.

dev. lgth *Abbreviation for:* developed length.

DEWAXING Refining process for removal of wax from a base oil to reduce its cloud and pour point.

DEWATERING 1. Removal of surface water from an area; **2.** Lowering of the groundwater table to produce a 'dry' area in the vicinity of an excavation that would otherwise extend below water; **3.** Removal of water by filtration, centrifugation, pressing, open-air drying, or other methods.

DEWATERING PUMP *See* **Pump.**

DEW POINT Temperature at which air becomes saturated with moisture and below which condensation occurs.

DF *Abbreviation for:* damage free; decimal fraction; direction finder.

dflct. *Abbreviation for:* deflection.

D4S Lumber that has been dressed four sides.

dfrm *Abbreviation for:* deform; deformed.

dfs. *Abbreviation for:* diffuse.

dfsn *Abbreviation for:* diffusion.

dft *Abbreviation for:* draft; drift.

dftg *Abbreviation for:* drafting.

dftsmn *Abbreviation for:* draftsman.

DG *Abbreviation for:* double glaze; double glazed; dust guard.

Dgl *Abbreviation for:* diameter at ground line.

dgr *Abbreviation for:* danger.

dgr. *Abbreviation for:* degrease.

d-h *Abbreviation for:* double hung.

DH *Abbreviation for:* double hung.

DHW *Abbreviation for:* double-hung window(s).

di *Abbreviation for:* drop inlet.

dia. *Abbreviation for:* diameter; diamond.

diag. *Abbreviation for:* diagonal; diagram.

DIAGNOSTICS Component-by-component self-check performed on a mechanical, electrical, or electronic device, or series of connected or interdependent components, by a microprocessor programmed to detect irregularities or abnormalities.

DIAGONAL Slanting line, row, course, etc.

DIAGONAL BOND *See* **Bond.**

DIAGONAL BRACING Inclined structural members carrying primarily axial loads, employed to enable a structural frame to act as a truss to resist horizontal loads.

DIAGONAL BREAK Fabric break in a tire that follows the path of the ply cords.

DIAGONAL BUTTRESS *See* **Buttress.**

DIAGONAL CRACK In a flexural member, an inclined crack caused by shear stress, usually at about 45° to the axis, or a crack in a concrete slab, not parallel to either the lateral or longitudinal directions.

DIAGONAL CRACKING Development of diagonal cracks. *See also* **Diagonal tension.**

DIAGONAL EYEPIECE Eyepiece of a prismatic telescope used for survey.

DIAGONAL GRAIN Faulty conversion of lumber resulting in the wood fibers being at an angle to the length of the piece.

DIAGONAL-GRAINED WOOD *See* **Grain.**

DIAGONAL RING Pair of mating packing rings, one of which has one face convex-conical and the other has one face concave-conical.

DIAGONAL SIDE-CUTTING PLIERS *See* **Pliers.**

DIAGONAL SLATING Laying of roofing slates with one diagonal horizontal (the head is nailed and the horizontal corners are cut off).

DIAGONAL SPACER Spacer bar, or fork, of cast iron, pressed steel, or tube, with slotted ends to fit the main reinforcing bars of a precast pile, used in pairs across opposite diagonals of the cross section of the pile.

DIAGONAL SYSTEM Divides an anti-lock braking system into two circuits (front wheel one side with rear on the other side) to allow partial system function should one diagonal malfunction.

DIAGONAL TENSION Principal stress resulting from the combination of normal and shear stresses acting upon a structural member. *See also* **Diagonal cracking.**

DIAGONAL TIE 1. Braces or ties that help stiffen a roof truss; **2.** Braces attached to an angle to tie framing members together.

DIAGRAM Drawing that illustrates pertinent characteristics, component positions, sizes, interconnections, controls, and actuation of components. There are many types of presentation, including:

Cutaway: Drawing showing principal internal parts of a component, controls and actuating mechanisms, etc., all interconnecting lines and functions of individual components.

Graphical: Drawing or drawings showing each piece of apparatus including all interconnecting lines by means of standard symbols.

Line drawing: Scale drawing in which objects are represented by single lines with no attempt made to show the relative thickness of what is being shown. In the case of a building, for instance, the lines might indicate the centerlines of walls, or the outside faces of walls. *Also called* **Single-line diagram.**

Network: Graphical display of the sequence, timing and interrelationships of the activities comprising a project.

Pictorial: Drawing showing each component in its actual shape according to the manufacturer's installation drawings.

Schematic: *See* **Graphical,** above.

Schematic wiring: Drawing that shows a wiring layout using graphic symbols, the electrical connections and functions of a specific circuit arrangement.

Single line: *See* **Line drawing,** above.

DIAL GAUGE Instrument that shows, by a needle indication on a circular graduated dial, very small displacements of the plunger, indicating pressure, or movement, or temperature.

DIAL INDICATOR 1. Measuring instrument, commonly calibrated in 0.001 in. or 0.01 mm, that shows the reading on a dial; **2.** Device used to determine when the axis of two shafts are in line.

diam. *Abbreviation for:* diameter.

DIAMETER Length of a line passing from one side of a circle to the other and passing through the center. *See also* **Inferior diameter,** and **Superior diameter.**

DIAMETER AT GROUND LINE Diameter measure of a standing tree at the estimated cutting height.

DIAMETER BREAST HIGHT Measurement of trees by their diameter at the chest height of a man, about 1.5 m (5 ft). Also called **dbh.**

DIAMETER CLASSES Classification of trees based on diameter outside bark measured at dbh (diameter at breast height). In forest surveys, each diameter class encompasses approximately 50 mm (2 in.): the 150 mm (6 in.) class would include trees 125 mm (5 in.) through 175 mm (6.9 in.) in dbh.

DIAMETER INSIDE BARK Diameter of a tree or log excluding bark thickness.

DIAMETER LIMIT Minimum and/or maximum diameter of trees to be cut, measured, or used, as in a timber sale contract.

DIAMETER OUTSIDE BARK Measurement of tree diameter in which the bark is included.

DIAMETER TAPE Graduated tape based on the relationship of circumference to diameter that provides direct measure of tree diameter when stretched around the outside of a tree, usually at breast height.

DIAMETRAL COMPRESSION TEST *See* **Splitting tensile test.**

DIAMICTON Walls constructed with an exterior of masonry and interior of rubble.

DIAMOND CUT Inclined, shallow drill holes near the center of a tunnel face, arranged so that when the first shots are exploded in a round, a diamond-shaped wedge of rock is removed, allowing relief for the remaining rock when the main charge detonates in the surrounding holes.

DIAMOND DRILL *See* **Bit,** and **Drill.**

DIAMOND INTERCHANGE *See* **Interchange types.**

DIAMOND LEAD Yarding from the back of a square lead.

DIAMOND MESH Metallic fabric having rhomboidal openings in a geometric pattern. *See also* **Metal lath.**

DIAMOND-POINT CHISEL *See* **Chisel.**

DIAMOND POINTED *See* **Rustication.**

DIAMOND SAW Circular saw blade with industrial diamonds embedded in its cutting edge, used to cut concrete, stone, and other hard materials.

DIAMOND TOOL Diamond dresser (for dressing grinding wheels).

DIAMOND WHEEL Grinding wheel in which the abrasive is natural bort diamond.

DIAPER Any continuous pattern in brickwork, of which the various bonds are examples. The term is more commonly applied to diamond or other diagonal patterns.

diaph. *Abbreviation for:* diaphragm.

DIAPHRAGM 1. Floor slab, metal wall or roof panel possessing a large in-plane shear stiffness and strength adequate to transmit horizontal forces to resisting systems; **2.** Flexible partition between two chambers; **3.** Flexible sheet used to enclose a tire during precure retreading.

DIAPHRAGM ACTION In-plane action of a floor system (also roofs and walls) such that all columns framing into the floor from above and below are maintained in their same position relative to each other.

DIAPHRAGM AND SHEAR WALLS Wall system constructed in such a manner as to resist lateral forces, e.g. wind or earthquake.

DIAPHRAGM SPRING CLUTCH *See* **Clutch.**

DIAPHRAGM PACKING Packing between rigid members in relative motion that is attached to both members and absorbs the motion through its own deformation.

DIAPHRAGM PUMP *See* **Pump.**

DIAPHRAGM-TYPE CELLULAR COFFERDAM Structure made of steel sheet piles with each of the inner and outer walls consisting of a series of arc segments that are connected at their intersections with diaphragms that extend through the cofferdam to form a series of cells. The cells are filled with earth, sand, gravel, or rock. *See also* **Circular-type cellular cofferdam.**

DIAPHRAGM WALL Technique of constructing, *in situ*, a separating wall in the ground. *See also* **Slurry wall.**

DIASTYLE Columnar spacing of three diameters.

DIATOMACEOUS EARTH Friable earthy material composed of nearly pure hydrous amorphous silica (opal) and consisting essentially of the frusticles of the microscopic plants called diatoms. *Also called*

Kieselguhr.

DIBBLE Spade-like tool used to prepare planting holes for seedlings.

DICALCIUM SILICATE Compound having the composition $2CaO \cdot SiO_2$, and impure form of which occurs in portland cement clinker. *See also* **Calcium-silicate hydrate.**

dict. *Abbreviation for:* dictionary.

DIE Tool used to cut external threads on pipe or bar stock.

DIE CASTING Fabrication of parts by forcing molten metal into a metal die or mold.

DIE-FORMED RING Packing ring, cut to length and mechanically compacted to remove voids prior to being installed in a stuffing box.

DIELDRIN Poisonous powder, used for killing wood-destroying insects.

DIELECTRIC Material that will resist the passage of an electric current.

DIELECTRIC UNION Nonconducting plumbing fitting, used to join pipes manufactured of different materials to prevent electrolysis.

DIELECTRIC STRENGTH Ability of insulation to withstand voltage without breaking down: expressed in volts per mil.

DIESEL-ELECTRIC TRUCK An electric truck in which the power source is a diesel-driven generator.

DIESEL ENGINE Internal combustion engine that burns crude oil, ignition being brought about by heat compression.

DIESEL HAMMER *See* **Pile hammer.**

DIESELING Explosions of mixtures of lubricating oil and air in the compression chambers or other parts of the air system of a compressor. Also called **Prefiring,** and **Preignition.**

DIESEL 1 PROGRAM Computer program for solving wave equations for diesel hammers. *See also* **Computer wave equation program.**

DIE SIZE Coded description of tire tread rubber dimensions.

DIE STOCK Tool used to turn dies when cutting external threads.

diff. *Abbreviation for:* difference; different; differential.

DIFFERENCE IN ELEVATION Vertical dimension between two level surfaces, not withstanding their horizontal displacement.

DIFFERENTIAL Device that permits a drive axle to transmit power to the wheels while allowing each drive wheel to rotate at a different speed when necessary.There are several types, including:

> **Antislip:** Differential having two spring-loaded jaw-toothed clutches. When traveling straight ahead, each axle clutch is engaged to the drive clutch, creating a solid axle with both wheels driving. In a turn, the outside axle overruns the inside, unlocking the clutches. The inside wheel turns at the ring gear speed and provides 100%

of driving torque while allowing the outside wheel to turn faster and prevent tire scuffing.

> **Conventional:** Differential that divides torque equally between two drive wheels, even when one is traveling faster in a turn.

> **Driver-controlled, full-locking main:** Option that provides greater traction to each driving axle for on-/off- and slippery-road conditions. The mechanism, actuated through an in-cab switch, locks in the main differential to control slip on the left and right sides (versus the inter-axle differential that controls front-to-back axle power).

> **Locking:** Device that permits an axle to transmit driving force to the rear wheels of a vehicle with better traction by reducing the possibility of a vehicle becoming immobile when one driving wheel loses traction.

> **Nonspin:** Limited-action differential that will turn both axles, even if one offers no resistance.

> **Torque proportioning:** Differential that partially locks both axles together whenever one wheel starts to spin. A helical gear design permits differential action

> **Two-speed:** Differential having a high/low gearshift between the driveshaft and the ring gear.

DIFFERENTIAL-ACTING HAMMER *See* **Pile hammer.**

DIFFERENTIAL AMPLIFIER Device that amplifies the difference between two inputs, or the difference between an input and an output.

DIFFERENTIAL CROSS Often called a spider, it contains four side differential pinions that can rotate with the cross and/or on the cross at the same time to provide differential movement to both driving sides, depending on the driving needs. Side gears are meshed to the pinions. This splits the drive effort to the wheels evenly through the rotating differential assembly.

DIFFERENTIAL DIAPHRAGM Diaphragm between two fluids at substantially different pressures.

DIFFERENTIAL LEVELING Process of determining the difference in elevation between two points using a bench mark of known elevation as a reference. *See also* **Profile leveling.**

DIFFERENTIAL LOCK (LOCKOUT) Movable tooth or jaw that can lock together a pair of differential-driven axles.

DIFFERENTIAL PINION Gear that meshes with the differential ring gear. Also called **First reduction gearing.**

DIFFERENTIAL PRESSURE *See* **Pressure.**

DIFFERENTIAL PRESSURE INDICATOR *See* **Indicator.**

DIFFERENTIAL PRESSURE VALVE *See* **Valve.**

DIFFERENTIAL PULLEY BLOCK *See* **Chainblock.**

DIFFERENTIAL SETTLEMENT 1. Where foundations of different parts of the same structure settle at different rates; **2.** In a landfill, the nonuniform subsidence of the landfill surface due to the variety in waste types (biodegradable *vs* inert), decomposition rates, compactive effort, and any voids when the waste was placed.

DIFFERENTIAL SHRINKAGE Different rates of shrinkage in the same material or member, due to different ages or moisture content or other factors.

DIFFERENTIAL THERMAL ANALYSIS Indication of thermal reaction by differential thermocouple recording of temperature changes in a sample under investigation compared with those of a thermally passive control sample that is heated uniformly and simultaneously.

DIFFERENTIAL TRANSFORMER Variable transformer having opposed secondary windings arranged so that the output voltage varies with the motion of its core.

DIFFERENTIAL WEATHERING Results of variations in the rate of weathering on different parts of a material, structure, or rock body.

DIFFERENTIAL WINDLASS Windlass with two drums of different diameters, used to increase the lifting power.

DIFFERENTIATOR Electrical circuit that responds only to differences, or changes in input signal.

diffr. *Abbreviation for:* diffraction.

DIFFRACTION 1. Breaking up of a ray of light into a visible spectrum; **2.** Deflection of a ray of light when passing through a transparent medium such as water or glass; **3.** Tendency of sound waves to flow around obstacles that are small in relation to the wave length of the sound.

DIFFUSE Spread out.

DIFFUSE ENERGY One of two components of solar energy arriving at the earth's surface. *See also* **Direct energy.**

DIFFUSER 1. Inner shell and water passages of a centrifugal pump; **2.** Device through which compressed air is injected in the form of fine bubbles into a liquid waste.

DIFFUSE REFLECTION Light reflected from a smooth surface.

DIFFUSE-REFLECTION RATIO Ratio of light diffusely reflected from a surface, to the light that falls upon it.

DIFFUSION 1. Spreading or scattering; **2.** Movement of dissolved elements in water or air from areas of higher concentrations to areas of lower concentration.

DIFFUSIVITY Measure of the rate with which heat diffuses through a material.

DIGESTER Vessel in which sewage sludge is processed under anaerobic conditions to produce, following further processing, an inert cake, with methane gas being produced as a by-product.

DIGESTION In waste treatment, the biochemical decomposition of organic matter under aerobic or anaerobic conditions.

DIGGING CYCLE Series of motions completed by a machine during the continuous process of digging, typically involving filling the bucket, swinging and

dumping the bucket contents, and returning the bucket to the original position.

DIGGING DEPTH Vertical distance from the ground line to the lowest possible bucket cutting edge position with the bucket in level position.

DIGGING LINE Cable on a shovel that forces the bucket into the material being excavated. Called **Crowd** in a dipper shovel; **Drag** in a pull shovel; and **Dragline** and **Closing line** in a clamshell.

DIGIT 1. Finger or other extremity; **2.** Numeral from 0 to 9.

DIGITAL Of or pertaining to the general class of devices or circuits whose output varies in discrete steps (i.e., pulses or 'on-off' characteristics).

DIGITAL CIRCUIT *See* **Binary circuit.**

DIGITAL INDICATION SYSTEM System that monitors performance by using digital display calibrators.

DIGITAL PROCESSING Electronic data processing technology in which information is expressed in numerical form. It is many times faster than analog processing and requires less space and provides greater reliability.

DIGITAL-TO-ANALOG CONVERTER Device capable of converting a digital signal to an analog voltage.

DIKE, or DYKE 1. Obstacle used to protect land from inundation by water from the sea or a river. *See also* **Seawall; 2.** Raised section built onto the sides of roads to control water runoff and erosion; **3.** Slender rock formation that cuts across the structure of surrounding rock.

dil. *Abbreviation for:* dilute; diluted; dilution.

DILATION Expansion of concrete during cooling or freezing, generally calculated as the maximum deviation from the normal thermal contraction predicted from the length-change-temperature curve or length-change-time curve established at temperatures before initial freezing.

DILATOMETER Apparatus for measuring the coefficient of thermal expansion of a particle of aggregate.

DILUTENT Substance, liquid or solid, mixed with the active constituents of a formulation to increase bulk or lower concentration.

DILUTION Reduction of the concentration of a soluble material by the addition of a solvent or water.

dim. *Abbreviation for:* dimension; dimensioned.

DIMENSION Distance in a given direction or along a given line.

DIMENSIONAL STABILITY Quality of not changing in size due to any cause: moisture change, heat, cold, pressure, etc.

DIMENSIONAL TOLERANCES Allowable differences in dimensions, squareness or thickness.

DIMENSIONED STONE Stone precut and shaped to specified sizes.

DIMENSION LINE On drawings, a line with arrowheads or other limiters at either end, marked with a written dimension, used to show the distance between the two points.

DIMENSION LUMBER Lumber with a nominal thickness of from 50 mm (2 in.) up to, but not including, 127 mm (5 in.) and up to 300 mm (12 in.) wide.

DIMENSION SHINGLES Shingles cut to regular rather than random widths.

DIMINISHING COURSES Roofing shingles, slates, or tiles laid to a gauge that lessens from eaves to ridge.

DIMINISHING PIPE Tapered pipe.

DIMINUTION Gradual reduction: of size, intensity, etc.

DIMPLE Depression made in the surface of drywall as the fastener is set slightly below the finished plane to permit concealment with joint compound.

DIODE Two-element device capable of rectification of an electrical current or of blocking it in one direction.

DIOPTER ADJUSTMENT Adjustment made through the eyepiece of an instrument that brings the reticle into sharp focus for the user's eye.

DIOXINS Class of organic compounds that form as a result of incomplete or inefficient combustion of carbon compounds.

DIP Angle at which strata, beds, or veins are inclined from the horizontal.

dip. *Abbreviation for:* dipped.

DIP COATING 1. Application of a coating by dipping an object into a vessel containing another material in suspension; **2.** Applying a plastic coating by dipping an article into a tank of melted or liquid resin, and then chilling or rapidly curing the adhering plastic.

DIP COMPASS Magnetic needle with a horizontal pivot that is set in the magnetic meridian. When the inclination to the horizontal of the needle is read, the resulting angle represents the dip of the earth's magnetic field at that point.

DIPMETER Instrument to record the depth below ground level of the surface of the water in a borehole or piezometric tube.

DIPPER Digging bucket rigidly attached to a dipper stick, arm, or handle.

DIPPER SHOVEL *See* **Shovel.**

DIPPER STICK 1. Standard revolving shovel; **2.** Straight shaft that connects the bucket with the boom of an excavator or hoe.

DIPPER TRIP Device that unlatches the door of a shovel bucket to dump the load.

DIP STICK Measuring rod used to gauge the level of liquids within a vessel; of fluids in the reservoirs of machines.

dir *Abbreviation for:* director.

dir. *Abbreviation for:* direct; direction.

dir. conn. *Abbreviation for:* direct connected.

DIRECT-ACTING PUMP *See* **Pump.**

DIRECT CHARGES Costs that can be identified

specifically with a product, service, or activity. *See also* **Charges,** and **Indirect charges.**

DIRECT CIRCULATION Circulation of drilling fluid by pumping it down a hollow drill pipe, through the drill bit, and back to the surface in the annular space around the drill pipe. *See also* **Reverse circulation.**

DIRECT COMBUSTION Combustion that occurs in the main chamber of an engine.

DIRECT COMPENSATION Payment for land or interest in land and improvements actually acquired for highway purposes. *Also called* **Direct damages.**

DIRECT CONNECTION *See* **Interchange elements.**

DIRECT COST Cost that can be reasonably identified with a specific unit of production or with a specific operation or other cost center.

DIRECT CURRENT (DC) Electrical current that flows only in one direction.

DIRECT DAMAGES *See* **Direct compensation.**

DIRECT DRIVE *See* **Transmission.** *See also* **Driving machine.**

DIRECT DUMPING Discharge of concrete directly into place from a crane bucket or mixer.

DIRECT-DUMP TRANSFER SYSTEM Unloading of solid waste directly from a collection vehicle into an open-top transfer trailer or container.

DIRECT ENERGY One of two components of solar energy arriving at the earth's surface. *See also* **Diffuse energy.**

DIRECT EXPENSES All expenditures incurred by or attributable to a specific project.

DIRECT-FEED INCINERATOR *See* **Incinerator.**

DIRECT-FINANCING LEASE *See* **Lease.**

DIRECT HEATING Heating a space by a heat source located within the area.

DIRECT INJECTION Injection of fuel into the combustion chamber formed by the cylinder head and piston crown at top dead center.

DIRECT IN-LINE MECHANICAL LINKAGE Metal control rods linked directly to valve sections. This type of linkage provides the operator with positive control as well as parts durability.

DIRECTIONAL ARROWS 1. Arrows, usually illuminated and frequently mounted on a vehicle, that indicate the need for approaching traffic to change lanes; **2.** Indicator lights that signal the direction in which an elevator car will normally travel on the next start.

DIRECTIONAL BORING SYSTEM Equipment designed to bore a hole through the earth, with means to change direction of the cutting head. *See also* **Boring system.**

DIRECTIONAL CONTROL VALVE *See* **Valve.**

DIRECTIONAL DISTRIBUTION Directional split of traffic during the peak or design hour, commonly expressed as percent in the peak and off-peak flow directions.

DIRECTIONAL DRILLING 1. Curving a rotary drill

hole to avoid obstacles or to reach a side area; **2.** Technique for conduit installation involving drilling in a shallow arc using a guided steerable drilling head, generally for long spans.

DIRECTIONAL FELLING Predetermining the way a tree will land when it hits the ground. When shears are used, the wedge-shaped blade provides a lever that directs the tree into its lay.

DIRECTIONAL GROWTH Geographic direction toward which a community is expanding.

DIRECTIONAL INTERCHANGE *See* **Interchange types.**

DIRECTIONAL LIMIT SWITCH *See* **Switch.**

DIRECTIONAL START SWITCH *See* **Switch.**

DIRECTIONAL TREAD Tire tread design that is effective in only one direction of rotation.

DIRECTION OF GRAIN As applied to plastic laminates, a sanded grit pattern that can be seen on the laminate back and that is usually parallel with a printed wood-grain pattern.

DIRECTION OF IRRIGATION Direction of flow of irrigation water; usually perpendicular to the supply ditch or pipe.

DIRECTION OF ORIENTATION Predominate direction of a layer or layers.

DIRECTION SELECTING CIRCUIT That portion of a wiring diagram that determines the direction an elevator should travel to answer a call.

DIRECTION SWITCH *See* **Switch.**

DIRECT-LIFT HOIST *See* **Truck hoist.**

DIRECT LIGHTING Lights that direct all or most of their illumination downward.

DIRECT MATERIAL *See* **Material.**

DIRECT NAILING Nailing perpendicular to the face of the work.

DIRECT PERSONNEL EXPENSES Salaries, wages, and benefits of those employees directly engaged on a project.

DIRECT-PLUNGER DRIVING MACHINE *See* **Driving machine.**

DIRECT-PLUNGER ELEVATOR *See* **Elevator.**

DIRECT POLARITY Direct electrical current flowing from base metal (anode) to electrode (cathode).

DIRECT-REDUCTION MORTGAGE *See* **Mortgage.**

DIRECT SEEDING Spreading seeds over the forest seedbed by hand or by machine.

DIRECT SOLAR GAIN Warming of an area by solar energy directly entering the area.

DIRECT STRESS Stress created by compression or tension only, not involving bending or shear.

DIRECT TENSION INDICATOR Compressible-washer-type indicator capable of indicating, through the degree of plastic deformation, the achievement of a specified minimum bolt tension in a tightened structural bolt.

DIRT CAPACITY Weight of a specified artificial contaminant that must be added to the fluid to produce a given differential pressure across a filter at specified conditions. Used as an indication of relative service life. *Also called* **Contaminant capacity,** and **Dust capacity.**

DIRT WIPER Mechanism on a hydraulic cylinder that cleans the cylinder shaft as it moves in and out.

DIRTY MONEY Additional pay to a construction worker for working in difficult or dirty conditions.

DISAPPEARING STAIR Attic ladder that is hinged for raising and that folds into a compartment at the raised level.

disassmbl. *Abbreviation for:* disassemble.

disassy *Abbreviation for:* disassembly.

disb. *Abbreviation for:* disburse; disbursements; disbursing.

DISBURSEMENT *See* **Expenditure.**

DISC One or more rows of plate-shaped steel wheels that cut into the earth and roll over it to fragment and mix the soil.

disc. *Abbreviation for:* disconnect; discontinue; discontinued.

DISC BRAKE *See* **Brake.**

DISC BRAKE CALIPER Nonrotational component, including its actuating mechanism and pads, that generates frictional force on the disc.

DISC BRAKE ROTOR Flat, circular rotating member attached to the spoke wheel or hub that is contacted by the friction pads.

disch. *Abbreviation for:* discharge.

DISCHARGE 1. Flow from a pipe, culvert, channel, etc.; **2.** Rated output of a pump; **3.** Legal effect of the repayment of a debt or release from an obligation.

DISCHARGE APRON Structural member of an open-end loader assembly that overlaps the drum opening and conveys the mixed concrete from the drum to the main discharge chute during discharge.

DISCHARGE COEFFICIENT Factor used in figuring flow through an orifice. It takes into account the fact that a fluid flowing through an orifice will contract to a cross-sectional area that is even smaller than that of the orifice, and that there is some dissipation of energy due to turbulence.

DISCHARGE CONVEYOR Conveyor that transports and deposits material away from the machine.

DISCHARGED AND WET Condition of a storage battery when it is filled with electrolyte and is discharged. An initial charge is required before it is ready for use.

DISCHARGE HOSE Temporary connection between the discharge outlet of a pump and the point of delivery.

DISCHARGE PIPE Pipe connecting the discharge outlet of a pump to the point of delivery.

DISCHARGE SPOUT Outfeed portion of a portable chipper. Also called **Chute.**

DISCHARGE VALVE *See* **Valve.**

DISCHARGING ARCH *See* **Arch.**

dischgd *Abbreviation for:* discharged.

DISCOLORATION Difference of color from that which is normal or desired.

DISCONNECTING MEANS Device, or group of devices, or other means by which electrical conductors of a circuit can be disconnected from their source of supply.

DISCONNECTING SWITCH *See* **Switch.**

DISCONNECTING TRAP Intercept trap.

DISCONTINUITY 1. Interruption in the normal physical structure or configuration of a part, such as laps, cracks, seams, inclusions, or porosity; **2.** Abrupt change or break in the shape or structure of a part.

DISCONTINUOUS CONSTRUCTION Technique of sound transmission control using minimum insulating material but including construction systems involving continuity breaks in walls, floors, and ceilings.

DISCOUNT Reduction from a list price by a stated amount. There are various forms, including:

Cash: Reduction of a quoted price in consideration of a cash payment within a prescribed period.

Quantity: Reduction in the unit price of goods or services in recognition of the quantity purchased in an individual transaction.

Trade: Deduction from a listed price, applicable to purchases made by a defined class or group of purchasers.

Volume: Reduction in the selling price of goods or services in consideration of the volume of units purchased over a stated period.

DISCOUNTED CASH FLOW In evaluating investment opportunities, the various costs and benefits anticipated in future years, discounted to the present. These values are expressed by either (a) their difference, giving a net present value; (b) the benefit-cost ratio; or (c) calculating the discount rate that equated them, giving the internal rate of return.

discrp. *Abbreviation for:* discrepancy.

DISH 1. Angular clearance on the sides of a saw tooth; **2.** Shape assumed by a 'dished' or open saw; which is a saw that does not stand up straight because it has been stretched too much in the inner area for the speed at which it is run.

DISHED DIAPHRAGM Molded diaphragm in which the entire center is depressed below the plane of the rim.

DISH WHEEL Grinding wheel shaped like a dish.

disinf. *Abbreviation for:* disinfectant; disinfection.

DISINTEGRATION Reduction into small fragments and subsequently into particles. *See also* **Deterioration,** and **Weathering.**

DISK CLUTCH *See* **Clutch.**

DISK GRINDER Machine on which abrasive disks are used for grinding.

DISK LOAD CELL Mechanical load cell that has an elastic element consisting of a cup spring. The element, fixed between an abutment plate and a top yoke plate, deflects when loaded, and a dial gauge measures the distance between the abutment and yoke plates.

DISK SANDER *See* **Sander.**

DISK SCREEN Flat, bed-like sizing device consisting of rows of disks on driven shafts that rotate the disks to transport material along its length. The up-and-down motion of the material as it travels over the disks separates it, causing smaller portions to fall between the disks to a collector or conveyor below. The spacing of the disks determines the size of the material falling through.

DISK TRENCHER Machine designed for mechanical site preparation by providing continuous rows of planting spots rather than intermittent patches, as provided by patch scarifiers. Consists of scarifying steel disks equipped with teeth.

DISK TUMBLER Circular or oval-shaped disk having a rectangular hole and one or more side projections, used side-by-side in a disk-tumbler lock.

DISK WHEEL 1. Road wheel consisting of a rim permanently affixed to a round metal plate and designed to be attached to an axle and hub assembly; **2.** Grinding wheel shaped similar to a straight wheel, but usually mounted on a plate and using the side of the wheel for grinding.

dism. *Abbreviation for:* dismantle.

disp. *Abbreviation for:* displacement.

DISPATCHER Operating control of an integrated electric generating system involving operations such as:

(a) Assignment of a load to a specific generating station and other sources of supply to effect the most economical supply as the total or the significant area loads rise and fall.

(b) Control of operations and maintenance of high-voltage lines, substations and equipment, including administration of safety procedures.

(c) Operation of principal tie lines and switching.

(d) Scheduling of energy transactions with connecting utilities.

DISPATCHING DEVICE Device to either operate a signal in an elevator car to indicate when the car should leave a designated landing, or to actuate the elevator starting mechanism when the car is at a designated landing.

DISPENSABILITY Property of a grease that governs the ease with which it may be transferred from its container to its point of application. Mostly used in reference to grease dispensing systems, where it includes both the properties of pumpability and feedability.

DISPERSANT Material that deflocculates or disperses finely ground materials by satisfying the surface energy requirements of the particles; used as a slurry thinner or grinding aid.

DISPERSANT ADDITIVE Additive that helps prevent deposits by holding the insoluble products of oil oxidation and fuel combustion in suspension in an oil.

DISPERSING AGENT Agent capable of increasing

the fluidity of pastes, mortars, or coelative concretes by reduction of interparticle attraction.

DISPERSION 1. Finely divided particles of a material in suspension in another substance; **2.** Distribute or spread material in an orderly manner.

DISPLACEMENT 1. Of an engine, the total volume displaced by one piston when it moves from bottom dead center to top dead center, multiplied by the number of cylinders, measured in cubic inches or liters; **2.** Quantity of fluid that can pass through a pump, motor, or cylinder in a single revolution or stroke; **3.** Amount of motion associated with waves of vibration, measured in millimeters or inches.

DISPLACEMENT PILE *See* **Pile.**

DISPLACEMENT PUMP *See* **Pump.**

DISPLACEMENT TRANSDUCER Transducer that converts displacement to a proportional electrical signal.

DISPLAY DRAWING *See* **Presentation drawing.**

DISPLAY INDICATOR *See* **Indicator.**

dispn *Abbreviation for:* disposition.

DISPOSABLE BIT *See* **Bit.**

DISPOSABLE CONTAINER *See* **Waste container.**

DISPOSABLE ELEMENT Filter that is discarded and replaced at the end of its service life.

DISPOSABLE FILTER Filter consisting of a filter element encased in a housing that is discarded and replaced in its entirety at the end of the service life of the element.

DISPOSAL AREA Area where excavated material can be dumped.

DISPOSAL FIELD *See* **Leach field.** *Also called* **Tile bed.**

DISPOSSESS PROCEEDINGS Summary process by an owner to evict a tenant and regain possession of premises for nonpayment of rent or other breach of the conditions of lease of occupancy.

disre. *Abbreviation for:* disregard.

DISSOCIATION Breaking up of combustion products into combustibles and oxygen, accompanied by an absorption of heat. This usually occurs at high temperatures, and is one of the factors limiting the maximum temperature of the flame.

DISSOLVED AIR Air that is dispersed in a fluid to form a mixture.

DISSOLVED WATER Water that is dispersed in a fluid to form a mixture.

dist. *Abbreviation for:* distance; distant; distort; distortion; distribution; distributor; district.

DISTANCE Measure of a space or interval.

DISTANCE BETWEEN GUIDES Distance between the faces of a pair of guide rails.

DISTANCE GAUGE Gauge used to measure a repetitive dimension.

DISTANCE PIECE Small piece of nonabsorbent material placed in the glazing compound of a light to

prevent dislodgement of the compound by pressure.

DISTILLATE Fraction of crude petroleum obtained by applying heat to crude oil, and condensing and collecting the vapor driven off by the heating process.

DISTILLATE OIL Oil that has been separated from crude oil by fractional distillation.

DISTILLATION 1. Separation of fractions of a material according to their boiling point: a basic refining process in fuel and lubricant manufacture; **2.** Various test methods used to characterize the volatility and boiling range of fuels and lubricants.

disting. *Abbreviation for:* distinguish.

DISTORTION 1. Change in a normal shape or appearance; **2.** Physical manifestation of cracking and distortion in a concrete structure as the result of stress, chemical action, or both; **3.** Any change in the geometry of a pavement during compaction or under traffic; **4.** Optical effect due to variation of thickness of sheet glass.

DISTRESS Soil that is in a condition where a cave-in is imminent or is likely to occur.

DISTRIBUTED LOAD Load spread over an entire surface or along the length of a beam.

DISTRIBUTED SAMPLE Soil sample that has been thoroughly mixed and is therefore not representative of its original characteristics.

DISTRIBUTION-BAR REINFORCEMENT Small-diameter bars, usually at right angles to the main reinforcement, intended to spread a concentrated load on a slab and to prevent cracking.

DISTRIBUTION BOX 1. Box in an electrical circuit in which the main feed terminates, and from which branch circuits originate; **2.** Buried chamber that receives the effluent from a septic tank and distributes it equally to the filter drains.

DISTRIBUTION LINE Primary electrical circuit to which branch circuits are connected.

DISTRIBUTION PANEL Insulated and enclosed board from which connections are made between the main feed line and branch lines, each isolated by a fuse or circuit breaker.

DISTRIBUTION PIPE Pipe carrying water from a storage tank to a point of use.

DISTRIBUTION RESERVOIR Water storage facility that is part of a distribution system.

DISTRIBUTION STEEL Subsidiary reinforcement in a concrete slab placed at right angles to the main steel to hold it in place during concrete placement and to distribute the loads over a wider area.

DISTRIBUTION SYSTEM Electrical connections that permit the distribution of electrical energy from convenient points on a transmission or bulk power system to consumers.

DISTRIBUTION TILE Agricultural tiles, laid in rows to form a drainage bed, that receive the effluent from a septic tank.

DISTRIBUTOR 1. Device, driven at one-half engine speed (in a four-stroke cycle engine) to time and deliver high voltage surges to the spark plug(s) in proper sequence to ignite the air-fuel mixture; **2.** Means of dividing fluid flows from a single path to two or more parallel paths.

DISTRIBUTOR TRUCK Truck equipped with a heated and insulated tank from which asphaltic products can be pumped to a spray bar attached to the rear of the vehicle. *Also called* **Boot truck.**

DISTRICT *See* **Zone.**

DISTRICT MAP Map showing the boundaries of the districts into which a planning area is divided.

DISTURBANCE TIME *See* **Machine time.**

DISTURBED SAMPLE Sample of soil taken without effectively minimizing disturbance of the soil mass.

DISUSE OF LAND Authority granted some public authorities to enforce actions such as removal of debris, basic site development, clearing of unmarketable titles, etc., by the owners of disused land.

DITCH Long narrow excavation, cut or channel in earth, deeper than it is wide, often for conveying drainage or other water but also to permit installation or construction. *See also* **Trench.**

DITCH CHECK Barrier placed in a ditch to reduce the stream velocity of flowing water.

DITCHER Machine used to excavate a ditch. *See also* **Ladder ditcher,** and **Wheel ditcher.**

DITCH-LINE Location of a ditch in sidehill road construction. When working in rock, it is often necessary to drill and blast a row of short downholes after the subgrade is built to produce the ditch.

DITHER Low-amplitude, relatively high-frequency, periodic electrical signal, sometimes superimposed on a servovalve electrical input to improve system resolution; expressed by the dither frequency (Hz) and the peak-to-peak dither current amplitude.

DIURNAL Temperature, relative humidity, wind, and stability changes between daytime and nighttime.

div. *Abbreviation for:* divide; divided; divider; division.

DIVER Underwater worker supplied with air, usually under pressure by pipeline from the surface, but also from self-contained equipment that is part of the diver's equipment.

DIVERGING Dividing of a single stream of traffic into separate streams.

DIVERSION 1. Route by which traffic is made to bypass a section of regularly traveled road; **2.** Channel excavated to make a stream or river bypass, either permanently or during construction.

DIVERSION CHAMBER Chamber on a channel containing means of diverting all or part of the flow to another channel or channels.

DIVERSION DAM Barrier across flowing water to divert all or part of the water.

DIVERSION RATE Measure of the amount of waste material being diverted for recycling compared with the total amount that was previously thrown away.

DIVERSION VALVE *See* **Valve.**

DIVERTER VALVE *See* **Valve.**

DIVIDE Separate into parts, equally or unequally, temporarily or permanently, wholly separately or by partitions or dividers.

DIVIDED HIGHWAY *See* **Highway.**

DIVIDER BEAM Horizontal structural member between adjacent hoistways of a multiple-car system.

DIVIDERS Hinged instrument with two sharpened legs used to mark off equal spaces, or to mark or scribe a circle or arc.

DIVIDER STRIPS In terrazzo work, nonferrous metal or plastic strips of different thicknesses, usually embedded, to form panels in the topping.

DIVIDING WALL *See* **Wall.**

DIVISION (OF A SPECIFICATION) Organizational subdivision of the text, usually annotated for quick and individual reference.

DIVISION WALL *See* **Wall.**

dk *Abbreviation for:* dark; deck.

dkg *Abbreviation for:* decking.

dl *Abbreviation for:* deadlight.

DL *Abbreviation for:* dead load; drawing list.

D-LINE CRACKS *See* **D-cracks.**

D-LOAD Constant load that in a structure is due to the mass of the members, the support structure, and permanent attachments or accessories.

dlvr *Abbreviation for:* deliver.

dlvy *Abbreviation for:* delivery.

dly *Abbreviation for:* daily; delay; dolly.

DM *Abbreviation for:* design manual.

dmh *Abbreviation for:* demolish

dmh. *Abbreviation for:* demolished; demolition; drop manhole.

dmpr *Abbreviation for:* damper.

dn *Abbreviation for:* down.

do *Abbreviation for:* ditto.

DO *Abbreviation for:* dissolved oxygen.

DOB *Abbreviation for:* diameter outside bark.

doc. *Abbreviation for:* document; documentation.

DOC *Abbreviation for:* date of change.

DOCK 1. Platform where trucks and trailers are loaded and unloaded; **2.** Sheltered basin for landing and mooring ships. *See also* **Landing.**

DOCKBOARD Portable or fixed device for spanning the gap or compensating for difference in level between loading platforms and carriers and having adequate strength to support the passage of personnel, equipment and materials.

DOCKBUILDERS *See* **Pile crew.**

DOCK LEVELER Adjustable ramp, either part of a vehicle or built into the lip of a loading dock, that compensates for differences in height between the dock and vehicle bed.

DOCK WALL Marginal wall on a wharf against which vessels can lie and to which they are tied.

DOCUMENTARY STAMPS Tax, in the form of stamps, required in some jurisdictions on deeds and mortgages when real property title passes from one owner to another.

DOCUMENT DEPOSIT Monetary sum deposited as security when taking a set of contract documents and their bidding requirements, usually refunded to *bona fide* bidders on return of the materials in good condition within a specified time.

DOG 1. Heavy-duty latch that can stop relative motion between two parts; **2.** Pawl used as a stop on a ratchet wheel; **3.** Pointed teeth located on a chain saw body against which pressure is applied, causing them to dig into a tree or log to aid in cutting; **4.** Device attached to a work piece by means of which the work is revolved.

DOG ANCHOR Iron rod or bar with the ends bent to a right angle, used for holding pieces of timber together. *Also called* **Bitch.**

DOG BAR Vertical member, several of which are spaced across the lower part of a gate.

DOGBONE PLUG *See* **Repairs.**

DOG COCK Choker knob and hook that has pulled tight after the log has slipped out of the loop.

DOG EAR Fold made in metal roofing to form an external corner without cutting.

DOGGED OFF Condition of a rope drum when its dog is engaged.

DOGHOUSE Engine tunnel or compartment that partially or entirely houses a vehicle's engine.

DOG IRON Piece of iron rod turned at each end in a right angle. The ends are pointed so one can be driven in to one edge of the block being sawed from a log and the other can be driven into the log itself to steady the log when sawing.

DOG IT To hold the load line secure, or to secure the boom; to stop movement, secure a machine winch or other equipment, apply brakes, or do not initiate action.

DOG KINK Empty choker noose from which a log has slipped out of or broken off.

DOGLEG 1. Angle away from a straight line; crooked; **2.** (In wire rope) a permanent bend or kink caused by improper use or handling.

DOGLEG PILE Pile curved or bent in driving.

DOGLEG STAIR Stair that makes a right-angle turn about a half-landing.

DOG'S TOOTH Bricks laid with their corners projecting from the face of the wall.

dol. *Abbreviation for:* dollar.

DOLLY 1. Small wheeled carriage used to support and transport heavy components or parts around a workshop or warehouse. *See also* **Bummer; 2.** Block of hardwood, or other suitable material, placed on top of a concrete or timber pile to cushion the blows from a pile-driving hammer. *See also* **Cushion block; 3.** Unit consisting of a draw tongue, an axle with wheels, and a turntable platform to support a trailer. There are sev-

eral types, including:

> **Converter:** Auxiliary undercarriage assembly comprising a chassis, fifth wheel, and towbar, used to convert a semi-trailer to an independent trailer.

> **Load divider:** Short, frame-type trailer complete with upper coupler, fifth wheel, and undercarriage that, when coupled to a semi-trailer and tractor, carries a portion of the trailer kingpin load while transferring the remainder to the tractor fifth wheel.

> **Permanent:** Undercarriage assembly with a permanently attached turntable and towbar.

When used in conjunction with a trailing boom kit, the dolly is attached directly to the rear overhanging boom with metallic pins. A dolly distributes a crane's weight over more axles, but through the use of pressurized cylinders does not force reallocation of weights.

DOLLY BAR Steel bar used for bucking up rivets.

DOLLY ROLLER Roller used to move steel.

DOLOMITE 1. Mineral having a specific crystal structure and consisting of calcium carbonate and magnesium carbonate in equivalent chemical amounts, that are 54.27% and 45.73% by mass, respectively; **2.** Rock containing dolomite as the principal constituent.

DOLPHIN Group of piles driven close together through water and tied together so that the group is capable of withstanding lateral forces from vessels and other floating objects.

dom. *Abbreviation for:* domestic.

DOME 1. Vault of even curvature erected on a circular base. In section, segmental, semicircular, pointed, or bulbous. (A dome may also be erected on a square or polygonal base.); **2.** Square prefabricated pan form used in two-way (waffle) concrete joist floor construction.

DOME FOUNDATION Foundation to support a column in which the underlying soil is first shaped into a dome and then covered by a uniform thickness of reinforced concrete.

DOME LIGHT One-piece spherical or rectangular glass dome.

DOMESTIC WASTE *See* **Domestic refuse.**

DOMINANT Trees with crowns extending above the general level of the canopy and receiving full light from above and partly from the side; taller than the average trees in the stand with crowns well developed.

DOMINANT TREES The most numerous and vigorous species in a mixed forest.

DOMINO FALLING Placing undercuts and backcuts in a series of trees, then 'pushing' them with another tree to cause them all to fall in sequence.

DONKEY DOCTOR Heavy-duty mechanic.

DONKEY PUNCHER Spar or yarder operator.

DONKEY WINCH *See* **Winch.**

DOODLE Empty closed-end tubular section, driven into the ground within operating radius of a pile rig. A corrugated shell is lowered into the hole to facilitate inserting a mandrel, thereby allowing the use of shorter leads or longer lengths of shell. The shell and mandrel are raised together out of the doodle hole and moved to the required pile location. *Also called* **Dummy hole, Makeup pile,** and **Rat hole.**

DOODLE HOLE *See* **Shelling-up.**

DOOR Movable element for closing an opening. Doors may be solid or made up of several components or materials. A six-panel door, for instance, comprises the following:

> **Hanging stile:** Vertical member running the full height of the door, to which the hinges are attached.
>
> **Muntin:** Vertical posts separating the panels, tenoned into the rails.
>
> **Panels:** The six panels that fill the openings between the stile, muntin, and rails.
>
> **Rails:** Top, frieze, lock, and bottom horizontal members that are tenoned into the stiles.
>
> **Shutting stile:** Vertical member running full height of the door and containing the closing or locking mechanism.

DOOR ARM Metal bar extending from an elevator door operator mechanism to transmit the door opening and closing force, through one or more connecting links, to the door.

DOOR BUCK Rough door frame set in a partition or wall, especially in a masonry wall, to which the door frame is attached.

DOOR CAM Device mounted on an elevator car door that unlocks and drives the hoistway doors.

DOORCASE The visible or inner frame of a door, including the finished trim with the two jamb pieces.

DOOR CASING Finish material around a door opening.

DOOR CHECK Device to retard the movement of a closing door and to guard against its slamming or banging against the wall behind.

DOOR CHEEK Vertical member of a door frame.

DOOR CLOSER Device containing a heavy spring and arm coupled to an air- or oil-operated cylinder that closes a door at a controlled rate.

DOOR CONTACT Electric switch operated by an elevator car door in the closed position that permits the car to run.

DOOR CONTROL CIRCUIT Part of a wiring diagram that controls the operation of elevator and hoistway doors.

DOOR FRAME The case that surrounds a door, into which the door closes and out of which it opens. The frame consists of two upright pieces called jambs and the lintel or horizontal piece over the opening for the door.

DOOR FURNITURE Metal or other accessories attached to the door (hinges, closures, knobs, handles, grilles, etc.).

DOOR GIB Device at the bottom of a horizontal sliding door panel that sticks into the sill grooves and prevents the door panel from swinging in or out.

DOOR GUIDE Angle or channel that stabilizes and keeps plumb a sliding or rolling door during operation.

DOOR GUIDE RAILS Vertical tracks that guide and direct the travel of biparting or vertically sliding door panels.

DOOR GUIDE SHOE Casting mounted on the edge of a vertical sliding door panel that rides on the door guide rails.

DOOR HANGER Assembly fastened to the top of a door that supports it and allows it to slide sideways.

DOOR HEAD Upper part of the frame of a door.

DOOR HOLDBACK Device used to hold trailer doors in the open position.

DOOR JACK Frame used by carpenters to hold a door while it is being worked on prior to final installation.

DOOR JAMB Two upright pieces fitted and held together by a head to form the lining for a door opening.

DOOR LOCK Mechanical device that prevents a door from being opened.

DOOR OPEN BUTTON Electrical contact that causes elevator doors to reopen.

DOOR OPERATOR Machine mounted on an elevator car, directly above the opening, that drives the car door open and closed.

DOOR PANEL Portion of a door or gate that covers the opening and moves to uncover the opening.

DOOR POST Vertical member of a door frame.

DOOR SCHEDULE A table that lists each type of door to be included in a project by quantity and size.

DOOR SILL Horizontal, weathertight member forming the bottom of an external door frame, over which the door closes.

DOOR STOP 1. Device fitted to the door, or on the floor near the door, to hold it open as far as may be required, or to prevent the door from being opened beyond a certain amount; **2.** Strip, part of the door frame, against which a door closes.

DOOR SWITCH *See* **Switch.**

DOOR TRACK Steel guide on a door opening on which the hanger roller or door guide runs.

DOOR TRIM Casing around an interior door opening that conceals the break between the plaster or other wall covering and the door frame or jamb.

DOOR TYPES Following are some of the many ways in which doors are fashioned and configured:

> **Accordion:** *See* **Folding.**
>
> **Bifold:** Flexible door, mounted in an overhead track, that folds into a series of relatively narrow panels when opened.
>
> **Biparting:** Protective devices for hoistway openings of freight elevators consisting of two steel panels that move vertically and counterweight each other.
>
> **Calamine:** Wood core door faced on each side with sheet metal.

Casement: Side-hung doors, mounted in other than a vertical plane.

Center-opening: Door type consisting of two horizontal sliding panels that move in opposite directions.

Combination: Door with a fixed outer frame that remains mounted in a casing, and having an interchangeable inner section; typically a solid section for winter use and a screened section for summer.

Double: Matched pair of doors which, when closed, meet in the center of a single opening.

Dutch: Door divided horizontally into two roughly-equal parts, either of which may be opened independently of the other, the upper section of which is often glazed.

Emergency: Door fitted on its interior face with a crush bar that causes the door to unlock and open when pressed or struck.

Fireproof: Door fabricated of materials, and constructed in a manner conforming to the requirements of a fire rating agency.

Flush: Door having a flush face on each side; can be either solid-core or hollow-core.

Folding: Door divided into two or more panels that fold against each other when opened. Also called **Accordion doors,** and **Bifold doors.**

French: Pairs of multi-panel glazed doors side hung to an opening.

Glazed: Door having one or more glass-filled panels.

Hollow-core: Door construction that leaves voids within the interior space between the skins.

Horizontal sliding: Entrance protection for both car and hoistway (usually for passenger elevators) that move sideways.

Overhead: Door that swings vertically through an arc to store in a horizontal plane above and behind the opening.

Pocket: Sliding door built within the thickness of a wall.

Power: Door that is propelled by electric motor(s).

Revolving: Door panels mounted on a common vertical rotating axis within a circular enclosure having an opening to the interior of the building and an opening to the exterior.

Single-speed: Type of door consisting of one horizontal sliding panel.

Six-panel: Door having six panels framed by stiles, rails, and muntins.

Sliding: One or more panels that slide sideways, either over each other or into a prepared storage area.

Solid-core: Flush door of solid construction.

Swing: One or more panels hinged to swing through an arc up to 180°.

Two-speed: Type of door consisting of two horizontal sliding panels that move in the same direction.

DOOR WARNING SIGNAL Buzzer, bell, or other audible alarm used with automatic door operation to warn of imminent activation.

DOORWAY Opening for passage through a wall.

DOPE 1. Thick liquid or pasty substance; **2.** Substance used by plumbers to coat the threads of pipes before joining to ensure a tight fitting.

DOPE COAT Neat cement applied to a setting bed.

DOPE GANG Workers who apply bitumen and wrap to pipes.

DOPE MACHINE Machine that applies bitumen to coat the exterior of pipe and then wraps the pipe.

DORIC An order of architecture characterized by fluted, heavy columns with simple capitals. The Greek Doric lacks a base.

DORMER Projection in a sloping roof. Its framing forms a vertical wall suitable for windows or other opening.

DORMER CHEEK Vertical side of a dormer, gable, or window.

DORMER WINDOW Projecting upright window in a sloped roof, complete with its own roof.

DORMITORY Building or part of a building containing a number of rooms designed for people to sleep in, usually several to a room.

DOSING SIPHON Automatic siphon that discharges the contents of a dosing tank.

DOSING TANK Storage tank into which an effluent is fed and held until a specific volume has been accumulated, at which point it is automatically siphoned off.

DOT Small mound of gypsum plaster on a surface between grounds as an aid in obtaining a uniform plaster thickness.

DOTE General term used to denote decay or rot in timber.

DOT GRID Transparent sheet of film (overlay) with systematically arranged dots, each dot representing a number of area units. Used to determine areas on maps, aerial photos, plans, etc.

DOT-MOUNTED TILE Wall tile packaged in sheet form with plastic or rubber dot-shaped spacers between individual tiles. The dots are removed after installation but prior to grouting.

DOT RECORDING SYSTEM Unit, used in conjunction with an analog or digital indicating system, that can directly produce a graph showing any required linear displacements.

DOT-TYPE LP GAS CYLINDER Fuel container for liquefied petroleum gas made and inspected under the Department of Transportation regulations. When used for storage of an industrial truck's fuel supply, it may be permanently or removably affixed to the vehicle.

DOUBLE Two pieces of rotary drill rod left fastened together during raising and lowering.

DOUBLE-ACTING Working in either direction; reciprocating.

DOUBLE-ACTING BUTT Butt hinges that allow a door to swing through a minimum of 180°.

DOUBLE-ACTING CYLINDER *See* **Cylinder.**

DOUBLE-ACTING GATE Dump body gate that is hinged at both top and bottom to provide the option of opening the gate for dumping and/or spreading operations.

DOUBLE-ACTING HAMMER *See* **Pile hammer.**

DOUBLE-ACTING HOIST *See* **Truck hoist.**

DOUBLE-ACTING PISTON/SEAL Movable element in a cylinder or fluid system in which the fluid force can be applied in either direction.

DOUBLE-ACTING RAM *See* **Ram.**

DOUBLE ACTION SHEAR Mechanized cutting tool for felling trees; works like a pair of scissors: one blade is slightly offset, but both work against the other. Some work edge-to-edge.

DOUBLE BEAD Two parallel beads separated by a quirk.

DOUBLE BEND *See* **Bend.**

DOUBLE-BITTED KEY Key with cuts on both edges of its blade that can be inserted into a lock with either face engaging the tumblers.

DOUBLE-BLADE GATE Vertical sliding, counterweighted device used to provide protection on freight elevators. It consists of two panels, usually made of expanded metal. The blades telescope to reduce overhead space requirements. *See also* **Single-blade gate.**

DOUBLE BLOCK AND BLEED Two safety shutoff valves separated by a vent valve.

DOUBLE BOTTOM TRAILER *See* **Trailer.**

DOUBLE-BOX-BEAM OUTRIGGER *See* **Outrigger.**

DOUBLE BRIDGING Two rows of herringbone strutting, dividing the span of each joist into three parts.

DOUBLE BULLNOSE *See* **Brick.**

DOUBLE-CLEAT LADDER Similar in construction to a single-cleat ladder, but with a center rail to allow simultaneous two-way traffic of workers ascending and descending.

DOUBLE CLUTCHING Disengaging and engaging an engine clutch two times during a single gear shift, so as to synchronize gear speeds.

DOUBLE CORNER BLOCK Concrete masonry unit having two finished ends in addition to two finished faces.

DOUBLE-COVERAGE ROOFING Roll roofing produced with 483 mm (19 in.) or slightly more than half its surface covered with mineral granules.

DOUBLE CURTAIN Two layers of vertical and horizontal reinforcing bars in a concrete wall, one at each face. *See also* **Curtain.**

DOUBLE CURVATURE Bending condition in which end moments on a steel member cause the member to assume an S-shape.

DOUBLE-CUT *See* **File.**

DOUBLE DOOR *See* **Door.**

DOUBLE-DOVETAIL KEY Wooden key resembling two dovetail pins joined at their narrow ends, driven into a butt joint between two timbers. *Also called* **Dovetail feather.**

DOUBLE-DROP-FRAME TRAILER *See* **Trailer.**

DOUBLE-DRUM WINCH *See* **Winch.**

DOUBLE DUPLEX Two duplex dwellings joined by a dividing wall and containing a total of four dwelling units.

DOUBLE EAVES COURSE Double layer of shingles laid at the lower edge of a roof slope or vertical face. *Also called* **Doubling course.**

DOUBLE EXPANSION JOINT *See* **Joint.**

DOUBLE FACED HAMMER *See* **Hammer.**

DOUBLE-FACED SAW *See* **Saw.**

DOUBLE FACED SLEDGEHAMMER *See* **Hammer.**

DOUBLE-FLEMISH BOND *See* **Bond**

DOUBLE-FRONTAGE LOT *See* **Through lot.**

DOUBLE GLAZING Light, fabricated of two panes of glass separated by a spacer and hermetically sealed. *See also* **Triple glazing.**

DOUBLE-HANDED SAW *See* **Saw.**

DOUBLE-HEADED NAIL *See* **Nail.**

DOUBLE HEADER Structural member made by nailing or bolting two or more timbers together, used where extra strength is required in the header, as around stair openings.

DOUBLE HEADING Two headings in a tunnel that can be driven from one plant setup.

DOUBLE HUB Cast iron sewer pipe having a bell on both ends.

DOUBLE-HUNG *See* **Sash.**

DOUBLE-HUNG WINDOW *See* **Window.**

DOUBLE-LAP TILE Tile that overlaps not only the next course but the next but one below it.

DOUBLE LAYER Two layers of drywall board, used to improve fire and sound resistance or for structural considerations.

DOUBLE-LINK SYSTEM Using half-width links, in pairs, across the width of a reinforced concrete beam so as to introduce more steel, at the section, to resist shear stress.

DOUBLE LOCK Parallel canal-lock chambers separated by a sluice.

DOUBLE NAILING Technique of applying drywall board using two nails spaced approximately 50 mm (2 in.) apart every 300 mm (12 in.) to insure more firm attachment to framing.

DOUBLE PARTITION Partition constructed with two rows of studding; for sound proofing or to accommodate a sliding door.

DOUBLE-PITCH ROOF Roof having two slopes. *See also* **Mansard roof,** and **Pitched roof.**

DOUBLE-PITCH SKYLIGHT Skylight that slopes in two directions.

DOUBLE POLE SCAFFOLD Scaffold supported from the base by a double row of uprights. This scaffold is independent of support from the walls and is constructed of uprights, runners, horizontal unit bearers, and diagonal bracing. *Also called* **Independent pole scaffold.**

DOUBLE-POLE SWITCH *See* **Switch.**

DOUBLE PRESTRESSING Prestressing a concrete member in two directions, usually at right angles to each other.

DOUBLE PRIMING Blasthole containing two priming units, usually on the same time delay, placed one near the top and one near the bottom of the blasthole.

DOUBLE REBATED Door jamb rebated on both edges to permit hanging a door to open either in or out, or to mount two doors.

DOUBLE-REDUCTION AXLE *See* **Axle.**

DOUBLE-RETURN STAIR Stair with a single flight rising to a landing, from which two flights rise in opposite directions to the next floor.

DOUBLE-ROD CYLINDER *See* **Cylinder.**

DOUBLE ROOF Roof in which common rafters are carried on purlins supported by a roof truss.

DOUBLE-ROPE TRAMWAY Aerial tramway having two track cables and one continuous traction rope.

DOUBLES *See* **Trailer.**

DOUBLE-SHAFT PIER Twin piers separated by a joining structure.

DOUBLE SHEETS Two sheet piles, interlocked and tack welded, then handled together for installation and pulling.

DOUBLE SHELL HOUSE Method of house construction that creates a space in the wall for a continuous convection current between exterior and interior house envelopes.

DOUBLE-SKIN ROOF Roof consisting of an upper, weathering layer and a lower, flat layer that forms a ceiling.

DOUBLE SLOPE Surface that is sloped along both axes.

DOUBLE STEP W-shaped notch between a rafter and tie beam to reduce the possibility of horizontal shear in the tie.

DOUBLE-STRENGTH GLASS Sheet glass 3 mm (0.125 in.) or thicker.

DOUBLE SURFACE TREATMENTS Successive applications of asphalt and mineral aggregate to a roof area.

DOUBLE-TEE BEAM Precast concrete member composed of two stems and a combined top flange, commonly used as a beam but also used in exterior walls.

DOUBLE-THROW SWITCH *See* **Switch.**

DOUBLE-TIER PARTITION Framed partition two storys high.

DOUBLE TIME Contractual payment that applies once a worker has completed a stipulated number of hours, or under defined circumstances.

DOUBLE-UP Method of plastering characterized by application in successive operations with no setting or drying time between coats.

DOUBLE-WALL COFFERDAM Cofferdam enclosed by a wall consisting of two parallel lines of sheeting tied together, and with filling between them, that is usually self-supporting against external pressure.

DOUBLE-WALL CONSTRUCTION *See* **Construction types.**

DOUBLE-WALLED HEAT EXCHANGER Heat exchanger in which the collector fluid is separated from potable water by two surfaces.

DOUBLE WINDOW Two complete windows filling a single opening, one on each face, the outer unit acting as a storm window.

DOUBLE WRAP Roping arrangement whereby hoist ropes pass around a secondary sheave and back over the main drive sheave.

DOUBLING AN ANGLE Survey technique designed to eliminate instrument error; it involves taking the same angular reading twice with the telescope and base plate rotated through 180° between sightings.

DOUBLING COURSE *See* **Double eaves course.**

DOUBLING PIECE In carpentry, a tilting fillet.

DOUBLY REINFORCED BEAM Reinforced concrete beam containing extra reinforcement in the lower tensile portion, plus balancing reinforcing in the upper compression portion, so as to reduce its depth or breadth below normal design values.

DOUGHNUT 1. Large washer of any shape used to increase the bearing area of bolts and ties; **2.** Round concrete spacer with a hole in the center, used to hold bars the desired distance from the forms.

DOUGLAS FIR/LARCH Group of coniferous trees generally considered to give the highest lumber strength.

DOVETAIL Wedge-shaped part (tenon) that fits into a corresponding indentation (mortise) to form a joint.

DOVETAIL CUTTER Tool used to cut the inner and outer dovetails for joints.

DOVETAIL FEATHER *See* **Double-dovetail key.**

DOVETAIL SAW *See* **Saw.**

DOWEL 1. Pin or peg of wood, metal, etc., usually fitted into corresponding holes in two pieces to fasten them together; **2.** Steel pin, commonly a plain round steel bar, that extends into adjoining portions of a concrete construction, as at a joint in a pavement slab; **3.** Deformed reinforcing bar intended to transmit tension, compression, or shear through a construction joint.

DOWEL BAR Short length of reinforcing steel, used to connect and align slabs, and usually to transmit loads from one slab to another.

DOWEL BAR PLACER Machine, forming part of a concrete paving train, consisting of an aligning jig and a vibrating beam fitted with forked inserters.

DOWEL CAP Cardboard or plastic tube placed over one half of a dowel bar in a doweled concrete slab joint to allow the dowel bar to slide freely.

DOWEL DEFLECTION Deflection caused by the transverse load imposed on a dowel.

DOWELED JOINT Joint in a concrete slab that uses dowel bars or dowel plates to provide load transmission from one slab to the next, and that allows horizontal movement of the slab.

DOWELING JIG Adjustable clamp used to guide bits when boring holes to receive dowels. The clamp can be adjusted to varying thicknesses of stock, and the position of the bit guide can be adjusted across the width of the stock to be drilled.

DOWEL LUBRICANT Material applied to part of the surface of a dowel to reduce its bond with the concrete and permit axial movement.

DOWEL PIN 1. Pins used to align two objects together; **2.** Metal pin, usually made of brass or galvanized metal, inserted into the top of a masonry wall with the end projecting to fit into a prepared hole in the coping stone.

DOWELS Two-piece tool for lifting stones.

DOWEL SCREW Metal screw having a coarse wood thread at both ends.

DOWEL SLEEVE Cap of light metal or cardboard on one end of a dowel bar to allow free movement of an expansion joint.

DOWEL TEMPLATE Frame that outlines the dimensions for setting dowel bars into footings for columns and walls.

DOWN ANGLE Ceramic trim tile with two rounded or curved edges, used to complete an inside corner.

DOWN CAB *See* **Fixed cab.**

DOWN-CUTTING/UP-CUTTING Direction of cutter drum rotation in relation to the direction of travel and the pavement surface.

DOWN-DRAFT Draft created in a chimney when air currents enter at the top and travel down; sometimes caused by the chimney not being carried high enough in relation to an adjacent or nearby roof ridge.

DOWNDRAG *See* **Negative skin friction.**

DOWN FEED Surface grinding. The rate at which an abrasive wheel is fed into the work. *See also* **Cross feed,** and **Feed.**

DOWNGRADIENT Any point that is downslope from a facility in the direction of water flow.

DOWNHOLE Borehole drilled in a near-vertical direction.

DOWNLINE Detonating cord line running from the top of a blasthole.

DOWNPASS Chamber or gas passage placed between two combustion chambers to carry the products of combustion downward.

DOWN PAYMENT Sum, a stipulated percentage of the purchase price, paid by a purchaser to the seller on signing of an agreement of sale.

DOWN PIPE Spout or pipe to carry rainwater from a roof.

DOWNSPOUT Any connector, such as a pipe, for carrying rainwater from the roof of a building to the ground or drainage system. *Also called* **Conductor,** and **Leader.**

DOWNSTREAM In the direction of flow.

DOWNSTREAM FACE Dry side of a dam.

DOWN-THE-HOLE DRILLING *See* **Rock drilling.**

DOWNTIME Operating time lost when a vehicle or item of equipment is not available for any reason, or work is stopped on the project.

DOWNTOWN *See* **Central business district.**

DOYLE RULE Log rule that underestimates board footage in small logs and overestimates in large logs.

doz. *Abbreviation for:* dozen.

dozer *Abbreviation for:* bulldozer, angle-dozer, or shovel dozer.

DOZER SHOVEL *See* **Shovel.**

dp. *Abbreviation for:* dampproof; dampproofing; double-pole switch.

DP *Abbreviation for:* double pitched; dripproof; dual purpose; dustproof.

dpc *Abbreviation for:* dampproof course.

dpdt *Abbreviation for:* double-pole, double throw switch.

DPS *Abbreviation for:* dustproof strike.

dptrk *Abbreviation for:* dumptruck.

DPV *Abbreviation for:* dry pipe valve.

DQ *Abbreviation for:* drawing quality.

dr *Abbreviation for:* dining room; door.

dr. *Abbreviation for:* drain; drawn; drill; driveway; drum.

DRAFT 1. A margin on the surface of a stone cut approximately to the width of the chisel; **2.** Current of air created by the variation in pressure resulting from difference in weight between cooler air and warmer air in a fireplace, usually between the cooler air outside the chimney and hot gases in the flue; **3.** Resistance to movement of a towed vehicle; **4.** Degree of taper; **5.** Angle of clearance in a mold to facilitate removal of parts; **6.** Pressure differential between a furnace and the atmosphere that causes the products of combustion to flow from the furnace to the atmosphere. There are several measures, including:

> **Forced:** The positive pressure created by the action of a fan or blower that supplies the primary or secondary air for combustion.

> **Induced:** The negative pressure created by the action of a fan, blower, or ejector that is located between the furnace and the stack.

> **Natural:** The negative pressure created by the height of a stack or chimney and difference in temperature between flue gases and the atmosphere.

DRAFT CHISEL Stonemason's chisel used to cut a draft on the face of a stone.

DRAFT CONTROLLER Automatic device that maintains a uniform draft at a particular point in a system by regulating a damper to change the system resistance.

DRAFTED MARGIN Smooth, uniform recess worked into the edges of the face of a masonry block.

DRAFTSMAN One who draws plans.

DRAFT SPECIFICATIONS Specifications that establish the general quality and standards to be maintained but that lack detail as to particular application; outline model against which detailed specifications are measured. *Also called* **Outline specifications,** and **Preliminary specifications.**

DRAFT STOP Obstruction placed in a concealed space to block the passage of flame or air currents upwards, or across a building. Also called **Fire stop.**

DRAFT TUBE Metal casing by which water exits a turbine.

DRAG 1. Pulling a bucket into the digging, or the mechanism by which the pulling is done or controlled; **2.** Ground skidding term for a turn of logs; **3.** Single sled used in dragging logs; one end of the log rests on the sled, the other drags on the ground; **4.** Device for leveling roads; **5.** Application of slight pressure on drive frictions to maintain minimal line tension; **6.** Negative skin friction; **7.** *See* **Dragline.**

DRAG BIT *See* **Bit.**

DRAG BRAKE *See* **Brake.**

DRAG CABLE Line in a dragline hoe that pulls the bucket toward the shovel.

DRAG COEFFICIENT Indication of the relative aerodynamic efficiency of a vehicle.

DRAGGED WORK Rock surface that has been smoothed with a drag.

DRAGGING Glazing technique for achieving a subtle mixture of fine stripes by pulling a wide, long-bristled brush through wet glaze.

DRAGGING OFF Straightedging or rodding concrete to a grade.

DRAGLINE 1. Long boom mounted on an excavator and fitted with a specially designed bucket that can be swung out and down onto the material to be excavated, then hauled in and back toward the machine, excavating material into the bucket as it comes in; **2.** Wire rope used for pulling excavating or drag buckets. *See also* **Closing line, Crowd, Digging line,** and **Drag.**

DRAGON BEAM Horizontal timber into which the end of a hip rafter is framed, with the outer end carried on the corner of the building and the inner end placed at the angle tie.

DRAGON'S BLOOD Any of several organic materials used for coloring varnishes.

DRAGON TIE An angle tie.

DRAG ROPE Rope for pulling in the bucket during dragline operations. Also called an **Inhaul line.**

DRAG SCARIFICATION Method of site preparation that disturbs the forest floor and prepares logged areas for regeneration. Often carried out by dragging chains or drums behind a skidder or tractor.

DRAG SCRAPER *See* **Scraper.**

DRAG SHOVEL Shovel that digs by pulling toward

itself, equipped with a jack boom, live boom, hinged stick and rigidly attached or hinged bucket. *Also called* **Backacter.**

DRAG TRENCHER Ditcher whose cutters drag dirt to the surface, rather than lifting it in buckets.

DRAIN Pipe and its associated supports and appurtenances, used to conduct fluids from one point to another. *See also* **Combined drain.**

DRAINAGE 1. System for the collection and removal of water, from the ground, or from appliances; **2.** Interception and removal of water from, on, or under an area or roadway.

DRAINAGE AREA Catchment area.

DRAINAGE BASIN Total area that contributes water to a stream.

DRAINAGE CANAL Canal that drains water from an area having no natural outlet for precipitation or runoff.

DRAINAGE CHANNEL Channel sloped downstream and leading to a side-ditch formed across the path normally taken by surface water to impede and slow velocity and to divert a portion of the flow.

DRAINAGE DITCH Excavation made along the perimeter of an area to which surface water is directed, and that is graded to a slope sufficient to conduct the collected water away from the area. *See also* **Culvert,** and **Drainage swale.**

DRAINAGE EASEMENT *See* **Easement.**

DRAINAGE FACILITY Mechanical device or structure (channel, culvert, ditch, pipe, sewer, etc.) designed and placed to intercept, divert, carry, or contain surface water runoff.

DRAINAGE FILL 1. Base course of granular material placed between a floor slab and subgrade to impede capillary rise of moisture. *Also called* **Porous fill; 2.** Lightweight concrete placed on floors or roofs to promote drainage.

DRAINAGE HEAD Highest elevation in a drainage area.

DRAINAGE PATTERN Configuration in which surface runoff accumulates and disperses downstream.

DRAINAGE PIPING 1. All or any portion of a drainage piping system; **2.** Pipe that, when fitted together as part of a system, produces a smooth internal surface, especially at joints.

DRAINAGE RIGHT-OF-WAY Land assumed by a public administration for the installation of stormwater channels, sewers, or other facilities for the collection and disposal of runoff.

DRAINAGE SWALE Excavation that is shallow in proportion to its width, intended to convey surface water away from an area. *Also called* **Drainage ditch.**

DRAINAGE SYSTEM Assembly of pipes, fittings, fixtures, traps, and appurtenances used to convey sewage, clear water waste, or storm water to a public sewer or a private sewage disposal system, but not including subsoil drainage pipes. *Also called* **Building drainage system.**

DRAINAGE WELL Borehole downstream of a protected area, used to collect and control seepage so as to reduce uplift pressure.

DRAIN-BACK SYSTEM Type of freeze protection for a solar water-heating system that circulates an antifreeze solution through the collector and into a heat exchanger in the water tank, where heat is transferred from the antifreeze to the water.

DRAINBOARD Work surface adjacent to a kitchen sink and sloping into it.

DRAIN COCK Faucet at the lowest point of a water system.

DRAIN-DOWN SYSTEM Freeze protection for a solar water-heating system designed with controls that drain the collector of water when subfreezing temperatures occur.

DRAIN LINE Line returning drain fluid independently to a reservoir or vented manifold.

DRAIN PIPE Pipe, manufactured from a variety of materials and in a wide range of sizes with accompanying fittings, used underground to convey sewage and stormwater.

DRAIN TILE Terra-cotta, plastic, or concrete pipe, sometimes perforated, laid on a bed of crushed stone to drain water either from waterlogged soil to a filter material, from a drainage system, or from around the footings of a building.

DRAPED TENDONS See **Deflected tendons.**

DRAW Natural depression or swale; a small watercourse. See also **Fuel tank capacity.**

DRAWBAR Fixed or hinged bar extending to the rear of a tractor and used as a fastening for lines and towed loads, machines, or attachments.

DRAWBAR HORSEPOWER See **Horsepower.**

DRAWBAR PULL Towing force of a machine or vehicle, exerted at its coupler or equivalent in the direction of motion of the coupling point. When the towing vehicle is moving in a straight line, the drawbar pull is equal to tractive effort minus (a) rolling resistance of the tractor, (b) grade resistance of the tractor, and (c) acceleration resistance of the tractor. See also **Maximum drawbar pull,** and **Rated normal drawbar pull.**

DRAWBOARD A type of mortise-and-tenon joint made with holes (mortises) that are shaped so that the joint becomes tighter as the pins (tenons) are driven home.

DRAW BOLT Barrel bolt.

DRAWBORE Holes drilled through a tenon and a mortised piece, about 3 mm (0.125 in.) out of line, through which a tapered pin is driven to draw the pieces more closely together.

DRAW-DOOR WEIR Weir with vertical gates.

DRAWDOWN Lowering of the level of groundwater, as in dewatering an area in which an excavation is to be dug.

DRAW FILING Using a file sideways along an edge to remove cross-filing marks.

DRAWING QUALITY PLATE Plate produced from low-carbon steel, suitable for drawing into structural shapes.

DRAWKNIFE 1. U-shaped metal blade sharpened on its inner, flat face, and with handles on either end facing in the same direction as the blade; **2.** Planing or scraping tool; **3.** Curved, two-handled knife used in digging clay.

DRAW PIN Removable pin used to attach a load to a drawbar.

DRAWPOINT Location where ore, fed by gravity from a higher elevation, is discharged into a hauling unit.

DRAW RATIO Measure of the extension applied to extruded and drawn film, used in the production of a reinforcing fiber.

DRAW SHEAR Carrier-mounted, single action, anvil shear used in mechanized forest cutting operations. The blade is drawn through the tree, toward the carrier.

DRAW TONGUE Bar hinged to a towed machine with a means for attaching it to a tractor.

DRAW WORKS Power distribution and control machinery of a rotary drill.

DRAY Sled used for yarding logs.

DRAYAGE Shipment of goods between point of manufacture and place of use.

DREDGE Machine for excavating material at the bottom, or at the banks, of a body of water, the excavated material being discharged on the bank or to a scow for transport.

DREDGE BULKHEAD Bulkhead that, after being anchored, has soil excavated or dredged from the front of its base.

DREDGE LEVEL Level to which the ground on the water side of a bulkhead, wharf, or quay has been dredged. Also called **Fill bulkhead.**

DREDGING Excavating underwater, usually with floating equipment; it may be an elevator ladder, hydraulic suction, grab or dipper bucket, scraper, dragline, clam shell, or backhoe.

DREDGING WELL Opening in a dredge through which the ladder or cutter passes.

DRESS To plane one or more sides of a piece of sawn lumber.

DRESSED AND MATCHED Boards or planks machined so there is a groove on one edge and a corresponding tongue on the other, the moldings positioned so that when laid together or matched, one mating surface is exactly flush and level.

DRESSED LUMBER See **Lumber.**

DRESSED SIZE Dimensions of lumber after being surfaced with a planing machine. The dressed size is usually 13 mm to 19 mm (0.5 in. to 0.75 in.) less than the nominal or rough size.

DRESSER Tool used for dressing a grinding wheel.

DRESSING 1. Operation of squaring or smoothing stones or lumber for building purposes; **2.** Process of modifying the shape and/or texture of a grinding wheel to improve or alter its cutting action; **3.** Stone worked to a smooth face, sometimes containing a molding, used at angles and around doors and windows or as a feature.

DRESSING COMPOUND Hot or cold bitumens, used for dressing the exposed surface of roofing felt.

DRESS UP Fasten structural members into sub-assemblies prior to erection.

DRF *Abbreviation for:* data reporting form.

DRIED STRENGTH Compressive or flexural strength of refractory concrete after being dried in an oven at 105°C to 110°C (220°F to 230°F) for a specified time.

DRIER Chemical, a volatile liquid, that promotes oxidation or drying of a paint or adhesive.

DRIFT 1. Horizontal or nearly horizontal tunnel, generally following the vein of a material. *See also* **Adit**; **2.** Continued deformation under strain. *See also* **Cold flow**, and **Creep**; **3.** Lateral deflection of a building; **4.** Change in a given hardness value after a specified period of time; **5.** Round steel pin cut square at both ends, used for fastening post or frame bents; **6.** Percentage above and below an operating pressure at a constant flow rate over a specified length of time; **7.** Deposit of loose detrital material, fragments of rock, boulders, sand, gravel, clay, or other soils driven together by ice, wind, or water. *See also* **Glacial till**; **8.** *See* **Punch**.

DRIFT BARRIER Wire-rope or chain barrier across a waterway to catch driftwood.

DRIFT BOLT Metal rod driven into a smaller diameter hole in timber, used to hold two or more timbers together.

DRIFTER Air-percussion drill mounted on a column or crossbar, used for drilling underground.

DRIFTER DRILLING *See* **Rock drilling**.

DRIFT INDEX Ratio of lateral deflection to the height of the building.

DRIFTING Pulling a suspended load laterally to change its horizontal position.

DRIFT PIN Pin tapered at both ends, used for lining up holes.

DRIFT PLUG Plug driven through a soft metal pipe to remove dents and distortions.

DRILL Mechanical device for gripping, rotating, and sometimes hammering a cutting tool. There are several types, including:

Blasthole: Air percussion, cable, fusion-type, or rotary drill designed to drill blastholes.

Churn: Machine that drills a hole in ground by raising and dropping a string of drilling tools suspended by a reciprocating cable. *Also called* **Cable drill**.

Core: Rotary drill equipped with a hollow bit and core lifter, used to obtain samples of the material being drilled through.

Diamond: Rotary drill that uses a coring bit studded with industrial diamonds, used chiefly in exploratory drilling.

Earth: Auger-type drill with a bucket, used for exploratory drilling in advance of rock-earth excavation.

Fusion: Drill that burns out a blasthole by means of fuel, oil, oxygen, and cooling water delivered to a blowpipe within the hole. *Also called* **Jet piercing drill**.

Hammer: Pneumatic, compressed air-driven reciprocating drill, commonly used to drill holes in hard rock.

Jet piercing: *See* **Fusion**, above.

Percussion: Drill that hammers and rotates a steel and bit.

Press: Chuck mounted on a power-driven, variable-speed, geared shaft that can be raised or lowered vertically above a base.

Reciprocating: Drill that operates by continuously thrusting a bit against a surface. The bit is held in a chuck that allows it to rotate between blows.

Rock: Pneumatic or electric drill for making holes in rock for blasting or other purposes.

Rotary: 1. Drilling machine that bores holes by the rotation of a kelly bar driving continuous-flight augers or a rotary table, etc.; **2.** A wet rotary drill rig using high-pressure water to open a hole for installation of a mandrel-driven pile.

Spudding: Drill that makes holes by lifting and dropping a chisel bit.

Track: Self-contained rock drill mounted on a self-propelled, tracked chassis.

Turbodrill: Rock drill used at depths below two miles.

Twist: Drilling tool used to bore holes in wood and metal.

Well: Churn drill used to develop a water well, usually truck-mounted.

DRILLABILITY Relative ease with which a hole can be drilled in a certain rock type; factors such as the rock's mineral composition, hardness or compressive strength, and degree of fracturing and weathering all have influence.

DRILL-BLASTER Rock-drill operator who also carries out loading and blasting duties.

DRILL COLLAR Thick-walled drill pipe, used immediately above a rotary bit to provide additional weight.

DRILLED SHAFT *See* **Caissoon pile**.

DRILL FEED 1. Mechanism that pushes a drill tool into a hole; **2.** Measure of the speed of advance of a drilling tool.

DRILLED-IN-CAISSON *See* **Pile**.

DRILLED PIER/PILE *See* **Pile**.

DRILLER'S STROKE Height of fall of a drop hammer weight, 63.5 kg (140 lb) or more, used to drive a casing or soil sampling tool. The weight is usually raised by wrapping a rope around a powered and continuously rotating spool or cat-head. The weight is raised by holding the rope taut, thus raising the hammer, then allowing slack in the rope to let the hammer fall to hit either the casing or the sampler rod. This distance from the top of the casing to the height before release is called the fall or driller's stroke. *See also* **N Value**.

DRILLING BUCKET *See* **Bucket auger.**

DRILLING CORE Exploratory drilling that includes cutting cylinders and bringing them to the surface for analysis.

DRILLING FLUID Water- or air-based fluid used to decrease friction between the drill string and the hole, cool the bit, and remove cuttings from a drill hole.

DRILLING LINE Cable that supports and manipulates the tools of a churn drill.

DRILLING LOG Detailed daily record of rocks passed through in the drilling of exploratory holes and blastholes.

DRILLING MUD Fluid mixture of clayey soil and water, or commercial mixture of sodium montmorillonite (bentonite) and a clay mineral.

DRILLING RIG Machine for drilling holes in earth or rock.

DRILLINGS Cuttings produced from the process of drilling.

DRILL JUMBO Movable frame on which drill positioners, jibs, and drills are mounted.

DRILL PIN Round pin protruding from the inside of a lock case to receive a barrel key. *Also called* **Barrel post.**

DRILL PIPE One or several sections of rotary drilling string that connect the kelly with the bit or collars.

DRILL POSITIONER Mechanical control for moving, rotating, and controlling jibs.

DRILL PRESS *See* **Drill.**

DRILL ROD Lengths of steel used to transmit impact energy and rotation from the striker bar to the drill bit; typically 3 m (10 ft) or 3.7 m (12 ft) long and hollow for the flow of flushing air (or water).

DRILL STEEL Hollow steel that connects a percussion drill with the bit.

DRILL STRING All revolving parts below ground of a rotary drill; or the tools hanging from the drilling cable of a churn drill.

D-RING D-shaped ring, often mounted on the periphery of a stake body or trailer, hinged on the straight leg, used as a tie down for ropes, chains, or straps securing a load.

DRIP Projection to shed water, preventing it from reaching the part below. *Also called* **Beak.**

DRIP CAP 1. Molding placed on the exterior top side of a door or window so as to cause rainwater to run off, or drip; **2.** Groove on the underside of a projecting part, such as a sill, to cause moisture to drop from a surface; **3.** Molding placed where the exterior sheathing and foundation meet, that throws water away from the foundation.

DRIP CHANNEL Throat of a drip.

DRIP EDGE L-shaped aluminum or galvanized metal molding fastened around the perimeter of a roof to protect the edges of the roof sheathing from weathering.

DRIP LINE Line projected on the ground that corresponds to the limit of a tree's foliage.

DRIP MOLD Projecting molding arranged to throw off rain water from the face of a wall.

DRIP PAN Pan below a piece of machinery, placed to catch oil drips.

DRIPSTONE Projecting molding over the heads of doorways, windows and archways, to throw off rain. *Also called* **Hood-mold**, and, when rectangular, **Label.**

DRIVE 1. Digging or making a tunnel; **2.** To hammer a pile.

DRIVE A TREE To push one tree to the ground by felling a second tree into it.

DRIVE AXLE *See* **Axle.**

DRIVE CAP Steel accessory placed over a pile to prevent damage from driving. It is suspended beneath a hammer by cables and contains a well or recess on top for cushion material and for seating the anvil, if used. The bottom is formed to accept a specific shape pile, along with it's cushion, if used. The outside incorporates lugs or an insert slot for attaching to the lead system. *Also called* **Anvil block, Bonnet, Cap block, Driving head, Follow block, Helmet, Rider cap,** and **Shield.**

DRIVE CHAIN Chain used to transfer power from a horizontal traction shaft to a track-drive sprocket.

DRIVE CHAIN SPROCKET Driving or driven sprocket in a chain drive system.

DRIVE CLAMP Collar fitted to a churn drill enabling it to be used as a hammer to drive pipe casing.

DRIVE-FIT SLEEVE Steel collar used for splicing pipe piles.

DRIVELINE Total system of drive components between the transmission and axles, including flanges, weld yokes, slip yokes, U-joints, and propeller shafts.

DRIVE MACHINE Power unit that raises or lowers a hoisting device.

DRIVEN DISK Clutch disk that contains the friction material and which is splined to the transmission's input shaft, connecting the flywheel to the transmission.

DRIVEN PILE Pile that is hammered or driven into the ground.

DRIVER-CONTROLLED, FULL-LOCKING MAIN DIFFERENTIAL *See* **Differential.**

DRIVE ROLL *See* **Compression roll.**

DRIVER One of the upper set of spring-activated pins in a pin-tumbler cylinder lock.

DRIVE SCREW Type of wood screw that can be driven with a hammer.

DRIVE SHAFT 1. Shaft or shafts leading from the transmission that connect the axle differential to supply power from the engine to the drive wheels; **2.** Excavation from which trenchless technology can be launched for the installation of conduits.

DRIVE SHEAVE *See* **Sheave.**

DRIVE SHOE *See* **Pile point.**

DRIVE SLEEVE Splice for joining two lengths of pipe pile, consisting of an outside sleeve with an inner flange, shaped with tapered shoulders to compress the pipe ends and hold the pile lengths together by friction.

DRIVETRAIN All components used to propel a vehicle.

DRIVE UNIT Motor and gear reduction unit that forms a drive mechanism.

DRIVEWAY Private road giving access from a public way to a building on abutting grounds.

DRIVING BAND Steel band placed around the head of a timber pile to prevent it from brooming while being driven.

DRIVING CLEARANCE Minimum distance from the center of a rivet to any other feature.

DRIVING CRITERIA 1. Requirement for resistance of a pile to penetration, stated in blows per increment of depth; **2.** Required tip elevation of a pile. *See also* **Refusal.**

DRIVING FORCE Those forces in a system that tend to cause failure.

DRIVING FORMULA *See* **Dynamic formula.**

DRIVING HEAD *See* **Drive cap.**

DRIVING HOME Placing a fastener into its final and finished position.

DRIVING LOG Field record of each driven pile, including location number, hammer model, pile type, blow count per unit of penetration, final resistance, driven length, etc. *Also called* **Log,** and **Pile log.**

DRIVING MACHINE Power unit that raises and lowers an elevator, material lift, dumbwaiter car, or that drives an escalator, inclined lift or moving walk. *Also called* **Elevator driving machine.** There are several types:

   Belt driven: Indirect-drive machine connected by a belt or belts to its electric drive motor.

   Chain drive: Indirect-drive machine connected to its electric drive motor by a chain.

   Direct drive: Electrically driven machine, the motor of which is directly connected to the drive sheave, drum, or shaft.

   Direct plunger: Hydraulic driving machine in which the plunger or cylinder is directly attached to the car frame or body.

   Electric drive: Machine where the energy is applied by an electric motor that incorporates, in addition to the motor, the brake, driving sheave, or drum and any connecting gears, belt, or chain.

   Elevator: Geared traction machine, used for elevators of 450 fpm capacity or less in which the drive sheave is connected to the motor through a gear train; a gearless traction machine is used for elevators of up to 1,200 fpm capacity in which the drive sheave is directly connected to the motor, which is controlled by variable voltage.

   Geared drive: Direct-drive machine in which the energy is transmitted from the motor to the driving sheave, drum, or shaft through gearing.

   Geared traction: Geared-driven traction machine.

   Gearless traction: Traction machine, without intermediate gearing, that has the traction sheave and the brake drum mounted directly on the motor shaft.

   Hydraulic drive: Machine in which the energy is applied by means of liquid under pressure in a cylinder equipped with a plunger or piston.

   Indirect drive: Electric driving machine, the motor of which is connected indirectly to the driving sheave, drum, or shaft by means of a belt or chain through intermediate gears.

   Roped hydraulic drive: Hydraulic driving machine in which the piston is connected to the car with wire ropes.

   Screw drive: Direct-drive or indirect-drive electric machine the motor of which raises and lowers a vertical screw through a nut, with or without gearing. The upper end of the screw is connected directly to the car frame or platform.

   Traction drive: Direct-drive machine in which movement of the car is obtained through friction between the suspension ropes and a traction sheave.

   Winding drum: Geared-drive machine in which the suspension ropes are fastened to, and wind on, a drum.

   Worm-gear: Direct-drive machine in which the motor energy is transmitted to the driving sheave or drum through worm gearing.

DRIVING POINT *See* **Pile point.**

DRIVING SHOE *See* **Pile point.**

drn *Abbreviation for:* drain.

drnbd *Abbreviation for:* drainboard.

drng. *Abbreviation for:* drainage.

DROOP ENGINE SPEED Difference between the speed of an engine, when rated load is applied, and the speed of the engine running at no load, with a fixed governor setting.

DROOP SCREW Screw used to control the volume of fuel delivered to an engine under overload.

DROOP-TYPE GOVERNOR *See* **Governor.**

DROP 1. Vertical pipe connected to a supply above and feeding an automatic sprinkler system piping below; **2.** Structure for vertically dropping water in a conduit to a lower level and simultaneously dissipating energy; **3.** Vertical measure of a cliff face; **4.** One of five basic conditions characterizing the lay of a tree. It exists when one end of a log to be cut is supported and the other is not. When the run has been completed the supported end will remain stationary; the unsupported end will drop to, or part pay to the ground. *See also* **End pressure, Bottom bind, Side bind,** and **Top bind.**

DROP APRON Metal strip fixed vertically at roof eaves, verges, and gutters and held by a lining plate.

DROP ARCH *See* **Arch.**

DROP AXLE *See* **Axle.**

DROP BALL *See* **Headache ball.**

DROP-BOTTOM BUCKET Container, moved around a site suspended from the boom of a crane, by which concrete is delivered to the point of placement, and that discharges its contents when its lower end comes in contact with the surface on which the concrete is to be deposited or has previously been deposited.

DROP CEILING *See* **Suspended ceiling.**

DROP-CENTER LOWBOY AND REMOVABLE GOOSENECK *See* **Trailer.**

DROP-CENTER TRAILER *See* **Trailer.**

DROPCHUTE Device used to confine or direct the flow of a falling stream of fresh concrete. *See also* **Articulated dropchute,** and **Flexible dropchute.**

DROP CLOTH Protective sheet used during interior decorating.

DROP CONNECTION *See* **Drop manhole.**

DROP ELBOW Pipe elbow with ears to allow fastening to a surface.

DROP ESCUTCHEON Pivoted metal plate that covers a keyhole when a key is not in the lock.

DROP-FORGING Process of forming metal by taking a heated blank, placing it in a suitable mold, and forming it under high impact into an unfinished (that is, unmachined) piece or part.

DROP-FRAME TRAILER *See* **Trailer.**

DROP HAMMER *See* **Pile hammer.**

DROP-IN ANCHOR Attachment device that is installed by drilling a hole in concrete, installing the anchor, and securing it using a 'setting tool.'

DROP-IN BEAM Precast concrete element simply supported on adjacent cantilevered elements.

DROP INLET Catch basin with the top set lower than the surrounding pavement.

DROP L L-shaped pipe fitting fabricated with ears that permit it to be attached directly to the building frame.

DROP LINE Suspended line, independent of the work platform, provided for direct attachment to the user's body belt, lanyard, or deceleration device.

DROP MANHOLE Vertical branch drain connection at a manhole. *Also called* **Back drop,** and **Drop connection.**

DROP NIPPLE Vertical nipple connecting the branch line above an automatic sprinkler system to a pendent sprinkler below.

DROP-OFF BOX *See* **Waste container.**

DROP PANEL Thickened structural portion of a flat slab in the area surrounding a column, column capital, or bracket, in order to reduce the intensity of stresses.

DROP-PANEL FORM Retainer or mold so erected as to give the necessary shape, support, and finish to a drop panel.

DROPPING POINT Temperature at which a drop of material falls from the orifice of the test apparatus.

DROP-SIDE DECK Portion of a lowbed or drop deck trailer loading deck outside the main frames that is lower than their tops.

DROP SIDING Weather-boarding that is rabbeted and overlapped.

DROP SYSTEM Plumbing heating circuit where the flow pipe rises to its maximum elevation, feeding downward through branches as it travels vertically to a return main.

DROP TEE Tee-shaped pipe fitting fabricated with ears that permit it to be attached directly to the building frame.

DROP WIRE Electrical cable dropped from the nearest pole to provide electrical service to a building.

DROP ZONE Wet or dry area where a helicopter delivers logs from the logging site.

DR PT *Abbreviation for:* door part.

drs. *Abbreviation for:* dressed.

DRUM 1. Wall, circular in plan, carrying a dome; **2.** Round stone block, part of a column; **3.** Truck-mounted concrete mixing cylinder; **4.** Cylindrical flanged barrel, either of uniform or tapering diameter, on which a rope is wound either for operation or storage; its surface may be smooth or grooved. *See also* **Winch.** *Also called* **Barrel; 5.** Cylindrical container, usually with one or more ridges deformed into its perimeter and spaced along its length as reinforcing; **6.** Container having a capacity of less than 230 L (60.7 gal) but more than 30 L (7.92 gal); **6.** Rotating cylindrical member used to transmit compaction forces to soil or other surface materials. *Also called* **Roll.**

DRUM BARREL Spool around which cable is wound.

DRUM BRAKE *See* **Brake.**

DRUM COUNTERWEIGHT On winding-drum machines, a counterweight that balances all or part of an elevator car. Also called a **Machine counterweight.**

DRUM DEBARKER *See* **Harvesting machines (single function).**

DRUM GATE Arc-shaped spillway gate.

DRUM HANDLER Equipment attachment to engage one or more drums. Certain models may also manipulate drums.

DRUM HOIST *See* **Winch.**

DRUM LAGGING *See* **Lagging.**

DRUM MACHINE *See* **Winch.**

DRUM MILL Long, inclined steel drum that rotates and grinds solids in its rough interior; smaller ground material drops through holes near the end of the drum and larger material drops out of the end.

DRUM MIX PLANT Asphalt mixing plant that combines both drying and mixing functions in a single drum.

DRUM ROTATION INDICATOR Device, used to indicate winch drum motion, that can also be used to monitor drum speed.

DRUM SANDER *See* **Sander.**

DRUM SCREEN Rotating cylindrical screen set in a channel through which raw sewage flows, used to trap solid material, which is then scraped off above the water line.

DRUM TRAP Trap consisting of a cylinder with its axle vertical.

drwl *Abbreviation for:* drywall.

drwn *Abbreviation for:* drawn.

DRY 1. Transition from a liquid to a solid that occurs after paint is deposited as a film and which includes the evaporation of solvents and any chemical changes; **2.** Absence of tack; no adhering properties.

DRY AREA Roofed space between the basement wall of a building and a facing retaining wall.

DRY (AUTOMATIC SPRINKLER) SYSTEM *See* **Automatic sprinkler system.**

DRY-BATCH WEIGHT Weight of the materials, excluding water, used to make a batch of concrete.

DRY BONDING Laying ashlar or masonry units without mortar.

DRY-BULB TEMPERATURE Air temperature as indicated by any type of thermometer not affected by the water vapor content or relative humidity of the air.

DRY-CAST PROCESS Process for producing concrete products such as pipe that uses low-frequency, high-amplitude vibration to consolidate dry-mix concrete in the form. *See also* **Centrifugal process, Packerhead process, Tamp process,** and **Wet-cast process.**

DRY CONSTRUCTION Building without plaster or mortar, including the use of prefabricated units.

DRY DENSITY Weight of dry material in a soil sample after drying at 105°C (221°F).

DRY-DENSITY/MOISTURE RELATIONSHIP Relationship between the dry density and the moisture content of soil for a specific amount of compaction.

DRY DOCK Dock into which a vessel can be floated, equipped with gates, and from which the water can be pumped to leave the vessel in a dry basin.

DRY EDGING Rough edges and corners of glazed ceramic ware due to insufficient glaze coating.

DRYER Machine that dries material to a specified moisture content using hot air.

DRY FACE Condition where the outer face fines or flakes of a composite board readily fall or flake off.

DRY FILM LUBRICANT Any class of lubricants wherein the reduction of friction and wear during sliding is caused by making the shearing take place within the crystal structure of a material with low shear strength in one particular plane. Examples include graphite, molybdenum disulfide, and certain soaps.

DRY FILM THICKNESS Thickness of a cured membrane.

DRY FLUE GAS Gaseous products of combustion exclusive of water vapor. Separation of the vapor from the flue gas (a practical impossibility) is a theoretical concept used in combustion calculations.

DRY GALVANIZING Process in which steel is first fluxed in hot ammonium chloride, hot-air dried, and passed through a bath of molten zinc.

DRYING Removal by evaporation of uncombined water or other volatile substance.

DRYING CREEP Creep caused by drying. *See also* **Basic creep.**

DRYING DISCOLORATION Unwanted variation in color tone of concrete visible some time after striking the formwork.

DRYING OIL In painting, an oil that forms a tough, elastic film when exposed to air.

DRYING SHRINKAGE Shrinkage resulting from loss of moisture.

DRY JOINT Joint where there is no actual connection between two structures or members so as to allow relative movement.

DRY JOINTED Laid without mortar.

DRY KILN Heated, sealed chamber in which wood is artificially seasoned.

DRY LINER *See* **Cylinder liner.**

DRY LOCATION Location not normally subject to dampness or wetness. A location classified as dry may be temporarily subject to dampness or wetness, as in the case of a building under construction.

DRY MASONRY Block or brick laid without mortar.

DRY MIX CONCRETE 1. Concrete, mortar, or plaster mixture, commonly sold in bags, containing all components except water; **2.** Concrete of near-zero slump; **3.** Concrete of very low water content, used in the dry-cast process. *See* **Dry-cast process.**

DRY MIXING Blending of the solid materials for mortar or concrete prior to adding the mixing water.

DRY-MIX SHOTCRETE Shotcrete in which most of the mixing water is added at the nozzle.

DRY PACKING 1. Placing of zero-slump, or near zero-slump concrete, mortar, or grout by ramming into a confined space; **2.** Pea gravel or similar material blown in between the lagging and excavated surface to provide support.

DRY-PENDENT SPRINKLER Sprinkler used in a pendent position on a dry-pipe system and designed to prevent water from being pocketed in the drop from branch line to sprinkler.

DRY-PIPE SPRINKLER SYSTEM *See* **Fire sprinkler system.**

DRY-PIPE VALVE *See* **Valve.**

DRY-PRESS Casting artificial stone with a very dry mix that permits early removal of the mold.

DRY-PRESS BRICK *See* **Brick.**

DRY PROCESS In the manufacture of cement, the process in which the raw materials are ground, conveyed, blended, and stored in a dry condition. *See also* **Wet process.**

DRY RETURN Return pipe in a steam heating system that carries both condensed water and air. *See also* **Wet return.**

DRY-RODDED VOLUME Bulk volume occupied by a dry aggregate compacted by rodding under standardized conditions, used in measuring the unit weight of aggregate.

DRY-RODDED WEIGHT Weight per unit volume of dry aggregate compacted by rodding under standardized conditions, used in measuring the unit weight of aggregate.

DRY RODDING In measurement of the weight per unit volume of coarse aggregate, the process of consolidating dry material in a calibrated container by rodding under standardized conditions.

DRY ROT Decay of lumber due to the attack of certain fungi.

DRY-RUNNING Running with liquid present at the seal surfaces.

DRY SAMPLING Method of sampling soil by augering a hole in the ground with a sampler or sample spoon attached to the end of an auger. The object is to obtain a complete and undisturbed sample of the natural soil for analysis. *Also called* **Core boring.**

DRY-SAND METHOD Technique for determining the field density of soils.

DRY SCRUBBER Emission control device used to neutralize acid gases. A mixture of lime and water is evaporated in a large cylinder through which the combustion gases pass.

DRY-SEAL PIPE THREAD Pipe threads in which sealing is a function of root and crest interference. *See also* **Pipe thread,** and **Tapered pipe thread.**

DRY-SET MORTAR Water-retentive hydraulic cement mortar, usable with or without sand.

DRY SHAKE Dry mixture of hydraulic cement and fine aggregate (either natural or special metallic) that is worked evenly into the surface of concrete flatwork before the time of final setting in order to impart a required surface, and then floated. The mixture may or may not contain pigment. *Also called* **Dry topping,** and **Monolithic surface treatment.**

DRY SLEEVE Cylinder wearing surface inserted into an engine block that makes no contact with the engine coolant.

DRY SOIL *See* **Soil types.**

DRY SPOT Small area on the face of a ceramic tile that has been insufficiently glazed.

DRY-STONE WALL Wall built without mortar, the stones being dry fitted together.

DRY TACK Property of some adhesives to adhere on contact to themselves.

DRY TAMP PROCESS Final placing of concrete or mortar by hammering or ramming a relatively dry mix into place.

DRY TAPE Application of drywall tape with adhesives (other than with jointing compound).

DRY TOPPING *See* **Dry shake.**

DRY VALVE *See* **Valve.**

DRY VENT Any vent that does not carry waste water.

DRY-VOLUME MEASUREMENT Measurement of the ingredients of grout, mortar, or concrete by their bulk volume.

DRYWALL Fabricated panel of gypsum plaster faced with paper on both sides. *Also called* **Gypsum board.**

DRYWALL CONSTRUCTION *See* **Construction types.**

DRYWALLL NAIL *See* **Nail.**

DRYWALL SCREWS Special fasteners, characterized by a very coarse thread about a slender shank, used for attaching drywall panels.

DRY WEATHER FLOW Sewer flow over 24 hours of dry weather.

DRY WELL 1. Covered pit with open-jointed linings, often filled with rocks, through which drainage from roofs, basement floors, or areaways may seep or leach into the surrounding soil; **2.** Well that was drilled to obtain water, but that did not penetrate an aquifer.

DRY-WOOD TERMITES *Kalotermes:* termites that live in wood.

ds. *Abbreviation for:* downspout.

DS *Abbreviation for:* disconnecting switch; double strength; drop siding.

DSA *Abbreviation for:* double strength A quality.

DSA-B *Abbreviation for:* double strength A and B quality.

DSB *Abbreviation for:* double strength B quality.

D SELECT *See* **Lumber grades.**

dspl *Abbreviation for:* disposal.

dsw. *Abbreviation for:* door switch.

dt *Abbreviation for:* double throw switch.

DT *Abbreviation for:* drain tile; dummy trim; dust tight.

DTA *Abbreviation for:* differential thermal analysis.

DT&G *Abbreviation for:* double tongue and groove.

dtd *Abbreviation for:* dated.

DUAL-BEAD TIRES Heavy-service tires using two or more sets of bead wires in each bead.

DUAL-DRIVE TANDEM AXLE *See* **Axle.**

DUAL FUEL SYSTEM Engine design where the fuel used can be changed from one type to another. For instance, an engine may use diesel fuel in one mode, and LNG (liquid natural gas) in another.

DUAL PEDAL BRAKING System used on some heavy equipment (wheel loaders, for instance) and consisting of two pedals: a brake only pedal and a transmission neutralizer/brake pedal. By depressing the brake-only pedal the operator stops the machine with the transmission engaged. By using the combined transmission neutralizer and brake pedal, the transmission is disengaged and brakes are applied. This allows the operator to maintain high engine rpm for good hydraulic response.

DUAL POROSITY ELEMENT Element that contains two media of different porosity in parallel.

DUAL POROSITY FILTER Filter that contains two media of different porosity offering parallel flow paths to the fluid.

DUAL-RATE CHARGER Automatic battery charger that is capable of maintaining starting batteries at a reduced rate, and then switching to a high-charge rate to rapidly recharge discharged batteries.

DUAL SETTING THERMOSTAT Thermostat that offers two temperature settings, usually for day and night, respectively.

DUAL SPACING Lateral distance between tire centerlines in a dual tire mounting.

DUAL-STROKE CYLINDER *See* **Cylinder.**

DUAL SYSTEM Two-pipe plumbing system.

DUAL VENT Vent connecting at the junction of two fixture branches, serving as a back vent for both.

DUAL VIBRATION Application of synchronous vibrators to each end of a precast concrete block-making mold to increase production speed.

DUBBING OFF THE END Narrow tapered condition on one end of a sanded particleboard panel.

DUBBING OUT Rough plaster formation of a cornice or other elaborate form before the finishing coat is run.

DUCK Term applied to a wide range of medium- and heavy-weight woven fabrics.

DUCKBILL SNIPS *See* **Tin snips.**

DUCKBOARD *See* **Cat ladder.**

DUCT 1. Tube, channel, or canal through which a gas, liquid, etc., moves; **2.** Pipe or conduit with wires or cables running through it; **3.** Hole formed in a concrete member to accommodate a tendon for post-tensioning.

DUCT FORMER Plastic or steel former to make circular, oval, or rectangular ducts in precast concrete units.

DUCTILE Refers to metals that are relatively easily formed into shapes and that can sustain not less than 5% elongation before fracturing. Especially refers to cast ductile iron that has less tendency to crack when subjected to high stress.

DUCTILE IRON PIPE *See* **Pipe.**

DUCTILITY Property of a material by virtue of which it may undergo large permanent deformation without rupture.

DUCTILITY FACTOR Ratio of the total deformation at maximum load to the elastic-limit deformation.

DUFF Partially decomposed organic material of the forest floor beneath the litter of freshly fallen twigs, needles, and leaves.

DULLING EFFECT Result of incompatible solvents, it shows up on a finished particleboard panel.

dum. *Abbreviation for:* dummy.

DUMBWAITER Internal, nonpassenger elevators designed to convey freight and other commodities between floors of a multistory building.

DUMMY 1. Tree rigged to raise a spar tree for use in yarding; **2.** Shapes, generally of wood, employed with lead batteries to fill the space left in a tray when the specified number of cells is less than the tray's capacity; **3.** Zinc or lead hammer, used for working soft stone.

DUMMY CYLINDER Lock cylinder containing one operating mechanism.

DUMMY FLUE Flue lining installed at the top of a chimney that is not connected to any source of heat, used to balance the design of a chimney stack.

DUMMY HOLE Empty tubular section that is driven into the ground within the operating radius of a pile rig, into which a shell or thin-wall pipe requiring a full-length mandrel for driving is lowered. Following insertion of the mandrel, the pile and mandrel are raised together into the leads of the driver and moved to the required location. *See also* **Doodle, Makeup pile,** and **Rat hole.**

DUMMY JOINT Joint extending to one-third of the depth of a concrete slab to form a plane of weakness that predetermines the position of possible cracks.

DUMP 1. Area where logs are off-loaded on land or into water; **2.** Unmonitored, unprepared land area where unrestricted unloading of refuse is illegally done.

DUMP ANGLE Degree of slope attained by a dump body at the top of its lift, usually measured in degrees from the horizontal plane of the truck axis.

DUMP BODY *See* **Truck.**

DUMPER 1. Rubber-tired vehicle with two large wheels in front and two smaller wheels behind, with a front, or sideways dumping hopper positioned over the front wheels. *Also called* **Dumpster; 2.** Attachment for dumping containers. Various types are available to suit container and operational requirements.

DUMP HEIGHT Vertical distance from the ground to the lowest point on a bucket cutting edge or teeth when the bucket is at a 45° dump angle.

DUMPLING Ground temporarily left in the middle of an excavation to serve as an abutment for timbering to surrounding trenches.

DUMPSTER Large metal box-like container used to hold garbage and other refuse at its point of generation, and equipped with means that enable it to be seized, held, lifted, rotated, and emptied into a mobile refuse collection vehicle, itself equipped with the means to effect transfer of the waste. *Also called* **Dumper.**

DUMP TIME Time taken to empty the body, bucket, bowl, etc., of mechanical equipment. *See also* **Hydraulic cycle time.**

DUMP TRAILER *See* **Trailer.**

DUMPY LEVEL Surveying telescope and level tube rigidly attached to a vertical spindle, complete with a means to level the instrument, and free to rotate on a horizontal plane.

DUNAGAN ANALYSIS Method of separating the ingredients of freshly mixed concrete or mortar to determine the proportions of the mixture.

DUNE Mound or rise of sand or other fine-grain material deposited by wind action.

DUNNAGE Loose packing put around a cargo for protection.

DUNTER 1. Monumental mason who prepares large surfaces for polishing using a pneumatic surfacing machine; **2.** Pneumatic stone surfacing machine.

DUNTING Cracking that occurs in fired ceramic bodies due to thermally induced stress.

dup. *Abbreviation for:* duplicate; duplication.

DUPLEX 1. Building containing two dwelling units, separated vertically or horizontally by party walls/floors, each having independent access; **2.** Assembly of two filters with valving for selection of either or both filters.

DUPLEX APARTMENT Apartment on two floors connected by a private stair.

DUPLEX CABLE Two electrical conductors, both insulated, in a common insulating cover that may also contain an uninsulated conductor. Also called **Twin cable.**

DUPLEX FUEL FILTER Second filter, in addition to a primary filter.

DUPLEX-HEAD NAIL *See* **Nail.**

DUPLEX OUTLET Electrical wall outlet having two plug receptacles.

DUPLICATION MOLD Mold made by casting over or duplicating another article.

DURABILITY Ability of a material to withstand conditions of service.

DURABILITY FACTOR Measure of the change in a material property over a period of time as a response to exposure to an influence that can cause deterioration, usually expressed as a percentage of the value of the property before exposure.

DURALUMIN Very high-strength, but lightweight alloy of aluminum, copper, magnesium and manganese, iron, and silicon.

DURATION Estimated time to perform an activity.

DUROMETER Instrument for measuring the hardness of rubber and plastic compounds.

DUROMETER HARDNESS Arbitrary numerical value that measures the resistance to indentation of the blunt indentor point of the durometer.

DUST 1. Particles of solid material, usually 100 microns or less, suspended in air; **2.** Particles formed by other than combustion processes.

DUST CAPACITY *See* **Dirt capacity.**

DUST FREE Stage in paint drying after which dust will not stick to the finish.

DUSTING 1. Development of a powdered material at the surface of hardened concrete; **2.** Application of dry portland cement to a wet floor or deck mortar surface.

DUSTING BRUSH Soft-hair brush used to remove dust from a surface prior to painting.

DUST LAYING OIL Low viscosity oil that seals a surface and inhibits the generation of surface dust.

DUST LOADING Particulate content of air or flue emissions.

DUST OF FRACTURE Rock dust created during the production process or handling of aggregate.

DUST PALLIATIVE Asphalt emulsion sprayed onto gravel roads to control dust generation and inhibit erosion.

DUST SHIELD Disk or plate placed to keep debris from a brake assembly.

DUTCH AUCTION Bidding process in which the asking price is gradually reduced until someone makes a qualifying bid.

DUTCH BLOCK Y-type rigging system, used to operate a diesel hammer with a single line that supports both the lead and hammer.

DUTCH BOND *See* **Bond.**

DUTCH CONE *See* **Penetrometer.**

DUTCH DOOR *See* **Door types.**

DUTCH GABLE Gable of curved form, sometimes crowned by a pediment.

DUTCH LAP Application of shingles so that each has a side lap and a head lap.

DUTCH LEADERS Pile hammer leads composed of two pipes loosely coupled at the boom point where the base of the hammer is guided between the pipes.

DUTCHMAN 1. Block arrangement used to alter the lateral placement of a line or pull the bight of a line to assist in landing logs; **2.** Flat area produced when the two horizontal cuts of an undercut do not meet at a point, i.e., an undercut that is not cut or cleaned out properly; **3.** Cut tile used as a filler in the run of a wall or floor area.

DUTCH MATTRESS Mattress of timber and reeds, used to protect the river or sea bed from scour.

DUTY Load and speed criteria to which an elevator is designed.

dvtl *Abbreviation for:* dovetail.

dw *Abbreviation for:* dry well

DW *Abbreviation for:* distilled water; double width; dumb waiter.

DW. *Abbreviation for:* dishwasher.

DWARF PARTITION Partition that ends short of the ceiling.

DWARF WALL 1. Wall enclosing a court, topped by railings; **2.** Wall between the upper ceiling level and finished roof level. *See also* **Wall.**

DWDI *Abbreviation for:* double-width double-inlet.

dwel. *Abbreviation for:* dwelling.

DWELLING Unit of housing, usually designed for a single family.

DWELLING UNIT One or more rooms, used or intended for the domestic use of one or more individuals living as a single housekeeping unit, with cooking, eating, living, sleeping, and sanitary facilities. *Also called* **Housing unit,** and **Living unit.**

DWELL TIME Time required for material to pass through a dryer or drum mixer.

dwg *Abbreviation for:* drawing.

dwl *Abbreviation for:* dowel.

dwl pn *Abbreviation for:* dowel pin.

dwt *Abbreviation for:* dead weight; deadweight ton.

DWV *Abbreviation for:* drain, waste, and vent.

DX *Abbreviation for:* distance.

DYKE *See* **Dike.**

dyn. *Abbreviation for:* dynamic.

DYNAMIC ANALYSIS Analysis of stresses in framing as functions of displacement under transient loading.

DYNAMIC BALANCE 1. Condition of rest or equilibrium created by forces of equal strength tending to move in opposite directions. *See also* **Balance,** and **Static balance**; **2.** Balancing a tire (and supporting rotating accessories: rim, brake assembly, mounting lugs, etc.) while it is spinning.

DYNAMIC BRAKING Use of counter voltage of a turning electric motor armature to oppose rotation.

DYNAMIC CREEP Creep that occurs under conditions of fluctuating load or fluctuating temperature.

DYNAMIC DISCHARGE HEAD Total of the static discharge head plus the friction in the discharge line.

DYNAMIC FORCE Force tending to produce motion.

DYNAMIC FORCE APPLIED Vectorial resolution of all the generated forces and the static forces at the interface of the drum and the material being compacted.

DYNAMIC FORMULA Mathematical equation that translates the energy per blow by a pile hammer and the observed set of the pile into an estimated static load-carrying capacity the pile will safely support. *Also called* **Driving formula.**

DYNAMIC HEAD *See* **Head.**

DYNAMIC JOINT Joint intended to accommodate expansion and contraction movements of a structure.

DYNAMIC LOAD Load that is variable, i.e., not static, such as a moving live load, earthquake, or wind.

DYNAMIC LOADING Loading from units (particularly machinery) that, by virtue of their movement or vibration, impose stresses in excess of those imposed by their dead load.

DYNAMIC MODULUS OF ELASTICITY Modulus of elasticity computed from the size, weight, shape, and fundamental frequency of vibration of a test specimen, or from pulse velocity.

DYNAMIC PENETRATION TEST Test involving moving parts, as distinguished from a static test.

DYNAMIC PILE FORMULA Formula for the safe load on a pile, calculated from the energy of the hammer blow and the penetration of the pile from each blow.

DYNAMIC ROCK STRENGTH Amount of stress that a rock can withstand without failing, under changing loading conditions.

DYNAMIC SEAL Seal that has rotating, oscillating, or reciprocating motion between its components. *See also* **Static seal.**

DYNAMIC STRENGTH Resistance to loads applied suddenly.

DYNAMIC SUCTION HEAD Total of the static suction head and suction line friction.

DYNAMITE High explosive, made in three basic types:

> **Extra dynamite:** Grade-for-grade is less sensitive to shock and friction than straight dynamite;

> **Gelatin dynamite:** Explosive with a base of nitrocotton-nitroglycerine gel, that is insoluble in water, used for blasting under wet conditions.

> **Straight dynamite:** Contains nitroglycerine as the principal or only explosive;

DYNAMO Machine that transforms mechanical work into electric current.

DYNAMOMETER Device used to measure the horsepower output of an engine; can be mechanical, hydraulic, or electrical in operation.

DYNE Obsolete unit of force, equal to 10 μN, that should not be used with the SI system of measurement. Symbol: dyn. *See also the appendix*: **Metric and nonmetric measurement.**

E

e *Abbreviation for:* coefficient of resitition.

E *Abbreviation for:* edge.

ea. *Abbreviation for:* each.

EA *Abbreviation for:* exhaust air.

EAGLE BEAK Outside corner tile trim shape measuring 150 by 19 mm (6 by 0.75 in.).

EAL *Abbreviation for:* Equivalent axle loading.

E&CB1S *Abbreviation for:* edge and center bead one side.

E&CB2S *Abbreviation for:* edge and center bead two sides.

E&CV1S *Abbreviation for:* edge and center vee one side.

E&CV2S *Abbreviation for:* edge and center vee two sides.

E&I *Abbreviation for:* equip and install.

E&OE *Abbreviation for:* errors and omissions omitted.

EAR Projection on a plumbing fitting, or on a pipe, by which it may be attached to a wall.

EARLIEST EVEN OCCURRENCE TIME Earliest time that all activities that precede an event will be completed.

EARLIEST FINISH Earliest day that the work item can finish if it starts at its earliest start and is completed in its expected time.

EARLIEST START Earliest day that the work item can start provided every preceding work item starts at its earliest start day and is completed in its expected time.

EARLY-AGE GRINDING Using a low-speed grinder to take off 1.0 to 2.0 mm (0.04 to 0.08 in.) from the surface laitance of fully compacted, leveled, floated, and hardened concrete to provide a low-cost finished floor surface.

EARLY STIFFENING Early development of an abnormal reduction in the working characteristics of a hydraulic-cement paste, mortar or concrete, that may be further described as **False set, Flash set,** or **Quick set.**

EARLY STRENGTH Strength of concrete or mortar, usually as developed at various times during the first 72 hours after placement.

EARNED HARVEST Timber management concept. It allows the timber manager an immediate increase in the allowable cut with manager-applied intensive management techniques that will accelerate future timber growth.

EARNEST MONEY Deposit given by a potential purchaser to a seller or his agent upon signing of an agreement of sale. On completion of the sale the deposit is applied against the purchase price; if the sale is not completed the deposit is forfeit unless otherwise specified in the agreement of sale.

EARTH Nonstatic particulate material lying above bedrock and around rocky outcrops, consisting mostly of the products of rock disintegration such as sand, clay, pebbles, cobbles and boulders, plus organic constituents. *See also* **Material density.**

EARTH ANCHOR Rod driven into the ground at an angle of approximately 30° in a direction opposite to that of the line of force, to which anchor cables are attached.

EARTH AUGER Large-diameter earth drill with slow-pitch flights. May be an attachment or self-powered.

EARTH BERM Mound of dirt placed to achieve some effect: diversion of water or wind, insulation, impoundment, etc.

EARTH BORER Truck-mounted drill rig.

EARTH DAM Water barrier made of earth, clay, sand, and gravel and having an impervious core in the form of a vertical curtain. *Also called* **Fill dam.**

EARTH DRILL *See* **Drill.**

EARTHENWARE Pottery made from brick earth, softer than stoneware.

EARTH-MOVING PLANT Machinery and equipment designed to excavate, transport, and place earth and granular material.

EARTH PIGMENTS Class of pigments produced by physical processing of materials mined from the earth. *Also called* **Mineral pigments,** and **Colors.**

EARTH PRESSURE Thrust from retained soil that varies between two extremes: the minimum or active earth pressure (the force from soil tending to overturn a free retaining wall), and the maximum passive earth pressure (resistance of an earth outface to deformation by other forces). *Also called* **Lateral pressure.**

EARTH PRESSURE BALANCE MACHINE Tunnel boring machine that applies mechanical pressure to the material to be excavated to provide temporary support to the surrounding ground.

EARTH PRESSURE CELL Large, nonferrous flat cell for the *in situ* measurement of soil pressure.

EARTHQUAKE Vibration or shaking of the ground due to some natural phenomenon within the earth.

EARTHQUAKE LOADING Load or stress placed on a wall or floor system due to the earthquake shaking action.

EARTHQUAKE POWDER Explosive mixture containing 79% nitrate and 21% charcoal.

EARTHQUAKE PROTECTION DEVICE Device designed to regulate operation of an elevator during or after an earthquake in excess of a defined magnitude.

EARTHWORK 1. Moving of surface materials to create a change of landform during site works; **2.** Embankment formed by heaping soil; **3.** Work of excavation.

EASED EDGE Tapered and slightly rounded factory edge of drywall boards.

EASEMENT 1. Curved member used to prevent abrupt changes in direction, as in a baseboard or handrail; **2.** Triangular piece in a stairway to match the inside string and the wall base where these join at the bottom of the stairs; **3.** Strips of land within a larger area over which right of access by a person or public authority other than the owner has been granted for specific purposes, such as the passage of services. There are several distinctions, including:

　Drainage: Easement for directing the flow of water.

　Planting: Easement for reshaping roadside areas and establishing, maintaining, and controlling plant growth thereon.

　Scenic: Easement for the conservation and development of roadside views and natural features.

　Sight-line: Easement for maintaining or improving the sight distance.

　Slope: Easement for cuts or fills.

EASEMENT CURVE Transition curve.

EASING THE WEDGES Loosening the wedges that hold firm such temporary erections as arch centering, shoring for underpinning, etc.

EASTING Measurement in compass degrees to the east from a survey instrument location to the point being measured.

EA STRENGTH Strength of a panel, determined by multiplying the modulus of elasticity by the cross-sectional area of the panel.

EAVES Overhanging of a roof that projects beyond the supporting exterior walls.

EAVES COURSE First course of tiles or shingles on a roof, including the underlying starter course.

EAVES FASCIA *See* **Fascia.**

EAVES FLASHING Drop apron from an asphalt roof dressed into the eaves gutter.

EAVES PLATE Plate reaching between vertical uprights at the eaves, on which the rafters rest when there is no wall to carry them.

EAVES STRUT Structural member at the eaves that transmits forces from roof brace rods to wall brace rods.

EAVES TROUGH *See* **Gutter.**

EBB *Abbreviation for:* extra best best.

EBC *Abbreviation for:* electric blasting cap.

EB1S *Abbreviation for:* edge bead one side.

EB2S *Abbreviation for:* edge bead two sides.

ec *Abbreviation for:* end of curve.

EC *Abbreviation for:* electrical conductor.

ecc. *Abbreviation for:* eccentric.

ECCENTRIC 1. Degree of being off-center; the distance between the center of an eccentric and its axis; **2.** Wheel or cam with an off-center axis of revolution.

ECCENTRIC FITTING Pipe fitting in which the centerline of the opening is offset.

ECCENTRICITY State, quality, or amount of being off-center or not concentric. *Also called* **Off center.**

ECCENTRIC LOADING Loads that do not bear axially upon the member resisting them, as on a pile that is driven out of plumb.

ECCENTRIC MOMENT Product of the unbalanced mass/unbalanced weight times the distance from the center of gravity of the unbalanced mass to the bearing center.

ECCENTRIC STUD Stud with an eccentric to its diameter, used as an adjusting screw; when turned it provides a cam-like action.

ECCENTRIC TENDON Prestressing tendon that follows a trajectory not coincident with the gravity axis of the concrete member.

ECCENTRIC WALL In hose or tubing, a wall of varying thickness.

ECHELON PAVING Use of two or more asphalt pavers operating one ahead of the other and offset to one side to obtain a hot joint between the mats.

ECHO Repetition of a sound caused by its reflection from a surface.

ecol. *Abbreviation for:* ecological; ecology.

ECOLOGY Study of plants and animals in relation to their physical and biological surrounds.

econ. *Abbreviation for:* economic; economy.

ECONOMIC LIFE Period for which a product or structure is designed to remain viable, useful, and/or profitable: usually shorter than physical life.

ECONOMIC OBSOLESCENCE Loss of value due to changing economic circumstances beyond the bounds of a structure or development.

ECONOMIZER Final convection heat exchanger in a waterfall furnace, used to preheat feedwater.

ECONOMY BRICK *See* **Brick.**

ECONOMY GRADE *See* **Lumber grades.**

ECONOMY OF SCALE Scale of operation where the benefits per production unit exceed the unit cost.

ECOSYSTEM Complex ecological community and environment forming a functional whole in nature.

ECT *Abbreviation for:* estimated completion time.

ECU *Abbreviation for:* electronic control unit.

ed. *Abbreviation for:* edition.

EDC *Abbreviation for:* estimated date of completion.

EDDY CURRENT CLUTCH *See* **Clutch.**

EDDY-CURRENT SEPARATOR Device that passes a varying magnetic field through material, thereby inducing eddy currents in the ferrous metals present in the feed. The eddy currents counteract the magnetic field and exert a repelling force on the metals, separating them from the field and the remainder of the feed.

edg. *Abbreviation for:* edge; edged; edging.

EDGE BANDING Any piece of wood, metal, plastic, or other material affixed to a panel edge to provide protection or improve appearance.

EDGE-BAR REINFORCEMENT Tension steel, sometimes used to strengthen otherwise inadequate edges in a concrete slab without resorting to thickening.

EDGE BEAM Stiffening beam at the edge of a slab.

EDGE CRACKS Longitudinal cracks near the edge of pavement, sometimes branching toward the shoulder, that are usually caused by insufficient shoulder support.

EDGE FIRING Technique of broadcast burning in which fires are set along the edges of an area and allowed to spread to the center.

EDGE FORM Formwork used to limit the horizontal spread of fresh concrete on flat surfaces such as pavement or floors.

EDGE-GRAINED *See* **Grain.**

EDGE HARDNESS Filter medium whose passages are formed by the adjacent surfaces of stacked disks, edge-wound ribbons, or by single-layer filaments.

EDGE JOINT Where the edges of two boards are glued together under pressure, in the same plane. The resulting joint can be reinforced through the use of dowels.

EDGELINE MARKING *See* **Pavement marking.**

EDGE MEDIUM Filter medium whose passages are formed by the adjacent surfaces of stacked discs, edgewound ribbons, or single-layer filaments.

EDGE-MOUNTED TILE Type of mounted tile wherein the tiles are assembled into units or sheets and are bonded to each other at the edges or corners or the back of the tiles by an elastomeric or resinous material that becomes an integral part of the tile installation.

EDGE NAILING Blind or secret nailing, a method used in laying and fastening hardwood flooring. *See also* **Edge toe nailing.**

EDGER Finishing tool used on the edges of fresh concrete to provide a rounded edge.

EDGE SEALING Application of a coating (e.g., sealant paint) to the edges of a structural wood panel to reduce its water absorption.

EDGESET Brick set on its narrow side instead of on its flat side.

EDGE SNIPE Narrow, tapered condition along the long edge of a sanded particleboard panel.

EDGE SPACING *See* **Panel spacing.**

EDGE STRIP Strip (in plan view) at each edge of a suspended concrete slab: each strip having a width of one-eighth the appropriate span of the slab. *See also* **Column strip,** and **Middle strip.**

EDGE SUPPORT Structural element located so as to support a panel from bending.

DGE-SUPPORTED *See* **Slab-on-grade.**

EDGE-SUPPORTED BELT-TYPE MOVING

WALKWAY *See* **Moving walkway.**

EDGE TAPE EASE Amount of bevel along the edge of edge-taped shelving.

EDGE TOENAILING Fastening boards by driving nails at a slant through the vertical edge so that their heads will be concealed by the adjacent board. *See also* **Edge nailing.**

EDGE TREATMENT Edge finishing method for structural wood panels, such as banding with wood or plastic, or filling with putty or spackle.

EDGE VOID Wood panel defect in which the edge or end of an inner ply has split or broken away during manufacture, leaving a gap in the edge of the plywood panel.

EDGING 1. Finishing of the outside edges of concrete slabs, sidewalks, steps, etc., into a concave arc; **2.** Small solid squares used in cabinet making, at the edge of a table top, for instance, to protect a veneer; **3.** Linear barrier, often of paving stone or preservative-treated lumber, between two surface materials; commonly used between a lawn and gravel.

EDGING TROWEL Rectangular hand trowel with one edge curved down, used to trim and smooth the edges of concrete slabs, curbs, etc.

edg. str. *Abbreviation for:* edging strip.

edgw. *Abbreviation for:* edgewise.

EDIFICE Imposing and distinguished building.

EDM *Abbreviation for:* electronic distance measuring.

EDP *Abbreviation for:* electronic data processing.

EDR *Abbreviation for:* equivalent direct radiation.

educ. *Abbreviation for:* education; eductor.

EDUCTOR Device for lifting water to a higher level, consisting of concentric tubes that are inserted below the surface, with air under pressure being injected through the central tube.

EE *Abbreviation for:* eased edge(s); electrical engineer; errors excepted.

E18KSAL *Abbreviation for:* equivalent 18 kip single axle (wheel) load.

EER Energy efficiency ratio: the refrigeration effect in Btu/hr divided by the power input in watts.

ef *Abbreviation for:* each face.

EF *Abbreviation for:* each face.

eff. *Abbreviation for:* effect; effective; efficiency.

EFFECTIVE AGE Estimated age of a structure based on the use and care it has received.

EFFECTIVE AREA Area of a filter medium through which fluid flows.

EFFECTIVE AREA OF CONCRETE Area of a concrete section assumed to resist shear or flexural stresses.

EFFECTIVE AREA OF FURNACE OPENING Area of an opening in an infinitely thin furnace wall that would permit a radiation loss equal to that occurring through an actual opening in a wall of infinite thickness. The effective area is always less than the actual area because some radiation always strikes the sides of the opening and is reflected back into the furnace.

EFFECTIVE AREA OF REINFORCEMENT Area obtained by multiplying the right-angle cross-sectional area of the metal reinforcement by the cosine or the angle between its centroidal axis and the direction for which its effectiveness is considered. *See also* **Area of steel.**

EFFECTIVE COLUMN HEIGHT Height of a concrete column to be used in its design.

EFFECTIVE DATE Date indicated as 'the effective date of this agreement' in contract documents. If no date is indicated, it is the date the agreement was signed by the last of the parties to the agreement.

EFFECTIVE DEPTH Depth of a beam or slab section measured from its compression face to the centroid of the tensile reinforcement.

EFFECTIVE DIAMETER Diameter of a sphere having the same volume as that of an irregular-shaped particle.

EFFECTIVE FLANGE WIDTH 1. Width of a slab adjoining a beam stem where the slab is assumed to function as the flange element of a T-beam section; **2.** For a concrete T-beam, the lesser of the width of the beam web plus one-fifth of the distance between points of zero moment, or the actual width of the flange; **3.** For a concrete L-beam, the lesser of the width of the beam plus one-tenth of the distance between points of zero moment, or the actual width of the flange.

EFFECTIVE FRONTAL AREA Area of a building or structure normal to the direction of the wind or 'shadow area.'

EFFECTIVE HEIGHT Height of a member to be assumed for calculating the slenderness ratio.

EFFECTIVE LEAKAGE AREA Orifice-flow area that will result in the same calculated flow for a given pressure drop as is measured for the seal in question.

EFFECTIVE LENGTH Equivalent length *KL* used in compression formulas and determined by a bifurcation analysis.

EFFECTIVE LENGTH (FACTOR *K*) Ratio between the effective length and the unbraced length of the member measured between the centers of gravity of the bracing members.

EFFECTIVELY GROUNDED Permanently connected to earth through a ground connection of sufficiently low impedance, and having sufficient ampacity, that ground fault current that may occur cannot build up to voltages dangerous to personnel. *See also* **Grounded.**

EFFECTIVE MODULUS OF ELASTICITY Elastic and plastic effects in an overall stress-strain relationship in a structure.

EFFECTIVE MOMENT OF INERTIA 1. Moment of inertia of the cross section of a member that remains elastic when partial plastification of the cross section takes place, usually under the combination of residual stress and applied stress; **2.** Moment of inertia based on effective widths of elements that buckle locally; **3.** Moment of inertia used in the design of partially com-

posite members.

EFFECTIVE OPENING Cross-sectional area of an opening from which water is discharged from a water supply pipe.

EFFECTIVE PERIMETER 1. Of a single reinforcing bar: 3.14 times its nominal size; **2.** Of a group of reinforcing bars in contact with one another: the sum of the effective perimeters of the individual bars, multiplied by a reduction factor, which is 0.8 for two bars, 0.6 for three bars, and 0.4 for four bars in a bundle.

EFFECTIVE PRESTRESS Prestressing force at a specific location in a prestressed concrete member under the effects of service dead load or total service load after all losses of prestress have occurred.

EFFECTIVE REINFORCEMENT Steel reinforcement of a concrete section assumed to be effective in resisting applied stresses.

EFFECTIVE SCREENING Full or partial concealment of unsightly views to render them unnoticeable from the through traffic lanes, by means of natural objects, plantings, fences, or other appropriate means.

EFFECTIVE SPAN Lesser of (a) the distance between supports or (b) the clear distance between supports plus the effective depth of the beam or slab.

EFFECTIVE STIFFNESS Stiffness of a member computed using the effective moment of inertia of its cross section.

EFFECTIVE STRESS Net stress across points of contact of soil particles, generally considered as equivalent to the total stress, minus the pore water. *See also* **Effective prestress.**

EFFECTIVE STROKE *See* **Stroke.**

EFFECTIVE TEMPERATURE Actual temperature at a location within a heated or cooled complex, regardless of the calculated temperature or the equipment setting.

EFFECTIVE THICKNESS Thickness of a member to be assumed for calculating the slenderness ratio.

EFFECTIVE THREAD Includes the complete thread, as well as that portion of the incomplete thread having fully formed roots, but having crests not fully formed.

EFFECTIVE WALL HEIGHT (a) Of an unbraced, plain concrete wall: 1.5 times the vertical height of the wall from base to top if, at the top, a floor or roof spans at right angles; (b) Of an unbraced plain concrete wall: twice the vertical height of the wall if there is no restraint at right angles to the top; (c) Of a reinforced concrete wall constructed monolithically with adjacent construction: the same as the effective height of a column being bent at right angles to the plane of the wall.

EFFECTIVE WIDTH Reduced width of a plate or slab that, with an assumed uniform stress distribution, produces the same effect on the behavior of a structural member as the actual plate width with its nonuniform stress distribution.

EFFECTIVE WIDTH OF SLAB That part of the width of a slab taken into account when designing T- or L-beams.

EFFICIENCY 1. Measurement of the amount of fuel energy put into an engine as opposed to the amount of usable energy put out by the engine as horsepower. A typical measurement value is brake specific fuel consumption; **2.** Ratio of a wire rope's actual breaking strength and the aggregate strength of all individual wires tested separately, usually expressed as a percentage; **3.** Ability, expressed as a percent, of a filter to remove specified artificial contaminants, at a given concentration, under specified test conditions.

EFFICIENCY APARTMENT *See* **Bachelor dwelling unit.**

EFFICIENCY FACTOR Percentage of theoretical production obtainable under actual conditions.

EFFICIENCY VARIANCE *See* **Variance.**

effl. *Abbreviation for:* efflorescence; effluent.

EFFLORESCENCE Deposit of white powder or crust on the surface of brickwork due to soluble salts in the mortar or the brick being drawn to the surface by moisture. *See also* **Blemish.**

EFFLUENT 1. Any solid, liquid, or gas that enters the environment as a by-product of a man-originated process; the substances that flow out of a designated source; **2.** Fluid leaving a component.

EFFLUENT SEEPAGE Diffuse discharge onto the ground of liquids that have percolated through a mass; may contain dissolved or suspended solids.

EFFORT-EXPENDED METHOD OF ACCOUNTING *See* **Method of accounting.**

EFL *Abbreviation for:* environment for living.

eg *Abbreviation for:* existing grade.

e.g. *Abbreviation for:* exempli gratia (for example).

EG *Abbreviation for:* edge grain.

EGG-AND-(ANCHOR, DART, TONGUE) Molding used in classical architecture consisting of an egg shape separated by another form.

EGG CRATE Baffle diffuser used below light fittings, particularly fluorescent fixtures.

EGG-CRATE FOUNDATION Reinforced concrete raft foundation stiffened underneath with a grid of ground beams.

EGG-SHAPED SEWER Sewer pipe with an egg-shaped cross section, installed small end down, used to improve the self-cleaning effect at low flows.

EGGSHELL Paint finish with a low degree of gloss, resembling that of an eggshell. It is between flat and semigloss.

EGRESS Way out; a place of exit.

EGRESS LIGHTING Illumination of the means of exiting a building or structure.

ehf *Abbreviation for:* extremely high frequency.

EJECTMENT Action taken when there is no contractual relationship between an owner and tenant to regain possession of real property, with damages for unlawful retention.

EJECTOR Cleanout device, usually a sliding plate.

EJECTOR GRILL Ventilation grill shaped to distribute the vented air in many directions.

EJECTOR WELL POINT Well-point dewatering system whereby a vacuum is created at the well-point tip by means of a high-velocity water flow through a jet nozzle, sucking water from the soil and forcing it up the return pipe.

el *Abbreviation for:* existing level.

el. *Abbreviation for:* elevation.

EL *Abbreviation for:* elastic limit; equipment list.

ELAPSED TIME Time used to complete a defined stage or action, or between stages or actions.

elas. *Abbreviation for:* elastic; elasticity.

ELASTIC ANALYSIS Determination of load effects (force, moment, stress as appropriate) on steel members and connections based on the assumption that material deformation disappears on removal of the force that produced it.

ELASTIC COMPRESSION Movement of soil particles due to the imposition of a load. *See also* **Plastic creep** and **Soil consolidation.**

ELASTIC CURVE Curve depicting the deflected shape of the neutral surface of a bent beam.

ELASTIC DEFORMATION Temporary deformation, proportional to the applied stress.

ELASTIC DESIGN Method of analysis in which the design of a member is based on a linear stress-strain relationship and corresponding limiting elastic properties of the material.

ELASTICITY 1. Characteristic of a material or combination of materials that enables it to react without damage to an imposed force that causes it to change its shape and dimension, regaining its original shape or dimension when the force is removed; **2.** Soil property that allows it to return to its approximate original shape when a compressing load is removed.

ELASTIC LIMIT Stress limit above which permanent deformation will take place within a material.

ELASTIC LOSS Reduction in prestressing load resulting from the elastic shortening of the member.

ELASTIC MOVEMENT Movement under load that is recoverable when the load is removed.

ELASTIC-PERFECTLY PLASTIC Material that has an idealized stress-strain curve that varies linearly from the point of zero strain and zero stress up to the yield point of the material, and then increases in strain at the value of the yield stress without further increases in stress.

ELASTIC SHORTENING 1. Reduction in length of a pile or structural member due to an imposed load; **2.** Shortening of a member that occurs immediately upon the application of forces induced by prestressing.

ELASTIC STRETCH Wire rope stretch that occurs when a load or loads are initially applied, due to individual strands in the rope seating themselves.

ELASTOHYDRODYNAMIC LUBRICATION *See* **Lubrication.**

ELASTOMER Elastic, rubber-like substance, such as natural or synthetic rubber, capable of returning to its original dimensions after tensile or compressive forces are applied that are within limits of its yield strength.

ELASTOMERIC Elastic substance resembling rubber.

ELBOW Pipe fitting made to allow a turn in direction of a pipe line.

ELDERLY HOUSING *See* **Senior citizen housing.**

elec *Abbreviation for:* electric.

elec. *Abbreviation for:* electrical; electrician.

elect. *Abbreviation for:* electrolyte.

ELECTRICAL CABINET Enclosure designed either for surface or flush mounting, and provided with a frame, mat, or trim in which a swinging door or doors are or may be hung.

ELECTRICAL CONDUCTOR Metal strip designed to conduct electrical energy, classified by the following types:

Bare: Conductor having no covering or electrical insulation whatsoever.

Covered: Conductor encased within material of composition or thickness that is not recognized as electrical insulation.

Insulated: Conductor encased within material of composition and thickness that is recognized as electrical insulation.

ELECTRICAL ENCLOSURE Case or housing of apparatus, or the fence or walls surrounding an installation, to prevent personnel from accidentally contacting energized parts, or to protect the equipment from physical damage.

ELECTRICAL EQUIPMENT General term including material, fittings, devices, appliances, fixtures, apparatus, and the like, used as a part of, or in connection with, an electrical installation.

ELECTRICAL GROUND Conducting connection between an electrical circuit or equipment and the earth, or some conducting body that serves in place of the earth.

ELECTRICAL INSULATION Nonconducting covering applied to wire or equipment to prevent short circuiting.

ELECTRICAL PORCELAIN Vitrified whiteware having an electrical insulating function.

ELECTRICAL QUIESCENT POWER Power required for differential operation of a servovalve when the current through each coil is equal and opposite in polarity.

ELECTRICAL SPECIFICATIONS Detailed drawings and descriptions of the work to be done by an electrical contractor.

ELECTRICAL STARTING SYSTEM Starting system that utilizes electrical energy (battery) through a motor.

ELECTRIC BLASTING CAP Cap used to initiate primers or detonating cord; it may be instantaneous or delayed. *Also called* **EBC.**

ELECTRIC BLASTING CIRCUITRY Wiring connecting a source of electric energy and the blasting cap(s) used to detonate an explosive, including:

Bus wire: An expendable wire, used in parallel or series circuits, to which are connected the leg wires of electric blasting caps.

Connecting wire: Insulated, expendable wire between electric blasting caps and the leading wires or between the bus wire and the leading wires.

Leading wire: Insulated wire used between the electric power source and the electric blasting cap circuit.

Permanent blasting wire: Permanently mounted insulated wire used between the electric power source and the electric blasting cap circuit.

ELECTRIC BRAKE *See* **Brake.**

ELECTRIC CABLE Bundle of insulated wires, used to conduct electricity.

ELECTRIC CURING Curing concrete by passing an alternating electric current through it to heat and early harden it so that formwork and molds can be stripped. *See also* **Curing.**

ELECTRIC DELAY BLASTING Caps designed to detonate at a predetermined period of time after energy is applied to the ignition system.

ELECTRIC DRIVE *See* **Driving machine.**

ELECTRIC ELEVATOR *See* **Elevator.**

ELECTRIC EYE 1. Phototube capable of detecting the presence or absence of light; **2.** Light beam (or beams) that spans an elevator door opening and, when interrupted, causes the door to reopen.

ELECTRIC FIELD Force of attraction between opposite charges that are close, but not touching.

ELECTRIC-FUSION-WELDED PIPE *See* **Pipe.**

ELECTRIC GAUGE Gauge that reads a value, such as psi, via a sending unit.

ELECTRIC GOVERNOR *See* **Governor.**

ELECTRIC HEAT *See* **Heating system.**

ELECTRICIAN Qualified worker who installs or repairs electrical circuits, wiring, or machines.

ELECTRIC LINE TRUCK Truck used to transport men, tools, and material, and to serve as a traveling workshop for electric power line construction and maintenance work.

ELECTRIC, LOAD-SENSING GOVERNOR *See* **Governor.**

ELECTRIC/OIL SYSTEM Motive or pressure system that comprises an electric motor driving a pumping unit that supplies oil under pressure.

ELECTRIC OVERHEAD CRANE Mobile lifting mechanism that travels along the upper side rails of a gantry. Can often be remotely controlled.

ELECTRIC PRECIPITATOR Device that causes dust particles to fall out of the air by inducing an electric charge on them.

ELECTRIC PUMP *See* **Pump.**

ELECTRIC-RESISTANCE-WELDED PIPE *See* **Pipe.**

ELECTRIC SET Equipment, fixed or portable, designed to generate electricity. There are several distinctions, including:

Peaking power: Electric set that assumes part of the load during peak-load periods.

Prime power: Electric set that is operated by the primary source of power. It may be primary because it is the sole source or because it provides a special type of power.

Standby power: Emergency electric power system that is on 'standby alert,' ready to assume the load when the normal power source fails.

ELECTRIC STAIRWAY *See* **Escalator.**

ELECTRIC SUBMERSIBLE PUMP *See* **Pump.**

ELECTRIC TRACTOR Industrial tractor in which the principal energy is transmitted from a power source to motor(s) in the form of electricity; the power can be either battery, gas-electric, diesel-electric, or tethered-electric.

ELECTRIC TRUCK Truck in which the principal energy is transmitted from power source to motor(s) in the form of electricity.

ELECTRIC UTILITIES All enterprises engaged in the production and/or distribution of electricity for use by the public.

ELECTRIC WALK *See* **Moving walk system.**

ELECTRIC WELDING Joining metal by arc or resistance welding.

ELECTRODE 1. Electric conductor through which a current enters or leaves a medium; 2. Terminal point to which electricity is brought in welding and from which an arc is produced to do the welding; in electric-arc welding, the electrode is usually melted and becomes a part of the weld.

ELECTRODYNAMIC SEPARATOR Device that incorporates a rotating drum or other moving poles in place of one or more fixed charged plates (poles). Can be used to separate electrically conductive material from nonconductive material.

ELECTROHYDRAULIC ELEVATOR *See* **Elevator.**

ELECTROHYDRAULIC SERVO-VALVE Servo-valve that is capable of continuously controlling hydraulic output as a function of electrical input.

ELECTROLYSIS Production of chemical changes by the passage of electrical current through an electrolyte.

ELECTROLYSIS/ELECTROLYTIC PITTING Migration (corrosion, pitting) of metal particles between two dissimilar metals.

ELECTROLYTE Conducting medium in which the flow of electric current takes place by migration of ions. The electrolyte for a lead storage cell is an aqueous solution of sulfuric acid; for alkaline storage cells an aqueous solution of certain hydroxides.

ELECTROMAGNET Magnet that derives its magnetic energy from a current of electricity passing through a coil of wire wound round a central core of iron.

ELECTROMAGNETISM Magnetic force produced by a coil surrounding a mass of ferrous material and carrying an electric current.

ELECTROMOTIVE FORCE Energy, measured in volts, that causes the flow of electric current.

ELECTRONIC CONTROL UNIT Central component or 'brain' of a microprocessor-based system that receives sensor signals, interprets transmitted data, and generates signals or data in response.

ELECTRONIC DISTANCE MEASUREMENT Measurements made with an instrument that compares the phase difference between transmitted and returned electromagnetic waves, of known speed and frequency, or the round-trip travel time of a pulsed signal, from which the distance is calculated.

ELECTRONIC FILTER Filter that uses an electrically charged plate to attract and retain dust particles and pollen. *Also called* an **Electrostatic filter.**

ELECTRO-OSMOSIS Groundwater lowering process in which the flow of water is induced by an electrical current flowing from a positive anode to a negative cathode.

ELECTROPLATING Deposition of one metal upon another by electrolysis.

ELECTROSTATIC FILTER *See* **Electronic filter.**

ELECTROSTATIC PRECIPITATOR Device that collects particulates by placing an electrical charge on them and attracting them onto a collecting electrode.

ELECTROSTATIC SEPARATOR *See* **Separator.**

elem. *Abbreviation for:* element; elementary.

ELEMENT 1. Distinctive construction item; 2. Substance that cannot be divided into substances different from itself; 3. Electrical resistance coil used for heating; 4. Porous device which performs the actual process of filtrtion. *Also called* **Cartridge**; 5. *See* **Functional element.**

ELEPHANT TRUNK Articulated tube or chute used in concrete placement to keep the concrete from falling freely into forms, which could cause segregation of cement and aggregate.

elev. *Abbreviation for:* elevate; elevator.

ELEVATING-BELT GRADER *See* **Loader.**

ELEVATING GATE Form of truck endgate, used in conjunction with a hydraulic or mechanical hoisting mechanism, to allow the gate to descend to ground level. Power elevation allows the gate to be used to raise freight to the truck floor level for loading. *Also called* **Lift gate, Load gate, Power gate,** and **Tailgate lift.**

ELEVATING SCRAPER Scraper in which the apron is replaced by a chain-driven elevator that lifts soil into the bowl.

ELEVATION 1. External faces of a building or structure; 2. Vertical projection drawing; 3. Established point above a known datum (bench mark or other reference).

ELEVATION LOSS Loss of pressure caused by raising water through hose or pipe to a higher elevation.

ELEVATOR 1. Machine that lifts material to a higher elevation on a belt or chain fitted with small buckets. 2. Lifting equipment serving defined levels and platforms, equipped with a car or platform that moves in vertical guides. There are several types:

Direct-plunger: Hydraulic elevator having a plunger or piston directly attached to the car frame or platform.

Electric: Power elevator where the energy is applied by means of an electric driving machine.

Electrohydraulic: Direct-plunger elevator where liquid is pumped under pressure directly into the cylinder by a pump driven by an electric motor.

Freight: Elevator used to carry material rather than people. There are several classifications:

Class A: General freight.

Class B: Motor vehicles.

Class C (1): Industrial truck loading where the truck is carried on the elevator.

Class C (2): Industrial truck loading where the truck is not carried on the elevator.

Class C (3): Loading with heavy concentrates where a truck is not used.

Gravity: Elevator utilizing gravity to move the car.

Hand: Elevator utilizing manual energy to move the car.

Hydraulic: Power elevator where the energy is applied, by means of liquid under pressure, in a cylinder equipped with a plunger or piston.

Multideck: Elevator having two or more compartments located one immediately above the other.

Observation: Elevator that gives its passengers a direct view of the exterior while the car is traveling.

Passenger: Elevator used primarily to carry persons other than an operator and persons necessary for unloading and loading.

Power: Elevator utilizing energy, other than gravitational or manual, to move the car.

Roped hydraulic: Hydraulic elevator having its piston connected to the car with wire rope.

Sidewalk: Freight elevator for carrying material between a landing in the sidewalk or other area exterior to a building and floors below the sidewalk grade level.

ELEVATOR CAR Cab, vehicle, or enclosure that conveys passengers and loads between levels in an elevator shaft.

ELEVATOR CAR LEVELING DEVICE Any mechanism that will, either automatically or under the control of the operator, move the car within the leveling zone toward the landing only, and automatically stop it at the landing. Where controlled by an operator by means of up-and-down continuous-pressure switches in the car, this device is known as an inching device; where used with a hydraulic elevator to correct automatically a change in car level caused by leakage in the hydraulic system, this device is known as an anti-creep device. There are several types:

Leveling zone: Limited distance above or below an elevator landing within which the leveling device is permitted to cause movement of the car toward the landing.

One-way automatic: Device that corrects the car level only in the case of underrun of the car, but will not maintain the level during loading and unloading.

Two-way automatic maintaining: Device that corrects the car level on both underrun and overrun, and maintains the level during loading and unloading.

Two-way automatic nonmaintaining: Device that corrects the car level on both underrun and overrun, but will not maintain the level during loading or unloading.

ELEVATOR DRIVING MACHINE *See* **Driving machine.**

ELEVATOR LANDING Portion of a floor, balcony, or platform used to receive and discharge passengers or freight.

ELEVATOR LIABILITY Responsibility for property damage or personal injury resulting from the ownership, maintenance, and use of elevators.

ELEVATOR PIT Portion of a hoistway extending from the threshold of the lowest landing door to the floor at the bottom of the hoistway.

ELEVATOR SHAFT Space in which an elevator and its counterweight move.

elig. *Abbreviation for:* eligible.

elim. *Abbreviation for:* eliminate; elimination.

ELITE TREE One that has been shown by testing to be capable of producing progeny with superior qualities.

ELL Pipe fitting shaped like a bent elbow or L.

ell. *Abbreviation for:* elbow; ellipse.

ELLIPSE Curve that is longer than it is wide.

ELLIPSOID Solid of which every plane section is an ellipse or a circle.

ELLIPTICAL ARCH *See* **Arch.**

ELLIPTICAL PIPE Pipe having x and y axes of different dimensions.

ELLIPTICAL STAIR Stair that follows a well having an elliptical shape in plan.

elon *Abbreviation for:* elongation.

elon. *Abbreviation for:* elongate; elongated.

ELONGATED PIECE Particle of aggregate for which the ratio of the length to the width of its circumscribing rectangular prism is greater than a specified value. *See also* **Flat piece.**

ELONGATION Increase in length. *See also* **Expansion,** and **Swelling.**

ELONGATION INDEX Index given by the mass of particles left following the quartering of a representative sample of aggregate, between 5 and 63 mm (0.2 and 2.5 in.), and passing the residue through a bank of standard sieves to leave between 100 and 200 particles on each sieve. The index is given by the mass of particles whose maximum dimension or length is greater than appropriate to the fraction to which they belong: to the mass of the total sample.

ELUTRAITION Process wherein materials are separated according to differences in their densities and/or shapes in a counter-current stream of a fluid (usually water), gas, or air.

ELUTRAITOR Classifier used in soil analysis in which large grains sink faster through liquid than small grains of the same material.

ELUVIUM Deposit of disintegrated material.

el./w. *Abbreviation for:* electric weld.

EM *Abbreviation for:* end matched; electromagnetic; elsewhere mentioned; expanded metal.

emb. *Abbreviation for:* embankment; emboss.

EMBANKMENT 1. Area of fill, the top of which is higher than the surrounding surface; **2.** Structure of soil, soil-aggregate, or broken rock between the embankment foundation and the subgrade.

EMBANKMENT DAM Structure of excavated or imported materials, used to retain water.

EMBANKMENT FOUNDATION Material below the original ground surface, the physical characteristics of which affect the support of the embankment.

EMBANKMENT WALL Retaining wall at the foot of a bank.

EMBAYMENT Deep depression in a shoreline forming a large bay.

EMBED To apply and cover drywall tape with joint compound.

EMBEDDED GRIT Grit that becomes embedded in wood chips in the process of whole-tree chipping.

EMBEDMENT 1. Length of a pile from the surface of the ground, or from the cutoff below the ground, to its tip; **2.** Extent of penetration of the top of the pile into the pile cap; **3.** Enclosure of an object in resin for presentation and display; the resin is always colorless and transparent; **4.** Steel component cast in a concrete structure that is used to transmit externally applied loads to the concrete structure by means of bearing, shear, bond, friction, or any combination. The embedment may be fabricated of structural-steel plates, shapes, bars, bolts, pipe, studs, concrete reinforcing bars, shear connectors, or any combination.

EMBEDMENT LENGTH Length of embedded reinforcement provided beyond a critical section.

EMBEDMENT-LENGTH EQUIVALENT Length of embedded reinforcement that can develop the same stress as that which can be developed by a hook or mechanical anchorage.

EMBOSSED 1. Decoration in relief, or excised; **2.** Structural panel surface treatment in which heat and pressure against a master pattern impress a variety of textured effects into panel surfaces, which remain smooth and paintable.

EMBRITTLEMENT Reduction in the normal ductility of a metal due to a physical or chemical change.

EMC *Abbreviation for:* equilibrium moisture content.

emer. *Abbreviation for:* emergency.

EMERGENCY BRAKE -*See* **Brake.**

EMERGENCY CIRCUIT Building load circuit separated from the normal circuits and operated separately only during emergencies.

EMERGENCY DOOR *See* **Door types.**

EMERGENCY EXIT 1. Exterior, outward-opening door or doors, the lock for which is actuated by an inside-mounted push bar; **2.** Removable or openable top or side panels of an elevator car.

EMERGENCY FIELD ORDER *See* **Extra work order.**

EMERGENCY KEY Key capable of opening any lock in a building, whether the door is locked from the inside or not.

EMERGENCY LOCK Air lock designed to hold and permit the quick passage of an entire shift of workers.

EMERGENCY POWER Independent reserve source of electric energy that, upon failure or outage of the normal source, provides electric power. *Also called* **Alternate source of power.**

EMERGENCY SIGNAL 1. Visual or audible signal that warns of a condition beyond that for which a system is designed or that the local environment can tolerate; **2.** *See* **Traffic signal.**

EMERGENCY STOP SWITCH *See* **Switch.**

EMERGENCY WORK ORDER *See* **Extra work order.**

EMERY Rock consisting of an intercrystalline mixture of corundum and either magnetite or hematite; a natural abrasive of the aluminum oxide type.

EMERY CLOTH Fabricized paper coated with metal particles, used to clean or polish metal.

EMERY WHEEL High-speed grinding wheel composed mostly of emery.

EMF *Abbreviation for:* electromotive force.

EMINENT DOMAIN Right to expropriate or condemn private property for public use. *See also* **Acquisition, Conveyance, Dedication, Expropriation, Negotiation, Option, Remainder,** and **Severality.**

EMISSION Air or gas expelled from an area or mechanical device.

EMISSION STANDARD Rule or measurement established to regulate or control the amount of a given constituent that may be discharged into the outdoor atmosphere.

EMISSIVITY Ability of a surface to radiate energy, as compared with that of a 'black body,' which emits radiation at the maximum rate possible at any given temperature and that has an emissivity of 1.0.

EMITTANCE Amount of heat radiated back from a solar collector.

Em. K. *Abbreviation for:* emergency master key.

EMkd *Abbreviation for:* emergency master-keyed.

EMPHYTEUTIC LEASE Contract, usually for the development of commercial property, for a maximum of ninety-nine years and a minimum of nine years, in which the owner of an immovable conveys it for a time to another, the lessee being required to make improvements, to pay an annual rent, and be responsible for such other charges as are agreed upon. During the term of the contract, the lessee enjoys all the rights of ownership including alienation, transfer, and hypothecation, without prejudice to the rights of the lessor. At the end of the contract period, the lessee must give up, in good condition, the property received from the lessor, as well as the structures he obliged himself to build.

EMPIRICAL FORMULA/RULE Formula or rule based on practical experiments, as against theoretical calculation.

empl. *Abbreviation for:* employee.

EMPLOYEE INJURIES LIABILITY INSURANCE Indemnity against injury to employees unless they are covered under workmen's compensation, unemployment compensation, or disability benefit laws.

EMPLOYER Contractor or subcontractor; one who contracts with another, orally or in writing, to pay for the performance of specified acts or duties.

EMPLOYER'S LIABILITY INSURANCE Protection for an employer against claims or common lawsuits by employees for damages arising out of their employment.

emu *Abbreviation for:* electromagnetic unit.

emul. *Abbreviation for:* emulsify; emulsion.

EMULSIFIED ASPHALT *See* **Asphalt.**

EMULSIFIER *See* **Surfactant.**

EMULSION 1. Colloidal dispersion of a liquid in another liquid; **2.** Agent used with water to liquefy an asphalt for construction operations, producing an emulsified asphalt. *See also* **Water-in-oil emulsion.**

EMULSION PAINT Paint, the vehicle of which is an emulsion binder in water. The binder may be oil, oleoresinous varnish, resin, or other emulsifiable binder.

enam. *Abbreviation for:* enamel; enameled.

ENAMEL Hard, vitreous material baked on the surface of metal, porcelain, or brick; a form of paint that dries with a hard, glossy surface. An enamel paint may be of either the lacquer or varnish variety.

ENATHALPY Total heat content above that at an arbitrary set of conditions chosen as the base or zero point.

enc. *Abbreviation for:* enclosure.

ENCASE Enclose; cover entirely; seal.

ENCASED KNOT Knot whose annual growth rings are separate from those of the surrounding wood.

ENCASED STEEL STRUCTURE Steel-framed structure in which all of the individual frame members are completely encased in cast-in-place concrete.

ENCASEMENT *See* **Pile encasement.**

ENCASTRÉ End fixing of a built-in beam.

ENCAUSTIC Color that is fused to or into an object.

ENCAUSTIC DECORATION Decoration fused onto such materials as brick, tiles, glass or porcelain.

encl. *Abbreviation for:* enclose; enclosing; enclosure.

ENCLOSED Surrounded by a case, cage, or fence, that will protect the contained equipment and prevent accidental contact of a person with live parts.

ENCLOSED CAB Equipment cab suitable for an all-weather operation; may be heated and/or air-conditioned.

ENCLOSED HYDRAULIC HOIST BRAKE *See* **Brake.**

ENCLOSED KNOT Knot that does not appear on the surface of a board.

ENCLOSED STAIRWAY Stairway enclosed by, and separated from, hallways and living units by means of walls or partitions and made accessible to such hallways or living units by means of a door or doors.

ENCLOSURE 1. Housing for components. *See also* **Electrical enclosure**; **2.** Ruggedly built room on a freight elevator in which the material being carried is located.

ENCLOSURE WALL *See* **Wall.**

ENCODER Device that converts an analog signal into digital information.

ENCROACHMENT 1. Building, part of a building, or obstruction that intrudes upon the property of another; **2.** Unauthorized use of highway right of way or easements, as for signs, fences, buildings, etc.; **3.** Any man-made works or facilities that interrupt, impede, or divert the natural flow of surface water or runoff.

ENCUMBRANCE Right to or interest in land that affects its value. *Also called* **Incumbrance.**

END 1. The junction of the ends of two boards of the same dimension, in the same plane at the same elevation. This joint has no structural strength; **2.** One of a bundle of essentially parallel fibers.

END ANCHORAGE 1. Length of reinforcement, or a mechanical anchor, or a hook, or combination beyond the point of nominal zero stress in the reinforcement of cast-in-place concrete; **2.** Mechanical device to transmit prestressing force to the concrete in a post-tensioned member. *See also* **Anchorage.**

END-BEARING PILE *See* **Pile.**

END-BEARING SLEEVE Device fitting over the abutting ends of two reinforcing bars for the purpose of assuring transfer of axial compression only, from one bar to the other. *See also* **Coupler, Coupling sleeve,** and **Mechanical connection.**

END BLOCK Enlarged section of a concrete member, intended to reduce anchorage stresses to allowable values and provide space needed for post-tensioning anchorages.*See also* **End of reinforcement.**

END-BLOCK REINFORCEMENT Mild steel reinforcement placed in the end block of a prestressed concrete I- or T-beam to distribute the anchorage forces.

END CAP Ported or closed cover for the end of a filter element.

END CHECK Surface check at the end of a piece of lumber.

END CLOSURE Plate or point attached to the tip of a pipe or shell pile.

END-CONSTRUCTION TILE Tile designed to be laid with axes of the cells vertical.

END CONTROL TRUCK Powered industrial truck in which the operator's position is located at the end opposite the load end of the vehicle.

END CRIMPING Crimping a dead-end anchor wire for a short distance at the end of its length.

END-CUTTING NIPPERS *See* **Pliers.**

END FRAME Frame at the end wall of a building that supports the roof load from cne half of the end bay.

END GATE Device attached to the end of an asphalt paver screed to prevent material from spilling beyond the paving width and to provide a smooth edge to the mat.

END GRAIN Face of a piece of lumber that is exposed when the fibers are cut transversely.

END HONEYCOMB End grain check of a piece of lumber that does not extend to any lateral face.

END HOOK Pointed hook, placed at the end of a log and used for loading.

END JOINT *See* **Butt joint.**

END-LAP JOINT Joint formed at a corner where two boards lap. The boards are cut away to half their thickness so they fit into each other, and are halved to a distance equal to their width. When fitted together the outer surfaces are flush.

ENDLESS ROPE Rope with ends spliced together to form a single continuous loop.

END LINE *See* **Brush line.**

END MANHOLE Access chamber at the upstream end of a sewer.

END MARK *See* **Mark.**

END MATCHED Lumber with tongued-and-grooved ends.

ENDO Dimension from the end of a reinforcing bar to a point of reference along its longitudinal axis; i.e., any bar is positioned in the forms transversely by 'cover' or 'spacing' and longitudinally by 'cover' or 'endo.'

END OF REINFORCEMENT *See* **End block.**

ENDOTHERMIC REACTION Chemical reaction that occurs with the absorption of heat.

END PLATE Plate that holds the nonrotating assembly of a mechanical seal and connects it to the seal chamber. *See also* **Gland.**

END PLAY Amount of axial (lengthwise) movement between two parts. *See also* **End thrust.**

END POST Method of mounting two stacks of rails on a common wall at the end of a hoistway.

END PREPARATION Treatment of the end of a length of wire rope, designed primarily as an aid for pulling the rope through a reeving system or tight drum opening. Unlike end terminations, these are not designed for use as a method for making a permanent connection.

END PRESSURE One of five basic conditions characterizing the lay of a tree. It occurs when a log to be bucked is lying straight up and down a slope. When a run in made, the upper section will slide downhill, keeping a continuous pressure on the end of the adjacent log, that will pinch out the saw bar, or if not withdrawn in time, will cause the bar to be hung up. *See also* **Drop, Bottom bind, Side bind,** and **Top bind.**

END REINFORCEMENT Extra reinforcing material applied to the end of a hose product to provide additional strength or stiffening. *Also called* **Block end.**

END RESISTANCE Static soil resistance at the pile end to loading. *Also called* **Tip resistance.**

END-RESULT SPECIFICATION Specification that gives parameters for the completed work but which does not describe or require how the work will be done.

END RING Ring at one end of a packing sct, generally an anti-extrusion ring.

ENDS *See* **Fabric count.**

END SEAL 1. Bond between the end cap and a filter medium; **2.** Sealing device that seals against the end cap by axial contact pressure.

END SECTION Flared metal end attachment to a culvert to prevent erosion and improve hydraulic efficiency.

END SPACING *See* **Panel spacing.**

END SPAN Beam or slab that is continuous only at its interior support.

END SPIDER Cover plate to the upper ends of penstocks, used to prevent logs and debris.

END SPLIT Lengthwisc separation of wood fibers at the end of a piece of lumber.

END TERMINATION Treatment at the end or ends of a length of wire rope, usually made by forming an eye or attaching a fitting, and designed to be the permanent end termination on the wire rope that connects it to the load.

END THREADING Process of making a helical thread on the end of a reinforcing bar or prestressing wire.

END THRUST Pressure exerted in the direction of the ends of a structural member, such as a girder, beam, truss or rafter. *Also called* **End play.**

ENDURANCE LIMIT Limiting stress below which metal will withstand an indefinite number of applications of such stress without fracturing.

ENDURANCE TEST Service or laboratory test, conducted to product failure, usually under normal use conditions.

END USE Proposed or potential final use of landfill property after closure.

ENERGIZED *See* **Alive.**

ENERGY Capacity to do work. There are several definitions, including:

> **Kinetic:** Energy due to motion.

> **Potential:** Energy due to position or condition.

> **Total:** Sum of kinetic energy and potential energy.

ENERGY AUDIT Accounting of all forms of energy inputs over a period.

ENERGY CHIPS Whole-tree chips used for energy

generation.

ENERGY EFFICIENCY RATIO Ratio of net cooling/heating capacity to total rate of electric input under designated operating conditions.

ENERGY HEAD Elevation of the hydraulic gradient, at any section, plus the head.

ENERGY RATIO Standard for damage caused by vibration from blasting, defined as acceleration in m(ft)/sec/frequency.

ENERGY RECOVERY Energy resource recovery where a part, or all, of a waste stream is processed to utilize its heat content to produce hot air, hot water, steam, electricity, synthetic fuel or other useful energy forms.

ENERGY RECOVERY PROCESS Process that recovers the energy content of combustible wastes directly by burning, or indirectly by converting the waste to another fuel form, such as gas or oil.

ENERGY-RELEASE COMPONENT Number related to the rate of heat release per unit area within the flaming front at the head of a moving forest fire.

ENERGY WOOD Wood that has been delivered specifically for burning in boilers. Includes forest, industrial, urban, and other wood waste, as well as whole-tree chips.

ENFILADE Doors aligned on an axis extending through a series of rooms.

eng. *Abbreviation for:* engine.

ENGAGE 1. To mesh with; to interlock; **2.** To attach one element to a simpler and more extensive one so that the former seems to be partly embedded, as an engaged column.

ENGAGED COLUMN Column of less than circular cross section, set against a vertical face.

ENGINE 1. Prime mover; **2.** Machine that uses energy to develop mechanical power. There are several basic types, including:

   Internal combustion: Petroleum- or diesel-driven power plant that converts chemical energy into mechanical energy.

   Naturally aspirated: Engine in which air is introduced to the combustion process by means of atmospheric pressure without the aid of fans or blowers.

   Turbocharged: Engine in which air is injected into the combustion chambers at pressures higher than the atmosphere to increase engine horsepower and performance.

ENGINE ADVANCE MECHANISM System, operated mechanically or by sensing pressure or vacuum, that changes the instant at which fuel injection takes place. Used to cause fuel injection to occur earlier as the engine speed increases.

ENGINE BRAKE *See* **Brake.**

ENGINE CHARGE AIR COOLER Heat exchanger used to cool the charge air of an internal combustion engine after is has been compressed by an exhaust-driven turbocharger and/or mechanically-driven blower. *Also called* **Aftercooler,** and **Intercooler** depending on its location, relative to the final compression stage,

in the air induction system.

ENGINE CLUTCH *See* **Clutch.**

ENGINE CONTROL VALVE When automatic traction control (ATC) is active, the engine control valve works with ATC to maximize traction to the driven wheels by regulating the amount of fuel entering the injection pump.

ENGINE DERATING Reduction in engine output by reducing the maximum flow of fuel.

ENGINE DISPLACEMENT Engine size when all cylinder displacements are added together.

ENGINE-DRIVEN BATTERY CHARGER Battery-charging alternator, or generator, driven by the engine.

ENGINE-DRIVEN GENERATOR Electrical generator driven by a prime mover.

ENGINEER Registered professional responsible for the design of structural elements within a constructed whole.

ENGINEER-ARCHITECT *See* **Architect-engineer.**

ENGINEERED BRICK *See* **Brick.**

ENGINEERED BRICK MASONRY 1. Masonry in which design is based on rational structural analysis; **2.** Load-bearing brick masonry wall.

ENGINEERED 24-IN. FRAMING Building system using structural wood panels over lumber framing spaced 24 in. (620 mm) on center in walls, floor, and roof.

ENGINEERING DESIGN Design system that uses the strength or the load resistant properties of building materials in the design of a structure.

ENGINEERING DRAWING *See* **Structural Drawing.**

ENGINEERING GEOLOGIST Geologist, who may also have engineering training, specializing in the application of geology to engineering problems.

ENGINEERING STUDY Detailed study of the loadings to be accommodated, the structural system to be employed, and the materials suggested for that purpose.

ENGINEER-IN-TRAINING Person qualified for registration as a professional engineer, but lacking the required professional experience.

ENGINEER'S CHAIN *See* **Chain.**

ENGINEER'S HAMMER *See* **Hammer.**

ENGINEER'S LEVEL *See* **Level.**

ENGINE GOVERNOR Device that limits the number of revolutions per minute that a reciprocating engine can develop.

ENGINE PREHEATER Device utilizing an external power source to keep the engine warm in a stationary vehicle.

ENGINE RATING Value of engine power output assigned by the manufacturer, to indicate the maximum power level at which the engine should be applied in a given application.

ENGINE SAFETY CONTROL Device that protects

against catastrophic damage by shutting the engine down in the event of high coolant temperature, low lube oil pressure, low coolant level, or overspeed.

ENGINE SIDESCREEN Rugged screen that fits on the engine housing of a vehicle used at a sanitary landfill to keep paper and other objects from accumulating and impairing engine performance.

ENGINE SPEED Rotating velocity of an engine flywheel, measured in revolutions per minute.

ENGINE TORQUE *See* **Torque.**

ENGLISH BOND *See* **Bond.**

ENGLISH GARDEN WALL BOND *See* **Bond.**

ENGLISH TILE *See* **Roof tile.**

ENGOBE Slip coating applied to a ceramic body for imparting color, opacity, or other characteristics, and subsequently covered with a glaze.

engr *Abbreviation for:* engineer.

ENGRAILED Indented with curved lines or small concave scallops.

engrv. *Abbreviation for:* engrave.

ENLARGED BASE Enlargement of the base area of a pile, formed (a) with a base larger than the shaft of a preformed pile; (b) *in situ*, by driving a plug of concrete into the surrounding ground; or (c) *in situ*, by undercutting soil at the base of a bored pile. *Also called* **Expanded base,** and **Pressure injected footing.**

ENLARGED END In hose, an end having a bore diameter greater than that of the main body of the hose in order to accommodate a larger fitting.

enlg. *Abbreviation for:* enlarge.

enlgd *Abbreviation for:* enlarged.

enr. *Abbreviation for:* en route.

ENRICHMENT Decoration added to an otherwise plain surface or detail.

ent. *Abbreviation for:* enter; entrance.

ENTABLATURE Upper part of an order, consisting of the architrave, frieze, and cornice. *See also* **Architrave.**

ENTAIL Require to be done as part of a larger whole.

ENTASIS The very slight convex curve applied to the shaft of a column, spire, or similar structure to correct the optical illusion of concavity that results if the sides are straight.

entd *Abbreviation for:* entered.

ENTOURAGE Surrounding environment; grounds surrounding a building.

ENTRAINED AIR 1. Air induced into a room by the primary air flow, creating a mixed air path; **2.** Microscopic air bubbles, typically betwen 10 and 1000 mm in diameter and spherical, or nearly so, intentionally incorporated in mortar or concrete during mixing, usually by use of a surface-active agent. *See also* **Air entrainment.**

ENTRANCE FRAME Door frame through which passengers enter or leave an elevator; consists of two side

jambs and one head jamb.

ENTRANCE HALL Circulation space immediately within a building at its entrance.

ENTRANCE HEAD Fluid head required to cause flow into a conduit.

ENTRANCE LOSS Fluid head lost from eddies and friction at the inlet to a conduit.

ENTRANCE SWITCH *See* **Mains breaker.**

ENTRAPPED AIR *See* **Accidental air.**

env. *Abbreviation for:* envelope.

ENVELOPE Total volumetric space occupied by a building.

envir. *Abbreviation for:* environment.

ENVIRONMENT Conditions, circumstances, and influences surrounding and affecting the development of an organism(s).

ENVIRONMENTAL DESIGN Location and design of a facility that includes consideration of the impact of the facility on the community or region based on aesthetic, ecological, cultural, sociological, economic, historical, conservation, and other factors.

ENVIRONMENTAL EVALUATION *See* **Environmental impact assessment.**

ENVIRONMENTAL IMPACT ASSESSMENT Assessment of the impact a development will have on the environment of the site and/or area on which it is to be constructed. *Also called* **Environmental evaluation.**

ENVIRONMENTAL IMPACT STATEMENT Document that identifies and analyzes in detail the environmental impacts of a proposed action.

ENVIRONMENTALLY SENSITIVE AREA In forestry, an area that includes potentially fragile or unstable soils that may deteriorate unacceptably after forest harvesting, and areas of high value to nontimber resources such as fisheries, wildlife, water, and recreation.

EOA *Abbreviation for:* effective on or about.

EOE *Abbreviation for:* equal opportunity employer.

EOM *Abbreviation for:* end of month.

ep *Abbreviation for:* edge of pavement.

EP *Abbreviation for:* explosion proof.

EPD *Abbreviation for:* earliest practical date.

EPICENTER Point within the earth's crust at which an earthquake or earth tremor is reckoned to have begun.

EPOXY Straight-chain thermosetting plastic.

EPOXY CONCRETE Mixture of epoxy resin, catalyst, fine aggregate, and coarse aggregate. *See also* **Epoxy mortar.**

EPOXY ESTER Chemical constituent of a coating formulated to be chemical resistant.

EPOXY GROUT Two-part grout system consisting of epoxy resin and epoxy hardener.

EPOXY MORTAR Mortar containing a hardener, resin, and powder that sets very hard with a durable surface, used for pointing up head and bed joints in glazed tile work where sanitary conditions require. *Also called* **Epoxy thinset adhesive.** *See also* **Epoxy concrete.**

EPOXY PLASTIC Thermosetting plastic having excellent adhesion properties, resistance to chemicals, heat, and weather, plus a slow curing rate.

EPOXY POLYESTER Two-part system used as an interior or exterior coating.

EPOXY RESIN Any of a class of organic chemical bonding systems used in the preparation of special coatings or adhesives.

EPOXY-RESIN GLUE *See* **Adhesive.**

EPOXY THINSET ADHESIVE *See* **Epoxy mortar.**

EPS *Abbreviation for:* emergency power supply.

EPT *Abbreviation for:* external pipe thread.

eq. *Abbreviation for:* equal; equation.

EQUALIZATION OF BOUNDARIES Method of calculating areas or irregular shapes by drawing straight lines that cut off on one side an amount equal to what they add on the other side, forming a conjoined series of triangles.

EQUALIZER 1. Piping arranged to maintain a common liquid level or pressure between two or more vessels; **2.** Suspension device used to transfer and maintain equal load distribution between two or more axles of a suspension; formerly called a rocker beam; **3.** Device used to equalize rope tension in a lifting system; **4.** Culvert that allows standing water to rest at a common elevation about a bank.

EQUALIZER BRACKET Bracket for mounting the equalizer beam of a multiple-axle suspension to a truck or trailer frame, that allows for the beam's pivotal movement. Formerly called the center hanger. Normally of three basic types: flange-mount, straddle-mount, and under- or side-mount.

EQUALIZING BED Layer of ballast or concrete on which pipes are laid in the bottom of a trench.

EQUALIZING CHARGE Extended charge to a measured end point given a storage battery to insure complete restoration of the active materials in all the plates of all the cells.

EQUALIZING SHEAVE Sheave at the center of a rope system over which no rope movement occurs other than equalizing movement.

EQUALIZING TIMER Used in conjunction with an automatic battery charger to insure all cells are charged.

EQUATION OF MOTION Mathematical expression of the relationship between displacement and the initiating dynamic excitation as a function of time.

equil. *Abbreviation for:* equilibrium.

EQUILATERAL ARCH *See* **Arch.**

EQUILIBRIUM Balance between two opposing forces.

EQUILIBRIUM MOISTURE CONTENT Moisture content at which wood neither gains nor loses moisture when surrounded by air at a given relative humidity and temperature.

equip. *Abbreviation for:* equipment.

EQUIPMENT All machinery and equipment, together with the necessary supplies for upkeep and maintenance and also tools and apparatus necessary for the proper construction and acceptable completion of the work.

EQUIPMENT FLOATER Insurance covering equipment and machinery, including mobile equipment, used by a contractor.

EQUIPMENT GROUNDING CONDUCTOR Conductor used to connect the non-current-carrying metal parts of equipment, raceways, and other enclosures to the system grounded conductor and/or the grounding electrode conductor at the service equipment or at the source of a separately derived system.

EQUIPMENT RENTAL RATE Equipment usage charges usually established on a time or mileage basis, including direct costs, indirect costs, and depreciation.

EQUIPMENT TRAIN Assembly of mobile equipment and machinery necessary to complete a specific task, or complete a task in a particular manner. *Also called* **Train.**

EQUIPOTENTIAL LINES Contours of equal water pressure in the soil mass around a water-retaining structure. *Also called* **Flow lines.**

EQUITY Value of real property in excess of any liens against it.

EQUITY OF REDEMPTION Right of an owner to reclaim property after legal right to it has been lost through foreclosure proceedings for nonpayment on a mortgage, by the payment of the debt, interest and costs.

equiv. *Abbreviation for:* equivalent.

EQUIVALENT Alternative designs, materials, or methods that can be demonstrated to provide an equal or greater degree of safety and performance as the item specified.

EQUIVALENT AXLE LOADING Means of calculating loads imposed on an asphalt pavement that takes into account both traffic density and the weight imposed by each axle.

EQUIVALENT DIRECT RADIATION Rate of heat transfer (by both radiation and convection) from a radiator or convector. The equivalent direct radiation is expressed in terms of the surface area of an imaginary standard radiator that would be required to transfer heat at the same rate as does the unit in question. 0.092 m^2 (1.0 ft^2) of EDR gives off 70.32 W (240 Btu/hr) for steam heating units, or 43.9 W (150 Btu/hr) for hot water heating.

EQUIVALENT 18 KIP SINGLE AXLE (WHEEL) LOAD Factor used to express the number of repetitive, but varied wheel loads on an equivalent basis.

EQUIVALENT FLUID PRESSURE Horizontal pressures of soil, or soil and water, in combination that increase linearly with depth, equivalent to that which would be produced by a heavy fluid of a selected unit weight.

EQUIVALENT FOOTING ANALOGY Analogy for replacing the effective bearing of a pile or piles by a simple mass foundation.

EQUIVALENT MASS Single concentrated mass at a

defined radius, having the same rotational charactistics as the distributed and/or discrete masses it represents.

EQUIVALENT RECTANGULAR STRESS DISTRIBUTION Assumption of uniform stress on the compression side of the neutral axis in the strength method of design to determine flexural capacity.

eq. sp. *Abbreviation for:* equally spaced.

ERECT Procedure of building.

ERECTING Positioning and fastening together of structural components.

ERECTING BILL Bill of materials for a structure, arranged so as to facilitate the finding and placing of members during erection.

ERECTION Assembling and connecting on site the members of a structure.

ERECTION SHOP Area where steel frames are joined after fabrication before being shipped as separate pieces to site.

ERECTOR Contractor responsible for the erection of the structural steel.

ERECTOR ARM Swing arm on a boring machine or shield, used for picking up supports or liner segments and setting them in position.

ERG Obsolete unit of energy, equal to 0.1 μJ, that should not be used with the SI system of measurement. Symbol: erg. *See also the appendix:* **Metric and nonmetric measurement.**

ERGONOMIC DESIGN Design of machines, equipment, and parts to best suit the human form and in a manner to reduce fatigue.

ERGONOMICS Study of the interactions between people and work, especially machines and their component parts.

EROSION 1. Progressive wearing away of land through natural actions of streams, wind, etc.; **2.** Progressive disintegration of a solid by the abrasive or cavitation action of gases, fluids, or solids in motion. *See also* **Cavitation damage.**

err. *Abbreviation for:* error.

ERROR OF CLOSURE Difference between a value of a quantity determined by surveying and the theoretical value for the same quantity. There are several distinctions, including:

> **Angles:** Difference between the sum of a succession of angles and the theoretically correct total.
>
> **Azimuth:** Amount by which two values of the azimuth of a line, deduced by different surveys, do not agree with each other.
>
> **Traverse:** Amount by which a value of a position of a traverse location, determined by calculation, fails to equal another value of the same position as established by a different set of observations.

ERROR OF COLLIMATION Angle between the line of sight of a surveying telescope and its collimation axis. *See also* **Collimation adjustment.**

ERRORS AND OMISSIONS INSURANCE Indemnity against loss sustained because of an error or over-

sight by the insured.

ERW *Abbreviation for:* electrical resistance weld.

ES *Abbreviation for:* expansion shield.

esc. *Abbreviation for:* escape; escutcheon.

ESCALATED BASE PRICE Estimate of the base price of a project at the time of tender.

ESCALATION 1. Anticipated increase in uncommitted costs due to the inflation of prices for resources (labor, materials, equipment); **2.** Component of a cost type; the allowance for escalation is a component within the anticipated award of a cost class.

ESCALATOR Power-driven, reversible, endless stairway. *Also called* **Electric stairway.**

ESCALATOR CLAUSE Provision in a contract that recognizes anticipated or unexpected cost increases and makes allowance for them.

ESCAPE Watercourse for discharging an entire stream flow.

ESCAPE CLAUSE Contract provision that allows one or more parties to cancel all or part of a contract under defined conditions, typically if certain events do or do not take place.

ESCAPE ROUTE 1. Planned and brushed-out path used by fallers to make their way clear when the backcut is completed; **2.** Path, clear of obstructions and overhead hazards, users by hookers and buckers working in helicopter logging to move to a predetermined safe position.

ESCARPMENT Steep slope or cliff.

ESCHEAT Reversion to the state (crown) of property when an owner dies intestate and without legal heirs.

ESCROW Use of a disinterested third party to complete a transaction between parties to a written agreement or contract.

ESCUTCHEON Perforated plate around an opening, typically a keyhole plate or the plate to which a door knob is attached. *Also called* **Key plate.**

ESCUTCHEON PIN Decorative nail used to fasten ornamental and/or decorative metal plates to wood.

ESKER Long, narrow ridge formed by glacial meltwater.

esntl *Abbreviation for:* essential.

esp. *Abbreviation for:* especially.

ESPLANADE Level open space used as a public walk.

ess. *Abbreviation for:* essence.

est. *Abbreviation for:* estimate; estimated.

estab. *Abbreviation for:* established; establishment.

ESTATE Extent of interest a person has in real property.

ESTATE AT WILL Indefinite occupation of land and property by a tenant, terminable by either or both parties at will.

ESTER GUM Hard, brittle resin, used in the manufacture of lacquers.

ESTIMATE A calculated prediction. *See also* **Cruise.**

ESTIMATED COST Prediction of cost based on past experience and/or on information supplied.

ESTIMATED *IN SITU* **CUBE STRENGTH** Strength of concrete at a location in a structural member estimated from indirect means and expressed in terms of cubic specimens.

ESTIMATES Predictions or forecasts of costs that will occur in the future, but that are not yet committed; a cost type. Estimates are described or qualified on the basis of the supporting design information available at the time the estimate is prepared. In ascending order of precision and certainty, the following descriptions are used: (a) order of magnitude, (b) concept, (c) design presentation or appropriation grade, and (d) pre-tender.

ESTIMATE TO COMPLETE *See* **Cost types.**

ESTIMATING Judging or calculating the amount of material required for a given item of work, including labor content, and extending it by the cost per unit of measurement to give an approximation of the value of the finished product.

ESTIMATOR 1. One who calculates quantities of materials and labor, and their costs, and the costs of construction. **2.** *See* **Cruiser.**

ESTIMATOR'S CONTINGENCY Estimated allowance for price variances due to the inability of the estimator to price any given item.

ESTOPPEL CERTIFICATE Certificate showing the unpaid principal and interest due on a mortgage as of that date.

esu *Abbreviation for:* electrostatic unit.

ET *Abbreviation for:* edge thickness.

ET AL *Abbreviation for:* et alibi (and elsewhere); *et alii* (and others).

etc. *Abbreviation for:* etcetera.

ETC *Abbreviation for:* estimated time of completion.

ETCH Process of using an acid or caustic to clean the surface of metal or glass, or to put a mark on them.

ETCHED NAIL Nail that has been chemically treated to improve its holding power.

ETCHING PRIMER Primer coating, generally a two-part system, that strips the surface of films, grease, and dirt, lightly etches the surface to provide a key, then produces a corrosion-inhibiting film on the surface to which they are applied. *Also called* a **Wash primer.**

ETHYL ACETATE Lacquer solvent.

e to e *Abbreviation for:* end to end.

et seq *Abbreviation for:* et sequens (and the following).

ETTRINGITE Sulfoaluminate hydrate, which when added in small quantities to cement will make it expand. *Also called* **Cement bacillus.**

EULER FORMULA Mathematical relationship expressing the value of the Euler load in terms of the modulus of elasticity, the moment of inertia of the cross section, and the length of a column.

EULER LOAD Critical load of a perfectly straight, centrally loaded, pin-ended column.

EUROPEAN LEAD *See* **Lead.**

EUTECTIC Alloy or solution of materials that has the lowest possible constant melting temperature.

EUTECTIC DEFORMATION Composition within a system of two or more components that, on heating under specific conditions, develops sufficient liquid to cause deformation at minimum temperature.

EUTECTIC SOLUTION Mixture that melts or freezes at constant temperature and with constant composition. *See also* **Solution.**

ev *Abbreviation for:* electron volt.

ev. *Abbreviation for:* every.

eval. *Abbreviation for:* evaluate; evaluated; evaluation; evaluator.

EVALUATION Study of potential property use, but not to establish its present value.

evap. *Abbreviation for:* evaporate; evaporation; evaporative; evaporator.

EVAPORABLE WATER Water present in set cement paste in capillaries or held by surface forces, measured as that removable by drying under specified conditions. *See also* **Nonevaporable water.**

EVAPORATION Change of state from liquid to vapor.

EVAPORATION LOSS Portion of a lubricant that volatilizes in use or in storage.

EVAPORATION RETARDANT Long-chain organic material such as cetyl alcohol which, when spread on a water film, retards evaporation.

EVAPORATIVE COOLER Air conditioning air by the effect of water evaporation.

EVAPORATOR Part of a refrigerating system in which refrigerant is vaporized to produce refrigeration.

EVAPOTRANSPIRATION Portion of precipitation that is returned to the atmosphere through direct evaporation and vegetative transpiration.

EVASE DUCT Expanding duct connection on the outlet of a fan in an air/gas flow passage; its purpose is to convert kinetic energy into static pressure.

EVEN AGED Forest stand or forest type in which relatively small (10 to 20 year) age differences exist between individual trees.

EVEN-AGED MANAGEMENT Silvicultural system in which the individual trees originate at about the same time and are removed in one or more harvest cuts, after which a new stand is established.

EVEN FLOW Same amount of timber produced annually for an indefinite, extended period of time from a natural forest or other unit of land.

EVENT Point in time where certain conditions have been fulfilled, such as the start or completion of one or more activities.

EVENT-ORIENTED SYSTEMS Systems that present information in terms of events (i.e., points in time).

EVICTION Court directed action to recover possession of real property. There are several qualifications, including:

Actual: Where a tenant or occupier is put out of

possession by process of law or by force.

Constructive: Any interruption of a tenant's possession by the owner making the premises unfit or unsuitable for the leased purpose.

Partial: Where a possessor is deprived of a part of the premises.

evol. *Abbreviation for:* evolution; evolve.

EV1S *Abbreviation for:* edge vee one side.

EV2S *Abbreviation for:* edge vee two sides.

EW *Abbreviation for:* each way.

EWO *See* **Extra-work order.**

ex. *Abbreviation for:* example; excess; extra.

exam. *Abbreviation for:* examination; examine; examiner.

exc. *Abbreviation for:* excavate; excavation; excavator; excellent.

EXCAVATION 1. Any digging operation involving the removal of earth; **2.** Space formed by the removal of earth or rock. There are several classifications, including:

Classified: Excavation paid for at a unit price for common excavation plus a unit price for rock excavation.

Common: General description of 'soft' excavation, such as earth and residual materials.

Rock: Excavation in rocky materials requiring blasting.

Unclassified excavation: Excavation paid for at one unit price, whether common or rock excavation.

EXCAVATION LINE One of a series of lines laid out on the surface of the ground, usually with lime, to indicate where trench excavation should be done.

EXCAVATOR General term for an excavating machine that digs its load by means of a bucket mounted on, or suspended from a boom. May be mounted on a wheeled or tracked chassis. *See also* **Hydraulic excavator.**

EXCESS AIR 1. Air remaining after a fuel has been completely burned; **2.** Air supplied in addition to the quantity required for complete combustion.

EXCESS FUEL DEVICE Any device provided for giving an increased fuel setting for starting only, generally designed to automatically restore action of the normal full-load stop after starting.

EXCESSIVE I/I Quantities of infiltration and inflow that can be economically eliminated from a sewer system by rehabilitation, as determined by cost-effectiveness analysis that compares the cost for correcting the condition with the total cost for transportation and treatment of the infiltration and inflow. *See also* **I/I analysis.**

EXCESS PORE PRESSURE Increment of pore water pressure greater than hydrostatic values, produced by consolidation stresses in compressible materials or by shear strain.

EXCESS PRESSURE PUMP *See* **Pump.**

exch. *Abbreviation for:* exchange.

EXCITATION CURRENT Direct current (DC) that flows through the field coil windings of a DC generator and produces a magnetic field.

EXCITER Direct-current generator that energizes the field magnets of an alternator.

excl. *Abbreviation for:* exclude; excluding; exclusion; exclusive.

EXCLUSION DEVICE *See* **Wiper ring.**

EXCLUSIVE AGENCY Agreement to use a broker or agency for a specified purpose and/or time to the exclusion of all others.

excp. *Abbreviation for:* except.

excpn *Abbreviation for:* exception.

EXDUCER Fluid exit portion of a radial turbine wheel.

exec. *Abbreviation for:* execute; executive.

exec. v-p *Abbreviation for:* executive vice president.

EXFILTRATION Uncontrolled escape of air or water from a structure, system, or vessel.

EXFOLIATED VERMICULITE Very lightweight aggregate (hydrous silicate) used in the production of ultra-lightweight concrete having little compressive strength.

EXFOLIATION 1. Disintegration occurring by peeling off in successive layers; **2.** Swelling up and opening into leaves or plates like a partly opened book.

exh. *Abbreviation for:* exhaust.

EXHAUST BRAKE Device for plugging the exhaust system during engine coast to use exhaust back pressure as an assist to engine braking.

EXHAUST DUCT Duct through which air is conveyed from a room or space to the outdoors.

EXHAUST EMISSIONS Content of the waste gas that leaves a prime mover through its exhaust system.

EXHAUST FAN Fan that withdraws air under suction.

EXHAUST GAS ANALYZER Instrument that measures the quantities of different gases present in an exhaust stream. The result is a measure of the engine's efficiency.

EXHAUST LINE Line returning power or control fluid back to a reservoir or atmosphere.

EXHAUST OPENING Port or void through which air is removed from a space that is being air conditioned.

EXHAUST PORT 1. Outlet fluid path of all fluid-powered hammers; **2.** Combined air inlet and exhaust port of a diesel hammer. *Also called* **Air port.**

EXHAUST PYROMETER Temperature-measuring device that is inserted into one or more of the exhaust passages of an engine allowing the engine operator to gauge the efficiency of the engine.

EXHAUST SHAFT Ventilating passage used to convey air away from rooms.

EXHAUST SYSTEM System of pipes and muffling devices that channels the products of combustion (exhaust gases) from an engine into the atmosphere at a desired location.

exhib. *Abbreviation for:* exhibit; exhibition.

exh. v. *Abbreviation for:* exhaust vent.

exist. *Abbreviation for:* existence; existing.

EXISTING BUILDING Detailed record of a building prior to renovation, remodeling, alteration, or demolition.

EXISTING LAND-USE MAP Map that displays the development and use of an area of land, current at a point in time. It will contain a variety of types of information that may be displayed pictorally, graphically, or texturally, and may include the growth rate of a community, available land, soil types, zoning, community infrastructure (roads, water supply, sewer, electrical distribution, etc.), vegetation, land contours, among others.

EXISTING WORK That part of a structure or system that exists prior to new work being done.

EXIT That part of a means of egress from a structure that leads from the floor area it serves, including any doorway leading directly from a floor area, to another floor area, to a public thoroughfare, or to an approved open space. *See also* **Horizontal exit.**

EXIT GRADIENT Hydraulic gradient (difference in piezometric levels at two points, divided by the distance between them) near to an exposed surface through which seepage is moving.

EXIT LEVEL Lowest level in an enclosed exit stairway from which an exterior door provides access to a public thoroughfare or to an approved open space with access to a public thoroughfare at approximately the same level either directly or through a vestibule or exit corridor.

EXIT SHAFT *See* **Reception shaft.**

EXIT STORY Story from which an exterior door provides direct access at approximately the same level to a public thoroughfare or to an approved open space with access to a public thoroughfare.

EXIT TEMPERATURE Temperature of flue gases as they leave a furnace.

EXOTHERM CURVE Graph of temperature plotted against time during the curing cycle.

EXOTHERMIC HEAT Heat given off by a chemical process.

EXOTHERMIC REACTION Chemical reaction that occurs with the evolution of heat.

EXOTIC SPECIES Plant or animal that is not native to a region but which has been introduced; reciprocal of **Indigenous species.**

exp. *Abbreviation for:* expand; expansion; expense; experiment; export; express.

EXPANDABLE PILOTS Valve-seat grinding pilot that is expanded into the valve guide to hold it in position.

EXPANDED BASE *See* **Enlarged base.**

EXPANDED BLAST-FURNACE SLAG *See* **Blast-furnace slag.**

EXPANDED CLAY Lightweight concrete aggregate produced by heating clay and then crushing and screening it to a suitable size.

EXPANDED METAL Metal sheets that are perforated with narrow slits that are then drawn out to form openings. Used as reinforcing in concrete construction and as a key for plaster or other types of screed. *See also* **Metal lath.**

EXPANDED SHALE Lightweight vesicular aggregate obtained by firing suitable raw materials in a kiln or on a sintering grate under controlled conditions.

EXPANDER SPRING Radial spring employed to drive a piston ring outwards against the cylinder wall.

EXPANDER TUBE BRAKE *See* **Brake.**

EXPANDING CONCRETE Cement, manufactured by adding a proprietary chemical to portland cement, that will produce a concrete that expands by a predetermined amount prior to setting up.

EXPANDING PILE Pile with a mechanical device for expanding the bottom for greater bearing or resistance to uplift.

EXPANDING SHOE BRAKE *See* **Brake.**

EXPANSION *See* **Deformation, Elongation,** and **Swelling.**

EXPANSION ANCHOR A metal expandable unit inserted into a drilled hole that grips by expansion.

EXPANSION BEND Loop in a pipe that permits the expansion and contraction of the pipe.

EXPANSION BOLT Used in conjunction with an expansion shield as a holdfast or anchor in materials that cannot accept a threaded anchor.

EXPANSION CHAMBER Chamber designed to reduce the velocity of the products of combustion and promote the settling of fly ash from a gas stream.

EXPANSION FIT Fit easily made by placing a cold (subzero) inside component within a warmer outside component and allowing an equalization of temperature.

EXPANSION GAUGE Gauge that is connected directly to an engine via an oil line and that registers oil pressure.

EXPANSION JOINT *See* **Joint.**

EXPANSION JOINT FILLER Pliable material used to fill gaps left for expansion so as to exclude foreign matter.

EXPANSION PLUG 1. Soft metal plug that is used to close designed openings in the water jacket of an engine and that will blow out in case of freezing of the engine coolant. *Also called* **Core plug, Freeze-out plug, or Frost plug,** and **Welch plug; 2.** Fiber, plastic, or lead sheath that, when inserted into a pre-drilled hole, expands upon insertion of a screw, providing purchase against the surrounding material.

EXPANSION SHIELD *See* **Wall anchor.**

EXPANSION SLEEVE Tubular metal covering for a dowel bar to allow its free longitudinal movement at a joint.

EXPANSION SLOT Slot extending from the rim of a circular saw in toward the eye to relieve the stresses developed when the saw expands.

EXPANSION STRIP Material used in discontinuous construction to fill the joint between adjacent components.

EXPANSION TANK In a hot water system, a tank designed to allow expansion of the water on heating.

EXPANSION TAPE Strip of resilient material to insulate the edge of glass against rigid contact with nonresilient material.

EXPANSIVE CEMENT Cement that, when mixed with water, produces a paste which, after setting, tends to increase in volume to a significantly greater degree than does portland cement paste, used to compensate for volume decrease due to shrinkage or to induce tensile stress in reinforcement (post-tensioning). It is available in three types:

> **Type K:** A mixture of portland cement, anhydrous tetracalcium trialuminate sulfate, calcium sulfate, and lime.

> **Type M:** Interground or blended mixtures of portland cement, calcium-aluminate cement, and calcium sulfate suitably proportioned.

> **Type S:** A portland cement containing a high computed tricalcium aluminate content and an amount of calcium sulfate above the usual amount found in portland cement.

Also called **Self-stressing cement, Shrinkage-compensating cement,** and **Sulfoaluminate cement.**

EXPANSIVE-CEMENT CONCRETE (MORTAR OR GROUT) Concrete (mortar or grout) made with expansive cement.

EXPANSIVE COMPONENT Portion of an expansive cement that is responsible for the expansion, generally one of several anhydrous calcium aluminate or sulfoaluminate compounds and a source of sulfate, with or without free lime.

EXPANSIVE SOILS Active clay soils that expand on wetting and shrink on drying.

exp.bt *Abbreviation for:* expansion bolt.

EXPECTED BRIDGE LIFE Useful life of a bridge for which it is designed and maintained.

EXPEDITE To carry out in the most efficient manner possible.

EXPENDABLE BASE Type of bottom mast section on static-mounted tower cranes that is cast into the concrete footing of the structure and abandoned following disassembly of the crane.

EXPENDITURE 1. *See* **Cost types**; **2.** Accrual accounting, meaning total charges incurred, including expenses, provisions for retirement of debt, and capital outlays. The making of a payment is a disbursement.

exper. *Abbreviation for:* experience; expert.

EXPERIMENTAL PLOT Area of ground laid out to determine the effects of a certain method of treatment or condition, often divided into subplots.

exp. jt *Abbreviation for:* expansion joint.

expl. *Abbreviation for:* explanation.

EXPLETIVE Stone used to fill a cavity in masonry construction.

EXPLOIT Excavation technique that uses defined materials from a vein or layer, discarding or wasting the remainder.

EXPLORATORY DRILLING Drilling performed to obtain data about subsurface materials and conditions.

EXPLORATORY PROGRAM Series of actions or tests carried out to an established plan and designed to reveal information relevant to and typical of an area or situation.

EXPLOSION Thermochemical process whereby mixtures of gases, solids, or liquids react with the almost instantaneous formation of gaseous pressures and near-sudden heat release.

EXPLOSION, COLLAPSE, OR UNDERGROUND DAMAGE Insurance clause that excludes certain classifications of property damage arising out of explosion, collapse, or underground damage

EXPLOSION-PROOF APPARATUS Electrical apparatus enclosed in a case that is capable of withstanding an explosion of a specified gas or vapor that may occur within it, and of preventing the ignition of a specified gas or vapor surrounding the enclosure by sparks, flashes, or explosion of the gas or vapor within, and that operates at such an external temperature that it will not ignite a surrounding flammable atmosphere.

EXPLOSION SUPPRESSION Technique by which burning in a confined space is detected and arrested during incipient stages, preventing development of pressure that could result in an explosion.

EXPLOSION VENTING Provision of an opening for the release of pressure and heated (explosive) gases, thus preventing the development of destructive pressure.

EXPLOSIVE Any solid, liquid or gaseous mixture or chemical compounds that, by chemical action, suddenly generates large volumes of heated gas, included in the following classifications:

> **Class A Explosives:** Possessing detonating hazard, such as dynamite, nitroglycerin, picric acid, lead azide, fulminate of mercury, black powder, blasting caps, and detonating primers.

> **Class B Explosives:** Possessing flammable hazard, such as propellant explosives, including some smokeless propellants.

> **Class C Explosives:** Including certain types of manufactured articles that contain Class A or Class B explosives, or both, as components, but in restricted quantities.

See also **High explosive** and **Low explosive.**

EXPLOSIVE-ACTUATED FASTENING TOOL Device that uses an explosive charge to propel a fastener or stud. There are several classes, including:

> **High-velocity:** Tool of machine that propels a stud, pin or fastener at velocities in excess of 91 m/s (300 ft/s) measured 2 m (6.5 ft) from the muzzel end.

> **Low velocity:** Tool with a heavy mass hammer supplemented by an explosive load.

> **Low velocity, pistol type:** Captive-type tool that uses an explosive charge to create a muzzzle velocity of less than 91 m/s (300 ft/s) at a

distance of 2 m (6.5 ft).

EXPLOSIVE CHARGE Quantity of explosive that is to be detonated.

EXPLOSIVE DECKS Explosives placed in certain areas of a blasthole, separated by drill cuttings.

EXPLOSIVE FORCE Force exerted on a pile by the explosion of the diesel fuel, that is equal to the gas pressure created by the explosion times the area of the cylinder bore.

EXPLOSIVE WELDING Joining of metals when powerful shock waves create pressure and cause metal to flow with resulting fusion.

expos. *Abbreviation for:* exposed; exposure.

EXPOSED 1. As applied to live electrical parts, capable of being inadvertently touched or approached nearer than a safe distance by a person. It is applied to parts not suitably guarded, isolated, or insulated. *See also* **Accessible** and **Concealed; 2.** As applied to wiring methods, on or attached to the surface or behind panels designed to allow access. *See* **Accessible; 3.** For the purposes of communications systems, where the circuit is in such a position that in case of failure of supports or insulation, contact with another circuit may result.

EXPOSED-AGGREGATE FINISH Decorative finish for concrete work achieved by removing, generally before the concrete has fully hardened, the outer skin of mortar and exposing the coarse aggregate.

EXPOSED AREA That portion of a building material not covered by similar material or other materials, such as roofing shingles or siding, and that bears the full effects of weather.

EXPOSED CONCRETE Concrete surfaces formed so as to yield an acceptable texture and finish for permanent exposure to view. *See also* **Architectural concrete.**

EXPOSED MASONRY Masonry constructed to have no surface finish other than paint.

EXPOSING BUILDING FACE That part of the exterior wall of a building that faces one direction and is located between ground level and the ceiling of its top story, or where a building is divided into fire compartments, the exterior wall of a fire compartment that faces one direction.

EXPOSURE 1. Location of a building or structure in relation to the sun, winds, etc.; **2.** Body of bedrock not covered by the regolith. *See also* **Aspect; 3.** Property that may be endangered by a fire in another structure or by a forest fire; **4.** Ability of a panel to maintain its manufactured properties after prolonged periods of outdoor use or long delays in construction.

EXPOSURE DURABILITY CLASSIFICATION Exposure ratings for American Plywood Association structural wood panels:

Exterior: Panels having a fully waterproofed bond and designed for applications subject to permanent exposure to the weather or to moisture.

Exposure 1: Panels having a fully waterproofed bond and designed for applications where long construction delays may be expected prior to providing protection, or where high moisture conditions may be encountered in service.

Exposure 2: Panels (identified as interior type with intermediate glue) intended for protected construction applications where only moderate delays in providing protection from moisture may be expected.

Interior: Panels that lack further glueline information in their trademarks, manufactured with interior glue, and intended for interior applications only.

expr. *Abbreviation for:* expire; expires; expression.

EXPRESSED WARRANTY *See* **Warranty.**

EXPRESSION OF SLOPE Means by which the rise (or fall) of a slope may be defined relative to the horizontal distance. There are several ways of expressing this, including:

Degree of slope: Angle of the face of a slope above or below the horizontal, i.e., a 5° indicates a slope that measures 5 degrees above a horizontal plane.

Gradient: 1. Ascending or descending with a uniform slope; **2.** Degree of slope;

Percent grade: Measure of the rate of ascent of an inclined plane numerically equal to the vertical rise divided by the horizontal length, multiplied by 100.

Percent of slope: Rise of fall of a slope in m/100 m, i.e., a 3% slope is a rise or fall of 3 m for every 100 meters of horizontal run.

Slope distance: Measured distance between the top and bottom of a slope.

Slope ratio: Expression of the total rise of a slope to its total horizontal distance, i.e., a ratio of 0.05:1 gives a vertical rise of 0.05 m for every 1 m of horizontal distance.

EXPRESSWAY Limited access highway with two or more lanes in each direction, separated by a barrier or median strip, not crossed on the same level by other traffic lanes.

exprn *Abbreviation for:* expiration.

EXPROPRIATION Public authority to take or modify the property rights of an individual or land owner. (In the U.S., the term 'eminent domain' refers to the right to take private property for public use, a right exercised through the process of 'condemnation'). *See also* **Acquisition, Conveyance, Dedication, Eminent domain, Negotiation, Option, Remainder,** and **Severality.**

expt *Abbreviation for:* experiment.

expul. *Abbreviation for:* expulsion.

expwy *Abbreviation for:* expressway.

exsec. *Abbreviation for:* exsecant.

ext *Abbreviation for:* extent; extract.

ext. *Abbreviation for:* extension; exterior; external; extinguisher.

EXTENDABLE-BOOM AERIAL PLATFORM Aerial work platform mounted on a telescopic or extendable boom.

EXTENDABLE FLATBED *See* **Trailer.**

EXTENDED LEAD *See* **Lead.**

EXTENDED LOAD MOUNTING Mounting, including a tag axle, that extends the bridge, permitting greater legal weight capability of a hauler or mixer truck.

EXTENDED OVERHEAD Overhead costs beyond those estimated when bidding, due to completion delays and extended work schedule.

EXTENDED SURFACE Addition of fins, disks, etc., to a heat transfer element to as to increase their heat transfer area.

EXTENDER Finely divided, inert mineral added to provide economical bulk in paints, synthetic resins and adhesives, or other products.

EXTENDER PIGMENT Inert, usually colorless and semitransparent pigment, used in paints as a fortifier.

EXTENDING DIPPER Variable length backhoe dipper that gives the operator more extension of the backhoe without necessarily breaking the digging cycle.

EXTENDING LADDER Ladder that can telescope to 10 m (30 ft) or more.

EXTENSIBILITY Maximum tensile strain that hardened cement paste, mortar, or concrete can sustain before cracking occurs.

EXTENSION 1. Line added to another to increase its length; **2.** 76 m (250 ft) piece of strawline.

EXTENSION AGREEMENT Agreement which extends the terms and/or conditions of a contract, signed by the parties to the original document.

EXTENSION (BOOM OR JIB) Section of a boom or jib that can come in various lengths, one or more of which are and used to increase the overall length of the basic boom or jib.

EXTENSION CYLINDER Hydraulic ram used to extend a section of a telescopic boom.

EXTENSION DEVICE Any device, other than an adjustment screw, used to obtain vertical adjustment of shoring towers, scaffolding, etc.

EXTENSION OF TIME Additional time granted for the completion of something beyond that stipulated in the contract documents.

EXTENSION PAN Accessory that adds length to the floor of a truck dump body and directs material away from the truck during dumping.

EXTENSIONS 1. Extended prices on bid sheets or estimates of cost; **2.** Fixed extension of any given size that is added to an existing screed to add to a paver width. Extensions duplicate the parent screed shape and size for continuity of screed design and performance.

EXTENSION SERVICES Assistance provided by a regulatory authority. In the case of woodland operators, for instance, this may include help with the preparation of forest management plans, cutting permits, marking trees for selective cutting, and guidance in carrying out slash disposal, site preparation, planting, etc.

EXTENSION SPRING *See* **Spring.**

EXTENSION TRESTLE LADDER Self-supporting, portable ladder, adjustable in length.

EXTENSOMETER Apparatus that measures thermal or tensile expansion or contraction in metal bars.

EXTERIOR 1. Not within a shelter or structure; open to the elements; **2.** On some materials, the side or face that should be oriented toward the outside of the building; **3.** The exposed face; **4.** One of four possible Exposure Durability classifications for American Plywood Association panels. Exterior panels that have a fully waterproof bond and are designed for applications subject to permanent exposure to water or moisture.

EXTERIOR ADHESIVE *See* **Adhesive.**

EXTERIOR BOND Gluing together of fibers or wafers with thermal-setting resin to give a bond that cannot be dissolved by water.

EXTERIOR CLADDING Those components of a building that are exposed to the outdoor environment and are intended to provide protection against wind, water, or vapor.

EXTERIOR GLAZED Glass set from the outside of a building using an appropriate glazing compound.

EXTERIOR GLUE *See* **Adhesive.**

EXTERIOR PANEL In a flat slab, a panel having at least one edge that is not common with another panel.

EXTERIOR SIDING Material, usually either panels or strips, that protects the outside of a building.

EXTERIOR STOP Removable glazing molding or bed.

EXTERIOR TRIM That part of an exterior finish other than the wall covering.

EXTERIOR TYPE Plywood manufactured for permanent outdoor or marine use and bonded with 100% waterproof adhesives.

EXTERIOR WALL *See* **Wall.**

EXTERNAL AUDIT Examination of fiscal and other source records maintained by those making claims to an agency.

EXTERNAL GLAZING Glazing in openings in external walls.

EXTERNAL GRINDING Grinding on the outside surface of an object, as distinguished from internal grinding.

EXTERNALLY OPERABLE Electrical system capable of being operated without exposing the operator to contact with live parts.

EXTERNAL PANEL WALL Part of an external wall infilling between structural members.

EXTERNAL THREAD Thread on the outside of a part.

EXTERNAL-TOOTH WASHER *See* **Washer.**

EXTERNAL VIBRATION Vibration device attached to strategic positions on forms to assist in consolidating the concrete during and immediately after placement. *See also* **Internal vibration, Surface vibration** and **Vibration.**

EXTERNAL VIBRATOR *See* **Vibrator.**

EXTERNAL WALL *See* **Wall.**

EXTERNAL YARDING DISTANCE Slope distance from a landing to the farthest point within the cutting unit boundary.

extg *Abbreviation for:* extracting.

EXTINGUISHING MEDIA Chemical mixture designed to combat specific types of fire: electrical, oil and grease, organic combustibles, etc.

extr. *Abbreviation for:* extreme; extrude; extrusion.

EXTRA Addition to a work contract after the contract has been awarded.

EXTRACTION 1. Separation of specific constituents from a matrix of solids or a solution, employing mechanical and/or chemical methods; **2.** Pulling to withdraw a previously installed pile from the ground, usually done with an impact pile extractor or a vibratory pile driver/extractor; **3.** Refining process used in lubricant manufacture to selectively remove unstable components and improve the oil's stability, response to additives and viscosity index.

EXTRACTOR Device for pulling piles or casings out of the ground. *See also* **Pile extractor.**

EXTRADOS *See* Arch.

EXTRA DYNAMITE *See* **Dynamite.**

EXTRA HAZARD *See* **Fire hazard.**

EXTRANEOUS ASH After combustion, that portion of the residue (ash and noncombustibles) that is derived from entrained materials which were mixed with the combustible materials.

EXTRA-RAPID-HARDENING CEMENT Rapid-hardening portland cement containing ground calcium chloride.

EXTRA WORK Item of work not provided for in the contract as awarded but found by the engineer (or other senior responsible person) to be essential for the satisfactory completion of the contract within its intended scope.

EXTRA WORK ORDER Unplanned authorization for a construction contractor to perform work beyond the scope of his contract, and hence a commitment document. An extra work order is generally issued in response to an emergency situation encountered during construction. It constitutes an extension to the contract. *Also called* **Emergency work order, Emergency field order** and **EWO.**

EXTREME COMPRESSION FIBER Farthest fiber from the neutral axis on the compression side of a member subjected to bending.

EXTREME PRESSURE ADDITIVE Additive designed to minimize the tendency of metal surfaces to weld and sieze under conditions of extreme local loads and boundary lubrication. Usually act by forming metal compounds which allow sliding at the contact points or asperities.

EXTREME PRESSURE LUBRICANT Lubricant containing additives designed to increase their capacity to withstand high pressure.

EXTREME PRESSURE PROPERTY Ability of a lubricant to reduce scuffing, scoring, and seizure of contacting bearing surfaces when applied loads are high.

EXTREME SERVICE BUCKET *See* **Bucket.**

EXTREME TENSION FIBER Farthest fiber from the neutral axis on the tension side of a member subjected to bending.

EXTRUDED Forced through the shaping die of an extruder. The extrusion may be solid or hollow in cross section.

EXTRUDED CONCRETE Premixed concrete extruded from the nozzle of a traveling hopper.

EXTRUDED SECTION Particular shape formed by forcing a molten, ductile, or thermosetting material through a die.

EXTRUDER Machine, generally with a driven screw, for continuous forming of rubber, plastic, or other material, through a die. *Also called* **Tubing machine.**

EXTRUSION 1. Operation in which material, while in a plastic state, is forced through a shaping orifice and subsequently sets or hardens to that profile; **2.** Permanent displacement of part of a seal into a gap, under the action of fluid pressure.

EXTRUSION GAP Clearance on the low-pressure side between components that confine a seal.

EXUDATION Liquid or viscous gel-like material discharged through a pore, crack, or opening in the surface of concrete.

EYE 1. Loop at the end of a wire or fiber rope, spliced or press-fitted; **2.** Hole in the center of a circular saw blade so it can be fitted on an arbor.

EYE BAR Bar with an eye formed at one or both ends.

EYEBOLT Bolt with a large eye fabricated to replace the head. The hole or eye is bigger than the diameter of the bolt system. Used to fasten wire rope to wood or steel.

EYEBROW Short curved roof extension, commonly used above wall openings such as windows.

EYEPIECE Lens of a telescope or other optical instrument nearest the observer; the part of the telescope that relays the focused image of the reticle and object to the eye. *Also called* the **Ocular.**

EYE RELIEF Distance at which the eye is placed from the end of the telescope.

EYE SPLICE Loop, with or without a thimble, formed at the end of a wire rope by interweaving the end of the wire rope back into itself.

EZE-SKI Lightweight leveling system with two articulated skis, mounted one at the front and one at the rear of a paver, and a stringline between the two units.

F

f *Abbreviation for:* fine; focal length; frequency; friction factor.

F *Abbreviation for:* Fahrenheit; failed; false; force; front; face.

°F *Abbreviation for:* degree Fahrenheit.

FA *Abbreviation for:* fire alarm.

fab. *Abbreviation for:* fabric; fabricate; fabricated; free aboard.

FABRIC Planar structure produced by interlaced yarns, fibers, or filaments.

FABRICATE Manufacture, form, assemble, construct, etc.

FABRICATED-FRAME SCAFFOLD Supported scaffold consisting of a platform(s) supported on reusable frames fabricated from wood and/or metal components.

FABRICATED PLANK Manufactured platform unit using metal and nonmetal structural components. *Also called a* **Fabricated platform.**

FABRICATED PLATFORM *See* **Fabricated plank.**

FABRICATED SCAFFOLD DECK Work unit equipped with end hooks that engage a scaffold bearer.

FABRICATED SCAFFOLD PLANK Plank used for supporting two users and material. These planks are usually employed with ladder-jack scaffolds, trestles, extension trestles, platforms, or stepladders.

FABRICATED SINGLE-POST SHORE Manufactured post shore. There are two types:

Type I: Single all-metal post with a fine-adjustment screw or other device in combination with pin-and-hole adjustment or clamp. *Also called* **House jack.**

Type II: Single or double wooden post members adjustable by a metal clamp or screw and usually manufactured as a complete unit.

See also **Adjustable timber single-post shore, Post shore,** and **Timber single-post shore.**

FABRICATED TUBULAR FRAME SCAFFOLD Supported scaffold consisting of a platform(s) supported on fabricated end frames with integral posts, horizontal bearers, and intermediate members.

FABRICATION 1. Preparing, making or assembling materials into elements and components, ready for use in a designed whole, and bringing those elements into use; 2. Actual work on reinforcing bars such as cutting, bending, bundling, and tagging.

FABRICATOR Contractor responsible for furnishing fabricated structural steel.

FABRIC COUNT Number of warp ends per mm (in.), and the number of filling picks per mm (in.). *Also called* **Ends.**

FABRIC FATIGUE Fabric degradation and resultant tire cord breakdown due to repeated flexing.

FABRIC FILTER COLLECTOR Asphalt mixing plant component that captures fine material contained in dryer and drum mixer exhaust gases as they pass through bags made of special filter material. The captured dust is generally returned to the mix. *See also* **Baghouse.**

FABRIC FINISH *See* **Fabric impression.**

FABRIC IMPRESSION Pattern in the rubber surface formed by contact with fabric during vulcanization. *Also called* **Cloth impression** and **Fabric finish.** *See also* **Impression.**

FABRIC PICKS/MM Number of filling (weft) yarns per mm.

fac. *Abbreviation for:* facade; facility.

FACADE Whole exterior of a building that can be seen at one view; the principal front.

FACE 1. Exposed surface of a wall or masonry unit; 2. Surface of a unit or material designed to be exposed in the finished structure; 3. End of an excavation toward which work is progressing or that which was last done. *See also* **Faces;** 4. More-or-less vertical surface of rock exposed by excavation or blasting; 5. Any rock surface exposed to air; 6. Edge of the timber stand along which fallers work; 7. Section of wood removed from a tree's base. *See also* **Undercut;** 8. Highest-grade side of any veneer-faced panel that has outer plies of different veneer grades; 9. Either side of a structural wood panel where grading rules draw no distinction between faces; 10. That part of a straight wheel on which cylindrical and surface grinding is usually done; 11. Width of a crusher roll; 12. To overlay one material with another, as to face a block wall with brick; 13. Wider surface of a piece of lumber; 14. Front cutting surface of a drilling, sawing, or cutting tool; 15. Working surface of a hand tool.

FACE-BEDDED STONE Stone laid so that the natural bed is vertical.

FACE BEVEL Angle to which a saw is filed with respect to the saw body. Teeth filed at 90° to the saw body are designated straight, filed at any other angle they are beveled.

FACE-BORING BACKCUT Variation on the standard backcutting procedure, used for particularly large trees or those with a heavy lean. Consists of reducing the amount of wood remaining to be cut in the final backcut by boring in through the already-cut face.

FACE BRICK A well-burned brick especially prepared, selected, and handled to provide an attractive appearance in the face of a wall.

FACE-CHECKING Partial separation of wood fibers parallel to the grain in wood or veneer surfaces of structural wood panels, caused chiefly by the strains of weathering and seasoning. *Also called* **Checking.**

FACE CORD Sometimes used in measuring firewod, a face cord is 1.22 m (4 ft) high by 2.44 m (8 ft) long but only as deep as the length of the individual fireplace pieces.

FACED BLOCK Concrete masonry unit having a face finished in a special manner.

FACED PLYWOOD Plywood faced with a material other than wood.

FACED WALL *See* **Wall.**

FACE GRAIN Direction of the grain of the outer ply (face) of a veneer-faced panel in relation to its supports.

FACE HAMMER *See* **Hammer.**

FACE LAYER Outer layer of drywall board in multilayer applications.

FACE LEFT Position of a theodolite when the vertical circle is to the left of the telescope when viewed from the eyepiece.

FACE LOG Logs placed in a crib parallel to the stream flow.

FACE MARK Mark put on a piece of finished carpentry indicating the face to which all sides are dressed true.

FACE MEASURE Measurement of the area of a board (not the same as board measure, except for 25-mm (1-in.)- thick boards.

FACE MIX Mixture of stone dust and cement used as a facing to concrete blocks to simulate stone.

FACE-MOUNTED TILE Ceramic tile packaged in sheet form with the mounting paper applied to the face of the tiles.

FACE NAILING To nail perpendicular to the initial surface or to the junction of the pieces joined.

FACE PIECE Face waling.

FACE PLATE Part of a marking gauge that presses against the face of a board being scored.

FACE PUTTY Triangular fillet of glazing compound on the exposed face of a light.

FACER BOARD *See* **Barge board.**

FACE RIGHT Position of a theodolite when the vertical circle is to the right of the telescope when seen from the eyepiece.

FACES Vertical or inclined earth surfaces formed as a result of excavation work. *Also called* **Sides.** *See also* **Face.**

FACE SHELL Side wall of a hollow concrete masonry unit.

FACE SHELL BEDDING Mortar applied to face shells for the bed joint, equal in depth to the face shell thickness for the head joint.

FACE SHOVEL Bucket-type shovel that excavates radially forward and upward.

FACE SIDE That side of a piece showing the best quality, or that carries the face mark.

FACE STRING String of a wood stair farthest from the wall.

FACE VENEER Decorative veneer.

FACE WALING Waling across the end of a trench, supported by the ends of the side walings that, together with the end strut also acting as a waling, supports the end face of a trench.

FACING 1. Material used as a finishing surface; **2.** External layer of a wall that is visible and exposed to the weather, supported by a structural wall behind it.

FACING BLOCK Concrete block having an architectural finish on at least one exposed face.

FACING BRICK *See* **Brick.**

FACING MIX Applied finish to a concrete surface employing a special aggregate or cement; the remainder of the structure employing normal aggregate and cement. Site construction consists of a sliding plate inside the normal formwork, distanced a set amount from the exterior face. The two concrete mixes are placed into the combination formwork in their respective areas. After the mixes have stiffened, the plate is slowly removed, allowing the facing mix to bond with the structural backup.

FACING TILE 1. Tile for exterior and interior masonry having exposed face work. **2.** *See* **Tile.**

FACING WALL Lining, usually of reinforced concrete, either precast or *in situ*, constructed against the exposed face of an excavation, used in place of sheeting, supported by the principal timbering, and usually left in place following construction.

fact. *Abbreviation for:* factor; factory.

FACTORED LOAD Load, multiplied by appropriate load factors, used to proportion members by the strength design method.

FACTOR OF SAFETY 1. Ratio of the forces tending to resist failure to those forces tending to cause failure; **2.** Ratio of load, moment, or shear of a structural member at the ultimate to that at the service level; **3.** For wire rope, a term originally used to express the ratio of nominal strength to the total working load but no longer used since it implies a permanent existence for this ratio when, in actuality, the rope strength begins to reduce the moment it is placed in service. For wire rope, *See* **Design factor.**

FACTORY Building used mainly for manufacturing or assembly of goods.

FACTORY AND SHOP LUMBER Lumber intended to be cut up for use in further manufacture, graded on the basis of the percentage of the area that will produce a limited number of cuttings of a specified minimum size and quality, especially in hardwoods.

FACTORY-BUILT CHIMNEY Chimney consisting entirely of factory-made parts, each designed to be assembled with the other without requiring fabrication on site.

FACTORY-BUILT HOUSING *See* **Construction types.**

FACTORY MUTUAL LISTING List of truck models complying with the requirements of and used by the Mutual Group of Insurance Companies to determine the insurance premium rate for various areas of operation.

FAHRENHEIT Non-SI temperature measurement

scale in which water freezes at 32°F (0°C) and boils at 212°F (100°C).

FAIENCE Rough-grade enameled clay products, produced by firing twice: once without and once with the glaze.

FAIENCE MOSAICS Faience tile less than 3870 mm² (6 in.²) in facial area, usually 8 to 9.5 mm (0.3 to 0.365 in.) thick, usually mounted to facilitate installation.

FAIENCE TILE Glazed tile, generally made by the plastic process, showing characteristic variations in the face, edges, and glaze that give a hand-crafted, nonmechanical, decorative effect.

fail. *Abbreviation for:* failure.

FAILURE Breakage, displacement, or permanent deformation of a structural member or connection so as to reduce its structural integrity and its supportive capabilities.

FAILURE ANALYSIS Step-by-step process of examination and analysis that allows the determination of the reason for failure of a part, component or system.

FAILURE OF A PILE FOUNDATION 1. Movement of the pile foundation or any part thereof, either as vertical settlement or laterally, to such an extent that objectionable damage results to the structure supported by the foundation; **2.** Failure of a pile or piles to pass a load test.

FAIR CASH VALUE *See* **Market value.**

FAIR-FACED BRICKWORK Brick surface built neatly and smoothly.

FAIR-FACED CONCRETE Concrete surface that, on completion of the forming process, requires no further (concrete) treatment other than curing. *See also* **Architectural concrete.**

FAIRLEAD 1. Device that enables cable to wind smoothly onto a drum; **2.** Permanently mounted, swiveling roller or sheave arrangement used to permit reeling in a cable from any direction; **3.** Area between the two front quarter guylines.

FAIR MARKET VALUE Price that would be agreed upon in an open and unrestricted market between fully informed, knowledgeable and willing parties dealing at arm's length without constraint.

FALL 1. Amount of slope given to horizontal pipe runs; **2.** Cut down trees in a predetermined and controlled manner.

FALL BLOCK Long, narrow block with a thick shell, a small sheave at one end, and a gooseneck at the other, used in north and south bend systems to add mechanical advantage for lifting the turn to the skyline.

FALLER Worker whose primary purpose is to fell trees and buck them to length for yarding. In western Canada, **Timber faller-bucker** (coastal) or **Tree faller** (interior).

FALLING To cut down a tree by hand.

FALLING HEAD PERMEAMETER Apparatus for determining the permeability of soils.

FALLING MOLD Developed elevation of the centerline of a handrail.

FALLING OBJECT PROTECTION STRUCTURE Attachment to mobile equipment designed to protect operators against falling objects.

FALLING WEDGE Wedge used to throw a tree in the desired direction.

FALL PIPE Downspout.

FALLS Set of blocks with line, used for hoisting.

FALSE BODY Natural stiffness of a thixotropic paint at rest. The paint becomes less stiff when mixed, but will return to its natural viscosity.

FALSE BRINELLING *See* **Fretting.**

FALSE CAR Movable hoistway working platform, used to install brackets and guide rails, etc.

FALSE CEILING Ceiling suspended below another.

FALSE HEADER *See* **Clipped header, Header,** and **Snap header.**

FALSE PILE 1. Temporary pile used to support falsework in construction of cast-in-place concrete bridges or to support superstructure units until they are bolted, riveted, or welded in place or are otherwise made self-supporting in the permanent structure; **2.** Pile used for a temporary bridge, cofferdam or pile template.

FALSE RAFTER Short extension to a main rafter, over a cornice or where there is a change in the roof line.

FALSE SET 1. Rapid development of rigidity in a freshly-mixed portland-cement paste, mortar or concrete without the evolution of much heat, which rigidity can be dispelled and plasticity regained by further mixing without addition of water. *Also called* **Early stiffening, Hesitation set, Premature stiffening,** and **Rubber set.** *See also* **Flash set; 2.** Temporary support for forepoles used in driving a tunnel in soft ground. *Also called* **Horsehead.**

FALSE TONGUE Tongue in a feather joint. *Also called* **Slip tongue.**

FALSEWORK 1. Temporary structure erected to support work in the process of construction, composed of shoring or vertical posting, formwork for beams and slabs, and lateral bracing. *See also* **Formwork; 2.** Steel or wooden supports upon which steel is erected, as in bridges.

fam. *Abbreviation for:* familiar.

FAN 1. Part of a portable chipper that creates an air stream, moving the chips out of the chipper housing; **2.** Floor of scaffold boards projecting over a pedestrian area; **3.** *See* **Alluvial fan; 4.** Power-driven device that blows or sucks air. There are many types, including:

   Attic: Exhaust fan to discharge air from near the top of an attic space while fresh air is drawn in through a louver at a lower level.

   Centrifugal: Fan rotor wheel within a scroll-type housing.

   Forced draft: Device that pushes air or gases.

   Induced draft: Device that draws or exhausts air or gases.

   Overfire air: Device that provides air to the combustion chamber of an incinerator, above the fuel bed.

Propeller: Propeller mounted on an axle or directly-mounted to the shaft of a motor.

Tubeaxial: Propeller or disk-type wheel within a cylinder.

Vaneaxial: Disk-type wheel within a cylinder, usually with air-guiding vanes mounted behind or ahead.

FAN COIL Fan and heat exchanger, usually mounted in a duct of a forced-air heating/cooling system.

FAN ECONOMIZER Means by which a fan motor can be operated as a cold diffuser during the shutdown phase after the coil of a refrigeration unit has been defrosted.

FANG 1. Fishtailed end of a metal bar or railing built into a wall; **2.** Shank of a tool.

FANLIGHT Window over a door, enclosed within the same frame, usually top hinged, sometimes semicircular, thus 'fan-.'

FAN MIXER Air blower in which gas is admitted to the inlet to be mixed with air.

FANNING Spacing tile joints to widen certain areas so they will conform to a section that is not parallel.

FAN VAULT *See* **Vault.**

FAO *Abbreviation for:* finish all over.

FARAD One of 17 derived units with special names of the SI system of measurement: a unit of electric capacitance of a condenser having a charge of 1 C, across the plates of which the potential difference is 1 V. (1 F=1 C/V). Symbol: F . *See also the appendix:* **Metric and nonmetric measurement.**

FAR FACE Face (as of wall) farthest from the viewer; may be the outside or inside face depending on whether one is inside looking out or outside looking in.

FAR FIELD Region sufficiently far from a sound source in which direct transmission of sound waves is negligible.

FARM DRAIN System of draining water from the surface of fields or grass areas by the use of ditches filled with gravel; perforated pipes may also be used. *Also called* **French drain.**

FARMER'S EYE Eye splice formed by unraveling three strands of wire rope, forming a loop, then wrapping the strands together to form a fast, temporary eye. *Also called* **Flemish eye,** and **Flemish splice.**

FAS *Abbreviation for:* first and seconds.

FASCIA 1. Plain, horizontal band, usually in an architrave, often consisting of two or three bands overlapping each other; **2.** Facing to eaves of a roof, covering the structural members. *Also called* **Eaves fascia.**

FASCINE 1. Bundle of long sticks bound together and used for such purposes as filling ditches and making parapets; **2.** Woven willow mattress used along river banks and during river pier construction to minimize scour.

FASTENER Generic term for welds, bolts, rivets, or other connecting devices; a mechanical device designed specifically to hold, join, or assemble single or multiple components. There are three principal types:

Adhesive wall clip: Fasteners that adhere to an appropriate surface without penetrating it.

Hollow-surface: Fasteners designed to afford purchase when inserted through the skin of a hollow wall.

Solid-surface: Fasteners designed to afford purchase when inserted into the face of a solid wall.

fastn *Abbreviation for:* fasten.

fastnr *Abbreviation for:* fastener.

FAST POWDER Explosive having a high speed of detonation.

FAST SAW Circular saw blade that wobbles, weaves, or snakes due to the fact that the rim is too long for the speed at which it runs.

FAST SET *See* **Quick set.**

FAST-TO-LIGHT Paint, the color of which is unaffected by exposure to natural light.

FAST-TRACK CONSTRUCTION *See* **Accelerated design and construction.**

FAST TRACKING *See* **Accelerated design and construction.**

FAT Highly cementitous material worked out of mortar; the working characteristics of highly plastic mortars.

FAT BOARD Mortar board used by bricklayers when pointing.

FAT CONCRETE Concrete containing a relatively large amount of plastic and cohesive mortar.

FAT EDGE Ridge of paint that collects at the bottom of a sloped or vertical surface when too much material has been applied.

FATIGUE 1. Weakening of a material caused by repeated or alternating loads, or from vibration; **2.** Process of progressive fracture resulting from the bending of individual wires in a wire rope. These fractures may, and usually do, occur at bending stresses well below the ultimate strength of the material; it is not an abnormality although it may be accelerated due to conditions in the rope such as rust or lack of lubrication; **3.** Ability of a filter element to resist structural failure of the filter medium due to flexing caused by cylic differential pressure.

FATIGUED Structural failure of a filter medium due to flexing caused by cylic differential pressure.

FATIGUE FAILURE Cracking or separation of a material when subjected to repeated loadings at a stress substantially less than the static strength of the material.

FATIGUE LIMIT Stress limit below which a material can be expected to withstand any number of stress cycles.

FATIGUE RESISTANCE Ability of a material to withstand repeated or alternating stresses.

FATIGUE STRENGTH Greatest stress that can be sustained for a given number of stress cycles without failure.

FATIGUE TEST Test involving repeated fluctuations

or reversals of stress to determine the endurance limit of a material.

FAT LIME Quicklime made by burning pure, or nearly pure limestone, used for plastering and masonry work.

FAT MIX Cementitious mix containing a higher proportion than normal of cement, lime, or some other binder.

FAT MORTAR Mortar containing a high percentage of cementitious elements. It is a sticky mortar that adheres to the trowel.

FATTENING Process by which the viscosity of paint tends to increase the longer it is stored.

FATTENING UP Increasing the plasticity of lime putty by storing it in excess water for approximately 30 days after slaking.

FATTY ACID Acid derived from a natural oil.

FATTY-OIL FLUID Fluid composed of fats derived from animal, marine, or vegetable origin (it may contain additives). *See also* **Fluid.**

FAUCET *See* **Valve.**

FAUCET EAR Projection from a faucet permitting nailing of the fixture to a wall.

FAULT Abrupt break in the continuity of strata or formation of rock by its elevation or depression on one side of a plane of fault.

FAULTING Differential vertical displacement.

FAUX French term loosely defined as a simulation of a real substance. Simulating wood is called **Faux bois**; simulating marble **Faux marbre.**

FAUX BOIS *See* **Faux.**

FAUX MARBRE *See* **Faux.**

fav. *Abbreviation for:* favorable.

FAVORABLE SLOPE (OR GRADE) Downhill (i.e., gravity-assisted) incline for hauling or skidding logs or other loads.

fax *Abbreviation for:* facsimile.

FAYING SURFACE Mating surfaces that fit so closely as to leave no space between them.

FB *Abbreviation for:* fuse block.

f.bk *Abbreviation for:* flat back.

fbm *Abbreviation for:* foot board measure (board foot).

fbr *Abbreviation for:* fiber.

fbrbd *Abbreviation for:* fiberboard.

fbrs *Abbreviation for:* fibrous.

FBT *Abbreviation for:* flat button tip.

fc *Abbreviation for:* foot-candle.

FC *Abbreviation for:* fire code.

fcg *Abbreviation for:* facing.

fclay *Abbreviation for:* fireclay.

fd *Abbreviation for:* feed.

FD *Abbreviation for:* fire department; fire door; floor drain; forced draft.

FDB *Abbreviation for:* forced draft blower.

FDC *Abbreviation for:* fire department connection.

fdn *Abbreviation for:* foundation.

fdr *Abbreviation for:* fire door.

fdry *Abbreviation for:* foundry.

FE *Abbreviation for:* fire escape.

fea. *Abbreviation for:* feature.

FEASIBILITY BUDGET Authorized appropriation of funds, based on an estimate, for the scope of work in the feasibility phase of a project. *See also* **Appropriation.**

FEASIBILITY PHASE That part of a project life during which the concept design is prepared. *See also* **Construction phase,** *and* **Development objectives.**

FEASIBILITY STUDY/REPORT Investigation of the probability that a project could be completed as proposed, within the terms of reference of the study, that may be limited to finances, or materials, or structural competence, or scheduling, or any combination of these and other considerations.

FEATHER 1. Blending the edge of one material into another; **2.** Wood slip separating the sash weights in a sash window; **3.** Cross tongue that joins matched boards.

FEATHERED Sloped off to a very thin edge.

FEATHERED EDGE 1. Junction of jointing compound and the surface of drywall panels when used to conceal joints between panels; **2.** Skived edge of drywall jointing tape.

FEATHER EDGE 1. Wood or metal tool having a beveled edge and used to straighten reentrant angles in finish plaster coat; **2.** Edge of a concrete or mortar patch or topping that is beveled at an acute angle.

FEATHER-EDGE BOARD Board that tapers across its wider face, used as weather boarding and close-boarding.

FEATHER-EDGED BRICK *See* **Brick.**

FEATHER-EDGED COPING Coping stone having one edge thicker than the other so that moisture is shed from its upper face.

FEATHER EDGE RULE Rule used in plastering, after a floating rule or for working angles.

FEATHEREDGING TILE Tile from which the body has been chipped beneath the facial edge in order to form a miter.

FEATHER JOINT Carpentry joint made between two boards in which a tongue is inserted into a groove plowed into the squared and butting edge of each.

FEATHERING Process of manipulating hydraulic controls so that a smooth, flowing action is created.

FEATURE STRIP Narrow strip of tile or other material that has a contrasting color, texture, or design.

fed. *Abbreviation for:* federal.

FEE Payment for a professional service.

FEE CURVE Means of determining a graduated fee percentage based on estimated construction costs.

FEED 1. Process of supplying material to a processor or conveyor; **2.** Mechanism that pushes a drill into the work. *See also* **Cross feed,** and **Down feed.**

FEEDABILITY Ability of a lubricating grease to flow to the suction of a dispensing pump at a rate at least equal to pump delivery capacity. *Also called* **Slumpability.**

FEEDBACK Transmission of current or voltage from the output of a circuit or device back to the input, where it interacts to modify operation of the circuit or device.

FEED BEAM Beam along which a rock-drill hammer slides. A motor is mounted on the structure to provide feed force through a chain or cable. *Also called* **Feed rails.**

FEEDER All circuit conductors between the service equipment, or the generator switchboard of an isolated plant, and the final branch-circuit overcurrent device.

FEEDER CONVEYOR *See* **Conveyor.**

FEED INDEX Cylindrical grinding. Measurement indicated by the index of the machine (on most machines this measurement refers to the diameter of the work; on a few to the radius).

FEED LEGS *See* **Air legs.**

FEED LINES Pattern on the work produced by grinding.

FEED MAIN Horizontal pipe in an automatic sprinkler system that connects: water supply to sprinkler riser; sprinkler riser to another sprinkler riser; sprinkler riser to cross main; or cross main to another cross main.

FEED PIPE 1. Main-line water pipe that carries a supply directly to the point of use, or to secondary lines; **2.** Pipe supplying feed water to a boiler.

FEED PLATE Vertical steel plate that prevents a tree in a portable chipper from passing beyond the disk.

FEED PUMP *See* **Pump.**

FEED RAILS *See* **Feed beam.**

FEED RATE Rate at which a material passes a cutting tool, measured in meters per minute or feet per minute.

FEED TRAVEL Distance a drilling machine moves the steel shank in traveling the length of its feeding range.

FEEDWATER Heated water pumped into a water-wall furnace to compensate for steam leaving the furnace.

FEEDWATER HEATERS Heat exchangers used as part of the feedwater system.

FEEDWATER SYSTEM System of pumps, deaerating feedwater heater, piping, and heat exchangers used to optimize the thermal efficiency of a steam cycle.

FEED WHEEL Material distributor or regulator in certain types of shotcrete equipment.

FEEL Working qualities of paint.

FEELER HOLE Hole driven ahead of an excavation face for exploratory purposes.

FEE SIMPLE Largest and most extensive estate, or full ownership, in property.

FEINT Bent edge made to cappings and flashings of flexible-metal roofing to form a capillary break.

FELL *See* **Harvest functions.**

FELLED AND BUCKED Timber that is felled and bucked, ready for yarding.

FELLER *See* **Faller,** and **Harvesting machines (single function).**

FLLER-BUNCHER *See* **Harvesting machines (multifunction).**

FELLER-CHIPPER *See* **Harvesting machines (multifunction).**

FELLER-DELIMBER *See* **Harvesting machines (multifunction).**

FELLER-DELIMBER-BUNCHER *See* **Harvesting machines (multifunction).**

FELLER-DELIMBER-SLASHER-BUNCHER *See* **Harvesting machines (multifunction).**

FELLER-DELIMBER-SLASHER-FORWARDER *See* **Harvesting machines (multifunction).**

FELLER-FORWARDER *See* **Harvesting machines (multifunction).**

FELLER-SKIDDER *See* **Harvesting machines (multifunction).**

FELLING Act of cutting down a standing tree. *Also called* **Falling** when done by hand.

FELT 1. Matted organic fibers used as a basis in the manufacture of composition sheet roofing; **2.** Compacted fibers of various materials used in flexible sheet form as insulation.

FELT-AND-GRAVEL ROOF Flat roof covered with bitumen felt topped by gravel.

FELTING DOWN Flatting a dry varnish or paint surface with a felt pad loaded with abrasive powder and lubricated with water.

FELT NAIL Clout nail.

FELT PAPER Building paper of strong, tough paper base saturated with hot bitumen and rolled smooth; used under roofing and siding materials as a protection against moisture and air infiltration.

fem. *Abbreviation for:* female.

FEMALE Item having an end that encloses a like item.

FEMALE COUPLING Internally threaded pipe.

FEMALE THREAD Internal thread.

FEMTO Prefix representing 10^{-15}. Symbol: f. Used in the SI system of measurement. *See also the appendix:* **Metric and nonmetric measurement.**

fen. *Abbreviation for:* fender.

FENCE 1. Barrier separating one area from another; **2.** Adjustable strip mounted on a table or machine to guide work in cutting, milling, or sanding; **3.** Molding-sand wall placed around a mold prior to pouring; **4.** Post in a lever-tumbler lock.

FENCE PLIERS *See* **Pliers.**

FENDER 1. Deflector that controls spray thrown from a rotating tire; **2.** Protective system built to prevent marine traffic from damaging bridge piers and docks.

FENDER CAP Horizontal member resting on, and framed to the top of fender piles.

FENDER CLUSTER Cluster of fender piles.

FENDER LOG Log placed at the end of a bridge on the road shoulder as a continuation of a guardrail to guide traffic onto the bridge, or used as a floating log protecting a pier or abutment from damage from floating objects.

FENDER PILE *See* **Pile.**

FENDER SYSTEM Piles and/or timbers used as guides to a bridge opening or dock and along the face of a wharf, sea wall, or other waterfront structure to absorb shock from impact and to minimize damage to structures and vessels.

FENDER WALL *See* **Wall.**

fenes. *Abbreviation for:* fenestration.

FENESTRATION Arrangement of windows in a building.

FERROCEMENT Composite structural material comprising thin sections consisting of cement mortar reinforced by a number of very closely spaced layers of steel wire mesh.

FERROUS Containing, or derived from iron.

FERROUS AGGREGATE CONCRETE Hard-wearing concrete floor surface finish where granulated ferrous aggregate replaces part of the aggregate in a normal concrete mix.

FERROUS METAL PRIMER Primer coating formulated for application on iron and steel to resist oxidation.

FERRULE 1. Short tube or bushing sweated or soldered onto a tube or shaft; **2.** Metallic collar placed over a hose end to anchor a coupling against the hose. The ferrule may be crimped, forcing the hose in against the shank of the coupling, or the shank may be expanded, forcing the hose out against the ferrule, or both. *See also* **Clamp,** and **Hose clamp.**

FERRULE BUTT-HOOK Socket on the butt-rigging into which the choker knob is fitted.

FERTILIZER An artificial or natural substance added to soil to provide one or more of the nutrients essential to the growth of plants; the principal ones being nitrogen, phosphorus, and potassium.

FESCUE *See* **Grass.**

FESTOON LIGHTING String of outdoor lights suspended between two points more than 4.57 m (15 ft) apart.

FETCH Straight-line distance between the rim of a dam and the farthest reservoir shore.

FETTLE Finishing-off work in a trade.

FF *Abbreviation for:* far face.

FFA *Abbreviation for:* full freight allowance.

ffh *Abbreviation for:* free fork height.

ffl *Abbreviation for:* finished floor level.

ffm *Abbreviation for:* full free mast.

ffqm *Abbreviation for:* full free quadruple mast.

fftg *Abbreviation for:* firefighting.

fftm *Abbreviation for:* full free triple mast.

FG *Abbreviation for:* fine grain; finish grade; flat grain.

fgr *Abbreviation for:* finger.

fh *Abbreviation for:* fire hydrant; fork height; frame height.

FH *Abbreviation for:* fire hose; flat head; full hard.

FHC *Abbreviation for:* fire-hose cabinet.

FHR *Abbreviation for:* fire-hose rack.

FHT *Abbreviation for:* fully heat-treated.

fhy. *Abbreviation for:* fire hydrant.

FIBER 1. For the purpose of microscopic particle counting, a particle whose length is greater than 100 micro-meters but at least ten times its width; **2.** General term for any long, narrow cell of wood; **3.** Wood volume produced by a tree or trees that can be converted into wood products, such as lumber, paper, or cardboard. *Also called* **Pulpwood;** **4.** In lubricating grease, form in which soap thickeners occur.

FIBER BALL Unwanted ball of chopped fibers collecting together in a concrete mixer.

FIBERBOARD Any of a number of sheet materials manufactured of refined or partially refined wood fibers.

FIBER CEMENT Hardened cement paste containing fibers.

FIBER CENTER Cord or rope of vegetable or synthetic fiber used as the axial member of a strand.

FIBER CONCRETE Cement-rich fine-particled concrete containing fibrous aggregate such as asbestos or sawdust.

FIBER CORE Cord or rope of vegetable or synthetic fiber used as the axial member of a rope.

FIBER FUEL Process where the combustible fraction of a solid waste is extracted, shredded, and used as a fuel.

FIBERGLASS Material made of spun, woven, matted, or chopped fibers of glass made by spinning melted glass into filaments.

FIBERGLASS-REINFORCED PLASTICS Polyester resins reinforced with glass fibers, the basic material from which a wide range of sheet and formed products is manufactured.

FIBERGLASS SHINGLE *See* **Shingle.**

FIBER OPTICS Method of transmitting light through continuous glass fibers.

FIBER-REINFORCED CONCRETE Concrete containing dispersed, randomly oriented fibers.

FIBER ROPE Rope fabricated from nonmetallic materials such as vegetable, animal, or synthetic fibers.

FIBER SATURATION POINT Stage in the drying

and wetting of wood at which the cell walls are saturated and the cell cavities are free from water, usually taken as approximately 30% moisture content based on weight when oven-dry.

FIBRIL Extremely small fiber, usually barely visible even at maximum magnification of the electron microscope. Fibrils may collect in bundles to form larger fibers.

FIBROUS PLASTER Plaster slabs consisting of canvas coated with thin layers of gypsum plaster.

FICTILE Formed or molded of clay.

FIDDLE BLOCK Block made up of two sheaves in the same plane, held in place by the same cheek plates.

FIDELITY BOND *See* **Bond.**

FIDELITY INSURANCE *See* **Insurance.**

FIDUCIAL LINE/POINT Line or point of established reference.

FIDUCIARY Person who transacts business or handles money or property not their own on behalf of another.

FIELD 1. Expanse of wall between openings, corners, etc.; **2.** Surface area of a piece of drywall panel; **3.** Space near a field magnet in which magnetic lines of force exist.

FIELD BENDING Bending of reinforcing bars on the job rather than in a fabricating shop.

FIELD BOOK Surveyor's notebook; the permanent record of field measurements and readings.

FIELD CAPACITY 1. Water that a soil in place retains, when drainage has become negligible. *Also called* **Field moisture capacity**; **2.** Amount of water retained in solid waste after it has been saturated and has drained freely.

FIELD CHECK On-site inspection of work, conditions, and/or materials.

FIELD CONCRETE Concrete delivered or mixed, placed, and cured on the job site.

FIELD-CURED CYLINDERS Test cylinders that are left at the jobsite for curing, as nearly as practicable in the same manner as the concrete in the structure, to indicate when supporting forms may be removed, additional construction loads may be imposed, or the structure may be placed in service.

FIELD MOISTURE CAPACITY *See* **Field capacity.**

FIELD NOTES Permanent record showing details of a survey, such as title, instrument used, date, etc. May be used as legal evidence in court cases.

FIELD OFFICE Satellite office responsible for day-to-day on-site organization and operations.

FIELD OF VIEW Width that can be sighted in relation to the distance from a telescope, expressed in angular units in surveying instruments.

FIELD ORDER Written order, issued by the architect or owner's agent, that orders a minor change in the work not involving a change in the contact sum or guaranteed maximum price, or the contract time.

FIELD REPRESENTATIVE *See* **Owner's representative.**

FIELDSTONE 1. Rubble; **2.** Stone used for a structural or decorative purpose without dressing or working.

FIELD SUPERVISION *See* **Site supervision.**

FIELD TEST Experiment conducted under field conditions. Ordinarily less subject to control than a formal experiment; it may also be less precise. *Also called* a **Field trial.**

FIELD TILE 1. Pipe installed as a subsurface drain to collect groundwater; may be perforated along its length or laid in short lengths with a gap at each joint to allow infiltration; **2.** Area of similar tile covering a wall or floor, the field being covered by tile trim. *See also* **Accent tile.**

FIELD TREATMENT On-site application of wood preservative.

FIELD TRIAL *See* **Field test.**

FIELD WELD Weld done in the field during the assembly of parts. *See also* **Shop weld.**

FIELD WIRING DIAGRAM Drawing showing the physical relationship between components and where the wires of various circuits are connected.

FIFO *Abbreviation for:* first-in first-out.

FIFTH OUTRIGGER *See* **Bumper outrigger.**

FIFTH WHEEL Weight-bearing swivel connection between haulage tractors and the trailer or equipment they are hauling. There are several types:

   Angle: Where the fifth-wheel pivot brackets are attached to two parallel lengths of angle iron, that fit over and are bolted to the frame rails.

   Oscillating: Providing side-to-side pivoting in addition to normal front-to-back.

   Plate: Where the fifth-wheel is attached to a rigid rectangular platform that, when bolted to the frame, acts as an extra crossmember to absorb stress and inhibit flexing.

   Sliding: Permitting movement of the fifth wheel forward or rearward to accommodate a variety of trailer kingpin settings, to vary load distribution between the tractor's front and rear axles, or to increase or decrease swing clearance as needed.

   Stationary: Installed on the frame at an operator-specified location ahead or over the centerline of the rear axle, as determined by the trailer kingpin setting and optimum weight distribution between the tractor's front and rear axles.

FIFTH-WHEEL PICKUP RAMP Sloping structure behind a tractor's fifth wheel that lifts and guides the upper coupler assembly of a trailer.

FIFTY-YEAR FLOOD Flood that has a 2% chance of occurring in a given year, based on historical data. *See also* **One-hundred year flood.**

fig. *Abbreviation for:* figure.

FIGHT HANG-UPS To change or reset the chokers to clear the turn being yarded when it gets hung up behind a stump or boulders.

FIGURE EIGHT Method of setting a choker on two logs when one is crossed over the other.

FIL *See* **Shoot wire.**

fil. *Abbreviation for:* filament; fillet; fillister.

FILAMENT Long, thin, threadlike material; often extruded plastic such as nylon.

FILAMENT REINFORCEMENT Continuous fibers of boron, carbon, crystalline ceramic, fused silica, glass, metal, nylon, etc., used as reinforcement for cement. *Also called* **A-glass.** *See also* **Aligned fibers.**

FILE Metal hand tool consisting of a bar or rod of various sizes and cross sections including flat (mill, flat and hand, in order of thickness), pillar (thick and flat), half-round, round, square, and taper (can be round or triangular). The face or faces of the bar are covered with small teeth or sharpened ridges, used to abrade wood and scrape away metal. *See also* **Riffler.** There are several classifications, each available in various graduations including smooth, dead smooth, second cut, bastard, and rough:

> **Curved cut:** Having slightly curved rows of parallel lines.
>
> **Double cut:** Having two sets of parallel lines crossing each other, each set at an angle to the line of the blade.
>
> **Rasp cut:** Short, triangular, separate teeth, sometimes slightly curved in their height.
>
> **Single cut:** Having one set of parallel lines set at an angle to the line of the blade.

FILE CARD Wire brush used for cleaning debris from between the teeth of files.

fil.hd *Abbreviation for:* fillister head.

FILIGREE Delicate and ornamental fretwork in metal or wood.

FILL 1. Any material that is moved or added to the existing terrain to raise its elevation; **2.** Height to which material is to be placed; **3.** Earthwork resulting from the placement of fill; **4.** Coarse grade of earth having low compressibility and good stability, used to fill in around walls or foundations, or to raise the level of ground. *Also called* **Soil fill.** *See also* **Backfill, Compacted fill,** and **Structural fill. 5.** *See* **Fuel tank capacity.**

fill. *Abbreviation for:* filler; filling.

FILL BANK Fill material used to shape a road from the outer edge of the traveled portion to its intersection with the existing ground profile.

FILL BULKHEAD Bulkhead that, after being driven and anchorages constructed, has backfill placed behind it so as to form a wharf. *Also called* a **Dredge bulkhead.**

FILL CAP FILTER Filter that covers the fill opening to a reservoir tank and which filters makeup fluid.

FILL CAPACITY Of a fuel tank, 5% less than the nominal capacity, which allows for fuel expansion.

FILL DAM *See* **Earth dam.**

FILLER 1. Substance used to fill holes, particularly in woodworking and plasterwork, that can usually be finished by cutting, planing, or sanding once it has set

up and dried; **2.** Finely divided inert material such as pulverized limestone, silica, or colloidal substances sometimes added to portland-cement paint or other materials to reduce shrinkage, improve workability or act as an extender; **3.** Material used to fill an opening in a form; **4.** Metal added in making welded, brazed, or soldered joints; **5.** Inert substance added to a plastic to make it less costly, and/or improve physical properties such as hardness, stiffness, and impact strength; **6.** Drying vehicle and an inert filler, used to level out the pores or cells of coarse-grained wood such as oak; **7.** Material added to a lubricant to increase bulk or density. Depending on type and amount, a filler may contribute to, detract from, or have no effect on lubricating properties of the grease. **8.** *See also* **Adapter; 9.** Heavily pigmented paint, used to fill imperfections or pores in a substrate; **10.** *See* **Shoot wire; 11.** Piece of steel plate used between two members to make the proper connection.

FILLER BLOCK Concrete masonry unit used in conjunction with concrete joists in floor and roof construction.

FILLER BOARD Material generally used in expansion joints, primarily to provide support to a sealant so that it can resist pressure from the surface. Filler boards are used to form the joint when one structural component is placed or cast against another. Commonly used materials are impregnated fiber or cork.

FILLER COAT First coat of paint applied to raw wood to fill the pores and provide a smooth surface for finishing coats.

FILLER PLATE Steel plate or shim used to fill space between compression members.

FILLER RING Ring of rubber or other material used to fill the hinge of a V-ring or U-ring.

FILLER STRIP Free-flowing rubber used under the tread of a tire when added thickness is needed.

FILLER WIRE 1. Small spacer wires within a strand that help position and support other wires; **2.** The type of strand pattern utilizing filler wires.

FILLET 1. Narrow band separating two moldings; **2.** Small square molding crowning a larger one. *Also called* a **List; 3.** Narrow band between flutes of a column; **4.** Weld metal in the internal vertex, or corner of an angle formed by two pieces of metal, giving the joint additional strength; **5.** Beveled inside corner, usually 45°, to avoid a sharp right-angle change in direction at the intersection of two concrete members. *See also* **Chamfer strip.**

FILLET AREA Angled, ground area between a crank journal and crankshaft.

FILLET BEAD *See* **Fillet seal.**

FILLET SEAL Triangular-sectioned material used to seal a right-angled joint. *Also called* a **Fillet bead.** *See also* **Sliding joint.**

FILLET WELD Weld made in the interior angle where the members meet at a right angle.

FILL FACE Active area of a landfill, where solid waste is being unloaded and then compacted into the daily cell.

FILLING IN Process of building in the center of a wall between the face and the back.

FILLING-IN GANG Crew of workers that follow the raising gang, 'filling in' or completing the erection of miscellaneous pieces of a structural steel frame. *See also* **Raising gang.**

FILLING-IN PIECE Piece of lumber shorter than its neighbors, such as a jack rafter.

FILLING PIECE Piece of lumber placed on another to make a plane surface.

FILLING THREAD Thread or yarn running at a right angle to the warp. *Also called* **Weft.**

FILL-IN PLANTING Tree planting required to supplement poorly stocked natural regeneration or to replace seedlings that have died on previously planted sites.

FILLISTER 1. Hand plane used for cutting grooves and rabbets; **2.** Rounded head of a machine screw.

FILL-TYPE INSULATION Loose-fill thermal insulation poured from sacks, hand packed, or blown in.

FILM 1. Layer of material not thicker than 0.25 mm (0.01 in.); **2.** Thin, continuous layer of liquid applied to a surface.

FILM FORMER Protective treatment that fills masonry pores, forming a continuous film on the surface.

FILM STRENGTH Ability of a film of lubricant to resist rupture due to load, speed, or temperature.

FILM THICKNESS In a dynamic seal, the distance separating the two surfaces that form the primary seal.

FILTER 1. Device having a porous medium, whose primary function is the separation and retention of particulate contaminates from a fluid or air stream. *See also* **Bypass filter; 2.** Device that permits the passage of waves and currents of certain frequencies only.

FILTERABLE SOLIDS Solids retained on a membrane for analysis by weight, count, or observation as a measurement of contamination.

FILTER BAG Device designed to remove particles from a carrier gas or air by passage of the gas through a porous fabric medium.

FILTER BED Fill of rock and pervious soil that receives the effluent from a septic tank.

FILTER BLOCK Hollow, vitrified clay masonry unit, sometimes salt-glazed, designed for trickling filter floors in sewage treatment plants.

FILTER CAKE 1. Suspended solids deposited and compacted on a porous medium during filtration; **2.** Mud deposited on the walls of a drill hole. *See also* **Cake.**

FILTER EFFICIENCY Ability, expressed as percent, of a filter to remove specified artificial contaminants from a specified fluid under specified test conditions.

FILTER ELEMENT Subassembly of a filter that contains the filter medium or media.

FILTER HOUSING Ported enclosure that contains the filter element and directs fluid or air flow.

FILTER MEDIUM (MEDIA) Porous material that performs the process of particle separation and retention.

FILTER PACK Clean graded sand and gravel placed in the annulus of a drilled water well to prevent formation material from entering the screen.

FILTER PERFORMANCE Those factors which describe the functions and attributes of a filter element.

FILTER PRESS Device for extracting moisture from a slurry to leave a semisolid cake.

FILTER PRESSURE DIFFERENTIAL Drop in pressure due to flow across a filter or element at any time. The term may be qualified by adding 'initial,' 'final,' or 'mean.'

FILTER RATED FLOW Maximum flow rate of a fluid of specified viscosity for which a filter is designed.

FILTER RATING Measure by which filter performance may be evaluated and rated, including:

> **Absolute:** The diameter of the largest hard spherical particle that will pass through a filter under specified test conditions. This is an indication of the largest opening in the filter element.
>
> **Filtration (b$_x$):** The ratio of the number of particles equal to and greater than a given size (x) in the influent fluid to the number of particles equal to and greater than the same size (x) in the effluent fluid.
>
> **Nominal:** An arbitrary micrometer value, based on weight percent removal, indicated by a filter manufacturer. Due to lack of reproducibility, this rating is depreciated.

FILTRATION RATIO Ratio of the number of particles of a given size entering a filter to the number of particles of the same size leaving a filter.

FIN 1. Narrow linear projection on a formed concrete surface resulting from mortar flowing into spaces in the formwork; **2.** Type of blade in a concrete mixer drum; **3.** Thin projection on a casting. *See also* **Flashing.**

fin. *Abbreviation for:* finance; financial; finish.

FINA *Abbreviation for:* following items not available.

FINAL ACCEPTANCE Acceptance by the owner or his agent of a completed construction project from the contractor with an obligation for final payment for the work done. *See also* **Acceptance of work, Interim acceptance,** and **Partial acceptance.**

FINAL BLOW COUNT Number of blows per inch, foot or other unit length of measure at which the driving of the pile or soil sampling device was stopped.

FINAL COMPLETION Stage when all work is completed and all contract requirements are fulfilled by the contractor.

FINAL DATE *See* **Closing date.**

FINAL DESIGN Project design that is approved by all responsible parties, including the owner, and from which working drawings and other contract documents are prepared.

FINAL DRIVE Reduction gearing close to or inside of a drive wheel. There are four common designs: single reduction, double reduction, triple reduction, and planetary.

FINAL FILTER Last stage of a multistage filter system.

FINAL FINISHER Equipment, forming part of a concrete paving train, consisting of an oscillating surface-finishing unit set at 60° to the center line of the pavement.

FINAL GRADE Height and shape of a finished landfill after the cap has been installed, or after settlement is complete.

FINAL INSPECTION Final review of a project by the owner's representative that leads to issuance of the final certificate for payment.

FINAL LAYOUT Drawing showing the general arrangement of related components supplied by a subcontractor and approved by the owner's representative.

FINAL PAYMENT Payment by the owner to the contractor, upon issuance of the final certificate for payment by his appointed agent, of the entire unpaid balance of the adjusted contract sum.

FINAL SET 1. Net penetration a pile moves under one blow or a specified series of blows at the end of driving (reciprocal of final blow count); **2.** Degree of stiffening of a mixture of cement and water greater than initial set, generally stated as an empirical value indicating the time in hours and minutes required for a cement paste to stiffen sufficiently to resist to an established degree the penetration of a weighted test needle.

FINAL SETTING TIME Time required for a freshly mixed cement paste, mortar, or concrete to achieve final set. *Also called* **Setting time.**

FINAL SETTLING BASIN Tank through which the effluent from a waste effluent oxidizing process is passed to remove settleable solids.

FINAL-STAGE MANUFACTURER Manufacturer that performs such manufacturing operations on an incomplete machine or vehicle that it becomes complete and ready for its intended purpose.

FINAL STRESS Stress that exists after substantially all losses have occurred in prestressed concrete.

FINANCE CHARGE Amount included in an installment sales contract to cover the cost of interest on the funds advanced and the costs of servicing the account, including a risk factor.

FINANCIAL EXPENSE Expenses relating to the cost of financing an operation.

FINANCIAL MATURITY Age at which a stand of timber offers the maximum return on investment in terms of volume and grade yield.

FINANCIAL ROTATION Rotation of tree crops determined solely by financial considerations (that are related to biological production potential) in order to obtain the highest monetary values over time, in terms of optimum net present value or return on investment.

FINANCING LEASE *See* **Lease.**

FINE ADJUSTMENT SCREW 1. Screw having a very fine pitch thread; **2.** Tangent screw on a surveyor's telescope.

FINE AGGREGATE *See* **Aggregate,** and **Soil types (sand).**

FINE COLD ASPHALT Wearing course of aggregate and bitumen, spread and compacted either hot or cold.

FINE GRADING Finish grading that follows rough grading in order to bring a profile to the necessary elevation.

FINE-GRAINED Rocks in which the minerals are less than 1 mm (0.04 in.) in diameter.

FINE-GRAINED SOIL Soil in which the smaller grain sizes predominate, such as fine sand, silt, and clay.

FINE GRAVEL Rock fragments ranging in size from 2 to 6 mm (0.08 to 0.25 in.)

FINE INCHING CONTROL Machine operator's control function that permits selection of a range of preselected engine speed settings. *See also* **One-touch deceleration/acceleration.**

FINELY GRADED AGGREGATE Sample of aggregate containing a high proportion of small particles.

FINENESS Measure of particle size.

FINENESS MODULUS Factor obtained by adding the total percentages of material in a sample that are coarser than each of a specified series of sieves (cumulative percentages retained) and dividing the sum by 100.

FINE RESIDUE Residue not suitable for chipping, such as sawdust, shavings, and veneer clippings.

FINES Finer-grained particles of soil, clay, silt and sand.

FINE SAND Rock fragment ranging in size from 0.06 to 0.2 mm (0.008 to 0.002 in.)

FINE SOLDER Solder composed of 50% tin and 50% lead.

FINE STUFF Plaster finishing coat.

FINGER JOINT Glued joint consisting of a series of interlocking fingers, precision machined on the ends of two pieces of wood to be joined.

FINGER LINK Substantial temporary connector used to secure pass chains or line stringing equipment that must be disconnected when under tension. *Also called* **Pelican hook.**

FINGER PLATE Metal plate fastened near the doorknob or doorpush to protect the surrounding finish.

FINGERS 1. Comb or drag used in plastering to roughen and key a preliminary coat; **2.** *See also* **Pants.**

FINIAL Ornament at the apex of a roof, pediment, canopy, tower corner, etc.

FINISH Texture of a surface after finishing operations have been performed.

FINISH CARPENTRY Installation of doors, trim, baseboard, etc.

FINISH COAT 1. Exposed coat of plaster or stucco; **2.** Final thin coat of shotcrete preparatory to hand finishing; **3.** Protective coat, generally applied last in a decorative painting process. Finish coats may be water- or paint-thinner-soluble.

FINISHED FLOORING Surface material used as a

finish to a constructed floor. *See also* **Floor.**

FINISHED INTERIOR Interior space ready to occupy in all respects: painted and wallpapered as required, with permanent floor coverings and all hardware and plumbing fixtures installed, plus lighting fixtures where specified.

FINISHED SIZE Overall measurements of any object completely finished and ready for use.

FINISHED STRING End string of a stair, fastened to the rough carriage.

FINISHER Tradesperson skilled in the finishing aspects of the trade and/or materials.

FINISH FLOOR Top or wearing surface of a floor system.

FINISH GRADE Surface elevation of an improved surface following grading.

FINISH GRADING Earthmoving work necessary to bring the site grades to those levels necessary for final drainage patterns and landscaping shapes.

FINISH GRINDING 1. Final grinding of clinker into cement, with calcium sulfate in the form of gypsum or anhydrite generally being added; **2.** Final grinding operation required for a finished concrete surface, e.g., bump cutting of pavement, fin removal from structural concrete, terrazzo floor grinding, etc.

FINISH HARDWARE Hardware that is visible on the final work; usually of better quality and finish. *See also* **Hardware.**

FINISHING 1. Leveling, smoothing, consolidating and otherwise treating surfaces of fresh or recently placed concrete or mortar to produce the desired appearance and service; **2.** Final cuts taken with a grinding wheel to obtain accuracy and the surface desired; **3.** Third major stage in plumbing: includes installation of fixtures and other exposed components. *See also* **Rough in; 4.** Final covering and treatment of surfaces and their intersections.

FINISHING MACHINE Power-operated machine used to produce the desired surface texture on a concrete slab.

FINISHING NAIL *See* **Nail.**

FINISHING OFF Preparing the finished surface of cabinet work or carpentry.

FINISHING TOOL Tool specially designed and used to impart a final finish on material being worked, applied, or fitted.

FINISHING TROWEL *See* **Trowel.**

FINISH JUMBO Traveling support for workers when repairing concrete or other tasks on the arch of a tunnel.

FINISH LUMBER Superior grade, as opposed to common or lower grades.

FINISH OUT ALLOWANCE Specific sum in a lease for office or retail space for the tenant to customize the space being occupied.

FINISH PLASTER Final or white coat of plaster.

FINISH SCREEN One of a series of vibrating screens (preferably horizontal) operated at a concrete batching plant so that excessive amounts of significantly undersized material are removed and delivered directly to their appropriate batcher bin without intermediate storage.

FINISH WEARING SURFACE *See* **Traffic surface.**

FINITE ELEMENT ANALYSIS Numerical analysis that allows complicated boundary shapes and variations in material properties to be introduced into the problem to be solved.

FINS 1. Thin metal plates spaced perpendicular about a heating element to help disperse warmth. *See also* **Fin.**

FIRE 1. Rapid oxidation of combustible materials resulting in light and heat; **2.** Act of initiating an explosive reaction in blasting.

FIREBACK 1. Firebrick construction above a hearth that reflects heat into the room and conducts smoke up and through the throat above; **2.** Plate of cast or wrought iron, often ornamented, at the back of a fireplace above hearth level to protect masonry from direct heat and flame.

FIRE BARRIER Fire resistant walls, doors, and similar construction to prevent spread of a fire in a building. *See also* **Fire stop.**

FIRE BEHAVIOR Manner in which fuel ignites, flame develops, and fire spreads, etc.

FIRE BLOCK Short pieces of wood, or blocks, nailed between studding to serve as braces and, in case of fire, to stop drafts and prevent the spread of fire to other parts of the building.

FIRE BOSS Person responsible for all suppression and service activities on a forest fire.

FIREBOX Fireplace area, built of firebricks and other masonry elements, where the fire is built.

FIREBRAND Source of heat, natural or man-made, capable of igniting a forest fire.

FIREBREAK 1. Area or strip of less flammable fuels that is either natural (standing timber or landslide) or is made in advance (cat trail or road) as a precautionary measure separating areas of greater fire hazard. *Also called* **Fuelbreak; 2.** Any natural or constructed barrier utilized to segregate, stop, and control the spread of fire or to provide a control line from which to work.

FIRE BREAK Any fire resistant fitting or construction: fire door, closed stairwell, concrete floor, division wall, etc.

FIREBRICK *See* **Brick.**

FIRE CAMP Temporary camp used to accommodate workers and equipment while suppressing a forest fire.

FIRE CLAY Earthy or stony mineral aggregate that has, as an essential constituent, hydrous silicates of aluminum with or without free silica, plastic when sufficiently pulverized and wetted, rigid when subsequently dried, and of suitable refractoriness for use in commercial refractory products, used for the manufacture of firebrick, furnace linings, etc.

FIRE COMPARTMENT Enclosed interior space in a building that is separated from all other parts of the building by enclosing construction providing a fire separation having a required fire resistance rating.

FIRE CONTROL Process of controlling an unwanted forest fire.

FIRE CONTROL EQUIPMENT Range of tools, machinery, vehicles, etc. used in forest fire control.

FIRE CONTROL IMPROVEMENTS Structures primarily used for forest fire control, such as lookout towers, housing, telephone lines, radio station, roads, etc.

FIRE CONTROL PLANNING Design of organizations, facilities, and procedures to protect forests and similar land areas from fire.

FIRECUT Angular cut at the end of a joist that is anchored in masonry. In the event of a fire the joist will collapse without forcing the wall to fall outward.

FIRE DAMAGE Loss caused by fire, present and estimated future.

FIRE DAMPER Closure that consists of a normally-held-open damper installed in an air distribution system or in a wall or floor assembly, and designed to close automatically in the event of a fire in order to maintain the integrity of the fire separation.

FIRE DANGER Measure of the likelihood of a forest fire, based on temperature, relative humidity, wind force and direction, and the dryness of the woods.

FIRE DANGER RATING Management system that integrates a range of fire-related factors into a single index.

FIRE DANGER RATING AREA Geographical area throughout which the fire danger is adequately represented by that measured at a single fire danger station; area relatively homogeneous in climate, fuels, and topography.

FIRE DECK Parting line of the block and head of an engine, surrounding the cylinder bore, where the hottest temperatures and highest pressures of combustion are produced.

FIRE DEVIL Small-diameter and low-energy cyclone of short duration that develops when heated air or hot gasses (as from a fire) rise at a rate faster than the surrounding atmosphere, and that can carry sparks, ash, and burning debris away from the area of an active fire.

FIRE DISTRICT Fire prevention/fighting organization organized and equipped to serve a defined area.

FIRE DIVISION PARTY WALL See **Wall.**

FIRE DIVISION WALL See **Wall.**

FIREDOG Metal support for logs in a fireplace; an andiron.

FIRE DOOR Comprises two types:

> **Heat-actuated fire door:** One in which a mechanism operates under the action of heat, causing the door to close automatically;

> **Self-closing fire door:** Fire door normally closed and designed to close automatically upon being opened.

FIRED STRENGTH Compressive or flexural strength of refractory concrete determined upon cooling after

first firing to a specific temperature for a specified time.

FIRED UNIT WEIGHT Unit weight of refractory concrete, upon cooling, after having been exposed to a specified firing temperature for a specified time.

FIRE EDGE Boundary of a fire at a point in time.

FIRE ENDURANCE Ability of a material or assembly to meet defined fire test criteria.

FIRE ESCAPE Device used for escape from a building: chute, exterior stairway, ladder, etc.

FIRE EXTINGUISHER Portable device containing chemicals that can be sprayed on a fire to extinguish it.

FIRE GUARD 1. Metal screen that can be placed in front of an open fire; **2.** Man-made barrier (often an area cleared of fuels) constructed at the time of a fire to control it and provide a point from which to carry out fire suppression.

FIRE HAZARD Degree of hazard from fire according to the type of occupancy and contents of a structure, classified as:

> **Light hazard:** Where the quantity and/or combustibility of contents is low and fires with relatively low rates of heat release are expected.

> **Ordinary hazard, Group 1:** Where combustibility is low, quantity of combustibles is moderate, stock piles of combustibles do not exceed 2.4 m (8 ft), and fires with moderate rates of heat release are expected.

> **Ordinary hazard, Group 2:** Where quantity and combustibility of contents is moderate, stock piles do not exceed 3.6 m (12 ft), and fires with moderate rates of heat release are expected.

> **Ordinary hazard, Group 3:** Where quantity and/or combustibility of contents is high, and fires of high rates of heat release are expected.

> **Extra hazard:** Where quantity and combustibility of contents is very high, where flammable liquids, dust, lint, or other materials are present, introducing the probability of rapidly developing fires with high rates of heat release.

FIRE HEADQUARTERS Control center for operations against a particular forest fire.

FIRE HYDRANT Water supply outlet with a wrench-actuated valve and connection for a fire hose.

FIRE LIABILITY Legal responsibility for loss or damage to the property of others caused by fire.

FIRE LINE Cleared area extending down to mineral soil that surrounds a forest fire to prevent the fire from reaching fresh fuels.

FIRE LINTEL Horizontal flap attached to an elevator cab upper door panel to close the space between the panel and the hoistway when the doors are closed, required to seal the hoistway in case of fire.

FIRE LOAD As applied to occupancy, means the combustible contents of a room or floor area expressed in terms of the average weight of combustible materials per square meter (square foot), from which the potential heat liberation may be calculated based on the calorific value of the materials. It includes the furnishings, finished floor, wall and ceiling finishes, and temporary and movable partitions.

FIREMAN'S SERVICE Devices that provide (a) a signal for the immediate recall of an elevator car to a designated landing, to remove such cars from normal use, and (b) to permit special operation of an elevator system for fire fighters.

FIRE PACK Tools, equipment, and supplies, readied in advance, to be carried by one person to fight a forest fire.

FIRE PARTITION *See* **Wall.**

FIREPLACE Open place at the bottom of a chimney.

FIREPLACE INSERT Device that increases the efficiency of a fireplace by circulating room-temperature air around the firebox, where it is heated, and then moving it back into the room. May be convection driven or use electric fans to increase circulation.

FIRE POINT Lowest temperature at which vapor from asphalt will ignite and remain burning. *Also called* **Flash point.**

FIRE POLISH Making glass smooth by the action of flame.

FIRE PREVENTION Planning and activities taken in advance to prevent the outbreak of fire and/or to minimize loss if fire occurs.

FIRE PROGRESS MAP Map that shows the state of development of a large fire and the disposition of forces deployed to fight it.

FIREPROOF Something that is not burnable (a redundant term). *See* **Fire resistance.**

FIREPROOF DOOR *See* **Door types.**

FIREPROOFING Material or combination of materials protecting structural members to increase their fire resistance. *See also* **Concrete cover.**

FIREPROOFING TILE Tile designed for protecting structural members against fire.

FIREPROOF WOOD Chemically treated wood, used where fire resistive materials are required.

FIRE PROTECTION Procedures, processes, materials and resources developed and used to prevent the outbreak of fire or for its containment and extinguishment.

FIRE PROTECTION DISTRICT Rural or suburban fire protection area that maintains its own fire apparatus or contracts with an adjacent fire department for fire protection.

FIRE PROTECTION RATING Time in hours or fraction of an hour that a closure, window assembly, or glass block assembly will withstand the passage of flame when exposed to fire under specified conditions of test and performance criteria.

FIRE PUMP Water pump having certain performance characteristics, complete with ancillary equipment such as hoses, nozzles, etc., reserved for and available for use to combat fire.

FIRE-RATED DOOR Door designed to resist standard fire tests.

FIRE-RATED SYSTEM Wall, floor, and roof construction, of specific materials and designs, that has

been tested and rated according to fire safety criteria (e.g., flame spread and fire resistance).

FIRE RESISTANCE Time that a structural element will perform its normal functions when subjected to standard heating conditions under laboratory test conditions. *See also* **Fireproof.**

FIRE RESISTANT COATING Paint or coating that will not support combustion and that will provide an effective fire barrier to the surface on which it is applied. *See also* **Fire-retardant coating.**

FIRE RESISTANT FLUID Fluid, difficult to ignite, that shows little tendency to propagate flame. *Also called* **Nonflammable fluid.** *See also* **Fluid.**

FIRE RESISTANT RATING Time in hours (or major fraction of an hour) that a material will withstand the passage of flame and the transmission of heat at a set distance and temperature.

FIRE RESISTING FINISH Paint formulation based on silicones, polyvinyl chloride, chlorinated waxes, urea formaldehyde resins, casein, borax, or other noncombustible substances that reduces the spread of fire on combustible materials.

FIRE RESISTING FLOOR/WALL Floor or wall construction that has a fire rating appropriate to the occupancy and fire load of the building.

FIRE-RESISTIVE CONSTRUCTION *See* **Construction types.**

FIRE RETARDANT Any substance that, by chemical or physical action, reduces the flammability of forest fuels, usually added to water.

FIRE RETARDANT COATING Paint or coating that will not support combustion, but that will burn, depending on the material to which it is applied. *See also* **Fire-resistant coating.**

FIRE RETARDANT RATING Rating, based on standard tests of fire-resistive and fire-protective characteristics, of building materials and assemblies.

FIRE-RETARDANT-TREATED Chemical treatment of wood and plywood to retard combustion.

FIRE RISK Chance of a fire starting.

FIRE SAFETY OFFICER Person responsible for identifying the accident and health hazards to forest fire suppression forces and for advising the fire boss on means of keeping the hazards to a minimum.

FIRE SCAR Fresh or healing injury of the cambium of a woody plant, caused by fire.

FIRE SCREEN Fine mesh metal screen that can be placed in front of an open fire to contain sparks.

FIRE SEASON Time(s) of year when fires are most likely to occur, rated according to potential hazard.

FIRE SEPARATION Construction assembly that acts as a barrier against the spread of fire, and that may or may not be required to have a fire-resistance rating.

FIRE SERVICE Fire prevention and fighting organization.

FIRE SETTING Starting of a forest fire, usually with malicious intent.

FIRE SPRINKLER SYSTEM System of water supply pipes, heat detectors, and sprinkler heads, mounted on or near the ceilings of a building, that automatically detects heat or the products of combustion above a preset limit and causes a continuous spray of water to cascade over the area below. There are several types, including:

Dense system: Type of dry-pipe sprinkler system in which all sprinklers are open and where pressure-sensitive devices, rate-of-temperature-rise releases, and/or heat sensitive controls activate the main water supply control valve, causing the entire system to activate flooding of the protected area. *Also called* **Deluge system.**

Dry-pipe system: Automatic sprinkler system that is normally dry, with no water in the sprinkler pipes above the dry-pipe valve; installed where there is a danger of freezing if water were in the pipes.

Wet-pipe system: Automatic sprinkler system that has water in the pipes right up to each sprinkler at all times.

FIRE STOP 1. Projection of brickwork on the walls between the joists to prevent the spread of fire; **2.** Blocking of incombustible material used to fill air passages through which flames might travel. *See also* **Draft stop; 3.** Horizontal block nailed between studs to deter the spread of fire.

FIRE STOP FLAP Device intended for use in horizontal assemblies required to have a fire resistance rating and incorporating protective ceiling membranes that operate to close off a duct opening through the membrane in the event of a fire.

FIRESTORM Violent convection caused by a large continuous area of intense fire, often characterized by destructively violent surface indrafts near and beyond the perimeter, and sometimes by tornado-like whirls.

FIRE SUPPRESSION ORGANIZATION Management structure that enables the line and staff duties of the fire boss to be carried out; supervisory and facilitating personnel assigned to forest fire suppression under the direction of a fire boss.

FIRE SURROUND Facing to the chimney breast, either side and above the fire opening or fireplace.

FIRE TAPING Taping of drywall joints without subsequent finishing coats in areas where final appearance is not important.

FIRE TOOL CACHE Placing of a supply of fire tools and equipment at a strategic location as part of a forest fire prevention or fire fighting plan.

FIRE TOWER Stairway serving all floors of a building, plus the exterior at ground level, accessible only through fire doors, and enclosed with fire-resistant construction. *Also called* **Lookout.**

FIRE TRAIL Cleared area constructed around logging slash or other fire hazards in order to prevent the spread of fire to this hazardous material.

FIRE TRIANGLE Oxygen, heat, and fuel: the three factors necessary for combustion and flame production.

FIRE TRUCK Any motorized vehicle carrying or fitted with fire-fighting equipment.

FIREWALL *See* **Wall.**

FIRE WARDEN Person responsible for fire control within a defined area.

FIRE WATCH Worker who remains at a logging site for approximately two hours at the end of the day to watch for possible fires caused by logging activities.

FIRE WEATHER Weather conducive to the start and rapid spread of fire.

FIRE WEATHER STATION Meteorological station specially equipped to measure weather elements important to forest fire prevention and control.

FIRE WHIRLWIND Revolving mass of air caused by a forest fire; may have sufficient intensity to snap off large trees.

FIRE WINDOW Window with its frame, sash, and glazing that, under standard test conditions, meets the fire protection requirements for the location in which it is to be used.

FIRING 1. Igniting prepared material; **2.** Ignition of combustible gases within an internal-combustion engine, either from compression (diesel) or by a spark plug (gasoline); **3.** Charging of fuel into a furnace; **4.** Exposure to heat of clay products in a kiln.

FIRING ORDER Order or sequence in which the cylinders of an internal combustion engine fire on the power stroke.

FIRING OUT *See* **Burning out.**

FIRING RATE Rate at which air, fuel, or an air-fuel mixture is supplied to a burner, or furnace.

FIRM PRICE CONTRACT Contract in which no variation of bid price is allowed.

FIRMER *See* **Wood chisel.**

FIRMER TOOLS Woodworking tools commonly used on a workbench, typically chisels, gouges.

FIRMWOOD Sound or solid wood in a log, suitable for either chips or solid wood products such as lumber or veneer.

FIRST ATTACK Initial suppression work on a forest fire.

FIRST-CLASS LEVER *See* **Lever.**

FIRST FLOOR Floor of a structure most nearly level with the ground. *Also called* **First story.** *See also* **Ground floor.**

FIRST GROWTH Timber from a forest that has not been previously logged. *Also called* **Virgin timber.**

FIRST-IN FIRST-OUT *Also called* **FIFO.** *See* **Cost flow.**

FIRST MORTGAGE *See* **Mortgage.**

FIRST-ORDER ANALYSIS Analysis based on first-order deformations in which equilibrium conditions are formulated on the undeformed steel structure.

FIRST REDUCTION GEARING *See* **Differential pinion.**

FIRST STORY *See* **First floor.**

FISH Coil of lightly tempered wire, used to thread through conduit, including through bends and around corners, when pulling wires. *Also called* a **Snake.**

FISHED A butt joint strengthened by wood nailed to the sides.

FISHING Recovering an object left, lost, or dropped in a confined or closed-in space, such as a drill hole or wall.

FISH JOINT Splice of two pieces placed end-to-end and secured by plates placed on each side and covering the joint, securely bolted.

FISH LADDER Channel containing a series of low weirs or ramps through which water continuously flows, used to help fish bypass a dam or waterfall. *Also called* a **Fishway**.

FISH MOUTH Crescent-shaped wrinkle in roofing felt.

FISH OIL Oil extracted from some species of fish and used as a vehicle in some specialized paints and varnishes.

FISHPLATE 1. Plate attached to the web of a chassis frame side member, running along the frame length to resist vertical loads; **2.** Plate or plank fastened to the sides of and joining two members laced end-to-end, forming a splice; **3.** Steel plate that spans the joint where two lengths of guide rail in a stack meet.

FISHTAIL Wedge-shaped piece of wood used as part of the support form between tapered pans in concrete joist construction. *See also* **Bit**.

FISHTAIL BIT 1. Device shaped like a fish tail, used to better utilize jetting water; **2.** Reaming bit used on the end of a flight auger.

FISHTAIL BOLT Anchor bolt with a split tail, cast into concrete or masonry.

FISHTAILS *See* **Pants**.

FISH TAPE Flexible steel wire or strip, used to push or thread through a conduit or space so that a wire or group of wires may be attached and pulled back to the originating point.

FISHWAY *See* **Fish ladder**.

FISSILE Capable of being split, as schist, slate, and shale.

FISSURE Fracture of rocks along which the opposite walls have pulled apart.

FISSURED SOIL *See* **Soil types**.

FIT Clearance or interference between mating parts.

FIT CLASSIFICATIONS Relative tightness of an assembly, classified as:

Class 1: Loose,:a large allowance, used where accuracy is not essential.

Class 2: Free: a liberal allowance, used for running fit at 600 rpm or over, and shaft pressure over 4137 kP (600 psi).

Class 3: Medium allowance: used for running fits under 600 rpm, journal pressures under 4137 kP (600 psi), and for sliding fits.

Class 4: Snug, zero allowance: the closest fit that can be assembled by hand. Used when moving parts are not intended to move freely.

Class 5: Wringing, zero, or negative allowance: commonly used where an assembly is not to be interchanged.

Class 6: Tight, slight negative allowance: where light pressure is required for assembly of parts.

Class 7: Medium force, negative allowance: where considerable pressure is required for assembly.

Class 8: Heavy force and shrink, considerable negative allowance: used where material can be highly stressed.

FIT-UP Formwork that can be struck and reused.

FITCH Small paint brush with a long handle, used to reach nearly inaccessible areas.

FITMENT An attachment.

FITTING 1. Any functional accessory, including:

Bushing fitting: Short externally threaded connector with a smaller size internal thread.

Closure fitting: Cap or plug.

Compression fitting: Fitting that seals and grips by manual adjustment and deformation.

Connector fitting: Fitting for joining a conductor to a component port or to one or more other conductors.

Flared fitting: Fitting that seals and grips by a preformed flare at the end of a tube.

Flareless fitting: Fitting that seals and grips by means other than a flare.

Reusable hose fitting: Hose fitting that can be removed from a hose and reused.

2. Electrical accessory, such as a lock nut, bushing, or other part of the wiring system, that is intended primarily to perform a mechanical rather than an electrical function; **3.** Part of a piping system, excluding valves, that joins lengths of pipe.

FITTING GAIN Amount of space inside a fitting for which pipe allowance must be made.

FIVE-QUARTER CUT Lumber that measures a true 31.75 mm (1.25 in.) thick, frequently used for stair treads.

FIVE-YEAR LIABILITY Implied or actual warranty effective for five years from date of purchase or acceptance or other defined date.

fix. *Abbreviation for:* fixture.

FIXED ARCH Arch not provided with hinges.

FIXED-AREA PLOT SAMPLING METHOD Controlled cruise method where small plots of a fixed size are used to sample a portion of a forest area to obtain information (such as tree volume) that can be used to describe the whole area.

FIXED ASSET Tangible, noncurrent asset, such as land, buildings, equipment, etc., held for use rather than for sale.

FIXED BASE Column base designed to resist rotation in addition to vertical or horizontal movement.

FIXED BEAM Beam fixed at one end.

FIXED BUDGET *See* **Budget**.

FIXED CAB Crane with a nonrotating operator's cab. *Also called* **Down cab**.

FIXED CAPITAL Investment in capital assets.

FIXED CHARGE Unavoidable expense, e.g., interest, rent, etc.

FIXED COSTS Operation costs that will remain relatively constant for all levels of output.

FIXED DELAY Delay caused by traffic controls. *See also* **Delay,** and **Operational delay.**

FIXED DISPLACEMENT HYDRAULIC SYSTEM Hydraulic system that produces a constant flow rate regardless of the actuators engaged at a particular time. Excess hydraulic flow is returned to the reservoir through return lines. Low line pressures provide just enough force to allow the oil to overcome the friction of movement, but are not sufficient to engage hydraulic actuators.

FIXED EARTH SUPPORT Pressure distribution assumed in the design of anchored sheet-pile walls where a point of contraflexure is assumed due to the fixed bottom-end support of the wall.

FIXED END Anchored end of a beam, girder, or strut.

FIXED EXPENSES Property expenses that remain the same regardless of occupancy.

FIXED FEE Provision of services for a stipulated sum.

FIXED-FORM PAVING TRAIN Procession of different types of special machinery that performs continuous concreting of road or airfield pavements. The train is supported on flat-bottomed rails or on previously cast concrete strips that also control the horizontal and vertical alignment. A typical train comprises side feeders into which mixed concrete is placed, spreaders to distribute the concrete over the width of the forms, compactors and finishers, a placer for positioning dowel bars or other joint steel, joint groove-forming and finishing equipment, final finish machine, surface texturing equipment, cure application, and protective tenting equipment.

FIXED GRATE *See* **Grate.**

FIXED-HEAD PIER Pier restrained at its head against lateral movement.

FIXED LADDER Ladder that cannot be readily moved or carried because of being an integral part of a building or structure.

FIXED LEAD *See* **Lead.**

FIXED LIGHT Window that is not made to open.

FIXED PACKER Adjunct of a refuse container system that compacts refuse at the site of generation into a detachable container.

FIXED PAYMENT MORTGAGE *See* **Mortgage.**

FIXED-PLATFORM TRUCK Industrial truck equipped with a load platform and not capable of self loading.

FIXED PRICE CONTRACT *See* **Stipulated price contract.**

FIXED RATE MORTGAGE *See* **Mortgage.**

FIXED SASH *See* **Sash.**

FIXED TIME Load, dump, and maneuver times that are relatively constant in a cycle.

FIXING Securing or fastening piece.

FIXING BRACKET Fitting on a glazing bar, used to fasten it to a structural member.

FIXING FILLET Piece of wood, the thickness of a mortar joint, used as a fixing for carpentry.

FIXTURE 1. Device designed for a particular purpose and fastened securely in place; **2.** Receptacle in which water or other waste may be collected for ultimate discharge into a plumbing system; **3.** Means for illumination attached to an electrical distribution system; **4.** Assembly that serves to hold pieces in place while they are worked upon; **5.** Improvement or personal property attached to land that becomes part of the real estate; **6.** Part of fixed assets, usually consisting of machinery or equipment attached to or forming a normal part of a building; **7.** Personal property so permanently attached or part of a structure as to become part of the real property.

FIXTURE BRANCH Drain from a fixture trap to the junction of the drain with a vent.

FIXTURE DRAIN Drain from a fixture branch to the junction of any other drain pipe.

FIXTURE SUPPLY PIPE Water supply pipe connecting the fixture to the supply system.

FIXTURE UNIT Unit flow from a fixture, usually established as $0.028m^3$ (1 ft^3) or 28 L (7.5 gal) of water per minute.

FJ *Abbreviation for:* finger jointing.

fk *Abbreviation for:* fork.

fl. *Abbreviation for:* flashing; floor; fluid; floor line; flush; foot-lambert; free lift.

FL *Abbreviation for:* floor line; fusible link.

FLAG 1. Split end of a bristle in a paint brush; **2.** *See* **Flagstone; 3.** Thin metal strip, attached in groups to a stationary structure in a hoistway, that actuates magnetically operated switches on an elevator car.

FLAGGING Flat stones or cast slabs used for walkways.

FLAGGING TAPE Colored plastic tape, produced in rolls, used to mark (flag) boundaries or identify objects.

FLAGMAN Worker who holds the flagpole or range pole at the points designated by the transitman.

FLAGPERSON Worker who controls pedestrian and vehicular traffic in the vicinity of road works.

FLAGSTONE Hard, flat-faced stone, from 25 to 75 mm (1 to 3 in.) thick. It is laid either in mortar or sand for walks, patios and floors. Can be laid in squared section bond or a random pattern with irregular pieces. *Also called* **Crazy Paving,** and **Flag.**

FLAIL 1. Hammer hinged to an axle so that it can be used to break or crush material; **2.** Swatter for smothering and beating out grass fires.

FLAIL MILL Rotating equipment that uses chains to open or spread solid waste and to shred or crush such material as paper, cardboard, or glass.

FLAKE RAISE Face flakes that are raised above surrounding flakes in a particleboard, appearing as a rough surface. Usually caused by excessive absorption of moisture.

FLAKING Detachment of a uniform layer (of a coating or substratum), usually related to internal movement of water thickness or paint film or a film forming a clear coating.

FLAME Light from burning gases and incandescent particles.

FLAME BLOW-OFF Phenomenon that occurs when a flame moves away from a burner, often resulting in the flame being extinguished.

FLAME CLEANING Use of a very hot flame to remove mill scale and moisture from weathered structural steel prior to painting.

FLAME-CUT PLATE Steel plate in which the longitudinal edges have been prepared by oxygen cutting from a large plate.

FLAME CUTTING Cutting performed by an oxygen-fuel gas torch.

FLAME FRONT Plane along which combustion starts, or the base of the flame.

FLAME GUN Large blowtorch.

FLAME PHOTOMETER Instrument used to determine the physical elements present in a sample, by the color intensity of their unique flame spectra. *Also called* a **Photometer.**

FLAME RETARDANT Chemical used to reduce or eliminate a material's tendency to burn.

FLAME RETENTION BURNER Burner whose nozzle is surrounded with small ports that act as a pilot to relight the main flame if it blows off.

FLAME SPREAD Spread of fire along the surface of a material.

FLAME SPREAD CLASSIFICATION *See* **Surface burning characteristics.**

FLAME SPREAD RATING Measurement of flame spread on the surface of a material or an assembly of materials as determined in a standard fire test.

FLAME VELOCITY Speed at which a flame progresses into a mixture relative to the speed of the mixture.

FLAMING FRONT Zone of a moving forest fire within which the combustion is primarily flaming and behind which the combustion is mainly glowing.

FLAMMABILITY Relative ease with which fuels and materials ignite, regardless of quantity.

FLAMMABILITY LIMITS Maximum and minimum percentages of a fuel in an air-fuel mixture that will burn.

FLAMMABLE LIQUID Any liquid having a flash point below 60°C (140°F) and having a vapor pressure not exceeding 275.8 kP at 37.7°C (40 psi at 100°F).

FLANGE 1. Top or bottom member of an I-beam, separated by the web; **2.** Rim-like end on a valve, or pipe fitting for bolting another flanged fitting, usually for large diameter or pressurized pipe; **3.** Ridge that prevents a sliding motion; **4.** Mounting device, usually consisting of a plate or collar; **5.** Horizontal top or bottom portion of a frame siderail; **6.** Circular metal plate on a grinding machine, one or more of which drive the grinding wheel. *Also called* **Collet.** *See also* **Wheel sleeves.**

FLANGE BRACE Brace used to provide lateral support to the flange of a beam, girder, or column.

FLANGE PACKING *See* **Packing.**

FLANGE POINT Point of contact between the rope and rope drum flange where the rope changes layers.

FLANGE UNIT Union secured with nuts and bolts.

FLANK FIRE Forest fire set along a control line parallel to the prevailing wind and allowed to spread at right angles to it.

FLANKING Attacking a forest fire by working along its flanks, either simultaneously or successively from a less active, or anchor point, and endeavoring to connect the two lines at the head.

FLANKING PATH Route taken by sound waves around an element designed to impede their passage.

FLANKING WINDOW Window adjacent to (and usually part of) an exterior door.

FLANK WALL *See* **Wall**

FLAP *See* **Tire construction.**

FLAPPER ACTION *See* **Valve.**

FLAP VALVE *See* **Valve.**

FLARE 1. Widening of a tube; **2.** Stack for burning excess quantities of waste combustible gases.

FLAREBACK Burst of flame from a furnace in a direction opposed to normal gas flows; occurs when accumulated combustible gases ignite or sudden momentary furnace overpressure occurs.

FLARED FITTING *See* **Fitting.**

FLARED HEAD Increase in the size of the top portion of a column, at a maximum angle of 45° to the vertical.

FLARED INTERSECTION *See* **Intersection types.**

FLARE HEADER *See* **Brick.**

FLARELESS FITTING *See* **Fitting.**

FLARE UP Sudden acceleration of forest fire spread or intensification of the fire. *See also* **Blow up.**

FLARING CUP Grinding cup wheel with the rim extending from the back at an angle so that the diameter at the outer edge is greater than at the back.

FLARING INLET Funnel-shaped inlet to a pipe or conduit that facilitates flow.

FLASH 1. Eye injury caused by rays from arc welding; **2.** Surplus metal formed at the seam of a resistance weld; **3.** Excess material extruded during casting or molding; **4.** To make a joint weathertight; **5.** Variation in the color of paint due to wall suction; **6.** Movement of solder as it melts around a joint.

flash. *Abbreviation for:* flashing.

FLASHBACK Phenomenon that occurs when a flame front moves back through a burner nozzle (and possibly back to the mixing point).

FLASHBACK ARRESTERS Check valves, usually installed between torch and welding hose, to prevent the flow of burning fuel gas and the oxygen mixture back into the hose and regulators.

FLASHBOARD *See* **Stop log.**

FLASH COAT Light coat of shotcrete used to cover minor blemishes on a concrete surface.

FLASH DRYING Rapid drying of an applied coat of material containing volatile solvents, through exposure to radiant heat.

FLASH FUEL Material, such as grass, leaves, draped pine needles, fern, moss, or some kinds of slash that ignites readily and is consumed rapidly when it is dry.

FLASHING 1. Sheet metal or asphalted felt used to make the junctions of materials (particularly dissimilar materials, and on external surfaces) watertight; **2.** Rapid change in fluid state, from liquid to gaseous; **3.** Manufacturing method to produce specific color tones; **4.** Material that is extruded between the faces of a mold, and that forms part of the casting or extrusion on setting or cooling. *See also* **Fin.**

FLASHOVER Rapid combustion and/or explosion of unburned gases trapped at some distance from a main forest fire front.

FLASH POINT Lowest temperature at which a sufficient quantity of vapors are given off to form an ignitable mixture of vapor and air immediately above the surface of the liquid.

FLASH SET Rapid development of rigidity in a freshly mixed portland-cement paste, mortar, or concrete, usually with the evolution of considerable heat, that cannot be dispelled without addition of water. *Also called* **Early stiffening, False set, Grab set, Premature stiffening, Quick set,** and **Rubber set.**

FLASH WELDING Process using an electric arc in combination with resistance and pressure welding.

FLAT 1. Thin, rectangular bar of iron or steel; **2.** Finish with no sheen. *Also called* **Matte finish.**

FLAT ARCH *See* **Arch.**

FLAT BACK COWL Cowl or chassis-cowl: a conventional chassis with hood, front fenders, and instrument panel, designed for use with a body that includes windshield, doors, and driver's seat.

FLATBED *See* **Trailer.**

FLATBED TRUCK *See* **Truck.**

FLAT CARVING Type of carving that leaves the design flat and cuts away only the background.

FLAT CHISEL *See* **Cold chisel.**

FLAT COAT Coat of filler.

FLAT CORNER BRACE *See* **Braces and plates**

FLAT COST Cost only of labor and materials.

FLAT CURE Method of curing fire hose in a flat form.

FLAT DRAWN GLASS Process in which a sheet of glass is drawn vertically from a batch of molten glass and is passed between coolers, causing the sheet to solidify when it attains a predetermined thickness.

FLAT FLOOR SLAB Concrete floor slab construction in which the slab is carried directly onto columns (that may have enlarged heads) without the use of beams, except at the extreme edges.

FLAT-GRAINED LUMBER *See* **Grain.**

FLAT JACK Hydraulic jack consisting of light-gauge metal that is bent and welded to a flat shape and that expands under internal pressure.

FLAT JOINT *See* **Pointing, flush.**

FLATNESS Condition of a surface having all elements in one plane.

FLAT PAINT Paint that dries without luster.

FLAT PIECE OF AGGREGATE One in which the ratio of the width to thickness of its circumscribing rectangular prism is greater than a specified value.

FLAT-PITCH ROOF Roof with only a moderately sloping surface.

FLAT PLATE SLAB Flat concrete slab without column capitals or drop panels. *See also* **Flat slab.**

FLAT-PLATE COLLECTOR *See* **Solar collector.**

FLAT POSITION WELD Horizontal weld on the upper side of a horizontal surface.

FLAT ROOF Roof that is horizontal or given a slope sufficient only to provide drainage.

FLAT ROPE Wire rope that is made of a series of parallel, alternating right-lay and left-lay ropes, sewn together with relatively soft wires.

FLAT SLAB Concrete slab reinforced in two or more directions and having drop panels or column capitals or both. *See also* **Flat plate slab.**

FLAT SLAB CONSTRUCTION Reinforced concrete floor and roof construction of uniform thickness that eliminates drops for beams and girders.

FLAT SPOTS 1. Areas of a painted surface lacking gloss; **2.** Flat areas on the surface of mandrel-cured hose, caused by deformation during vulcanization.

FLAT STRETCHER COURSE Brick course consisting of stretchers set on edge and exposing their flat sides on the surface of the wall. Often done with brick finished for the purpose of exposing the flat side, such as enameled or glazed brick.

FLATTENED STRAND ROPE Wire rope that is made either of oval or triangular-shaped strands in order to form a flattened rope surface.

FLATTING 1. Process of flattening out buckled face veneers; **2.** Paint finish that leaves no gloss.

FLATTING AGENT Compound added to paints, varnishes, and other finishing materials to reduce the gloss of the dried film.

FLAT TOP GRIND Saw tooth filed square on top.

FLAT TROWEL Trowel used to transfer mortar from a mortarboard to a wall or other vertical surface, or for spreading mortar on floor surfaces before tiles are set.

FLAT TRUSS *See* **Truss.**

FLAT WASHER *See* **Washer.**

FLAT WIDTH For a rectangular steel tube, the nominal width minus twice the outside corner radius. In absence of knowledge of the corner radius, the flat width may be taken as the total section minus three times the thickness.

FLAT WIRE Rectangular cross section wire commonly used as the inner element of rough bore suction hose.

FLAUNCHING Cement mortar fillet around the top of a chimney stack that acts to throw off water.

FLB *Abbreviation for:* full-length bundle.

flbd *Abbreviation for:* floorboard.

fld *Abbreviation for:* field.

fld engr *Abbreviation for:* field engineer.

FLECHE Slender spire rising from the ridge of a roof.

FLEET ANGLE Angle between a wire rope's position at the extreme end wrap on a drum, and a line drawn perpendicular to the axis of the drum through the center of the nearest fixed sheave.

FLEETING SHEAVE Shaft-mounted sheave parallel to the rope-drum shaft, arranged so that it can slide laterally as the rope spools, permitting close sheave placement without excessive fleet angle.

FLEMISH BOND *See* **Bond.**

FLEMISH CROSS BOND *See* **Bond.**

FLEMISH DOUBLE BOND *See* **Bond.**

FLEMISH DOUBLE CROSS BOND *See* **Bond.**

FLEMISH EYE *See* **Farmer's eye.**

FLEMISH SPLICE *See* **Farmer's eye.**

FLEMISH GARDEN BOND *See* **Bond.**

flex. *Abbreviation for:* flexibility; flexible.

FLEX CRACKING Surface cracking induced by repeated bending and straightening.

FLEX DUCT Duct made of spiral wire covered with plastic, used in air conditioning and forced-air heating systems.

FLEXIBLE BUDGET *See* **Budget.**

FLEXIBLE CONNECTION Connection permitting a portion, but not all, of the simple beam rotation of a member end.

FLEXIBLE DROPCHUTE Device consisting of a heavy rubberized canvas or plastic collapsible tube, used to confine or direct the flow of a falling stream of fresh concrete. *See also* **Articulated dropchute,** and **Dropchute.**

FLEXIBLE EXHAUST CONNECTION Flexible section between an exhaust manifold and exhaust pipe.

FLEXIBLE FUEL LINE Pliant coupler line used between engine and supply lines.

FLEXIBLE JOINT FILLER Joint filler that is readily compressible yet capable of recovery when the joint opens out again.

FLEXIBLE LEAD Flexible conductor connected to a source of current.

FLEXIBLE MANDREL Long, round, smooth rod capable of being coiled in a small diameter, used for support during the manufacture of certain types of hose.

FLEXIBLE METAL CONDUIT Electrical conduit fabricated from spirally wound steel strip. *See also* **Conduit.**

FLEXIBLE METAL ROOFING Flat, flexible metal sheet used for roof covering and flashings.

FLEXIBLE PAVEMENT Pavement structure that maintains intimate contact with, and distributes loads to the subgrade and depends on aggregate interlock, particle friction, and cohesion for stability. Cementing agents, where used, are generally bituminous materials, as contrasted to hydraulic cement (as in the case of rigid pavement). *See also* **Rigid pavement.**

FLEXIBLE WALLS Sheet-piled and thin-stemmed reinforced concrete retaining walls.

FLEXIBILITY Degree to which a material will bend without cracking, breaking, or becoming permanently deformed.

FLEXIBILITY FACTOR Degree to which a material or product can be flexed without permanent deformation.

FLEXIBILIZER Additive that makes a resin or rubber more flexible.

FLEX LIFE Relative ability of rubber to withstand cyclical bending stresses.

FLEX LIFE TEST Laboratory method used to determine the life of a rubber product when subjected to dynamic bending stresses.

FLEXURAL BOND STRESS Stress in structural concrete members between the concrete and the reinforcing element that results from the application of external load.

FLEXURAL MODULUS OF STRENGTH Strength of a material in bending, expressed in tensile stress of the outermost fibers of a bent test sample at the instant of failure.

FLEXURAL MOMENT *See* **Positive moment.**

FLEXURAL RIGIDITY Measure of stiffness of a member, indicated by the product of modulus of elasticity and moment of inertia divided by the length of the member.

FLEXURAL STRENGTH Property of a material or a structural member that indicates its ability to resist failure in bending.

FLEXURE *See* **Bending.**

flg *Abbreviation for:* Flooring.

flg. *Abbreviation for:* flange.

FLIGHT 1. Slow-pitch thread of an auger; **2.** Continuous series of stairs, more than 3 and less than 16; **3.** Metal plate in any of various shapes, placed longitudinally inside the shell of a dryer or drum mixer drum to lift the aggregate through the flame and hot gases as it travels through the drum.

FLIGHTING Assembly attached to the cutter drum shell to which the working tool(s)/holders are attached.

FLIGHT PATH Helicopter's path of operation while

flying between the logging area and drop zone.

FLINT A variety of chert.

FLINT WALL Wall built with an outer shell of knapped (split) flints set in lime mortar.

FLIP AXLE *See* **Axle.**

FLIP-FLOP Digital component or circuit with two stable states and sufficient hysteresis so that it has 'memory.' The state is changed with a control pulse, a continuous control signal is now necessary for it to remain in that state.

FLIP-OVER PALLET FORKS Pallet forks, one of the fittings of a multipurpose bucket, that are normally stored behind the bucket when not in use. The operator is able to flip the forks over into the front-facing position, allowing the front loader to be used as a forklift.

FLIP-UP RAMPS Loading ramps that are mounted at the trailing edge of an equipment trailer and that are hinged to rotate through 180° to store on the trailer deck.

FLITCH 1. Portion of a log sawed on two or more sides and intended for remanufacture into lumber. *Also called* **Cant**; **2.** Frame reinforcement.

FLITCH BEAM Built-up beam formed by a metal plate sandwiched between two wooden members and bolted together.

FLITCH PLATE Thin strip of steel bolted between two planks to strengthen them.

flm *Abbreviation for:* film.

flm. *Abbreviation for:* flame.

flmprf *Abbreviation for:* flameproof.

FLOAT 1. Position in a hydraulic control valve that allows loader arms to raise and lower freely; **2.** Dozer blade resting by its own weight, or held from digging by the upward pressure of a load of dirt against its moldboard; **3.** Tool (not a **Darby**), usually of wood, aluminum, or magnesium, used in finishing operations to impart relatively even but still open texture to an unformed, fresh concrete surface; **4.** Suspended scaffold hung from overhead supports by means of ropes and consisting of a substantial platform having diagonal bracing underneath, resting on and securely fastened to two parallel plank bearers at right angles to the span; **5.** Loose fragments or particles of rock, as distinguished from the outcrop or bedrock; **6.** Additional time available to complete noncritical activities or work items without affecting the critical path (i.e., the overall project duration); **7.** Object floating on water that opens and closes a valve in a water tank based on water level; **8.** Funds available for immediate use; **9.** *See* **Tolerance.**

FLOAT ARM Rod connecting the float ball to the inlet valve in a water closet tank.

FLOATATION 1. Ability to float, as with large, low-pressure tires that allow a vehicle or machine to be supported over soft ground; **2.** Separating minerals by floating the lighter components in a fluid.

FLOAT BALL Ball used to control the inlet valve in water closet tanks.

FLOAT COAT Final mortar coat over which the neat coat, pure coat, or skim coat is applied.

FLOATED BED Bed of mortar applied to a surface and floated to a uniform level to serve as a setting surface for ceramic tile.

FLOATER Tool for finishing mortar screeds.

FLOAT FINISH Rather rough concrete surface texture obtained by finishing with a float.

FLOAT GLASS Transparent glass, the two sides of which are flat, parallel, and fire-polished giving clear, undistorted vision and reflection.

FLOATING 1. Leveling a freshly placed concrete or plaster surface with a float; **2.** Creating sufficient surface paste needed for troweling the concrete. Floating is done immediately after the concrete surface has been consolidated and struck off.

FLOATING ANGLE Technique of installing drywall panels at interior corners so as to allow structural movement.

FLOATING CRANE Crane mounted on a barge or pontoon.

FLOATING EDGE Drywall panel installed in such a manner that the beveled edge does not lie over a framing member or other support.

FLOATING FLOOR Floor supported upon another in a way to provide insulation against sound or vibration, or to provide a less rigid structure (as for a dance floor).

FLOATING FOUNDATION Foundation that distributes loads imposed upon it rather than transmitting them directly to the ground below, used on unstable soil or where there is a high water table.

FLOATING HARBOR Breakwater of pontoons attached end-to-end.

FLOATING HARNESS Frame, part of the boom suspension, supporting sheaves for the live suspension ropes and attached to the fixed suspension ropes (pendants). *Also called* **Bridle**, **Live spreader**, **Spreader bar**, and **Upper spreader.**

FLOATING JOINT Condition where the butt joint formed by the meeting of unbeveled edges of drywall does not lie over backing or other support.

FLOATING MAINSHAFT When engaged, a floating mainshaft transfers engine torque evenly through its gears to the rest of the transmission and ultimately to the rear axle. *See also* **Mainshaft.**

FLOATING PILE DRIVER Pile driver mounted on a barge or scow.

FLOATING PIPELINE Pipeline supported on or by pontoons, used between a floating dredge and land to transport dredged material to be used as hydraulic fill.

FLOATING RULE Long wooden rule with which a plaster float coat is leveled to a plane surface between screeds.

FLOATING SCREED Asphalt paver screed that is attached to a tractor unit by a towpoint on each side and that is free to float on an asphalt mix, the design permitting gradual thickness changes and automatic leveling.

FLOATING SLAB *See* **Slab-on-grade.**

FLOATING ZONE Land area described in the text of zoning regulations but not included in the zoning map until a developer applies for rezoning.

FLOAT SCAFFOLD Suspension scaffold consisting of a braced platform resting upon two parallel bearers and hung from overhead supports by ropes of fixed length. *Also called* **Ship scaffold.**

FLOAT STRIP Strip of wood, about 6 mm (0.25 in.) thick and 32 mm (1.25 in.) wide, used as a guide to align mortar surfaces.

FLOAT SWITCH *See* **Switch.**

FLOAT VALVE *See* **Valve.**

FLOC Porous mass formed by the agglomeration of suspended particles, in water or in air, often following application of a flocculant.

FLOCCULATION Induced aggregation of colloids in suspension in order to increase their rate of settlement.

FLOGGING Technique for creating a small repetitive pattern (usually of pores in graining) by slapping the flat side of a brush into a wet medium while moving along a surface. A wide, long-bristled brush called a flogger usually is used, although other brushes may be employed.

FLOOD Condition where the volume of surface water runoff is greater than the capacity of natural and artificial channels and conduits to adequately and safely contain and conduct the resulting flow, resulting an an accumulation of water over otherwise dry land.

FLOOD BASE ELEVATION Elevation of the highest recorded flood for an area, specific to a location and to a point in time.

FLOOD CHANNEL Wetted area that the flood occupies at its peak. *See also* **Channel.**

FLOOD CONTROL Construction (or the prevention of construction) intended to assist in the conveying of excess surface water runoff.

FLOOD CURRENT Current in water associated with tidal action.

FLOODGATE Mechanical means by which to regulate or restrain the flow of floodwater through a channel.

FLOOD HAZARD Potential risk due to a sudden and temporary increase of surface water flow.

FLOOD LEVEL Highest level that flowing water will reach, usually rated as annual, 10-year, 25-year, etc.

FLOOD LEVEL RIM Top edge of a plumbing fixture from which water will overflow.

FLOODLIGHTING Use of high-intensity lamps to bathe an exterior area, or the exterior of a structure with light, at night.

FLOODPLAIN That part of a stream valley that is inundated during floods.

FLOODPLAIN REGULATIONS Building regulations that stipulate the type of construction, or prevent construction, on land within a floodplain.

FLOOD PROFILE Graphic representation of data showing the water surface elevation resulting from a calculated flood condition, relevant to the existing topography.

FLOOD RESERVOIR Ponding area for the temporary impounding of excess surface runoff.

FLOODWAY Total complex of floodplain areas, streams, and channels, both natural and constructed, that serve to contain and conduct the surface runoff due to excessive inundation.

FLOOR 1. Inside bottom surface of a room or area; **2.** Level or story in a building; **3.** Bottom horizontal, or nearly so, part of an excavation upon which haulage or walking is done; **4.** Self-propelled platform used in tunnel excavation.

FLOOR ARCH *See* **Arch.**

FLOOR AREA Space on any story of a building between exterior walls and required firewalls, including the space occupied by interior walls and partitions, but not including exits and vertical service spaces that pierce the story. *See also* **Area, Area of building, Gross area, Net area,** and **Net room area.**

FLOOR AREA RATIO Relationship of the total area of a building to the area of the land upon which it sits.

FLOOR ARMOR Heavy steel-plate covering, usually laid over the regular floor of a dump body, with a wood or corrugated-steel filler between the plate and body, as a protection against impact.

FLOOR ASSEMBLY Combination of materials providing the horizontal division of a building into stories.

FLOOR BEAM Transverse beam or girder, placed at the panel points of a span, that supports stringers carrying the floor.

FLOOR BRICK *See* **Brick.**

FLOOR CENTER Adjustable-length beam used to support formwork for a floor slab.

FLOOR CLIP Strips of steel anchored into a concrete slab after leveling that, after the concrete has hardened, are bent up and nailed to floor battens to which a finished wood floor is fixed.

FLOOR CRAMP Cramp (reverse clamp) used to force floorboards together as they are nailed down.

FLOOR DRAIN Wastewater outlet and trap, usually placed at a low point in a sloping concrete floor.

FLOOR FLANGE Fitting attached to a plumbing system at floor level so that a fixture can be bolted to the drainage piping.

FLOOR FRAMING Common joists, strutting, and supports.

FLOOR GRILLE *See* **Floor register.**

FLOOR GRINDER Motorized grind machine, usually hand-propelled, used for removing trowel marks, mortar droppings, etc. from hardened concrete.

FLOOR HOLE Opening measuring less than 300 mm (12 in.), but more than 25 mm (1 in.) in its least dimension in any floor, roof, or platform through which materials, but not persons, may fall. *See also* **Floor opening.**

FLOORING Material used in the construction of floors. The surface material is known as **Finished**

flooring while the base material is called **Subflooring**.

FLOORING SAW *See* **Saw.**

FLOOR JOIST Wood or metal supports that span an opening and support a floor.

FLOOR LAYOUT *See* **Floor plan.**

FLOOR LINE Mark made on a foundation wall, stanchion, etc., denoting finished floor level.

FLOOR LOAD Total weight on a floor, including self weight of the floor itself and any live or transient load.

FLOOR-LOADING DUMBWAITER Dumbwaiter with the sill of the entrance at floor level.

FLOOR LOADING HEIGHT Height above ground of the floor of an unladen cargo-carrying vehicle.

FLOOR OPENING Opening measuring 300 mm (12 in.) or more in its least dimension in any floor, roof, or platform, through which persons may fall. *See also* **Floor hole.**

FLOOR PLAN Drawing showing the length and breadth of a building and the location of the rooms and other areas, window and door openings, and such structural and servicing details as are appropriate to the scale of the drawing. *Also called* **Floor layout.**

FLOOR PLATE 1. Bottom horizontal member of a framed wall or partition that rests directly upon the subfloor; **2.** Removable cover plate at top and bottom landings of an escalator; **3.** Deck plate, usually removable, set into a structural grid to produce a walkway or complete floor. May be solid, with or without a raised surface pattern, or open pattern.

FLOOR PLUG Socket outlet mounted in the floor.

FLOOR REGISTER Grille or other vented outlet that fits over a forced hot air heating register box. *Also called* **Floor grille.**

FLOOR/ROOF INSULATION (SOLID) Low-density concrete used for insulating purposes only and placed over a structural system.

FLOOR SANDER Power-operated sanding machine used to finish wood floors.

FLOOR SAW *See* **Saw.**

FLOOR SCABBLER Motorized, multihead, hand-propelled scabbling machine used to reduce the level of a hardened concrete slab, for keying a concrete surface to take a topping, and roughing walkways and other surfaces.

FLOOR SHEATHING Group of panels covering a floor structure.

FLOOR SLAB Reinforced or prestressed concrete floor, particularly that portion between supporting beams.

FLOOR-STAND GRINDER Off-hand grinder, mounting one or two wheels running on a horizontal spindle, fixed to a metal base attached to the floor.

FLOOR STRUTTING *See* **Herringbone strutting.**

FLOOR SYSTEM System of structural steel components separating the floors of a building.

FLOOR TILE 1. Structural units for floor and roof slab construction; **2.** Decorative ceramic tiles for floor finishing.

FLOOR-TO-FLOOR TIME Time required for an elevator to run from one floor to the next.

flot. *Abbreviation for:* flotation.

FLOTATION Ability of a tire to support a load on soft, yielding terrain.

FLOURY SOIL Soil that looks like clay when wet but which becomes a powder when dry.

FLOW 1. To move as a liquid does. *See also* **Laminar flow, Metered flow,** and **Turbulent flow; 2.** Rate of movement of water in a conduit; **3.** Time-dependent irrecoverable deformation of concrete. *See also* **Creep** and **Rheology; 4.** Measure of the consistency of freshly mixed concrete, mortar, or cement paste in terms of the increase in diameter of a molded, truncated cone specimen after jigging a specified number of times; **5.** Fluidity of plastic material; **6.** Movement of traffic.

FLOWBACK Degree a material will compress before penetration of a cutting tool occurs.

FLOW COEFFICIENT Correction factor used for figuring the volume flow rate of a fluid through an orifice.

FLOW CONE 1. Device for measurement of grout consistency in which a predetermined volume of grout is permitted to escape through a precisely sized orifice, the time of efflux (flow factor) being used as the indication of consistency; **2.** Mold used to prepare a specimen for the flow test.

FLOW CONTROL (DECELERATION) VALVE *See* **Valve.**

FLOW CONTROL (FLOW METERING) VALVE *See* **Valve.**

FLOW CRACK Surface imperfection caused by improper flow and failure of stock to knit or blend with itself during the molding operation.

FLOW DIAGRAM 1. Graphic representation of a program or routine, consisting of procedure, question, and statement boxes; **2.** The directions taken by a fluid showing any processes undertaken and noting special equipment.

FLOW DIVIDER *See* **Valve.**

FLOW GRADIENT 1. Slope to which a pipe must be laid to ensure a self-cleansing gradient; **2.** Drainage slope determined by the elevational differences between the inlet and outlet, the distance between those two points, and the required volume and velocity.

FLOW IMPROVER Additive designed to modify wax crystal growth in a fuel or lubricant, thereby lowering the pour point and improving its low temperature fluidity.

FLOW INDEX Slope of a flow curve.

FLOW LINE Line that indicates the direction followed by groundwater toward points of discharge. *See also* **Equipotential line,** and **Flow mark.**

FLOW MARK Surface imperfection similar to a flow crack, but the depression is not quite as deep.

FLOW METER 1. Device that indicates either flow rate, total flow, or a combination of both; **2.** Instrument designed to measure the quantity of fuel used in an engine.

FLOWNET 1. Graphical method used to study the hypothetical flow of water through a soil. It is used to indicate the paths of travel followed by moving water and the hydraulic pressures resulting from such water flow; **2.** Pictorial description of the path of water through a dam.

FLOW PROMOTER Substance added to a coating to enhance brushability, flow, and leveling.

FLOW RATE Volume of fluid per unit of time passing a given cross section of a flow passage in a given direction.

FLOW RESISTANCE Measure of the ability of a material to impede the flow of air or water through it.

FLOW SLIDE Shear failure in which a soil mass moves over a relatively long distance in a fluid-like manner, occurring rapidly on flat slopes in loose, saturated, uniform sands, or in highly sensitive clays.

FLOW SWITCH *See* **Switch.**

FLOW TABLE Flat, circular jigging device used in making flow tests for consistency of cement paste, mortar, or concrete.

FLOW-THROUGH CHAMBER Upper compartment of a two-story sedimentation chamber.

FLOW TROUGH Sloping trough used to convey concrete by gravity flow from either a truck mixer or a receiving hopper to the point of placement. *See also* **Chute.**

fl. oz *Abbreviation for:* fluid ounce.

flr *Abbreviation for:* floor.

flr. *Abbreviation for:* flooring.

flt *Abbreviation for:* flight; float.

flt. *Abbreviation for:* filter.

fltg *Abbreviation for:* floating.

fltr *Abbreviation for:* floater.

flu. *Abbreviation for:* flute; fluted.

FLUE Space or passage in a chimney through which smoke, gas, or fumes are conducted to the outside atmosphere.

FLUE COLLAR Portion of a fuel-fired appliance designed for the attachment of the flue pipe or breeching.

FLUE GAS Gases that leave a furnace by way of the flue, including gaseous products of combustion, water vapor, excess oxygen, and nitrogen.

FLUE-GAS ANALYSIS Statement of the quantities of the various components of a sample of flue gas, usually expressed in percentages by volume.

FLUE-GAS LOSS Sensible heat carried away by the dry flue gas, and the sensible and latent heat carried away by water vapor in the flue gas.

FLUE GAS WASHER/SCRUBBER Equipment for removing objectionable constituents from the products of combustion by means of spray, wet baffles, etc.

FLUE LINING Smooth, one-celled hollow tile of fireclay or terra cotta, used to protect the masonry of a chimney and to provide a smoother duct through which to exhaust smoke and gases. May be rectangular or round in shape; usually 0.6 m (2 ft) long. *See also* **Chimney lining.**

FLUE PIPE Pipe connecting the flue collar of an appliance to a chimney.

FLUE TILE Glazed or unglazed tile about 600 mm (24 in.) long, either round, oblong, or square in section, used to line a chimney flue.

FLUID Liquid, gas, or a combination thereof. *See also* **Fatty-oil fluid, Fire-resistant fluid, Hydraulic fluid, and Pneumatic fluid.**

FLUID ANALYSIS Method by which the contamination level of a lubricating or hydraulic fluid may be determined. There are a a number of methods, including:

> **Automatic particle count:** Technique that determines the number of particles per milliliter of fluid. It is fast and repeatable, limited to being sensitive to particle concentration and to nonparticulate contaminants, e.g., H_2O, air, gels.

> **Ferrography:** Technique that produces a scaled number of the large-to-small particles present in a sample. It provides basic information that will indicate the need for more sophisticated testing upon abnormal results. It is limited by not being capable of detecting nonferrous particles (eg, brass, copper, silica, etc.).

> **Gravimetric:** Test that determines the milligrams of contaminants per liter of sample. It indicates the total amount of contaminant but cannot distinguish particulate size.

> **Optical particle count:** Technique that determines the number of particulate contaminants per milliliter of sample and that provides an accurate measure of size and quantity distribution.

> **Patch test and fluid contamination comparator:** A field-conducted, visual comparison against a cleanliness code that produces a rapid approximation of system fluid cleanliness levels and that also helps to identify the types of contaminants.

> **Spectrometry:** Technique that identifies and quantifies contaminant material as parts-per-million of the sample being tested. However, it cannot size contaminants, and has limited sensitivity above 5 mm.

FLUID CLUTCH *See* **Clutch.**

FLUID COUPLING Hydrodynamic drive that transmits power without the ability to change torque (torque ratio is unity for all speed ratios).

FLUID DRIVE Connection between two shafts that transmits torque through a fluid.

FLUID FILM LUBRICATION *See* **Lubrication.**

FLUID FRICTION Friction due to viscosity of fluid.

FLUIDIC Of or pertaining to devices, systems, assemblies, etc., utilizing fluidic components.

FLUIDIC AMPLIFIER Device that enables one or more fluid dynamic signals to control a source of power, and thus is capable of delivering at its output an amplification of the essential characteristics of the input signal.

FLUIDIC NOISE Root mean square of random pressure variations with respect to the operating pressure, defined in terms of a signal-to-noise ratio.

FLUIDICS Engineering science pertaining to the use of fluid dynamic phenomenon to sense, control, process information, and/or actuate.

FLUIDIFIER Admixture employed in grout to increase or decrease the flow factor without changing water content. *See also* **Water-reducing admixture.**

FLUIDIZED BED REACTOR Process in which heat is transferred from a churning suspension of finely divided particles, such as sand, to the suspended particles of introduced materials for calcining, heat treatment, regeneration, or combustion.

FLUID LEVEL GAUGE *See* **Gauge.**

FLUID POWER Energy transmitted and controlled through use of a pressurized fluid within an enclosed circuit. *Also called* **Motive fluid.**

FLUID POWER SYSTEM System that transmits and controls power through use of a pressurized fluid within an enclosed circuit.

FLUID SAMPLE Small portion of a fluid taken from a system or test apparatus, usually obtained for analyzing fluid properties and/or amount of contamination.

FLUME Artificial channel, often set at an angle, used for measuring the flow of water, or for carrying materials conveyed by water, such as logs. *Also called* **Control flume.**

fluor. *Abbreviation for:* fluorescent; fluoridation.

FLUORESCENT LAMP Light source in which fluorescent powder and phosphor, coated on the inner surface of a glass tube, is activated by electrical energy. They are manufactured to produce several types of light:

    Daylight: 6700 kelvin.

    Deluxe cool white: 4300 kelvin.

    Deluxe warn white: 3050 kelvin.

    Design white: 5350 kelvin.

    Incandescent fluorescent: 2700 kelvin.

    Natural: 3900 kelvin.

    Standard cool white: 4300 kelvin.

    Standard warn white: 3000 kelvin.

    Standard white: 3560 kelvin.

FLUORESCENT PAINT Paint containing pigments that absorb energy from the UV end of the spectrum and reemit it as light in a visible wavelength. *See also* **Luminous paint,** and **Phosphorescent paint.**

FLUOSILICATE Magnesium or zinc silico-fluoride used to prepare aqueous solutions, sometimes applied to concrete as surface-hardening agents.

FLUSH 1. Release of a body of water in a controlled manner; **2.** Having the surface even with the adjoining surface.

FLUSH BALL Ball-shaped closure that controls the flow of water into the bowl of a water closet.

FLUSH BEAD MOLDING Inset bead or convex molding where the apex of the curve is flush with the adjacent surfaces.

FLUSH BOLT Fastening bolt flush with the surface.

FLUSH BUSHING Pipe fitting that reduces the diameter of a female-threaded pipe fitting.

FLUSH DOOR *See* **Door types.**

FLUSHED Filled up to the surface.

FLUSHING 1. Process of expelling drill cuttings from a borehole, by directing compressed air (or compressed air and water) through the drill string to the drill bit face. **2.** *See also* **Bleeding; 3.** Release of water in a controlled manner so as to cause the resulting overflow to cleanse the downstream channel.

FLUSHING MANHOLE Manhole equipped with an opening gate so that sewage flows may be accumulated and then released to flush the downstream section of pipe.

FLUSH JOINT Mortar joint of which the surface is in the same plane as the surface of the masonry wall of which it forms a part.

FLUSHOMETER Device that allows a predetermined quantity of flushing water to be accumulated and then released, actuated by water pressure.

FLUSH PANEL Panel flush with its framing.

FLUSH POINTING *See* **Pointing.**

FLUSH SEAL Application of asphaltic material to a road surface without a covering course of aggregate.

FLUSH VALVE *See* **Valve.**

FLUSH WATER *See* **Wash water.**

FLUSH WELD Weld that is flush to the level of the pieces being joined; there is no buildup for reinforcing or strengthening of the weld.

FLUSHWORK Decorative use of flint in conjunction with dressed stone to form patterns, monograms, inscriptions, etc.

FLUTE Concave channel or groove.

FLUTED BIT *See* **Bit.**

FLUTING 1. Shallow, concave grooves, vertically on a column or pilaster, for instance; **2.** Grinding the grooves of a twist drill or tap.

FLUTTER Small, rapid vibrations or oscillations.

FLUVIAL Of or pertaining to rivers, or produced by river action.

FLUVIAL SOIL Soil whose properties have been the subject of water action; characterized by the roundness of individual particles.

FLUX Chemical used to promote fusion of metals during welding, brazing, or soldering.

FLUXING OIL 1. Thick, relatively nonvolatile fraction of petroleum that may be used to soften asphalt to a desired consistency; often used as base stock for the manufacture of roofing asphalt; **2.** Asphalt used in the production of powdered asphalt mixes for low-cost construction, generally an SC-250 or SC-800 liquid asphalt.

FLY 1. Number of chokers being used at one time on any equipment (flying one or more chokers); **2.** Space in a theater, above and behind the proscenium arch, in which scenery is hung.

FLY ASH Finely divided residue that results from the combustion of ground or powdered coal and that is transported from the firebox through the boiler by flue gases.

FLY ASH COLLECTOR Equipment for removing fly ash (particulates) from combustion gases prior to their discharge to the atmosphere.

FLY-ASH STABILIZATION *See* **Soil stabilization.**

FLYBALL GOVERNOR *See* **Governor.**

FLYER 1. Flying shore; **2.** Rectangular stair tread.

FLYING BUTTRESS *See* **Buttress.**

FLYING CHOKER On a grapple rigging, yarding with a choker.

FLYING FORM Large, prefabricated unit of formwork incorporating support, designed to be moved from place to place. They are commonly transported by crane, particularly a tower crane, thus the term 'flying.'

FLYING HAMMER *See* **Pile hammer.**

FLYING SCAFFOLD Scaffold hung from an outrigger beam.

FLYING SHORE Horizontal strut between two walls above ground level.

FLY JIB *See* **Jib.**

FLY RAFTER Decorated barge board.

FLYROCK Rock that is propelled into the air by the force of an explosion, usually originating from prebroken material on the surface or upper open face.

FLY SECTION Lightest boom section of a crane, located at the end farthest from the boom pivot pin. The section may be actuated by either a manual power pin or full-power hydraulic system. The boom nose is attached to this section.

FLYWEIGHT Integral part of a mechanical governor that senses speed changes.

FLYWHEEL Large, heavy wheel, connected to the crankshaft of an engine, that absorbs power impulses.

FLYWHEEL CLUTCH *See* **Clutch,** and **Transmission.**

FLYWHEEL-DRIVEN TRANSMISSION *See* **Power take-off.**

FLYWHEEL RING GEAR External gear mounted on an engine's flywheel and driven by the starter pinion when starting the engine.

fm *Abbreviation for:* fathom.

FM *Abbreviation for:* field manual; fire main; frequency modulation.

fman *Abbreviation for:* foreman.

FMV *Abbreviation for:* fair market value.

fndry *Abbreviation for:* foundry.

FO *Abbreviation for:* fuel oil.

FOAM Chemical fire-extinguishing mixture that forms bubbles on application, greatly increasing the mixture volume.

FOAM APPLICATOR *See* **Applicator.**

FOAM CORE 1. Center of a plywood 'sandwich' panel in which liquid plastic is foamed into all spaces between the plywood panels to serve both as insulation and support for component skins; **2.** Plywood skins pressure-glued to both sides of rigid plastic foam boards or billets.

FOAM COVER Material that is sprayed on solid waste in lieu of daily soil cover.

FOAMED ADHESIVE *See* **Adhesive.**

FOAMED CONCRETE Concrete made very light and cellular by the addition of a prepared foam or by the generation of gas within the unhardened mixture. *See also* **Aerated concrete, Cellular concrete,** and **Gas concrete.**

FOAMED-IN-PLACE INSULATION Two-component liquid resin product that is injected into building cavities. As the component chemicals combine they produce a foam that expands to fill the cavity. The density of the foam and the resulting insulating capability is determined by the relative proportions and types of chemicals used.

FOAMING AGENTS Chemical additives that generate inert gases, causing the host material to assume a cellular structure.

FOAM-IN-PLACE Use of a foaming machine at the work location.

FOB *Abbreviation for:* free on board.

FOC *Abbreviation for:* free of charge.

FOG Jet of fine water spray discharged by spray nozzles, used to extinguish fires.

FOG APPLICATOR *See* **Applicator.**

FOG CURING 1. Storage of concrete in a moist room in which the desired high humidity is achieved by the atomization of fresh water. *See also* **Moist cabinet; 2.** Application of atomized fresh water to concrete, stucco, mortar, or plaster.

FOG SEAL *See* **Bituminous coating.**

FOHC *Abbreviation for:* free of heart centers.

FOIL 1. Leaf-shaped curve formed by cusping a circle or an arc; the number of foils is indicated by the prefix, e.g., trefoil, multifoil; **2.** Metal, hammered or rolled into a thin sheet; **3.** Cellulose papers weighing between 40 gm and 140 gm per square meter untreated. The papers may be impregnated with malamine thermoplastic resins or left untreated.

FOIL BACKED Drywall panel backed with aluminum foil laminated to its surface.

fok *Abbreviation for:* free of knots.

fol. *Abbreviation for:* follow; following.

FOLD Pronounced bend in layers of rock.

FOLDED PLATE 1. Framing assembly composed of sloping slabs in a hipped or gabled arrangement; **2.** Prismatic shell with an open polygonal section.

FOLDING DOORS *See* **Door types.**

FOLDING GOOSENECK *See* **Gooseneck.**

FOLDING LINE Image plane line that intersects with another in orthographic projection.

FOLDING STAIR Attic ladder, hinged so that it can be raised and lowered and stored in a folded state within an enclosure.

FOLDING WEDGE Wedge, one of a pair that overlap each other, and that are driven in opposite directions to hold or force apart two parallel surfaces.

FOLIATED Carved with leaf-like ornamentation.

FOLIATION Crystalline segregation of certain minerals in rock in a dominant plane due to metamorphism; schistosity, flow cleavage, and fracture cleavage are considered types of foliation.

FOLLOW BLOCK *See* **Drive cap.**

FOLLOWER 1. Member interposed between the hammer of a pile driver and the head of the pile, used to transmit blows when the top of the pile is below the reach of the hammer; **2.** Piston that maintains a light pressure against a variable quantity of fluid in a container; **3.** Sleeve on a pipe die that aligns the die with the pipe.

FOLLOWER PLATE Plate fitted to the top surface of a lubricating grease container and so designed that as lubricating grease is dispensed, atmospheric pressure assists gravitational forces in delivering grease to the inlet of the dispensing system.

FOLLOWING TOOL Tool used to hold the pin tumblers and springs in place while the pin-tumbler cylinder of a lock is being worked on.

FOLLY Useless structure built only to satisfy an oblique requirement; often a completely detailed Gothic or classical building or ruin, frequently to scale, placed so as to enhance a view or complement landscaping.

FOOD WASTE *See* **Solid waste.**

FOOT 1. Lower end of a pile; **2.** One of a number of projections from the drum of a compactor. *Also called* a **Sheep's foot**; **3.** Nonmetric unit of length equal to 12 inches. Symbol: ft. Multiply by 0.3048 to obtain meters, symbol: m; by 304.8 to obtain millimeters, symbol: mm. *See also the appendix:* **Metric and nonmetric measurement.**

FOOTAGE Common measure of rock-drill production; the number of linear feet drilled, usually per shift.

FOOTBLOCK 1. Timber pad used to spread a load from a prop or side tree; **2.** Steel wedge or assembly used as the base mounting for a guy derrick, gin pole, or Chicago-boom derrick.

FOOTBOARD Wood board laid on concrete blocks or bricks (a simple type of scaffold) used by a mason to stand on when setting stone or laying brick. *Also called* **Hopping board.**

FOOT BOLT Vertically set tower bolt.

FOOTBRAKE VALVE *See* **Valve.**

FOOT CANDLE Non-SI unit of illuminance, equal to the amount of direct light thrown by one international candle on a square foot of surface, every part of which is one foot away. Symbol: ft/candle. Multiply by 10.763

91 to obtain lux, symbol: lx. *See also the appendix:* **Metric and nonmetric measurement.**

FOOT CUT *See* **Plate cut.**

FOOTER *See* **Footing.**

FOOTING Support, usually rectangular in section, and of concrete, wider than the bottom of the foundation wall or pier that it supports and resting on earth. *Also called* **Footer.**

FOOTING BEAM Tie beam for a roof.

FOOTING FORM Formwork of wood or steel for shaping and holding concrete for the footing of a wall.

FOOTLAMBERT Nonmetric unit of photometric brightness (luminance) equal to 1 mcd/ft²; it is the luminance of a theoretically perfect diffusing surface emitting or reflecting flux at the rate of 1 lumen. *See also the appendix:* **Metric and nonmetric measurement.**

FOOT OF A PILE Lower end of a driven pile.

FOOTPATH Constructed walk for pedestrians.

FOOT PER HOUR Nonmetric unit of velocity. Symbol: ft/h. Multiply by 0.084 666 7 to obtain millimeters per second, symbol: mm/s; by 304.8 to obtain millimeters per hour, symbol: mm/h. *See also the appendix:* **Metric and nonmetric measurement.**

FOOT PER MINUTE Nonmetric unit of velocity. Symbol: ft/min. Multiply by 0.005 08 to obtain meters per second, symbol: m/s; by 5.08 to obtain millimeters per second, symbol: mm/s. *See also the appendix:* **Metric and nonmetric measurement.**

FOOT PER SECOND Nonmetric unit of velocity. Symbol: ft/s. Multiply by 0.3048 to obtain meters per second, symbol: m/s. *See also the appendix:* **Metric and nonmetric measurement.**

FOOT PIN Hinge attaching the boom to a revolving shovel.

FOOT PLATE Horizontal board laid over and crossing a wall plate.

FOOT-POUND Nonmetric unit of energy, equal to the work done when a mass of one pound, accelerated at a rate of one foot per second per second, has moved one foot. Symbol: ft/lb. Multiply by 1.355 818 to obtain joules, symbol: J. *See also the appendix:* **Metric and nonmetric measurement.**

FOOT-POUND BLOW Energy in foot-pounds delivered per blow by the ram of a pile hammer.

FOOTPRINT 1. Surface area occupied by an item of equipment or machinery; **2.** Tread area of a tire when actually touching a flat surface.

FOOT SCREW One of a set of three adjusting screws connecting the tribranch of a theodolite or level to the plate that is screwed onto the head of a tripod, used to level the instrument.

FOOTSTONE Stone built into the lower end of a gable to support its coping.

FOOT TIE Structural member that stabilizes two or more formwork standards.

FOOT VALVE *See* **Valve.**

FOPS *Abbreviation for:* falling object protective structure.

FORCE 1. Resultant of distribution of stress of a prescribed area. A reaction that develops in a member as a result of load (formerly called total stress or stress); **2.** That which tends to produce or modify motion, including:

Compression force: Force acting on a body, tending to compress it.

Shear force: Force acting on a body that tends to slide one portion of the body against its other side.

Tensile force: Force acting on a body tending to elongate it.

Torsion: Force acting on a body that tends to twist it.

FORCE ACCOUNT WORK Prescribed work paid for on the basis of actual costs and appropriate additives.

FORCE CUP Flexible cup attached to a pole-like handle, used for clearing clogged drains and water closets.

FORCED-ACTION MIXER Concrete mixer consisting of a rotating open pan and mixing 'star,' carrying steel blades mounted eccentrically to the pan. The 'star' blades either revolve in the same direction as the pan or in the opposite direction.

FORCED AIR Hot or cold air blown from a heating or air-conditioning unit by means of a fan or blower.

FORCED AIR FURNACE Heating unit with an attached fan or blower.

FORCED-CIRCULATION AIR COOLER Cooler that includes a fan or blower for positive air circulation.

FORCED CONVECTION Convection heat transfer by artificial fluid agitation.

FORCED DRAFT *See* **Draft.**

FORCED-DRAFT FAN *See* **Fan.**

FORCE DRY Causing evaporation of a solvent at a rate faster than that due to exposure to ambient temperature; e.g., by exposing the coated surface to the passage of warm or hot air.

FORCED WARM-AIR HEATING SYSTEM System in which the circulation of heated air is effected by a fan. Such a system includes air filtering or cleaning devices.

FORCE FIT Various interference fits between parts assembled under various amounts of force.

FORCE MAJEURE Unavoidable circumstance, to perform an obligation, for instance, or that is the cause of an event.

FORCE MOTOR Type of electromechanical transducer having linear motion ,used in the input stages of servovalves.

FORCE PUMP Pump used to deliver fluids to a level considerably higher than the cylinder.

FORD Road crossing a stream under water.

FORE-AND-AFT Direction parallel to the travel direction of a bridge.

FORE-AND-AFT PLANKING Bridge planking laid parallel to the stringers.

FORE BATTER Batter where the bottom of the pile is inclined away from the pile driver. *Also called* **Forward batter,** and **Positive batter.**

FOREBAY Reservoir or pond at the head of a penstock or sluice.

FORECAST Estimate of future events or conditions.

FORECAST TO COMPLETE Estimate of the value that remains to be committed within a cost class as of a specified date. Adding this estimate to the committed cost yields the estimated final cost of the scope of work in the class. *See also* **Cost reviews,** and **Cost types.**

FORECLOSURE Termination of the equity of redemption of a lender in property covered by a mortgage.

FOREIGN MATERIAL Any extraneous material, such as wood, paper, metal, sand, dirt, or pigment, that should not normally be present in the tube or cover of a hose.

FOREMAN Person in charge of a specific trade, group of other workers, or part of a project, responsible to a project or site superintendent.

FORE PLANE *See* **Plane.**

FOREPOLE Plank driven in advance of a tunnel face to support the roof or wall during excavation. *Also called* **Spile.**

FORESIGHT Observation made and data read in a direction other than that of a backsight; sight taken on a point, the elevation of which is to be determined with a transit or level; sight taken on a point with a transit, the direction of which is to be determined.

FOREST Plant community, predominantly of trees and other woody vegetation, growing more-or-less closely together.

FORESTATION Establishment of a forest, naturally or artificially, on an area, whether previously forested or not.

FOREST COVER MAP Map showing relatively homogeneous forest stands or cover types, produced from the interpretation of aerial photos and information collected in field surveys. Commonly includes information on species, age class, height class, site, and stocking level.

FOREST ECOLOGY Relationship between forest organisms and their environment.

FOREST ECONOMICS Generally, that branch of forestry concerned with the forest as a productive asset subject to economic principles.

FOREST FIRE Any wildland fire not prescribed for the area by an authorized plan. *Also called* **Wildfire.**

FOREST FLOOR General term for the surface layer of soil supporting the forest vegetation; includes all dead vegetation on the mineral soil surface in the forest as well as litter and unincorporated humus.

FOREST INVENTORY Survey of a forest area to collect such data as area condition, timber volume, and species, for specific purposes such as planning, purchase, evaluation, management, or harvesting.

FOREST LAND See **Land use classes (forest management).**

FOREST MANAGEMENT Generally, the practical application of scientific, economic, and social principles to the administration and working of a specific forest area for specified objectives.

FOREST MANAGEMENT CYCLE Phases that occur in the management of a forest, including harvesting, site preparation, reforesting, and stand tending.

FOREST MANAGEMENT PLAN General plan for the management of a forest area, usually for a full rotation cycle, including the objectives, prescribed management activities, and standards to be employed to achieve specific goals. Commonly supported with more detailed development plans. See also **Managed forest land.**

FOREST PRACTICE 1. Any activity that enhances and/or recovers forest growth or harvest yield, such as site preparation, planting, thinning, fertilization, and harvesting; **2.** Road construction or reconstruction within forest lands for the purpose of facilitating harvest or forest management; **3.** Any management of slash resulting from the harvest or improvement of tree species.

FOREST PROTECTION Prevention and control of any cause of potential forest damage.

FOREST REGENERATION See **Reforestation.**

FOREST RENEWAL Renewal of a tree crop by either natural or artificial means.

FOREST RESIDUALS 1. Sum of wasted and unused wood in the forest, including logging residues, rough, rotten and dead trees, and annual mortality; **2.** Unmerchantable material normally left following conventional logging operations other than whole-tree harvesting.

FORESTRY Generally, a profession embracing the science, business, and art of creating, conserving, and managing forest, and forest lands for the continuing use of their resources, materials, and other forest products.

FOREST TYPE 1. Group of forested areas or stands of similar composition (species, age, height, and stocking) that differentiates it from other such groups; **2.** Classification of forest land in terms of potential cubic-measurement volume growth per ha (acre) at the culmination of mean annual increments in fully stocked natural stands.

FOREST TYPE LABEL Symbol that is used to code information about the forest type on a forest cover map, e.g., site, disturbance, age and height class, species, stocking.

FOREST TYPE LINE Line on a map or aerial photo outlining a forest type.

FOREST UTILIZATION FIRE Fire resulting directly from timber harvesting, harvesting other products, and forest and range management, except related use of equipment, smoking, and recreation.

forf. Abbreviation for: forfeit.

FORFEITURE Loss of something of value by way of penalty for failure to perform.

forg. Abbreviation for: forging.

FORGE Open-hearth furnace to which air is fed under pressure to produce elevated temperatures, used to heat metal to the point where it can be worked by beating on an anvil.

FORGED Metal shape made by either hammering or squeezing the original piece of metal.

FORGING Shaping metal by beating or hammering while softened by heat.

FORGING-QUALITY PLATE Structural steel plate intended for forging, heat treating or similar application.

FORGING RANGE Temperature range in which metal can be forged successfully.

FORGING STRAIN Differential strain resulting from forging or from cooling from the forging temperature.

FORK 1. Two-pronged rod or yoke used to slide shifting collars along their shafts; **2.** Development of a tree when two or more leaders develop following the death or removal of the original leader;

FORK ADAPTER Quickly detachable lift truck accessory that mounts on forks to adapt the industrial truck to handle specific loads.

FORK ADJUSTER Manual- or power-actuated lift truck accessory to facilitate lateral positioning of the forks.

FORK EXTENSION Attachable lift truck fork accessory that increases the load-carrying surface of the forks.

FORK HEAD See **Caster frame.**

FORK HEIGHT Vertical distance from the floor to the load-carrying surface of a lift truck, adjacent to the heel of the forks with the mast vertical, and in the case of reach trucks, with the forks extended.

FORK LIFT TRUCK High-lift, self-loading industrial truck, equipped with a load carriage and forks for transporting and/or tiering loads.

FORKS Horizontal tine-like projections, typically suspended from the carriage of a lift truck but also available as an attachment for other types of mechanical equipment, for engaging and supporting loads. There are many types, including:

> **Log or lumber:** Single, double, or full-width forks with optional clamps.
>
> **Pulpwood:** Forks with either single or double top clamps giving positive control of the material being carried.
>
> **Sorting:** Heavy-duty forks designed for durability and efficiency in stacking operations.
>
> **Stinger:** Single-shaft fork to penetrate junked autos, round hay bales, carpet, etc.
>
> **Utility pallet:** Double-tine, general-purpose forks available in a range of tine widths and lengths.
>
> **Wide frame:** Forks that are adjustable for control of long pipes, culverts, etc.

FORM Temporary structure or mold for the support of concrete while it is setting and gaining sufficient strength to be self-supporting. See also **Formwork** and

Plaster mold.

form. *Abbreviation for:* formation; forming; formula.

FORMALDEHYDE Reactive organic compound cH_2O.

FORM ANCHOR Device used to secure formwork to previously placed concrete of adequate strength, the device being normally embedded in the concrete during placement. *See also* **Anchor.**

FORMATION Ordinary unit of geologic mapping consisting of large and persistent strata of some kind of rock.

FORM CLASS Any of the intervals into which a numerical expression of the taper of a tree or log may be divided for classification or use.

FORMATION STABILIZER Sand or gravel placed in the annulus of a drilled well, between the borehole wall and the well screen, to provide temporary or long-term support for the borehole.

FORM COATING Liquid applied to formwork surfaces for a specific purpose, such as to promote easy release from the concrete and to preserve the form material, or to retard setting of the near-surface matrix for preparation of exposed-aggregate finishes.

FORMED-STEEL CONSTRUCTION *See* **Construction types.**

FORMER Machine that forms the furnish into a particleboard mat prior to hot pressing. May use mechanical or air classification methods of forming the mat.

FORM HANGER Device used to support formwork from a structural framework.

FORMING Changing the shape of a metal part without changing its thickness.

FORM INSULATION Insulating material, applied to the outside of forms between studs and over the top in sufficient thickness and air tightness to conserve the heat of hydration in order to maintain the concrete at required temperatures in cold weather.

FORM LINING Materials used to line the concreting face of formwork in order to impart a smooth or patterned finish to the concrete surface, to absorb moisture from the concrete, or to apply a set-retarding chemical to the formed surface. *See also* **Sheathing.**

FORM OIL Oil applied to the interior surfaces of forms to promote easy release from the concrete when the forms are removed. *See also* **Bond breaker,** and **Release agent.**

FORM PANEL LAYOUT Plan showing how the various foundation form panels are to be located and what size panels are to be used.

FORM PRESSURE Lateral pressure acting on vertical or inclined formed surfaces, resulting from the fluid-like behavior of the unhardened concrete confined by the forms.

FORM RATIO Area of formed surface required per volumetric measure of concrete; m^2 of formwork per m^3 of concrete (ft^2 of formwork per ft^3 of concrete).

FORM Enclosure made of boards, plywood, 'Plyform', or metal for holding green concrete to the desired shape until it has set and thoroughly dried.

FORM SCABBING Inadvertent removal of the surface of concrete because of adhesion to the form.

FORM TIE Mechanical connection in tension used to prevent concrete forms from spreading due to the fluid pressure of fresh, unhardened concrete.

FORM TRAVELER Traveling frame used to strip, collapse, transport, and erect full-circle or arch forms in tunnel construction.

FORMULA 1. List of ingredients and their amount, used in the preparation of a compound; **2.** Any general equation; a rule of principle expressed in algebraic symbols; a method of reasoning stated in the form of an equation.

FORMWORK Total system of support for freshly-placed concrete, including the mold or sheathing that contacts the concrete as well as all supporting members, hardware, and necessary bracing. The concrete, when set, will permanently adopt this shape once the form is removed, or stripped. *See also* **Centering, Falsework,** and **Form.** *Also called* **Shuttering.**

FORSTNER BIT *See* **Bit.**

FORUM Place intended for the conduct of public business and discussion. Originally, an open area, often surrounded by public buildings. More recently, an open, roofed or enclosed area for similar purpose.

FORWARD *See* **Harvest functions.**

FORWARD-ACTING CIRCUIT Circuit in which a relay is operated by an increase of activity: light, pressure, voltage, etc.

FORWARD BATTER *See* **Fore batter.**

FORWARD CONTROL Configuration in which more than half of the engine length is rearward of the foremost point of the windshield base and the steering wheel hub is in the forward quarter of the vehicle length. *Also called* **Cab forward.**

FORWARD CONTROL CHASSIS Vehicle with driver controls (pedals, steering wheel, instruments) located as far forward as possible.

FORWARDER *See* **Harvesting machines (single function).**

FORWARDING Transporting trees or parts of trees by carrying them completely off the ground rather than by pulling or dragging them along the ground. *Also called* **Prehauling.**

FORWARD PASS Calculation of the earliest start and earliest finish dates on a precedence network.

FORWARD PRICE Price of a commodity for delivery at a specified future date.

FORWARD VOLTAGE Minimum voltage necessary to allow a device to continue operation.

FORWARD WELDING Fusing metal in the same direction as the torch flame.

FOS *Abbreviation for:* factor of safety.

FOSSIL Naturally preserved remains or traces of animal or plant life, generally associated with sedimentary rocks.

FOSSILIFEROUS Containing organic remains as fossils.

FOUL AIR DUCT Suction line of a tunnel ventilating system.

FOULED Anything that hangs up, jams; anywhere the intended movement is restricted.

FOULING Material accumulating in gas passages or on heat absorbing surfaces. *See also* **Slag.**

FOUNDATION 1. Solid mass designed to support a superstructure; **2.** Material that supports a structure, cut or fill, whether strengthened or not by piles, mats, or other means, to secure adequate bearing.

FOUNDATION AREA Area required to support the bridge traffic and the load of the structure. *See also* **Cribbing foundation area.**

FOUNDATION BED Rock or soil on which a foundation rests.

FOUNDATION BOLT *See* **Anchor bolt.**

FOUNDATION DRAIN Subsoil drain adjacent to the external, and sometimes internal, face of foundations that permits the infiltration of groundwater and conveys it to a sump or tile bed.

FOUNDATION GROUTING Pressure injection of grout into deposits of rock containing fissures, cavities, seams, etc.; to solidify and strengthen the formation, to reduce or eliminate a flow of water through the formation, and/or to reduce the hydrostatic uplift under a structure such as a dam. *Also called* **Pressure grouting.** *See also* **Grout.**

FOUNDATION LOG Log placed under the mudsill in a cribbing to spread the load over a larger area.

FOUNDATION PILE *See* **Pile.**

FOUNDATION SLOPE CORRECTION Excavation of the rock surface under the impervious core of an earth- or rock-filled dam, necessary to create an upstream slope in relation to the dam axis

FOUNDATION SOIL *See* **Subgrade.**

FOUNDATION UNIT One of the structural members of the foundation of a building, such as a footing, raft, or pile.

FOUNDATION WALL *See* **Wall.**

FOUNDED Caisson settled on its bed.

fount. *Abbreviation for:* fountain.

FOURBLE Unit of four drill pipes left coupled together in a rotary drilling unit.

FOUR-BY-FOUR (4x4) *See* **Four wheel drive.**

4/c *Abbreviation for:* four conductor.

FOUR-CENTERED ARCH *See* **Arch.**

FOUR-CHANNEL ABS Use of four wheel speed sensors and four brake control valves. On a 4x2, 6x2, or 6x4 tractor, 4-channel ABS monitors four of the six wheel ends, though it controls left and right side drive axle wheel ends in pairs so that both receive the benefits of ABS. *See also* **Six-channel ABS.**

FOUR CYCLE Refers to an engine where intake, compression, power, and exhaust each take place, in sequence, as the piston moves through its cyclical travel during two crankshaft revolutions. *Also called* **Four stroke.**

FOUR-IN-ONE BUCKET *See* **Bucket.**

FOUR-LEG INTERSECTION *See* **Intersection types.**

FOUR-LEG SLING Chain or rope sling with four hooks hung from one link or thimble.

FOUR-PART LINE *See* **Parts of line.**

FOUR STROKE *See* **Four cycle.**

FOURTH DRUM Fourth hoist drum, in addition to two main hoist drums and a third hoist drum.

FOUR-WAY REINFORCEMENT System of reinforcement in flat-slab concrete construction comprising bands of bars parallel to two adjacent edges and also to both diagonals of a rectangular slab.

FOUR-WAY SWITCH *See* **Switch.**

FOUR-WAY SWIVEL Universal-type joint used for hanging blocks on a machine to eliminate wear of block components.

FOUR-WAY VALVE *See* **Valve.**

FOUR-WHEEL DRIVE Motive system that provides power to two axles and the wheels attached to them. The drive wheel is designated as the wheel meeting the least resistance. Therefore, conditions such as steering mode and ground condition will influence which wheel will receive power. If a no-spin differential is used, the axle using it will distribute power equally to both wheels. *Also called* **Four-by-four.**

FOYER Entrance hall or vestibule.

fp *Abbreviation for:* freezing point.

fp. *Abbreviation for:* faceplate; fireplace; flameproof.

FP *Abbreviation for:* fixed pin; fixed price; full period.

fpm *Abbreviation for:* feet per minute.

fprf *Abbreviation for:* fireproof.

fps *Abbreviation for:* foot-pound per second; feet per second.

fpt *Abbreviation for:* female pipe thread.

fr. *Abbreviation for:* frame; framing; front.

frac. *Abbreviation for:* fraction; fractional.

FRACTIONAL-HORSEPOWER MOTOR Motor, commonly electrical, with a rating of less than one horsepower.

FRACTIONAL SAMPLING Mechanical sampling of granular material using equipment that divides or decimates a sample without segregation.

FRACTURE 1. Crack or break, as of concrete or masonry; **2.** Configuration of a broken surface; **3.** Action of cracking or breaking; **4.** Breaking of rock without movement of the broken pieces. *See also* **Crack.**

FRACTURED FACES Freshly broken surface of crushed aggregate.

FRACTURE RADIUS *See* **Radius of rupture.**

FRACTURE TOUGHNESS Measurement of the ability to absorb energy without fracture. Generally determined by impact loading of specimens containing a notch having a prescribed geometry.

frag. *Abbreviation for:* fragment.

FRAGMENT Rock or mineral particle 3 mm (0.125 in.) or larger.

FRAGMENTATION Extent to which rock is broken into small pieces by primary blasting.

FRAME 1. Skeleton structure supporting cladding; **2.** Entire framework, including joists, studs, plates, sills, partitions, and roofing; all the parts that make up the skeleton of a building; **3.** Timber work supporting the various structural parts, such as windows, doors, floors, and roofs; **4.** Woodwork of doors and windows; **5.** Lintel or header, bucks, and sill assembly forming the entrance of an elevator hoistway; **6.** Principal prefabricated, structural unit in a fabricated tubular-frame scaffold. *Also called* **Panel**; **7.** Any pair of walings on opposite faces of a trench, together with the struts that support them; **8.** Central chassis member, comprised of side-members, cross-members, gussets, and reinforcements (if required) that provides a mounting place for suspension, steering, power-train components, cab, and body, and that also supports the load; **9.** Structure on which either upper or carrier machinery is installed; **10.** Foundation of a machine, supporting other machine components.

FRAME BENT Bent in a pier or abutment made from sawn timber.

FRAME BUCKLING Condition under which bifurcation may occur in a frame.

FRAME CONSTRUCTION Building of houses, apartments, etc., with wood or metal framing members. There are two basic systems:

> **Balloon:** Framing system in which all vertical structural elements of the bearing walls and partitions consist of single pieces extending from the top of the soleplate to the roofplate; all floor joists are fastened to these vertical elements.

> **Platform:** Framing system in which the floor joists of each story rest on the top plates of the story below (or on the foundation sill for the first story), and the bearing walls and partitions rest on the subfloor of each story.

FRAME CUTOFF Where the frame extension behind the rear axle of a vehicle may be shortened, or shortened to create a special-purpose body that may be unusually short for the wheelbase on which it is mounted. The shortest allowable extension for each vehicle is referred to as the **Maximum frame cutoff.**

FRAMED BUILDING Structure whose weight is carried by a framework, as against by structural walls.

FRAMED OPENING Framework and flashing that surround an opening in the wall or roof of a building.

FRAMED PARTITION Partition built up of studs and plates.

FRAMED STRUT Replacement for a needle beam.

FRAME HIGH 1. Height of the top of window and door frames; **2.** Level at which the lintel is to be placed.

FRAME INSTABILITY Condition under which a frame deforms with increasing lateral deflection under a system of increasing applied monotonic loads until a maximum value of the load, called the stability limit, is reached, after which the frame will continue to deflect without further increase in load.

FRAMER Tradesman skilled in framing.

FRAME REINFORCEMENT Extra thickness of side rail material attached over, or inside the main side rails for increased frame strength.

FRAMER'S HAMMER *See* **Hammer.**

FRAME SECTION MODULUS Relative strength of frames as it relates to shape, taking into account frame depth, flange width, and material thickness. All other things being equal, the frame with the largest section modulus will have the greatest strength and stiffness, i.e., the ability to more effectively resist sagging under loads.

FRAMEWORK 1. Load-carrying frame of a structure; **2.** Carpentry work consisting entirely of framing or rough work.

FRAMING The structural components that form the rough structure of a building, including the flooring, roofing, partitioning, ceiling, and beams.

FRAMING CHISEL *See* **Wood chisel.**

FRAMING CONTRACTOR Contractor who subcontracts to complete the framing of a house or structure.

FRAMING MEMBER Structural member to which exterior and interior surfacing materials can be attached.

FRAMING SQUARE An L-shaped device with a 600-mm (24-in.)-long, 50-mm (2-in.)-wide blade and a 400- or 460-mm (16- or 18-in.)-long, 38-mm (1.5-in.)-wide tongue. The edges of each leg are engraved with dimensional units. The square can be used to calculate and lay out a wide range of functions.

FRANCHISE Exclusive right to something granted a contractor, typically to collect and/or dispose of solid wastes from a district or community.

FRANCHISE COLLECTION *See* **Collection.**

FRANGIBLE-BULB-TYPE SPRINKLER *See* **Automatic sprinkler.**

FRANGIBLE-PELLET-TYPE SPRINKLER *See* **Automatic sprinkler.**

FRANKLIN STOVE Cast iron stove resembling a fireplace; invented by Benjamin Franklin.

FRAZIL ICE Granular or laminar ice that forms on fast-flowing water during long cold spells.

frbk *Abbreviation for:* firebrick.

FREE *See* **Fit classification.**

FREE AIR Air at normal atmospheric condition for the elevation above sea level. *See also* **Compressed air,** and **Standard air.**

FREE AREA Total area across the face of a pipe or duct.

FREEBOARD Brim of a dam, etc., above maximum water level.

FREE BURNING Condition of a fire or part of a fire that has not been checked by natural barriers or by control measures.

FREE CAR In a duplex or triplex elevator system, a car that remains where last used during periods of no traffic or when calls to the system are behind it and ahead of the home car.

FREE CAPPING One-half the overall diameter of an unloaded new tire.

FREE CONTRACTION JOINT Contraction joint placed in a concrete slab-on-grade.

FREE EARTH SUPPORT Pressure distribution assumed in the design of anchored sheet pile walls where the wall is considered to be free to rotate about its base.

FREE END End of a beam that is unsupported, as of a cantilever.

FREE FACE Area of least resistance toward which the rock will move following detonation of an explosive; usually an open face.

FREEFALL 1. Lowering of the hook, or lowering of the boom by gravity, without it being coupled to the power train. The lowering speed is controlled by a brake or clutch; **2.** Descent of freshly mixed concrete into forms without dropchutes or other means of confinement; **3.** Uncontrolled fall of aggregate.

FREE FLOAT Extra time available for an activity if every activity in the project starts as early as possible: the amount of float that can be allocated to an activity without interfering with subsequent work.

FREE FLOW Flow that is not hindered by obstruction or constriction; the maximum velocity possible for the condition.

FREE FORK HEIGHT Attainable fork height of a lift truck before the stated overall lowered height of the mast is exceeded by any standard part of the forks, mast, or carriage assemblies, when loaded.

FREE FORM Design or shape that is unconstrained; without constraint; flowing.

FREE FORMALDEHYDE Uncombined or unreleased formaldehyde available for release or emission from a particleboard panel.

FREE GROWING Young trees that are as high, or higher than competing brush vegetation with 1 m (3.3 ft) of free-growing space surrounding their leaders.

FREE GROUNDWATER See **Groundwater.**

FREEHAND GRINDING Grinding by holding the work against the wheel by hand, usually called **Off-hand grinding.**

FREEHAND SKETCH Depiction that does not necessarily represent a three-dimensional object to scale or in correct perspective.

FREE-HANGING HAMMER See **Pile hammer.**

FREE-HANGING LEAD See **Lead.**

FREEHAUL Distance within which material is moved without extra compensation, usually 300 m (1000 ft) or less. See also **Average haul, Haul, Overhaul,** and **Station yards.**

FREEHOLD Tenure of land in fee simple (unencumbered).

FREE LENGTH Lineal measurement of hose between fittings or couplings.

FREE LIFT Attainable lift of a lift truck from the extreme lowered section of the carriage before the stated overall lowered height of the mast is exceeded by any standard part of the forks, mast or carriage assemblies.

FREE LIQUID Liquid that readily separates from the solid portion of a waste under ambient temperature and pressure.

FREE MOISTURE 1. Moisture having essentially the properties of pure water in bulk; **2.** Moisture not absorbed by aggregate. See also **Surface moisture.**

FREE ON BOARD Point at which transfer of title takes place: There are several variations, including:

> **F.O.B. point of origin:** freight collect from the shipping point.
>
> **F.O.B. point of destination:** freight prepaid from the original shipping point to its destination.

FREE-RUNNING PISTON Piston not connected with a rod, working with hammer-like blows. See also **Piston,** and **Slave piston.**

FREE STANDING Independent of other means of support.

FREE TIME Period of time that freight will be held before storage charges are applied.

FREE WATER 1. Non-chemically combined water; **2.** Water droplets or globules in a system fluid that tend to accumulate at the bottom or top of the system fluid, depending on the fluid's specific gravity.

FREEWAY Limited access highway, devoted to the use of motor vehicles, in which all intersections are separated so that no traffic crosses the grade and opposing streams of traffic are separated.

FREEWHEELING CLUTCH See **Clutch.**

FREEZE Condition where some piles increase their load-carrying capacity after being driven. See also **Set-up.**

FREEZELESS WATER FAUCET Water faucet that discharges on the exterior face of a wall but which has its valve seat on the interior face of the wall.

FREEZE LINE See **Frost line.**

FREEZE-OUT PLUG See **Expansion plug.**

FREEZE RESISTANCE Ability to resist the effects of temperatures below the freezing point.

FREEZE-THAW CYCLE Cycle from completely frozen to completely thawed and back to completely frozen.

FREEZE-UP 1. Date when outdoor conditions in an area mean that open water can be expected to be frozen, the soil will contain ice crystals, exposed surfaces will be frosted in the early morning; full winter conditions. See also **Break-up; 2.** Condition where a moving part refuses to operate, for whatever reason.

FREEZING INDEX Total number of degree days below freezing for a winter, calculated from the mean night and day air temperatures.

FREEZING-POINT DEPRESSANT Admixture

added to a concrete mix to act as an antifreeze.

FREEZING PROCESS Technique of temporarily freezing a water-bearing soil to increase its strength, as well as to eliminate passage of water through the soil.

FREEZING TIME Time for a freezing process to be complete.

FREIGHT Any commodity being transported.

FREIGHT BILL Document giving a description of the freight, its weight, amount of charges, rate for charges, taxes, and whether collect or prepaid.

FREIGHT ELEVATOR *See* **Elevator.**

FREIGHT FORWARDER Company that assembles small shipments from various shippers into larger shipments, usually full truck or car loads.

FREIGHT IN Freight charge paid by the purchaser on incoming shipments of goods and materials.

FREIGHT OUT Freight charge paid or allowed for by the seller of goods and materials.

FRENCH CONCRETE Concrete containing large stones or rock in order to save cement. Usually used as fill for leveling out rock surfaces where it is not subject to tension or bending.

FRENCH DOOR *See* **Door types.**

FRENCH DRAIN *See* **Farm drain.**

FRENCH POLISH Highly polished finish for wood obtained by repeated rubbing with shellac, or a varnish gum, dissolved in an abundance of alcohol.

FRENCH WINDOW Window that reaches to floor level and that opens in two leaves.

freq. *Abbreviation for:* frequency; frequent.

FREQUENCY Number of vibrations or complete oscillations occurring in one second (designated Hertz or cycles/second).

FREQUENCY RESPONSE Range of frequencies that can be sensed (within certain acceptable limits of error) by a device.

fres. *Abbreviation for:* fire-resistant.

FRESH-AIR INLET Opening for the introduction of atmospheric air.

FRESH CONCRETE Unhardened concrete that can be consolidated by the intended method.

FRESHENING CHARGE Charge given to a charged and wet battery in storage or during inactive periods to replace losses due to local action and to insure that every cell is brought periodically to a full state of charge.

FRET Ornamentation comprising continuous combinations of straight lines, joined usually at right angles.

FRET SAW *See* **Saw.**

FRETTING Erosion of the surface of a material caused by contact and repeated movement between the parts, that tends to wear away small gouges of material from the surface. *Also called* **False brinelling,** and **Friction oxidation.**

FRETTING CORROSION Oxidation of fretting wear debris.

FRETWORK Ornamental openwork, not necessarily completed with a fret saw.

FRIABLE PARTICLE Particle that crumbles easily under slight rolling pressure.

fric. *Abbreviation for:* friction.

FRICTION Resistance to motion when one body is sliding or tending to slide over another.

FRICTION MODIFIER Additive designed to affect the frictional properties of rubbing surfaces, used to prevent stick-slip oscillations or noise in, for example, wet-clutch operation or reduce energy consumption in, for example, engine oils. Act by forming thin layers on the surfaces by physical adsorption.

FRICTIONAL SOIL Silt, sand, or gravel whose shearing strength is mainly decided by the friction between particles.

FRICTION BLOCK Block or pad, employing the same principle as a brake shoe, used to apply friction to a drum to transmit rotation energy.

FRICTION BRAKE *See* **Brake.**

FRICTION CLUTCH POWER TAKE OFF *See* **Clutch.**

FRICTION COATING Rubber covering applied to the weave of a fabric, simultaneously with impregnation.

FRICTION COURSE *See* **Hot mix seal coat.**

FRICTIONED FABRIC Fabric impregnated with a rubber compound by friction motion (calender rolls running at different surface speeds).

FRICTION/END-BEARING PILE *See* **Pile.**

FRICTION FACTOR Factor used in calculating loss of pressure due to the friction of a fluid flowing through a pipe.

FRICTION-FIT BATT Batt insulation without vapor barrier that is held secure within framing members by friction. *Also called* **Batt.**

FRICTION HEAD *See* **Friction loss.**

FRICTION HORSEPOWER *See* **Horsepower.**

FRICTION LOSS 1. Loss of pressure created by movement of water in a pipe, hose, or fitting. *Also called* **Friction head**; **2.** Stress loss in a prestressing tendon resulting from friction between the tendon and duct or other device during stressing.

FRICTION OXIDATION *See* **Fretting.**

FRICTION PILE *See* **Pile.**

FRICTIONS Any friction block and drum-drive assembly.

FRICTION SURFACE Exposed portion of a hose formed by a layer of rubber-impregnated fabric, as distinguished from a product having the fabric completely covered with a layer of rubber.

FRICTION-TYPE BEARING Bushing; bearing that does not employ balls or rollers.

FRIEZE 1. Decorated band along the upper part of an internal wall; **2.** Middle division of an entablature, between the architrave and cornice.

FRIEZE BLOCK Piece of lumber between rafters at the eaves, used to close the attic space.

FRIEZE PANEL Upper panel in a door having more than four panels.

FRINGE BENEFITS Those amounts that are added to a basic wage for work done; to cover such benefits as an extended health plan, company pension, unemployment insurance, vacations, education, etc.

FRITTED GLAZE *See* **Glaze.**

frm. *Abbreviation for:* formed; framer; framing.

FROE Long, wedge-shaped blade for splitting wood into shingles, shakes, and other split products.

FROG 1. Depression in the bed surface of a brick to receive mortar. *Also called* **Key,** and **Panel; 2.** Piece of material with two holes, used on a worker's belt to carry spud wrenches.

FRONT Working attachment, such as the boom of a dragline or the boom, handle, and dipper of a shovel.

FRONTAGE Length of a site in contact with a public road.

FRONTAGE STREET *See* **Street.**

FRONT AXLE *See* **Axle.**

FRONT CONNECTED Where connections are made to normally exposed surfaces of components.

FRONT ELEVATION View of the face of a building showing the main entrance and type of architecture.

FRONT END Logging area closest to a yarder.

FRONT-END ATTACHMENT Optional load-supporting member for use on a mobile crane.

FRONT-END LOADER *See* **Loader.**

FRONT-END STABILIZER Fifth hydraulic jack mounted in the front end of truck-mounted cranes to provide a 360° area of operation.

FRONT-END PROCESS Size reduction, separation and/or physical modification of solid wastes to afford practical use or reuse.

FRONT FOOT One foot (0.3 m) length of land measured along the frontage of a lot. When used as the basis for the sale of commercial property it implies a square measure and includes the ground lying back of the frontage to the rear boundary of the property. *See also* **Front meter.**

FRONT HANGER Bracket for mounting the equalizer beam of a multiple-axle suspension to a truck or trailer frame that allows for the beam's pivotal movement. Formerly called a center hanger. Normally of three basic types: flange-mount; straddle-mount, and under- or side-mount.

FRONTING Front glazing or front pointing, done where two parallel edges have a recess between them, such as the shallow gap between two adjoining window frames, that requires a flexible filler.

FRONTISPIECE Main facade or principal entrance bay of a building.

FRONT LINTEL Lintel supporting the wall above an opening, as seen from the exterior.

FRONT-LOADING REFUSE TRUCK *See* **Refuse truck.**

FRONT LOT LINE Boundary line of a lot along a street; in the case of a corner lot, either of the boundary lines along a street, the other being considered a side lot line; in the case of a through lot, each of the two shorter boundary lines along streets.

FRONT METER One meter (3..28 ft) of land measured along the frontage of a lot. When used as the basis for the sale of commercial property it implies a square measure and includes the ground lying back of the frontage to the rear boundary of the property. *See also* **Front foot.**

FRONT-MOUNT HOIST Hydraulic cylinder, usually long-stroke telescopic, mounted vertically at the extreme front of a dump body, commonly enclosed in a housing that extends into the front portion of the body. *Also called* **Head lift, Head mount,** and **Vertical hoist.**

FRONT PUTTY Glazing compound forming a triangular fillet between the surface of the glass and the front edge of the rebate, or, when beads are used, forming the ribbon between the surface of the glass and the bead.

FRONT SHOVEL Tracked equipment equipped with a boom and stick on which is mounted a bucket that is pushed into the dirt and lifted up and away from the machine to load, and that discharges by rotating forward and down, or discharges through the bottom or front by releasing hinged sections.

FRONT STAND Structural, front-mounted supporting member that attaches to the main frame rails of a mixer truck and that furnishes support to the trunnion bearing located at the head (front) end of the mixer.

FRONT YARD Area of ground between the front lot line of a property and the setback.

FROST Weather condition during which dew turns to ice.

FROST ACTION Phenomenon that occurs when water in soil is subjected to freezing which, because of the water ice phase change or ice lens growth, results in a total volume increase or the buildup of expansive forces under confined conditions, or both, and the subsequent thawing that leads to loss of soil strength and increased compressibility.

FROST BOIL Softness of soil that has thawed after frost heave.

FROST CRACK Radial, longitudinal split in the wood of a tree, generally near the base of the bole, caused by internal stresses due to extremely cold weather.

FROST GRIP *See* **Adfreezing.**

FROSTED *See* **Rustication.**

FROSTED GLASS Surface treatment for glass making it translucent in which one or both surfaces are acid etched or sand blasted, either wholly, or in areas to form a decoration.

FROST HEAVE Lifting action of structures, waste material, or surface soil, caused by the expansion of material during the freeze/thaw cycle of water contained in the top soil.

FROSTING *See* **Chalking.**

FROSTLINE Maximum depth to which frost penetrates the ground. This varies in different parts of the country and from year to year. For design purposes it is a figure established for a locality as determined from historical records. It is the depth below which foundations should be placed so as to prevent movement due to frost heave. *Also called* **Freeze line.**

FROST PLUG *See* **Expansion Plug.**

FROST-PROOF TILE Tile produced for use where freezing and thawing conditions occur.

FROST WEDGING Pushing up or apart of rock particles due to the action of ice formation.

FROZEN Condition where a part normally free to rotate cannot move.

FROZEN SOIL Soil below 0°C (32°F) in which part of the pore water has frozen.

FRP *Abbreviation for:* fiberglass-reinforced plastic.

frst *Abbreviation for:* frost.

frstd *Abbreviation for:* frosted.

frt *Abbreviation for:* freight; front.

frwk *Abbreviation for:* framework.

frwy *Abbreviation for:* freeway.

frz. *Abbreviation for:* freeze.

fs *Abbreviation for:* finished surface.

FS *Abbreviation for:* full scale.

fsbl. *Abbreviation for:* fusible.

fsp *Abbreviation for:* fiber saturation point.

fst. *Abbreviation for:* forged steel.

FST *Abbreviation for:* flat seamed tin.

ft *Abbreviation for:* foot; fork thickness.

FT *Abbreviation for:* flush threshold; fume-tight.

ft² *Abbreviation for:* square foot.

ft³ *Abbreviation for:* cubic foot.

ftbd *Abbreviation for:* footboard.

ft-c *Abbreviation for:* foot-candle.

ftg *Abbreviation for:* fitting; footing.

ft-L *Abbreviation for:* foot-lambert.

ft-lb *Abbreviation for:* foot-pound.

f to f *Abbreviation for:* face to face.

FTR *Abbreviation for:* flat tile roof.

fu. *Abbreviation for:* fuse.

FUEL 1. Substance or combination of substances that can be burned; **2.** In explosive calculations, the chemical compound used to combine with oxygen to form gaseous products and cause a release of heat.

FUEL-AIR RATIO Ratio of fuel supply flow rate to the air supply flow rate when both rates are measured in the same units under the same conditions.

FUEL BED Layer of solid fuel or solid waste on a furnace grate or hearth.

FUELBREAK *See* **Firebreak.**

FUEL CLASS Group of forest floor fuels possessing common characteristics; dead fuels are grouped according to their time lag (1-, 10-, and 100-hour) and living fuels as to whether they are herbaceous or woody.

FUEL EFFICIENCY Relationship between fuel consumption and machine productivity. It is expressed in units of weight carried or material moved per volume of fuel consumed.

FUEL EVAPORATIVE EMISSION Vaporized fuel released into the atmosphere from the fuel system of an engine.

FUEL HEATER Device used to heat fuel at cold ambient temperatures.

FUEL INJECTION SYSTEM System that injects a combustible fuel into the cylinders of an engine.

FUEL INJECTION TUBING Tube connecting the injection pump to the nozzle holder assembly.

FUEL LINE Tube used to convey fuel to an engine.

FUEL MANAGEMENT Forest activities carried out to modify fuel accumulations (slash) to reduce the chance of ignition and rate of fire spread.

FUEL MODEL Simulated forest floor fuel complex for which all the fuel descriptors required for solution of a mathematical fire-spread model have been specified.

FUEL MOISTURE ANALOG Device that simulates the response of the moisture content of specific classes of dead fuels when exposed in the same environment.

FUEL MOISTURE CONTENT Quantity of moisture in forest floor fuel, expressed as a percentage of the weight when thoroughly dried at 100°C (212°F).

FUEL MOISTURE STICK Wooden stick of known dry weight used to determine changes in moisture content and flammability of forest fuels.

FUEL STORAGE TANK Container used to store the fuel used by the prime mover.

FUEL STRAINER Coarse wire-mesh strainer, usually used in conjunction with gas lines and heavy fuels.

FUEL SYSTEM A basic (diesel) fuel system consists of a tank, lines, transfer pump, filters, water separator, injection pump, and injectors.

FUEL TANK CAPACITY

> **Draw:** Determined by the location of the fuel-supply line; the amount of fuel that can be drawn from a tank.
>
> **Fill:** Usually, 95% of nominal capacity, to allow for heat expansion.
>
> **Nominal:** Capacity based on the maximum volume of the tank.

FUEL TRANSFER PUMP *See* **Pump.**

FUEL TYPE Identifiable association of forest floor fuel elements of distinctive species, form, size, arrangement, or other characteristics that will cause a predictable rate of fire spread or control difficulty under specified weather conditions.

FUELWOOD Plant that obtains its nourishment through the organic matter of other plants, causing decays.

FUGITIVE DYE Dye whose color fades to neutral after a few days on exposure, usually due to ultraviolet rays in sunlight; used to temporarily color membrane-curing compounds so that coverage of the concrete surface can be observed.

ful. *Abbreviation for:* fulcrum.

FULCRUM Pivot for a lever.

FULL AIR BRAKE *See* **Brake.**

FULLBACK Concrete block that has all of the cell webs cut out except the last one, which holds the block together.

FULL CAPPING Of a tire, application of new rubber to the tread area plus some distance down the sidewall of the casing.

FULL-CELL PROCESS Process for impregnating wood with preservative or chemical in which a vacuum is drawn to remove air from the wood before admitting the preservative.

FULL-CIRCLE POUR Process whereby the complete concrete lining in a tunnel is poured in one operation between bulkheads. *See also* **Arch pour** and **Invert pour.**

FULL COAT As thick a coat of paint, varnish, or lacquer as can be applied in one operation and produce a final film of uniform appearance with satisfactory hardness. *See also* **Base coat, Glaze coat,** and **Mist coat.**

FULL CONTROL OF ACCESS Authority to control access, exercised to give preference to through traffic by providing access connections with selected public roads only, by prohibiting crossings at grade or direct private driveway connections. *See also* **Control of access** and **Partial control of access.**

FULL-CUT LUMBER Lumber cut so that its thickness and width measure fully to the specified dimensions.

FULL-DEPTH ASPHALT PAVEMENT Pavement in which asphalt mixtures are employed for all courses above subgrade, laid directly on the prepared subgrade.

FULL DISC BRAKE *See* **Brake.**

FULLER'S CURVE Empirical curve for gradation of aggregates, also known as the Fuller-Thompson ideal grading curve. The curve is designed by fitting either a parabola or an ellipse to a tangent at the point where the aggregate fraction is one-tenth of the maximum size fraction. *See also* **Grading curve.**

FULLER'S EARTH Fine-grained earth resembling clay but lacking plasticity, used as a filtering agent.

FULL FACE Technique of tunnel blasting to the full bore size with each round.

FULL-FLOATING AXLE *See* **Axle.**

FULL FLOW OIL FILTER Filter through which all of the system's oil flows, typically trapping all particles larger than 10 microns. A magnetized screen is used to trap metal particles.

FULL GLOSS Highest possible gloss obtainable with paint.

FULL HEADER A course consisting of all headers.

FULL-LENGTH LUMBER Lumber cut so that the ends can be squared to exact length.

FULL-LOAD CURRENT Greatest load that a circuit or device is designed to carry under specific conditions: any additional load is an overload.

FULL LOAD SPEED Speed at which an engine runs when it is delivering its full rated horsepower.

FULL LOAD STOP Device that limits the maximum amount of fuel injected into the engine cylinders at the rated load and speed specified by the engine manufacturer.

FULL-PENETRATION JOINT *See* **Pile joint.**

FULL POWER BOOM Where all boom sections of a lifting device are powered by hydraulic telescoping cylinders that are controlled by the operator inside the machine's cab.

FULL-SCALE LOAD TEST Load test made on an above- or below-ground structural element, with the load carried being at least that of the structural design, or more, according to the factor of safety of the design or test.

FULL SPEED Contract speed at which an elevator should run.

FULL-TIDE COFFERDAM Cofferdam built high enough to exclude water at all tides.

FULL TRAILER *See* **Trailer.**

FULL TREE *See* **Whole tree.**

FULL-WIDTH GOOSENECK *See* **Gooseneck.**

FULLY COMPOSITE BEAM Composite beam with sufficient shear connectors to develop the full flexural strength of the composite section.

FULLY EQUIPPED ENGINE One that is equipped with all the accessories necessary to perform its intended functions unaided. This includes, but is not restricted to, intake air system, exhaust system, cooling system, generator or alternator, starter, and emission control equipment.

FULLY FIXED Member in a structural frame that has a fixed end.

FULLY SANDED Panel that has had the entire surface smoothed by sanding (in comparison to skip sanding, that removes only the high spots).

FULLY SERVICED LOT *See* **Serviced lot.**

FULLY STOCKED STANDS *See* **Stocking classes.**

fum. *Abbreviation for:* fumigate.

FUME 1. Gas, smoke, or vapor, usually offensive and sometimes suffocating; **2.** Treating finished wood with ammonia fumes to fix certain colors.

FUMIGATION Disinfection with a gas for the destruction of germs, insects, or animal life.

FUNCTIONAL ADDITION Material or agent added to modify the use properties of a finished product.

FUNCTIONAL ELEMENT Part of a building, classified according to the purpose or function it serves rather than its composition. Elemental cost analysis is a technique for estimating a building cost before the specific design is known. *Also called* **Element.**

FUNCTIONALISM Establishing form and structure based on the most economical satisfaction of physical need.

FUNCTIONAL OBSOLESCENCE Decreased capacity of a structure to perform the function for which it was designed due to changing standards and/or needs.

FUNCTIONAL PROGRAM Written description of the client needs that must be fulfilled by a proposed building. The description generally consists of a statement of the functions and approximate space requirements. The architectural program is, in turn, based on the functional program. Both are components of the conceptual design. *See also* **Architectural program, Conceptual program,** and **Requirements.**

FUNCTION LIMITER Device incorporated into the anti-two block system or rated capacity indicator system that will disable such crane functions as winch up, telescope out, and/or boom down (as applicable) as two block or overload situations approach. *Also called* **Function lockout, Hydraulic cutout,** and **Hydraulic kickout.**

FUNCTION LOCKOUT *See* **Function limiter.**

fungi. *Abbreviation for:* fungicide.

FUNGICIDAL ADMIXTURE Toxic chemical added to a concrete mix to deter the growth of algae and lichen on the hardened concrete.

FUNGICIDE Poison, used to kill or retard the growth of fungus.

FUNGUS Group of living plants, like mildew and mold, that feed on plant ingredients and that can attack most organic material under suitable conditions of moisture and temperature.

fur *Abbreviation for:* further.

fur. *Abbreviation for:* furnace; furred; furring.

FURAN MORTAR Two-part mortar system of furan resin and furan hardener, used for bonding tile to back-up material where chemical resistance of floors is important.

furn. *Abbreviation for:* furnish; furnished; furniture.

FURNACE Enclosed place in which heat is intentionally released by combustion, electrical device, or nuclear reaction. There are several types, including:

Arc: Heat produced by means of an electric arc between carbon electrodes and the furnace charge.

Crucible: Furnace fired with coal, coke, oil, or gas, in which metal contained in crucibles is melted.

Induction: Where heat is produced by electric currents induced in the charge itself.

Muffle: Where heat is applied to the outside of a refractory chamber containing the hearth.

Multiple hearth: Combustion unit that transfers burning material down a series of refractory hearths using a rabble arm between each two hearths.

Waterwall: Field-erected incineration equipment, the side of which consists of water-carrying tubes. As water circulates through the tubes it extracts the energy produced by combustion and turns it into steam.

FURNACE PRESSURE Gauge pressure that exists within a furnace combustion chamber.

FURNACE VOLUME Total internal volume of a combustion chamber.

FURNACE WALL *See* **Wall.**

FURNISH 1. Provide, supply; **2.** Blended particles, binders, and additives ready for the board-forming process.

FURNISHED BY OTHERS Materials or apparatus to be installed by the contractor but supplied to the project by someone else, named or not named in the contract documents.

FURNISHINGS Curtains, carpets, furniture and similar soft and/or movable items that finish a place for people's use.

FURNITURE 1. Personal effects that make a dwelling useful and habitable; **2.** Hardware that is fitted to something to make it functional: door furniture, for instance, includes hinges, handsets, locks, catches, push plates, etc.

FURNITURE STEEL Grade of sheet steel used for elevator cabs and entrances.

FURR DOWN Lowering the level of a ceiling to conceal a duct.

FURRED Pipes and boilers that have become encrusted internally with salts and minerals deposited from the water, usually at elevated temperature.

FURRED CEILING Ceiling having spacer elements between it and the supporting structure above.

FURRING 1. Process of leveling up part of a wall, ceiling, or floor by the use of wood or metal strips; **2.** Use of thin strips spaced at intervals behind a wall and the plaster to provide an air space, to give appearance of greater thickness, or for the application of an interior finish such as plaster.

FURRING CHANNEL 1. Steel member used to support interior finish; **2.** Smallest horizontal member of a suspended ceiling system.

FURRING STRIP Thin strip of any material applied to a surface to even it and to serve as a fastening base for finish material.

FURRING TILE Tile designed for lining the inside of exterior walls, and carrying no superimposed loads.

FURROWING Striking a V-shaped trough in a bed of mortar.

fus. *Abbreviation for:* fusible.

FUSE 1. Attachment to an electrical circuit designed to interrupt the circuit under conditions of overload or short-circuiting; **2.** Length of wire that will melt when a specified amount of electric current is passed through it; **3.** Device for initiating an explosive. *See also* **Blasting fuse.**

FUSE BLOCK Insulated panel holding the fuses that protect an electrical circuit.

FUSED SWITCH *See* **Switch.**

FUSE LIGHTER Special device used to ignite a safety fuse.

FUSE PLUG 1. Fuse encapsulated in a screw plug that is screwed into a fuse block to complete and protect an electric circuit. **2.** *See* **Spillway.**

FUSE WIRE Wire made of an alloy that melts at a predetermined and low temperature; each alloy selected will melt at a different temperature and is rated in terms of electrical amperage it will pass.

FUSIBLE LINK Component comprising a short length of metal having a low melting point; used to actuate a fire door, sprinkler head, etc., under conditions of elevated temperature.

FUSION Process of melting; usually the result of interaction of two or more materials.

FUSION DRILL *See* **Drill.**

FUSION POINT Temperature at which the solid and liquid states of a substance can exist together in equilibrium (often thought of as the melting or freezing point of a substance).

FUSION WELDING Process where metal is joined by fusion, with or without filler-metal at the same time.

fut. *Abbreviation for:* future.

FUTURE DEPRECIATION Loss in value that will occur in the future: the basis for the recapture of capital now represented by the machine, equipment, or structure, used in the income approach to value.

fv *Abbreviation for:* face velocity.

FV *Abbreviation for:* front view.

FVF *Abbreviation for:* facility value formula.

fw *Abbreviation for:* frame width.

FW *Abbreviation for:* feed water.

fwd *Abbreviation for:* forward.

fwdg *Abbreviation for:* forwarding.

fxd *Abbreviation for:* fixed.

FY *Abbreviation for:* fiscal year.

FYA *Abbreviation for:* for your attention.

FYI *Abbreviation for:* for your information.

g *Abbreviation for:* gram; granite; gravitational acceleration; gravity.

ga *Abbreviation for:* gauge.

GABION Medium-sized, 200 to 600 mm (8 to 24 in.), rocks confined in a rectangular wire cage, used to restrain the toe of an excavation, as a retaining wall, or for other pressure-resisting purposes.

GABLE Triangular vertical end of a building, above the ceiling line, supporting the roof.

GABLE END Entire end wall of a house having a gable roof. *Also called* **Gable wall.**

GABLE-END TRUSS *See* **Truss.**

GABLE MOLDING Molding on the upper exterior face of a gable, applied to the barge board where it meets the tiles or shingles.

GABLE POST Short vertical post at the apex of a gable into which the barge boards are dressed.

GABLE ROOF Ridged roof having a gable at one or both ends.

GABLE WALL Wall of which the gable forms a part. *Also called* **Gable end.**

GABLE WINDOW Window placed in the gable end of a roof, sometimes with its sides conforming to the angles of the roof.

GAD Chisel or pointed bar, used for loosening rock.

GADROONED Decorated with convex curves; the opposite of fluting.

GAGE *See* **Gauge.**

GAG PROCESS Bending structural steel shapes in a gag press.

GAIN Notch or mortise cut to receive the end of another structural member or piece of hardware.

gal. *Abbreviation for:* gallon.

GALL Small, local swelling on the stem or branch of a tree, caused by fungi, bacteria, insects, or physiological disorders.

gall. *Abbreviation for:* gallery.

GALLERY 1. Covered promenade, partly open at one side; **2.** Railed platform projecting into and overlooking a room; **3.** Long narrow room designed for specific purposes: display of art works, indoor target practice, etc.; **4.** In underground mining, a horizontal passage; **5.** Passage carved out under the bark or in the wood of a

tree by insects feeding or laying eggs.

GALLET Stone chip.

GALLETING Inserting into mortar courses, while still soft, small pieces of stone, chips of flint, etc., sometimes for structural but usually for decorative reasons.

GALLING 1. Material failure, usually characterized by surface loss due to very high pressure; **2.** Polished depression in a metal surface.

GALLON (IMPERIAL) Nonmetric unit of volume, equal to 277.42 cubic inches; 160 fluid ounces; 10.02 pounds of water. Symbol: gal. Multiply by 0.004 546 09 to obtain cubic meters, symbol m³; by 4.546 09 to obtain liters, symbol: L. *See also the appendix:* **Metric and nonmetric measurement.**

GALLON (IMPERIAL) PER DAY Nonmetric unit of flow. Symbol: gpd. Multiply by 0.004 546 09 to obtain cubic meters per day, symbol: m³/d; by 0.000 052 618 8 to obtain liters per second, symbol: L/s. *See also the appendix:* **Metric and nonmetric measurement.**

GALLON (IMPERIAL) PER MINUTE Nonmetric unit of flow. Symbol: gpm. Multiply by 0.007 768 to obtain liters per second, symbol: L/s. *See also the appendix:* **Metric and nonmetric measurement.**

GALLON (U.S.) Nonmetric unit of volume, equal to 231 cubic inches; 128 fluid ounces; 8.34 pounds of water. Symbol: gal. Multiply by 0.003 785 412 to obtain cubic meters, symbol: m³; by 3.785 412 to obtain liters, symbol: L. *See also the appendix:* **Metric and nonmetric measurement.**

GALLON (U.S.) PER MINUTE Nonmetric unit of flow. Symbol: gpm. Multiply by 0.063 090 2 to obtain liters per second, symbol: L/s. *See also the appendix:* **Metric and nonmetric measurement.**

GALLOWS BRACKET Triangular-shaped framing projecting from a wall with its horizontal member at the top.

galv. *Abbreviation for:* galvanize; galvanized.

GALVALUME SHEET *See* **Steel.**

GALVANIC CORROSION Corrosive action occasioned when two different metals touch each other in an environment sufficiently moist to permit the passage of an electric current between them, causing one metal to corrode at the expense of the other.

GALVANIZED IRON Steel or iron, sheets or parts, coated with zinc.

GALVANIZED PIPE *See* **Pipe.**

GALVANIZED STEEL PIPE *See* **Pipe.**

GALVANIZING To coat a metal with zinc, either by hot dipping or by electro-deposition. *Also called* **Hot-dipped galvanizing.**

GALVANOMETER Device containing a silver chloride cell, used to measure resistance in an electric blasting circuit.

GAMBREL ROOF Pitched roof having a double slope between the eaves and ridge, the lower slope being steeper than the upper slope. Also known as a **Double-pitch roof,** and **Mansard roof.**

GANG 1. Number of tools, machines, etc., that operate together, usually powered from a single source; **2.** Crew of workers.

GANGED FORMS Prefabricated panels joined to make a much larger unit (up to 10 x 15 m (30 x 50 ft)) for convenience in erection, stripping, and reuse, usually braced with wales, strongbacks, or special lifting hardware.

GANG MOLD Mold used to cast several identical items at the same time.

GANG SAW *See* **Saw.**

GANG SAWING Use of multiple fixed saws on a common arbor.

GANGWAY Temporary or movable footbridge.

GANISTER Highly refractory siliceous sedimentary rock used for furnace linings.

GANTRY 1. Framework, supported at each end so that it spans a distance, used for carrying a traveling crane; **2.** Overhead structure of an excavator that supports operating parts; **3.** Structural part of a heel-boom log-loader that supports the boom; **4.** Part of the upper structure of a crane, fixed or adjustable in height, to which the lower spreader is anchored, used to support the boom and hoist system. *Also called* **A-frame, A-frame gantry,** and **Back-hitch gantry.**

GANTRY BOOM *See* **Topping line.**

GANTRY JUMBO Drill jumbo that has an open space in the center large enough for muckers, cars, and locomotives to pass through.

GANTRY LINE Line that is used to raise and lower the boom of a yarding crane.

GAP-GRADED AGGREGATE *See* **Aggregate.**

GAP-GRADED CONCRETE Concrete containing a gap-graded aggregate.

GAP GRADING Distribution of aggregate in which particles of certain intermediate sizes are wholly or substantially excluded.

gar. *Abbreviation for:* garage.

GARAGE Enclosed space for the storage, servicing, or repair of vehicles and equipment.

garb. *Abbreviation for:* garbage; garburator.

GARBAGE *See* **Solid waste.**

GARBURATOR *See* **Waste disposal unit.**

GARDEN APARTMENT Dwelling, part of a multidwelling, and possibly multistory structure, having direct access to a landscaped area.

GARDEN CITY Planned community in which single-family detached houses and open spaces predominate.

GARDEN WALL BOND *See* **Bond.**

GARGOYLE Water spout projecting from a roof or the parapet of a wall or tower, carved into grotesque shapes, usually human, animal, or mythical.

GARNET PAPER Generic name for a paper sheet with an abrasive grit glued on one side. The paper is graded according to the size of the grit used, and the grade represents its cutting effect.

GARRET Room, often unfinished, within the space of a pitched roof.

GARRETING Process of pressing small slivers of stone into the joints between coarse masonry.

gas. *Abbreviation for:* gasoline.

GAS COCK *See* **Valve.**

GAS CONCRETE Lightweight concrete produced by developing voids with gas generated within the unhardened mixture (usually from the action of cement alkalis or aluminum powder used as an admixture). *See also* **Foamed concrete.**

GAS-ELECTRIC TRUCK Electric truck in which the power source is a gasoline- or LP gas engine-driven generator.

GAS ENGINE Internal combustion engine that uses gas as a fuel.

GAS HOLE Hole created by gas escaping from molten metal.

GASIFICATION Process whereby carbonaceous solid or liquid matter is converted to gases, e.g., carbon dioxide, methane, or ammonia.

GASIFIER That part of an engine that supplies heated, pressurized gas to a power turbine. Typically the compressor/turbine/combustor setion of a two-shaft, free-power turbine engine.

GASKET Pliable material placed between two surfaces to create a water-, air-, or oil-tight seal when the surfaces are drawn together under pressure.

GASKIN Rope or hemp gasket placed in a bell-and-socket joint prior to sealing with mortar or lead.

GAS MAIN Principal gas supply pipe from which service feeds are taken.

GAS METAL ARC WELDING Welding using a continuously fed consumable electrode and a shielding gas.

GAS METER Device that measures and records the volume of gas consumed by an individual location or appliance.

GAS POCKET Cavity in a weld caused by entrapped gas.

GAS TUNGSTEN ARC WELDING Welding using a tungsten electrode and a shielding gas.

GAS TURBINE ENGINE Rotary prime mover that uses an essentially continuous process to compress, heat, and expand a gaseous working fluid.

GAS VENT Opening to the atmosphere from a system or vessel generating or containing lighter-than-air gases.

gas/w *Abbreviation for:* gas weld.

GATE 1. Hinged barrier, usually of open construction. *See also* **Double-blade gate,** and **Single-blade gate;** **2.** Hinged metal flap used to control the flow of water; **3.** Part of a casting formed by the opening in the mold through which the metal is poured. *See also* **Pile gate;** **4.** Opening in a lever tumbler that allows them to pass the fence in a lever-tumbler lock.

GATE BLOCK *See* **Snatch block.**

GATE HOOK Right-angle-shaped fitting, one leg of

which is spiked and is driven into a wooden gate post or built into a masonry pier, the other leg of which forms an upright pin on which the hinge of a gate pivots.

GATEHOUSE Small building, built as part of a perimeter wall, located at a principal entrance, often serving as a guard post or security location.

GATEPOST Post on which a gate hinges, or against which it latches.

GATE VALVE *See* **Valve.**

GATE YARD Solid waste volumes as calculated at the landfill entrance in the incoming trucks, generally compacted into a space in the landfill equivalent to 50% of that occupied in the truck.

GATHER To bring together, as flues in a stack.

GATHERED-RATIO TRANSMISSION *See* **Transmission.**

GAUGE (ALSO SPELLED GAGE) 1. Standard measure; **2.** An instrument for measuring, indicating, or comparing a physical characteristic. There are many types, including:

   Fluid level: Gauge that indicates a fluid level.

   Manometer: Differential pressure gauge in which pressure is indicated by the height of a liquid column of known density. Pressure is equal to the difference in vertical height between two connected columns multiplied by the density of the manometer liquid.

   Pressure: Gauge that indicates the pressure in the system to which it is connected.

   Vacuum: Pressure gauge for pressures less than atmospheric.

   Water: Glass U-shaped tube half filled with water, the other end connected by a flexible tube to a system of drains, gas pipes, etc., being tested. It shows whether the pipes are gas tight. *Also called* a **U-gauge.**

3. To measure for a particular purpose; **4.** Of a roofing slate or tile, that portion exposed to view; **5.** Tool for scoring a line parallel with the edge of a board; **6.** Amount of plaster of paris used with common plaster to quicken setting of the mix; **7.** Designation for specific thickness of sheet metal (i.e., 4-gauge, 10-gauge, etc.) or wire; **8.** Thickness of a saw blade measured in decimals of an inch, mm, or Birmingham gauge; **9.** Spacing of tracks or wheels.

GAUGE BOX Box of specific volume, used for batching.

GAUGED ARCH *See* **Arch.**

GAUGED BRICK *See* **Brick.**

GAUGED MORTAR Mortar mixed with plaster of paris to make it set more quickly.

GAUGE GLASS Vertical glass tube that shows a liquid level.

GAUGE PRESSURE *See* **Pressure.**

GAUGE SIZE Width of a drill bit along its cutting edge.

GAUGE STICK 1. Lath marked to represent the intervals between repeated elements, such as courses,

or center-to-center spacing; **2.** Piece of material sized to represent a specific dimension or spacing.

GAUGING 1. Adding gauging material; **2.** Cementitious material added to lime putty to provide and control set.

GAUGING BOARD Portable, flat surface on which mortar, plaster, etc., is mixed.

GAUGING PLASTER Gypsum plaster or plaster of paris.

GAUGING TROWEL Larger than a pointing trowel, smaller than the buttering trowel, usually 83 x 178 mm (3.25 x 7 in.).

GAUL Hollow in a mortar finishing coat.

GAUSS Non-SI unit of magnetic flux density, equal to one line of magnetic force per square centimeter. Symbol: Gs or G. Not permitted for use with the SI system of measurement. *See also the appendix*: **Metric and nonmetric measurement.**

GAWR *Abbreviation for:* gross axle weight rating.

GAZEBO Traditionally, a structure offering a view; small look-out tower, summerhouse, belvedere, lantern, turret, balcony, etc. More recently, an ornamental garden house offering shade and views.

gb *Abbreviation for:* grade break.

GB *Abbreviation for:* glass block.

g-cal *Abbreviation for:* gram-calorie.

GCF *Abbreviation for:* greatest common factor.

GCI *Abbreviation for:* gray cast iron.

G-CLAMP G-shaped clamp having an adjustment screw as its vertical leg, used when gluing wood or holding components.

GCW *Abbreviation for:* gross combination weight.

GCWR *Abbreviation for:* gross combination weight rating.

gd *Abbreviation for:* guard; gutter drain.

gd. *Abbreviation for:* guide.

GEAR Two or more toothed wheels meshed together so that the motion of one is transmitted to the other. There are many types, including:

   Bevel: Gear made of teeth cut in the surface of a truncated cone, used to transmit power at right angles.

   Cluster: Two or more gears of different sizes made in one piece.

   Helical: Type of gear where the straight or curved teeth are positioned diagonally across the face of the gear wheel at an angle of less than 90° to the direction of rotation.

   Herringbone: Gear with V-teeth.

   Idler: Gear meshed with two others that does not transmit power to its shaft, used to reverse the direction of rotation in a transmission.

   Pinion: Small gear that drives a larger gear.

   Planetary set: Set of gears consisting of an inner (sun) gear, and outer ring with internal teeth, and two or more small (planetary) gears meshed with both the sun gear and the ring.Used to either speed up or slow down the input *vs* output to gain speed or power.

Rack: Toothed bar.

Ring: Large gear, driven by a worm gear, that is actuated by the drive sheave.

Spiral bevel: Gear with spiral-shaped teeth, used primarily to change the direction of transmitted power.

Sprocket: Metal disk with projecting teeth about its periphery, usually employed in combination with a large chain to transmit rotary power.

Spur: Gear on which the teeth are cut parallel to the axis of the shaft.

Sun: 1. Central gear in a planetary set. 2. Planetary gear set consisting of a central gear and an internal-toothed ring gear, and two or more planet gears meshed with both of them.

GEAR BACKLASH Clearance between meshed gear teeth

GEAR CASE Enclosure containing a gear train, the shafts on which they are mounted, plus lubricant.

GEARED DRIVE *See* **Driving machine.**

GEARED MPH Theoretical speed in miles per hour based on engine speed, combined gear ratio and tire revolutions per mile.

GEARED SPEED Maximum attainable road speed based on engine-governed rpm, transmission gear ratio, rear axle ratio, and tire size.

GEARED TRACTION MACHINE *See* **Driving machine.**

GEAR HYDRAULIC PUMP *See* **Pump, hydraulic.**

GEARLESS TRACTION MACHINE Type of elevator hoisting machine on which the hoist ropes pass over a traction drive sheave that is an integral part of the armature. *See also* **Driving machine.**

GEAR PUMP *See* **Pump.**

GEAR RATIO Number of revolutions a driving gear requires to turn a driven gear through one complete revolution. For a pair of gears, the ratio is found by dividing the number of teeth on the driven gear by the number of teeth on the driving gear.

GEAR REDUCTION UNIT Assembly of meshed gears in which the rpm of the input shaft is higher than that of the output shaft.

GEAR SEGMENT Part of a gear that is segmented to provide a reference for timing, such as combining two teeth together.

GEAR STEP Transmission gear step measured by the proportion of change between successive gears. It is determined by dividing the ratio of the lower gear by the ratio of the next higher gear. The larger the step the greater the rpm drop upon downshifting. Gear steps are usually expressed as percentages.

GEAR TRAIN Assembly of meshed gears in which the related shafts revolve at the same, or different, rpm than that of the input shaft.

GEL Matter in a colloidal state that does not dissolve but remains suspended in a solvent from which it fails

to precipitate without the intervention of heat or an electrolyte. *See also* **Cement gel.**

GELATIN Jelly-like high explosive made by dissolving nitrocotton in nitroglycerine.

GELATIN DYNAMITE *See* **Dynamite.**

GEL COAT Thin outer layer applied to a molded shape, usually as a cosmetic.

GELLING Transformation of a liquid into a jelly-like consistency.

GEL PORE Void, finely divided from other such voids, left when a cement gel is dried, by heating to 105°C (221°F), and the water gel is driven off.

GEL SPACE RATIO Ratio of the concentration of newly formed hydration products to the original capillary pore space available.

GEL STRENGTH Stress required to break up the gel structure of a bentonite slurry formed by thixotropic buildup, under static conditions.

GEL TIME Time required to change a flowable liquid resin into a nonflowing gel.

GEL WATER Excess water that is bound to a gel but that can be driven off by a strong drying action.

gen. *Abbreviation for:* generation; generator.

GENERAL BENEFIT Advantage accruing from a given public works, typically a highway improvement, to a community as a whole, applying to all property similarly situated. *See also* **Benefit,** and **Special benefit.**

GENERAL CONDITION One of the standard clauses within a contract that pertain to accepted practices of construction and project management, etc.

GENERAL CONTRACT Contract between an owner and a contractor covering the entire work to be done; one not limited in its scope.

GENERAL CONTRACTOR Owner's designated representative with full responsibility for the completion of the contracted project.

GENERAL DRAWING Drawing showing the plan, elevations, and a cross section of a structure or works, and including overall dimensions, etc.

GENERAL ESTIMATE METHOD Simple estimating method based on a square foot (or equivalent) cost for typical construction types (residential, warehouse, commercial, etc.).

GENERAL FOREMAN Contractor's representative in charge of all labor who coordinates the work of trades foremen, whether directly employed or subcontracted.

GENERAL INSURANCE *See* **Insurance.**

GENERAL LIEN Lien that includes all of the property owned by the debtor, rather than a specific property.

GENERAL LIGHTING Area lighting designed to provide a minimum level of illumination.

GENERAL-PURPOSE BUCKET *See* **Bucket.**

GENERAL RATE Rate that is levied annually by a municipal taxing authority on that part of the tax rate that includes real property, or as a business assessment, to fund municipal expenditures.

GENERAL-USE SNAP SWITCH *See* **Switch.**

GENERAL-USE SWITCH *See* **Switch.**

GENERAL WARRANTY Clause in a deed whereby a grantor agrees to protect the grantee against any defect in the title being transferred.

GENERATED CONTAMINANT Contamination created by the operation of a fluid system or component. *See also* **Artificial contaminant, Built-in contaminant,** and **Contaminant.**

GENERATED HEAT Heat resulting from the removal of metal by a grinding wheel.

GENERATOR 1. Mechanism consisting of an armature, field coils, etc., which generates electricity when the armature coils are rotated in the magnetic field produced by the field windings; **2.** Engine-driven electrical device that produces direct or alternating current.

GENERATOR SATURATION Point at which an increase in excitation current produces little or no increase in flux density.

GENERATOR SET Integrated diesel- or gasoline-powered engine and electrical generator. They are rated according to:

Prime power: For continuous electrical service with 10% overload capability for one hour in 12.

Standby power: For continuous electrical service during interruption of normal power.

genl *Abbreviation for:* general.

genly *Abbreviation for:* generally.

geo. *Abbreviation for:* geometric.

geod. *Abbreviation for:* geodetic.

GEODESIC DOME A structure of 60, 90, or 120 triangles that closely follows the shape of a sphere and whose edge lengths closely follow the path of great circles on the sphere.

GEODETIC CONSTRUCTION Stressed-skin construction in which the principal joints follow the perimeter of a sphere.

GEODETIC CONTROL System of horizontal and/or vertical control points determined by geodetic survey.

GEODETIC SURVEYING *See* **Surveying.**

GEODIMETER Surveying instrument employing the length of a generated wave of pulse light or modulated wave to determine the distance between the transmitter and a reflector.

GEOHYDROLOGIC Having to do with subsurface water and related geologic aspects of surface waters.

geol *Abbreviation for:* geological.

geol. *Abbreviation for:* geologist.

GEOLOGICAL ENGINEER Professional engineer who specializes in studies of the earth's structure.

GEOLOGICAL MAP Map showing the significant geological features of an area identifying the salient materials.

GEOLOGICAL REPORT Report on a specific area or site identifying and locating the materials, and their interrelationship, to prescribed depths and in specified detail, primarily intended to determine load-bearing capabilities, susceptibility to movement, and seismic potential.

GEOLOGICAL SURVEY Investigation of subsurface materials and conditions to meet a requirement and/or specification.

GEOMEMBRANE Any synthetic, hydraulically impervious material intended to be used with soil as part of a man-made fluid containment system.

GEOMETRY OF CIRCULAR CURVES The following geometric terms and symbols are used in the computation and recording of circular curve data for roads and highways:

Arc (L): L, or Arc, is used to denote the length of a circular curve.

Chord (C): Line drawn from point of curvature to point of tangency.

Deflection angle: Angle between the chord and the tangent: one half of the angle subtended by the arc.

Degree of curve: Angle required for a curve of a given radius to subtend an arc of 100 m

Delta (Δ): Used to designate the angle subtended by the circular arc, shown in degrees, minutes and seconds.

Point of curvature (PC): Point where an arc begins.

Point of intersection (PI): Point where two connected tangent lines intersect.

Point of tangency (PT): Point where the arc becomes tangental with the departing line.

Radius (R): Length of the radius.

Tangent (T): Distance from the beginning of the circular arc (point of curvature); the point of intersection of the two lines being joined.

GEOMETRICAL STAIR Winding stair that returns upon itself with winders built around a well.

GEOPHYSICAL EXPLORATION Preliminary subsurface survey using devices located on or above the surface to find the change in wave velocity, electrical resistivity, gravity, magnetism, or other physical variables.

GEORGIAN GLASS Thick glass with square-mesh wire embedded in it.

GEOSYNTHETIC Any synthetic material used with soil as an integral part of a man-made system.

GEOTECHNICAL ENGINEER Engineer with specialized training and knowledge of soils and rocks, employed to do soil investigations, design of structure foundations, and provide field observation. *Also called* **Soils engineer.**

GEOTECHNICAL PROCESSES Processes that cause change in soil: freezing/thawing, groundwater levels, etc.

GEOTECHNOLOGY Study of geology, soil, and rock mechanics.

GEOTEXTILE Woven or nonwoven fabric manufactured from synthetic fibers or yarns that is designed to serve as a continuous membrane between soil and aggregate in a variety of earth structures.

GEOTHERMAL ENERGY Energy generated within the earth's interior.

GF *Abbreviation for:* glass fiber.

GGMK *Abbreviation for:* great grand master key.

GGMkd *Abbreviation for:* great grand master keyed.

GHOSTING Paint applied so thin that previously applied colors appear through.

GI *Abbreviation for:* galvanized iron.

GIANT Large-diameter nozzle, manually or mechanically controlled, for directing a jet of water under high pressure for hydraulic excavating. *Also called* **Monitor.**

GIBS 1. Channel-shaped lengths of metal, used as a clamp; **2.** Guide that attaches a diesel hammer and its tripping device to a spud lead. *See also* **Jaws,** and **Side channel.** *Also called* **Guide channels,** and **Spud clip.**

GIGA Prefix representing 10⁹. Symbol: G. Used in the SI system of measurement. *See also the appendix:* **Metric and nonmetric measurement.**

GILD Overlay with a thin layer of gold or coat with a gold color.

GILDING METAL Alloy of copper and zinc, with a greater proportion of copper than in brass.

GILLMORE NEEDLE Device used in determining the time of setting of hydraulic cement.

GILSONITE Form of natural asphalt, hard and brittle, occurring in rock crevices or veins from which it is mined.

GIMBAL EXPANSION JOINT *See* **Joint.**

GIMLET Small hand tool having a handle that forms a tee to the shaft, and with a point fashioned like an auger with a threaded tip; used to bore a starter hole with minimal stress to the work.

GIN Vertical pole or frame used to raise objects.

GIN BLOCK Single-sheave pulley, carried in a steel cage, having a hook at one end by which it can be hung.

GINGERBREAD Ornate decoration, especially in the exterior trim of a house.

GIN POLE Single mast with winch, held upright by guylines, over which a cable is run as a means of lifting objects to a height; a form of derrick.

GIRANDOLE 1. Revolving water jet used in a fountain; **2.** Lighting fixture in the form of a wall bracket with branching arms, often backed by a convex mirror.

GIRDER 1. Main horizontal support beam, usually supporting other beams. *See also* **Beam,** and **Pile cap; 2.** Main truss supporting secondary trusses framing into it.

GIRDER BRIDGE Bridge where the deck is supported by girders or large beams.

GIRDER CASING Material used to enclose and protect a steel girder.

GIRDLING Killing of a tree by severing or damaging the cambium layer and interrupting the flow of food between the leaves and the rest of the tree.

GIRT Small beam spanning between columns, generally used in industrial buildings to support outside walls. *See also* **Beam.**

GIRTH Circumference of any circular or near-circular object.

GIRT STRIP Boards fastened to studding to carry floor joists.

GIVE LINE To direct the placing of a flagpole, pin or other object in line. *See also* **Line.**

gl. *Abbreviation for:* glass; glaze; gloss.

GL *Abbreviation for:* grade line.

GLACIAL DRIFT Any rock material, such as alluvia, transported by a glacier and deposited by the ice or by the meltwater from the glacier.

GLACIAL TILL Material deposited by glaciation, usually a wide range of particle sizes, not subjected to the sorting action of flowing water. *See also* **Drift.** *Also called* **Moraine.**

GLADHAND Quick-release coupling, attached to the cab or chassis, that connects tractor brake hoses to the semi-trailer or trailer. *See also* **Polarized gladhand.** *Also called* **Brake-hose coupler.**

GLAND 1. Cavity of a stuffing box; **2.** Compressible copper or brass sleeve. *Also called* **End plate; 3.** Waterproof seal at the end of an electrical cable.

GLAND BOLT Bolt used for tightening an unthreaded gland.

GLAND JOINT Pipe joint that allows for movement due to thermal expansion or contraction.

GLAND RING Part of a stuffing box assembly that exerts a uniform pressure on the packing.

GLARE Brightness that causes visual distortion or lack of clarity.

GLASPHALT Asphalt mix in which a portion of the aggregate consists of crushed glass (cullet). *See also* **Glassphalt.**

GLASS Hard, brittle substance made by fusing silicates with soda or potash, lime, and sometimes metallic oxides. Used in sheet form for glazing, and as hollow blocks in nonstructural walls and partitions. Sheet glass is drawn or rolled, in several thicknesses or weights, with or without ornamentation to the surface, and in a wide range of translucency and pigmentation.

GLASS BEADING Surface treatment (for cleaning and/or polishing) in which tiny glass beads are blown, under high pressure, against a metallic surface.

GLASS BLOCK Hollow, translucent glass units with patterns molded on their interior or exterior, or both.

GLASS-CONCRETE CONSTRUCTION Composite construction where loads are carried by reinforced concrete columns and beams with glass blocks used for infilling. *See also* **Pavement light.**

GLASS CUTTER Industrial diamond or tungsten carbide wheel mounted at the tip of a stylus. When drawn across the surface of a sheet of glass, the diamond or wheel scores the surface. Subsequent breaking/downward/outward pressure exerted on the underside of the scored line will induce a clean break along the line.

GLASS (DRILLING) BIT *See* **Bit.**

GLASS FIBER Drawn fiber of glass, used as reinforcement for concrete.

GLASS-FIBER FELT Asphalt-coated sheet of glass fiber, generally used in built-up roofing. *See also* **Built-up roof, Organic felt,** and **Tarred felt.**

GLASS-FIBER REINFORCED CEMENT Composite material consisting essentially of a matrix of hydraulic cement paste or mortar reinforced with glass fibers, typically precast into units less than 25 mm (1 in.) thick.

GLASS MAT Glass fiber sheet containing random textile fibers, used in the manufacture of roofing products.

GLASS MOSAIC TILE Tile made of glass, usually no more than 50 mm (2 in.) square and 6 mm (0.25 in.) thick, they come mounted on sheets of paper about 300 mm (12 in.) square.

GLASSPAPER Abrasive paper where glass granules, flint, garnet, corundum or similar materials are glued to paper and graded according to their coarseness.

GLASSPHALT Proprietary name for **Glasphalt.**

GLASS STOP Glazing bead; a device to stop a sheet of glass from sliding vertically within a frame or mounting.

GLASS-TRANSITION TEMPERATURE Midpoint of the temperature range over which an amorphous material (such as glass or a high polymer) changes from (or to) a brittle, vitreous state to (or from) a plastic state.

GLASS WOOL Thermal insulation made from glass fibers.

glaz. *Abbreviation for:* glazed; glazing.

GLAZE 1. To put panes of glass in a sash, frame, or prepared opening. *See also* **Double glazing,** and **Single glazing; 2.** Translucent film of color made from paint media. Also used to refer to the medium itself; **3.** Transparent liquid applied to clay tiles before firing in order to produce a glossy and impermeable surface; of various types:

> **Bright glaze:** High-gloss coating with or without color.
>
> **Clear glaze:** Transparent glaze with or without color.
>
> **Crystalline glaze:** Glaze that contains microscopic crystals.
>
> **Fritted glaze:** Glaze in which a part or all of the fluxing constituents are prefused.
>
> **Mat glaze:** Low-gloss ceramic glaze with or without color.
>
> **Opaque glaze:** Nontransparent glaze with or without color.
>
> **Raw glaze:** Glaze compounded primarily from raw constituents. It contains no prefused materials.
>
> **Semimat glaze:** Medium-gloss ceramic glaze with or without color.

> **Speckled glaze:** Glaze containing granules of oxides or ceramic stains that are of contrasting colors.

GLAZE COAT Thin application of paint or varnish, sufficient to seal and protect the substrate. *See also* **Base coat, Full coat,** and **Mist coat.**

GLAZE-COAT SURFACING Optional final coating to a built-up roof construction consisting of a final coat of asphalt flooded over the top layer of felt to protect it from ultraviolet rays; may be finished in a light-colored surfacing to help reflect heat.

GLAZED BRICK Building brick having one or more faces with a glaze fused to its surface.

GLAZED CERAMIC MOSAIC TILE Ceramic mosaic tile with glazed faces.

GLAZED DOOR *See* **Door types.**

GLAZED FINISH Finish applied to concrete or brick produced by mixing, to a thick paste, colored portland cement with a gauging liquid that controls the set. The mixture is then trowelled onto a wall surface in two or three layers, followed by two sprayed coats of clear plastic.

GLAZED INTERIOR TILE Glazed tile with a body that is suitable for interior use and that is usually nonvitreous, and is not expected to withstand excessive impact or be subject to freezing and thawing conditions.

GLAZED TILE Fired clay tile finished with a glasslike surface.

GLAZED WARE Stoneware pipe and fittings, glazed from salt thrown into the kiln during firing.

GLAZE FIT Stress relationship between the glaze and body of a fired ceramic body.

GLAZE MEDIUM Paint thinner or water-soluble medium used to make opaque pigments translucent.

GLAZIER Specialist who cuts and installs glass.

GLAZIER'S CHISEL Chisel-shaped putty knife.

GLAZING 1. Panes of glass fitted into doors and windows; **2.** Application of sealant in the process of installing glass in prepared openings in windows, doors, panels, screens, and partitions; **3.** Thin or transparent painting allowing dry underlayers to show through; **4.** Clogging of the surface particles of a grinding wheel resulting in a decreased rate of cutting.

GLAZING BAR Intermediate member of a window sash, dividing the glass into panes or squares.

GLAZING BEAD Molding or stop around the inside of a frame to hold the glass in place.

GLAZING COMPOUND Putty-like compound that retains its elasticity when exposed to weather, used to seal the joints between panes of glass and the sashes or frames into which they are mounted. *Also called* **Glazing putty.**

GLAZING GROOVE Groove made to receive glass.

GLAZING POINT Small, often triangular piece of galvanized metal used to 'pin' a pane of glass in position on the back putty in a sash or frame prior to face puttying. *Also called* **Point.**

GLAZING PUTTY *See* **Glazing compound.**

gld *Abbreviation for:* gold.

GLEN Small, narrow valley, usually bounded by gently sloped concave sides. *Also called* **Dale**

GLIDER Cab and chassis less drivetrain components.

GLIDE SWING System designed to allow free swing of a crane's superstructure, allowing it to come to a feathered stop, thereby reducing the swinging of loads during craning operations.

glob. *Abbreviation for:* globular.

GLOBE VALVE *See* **Valve.**

GLORY HOLE Type of open-pit mine that exploits a large, vertical-sided pit where material is gravity fed through a funnel-shaped hole in the floor of the pit, discharging to hauling units located in an adit beneath the pit.

GLOSS Degree of brightness, reflectiveness, or luster of a surface.

GLOW PLUG Electrically heated element, typically located in the precombustion chamber of a diesel engine, to aid in fuel burning during the cold start-up phase of engine operation.

glr. *Abbreviation for:* glare.

GLUE *See* **Adhesive.**

GLUE BLOCK Small piece of wood, usually triangular in cross section, coated with glue on the right-angled face, and used to reinforce a joint.

GLUED-LAMINATED TIMBER Fabrication of relatively thin strips of wood glued together at their faces, either as a flat beam or to a designed shape.

GLUE GUN Hand-held tool for the application of cartridge- or stick-type adhesives.

GLUELINE Adhesive joint formed between veneers in a plywood panel or between face veneers and the core in a composite panel, or between lumber and structural wood panel parts in an assembly such as a component.

GLUE-NAILED Gluing structural wood panel joints and connections with pressure provided by nailing.

GLUING Act of joining two surfaces together with an adhesive.

GLULAM Short for glued-laminated structural timber.

GLUT Short piece of timber placed between bracing and a pile as a filler.

glv. *Abbreviation for:* globe valve.

GLYCERIN Colorless, syrupy liquid prepared from fats and oils. Extends media drying time in water-based media.

GM *Abbreviation for:* grade marked.

G-MANDREL Two-segment mandrel actuated by a hydraulic cylinder pulling on cables that pass over sheaves in a manner to force the two segments apart, used to hold a corrugated shell for driving.

GMAW *Abbreviation for:* gas metal-arc weld.

gmd *Abbreviation for:* geometric mean distance.

GMK *Abbreviation for:* grand master key.

GMKd *Abbreviation for:* grand master keyed.

gmr *Abbreviation for:* geometric mean radius.

GN *Abbreviation for:* grommet nut.

gnd *Abbreviation for:* ground.

GOB HOPPER Hopper under the mixer of a concrete-mixing plant into which mixed concrete is dumped, and from which concrete buckets are filled.

GO-DEVIL 1. Ball of rolled-up burlap or paper or a specially fabricated device put into the pump end of a pipeline and forced through the pipe by water pressure in order to clean the pipeline; **2.** Device used with tremie concrete operations; **3.** Movable hoistway working platform, used when installing elevator brackets and guide rails.

GOING Horizontal distance between the front face of one step and the next step above or below.

G1S *Abbreviation for:* good one side.

GOOD ONE SIDE (G1S) Exterior plywood that is sanded for best appearance on one side only and that may contain neat wood patches, inlays, or synthetic patching material.

GOOD PRACTICE Work done to a level of expertise and skill that is between the minimum necessary for qualification within that speciality and the best that can be produced.

G2S *Abbreviation for:* good two sides.

GOOD TWO SIDES (G2S) Exterior plywood that is sanded for best appearance on both faces and that may contain neat wood patches, inlays, or synthetic patching material.

GOOSENECK 1. Yoke of a block; **2.** Curved or bent section of a stair handrail; **3.** Arched connection between tractor and trailer. There are several designs, including:

> **Folding:** Gooseneck articulated so that it will extend flat upon the ground when the trailer is disconnected from the tractor.
>
> **Full-width:** Gooseneck of the same width as the trailer deck.
>
> **Removable:** Gooseneck that can be removed and reconnected.

GOOSENECK BOOM Boom on a shovel log loader constructed in an arc to heel logs.

GORE LOT Small triangular building lot.

GOTHIC ARCH *See* **Arch.**

GOUGE 1. *See* **Wood chisel;** **2.** Finely ground-up material found in fault areas.

GOUGE SLIP Oilstone slip.

GOUGING Cutting a groove in the surface of a metal.

gov. *Abbreviation for:* governed; governor.

GOVERNMENT RECTANGULAR SURVEY Rectangular system of land survey that divides a district into 62-km² (24-mile²) quadrangles from a meridian and baseline. These areas are further divided into 16 km^2 (6-mile^2) townships, that are further subdivided into 2.58-km^2 (1-mile^2) sections.

GOVERNOR 1. Device that senses and regulates the speed of a rotating shaft; **2.** Speed-monitoring device (on traction elevators) that triggers the safety when the elevator overspeeds; **3.** In a reciprocating engine, a device that senses and regulates engine speed. The governor acts upon any fuel delivery mechanism (such as the rack or control rods) to control fuel delivery and cutoff. There are many governor types in use, including:

Dead-beat: Governor having a high degree of precision and stability.

Droop-type: Governor that regulates speed so that steady-state speed increases slightly as load is removed. Speed is highest at no-load and lowest at full-load.

Electric: Governor that senses speed electrically by means of a magnetic pick-up or tachometer generator or by sensing the frequency of the electric-set generator. Fuel level actuation is usually by a hydraulic actuator.

Electric, load-sensing: Electric governor that senses load as well as speed. Load is sensed by monitoring the electric-set generator output current. Sensing load changes quickens response by causing the governor to respond immediately before a significant speed change occurs. In parallel electric sets operating isochronously (no variation of steady speed with load), each governor senses differences in load between its electric set and the others and adjusts its engine fuel lever to balance the load.

Flyball: Device that limits the speed at which a vertical spindle rotates by causing balls, attached to the spindle on hinged arms, to fly outwards by centrifugal force.

Hydraulic: Governor that achieves prime mover speed control by balancing a hydraulic force against a spring force.

Isochronous: Governor that can be adjusted to zero droop so that steady-state speed is the same at all loads.

Overspeed: Governor that limits the maximum rotational speed of an engine by limiting or stopping the fuel supply above a specified rpm.

GOVERNOR DROOP Difference between engine speed at full load and at no load. Percentage droop is this difference divided by the full-load rpm and multiplied by 100.

GOVERNOR REGULATION Difference between the steady-state engine speed at rated load and the steady-state engine speed at no load, expressed as a percentage of the rated load speed.

GOVERNOR ROPE Wire rope attached to an elevator safety-activating means or releasing carrier that drives a governor, and that, when stopped by a governor, initiates setting of the car or counterweight safety.

GOVERNOR SWITCH *See* **Switch.**

GOVERNOR TENSION SHEAVE ASSEMBLY Sheave mounted in an elevator pit to guide the governor rope and keep it under tension.

govt *Abbreviation for:* government.

GOW CAISSON Short, steel-plate cylinder, large enough for a worker to work inside. As excavation proceeds, successively smaller cylinders are set inside until the bearing stratum is reached, at which point the cylinders are withdrawn and concrete placed in the hole.

gp *Abbreviation for:* group; grade plain.

GP *Abbreviation for:* general purpose.

gpd *Abbreviation for:* gallons per day.

gpf *Abbreviation for:* gasproof.

gph *Abbreviation for:* gallons per hour; graphite.

gp hd *Abbreviation for:* group head.

gp ldr *Abbreviation for:* group leader.

gpm *Abbreviation for:* gallons per minute.

gr. *Abbreviation for:* gear; grade; grains; gravity; gross.

GRAB Crane fitted with a grabbing bucket that is lowered vertically and digs into, closes round, and picks up material.

GRAB HOOK Chain hook that will slide over any single link, but not along the chain. *See also* **Cable hook, Hook, Pintle hook, Round Hook, Safety hook,** and **Swivel hook.**

GRAB SAMPLE Random sample of a material, obtained for laboratory diagnosis.

GRAB SET *See* **Flash set.**

GRAB TEST Tensile test for woven fabric using specimens considerably wider than the jaws holding the ends of the test specimen.

GRACE PERIOD Time beyond the due date, allowed to perform an act or make a payment before default occurs.

GRAD Hundredth part of a right angle.

grad. *Abbreviation for:* gradient; graduate; graduation.

GRADABILITY Slope that a machine can climb, expressed as a percentage (45° = 100% slope).

GRADATION Amount, expressed in percentages or weights, of various material sizes present in an aggregate.

GRADE 1. Ground level; **2.** Percentage of rise or fall to the horizontal distance. *See also* **Profile grade**; **3.** Finished design profile of a roadway or other worked ground; **4.** To cut, fill, and smooth a given area to a given profile; **5.** Designation of quality when assigned a letter, number, or description; **6.** Designation of the quality of a manufactured piece of wood or logs; **7.** Strength of bonding of a grinding wheel, frequently referred to as its hardness; **8.** Cut-off elevation of a pile.

GRADE AND SLOPE CONTROL Automatic system that controls a machine for longitudinal grade and transverse slope to a consistent profile during a planing/milling pass.

GRADE AND SLOPE EQUIPMENT 1. Instruments that help equipment operators maintain the required slope by either noting the plus-or-minus from a predetermined level or by automatically adjusting the equipment so as to compensate grading error; **2.** Attachments that enable a machine to perform fine grading tasks for which it was not primarily designed.

GRADE ASSISTANCE Assistance given a machine or body when descending a grade, caused by the force of the inclined component of its weight.

GRADE BEAM 1. Beam, usually reinforced concrete, placed just below the finish grade line for the purpose of supporting a masonry wall or the walls of a building; **2.** Structural member installed at or near ground level, usually acting as a foundation.

GRADE BREAK Change in slope from one incline ratio to another.

GRADE CLEARANCE Maximum grade change that a vehicle will clear empty, or loaded, without abnormal mechanical contact between the ground and the vehicle.

GRADED AGGREGATE Sample of aggregate made up from particles of different, specified sizes.

GRADED STANDARD SAND Ottawa sand accurately graded between No. 30 (0.59 mm, 0.0232 in.) and No. 100 (0.149 mm, 0.0059 in.) sieves for use in the testing of cements. *See also* **Ottawa sand,** and **Standard sand.**

GRADE LATH Lath that the grade setter has marked to show equipment operators the correct grade.

GRADE LINE 1. Predetermined line indicating a proposed elevation; **2.** Erected string or wire line used to establish a reference for a paver's automatic screed controls. *Also called* **String line.**

GRADE MARK Marking rolled onto a reinforcing bar to identify the grade of steel.

GRADE OF STEEL Means by which a design engineer specifies the strength properties of the steel required in each part of a structure.

GRADE PERCENTAGE Steepness of a grade, measured by dividing the change in elevation by the horizontal distance.

GRADE PIN Steel rod driven into the ground at a surveyor's hub. A string is stretched between the grade pins at the grade indicated on the survey stakes.

GRADER Power excavating machine with a central blade that can be angled to cast soil to either side and that has an independent hoist control on either side.

GRADE REFERENCE Erected stringline, curb, gutter, adjacent mat, or mobile averaging device used to provide input to the automatic control system of a paver, cold planer, or other equipment.

GRADE RESISTANCE 1. Gravity's force that must be overcome when going up (or down) a grade; **2.** Component of gravitational force affecting movement up, or down, an inclined plane and acting at the center of gravity of a vehicle in a direction parallel to the surface on which the wheels are supported. It is numerically equal to the product of gross vehicle weight and sine of the angle of incline or decline.

GRADE ROCK Rock within the road right-of-way that must be drilled, blasted, and excavated to produce a finished road.

GRADE ROD Small length of round or rectangular wood or metal used to check grades.

GRADE SECTION Portion of the cross section consisting of cut-and-filled sections graded or shaped to the specified elevations.

GRADE SENSOR Portion of an automation system that controls grade.

GRADE SEPARATION Crossing of two highways, or a highway and a railroad, at different levels.

GRADE SHOE *See* **Matching shoe.**

GRADE STAKE Stake that indicates the amount of cut (-) or fill (+) at a stated point.

GRADE STRESS Stress that can be safely sustained by formwork timber of a particular strength class, or species and grade.

GRADE STRIP Usually a thin strip of wood tacked to the inside surface of forms at the elevation to which the top of the concrete lift is to rise, either at a construction joint or the top of the structure.

GRADIENT 1. *See* **Expressions of slope; 2.** Rate of change in a variable over time or a distance, as of temperature or moisture.

GRADIENT OF SLOPE Decimal form of rise and fall in m/m, i.e., 0.05 is the rise or fall of a slope that has 0.05 m of rise of fall for every horizontal meter of run. *See also* **Expression of slope.**

GRADING 1. Process of changing the lay of the ground, usually to direct the flow of surface water; **2.** Distribution of particles of granular material among various sizes, usually expressed in terms of cumulative percentages larger or smaller than each of a series of sizes (sieve openings) or the percentages between certain ranges of sizes (sieve openings); **3.** Classifying timber, lumber or logs, according to quality or end-use.

GRADING BUCKET *See* **Bucket.**

GRADING CURVE Graphical representation of the proportions of different particle sizes in a granular material. *See also* **Fuller's curve.**

GRADING LIMITS Allowable maximum and minimum limits, in a sieve analysis test, for a sample of coarse aggregate.

GRADING PLAN Working drawing showing the existing and proposed vertical dimensions of a site layout, by means of contour lines and spot elevations at high and low points.

GRADUAL LOADING Gradual application of loads anticipated to apply to a structure.

GRADUATED ACTING Control instrument that gives throttling control, working between fully on or open and fully off or closed.

GRADUATED COURSES Diminishing courses of brick, block, etc.

GRADUATED LEASE Lease that provides for variations in the amount of rent paid, more of less, based on time expired on the life of the lease.

GRADUATED PAYMENT MORTGAGE *See* **Mortgage.**

GRADUATED TAX *See* **Progressive tax.**

GRAIN 1. Direction, size, arrangement, appearance, or quality of fibers in wood. There are many definitions, including:

Close-grained: Wood with narrowly spaced annual rings.

Coarse-grained: Wood with wide and conspicuous annual rings.

Cross-grained: Wood in which the fibers deviate from a line parallel to the sides of the piece. *Also called* **Slash grain.**

Curly-grained: Wood in which the fibers are distorted so that they have a curled appearance, as in 'birds-eye' wood.

Diagonal-grained: Wood in which the annual rings are at an angle with the bark of the tree or log.

Edge-grained: Lumber that has been sawed so that the wide surfaces extend approximately at right angles to the annual growth rings. (Lumber is considered edge grained when the rings form an angle of 45° to 90° with the wide surfaces of the piece.)

Flat-grained: Lumber sawed parallel to the pith and approximately tangential to the growth rings. (Lumber is considered flat grained when the annual growth rings make an angle less than 45° with the surface of the piece.)

Interlocked-grained: Grain in which the fibers put on for several years may slope in a right-handed direction and then, for a number of years, slope in an opposite direction, and so on.

Mixed-grain: Unsorted lumber containing a mixture of pieces with flat and vertical grain.

Open-grained: Common classification for woods with large pores.

Spiral-grained: Wood in which the fibers take a spiral course about the trunk of a tree.

Straight-grained: Wood in which the fibers run parallel to the axis of the piece.

Vertical-grained: Edge grain of quarter-sawn lumber.

Wavy-grained: Wood in which the fibers collectively take the form of waves or undulations.

2. Mineral or rock particle having a diameter less than 3 mm (0.125 in.); **3.** Unidirectional orientation of rubber or filler particles resulting in anisotropy of rubber compounds; **4.** Abrasive classified into predetermined sizes for use in polishing, in grinding wheels, and in coated abrasive.

GRAINING Decorative painting technique used to simulate woods.

GRAIN RAISE Condition on the surface of a plywood panel resulting from harder or denser wood fibers swelling and rising above softer surrounding wood.

GRAIN SIZE Size of the cutting particles of a grinding wheel or polishing abrasive.

GRAIN SPACING Relative position of the cutting particles in a grinding wheel.

GRAINY *See* **Bulk appearance.**

GRAM Unit of mass, equal to 1/1000 kilogram. Symbol: g. Used in the SI system of measurement. Multiply by 0.035 273 96 to obtain ounces, symbol: oz; by 0.002 679 2 to obtain pounds, symbol: lb. *See also the appendix*: **Metric and nonmetric measurement.**

GRAM PER CUBIC CENTIMETER Unit of density. Symbol: g/cm³. Used in the SI system of measurement. *See also the appendix*: **Metric and nonmetric measurement.**

GRAM PER LITER Unit of density. Symbol: g/L. Permitted for use in the SI system of measurement. *See also the appendix*: **Metric and nonmetric measurement.**

gran. *Abbreviation for:* granular.

GRANITE Coarse-grained, light-to-dark colored, intrusive igneous rock, soft to hard, and medium to heavy in weight. *See also* **Material density.**

GRANITE SETTS Granite blocks of rectangular shape and of approximately brick-size dimensions.

GRANITIC FINISH Face mix, resembling granite, on precast concrete.

GRANOLITHIC CONCRETE Concrete suitable for use as a wearing surface finish to floors, made with specially selected aggregate of suitable hardness, surface texture, and particle shape.

GRANOLITHIC FINISH Surface layer of granolithic concrete that can be laid on a base of either fresh or hardened concrete.

GRANT Term in deeds of land conveyance indicating the transfer from one party to another.

GRANTEE Person to whom a grant is made.

GRANTOR Person who makes a grant.

GRANULAR Uniformly sized grains or crystals of rock.

GRANULAR DUST Mineral filler consisting of finely powdered rock dust.

GRANULAR SOIL *See* **Soil types.**

GRANULAR STRUCTURE Nonuniform appearance of a finished material due to incomplete fusion of particles within the mass.

GRANULATED BLAST-FURNACE SLAG *See* **Blast-furnace slag.**

GRANULE Natural colored or artificially colored particles of a silicious material used on the exposed surface of roofing material.

GRAPHICAL (SCHEMATIC) DIAGRAM *See* **Diagram.**

GRAPHIC LANGUAGE Diagram employing a language comprised of symbols to represent an object or objects.

GRAPHITE Soft form of carbon that, as a powder, can be used as a dry lubricant.

GRAP HOOK Hook with a narrow throat adapted to cover a link on a chain and not slip.

GRAP LINK Pear-shaped link used to connect chain.

GRAPPLE Clamshell-type bucket, sometimes with heavy tongs, having three or more jaws and generally used to handle rock or large-size demolition rubble.

GRAPPLE LEG Either of the two main legs of a grapple.

GRAPPLE LINE *See* **Holding line.**

GRAPPLER Wedge-shaped spike at the top end of a bracket scaffold that is driven into a brick joint and to which the scaffold is attached.

GRAPPLE SKIDDER *See* **Harvesting machines (single function).**

GRAPPLE YARDER Machine used in harvesting to bring logs into a landing. The grapple closes like teeth around the log and is controlled by the machine operator.

GRASS Category of plants related to cereals that represents about 10% of the world's flora. A grass leaf typically consists of a sheath and a blade. There are several types having significance in landscaping and construction, including:

Bentgrass: Strain of grasses normally used on golf courses.

Bluegrass: Grass that spreads by underground stems and makes a thick sod, preferring sunny locations and well-drained soils.

Fescue: Shade-tolerant grass, useful for lawns that receive limited maintenance; used on poor and dry soils.

Ryegrass: Fast-growing, rough grass, either annual or perennial, used to establish a quick grass cover where appearance is not a major factor.

GRASSHOPPER Traveling frame with hinged ends, riding on a separate wide-gauge track, used for passing cars in tunnel work.

GRATE Device used to support solid fuel or solid waste in a furnace during drying, ignition, and combustion. Openings (tuyeres) are provided for passage of combustion air. There are several types, including:

Chain: Stoker that has a massive moving chain as a grate surface; the grate consists of links mounted on rods to form a continuous belt-like surface that is generally pulled by sprockets on the front shaft.

Dead plate: A stationary grate through which no air passes.

Fixed: A stationary grate.

Movable: A grate designed to feed a solid fuel or solid wastes into a furnace and discharge combustion residue. *Also called* **Stoker.**

Oscillating: Where the grate surface oscillates to move the fuel and residue from feed end to discharge.

Reciprocating: Stoker grate surface having alternate lateral stationary and moving rows that reciprocate continuously and slowly, forward and backward, for the purpose of stirring the combustible bed of material while conveying it and the resulting residue from the infeeding end to the discharge end of the furnace.

Rocking: An incinerator stoker with moving and stationary grate bars that are trunnion supported. In operation, the moving bars oscillate on the trunnions, imparting a rocking motion to the bars, thus agitating and conveying the solid fuel and resulting residue through the furnace.

Traveling: A traveling grate stoker consisting of a belt-like arrangement of air-admitting grate bars, similar to a chain grate but with grate bars mounted on transverse beams, usually pulled by chains and sprockets through the furnace.

GRATICULATE System of squares superimposed over an existing drawing or sketch, used to subdivide the original for the purpose of making a larger or smaller version.

GRATING Open screen within an opening in a vertical or horizontal surface.

grav. *Abbreviation for:* gravity.

GRAVEL Granular material, from 2 mm (0.078 in.) to 64 mm (2.5 in.) in diameter, predominantly retained on the No. 4 (4.76 mm, 0.187 in.) sieve, and resulting either from natural disintegration and abrasion of rock or the processing of weakly bound conglomerate. *See also* **Pit run gravel.**

GRAVEL FILL Layer of crushed rock or gravel, placed to a specific thickness at the bottom of an excavation.

GRAVEL FRACTION Fraction of a soil composed of particles between 60 and 2 mm (2.4 and 0.078 in.).

GRAVEL MACADAM *See* **Stone macadam.**

GRAVEL PUMP Centrifugal pump used to move hydraulically loosened gravel.

GRAVEL ROOF *See* **Gravel surfacing.**

GRAVEL STOP Transition member between a roof surface and the fascia, used to contain gravel used on built-up and gravel-surfaced roofing.

GRAVEL SURFACING Final layer of a built-up roof construction consisting of a flood coat of bitumen to which an aggregate such as rock chips, gravel, crushed slag, or ceramic granules is applied while still mastic. *Also called* **Gravel roof.**

GRAVER Sheet metal cutting tool with a diamond-shaped cutting point.

GRAVIMETRIC VALUE Weight of suspended solids per unit volume of fluid.

GRAVITATIONAL WATER Water that enters and flows through the ground by gravity. This water is free-flowing and can be removed from the ground by pumping. *See also* **Absorbed water, Capillary water,** and **Moisture.**

GRAVITY ABUTMENT Central pier supporting a bridge, having two wing walls to retain fill, all supported on a single footing.

GRAVITY-ARCH DAM Dam that obtains its resistance to thrust from arch action and from self-weight.

GRAVITY CIRCULATION Hot water plumbing circuit that uses temperature differential to move hot water around the piping system.

GRAVITY DAM Dam that relies solely on its own mass to resist lateral pressure.

GRAVITY DRAINAGE Drainage of water by gravity.

GRAVITY ELEVATOR *See* **Elevator.**

GRAVITY FAULT Rock fracture caused by forces that pull the rocks apart. *Also called* **Normal fault.**

GRAVITY FURNACE Hot air furnace in which hot air rises through piping and ducts.

GRAVITY FURNACE WALL Furnace wall supported directly by the foundation or floor of a structure.

GRAVITY HAMMER Weight configured to slide in the pile hammer leads or within a hollow pile that has a formed bale or swivel at its top by which it can be mechanically lifted and dropped to drive a pile. *See also* **Drop hammer.**

GRAVITY LOGGING Any cable system that depends on the force of gravity for downhill travel of the carriage.

GRAVITY MAIN Pipeline in which water flows downhill from an impounding reservoir to a service reservoir.

GRAVITY SEPARATION Concentration or separation of a mix of materials based on difference in specific gravity and sizes of materials.

GRAVITY SUPPLY Water distribution system that relies on gravity fall to provide water to all outlets.

GRAVITY TANK Water storage tank that supplies a flow by gravity pressure.

GRAVITY WALL Mass concrete, unreinforced retaining wall that depends for its stability entirely on its own mass.

GRAVITY WARM-AIR HEATING SYSTEM System in which the heat flow depends on the difference in weight between the heated air leaving the casing and the cooler air entering the bottom of the casing.

GRAY One of 17 derived units with special names of the SI system of measurement: a unit of absorbed dose of ionizing radiation equal to one joule per kilogram. (1 Gy=1 J/kg). Symbol: Gy. *See also the appendix*: **Metric and nonmetric measurement.**

GRAY CAST IRON Cast iron containing gray, crystalline, coarse flakes of graphite either embedded in a matrix of iron or chemically combined with the iron.

GRAY PAPER Unsized, uncalendered paper used to back drywall panels, and as the face and back paper on backing board.

grd *Abbreviation for:* grind; ground.

GREASE Thick oil used to lubricate bearings and other working parts.

GREASE NIPPLE *See* **Nipple.**

GREASE TRAP Drum-type trap installed in a drainage line to separate and collect grease from the flow being conveyed.

GREEN 1. Brickwork in which the mortar has not yet set; **2.** Concrete that has set but not appreciably hardened; **3.** Freshly sawed or undried wood that has a moisture content above the fiber-saturation point (approximately 25% to 30%).

GREEN BELT Strip of undisturbed soil and vegetation left along waterways or access routes to minimize the environmental impact from development. *See also* Buffer strip.

GREEN BOARD 1. Drywall panel finished with a face paper tinted to a color to signify a special board type; **2.** Drywall panel having a higher-than-optimum moisture content.

GREEN FUEL Living vegetation of high moisture content that ordinarily will not burn unless it is first dried out by excessive heating or lack of rain.

GREENHEART Very hard, naturally durable wood sometimes used for piling; native to the Amazon area. Less durable than pressure-treated wood, especially in resistance to marine borers.

GREENHOUSE EFFECT Solar radiation waves admitted to the earth's surface that cannot then exit the atmosphere for any reason; within a building, the solar radiation waves admitted through glass that are transformed to heat waves that cannot pass back through the glass.

GREEN LUMBER Lumber that has not been adequately dried and which has a tendency to warp.

GREEN STRIP Uncut strip of timber left along streams and roads. *Also called* **Buffer strip, Leave strip,** and **Streamside management zone.**

GREEN TIMBER Uncut forest.

GREEN TIME *See* **Traffic signal.**

GREENWARE Clay products, before burning, during and after drying.

gr. fl. *Abbreviation for:* ground floor.

GRID 1. Template for spotting piles in prescribed locations. *See also* **Pile template; 2.** Set of surveyor's closely-spaced reference lines, laid out at right angles, with elevations taken at intersections of the lines, used to calculate quantities of fill or excavation.

GRID BEARING Angle between a required direction and that of a line in a grid, the verticals of which are generally north-south.

GRID FOUNDATION Combined footing formed by intersecting continuous footings, loaded at the intersection points and covering much of the total area within the outer limits of the assembly.

GRIDIRON Rectangular plotting of city streets.

GRID LINE Line in a reference pattern.

GRID PLAN Plan in which grid lines coincide with the principal walls and building components.

GRID POWER *See* **Commercial power.**

GRID ROLLER Soil compaction roller consisting of rolls made up of 38 mm (1.5 in.) diameter steel bars at 130 mm (5 in.) centers.

GRILLAGE Framework of structural horizontal members crossed in layers, used to support loads.

GRILLE 1. Open screen within a relatively large space in a wall or ceiling; **2.** Grating or screen through which air is drawn into a return-air duct.

GRIND 1. Reduce a substance to powder by crushing or friction; **2.** Abrade or wear away a surface.

GRINDER Equipment for the size reduction of a material. *Also called* **Cold planer.**

GRINDING 1. Particle-size reduction by attrition and/or high-speed impact; **2.** Removing material with a grinding wheel.

GRINDING ACTION Cutting ability of, and the finish produced by a grinding wheel.

GRINDING AID Material used to expedite the process of grinding by eliminating ball coating or by dispersing the finely ground product, or both.

GRINDING MACHINE Any machine on which a grinding wheel is operated.

GRINDING MEDIUM Hard, free-moving charge in a ball or tube mill, used to reduce the particle size of introduced materials by attrition or impact.

GRINDING WHEEL Cutting tool of circular shape made of abrasive grains bonded together.

GRINDSTONE Flat, circular wheel cut from natural sandstone, sometimes used for sharpening tools.

GRIP 1. Ability of an adhesive or mortar to hold an applied article in place during and after the setting and/or curing process. *See also* **Hang**; **2.** Internally tapered steel barrel and separate concentric annular-split wedge that grips the end of a stressed tendon.

GRIT 1. Coarse sand or sandstone, formed mostly of hard, angular quartz grains; **2.** Abrasive materials of various degrees of fineness; **3.** Nomenclature used to distinguish abrasives according to their grit size; **4.** Conglomeration of mineral particles of many sources and a wide range of sizes, usually deposited from flowing water; **5.** Contaminant that may be found in whole-tree chips.

GRIT BLASTING Technique for exposing the aggregate in hardened concrete for architectural effect.

GRIT CHAMBER Chamber in a storm sewer below the invert and discharge levels, used to reduce flow velocity and permit settling of particulate matter.

GRIT CONTROL FILTER *See* **Chip control filter.**

GRIZZLY Simple, stationary screen, or series of equally spaced parallel bars, set at an angle to remove oversize particles in processing aggregate or other material.

grl. *Abbreviation for:* grille.

grn *Abbreviation for:* green.

grnhse *Abbreviation for:* greenhouse.

GROG Burned refractory material, usually calcined clay or crushed brick bats.

GROIN, or GROYNE 1. Framework or low broad wall run out from a shore to check lateral drifting of the beach and to deter erosion by wave and tidal action. *Also called* **Spur dike**; **2.** Edge formed by intersecting vaults; **3.** Fillet covering this.

GROIN ARCH *See* **Arch.**

GROMMET 1. Reinforcing device placed around an orifice in a relatively thin component; **2.** Endless circle or ring fabricated from one continuous length or strand or rope.

GROOVE 1. Groove in a drive sheave: a vee-groove produces a pinching or wedging effect on the belt or cable, and is used in geared machines; a U-groove relies on friction between groove and belt or cable and is used on gearless machines; **2.** Space between two adjacent tread ribs of a tire; **3.** Depression, helical or parallel, in the surface of a sheave or drum that is shaped to position and support wire rope. *See also* **Gib.**

GROOVE CRACKING Cracking that occurs at the bottom of the tread of a tire.

GROOVED COUPLING Clamp-type gasketed pipe coupling having a groove machined or cast into its internal circumference; it fits over flanges cast on the exterior circumference of pipe ends.

GROOVED DRUM Drum with a grooved surface that accommodates a rope and guides it for proper winding.

GROOVED SURFACE FINISH Concrete road surface-finish produced by cutting grooves in plastic concrete with a grooving machine.

GROOVE JOINT Joint created by forming a groove in the surface of a concrete pavement, floor slab, or wall to control random cracking.

GROOVER Tool used to form grooves or weakened-plane joints in a concrete slab before hardening, to control crack location or provide pattern.

GROOVE WELD Welding rod fused into a joint that has the base metal removed to form a trough at the edge of the pieces to be joined.

GROSS AREA 1. Enclosed floor area of a building, measured from the inside surface of the exterior walls. *See also* **Area, Area of building, Floor area, Land use classes (forest management),** and **Net area**; **2.** Total area within the outer periphery of any section of a structural unit, perpendicular to the stress to be resisted.

GROSS AXLE WEIGHT RATING Rated capacity for a particular axle system, including springs, brakes, wheels, and tires. *Also called* **Axle rating.**

GROSS COMBINATION WEIGHT Combined weight of a truck and trailer, or tractor and trailer or semitrailer.

GROSS CUT Total amount of excavation, without regard to fill requirements. *See also* **Cut**, and **Net cut.**

GROSS ERROR Measuring mistake so large that it is easily detected by reason of its size in proportion to related, accurate measurements.

GROSS HEATING VALUE Total heat obtained from combustion of a specified amount of fuel that is at 15.5°C (60°F) when combustion starts, and the combustion products of which are cooled to 15.5°C (60°F) before the quantity of heat released is measured.

GROSS HORSEPOWER *See* **Horsepower.**

GROSS LEASABLE AREA Floor area that can be used by tenants.

GROSS LEASE Property lease in which the landlord is responsible for payment of all property expenses (tax, insurance, utilities, repairs, etc.).

GROSS LOADING INTENSITY Intensity of vertical loading at the base of a foundation due to all loads above that level.

GROSS MEASURE Board measure of lumber calculated for measurements of named sizes.

GROSS MERCHANTABLE VOLUME OF WOOD Volume of a main stem, excluding stump and top but including defective and decayed wood, of trees or stands of trees.

GROSS PENETRATION *See* **Penetration.**

GROSS POWER Power output of a 'basic' engine.

GROSS SAMPLE Sample representing one lot and composed of a number of increments on which neither reduction nor division has been performed.

GROSS SALE Where the buyer purchases, on an area basis, all timber for a fixed price.

GROSS SCALE Measurement of log volume in which no deduction is made for defect.

GROSS SETTLEMENT Total downward movement of a structure, pile, or group of piles that occurs under an applied load. *See also* **Net settlement,** and **Settlement.**

GROSS TORQUE *See* **Torque.**

GROSS TOTAL VOLUME 1. Volume of the main stem of a tree, including stump and top; **2.** Volume of a stand including all trees.

GROSS VEHICLE LOAD *See* **Gross vehicle weight.**

GROSS VEHICLE WEIGHT Total weight of machine and payload. *Also called* **Gross vehicle load.**

GROSS VEHICLE WEIGHT RATING Manufacturer's rating for the vehicle: the maximum amount that the loaded truck should weigh.

GROSS VOLUME OF CONCRETE MIXERS 1. Total interior volume of the revolving portion of the mixer drum of a revolving-drum mixer; **2.** Total volume of the trough or pan of an open-top mixer, calculated on the basis that no vertical dimension of the container exceeds twice the radius of the circular section below the axis of the central shaft.

GROTTO 1. Picturesque natural seashore cave; **2.** Artificial cave decorated with sea shells, etc.

GROUND 1. Connection, either intentional or accidental, between an electric circuit and the earth, or to some conducting body serving in place of the earth; **2.** Surface upon which paint media is first applied; **3.** Strip of wood to which the thickness of plaster or cement is worked. *Also called* **Grounds**; **4.** Territory on which a logging operation is being conducted.

GROUND ANCHOR Structural member that transmits an applied tensile force to 'competent' ground; the anchor may be a tension pile, rock bolt, or high-strength steel tendon installed and grouted-in at a calculated inclination.

GROUND AREA Area computed from the exterior dimensions of the ground floor.

GROUND BEAM Beam in a substructure transmitting a load to a pile, pad, or other foundation.

GROUND CALIBRATION Determination of the vibration transmission characteristics of a region.

GROUND CASING In carpentry, a blind casing.

GROUND CLEARANCE 1. Distance by which a vehicle's or machine's lowest point, exclusive of the wheel assembly, clears the ground, measured perpendicularly from that point to a plane surface on which the vehicle rests; **2.** General term for removing unwanted vegetation, slash stumps, roots and stones from a site before forestation or reforestation.

GROUND COAT *See* **Base coat.**

GROUND COLOR Background color against which the top colors create a pattern, design, or representation.

GROUND CONTROL Points of known position and elevation that are marked on the ground, and that will be visible in an aerial photograph, used as reference points in photo-interpretation.

GROUND COVER Mat of vegetation produced by low, spreading or creeping plants.

GROUNDED Connected to earth or to some conducting body that serves in place of the earth. *See also* **Effectively grounded.**

GROUNDED CONDUCTOR System or circuit conductor that is intentionally grounded.

GROUNDED NEUTRAL Point of an electrical system that is intentionally connected to ground.

GROUND-FAULT CIRCUIT INTERRUPTER Device for the protection of personnel that functions to de-energize a circuit or portion thereof within an established period of time when a current to ground exceeds some predetermined value that is less than that required to operate the overcurrent protective device of the supply circuit.

GROUND FINISH Surface produced by grinding or buffing.

GROUND FIRE Forest fire at ground level, as contrasted to a crown fire or structural fire at upper levels; a forest fire that consumes organic material such as peat or muck below the surface litter or duff.

GROUND FLOOR Floor nearest the ground level adjacent to the main entrance of a building. *See also* **First floor.**

GROUND FRAME Timber frame of walings and struts laid 300 mm (12 in.) or more below ground level, used as a guide for the first setting of runners or trench sheeting.

GROUNDING CONDUCTOR Conductor used to connect equipment, or the grounded circuit of a wiring system, to a grounding electrode or electrodes.

GROUNDING ELECTRODE CONDUCTOR Conductor used to connect the grounding electrode to the equipment grounding conductor and/or to the grounded conductor of the circuit at the service equipment, or at the source of a separately derived system.

GROUNDING ELECTRODE RESISTANCE Resistance of a grounding electrode to earth.

GROUND JOINT 1. Closely mating joint between masonry blocks; **2.** Joint between metal surfaces that has been machined to a close tolerance.

GROUND-LEAD LOGGING Yarding with no lift for either the rigging or logs.

GROUND LENGTH Extent to which the ground around a tree is broken by ridges, gullies or swells, rock outcrops, and sharp slope changes. *See also* **Length**

of ground.

GROUND LINE 1. Natural grade level from which measurements for excavation are taken; **2.** Line about a structure to which fill will be placed to form a finished grade.

GROUND LOSS Subsidence of the surface of the ground in the vicinity of a shaft excavation, caused by soil moving laterally into the excavation during drilling, or during dewatering.

GROUND MODIFICATION Technique of dropping a heavy weight from considerable height to dynamically compact loose soil.

GROUND PLATE *See* **Soleplate.**

GROUND PRESSURE Weight of a structure or machine divided by the total area of the surfaces supporting it (i.e., foundations, total pad area, or total footprint area).

GROUND PROP Prop or puncheon placed between the lowest frame and a foot block on the bottom surface of an excavation, used to support the mass of timbering.

GROUND RENT Amount paid for the use of ground for specified purposes.

GROUNDS *See* **Ground.**

GROUND SKIDDING Pulling logs parallel to the ground without using an arch or fairlead to raise the forward end.

GROUNDWATER Underground water that has collected due to porosity and fissuring of rocks and that is contained by an underlying impervious strata. There are several classifications, including:

Confined: Groundwater that is under pressure substantially greater than atmospheric.

Free: Groundwater in aquifers not bounded or confined by impervious strata.

Juvenile: Geologic water confined in voids in the earth that are not subject to recharge.

Perched: Unconfined groundwater separated from an underlying body of groundwater by an unsaturated zone.

Unconfined: Water in an aquifer that has a water table.

GROUNDWATER DISCHARGE Rate at which water flows from a saturated zone to an unsaturated zone or is lost at the surface.

GROUNDWATER DRAIN Drain through which groundwater is taken from an area.

GROUNDWATER FLOW Rate of flow of water in an aquifer or soil; that portion of the discharge of a stream that is derived from groundwater.

GROUNDWATER LOWERING Lowering the level of groundwater by pumping to create a dry excavation.

GROUNDWATER MINING Withdrawal of groundwater at a rate exceeding that of recharge.

GROUNDWATER RECHARGE Rate at which water is added to the water table from an unsaturated zone.

GROUNDWATER RUNOFF That part of groundwater that is discharged into a stream channel as spring or seepage water.

GROUNDWATER TABLE Upper surface of the zone of saturation in permeable rock or soil.

GROUND WAVES Vibrations of soil or rock.

GROUND WIRE Small-gauge, high-strength steel wire used to establish line and grade, as in shotcrete work. *Also called* **Alignment wire,** and **Screed wire.**

GROUP FELLING Felling method used to orient the butts of small-diameter, tree-length timber in one direction for skidding.

GROUP TASK *See* **Collection method.**

GROUP VENT Branch vent to a drain that serves two or more traps.

GROUSER Heavy lug attached to crawler tread plates and tractor wheels to obtain better traction in soft ground. Single, double, and triple grousers are used: single-grouser shoes are used on track-type tractors; double and triple grousers on track loaders and excavators to ease turning and provide less ground disturbance.

GROUT A cementitious compound of high water-to-cement ratio, permitting it to be poured into spaces within masonry walls, between tiles, and other crevices. Grout consists of portland cement, lime, and aggregate. It is sometimes formed by adding water to mortar. *See also* **Foundation grouting.**

GROUT BOX Cone-shaped metal box, set in concrete with an anchor plate at its narrow end, through which an anchor bolt passes.

GROUT CURTAIN Wall of injected material formed below ground level.

GROUTED-AGGREGATE CONCRETE Concrete formed by injecting grout into previously placed coarse aggregate.

GROUTED MACADAM Road constructed by pouring a bituminous or cement grout into a bed of coarse aggregate.

GROUTED MASONRY Unit masonry composed of either hollow units, wherein the cells are filled with grout, or multiple wythes, where spaces between the wythes are filled with grout.

GROUTING Process of filling with grout. *See also* **High-lift grouting,** and **Low-lift grouting.**

GROUT LIFT Increment of a grout pour.

GROUT LOSS Unwanted, sand-textured areas, lacking cement paste, on the surface of concrete.

GROUT POUR Height of masonry to be filled with grout prior to laying of additional masonry courses: a pour consists of one or more lifts.

GROUT SAW *See* **Saw.**

GROUT SCRUBBING PAD Nonscratch nylon pad impregnated with abrasive, used for cleaning grout off tile.

GROUT SLOPE Natural slope of fluid grout injected into preplaced-aggregate concrete.

GROUT VENT Pipe provided at intervals along the length of a tendon to be completely grouted-in following stressing.

GROWING-EQUITY MORTGAGE *See* **Mortgage.**

GROWING-STOCK TREE Live tree of any size, except a rough or rotten tree. *See* **Tree classes.** *See also* **Timber volume, Volume of growing stock,** and **Volume of saw timber.**

GROWLER Electrical test instrument, used to check generator armatures for opens, shorts, and grounded circuits.

GROWTH Increase in diameter, basal area, height, and volume of individual trees or stands during a given period of time. *Also called* **Increment.**

GROWTH RING *See* **Annual growth ring.**

GROYN *See* **Groin.**

GRP *Abbreviation for:* glass reinforced plastic.

grs. *Abbreviation for:* grease.

grt *Abbreviation for:* grout.

grtg *Abbreviation for:* grating.

grtr *Abbreviation for:* grouter.

GRUB AXE Tool with an adze-like blade at one end for pulling roots, and an axe blade at the other.

GRUBBING Removing stumps and roots from an area.

GRUB SAW *See* **Saw.**

grv. *Abbreviation for:* groove; grooved.

grvl *Abbreviation for:* gravel.

gr.wt *Abbreviation for:* gross weight.

GS *Abbreviation for:* galvanized steel.

gsp *Abbreviation for:* galvanized steel pipe.

GSU *Abbreviation for:* glazed structural unit.

GSWR *Abbreviation for:* galvanized steel wire rope.

gt *Abbreviation for:* girt.

GT *Abbreviation for:* glazed tile; gross ton.

GT DIMENSION Distance from the ground to the top of the tailgate in the open position; or from the ground to the top of rear load floor and/or frame at the rear of the vehicle.

gt.v. *Abbreviation for:* gate valve.

G2S *Abbreviation for:* good two sides.

guar. *Abbreviation for:* guarantee.

GUARANTEE Assurance that something is as represented or will be done in the manner specified, with a remedy spelled out should that not occur. *See also* **Guaranty.**

GUARANTEE BOND *See* **Bond.**

GUARANTEED MAXIMUM COST Amount established in a contract between an owner and contractor setting out the maximum cost of completing specified work. *Also called* **Upset price.**

GUARANTEED MAXIMUM PRICE Maximum dollar amount for which the owner will be liable to the contractor for performance of all work.

GUARANTY Pledge by which a person commits to the payment of another's debt or the fulfillment of another's obligation in case of default. *See also* **Guarantee.**

GUARD Any device created or installed to protect a worker or equipment.

GUARD BOARD Board placed on edge at the perimeter of scaffolding to prevent objects from falling off.

GUARDED Covered, shielded, or otherwise protected by means of suitable covers, casings, barriers, rails, screens, mats, or platforms to remove the likelihood of approach to a point of danger or contact by persons of objects.

GUARD LOCK Lock separating a dock from tidal water.

GUARDLOG Use of logs in place of sawn timber as a guardrail.

GUARD PILE *See* **Pile.**

GUARDRAIL 1. Safety barrier at the edge of an elevated platform or flight of stairs; **2.** Traffic barriers, used to shield hazardous areas from errant vehicles.

GUDGEON 1. Reinforced bushing; a thrust-absorbing block; **2.** Stationary leaf of a hinge; **3.** Gate hook; **4.** Metal dowel locking adjacent masonry blocks.

GUDGEON PIN Pin at the top of a derrick mast forming a pivot for the spider or for the mast of a stiffleg derrick.

GUIDE BAR Support and guide for the chain of a chainsaw.

GUIDE BAR NOSE Tip of the guide bar of a chainsaw away from the powerhead.

GUIDE CHANNELS *See* **Gibs.**

GUIDE COAT Thin coat of paint applied to indicate high spots.

GUIDE FRAME Timber frame erected above ground level to act as a guide for runners or trench sheeting, and as a staging from which they may be driven.

GUIDE LINE Line placed at the side of an asphalt paver, parallel to the centerline of the road, to assist the operator in steering the machine.

GUIDE PILE *See* **Pile.**

GUIDE PIN Oil-soaked hardwood pin used to fasten a saw guide to the guide support or husk. *See also* **Saw guide.**

GUIDE RAIL Tee-shaped device, accurately mounted in a hoistway to positively locate an elevator as it moves through the hoistway.

GUIDE RIB Protruding rubber rib located over the bead. Serves as a guideline for mounting the tire on a rim.

GUIDE ROLL Front or steering wheel of a roller.

GUIDE ROLLER Guide shoe incorporating rollers that rotate on a guide rail.

GUIDE RUNNER Runner driven ahead to form a guide for driving intermediate runners in trench work.

GUIDES That part of pile leads forming a pathway for the hammer, and consisting of parallel members that mate with grooves on the hammer.

GUIDE SHOE Device on a sling that slides or rolls on the rails to guide an elevator through the hoistway.

GUIDE WALL Temporary, shallow-depth wall constructed in soft or unstable soil in advance of a deeper, permanent wall.

GUIDING RATE OF RETURN Rate attached to the use of capital that guides a firm in its choice of investments. In general, investments that promise a rate of return less than the guiding rate are rejected after due allowance for factors such as risk.

GUILLOTINE Piston-type line cutter, used to cut sheet metal and steel plate.

GUILLOTINE DAMPER See **Damper.**

GUILLOTINE SHEAR Type of carrier-mounted, single-action, anvil shear used in mechanical tree cutting where the blade is pushed through the stem and away from the carrier, instead of being pulled, as in a draw shear.

GUINEA Survey marker driven to grade and colored blue, used for finishing and fine trimming.

GUINEA HOPPER Member of a grading crew who uncovers the blue-topped stakes and signals the equipment operator to cut or fill as required.

GULLET Area of a saw tooth on which the chip is carried; the gap between the teeth.

GUM Solid resinous material that can be dissolved in an appropriate medium and that will form a film when such a solution is evenly spread or sprayed on a surface and the solvent allowed to evaporate.

GUMBO Very sticky, dark-colored, plastic clays distinguished by soapy or waxy appearance in the plastic state and by great toughness.

GUM COMPOUND Rubber compound containing only those ingredients necessary for vulcanization. Small amounts of other ingredients may be added for processability, coloring, and improving resistance to aging.

GUM VEIN Accumulation of resin in a streak in a tree.

GUN 1. Shotcrete material delivery equipment, usually consisting of double chambers under pressure, although equipment with a single pressure chamber is used to some extent; **2.** Pressure cylinder used to propel freshly mixed concrete pneumatically.

GUN FINISH Undisturbed final layer of shotcrete as applied from the nozzle, without hand finishing.

GUNITE Proprietary term for shotcrete.

GUNMAN Worker on a shotcreting crew who operates the delivery equipment.

GUNMETAL Alloy of copper and tin.

GUNNED CONCRETE See **Shotcrete.**

GUNNING 1. Act of applying shotcrete; **2.** Ejection of shotcrete material from a nozzle and its impingement on the surface to be gunned; **3.** Technique involving marks on a chain saw, used to estimate the direction the tree being cut will fall. See also **Sighting.**

GUNNING PATTERN 1. Conical outline of material

discharge stream in shotcrete operation; **2.** Sequence of shotcrete gunning operations to insure complete filling of the space, total encasement of reinforcing bars, easy removal of rebound, and thickness of shotcrete layers.

GUNTER'S CHAIN See **Chain.**

gup Abbreviation for: grading under pavement.

GUS Abbreviation for: guaranteed ultimate strength.

gus. Abbreviation for: gusset.

GUSSET Flat plate used to reinforce a connection at the intersection of members.

GUT Section of unsuitable foundation material that passes through the foundation of a dam.

gut. Abbreviation for: gutter.

GUT HOOK To grapple or choke a log in the middle.

GUT LINE Steel cable forming the load line on a cableway.

GUTTER 1. Shallow channel or conduit set below and parallel to the eaves to catch and carry rainwater from a roof. Also called **Eaves trough; 2.** Built or formed trough at the edge of pavement to collect and convey runoff.

GUTTER TOOL Tool used to give the desired shape and finish to concrete gutters.

GUTTER TRENCH Ditch dug on a slope below a forest fire, designed to catch rolling burning material. See also **Undercut line.**

GUY Rope, chain, or rod attached to a brace, steady or guide.

GUY DERRICK Boom operated from a mast held in an upright position by guylines, usually used on tall buildings.

GUYLINE Strand or rope, usually galvanized, for stabilizing or maintaining a structure in a fixed position. See also **Guy rope, Pendant, Standing line,** and **Stay rope.**

GUY ROPE Fixed-length supporting rope intended to maintain a nominally fixed distance between points of anchorage. Also called **Pendant,** and **Stay rope.** See also **Guyline.**

GV Abbreviation for: grid variation.

GVW Abbreviation for: gross vehicle weight.

GWT Abbreviation for: glazed wall tile.

gym. Abbreviation for: gymnasium.

gyp. Abbreviation for: gypsum.

GYPPO, or GYPO 1. Independent logger who usually runs a small-scale logging operation, sawmill, or other wood related operation; **2.** Personnel paid on piecework basis.

GYPSITE Dirty variety of gypsum with some impurities.

GYPSUM Mineral having the composition calcium sulfate dihydrate.

GYPSUM BOARD Wallboard made from gypsum plaster, with a covering of paper. See also **Drywall.**

GYPSUM CONCRETE Concrete in which the

cementitious constituent is partially dehydrated calcium sulfate (plaster).

GYPSUM LATH Gypsum board used as a foundation for the application of gypsum plaster.

GYPSUM PLASTER Plaster made using plaster of paris.

GYPSUM SHEATHING Type of gypsum board used as an exterior wall membrane and base for exterior siding.

GYPSY SPOOL Capstan winch.

GYRATORY CRUSHER *See* **Crusher.**

GYS *Abbreviation for:* guaranteed yield strength.

H

h *Abbreviation for:* harbor; hard; height; hour; hundred.

H *Abbreviation for:* horizontal.

ha *Abbreviation for:* hectare.

Ha *Abbreviation for:* abrasive hardness value.

HA *Abbreviation for:* hour angle.

HABENDUM CLAUSE 'Have and hold' clause defining the quantity of the estate granted in the premises of a deed.

HABITABLE ROOM OR SPACE Room or space intended primarily for human occupancy. *Also called* **Living space.**

HABITAT 1. Dwelling place; **2.** Place where an organism lives and/or the conditions of that environment including the soil, vegetation, water, and food.

HABITATION Place in which to live: dwelling, residence, home.

HABITAT MANAGEMENT Management of the forest to create environments that provide habitats (food, shelter) to meet the needs of particular species of wildlife, birds, etc.

HACK AND SQUIRT Method of conifer release and juvenile spacing where the bark of a tree is cut (hack) and herbicides are injected (squirt) to kill the tree.

HACKING 1. Stacking brick in a kiln or on a kiln car with the bottom edge set in from the surface of the wall; **2.** Unloading bricks from a truck by manual labor using brick tongs; **3.** Setting two masonry courses in the vertical space normally occupied by one; **4.** Roughening of a surface by striking with a tool.

HACKING KNIFE Glazier's knife used to remove old putty from a window frame.

HACKSAW *See* **Saw.**

HAFT Handle of a tool.

HAGEN-POISEUILLE LAW Friction factor of darcey's formula, or the ratio of 64 to the Reynold's numbers when flow is lamina.

HAIRLINE CRACK Crack in an exposed concrete surface having a width so small as to be barely perceptible.

HAIRPIN 1. Wedge used to tighten some types of form ties; **2.** Hairpin-shaped anchor set in place while concrete is unhardened; **3.** Light hairpin-shaped reinforcing bar used for shear reinforcement in beams, tie reinforcement in columns or prefabricated column shear heads or as a short hooked spacer bar in columns and walls and for special dowels; **4.** Gravity hammer in the shape of an inverted 'U' used without leads to start sheet piles into the ground. *Also called* **Pants.**

HAIRPIN LEAD *See* **Lead.**

HALFBACK Concrete block that has the inside of two cells cut out.

HALF BAT *See* **Brick.**

HALF BATH Room in a dwelling that contains a toilet and washbasin but no bathing facilities.

HALF CAB One-person cab, usually located beside the engine.

HALF-HIGH BLOCK Concrete block that is cut in half lengthwise.

HALF LANDING Platform separating two flights of stair, that may continue in the same direction or at 90° to left or right, or both.

½ld *Abbreviation for:* half load.

HALF-MOON STAKE Anvil with a curved edge, used for bending sheet metal.

HALF PACE 1. Platform separating two flights of stair that turn through 180°; **2.** Raised section of floor, as in a bay window.

HALF PRINCIPAL Rafter that does not reach the ridge.

½rd *Abbreviation for:* half round.

HALF-ROUND FILE Rasping tool having ONE flat surface and one that is an arc (less than a semicircle).

HALF-ROUND MOLDING *See* **Molding.**

HALF-ROUND SCREW Screw with a hemispherical head. *Also called* a **Button-headed screw.**

HALF SECTION Any two quarter sections within a section of land having a common border. *Also called* **Section half.**

HALF-SPAN ROOF Lean-to roof.

HALF-STORY A mezzanine.

HALF-TIMBERED Self-supporting, heavy wood frame to a structure, usually a residence, with the space filled in with non-load-bearing materials.

HALF-TRACK Heavy duty truck with steering wheels in the front and propelled by high-speed crawler tracks in the rear.

HALFWAY BOX Junction box, mounted in the hoistway just above the halfway point of car travel, from which the stationary ends of the traveling cables are hung.

HALL 1. Large public room; **2.** Corridor onto which rooms open; **3.** Corridor side of a hoistway.

HALOGENATED SYNTHETIC FLUID Fluid composed of halogenated organic materials. It may contain additives.

HALVED JOINT Made from two pieces of wood, each of which has had a section cut away so that they fit together flush.

HAMMER 1. Component of a drilling machine that provides impact energy and rotation for the drill string; the working principle centers on an oscillating piston driven by pneumatic or hydraulic pressures. *Also called* **Drifter; 2.** Hand tool consisting of a solid head set crosswise on a handle, used for beating or driving. There are many types, including:

Ball-peen: Double headed steel hammer, one end of which is a flat striking face, the other a domed striking face.

Beetle: Heavy wood-, plastic-, or leather-faced mallet, used for striking items that might be damaged from blows by a steel-headed implement such as a sledge hammer.

Blocking: Hammer with two large flat faces.

Carpenter's: Hammer with a striking face on one end of its head and a claw on the other.

Chipping: Lightweight hammer, available in a variety of sizes, the head and back of which may be tungsten carbide capped, used by a tilesetter to chip excess material from the backs and edges of quarry tiles.

Claw: Most common type of hammer, having a metal head with a striking surface, and an opposing notched, curved claw for pulling nails. The shaft may be wood or fiberglass but is often an integral construction with the head. Standard weight is 453 gr (16 oz), although models range up to 800 gr (28 oz) for framing and down to 225 gr (8 oz) for cabinetwork.

Club: Double-headed hammer weighing about 1.8 kg (4 lb).

Creasing: Hammer with two narrow cross peens, used in sheet metal work.

Deadblow: Type of soft-face hammer with the head filled with shot to eliminate rebound.

Double faced: Hammer having a striking surface at each end of the head.

Double faced sledgehammer: *See* **Sledgehammer,** below.

Engineer's: Hammer having a striking face and a ball peen, cross peen, or straight peen.

Face: Hammer with a striking face and a cutting peen.

Framer's: Hammer, weighing approximately 550 gr (20 oz) with a longer-than-usual shaft topped by a head having a striking face on one end and a claw on the other.

Hatchet: Type of hammer used by drywallers, having a convex head for setting and dimpling the fasteners, opposed by a blade used to adjust framing and use as a jacking wedge.

Hollowing: Hammer with two different-size ball peens.

Knapping: Hammer used for shaping stones, in particular for splitting flints.

Lath: Hammer having a striking face and a notched axe edge.

Magnetic: Hammer with a magnetized head, used to hold metal fasteners.

Mallet: Hammer faced with a nonbrittle material (plastic, rawhide, lead) used to drive parts without damaging or marking them. *See also* **Soft-faced hammer,** below.

Maul: Heavy wooden mallet. *See also* **Sledgehammer,** below.

Planishing: Hammer with a highly polished flat or domed face.

Raising: Hammer having a rounded face, used in lifting or raising metal.

Repoussé: Lightweight hammer with a broad-faced peen.

Rip: Similar to a claw hammer but with a straighter claw.

Single-face sledgehammer: *See* **Sledgehammer,** below.

Sledgehammer: Hammer with an oblong, faceted, heavy head mounted on a short handle. Weights range from 1.4 to 7.3 kg (3 to 16 lb). A double-face sledgehammer has identical faces; a single-face sledgehammer has one flat striking face and one wedge-shaped face.

Slide: Mechanical puller that uses a sliding weight to apply force.

Soft: Hammer made of mild steel having an 'S' designation.

Soft-faced: Large-diameter cylinder mounted on a shaft. The cylinder may be of virtually any pliable or resilient material: wood, many types of plastic, rubber, leather-faced, etc. Used where a metal hammer would tend to blemish the surface being struck.

Striking: Heavy steel hammer weighing from 0.9 to 1.8 kg (2 to 4 lb) used to settle stone into place or for driving stakes. *Also called* **Mash hammer.**

Tack: Light, double-headed, narrow, square-headed hammer with the head split and magnetized, used to drive tacks and small nails.

Upholsterer's: Lighter version of a tack hammer and with a rounded head.

HAMMER BEAM Beam on which rafters land.

HAMMER BRACE Member that braces a hammer beam against the pendant post.

HAMMER CRADLE Structure that slides on the rails of a pile driver leads and carries the hammer forward of the lead rails.

HAMMER CUSHION *See* **Pile cushion.**

HAMMER DRILL 1. *See* **Drill; 2.** Electrically powered hand drill that imparts a reciprocating and rotating action to a chuck-held tool or bit. *See also* **Drill.**

HAMMERED FINISH Finish to stone or concrete resulting from light hammering with a tool.

HAMMER EFFICIENCY Ratio of kinetic energy of the ram immediately prior to impact, divided by the rated energy.

HAMMER ENERGY Capacity of a pile-driving hammer to do work at impact.

HAMMER FORGING Deforming metal by repeated blows.

HAMMER FRACTURE Rupture of the face of a drywall panel and underlying gypsum core, caused by overdriving the fastener.

HAMMER GRAB Heavy tool used in breaking and removing obstructions in large-diameter cassions or other excavation.

HAMMERHEAD BOOM Boom tip arrangement in which both the boom suspension and the hoist ropes are offset from the boom longitudinal centerline, providing increased load clearance.

HAMMERHEAD PEAK *See* **Longitudinal centerline.**

HAMMERING Straightening, tensioning, or flattening a saw blade manually with a hammer.

HAMMER LINE Wire rope line of a crane assigned to raising and lowering a hammer.

HAMMER LOSS Test to determine the impact resistance of gypsum board.

HAMMERMILL *See* **Crusher.**

HAMMER POST Post with its base resting on a hammer beam.

HAMMER RACKING Severe bouncing of the casing of a power pile hammer during driving operations.

HAMMER SKIRT Extension fastened to the sides of a pile hammer to engage the leads of template-supported timber, pipe, concrete, or H-pile to keep the hammer centered upon and in vertical alignment with the pile to permit the hammer's free hanging operation.

HAMMER SPEED Number of complete strokes of a pile hammer achieved by the ram per minute.

HAMMER UPLIFT *See* **Cylinder lift.**

HAMM TIP Flared shotcrete nozzle having a larger diameter at midpoint than at either inlet or outlet. *Also called* **Premixing tip.**

HAND 1. Side, direction, or position indicated by one hand or the other (right- or left-hand); **2.** Swing of a door based on the side of the hinges as seen from the inside of the door; **3.** Side on which a log passes the sawyer as he faces the saw.

HAND BORING Site investigation technique of boring a hole in self-supporting soil using a shell and hand auger, down to a maximum depth of 4 m (13 ft).

HAND-BUILT HOSE Hose made by hand on a mandrel, reinforced by textile or wire or combination of both.

h&c *Abbreviation for:* hot and cold.

HAND DISTRIBUTOR Hand-directed spray for spreading road asphalts. *Also called* a **Hand sprayer.**

HAND DRILL Chuck mounted on a handle to which is attached a rotating geared crank.

HAND ELEVATOR *See* **Elevator.**

HAND FLOAT Hand tool for finishing plaster or concrete surfaces.

HAND GRINDING *See* **Off-hand grinding.**

HANDHOLD Provision made to grasp with the hand, typically when climbing onto a large vehicle or machine.

HAND HOLE 1. Small port allowing entry of a hand, typically to gain access for cleaning or adjustment; **2.** Opening in a disk wheel in order to gain access to the valve stem, access to the inside dual tire, or for chain installation.

HAND LEVEL Small, hand-held instrument consisting of a telescope and leveling-bubble tube, used for sighting a horizontal line.

HANDLING CAPACITY Maximum number of people that an elevator or escalator can service in a given period.

HANDLING STRESS Stress induced in a precast concrete unit by handling, storage, transport, or erection, that may be greater, though of a different type, than the design working load.

HANDLING TIGHT Pipe and couplings screwed together by hand so tight, that they require a wrench to loosen.

HAND SPRAYER Hand-operated device for distributing a liquid in the form of fine droplets.

H&M *Abbreviation for:* hit-and-miss.

H&P *Abbreviation for:* hasp and padlock.

HAND PRIMING Filling fuel lines with a hand pump prior to starting the engine.

HANDRAIL 1. Rail serving as a guide or support, as on a stairway or in an elevator car; **2.** Moving section over the top of the balustrade of an escalator.

HAND/RIDER TRUCK Dual purpose powered industrial truck that is designated to be controlled by a pedestrian but that may be controlled by a riding driver.

HAND SAW *See* **Saw.**

HAND SCREW Wooden clamp having parallel jaws connected by opposing screws, used by carpenters.

HAND SPIKE Short, slender pole used to position and hold a pile in the leads.

HANDSPLIT AND RESAWN *See* **Shake.**

HAND TAMPER Steel-shod wooden tamper fitted with a plough-type handle at each end.

HAND TOOL Any tool that is operated and guided by hand.

HAND TRUCK General group of nonpowered industrial trucks for operation by a pedestrian.

HANDYMAN Worker skilled at performing various jobs as the need arises.

HANG 1. Installation of something that is hinged: a door, window, etc.; **2.** Adhesive's ability to hold a wall tile in place on a vertical surface during its curing stage. *See also* **Grip.**

HANG A BLOCK Place a block in position when rigging up.

HANGER 1. Vertical support beam; **2.** Drop support, made of strap iron or steel, attached to the end of a joist or beam used to support another joist or beam; **3.** Wire hanger used to support a suspended ceiling; **4.** Support used to help hold a reinforced concrete joist; **5.** Support specially designed to support pipe lines; **6.** Worker who installs drywall.

HANGER BEARING Shaft support at the end of a hanger.

HANGER BENCH Low scaffold, used by drywall hangers to reach ceilings.

HANGER BOLT Double-ended fastener having a wood screw at one end and a machine screw at the other, used to attach hangers to woodwork.

HANGER ROLLER Roller, one of two per panel, from which horizontal door panels are suspended.

HANGING BUTTRESS *See* **Buttress.**

HANGING LEAD *See* **Lead.**

HANGING PENDULUM Device used to monitor movements in dams, abutments, foundations, etc.

HANGING STILE *See* **Door.**

HANG STRAP Wire rope strap choked onto a spar, used to support a jack strap.

HANG THE RIGGING Rig up an A-frame, head spar, or back spar.

HANG UP 1. Logs stuck behind a stump or other obstacle when yarding; **2.** Rigging fouled in some manner so as to prevent logging; **3.** Partially fallen tree supported by other standing timber.

HANK 1. Skein of yarn; **2.** Standard length of stubbing, roving, or yarn. The yarn is specified by the yarn numbering system in use, e.g., cotton hanks have a length of 840 yards; **3.** Stubbing or roving, indicating the yarn number (count), e.g., a 1.5 hank roving.

har. *Abbreviation for:* harbor.

HARD ASPHALT *See* **Asphalt.**

HARD-BITE Cast-steel driving tip with cutting teeth for H-piles.

HARDBOARD Dense building sheet made of softwood pulp formed into sheets under heat and pressure.

HARD-BURNED Nearly vitrified clay products that have been fired at high temperatures. They have relatively low absorptions and high compressive strengths.

HARD COMPACT SOIL Earth not classified as running or unstable.

HARDCORE Lumps of stone, brick, furnace slag, concrete rubble, etc., used for filling soft ground.

HARD DRY Stage in the drying of a paint film when it is sufficiently dried throughout its depth that another coat can be applied.

HARD EDGE Gypsum core formulation to protect the paperbound edges of drywall.

HARDENABILITY Loss of ductility between a weld and the parent material.

HARDENER 1. Chemical applied to concrete floors to reduce wear and dusting; **2.** Two-component adhesive or coating, the chemical component of which causes the resin component to cure.

HARDENING 1. Making metal harder through a process of heating and cooling; **2.** Increase in resistance to indentation.

HARDENING ACCELERATOR Compound added to a concrete mix to increase the rate of hardening.

HARDENING OFF Preparing seedlings or root cuttings in a nursery for transplanting or planting out by gradually reducing water, shade, and shelter, and thus making them more resistant to desiccation and temperature changes.

HARD EXTENSION *See* **Screed extension.**

HARD FACING Filler material used to toughen a surface to resist abrasion, erosion, wear, galling, and impact wear, placed on a surface by welding, spraying, or braze welding.

HARD FINISH Smooth finishing coat of fine plaster applied over rougher coats.

HARD HAT Protective head gear.

HARD-METAL SHEATHED CABLE Electrical conductor(s) sheathed in flexible metal tube.

HARDNESS Ability of a material to resist deformation.

HARDNESS TESTING Use of laboratory test apparatus to determine the surface hardness of materials.

HARDPAN *See* **Soil types.**

HARD SCREED Mortar screed that has become firm.

HARD SOLDER Solder that melts at red heat or hotter.

HARDSTANDING Natural or prepared surface suitable for parking heavy vehicles and equipment.

HARD STOPPING Stiff, quick-setting paste used to fill small holes and repair blemishes in a surface prior to painting.

HARD SUCTION Noncollapsible suction hose, used for drafting water from static sources lower than the pump inlet.

HARDWARE Fittings (metal, plastic, etc.) that are permanently incorporated into a structure; those that will be permanently visible are termed **Finish hardware.**

HARDWARE CLOTH Woven steel mesh, available in a range of gauges and mesh sizes, used as light reinforcement.

HARDWOOD Generally, a deciduous tree of the angiosperm class, having broad leaves (in contrast to a conifer, or softwood). The term has no reference to the actual hardness of the wood.

HARDY CROSS METHOD *See* **Moment distribution.**

HARMONIC 1. Series of numbers whose reciprocals are in arithmetical progression; **2.** Deviations from the fundamental frequency sine wave that can be expressed as additional sine waves of frequencies that are a multiple of the generated frequency. They are expressed as third, fifth, etc., harmonics, denoting their frequency as a multiple of the main frequency.

HARMONIC BALANCER Component, usually attached to the front of the crankshaft of an engine, that dampens unwanted vibrations in the rotating system.

HARMONIC MOTION Period motion having a single frequency or amplitude.

HARPED TENDONS *See* **Deflected tendons.**

HARSH MIX Concrete or asphalt mix that lacks desired workability and consistency.

HARVEST CUT Felling of the mature crop of trees either as a single clearcut or a series of regeneration cuttings.

HARVESTER *See* **Harvesting machines (multifunction).**

HARVEST FUNCTIONS Range of activities oriented toward cutting and removing trees from a forested area. This includes:

Buck: To saw a log into shorter pieces. *Also called* **Slash.**

Bunch: To gather trees or logs into small piles for subsequent skidding by other equipment.

Chip: Small piece of wood used to make pulp. Chips are made either from wood waste in a sawmill or pulpwood operation, or from pulpwood specifically cut for this purpose. Chips are larger and coarser than sawdust.

Debark: To remove the bark from a tree trunk or its limbs.

Delimb: To remove the limbs from a tree trunk, either before or after felling.

Fell: To cut down a tree and cause it to fall in a predetermined direction and position.

Forward: To move a tree that has been felled from its original position to a point where either further work will be carried out on it, or it will be loaded for transport off the site.

Load: To load a bucked and limbed log onto a transporter.

Pile: To stack limbed and bucked logs in a pile, usually with all of the but ends in one direction, in preparation for loading.

Skid: To move logs with a skidder or crawler tractor.

Slash: 1. Residue left on the ground after felling, including unused logs, uprooted stumps, broken tops, limbs, etc.; **2.** *See* **Buck,** above.

Top: To cut off the unmerchantable top of a tree.

Yard: To accumulate logs in a cleared forest area in preparation for further work or for transport.

HARVESTING Cutting and removal of trees from a forested area.

HARVESTING MACHINES Mobile machines used in the forest industry, generally classified as singlefunction and multifunction. They include the following:

SINGLE FUNCTION MACHINES

Bucker: Equipment that saws felled trees into required lengths, such as logs, bolts, or sticks. *Also called* **Slasher.**

Cable skidder: Log skidder that uses a main winch cable and cable chokers to assemble and hold a load.

Chipper: Machine designed to chip whole trees or parts of trees.

Clam-bunk skidder: Skidder that uses an integrally mounted loader to assemble the load, and a clam or top-opening jaws to hold it.

Debarker: Machine for removing bark from logs or bolts.

Delimber: Self-propelled or mobile machine designed to remove all limbs from trees with flailing chains or knives.

Drum debarker: Used primarily to remove bark from pulpwood. Bolts tumble together forcibly and repeatedly in their passage through a large drum, rubbing off the bark as they roll against each other and against the corrugated interior of the drum.

Feller: Machine designed to automatically cut trees of a defined size, either by sawing or through use of a guillotine.

Forwarder: Self-propelled machine, usually self-loading, designed to transport trees or parts of trees by carrying them completely off the ground.

Grapple skidder: Skidder fitted with a grapple rather than chokers to hold and handle logs.

Loader: In logging operations, a machine equipped with a grapple and supporting structure, designed to pick up and discharge trees or parts of trees for the purpose of piling or loading.

Mobile yarder: Self-propelled or mobile machine designed to perform cable logging with the use of a tower that may be integral to the machine or a separate structure.

Ring debarker: Single-function forest harvesting machine used primarily to remove bark from saw logs and veneer bolts. An infeed conveyor advances the log longitudinally into the feed rollers, which automatically center the log in the rotating mechanical ring. The ring has five crescent-shaped fingers that open automatically as the feed rollers force the log against them and the log advances through the rotating mechanical ring.

Skidder: Wheeled or tracked vehicle used for sliding/dragging logs from the stump to a landing.

Slasher: Mobile machine designed to cut felled trees to a predetermined length with a shear or saw. *Also called* **Bucker.**

Swath cutter: Self-propelled forest harvesting machine capable of continuous movement while simultaneously felling multiple stems across a 2-m- to 2.5-m- (6-ft- to 8-ft-) wide swath.

MULTIFUNCTION MACHINES

Delimber-bucker: *See* **Delimber-slasher,** below.

Delimber-buncher: Used to delimb trees and arrange logs in piles on the ground.

Delimber-slasher: Used to delimb and slash trees. *Also called* **Delimber-bucker.**

Delimber-slasher-buncher: Used to delimb and slash trees and arrange logs in piles on the ground.

Feller-buncher: Forest harvesting machine that cuts a tree with shears or a saw and then piles it ready for yarding or skidding.

Feller-chipper: Multifunction machine used to

fell and chip whole trees.

Feller-delimber: Self-propelled multifunction machine designed to fell and delimb trees.

Feller-delimber-buncher: Self-propelled multifunction machine designed to fell, delimb, and arrange the trees in bunches.

Feller-delimber-slasher-buncher: Self-propelled multifunction machine designed to fell, delimb, and slash trees and arrange tree parts in piles on the ground.

Feller-delimber-slasher-forwarder: Self-propelled multifunctional machine designed to fell, delimb, and slash trees and carry tree parts to a landing.

Feller-forwarder: Self-propelled, self-loading machine designed to fell standing trees and transport the stems by carrying them completely off the ground.

Feller-skidder: Self-propelled, self-loading machine designed to fell standing trees and transport them by skidding.

Harvester: Self-propelled multifunction machine that may be capable of operating as a swath cutter but also performs chipping and/or forwarding functions in addition to felling.

Limited-area feller-buncher: Feller-buncher with a shear mounted on a knuckleboom, allowing the machine to reach and fell several trees while remaining stationary.

Processor: Multifunction machine that does not fell trees but handles two or more subsequent functions.

Slasher-buncher: Used to cut logs to predetermined lengths and arrange them on the ground.

HASP Fastening device in which a slotted plate fits over a staple and is secured to it by means of a padlock or peg.

HASS Inside curve of a bent pipe.

HAT BRACKET *See* **Chair bracket.**

HATCH Covered opening that provides access to an area: attic, roof, crawl space, etc.

HATCHET 1. Small hand axe with a short handle; **2.** *See* **Hammer.**

HATCHET IRON Hatchet-shaped soldering iron.

HATCHET STAKE Anvil with a straight, sharp edge, used for bending sheet metal.

HATCHING Parallel lines placed close together. *See also* **Cross hatching.**

HAUL 1. Distance material is moved from the loading point to the point of disposition. *See also* **Average haul, Free haul, Over haul,** and **Station yards; 2.** Conveying wood from a loading point to an unloading point.

HAULAGE ROPE Wire rope used for pulling movable devices.

HAULAGEWAY Main tunnel connecting an underground excavation with an exit.

HAULAWAY Technique of excavation involving hauling spoil away from the hole.

HAULBACK Line used to outhaul the rigging or grapple.

HAULBACK BLOCK Block through which the haulback runs.

HAULBACK DRUM Winch drum on a yarder that holds the haulback.

HAULBACK LINE Rope used in cable logging to haul the main line and its fittings back to the point where the logs are to be attached.

HAUL DISTANCE 1. Distance along the most direct and/or practical route between the center of the area to be excavated and the center of the area to be filled; **2.** For refuse collection:

(a) The distance a collection vehicle travels from its last pickup stop to the solid waste transfer station, processing facility, or sanitary landfill.

(b) The distance a vehicle travels from a solid waste transfer station or processing facility to a point of final disposal.

(c) The distance that cover material must be transported from an excavation or stockpile to the working face of a sanitary landfill.

HAULING General term for the transport of logs from one point to another, usually from a landing to the mill or shipping point.

HAUL ROAD Road dedicated to the movement of vehicles and materials. Typically between a working face and a crushing plant.

HAUL TIME Time it takes to travel from the load area to the dump area.

HAUNCH 1. Deepened portion of a beam in the vicinity of a support. *See also* **Arch; 2.** Sides and lower third of the circumference of a pipe.

HAUNCHING 1. Concrete support to the sides of a drain or sewer pipe above the bedding; **2.** Work done in strengthening or improving the outer strip of a roadway.

HAWK Tool comprising a platform of wood or metal, approximately 250 mm to 300 mm (10 in. to 12 in.) square, hand-held from below using a vertical handle in the center, used to carry plaster or mortar. *See also* **Hod,** and **Mortarboard.**

HAWK'S-BILL SNIPS *See* **Tin snips.**

HAWSER: Wire rope, usually galvanized, used for towing or mooring marine vessels.

HAWSER TWIST Cord or rope construction in which the first and second twists are in the same direction while the third twist is in the opposite direction.

HAYWIRE Any unsafe or slipshod work procedure.

haz. *Abbreviation for:* hazard; hazardous.

HAZARD Condition that presents a threat to normal expectations.

HAZARD INSURANCE Insurance that protects against property damages due to fire, windstorm, and other common hazards.

HAZARDOUS ATMOSPHERE Atmosphere that, by reason of being explosive, flammable, poisonous, corrosive, oxidizing, irritating, oxygen deficient, toxic, or otherwise harmful, may cause death, illness, or injury.

HAZARDOUS SUBSTANCE Substance that, by reason of being explosive, flammable, poisonous, corrosive, oxidizing, irritating, or otherwise harmful, is likely to cause death or injury.

HAZARDOUS WASTE *See* **Solid waste.**

HAZARD REDUCTION Treatment of a hazard that reduces or eliminates threat.

hb *Abbreviation for:* hose bib.

HB *Abbreviation for:* H-beam; hollow back.

HBB *Abbreviation for:* heavy beam bolster.

HBBU *Abbreviation for:* heavy beam bolster upper.

H-BEAM Structural steel beam, not unlike an I-beam but with wider flanges. *Also called* **H-pile.** *See* **Pile.**

H-BEAM LEAD *See* **Lead.**

hbr *Abbreviation for:* harbor.

HBS *Abbreviation for:* heavy beam spacer.

HBSU *Abbreviation for:* heavy beam spacer upper.

HC *Abbreviation for:* high carbon; high chairs; hollow core.

hcg *Abbreviation for:* horizontal center of gravity.

hch *Abbreviation for:* hitch.

HCHC *Abbreviation for:* heavy continuous high chairs.

H-CLIP Metal clip in the form of an 'H,' used to join edges of panels 9.2 mm (0.365 in.) thick, or greater, when used as roof sheathing.

HCP *Abbreviation for:* hard chrome plate.

hd *Abbreviation for:* head.

HD *Abbreviation for:* heavy duty.

hdbk *Abbreviation for:* handbook.

hd fz. *Abbreviation for:* hard freeze.

hdg *Abbreviation for:* heading.

HDG *Abbreviation for:* high-density graphite.

hdls *Abbreviation for:* headless.

HDPE *Abbreviation for:* high density polyethylene.

hdw. *Abbreviation for:* hardware.

hdwd *Abbreviation for:* hardwood.

HE *Abbreviation for:* heat exchanger; heavy equipment; high explosive.

HEAD 1. Topmost member of a door or window frame, used as a lintel; **2.** Upper end of a vertical timber or pile; **3.** Capital of a column; **4.** End closure for a filter case or bowl that contains one or more parts; **5.** Pressure resulting from a column of water or an elevated supply of water, within a plumbing system or in the ground, consisting of the sum of static head, velocity head, and friction head. There are several definitions, including:

> **Dynamic:** Maximum elevation that water within a system will reach due to pump pressure, or to induced pressure from a contained source (reservoir or header tank).

> **Static:** Height of a column or body of fluid above a given point.

> **Static discharge:** Static head from the centerline of a pump to the free discharge surface.

> **Total:** Sum of the static and velocity heads of a flowing fluid at the point of measurement. *Also called* **Dynamic head.**

> **Velocity:** Equivalent head through which a liquid would have to fall to attain a given velocity.

5. Back-pressure against a pump from an elevated outlet.

HEADACHE BALL 1. Heavy metal ball fastened to the end of a single-part line (whipline), the size of which is determined by the amount of weight needed to overhaul the wire rope and the load to be lifted; **2.** Iron or steel weight held on a wire rope that is dropped on or swung against objects or material to be broken. *Also called* **Drop ball,** and **Wrecking ball.**

HEAD BAY Widened part of a canal, upstream from the lock gates.

HEAD BLOCK 1. Main block of a crane, equipped with multiple sheaves and used for heavy lifts. *Also called* **Top block; 2.** Top section of a fixed, or semifixed or extended pile drive lead with sheaves for carrying lines holding the pile and hammer over the top of the leads. *Also called* **Sheave head assembly.**

HEAD BOARD Board at the roof of a heading, in contact with the ground above.

HEAD BUTTON Formed head of a bar, bolt, or rivet, shaped like a button.

HEAD CASING Outside casing or trim over a window opening that serves as a stop for the wall covering.

HEADER 1. Narrower of the two faces of a brick; **2.** Masonry unit that overlaps two or more adjoining wythes of masonry to tie them together. *Also called* **Bonder;** *See also* **Blind header, Brick, Clipped header,** and **False header; 3.** Brick laid on its flat side across the thickness of a wall so as to show the end of the brick on the surface of the wall; **4.** Stone having its greatest dimension at right angles to the face of the wall; **5.** Beam placed perpendicular to joists and to which joists are fastened in framing for an opening (stairway, chimney, etc.); **6.** Wood lintel; **7.** Structural member placed at right angles to the majority of framing members in a wall, floor, or roof; **8.** End of a hydraulic cylinder through which the piston rod extends. *Also called* **Cylinder head; 9.** Water supply pipe to which two or more branch pipes are connected to service fixtures; **10.** Manifold or supply pipe to which a number of branch pipes are connected.

HEADER BLOCK Concrete masonry units made with part of one side of the height removed to provide space for bonding with adjoining units, such as brick.

HEADER BOARD Board used to contain concrete at the end of a pour.

HEADER BOND *See* **Bond.**

HEADER COURSE A course composed entirely of headers.

HEADER CUT *See* **Butt cut.**

HEADER HIGH The height up to the top of the course directly under a header course.

HEADER JOINT 1. Joint between the ends of two bricks in the same course; **2.** Vertical joint.

HEADER JOIST The large beam or timber into which the common joists are fitted when framing around openings for stairs, chimneys, or any openings in a floor or roof, placed so as to fit between two long beams and support the ends of short timbers.

HEADER TILE Tile containing recesses for brick headers in masonry-faced walls.

HEAD FIRE Forest fire spreading, or set to spread with the wind.

HEADFRAME Tower over a shaft that supports the ropes used for raising and lowering workers, equipment, and materials.

HEAD GATE Upstream gate of a lock, irrigation system, or conduit.

HEADING 1. Direction of travel according to compass degrees; **2.** Distance worked over time of a tunneling operation; **3.** Collection of joints in a rock formation; **4.** Area adjacent to the face of a tunnel in which the excavation crew works.

HEADING COURSE 1. Continuous bonding course of header brick; **2.** Header course.

HEAD JAMB Term sometimes applied to the horizontal top member of a door or window frame. *Also called* **Yoke.**

HEAD JOINT Vertical joint between the ends of masonry units. *Also called* **Cross joint.**

HEADLAND *See* **Bay,** and **Promontory.**

HEAD LAP Portion of a shingle, shake, or tile that is covered by the succeeding course and that is not exposed to weathering.

HEAD LEAN Downhill or outward lean of a standing tree. *See also* **Lean.**

HEAD LIFT *See* **Front-mount hoist.**

HEAD MACHINERY Arrangement of sheaves on the end of an attachment used to reeve wire rope.

HEAD MAST Tower carrying the working lines of a cable excavator.

HEAD MOLDING Molding over an opening.

HEAD MOUNT *See* **Front-mount hoist.**

HEAD OF A FIRE Most rapidly spreading portion of a forest fire's perimeter, usually to the leeward or up slope.

HEAD OF MATERIAL Amount of mix spread by the spreading screws in front of an asphalt paver screed.

HEAD PLATE *See* **Wall plate.**

HEAD PRESSURE Pressure developed by a standing column of water; the pressure a pump must overcome before raising water to a higher level.

HEAD RACE Tunnel or open channel connecting a head pond to a penstock.

HEAD RIG Principal machine in a sawmill, used for the initial breakdown of logs by sawing along the grain, converting the logs into cants. *Also called* **Head saw.**

HEADROOM Vertical distance measured above a defined position, often prescribed by regulation; clear space above a stair tread, for instance; or above machinery.

HEAD SAW *See* **Head rig.**

HEAD SECTION *See* **Boom tip section.**

HEAD SPAR Spar to which logs are yarded. *Also called* **Head tree,** and **Home spar.**

HEAD STEEL Extra secondary (or lateral) reinforcement for a distance from the top of a precast concrete pile, of about three times the width of the pile, to prevent damage during driving.

HEAD TOWER Tower of a cableway, containing (or near) the drums and hoist.

HEAD TREE *See* **Head spar.**

HEADWALL Wall at the end and sides of a culvert (occasionally only the upstream wall) that prevents scour, acts as a retaining wall, and sometimes diverts direction of flow.

HEADWAY 1. Clear space or height under an arch or lintel, or over a stairway and the like; **2.** Spatial distance or time interval between the front ends of vehicles moving along the same lane or track in the same direction.

HEALTH AND WELFARE BOND Bond that guarantees the payment of health, welfare, pension, vacation funds, and, in some cases, wages up to a defined limit.

HEALTH RISK ASSESSMENT Scientific evaluation of the risk to a population's health resulting from certain actions or conditions.

HEAPED 1. To pile loosely **2.** Load of material being carried above the sides of a hauler or the bucket of an excavator. *See also* **Bucket capacity.**

HEAPED CAPACITY *See* **Bucket capacity.**

HEART BOND When two headers meet in the middle of a wall and the joint between them is covered by another header.

HEART ROT Decay in the heartwood of a tree.

HEARTH That portion of a fireplace upon which the fire is built (the rear portion extending into the fire opening is known as the **Back hearth**).

HEARTH TILE Unglazed, machine-made tile about 12 mm (0.5 in.) thick, used to surface a fireplace hearth.

HEARTWOOD Wood extending from the pith to the sapwood, the cells of which no longer participate in the life process of the tree. Heartwood may contain phenolic compounds, gums, resins, and other materials that usually make it darker and more decay resistant than sapwood.

HEAT Temperatures above ambient, as produced by burning or oxidation. There are a number of measures, including:

Humid heat: Ratio increase of enthalpy per pound of dry air with its associated moisture to rise of temperature, under conditions of constant pressure and constant specific humidity.

Latent heat: Heat absorbed or given off by a substance without changing its temperature.

Sensible heat: Heat, the addition or removal of which results in a change in temperature, as opposed to latent heat.

Specific heat: Amount of heat required per unit mass to cause a unit rise of temperature over a small range.

HEAT-ABSORBING GLASS Glass that is substantially opaque to infrared radiations, including short-wave infrared, but having a moderately high degree of transparency to most visible radiations.

HEAT-ACTIVATED ADHESIVE See **Adhesive.**

HEAT-ACTUATED FIRE DOOR See **Fire door.**

HEAT CAPACITY Heat required to raise the temperature of a given mass one degree.

HEAT CHECK Pattern of surface checks formed on surfaces that are alternately subjected to rapid heating and cooling.

HEAT CONDUCTOR Material capable of readily conducting heat at a constant rate.

HEAT CONTENT Sum total of latent and sensible heat stored in a substance minus that contained at an arbitrary set of conditions chosen as the base or zero point.

HEAT DEFLECTION TEMPERATURE 1. Temperature at which a plastic material has an arbitrary deflection when subjected to an arbitrary load and test condition; **2.** An indication of the glass-transition temperature.

HEAT DETECTOR Device for sensing an abnormally high air temperature or an abnormal rate of heat rise and automatically initiating a signal indicating this condition.

HEAT ENDURANCE Time that a sample can retain its original characteristics and physical properties at a specific temperature.

HEATER 1. Device that produces heat; **2.** Worker who heats rivets in a forge; **3.** Device used to preheat the screed plates to approximately the same temperature as the asphalt mix being applied. Fuels used to heat the screed can be: diesel, propane, or electric energy.

HEAT EXCHANGER Device that transfers heat from a warmer to a cooler medium, normally by conduction.

HEAT EXCHANGER COOLING Engine coolant heat dissipated to water through a liquid-to-liquid heat exchanger.

HEAT FLUX Rate of flow of heat through a unit area.

HEATING ELEMENT Electrical resistance coil used to heat water or air.

HEATING LOAD Amount of heat required to maintain a desired temperature within the warmed area.

HEATING RATE Rate, expressed in degrees per hour, at which the temperature of an object or area being heated increases.

HEATING SEASON Period of the year when it is necessary to operate heating equipment for indoor comfort, sometimes established by local regulation.

HEATING SYSTEM Complete works and installation necessary to generate heat, regulate and deliver it to the areas required to be warmed. There are many types, including:

Buried-cable: Heating system in which electrical resistance cable is buried in a ground-floor concrete slab.

Electric: System where heat is provided from the installation of electric resistance heaters at strategic locations.

High-pressure steam: Steam heating system employing steam at pressures above 1.05 kg/cm² (15 psig).

High-pressure water: Heating system in which water, having a supply temperature above 176°C (350°F), is used as a medium to convey heat from a central boiler, through a piping system, to radiators.

Hot water: Heating system in which water, having supply temperatures less than 121 °C (250°F) is used as a medium to convey heat from a central boiler, through a piping system, to radiators, either by gravity or by a circulating pump.

Low-pressure steam: Steam heating system employing steam at pressures between 0 and 1.05 kg.m² (0 and 15 psig).

Medium-temperature water: Heating system in which water having supply temperatures between 121°C and 176°C (250°F and 350°F) is used as a heating medium.

Off-peak, electric: System where electrical resistance cable is buried within a ground-floor concrete slab and operated by a separately metered circuit that operates only during off-peak load hours of the distribution system.

Panel: Heating system consisting of coils or ducts installed in wall, floor, or ceiling panels to provide a large surface of low-intensity heat in which heat is transmitted by both radiation and convection from the panel surfaces to both air and surrounding surfaces.

Perimeter: Heating system where the radiators are placed adjacent to the exterior walls, usually under windows.

Perimeter warm air: Heating system in which warmed air is conducted via ducts cast in a ground slab to registers placed at intervals around the perimeter of the area to be warmed.

Radiant: Heating method usually consisting of coils, pipes, or electric heating elements placed in or mounted on the floor, wall, or ceiling in which only the radiated heat is effective in providing the heating requirements.

Split: System in which heating is accomplished by radiators or convectors supplemented by mechanical air circulation.

Steam: Heating system in which steam is used as a heat transfer medium, at or above atmospheric pressure.

Vacuum: Method of building heating in which a vacuum pump connected to the return main removes moisture and air from the radiators and returns water to the boiler feed tank; a two-pipe heating system that operates below atmospheric pressure.

Vapor: Steam heating system in which the primary supply is at or near atmospheric pressure and the return system conveys the condensate to the boiler by gravity.

Warm air: System in which warmed air is ducted to points of application and vented to the atmosphere.

HEATING VALUE Heat released by combustion of a unit quantity of fuel, measured either in calories or Btu.

HEAT INSULATION Ability of a material to impede heat flow.

HEAT LOSS Heat lost through openings around doors, windows, etc. and due to the lack of insulation or the inefficiency of insulation.

HEAT MIRROR Plastic film, fully or partially reflective, placed on or between panes of glass to increase their insulating properties.

HEAT OF COMBUSTION Heat released by combustion of a unit quantity of a fuel, measured either in calories or Btu.

HEAT OF HYDRATION Heat evolved by chemical reactions with water, such as that evolved during the setting and hardening of portland cement, or the difference between the heat of solution of dry cement and that of partially hydrated cement. *See also* **Heat of solution.**

HEAT OF SOLUTION Heat evolved or absorbed when a substance is dissolved in a solvent. *See also* **Heat of hydration.**

HEAT PUMP Heating device that extracts usable heat from a medium like air or water by raising (pumping) its temperature. In its reverse mode it can be used for cooling.

HEAT RELEASE RATE Amount of heat liberated in a chamber during complete combustion.

HEAT RESISTANCE Property or ability to resist the deteriorating effects of elevated temperatures.

HEAT-RESISTANT CONCRETE *See* **Refractory concrete.**

HEAT-RESISTANT GLASS Transparent glass having a low coefficient of expansion.

HEAT-RESISTANT PAINT Thermally tolerant paint for use on such surfaces as radiators or hot-water piping, etc.

HEAT-SETTING ADHESIVE *See* **Adhesive.**

HEAT SINK Medium that conducts heat away from electronic devices.

HEAT SOFTENING POINT Temperature at which a standard test bar deflects 0.25 mm (0.010 in.) under a stated load.

HEAT TRANSFER Movement and dispersion of heat from its source, by convection, radiation, conduction, or transfer.

HEAT TRANSFER FOIL Panel coating system that involves the transfer of a complete coating system from a carrier film to a substrate by means of heat and pressure.

HEAT TRANSFER LIQUID Liquid used to transport thermal energy.

HEAT TRANSMISSION Heat flow over time.

HEAT TRANSMISSION COEFFICIENT Coefficient used in the calculation of heat transmission by conduction, convection, or radiation.

HEAT TREATMENT Application of heat in order to change the state or condition of something.

HEAVE 1. Upward movement of soil and/or structures supported on soil caused by swelling of the soil from the freeze/thaw cycle; **2.** Swelling due to increased moisture content, release of overburden pressures, or oxidation of sulfite soils exposed to air; **3.** Displacement of earth adjacent to structures driven into the earth; **4.** Uplift of earth adjacent to a previously driven pile, caused by the driving of an adjacent pile.

HEAVED PILE *In situ* pile that heaves due to ground displacement resulting from the driving of adjacent piles.

HEAVE FORCE Force producing soil uplift due to frost, and to swell of the ground.

HEAVE GAUGE Fabricated steel stud used to assist in the monitoring of ground heave due the development of an excavation.

HEAVY BENDING Reinforcing bar sizes #4 through #18, that are bent at not more than six points in one plane (unless classified as light bending or special bending) and single radius bending. *See also* **Light bending,** and **Special bending.**

HEAVY CONSTRUCTION Construction of large projects, usually involving earth moving and possibly rock blasting, and requiring the use of heavy equipment.

HEAVY-DUTY AIR CLEANER Engine air cleaner with greater dust-holding capacity for applications where operations will be in heavy dust concentrations for sustained periods.

HEAVY-DUTY BUCKET *See* **Bucket.**

HEAVY-DUTY DUMP BODY *See* **Truck.**

HEAVY-DUTY SCAFFOLD Scaffold designed and constructed to carry a working load not to exceed 34 kg/0.093m² (75 lb/ft²).

HEAVY-DUTY TILE Tile suitable for areas where heavy pedestrian traffic is prevalent.

HEAVY-DUTY TRUCK Class 8 and above vehicle.

HEAVY-DUTY VEHICLE Motor vehicle rated at more than 3056 kg (8500 lb) GVRW or that has a curb weight of more than 2712 kg (6,000 lb) or a basic frontal area in excess of 4.18 m² (45 ft²).

HEAVY-EDGE REINFORCEMENT Wire fabric reinforcement for highway pavement slabs having one to four edge wires heavier than the other longitudinal wires.

HEAVY EQUIPMENT Mobile and static equipment used on heavy construction and in operations such as forestry, road and highway construction and maintenance, open-pit mining, etc.

HEAVY FORCE AND SHRINK *See* **Fit classifications.**

HEAVY FUEL OIL Grades 5 and 6 Bunker C.

HEAVY FUEL Forest fuel of large diameter, such as logs and large limbwood, that ignites and burns more slowly than a flash fuel. *Also called* **Coarse fuel.**

HEAVY JOIST Timber measuring between 100 and 150 mm (4 and 6 in.) in thickness and 200 mm (8 in.) or over in width.

HEAVY LOAD RATING *See* **Load rating.**

HEAVY-MEDIA SEPARATION Method in which a liquid or suspension of given specific gravity is used to separate particles of a portion lighter than those that float and a portion heavier than those that sink from the medium.

HEAVY METALS Trace emissions that make up the solid particulate matter emissions and residue from combustion processes of incineration. The emissions are dependent on the composition of the incoming waste stream but can include silver, chromium, lead, tin, zinc, cadmium, mercury, and nickel.

HEAVY SOIL Soil largely composed of clay.

HEAVY SPECIALIZED CARRIER Trucking company equipped and authorized to transport equipment and goods that, because of inherent characteristics of size, shape, weight, etc., require special equipment for loading, unloading, or transporting.

HEAVY TIMBER CONSTRUCTION Construction system with good fire endurance (widely recognized as comparable to one-hour rating), comprising a panel roof deck of 28-mm (1.125-in.) tongue-and-groove plywood with exterior glue, over 100-mm (4-in.) wide supports.

HEAVYWEIGHT AGGREGATE *See* **Aggregate.**

HEAVYWEIGHT CONCRETE Concrete of substantially higher density than that made using normal-weight aggregates, usually obtained by use of heavy-weight aggregates and used especially for radiation shielding.

HECTARE Non-SI unit of area, equal to 10 000 m², permitted for use with the SI system of measurement. Symbol: ha. Multiply by 2.471 054 to obtain acres, symbol: ac. *See also the appendix:* **Metric and nonmetric measurement.**

HECTO Prefix representing 10². Symbol: h. Used in the SI system of measurement. *See also the appendix:* **Metric and nonmetric measurement.**

HEEL 1. Trailing edge of an angled blade; **2.** Socket or floor brace for wall-bracing timbers; **3.** Back of the base of a retaining wall; **4.** Opposite end of an anvil to the beak; **5.** Joint in a pitched truss where the top and bottom chords meet.

HEEL BEAD Compound applied at the base of a glazing channel after setting the light and before the removable stop is installed.

HEEL BOOM 1. Loading boom that uses tongs to heel or force the end of a log against the underside of the boom; **2.** Type of loader that braces one end of a tree length or long log against a plate on the boom to control and carry it.

HEELING IN Temporary planting of saplings and shrubs.

HEEL TACKLE System of lines and blocks used to tighten a skyline.

HEIGHT CLASS Any interval into which the range of tree heights is divided for classification and use, usually 3 m, 5 m, or 10 m (10 ft, 16 ft, or 33 ft) classes.

HEIGHT/DIAMETER CURVE Graphic representation of the relationship between individual tree heights and diameters used to determine tree volumes in localized areas.

HEIGHT OF BUILDING Vertical distance between a horizontal plane through the average grade level and: (a) the highest point of the roof assembly, in the case of a building with a flat roof or a deck roof (a roof having a slope of less than 20° is considered a flat roof for this definition); or (b) the average level of a sloping roof between the highest ceiling level and the highest point of the roof.

HEIGHT OF BUILDING IN STORIES Number of stories contained between the highest roof of a building (except for a penthouse containing no dwelling units) and the floor of its first story.

HEIGHT OF INSTRUMENT Elevation of the line of sight of the telescope when the instrument is leveled. *See also* **Collimation method.**

HELD LINE Forest fire control line that still contains the fire when mop-up is completed, excluding lost line, natural barriers not backfired, and unused secondary lines.

HELD WATER Water above the standing water table. *Also called* **Water of capillarity.**

HELICAL BINDERS Mild-steel rods spirally arranged about the main reinforcing steel to bind the latter to form a cage.

HELICAL CONVEYOR Large-diameter, slow-pitch screw set about a central axle that, when rotated, moves materials within its flight. An Archimedes screw.

HELICAL CORD In hose, a reinforcement formed by a cord or cords wound spirally around the body of a hose.

HELICAL GEAR *See* **Gear.**

HELICAL HINGE *See* **Rising butt hinge.**

HELICALLY CRIMPED WIRE Prestressing wire that has been manufactured with a helical crimp along its length.

HELICAL REINFORCEMENT Steel reinforcement of hot rolled bar or cold-drawn wire, fabricated into a helix. *Also called* **Spiral reinforcement.**

HELICAL SHELL Corrugated (usually 12 to 18 gauge) steel sheet wound into a pipe, installed as a pile with an inside mandrel or a removable outside casing.

HELICAL STAIR *See* **Spiral stair.**

HELICOIL Threaded steel insert.

HELIOGRAPH Survey instrument that reflects sunlight to create a flashing light, used to identify a distant station.

HELIOSTAT Mirror used to reflect the sun's rays into a collector.

HELIPAD Structure, or area, for specific use of individual helicopter operations. *Also called* **Pad.**

HELIPORT Permanent facility for specific use of helicopter operations, equipped and manned for multiple aircraft use.

HELIUM Inert, colorless, gaseous element used as a shielding gas in welding.

HELIX 1. Any spiral, either lying in a single plane or, especially, moving around a cylinder, cone, etc., as the thread around a screw; **2.** In hose, a shape formed by spirally winding a wire or other reinforcement around the cylindrical body of the hose.

HELMET 1. Protecting hood equipped with a lens of safety glass that fits over an arc welder's head to permit safe observation of an electric or gas arc; **2.** Cast steel driving cap. *See also* **Driving cap.**

HELPER SPRING *See* **Springs, auxiliary.**

HELVE Handle of a tool, typically an axe or hatchet.

hem. *Abbreviation for:* hemlock.

HEMATITE Mineral, iron oxide, used as aggregate in high-density concrete and in finely divided form as a red pigment in colored concrete.

HEMIHYDRATE Hydrate containing one-half molecule of water to one molecule of compound, the most common form of which is partially dehydrated gypsum. *Also called* **Bassanite,** and **Plaster of paris.**

HEMMING MACHINE Machine used for grinding flat surfaces.

HEMP Tough, organic fiber used to make rope.

HENRY One of 17 derived units with special names of the SI system of measurement: a unit of inductance of a closed circuit which gives rise to a magnetic flux of 1 Wb/A. (1 H=1 Wb/A). Symbol: H. *See also the appendix*: **Metric and nonmetric measurement.**

HERBACEOUS FUEL Forest floor undecomposed material, living or dead, derived from herbaceous plants (one that does not develop woody, persistent tissue but is relatively soft and succulent, including grasses and ferns).

HERBICIDE Chemical used to kill or retard the growth of plants; a weed killer.

HEREDITAMENTS Any property that can be inherited.

HERRINGBONE Arrangement of stones, bricks, tiles or other elements that are laid diagonally with alternate courses in opposite directions, forming a zig-zag pattern.

HERRINGBONE BOND *See* **Bond.**

HERRINGBONE DRAIN Stone- and gravel-filled trenches laid out in alternating diagonals. *Also called* **Chevron drain.**

HERRINGBONE GEAR *See* **Gear.**

HERRINGBONE STRUTTING Solid or other strutting between floor joists. *Also called* **Floor strutting.**

HERRINGBONE WRAP Narrow herringbone-woven tape spiraled circumferentially over the outside of a hose product to apply external pressure during vulcanization. *See also* **Cross wrap,** and **Wrapped cure.**

HERTZ One of 17 derived units with special names of the SI system of measurement: a unit of frequency equal to a periodic occurrence which has a period of one second. (1 Hz=1 s^{-1}). Symbol: Hz. *See also the appendix*: **Metric and nonmetric measurement.**

HESITATION SET *See* **False set.**

HESSIAN *See* **Burlap.**

HETEROGENEOUS Nonuniform in structure or composition throughout.

HEWN Any material fashioned through the processing of chopping with a tool; hewn stone, hewn wood, etc.

hex. *Abbreviation for:* hexagon; hexagonal.

HEXADECIMAL Alphanumeric numbering system used in digital electronics. It has a base of 16 (as compared to a base of 10 for the decimal system, or a base of two for a binary system) using the digits 0 through 9 and the letters A through F (to represent the numbers 10 through 15).

HEXADOME Dome structure of 24 triangles.

hf *Abbreviation for:* half.

HF *Abbreviation for:* high frequency; hot finished.

HFMS *Abbreviation for:* high-float medium-settling.

hgr *Abbreviation for:* hanger.

hgt *Abbreviation for:* height.

HH *Abbreviation for:* high humidity.

H-HINGE Hinge shaped like the letter H.

HI *Abbreviation for:* height of instrument; heat input.

HICKEY Hand tool with side-opening jaws, used in developing leverage for making bends in bars or pipes in place.

HIDDEN LINES Broken lines on a drawing that represent hidden edges and outlines.

HIDING POWER *See* **Opacity.**

HIGBEE INDICATOR Raised, or indented, in-line marking on a fire hose coupling or nipple that, when placed in line with a corresponding indicator, will ensure proper mating of threads, reducing the possibility of cross-threading when completing a joint.

HIGH AIR High-pressure air used to power pneumatic tools and devices. *See also* **Low air.**

HIGH-ALUMINA CEMENT *See* **Calcium-aluminate cement.**

HIGHBALL To do something swiftly, as to 'highball a load.'

HIGH-BOND BAR *See* **Deformed bar.**

HIGH-BOND MORTAR Portland cement mortar with a special additive for extra strength. High-bond mortar makes possible the factory production of 100-mm (4-in.) wythe masonry panels.

HIGH CENTERED Condition in which the tracks of mechanical equipment sink into soft soil allowing the undercarriage of the machine to rest on the ground;

often prevents further movement.

HIGH CHAIR *See* **Continuous high chair,** and **Individual high chair.**

HIGH-CYCLE FAILURE Failure resulting from more than 20,000 applications of cyclic stress.

HIGH-DENSITY CONCRETE Concrete of exceptionally high unit weight, obtained through the use of heavyweight aggregate.

HIGH-DENSITY OVERLAY Exterior-type plywood finished with a resin-impregnated fiber overlay to provide extremely hard surfaces that need no additional finishing and that have high resistance to chemicals and abrasion.

HIGH-DENSITY POLYETHYLENE Thermoplastic material composed of many molecules of ethylene.

HIGH-DENSITY SCREED Any asphalt paver screed designed to achieve greater compaction than can be obtained with a standard screed.

HIGH-DISCHARGE MIXER *See* **Inclined-axis mixer.**

HIGH-DUMP BUCKET *See* **Bucket.**

HIGH DUMPER Dump body and hoist combination giving a high angle of elevation, to the extent that retractive power is required for lowering the body rather than the usual gravity return.

HIGH-DUTY FIRECLAY BRICK *See* **Brick.**

HIGH-EARLY-STRENGTH CEMENT Cement characterized by attaining a given level of strength in mortar or concrete earlier than normal cement.

HIGH-EARLY-STRENGTH CONCRETE Concrete that, through the use of high-early-strength cement or admixtures, attains a given level of strength earlier than normal concrete.

HIGH EFFICIENCY PARTICULATE AIR (HEPA) FILTER Filter capable of trapping and retaining at least 99.97% of all monodispersed particles of 0.3 micrometers in diameter or larger.

HIGHEST AND BEST USE Most productive use, reasonable but not speculative or conjectural, to which property may be put in the near future.

HIGH EXPLOSIVE Explosive designed to detonate, and containing at least one high-explosive ingredient; an explosive that decomposes with extreme rapidity. *See also* **Explosive,** and **Low explosive.**

HIGH-FLOAT MEDIUM SETTING Anionic asphalt emulsion, used primarily in plant mixes, seal coats, and road mixes.

HIGH GLOSS Highest sheen of finish.

HIGH-GRADING Removal of only the best trees from a stand, often resulting in a poor-quality residual stand.

HIGH HAZARD INDUSTRIAL OCCUPANCY *See* **Industrial occupancy.**

HIGH IDLE Governed engine speed at full throttle and no load.

HIGHLEAD SYSTEM Logging system that uses cables rigged to a spar high enough above the ground so that one end of the logs can be lifted during yarding.

HIGH-LIFT GROUTING Technique in concrete masonry wall construction in which the grouting operation is delayed until the wall has been laid up to a full story height. *See also* **Grouting,** and **Low-lift grouting.**

HIGH-LIFT PLATFORM TRUCK Self-loading outrigger-type industrial truck equipped with a load platform, intended primarily for transporting and tiering loaded skid platforms.

HIGH-LIFT TRUCK Self-loading powered industrial truck equipped with an elevating mechanism designed to permit tiering of one load upon another. Popular types are high-lift fork truck, high-lift ram truck, high-lift boom truck, high-lift clamp truck, and high-lift platform truck.

HIGHLIGHT 1. To emphasize, through design, finish, etc.; **2.** Area of a finish distinct by its sheen from its surroundings.

HIGH LINE High-tension electrical distribution line.

HIGH LOAD Expression meaning 'high enough.'

HIGH OCCUPANT LOAD Condition where the number of persons in a room or floor area is such that the area of floor per person is not more than 1.11 m² (12 ft²).

HIGH-PERFORMANCE SEALANTS Sealants formulated to exhibit movement capability of ±50% joint movement.

HIGH-PRESSURE FOG Small-capacity spray jet produced at very high pressures and discharged through a small hose having a gun-type nozzle.

HIGH-PRESSURE GROUTING Consolidation grouting used to strengthen rock or to cut off water inflows.

HIGH-PRESSURE LAMINATE Sheet material formed from multiple layers of kraft paper saturated with phenolic resin; a decorative layer of paper saturated with melamine resin; and a very thin top sheet of paper heavily saturated with a melamine resin. Fused together in a hot press under high temperature and pressure to produce a stiff plastic sheet.

HIGH-PRESSURE STEAM CURING *See* **Autoclave curing.**

HIGH-PRESSURE STEAM HEATING SYSTEM *See* **Heating system.**

HIGH-PRESSURE WATER HEATING SYSTEM *See* **Heating system.**

HIGH-RATE, WATER-REDUCING ADMIXTURE *See* **Admixture.**

HIGH RIGGER Logger who tops trees and rigs them with guys, blocks, and lines.

HIGH-RISE Structure, usually for commercial or residential purposes, having more than six stories and equipped with elevators.

HIGH-SPEED BIT *See* **Bit.**

HIGH-STRENGTH CONCRETE Concrete that has a specified compressive strength for design of 41 370 kP (6,000 psi), or greater.

HIGH-STRENGTH LOW-ALLOY STEEL *See* **Steel.**

HIGH-STRENGTH REINFORCEMENT *See* **High-strength steel.**

HIGH-STRENGTH STEEL Steel with a high yield point, in the case of reinforcing bars, 60,000 psi and greater. *Also called* **High-strength reinforcement.**

HIGH-STRENGTH STRUCTURAL BOLTS High tensile strength bolts and nuts used for assembling structural steel.

HIGH STUMP Stump that is higher than a specified standard.

HIGH-TEMPERATURE INDICATING CRAYONS Wax-like indicating sticks that change color at rated temperatures from 120°C to 600°C (248°F to 1,112°F). Used as a troubleshooting in-field maintenance tool on off-highway equipment to indicate overheating of components.

HIGH-TEMPERATURE STEAM CURING *See* **Atmospheric-pressure steam curing,** and **Autoclave curing.**

HIGH-TENSION BOLTS High-tension steel bolts designed to be tightened with a torque wrench to a predetermined tension, used as a substitute for rivets in structural steel framing.

HIGH-VELOCITY TOOL *See* **Explosive-actuated fastening tool.**

HIGHWALL 1. Bench, bluff, or ledge on the edge of a surface excavation, commonly used in coal strip mining; **2.** Face that is being excavated, as distinguished from a spoil pile.

HIGH WATER MARK Highest visible mark left by a flood. It is usually the sign of the 10-year (or other arbitrary period) peak flow. *See also* **Low water mark.**

HIGHWAY Public road maintained at public expense, generally connecting principal centers, with minor roads directly connecting. There are several definitions, including:

- **Arterial:** General term denoting a highway primarily for through traffic, usually on a continuous route.
- **Belt:** Arterial highway for carrying traffic around an urban area.
- **Bypass:** Highway that circumnavigates a location, usually the core of a community or a residential area.
- **Divided:** Highway with separate roadways for traffic in opposite directions.
- **Major:** Arterial highway with intersections at grade and direct access to abutting property, and on which geometric design and traffic control measures are used to expedite the safe movement of through traffic.
- **Radial:** Arterial road leading to or from an urban center.
- **Through:** Highway that passes through any number of adjacent or contiguous communities or locations, giving access to all or some of them.

HIGHWAY SPEED Speed up to the legal maximum, in excess of the minimum (where applicable) posted.

HIGHWAY TRUCK *See* **Truck.**

HILEY FORMULA Dynamic pile driving formula for estimating the static load-bearing capacity of a pile driven from its penetration resistance. It includes terms for pile weight to ram weight ratio, pile rebound, and cushion properties.

HINGE Connection permiting motion about an axis in one plane.

HINGED EXPANSION JOINT *See* **Joint.**

HINGE JOINT Any joint that permits rotation with no appreciable moment developed in the members at the joint. *See also* **Mesnager hinge,** and **Semiflexible joint.**

HINGE POINT Point indicating where a fill slope stops and a road or shoulder grade begins. *Also called* **Catch point.**

HINGE-TYPE RIPPER *See* **Ripper.**

HINGE WOOD Holding wood between the undercut and backcut that acts as a hinge to control the direction of fall.

HIP External angle formed by the meeting of two sloping roof surfaces.

hip. *Abbreviation for:* hipped.

HIP CHAIN Device used to measure distance by means of an anchored filament wrapped around a wheel that revolves as the user walks.

HIP JACK Roofing member that runs from the rafter plate to the hip rafter.

HIP RAFTER *See* **Rafter.**

HIP ROLL Metal roofing trim consisting of a half-cylinder with a flat extension to each side, used for finishing the hip of a roof.

HIP ROOF Roof that rises by inclined planes from all four sides of a rectangular building.

HIP TILE *See* **Roof tile.**

HIP TRUSS *See* **Truss.**

HIP VALLEY CRIPPLE Strut that extends between the valley and the hip rafters.

hist. *Abbreviation for:* historical.

HISTORIC SITE Building, monument, park, cemetery or other site having public interest and national, regional, or local significance.

HISTORICAL AGE *See* **Actual age.**

HITCH 1. Connection between two machines; **2.** Horizontal shelf along the side of a rock tunnel, suporting roof timbers; **3.** To attach a rope to a post, rail, or attachment.

HITCHHIKER *See* **Hobo,** and **Pogo stick.**

HITCH PLATE Plate (on traction elevators) clamped to the underside of the crosshead and to which the shackles are attached.

h/l *Abbreviation for:* headlight.

hld *Abbreviation for:* hold.

hldg *Abbreviation for:* holding.

hldr *Abbreviation for:* holder.

hlp. *Abbreviation for:* helper.

hltp *Abbreviation for:* hilltop.

HM *Abbreviation for:* hollow metal.

hmr *Abbreviation for:* hammer.

hn. *Abbreviation for:* hone.

HNA *Abbreviation for:* high-nickel alloy.

hnd *Abbreviation for:* honed.

hndl. *Abbreviation for:* handle.

hndlg *Abbreviation for:* handling.

hng *Abbreviation for:* hang; hanging; honing.

hng *Abbreviation for:* hinge.

hnycmb *Abbreviation for:* honeycomb.

HO *Abbreviation for:* hold open.

HOA *Abbreviation for:* hold-open arm.

HOARDING Solid enclosure around a construction site.

HOBO Unchoked log that is carried to the landing by the choked turn. *Also called* **Hitchhiker.**

HOD V-shaped trough or tray, supported by a pole handle that is borne on the carrier's shoulder, for carrying small quantities of brick, tile, mortar, or similar load. *See also* **Hawk,** and **Mortarboard.**

HOE Construction shovel that digs by pulling a boom-and-stick-mounted bucket toward itself. *Also called* **Backhoe,** and **Pullshovel.**

HOE CHUCKING Moving logs to a landing or road system by a hydraulic grapple loader.

HOEDAG Hoe-like tool with an elongated blade used for planting trees.

HOE SHOVEL *See* **Shovel.**

HOG 1. Coarse wood chips to be burned as a fuel; 2. Machine used to grind wood into chips for use as a fuel or for other purposes.

HOGBACK Ridge formed by the outcropping edge of a tilted strata of rock, characterized by concave slopes at the sides.

HOG BOX Concrete box in which water and dirt are mixed before being pumped to a hydraulic fill.

HOGGED FUEL 1. Fuel made by grinding waste wood in a hog; 2. Mix of wood residues such as sawdust, planer shavings, and sometimes coarsely broken-down bark and solid wood chunks produced and manufactured of wood products and normally used as a fuel.

HOGGIN Natural deposit of stony sand and gravel containing a small amount of clay, sufficient to hold the mass together.

HOG LINE *See* **Intermediate suspension line.**

HOG ROD Turnbuckle-equipped rod installed horizontally across a shield-driven tunnel, fastened to the liner plates. Used to prevent deflection of the liner plate until the shield tail void is filled.

HOG WOOD Pulpwood logs to be chipped in a hog.

HOIST *See* **Winch.**

HOIST DRUM Rotating cylindrical spool with side flanges, used to wrap the winch rope during the raising and lowering of the load with the winch.

HOIST FRAME SIDE RAIL Main channel or formed rails enclosing the sides of hoist mechanisms and acting as supports for the dump body to rest upon when in a lowered position. *Also called* **Sill,** and **Channel.**

HOISTING ENGINE *See* **Winch.**

HOIST LINE Line that controls the elevation of a crane hook or cableway.

HOIST ROPE Wire rope used to reeve the winch and fastened to the attachment for lifting loads.

HOISTWAY Any shaftway, hatchway, well hole, or other vertical opening or space in which an elevator or dumb-waiter is designed to operate.

HOISTWAY ACCESS SWITCH *See* **Switch.**

HOISTWAY ENCLOSURE Fixed structure of vertical walls or partitions isolating a hoistway from all other parts of a building, or from adjacent hoistways, and in which the hoistway doors and assemblies are installed.

hol. *Abbreviation for:* hollow.

HOLD Neutral position in the control of hydraulically-operated devices in which the movable part is locked in position.

HOLDBACK 1. Percent of a contract sum held back on completion of the project for a contractual period to cover outstanding liabilities, contingencies, etc. *Also called* **Retention money;** 2. Automatic safety device that prevents a conveyor belt from running backward.

HOLD-DOWN *See* **Clamp.**

HOLD-DOWN BOLT Anchor bolt provided near the ends of shear walls for transferring boundary member loads from the shear wall to the foundation. *See also* **Anchor bolt.**

HOLDING LINE 1. Wire rope on a grapple, clamshell, or orange-peel bucket that suspends the bucket while the closing line is released to dump its load; 2. One of the two lines used for loading logs on line grapple loaders. *Also called* the **Grapple line.**

HOLDING PERIOD *See* **Presteaming period.**

HOLDING VALVE *See* **Valve.**

HOLDING VALVE (INTEGRAL) *See* **Valve.**

HOLDING WOOD Hinge of wood left uncut between the back of the undercut and the backcut.

HOLDING ZONE Area of land within a planning region, zoned for a type of development, but 'held' pending provision of services or other infrastructure.

HOLDOVER FIRE Forest fire that remains dormant for a considerable time.

HOLDOVER TENANT Tenant remaining in possession after expiration of the lease term.

HOLD TEST Hydrostatic pressure test in which hose is subjected to a specified internal pressure for a specified period of time.

HOLE BURDEN Horizontal distance from the rock face of an excavated area or quarry to the blast holes.

HOLE PATTERN Horizontal spacing of drill holes.

HOLE SAW *See* **Saw.**

HOLE SPACING Horizontal distance between drill holes, measured parallel to the rock face.

HOLIDAY Area that is inadvertently skipped during the application of a liquid film.

HOLING THROUGH Point in a tunnel excavation where the face daylights at a portal or meets another face.

HOLLOW BRICK *See* **Brick.**

HOLLOW CORE DOOR *See* **Door types.**

HOLLOW GRIND 1. Reducing the face of a material by grinding so as to form a plane or curve; 2. Grinding a saw plate on both sides so that the blade is thinner toward the eye than at the rim but leaving a hub in the center the same thickness as the rim.

HOLLOWING HAMMER *See* **Hammer.**

HOLLOW MASONRY UNITS One whose net cross-sectional area in any plane parallel to the bearing surface is less than 75% of the gross. *See also* **Masonry unit.**

HOLLOW PARTITION 1. Partition built of hollow masonry units; 2. Partition built in two sections with a gap between, used for sound and/or thermal insulation or to accommodate a sliding door.

HOLLOW-POST KEY *See* **Barrel key.**

HOLLOW-STEM AUGER Earth auger with an end bit on a hollow center shaft.

HOLLOW-SURFACE FASTENER *See* **Fastener.**

HOLLOW TILE Fired clay tile, similar in shape and cross section to a hollow masonry unit.

HOLLOW-UNIT MASONRY Masonry consisting either entirely or partially of hollow masonry units laid in mortar.

HOLLOW WALL *See* **Wall.**

HOLLOW-WEB GIRDER Box girder.

HOLT Anchor for rigging, or the system of attaching a choker, or other rigging to an object.

HOME Place where a person or family lives; a dwelling.

HOME INSPECTOR Person qualified or experienced in the inspection and evaluation of the structural and mechanical condition of a dwelling.

HOME SPAR *See* **Head spar.**

homo. *Abbreviation for:* homogeneous.

HOMOGENEOUS Uniform throughout.

HOMOGENEOUS BOARD Particleboard product manufactured with the same kind, size, and quality of particle throughout its thickness.

HOMOGENIZATION Process of subjecting a substance to intimate mixing resulting in a more uniform dispersion of components.

HONE 1. Material, such as a whetstone, used to sharpen a cutting edge; 2. Process of sharpening a cutting edge.

HONEYCOMB 1. Cell-like structure. *See also* **Rock pocket**; 2. Voids left in concrete due to failure of the mortar to effectively fill the spaces among coarse aggregate particles. *See also* **Blemish.**

HONEYCOMB CORE Structure, often made of heavyweight treated paper, arranged to form large cells, used to separate facing panels of hollow-core doors, partitioning panels, etc.

HONEYCOMBING Checks, often not visible at the surface, that occur in the interior of a piece of wood, usually along the wood rays.

HONING Abrasive operation typically performed on internal cylindrical surfaces and employing bonded abrasive sticks in a special holder to remove stock and obtain surface accuracy.

HOOD 1. Hinged covering to an engine compartment; 2. Curved baffle that prevents separation and scattering of material being discharged by a belt conveyor; 3. Casing on the end of an underwater suction line that causes it to pick up material from the bottom only; 4. Metal guard used for protection against abrasive wheel breakage. *Also called* **Protection hood**; 5. *See also* **Pile driving cap.**

HOOD ACCESS HATCHES Hatches on the covering to an engine compartment that allow minor maintenance checks to be performed without having to tilt or open the entire hood or cab.

HOOD-MOLD Projecting molding on the face of a wall above an opening or recess to throw off rain. *See also* **Dripstone**, and **Label.**

HOOK 1. Angle at which the face of a saw tooth contacts the material to be cut; 2. Bend (90° or 180°) in the end of a reinforcing bar to provide anchorage in concrete. For stirrups and column ties only, turns of either 90° or 135° are used. *See also* **Bent bar, Cable hook, Grab hook, Pintle hook, Round hook, Safety hook, Stirrup, Swivel hook,** and **Tie**; 3. Attachment to mechanical equipment, particularly toolcarriers, designed to mate with dumpsters, bins, troughs, etc., to permit quick, easy movement.

HOOK ARM *See* **Roll-off bodies.**

HOOKBLOCK Block with hook attached, used in lifting operations. It may have a single sheave for double- or triple-line work, or multiple sheaves for increased multiple reeving.

HOOK CARRIAGE Carriage of solid plate or two-bar construction upon which forks or other attachments can be mounted on a lift truck by means of hook-like projections.

HOOKED BAR Reinforcing bar with the end bent into a hook to provide anchorage. *See also* **Bent bar, Hook, Stirrup,** and **Tie.**

HOOKER 1. In logging, the foreman of a yarding crew. *Also called* a **Hooktender**; 2. Head man of the choker crew on a skidder; 3. Worker who directs the helicopter pilot in helicopter logging to the load site and hooks up the load to the helicopter load hook. *See also* **Hooktender.**

HOOKE'S LAW Law that holds practical for strains within the elastic limit that the strain is proportional to the stress producing it. *See also* **Proportional limit** and **Modulus of elasticity.**

HOOK KNIFE Hand knife with a short, sharply curved blade.

HOOK ON AND/OR HOOK OFF Act of placing or removing chokers or slings on or off a bundle of bars and connecting or disconnecting the crane hook.

HOOK ROLLER Roller attached by a bracket to the revolving section of a revolving shovel and contacting the lower face of a circular track on the travel unit.

HOOK SCALE Load measuring device that attaches to the hook of a crane or other lifting device, and to which the load is attached.

HOOKTENDER Foreman of a skidding or yarding crew. *Also called* **Hooker.**

HOOK UP Fastening rigging to something; fastening a choker to logs.

HOOPING Curved reinforcement in a circular vessel.

HOOP REINFORCEMENT One-piece closed tie or continuously wound tie not less than No. 3 in size, the ends of which have a standard 135° bend with a 10-bar diameter extension, that encloses the longitudinal reinforcement.

HOPPED-UP MUD Mortar mixed with an accelerator.

HOPPER 1. Square or conical-shaped storage structure with a spout extension at its lower end; **2.** Receptacle for receiving and holding a supply of material for the paving process. Each hopper side is hinged and is capable of being raised hydraulically to stimulate the flow or discharge of material toward the conveyor(s) at the center of the hopper area.

HOPPER BODY Body of a hauler that permits dumping the load through the bottom hopper doors or gates. *Also called* **Bottom dump.**

HOPPER/CONVEYOR *See* **Spreader.**

HOPPER/CONVEYOR SPREADER DUMP BODY *See* **Truck.**

HOPPER HEAD Rainwater head at the head of a downspout and into which eaves gutters discharge.

HOPPER SASH *See* **Sash.**

HOPPER-TYPE DUMP TRAILER *See* **Trailer.**

HOPPER WINDOW Frame containing one or more sashes, each of which is installed in a vertical plane and hinged to permit the top of the sash to open inward.

HOPPING BOARD *See* **Footboard.**

hor. *Abbreviation for:* horizon; horizontal.

HORIZON 1. Line where the earth meets the sky; **2.** In perspective drawing, a line (usually horizontal) that establishes the viewing height, to which all lines above or below converge; **3.** Division between two layers of rock denoting different geological ages.

HORIZON GLASS Half silvered, half clear glass in a sextant.

HORIZONTAL 1. Parallel to the plane of the horizon; **2.** Reinforcing bars running horizontally.

HORIZONTAL ALIGNMENT Configuration of a roadway in the horizontal plane.

HORIZONTAL ANGLE Angle formed by the intersection of two lines on a horizontal plane.

HORIZONTAL APPLICATION Installation of drywall panels with the long edge perpendicular to framing members.

HORIZONTAL-AXIS MIXER Revolving drum-type concrete mixer in which the drum rotates about a horizontal axis.

HORIZONTAL BRACE *See* **Ledger.**

HORIZONTAL BRANCH Drain branch extending laterally from a soil or waste stack or building drain.

HORIZONTAL BROKEN JOINTS Style of laying tile with each course offset one-half of its length.

HORIZONTAL CENTER OF GRAVITY *See* **Center of gravity.**

HORIZONTAL CIRCLE Circular, graduated plate under the telescope of a theodolite, used to measure horizontal angles.

HORIZONTAL DIAGONAL BRACING Diagonal braces running horizontally between vertical uprights of scaffolding.

HORIZONTAL DISTANCE In plane surveying, a distance measured along a level line.

HORIZONTAL EXIT Connection by a bridge, balcony, vestibule, or doorway of two floor areas at substantially the same level; such floor areas being located either in different buildings or located in the same building and fully separated from each other by a firewall. *See also* **Exit.**

HORIZONTAL FACE CUT First of two cuts required to face or undercut a tree. Its depth is approximately one-third the diameter of the tree and horizontal. *See also* **Conventional face, Humboldt face,** and **Undercut.**

HORIZONTAL JOINT Joint whose central axis lies primarily in the horizontal plane.

HORIZONTAL LEDGE *See* **Ledge.**

HORIZONTAL LINE Line tangent to a level surface. To all intents and purposes, a straight line.

HORIZONTAL PIPE Any fitting or pipe that makes 45° or more with the vertical.

HORIZONTAL PLANE Plane perpendicular to plumb, or vertical.

HORIZONTAL SERVICE SPACE Space such as an attic, duct, ceiling, roof, or crawl space oriented essentially in a horizontal plane, concealed and generally inaccessible, through which building service facilities such as pipes, ducts, and wiring may pass.

HORIZONTAL-SHAFT MIXER Concrete mixer having a stationary cylindrical mixing compartment, with the axis of the cylinder horizontal, and one or more rotating horizontal shafts to which mixing blades or paddles are attached. *See also* **Pugmill.**

HORIZONTAL SHORING Load-carrying beam or trussed section, used to carry a shoring load from one bearing point, column, frame, post, or wall to another.

HORIZONTAL SLIDING DOOR *See* **Door types.**

HorM *Abbreviation for:* hit-or-miss.

HORN *See* **Beak.**

HORSE Simple frame used, mostly in pairs, to support material being worked on. *Also called* **Sawhorse.**

HORSEHEAD Temporary support for forepoles, used in tunneling soft ground. *Also called* a **False set.**

HORSEPOWER 1. Nonmetric unit of power equal to a rate of 33,000 foot-pounds per minute. Symbol: hp. Multiply by 0.745 699 9 to obtain kilowatts, symbol: kW. *See also the appendix*: **Metric and nonmetric measurement; 2.** There are many other ways of measuring horsepower, including:

Boiler: Equivalent of the heat required to change 16 kg (34.5 lb) per hour of water at 100°C (212°F) to steam. It is equal to a boiler heat output of 33,475 Btu per hour.

Brake: Amount of horsepower of an engine, motor, or mechanical device produced as indicated using a brake dynamometer, measured at the end of the crankshaft.

Continuous: 1. Amount of horsepower the engine is capable of developing, at a stated speed and under full load, for more than 24 hours; **2.** Power recommended by a manufacturer for satisfactory operation under specified continuous-duty conditions.

Demand: Amount of horsepower a powertrain needs to propel a specified gross weight at a specified rate of speed; the sum of the power absorbed by the rolling resistance, air resistance, and grade resistance.

Drawbar: Horsepower of an engine, minus friction and slippage loss in the drive mechanisms and in the tracks or tires.

Friction: Power needed to overcome friction within an engine resulting from the pressure of the piston and rings against the cylinder walls, rotation of the crankshaft and camshaft in their bearings, and friction developed by other moving parts.

Gross: Power rating obtained by a dynamometer test of an engine without allowance for power absorbed by engine accessories.

Horsepower: Torque or twisting force produced by an engine over a period of time. A measurement of work. Equation: 550 lb/ft/sec = 1 horsepower (76 m-kg/sec), or 746 watts.

Horsepower hour: The equivalent of developing one horsepower for one hour. A heat equivalent of approximately 2454 btu.

Hydraulic: Horsepower computed from flow rate and pressure differential.

Indicated: Calculated horsepower value using a cylinder pressure gauge and known engine dimensions to provide a reference horsepower value.

Intermittent: Amount of full-load horsepower, at a stated speed, that an engine can maintain on a cycle of 30 minutes under full load, and 60 minutes under no (or reduced) load.

Peak: Horsepower an engine can maintain at maximum load for one minute without speed loss, with a reasonably clean exhaust and with the engine in proper adjustment.

Rated: Horsepower an engine can develop under full load, at a speed recommended by the engine manufacturer, at a measured altitude and temperature.

SAE net: Brake horsepower remaining at the flywheel of an engine to do useful work after the power required by the engine accessories has been provided.

Shaft: Actual horsepower produced by an engine, after deducting the drag of accessories.

Taxable: Arbitrary formula for estimating horsepower that assumes that engines deliver their rated power at a piston speed of 304 m (1,000 ft) per min. and that mechanical efficiency will average 75%.

Wheel: Average horsepower delivered at the points of contact between the driving wheels and the ground surface, equal to the flywheel horsepower less the power lost through the power transmission assemblies.

Working: Horsepower delivered at the flywheel with full engine accessories for the working conditions.

HORSEPOWER HOUR *See* **Horsepower.**

HORSE SCAFFOLD Scaffold consisting of a platform supported by construction horses.

HORSESHOE ARCH *See* **Arch.**

hort. *Abbreviation for:* horticultural; horticulture.

HOSE Flexible conduit consisting of a tube, reinforcement, and usually the outer cover.

HOSE ASSEMBLY Length of hose with a coupling attached to one or each end. *Also called* **Coupled length.**

HOSE BIB *See* **Bib.**

HOSE CLAMP Collar, band or wire used to hold hose on to a fitting. *See also* **Clamp,** and **Ferrule.**

HOSE COUPLING Joint or connection between a hose and a pipe, or a hose and another hose.

HOSE DUCK Woven fabric made from plied yarns with approximately equal strength in the warp and filling directions.

HOSE REEL Drum on which a hose may be reeled for storage and/or transport, often wall mounted in the area where the hose is to be used to dispense compressed air, lubricants, oil, etc., or on mobile equipment, and equipped with a spring- or power-operated means to rewind the hose(s).

HOSING Synthetic flexible tubing used in hydraulic plumbing, having the distinct advantage of providing flexibility for easy routing, noise and vibration absorption, and ease of replacement, and having the disadvantage that it acts as an insulator, decreasing heat dissipation.

hosp. *Abbreviation for:* hospital.

HOSPITAL SIDE The unsafe side of a load being lifted or moved.

HOT AIR CURE Vulcanization using heated air, with or without pressure. *See also* **Air cure,** and **Vulcanization.**

HOT AIR HEATING SYSTEM Heating system in which air is warmed and then pumped through ducts to vents.

HOT APPLIED Products supplied in solid form that must be heated until they become liquid prior to application. *See also* **Cold applied.**

HOT BIN Bin used to store any of various sizes of recently dried material in an asphalt mixing plant prior to proportioning and mixing.

HOT CATHODE Instant-starting fluorescent lamp using electrodes of coiled tungsten filaments. *See also* **Cold cathode.**

HOT CEMENT Newly manufactured cement that has not had an opportunity to cool after burning and grinding of the component materials.

HOTCHKISS DRIVE Type of chassis design where the rear springs are mounted at the forward end in a stationary bracket (not shackled at the rear end) and all driving and braking forces are cushioned by the springs and transferred directly to the frame side members. Open-type universal joints and propeller shafts are used in this design.

HOT DECK Pile of logs from which logs are hauled as soon as they are yarded. *Also called* **Hot landing.**

HOT-DIP GALVANIZING *See* **Galvanizing.**

HOT DRAWN Forming a metal product while the material is heated to the point of being malleable.

HOTEL Building or part of a building in which rooms are provided for rent as temporary dwellings with no provision for cooking, but with a public dining room.

HOT ELEVATOR Bucket elevator used to carry hot, dried aggregate from the dryer to the weight hopper or gradation unit of an asphalt mixing plant.

HOT FACE Surface of a refractory section exposed to the source of heat.

HOT FORMING Operations such as bending, drawing, forging, heading, piercing or pressing, performed on metal while it is above its recrystallization temperature.

HOT-LAID MIXTURE Any asphaltic mixture that must be spread and compacted while in a heated condition.

HOT LANDING *See* **Hot deck.**

HOTLINE EQUIPMENT Tools and ropes especially designed for work on energized high-voltage lines and equipment.

HOT LOADING 1. Loading out logs as they are landed by the yarder; **2.** Practice in helicopter logging of landing sling loads of shake blocks directly to truck trailer decks.

HOT LOAD TEST Test for determining the resistance to deformation or shear of a refractory material when subjected to a specified compressive load at a specified temperature for a specified time.

HOT LOGGING Logging operation in which the logs are not stored or decked, but loaded onto a truck as soon as they are skidded to a landing.

HOT-MELT ADHESIVE *See* **Adhesive.**

HOT MILL Heating metal, then shaping it.

HOT-MIX ASPHALT Asphalt mix, usually produced in a central mixing plant, in which the aggregates have been heated and dried for placement at high temperatures.

HOT-MIX SEAL COAT Layer of hot mix, usually less than 25 mm (1 in.) thick, placed on a structurally sound pavement to improve skid resistance and smoothness. *Also called* **Carpet course, Friction coat, Light asphalt resurfacing, Plant-mixed surface treatment, Plant-mix seal,** and **Thin overlay.**

HOT-OIL HEATER Heating system designed to increase or maintain the temperature of liquid asphalt stored at an asphalt mixing plant and to provide heat transfer oil that heats many plant process operations.

HOT-POURED SEALING COMPOUND Rubberized or bitumen compound, poured into a horizontal concrete joint at higher than its maximum service temperature.

HOT REFUELING Refueling a helicopter while the engine is still running.

HOT ROLLING Manufacture of steel sections by passing hot steel bars through pairs of massive rolls.

HOT SLING RECOVERY Practice in helicopter logging of recovering slings from an active load landing area.

HOT WALL Freshly plastered wall containing an unusually high proportion of free lime, that burns out a paint film.

HOT WATER HEATING *See* **Heating system.**

HOT WATER SUPPLY SYSTEM Storage tank (including means of heating stored water) plus connected distribution piping and, in the case of a circulating hot water system, a pump. *See also* **Circulating hot water system,** and **Noncirculating hot water system.**

HOT WIRE Wire that carries power in an electrical distribution system, usually black or red (as distinguished from the neutral, white wires).

HOT WORKING Shaping of metal at a temperature and rate that does not cause strain hardening.

HOUR Non-SI unit of time, equal to 3600 seconds or 60 minutes, permitted for use with the SI system of measurement. Symbol: h. *See also the appendix*: **Metric and nonmetric measurement.**

HOUSE 1. Building designed as a single dwelling; **2.** Deck, deck machinery, and cab of revolving excavators and cranes.

HOUSE ASSEMBLY Housing that covers the machinery mounted on an upper revolving frame.

HOUSED JOINT Piece of lumber fitted into a second piece.

HOUSE DRAIN *See* **Building drain.**

HOUSED STAIR Staircase in which the stringers are grooved, or housed, to receive the ends of the treads and risers.

HOUSED STRING Stair string with horizontal and vertical grooves cut on the inner face to receive the ends of the risers and treads. Wedges covered with glue often are used to hold the risers and treads in place in the grooves.

HOUSEHOLD SOLID WASTE *See* **Solid waste.**

HOUSE JACK *See* **Jack post.**

HOUSE LOCK Mechanical lock engaged to prevent undesired swinging of a crane's superstructure, normally used during transport or pick-and-carry work.

HOUSE SEWER Pipe conveying sewage from a single residence to a common sewer.

HOUSE TRAP Sanitary seal provided between the building drain and the building sewer. *See also* **Running trap.**

HOUSING 1. Part cut out of one member to receive another; **2.** Jointing of two timbers by fitting the entire end of one piece into a blind mortise cut into the other; **3.** Buildings for residential use; **4.** Case or enclosure for rotating parts; **5.** Ported enclosure which directs a flow through a filter element.

HOUSING CODE (CANADA) *See* **Building code, National building code.** *Also called* **Minimum standards bylaw.**

HOUSING CODE (US) *See* **Building code, housing code (U.S.),** and **Building code, National building code.** *Also called* **Minimum standards bylaw.**

HOUSING COMPLETIONS Number of dwelling units completed within a given period according to required codes and standards and ready for occupancy.

HOUSING COOPERATIVE Voluntary organization of nongovernment people or association responsible for the financing, operation, and management of housing accommodation. *See also* **Cooperative housing.**

HOUSING POLICY Policy statement for existing and future development of the housing stock of a community.

HOUSING START Following issuance of a building permit, completion of excavation and commencement of building.

HOUSING STARTS Number of houses on which excavation is completed and building has begun within a given period.

HOUSING STOCK Total number of dwelling units within a geographic area.

HOUSING SUPPLY Number of dwelling units available for sale or rent within a geographic area at a point in time.

HOUSING TYPE Architectural or building style of a house, i.e., high-rise, condominium, garden, attached, row, etc.

HOUSING UNIT *See* **Dwelling unit.**

HOYER EFFECT Frictional forces in pretensioned, prestressed concrete that result from the tendency of the tendons to regain the diameter that they had before they were stressed.

hp *Abbreviation for:* high pressure; hinge point; horsepower.

h-p *Abbreviation for:* high pressure.

HP *Abbreviation for:* high pressure.

H-PILE *See* **Pile.**

HPL *Abbreviation for:* high pressure laminate.

HPP *Abbreviation for:* high-pressure pump.

h.pt *Abbreviation for:* high point.

HQ *Abbreviation for:* headquarters.

hr *Abbreviation for:* hour.

HR *Abbreviation for:* hot rolled.

HRA *Abbreviation for:* hot rolled and annealed.

hrdbd *Abbreviation for:* hardboard.

hrdn *Abbreviation for:* harden.

hrdwd *Abbreviation for:* hardwood.

HRM *Abbreviation for:* high-resistant multiplate.

HRS *Abbreviation for:* hot-rolled steel.

Hrt *Abbreviation for:* heart.

Hrt CC *Abbreviation for:* heart cubic content.

Hrt FA *Abbreviation for:* heart facial area.

Hrt G *Abbreviation for:* heart girth.

hrtwd *Abbreviation for:* heartwood.

HS *Abbreviation for:* hermetically sealed; high school; high speed; horizontal sliding.

hse *Abbreviation for:* house.

hsg *Abbreviation for:* housing.

hshld *Abbreviation for:* household.

HSP *Abbreviation for:* hydraulic self-propelled.

ht *Abbreviation for:* heat; height.

HT *Abbreviation for:* hard top; heat treated; high tension; highway truck; hospital tip.

HTCI *Abbreviation for:* high-tensile cast iron.

htl *Abbreviation for:* hotel.

HTQ *Abbreviation for:* high-tensile quality.

htr *Abbreviation for:* heater.

HTS *Abbreviation for:* high-tensile steel; high-tensile strength.

HUB 1. Strengthenedpart or mounting of a wheel or gear; **2.** Transit station, or point over which the transit is set, in the form of a heavy, stable reference set nearly flush with the ground, with a tack in the top marking the point; **3.** Enlarged end of a hub-and-spigot cast iron pipe; **4.** Point-of-origin stake that identifies a point on the ground. The top of the hub establishes the point from which soil elevations and distances are computed.

HUB END Pipe end connections that are leaded and caulked, such as on cast-iron sewer pipe.

HUBLESS PIPE *See* **Pipe.**

HUBODOMETER Mile-counting device mounted in the hub of an axle.

HUE Attribute by which a perceived color is distinguished as red, yellow, green, blue, purple, or a combination of these. White, gray, and black colors possess no hue. *See also* **Color, Lightness,** and **Saturation.**

HUMBOLDT FACE One of the two types of face or undercut commonly used to fall a tree. The face section is removed from the stump of the tree and consists of a horizontal cut made to approximately one-third of the diameter of the tree with a second cut angling down to meet the point of maximum penetration of the first. *See also* **Conventional face, Horizontal face cut,** and **Undercut.**

humid. *Abbreviation for:* humidity.

HUMID HEAT *See* **Heat.**

HUMIDIFICATION Process of adding moisture, to the atmosphere, or to a product.

HUMIDIFIED BOND Ability of the surfacing paper of drywall to resist delamination under conditions of high humidity.

HUMIDIFIER Device that adds moisture vapor to the atmosphere within a room or building or heating/ventilating system.

HUMIDISTAT Control mechanism that regulates the amount of humidity in the atmosphere of a room, building, or heating/ventilating system.

HUMIDITY Amount or degree of moisture in the air. There are several distinctions, including:

  Absolute: Mass of water vapor per unit volume of air.

  Percent: Ratio of weight of water vapor in a given weight of dry air to the weight of water vapor that would be present if the same weight of air were saturated, expressed as a percent.

  Relative: Ratio of the amount of water vapor present in the air to that which the air would hold at saturation at the same temperature.

  Specific: Ratio of the mass of water vapor to the mass of the system: weight of water vapor relative to a given volume of air.

HUMUS Decomposed residue of plant and animal tissue.

HUNDREDWEIGHT Nonmetric unit of mass, equal to 100 pounds. Symbol: cwt. Multiply by 43.359 237 to obtain kilograms, symbol: kg. *See also the appendix:* **Metric and nonmetric measurement.**

HUNGRY *See* **Starved.**

HUNTING 1. In a diesel or gasoline engine, a perceptible, varying change in engine rpm resulting from the inability of a governor to match injection or fuel supply demands with load demands. Sensed at or near idle rpm; 2. In an electric generating set, the oscillation of voltage or frequency above and below the mean value; an unstable condition.

HUNTINGTON DRESSER Tool using star-shaped cutters for truing and dressing grinding wheels.

HUNTING TOOTH Sprocket and roller chain combination in which one has an odd number of contacts and the other an even number, so that no tooth will contact the same pin twice in succession.

hurcn. *Abbreviation for:* hurricane.

HURRICANE ANCHOR Metal fastener nailed to a rafter and the top plate.

HUSK Parts of a sawing system comprising the arbor, saw, saw guide, and splitter, usually on a round headrig.

HV *Abbreviation for:* heating-ventilating; high voltage.

HVAC Complete heating, ventilating, and air-conditioning system.

HVEEM Method of asphalt mix design based on the cohesion and friction of a compacted specimen .

hvy *Abbreviation for:* heavy.

HW *Abbreviation for:* high water; hot water.

HWC *Abbreviation for:* hot water circulating.

h x w x l *Abbreviation for:* height by width by length.

HWM *Abbreviation for:* high water mark.

HWT *Abbreviation for:* hot water temperature.

hwy *Abbreviation for:* highway.

hyb. *Abbreviation for:* hybrid.

HYBRID BEAM Fabricated steel beam composed of flanges with a greater yield strength than that of the web. Whenever the maximum flange stress is less than or equal to the web yield stress, the girder is considered homogeneous.

HYBRID SOLAR SYSTEM Passive solar system that uses only a few mechanical devices to utilize the collected energy. *See also* **Active solar system.**

hyd. *Abbreviation for:* hydrated; hydraulic.

HYDRANT Water supply outlet, with its valve located below ground, usually below the frost line, that permits the discharge of water at full bore and full available pressure.

HYDRATE Chemical combination of water with another compound or an element.

HYDRATED LIME Quicklime treated with sufficient water to satisfy its chemical needs and then processed for use.

HYDRATED SOAP Soap that has water associated with its structure. A typical example is a water-stabilized calcium soap grease that owes its stability to hydrated calcium soap.

HYDRATION 1. Formation of a compound by the combining of water with some other substance; 2. In concrete, the chemical reaction between hydraulic cement and water.

HYDRAULIC 1. Operated by the movement and force of liquids; 2. One of two methods by which an elevator is moved, whereby the elevator is 'pushed' up by oil. *See also* **Traction.**

HYDRAULIC ACTUATOR Mechanism that converts hydraulic pressure to mechanical motion.

HYDRAULICALLY EXTENDABLE COUNTERWEIGHT Counterweight that is extended and retracted by means of a hydraulically powered cylinder to increase the crane moment so as to increase crane capacities and stability. *See also* **Counterweight,** and **Power removable counterweight.**

HYDRAULIC BARKING Removal of bark from round timber, such as logs, bolts, or billets, by high-pressure jets of water as the pieces are mechanically rotated in a closed chamber.

HYDRAULIC CEMENT Cement that sets and hardens by chemical interaction with water and that is capable of doing so under water. *See also* **Cementitious materials.**

HYDRAULIC COLLAPSE Hydrostatic pressure in the ground (usually below a clay strata) that will cause the collapse of thin pile casing.

HYDRAULIC CONDUCTIVITY Rate of groundwater flow per unit of time through a cross section of unit area in a unit hydraulic gradient at ambient temperature.

HYDRAULIC CONTROL VALVE *See* **Valve.**

HYDRAULIC CUTOUT *See* **Function limiter.**

HYDRAULIC CUTTER Manually operated, hydraulic-shear, piston-type line cutter.

HYDRAULIC CYCLE TIME Time required for a related series of hydraulic units, and the devices they power, to complete one complete set of operations. Typically, for wheel loader bucket operations, this includes:

> **Raise time:** Time in seconds required to raise an empty bucket to its full height from a level position on the ground.
>
> **Lower time:** Time in seconds required to lower the empty bucket from the full height to a level position on the ground.
>
> **Dump time:** Time in seconds required to move the bucket from the load-carrying position at maximum height to the full dump position.

HYDRAULIC DASHPOT Device used with an engine governor to prevent low idle underrun and give a more stable idle speed.

HYDRAULIC DREDGE Floating pump used to suck up a mixture of water and sediment, usually for discharge to land, through pipes, but also to another underwater location.

HYDRAULIC DRIVE *See* **Driving machine.**

HYDRAULIC EJECTOR Pipe for removing sand, mud, or small gravel from the working chamber of a pneumatic caisson.

HYDRAULIC ELEVATOR *See* **Elevator.**

HYDRAULIC EXCAVATION Use of water under high pressure directed against the surface of a material to be moved, the dislodged material being carried away by the resulting flow.

HYDRAULIC EXCAVATOR Wheeled or tracked digging machine with a revolving superstructure having an engine mounted at one end and a hydraulically powered arm and stick at the other. A wide range of buckets and other attachments can be mounted on the stick. *See also* **Excavator.**

HYDRAULIC FILL Fill built by transporting material by water, either through an open channel or by pumping from a dredge.

HYDRAULIC FILL DAM Dam developed by the deposition of hydraulic fill.

HYDRAULIC FLUID Fluid suitable for use in a hydraulic system. *See also* **Fluid.**

HYDRAULIC FRICTION Flow resistance in a pipe or channel caused by surface drag, roughness, or internal obstructions.

HYDRAULIC GOVERNOR *See* **Governor.**

HYDRAULIC GRADIENT Slope of the surface of open or underground water.

HYDRAULIC HAMMER Attachment used to break rock or concrete.

HYDRAULIC HORSEPOWER *See* **Horsepower.**

HYDRAULIC HOSE Flexible oil lines used to transmit fluid.

HYDRAULIC HYDRATED LIME Hydrated dry cementitious product obtained by calcining a limestone containing silica and alumina to a temperature short of incipient fusion so as to form sufficient free calcium oxide to permit hydration and, at the same time, leaving unhydrated sufficient calcium silicates to give the dry powder its hydraulic properties.

HYDRAULIC JACK Lever-actuated lifting device.

HYDRAULIC JACK PAD Thick steel pad that is placed between the hydraulic jack plunger and butt of a tree to distribute the upward push over a larger area.

HYDRAULIC KICKOUT *See* **Function limiter.**

HYDRAULIC LIFTING CAPACITY Measure of hydraulic power, rated with the rear of the machine tied down, not taking into account the static tipping load or normal operating conditions.

HYDRAULIC LIME Lime having the property of hardening under water.

HYDRAULIC LOADER *See* **Loader.**

HYDRAULIC MONITOR *See* **Monitor.**

HYDRAULIC MORTAR Mortar that will harden under water, used for foundation work or any masonry constructed under water.

HYDRAULIC MOTOR 1. Rotary motion device that changes hydraulic energy into mechanical energy; **2.** Rotary actuator.

HYDRAULIC OIL Fluid used to transmit power for operation of hydraulic systems.

HYDRAULIC PILE DRIVING Method of driving piles using hydraulic force.

HYDRAULIC POWER Transmission of power through fluid under pressure.

HYDRAULIC POWER UNIT Combination of componentry to facilitate fluid storage and conditioning, and delivery of the fluid under conditions of controlled pressure and flow to the discharge port of a pump, including maximum pressure controls and sensing devices where applicable. Circuitry components, although sometimes mounted on the reservoir, are not considered part of the power unit.

HYDRAULIC PUMP *See* **Pump.**

HYDRAULIC RADIUS Cross-sectional area of a stream of water divided by the length of that part of its periphery in contact with its containing conduit.

HYDRAULIC REACH Section of a river or stream between significant changes in hydraulic character: increase or decrease in flow, bottom, bank height, volume, etc.

HYDRAULIC RELIEF VALVE *See* **Valve.**

HYDRAULIC RESERVOIR Vessel or tank for storing and conditioning liquid in a hydraulic system.

HYDRAULICS Science of transmitting force and/or motion through the medium of a liquid.

HYDRAULIC SCOOPER Self-propelled crawler vehicle equipped with hydraulically operated arms that lift, empty, and replace containers carried on a transfer trailer bed.

HYDRAULIC SHOVEL *See* **Shovel.**

HYDRAULIC SLUICING Moving materials by water pressure.

HYDRAULIC STARTING SYSTEM Engine starting system that utilizes pressurized hydraulic oil through a motor.

HYDRAULIC STRIKE-OFF A strike-off only (no screed plate, heat, or vibration).

HYDRAULIC TEST Test for piping systems in which all conduits are sealed and filled with water to, or slightly above, design pressure.

HYDRAULIC TESTER Device that can measure flow, pressure, and temperature of an active hydraulic circuit, in both forward and reverse flow.

HYDRAULIC TIPPER Device that unloads a transfer trailer by raising its front end to a 70° angle.

HYDRAULIC VOLUME Rate of hydraulic flow generated by a system's pumps.

HYDRIC Soil that is saturated for sufficient periods of time to produce anaerobic conditions.

hydrl. *Abbreviation for:* hydroelectric.

HYDROCRACKING Process used in petroleum refining involving the cracking or thermal degradation of petroleum in the presence of hydrogen and a catalyst to produce petroleum products. *Also called* **Hydrotreating.**

HYDRODYNAMIC Hydraulic device that uses the impact or kinetic energy in a liquid to transmit force.

HYDRODYNAMIC LUBRICATION *See* **Lubrication.**

HYDRODYNAMICS Engineering science pertaining to the energy of liquid flow and pressure.

HYDROELECTRIC POWER Electrical energy generated from the force of falling water, used to drive turbines that, in turn, drive generators.

HYDROFINISHING Process used in the refining of lubricants involving treatment with hydrogen in the presence of a catalyst to remove unstable components and improve the oxidation stability of the resultant base oil.

HYDROGEN SULFIDE Poisonous gas with the odor of rotten eggs that is produced from the putrefaction of sulfur-containing organic material.

HYDROGEOLOGY Science dealing with the occurrence of surface and groundwater, its utilization, and its functions in modifying the earth, primarily by erosion and deposition.

HYDROGRAPH Graph showing various conditions of the properties of water with respect to time.

HYDROGRAPHIC SURVEY Determination of the configuration of the bottom of a body of water.

HYDROKINETICS Engineering science pertaining to the energy of liquids in motion.

HYDROLOGICAL CYCLE General pattern of water movement at, and near the earth's surface.

HYDROLOGY Science concerned with the properties, occurrences, distribution, and circulation of water, particularly underground water.

HYDROLYSIS Chemical reaction between a material and the ions of water.

HYDROMETER Device that measures the specific gravity of a fluid, such as engine antifreeze or battery electrolyte.

HYDRONICS Practice of heating or cooling with water.

HYDROPHILIC Substance that absorbs or has exhibited affinity for water.

HYDROPHOBIC CEMENT Unhydrated cement treated so as to have a reduced tendency to take up moisture.

HYDROPHYTE Plant typically found in wet areas or in water where oxygen deficiencies occur periodically.

HYDROPLANING Condition where one or more tires of a moving vehicle are separated from the pavement by a film of water, usually due to a combination of depth of water, pavement surface texture, vehicle speed, tread pattern, tire condition, and other factors.

HYDROPNEUMATIC ACCUMULATOR Accumulator in which compressed gas applies pressure to a stored liquid in the same vessel. *See also* **Accumulator.**

HYDROPNEUMATICS Pertaining to the combination of hydraulic and pneumatic fluid power.

HYDROSEEDING Process of spraying a combination of fertilizer, grass seed, water, and fibrous binder onto prepared ground in the form of a slurry; often used on steep slopes and hard-to-reach places.

HYDROSTATIC Pressure or equilibrium of fluids.

HYDROSTATIC DRIVE Drive system that is completely hydraulic with no mechanical connection; allows an easily variable speed control.

HYDROSTATIC LUBRICATION *See* **Lubrication.**

HYDROSTATIC PORE PRESSURES Pore water pressures or groundwater pressures exerted under conditions of no flow where the magnitude of pore pressures increases linearly with depth below the ground surface.

HYDROSTATIC PRESSURE *See* **Pressure.**

HYDROSTATICS Branch of physics devoted to the study of confined liquids and the resultant transmission of force through exerted pressure. Energy is transferred by the fluid itself in a closed circuit between a pump and an actuator. When the fluid moves through hydraulic lines, the system is still considered to be hydrostatic because work is performed as a result of the pressure differential on each side of the actuator, not as a result of the kinetic energy or momentum of the fluid.

HYDROSTATIC TRANSMISSION *See* **Transmission.**

HYDROTREATING *See* **Hydrocracking.**

hyg. *Abbreviation for:* hygiene.

HYGROMETER Instrument that measures the degree of moisture suspended in the air.

HYGROSCOPIC Water absorbed from the atmosphere.

HYGROSCOPIC COEFFICIENT Percent moisture that a dry material will absorb in saturated air at a given temperature.

HYGROSTAT Automatic control responsive to humidity.

hyp. *Abbreviation for:* hypotenuse; hypothesis.

hyperb. *Abbreviation for:* hyperbola.

hyperbol. *Abbreviation for:* hyperbolic.

HYPERBOLIC PARABOLOID ROOF Roof in the shape of a double-curved shell, the geometry of which is generated by straight lines. The shape consists of a continuous plane developing from a parabolic arch in one direction to a similar inverted parabola in another.

HYPOID GEAR 1. Pinion-and-ring gear set transmitting rotation through a right angle; **2.** Design that puts the pinion below-center of the ring gear.

HYPOTHECATE Obligation, right, or security given by contract or under law to a creditor over property of the debtor without transfer of possession or title to the creditor.

HYPSOMETER Simple instrument, often a stick or other straightedge, used to measure the heights of trees on the basis of similar angles.

HYSTERESIS Loss of energy due to successive deformation and relaxation. It is measured by the area between the deformation and relaxation stress-strain curves.

HYSTERESIS LOOP Plot of force versus displacement of a structure or member subjected to reversed, repeated loads into the inelastic range, in which the path followed during release and removal of load is different from the path for the addition of load over the same range of displacement.

Hz *Abbreviation for:* hertz.

I

i *Abbreviation for:* inclination; intensity; moment of intertia.

I *Abbreviation for:* current; I-beam.

I&M *Abbreviation for:* installation and maintenance.

I-BEAM Structural beam with a cross section similar to the letter I, consisting of an upper and lower flange separated by a web.

i-c *Abbreviation for:* interconnect.

ICE APRON Ramp upstream of a bridge pier, sloping up from below minimum water level, used to lift floating ice, forcing it to break.

ICE LUG Piece of metal welded onto a grouser to provide better traction on ice and frozen ground.

ID *Abbreviation for:* induced draft; inside diameter; inside dimension; internal diameter; item description.

ID card *Abbreviation for:* identification card.

IDEAL CAPACITY *See* Capacity.

ident. *Abbreviation for:* identical; identification; identify.

IDENTIFICATION INDEX Series of numbers and/or letters, used to indicate a sequence of grades, blends, etc.

IDENTIFICATION LIGHT One of a cluster of three lights spaced 150 to 300 mm (6 to 12 in.) apart and mounted in a horizontal row near the top center of the front (amber lights) or rear (red lights) of a vehicle 2.032 m (80 in.) or more in width.

IDENTIFIED CONDUCTOR OR TERMINAL Identified, as used in reference to a conductor or its terminal, means that such conductor or terminal can be recognized as grounded.

IDENTIFIED FOR USE Recognized as suitable for the specific purpose, function, use, environment, application, etc., where described as a requirement.

IDLE CAPACITY *See* **Capacity.**

IDLER 1. Wheel that changes the direction of rotation of belt-driven shafts; **2.** Sheave or roller used to guide or support a rope.

IDLER GEAR *See* **Gear.**

IDLER ROLLER Roller of a track mechanism that is not power driven but is are used to maintain proper tension on the track.

IDLER SHAFT Shaft that carries a gear that reverses the direction of rotation.

IDLER SHEAVE Sheave in a running line tensiometer or other assembly that acts simply as a guide for the running line.

IDLE TIME *See* **Machine time.**

i.e. *Abbreviation for: id est* (that is).

IE *Abbreviation for:* industrial engineer; industrial exhaust; invert elevation.

IF *Abbreviation for:* inside face; internal friction.

IFB *Abbreviation for:* invitation for bid.

IG *Abbreviation for:* insulating glass.

ign *Abbreviation for:* ignition.

IGNEOUS Classification of rock that is the result of cooling of hot molten materials from beneath the surface of the earth.

IGNITABLE Mixture within lower and upper limits of the flammable range that is capable of propagation of flame away from the source of ignition when lighted.

IGNITION 1. Catching on fire of something combustible; **2.** Process of igniting the air/fuel mixture within the cylinder of an engine so that it will burn.

IGNITION COMPONENT Number related to the probability that a spreading fire will result if a firebrand encounters fine fuel.

IGNITION DELAY Period between injection of a fuel and air mixture into an engine cylinder and the beginning of ignition.

IGNITION KEY Key used to operate a lock that, in its closed position, completes the ignition circuit of an engine. *See also* **Ignition switch.**

IGNITION LAG Time between a fuel being ignited and is starting to burn.

IGNITION LOSS *See* **Loss of ignition.**

IGNITION POWDER Mixture, usually powdered aluminum and an oxidant, used to start the reaction in fusion welding.

IGNITION SWITCH *See* **Switch.**

IGNITION TEMPERATURE Lowest temperature at which a fuel ignites and flame is self-propagating.

ihp *Abbreviation for:* indicated horsepower.

I/I *See* **Excessive I/I; I/I analysis;** and **Infiltration/inflow.**

I/I ANALYSIS Analysis demonstrating possible excessive or nonexcessive infiltration and/or inflow to a sewer system. *See also* **Excessive I/I.**

ild *Abbreviation for:* inside length dimension.

illum. *Abbreviation for:* illumination.

illum. engr *Abbreviation for:* illumination engineer.

ILLUMINATION Amount of light bearing upon a surface.

illus. *Abbreviation for:* illustration.

imag. *Abbreviation for:* imaginary.

IMBRECATE 1. Cover with gutter tiles; **2.** Overlap

and break joint of roofing tiles.

IMHOFF TANK Two-storied, anaerobic, sewage treatment tank comprising an upper, continuous sedimentation chamber and a lower, sludge-digestion chamber.

imit. *Abbreviation for:* imitation.

immed. *Abbreviation for:* immediate.

IMMEDIATE SET Amount of deformation measured immediately after removal of the load causing the deformation.

IMMEDIATE SETTLEMENT Settlement of a foundation that takes place as the initial construction load is applied.

immer. *Abbreviation for:* immerse; immersed; immersion.

IMMERSE To dip into or cover with a fluid.

IMMERSION CLEANING Cleaning a workpiece by immersion in a liquid solution.

IMMERSION-COMPRESSION TEST Means of determining the resistance of an aggregate to stripping.

IMMERSION HEATER Electric resistance heater that is submerged in water.

imp. *Abbreviation for:* impact; imperial; import.

IMPACT-ALLOWANCE LOAD Percentage allowance for impact applied to the equivalent of the uniform live load.

IMPACT ATOMIZATION Type of fuel atomization used in a diesel hammer in which the raw fuel injected and trapped between the ram and the anvil is atomized by impact of the ram on the impact block. The hammer normally will have a time delay between impact and combustion. *See also* **Air atomization, Atomization,** and **Injection atomization.**

IMPACT BLOCK Normally used with a diesel hammer. *See also* **Anvil.**

IMPACT BLOW Transmission of energy during a short interval of time from one moving body to another.

IMPACT ENERGY 1. Amount of energy that must be exerted to fracture a part; **2.** Kinetic energy delivered by a pile hammer to the drive cap.

IMPACT FACTOR Factor used in the design of bridges and structural foundations equal to unity when a supported structure is static; for moving loads the factor depends upon the speed of a vehicle and its vibratory action.

IMPACT HAMMER *See* **Rebound hammer.**

IMPACT INSULATION CLASS Values that rate the capacity of floor assemblies to control impact noise, such as footfalls.

IMPACT LOAD Load carried by a member, panel, or sub-assembly when a load is dropped on it from a height.

IMPACT MILL Machine that grinds material by throwing it against heavy metal projections rigidly attached to a rapidly rotating shaft.

IMPACT MOLING Technique for conduit installation using a percussive soil displacement tool to form a bore.

IMPACT NOISE RATING Values for floor assembly impact sound transmission.

IMPACT PLY *See* **Breaker strip/ply.**

IMPACT RESISTANCE Ability of a material to resist cracking or breaking when subjected to a blow or knock of predetermined force.

IMPACT SPANNER *See* **Impact wrench.**

IMPACT STRENGTH Material's ability to resist shock.

IMPACT TEST Test in which one or more blows of specific force are suddenly applied to a specimen.

IMPACT TRANSMISSION Sound or vibration resulting from a blow transferred through a structure or assembly.

IMPACT VELOCITY Velocity of the ram of a gravity or single-acting steam/air hammer when it strikes the anvil.

IMPACT WRENCH Air-driven wrench used for tightening bolts. *Also called* **Impact spanner.**

IMPASTO PAINTING Application of paint with brush or knife to create thick textural effects.

IMPEDANCE Total resistance to flow of alternating current as a result of resistance.

IMPELLER Component of a centrifugal pump that uses centrifugal force to discharge a fluid into the outlet passages.

IMPENDING SLOUGH Consistency of a shotcrete mixture containing the maximum amount of water such that the product will not flow or sag after placement.

imperf. *Abbreviation for:* imperfect.

IMPERIAL GALLON Nonmetric measure of a fluid gallon equal to approximately 1.2 U.S. gallons. *Also called* **British imperial gallon.** *See also* **Gallon (imperial).**

imperm. *Abbreviation for:* impermeability; impermeablie.

IMPERMEABLE Material that will not permit the passage of water.

IMPERMEABILITY FACTOR Ratio of the amount of rain that runs off a surface to that which falls on it.

IMPERVIOUS 1. Incapable of being passed through or penetrated; **2.** Soil in which the spacing of the particles is such as to permit only extremely slow passage of water.

IMPERVIOUS CORE *See* **Impervious zone.**

IMPERVIOUS TILE Tile having water absorption of 0.5% or less.

IMPERVIOUS ZONE Clay or silt zone of an earth- or rock-filled dam that provides a water barrier. *Also called* **Impervious core.**

impf. *Abbreviation for:* imperfect.

imp. gal. *Abbreviation for:* imperial gallon.

IMPINGEMENT 1. Direct impact of fluid flow upon

or against a surface; **2.** Direct high-velocity impact of a fluid flow upon, or against any internal portion of a filter.

impl. *Abbreviation for:* implement.

IMPLIED WARRANTY *See* **Warranty.**

impos. *Abbreviation for:* impossible.

IMPOST Upper course of a pillar, or a member in a wall, usually in the shape of a molded bracket, on which the end of an arch rests. *See also* **Arch.**

IMPOUNDING RESERVOIR Reservoir to which surplus water is directed, and from which water is drawn through the same system.

impr. *Abbreviation for:* impractical; improvements.

impreg. *Abbreviation for:* impregnate.

IMPREGNATION Act of filling the interstices of an article with a compound, typically the treatment of textile fabrics and cords or a fibrous substrate.

IMPRESSED CURRENT Current of electricity introduced for cathodic protection.

IMPRESSION Design formed during vulcanization in the surface of a hose or tire by a method of transfer, such as fabric impression or molded impression. *See also* **Fabric impression.**

improv. *Abbreviation for:* improve; improvement.

IMPROVED LAND Land that has been partially or fully developed for a higher and better use.

IMPROVED SUBGRADE Subgrade, improved as a working platform (a) by the incorporation of granular materials or stabilizers such as asphalt cement, emulsion or cutback, lime, or portland cement, prepared to support a structure or a pavement system, or (b), any course or courses of select or improved material placed on the subgrade soil below the pavement structure. *See also* **Subgrade.**

IMPROVEMENT Addition to land or to a structure that increase its present or future value.

IMPROVEMENT CUTTING Removal of trees of undesirable species, form, or condition from the main canopy of a stand to improve the health, composition, and value of the stand.

IMPROVEMENT DISTRICT Area within defined boundaries formed by property owners for the purpose of undertaking improvements of communal benefit: paving, storm drainage, water supply, etc.

IMPROVEMENT LINE Building line; the line of an improved road.

IMPULSE Application of force in a manner to produce sudden strain or motion.

IMPULSE LINE Small-diameter pipe or tube used to convey pressure from a piping system to a diaphragm- or bellows-operated mechanism.

IMPULSE TURBINE Steam or water turbine driven by the speed of a fluid, rather than by its change in pressure.

IMPURITY Undesirable compound or element in a material.

in. *Abbreviation for:* inch; inside.

IN-ANTIS *See* **Portico.**

INBAND Header stone visible in a reveal.

INBARK Bark embedded or ingrown in the wood of a tree.

inc. *Abbreviation for:* income; incoming; incorporate; incorporated; increase.

incand. *Abbreviation for:* incandescent.

INCANDESCENT FLUORESCENT *See* **Fluorescent lamp.**

INCANDESCENT LIGHT Light source consisting of a metal filament suspended in a vacuum within a glass envelope, that emits light when excited by an electrical current.

INCENDIARY FIRE Fire willfully set to burn vegetation or property not owned or controlled by the instigator and without consent of the owner or his agent.

INCH Nonmetric unit of length, equal to 1/12 foot. Symbol: in. Multiply by 2.54 to obtain centimeters, symbol: cm; by 0.0254 to obtain meters, symbol: m; by 25.4 to obtain millimeters, symbol: mm. *See also the appendix:* **Metric and nonmetric measurement.**

INCIDENT RADIATION Energy arriving at the surface of a solar collector, including both direct and diffuse radiation.

INCINERATOR 1. Furnace capable of consuming all of the volatiles of a range of materials; **2.** Engineered apparatus used to burn waste substances and in which all the factors of combustion (temperature, retention time, turbulence, and combustion air) can be controlled. There are several types, including:

> **Batch fed:** Incinerator that is periodically charged with solid waste; one charge is allowed to burn down or burn out before another is added.
>
> **Cell-type:** Incinerator, usually batch fed, with grate areas divided into cells, each of which has its own underfire air control and ash dumping grate.
>
> **Central:** A conveniently located facility that burns solid waste collected from many different sources.
>
> **Chute-fed:** Incinerator charged through a chute that extends on one or more floors above the furnace.
>
> **Continuous-feed:** Incinerator into which solid waste is charged almost continuously to maintain a steady rate of burning.
>
> **Controlled-air:** Incinerator with two or more combustion areas in which the amounts and distribution of air are controlled. Partial combustion takes place in the first zone, and hydrocarbon gases are burned in a subsequent zone(s).
>
> **Direct-feed:** Incinerator charged through a chute that also functions as a flue to exhaust the products of combustion.
>
> **Industrial:** Incinerator specifically designed to burn a particular industrial waste.
>
> **Multiple-chamber:** Incinerator consisting of two or more chambers, arranged as in-line or with retorts, interconnected by gas passage ports or

ducts.

Multiple hearth: Incinerator consisting of a series of circular hearths stacked vertically. The combustible material is fed to the uppermost hearth and moved around it by a rotating scrabble arm until it reaches a port or opening where it discharged to the hearth below, where the process is repeated. The upper hearths progressively dry the charge, allowing the lower hearths to more fully oxidize the material before it is discharged as a mixture of ash and clinker.

On-site: Incinerator that burns waste on the property of the generator of the waste.

Open-pit: Burning device that has an open top and a system of closely spaced nozzles that place a stream of high-velocity air over the burning zone.

Residential: Predesigned, shop fabricated and assembled unit, shipped as a package for use in individual dwellings.

Single-chamber: Refractory-lined cylindrical furnace charged through a door in the upper part of the chamber. Refuse is batch fed periodically.

INCIPIENT EROSION Early stages of erosion, marked by such developments as gullying.

INCISE To cut inwardly, or engrave.

incl. *Abbreviation for:* inclination; include; including; inclusion; inclusive.

INCLINE Flat surface, one end of which is at a lower elevation that the other end.

INCLINED-AXIS MIXER Truck with a revolving drum that rotates about an axis inclined to the bed of the truck chassis. *Also called* **High-discharge mixer.**

INCLINED CABLEWAY Monocable cableway where the track cable is sloped along its entire length at an angle steep enough to cause the carrier to descend under its own weight.

INCLINED DRILLING Drilling so that the holes are perpendicular to the face angle of a quarry.

INCLINED GAUGE Sloping staff, graduated to read vertical distances above a certain datum.

INCLINED LIFT Powered, one-person passenger lift installed on a stairway.

INCLINED PLANE 1. Surface inclined to the plane of the horizon; **2.** Slope used to change the direction and the speed/power ratio of a force.

INCLINOMETER Instrument that measures a deviation from vertical.

INCLUDING When part of an agreement or the contract documents, the word 'including' is interpreted to read 'including without limitation.'

INCLUSION Foreign matter introduced to castings or welds.

incom. *Abbreviation for:* incomplete.

incomb. *Abbreviation for:* incombustible.

INCOMBUSTIBLE CONSTRUCTION Construction in which all structural elements are of incombustible materials having a fire-resistance of one hour or less.

INCOMBUSTIBLE MATERIAL Material that will not ignite or actively support combustion in a surrounding temperature of 648°C (1,200°F) for five minutes, and will not melt in a surrounding temperature of 482°C (900°F) for five minutes.

INCOMPATIBILITY 1. Successive paint coats of radically different composition causing premature failure of the final coat; **2.** Situation when two lubricating greases show incompatibility when a mixture of the products shows physical properties and/or service performance that is markedly inferior to that of either of the greases before mixing.

INCOMPATIBLE FLUIDS Fluids that when mixed in a system, will have a deleterious effect on that system, its components, or its operation.

INCOMPETENT GROUND Ground that must be supported when a tunnel is excavated through it.

INCOMPLETE COMBUSTION Combustion in which fuel is only partially burned and is capable of being further burned under proper conditions.

INCOMPLETE FUSION Less than complete fusion of weld material with base metal or with the preceding bead.

INCOMPLETE PENETRATION Failure of two bead welds to fuse through incomplete root penetration.

INCOMPLETE VEHICLE Assemblage consisting, as a minimum, of frame and chassis structure, powertrain, steering system, suspension system, and braking system, to the extent that those systems are to be part of the completed vehicle that requires further manufacturing operations other than the addition of readily attachable components.

INCOMPLETE VEHICLE MANUFACTURER Manufacturer that produces an incomplete vehicle (chassis) by assembling components, none of which, taken separately, constitutes an incomplete vehicle.

incr. *Abbreviation for:* increase; increasing; increment; incremental.

INCREASER Short pipe fitting with one end of a larger diameter.

INCREMENT *See* **Growth.**

INCREMENTAL ANALYSIS *See* **Analysis.**

INCREMENTAL ENCODER Angle-reading instrument that counts from a known location on the circle to determine the angle a surveying or other instrument has been rotated through.

INCREMENTAL LOADING Application of a load in equal units of weight, in a programmed manner.

INCREMENTAL SILVICULTURE *See* **Intensive silviculture.**

INCREMENTATION Steps of applying a test load to a pile. Generally the specifications will recite the steps as being 25%, 50%, 75%, 100%, 125%, 150%, 175% and 200% of design load, or some other increment dependent on design considerations.

INCREMENT BORER Tool used to extract a core of wood from a living tree for the purpose of studying the annual growth rings of the tree.

INCREMENT CORE That part of the cross section of

a tree extracted by an increment borer, used to determine tree age and growth pattern.

INCRUSTATION or ENCRUSTATION Crust or coating, generally hard, formed on the surface of concrete or masonry construction or on aggregate particles.

INCUMBRANCE *See* **Encumbrance.**

ind. *Abbreviation for:* independent; index; indicate; indicator; induction; industrial; industry.

indef. *Abbreviation for:* indefinite; indefinitely.

INDEMNIFICATION Protection (implied) or insurance (contractual) against loss or damage from specified liabilities.

INDEMNITY AGREEMENT Agreement given by a principal to the person or organization providing surety guaranteeing that no loss will result from the provision of a bond.

INDENT Tooth-like notches.

INDENTATION 1. Extent of deformation by the indentor point of any one of a number of standard hardness testing instruments; **2.** Recess in the surface of a hose.

INDENTED BAR Steel reinforcing bar having its surface deformed by a series of regular indentations to increase the mechanical bond with the surrounding concrete.

INDENTED BOLT Anchor bolt, cast in concrete, having a lower plain shank with indentations.

INDENTED STRAND Reinforcing strand having machine-made surface indentations to improve bond.

INDENTED WIRE Reinforcing wire having machine-made surface indentations intended to improve bond, used for either concrete reinforcement or pretensioning, depending on type. *See also* **Crimped wire.**

INDENTURE Contractual agreement between the issuer of a bond and the bondholder.

indep. *Abbreviation for:* independent.

INDEPENDENT One who logs and sells his output on the open market; not associated with a mill or under company or dealer contract.

INDEPENDENT CENTRAL STATION Continually supervised station under the control of a company independent of the owners of the building to be protected.

INDEPENDENT CONTRACTORS' INSURANCE *See* **Insurance.**

INDEPENDENT POLE SCAFFOLD *See* **Double pole scaffold.**

INDEPENDENT WIRE ROPE CORE A wire rope used as the axial member of a larger wire rope. *Also called* **IWRC**

indet. *Abbreviation for:* indeterminate.

INDEX 1. Pointer or indication, as the needle on a dial; **2.** Relation or proportion of one amount or dimension to another; **3.** Reference, of one object or article to another; a drill index is a holder for different diameters of twist drills, in order of increasing dimension, with each socket marked as to its diameter.

INDEX OF REFRACTION Ratio of light velocity in a vacuum to light velocity in a transparent medium; ratio of the sine of the angle of light refraction measured from perpendicular to the transparent medium's surface.

INDICATED HORSEPOWER *See* **Horsepower.**

INDICATED LOAD Value of a load that is displayed or otherwise shown on an indicator, meter or readout device.

INDICATING BOLT Door bolt that indicates on the exterior face of the door whether the lock is open or closed, typically used on toilet stalls.

INDICATOR Device or instrument that provides a visual indication or display of a measured parameter. There are many types, including:

> **Bypass:** Indicator that signals that an alternate flow path is being used.
>
> **Differential pressure: 1.** Indicator that signals the difference in pressure between two points; **2.** Device that indicates continuously during operation the differential pressure across a filter element.
>
> **Display:** Aid or instrument that provides a visual indication of a measured parameter.
>
> **Meter:** Type of indicator display or readout. It may be mechanical, hydraulic, electrical, or electronic.

INDICATOR PILE *See* **Pile.**

INDICATOR VALVE *See* **Valve.**

indig. *Abbreviation for:* indigenous.

INDIGENOUS SPECIES Plant or animal that occurs naturally in, or is native to a region; reciprocal of **Exotic species.**

INDIRECT CHARGES Costs that cannot be identified readily with a product, service or activity, the distribution of which must be made by proration. *See also* **Charges,** and **Direct charges.**

INDIRECT COMBUSTION Combustion that occurs in a small prechamber ahead of the combustion chamber.

INDIRECT DRAIN *See* **Indirect waste piping.**

INDIRECT DRIVE *See* **Driving machine.**

INDIRECT EXPENSE Overhead expense, not chargeable to a specific project or task.

INDIRECT HEATING System of heating by convection; heating of an area by heat generated outside of the specific area and brought to it.

INDIRECT LIGHTING Light that arrives at its intended place having been reflected off of one or more surfaces.

INDIRECT MATERIAL *See* **Material.**

INDIRECT METHOD Method of forest fire suppression in which the control line is located along natural fire breaks, along favorable topography, or at considerable distance from the fire and the intervening fuel, is burned out.

INDIRECT SERVICE WATER HEATER Service water heater that derives its heat from a heating medium such as warm air, steam, or hot water.

INDIRECT SOLAR Passive solar heating system in which the heat storage unit, a Trombe wall or liquid storage vessel, is physically located between the collector and distributor.

INDIRECT SOLAR GAIN Warming of an area by solar heat gained in an adjacent area: an adjacent sunspace, for instance, or via a Trombe wall.

INDIRECT WASTE PIPING Sanitary piping that does not connect directly with the drainage system but discharges liquid wastes into some other fixture or receptacle connected to the drainage system. *Also called* **Indirect drain.**

indiv. *Abbreviation for:* individual.

INDIVIDUAL ASSIGNMENT METHOD System for organizing workers to control a forest fire's perimeter in which each worker is assigned a specific section of the control line where he is responsible for all suppression jobs from hot-spotting to mop-up.

INDIVIDUAL FOOTING Footing that supports a single column.

INDIVIDUAL HIGH CHAIR Welded wire bar support used under a support bar to provide support for top bars in slabs, joists, or beams; also used to support upper mats of bars in slabs without support bars. *Also called* **High chair.** *See also* **Continuous high chair.**

INDIVIDUAL-RUNG STEP LADDER Ladder without a side rail or center rail support, made by mounting individual steps or rungs directly to the side or wall of the structure.

INDIVIDUAL STORAGE AREA Area occupied by piles, bin boxes, racks, or shelves, including subsidiary aisles providing access to the stored products, which is separated from adjacent storage by aisles not less than 2.4 m (7.9 ft) in width.

indl engr *Abbreviation for:* industrial engineer.

INDRAFT Movement of air into a fireplace or furnace.

INDUCED AIR 1. Air that flows into a furnace through openings because the furnace pressure is less than atmospheric pressure; **2.** Air brought into a furnace by entrainment in a high velocity stream.

INDUCED CURRENT Secondary current, set in motion when a second conductor in the shape of a closed loop is placed in the magnetic field around another current-carrying conductor.

INDUCED DRAFT *See* **Draft.**

INDUCED-DRAFT FAN *See* **Fan.**

INDUCED VOLTAGE Voltage that is produced in a conductor that has motion relative to a magnetic field, while under the influence of that field.

INDUCTANCE Property of an electric circuit that opposes any change in current flow.

INDUCTION FURNACE *See* **Furnace.**

INDUCTION GENERATOR Induction machine, when driven above synchronous speed by an external force of mechanical power, used to convert mechanical power to electric power.

INDUCTION HARDENING Process of heating a ferrous metal using electromagnetic induction, then cooling the metal in a controlled way to improve its surface characteristics.

INDUCTION PERIOD Period of time during which oxidation of a grease occurs at a relatively low rate.

INDUCTIVE SPEED SENSOR Magnet, a component of an antilock braking system (ABS), with a round pole pin and wire coil that monitors impulses from a tooth wheel and transmits data to the electronic control unit.

INDUSTRIAL INCINERATOR *See* **Incinerator.**

INDUSTRIALIZED BUILDING Integration of planning, design, programming, manufacturing, site operations, scheduling, financing, and management into a disciplined method of mechanized production of buildings. *Also called* **Systems building.**

INDUSTRIAL LIFT TRUCK *See* **Powered industrial truck.**

INDUSTRIAL OCCUPANCY Occupancy or use of a building for assembling, fabricating, manufacturing, processing, repairing, or storing of goods and materials. There are several subdefinitions, including:

> **High hazard:** Industrial occupancy containing sufficient quantities of highly combustible and flammable or explosive materials that, because of their inherent characteristics, constitute a special fire hazard.

> **Medium hazard:** Industrial occupancy in which the combustible content is more than 4.5 kg (10 lb) or 105 500 kJ (100,000 Btu) per sq ft of floor area and not classified as high hazard industrial occupancy.

> **Low hazard:** Industrial occupancy in which the combustible content is not more than 4.5 kg (10 lb) or 105 500 kJ (100,000 Btu) per sq ft of floor area.

INDUSTRIAL PARK Area zoned and dedicated for manufacturing and associated activities.

INDUSTRIAL SILENCER Exhaust muffler used to produce the silencing level normally associated within industrial areas.

INDUSTRIAL TOW TRACTOR (TRUCK) Powered industrial truck designed to tow one or more non-powered trucks, trailers, or other mobile loads on improved surfaces.

INDUSTRIAL TRUCK Wheeled vehicle, primarily intended for the movement of objects or materials and usually associated with manufacturing, processing, or warehousing, but not including vehicles intended primarily for earth-moving or over-the-road hauling.

INDUSTRIAL WASTE *See* **Solid waste.**

INDUSTRIAL WOOD All roundwood products except fuelwood.

INELASTIC ACTION Material deformation that does not disappear on removal of the force that produced it.

INELASTIC BEHAVIOR Deformation that does not disappear on removal of the force that produced it. *Also called* **Creep.**

INELASTIC DEFORMATION *See* **Deformation.**

ineq. *Abbreviation for:* inequality.

INERT Having inactive chemical properties.

INERT ATMOSPHERE Gas incapable of supporting combustion.

INERTIA Property of matter by which it will remain at rest or in uniform motion in a straight line unless acted upon by an external force.

INERTIAL SEPARATOR *See* **Separator.**

inf. *Abbreviation for:* infinite; infinity.

INFECTIOUS WASTE *See* **Solid waste.**

INFEED THROAT The somewhat funnel-shaped portion of the infeed of a portable chipper that causes the tree to move forward to the feed and downward to the anvil.

INFERIOR DIAMETER Diameter of the lower end of a column. *See also* **Diameter,** and **Superior diameter.**

INFILTRATION 1. Flow of water from the land surface into the subsurface; **2.** Water entering a sewer system and service connections from the ground through such means as defective pipes, pipe joints, connections, or manhole walls; **3.** Uncontrolled admittance of air through cracks and pores into a building.

INFILTRATION AIR Air that leaks into the chambers or ducts of a combustion system.

INFILTRATION/INFLOW Total quantity of water entering a sewer system from both infiltration and inflow, without distinguishing the source. *Also called* **I/I.**

infl. *Abbreviation for:* inflammable; inflection; inflow.

INFLATABLE TUBE Removable pneumatic rubber tube specially designed to form a duct in concrete as it is placed.

INFLECTURE Reversal of direction of a curve.

INFLOW Water discharged into a sewer system

INFLUENCE LINE Line drawn for any point, or section, in a structural system to show the effect at that location of a unit load placed successively at all other points in the system.

INFLUENT Fluid entering a component or system.

INFLUENT STREAM Stream or portion of a stream that contributes water to the groundwater supply.

info. *Abbreviation for:* information.

INFORMATION CENTER Building or portion of a building in a roadside area under the control of the highway agency, primarily to furnish travel, other information, and services to motorists.

INFORMATION STAKE Explains in surveyor's code what grades are to be established and the distances to them.

INFRARED Invisible rays just beyond the red of the visible spectrum; waves longer than those of the spectrum colors but shorter than radio waves, having a penetrating heating effect.

INFRARED CURING Use of infrared heat to cure freshly placed concrete.

INFRARED DETECTOR Sensor capable of sensing an infrared spectrum of light.

INFRARED SPECTROSCOPY Use of a spectrophotometer for determination of infrared absorption spectra of materials.

INFRASTRUCTURE Public and private services such as water, telephone, electricity, cablevision, gas, and sewage disposal.

INGLAZE DECORATION *See* **Decoration.**

INGLENOOK Bench or seat built into a fireplace or fire surround.

INGOT Metal block, other than lead or iron, cast at a smelting furnace. *See also* **Pig.**

INGRESS Entrance.

INHAUL Loaded segment of a yarding cycle: yarding the logs to a landing.

INHAUL CABLE Line device by which a cable excavator bucket is pulled toward the dumping point.

INHAUL CYCLE Ahead on the mainline; mainline coming in to the landing; bringing in a turn of logs.

INHAUL LINE *See* **Drag rope.**

INHERENT ASH Portion of the ash or other material found after combustion that is chemically bound to the molecules of the combustible, as distinguished from extraneous noncombustible materials, that comes from other sources or that may be mechanically entrained with the combustible.

INHERENT COLOR VARIATION Differences in color of the surface of hardened concrete due to (a) fine aggregates having come from different sources, (b) cement being supplied by different manufacturers, (c) batch variations, and/or (d) variations in mixing procedures or duration

INHERENT VOLTAGE DROOP Decrease in voltage produced by a gen-set from no load to full load with excitation fixed at 100% volts no load.

INHERENT VOLTAGE REGULATION Inherent voltage droop expressed as a percentage of no-load voltage.

inhib. *Abbreviation for:* inhibit; inhibiting.

INHIBITOR Substance that slows or prevents a chemical or organic reaction.

init *Abbreviation for:* initial; initiate.

INITIAL ATTACK First point of attack, and the hose lines employed to prevent further extension of a forest fire; safeguarding lives while additional lines are being laid and placed in position.

INITIAL CHARGE Charge given to a new storage battery before putting it into service.

INITIAL DRYING SHRINKAGE Difference between the length of a concrete specimen (molded and cured under stated conditions) when first dried and a constant length, expressed as a percentage of the moist length.

INITIAL PRESTRESS Prestressing force applied to concrete at the time of stressing.

INITIAL RATE OF ABSORPTION Weight of water absorbed, expressed in grams per 195 cm² (30 in.²) of contact surface, when a brick is partially immersed for one minute.

INITIAL SET Degree of stiffening of a mixture of cement and water, less than final set, generally stated as an empirical value indicating the time in hours and minutes required for cement paste to stiffen sufficiently to resist to an established degree the penetration of a weighted test needle.

INITIAL SETTING TIME Time required for a freshly mixed cement paste, mortar, or concrete to achieve initial set. *Also called* **Setting time.** *See also* **Final setting time.**

INITIAL STRESS Stress occurring in a prestressed concrete member before any losses occur.

INITIATING FIRE Forest fire exhibiting reasonably predictable behavior (no crowning or spotting).

INITIATION Act of detonating a high explosive by mechanical or other means.

INITIATOR Device or product used to transmit and/ or supply heat and/or shock to start an explosion.

INJECTION ATOMIZATION Type of fuel atomization used in a diesel hammer in which the fuel is atomized by an injector. Combustion will normally occur prior to impact. *See also* **Air atomization, Atomization,** and **Impact atomization.**

INJECTION LAG Time between activation of an injector and the presence in the passage or cavity of the substance being injected, created by the compressibility of the substance and elasticity of the line.

INJECTION PUMP *See* **Pump.**

INJECTOR Device that forces a substance into a passage or cavity.

INJECTOR NOZZLE That portion of an engine fuel injector that sprays the fuel directly into the combustion chamber or prechamber.

INJECTOR PLUNGER Rod on which a helix is cut, and that moves to force fuel downstream, under pressure, to the injector and its nozzle.

INJECTOR RACK Geared rod that rotates the plunger and helix within the injector barrel to vary the amount of fuel delivered to an engine's combustion chamber.

INJECTOR STROKE Effective distance the plunger moves between fuel delivery and fuel shut off. Injector stroke (variable) and injector bore (nonvariable) are what determine the total amount of fuel any one injector can deliver during any instance of delivery.

INJUNCTION Court issued writ or order that restrains one or more parties from some described act.

inl. *Abbreviation for:* inlet.

INLAND MARINE INSURANCE *See* **Insurance.**

INLAY Material placed into an area excised from the surface of another.

INLET 1. Port through which matter or material will pass in one direction; **2.** Connection to a closed drain.

INLET PORT 1. Opening in the cylinder of a two-cycle engine through which air enters; **2.** Inlet fluid path of fluid-powered hammers.

INLET PRESSURE Pressure of motive fluid at the inlet of an air/steam hammer.

INLET VALVE Valve that permits air, or fuel, to enter the cylinder of an engine.

INLET WELL Sump or surface opening to which water is conducted and that connects to a sewer.

INLINE ENGINE Where the engine cylinders are arranged in a straight line, one behind the other.

INNER CASING Metal partition in a heat exchanger.

INNER COURT Unroofed, internal area enclosed on all sides by walls. These walls may be specific to one building, or may be the external walls of several buildings and/or walls built on lot lines.

INNER HEARTH Flat bottom part of the firebox on which the fire is built.

INNER LINER *See* **Tire construction.**

INNER MIDSECTION Segment of a four-section telescoping boom that is attached to the base and outer midsections.

INNER PLY Any ply of a plywood panel except the face or the back.

INNER WIRE Any wire of a strand except the outer or cover wires.

inorg. *Abbreviation for:* inorganic.

INORGANIC COATING Coating based on silicates or phosphates and usually used pigmented with metallic zinc.

INORGANIC THICKENER *See* **Nonsoap thickener.**

IN-PLACE RECYCLING Rejuvenation and reuse of reclaimed asphalt pavement (RAP) on-site by planing, heating, scarifying, remixing, or upgrading by adding new aggregate, asphalt, or additives.

INPUT Driving force (current or voltage) applied to a circuit or device.

INPUT CLUTCH Friction clutch between a power source and torque converter.

INPUT/OUTPUT ANALYSIS *See* **Analysis.**

INPUT RATE Quantity of heat, fuel, or air supplied per unit, measured in volume, weight, or heat units.

INPUT SHAFT Shaft that delivers engine power to a transmission or clutch.

INPUT SYSTEM Ventilation method that relies on air drawn from the atmosphere and pumped into a vented space.

inq. *Abbreviation for:* inquiry.

IN REM Action taken directly against real property.

INRUSH CURRENT Maximum electrical current that flows after a machine or apparatus is suddenly and fully energized.

ins. *Abbreviation for:* inside; insulate; insulation; insurance; insure.

INSERT 1. Device buried in concrete to receive a bolt or screw to support a shelf angle, machinery, etc.; **2.** *See* **Adapter, Repairs,** and **Shim.**

INSERTED NUT Disk, segment, or cylinder wheel

having nuts embedded in a back surface for mounting on the machine.

INSERTED TOOTH 1. Replaceable tooth not formed from the saw body; **2.** Part of a replaceable tooth system fitted to the cutting edge of the bucket of an excavator, hoe, or similar equipment.

IN-SHIFT MOVING TIME *See* **Machine time.**

IN-SHIFT REPAIR TIME *See* **Machine time.**

IN-SHIFT SERVICE TIME *See* **Machine time.**

INSIDE-ANGLE TOOL Right-angle-shaped float, used for finishing inside angles.

INSIDE CALIPER Measuring instrument used to gauge an inside diameter or other dimension.

INSIDE CORNER BRACE *See* **Braces and plates.**

INSIDE GLAZING Glass placed in a frame from the inside face.

INSIDE HEIGHT Unobstructed inside loading height, measured at the side of a van body.

INSIDE LENGTH DIMENSION In the case of a truck or trailer body, the inside measurement front to rear of the body, usually at the floor.

INSIDE LINING Inner members of a cased frame.

INSIDE LOT Building lot surrounded on three sides by other building lots; bounded by a street on one side only. *Also called* **Interior lot.**

INSIDE SAFETY RELEASE Device mounted on the inside of a vehicle or machine door to allow emergency exit if the door is accidentally locked from the outside.

INSIDE STOP Bead used to hold a window sash in place.

INSIDE TRIM Architrave of a door or window.

INSIDE WIDTH Unobstructed inside loading width of a van body.

IN SITU In original or final place.

IN SITU **CONCRETE** Concrete that is deposited and allowed to harden in the place where it is required to be in the completed structure. *Also called* **Cast-in-place concrete.**

IN SITU **PILE** *See* **Pile.**

IN SITU **SOIL TEST** Test made on soil while in its natural setting.

INSOLATION Amount of solar radiation received on a surface.

INSOLUBLE Component of a lubricating grease that are insoluble in the prescribed reagents in an analytical procedure.

INSOLUBLE RESIDUE Portion of a cement or aggregate that is not soluble in dilute hydrochloric acid of stated concentration.

insp. *Abbreviation for:* inspect; inspection; inspector.

INSPECTION Examination of work in progress or completed to determine its compliance with contract requirements.

INSPECTION BLOCK Description on a drawing of the dimensional inspection to which a part, component or assembly will be subjected.

INSPECTION CHAMBER Shaft between grade and the invert level of a sewer, giving access for personnel.

INSPECTION FITTING Access eye.

INSPECTION LIST List, made near the completion of work, indicating items to be furnished or work to be done by the contractor or subcontractor to complete the work as specified. *Also called* **Punch list.**

INSPECTOR Authorized representative assigned to make detailed inspections of contract performance.

INSPECTOR'S TEST CONNECTION A 25-mm (1-in.) connection to an automatic sprinkler system, preferably at the highest point and farthest from the sprinkler valve, terminating in a 6-mm (0.5-in.) sprinkler orifice outside the building, and controlled by a 25-mm (1-in.) valve that permits observation of water flow causing a local alarm to sound.

inst. *Abbreviation for:* installation; instant; instantaneous; institute; institution; instruction; instrument.

INSTABILITY Condition reached in the loading of an element or structure in which continued deformation results in a decrease of load-resisting capacity.

instal. *Abbreviation for:* installation; installer.

INSTALL To place or fasten in position ready for use.

INSTALLATION FLOATER Insurance coverage of machinery and equipment being transported to the job site, installed and tested.

INSTALLMENT Part of a debt; installments are due at specified intervals.

INSTANTANEOUS MODULUS Slope of a stress-strain curve at a single point, employed when modulus varies from point to point.

INSTANTANEOUS PEAK (ELECTRIC) DEMAND *See* **Demand, (electric).**

INSTANTANEOUS SAFETY Safety device, limited to speeds up to 45 m/sec (150 fpm), that abruptly stops an elevator car.

INSTITUTIONAL OCCUPANCY Occupancy or use of a building by persons who are involuntarily detained, or detained for penal or correctional purposes, or whose liberty is restricted, or who require special care or treatment because of age, or mental or physical limitations.

instr. *Abbreviation for:* instruction; instructor; instrument; instrumentation.

INSTRUCTIONS TO BIDDERS Requirements contained in the bidding documents describing procedures for submitting bids.

INSTRUCTIONS TO CONTRACTOR *See* **Job-site instructions.**

INSTRUMENT 1. Surveying tool: theodolite, level, transit, sextant, planimeter, etc.; **2.** Written legal document that establishes the rights and liabilities of the parties named.

INSTRUMENT MAN Person who operates a survey instrument and records the readings, and who directs a rodman and/or flagman.

INSTRUMENT PANEL Board on which measuring

and recording instruments, switches, and controls are mounted.

INSTRUMENT SHELTER Naturally or artificially ventilated structure, used to shield temperature-measuring instruments from direct sunshine and precipitation.

insuf. *Abbreviation for:* insufficient.

insul. *Abbreviation for:* insulate; insulation.

INSULATED AERIAL WORK PLATFORM Boom-mounted aerial work platform or bucket, the boom of which either is wholly constructed of electrically insulating material, or contains a section wholly made of such material, designed to allow work on overhead energized power transmission lines.

INSULATED CONDUCTOR *See* **Electrical conductor.**

INSULATED FURNACE WALL *See* **Wall.**

INSULATED ROOM Room in which the temperature can be maintained above or below the ambient temperature through insulation and the provision of heating or cooling systems.

INSULATED METAL SHEETING Light-gauge flexible metal sheets permanently backed with insulation, used for wall and roof covering.

INSULATING BRICK *See* **Brick.**

INSULATING CONCRETE Concrete having low thermal conductivity. *See also* **Lightweight concrete,** and **Low-density concrete.**

INSULATING FOAM BOARD Rigid foamed-plastic board used as insulation. *See also* **Polystyrene insulating board,** and **Polyurethane insulating board.**

INSULATING PLASTERBOARD Drywall panel backed with polished aluminum foil.

INSULATION 1. Any nonconductive material; **2.** Material with above-average thermal or acoustic resistance, that inhibits the flow of heat or sound. **3.** Any material used in walls, floors, and ceilings to prevent heat or sound transmission.

INSULATION BOARD Structural building board made of wood or cane fiber. It is often finished on one face with a decorative treatment.

INSULATION RESISTANCE Resistance that an electrical insulating material has to the passage of current to ground or to another conductor.

INSULATION VALUE Ability of a material to resist heat flow. Stated as R or thermal resistance value.

INSULATOR Material or device used to isolate an active conductor.

INSURANCE Two-party risk transfer mechanism whereby one party pays to have another party protect it from certain well-defined risks. There are many types, including:

All-risk: Indemnity against loss from any cause other than those perils or causes specifically detailed.

Bid bond: Form of insurance providing indemnity in the event that a person fails to enter into a contract for which his bid has been accepted.

Business interruption: Insurance against continuing expenses and loss of earnings resulting from interruption of business caused by fire or other insured peril.

Buy-out: Cross insurance, usually by partners for the purpose of providing funds to purchase the deceased's share in a partnership on death of a partner.

Coinsurance: 1. Insurance contract clause requiring the insured to maintain a stipulated percentage of insurance to the value of the thing insured or, to contribute proportionally to his own loss; **2.** Joint underwriting of insurance.

Fidelity: Insurance providing indemnity in the event of embezzlement or other dishonesty by employees.

General: Insurance other than life; e.g., fire, accident, medical, business interruption, etc.

Independent contractors': Indemnity against claims resulting from the actions or operations of subcontractors.

Inland marine: Type of insurance designed to cover a contractor's equipment (other than vehicles designed for highway use). Because it applies to things that are mobile or 'floating,' it is called an equipment floater.

Life: Insurance in which the amount specified in the contract is payable on the death of the insured.

Performance: Insurance providing indemnity in the event that a contractor fails to meet specific contract conditions.

Reinsurance: Contract between insurers whereby one assumes all or part of the risk on an insurance contract issued by the other.

Self: Assumption by a person of a risk which otherwise might have been covered by insurance.

INSURANCE CARRIER Company that assumes financial responsibility for the risks of others.

INSURANCE COVERAGE Nature and amount of risk insured in an insurance contract.

INSURANCE FACTOR Factor used to average insurance costs over the life of a machine.

INSURANCE POLICY Written contract describing an agreement to insure.

int *Abbreviation for:* interest; intersect.

int. *Abbreviation for:* integer; integral; integrate; interior; internal; interval.

INTAGLIO Surface decoration created by incising a line or pattern.

INTAKE Portion of a pipe, pump, or structure through which water or air enters from the source of supply.

INTAKE MANIFOLD Pipes and fittings that connect a source to several destinations; in an internal combustion engine, connecting the air cleaner outlet air pipe to each cylinder inlet.

INTARSIA Surface decoration created by inlaying small pieces of contrasting and/or colored wood.

intchg. *Abbreviation for:* interchange; interchangeable.

intcp. *Abbreviation for:* intercept; interceptor.

INTEGRAL Components that form a complete unit.

INTEGRAL CURB Curbing formed together with a gutter and/or a roadway.

INTEGRAL GUIDE Guide that is part of the cylinder head of an engine and not removable.

INTEGRAL HOLDING VALVE *See* **Valve.**

INTEGRALLY CAST Elements such as concrete joists and top slab, cast in one piece. *See* **Monolithic concrete.**

INTEGRATED CEILING Suspended ceiling system that integrates facilities in addition to the ceiling surface: lighting, air-handling, communications, etc.

INTEGRATED CURBSIDE COLLECTION Residential area service combining collection of source-separated, recyclable materials and refuse simultaneously.

INTEGRATED (ELECTRIC) DEMAND *See* **Demand (electric).**

INTEGRATED FOREST COMPANY Forest company that both produces logs and manufactures them into lumber, pulp, and other wood products.

INTEGRATED LOGGING Logging operation that segregates and delivers a variety of products to mills and processors that will use them to best advantage.

INTEGRATED RESOURCE MANAGEMENT Management of two or more resources in the same general area: commonly includes water, soil, timber, range, fish, wildlife, and recreation.

INTEGRATED SOLID WASTE MANAGEMENT Solid waste management strategy that ranks the preferred alternatives, typically in the following order: source reduction, recycling, resource recovery, and landfill disposal.

INTEGRATING METER Device that measures and records the total quantity of a fluid, or of electricity that has passed it.

INTEGRITY TEST Manual or automatic check of a system to determine that it is operative prior to use. *Also called* **Operational test,** and **Self-test.**

INTENSIFIER Device that converts low-pressure fluid to a higher-pressure fluid power.

INTENSIVE FOREST MANAGEMENT Utilization of a wide variety of silvicultural practices, such as planting, thinning, fertilization, harvesting, and genetic improvement, to increase the capability of the forest to produce fiber.

INTENSIVE SILVICULTURE Treatments carried out to maintain or increase the yield and value of forest stands; includes treatments such as site rehabilitation, conifer release, spacing, pruning, and fertilization. *Also called* **Incremental silviculture.** *See also* **Silviculture.**

inter. *Abbreviation for:* intermediate.

INTER-AXLE DIFFERENTIAL *See* **Transmission.**

INTERCEPT Length of a staff seen between the sta-

dia hairs of a telescope.

INTERCEPTING DRAIN Drain constructed between a source of ground, or surface water and the area to be protected. *Also called* **Curtain drain.**

INTERCEPTING SEWER Sewer that receives the dry-weather flow from a number of transverse sewers or outlets, with or without a determined quantity of storm water.

INTERCEPTOR Receptacle that is installed to prevent oil, grease, sand, or other materials from passing into a drainage system.

INTERCEPTOR TRENCH Trench filled with stone or gravel that intercepts excess water runoff before it reaches a building.

INTERCHANGE System of interconnecting roadways in conjunction with one or more grade separations, providing for the movement of traffic between two or more roadways on different levels.

INTERCHANGE ELEMENTS Various elements associated with interconnecting roadways, including:

> **Direct connection:** One-way turning roadway that does not deviate greatly from the intended direction of travel.
>
> **Interchange ramp:** Turning roadway at an interchange for travel between intersection legs.
>
> **Loop:** One-way turning roadway that curves about 270° to the right to accommodate a left-turning movement. It may include provision for a left turn at a terminal to accommodate another turning movement.
>
> **Outer connection:** One-way turning roadway primarily for a right-turning movement. It may include provision for a left turn at a terminal to accommodate another turning movement.
>
> **Two-way ramp:** Ramp for travel in two directions. At a cloverleaf it serves as both an outer connection and a loop.

See also **Interchange types, Intersection elements,** and **Intersection types.**

INTERCHANGE RAMP *See* **Interchange elements.**

INTERCHANGE TYPES Various types of road interchange, including:

> **Cloverleaf:** Four-leg interchange with loops for left turns and outer connections for right turns, or two-way ramps in these turns. A full cloverleaf has ramps for two turning movements in each quadrant.
>
> **Diamond interchange:** Four-leg interchange with a single one-way ramp in each quadrant. All left turns are made directly on the minor highway.
>
> **Direct connection:** Intersection of two or more roads, usually without provision for feeding right-turning traffic, and with traffic movement generally controlled by signs and/or signals.
>
> **Directional interchange:** Interchange, generally having more than one highway grade separation, with direct connections for the major left-turning movements.

See also **Interchange elements, Intersection elements,** and **Intersection types.**

INTERCOLUMNATION Space between columns, measured in diameters. There are five main ratios: 1.5 diameters (Pycnostyle), 2 diameters (Systyle), 2.25 diameters (Eustyle), 3 diameters (Diastyle), and 4 diameters (Araeostyle).

INTERCOM Corruption of intercommunication. A device used to provide oral communication between two or more locations.

intercom. *Abbreviation for:* intercommunication.

INTERCONNECTED FLOOR SPACE Superimposed floor areas or parts of floor areas in which floor assemblies that are required to be fire separations are penetrated by openings that are not provided with closures.

INTERCONNECTOR Pipe that joins two or more water supply systems.

INTERCOOLER Device that cools a gas between the compressive steps of a multiple-stage compressor. *See also* **Aftercooler,** and **Engine charge air cooler.**

INTERDOME Space between the inner and outer shells of a dome.

INTEREST-ONLY MORTGAGE *See* **Mortgage.**

INTEREST RATE Percent of a sum of money charged for its use.

INTERFACE 1. Meeting of two otherwise separate surfaces; **2.** Joint or component where a transition is made between power levels, modes of operation, etc.; **3.** Boundary between two rocks or formations with different physical characteristics.

INTERFERENCE Overlap of influence between two wells pumping from the same aquifer.

INTERFERENCE ANGLE Difference in angle between a valve seat and its mating valve.

INTERFERENCE-BODY BOLT High-bearing and shear-strength bolt having a deformed threaded shank.

INTERFERENCE FIT Fit that results when a larger object is forced into a smaller hole.

INTERFERENCE SETTLEMENT Settlement of foundations due to imposed loads on adjacent ground.

INTERFLOW That portion of rainfall that infiltrates into the soil and moves laterally through the upper soil horizons until intercepted by a stream channel or until it returns to the surface at some point down slope from its point of infiltration.

intergovt *Abbreviation for:* intergovernment.

INTERIM ACCEPTANCE Acceptance of a construction project from a contractor subject to conditions that must be met before final acceptance. *See also* **Acceptance of work, Final acceptance,** and **Partial acceptance.**

INTERIM FINANCING Funds used to bridge the period between start of a project and completion of long-term financing.

INTERIM FOREST Forest that exists or will exist until conversion to a target forest is complete. An interim forest may develop under intensive forest man-agement, and may have excellent stocking, but does not necessarily represent the forest desired at some future time. *Also called* a **Transition forest.**

INTERIOR ADHESIVE *See* **Adhesive.**

INTERIOR DESIGN Detailed design of all decoration, applied fittings, and furnishings.

INTERIOR FINISH Completion of the interior of a building, including decoration and furnishings.

INTERIOR GLAZED Glass set from the inside of the building.

INTERIOR GLUE *See* **Adhesive.**

INTERIOR-HUNG SCAFFOLD Scaffold suspended from the ceiling or interior of the roof structure.

INTERIOR LOT *See* **Inside lot.**

INTERIOR PLYWOOD Plywood bonded with adhesives adequate for interior conditions but usually not waterproof.

INTERIOR SIDE YARD Side yard located adjacent to another zoning lot, or to a lane separating it from another zoning lot, or to the wall of a building in a planned building group.

INTERIOR SPAN Continuous beam or slab with supports that are continuous with adjacent spans.

INTERIOR STAIRWAY Stairway within the exterior walls of a building.

INTERIOR STOP Removable molding or bead that holds an interior light or panel in place.

INTERIOR TYPE Plywood manufactured for indoor use or construction subjected to only temporary moisture.

INTERLAMINAR SHEAR Shear strength at rupture in which the plane of the fracture is located between layers of reinforcement of a laminate.

INTERLAYER Plastic reinforcing material used in conjunction with glass in the manufacture of laminated safety glass.

INTERLAYER BARRIER *See* **Isolating coat.**

INTERLOCK 1. Device or system that connects two or more components, preventing action by one or both unless they are in their proper position or orientation. **2.** Meeting of two otherwise separate surfaces; **3.** Joint or component where a transition is made between power levels, modes of operation, etc.; **4.** Boundary between two rocks or formations with different physical characteristics.

INTERLOCKED-GRAIN LUMBER *See* **Grain.**

INTERLOCKING 1. Joined together so that no one part may act independently; **2.** Binding of particles one with another.

INTERLOCKING CONTACT Electrical contacts arranged to prevent two or more relays from being simultaneously energized.

INTERLOCKING PAVING STONES Precast concrete pavers of about brick-size dimensions, with shapes designed to interlock in order to provide mutual lateral support.

INTERLOCKING YARDER Incorporating a means

of coupling the mainline and haulback drums so as to maintain a consistent tension on the two lines while yarding, without the application of foot brakes by the operator.

INTERMEDIATE ADHESIVE *See* **Adhesive.**

INTERMEDIATE BEND Any bend between the end bends.

INTERMEDIATE COVER Soil material placed on a completed landfill lift to act as a layer between the completed lift and a planned lift on top of it.

INTERMEDIATE-DUTY FIRECLAY BRICK *See* **Brick.**

INTERMEDIATE FLOOR Landing or floor served by an elevator between its upper and lower terminal landings.

INTERMEDIATE GIRDER Any girder between two external girders.

INTERMEDIATE GLUE *See* **Adhesive.**

INTERMEDIATE HITCH *See* **Intermediate suspension line.**

INTERMEDIATE MANUFACTURER Manufacturer, other than an incomplete vehicle manufacturer, or final manufacturer, that performs manufacturing operations on an incomplete vehicle.

INTERMEDIATE PROCESSING CENTER Facility that separates mixed glass and metal containers, paper and plastic, and processes those materials for sale to end users.

INTERMEDIATE RAFTER Common rafter.

INTERMEDIATE RIPPER TIP *See* **Ripper tip.**

INTERMEDIATE SHAFT Shaft, driven by one shaft, that drives another.

INTERMEDIATE SIGHT In survey, a staff reading that is neither a backsight nor a foresight.

INTERMEDIATE SUPPORT Means of giving lift to a skyline in a multispan system at positions between the head spar and backspar by use of tree jacks.

INTERMEDIATE SUPPORT SPAR Spar tree located between the head spar and tail spar to support a multi-span skyline.

INTERMEDIATE SUSPENSION (LINE) Additional set of boom-suspension lines attached to the boom between the main suspension and the boom foot. *Also called* **Hog line, Intermediate hitch, Midpoint suspension,** and **Midpoint hitch.**

INTERMEDIATE-TEMPERATURE-SETTING ADHESIVE *See* **Adhesive.**

INTERMEDIATE TREE Tree with a small, crowded crown below (but extending into) the general canopy level; such trees receive a little light from above and none from the side.

INTERMITTENT BRAKE POWER Highest power recommended by the manufacturer for satisfactory operation within the manufacturer's specified conditions.

INTERMITTENT HORSEPOWER *See* **Horsepower.**

INTERMITTENT OVERLOAD Power in excess of

rated power that a generator is capable of delivering, without damage, for a specified period of time.

INTERMITTENT SAMPLING Sampling successively for limited periods throughout an operation or for a predetermined period of time.

INTERMITTENT WELD Joining two pieces but leaving sections unwelded.

INTERMODAL Shipment of goods involving more than one conventional mode: air, rail, sea, motor carrier, etc.

INTERNAL BOND Force applied perpendicular to a panel face required to pull the panel apart.

INTERNAL BOND STRENGTH Overall measure of a particleboard's integrity illustrating how well the core materials are bonded together; tested by applying tension perpendicular to the panel surface.

INTERNAL COMBUSTION Combustion that occurs within the cylinder(s) of an engine.

INTERNAL COMBUSTION ENGINE *See* **Engine.**

INTERNAL COMBUSTION TRACTOR Industrial tractor in which the power source is a gasoline, LP gas, or diesel engine and the power is transmitted to the driving wheels through a mechanical, hydrodynamic, or hydrostatic transmission.

INTERNAL COMPACTION TRANSFER SYSTEM Reciprocating action of a hydraulically powered bulkhead contained within an enclosed trailer that packs and compresses solid waste.

INTERNAL CONTROL Plan of organization and other coordinate methods and measures adopted to safeguard assets, produce accurate and reliable accounting data, promote operational efficiency, and encourage adherence to prescribed managerial policies.

INTERNAL DORMER Vertical window or door within the general line of a sloping roof.

INTERNAL ENERGY Energy of a body due to its conditions: pressure, temperature, etc.

INTERNAL FORCE Stress.

INTERNAL FRICTION 1. Resistance between mating parts; **2.** Resistance of rock or earth particles to sliding over each other; **3.** Power losses in the power train, from engine flywheel to the final drives.

INTERNAL GRINDING Grinding the inside surface of the hole in a piece of work.

INTERNAL HAMMER *See* **Pile hammer.**

INTERNAL INSPECTION Part of a sewer system evaluation survey that involves inspecting, by physical, photographic, and/or TV methods, sewer lines that have previously been cleaned.

INTERNALLY LUBRICATED Wire rope or strand having all of its wire components coated with lubricants.

INTERNAL MOISTURE Moisture contained within an aggregate particle.

INTERNAL SPUR-GEARED MACHINE Machine with a double-reduction gear, the first of which is a worm-and-gear, the second being a spur gear that meshes with an internal annular gear on the sheave or drum.

INTERNAL THREAD Thread on the inside of a pipe or part.

INTERNAL-TOOTH WASHER *See* **Washer.**

INTERNAL VIBRATION One or more vibrating elements inserted into fresh concrete at selected locations to assist in consolidation. *See also* **External vibration Surface vibration, and Vibration.**

INTERNATIONAL LOG RULE Formula that allows 12 mm (0.5 in.) of taper for each 1.2 m (4 ft) of length and allows for a 1.5-mm (0.06-in.) shrinkage for each 25 mm (1.0 in.) of board thickness. *See also* **International 1/4-in. scale.**

INTERNATIONAL SYSTEM OF UNITS International metric system (Système International d'Unités) or SI. *See the* appendix **Metric and nonmetric measurement** for complete details.

INTERNATIONAL 1/4-in. SCALE Log scale modification of an earlier rule using a 1/4-in. kerf, based on an analysis of the loss of wood fiber incurred in the conversion of saw logs to lumber. One of the few rules incorporating a basis for dealing with log taper. *See also* **International log rule.**

interp. *Abbreviation for:* interpolation; interpretation.

INTERPILE SHEATHING Horizontal sheeting installed between, and supported by piles as shoring.

INTERPOLATION Inferring the position of a point between two known points by assuming the variation between them is linear.

INTER-ROUTE RELIEF *See* **Collection method.**

INTERRUPTED ARCH *See* **Arch.**

INTERRUPTER Switch capable of making, carrying, and interrupting specified currents.

INTERRUPTION Secondary cutter in an auger drill.

INTERSECTION General area where two or more highways join or cross, within which are included the roadway and roadside facilities for traffic movements in that area.

INTERSECTION ANGLE *See* **Intersection elements.**

INTERSECTION ELEMENTS Component parts of the intersection of two highways, including:

Angle of turn: Angle through which a vehicle travels in making a turn.

Intersection angle: Angle between two intersection legs.

Intersection entrance: That part of an intersection leg for traffic entering the intersection.

Intersection exit: That part of an intersection leg for traffic leaving the intersection.

Intersection leg: Any one of the highways radiating from and forming part of an intersection. The common intersection of two highways crossing each other has four legs.

Island: Defined area between traffic lanes for control of vehicle movements or for pedestrian refuge. Within an intersection a median or an outer separation is considered an island.

Median opening: Gap in a median providing for crossing and turning traffic.

Merging end: End of an island, or area between converging roadways, beyond which traffic merges.

Minimum turning path: Path of a designated point on a vehicle making its sharpest turn.

Minimum turning radius: Radius of the minimum turning path of the outside of the outer front tire.

Turning movement: Traffic making a designated turn at an intersection.

Turning path: Path of a designated point on a vehicle making a specified turn.

Turning roadway: Connecting roadway for traffic turning between two intersection legs.

Turning roadway terminal: General area where a turning roadway connects with a through traffic roadway. 'Exit' used as a modifier refers to leaving the through traffic lanes and 'entrance' refers to entering the through traffic lanes.

Turning track width: Radial distance between the turning paths of the outside of the outer tire and the outside of the rear tire that is nearest the center of the turn.

See also **Interchange elements, Interchange types, and Intersection types.**

INTERSECTION ENTRANCE *See* **Intersection elements.**

INTERSECTION EXIT *See* **Intersection elements.**

INTERSECTION LEG *See* **Intersection elements.**

INTERSECTION TYPES Various configurations made by crossing highways, including:

At-grade intersection: Intersection where all roadways join or cross at the same level.

Channelized intersection: At-grade intersection in which traffic is directed into definite paths by islands.

Flared intersection: Unchannelized intersection, or a divided highway intersection without islands other than medians, where the traveled way of any intersection leg is widened or an auxiliary lane is added.

Four-leg intersection: Intersection with four legs, as where two highways cross.

Multileg intersection: Intersection with five or more legs.

Rotary intersection: Confluence of three or more intersection legs at which traffic merges into and emerges from a one-way roadway in a counterclockwise direction around a central area.

Three-leg intersection: Intersection with three legs, where two highways join.

T-intersection: Three-leg intersection in the general form of a 'T.'

Unchannelized intersection: At-grade intersection without islands for directing traffic into definite paths.

Y-intersection: Three-leg intersection in the general form of a 'Y.'

See also **Interchange elements, Interchange types,** and **Intersection elements.**

INTERSTICE Small opening, such as between fibers in a cord or threads in a woven or braided fabric.

INTERSTICE VISCOSITY Ratio of the difference of the viscosity of a solution at a given concentration and the viscosity of a pure solvent to the product of the viscosity of the pure solvent and the volume concentration of the solution.

INTERSTITIAL Of, forming, or occurring in interstices.

INTERTREE DISTANCE Distance between tree crowns, usually in the context of thinning.

INTERVAL 1. Discrete portion of a signal cycle during which signal indications do not change; **2.** *See* **Traffic signal.**

INTESTATE Death of an individual without a valid will.

intgl *Abbreviation for:* integral.

intk. *Abbreviation for:* intake.

intl *Abbreviation for:* international.

intlk *Abbreviation for:* interlock.

intmed. *Abbreviation for:* intermediate.

intmt *Abbreviation for:* intermittent.

INTOLERANT Tree relatively incapable of developing and growing normally in the shade of, and in competition with other trees.

INTRADOS Inner curve or underside of an arch. *Also called* the soffit. *See also* **Arch.**

INTRAVANCE Automatic timing device that regulates the injection timing of fuel while an engine is running.

INTRINSICALLY SAFE EQUIPMENT Equipment and associated wiring in which any spark or thermal effect, produced either normally or in specified fault conditions, is incapable, under certain prescribed conditions, of causing ignition of a mixture of flammable or combustible material in air in its most easily ignitable concentration.

intro. *Abbreviation for:* introduction.

INTUMESCENT PAINT Paint that, when subjected to excessive heat or flame, puffs out to form a thick, insulating crust to insulate the underlying strata and prevent the spread of fire.

intvl *Abbreviation for:* interval.

INUNDATE Cover with water, as in the case of flooding.

inv. *Abbreviation for:* inventory; inverse; invert.

invar. *Abbreviation for:* invariant.

INVENTORY *See* **Management-volume inventory.**

INVERSE CONDEMNATION Legal process that may be initiated by a property owner to compel payment of just compensation where a property has been taken, or damaged, for public purpose. *See also* **Con-**

demnation.

INVERSION Horizontal layer of air through which temperature increases with increasing height.

INVERT 1. To turn over; **2.** Inside bottom level of a pipe, trench, or tunnel.

INVERTED ARCH *See* **Arch**

INVERTED ASPHALT EMULSION Emulsified asphalt in which the continuous phase is asphalt, and the discontinuous phase is minute globules of water in relatively small quantities.

INVERTED CROWN Road, driveway, or path where the center is at a lower elevation than the sides.

INVERTED PENETRATION *See* **Single surface treatment.**

INVERTED T-BEAM Beam having a cross section in the shape of an inverted T. *See also* **Ledge.**

INVERTER Device that converts direct current to alternating current.

INVERT PAVEMENT Smooth bituminous layer on the lower segment of corrugated metal pipe, intended to improve flow, prevent deposition of silt in the invert corrugations, and provide resistance to scour and erosion.

INVERT POUR Process in which the lower half of a tunnel lining is poured separately from the upper half. *See also* **Arch pour,** and **Full-circle pour.**

INVERT STRUT Compression strut installed across the invert of a tunnel under construction to resist inward movement of the tunnel ribs.

invest. *Abbreviation for:* investment.

INVITATIONAL BIDDING *See* **Selective bidding.**

INVITATION FOR PROPOSALS *See* **Call for proposals.**

INVITATION TO BID Advertisement for proposals for all work or materials on which bids are required. Such advertisements will indicate, with reasonable accuracy, the quantity and location of the work to be done or the character and quantity of the material to be furnished and the time and place of the opening of proposals.

INVITED BIDDERS Bidders selected by the owner and his agent as the only ones from whom bids will be received. *Also called* **Closed list of bidders,** and **Selected bidders.**

INVOICE Document issued as a request for payment; by a contractor in accordance with the terms of his contract for works performed by him.

invol. *Abbreviation for:* involute.

INVOLUNTARY LIEN Lien imposed against property without consent of the owner.

INVOLUTE Spiral curve in one plane.

inx *Abbreviation for:* index.

IO *Abbreviation for:* input-output.

ION Electrically charged atom or group of atoms.

IONIC Style of classical Greek architecture characterized by ornamental scrolls or volutes on the capital.

i-p *Abbreviation for:* intermediate pressure.

IP *Abbreviation for:* initial point; instrument panel.

ipm *Abbreviation for:* inches per minute.

IPO *Abbreviation for:* installation planning order.

ips *Abbreviation for:* inches per second.

IPS *Abbreviation for:* inside pipe size; international pipe standard; iron pipe size.

IPT *Abbreviation for:* iron pipe thread.

IR *Abbreviation for:* infrared; inside radius.

irid. *Abbreviation for:* iridescent.

IRON Metallic chemical element. *See also* **Alloying elements, Cast iron, Malleable cast iron,** and **Wrought iron.**

IRONING *See* **Milking.**

IRON PIPE THREAD Standard system of threads for connecting various types of rigid piping.

IRONWORK Ornamental cast iron.

IRON WORKER Worker who handles and places steel and ornamental iron, including all types of reinforcing steel and bar supports. *Also called* **Bar placer, Metal lather,** and **Rebar contractor.**

irreg. *Abbreviation for:* irregular.

IRREGULAR COURSED RUBBLE *See* **Ashlar.**

IRREVOCABLE Not able to be changed, recalled, or cancelled.

IRRIGATION Artificial distribution of water to promote plant growth.

ISACOUSTIC Equal in acoustic correction.

I-SECTION Beam cross section consisting of top and bottom flanges connected by a vertical web.

isl. *Abbreviation for:* island.

ISLAND *See* **Intersection elements.**

ISLAND FIREPLACE Fireplace built within a room, clear of the surrounding walls, open on more than one side.

ISOCHRONOUS Maintaining constant engine speed independent of the imposed load.

ISOCHRONOUS GOVERNOR *See* **Governor.**

ISOCON Line of equal concentration of a specified chemical constituent or parameter.

isol. *Abbreviation for:* isolate.

ISOLATED Not readily accessible to persons unless special means of access are used.

ISOLATED POWER SYSTEM System comprised of an isolating transformer or its equivalent, a line isolation monitor, and its underground circuit conductors.

ISOLATED SOLAR GAIN Passive solar heating system in which heat is collected at one location for use at another.

ISOLATING COAT Coat used to isolate one paint layer from another to prevent interaction between them. *Also called* **Interlayer barrier.**

ISOLATING MEMBRANE An underlay.

ISOLATING SWITCH *See* **Switch.**

ISOLATION Process of reducing direct sound transmission paths.

ISOLATION JOINT *See* **Joint.**

ISOLATOR Any part of a whole designed to be removed so as to create separate, isolated parts; typically, in an electrical circuit.

isom. *Abbreviation for:* isometric.

ISOMETRIC PROJECTION Geometrical, three-dimensional drawing where the plan is projected with lines at an equal angle (usually 30°) to the horizontal, with verticals projected vertically. All diagonals and curves are distorted; other dimensions may be scaled.

ISOPROPYL ACETATE Fast-acting lacquer solvent.

isos. *Abbreviation for:* isosceles.

ISOSHEAR Having the same shear strength.

ISOSTATIC LINE Line tangential to the direction of one of the principal stresses at every point through which it passes.

ISOTHERMAL Having the same temperature.

ISOTROPIC SOIL Soil mass having essentially the same properties in all directions at any given point.

ISOTROPY Behavior of a medium having the same properties in all directions.

IST *Abbreviation for:* inside trim.

ital. *Abbreviation for:* Italian.

ITEM Nomenclature for a particular kind of work to be performed or material to be supplied under the contract documents.

IVORY PAPER *See* **Cream paper.**

IWRC *Abbreviation for:* independent wire rope core.

J

J *Abbreviation for:* jack; jamb; joist; journal.

jac. *Abbreviation for:* jacket.

JACK 1. Mechanical or hydraulic device used for raising objects small distances; **2.** Mechanism for applying force to prestressing tendons; **3.** Device for adjusting the elevation of concrete forms or form supports; **4.** Hanger device used to support a skyline; **5.** Plunger and cylinder of a direct-plunger hydraulic elevator.

JACK ARCH *See* **Arch.**

JACK BEAM Beam that supports another beam or truss, eliminating the need for a column.

JACKBLOCK CONSTRUCTION Technique where the roof of a multistory building is cast at ground level and raised by hydraulic jacks. The supporting walls of the slab and the floor or top story are constructed and finished under the roof and the whole again raised by jacks. This process is repeated until the full height of the building is completed.

JACK BOOM Boom on a pull shovel or dredge that supports the sheaves between the hoist drum and main boom. *See also* **Boom.**

JACK CLEARANCE Maximum clearance required behind, or adjacent to the anchorage of a prestressed concrete member for jacking.

JACKED PILE Pile forced into the ground by pressure exerted by a jack.

JACKET 1. Seamless tubular braided or woven ply, generally on the outside of a hose; **2.** Woven fabric used during vulcanization by the wrapped cure method. *Also called* **Woven jacket; 3.** Insulation that covers exposed heating and cooling pipes.

JACKING 1. Means of imposing a static driving force on a pile using jacks. Used extensively to install piles in underpinning of structures. A jacked pile; **2.** Means of precisely transferring load from another source (either dead weight or a reaction platform) in the performance of a pile load test.

JACKET-WATER AFTERCOOLING Aftercooling system in which intake air is cooled by engine coolant circulating through the intake manifold.

JACKETING Surrounding a prime flow with a confined bath or flow of fluid for temperature control or heat absorption.

JACK HAMMER 1. Air drill that hammers and rotates a hollow steel and bit. *See also* **Air hammer; 2.** Air drill that can be operated by one person.

JACK HOLE Cavity below ground level that contains the cylinder of a hydraulic elevator.

JACKING CONDUIT Orifice formed horizontally in the ground by jacking, into which a service pipe is then introduced.

JACKING DEVICE 1. Device used to stress the tendons for prestressing concrete; **2.** Device for raising a vertical slipform.

JACKING FORCE Temporary force exerted by the device that introduces tension into the tendons in prestressed concrete.

JACKING OPERATION Task of lifting a slab (or group of slabs) vertically from one location to another.

JACKING PIPE Pipe installed using pipe jacking techniques to form part of a conduit.

JACKING PLATE Steel plate placed on top of a pile during jacking to transmit the jacking force.

JACKING STRESS Maximum stress occurring in a prestressed tendon during stressing.

JACKKNIFE 1. Angle developed between a tractor and the trailer it is pulling so acute that the tractor cannot move forward; **2.** Accidental raising of a derrick boom by its load.

JACKKNIFE CLEARANCE *See* **Swing clearance.**

JACKLEG Outrigger post.

JACK LINE *See* **Cross support.**

JACK PLANE *See* **Plane.**

JACK POST Two heavy steel pipes, each about 1.2 m (4 ft) long, free to slide one within the other and drilled to receive a heavy metal pin, used to secure the tubes to a fixed length. A screw jack extends the combined tubes a short distance. The post and jack are used vertically between metal plates at top and bottom and jacked either to support a load, or to raise a beam to make it level. *Also called* **House jack.**

JACKPOT 1. Unstable logs crisscrossed or difficult to break free; **2.** Trees tied together or leaning into one another that create a hazard for fallers.

JACK RAFTER *See* **Rafter.**

JACK RIB Curved jack rafter, used in a dome.

JACK ROD Plain rod, usually 22 or 25 mm (0.875 or 1.0 in.) in diameter, with square-cut or threaded ends, used to support sliding forms in connection with a jack. In some cases, these jack rods are also used as a portion of the vertical reinforcement required.

JACK ROD SLEEVE Piece of pipe that joins two jack rods for an end-to-end butt connection.

JACKSCREW Adjustment screw used to lift weights and level plant and equipment.

JACKSHAFT Short drive shaft, usually connecting a clutch and transmission.

JACK SHORE Telescoping, or otherwise adjustable single-post metal shore.

JACK STUD Less than full length stud, commonly used above door and window openings between the lintel and plate.

JACK TEST High-capacity hydraulic jacking system

used to apply loads to a pile under test. *See also* **Pile.**

JACK TRUSS *See* **Truss.**

JACOB'S LADDER Ladder of rope or chain with wooden or metal rungs.

JAG Bundled scrap metal used as a wrecking ball.

JAGGER 1. Broken wire that juts out of a wire rope; **2.** Toothed chisel used to dress stone.

JAKE BRAKE Diesel engine compression release braking device that allows for recompression near the top of each compression stroke.

jal. *Abbreviation for:* jalousie

JALOUSIE Frame containing a number of movable, shutter-like, overlapping glass, wood, metal, or plastic panels, mounted horizontally or vertically, that may be adjusted to regulate the air and light coming through.

JAMB Vertical side of a door, window, archway, or other opening.

JAMB BLOCK Concrete block especially formed with a slot for holding the jambs of window or door frames.

JAMB POST Post or stone forming a door jamb.

JAM LINING Board facing to a jamb.

JAMMER 1. Lightweight, two-drum yarder, usually on a truck, with a spar and boom; may be used for both short distance yarding and loading; **2.** Frame mounted on a sled or vehicle for loading logs.

JAMMER LOGGING Cable logging system generally restricted to one skidding line and used for winching logs up to 90 m (300 ft) from the cutting area to a log collection point.

JAM NUT *See* **Nut.**

J&P *Abbreviation for:* joists and planks.

jap. *Abbreviation for:* japan; japanning.

JAPAN PAINT Paint-thinner- and lacquer-thinner-soluble paint with pigments dispersed in driers and flat varnish. These paints have no linseed oil.

JARS Tool in the string of tools of a cable drill that contains slack to allow upward hammering to free a stuck drill bit.

JAW 1. One of a pair of toothed rings in a clutch mechanism, the teeth of which face the other jaw; **2.** One of a pair of nearly flat faces separated by a wedge-shaped opening in a crusher.

JAW BRAKE *See* **Brake.**

JAW CLUTCH *See* **Clutch.**

JAW CRUSHER *See* **Crusher.**

JAWS 1. That part of a elevator overspeed safety device that grips the governor rope or guide rails; **2.** Device that interfaces with and is used to clamp a vibratory driver/extractor to a pile. *See also* **Gibs, Side channels,** and **Spud clip.**

JB *Abbreviation for:* junction box.

JC *Abbreviation for:* joist chair.

jct. *Abbreviation for:* junction.

JEEP Two-axle trailer that mounts between a tractor

and and load-carrying trailer to further disperse the axle loading.

JEMMY *See* **Jimmy.**

JENNY Machine for flanging and wiring sheet metal edges.

JERKINHEAD Pitched roof in which the top of the gable is cut off by a seconday slope that forms a hip.

JET Nozzle containing a small-diameter exit orifice to which the pipe or tube is fitted, having the effect of controlling the air or fluid being expelled.

JET ACTION Valve design type in which flow effect is controlled by the relative position of a nozzle and a receiver.

JET CEMENT Extra-rapid-hardening cement consisting of approximately 60% alite, 20% fluoraluminate, and gypsum.

JET DRILLING Technique of drilling rock by injecting a mixture of oxygen and fuel through a water-cooled pipe and then burning the mixture within the hole.

JET PIERCING DRILL *See* **Fusion drill.**

JET PUMP *See* **Pump.**

JETTING 1. Use of a water jet to aid in the placing or driving of a pile through hydraulic displacement of parts of the soil. *Also called* **Wash boring**; **2.** Process of injecting great amounts of water through a hose into soil and granular material to speed the process of compaction.

JETTY Structure or long fill extending into a body of water from the shore that serves to aid in access to deeper water so as to load and unload vessels, and to change the direction or velocity of the water flow.

JEWEL Boss of glass, often colored, in stained glass work.

J-HOOK Attachment device used for towing/recovery.

JIB 1. Pendant-supported attachment that connects to the head of a main boom for extra lift height or offset. *Also called* **Fly jib**; **2.** Horizontal support for drifter drills bolted to a jumbo that allows mechanical positioning of the drill at the working face.

JIB BOOM Extension hinged to the upper end of a crane boom.

JIB CRANE 1. Crane fitted with a jib; **2.** Cantilevered boom or horizontal beam with hoist and trolly, able to access loads from any position about the column to which it is attached.

JIB DOOR Concealed door, flush with the surface, the face of which is decorated to correspond with the adjacent surface.

JIB MAST Short strut or frame mounted on a boomhead. *Also called* **Jib strut,** and **Rooster.**

JIB STRUT *See* **Jib mast.**

JIG Fixture or template to accurately position and hold a part during fabrication or assembly.

JIGSAW *See* **Saw.**

JILLPOKE Log driven into position between two

anchor stumps, used to increase the stump's stability and holding power.

JIMMER Fixed hinge; one in which the leaves are inseparable.

JIMMY Short prybar, approximately 300–600 mm (12–24 in.) long. *Also called* **Jemmy.**

JITTERBUG Grate tamper for pushing coarse aggregate slightly below the surface of a slab to facilitate finishing. *See also* **Tamper.**

jmb *Abbreviation for:* jamb.

JOBBER Semiskilled worker able to do a variety of jobs.

JOB CAPTAIN Leader of a design team.

JOB MADE Fabricated on the building site.

JOB-MADE LADDER Ladder that is fabricated by workers, typically at the construction site, and is not commercially manufactured.

JOB SITE Location of the project; place where the work is to be done. *Also called* **Construction site, Project site,** and **Site.**

JOB-SITE INSTRUCTIONS Written clarification or interpretation of the terms, conditions, specifications, or drawings contained in a contract. Such an interpretation is rendered by the owner or his agent in response to a request for an urgent reply from the site of the work. *Also called* **Instructions to contractor.**

JOB SUPERINTENDENT Person, on site, responsible for the works.

JOE *See* **McGee.**

jog. *Abbreviation for:* joggle.

JOGGLE or JOGGING A particular way of fitting masonry blocks together consisting of a horizontal tongue and groove of significant proportions.

JOGGLE PIECE Post, shouldered to provide an abutment for a strut.

JOGGLE POST King post.

JOHNSON BAR Lever consisting of a long (approximately 1.5 m (5 ft)), leg with a short upturn at one end, having small wheels or rollers on the external angle, used to lift and move heavy loads.

JOINER Craftsman in woodworking who constructs joints; usually a term applied to a worker in a shop who constructs doors, windows, and other fitted parts of a house.

JOINER'S GAUGE Marking gauge.

JOINERY 1. Fabrication and assembly of worked timber components and panel products, other than structural timbers and cladding; 2. Various types of joints used in woodworking.

JOINT 1. Narrow space between adjacent stones, bricks, or other building blocks, usually filled with mortar; 2. Physical separation in concrete, whether precast or cast-in-place, including cracks if intentionally made to occur at specified locations; 3. Region where structural members intersect, such as a beam-column joint; 4. Area where different pieces of wood come together, or the various methods of fitting pieces of wood together; 5. Where two pipes are connected

either by bolting, welding, or by screwed connection; 6. Intersection of two or more members of a truss; 7. Rock fracture in which no appreciable movement parallel with the fracture has occurred; 8. Discontinuity in a surface at a predetermined position, that may be filled with a sealant or left unfilled; 9. Area where two or more ends, surfaces, or edges are attached, categorized by type of fastener or weld used and method of force transfer; 10. Point of intersection of a chord with the web or webs of a truss; 11. Designed vertical plane of separation or weakness. There are several types, including:

Construction: a. Rigid, immovable joint creating a single structural unit from two or more individual parts or materials, or between stages of construction, not necessarily intended to accommodate movement; **b.** Surface where two successive placements of concrete meet; **3.** Temporary joint employed when the placing of concrete must be interrupted for any reason.

Contraction: a. Control joint; **b.** Tooled groove made in wet concrete to permit a controlled crack to occur should the slab shrink from thermal contraction; **c.** Joint between building components where the only movement to be expected is due to the shrinkage of either or both components.

Expansion: a. Joint or space left in construction to allow for expansion and contraction of materials due to temperature changes; **b.** Joint located to provide for expansion of a rigid slab, without damage to itself, adjacent slabs, or structures; **c.** Any device containing one or more metal bellows used to absorb dimensional changes such as those caused by thermal expansion or contraction of a pipe-line, duct, or vessel. There are several types, including:

Double: Consisting of two bellows jointed by a common connector that is anchored to some rigid part of the installation by means of an anchor base. Each bellows functions independently as a single unit.

Gimbal: Expansion joint designed to permit angular rotation in any plane.

Hinged: Expansion joint containing one bellows designed to permit angular rotation in one plane only.

Isolation: Separation between adjoining parts of a concrete structure, usually a vertical plane, at a designed location so as to interfere least with performance of the structure, yet such as to allow relative movement in three directions and avoid formation of cracks elsewhere in the concrete and through which all or part of the bonded reinforcement is interrupted.

Pressure-balanced: Expansion joint designed to absorb axial movement and/or lateral deflection while restraining the bellows pressure thrust force.

Single: The simplest form of expansion joint; of single bellows construction, it absorbs all of the movement of the pipe section into which it is installed.

Swing: Expansion joint designed to absorb

lateral deflection and/or angular rotation in one plane only by use of swing bars, each of which is pinned at or near the ends of the unit.

Universal: Expansion joint containing two bellows joined by a common connector for the purpose of absorbing any combination of the three basic movements (axial motion, lateral deflection, or angular rotation).

Universal tied: Expansion joint used when it is necessary for the assembly to eliminate pressure thrust forces from the piping system (the joint will absorb lateral movement but not absorb axial movement external to the tied length).

Hinged: *See* **Warping,** below.

Longitudinal: Joint parallel to the length of a structure or pavement, in pavement, normally placed between traffic lanes.

Warping: Joint with the sole function of permitting warping of pavement slabs when moisture and temperature differentials occur between the top and bottom of the slabs, i.e., longitudinal or transverse joints with bonded or tie bars passing through them. *Also called* **Hinged.**

JOINT ASSEMBLY Prefabricated cradle, usually of steel, for supporting dowel or tie bars in their correct position at a joint during concreting.

JOINT COMPOUND Gypsum compound used for taping and finishing drywall panels.

JOINTED CORE Core veneer that has edges machined square. Gaps between pieces of core do not exceed 10 mm (0.375 in.), and the average of all gaps in the panel does not exceed 5 mm (0.188 in.).

JOINTED CROSSBAND Crossband veneer with square, machined edges. Gaps between pieces of crossband and inner ply veneer do not exceed 8.3 mm (0.325 in.) and the average of all gaps in the panel do not exceed 4.1 mm (0.1625 in.).

JOINT EFFICIENCY 1. Strength of joined materials; **2.** Strength of a welded joint given as a percentage of the strength of the base metal.

JOINTER 1. Tool used for smoothing or indenting the surface of a mortar joint or to cut a joint partly through fresh concrete; **2.** *See* **Plane.**

JOINT FILLER 1. Compressible material used to fill a joint to prevent infiltration of debris and to provide support for sealants; **2.** Filler compound of gypsum plaster and various additives, used to fill the joints between gypsum drywall panels. *Also called* **Mud.**

JOINT GROOVE FINISHER Equipment, part of a paving train, consisting of a manually guided horizontal vibrating plate that recompacts the concrete after it has been disturbed by formation of the groove.

JOINTING The process of facing or tooling mortar joints.

JOINTING MATERIAL Sheet material from which gaskets or washers are cut.

JOINT MATCHER Small grade-reference shoe that permits the paver automatic screed controls to match a curb, gutter, or other structure.

JOINT MOVEMENT Difference in the width of a joint opening between the fully open and fully closed positions.

JOINT PENETRATION Depth of weld metal and fusion in a welded joint.

JOINT PROTECTOR *See* **Sub.**

JOINT REINFORCEMENT Steel mesh, wire, or bar placed in or on mortar bed joints.

JOINT REINFORCING MESH Woven fiberglass mesh used with jointing compound in completing the joint between drywall panels. *Also called* **Mesh tape.**

JOINT RULE Plasterer's tool having one end forming an angle of 45°, used to form miters at the junctions of cornice moldings.

JOINT RUNNER Tool composed of asbestos rope and a clamp, used in leading joints in horizontal runs of bell-and-spigot cast iron pipe.

JOINTS Planes within rock masses along which there is no resistance to separation and along which there has been no relative movement of the material on each side of the break. *Also called* **Partings.**

JOINT SEALANT Compressible material used to exclude water and solid foreign materials from joints.

JOINT-SEALING COMPOUND Impervious material used to fill joints.

JOINT SPALL Spall adjacent to a joint.

JOINT TAPE Paper product used with jointing compound in completing the joint between drywall panels.

JOINT TENANCY Ownership of real property by two or more persons, each of whom has an undivided interest with the right of survivorship.

JOINT VENTURE Undertaking by two or more parties who share proportionately certain responsibilities (such as financing) and/or who bring to the project specific values (such as expertise).

JOIST 1. Horizontal floor or ceiling supporting member, laid parallel to other joists; *See also* **Beam; 2.** Small, T-shaped beam used in a parallel series in concrete joist floor construction.

JOIST ANCHOR Wall anchor.

JOIST BRIDGING *See* **Bridging.**

JOIST CHAIR Bent or welded wire support that holds and spaces reinforcing bars in the bottom of a concrete joist.

JOIST HANGER Steel section shaped like a stirrup, bent so it can be fastened to a beam to provide support for joists, headers, etc.

JOIST PLAN Drawing showing where each joist is located.

JOIST SCHEDULE 1. Table giving the quantity and mark of the joists; **2.** Quantity, size, length, and bending details of reinforcing bars, and usually the quantity of joist chairs in each concrete joist.

JOIST SPACING Distance between joists measured from center line to center line.

JOLT VIBRATOR Vibrator that repeatedly raises and drops a mold onto a fixed base.

JOULE One of 17 derived units with special names of the SI system of measurement: a unit of energy, work done when the point of application of a force of 1 N is displaced through a distance of 1 m in the direction of the force. (1 J=1 N•m). Symbol: J. Multiply by 0.000 947 8 to obtain British thermal units, symbol: Btu; by 0.737 562 to obtain foot-pounds, symbol ft/lb; by 0.3725 x 10^6 to obtain horsepower-hour, symbol: hp/hr; by 0.277 8 x 10^6 to obtain kilowatt hour, symbol: kW·h; and by 1.0 to obtain newton meters, symbol N·m. *See also the appendix*: **Metric and nonmetric measurement.**

JOULE PER KELVIN A derived unit of heat capacity with a compound name of the SI system of measurement. Symbol: J/K. *See also the appendix*: **Metric and nonmetric measurement.**

JOULE PER KILOGRAM A derived unit of specific latent heat with a compound name of the SI system of measurement. Symbol: J/kg. *See also the appendix*: **Metric and nonmetric measurement.**

JOULE PER KILOGRAM KELVIN A derived unit of specific heat capacity with a compound name of the SI system of measurement. Symbol: J/(kg•K). *See also the appendix*: **Metric and nonmetric measurement.**

JOULE PER MOLE A derived unit of molar internal energy with a compound name of the SI system of measurement. Symbol: J/mol. *See also the appendix*: **Metric and nonmetric measurement.**

JOUNCE LINES Flexible air or hydraulic hoses connecting brake lines between the chassis and wheels and/or axles, to permit axle articulation or wheel steering movement relative to a truck frame.

JOURNAL Portion of a rotating shaft or axle that turns in a load-supporting bearing.

JOURNEYMAN 1. Workman who has learned his trade by serving an apprenticeship; **2.** Term usually applied to a skilled workman who is able to command the standard wage rate in his particular trade.

JOYSTICK Control lever that combines two or more functions and which is manipulated in four directions instead of two.

JP *Abbreviation for:* jet pump.

jr *Abbreviation for:* junior.

J/S *Abbreviation for:* jigsaw.

J-SIDE Side or skin damping value used in wave equation analysis.

jst *Abbreviation for:* joist.

jt *Abbreviation for:* joint.

J-TIP Tip or toe damping value used in wave equation analysis.

JUBILEE RAILWAY Narrow-gauge railway suitable for transporting materials around a construction site.

JUDAS Hinged door or inspection panel within a door.

JUICER Hydraulic loader.

JUMBO 1. Machine designed to contain two or more mounted drilling units that may, or may not, be operated independently; **2.** Traveling support for forms, commonly used in tunnel work.

JUMBO BRICK *See* **Brick.**

JUMPER 1. Stretcher that covers more than one cross joint in squared or snecked rubble; **2.** Temporary electrical connection; **3.** Movable, inverted dome-shaped washer in a faucet.

JUMPER PIPE Small-diameter pipe used to connect the inlet and return fuel manifold to the injectors.

JUMPING 1. Raising a derrick from one operational mounting height to another; **2.** Moving a spar tree in an upright position to a better location at the landing.

JUMPING JACK Power driven hand tamper.

JUMPING SHOE Bracket on which the foot of the boom of a guy derrick is placed when jumping the unit to the next tier of steel.

JUMPOVER *See* **Return offset.**

JUMP SET Tunnel rib system installed between two previously placed supports.

JUNCTION BOX Box to which electrical conduit is fastened and in which electrical connections are made in domestic wiring.

JUNCTION CHAMBER Access manhole in which one or more branch sewers are joined to a main sewer.

JUNCTION POINT Point where a circular curve joins a noncircular curve.

JUNIOR MORTGAGE *See* **Mortgage.**

JUNKBUTT Badly splintered end of a felled tree that has been cut back to sound wood.

JUNKET RING *See* **Backup ring.**

JUNK FILL Fill that has been dumped over a long period and that contains refuse of all kinds.

juris. *Abbreviation for:* jurisdiction.

JURY RIG Any kind of improvised or expedient rigging.

JUST COMPENSATION Payment required by law for the loss sustained by an owner as a result of taking or damaging private property for highway purposes.

JUT Part of a building that projects abruptly outward from the main face.

JUVENILE SPACING Silviculture treatment to reduce the number of trees in young stands, often carried out before the stems removed are large enough to be used or sold as a forest product.

JUVENILE WATER *See* **Groundwater.**

JUVENILE WOOD Inner core of xylem surrounding the pith of a tree. The cells are smaller and less structurally developed than those of the outer xylem. The time during which juvenile wood is formed is termed the juvenile period, which varies among individuals with species, and with environmental conditions.

k Thermal conductivity: the amount of heat passing through one square foot of a homogenous material, one inch thick.

k *Abbreviation for:* key; kilo; kip.

K *Abbreviation for:* kelvin.

Ka *Abbreviation for:* cathode.

KA *Abbreviation for:* keyed alike.

KANAT Horizontal tunnel collection system developed within an aquifer at the bottom of a water well.

KAOLIN Rock, generally white, consisting primarily of clay minerals of the kaolinite group, composed principally of hydrous aluminum silicate, of low iron content, used as raw material in the manufacture of white cement.

KAPLAN TURBINE Water turbine, the propeller blade pitch of which is automatically varied according to the demand load.

KARST Geologic setting where cavities are developed in massive limestone beds by solution in flowing water. *Also called* **Karstic limestone.**

KARSTIC LIMESTONE *See* **Karst.**

KB *Abbreviation for:* knee brace.

K-BRACING System of struts used in a braced frame in which the pattern of the struts resembles the letter K, either normal or on its side.

kc *Abbreviation for:* kilocycle.

KD *Abbreviation for:* keyed differently; kiln dried; knocked down.

KEARBY METHOD Method of calculating quantities of asphalt and cover aggregate for standard seal coats.

KECKLE To cover or protect something by winding it with cable, rope, or wire.

KEEL 1. Longitudinal protrusion on the underside of a retaining wall foundation to prevent the wall from sliding away from its retained material; 2. Plumbers colored marking crayon.

KEEL ARCH *See* **Arch.**

KEEL MOLDING Molding formed of two ogee curves back-to-back.

KEENE'S CEMENT Cement composed of finely ground, anhydrous, calcined gypsum, the set of which is accelerated by the addition of other materials. Com-

monly used around showers and bathtubs.

KEEPER 1. Piece of lumber or steel used to hold a pile in the leads; 2. Receiving member of a lock, bolt, or latch. *Also called* **Striker plate.**

KEEPER PIN Similar to a cotter pin but with one leg deformed to pass over and grip the item being secured by the pin. Used where frequent removal is necessary.

KEEPING THE GAUGE Maintaining the vertical spacing of brick or masonry courses.

KEEPING THE PERPENDS Laying regularly-proportioned materials, such as brick, tile, slate, etc., so that the cross joints are vertically above those where they occur in preceeding courses.

KELLSTONE Stucco with crushed finish.

KELLY *See* **Kelly bar.**

KELLY BALL Apparatus used for indicating the consistency of fresh concrete, consisting of a cylindrical weight 150 mm (6 in.) in diameter and weighing 13.6 kg (30 lb) with a hemispherically shaped bottom, a handle consisting of a graduated rod, and a stirrup to guide the handle and serve as a reference for measuring the depth of penetration. *See also* **Ball test.**

KELLY BAR Square or splined shaft that can slide vertically through a square or splined opening in the driving head of a rotary drill to turn an auger or drill bit. *Also called* **Kelly.**

KELLY CROWD Mechanism that maintains down pressure on a kelly bar.

KELVIN One of the seven base units of the SI system of measurement: a unit of thermodynamic temperature equal to the fraction 1/273.16 of the thermodynamic temperature of the triple point of water. Symbol: K. The Kelvin scale has its origin, or zero point, at absolute zero (the point at which all atomic vibration ceases). Its fixed point of 273.16 K or 0°C is the temperature at which water exists in vapor, liquid, and solid state. The Celsius scale is derived from the Kelvin scale and has 0° for the freezing point of water and 100° as the boiling point. *See also the appendix:* **Metric and nonmetric measurement.**

KENTLEDGE Reaction weights for test loading. *Also called* **Caisson weights,** and **Load-test weight.**

KERF 1. Cut made by a saw blade (the thickness of the blade plus the left and right set given to the individual teeth); 2. To cut or notch, as a beam, transversely along the underside to curve it; 3. Cut or notch in a member such as a rustication strip to avoid damage from swelling of the wood and to permit easier removal.

KERF DUTCHMAN Special falling technique in which the constant relationship of the face (undercut), holding wood, and backcut are intentionally altered to solve a particular falling problem, i.e., to force a tree to jump off the stump.

KERFING Process of cutting grooves, or kerfs, across a board, usually to permit bending.

KERN AREA Area within a geometric shape in which a compressive force may be applied without tensile stresses resulting in any of the extreme fibers of the section.

KERN DISTANCE Distance from the centroid of a

section to the farthest point from the centroid at which resultant force can act without inducing a stress of opposite sign at the extreme fiber on the opposite side of the centroid.

Kes. *Abbreviation for:* Kessler.

KETTLE Movable tank used to store liquid asphalt during roofing operations.

KEVEL 1. Hinged bar that secures a door; 2. Cleat or peg to which a rope is secured.

KEY 1. Relative position of the headers of various courses of brickwork with reference to a vertical line; 2. Hard-steel strip inserted in a matching way (groove) in a shaft and a hub to form a positive connection and cause them to rotate as a unit; 3. Removable operating component of a lock; 4. Wedge-shaped piece of wood, metal, or plastic used to hold one or more components tight against each other, or in place; 5. Depression in a brick or tile to hold mortar. *See* **Frog**; 6. Plaster or mortar extruded between laths that serves to hold the mass in place; 5. Slotted joint in concrete, such as tongue-and-groove.

KEY BLANK Uncut key of a pattern to fit one or more lock designs.

KEY BLOCK *See* **Keystone.**

KEY DROP Pivoted escutcheon cover.

KEYED Fastened or fixed in position in a notch or other recess.

KEYED BEAM Beam with a lap joint into which joggles have been cut in each member, into which keys are driven.

KEYED DOWEL Dowel having grooves along its length to permit escape of air and excess glue.

KEYED JOINT *See* **Pointing.**

KEYED MORTISE AND TENON Tusk tenon.

KEYHOLE 1. Opening into which a key is fitted; 2. Welding technique in which concentrated heat penetrates the workpiece leaving a hole at the leading edge of the weld. As the heat source moves on, molten steel fills the hole forming the weld bead.

KEYHOLE SAW *See* **Saw.**

KEYHOLE SLOT Slot enlarged at one end to allow entrance of a chain or bolt shaft that can then be held by the narrow end.

KEYHOLE TIE-DOWN Keyhole-shaped holes through which chain can be inserted and held secure.

KEYING Use of a wedge-shaped piece to strengthen a joint.

KEYING-IN Bonding a brick or block wall to one already constructed.

KEY-OPERATED SWITCH *See* **Switch.**

KEY PLAN Small-scale plan that relates each part of the site to the whole; normally used in conjunction with a set of working drawings.

KEY PLATE *See* **Escutcheon.**

KEYSTONE Central stone of an arch or rib vault (sometimes called a **Lockstone**). *See also* **Arch.** *Also called* **Key block.**

KEY SWITCH *See* **Switch.**

KEY VALVE *See* **Valve.**

KEYWAY 1. Recess or groove in one lift or placement of concrete that is filled with concrete of the next lift, giving shear strength to the joint. *See also* **Tongue and groove**; 2. Square-edged lengthwise slot in a shaft or hub; 3. Opening in a lock into which a key may be inserted.

kg *Abbreviation for:* keg; kilogram.

KG BLADE Blade on a crawler, used to clear unwanted vegetation in preparation for planting tree seedlings.

kg-cal *Abbreviation for:* kilogram-calorie.

kHz *Abbreviation for:* kilohertz; cycles per second.

KIBBLE Bucket used to raise workers, tools, or material from a shaft.

KICK Holt set with a choker to get a log around a stump or tree.

KICK ANGLE Angle iron pieces fastened to the beams of a machine mounting to prevent its horizontal movement.

KICKBACK 1. Sudden and dangerous jump of the butt of a falling tree as it comes down; 2. Abrupt and dangerous backward movement of a chainsaw toward the operator, often caused by touching the moving chain at the tip of the bar to an object when starting to cut.

KICKER 1. Wood block or board attached to a formwork member in a building frame or formwork to make the structure more stable. *Also called* **Kick strip**; 2. Concrete plinth, between 50 and 75 mm (and 3 in.) high above a concrete floor, forming the start of a concrete column or wall to which formwork can be clamped.

KICKER BLOCK *See* **Thrust block.**

KICKING PIECE Short timber fastened to a waling to absorb the thrust of a raking shore.

KICK LIFT Jacking wedge used to elevate or shift sheet materials into proper position on a wall during installation.

KICKOUT 1. Lower section of a downspout that directs water away from a structure; 2. Accidental release or failure of a cross brace.

KICK PLATE Metal strip fastened to the lower edge of a door. *See also* **Plate.**

KICK STRIP *See* **Kicker.**

KIESELGUHR German name for diatomaceous earth, a filler or carrying agent in high explosives. *See also* **Diatomaceous earth.**

KILL 1. Stop an engine; 2. Cut off electric current from a circuit; 3. Coat an active substance to prevent it bleeding through.

KILN Chamber having controlled air flow, temperature, and relative humidity.

KILN-DRIED LUMBER Lumber that has been kiln dried with the use of artificial heat to a moisture content of between 6% and 12%. *See also* **Seasoning.**

KILN RUN Brick or tile from one kiln that has not been sorted or graded for size or color variation.

KILN STICK *See* **Sticker.**

KILO Prefix representing 10^3. Symbol: k. Used in the SI system of measurement. *See also the appendix:* **Metric and nonmetric measurement.**

KILOGRAM One of the seven base units of the SI system of measurement: a unit of mass equal to the mass of the international prototype of the kilogram, a cylinder of platinum-iridium alloy, kept by the International Bureau of Weights and Measures in Paris, France. Symbol: kg. Multiply by 35.273 96 to obtain ounces, symbol oz; by 2.204 62 to obtain pounds, symbol lb; by 0.000 984 21 to obtain long tons (2,240 lb); and by 0.001 102 3 to obtain short tons (2,000 lb). *See also the appendix:* **Metric and nonmetric measurement.**

KILOGRAM-FORCE Obsolete unit of force, equal to 9.8 Newtons, that should not be used with the SI system of measurement. Symbol: kg·f. *See also the appendix:* **Metric and nonmetric measurement.**

KILOGRAM METER PER SECOND A derived unit of momentum with a compound name of the SI system of measurement. Symbol: k·m/s. *See also the appendix:* **Metric and nonmetric measurement.**

KILOGRAM METER SQUARED A derived unit of the moment of inertia with a compound name of the SI system of measurement. Symbol: kg·m². *See also the appendix:* **Metric and nonmetric measurement.**

KILOGRAM METER SQUARED PER SECOND A derived unit of angular momentum with a compound name of the SI system of measurement. Symbol: kg·m²/s. *See also the appendix:* **Metric and nonmetric measurement.**

KILOGRAM PER CUBIC METER A derived unit of density (mass density) with a compound name of the SI system of measurement. Symbol: kg/m³. *See also the appendix:* **Metric and nonmetric measurement.**

KILOGRAM PER LITER A derived unit of density with a compound name of the SI system of measurement. Symbol: kg/L. *See also the appendix:* **Metric and nonmetric measurement.**

KILOGRAM PER MOLE A derived unit of molar mass with a compound name of the SI system of measurement. Symbol: kg/mol. *See also the appendix:* **Metric and nonmetric measurement.**

KILOGRAM PER SECOND A derived unit of mass flow rate with a compound name of the SI system of measurement. Symbol: kg/s. *See also the appendix:* **Metric and nonmetric measurement.**

KILOHERTZ Unit of frequency, equal to 1000 hertz, of the SI system of measurement. Symbol: kHz. *See also the appendix:* **Metric and nonmetric measurement.**

KILOKELVIN Unit of temperature, equal to 1000 kelvin, of the SI system of measurement. Symbol: kK. *See also the appendix:* **Metric and nonmetric measurement.**

KILOJOULE Unit of energy, equal to 1000 joules, of the SI system of measurement. Symbol: kJ. *See also the appendix:* **Metric and nonmetric measurement.**

KILOLITER Unit of volume, equal to 1000 liters. Permitted for use with the SI system of measurement. Symbol: kL. *See also the appendix:* **Metric and nonmetric measurement.**

KILOMETER Unit of length, equal to 1000 meters, of the SI system of measurement. Symbol: km. *See also the appendix:* **Metric and nonmetric measurement.**

KILOMETER PER HOUR Unit of speed, equal to 1000 meters covered in one hour. Permitted for use with the SI system of measurement. Symbol: km/h. *See also the appendix:* **Metric and nonmetric measurement.**

KILONEWTON Unit of force, equal to 1000 newtons, of the SI system of measurement. Symbol: kN. *See also the appendix:* **Metric and nonmetric measurement.**

KILOPASCAL Unit of pressure, equal to 1000 pascals, of the SI system of measurement. Symbol: kPa. *See also the appendix:* **Metric and nonmetric measurement.**

KILOVOLT-AMPERE Electrical unit of power, 1000 volt amperes (aparent power), equivalent to about 0.89 kW.

KILOVOLT AMPS REACTIVE 1000 volt amps reactive (reactive power).

KILOWATT Unit of power, equal to 1000 watts, of the SI system of measurement. Symbol: kW. *See also the appendix:* **Metric and nonmetric measurement.**

KILOWATT HOUR Non-SI unit of energy, equal to 3.6 mega joules, permitted for use with the SI system of measurement for a limited time. Symbol: kW·h. Multiply by 3412 to obtain British thermal units, symbol: Btu; by 1.340 5 to obtain horsepower hours, symbol: hp/hr. *See also the appendix:* **Metric and nonmetric measurement.**

kin. *Abbreviation for:* kinetic.

KINEMATIC VISCOSITY *See* **Viscosity.**

KINETIC ENERGY *See* **Energy.**

KING BOLT Vertical tie rod in a truss.

KING CLOSER *See* **Brick.**

KING PILE Center pile in a dolphin or cluster, usually with its top higher than the adjacent piles, used to hold lines from a ship.

KINGPIN 1. Vertical swivel or hinge pin, usually supported at top and bottom, for a front-wheel spindle; 2. Anchor pin at the center of a semitrailer coupling that engages with locking jaws at the center of the tractor's fifth wheel.

KINGPIN INCLINATION Tilt, inboard from vertical, of the top of the king pin of a steering mechanism.

KINGPIN PLATE Flat plate on the underside of an upper coupler through which a kingpin protrudes and that rests directly on a tractor fifth wheel.

KING POST Center, vertical member of a roof truss connecting the tie-beam or collar-beam with the ridge.

KING RAIL One rail of a set, used as a reference rail.

KINK A unique deformation caused by a loop of a product, typically a wire rope, being pulled down tight. It can represent irreparable damage to and an indeterminate loss of strength in the product or, in the case of a hose, prevention of flow.

KIOSK Traditionally, an open summerhouse or pavilion. Currently, a lightly built, often transportable, small enclosed structure used for retailing.

KIP Nonmetric unit of mass, equal to 1000 lb. Symbol: kip. Multiply by 4.448 222 to obtain kilonewtons, symbol: kN. *See also the appendix:* **Metric and nonmetric measurement.**

kip *Abbreviation for:* 1,000 pounds.

kip-ft *Abbreviation for:* 1,000 foot-pounds.

KIP PER SQUARE INCH Nonmetric unit of force. Symbol: kip/in.². Multiply by 6.894 757 to obtain megapascals, symbol: MPa. *See also the appendix:* **Metric and nonmetric measurement.**

KISS MARK Mark where bricks have touched during firing.

kit. *Abbreviation for:* kitchen.

KITE Sheet of Kraft paper applied to a sheet of coated roofing during manufacture, used to measure the weight of surfacing granules.

KLF *Abbreviation for:* kips per linear foot.

km *Abbreviation for:* kilometer.

KNAPP To split a rock, particularly a flint, revealing surfaces that can be exposed in face work without further working.

KNAPPED FLINT Large flints, split and laid so that the smooth, dark surfaces form the facing of a wall, or a decoration within the wall structure.

KNAPPING HAMMER *See* **Hammer.**

KNEE 1. Convex length of handrail; **2.** 90° bend in a pipe.

KNEE BRACE Brace between horizontal and vertical members in a building frame or formwork to make the structure more stable.

KNEELER Stone shaped to provide a change in direction.

KNEELING BOARD Board used to kneel on while finishing concrete.

KNEE WALL *See* **Wall.**

KNIFE 1. Cutting edge of a digging or excavating machine; **2.** Tree shear's cutting blade.

KNIFE-BLADE SWITCH *See* **Switch.**

KNIFE SWITCH *See* **Switch.**

KNITTED FABRIC Flat or tubular structure made from one or more yarns or filaments whose direction is generally transverse to the fabric axis but whose successive passes are united by a series of interlocking loops.

KNITTED PLY Layer of textile reinforcement in which the yarns are applied in an interlocking looped configuration in a continuous tubular structure.

KNITTER Machine for forming a fabric by the action of needles engaging threads in such a manner as to cause a sequence of interlaced loops. (Interlaced loops forming a continuous tubular structure are commonly used as hose reinforcement.)

knl *Abbreviation for:* knurl.

KNOB 1. Ferrule attached to the terminal end of a wire rope by babbitt, wedges, or hydraulic pressure; **2.** Abrupt hill with concave slopes and a rounded top; **3.** Rounded, projecting handle.

KNOB-AND-TUBE WIRING Obsolete method of exposed wiring using knobs and tubes of nonconducting materials to insulate the wiring from surfaces on which, or through which it is installed.

KNOBBING Rough dressing of quarried stones to remove humps and large projections.

KNOCK Uncontrolled explosion of the last portion of the burning fuel-air mixture in the combustion chamber of a spark-ignition engine. Knock or detonation results in the development of sudden pressure which reduces engine efficiency and can lead to engine damage.

KNOCK DOWN In fire fighting, to reduce the flame or heat on the more vigorously burning parts of a fire edge.

KNOCKED DOWN Structure or fitting, complete and finished in all respects, unassembled on delivery.

KNOCKER Hinged striker attached to a door.

KNOCKINGS Oversize residue obtained in screening a ceramic slip.

KNOCKING UP Working stiffened lime mortar to restore it to a plastic condition.

KNOCKOUT Stamped section on an electrical box designed to be knocked out so fittings can be attached or wires installed.

KNOCKOUT BOX Device used to slow the velocity of the exhaust gases from an asphalt plant dryer or drum mixer to remove the larger dust particles from the air stream and increase the efficiency of the dust collection system.

KNOCKOUT PUNCH Tool for cutting circular holes in ducts or boxes.

KNOCK OVER Fasten the end of one duct over the end of another.

KNOLL Hill, usually round to oval-shaped, with convex slopes.

KNOT 1. The portion of a branch or limb of a tree that appears on the edge or face of a piece of lumber; **2.** Fastening made by intertwining one or more pieces of cord, twine, etc.; **3.** Nonmetric unit of velocity, equal to one nautical mile per hour. Symbol: kn. Multiply by 1.852 to obtain kilometers per hour, symbol: km/h. *See also the appendix:* **Metric and nonmetric measurement.**

KNOTHOLE Void produced when a knot drops out of veneer.

KNOTTING Shellac dissolved in methyl hydrate, used to seal knots in new wood to prevent sap or resin from bleeding through.

knty *Abbreviation for:* knotty.

KNUCKLE Curved part of a hinge through which the pin passes.

KNUCKLEBOOM Hydraulically operated loading boom whose mechanical action imitates the human arm. *See also* **Loader,** and **Truck-mounted cranes.**

KNUCKLE JOINT Curb joint in a mansard roof.

ko *Abbreviation for:* knockout.

KP *Abbreviation for:* kick plate.

KP&D *Abbreviation for:* kick plate and drip.

kpr *Abbreviation for:* keeper.

KRAFT PAPER Strong brown paper, sometimes used as a building paper.

KRUGER H SYSTEM Electrically based system that provides an audiovisual warning to a crane operator to signify an approaching two-block situation. *See also* **Anti-two block system.**

KRUEGER HAP SYSTEM Electrically based system that provides a crane operator with an audiovisual anti-two block warning, boom angle indicator, and minimum/maximum boom angle presets.

KRUEGER HLAP SYSTEM Electrically based system that provides a crane operator with an audiovisual anti-two block warning, boom angle indicator, minimum/maximum boom angle presets, and boom length indicator.

KRUEGER LMI SYSTEM Electrically based load moment indicating system that provides a crane operator with an audiovisual anti-two block warning, boom angle indicator, minimum/maximum angle preset, boom length indicator, and a load measuring system that expresses the lifted load as a percentage of the machine's capacity.

ksf *Abbreviation for:* kips per square foot.

ksi *Abbreviation for:* kips per square inch.

kv *Abbreviation for:* kilovolt; thousand volts.

KV *Abbreviation for:* key valve.

kva *Abbreviation for:* kilovolt-ampere; thousand volt-amperes.

K-VALUE 1. Thermal conductivity of a material; **2.** Load-carrying capacity of a subgrade or pavement layer based on a static load (plate-bearing) test. *Also called* **Modulus of subgrade reaction.**

kw *Abbreviation for:* kilowatt; thousand watts.

kwhm *Abbreviation for:* kilowatt-hour meter.

kwhr *Abbreviation for:* kilowatt-hour; thousand watts per hour.

L

l *Abbreviation for:* line; lumen.

L *Abbreviation for:* angle; lambert; left; length; lip.

la. *Abbreviation for:* lane; load-to-axle

LA *Abbreviation for:* lightning arrester.

lab. *Abbreviation for:* labor; laboratory.

LABEL *See* **Dripstone**, and **Hood-mold.**

LABELED Equipment or materials to which has been attached a label, symbol, or other identifying mark of a qualified testing laboratory, indicating compliance with appropriate standards of performance in a specified manner.

LABEL STOP Ornamented terminal to a drip mold.

LABOR Cost of services of employees performing physical or mechanical tasks, usually paid as wages rather than salaries.

LABOR AND MATERIAL PAYMENT BOND *See* **Bond.**

LABORATORY SAMPLE Representative portion of a gross sample received by a laboratory for further analysis.

LABOR CONSTANT Amount of labor necessary to perform a defined unit of work.

LABORER One who performs manual labor or who labors at at an occupation requiring physical strength.

LAC Resinous excretion left by insects on East Indian trees; the basis of shellac, lacquers, and some varnishes.

LACE 1. To raise and slip an interlock of a pile, such as a steel sheet pile, into the interlock of an adjoining pile. *Also called* **Thread**; **2.** To raise and slip a pile hammer into pile hammer leads so that its side channels properly engage to slide upon the guides of the leads.

LACED FALL Line reeved through multisheaved pairs of blocks or pulleys.

LACED VALLEY Valley formed without a valley gutter, the tiles being laced or interleaved.

LACER BAR Small-diameter bar used in a strategic locations under top reinforcing bars, wired to the top bars to hold them in the correct lateral position.

LACING 1. Act of interlocking sections of sheet pile to form a wall. *Also called* **Threading**; **2.** Horizontal bracing between shoring members; **3.** Course of brick in a rubble wall; **4.** Bonding course between brick courses of an arch; **5.** Location of the working tools on a cutter drum.

LACK OF FUSION Weld defect caused by a lack of union between weld metal and base metal.

LACQUER Finish made of resins dissolved in ethyl alcohol.

LACUNA Depression where something has been omitted or left out.

LACUNARIA Coffers or panels in a ceiling or the underneath of a horizontal surface.

lad. *Abbreviation for:* ladder.

LADDER 1. Structure for climbing up or down, consisting of two side rails separated by a series of flat crosspieces at regular intervals of approximately 300 mm (12 in.); **2.** Digging-boom assembly of a hydraulic dredge or chain bucket-type excavator.

LADDER DITCHER Trench digging means that operates by means of buckets mounted on a pair of chains traveling on the outside of a boom. *See also* **Ditcher**, and **Wheel ditcher.**

LADDER-JACK SCAFFOLD Light-duty supported scaffold consisting of a platform supported by brackets attached to ladders.

LADDER SCRAPER Self-propelled excavating machine that incorporates a conveyor loading device to move earth scraped from the surface into its bowl.

LAG 1. Delay between consecutive actions; **2.** To wrap pipes either to reduce heat losses or prevent freezing; **3.** Wrap a cylinder so as to increase its diameter; **4.** Plate, block or structural section secured to a pile to increase its bearing capacity.

lag. *Abbreviation for:* lagging.

LAG BOLT/SCREW Heavy wood screw with a square or hex head. *Also called* **Coach screw.**

LAG EXPANSION INSERT Anchor used with a lag bolt. The anchor is inserted in a predrilled hole in concrete, masonry, etc., and expands to grip as the bolt is screwed into it.

LAGGED PILE Pile to which longitudinal pieces (lags) are fastened to increase its frictional area in soil.

LAGGING 1. Heavy sheathing used in underground work to withstand earth pressure; **2.** Planking placed horizontally between soldier piles in a shored or braced excavation; **3.** External wood covering on a reel to protect the wire rope or strand; **4.** Grooved shell of a drum. *See also* **Drum lagging**, and **Split lagging.**

LAGGING POWER FACTOR Condition caused by inductive loads, such as motors and transformers, in which the current lags behind the voltage in an alternating curent network.

LAITANCE An accumulation of fine particles on the surface of fresh concrete due to an upward movement of water. *See also* **Blemish.**

LAKE ASPHALT Naturally occurring asphalt deposits that form small lakes in natural depressions of the earth's crust.

LAKE DWELLING House built over water, on piles, and connected to the land by a bridge or walkway.

LAKE SAND Sand consisting mainly of fine, rounded granules.

LALLY COLUMN Metal pipe, sometimes filled with concrete, used to support girders or beams. *Also called* **Pipe column.**

lam. *Abbreviation for:* laminate; laminated

LAMBERT Unit of brightness: average brightness of a surface emitting or reflecting one lumen per square centimeter.

LAMELLA ROOF Large-span vault consisting of members joined in a diamond pattern.

LAMELLAR TEARING Separation in highly restrained base metal, caused by through-thickness strains induced by shrinkage of adjacent weld metal.

LAMINA Stratum less than 10 mm (0.375 in.) thick.

LAMINAR CONSTRUCTION Fabrication of a structural member by gluing or fastening layers of like material to a designed depth.

LAMINAR FLOW Water flow in which the stream lines remain distinct and in which the flow direction at every point remains unchanged with time. *See also* **Flow.**

LAMINAR (STREAMLINE) FLOW Flow situation in which fluid moves in parallel lamina or layers. *See also* **Reynold's number.**

LAMINAR VELOCITY Speed in a specific horizontal stream of water, distinct from those above or below it, in which streamline or turbulent flow may occur.

LAMINATE Building up in layers, as in plywood.

LAMINATED ARCH Arched rafter or beam formed by gluing and nailing thin strips of wood over a mold.

LAMINATED COVER Hose cover formed to desired thickness from thinner layers vulcanized together.

LAMINATED DECKING Bridge decking with laminations used in place of the crossties and planks.

LAMINATED GLASS Transparent safety glass composed of one or more layers of a plastic material between sheets of plate glass.

LAMINATED INSULATION Insulation mass comprised of different materials placed in layers against each other to take best advantage of the specific (structural, thermal, waterproof, etc.) characteristics of each.

LAMINATED JOINT Combed or finger joint.

LAMINATED PLANK Board of laminated wood veneers.

LAMINATED PLASTIC SHEET Semirigid decorative board formed by soaking sheets of paper with synthetic resin.

LAMINATED WALLBOARD Two or more layers of drywall bonded together with an adhesive.

LAMINATED WOOD Timber section built up of plies or laminations that have been glued and/or mechanically joined.

LAMINATING COMPOUND Cementitious material used to bond two or more layers of drywall together.

LAMINATION Banding of rocks caused by varia-

tions in different minerals, usually distinguishable by different colors.

LANCET ARCH *See* **Arch.**

LAND 1. That part of a grooved surface that is not indented; **2.** Metal left as a bevel on the flange of an H-pile for backup of a weld; **3.** A backfurrow; **4.** To place or drop logs in a landing during yarding.

LAND ABUSE *See* **Land-use control.**

LAND AREA *See* **Land-use classes (forest management).**

LAND ASSEMBLY Acquisition of, or option to acquire, contiguous parcels of land with the intent to create an area for development, or reserve against development.

LAND BANKING Acquisition of land, usually by a public authority, with the intent of reselling the whole or part of the area for designated types of development at some future date.

LAND BASE (FOREST MANAGEMENT) Areas of forest land that are actually available for forest management. This involves future trends not only in forest growth but also in deletions from the land base.

LAND CLASSIFICATION (FOREST MANAGEMENT SYSTEM) System that defines land according to ability to sustain specific types of forest growth, or with regard to the type of existing forest growth, including:

> **Basal area per ha (acre):** Land class based on total basal area per hectare (acre).
>
> **Site class:** Classification of forest land in terms of its inherent capacity to grow crops of industrial wood. Expressed in m^3 (cubic-foot) growth per hectare (acre) per year.
>
> **Site index:** Expression of the growing potential of a specific forest site based on the height of a free-growing dominant or codominant tree of a representative species in a forest of the same type at a specified age.
>
> **Stand age:** Age of trees of the dominant forest type and stand-size class.

L&CM *Abbreviation for:* lime-and-cement mortar.

LAND DEVELOPMENT Work done on land either to bring it to a stage of higher or better use or in anticipation of a future construction project. *See also* **Development.**

LAND DRAINAGE Process of removing water from on top of and within the ground.

LANDFILL *See* **Sanitary landfill.**

LANDFILL CELL Engineered volume in which waste is placed, compacted, and covered daily to form a portion of the landfill.

LANDFILL COMPACTOR *See* **Loader.**

LANDFILL GAS Natural by-product of the decomposition of organic material in the waste stream following disposal. Anaerobic decomposition produces the gas, which consists of approximately equal proportions of methane and carbon dioxide.

LANDFILL MACHINE Any machine that is used on a sanitary landfill; generally considered to be dozers, tractors, loaders, compactors, and/or scrapers.

LANDFORM Physical feature of the surface of the land, such as a hill or a valley, whether resulting from natural processes or man-made.

LAND IMPROVEMENT Work done and expenditures incurred in the process of putting land into usable condition, e.g., clearing, grading, landscaping, paving, installing utilities and services, etc.

LANDING 1. Platform between flights of stairs or at the termination of a flight of stairs; **2.** Area where logs are collected for loading. *Also called* **Brow; 3.** Area where logs are landed by a yarder; **4.** *See* **Dock.**

LANDING BOARD First board on a landing, immediately above the last riser.

LANDING BUCKER Worker who bucks logs at a landing.

LANDING GEAR 1. Dolly, or portion of a trailer, that holds the trailer upright when it is not supported by the truck tractor; **2.** Block or roller attached to a stationary object that guides the pull of a cable.

LANDING GEAR SAND SHOE Pivoted flat plate attached to the bottom of each leg to afford increased ground bearing surface.

LANDING GEAR WHEEL Wheel attached to the bottom of a landing gear lower leg.

LANDING NEWEL Post at the landing point of a stair supporting the handrail.

LANDING TREAD Front of a landing connecting to a flight of stairs, the front edge of which forms the uppermost tread, the nose of which is the same thickness as the remaining treads, even if this differs from that of the landing floor.

LANDING WORKER Worker who bucks, limbs, and trims log ends, unhooks chokers, and assists in hooking up trailers to log trucks. *Also called* the **Chaser.**

LANDING ZONE 1. Designated area where materials may be moved to, off-loaded, or stored; **2.** Area extending from 450 mm (18 in.) below a landing to 450 mm (18 in.) above it.

L&L *Abbreviation for:* latch and lock.

LANDLOCKED Building lot that has no access to a public thoroughfare except through an adjacent lot or lots.

LANDLORD One who rents property to another.

LANDLORD'S PROTECTIVE LIABILITY Insurance coverage for property owners who lease the entire premises to others who assume full control of the property.

L&M *Abbreviation for:* labor and material.

LANDMARK Fixed natural or man-made feature that serves as a boundary mark for a defined area of land.

LAND RECLAMATION Gaining usable land by removing surface water, preventing ingress of water (diking), lowering groundwater, or otherwise making stable land that was previously incapable of supporting loads.

LAND REGISTRAR Public office where deeds, mortgages, and other legal instruments affecting realty are recorded and maintained and are available for public inspection. (In the U.S., this function is more commonly handled by the **Recorder of Deeds** or **Registrar of Deeds**).

LAND REGISTRATION Recording of deeds, mortgages, and other instruments in a public office to evidence one's rights in real property.

landsc. *Abbreviation for:* landscape; landscaped.

LANDSCAPE ARCHITECTURE Integration of site layout and grading in addition to the use of plant and man-made materials into a designed whole.

LANDSCAPE CONTRACTOR One who contracts to modify landforms, supply and install materials as part of a design for the landscape of a defined area.

LANDSCAPING General use of plant and man-made materials on a site (as opposed to landscape architectural design).

landsc. arch. *Abbreviation for:* landscape architect.

LANDSLIDE Sudden downward movement of a mass of surface soil and/or rock, usually at a shear plane, often triggered by excess moisture content or seismic vibration, and frequently contributed to by changes in the ground profile in the immediate vicinity.

LAND SPECULATION Option on, or purchase of land in anticipation that its value will increase.

LAND SURVEY *See* **Surveying.**

LAND SURVEYOR Person qualified to measure land and buildings and prepare drawings of the results.

LAND THE WEIGHTS To set a counterweight on its buffer, or on supports placed for that purpose.

LAND TIE Tie rod or tendon holding sheet piling or a retaining wall to a buried deadman or stay pile.

LAND TILE *See* **Tile.**

LAND USE Division of available land into defined categories under such headings as use, class, ground cover, mineralization, etc.

LAND-USE CLASSES (FOREST MANAGEMENT)

> **Commercial forest land:** Forest land capable of producing crops of industrial wood and not withdrawn from timber utilization by regulation. Included are areas suitable for growing crops of industrial wood and generally capable of producing in excess of 0.7 m³ (25 ft³) per ha (acre)) of annual growth.

> **Forest land:** Land at least 16.7% stocked by forest trees of any size, or formerly having such tree cover, and not currently developed for nonforest use.

> **Gross area:** Entire area of land and water.

> **Land area:** Area of dry land and land temporarily or partially covered by water, such as marshes, flood plains, streams, sloughs, and estuaries. Canals less than 0.2 km (0.124 mile) wide, and lakes, reservoirs, and ponds smaller than 16 ha (40 acres).

> **Noncommercial forest land:** Unproductive forest land incapable of yielding crops of industrial wood because of adverse site conditions. Also,

productive forest land withdrawn from commercial timber use by regulation, or exclusively used for Christmas tree production.

Nonforest land: Land that has never supported forests, or land formerly forested where forest use is precluded by development for nonforest uses such as cropland, improved pasture, residential areas, and city parks.

Reserve forest land: Noncommercial forests that are productive but reserved for recreation or other nontimber uses.

LAND-USE CONTROL Public control over the manner in which land is used and/or developed. This control is exercised in several distinct ways:

Land abuse: Abortive subdivision, cut-over lands, etc.

Land use prescription: Zoning, building restrictions, etc.

Non-use or disuse: Taxing to enforce development and/or maintenance, clearing, etc.

Public protection: Building standards, density regulations, etc.

Reuse control: Urban redevelopment, slum clearing, etc.

LAND-USE DESIGNATION Type and extent of development established for an area of land defined on an Official Plan.

LAND-USE INTENSITY Expression of the intensity to which an area of land is developed in relation to an established zoning plan.

LAND-USE MAP Overall map of a community showing the character of and density of land use.

LAND-USE PLAN Official determination for the future uses of land contained within an Official Plan, showing the public and private improvements to be made on the land and assumptions and reasons for arriving at the indications.

LAND-USE PLANNING Development of plans intended to produce the highest and best use of land, together with proposals as to how such use can be achieved.

LAND-USE PRESCRIPTION *See* **Land-use control.**

LAND-USE REGULATION Zoning, official maps, and subdivision regulations to guide or control land development.

LANE 1. Any passageway or right-of-way, open from ground to sky, not constituting a street, but laid down upon a registered plan and dedicated to public use; **2.** Strip of roadway used for a single lane of traffic. There are several types, including:

Auxiliary lane: Portion of a roadway adjoining the through traveled way for parking, speed change, turning, storage for turning, weaving, truck climbing, or for other purposes supplementary to through traffic movement.

Median lane: Speed-change lane within the median to accommodate left-turning vehicles.

Parking lane: Auxiliary lane primarily for parking of vehicles.

Speed-change lane: Auxiliary lane, including tapered areas, primarily for the acceleration or deceleration of vehicles entering or leaving the through traveled way.

Traffic: Portion of the traveled way for the movement of a single line of vehicles.

LANE DELINEATOR Portable device used in traffic control to mark differences between lanes.

LANELINE MARKING *See* **Pavement marking.**

LANG LAY *See* **Lay types.**

LANGLEY Nonmetric unit of solar radiation measurement equal to one calorie of radiation energy per square centimeter per minute (220 Btu per square foot per hour).

LANOVA COMBUSTION CHAMBER Specific pre-combustion chamber design in which fuel is injected toward the chamber, burns turbulently inside the chamber, and is then reinjected into the main combustion chamber, as a flame, to complete its burning in the cylinder.

LANTERN 1. Small circular or polygonal turret with windows all round, surmounting a roof or dome; **2.** Hollow casing on the engine side of the body of a centrifugal pump.

LANTERN RING Ring in line with a port in a gland to introduce a lubricant or a coolant to the packing or stuffing box. *Also called* **Seal cage.**

LANYARD Safety line attached to a safety belt.

LAP 1. Part that extends over itself or like part, usually by a desired and predetermined amount; **2.** Distance one brick extends over another; **3.** Joint made when two pieces of wood overlap each other; **4.** Length by which one reinforcing bar or sheet of fabric reinforcement overlaps another; **5.** Folded-over section of metal that is then rolled or forged into the surface.

lap. *Abbreviation for:* lapped.

LAP CEMENT Asphaltic material used to seal the seams of sheet roofing.

LAP JOINT Joint in which the structural units being joined override each other so that, with movement, a sealant is in shear between the joint faces.

LAPPED Overlapped and fitted together.

LAPPED TENONS Two tenons that enter a mortise from opposite sides and lap within it.

LAPPING 1. Overlapping of reinforcing steel bars, welded wire fabric, or expanded metal so that there may be a continuity of stress in the reinforcing when the concrete member is subjected to loading; **2.** Finishing process typically employing loose abrasive grain, but more often including similar types of operation with bonded abrasive wheels or coated abrasives.

LAP SEAM Seam made by placing the edge of one piece of material extending flat over the edge of a second piece of material.

LAP SIDING Boards used to cover the sides of buildings, the lower edge of one board being lapped over the upper edge of the board below.

LAP SPLICE 1. Connection of reinforcing steel made by lapping the ends of the bars; similarly, the side and end overlap of sheets or rolls of welded wire fabric; **2.** Length of overlap of two reinforcing bars. *See also* **Contact splice.**

LAP SPLICE CONNECTOR SYSTEM Any of various proprietary prefabricated plastic or metal boxes or strips of foam acting as keyway forms and containing prefabricated anchorage and lap splice reinforcement. The lap splice reinforcement is field-straightened with a pipe and/or hickey when the keyway form is exposed for the next pour.

LAP WELD Weld along the edge of overlapping plates or pieces.

LAP WELD PIPE *See* **Pipe.**

laq. *Abbreviation for:* lacquer; lacquered.

LARGE-DIAMETER PILE One with a nominal diameter exceeding 600 mm (24 in.).

LARGE ROUTE *See* **Collection method.**

LASER Light Amplification by Stimulated Emission of Radiation (LASER), usually in the form of a small-diameter beam of intense light.

LASER LEVEL Instrument that generates a laser beam, either as a static ray or through a rotating lens which, mounted and leveled on a horizontal axis, will indicate all points of common elevation up to rated projection length of the instrument.

LASH To fasten together with cordage or wire rope.

LASHING RING Attachment to the underside of a trailer deck, used to fasten tiedown equipment.

LAST-CHANCE FILTER *See* **Chip control filter.**

LAST-IN FIRST-OUT *See* **Cost flow.**

lat. *Abbreviation for:* lateral; latitude; lattice.

LATCH Door catch consisting of a pivoted arm, the outer end of which engages a slot.

LATCHBOLT Beveled spring bolt.

LATENT HEAT *See* **Heat.**

LATERAL Anything that lies at an angle to the main component and that contributes to its function.

LATERAL BRACING MEMBER Member used individually, or as a component of a lateral bracing system, to prevent buckling of members or elements and/or to resist lateral loads and to prevent overturning.

LATERAL BUCKLING Buckling of a member involving lateral deflection and twist. *Also called* **Lateral-torsional buckling.**

LATERAL CENTER OF GRAVITY *See* **Center of gravity.**

LATERAL DEFLECTION Motion that occurs perpendicularly, or at right angles to the centerline of the object. In the case of an expansion joint, motion that is perpendicular to the centerline of the bellows. Lateral deflection can occur along one or more axes simultaneously. *See also* **Offset.** *Also called* **Side load.**

LATERAL DISPLACEMENT Sideways movement of asphaltic mix during compaction.

LATERAL EARTH PRESSURE Lateral forces interacting between an earth-retaining structure and the retained earth mass.

LATERAL FORCE Force generated across a moving vehicle while turning.

LATERAL LOAD Load that acts at 90° to the supporting structure.

LATERAL MOVEMENT Horizontal movement of a structure, earth, etc.

LATERAL NAIL RESISTANCE Amount of force required to rack a frame made of 1.2 x 2.4 m (4 x 8 ft) panels nailed to 50 x 100 mm (2 by 4 in.) studs so that the nails pull through the panel edges.

LATERAL PILE LOAD TEST *In situ* test to estimate the lateral resistance of a driven pile.

LATERAL PRESSURE *See* **Earth pressure.**

LATERAL REINFORCEMENT Usually applied to ties, hoops, and spirals in columns or column-like concrete members.

LATERAL SEWER Sewer that discharges into a branch or other sewer and has no other sewer tributary to it.

LATERAL STRAIN INDICATOR Device that measures lateral strain mechanically or electrically.

LATERAL SUPPORT 1. Batter piles or reinforcement to resist lateral forces on piles or footings; **2.** Means whereby walls are braced either vertically or horizontally by columns, pilasters, crosswalls, beams, floors, roofs, etc.

LATERAL THRUST Pressure of a load that extends to the sides.

LATERAL-TORSIONAL BUCKLING *See* **Lateral buckling.**

LATEST EVENT OCCURRENCE TIME Deadline by which an event must be completed so as not to delay the project.

LATEST FINISH DATE Latest day a work item can finish without affecting the project duration, assuming that all subsequent work items start as soon as they are able and are completed in their expected time.

LATEST START DATE Latest day that a work item can start without affecting the final project duration, assuming that it is completed in its expected time and that all subsequent work items start as soon as they are able and are completed in their expected times.

LATEWOOD *See* **Summerwood.**

LATEX Water emulsion of a high molecular-weight polymer, used especially in paints and finishes, coatings, adhesives, leveling compounds, and patching compounds.

LATEX EMULSION Dispersion in water of rubber or one of several synthetic resins. *Also called* **Plastic emulsion.**

LATH Narrow strips of wood, metal, gypsum, or fiberboard fastened to the frame of a building to serve as a base for plastering, slating, or tiling.

LATH AND PLASTER MEMBRANE Sheet of lath and plaster (including an integral support or stiffening) used as a structural unit.

LATH HAMMER *See* **Hammer.**

LATH, PLASTER, AND SET Two-coat work: a floating coat and finishing coat on lath.

LATH, PLASTER, FLOAT, AND SET Three-coat work: a floating coat, finishing coat, and set coat over lath.

LATHE Machine used to turn circular work from bar or cylinder stock.

LATHING Base for plaster.

LATITUDE Difference in northing from the survey instrument location to the point being observed.

LATRINE Multiseat toilet discharging to a single trough.

LATTICE Openwork structure of crossed or woven strips or bars. *Also called* **Trellis.**

LATTICE BOOM Lightweight boom fabricated of criss-crossed members, bar, tube, or angle, used on cranes and shovels, etc. *See also* **Boom.**

LATTICE TRUSS Truss in which the upper and lower members are connected by struts in a lattice-like pattern.

LATTICE WINDOW/DOOR Window or glazed door with diamond-shaped panes of glass set in lead-work, or decorated to resemble same.

lau. *Abbreviation for:* laundry.

lav. *Abbreviation for:* lavatory.

LAVATORY 1. Basin for washing hands and face; **2.** Place providing sanitary facilities.

LAWS AND REGULATIONS When included in the agreement or contract documents, the term 'laws and regulations' means all applicable laws, statutes, ordinances, building codes, rules, regulations, and lawful orders of any public authority.

LAY 1. Manner in which the wires in a strand, or the strands in a wire rope, are helically laid; **2.** Distance measured parallel to the axis of a rope (or strand) in which the strand (or wire) makes one complete helical convolution about the core (or center). *Also called* **Lay length,** and **Pitch; 3.** Position in which a felled tree is lying.

LAY BAR Horizontal glazing bar.

LAYDOWN MACHINE *See* **Asphalt paving machine.**

LAYER 1. Single thickness, coat, fold, or stratum; **2.** Wrap of wire rope around a rope drum barrel, extending fully from flange to flange; **3.** In plywood, one or more adjacent plies having the wood grain in the same direction; **4.** Single layer of rubber or fabric between adjacent parts of a hose.

LAYER BOARD Board on which the lowest sheet of a box gutter is placed. *Also called* **Lear board.**

LAYERED LAP Most common type of welded fabric lap where one sheet of fabric is just lapped over the top of another.

LAYERED SYSTEM (SOIL) *See* **Soil types.**

LAY IN Setting tiles or panels into prepared places, often within previously installed metal channels, as in suspended ceilings.

LAYING LINE Lines printed or scored on a material to indicate the amount of lap.

LAYING OVERHAND Building the face of a wall while positioned alongside the further face.

LAYING TO BOND Laying the brick of an entire course without a cut brick.

LAYING UP Hand placement of filament reinforcement mats in molds for the fabrication of fiber-reinforced products.

LAY LENGTH *See* **Lay.**

LAYLIGHT Glazed false ceiling, usually designed to diffuse artificial lighting or natural light from roof lights.

LAYOUT 1. Diagram or working plan, draw, to scale, showing the position of all components or members and their relationship to each other; may be specific to one or more aspects of the project: foundations, plumbing, wiring, framing, etc.; **2.** Marks or other physical indications made on the job to indicate position or direction; **3.** Logging plan.

LAYOUT BOARD Grooved board used by a locksmith to hold pin-tumbler parts in order.

LAYOUT OF CONSTRUCTION PLANT Disposition on site of the physical properties, fixed equipment, and materials necessary to complete a project.

LAYOUT PLAN Plan that shows the exact locations and horizontal dimensions of proposed site works, buildings, roads, and site features in relation to the existing site and structures.

LAYOUT STICK Long strip of wood marked at the appropriate joint intervals for the tile to be used. It is used to check the length, width, or height of tilework.

LAYOVER Nonworking time spent on location or away from base.

LAY PANEL Door or other panel with its length horizontal.

LAY SHAFT Fixed shaft supporting rotating drums or gears.

LAY TYPES In wire rope the various types are:

> **Alberts lay:** An old, rarely used term for **Lang lay,** below.
>
> **Alternate lay:** Lay of a wire rope in which the strands are alternately regular and lang lay.
>
> **Cross lay:** Rope or strand in which one or more operations are performed in opposite directions.
>
> **Lang lay:** The type of rope in which the lay of the wires in the strand is in the same direction as the lay of the strand in the rope. The crowns of the wires appear to be at an angle to the axis of the rope.
>
> **Left lay:** The direction of strand or wire helix corresponding to that of a left-hand screw thread.
>
> **Regular lay:** The type of rope wherein the lay of the wires in the strand is in the opposite direction to the lay of the strand in the rope. The

crowns of the wires appear to be parallel to the axis of the rope.

Reverse lay: Another term for **Alternate lay,** above.

Right lay: The direction of strand or wire helix corresponding to that of a right-hand screw thread.

Spiral lay: Manner in which a spiral reinforcement is applied with respect to the angularity and lead or pitch, as in a hose or cylindrical article.

Spring lay: This is not definable as a unique lay; more properly, it refers to a specific wire rope construction.

LAY UP 1. Process of placing reinforcing material into position in a mold; **2.** Step in structural wood panel manufacture in which veneers or reconstituted wood layers are 'stacked' in complete panel 'press loads' after gluing and before pressing.

LAY-UP Floor-to-ceiling full-scale representation of the various paint colors and/or techniques selected for large areas. A lay-up is examined by the client for approval prior to beginning a project.

LAZY AXLE See **Axle.**

LAZY GUY Front guy of any gin pole or simple derrick that takes no load whatsoever but serves primarily to prevent a kickback if the load is suddenly slackened.

LAZY JACK System of pivoted levers giving mechanical advantage.

lb Abbreviation for: pound.

LB Abbreviation for: lag bolt; linoleum base.

L-BEAM 1. Beam having a cross section in the shape of an L; **2.** Beam having a ledge on one side only. See also **Ledge.**

lb-ft Abbreviation for: pound-foot.

lbl Abbreviation for: label.

lbr Abbreviation for: labor lumber.

lbr. Abbreviation for: load backrest.

L-BRACKET Formed steel bracket used to attach and secure something: guide rails to a structure, for instance, a machine, etc.

lc Abbreviation for: load center.

LC Abbreviation for: laundry chute.

L/C Abbreviation for: less cylinder.

LCD Abbreviation for: liquid crystal display.

lcg Abbreviation for: lateral center of gravity.

lch Abbreviation for: latch.

l-cl. Abbreviation for: linen closet.

LCL Abbreviation for: less than carload.

LCM Abbreviation for: lead-coated metal; least common multiple.

L-COLUMN Portion of a precast-concrete frame composed of the column, haunch, and part of the girder.

ld Abbreviation for: lead; leader drain; load.

ldg Abbreviation for: landing; leading; loading.

ldr Abbreviation for: leader.

LE Abbreviation for: leading edge.

LEACH 1. To cause liquid to filter through a material; **2.** Extraction of a soluble material through immersion in water or by flowing water.

LEACHATE Liquid that has percolated through a material mass and contains dissolved or suspended microbial constituents.

LEACHATE RECIRCULATION Recycling or reintroduction of leachate into or on a disposal facility.

LEACH BED System of subsoil piping that permits absorption of fluids into the earth. Also called **Disposal field,** and **Tile bed.**

LEACHING Continuous removal of soluble matter by the dissolving action of water.

LEAD (pronounced leed) 1. An insulated conductor furnishing an electric path between the terminal of a battery and a point of further contact. This may be a terminal connector or a cable connector; **2.** Section of a wall built up, usually at corners, and racked back on successive courses (a line is attached to leads as a guide for constructing a wall between them); **3.** Nonparallel alignment of the guide track and circle saw plate that offsets carriage travel with respect to the saw plane to provide clearance for the back of the saw, and that counteracts the tendency of the saw to run out of the logs; **4.** Direction in which the lines run out from a yarder; **5.** Amount of lift above ground that the running lines have at a yarder, e.g., high-lead or ground-lead; **6.** Alignment of sheaves and winches; **7.** Position of logs relative to the yarding distance; **8.** Established direction in which all trees in a quarter or strip are to be felled, usually governed by the terrain of the area, or its general slope or skid road system; **9.** On a pile driver, the device that guides the pile and pile driver or hammer, including lead column, brace, and other components. The lead may be suspended at the top so as to permit vertical or lateral movement, or it may be held in a fixed position. There are several types, including:

Box: Lead configured in the shape of a 'U' with guiding rails for the hammer in the open portion of the 'U.' Also called **Steam lead,** and **U-lead.**

Cage: See **Offshore lead,** below.

Can: See **Offshore lead,** below.

Cardonic: Lead fixed to a crane, normally in an extended configuration, but with three directions of freedom about the boom point (i.e., in and out, left and right, and around) rather than just in and out, left and right movement. Also called **Swivel lead,** and **Three-way lead.**

Chuck: See **Offshore lead,** below.

Compression: Pile hammer lead designed to withstand the compressive forces from pile extraction operations (i.e., used in pulling a sand drain or mandrel).

European: System in which the hammer is mounted in front of the lead. Also called **Mast lead, Pipe lead,** and **Triangular lead.**

Extended: Pile hammer lead that extends above, and attaches to the boom point of a crane. Also *called* **Fixed lead,** and **Overhead lead.**

Fixed: *See* **Extended lead,** above.

Free-hanging: *See* **Swinging lead,** below.

Hairpin: Structure in which a pile hammer is mounted to adapt it to box leads with rails spaced wider than the hammer width. *Also called* **Pony lead,** and **Telescope lead.**

Hanging: *See* **Swinging lead,** below.

H-beam: *See* **Spud lead,** below.

Mast: *See* **European lead,** above.

Monkey stick: Spud lead system in which the lead column passes through a slide box at the tip of the crane boom. The lead column elevation is controlled with a crane line and a brace is not normally used, the lead base being set on the ground. The hammer is mounted on the back side of the lead with cables coming directly from the crane boom point sheaves. *Also called* **Pipe lead,** and **Spud lead.**

Offshore: Pile hammer lead that has an upper section containing the hammer and a lower section that closely fits about and guides the pile. The hammer is supported and aligned by the pile after engagement. The pile is usually supported by a template. *Also called* **Cage lead, Can lead, Chuck lead, Pipe lead,** and **Rope-suspended lead.**

Overhead: *See* **Extended lead,** above.

Pendulum: *See* **Swinging lead,** below.

Pipe: Lead, with the main member a pipe, to which a guide is attached for the hammer to slide on. *Also called* **European lead, Monkey stick, Pogo stick, Offshore lead,** and **Spud lead.**

Pogo stick: *See* **Pipe lead,** above.

Pony: *See* **Hairpin lead,** above, and **Telescope lead,** below.

Rope-suspended: *See* **Offshore lead,** above.

Semifixed: An extended pile hammer lead that may be raised or lowered independently of the boom by a cable from the crane. *Also called* **Vertical travel lead.**

Spud: Steel wide-flange or H-beam used in place of leads. The hammer engages and slides upon one flange of the beam through the use of spud clips that are bolted to the hammer. *Also called* an **H-beam lead, Monkey stick,** and **Pipe lead.**

Steam: *See* **Box lead,** above.

Swinging: Pile hammer lead topped with a bail and hung from one load line of a crane. *Also called* **Free-hanging lead, Hanging lead,** and **Pendulum lead.**

Swivel: *See* **Cardonic lead,** above.

Telescope: Structure that holds a pile hammer in wider-than-hammer-width box leads and permits extended driving by carrying the hammer beyond the bottom of the leads. *Also called* **Hairpin lead,** and **Pony lead.**

Three-way: *See* **Cardonic lead,** above.

Triangular: *See* **European lead,** above.

Truss: Pile hammer lead fabricated with four main chords in a generally rectangular shape, with the cords being connected by a diagonal bracing. May have internal diagonals to form a tetrahedral design. The pile hammer is mounted in front of, and rides on two of the main chords.

U-lead: *See* **Box lead,** above.

Underhung: Pile hammer lead that hangs from the boom point pin of a crane by straps or pendant cables.

Vertical-travel: *See* **Semifixed lead,** above.

LEAD (leed) BLOCK Block used to change the direction of a line pull.

LEAD (led) BURNING Welding lead without solder.

LEAD (led) BURST Leak in a lead press hose during vulcanization, caused by a rupture of the lead casing.

LEAD (led)-CAPPED NAIL Roof nail fitted with a lead washer to make a watertight joint.

LEAD (led) CASING Extruded lead tube or sheath that confines a hose during vulcanization.

LEAD (led) CHIP MARK Minor nick or mark in the surface of the cover of lead press hose caused by particles of lead flakes sloughing off the lead press die during the process of lead covering.

LEAD (leed) COMPANY Member company of a joint venture that takes principal responsibility under the contract.

LEAD (led) CURE FINISH Type of exterior surface, either ribbed, smooth, or longitudinally corrugated, obtained by the lead pipe mold method of vulcanization.

LEAD (led) DENT Indentation in the surface of lead press hose caused by deformations in the lead covering before vulcanization.

LEAD (led) DIE MARK Longitudinal line or mark in the cover of a lead press hose caused by a damaged lead press pin.

LEAD (led) DISCOLORATION Dark stain on the colored cover of lead press hose caused by a chemical reaction of the lead with the rubber compound.

LEADED (leded) LIGHT Window formed by fixing small pieces of glass together by means of lead cames.

LEADER (leeder) Any connector, such as a pipe, for carrying rainwater from the roof of a building to the ground. *See also* **Downspout.**

LEADER (leeder) HEAD 1. Rainwater head; **2.** Growing shoot or sprout of a plant or tree.

LEAD (led) FLAKE Particle of lead that remains on the cover of lead press hose after the lead covering has been stripped from the hose.

LEADING EDGE CROWN Condition where only the front or leading edge of a screed is bent or warped upward. This allows material to more easily flow under the leading edge of the screed, resulting in improve mat appearance.

LEADING WIRE *See* **Electric blasting circuit.**

LEAD (led) JOINT Molten lead poured into the joint

between sections of cast-iron pipe.

LEAD (leed) LINE That part of a rope tackle leading from the first, or fast, sheave to the drum.

LEAD (led) NAIL Copper-alloy nail used to fix lead sheet to a solid backing.

LEAD (led) PAINT Paint containing lead pigment.

LEAD (led) PLUG 1. Lead sleeve driven into a hole in masonry, brick, or concrete that acts as a tight fixing for a screw; **2.** Lead fillet placed into mating groves cut into adjacent stones to hold them together.

LEAD (led) POP Surface protrusion, the result of a rupture of lead sheath during vulcanization of hose.

LEAD (led) PRESS CURE Process wherein a lead sheath acts as a restraining member or mold during vulcanization of a hose.

LEAD (led) PRESS FINISH Type of exterior surface obtained by the lead press method of hose vulcanization.

LEAD (led) PRESS JOINT *See* **Lead stop.**

LEAD (led) SHEATH Lead covering to a power or communication cable.

LEAD (led) SOAKER *See* **Soaker.**

LEAD (led) STOP Mold mark in a lead press hose cover, caused by stopping the lead press to add another lead billet. *Also called* **Charge mark,** and **Lead press joint.**

LEAD (led) STORAGE BATTERY Storage battery, with the electrodes made of lead and the electrolyte consisting of a solution of sulfuric acid.

LEAD (leed) WIRE Wire connecting the electrodes of an electric blasting machine with the final leg wires of a blasting circuit.

LEAF One of a pair of similar components (door, wall, etc.)

LEAF CHAIN *See* **Chain.**

LEAK DETECTOR Device capable of determining the presence of water leaking from a pipe or vessel, pinpointing its source and estimating the rate of flow.

LEAKER 1. Crack or hole in a tube that allows fluid to escape; **2.** Hose assembly that allows fluid to escape at the fittings or couplings.

LEAN 1. Angle at which a tree naturally stands: can be referred to as heavy or slight. *See also* **Head lean,** and **Side lean; 2.** Mixture using less than the normal quantity of one or more of the constituents.

LEAN CONCRETE Concrete of low cement content.

LEANER Tree that leans excessively, not growing straight.

LEANING PIER Pier that is not vertical. It is usually located ashore and leans out over the stream to support the stringers.

LEAN MIXTURE Air-fuel mixture in which an excess of air is supplied in proportion to the amount of fuel. *See also* **Rich mixture.**

LEAN MORTAR Mortar that is deficient in cementitious components. It is usually harsh and difficult to spread.

LEAN-TO Secondary structure appended to a main building and covered with a single-slope roof.

LEAN-TO ROOF Single pitch roof, the upper edge of which leans against a wall. *See also* **Pent roof.**

LEAR BOARD *See* **Layer board.**

LEASE Contract by which a lessor (owner) conveys to the lessee (renter) the temporary right to use real or personal property for a specified purpose and for a specific time in exchange for payment or other consideration. There are several types, including:

Capital: Lease that transfers substantially all of the benefits and risks incidental to ownership of property to a lessee.

Direct finance: Lease that transfers substantially all the benefits and risks incident to ownership of property to the lessee and where, at the inception of the lease, the fair value of leased property is the same as its carrying amount to the lessor.

Financing: Any lease that is essentially a method of financing the purchase of property.

Lease-option agreement: Lease that gives the lessee the option to purchase the property at a specified date for a stipulated price.

Leveraged: Financial arrangement in which a purchaser acquires an asset, to be leased to a third party, and pays for it partly with his own funds and partly with funds raised from lenders on the security of the leased asset.

Long-term lease: Generally, a lease agreement extending for 10 years or more.

Maintenance: Lease that requires the lessor to assume all the costs associated with maintenance and upkeep.

Net: Lease where, in addition to an agreed rent, the lessee assumes responsibility for payment of such property charges as taxes, insurance, assessments, and maintenance. *Also called* **Net-net lease,** and **Net-net-net lease.**

Operating: Lease in which the lessor retains substantially all the benefits and risks incident to ownership of property.

Sales-type: Lease that transfers substantially all the benefits and risks incident to ownership of property to the lessee and where, at the inception of the lease, the fair value of the leased property is greater or less than its carrying amount, thus giving a profit or loss to the lessor.

LEASE BOND *See* **Bond.**

LEASEHOLD Interest in real estate conveyed by one person to another for a specified period in return for rent.

LEASEHOLD IMPROVEMENT Fixtures attached to real estate that are generally acquired or installed by the tenant and that can be removed on expiration of the lease provided such removal does not damage the property.

LEASEHOLD MORTGAGE *See* **Mortgage.**

LEASE OPTION AGREEMENT *See* **Lease.**

LEAST COUNT Smallest measurement made using a

vernier.

LEAVE STRIP Unfelled stand of timber left on purpose. Sometimes called a **Fire break.** *Also called* **Buffer strip, Green strip,** and **Streamside management zone.**

LEAVE TREE Tree left standing after timber has been felled in a cutting unit.

LED *Acronym for:* light-emitting diode.

LEDGE 1. Bedding, or several beddings of rock, as in a quarry; **2.** An outcropping of horizontal or nearly horizontal rock. *Also called* **Horizontal ledge; 3.** Shelf-like projection from the face of a wall; **4.** Any member with a protrusion (or protrusions) that supports other structural members. *See also* **L-beam,** and **Inverted T-beam.**

LEDGE BATTEN Top or bottom batten to which the vertical boards of a batten door are fastened.

LEDGED AND BRACED DOOR Batten door diagonally braced between the ledges.

LEDGE DRAIN Type of drainage system installed in the upstream face of a concrete dam.

LEDGER 1. Horizontal scaffold member that extends from post to post and which supports the putlogs or bearers, forming a tie between the posts. *See also* **Beam.** *Also called* **Horizontal brace,** and **Stringer; 2.** Horizontal formwork member supporting the joists; **3.** Horizontal stone slab covering a tomb.

LEDGER BOARD 1. Top or bottom horizontal board to which the vertical boards of a fence are fastened; **2.** Board let into the face of studding to support floor joists on its upper edge; **3.** Horizontal member in scaffolding.

LEDGER STRIP Strip of lumber nailed along the bottom or the side of a girder on which joists rest.

LEFT BANK Bank bounding flowing water that is on the left when looking downstream.

LEFT LAY *See* **Lay types.**

LEG 1. Side post in tunnel timbering; **2.** Tile wall running alongside a bathtub or abutment; a narrow strip of tile floor; **3.** Connector or wire in one side of an electrical circuit; **4.** Tension filament appearing when cemented or frictioned plies are pulled apart.

leg. *Abbreviation for:* legal.

LEGAL DESCRIPTION Description that will be upheld in a court of law.

legis. *Abbreviation for:* legislative; legislature; legislation.

LEG PROTECTOR Ballistic nylon pad attached to one or both pant legs to protect the leg(s) of workers using a chain saw from contact with the chain.

LEG VISE Bench vise having a leg that extends to the floor.

LENGTH 1. Measured distance or dimension; **2.** Longer or longest dimension of an object; **3.** Usable section of material; a length of lumber, length of steel, etc.

LENGTH CHANGE *See* **Deformation.**

LENGTH INDICATOR Device on a telescoping crane

boom that shows the boom length from minimum to maximum.

LENGTH OF ENGAGEMENT 1. Length of contact between two mating parts; **2.** Length along which two connected or attached parts act as one.

LENGTH OF GROUND Terrain factor affecting tree cutting efficiency. In short ground, the contour (at the felling end) is broken up with ridges, gulleys, rock outcrops, etc., and the lead must constantly be changed to follow the ground contour. *See also* **Ground length.**

LENO BREAKER Open-mesh fabric made from coarse ply yarns, with a leno weave. *See also* **Breaker ply.**

LENO WEAVE Fabric structure in which the warp yarns are bound in by the filling, resulting in an open perforated fabric.

LESSEE One who possesses by lease the right of use of another's property.

LESSOR One who holds title to and conveys the right to use or occupy their property or land.

LET-IN BRACES Diagonal braces notched into studding.

LETTER OF AGREEMENT Letter setting out the general terms of an agreement; when signed by the addressor and addressee, a form of contract.

LETTER OF CREDIT Conditional arrangement whereby a lender agrees to substitute its credit for a customer's.

LETTER OF INTENT Written expression of the intent to do something, that incurs a liability, but that is not wholly binding.

lev. *Abbreviation for:* level.

LEVEE 1. Rock or earth embankment to prevent inundation or erosion; **2.** Landing place on the bank of a river.

LEVEL 1. Horizontal plane; **2.** Instrument for determining a true horizontal by centering a bubble in a glass tube filled with spirit. The types of level can be simply divided into two classes:

    Builder's level: Wood or metal bar to which is fitted one or more alcohol- or ether-filled glass tubes whose angle is adjusted to correspond to that of the exerior long surface of the frame. The tubes can be mounted parallel to the working surface, perpendicular to it (to give a true vertical reading) or at an angle to it (usually 45°). Builders levels are rarely less than 600 mm (24 in.) long and may be up to 1.5 m (5 ft) or longer, or, an electronic device that illuminates a LED and sounds a tone when the casing in which it is housed is held against a truly horizontal or truly vertical face.

    Surveyor's level: Tripod-mounted, telescopic instrument that allows the transfer of its own level above datum to any number of distant points. The instrument consists of a telescope mounted on a rotating horizontal plate, both of which can be adjusted to cause the optical axis of the telescope to lie in a horizontal plane. The focusing telescope is fitted with hairlines that intersect on the optical axis. *Also called*

Engineer's level.

LEVEL BOOK Surveyor's field book, ruled vertically to permit recording of readings taken with an optical instrument.

LEVEL CIRCUIT Measurement of a bench mark elevation by bubble level in two different directions from one end of a circumference to the other.

LEVEL CONTROL Device that senses and maintains the water level in a boiler.

LEVELING 1. Process of establishing the difference in elevation between two points; **2.** Slow rate of speed used for the final approach to a floor of an elevator car to ensure an accurate floor stop.

LEVELING ARM Long arm extending forward from an asphalt paver screed and attached to a tow point on each side of the tractor, allowing the screed to float on the mix being placed.

LEVELING COURSE Layer of new asphalt placed over a distressed roadway to improve its geometry prior to resurfacing.

LEVELING INSTRUMENT 1. Device containing a means of determining when the surface it is placed on is truly horizontal; **2.** Pivoting telescope to which a spirit level is attached.

LEVELING PLATE Steel plate on which a structural column rests.

LEVELING ROD Telescoping or folding rod marked in gradations (meters, centimeters, and millimeters, or feet and inches), used with a surveyor's level or transit.

LEVELING RULE Straightedge used with a spirit level for bringing dots and screeds to a uniform level.

LEVELING SCREW Fine-thread adjustment screw; a foot screw.

LEVELING SKI *See* **Averaging ski.**

LEVELING WEDGE Patch of asphalt mix used to level a sag or depression in an old pavement prior to surfacing.

LEVELING WING *See* **Snow plow.**

LEVELING ZONE *See* **Elevator car leveling device.**

LEVEL LUFFING Automatic arrangement whereby a crane or derrick does not significantly change elevation as the boom derricks.

LEVEL-PAYMENT MORTGAGE *See* **Mortgage.**

LEVEL OF SERVICE Qualitative rating of the effectiveness of a highway in serving traffic, measured in terms of operating conditions. (NOTE: The Highway Capacity Manual identifies operating conditions ranging from 'A' for best operation (low volume, high speed) to 'E' for poor operations at possible load capacity.)

LEVEL RECORDER Pressure- or float-operated device that continuously senses and records the level of a fluid in a channel or vessel.

LEVEL SURFACE Surface that is parallel with the main spherical surface of the earth.

LEVEL TRIER Instrument that indicates the angle of a slope by the position of a bubble in a graduated tube.

LEVEL TUBE *See* **Bubble tube.**

LEVER 1. Tree limbing method in which the chainsaw rests on the tree trunk and is levered forward to cut limbs. *Also called* **Scandinavian technique**; **2.** Bar that pivots so that force applied at one part can do work at another. There are three distinct applications:

> **First-class lever:** Bar having a pivot point between the points where force is applied and where it is exerted.
>
> **Second-class lever:** Lever whose force is exerted between the fulcrum and the point where it is applied.
>
> **Third-class lever:** Lever to which force is applied between the fulcrum and the work point.

LEVERAGED LEASE *See* **Lease.**

LEVER ARM 1. Perpendicular distance of a transverse force from a point about which a moment is taken; **2.** In a concrete structural member, the distance from the center of the tensile reinforcement to the center of action of the compression zone.

LEVER CAP Cam-shaped metal piece above the back iron of a metal plane that holds the back and cutting irons in place.

LEVER HANDLE Horizontal handle for operating the latchbolt of a lock.

LEVER LOCK Lock in which a key must engage and move several levers in order to actuate the bolt.

LEVERMAN One who operates the controlling levers of a machine; an operator.

LEWIS ANCHOR Lead sleeve, inserted into a hole made in a masonry block, into which a lewis bolt is screwed.

LEWIS Attachment for lifting blocks of stone consisting of dovetail-shaped keys that fit into mortices cut into the stones. *Also called* **Lifting pin.**

LEWIS BOLT Anchor bolt held in place in masonry by a lead casing. *Also called* **Rag bolt.**

LEWIS HOLE Mortise cut into the nonexposed surface of a masonry block for engaging a lewis anchor.

LEWISING TOOL Chisel used to cut lewis holes.

LEWIS PIN Steel pin that is placed in a predrilled hole in a dressed stone block on an angle of approximately 60°. A second pin is angled in the opposite direction. The pair of pins is connected by a chain which is used for lifting the block.

lf *Abbreviation for:* left front; linear foot.

LF *Abbreviation for:* linoleum floor.

lg *Abbreviation for:* long.

lg. *Abbreviation for:* large; logarith.

LG *Abbreviation for:* liquid gas.

lgr *Abbreviation for:* longer.

lgt *Abbreviation for:* light.

lgth *Abbreviation for:* length.

LH *Abbreviation for:* left hand.

L-HEAD Top of a shore formed with a braced horizontal member projecting from one side forming an inverted L-shaped assembly. *Also called* **L-shore.**

LHR *Abbreviation for:* left hand, reversed.

LHRB *Abbreviation for:* left-hand reversed bevel.

LHS *Abbreviation for:* left-hand side.

LHT *Abbreviation for:* left hand thread.

li. *Abbreviation for:* link.

liab. *Abbreviation for:* liability; liable.

LIABILITY Debt or financial obligation.

LIABILITY BOND *See* **Bond.**

libr. *Abbreviation for:* librarian; library.

lic. *Abbreviation for:* license; licensed.

lic. appl. *Abbreviation for:* licensed applicator.

LICENSE BOND *See* **Bond.**

lic. inst. *Abbreviation for:* licensed installer.

LIEN Claim on the property of another against payment of a debt. *See also* **Mechanic's lien.**

LIEN BOND *See* **Bond.**

LIEN RELEASE Legal document that assures that materials and services furnished to a project have been paid for.

LIEN WAIVER Undertaking by a person or entity who has, or may have, a right of mechanic's lien against the property of another, to relinquish such right. *Also called* **Release of lien.**

LIFE CYCLE COST Total of the costs incurred during the design, or actual life of a product, typically including capital cost, maintenance costs, labor (operating) costs, running (fuel, supplies) costs, etc.

LIFE ESTATE Conveyance of title to property for the duration of the life of the grantee.

LIFE EXPECTANCY Length of time a device, material, construction or other fabrication may be expected to remain operating under the circumstances for which it was designed.

LIFE INSURANCE *See* **Insurance.**

LIFELINE 1. Rope, suitable for supporting one person, to which a lanyard or safety belt (or harness) is attached; 2. In elevator work, a wire rope of minimum 2450 kg (5,400 lb) breaking strength, hung the full length of a hoistway, having fittings attached at intervals of approximately 1.5 m (5 ft) to prevent a person whose belt lanyard is attached to the line from falling over 1.8 m (6 ft).

LIFE NET Rope net slung directly below an elevator work area.

LIFE TENANT Person whose interest in a property is limited to the duration of his life.

LIFE TEST Laboratory procedure used to determine the resistance of a product to a specific set of destructive forces or conditions. *See also* **Accelerated life test.**

LIFO *Acronym for:* last-in first-out.

LIFT 1. Concrete placed between two consecutive

horizontal construction joints, usually consisting of several layers or courses; 2. Vertical travel of the carriage of a lift truck with the mast vertical. 3. Depth of a cut taken out, or of a fill placed during a cycle of excavation; 4. Layer of asphalt mix placed and compacted separately; 5. Number of scaffolding frames stacked vertically one above each other. *Also called* **Tier;** 6. Units of reinforcing tied together for shop or field convenience, classified in two categories: (a) shop lifts, and (b) field lifts. Shop lifts are units of reinforcing loaded for shipment. Field lifts are units of reinforcing required for field handling by contractors. A field lift may consist of single bundles or two or more smaller bundles tied together. A shop lift may consist of one or more bundles, the same as field lifts, or consist of two or more field lifts. Straight and bent bars are not combined in the same lift. Maximum weight of a lift is dependent on regional practices and site conditions. *See also* **Bundle of bars;** 7. Vertical distance between a static water source and the suction chamber of a pump; 8. Height of a column or body of fluid below a given point, expressed in linear units; 9. In sanitary landfill, a compacted layer of solid wastes and the top layer of cover material.

LIFT-AND-CARRY *See* **Waste container.**

LIFT ANGLE 1. Angle between an imaginary vertical line and an imaginary line between the load and the hoist or lifting device; 2. *See* **Lift member.**

LIFT ARMS Front-end loader arms that carry an attachment used for lifting.

LIFT BAR Transverse, horizontally pivoting member attached to the boom of a wheel lift, or an underlift, for attaching frame or wheel-lift devices. *Also called* a **Cross bar.**

LIFT BEAM *See* **Lift member.**

LIFT BRIDGE Bridge, part of whose deck can be raised vertically within a fixed framework so as to provide increased clearance.

LIFT CAPACITY *See* **Rated load.**

LIFT CAPACITY TO MAXIMUM HEIGHT Maximum weight a loader can lift when not tied down: a measure of the ability of the hydraulic lift circuits to lift weight.

LIFT CONTROL SYSTEM Means to control the raising, lowering, and stopping of a lift or hoist.

LIFT DEPTH Vertical thickness of a compacted volume of solid wastes and the cover material immediately above it in a sanitary landfill.

LIFTER Wide, chisel-like drift, used for removing stone out of the rough.

LIFTER HOLE Borehole drilled in a near-horizontal position when constructing forest access roads.

LIFT GATE 1. Vertically rising lock gate; 2. Hydraulically powered, horizontal tailgate on a truck. *See also* **Elevating gate.**

LIFTING Softening and damage to a dry coat of paint or varnish caused by application of another coat.

LIFTING BLOCK Set of pulleys used with a line to gain mechanical advantage for raising heavy loads.

LIFTING CAPACITY Rated load for any given load radius and boom angle under specified operating conditions and machine configurations.

LIFTING MAGNET Electromagnet, typically suspended from the boom of a crane or the arm of an excavator, used to separate and lift ferrous metals to separate them from other materials.

LIFTING PIN *See* **Lewis.**

LIFT JOINT Surface at which two successive lifts of concrete meet. *See also* **Blemish.**

LIFT LATCH Thumb latch for a door.

LIFT MEMBER Frame rail along one side of a dump body understructure. The lift members contain the connecting holes for attachment to the hoist mechanism. *Also called* **Channel, Lift angle,** and **Lift beam.**

LIFT PUMP *See* **Pump.**

LIFT SHOE Device used by lifting equipment to transmit lifting forces from the equipment or vehicle to the ground.

LIFT SLAB 1. Method of concrete construction in which floor and roof slabs are cast on or at ground level and hoisted into position by jacking; **2.** Slab that is a component of such construction.

LIFT SPEED Average velocity in m/min (fpm) when raising a load carriage throughout its operating range, specified as empty and/or loaded.

LIFT-STARTER FORM Noncantilever form used after the first pour has been made on a dam foundation until sufficient height has been poured in each monolith. Such forms are used to provide clearance for cantilever forms.

LIFT TAIL GATE Power-operated tail gate capable of lifting or lowering a load between the ground and the level of a truck or trailer floor.

LIFT-TOW RATING Rating of a wrecker or recovery vehicle that gives the maximum weight of a vehicle to be towed.

LIFT TREE Single, unguyed tree, used to hang a block to hold the bight of the haulback a short distance from the ground.

LIGHT 1. Individual window between mullions and transoms (may be further subdivided by individual panes, as in a lattice window); **2.** Pane of window glass (may also be double, or triple glazed).

LIGHT ASPHALT RESURFACING *See* **Hot-mix seal coat.**

LIGHT BAR Array of lamps mounted on a machine or vehicle and used in compliance with local regulations.

LIGHT BENDING All #3 reinforcing bars, all stirrups and ties, and all bars #4 through #18 that are bent at more than six points in one plane, or bars that are bent in more than one plane (unless **Special bending**), all one-plane radius bending with more than one radius in any bar (three maximum), or a combination of radius and other type bending in one plane (radius bending being all bends having a radius of 250 mm (12 in.) or more to inside of bar). *See also* **Heavy bending,** and **Special bending.**

LIGHT BURNING Periodic broadcast burning to prevent the accumulation of forest floor fuels in quantities that would cause excessive damage, or difficulty in suppressing, in the event of accidental fire.

LIGHT-DUTY DUMP BODY *See* **Truck, dump body.**

LIGHT-DUTY SCAFFOLD Scaffold designed and constructed to carry a working load not exceeding 11.3 kg (25 lb) per square foot.

LIGHT-EMITTING DIODE Electronic device in which excitation of a diode will cause it to emit light of a specific color, used in instrumentation to signal something and to form alphanumeric characters. *Also called* **LED.**

LIGHT FASTNESS Ability of a finish to resist color changes caused by light.

LIGHT-GAUGE COPPER TUBE Thin-wall copper pipe, used for domestic plumbing (water distribution), connected with fittings and solder joints.

LIGHT-GAUGE STEEL CONSTRUCTION *See* **Construction types.**

LIGHT HARD Term applied to red bricks that are not the hardest in the kiln, although suitable for carrying moderate loads. Such brick may not be able to withstand alternate freeze and thaw cycles.

LIGHT HAZARD *See* **Fire hazard.**

LIGHTING FIXTURE Permanently installed electric light fixture.

LIGHTING OUTLET An outlet intended for the direct connection of a lampholder, a lighting fixture, or a pendant cord terminating in a lampholder.

LIGHT LOAD RATING *See* **Load rating.**

LIGHTLY LOADED VEHICLE WEIGHT 1. For vehicles with a GVWR of 4536 kg (10,000 lb) or less, unloaded vehicle weight plus 136 kg (300 lb), including driver and instrumentation; **2.** For vehicles with a GVWR greater than 4536 kg (10,000 lb), unloaded vehicle weight plus 227 kg (500 lb), including driver and instrumentation.

LIGHT MATERIAL BUCKET *See* **Bucket.**

LIGHTNESS Attribute by which a perceived color is judged to be equivalent to a member of a continuous series of grays ranging from black to white. *See also* **Color, Hue,** and **Saturation.**

LIGHTNING ACTIVITY LEVEL Number, on a scale of 1 to 5, reflecting the frequency and character of cloud-to-cloud lightning, either forecast or observed. The scale is exponential, based on the powers of 2.

LIGHTNING FIRE Forest fire caused directly or indirectly by lightning.

LIGHT PIPING Ability of optical glass or polished acrylic to pass light from one end to the other with little loss, even around bends.

LIGHT PLANT Generator unit on a loader or grapple yarder that supplies power to boom-mounted lights for night loading and yarding.

LIGHT PYLON Structure upon which a light and/or light bar is mounted.

LIGHT REFLECTIVE VALUE Measurable degree of light that is reflected back from every surface, expressed as a percentage.

LIGHT RESISTANCE Ability to retard the deleterious action of light.

LIGHT SAP STAIN Detectable discoloration that will not materially impair the appearance of naturally finished lumber.

LIGHT SCALING *See* **Scaling.**

LIGHT-TRUCK TIRE Tire designated by its manufacturer as primarily intended for use on lightweight trucks or multipurpose passenger vehicles.

LIGHTWEIGHT AGGREGATE *See* **Aggregate.**

LIGHTWEIGHT CONCRETE Concrete of substantially lower density than that made using aggregates of normal density. *See also* **Insulating concrete,** and **Low-density concrete.**

LIGHTWEIGHT CONCRETE BLOCK *See* **Concrete masonry unit.**

LIGHT WELL 1. Unroofed open shaft within the perimeter of a building that allows light to reach internal windows; **2.** Subsurface space in front of a basement window.

LIGNIN Resins that cement wood fibers together.

LIGNORUM *See* **Marine borer.**

LILY PAD Thin slice of wood, sometimes taken off the stump and used to cover a chain saw if it's to be left out in the open.

lim. *Abbreviation for:* limited; limiting; limits.

LIMB To cut branches off trees or logs.

LIMBWOOD Part of the tree above the stump that does not meet the requirements for saw logs or upper stem portions. Includes all live, sound branches to a 100-mm (4-in.) outside bark diameter minimum.

LIME 1. Limestone burned in a kiln until the carbon dioxide has been driven off; **2.** Base constituent of mortar.

LIME BLOOM Unwanted chalky-white powder appearing on the surface of hardened concrete, usually due to the formation of insoluble calcium carbonate formed from rainwater flowing over the concrete surface immediately after striking the formwork.

LIME MORTAR Lime, sand, and water, sometimes mixed with a quantity of cement equal to that of the lime, used for laying bricks.

LIME PLASTER Basecoat plaster made of lime and an aggregate.

LIME POWDER Powder obtained by air slaking lime.

LIME PUTTY Slaked lime in a soft, putty-like condition before sand or cement is added to produce mortar.

LIME ROCK Naturally occurring calcium carbonate containing varying quantities of silica, that hardens on exposure to the atmosphere.

LIME STABILIZATION Use of burned lime products, quick-lime, or hydrated lime, as an additive to plastic clayey soils and granular materials to improve water resistance and cohesive properties of the particles.

LIMESTONE Sedimentary rock consisting primarily of calcium carbonate. *See also* **Material density.**

LIMEWASH *See* **Whitewash.**

LIMIT 1. Greatest permissible dimensional tolerance of a part; **2.** Point where something ends.

LIMIT CONTROL Device that activates, or deactivates another mechanism within preset limits.

LIMIT DESIGN Method of proportioning reinforced concrete members based on calculations of their strength. *See also* **Strength-design method.**

LIMITED-AREA FELLER-BUNCHER *See* **Harvesting machines (multifunction).**

LIMITED TENDERING *See* **Selective bidding.**

LIMITING DISTANCE Distance from any exposing building face towards a property line, the center of a street, lane, public thoroughfare or an imaginary line between two buildings on the same property, measured at right angles to the exposing building face.

LIMITS Maximum permitted dimensions of a part.

LIMIT STATE 1. Condition in which a structure or component becomes unfit for service and is judged either to be no longer useful for its intended function (serviceability limit state) or to be unsafe (strength limit state); **2.** Limit of structural usefulness, such as brittle fracture, plastic collapse, excessive deformation, durability, fatigue, instability and serviceability.

LIMIT SWITCH *See* **Switch.**

LIMNOLOGY Study of the biological productivity and characteristics of fresh water bodies.

LIMNORIA TRIPUNCATE *See* **Marine borer.**

LIMONITE *See* **Brown oxide.**

LIMPET DAM Small open caisson shaped to fit against a dock wall that, when lowered into position and pumped out, is held against the wall by external water pressure.

LIMPET WASHER Dome-shaped washer used under a fastener to hold down a corrugated sheet.

lin. *Abbreviation for:* lineal; linear; lintel.

LINDERMANN JOINT Glued dovetail joint, shaped by a Lindermann jointer, joining two pieces of wood edge-to-edge longitudinally.

LINE 1. The string stretched taut from lead to lead as a guide for laying the top edge of a brick course; **2.** Cable, rope, chain, or other flexible medium for transmitting pull; **3.** Synonymous term for **Wire rope;** **4.** Survey line; path or route between points of control along which measurements are taken to determine distances or angles. *See also* **Give line;** **5.** Setting a boundary; **6.** Verbal stop signal when line is being pulled by hand; **7.** Fire hose; **8.** Pipe or hole for conducting fluid; **9.** Wires conducting electrical power to a switch, relay, or other device.

LINE AND PIN Chalk line attached to metal pins driven into brick courses at each end of a wall being built, to act as a guide to the true alignment of the intermediate bricks.

LINEAR 1. Pertaining to a line. Linear measurement is a measurement of a single dimension, as opposed to measurement of an area or volume; **2.** Having an output directly proportional to input.

LINEAR ACTUATOR *See* **Cylinder.**

LINEAR CENTRIFUGAL FORCE Centrifugal force generated per vibratory roller drum, divided by the drum width.

LINEAR EXPANSION 1. Measure of growth along the length and width of particleboard when exposed to conditions from low to high humidity, stated in percent; **2.** Increase in one dimension of a soil mass, expressed as a percentage of the shrinkage limit, when the water content is increased.

LINEAR FORCE Force exerted in a straight line.

LINEAR PERSPECTIVE Graphic representation in which the line of sight is perpendicular to the plane of projection. There are two types:

    Centrolinear: In which the viewer is assumed to be at a local fixed distance from the object. *Also called* **Station-point projection** and **Conic projection.**

    Orthogonal: In which the viewer is assumed to be at an infinite distance from the object. *Also called* **Cylindrical projection.**

LINEAR PRESTRESSING Prestressing applied to linear members such as beams, columns, etc.

LINEAR SCALE Sound-level measurement scale that is nonweighted so that there is little or no discrimination at low frequencies.

LINEAR SCALE RETICULE Optical glass marked with a straight scale to permit measurement or estimation of distance or length.

LINEAR SPEED-CONTROL LOGIC Circuitry that smooths a speed reference signal so an elevator is not abruptly started or stopped.

LINEAR STATIC FORCE Static weight of a vibratory roller, divided by the drum width.

LINEAR TRANSFORMATION Method of altering the path of a prestressing tendon in any statically indeterminate prestressed structure by changing the location of the tendon at one or more interior supports without altering its position at the end supports and without changing the basic shape of the path between any supports.

LINEAR-TRAVERSE METHOD Determination of the volumetric composition of a solid by integrating the distance traversed across areas of each component along a line or along regularly spaced lines in one or more planes intersecting a sample of a solid. *See also* **Point count.**

LINE BORE To cut or ream all bores or holes in a fuel line.

LINE BOSS Supervisory officer in a forest fire suppression organization, responsible for executing the fire suppression plan adopted by the fire boss.

LINE CHANGE Changing the rigging from one yarding road to another.

LINED BOLT HOLE Bolt hole that has been given a protective coating to cover the internal structure.

LINED HOSE Generally, fire hose having a seamless woven jacket or jackets and a tube.

LINE DRAWING *See* **Diagram.**

LINE DRILLING Series of holes drilled in rock along the line to which it is to be excavated.

LINE DROP 1. Electrical distribution power line connecting a service to a transformer; **2.** Drop or loss of voltage in an electrical conductor due to resistance.

LINE FIRING Setting fire to only the forest border fuel immediately adjacent to the control line. *See also* **Strip firing.**

LINE HORSE Line storage winch, usually mounted on a truck carrier, used to transfer lines.

LINE LEVEL Bubble tube that can be attached to a cord stretched between two points.

LINE LOCATOR In forest fires, the person responsible for on-the-ground location of fire lines to be constructed.

LINE MANHOLE Manhole in a sewer located for reasons other than permitting connection of a branch sewer, such as change of direction, service access, change of grade, etc.

LINE OF CREDIT Agreement whereby a lender promises to lend up to a certain amount without the need to seek detailed approval.

LINE OF FLIGHT Angle of ascent of a stair.

LINE OF LEAST RESISTANCE Shortest distance between the center of an explosive charge and the nearest free face.

LINE-OF-PLATE *See* **Struck measure.**

LINE OF SIGHT Unobstructed straight line between two points.

LINE OF THRUST Line containing the resultant thrusts on the back of a retaining wall.

LINE OILER Oil reservoir and metering unit in a compressed air line, used to lubricate air tools, or inserted in the pressure line to lubricate a hammer. *See also* **Lubricator, Oiler,** and **Sight-feed lubricator.**

LINE PIN Metal pin to which a line is attached for alignment of masonry units.

LINE PULL Pulling force developed by a hoist on a wire rope. This force varies according to the number of layers of rope on the drum.

LINE PULLING Felling a tree against its lean by securing a line to it and pulling it over.

LINER 1. Something placed inside another object and covering it's inside surface; **2.** Replaceable wearing part; **3.** Metal pipe used as a temporary form when placing concrete in a pier hole; **4.** Insulating fabric insert worn underneath safety caps or hats; **5.** Continuous layer of natural (e.g., earthen) or man-made material beneath, or on the sides of a surface impoundment, landfill, or landfill cell, that restricts the downward or lateral escape of hazardous waste, hazard waste constituents, or leachates; **6.** Layer of rubber that is laminated to the inside of a tubeless tire to insure the air-retention quality. *See also* **Tire construction, inner liner.**

LINE RESISTANCE METHOD Current-limiting technique for starting AC motors.

LINE RETRACTION Removing a measured amount of fuel from a line between fuel injection cycles.

LINE RIDER *See* **Running line tensiometer.**

LINER PANEL Panel applied as an interior finish.

LINER PLATE Steel plate with turned-back edges that contain bolt holes which, when bolted together, support the arch sides of a tunnel, and sometimes the invert.

LINER ROCK Condition in engine cylinder liners where the lower portion of the liner is not precisely secured in the engine, thus allowing movement or rocking of the liner during piston movement.

LINE RUNNING Locating, tracing, and marking of land ownership lines.

LINE SCOUT Person in a forest fire suppression organization assigned to scouting duties on the fire line.

LINE SPEED Rope velocity at a rope drum at a specified pitch diameter.

LINE SPINNING Method of rotating a metal pipe by wrapping a chain or line around it; commonly used to make and break a threaded connection.

LINE STRAINER Wire mesh placed in an oil line to catch particulate matter.

LINE-TO-LINE VOLTAGE Voltage existing between any two-phase conductors in a polyphase circuit. Equal to line-to-neutral voltage times 1.732.

LINE-TO-NEUTRAL VOLTAGE Voltage existing between any phase conductor and the neutral conductor. Line-to-neutral voltage equals line voltage divided by 1.732.

lin. ft *Abbreviation for:* linear foot; linear feet.

LINING Any sheet, plate, or layer of material attached directly to the inside face of an area; to the inside face of formwork to improve the surface texture and quality of the finished concrete; to the inside faces of exterior walls to act as a vapor barrier, etc. *See also* **Tube.**

LINING AREA Area of the friction material in a brake or clutch.

LINING PLATE Metal strip nailed to the eaves or verge and attached to flexible-metal roofing.

LINING TOOL Device used for painting lines.

LINK Small-diameter, loop-shaped steel bar used to hold together the main steel in a reinforced concrete element.

LINKAGE Any series of rods, yokes, levers, pedals, and other shapes used to transmit motion.

LINK CAGE Long, welded-fabric box with the main bars bent to form links.

LINK DIAGRAM Diagram used to work out the distribution of links in a reinforced concrete beam, similar to a shear-force diagram.

LINK DORMER Large dormer, sometimes incorporating side windows, used to join roof sections or incorporate roof projections such as chimneys.

LINKED SWITCH *See* **Switch.**

LINK FUSE Fuse not protected by a cover plate.

lino. *Abbreviation for:* linoleum.

LINOLEUM Sheet-form floor finishing material manufactured from linseed oil, ground cork, and other fillers, oxidized upon a fabric base; available in two basic finishes: plain or solid color, and inlaid, comprising different colored mixes extending throughout the entire wearing surface.

LINSEED OIL Vegetable oil made from flax plants. It is used in the manufacture of oil-base paints and finishes; blended with drying agents it is called **Boiled linseed oil** and can be used as a wood finish when cut with turpentine in mixtures ranging from 1:1 to 2:1.

LINTEL 1. Horizontal beam bridging an opening; **2.** In steel construction, a horizontal member spanning an opening and supporting an imposed load.

LINTEL BLOCK U- or W-shaped concrete block, used in construction of horizontal bond beams and lintels.

LIP 1. Cutting edge of a bucket or dipper; **2.** Short length of timber fixed to the top of a strut, and projecting sufficiently beyond its end so as to rest on a waling; supports the mass of the strut while wedges are driven.

LIP UNION In plumbing, a union with a ring-like inner projection that restrains the gasket in proper alignment.

liq. *Abbreviation for:* liquid.

LIQUEFACTION Change of state to liquid: can be from a previously gaseous state, or from a solid state. In the latter it is a loss of strength occurring in saturated, fine-grained, cohesionless soil when exposed to shock or vibrations. Under such circumstances, the soil particles momentarily lose contact due to pore pressure increase. The material then behaves as a fluid without shear strength. *See also* **Quicksand.**

LIQUEFIED PETROLEUM GAS Commonly called LPG and LP gas, it includes any material that is composed predominantly of any of the following hydrocarbons, or mixtures of them, such as propane, propylene, butane (normal butane or isobutane), and butylenes.

LIQUID ASPHALT Asphaltic cutback or road oil so soft that its consistency cannot be measured using the standard penetration test.

LIQUID ASSET Cash on hand or in the bank or temporary investments that are readily convertible into cash.

LIQUIDATED DAMAGES Sum established in a construction contract, usually as a per diem, to be levied as damages suffered by the owner due to failure to complete the work within a stated time.

LIQUIDATION 1. Payment of a debt; **2.** Conversion of assets into cash; **3.** Wind-up of the affairs of an organization by settling with its debtors and creditors and distributing any remaining assets.

LIQUIDATION VALUE Amount that might be realized through a forced sale or the wind-up of a business.

LIQUIDATOR Person appointed to wind up the affairs of an organization.

LIQUID COLLECTOR Solar collector that uses water or other liquid as the heat transfer medium.

LIQUID-COOLED ENGINE Internal combustion engine that is cooled by means of liquid coolant circulated about its heated parts. The coolant is then passed through a radiator or heat exchanger, where it is cooled and then re-circulated to the engine.

LIQUID CRYSTAL DISPLAY Instrument display that uses external light for contrast with dark-colored elements for display of characters and digits.

LIQUID DRIER Paint-thinner-soluble liquid chemical that promotes drying of paints and finish coats. Japan driers and cobalt driers are two examples.

LIQUID LIMIT Minimum moisture content that will cause soil to flow if jogged.

LIQUID-MEMBRANE CURING COMPOUND Liquid sealant.

LIQUID PETROLEUM Material composed of propane, propylene, butane (normal or isobutane), and butylenes, held in storage at pressures sufficient to render it liquid.

LIQUID SPECIFIC GRAVITY Ratio of the weight of a given volume of liquid to an equal volume of water. *See also* **Specific gravity.**

LIQUID-VOLUME MEASUREMENT Measurement of grout based on the total volume of solid and liquid constituents.

LIQUID WASTE Fluid discharge that must be contained (in a pipe or other vessel) during transport and treated before discharge or release to the environment.

LIQUID WEIGHT Water content, expressed as a percentage of the dry weight of soil, at which the soil passes from the plastic to the liquid state under standard test conditions.

LIS PENDENS Notice that an action or proceeding affecting the title to property is pending in the courts.

LIST *See* **Fillet.**

LISTED Equipment or materials included in a list published by a qualified testing laboratory whose listing states either that the equipment material meets appropriate standards, or has been tested and found suitable for use in a specified manner.

LISTING 1. Employment of a broker to sell real property; **2.** Record by a broker of properties he/she is authorized to sell.

LIST PRICE Price set out in a catalog.

lit. *Abbreviation for:* literature.

LITER Non-SI unit of volume, equal to a cubic decimeter, permitted for use with the SI system of measurement. Symbol: L. Multiply by 0.035 315 to obtain cubic feet, symbol: ft³; by 61.023 744 to obtain cubic inches, symbol: in.³; by 0.001 to obtain cubic meters, symbol: m³; by 0.219 969 to obtain (imp) gal, symbol: (I)gal; by 0.264 172 to obtain (U.S.) gal, symbol: (U.S.)gal; by 35.195 1 to obtain fluid ounces, symbol: fl oz; and by 0.879 877 to obtain quarts, symbol: qt. *See also the appendix:* **Metric and nonmetric measurement.**

LITHOLOGY Study of rocks. *See also* **Petrography,** and **Petrology.**

LITHOSPHERE Rocky portion of the earth's crust composed mainly of solid materials such as rock, clay, earth, gravel, etc.

LITTER LAYER Layer of organic debris, mainly bark, twigs, and leaves, on the forest floor.

LIVE *See* **Alive.**

LIVE AXLE *See* **Axle.**

LIVE BOOM Shovel boom that can be lifted and lowered without interrupting the digging cycle.*See also* **Boom.**

LIVE BOTTOM BIN Storage bin for shredded or granular material whereby controlled discharge is through a mechanical or vibrating device across the bin bottom.

LIVE BOTTOM PIT Storage pit, usually rectangular, receiving truck-unloaded material, and using a push platen or bulkhead, reciprocating rams or mechanical conveyor across the pit floor for controlled discharge (retrieval) of the material.

LIVE-BOTTOM TRAILER Transfer trailer whereby controlled discharge is by a mechanical or vibrating device across the bottom of the trailer.

LIVE BURNING Progressive burning of green slash as it is cut.

LIVE EDGE Edge of paint that can still be blended with newly applied paint without showing a lap or seam.

LIVE END Moving end of a rope or chain, used to perform a lifting or pulling operation, the other end being the dead end; **2.** Load side of an eye splice or other rope-grip-type fastening.

LIVE KNOT Knot whose fibers are intergrown with the living wood.

LIVE LOAD *See* **Load.**

LIVE MAST Frame hinged at or near the boom foot and extending above the cab, for use in connection with supporting a boom. The head of the mast is usually supported, and raised or lowered by boom hoist ropes.

LIVE REEL Reel that supplies air, electricity, or water to the fixed end of a hose or cable wound onto it. *See also* **Reel.**

LIVERING Gelling in (rubber) cement giving a liver-like consistency.

LIVE SKYLINE Skyline that can be raised and lowered during yarding to facilitate logging.

LIVE SPREADER *See* **Floating harness.**

LIVE WIRE Electrical conductor through which electrical energy is actually passing (as against a conductor capable of carrying such energy).

LIVE ZONE Volume of a pugmill below a line extending across the top of the paddle arc to the inside walls of the pugmill; the volume of the shafts, liners, arms, and tips are not included.

LIVING AREA Area contained within the exterior dimensions of all the habitable spaces of a residence, regardless of the number of floors, but excluding attic, basement, breezeway, or porch unless any such areas are built to accommodation standards for habitation, i.e., in compliance with building regulations as to insulation, heating and ventilation, services, etc.

LIVING SPACE *See* **Habitable room or space.**

LIVING UNIT *See* **Dwelling unit.**

liv.ld *Abbreviation for:* live load.

lk *Abbreviation for:* link; lock.

lkg *Abbreviation for:* linking.

lkg. *Abbreviation for:* leakage.

lkq *Abbreviation for:* like kind and quality.

lkr *Abbreviation for:* locker.

LL *Abbreviation for:* lead lined; live load.

LLC *Abbreviation for:* limited-life construction.

llf *Abbreviation for:* longleaf.

lm. *Abbreviation for:* lumen.

lmc *Abbreviation for:* load moment control.

lmp *Abbreviation for:* lamp.

ln *Abbreviation for:* linen.

lnd *Abbreviation for:* land.

lng *Abbreviation for:* lining.

l/o *Abbreviation for:* layout.

O *Abbreviation for:* lubricating oil.

LOA *Abbreviation for:* length overall.

LOAD 1. Capacity rating a piece of equipment is designated to safely handle; **2.** Sum of the dead load and the live load; **3.** Weight of the burden carried at one time or in one trip; **4.** Weight borne by a structure; **5.** In electricity, the amount of energy supplied or required; **6.** Amount of resistance offered to an engine by the machine it is operating; **7.** Placement of explosives in a blasthole; **8.** To apply additional weight (as in to preload a piece of ground to be developed); **9.** To transfer material to a haul unit or hopper; **10.** *See* **Harvest functions.**

LOAD Force or other action that arises on a structural system from the weight of all permanent construction, occupants and their possessions, environmental effects, differential settlement and restrained dimensional changes. Can be characterized as:

Concentrated: Load applied over a very small area.

Dead: Stationary, permanent load; that is, the weight of all the material used in construction.

Live: **a.** Planned load the structure must carry under normal conditions (such as human, furniture, or equipment) that would be moved across the structure's surface; **b.** Weight of moving traffic on a bridge deck.

Nominal: Magnitude of a load as specified by the applicable code.

Permanent: Load in which variations in time are rare or of small magnitude.

Uniform: Load that is evenly distributed over a large area.

LOAD AND CARRY *See* **Tire types.**

LOAD AND RESISTANCE FACTOR DESIGN Method of proportioning structural components (members, connectors, connecting elements and assemblages) such that no applicable limit state is exceeded when the structure is subjected to all appropriate load combinations. *Also called* **LRFD.**

LOAD AXLE Vehicle axle nearest the load.

LOAD BACKREST EXTENSION Removable structure extending vertically from the carriage of an industrial truck frame to provide increased support and stability for unusually high loads. *Also called* **Carriage backrest extension.**

LOAD BALANCING Technique used in the design of prestressed concrete members in which the amount and path of the prestressing is selected so that the forces imposed upon the member or structure by the prestressing counteract or balance a portion of the dead and live loads for which the member or structure must be designed.

LOAD-BEARING TILE *See* **Tile.**

LOAD-BEARING WALL *See* **Wall.**

LOAD BINDER Device consisting of a lever-operated toggle and lock, used to tighten chains or straps holding loads in place upon a truck bed.

LOAD BLOCK *See* **Load sensor.**

LOAD CAPACITY *See* **Load carrying capacity.**

LOAD CARRYING CAPACITY Capacity of a material to sustain an imposed load (structural, static, electrical, etc.). *Also called* **Load capacity.**

LOAD CELL Device that produces an output signal proportional to the applied weight or load. It may utilize any physical principal included in the fields of electricity, electronics, hydraulics, magnetics, mechanics, pneumatics, or combinations thereof. *See also* **Load sensor.**

LOAD CELL CAGE Mechanical arrangement of a load cell assembly, designed to retain the load in the event of load cell failure.

LOAD CELL LINK Mechanical assembly connecting a load cell to a crane's structure or rigging.

LOAD CENTER Horizontal longitudinal distance from the intersection of the horizontal load-carrying surface of an industrial truck and the vertical load engaging face of the forks (or equivalent load positioning structure) to the center of gravity of the load.

LOAD CHART Table depicting the maximum rated load a machine can lift at a given radii, boom length, and boom angle in a defined area of operation.

LOAD COMPENSATION Ability of an elevator system to automatically adjust to variations in weight to be carried.

LOAD CURRENT Total electrical current across a power source.

LOAD CURVE Curve on a chart showing power (kilowatts) supplied, plotted against time of occurrence, and illustrating the varying magnitude of the load during the period covered.

LOAD DECK Explosive charges spaced apart in a borehole and fired by separate primers or by detonating cord.

LOAD DISTRIBUTION Relationship of the total gross load on the front and rear axles of a vehicle or machine.

LOAD DIVERSITY Difference between the sum of the maximum of two or more individual loads and the coincident or combined maximum load, usually measured in kilowatts.

LOAD DIVIDER *See* **Dolly.**

LOAD-DIVIDING PRESSURE CONTROL VALVE *See* **Valve.**

LOADED When not otherwise specified, 'loaded' is understood to mean the condition when an industrial truck is handling the equivalent of a symmetrical cubic basic capacity load.

LOADED FILTER Filter at the foot of an earth dam. It stabilizes the toe of the dam by its weight and permeability since water cannot exist in it under pressure.

LOADED RADIUS Distance from the wheel axle centerline to the ground on a properly inflated, loaded tire.

LOADER Rear-engined, tracked or wheeled machine equipped with a bucket, or other moving/lifting device or special-purpose attachment at the front. Operation may be swing-to-load, or travel-to-load. *Also called* **Hydraulic loader,** and **Knuckleboom,** if it swings to load and has hydraulically activated boom members. *See also* **Harvesting machines (single function), Backhoe/loader,** and **Skid-steer loader.** The many designs include:

> **Bucket loader:** Machine equipped with a digging and gathering rotor, complete with a set of chain-mounted buckets to elevate the material to a dumping point.

> **Elevating-belt grader:** Machine whose forward motion cuts soil with a plowshare or disk and pushes it to a conveyor belt that elevates it to a dumping point.

> **Front-end loader:** Tractor/loader that both digs and dumps in front.

> **Landfill compactor:** Wheel loader used for dozing, filling, and compacting, particularly sanitary landfills, having wheels with tamping foot designs. These units are commonly equipped with a wide range of devices to prevent clogging of wheel treads, damage from debris, and buildup of blowing refuse.

> **Paddle loader:** Belt loader equipped with chain-driven paddles that move loose material to a belt.

> **Reversed loader:** Front-end loader having its driving wheels in front and steering wheels at the rear.

> **Soil compactor:** Wheel loader used for dozing, filling, and compacting, having wheels with tamping foot designs.

> **Swing loader:** Tractor/loader that digs in front and that can swing the bucket to dump to either side.

> **Tower loader:** Front-end loader whose bucket is lifted along tracks on a more-or-less vertical tower.

> **Tractor loader:** Tractor equipped with a digging bucket that can be raised and tilted forward to dump its contents.

LOAD FACTOR 1. Average load carried by an engine, machine or plant, expressed as a percentage of its maximum capacity; **2.** Factor by which a service load is multiplied to determine a factored load used in the strength design method. *See also* **Phi factor; 3.** Ratio that measures the decrease in density when a bank cubic meter/bank cubic yard of material is disturbed or removed from the bank state; **4.** Ratio of the collapse load to the working load on a structure or section; **5.** Percent of the total connected fixture unit flow rate likely to occur at any point of a drainage system; **6.** Ratio of the average load in kilowatts supplied during a designated period to the peak maximum load occurring in that period.

LOAD FREQUENCY DESIGN FACTOR Means of equating different traffic conditions, based on traffic analysis.

LOAD GATE *See* **Elevating gate.**

LOAD HEIGHT Height a bucket, attached to an excavator, loader or other equipment, can be raised to dump a load into a truck.

LOAD HOIST LINE The main hoist (the secondary hoist is the whip line).

LOAD HOOK In helicopter logging, that part of a helicopter's load rigging that is connected to the lower end of the long line.

LOAD-INDICATING BOLT High-strength, friction-type bolt that indicates when the specified minimum tension has been achieved through the compression of small projections on the underside of the head.

LOAD-INDICATING DEVICE Device that measures and displays the net load hoisted, or percentage of rated lifting capacity of a crane.

LOAD-INDICATING SYSTEM Load-indicating device applied to a crane. It includes all mounting and crane components.

LOADING 1. Picking up trees or parts of trees from the ground or from a vehicle, transporting them, and then piling them into another vehicle (such as a highway logging truck or rail car); **2.** Filling the pores of the grinding surface with the material being ground, usually resulting in a decrease in production and poor finish.

LOADING BOOM Overhanging structure from which material is loaded and unloaded.

LOADING HOPPER Hopper in which concrete or other free-flowing material is deposited for discharge into buggies or other conveyances, used for delivery to the forms or other place of processing, use, or storage.

LOADING JACK Rigging suspended from a spar tree guyline immediately above the line of haul, and terminating in a loading block.

LOADING PIPE Filling a pipe with a suitable material, or a device, to prevent distortion or collapse of the pipe wall during bending.

LOAD JIB *See* **Saddle jib.**

LOAD LIMIT ALARM Visual and/or audible indication that a specific load limit is being approached or has been reached.

LOAD LINE Rope used for hoisting and lowering loads.

LOAD LINE TENSIOMETER *See* **Running line tensiometer.**

LOAD MOMENT 1. Generally, the product of a force and its moment arm; specifically, the product of the load and the radius-of-load; **2.** Factor used in determining the lifting capacity of a crane; **3.** In counterbalanced trucks, reach trucks and sideloaders, the nominal force produced by the load tending to overturn the truck. Mathematically it is represented by the following formula: load moment = load x (load center + load moment constant).

LOAD MOMENT CONSTANT In a counterbalanced truck, the longitudinal dimension from the overturning axis to the vertical load-engaging face, with the mast vertical.

LOAD MOMENT SYSTEM System consisting of a means to sense crane loading, boom length, and operating radius, or functions of these, that automatically gives a signal when the loading condition approaches or exceeds rated values of radius or load.

LOAD PIN *See* **Load sensor.**

LOAD-PROPORTIONING BRAKE CONTROL System that modulates the input force to the brakes on an axle, proportional to the load on that axle.

LOAD PUSHER Lift truck attachment that holds the load in position while the truck moves in reverse, withdrawing the forks, or that pushes the load off the forks or platen(s). *Also called* **Unloader.**

LOAD RADIUS Nominally, the horizontal distance from the axis of rotation to the center of gravity of a lifted load. *Also called* **Operating radius**

LOAD RANGE A letter designation that indicates the maximum permissible load on a tire (also referred to as a ply rating, i.e., load rating G = 14 ply rating).

LOAD RATING 1. Weight or load required to compress a spring by an amount equal to its stroke; **2.** Crane rating in pounds or other comparable units of measure established by the manufacturer; is always referenced to a specific radius of load or boom angle to a specific crane configuration; **3.** Maximum loading, as applied to the following categories of scaffold:

Heavy: Scaffold designed and constructed to carry a working load of 34 kg/0.09 m² (75 lb/ft²), such as intended for stone masonry work, with storage of material on the platform.

Medium: Scaffold designed and constructed to carry a working load of 23 kg/0.09 m² (50 lb/ft²), such as intended for bricklayers or plasterers, with weight of material in addition to users.

Light: Scaffold designed and constructed to carry a working load of 11 kg/0.09 m² (25 lb/ft²), such as intended for users with no material storage other than weight of tools.

Special: Scaffold designed and constructed to carry weights of a specified nature, such as palletized materials.

LOAD RATING CHART Chart posted in the cab of a crane that specifies the relationship between the maximum allowable load that can be hoisted for a given crane configuration and radius of load and boom angle.

LOAD SENSOR Device that produces an output signal proportional to weight or force. The device may sense tension or compression, and it may be one of many types of mechanical, electrical, electronic or hydraulic devices. Special configurations may be referred to as **Load block, Load pin, Load strap,** etc. *Also called* **Load cell,** and **Transducer.**

LOAD SHARE BRIDGING Diagonal braces or continuous 50-mm- (2-in.-) by members fastened between floor and roof joists to make each joist act with those next to it in a load-sharing manner.

LOAD STRAP *See* **Load sensor.**

LOAD TEST Procedure in which a structure is subjected to a maximum imposed load that conforms to its design criteria.

LOAD TEST WEIGHT Method of testing capacity and relation of load to movement by putting a load on the bearing element before actually building upon the foundation. *See also* **Test load.**

LOAD TIME Time it takes to load a truck, fill a bucket, etc.

LOAD-TRANSFER ASSEMBLY Unit (basket or plate) designed to support or link dowel bars during concreting operations so as to hold them in place while in the desired alignment. *See also* **Basket.**

LOAD TRANSFER DEVICE Mechanical means designed to carry loads across a joint in pavement.

LOAM Earth having a relatively even mixture of clay, silt, and sand plus a considerable proportion of organic matter. *Also called* **Topsoil.** *See also* **Material density,** and **Soil types.**

LOBBY Public or common entrance space in a building

LOBE Oblong ground section on a camshaft, used to lift the cam followers.

loc. *Abbreviation for:* local; locality; locate; location; locus.

LOCAL ATTRACTION Deviation of a magnetic compass needle due to a ferrous metal mass in the immediate vicinity of the instrument.

LOCAL BOND FAILURE Failure of bond between reinforcement and the surrounding concrete.

LOCAL BUCKLING Buckling of a compression element that may precipitate the failure of the whole member.

LOCAL HOUSING AUTHORITY Government entity or public body that engages in the development and/or operation of public, low-rent housing.

LOCAL IMPROVEMENT RATE Development tax based on property frontage to a public road, assessed value, square footage, or other assessment, designed to meet the cost of designing and developing such municipal infrastructure (sewers, streets, curbs, sidewalks, etc.) as are not charged to the general funds of the corporation, but rather to the owners of the properties directly benefited.

LOCAL STREET *See* **Street.**

LOCATION Centerline and grade line of an engineered structure preparatory to its construction.

LOCK 1. Any special device or method of construction used to secure a bond in masonry; **2.** Chamber in a compressed air system that can be opened to pressurized air at one end, and to atmosphere at the other; **3.** Chamber in a canal or river with gates on each end through which vessels can pass up- or downstream and be raised or lowered to a new elevation.

l/o ck *Abbreviation for:* layout and check.

LOCK BLOCK Wood block in a hollow-core door into which a lock or handset can be fastened.

LOCKED COIL STRAND In wire rope, a smooth-surfaced strand ordinarily constructed of shaped, outer wires arranged in concentric layers around a center of round wires.

LOCKFACE Metal plate that shows in the edge of a door when a lock is installed.

LOCK GATE Gate at either end of a lock permitting passage of vessels into or out of the chamber, or within the length of the chamber, dividing the lock into two or more compartments.

LOCKING DEVICE Device used to secure a cross brace in scaffolding to the frame or panel.

LOCKING DIFFERENTIAL *See* **Differential.**

LOCKING DOG Part of a padlock that engages the shackle and holds it locked.

LOCKING PLATE Plate used with slotted brackets to lock a guide rail in place on a bracket after the rail has been aligned.

LOCKING PLIERS *See* **Pliers.**

LOCK NUT *See* **Nut.**

LOCKOUT RELAY Electrically reset or hand-reset auxiliary relay, used to hold associated devices inoperative until it is reset.

LOCK RAIL Rail of a multipanel door into which a lock or handset may be fastened.

LOCK RING Third piece of a three-piece wheel rim assembly that locks the side ring to the rim's base.

LOCK SEAM Spiral or longitudinal seam in corrugated or other pipe or sheet metal, formed by overlapping or folding the adjacent edges.

LOCK SILL Part of the floor of a lock chamber against which the gates rest when closed.

LOCK STILE Stile of a single-panel or framed door into which a lock or handset may be fastened.

LOCKSTONE *See* **Keystone.**

LOCKUP CLUTCH *See* **Clutch.**

LODGE 1. Small house, usually on the grounds of a larger building; **2.** Commercial building offering lodging and other facilities, usually in a rural setting.

LOESS Fine, porous mineral filler deposited by the wind.

LOFT Accessible space within the roof space of a building.

LOFT LADDER Folding ladder that, when lowered, gives access to an attic through a trapdoor that houses the ladder in its folded position.

LOG 1. Detailed record of an operation; of the rocks passed through during a drilling operation; **2.** To harvest trees on an area; **3.** Section of the trunk of a tree in suitable length for sawing into commercial lumber. *See also* **Driving log.**

log. *Abbreviation for:* logarithm.

LOGARITHM Exponent indicating the power to which a base number must be raised to produce a given number.

LOGARITHMIC SPIRAL Spiral that intersects all of its radiants at the same angle.

LOG BODY Truck or trailer body designed primarily for the transportation of logs or other loads, that may be boomed or chained in place.

LOG BOOK Driver's or operator's record of hours, routes, etc.

LOG CHUTE Channel through or beside a dam for passage of logs and driftwood.

LOG DECK Pile of logs.

LOG FORK *See* **Fork.**

LOGGER Worker employed in the production or maintenance phase of the logging industry. *See also* **Lumberjack.**

LOGGIA Gallery or arcade, open on one or more sides.

LOGGING All or any part of turning trees into logs and transporting them to an unloading area.

LOGGING CHAIN *See* **Chain.**

LOGGING PLAN Schedule of operations for a specific area that describes in words and on a map how and where harvesting will take place.

LOGGING RESIDUE Unused portion of pole timber and sawtimber trees killed by land clearing, cultural operations, or timber harvesting.

LOGGING SETTING Area to be logged; a block or strip.

LOGGING SYSTEM Method, such as tractor skidding, forwarding, etc., by which logs are placed on the landing.

LOGGING THE TREELINE Helicopter removal of logs right up to the treeline.

LOGGING TONGS Tongs with hooks shaped at the ends that dig in when the tongs are pulled.

LOGGING TRUCK Vehicle used to transport logs. A logging truck consists of a cab (including the engine compartment) and a trailer on which logs are carried. The trailer usually has an adjustable carriage to accommodate loads of various lengths, and can be carried by the tractor on the backhaul run or when running empty.

LOGGING WINCH *See* **Winch.**

LOGIC Result of planning a data-processing system.

LOGIC BOARD Assembly of decision-making circuits on a printed-circuit board.

LOGIC DEVICE One of a general category of components that perform logic functions; for example AND, NAND, OR, and NOR. They can permit or inhibit signal transmission with certain combinations of control signals.

LOGIC PROBE Instrument used to determine the logic level at a tested point.

LOGIC STATE Signal levels in logic devices characterized by two stable states, the logical 1 (one) state and the logical 0 (zero) state. The designation of the two states is chosen arbitrarily. Commonly, the logical 1 state represents an 'on' signal, and the 0 state represents an 'off' signal.

LOGIC SYMBOL Symbol used to represent a logic element.

LOG JACK Tool, similar to a peavey but with a flattened steel loop on the underside so that, when the hook fastens into a log on the ground and the handle is lowered, the log is jacked up and remains elevated on the jack; used to raise a log from the ground during bucking.

LOG JAM Congested group of logs and debris obstructing the flow of a stream.

LOGO Distinctive emblem, symbol or trademark that identifies a product or service.

LOG RULE Table intended to show amounts of lumber that may be sawed from logs of different sizes under various assumed conditions.

LOG SCALE Measure of the volume of wood in a log or logs, usually expressed in board feet and based on various log scaling rules.

LOI *Abbreviation for:* letter of instruction.

LONG BREAK LINE Line drawn at an angle through a drawing, broken in its length by a zig-zag, that, in pairs. indicate an artificial termination of the presentation for reasons of economy of space, convenience, etc.

LONG BUTT Short section cut off the butt end of a felled tree in order to remove cull or excess sweep.

LONG CAP Cap of a pile bent that usually holds the piles of a single bent together and distributes the load over them.

LONG COLUMN 1. Column whose effective length is 15 times greater than its least lateral dimension; **2.** Column whose load capacity is limited by buckling rather than strength. *See also* **Slender column.**

LONG DIRECTION Direction of a panel in the 2.4-m (8-ft) length.

LONG DOLLY Follower behind a trailer.

LONG FIBER *See* **Texture.**

LONG FLOAT Float requiring two workers to handle.

LONG-HOLE DRILLING Holes drilled at intervals along the course of a tunnel, following completion of a pilot drift, that are then loaded and exploded to form a working chamber.

LONGITUDINAL Body member attached to and running the length of an underframe. *Also called* **Longrail, Longsill, Riser,** and **Stringer.**

LONGITUDINAL AXIS *See* **Longitudinal centerline.**

LONGITUDINAL BAR Any bar placed in the long direction of the member. *See also* **Longitudinal reinforcement.**

LONGITUDINAL BEAD TEST Weldability test involving a steel plate on which a bead has been welded. The plate is dropped, then bent double. If the plate or weld metal cracks the material is not weldable.

LONGITUDINAL BOND *See* **Bond.**

LONGITUDINAL CENTERLINE Structural neutral axis of a crane boom. (Certain types of boom section may deviate from this: e.g., **Hammerhead peak, Pffset, Offset base,** etc.). *Also called* **Longitudinal axis.**

LONGITUDINAL COMPONENT Component of vibration that produces motion in the direction of a line joining the vibration source and the seismograph.

LONGITUDINAL CRACK 1. Crack that develops parallel to the length of a structural concrete member. **2.** Pavement crack that is roughly parallel to the centerline.

LONGITUDINAL FALL Fall given to a gutter to facilitate drainage; minimum fall is 1:250.

LONGITUDINAL JOINT *See* **Joint.**

LONGITUDINAL JOINT GROOVE FORMER Machine, part of a paving train, comprising a hollow vertical knife that travels submerged in the concrete and that is fitted to the underside of a flat plate holding a vibrator. The joint-former material is fed from a reel through the knife blade into the compacted concrete.

LONGITUDINAL PROFILE Vertical section through the centerline of a road or similar extended structure, showing the original and final grades.

LONGITUDINAL REINFORCEMENT Reinforcement parallel to the length of a concrete member or pavement. *See also* **Longitudinal bar.**

LONGITUDINAL SECTION Drawing showing a cut along the longitudinal axis of a body.

LONGITUDINAL TRACE Line on a vibration record that records the longitudinal component of motion.

LONGITUDINAL WAVE VELOCITY Speed of a wave travelling parallel to the direction of propagation.

LONG LINE Helicopter load line, attached to the belly hook.

LONG-LINE SKIDDING Cable system method of skidding logs to a landing from distances up to 365 m (1,200 ft) away.

LONG-NOSE PLIERS *See* **Pliers.**

LONG-OIL ALKYD Alkyd resin containing more than 60% oil as a modifying agent.

LONG-OIL VARNISH Varnish having a high ratio of oil to resin.

LONG QUARTER BEND 90° pipe fitting with one end longer than the other.

LONGRAIL *See* **Longitudinal.**

LONG RIPPER TIP *See* **Ripper tip.**

LONG SCREW In plumbing, a threaded connector.

LONGSILL *See* **Longitudinal.**

LONG SNORKEL One-piece wood or steel boom extension mounted on a loader to increase the distance that logs can be reached for yarding and loading purposes.

LONG SPAN SKIDDING Cable system capable of skidding logs for 900 m (3,000 ft) of more.

LONG SPLICE Wire rope splice, approximately 18 m (60 ft) long, that does not increase line diameter, used to join a broken line.

LONG-SWEEP FITTING Pipe fitting that has a longer than normal radius curve.

LONG-TERM FINANCING Secured funds obtained for the purpose of completing a project, and scheduled to be repaid over an extended period.

LONG-TERM LEASE *See* **Lease.**

LONG TON Nonmetric weight equal to 2,240 lb (1016 kg). *See also* **Metric ton** and **Short ton.**

LONG TRACK Version available of some tracked equipment consisting of an extended track frame. The resulting equipment is particularly suitable for fine grading.

LONGWOOD Pulpwood 3 m (10 ft) or more in length; stemwood delivered in lengths that exceed 4.6 m (15 ft).

LONGWOOD HARVESTING Timber harvesting method in which harvested trees are moved to the landing either as whole trees or as topped and limbed tree-length logs. At the landing, further processing such as limbing, topping, bucking, chipping, or loading is carried out as necessary.

LOOKOUT 1. Short member used to support the overhanging portion of a roof at a gable end; **2.** Person designated to detect and report forest fires, working from a vantage point; **3.** Lookout station. *Also called* **Fire tower.**

LOOKOUT LEDGER Timber band around the top of a wall to which lookouts are attached.

LOOKOUT TOWER Structure enabling a person to be above nearby obstructions and to sight and fix the location of forest fires.

LOOP 1. 360° change of direction in the course of a wire rope that, when pulled down tight, will result in a kink; **2.** *See* **Interchange elements.**

LOOP CIRCUIT Continuous circuit connecting the motor and generator armatures.

LOOP EDGE Selvage, formed by having the filling loop around a catch cord or wire, that is later withdrawn, leaving small loops along the edge of the cloth.

LOOP EDGE TAPE Tape woven with a selvage edge, formed by looping the filling threads to prevent raveling and allowing extensibility for even tension.

LOOPER That part of a roofing machine that saturates the sheet roofing, then forms and allows it to land in loops to facilitate cooling.

LOOPING IN Electrical wiring technique to reduce the number of T-joints in conduit by keeping one cable permanently connected to the socket, the other passing through the switch.

LOOP VENT Vent that loops back and connects with a waste stack vent.

LOOSE *See* **Fit classifications.**

LOOSE COVER Separation of the cover from the carcass or reinforcements.

LOOSE CUBIC METER/CUBIC YARD One cubic meter or cubic yard of material after it has been removed from natural conditions. *See also* **Bank cubic meter/cubic yard.**

LOOSE FILL Bulk insulation material, such as granulated cork, foamed slag, gypsum insulation, mineral wool, vermiculite, etc., used in walls and ceilings.

LOOSE FIT Fit between mating parts that allows considerable freedom.

LOOSE GROUND Material having a relative compaction less than 90%.

LOOSE-JOINT BUTT Hinge that can be taken apart by lifting one leaf from the other.

LOOSE KNOT Knot held in position by surrounding wood fibers.

LOOSE-LAY FLOORING Special type of flooring developed for use over wood subfloors where, rather than being fully glued down, it is stapled or glued around the perimeter of the room only.

LOOSE MATERIAL BUCKET *See* **Bucket.**

LOOSE-MEASURE VOLUME Volume of earth once it is moved from its original position and deposited in another location, either for transport or for storage.

LOOSE-PIN BUTT Butt hinge with a removable pin enabling the hinge to be separated.

LOOSE PLY Separation between adjacent plies.

LOOSE ROCK Rock that can be moved without blasting but that may still require ripping or digging.

LOOSE TONGUE Cross tongue.

LOOSE TUBE Tube separated from the carcass. *Also called* **Pulled-down tube.**

LOOSE YARD Cubic yard loose measurement, a unit of volume of excavation.

LOP To cut limbs from standing trees.

l or r *Abbreviation for:* left or right.

LOS ANGELES ABRASION TEST Test for abrasion resistance of aggregates to wear and abrasion. *Also called* **Aggregate attrition test,** and **Delaval test.**

LOSS FACTOR Reductions made to gross volumes to allow for decay, waste, and breakage.

LOSS OF HEAD Hydraulic friction. *Also called* **Lost head.**

LOSS OF PRESTRESS Reduction in the prestressing force that results from the combined effects of slip at anchorage, relaxation of steel stress, frictional loss due to curvature in the tendons, and the effects of elastic shortening, creep, and shrinkage in the concrete.

LOSS OF USE INSURANCE Protection against loss during the time required to repair or replace an insured's property damaged or destroyed.

LOSS ON HEATING *See* **Oven loss test.**

LOSS ON IGNITION Percentage loss in weight of a sample ignited to its constant weight at a specified temperature.

LOST CIRCULATION Drilling fluid that escapes from the borehole through crevices or porous media.

LOST GROUND Voids left by material that runs into an excavation under or through the shoring, or as a boil in the bottom.

LOST HEAD *See* **Loss of head.**

LOT 1. Defined quantity; 2. Plot of land; 3. Subdivision of a block in a town or city; 4. Parcel of land in a cemetery; 5. Uniquely defined quantity of material from a single source, or homogeneous segment of construction, on which decision is made for acceptance, rejection, or other disposition.

l/o temp. *Abbreviation for:* layout template.

LOT LINE Line that bounds a plot of ground legally described as a lot in the title of the property.

LOT WIDTH Average horizontal distance between the side lot lines, measured perpendicular to the lot depth.

LOUDNESS LEVEL Intensity level of a sound to that of a sound of 1,000 cycles frequency which sounds equally loud.

LOUVER Opening fitted with fixed or adjustable slats mounted at an angle to the horizontal or vertical, to control the passage of air and sunlight to the exclusion of rain, or vision.

LOW AIR Air supplied to pressurize working chambers and locks. *See also* **High air.**

LOW-ALKALI CEMENT Portland cement that contains a relatively small amount of sodium or potassium, or both.*See also* **Alkali.**

LOW BASIS WEIGHT Laminating papers, often referred to as 'micro-papers' or 'rice-papers,' that range in weight from 23 to 30 gm, sometimes impregnated with resin.

LOWBED *See* **Trailer.**

LOWBED, LEVEL-DECK TRAILER *See* **Trailer.**

LOW BID Bid for work that complies with all bidding requirements and that contains the lowest total price.

LOW BOY *See* **Trailer.**

LOW CAB FORWARD *See* **Cab.**

LOW CONSISTENCY PLASTER Neat gypsum base-coat plaster requiring less mixing water to produce workability than standard gypsum base-coat plaster.

LOW-COST HOUSING Housing of low capital cost per unit (including common amenities).

LOW-CYCLE FATIGUE Fracture resulting from a relatively high stress range resulting in a relatively small number of cycles to failure.

LOW-DENSITY CONCRETE Concrete having an oven-dry unit weight of less than 23 kg/0.028 m³ (50 lb/ft³). *See also* **Insulating concrete,** and **Lightweight concrete.**

LOW-DENSITY EXPLOSIVE Explosive that has less breaking power, due to its lower density; density may be decreased by loose packing, by altering the coarseness of the components, or by adding space-consuming materials such as gas, woodmeal, or microballoons.

LOW DISPLACEMENT PILE Pile with minimal soil displacement, such as H, open-end pipe, or sheet piles, *Also called* **Pile, nondisplacement.**

LOW EFFORT CONTROLS Improved control system with strengthened bell housings and more sensitive valve springs for improved hydraulic lever control that, in a crane, enables the operator to more precisely place a load with less personal fatigue.

LOW ENERGY DETONATING CORD Used to initiate nonelectric detonating caps at the bottom of boreholes.

LOWER BOUND LOAD Load computed on the basis of an assumed equilibrium moment diagram in which the moments are not greater than M_p, that is, less than or at best equal to the true ultimate load.

LOWER CARRIAGE Equipment at the bottom of an escalator truss that guides the step chain.

LOWERED CURB Section of curb that is lowered in order to permit step-free passage from pavement to road. *See also* **Curb.**

LOWER EXPLOSIVE LIMIT Minimum concentration of vapor in air at which the propagation of flame occurs on contact with a source of ignition.

LOWERING VALVE *See* **Valve.**

LOWER LEVEL Area to which a worker can fall from a stairway or ladder.

LOWER PLATE Fixed element of a telescope-equipped survey instrument, attached to the tripod, about which the rest of the instrument rotates.

LOWER RAIL Lower framing member of the front, sides, and occasionally rear sections of a van body.

LOWER SPREADER A-frame forming part of a boom suspension, supporting sheaves for the live suspension ropes and attached to the gantry or upper structure.

LOWER TIME *See* **Hydraulic cycle time.**

LOWEST BID Bid that offers to complete the specified work according to the terms of the contract documents for the least total sum.

LOW EXPLOSIVE Explosive designed to deflagrate and containing no ingredients that, by themselves, can be exploded. *See also* **Explosive,** and **High explosive.**

LOW FORWARD ENTRY CAB *See* **Cab.**

LOW FREQUENCY CURRENT Electric current having a relatively small number of cycles per second.

LOW GROUND PRESSURE Version of some tracked equipment that combines an increase in track length with wider-than-normal track shoes. A machine so equipped distributes its weight over a greater area with a resulting drop in pressure exerted on the ground.

LOW HAZARD INDUSTRIAL OCCUPANCY *See* **Industrial occupancy.**

LOW-HEAT CEMENT Portland cement that produces limited generation of heat during setting.

LOW-LIFT GROUTING Technique of concrete-masonry wall construction in which the wall sections are built to a height of not more than 1.5 m (5 ft) before the cells of the masonry units are filled with grout. *See also* **Grouting,** and **High-lift grouting.**

LOW-LIFT PLATFORM TRUCK Self-loading industrial truck equipped with a load platform intended primarily for transporting loaded skid platforms.

LOW LUSTER Finish with a low sheen. *Also called* **Satin finish.**

LOW-LYING AREA TECHNIQUE *See* **Sanitary landfilling, wet or low-lying area technique.**

LOW ORDER Condition of detonation that is not as rapid or complete as it should be.

LOW-PRESSURE GROUT Grout pumped through a concrete lining under low pressure to fill any voids between the lining and the tunnel surface or between tunnel liners and backfill concrete. *Also called* **Backfill grout.**

LOW-PRESSURE LAMINATE Preprinted or solid color decorative paper that has been saturated with resin. Under heat and pressure, it bonds to a board surface without need for additional adhesive.

LOW-PRESSURE STEAM CURING *See* **Atmospheric-pressure steam curing.**

LOW-PRESSURE STEAM HEATING *See* **Heating system.**

LOW-PRESSURE STORAGE TANK Storage tank designed to operate at pressures greater than 3.5 kPa (0.51 psi) to 100 kPa (14.5 psi) gauge.

LOW-PRESSURE TIRE Larger cross section tire for operation at lower pressure (increased air capacity permits lower pressure).

LOW-PROFILE TIRE Tire in which the cross section has a squat appearance. While most tires have cross section widths that are about the same as their heights, low profile tires have heights only about 85% of their width.

LOW-RENT HOUSING Public housing where the rent is subsidized in some form either by direct expenditure of public funds or by the foregoing of profits normally expected from an investment in rentable housing.

LOW SLUMP Very stiff concrete or mortar.

LOW-STRENGTH MATERIAL *See* **Controlled low-strength material.**

LOW-TEMPERATURE FLEXIBILITY Ability of a hose to be flexed, bent, or bowed at low temperatures without loss of serviceability.

LOW-TEMPERATURE FLEXING Act of bending or bowing a hose under conditions of cold environment. *Also called* **Cold flex.**

LOW VELOCITY TOOL *See* **Explosive-actuated fastening tool.**

LOW WATER MARK Average surface elevation of the expected low water. *See also* **High water mark.**

lox *Abbreviation for:* locks.

LOX *Abbreviation for:* liquid oxygen.

LP *Abbreviation for:* lighting panel; lightproof; lime putty.

L-P *Abbreviation for:* low pressure.

LPA *Abbreviation for:* local public agency.

LPG *Abbreviation for:* liquid propane gas.

LPL *Abbreviation for:* low-pressure laminate; lightproof louver.

LPP *Abbreviation for:* low-pressure pump.

LPS *Abbreviation for:* lightproof shade.

l pt *Abbreviation for:* low point.

LPV *Abbreviation for:* lightproof vent.

LR *Abbreviation for:* left rear; living room.

L-R *Abbreviation for:* left to right.

LRFD *Abbreviation for:* load and resistance factor.

LR.MCO *Abbreviation for:* log run, mill culls out.

LS *Abbreviation for:* lead shield; low speed.

ls *Abbreviation for:* left side; lump sum.

ls. *Abbreviation for:* limestone.

LSD *Abbreviation for:* legal subdivision.

lshd *Abbreviation for:* leasehold.

L-SHORE Shore with an L-head. *See also* **L-head.**

lt *Abbreviation for:* laundry tray; light.

LT *Abbreviation for:* long-turn.

L/T *Abbreviation for:* long ton.

ltb *Abbreviation for:* lime treated base.

ltd *Abbreviation for:* limited.

ltg *Abbreviation for:* lighting.

lth *Abbreviation for:* lath.

lthr *Abbreviation for:* leather.

ltprf *Abbreviation for:* lightproof.

ltr *Abbreviation for:* letter.

ltwt *Abbreviation for:* lightweight.

lub. *Abbreviation for:* lubricant; lubricate.

LUBE OIL HEATER Device used to heat the engine lubricating oil at cold ambient temperatures.

LUBRICANT Liquid, solution, or compound used to ease, lubricate, and cool surfaces moving about or across each other.

LUBRICATING GREASE Solid to semisolid product of dispersion of a thickening agent in a liquid lubricant. Additives imparting special properties may be included.

LUBRICATING GREASE PENETRATION Depth in tenths of a millimeter that a standard cone penetrates a sample under prescribed conditions of weight, time, and temperature. *See also* **Penetration.**

LUBRICATING GREASE STRUCTURE Physical arrangement of the component particles of a lubricating grease thickener, additive (if any), and liquid lubricant. It is the nature and stability of this arrangement that determines the appearance, texture, and physical properties of lubricating grease.

LUBRICATION Reduction of friction or wear between two load-bearing surfaces, in relative motion, by the application of a lubricant. There are four basic types of lubrication:

Boundary lubrication: Condition in which the lubricant film becomes too thin to give full fluid separation of the rubbing surfaces. As a result, surface asperities come in contact. Consequently, friction and wear protection properties are determined by the chemical nature of the lubricant rather than by its bulk properties.

Elastohydrodynamic lubrication: Condition in which surfaces of heavily loaded machine elements are either completely, or in part separated by a very thin lubricant film. The elastic deformation of the contacting surfaces traps the lubricant, subjecting it to high pressures that increase its viscosity and its load-carrying capacity.

Hydrodynamic lubrication: Condition in which the shape and relative motion of the sliding surfaces cause the formation of a continuous fluid film under sufficient pressure to prevent any contact between the surfaces. *Also called* **Fluid film lubrication.**

Hydrostatic lubrication: State of lubrication in which the lubricant is supplied to a plain bearing under sufficient external pressure to separate the opposing surfaces by a continuous lubricant film.

LUBRICATOR 1. Device that adds controlled or metered amounts of lubricant into a pneumatic system; **2.** Device (in elevators) that feeds oil to the rail and provides lubrication for sliding shoes. *Also called* **Oiler.** *See also* **Line oiler,** and **Sight-feed lubricator.**

LUFFING Changing boom angle in the vertical plane by varying the length of the boom suspension ropes or by extension of the boom cylinders. *Also called* **Booming in (and out), Derricking,** and **Topping.**

LUFFING ATTACHMENT Crane attachment adaptable to a basic cable crane, consisting of a vertical luffing boom that is capable of being offset, with a luffing jib (usually the basic crane boom) affixed to the upper part of the luffing boom.

LUFFING CABLEWAY Cableway with two fixed towers that provide transverse hook coverage by drifting the tops of the towers as the lengths of the side guys are adjusted.

LUFFING JIB Jib that can vary its angle relative to the main boom by means of reeving attached to one of the crane's hoists.

LUG 1. Mounting device consisting of a block extending past the basic cylinder profile. The block usually has a tapped or through-mounting hole at right angles to the cylinder axis; **2.** Aggressive tire tread pattern feature used to improve traction; **3.** Small projection

engaging an adjoining unit; **4.** Part of a door or window sill extending horizontally beyond the opening; **5.** Connector for fastening a wire to a terminal.

LUG DOWN Slowdown in engine speed of mechanical equipment caused by an increase in the load being hauled or carried beyond the capacity of the engine of the prime mover. Usually occurs when machinery is crossing soft or unstable soil or is pushing or pulling beyond its rated capacity.

LUGGER BODY Used for commercial/industrial waste collection systems, dropping off and retrieving trough-shaped containers of 4.6 to 15.3 m³ (6 to 20 yd³) capacity.

LUGGING 1. Operating an engine at extremely low rpm, below the efficient operating range of the engine; **2.** In reciprocating engines, continuous operation on the torque-rise end of the powerband.

LUG TEARING Rupture of the lug tread pattern of a tire resulting from violent operation or mechanical interference.

LUG TIRE Tire with a cross tread and relatively deep grooves.

LUMBER Product of the sawmill: unplaned, unmilled boards of varying widths and lengths. It can be grouped as follows:

Board: Yard lumber less than 50 mm (2 in.) thick and more than 50 mm (2 in.) wide.

Dimension: Yard lumber from 50 mm (2 in.) up to, but not including, 125 mm (5 in.) thick, and 50 mm (2 in.) or more inches wide. The actual size of such lumber after shrinking from its 'green' dimension and after machining to size is called **Dressed lumber.**

Dressed: Dimensions of lumber after shrinking from its 'green' size and after planing, usually 9.5 mm (0.375 in.) less than the nominal or rough size.

Matched: Lumber that is edge-dressed and shaped to make a close tongue-and-groove joint at the edges.

Shiplap: Lumber that is edge-dressed to make a close rabbeted or lapped joint.

Timber: Yard lumber 127 mm (5 in.) or more in least dimension.

lum. *Abbreviation for:* lumber; luminous.

LUMBER BODY Platform truck or trailer with transverse rollers, designed primarily for the transportation of sawed lumber.

LUMBER FORK *See* **Fork.**

LUMBER GRADES System of ranking lumber according to its physical strength and appearance. Lumber grades are divided into clear, and common:

CLEAR GRADES

A Select: Also known as **Clear.**

B Select: No knots or blemishes.

C Select: A few blemishes on one side only.

D Select: Slight blemishes on both sides.

COMMON GRADES

No. 1 Common: Quality overlaps with Clear grade

D select. *Also called* **1C,** and **Select merchantable.**

No. 2 Common: Some knots and blemishes. Also known as **2C.**

No. 3 Common: Contains some knots, blemishes, and small holes. Also known as **3C.**

No. 4 Common: Contains numerous imperfections; a relatively poor quality product. *Also called* **Utility Grade.**

No. 5 Common: Poor quality product, not suitable for construction. *Also called* **Economy grade.**

LUMBERJACK Traditional term for a forest worker, performing any of a variety of jobs related to timber harvesting; most commonly used in the eastern half of North America. *Also called* **Logger.**

LUMBER TALLY Record of lumber giving the number of boards or pieces by size, grade, and species; often expressed in MFB (million board feet).

LUMEN One of 17 derived units with special names of the SI system of measurement: a unit of luminous flux emitted in a cone of solid angle of 1 sr by a spherical point source of uniform luminous intensity of 1 cd. (1 lm=1 cd·sr). Symbol: lm. *See also the appendix:* **Metric and nonmetric measurement.**

LUMEN-HOUR Lighting output of one lumen for one hour. Symbol: lm·hr.

LUMEN SECOND A derived unit of the quantity of light with a compound name of the SI system of measurement. Symbol: lm·s. *See also the appendix:* **Metric and nonmetric measurement.**

LUMINAIRE Complete lighting unit: light source, holder, globe or shade, reflector (where provided), and means of attachment.

LUMINOUS PAINT Coating containing a substance that gives off light due to fluorescence, phosphorescence, or radioactivity. *See also* **Fluorescent paint,** and **Phosphorescent paint.**

LUMP SUM AGREEMENT *See* **Lump sum contract,** and **Stipulated price agreement.**

LUMP SUM BID Single sum bid without breakdown.

LUMP SUM CONTRACT Contract in which the owner will pay the contractor a specified sum of money for a project completed to the terms of the contract document. *Also called* **Lump sum agreement.** *See also* **Stipulated price agreement.**

LUNETTE EYE Steel eye mounted on the drawbar of a trailer or dolly, designed to couple with the pintle hook of a towing vehicle.

lust. *Abbreviation for:* lustrous.

LUSTER Reflecting quality and brilliance of a surface. *See also* **Appearance.**

LUTE Type of rake used to smooth out minor surface irregularities behind an asphalt paver.

LUX One of 17 derived units with special names of the SI system of measurement: a unit of illumination (or illuminance) of 1 lm uniformly over an area of 1 m². (1 lx = 1 lm/m²). Symbol: lx. *See also the appendix:* **Metric and nonmetric measurement.**

LUX-SECOND Derived unit of light exposure with a compound name of the SI system of measurement. Symbol: lx·s. *See also the appendix:* **Metric and nonmetric measurement.**

LUXURY VINYL TILE *See* **Solid vinyl tile.**

LV *Abbreviation for:* low voltage.

lvd *Abbreviation for:* louvered door.

lvr *Abbreviation for:* lever; louver.

l/w *Abbreviation for:* line weld.

LW *Abbreviation for:* lap weld; low water.

lw. *Abbreviation for:* lengthwise.

LWC *Abbreviation for:* lightweight concrete.

LWIC *Abbreviation for:* lightweight insulating concrete.

LWM *Abbreviation for:* low water mark.

lwr *Abbreviation for:* lower.

LWR *Abbreviation for:* local wage rate.

LYCEUM Building for an institution of learning.

LYNCH GATE Roofed wooden gateway. (Originally: at the entrance to a churchyard, providing a resting place for a coffin (lynch being Saxon for corpse). Part of the burial service would sometimes be read at this place.)

LYP *Abbreviation for:* lower yield point.

lyr *Abbreviation for:* layer.

LYSIMETER Structure containing a mass of soil, designed to permit measurement of water flowing through the soil.

M

m *Abbreviation for:* meter; milli.

M *Abbreviation for:* male; metal; thousand;.

ma. *Abbreviation for:* master.

mA *Abbreviation for:* milliampere.

MA *Abbreviation for:* moist air; mechanical advantage.

mac. *Abbreviation for:* macadam.

MACADAM Layer of coarse, graded, angular mineral aggregate with a filler of fine aggregate, interlocked by compaction.

mach. *Abbreviation for:* machine; machined; machinery; machinist.

MACHINE 1. Apparatus for applying mechanical power; **2.** *See* **Driving machine.**

MACHINE AVAILABILITY *See* **Machine time.**

MACHINE BEAM Horizontal steel beam that supports machinery, typically an elevator driving machine.

MACHINE BOLT Bolt with a square or hexagonal head and the upper portion of the shank not threaded.

MACHINE BURN Scorching of the surface of wood being planed due to overheating of the knives or rolls when the pieces are stopped in the machine.

MACHINE COUNTERWEIGHT *See* **Drum counterweight.**

MACHINED 1. Surface produced on metal by a machine tool, most frequently a smooth surface; **2.** Shape produced by cutting or grinding.

MACHINED END END Pipe that has been milled on each end and left rough in the center. *Also called* **M.E.E.**

MACHINE DIRECTION Particleboard or drywall orientation that corresponds with the direction that the product moved through the machine that manufactured or machined it. *Also called* **Parallel direction.**

MACHINED OVER ALL Pipe that has been milled end-to-end to allow easier joining of the pipe if the length must be cut to fit. *Also called* **M.O.A.**

MACHINE DOWNTIME *See* **Machine time.**

MACHINE DRAWING Drawing of a mechanical part, including dimensions and notes.

MACHINE EFFICIENCY Ratio of the rate of horsepower energy output to rate of horsepower energy input over a given period.

MACHINE LANGUAGE Series of binary numbers for interpretation by a computer.

MACHINE MADE Product made on or by a machine, as against being handmade.

MACHINE MOMENT *See* **Machine resisting moment.**

MACHINE OPERATING EFFICIENCY Ratio of actual work time to available work time, expressed as minutes worked per hour.

MACHINE RATE Cost per unit for owning and operating a machine or other piece of equipment. The rate is composed of fixed costs such as depreciation, interest, taxes, and license fees, and variable costs including fuel, lubricants, and repairs and replacement of components such as tires.

MACHINE RATING Amount of load, or power, an electrical machine can deliver without overheating.

MACHINE RESISTING MOMENT Moment of the dead weight of a crane or derrick, less boom weight, about the tipping fulcrum; the moment that resists overturning. *Also called* **Machine moment,** and **Stabilizing moment.**

MACHINE ROOM Room where the driving machine for an elevator is located.

MACHINE SCREW *See* **Screw.**

MACHINE STRESS RATED Lumber that has been nondestructively tested by mechanical stress rating equipment and marked to indicate its modulus of elasticity.

MACHINE TIME Time allocated to a machine or piece of equipment under a range of cost or profit headings, including:

Active repair time: Time during which actual repair work is carried out on the machine itself or a dismantled part of the machine.

Actual productive time: Time spent using a machine to carry out an actual task.

Delay time: Time when, for any reason, a machine, ready, available, and capable of performing its assigned task, is unable to do so.

Disturbance time: Examples are: time spent for securing a load, towing, detail planning, talking to supervisor, waiting for a load, and waiting for better weather.

Idle time: Scheduled nonoperating time during which a machine is not working, moving, under repair, or being serviced.

In-shift moving time: That part of nonmechanical delay time during which a machine is moving or being transported. Includes the time taken to move or transport the machine between operating sites or between base and site, assuming the machine is not under repair. It does not include time spent moving between adjacent working positions on any one site.

In-shift repair time: Part of mechanical delay time when a machine is actually undergoing repair plus the time during which a machine is waiting to be repaired or for repair parts, mechanics, or facilities.

In-shift service time: Part of mechanical delay

time when a machine is waiting for service parts, mechanics, or repair facilities.

Machine availability time: Percent of the scheduled operating time during which a machine is not under repair or service.

Machine downtime: Time during which a machine cannot be operated in production or auxiliary work because of breakdown, maintenance requirements, or power failure.

Machine utilization time: Percentage of the scheduled operating time that is productive time. It is computed by productive time divided the scheduled operating time multiplied by 100.

Mechanical delay time: Part of scheduled operating time spent in repair or service during which a machine cannot work. It docs not include replacement of oil filters and spark plugs as scheduled in a preventive maintenance program. Servicing is fueling, lubricating, and doing the work specified in a scheduled preventive maintenance program. When a machine is being serviced while under repair, the time involved is classified as repair time, not service time. Repair and service time occur in both scheduled operating and nonoperating times.

Operating time: Productive time.

Operational lost time: Time during which production is halted due to such factors as operating conditions, nonavailability of auxiliary equipment, or using the machine or equipment in a non-productive manner to assist other machines.

Other productive time: Time when a machine is carrying out tasks other than those for which it is intended.

Out-of-shift repair time: Part of nonoperating time when a machine is actually undergoing service time. Does not include waiting time.

Out-of-shift service time: Part of nonoperating time during which a machine is actually undergoing repair. Waiting time is not included here as an in-shift repair time element.

Personnel time: Part of nonmechanical delay time in which a machine lacks an operator or any other member of the machine crew.

Productive machine hour: Time during scheduled operating hours when a machine performs its designated function (time exclusive of such things as machine transport, operational or mechanical delays, and servicing or repair).

Productive time: Part of scheduled operating time in which a machine is performing a function for which it was intended. Also, time spent in carrying out the task; the sum of actual productive and other productive time.

Repair time: Sum of active repair time, waiting repair time, and time spent servicing the machine while undergoing repair.

Scheduled machine hour: Allocation of productive or nonproductive machine time to an accountable classification.

Scheduled nonoperating time: Time when no production is scheduled for a machine or item of equipment.

Scheduled operating time: Time when a machine is scheduled to do productive work. Time during which a machine is on standby as a replacement is not considered as scheduled operating time. When a machine is replaced, the scheduled operating time of the replaced machine is considered as ending when the replacement arrives on the job. Scheduled operating time of the replacement commences when it starts to move to the location of the machine it is replacing. Extension of the regular shift operation into overtime is considered as scheduled operating time.

Service time: Time for normal service and maintenance of machines and equipment.

Waiting repair time: Time during which the machine is waiting for a mechanic, spare parts, or repair equipment. Includes time for transporting the machine to and from the workshop.

MACHINE TOOL Hardened metal shape designed to cut or mark steel.

MACHINE UTILIZATION *See* **Machine time.**

MADE GROUND Usable land created by dumping and/or filling.

mag. *Abbreviation for:* magazine; magnetic; magneto; magnitude.

MAGAZINE Structure or container for the storage of explosives.

MAGNA-FLUX MACHINE Instrument that employs lines of magnetic force to check the soundness of metallic parts.

MAGNET Electromechanical device attached to the arm or jib, used in handling metal.

MAGNETIC BEARING Horizontal angle of a bearing or sight from magnetic north.

MAGNETIC DIP Angle that a magnetized needle, under artificial influence, will make with the horizontal plane.

MAGNETIC DRAIN PLUG Magnetized plug that collects metallic particles at the bottom of an axle housing or oil sump.

MAGNETIC DRIVER Magnetized chuck to hold fasteners (screws, nails, etc.) while they are being presented to the work and being driven.

MAGNETIC EFFECT Results when an electrical current produces a magnetic field.

MAGNETIC FIELD Region surrounding a magnet or conductor through which current is flowing.

MAGNETIC FLUX Total flow of magnetism.

MAGNETIC FRACTION That portion of municipal ferrous scrap remaining after the nonmagnetic contaminants have been removed and the magnetic fraction washed with water and dried at ambient temperature.

MAGNETIC HAMMER *See* **Hammer.**

MAGNETIC NORTH Direction indicated by a magnetized, free-floating or pivoting needle.

MAGNETIC PLUG Magnetized drain or inspection plug that will attract and hold ferrous particles in a lubricant. *See also* **Plug.**

MAGNETIC REPULSION Condition where like magnetic poles repel each other.

MAGNETIC SEPARATOR *See* **Separator.**

MAGNETO Small electric generator in which a permanent magnet carries the magnetic field.

MAGNIFICATION Apparent enlargement of an object by an optical instrument. A 25x telescope increases the apparent size of an object 25 times its real size.

mai *Abbreviation for:* mean annual increment.

MAIN Principal conductor or conduit: in electricity, the circuit from which smaller circuits are taken; in piping systems, the main supply pipe.

MAIN AIR Air supplied through a burner, but not including that used for atomizing. *Also called* **Combustion air.**

MAIN BAR *See* **Main reinforcement.**

MAIN BEAM Beam that supports other beams and is not itself supported by another beam.

MAIN BEARING BORE ALIGNMENT Having all main bearing bores in exact alignment from one end of an engine block to the other.

MAIN CIRCULATING LOOP Section of an oil handling system that delivers the oil from storage to the branch circuits and returns the unused oil to the storage tank.

MAIN CONTRACTOR *See* **Prime contractor.**

MAIN COUPLE Principal truss in a timber roof.

MAIN DIFFERENTIAL Mechanism that allows a vehicle's left- and right-side wheels to rotate at different speeds during turns for smoother vehicle operation.

MAIN FLOOR Floor providing normal or principal egress from a building.

MAINFRAME EXTENSION Extension(s) that bolts to either side of a paving tractor's rear bulkheads to help confine material when auger extensions are in place. Confining the material reduces the head of material that must be 'dragged down the road' and helps eliminate segregation problems.

MAIN GUT Track cable of a cableway.

MAIN HOIST Main load lifting device containing the primary lift line.

MAIN LINE In cable yarding, the line used to bring logs to the landing.

MAIN-LINE BLOCK Block on a spar through which the main line runs.

MAIN-LINE JUMBO Drill jumbo that travels on the main rail line in a tunnel.

MAIN RAFTER Rafter extending at right angles from the plate to the ridge.

MAIN REINFORCEMENT Steel reinforcement to concrete that resists stresses resulting from design loads and moments. *Also called* **Main bar.**

MAIN ROAD Road that supports a high level of traffic, usually well designed, built, and maintained.

MAIN RUNNER One of the principal support members in a suspended ceiling system, itself supported by hangers attached to the building structure and supporting furring channels or rods.

MAINS BREAKER Electrical switch that disconnects the incoming main feed to a distribution panel. *Also called* an **Entrance switch.**

MAIN SEWER Large-diameter sewer to which building drains (and, in a combined system, storm drains) are connected.

MAINSHAFT In a transmission, the mainshaft consists of a heavy-duty central shaft and several gears that turn freely when not engaged. The mainshaft can move to allow for equalization of the loading on the countershafts. *See also* **Floating mainshaft.**

maint. *Abbreviation for:* maintain; maintenance.

MAINTAINED LOAD TEST Loading test on a structural member, held constant for a specific period or until the rate of movement falls to a defined value.

MAINTENANCE Process of sustaining the level of physical quality of an existing building, machine or site, usually involving a program of inspection, cleaning, and repair activities.

MAINTENANCE BOND *See* **Bond.**

MAINTENANCE CONTRACT Contract, not necessarily with the supplier of equipment or builder of a structure, for a range of maintenance procedures over a specified period.

MAINTENANCE LEASE *See* **Lease.**

MAINTENANCE MANAGEMENT SYSTEM Formal procedure used to plan, organize, direct, control, and evaluate maintenance programs and maintenance management systems.

MAINTENANCE PERIOD Time, following completion of a project, during which the contractor is still responsible for his work.

MAINTENANCE SCHEDULE Detailed list of maintenance procedures required to be completed according to an established timetable.

MAINTENANCE STANDARD Formally established criterion for a specific operation that encompasses elements usually found in quality, quantity, and performance standards.

MAIN TRANSMISSION *See* **Transmission.**

MAIN VENT Principal soil or waste stack in a plumbing system that connects the system to the open sky.

MAIN WATER LINE *See* **Water main.**

MAISONETTE Dwelling of more than one story within a larger building.

maj. *Abbreviation for:* major.

MAJOR AND MINOR CONTRACT ITEMS Major contract items are listed as such in a bid schedule or in the special provision; all other original contract items are considered as minor items; or in cases where the major contract items are not listed as such, the original contract item of greatest cost, computed from the original contract price and estimated quantity, and such other original contract items next in sequence of lower quantities of not less than ...% (60% suggested) of the original contract cost are considered as a major item or items; or any item having an original contract amount is considered as a major item or items.

MAJOR ARCH *See* **Arch.**

MAJOR DIAMETER Largest diameter of a screw thread.

MAJOR HIGHWAY *See* **Highway.**

MAJOR OCCUPANCY Principal occupancy for which a building or a part of a building is used or intended to be used. *See also* **Occupancy.**

MAJOR STREET *See* **Street.**

MAKE GOOD Repair or otherwise correct a defect.

MAKEUP PILE *See* **Doodle.**

MAKEUP WATER Water that is added to a boiler, tank, or some other container to replace water that has been lost, thus maintaining the proper water level.

MAKING THE COUNT Totaling the number of masonry units laid by the mason in one day.

MALE COUPLING Threaded hose nipple that fits in the thread of a female swivel coupling of the same pitch and appropriate diameter.

MALE THREAD External pipe thread.

MALFUNCTION To function other than in the designed manner.

MALL 1. Shaded walk; **2.** Retail shopping complex either enclosed and roofed or with sufficient roof areas to provide shaded walks.

MALLEABILITY Ability of a metal to be deformed without breaking.

MALLEABLE CASTINGS Cast forms of metal that have been heat treated to reduce their brittleness.

MALLEABLE CAST IRON Cast iron made by annealing white cast iron while the metal undergoes decarburization, graphitization, or both, thus eliminating all or most of the cementite. *See also* **Iron,** and **Wrought iron.**

MALLEABLE IRON PIPE *See* **Pipe.**

MALLET *See* **Hammer.**

MALPRACTICE Alleged professional misconduct or lack of skill by a professional.

man. *Abbreviation for:* manage; managed; manual.

MANAGED FOREST LAND Forest land that is being managed under a **Forest management plan.**

MANAGED HARVEST Estimated volume of timber on commercial forest land that could be cut annually for the next 10 years while improving tree stocking and bringing about a more even distribution of age classes. Annual managed harvest is considered separate from harvest cuttings and thinnings and is determined by computer using an area control system that specifies the number of acres to be cut annually.

MANAGEMENT Process in the project management system that consists of direction and control to achieve results in accord with established performance stand-

ards for scope, time, and budget. The process includes participation in setting the performance standards and in monitoring performance. *See also* **Project management.**

MANAGEMENT AREA Stands or forest types that require similar management practices and that can be grouped for treatment as a management unit.

MANAGEMENT OBJECTIVES Long-term management goals, and the strategies to achieve them.

MANAGEMENT-VOLUME INVENTORY Computation of pertinent forest data, such as volume or basal area and increment and mortality of stands, to assess silvicultural opportunities. *Also called* **Inventory.**

MANDATORY Requiring absolute conformity.

MANDATORY AND CUSTOMARY BENEFITS Legislated personnel benefits (social security, workers' compensation, etc.); those that are customarily acknowledged (sick leave, holidays, etc.), plus those that are particular to each employer (company pensions, etc.).

MANDATORY RECYCLING Programs requiring that residents and/or businesses keep secondary materials from their solid wastes.

M&F *Abbreviation for:* male and female.

M&R *Abbreviation for:* maintenance and repair.

MANDREL 1. Tight-fitting metal rod inserted into something to hold it while it is being machined; **2.** Curved support inserted in a tire to prevent the casing from collapsing while under repair; **3.** Core that is inserted into a closed-end thin-shell tubular pile. There are two types:

> **Collapsible:** A core, the outer diameter of which can be changed by mechanical or other means, capable of transmitting the hammer energy to the bottom of the pile and the periphery of the thin-shell casing. It is inserted into the pile in a collapsed condition and then expanded to grip the inner surface of the pile with sufficient force to prevent slipping.

> **Solid or semisolid:** A heavy tubular section that will transmit the hammer energy to either the casing point or, for a step-tapered pile, to the transition rings or plow rings and the point. *Also called* **Pile core.**

MANDREL BUILT Hose fabricated and/or vulcanized on a mandrel.

MANDREL PILOT Machined rod, the same size as an inlet or exhaust valve, onto which a valve seat grinding stone mandrel is placed.

MANDREL WRAPPED Tubing, built up by wrapping a thick unvulcanized sheet around a mandrel.

MANEUVER TIME Part of cycle time, and for mobile equipment, includes basic travel and four changes of direction at full throttle.

MANGANESE *See* **Alloying elements.**

MANHOLE 1. Chamber giving access to an underground service; **2.** Covered access hole to a tank or boiler.

MANHOLE COVER Removable cast-iron plate mounted in a metal frame that is set flush with grade or the finished road surface.

MANHOLE HEAD Two-part cast-iron fixture: a frame that rests on the shaft of the manhole, and a removable cover. The frame may be equipped for height adjustment.

MAN HOUR Unit of work performed by one worker in one hour.

mani. *Abbreviation for:* manifold.

MANIFOLD 1. Chamber or tube having several inlets and one outlet, or one inlet and several outlets. *See also* **Vented manifold; 2.** Filter assembly containing multiple ports and integral related components which services more than one fluid circuit.

MANIPULATIVE JOINT *See* **Compression joint.**

MANLIFT 1. Mechanical equipment, often self-propelled, and including designs for rough terrain, designed to elevate a work platform and a limited weight of workers and their equipment; **2.** Power-driven device for the vertical transportation of people, consisting of an endless belt provided with steps or platforms and handholds.

MAN LOCK Chamber through which workers pass from one air pressure environment to another.

manom. *Abbreviation for:* manometer.

MANOMETER *See* **Gauge.**

MANSARD ROOF Pitched roof having a double slope between the eaves and ridge, the lower slope being steeper than the upper slope. Also known as **Double-pitch roof,** and **Gambrel roof.**

MANSION Large and often pretensious dwelling.

MANTELPIECE Covering or surround to a fireplace, usually decorative and mostly incorporating a projecting slab or shelf above.

MANTELSHELF Shelf or beam over a fire opening, usually projecting from the chimney breast.

MANTEL TREE Beam across a fireplace opening.

MANTLE Soil, sand, and other loose materials covering bedrock.

MANUAL Operated or used by hand.

MANUAL BATCHER *See* **Batcher.**

MANUAL BURNER Burner that is purged, started, ignited, modulated, and stopped manually.

MANUAL DRUM POSITIONING CONTROL Manual control of grade and slope.

MANUAL INSERT Optional, nonpowered section in a telescoping boom forming the outermost boom section.

MANUALLY PROPELLED MOBILE SCAFFOLD Scaffold assembly supported by casters and moved only manually.

MANUAL SCAFFOLD *See* **Suspended scaffold.**

MANUAL SECTION Outermost segment of a four- or five-section telescoping boom, attached to the outer mid-section. This boom section shares the telescope cylinder used to operate the outer mid boom section. This section also contains the head machinery of the attachment.

MANUAL SEPARATION Separation of materials from waste by hand sorting.

manuf. *Abbreviation for:* manufacturer.

MANUFACTURED Made, as distinct from assembled, supplied, retailed, sold, etc.

MANUFACTURED (MOBILE) HOME DECKING Particleboard product designed for use as a single-layer floor system in the construction of mobile (manufactured) homes. *Also called* **MHD.**

MANUFACTURERS' AND CONTRACTORS' LIABILITY INSURANCE Insurance covering manufacturing, contracting, and installation liability.

MANUFACTURERS' IDENTIFICATION Code symbol used on some products to indicate the manufacturer.

MAP CRACKING 1. Intersecting cracks that extend below the surface of hardened concrete, caused by shrinkage at the drying surface of concrete that is restrained by concrete at greater depths where little or no shrinkage occurs, and that vary in width from fine and barely visible to open and well-defined; **2.** Chief symptom of chemical reaction between alkalis in cement and mineral constituents in aggregate within hardened concrete.*Also called* **Pattern cracks.** *See also* **Craze cracking.**

mar. *Abbreviation for:* marine.

marb. *Abbreviation for:* marble; marbling; marbleized.

MARBLE Metamorphic rock composed essentially of re-crystallized calcite, dolomite, or both.

MARBLE MOSAIC TILE Tile made of small marble tesserae that vary slightly in size, usually about 13 mm (0.5 in.) square, and mounted on sheets of paper about 300 mm (12 in.) square to facilitate installation.

MARBLE TILES Marble cut into tile squares 300 mm (12 in.) square or less, usually 13 mm (0.5 in.) thick. May be polished, honed, split-faced, etc.

MARBLING Decorative painting technique used to simulate marble.

marg. *Abbreviation for:* margin.

MARGIN Edge or border. *See also* **Verge.**

MARGINAL WHARF Structure for mooring a vessel parallel to the shore.

MARINE ADHESIVE *See* **Adhesive.**

MARINE BORER One of several species of mollusks (*Teredo* (shipworm), *Bankia, Pholad*) and crustaceans (*Lignorum, Limnoria tripuncate, Quadripunctata*) that bore into untreated wood and concrete. *Also called* **Shipworm.**

MARINE GLUE *See* **Adhesive.**

MARINE GRADE Plywood panels manufactured with the same glueline durability requirements as other exterior-type panels but with more restrictive veneer quality and manufacturing requirements.

MARK 1. To select and indicate (usually with paint) trees to be felled in a logging operation. Trees to be left may also be marked. *Also called* **End mark; 2.** Panel identification number and/or symbol.

MARKED FACE Face side of timber.

MARKER Professional log scaler who indicates proper log lengths and correct trim for a log to be bucked in order to obtain the highest grade recovery.

MARKER LIGHT Amber or red light attached to a machine or vehicle body to indicate overall length.

MARKETABLE TITLE Title to real property free of defects.

MARKET PRICE Amount actually paid in a transaction.

MARKET VALUE Highest price for which property can be sold in the open market by a willing seller to a willing purchaser, neither acting under compulsion and both exercising reasonable judgement, both being fully aware of the highest and best use to which the property can be put. *Also called* **Actual value**, and **Fair cash value.**

MARKING Selecting and indicating, by a blaze or paint spot, a tree to be cut or left in a timber cutting operation.

MARKING GAUGE Tool used for scribing lines on wood parallel to the grain or face. Consists of a calibrated shaft with one or more sharpened pins piercing one end, along which a face plate rides. The face plate can be fixed in any position by means of a thumbscrew.

MARKING GUN Squirt gun used in applying marking paint for end-marking of logs for identification prior to transportation to the mill.

MARKING PAINT 1. Paint applied to identify an object according to some classification; **2.** Paint that marks the laying lines on saturated felt and similar roofing products.

MARK OUT 1. Indicating with keel (crayon) marks on the forms, the exact location of each reinforcing bar; usually done by a foreman from information derived from the detail sheets prior to starting actual placing operations in an area; **2.** Paint marking on road surface scheduled for reconstruction showing the location of buried services or underground structures.

MARL Calcareous clay found in the bottoms of shallow lakes, swamps, or extinct freshwater basins.

MARLIN SPIKE Tapered steel pin used in splicing rope.

marq. *Abbreviation for:* marquee.

MARQUEE Protective canopy projecting out over a principal entrance.

MARQUETRY Mosaic made of woods (and occasionally other materials) of different colors and textures, often to a finely detailed geometric pattern.

MARRYING Process by which ropes are joined together, end-to-end, forming a joint that is at least as strong as the lesser of the ropes being joined.

MARSHALL Method of asphalt mix design in which cylindrical test specimens are tested for density-voids and stability-flow to determine the optimum asphalt content.

MARSHALL STABILITY Maximum load resistance that an asphalt mix test specimen will develop at 60°C (140°F) under controlled conditions.

mas. *Abbreviation for:* masonry.

MASH HAMMER *See* **Hammer.**

MASKING Covering of part of a surface to exclude paint.

MASKING TAPE Tape, coated with a light adhesive, used to mask a surface so as to exclude paint.

MASON 1. Worker skilled in laying brick, block, or stone (as a brick mason, block mason, or stonemason); **2.** Name sometimes given a concrete finisher.

MASONRY Stone, brick, concrete, hollow-tile, concrete block, or similar building units or materials, or a combination of such materials, bonded together with mortar.

MASONRY BIT *See* **Bit.**

MASONRY CEMENT Hydraulic cement for use in mortars for masonry construction that contains one or more of the following materials: portland cement, portland blast-furnace slag cement, portland-pozzolan cement, natural cement, slag cement, or hydraulic limes, and in addition usually contains one or more materials such as hydrated lime, limestone, chalk, calcareous shell, talc, slag, or clay. *See also* **Cementitious materials.**

MASONRY FILL Insulating material used to fill the voids in hollow masonry units.

MASONRY FILLER UNIT Masonry unit used to fill in between joists or beams in order to provide a platform for a cast-in-place concrete slab.

MASONRY JOINT *See* **Pointing.**

MASONRY LIFT Height to which masonry is laid between periods of grouting.

MASONRY NAIL *See* **Nail.**

MASONRY PANEL WALL *See* **Wall.**

MASONRY REINFORCING Steel reinforcing rods, or mesh, used laterally between courses of masonry construction.

MASONRY UNIT 1. Natural or manufactured building unit of burned clay, stone, glass, gypsum, etc. *See also* **Hollow masonry unit, Modular masonry unit,** and **Solid masonry unit; 2.** Structural element consisting of concrete masonry units bonded by mortar, grout, or both.

MASONRY VENEER Surface shell or cladding of masonry units attached to a backing, not so bonded as to exert a common reaction under load.

MASONRY WALL *See* **Wall.**

MASON'S ADJUSTABLE MULTIPLE-POINT SUSPENSION SCAFFOLD Two-point or multipoint adjustable suspension scaffold designed and used for masonry operations.

MASON'S PUTTY Lime putty, portland cement, and stone dust mixture used for jointing ashlar.

MASON'S TROWEL *See* **Trowel.**

MASS 1. Physical property of matter that causes it to have weight; **2.** Quantity of matter in a body.

MASS BURN Municipal waste combustion in which solid waste is burned in a controlled system without prior processing.

MASS CONCRETE Any volume of concrete with dimensions large enough to require that measures be taken to cope with the generation of heat from hydration of the cement and attendant volume change to minimize cracking.

MASS CURING Adiabatic curing in sealed containers.

MASS DIAGRAM In earthwork calculations, a graphical representation of the algebraic cumulative quantities of cut and fill along the centerline, where cut is positive and fill is negative; used to calculate haul in terms of station yards.

MASS EXCAVATION BUCKET *See* **Bucket**.

MASS EXCAVATOR Excavator with a specially configured front end and bucket selection designed to dig and load large volumes of material considered as 'easy' digging. In general terms, the boom is shorter and both the boom and stick have bigger cross sections to handle the larger buckets and higher forces.

MASS HAUL CURVE Diagram showing the amount of excavation in a cut that is available for fill.

MASS PROFILE Road profile detailing the volume of cut and fill.

MASS SHOOTING Simultaneous exploding of charges in all of a large number of blastholes.

MAST 1. Essentially vertical, load-bearing component of a crane or derrick; the tower of a tower crane. *See also* **Boom mast**, and **Jib mast**; **2.** Support member of an industrial truck providing the guideways permitting vertical movement of the carriage. It is usually constructed in the form of channels or similar sections providing the supporting pathway for the carriage rollers.

mast. *Abbreviation for:* mastic.

MAST CAP *See* **Spider**.

MASTER BATCH Preliminary mixture of rubber and one or more compound ingredients, for such purposes as more thorough dispersion or better processing, that will later become part of the final compound in a subsequent mixing operation.

MASTER COUPLING LINK Alloy-steel-welded coupling link, used as an intermediate link to join alloy-steel chain to master links.

MASTER CYLINDER Main supply cylinder from which brake fluid is displaced to the wheel cylinders to actuate hydraulic brakes.

MASTER KEY Key that operates two or more locks requiring individually distinct keys. *Also called* **Passkey**.

MASTER LINK 1. Two-piece serrated link that bolts together and that can readily be unbolted allowing a track to be separated; **2.** Forged or welded-steel link used to support all members of a chain or wire-rope sling.

MASTER PIN The only pin in an integrated crawler track that, when driven out, will open the track. *See also* **Track pin**.

MASTER PLAN Overall plan that incorporates the concepts and principles in more detailed, supporting plans.

MASTER SPECIFICATIONS Specifications that serve as the principal reference and guide for more specific requirements particular to a location or circumstance, the intent of which may be amplified but may not be altered or diminished.

MASTER SWITCH *See* **Switch**.

MASTIC Sealant with putty-like properties.

MASTIC ADHESIVE *See* **Adhesive**.

MASTIC GROUT Chemical mixture of organic and inorganic ingredients forming a one-part grouting composition that is used directly from the manufacturer's container.

MASTIC SEALANT Soft, pliable and moldable sealing compound.

MAST LEAD *See* **Lead**.

MAT 1. Asphalt as it is produced from a spreader box or paving machine as a smooth, flat blanket; **2.** *See* **Bar mat**, and **Blasting mat**; **3.** Randomly distributed layer of glass fibers lightly bonded together by a polymeric binder; **4.** Wood platform, used in sets to support machinery on soft ground; **5.** Large footing or foundation slab used to support an entire structure; **6.** Grid of reinforcing bars; **7.** Log or lumber-type block placed under the foot of a hydraulic jack to increase surface area, give extra height, and compensate for ground irregularities for stability; **8.** Temporary roadway constructed of hardwood lumber; **9.** Heavy, flexible fabric of chain, wire rope, or other resilient material used to confine blasts.

mat. *Abbreviation for:* material; matrix.

MATCHED LUMBER *See* **Lumber**.

MATCHING Positioning veneer sheets to produce a particular pattern.

MATCHING SHOE Integral part of the automation system utilized to match a planing/milling pass to an existing profile. *Also called* **Grade shoe**.

MATCH MARK 1. Mark on each of two closely fitting parts, intended to ensure the same relative positioning when the two are reassembled. *Also called* **Reference mark**; **2.** Mark on sheet material that indicates a pattern repeat point.

MATERIAL Substance used to form products or construction works. There are several classes, including:

Direct: Material used in a manufacturing process that will form an integral part of the final product.

Indirect: Material in a manufacturing process that is necessary to the production of the final product but which does not form an integral part of it.

Raw: Goods acquired for the purpose of being consumed or changed in form in the manufacturing process.

MATERIAL BALANCE An accounting of the weights of material entering and leaving a process, usually made on a time-related basis.

MATERIAL DENSITY Mass of a unit volume of a substance (weight is synonymous with mass in the case of unit volume). Some typical material densities are:

Caliche: 1250 kg/m³ (2,100 lb/yd³).

Clay (natural bed): 1600 kg/m³ (2,800 lb/yd³).

Clay (dry): 1480 kg/m³ (2,500 lb/yd³).

Clay (wet): 1660 kg/m³ (2,800 lb/yd³).

Clay (with gravel, dry): 1420 kg/m³ (2,400 lb/yd³).

Clay (with gravel, wet): 1540 kg/m³ (1,600 lb/yd³).

Earth (dry, packed): 1510 kg/m³ (2,550 lb/yd³).

Earth (wet, excavated): 1600 kg/m³ (2,700 lb/yd³).

Granite (broken or large crushed): 1660 kg/m³ (2,800 lb/yd³).

Gravel (dry): 1510 kg/m³ (2,550 lb/yd³).

Gravel (pit run to graveled sand): 1930 kg/m³ (3,250 lb/yd³).

Gravel (dry to 13-50 mm (0.5-2 in.): 1690 kg/m³ (2,850 lb/yd³).

Gravel (wet to 13-50 mm (0.5-2 in.): 2020 kg/m³ (3,400 lb/yd³).

Limestone (broken or crushed): 1540 kg/m³ (2,600 lb/yd³).

Loam: 1250 kg/m³ (2,100 lb/yd³).

Sand (dry): 1420 kg/m³ (2,400 lb/yd³).

Sand (wet): 1840 kg/m³ (3,100 lb/yd³).

Sand (with gravel, dry): 1720 kg/m³ (2,900 lb/yd³).

Sand (with gravel, wet): 2020 kg/m³ (3,400 lb/yd³).

Sandstone (broken): 1510 kg/m³ (2,550 lb/yd³).

Shale: 1250 kg/m³ (2,100 lb/yd³).

Slag (broken): 1750 kg/m³ (2,950 lb/yd³).

Stone (crushed): 1600 kg/m³ (2,700 lb/yd³).

Topsoil: 950 kg/m³ (1,600 lb/yd³).

MATERIAL HANDLER Range of mechanical equipment designed to handle different types of materials (as distinct to processing material, or simply moving it by pushing (as with a dozer) or digging it (as with an excavator or backhoe)). Includes forklifts, telescopic boom equipment, etc.

MATERIAL HOSE *See* **Delivery hose.**

MATERIAL LIEN *See* **Mechanic's lien.**

MATERIAL LIFT Hoist designed to move materials vertically.

MATERIAL RELEASE Acceptance number for a product for use in mobile or manufactured homes.

MATERIAL SAFETY DATA SHEET Information data sheet listing the components of a product, their hazard level, the hazard level of the product when used, and how to extinguish fires should the material catch fire.

MATERIALS LOCK Chamber through which materials and equipment pass from one air pressure environment to another.

MATERIAL SPECIFICATION Stipulation of the character of certain materials to meet necessary performance criteria.

MATERIALS RECOVERY Concept of resource recovery where emphasis is on separating and processing waste materials for beneficial use or reuse.

MATERIAL TENDER Self-propelled, wheel-mounted bulk handling/delivery vehicle used in a hot-mix asphalt paving operation.

MATERIAL TRANSFER PAVER Asphalt paver that combines a windrow pickup device or truck dump hopper with large on-board mix storage capacity to achieve more continuous paving operation.

MATERIAL TRANSFER VEHICLE Asphalt paver that receives mix from haul trucks, provides on-site surge capacity, and feeds mix to the asphalt paver as required for continuous paver operation.

MAT-FORMED PARTICLEBOARD Particleboard in which the coated particles are formed into a mat before being hot pressed.

MAT FOUNDATION Continuous footing supporting an array of columns in several rows in each direction, having a slab-like shape, with or without depressions or openings, covering an area at least 75% of the total area within the outer limits of the assembly. *See also* **Raft foundation.**

MAT GLAZE *See* **Glaze.**

MATING PART Part fashioned to mate exactly with another, and only with that other part.

MATING CONNECTOR Mechanical or electrical connector that can only be coupled in a specified manner.

matl *Abbreviation for:* material.

matls engr *Abbreviation for:* materials engineer.

MATRIX 1. That which gives origin or form to something enclosed within it; **2.** Mold for a casting; **3.** Cement paste in which the fine aggregate particles are embedded to form a mortar; **4.** Mortar in which coarse aggregate particles are embedded to form concrete; **5.** Aluminum or steel rings or segments that form the cavity in which a tire is cured by hot capping and that form the tread design.

MAT SINKAGE *See* **Mat well.**

MATS METER Type of roughometer that gives a cumulative readout of pavement roughness in inches per mile or centimeters per kilometer.

MATTE FINISH Gloss-free finish. *See also* **Flat.**

MATTOCK Hand tool used for digging and grubbing, having a narrow hoeing surface at one end of the blade and either a pick or cutting blade at the other end.

MATTRESS Concrete slab on ground, used as a base for equipment or machinery.

MATURE TIMBER Stand of trees that has attained an age or size that satisfies the primary economic goal for which it was managed.

MATURING 1. Aging of material, during which time there will be change in physical state: **2.** Fattening up mortar.

MATURING TEMPERATURE Specific temperature, or temperature range, at which predictable changes in the state or cure of a material will take place, instantly or over a determined period.

MATURITY FACTOR Factor that is a function of the age of concrete (hours or days) multiplied by the difference between the average temperature of the concrete (degrees) during curing and a datum temperature below which hydration stops. *See also* **Degree-hour.**

MAT WELL Depression in a floor to receive a mat that, when placed in position, will be at approximately floor level. *Also called* **Mat sinkage.**

MAUL *See* **Hammer.**

MAUSOLEUM Large and imposing tomb.

max. *Abbreviation for:* maximum.

MAXIMUM AGGREGATE SIZE Smallest sieve opening through which an entire amount of aggregate is required to pass.

MAXIMUM ALLOWABLE COST Maximum cost for a project undertaken on a unit cost or cost-plus basis.

MAXIMUM ALLOWABLE SLOPE Steepest incline of an excavation face that is acceptable for the most favorable site condition as protection against cave-ins, and is expressed as the ratio of horizontal distance to vertical rise (H:V).

MAXIMUM BRAKE POWER Highest power developed at a given speed.

MAXIMUM DENSITY AND OPTIMUM MOISTURE Highest point on the moisture density curve, considered the best compaction of soil.

MAXIMUM DRAWBAR PULL Maximum towing force at a specified coupler height that a vehicle will develop on a level surface, based upon a prescribed coefficient of friction existing between the driving wheel(s) and the supporting surface when the load is being moved at a uniform rate of not less than 1% of the top level travel speed. *See also* **Drawbar pull,** and **Rated normal drawbar pull.**

MAXIMUM (ELECTRIC) DEMAND *See* **Demand (electric).**

MAXIMUM EMISSION CONCENTRATION Regulatory standard(s) for the maximum concentration of air pollutant emission from stationary or moving sources.

MAXIMUM EXCURSION Maximum pressure deviation from an operating pressure after an abrupt disturbance.

MAXIMUM FORK HEIGHT Loaded fork height attainable by a lift truck with the mast in a fully raised position.

MAXIMUM FRAME CUTOFF *See* **Frame cutoff.**

MAXIMUM GRADIENT Steepest gradient that the vehicles using a road or highway can be expected to negotiate in a normal operating mode.

MAXIMUM GROSS COMBINATION WEIGHT Manufacturer's gross weight limitation on how much a truck or tractor can pull, including weight of the truck and trailer.

MAXIMUM GROSS VEHICLE WEIGHT RAT-

ING Fixed weight rating applied to a truck by the manufacturer, determined by component sizes and ratings.

MAXIMUM HIGHWAY LOAD RATING Maximum load a disk wheel or demountable rim is rated to carry at highway speeds.

MAXIMUM INFLATION PRESSURE Maximum cold inflation pressure of a tire when fitted to a wheel.

MAXIMUM INLET PRESSURE Maximum rated pressure applied to the inlet port of a device.

MAXIMUM INSIDE FORK SPACING Nominal distance between the inside edges of adjustable forks at their greatest separated position.

MAXIMUM INTENDED LOAD Total load of all users, equipment, tools, materials, transmitted loads, wind loads, and other anticipated loads intended to be on or applied to a scaffold component at any one time. *Also called* **Design load.**

MAXIMUM LIFT Lift from the extreme lowered position of the carriage of an industrial truck to the fully elevated position when loaded.

MAXIMUM LOAD RATING Load rating at the maximum permissible inflation pressure for a specific tire.

MAXIMUM LOADED VEHICLE WEIGHT Sum of curb weight, passengers, and cargo.

MAXIMUM OUTSIDE FORK SPACING Nominal distance between the outside edges of the adjustable forks of an industrial truck when the forks are at their greatest separated position.

MAXIMUM PERMISSIBLE INFLATION PRESSURE Maximum cold inflation pressure to which a tire may be inflated.

MAXIMUM RATED LOAD Total of all loads including the working load, the weight of the scaffold, and such other loads as may be reasonably anticipated.

MAXIMUM ROLLING GRADE (GRADEABILITY) Greatest grade a vehicle or machine is able to climb while under motion.

MAXIMUM SAFE BEARING CAPACITY Maximum pressure to which soil can be subjected without risk of shear failure.

MAXIMUM SERVICE TEMPERATURE Temperature above which excessive shrinkage occurs in refractory concrete, usually between 66°C and 93°C (150°F and 200°F), below the actual temperature at which the refractory concrete softens.

MAXIMUM SIZE OF AGGREGATE Largest size aggregate particles present in sufficient quantity to affect the physical properties of the concrete of which they are a component.

MAXIMUM SPEED Speed attainable by accelerating at maximum rate from a standing start for one mile.

MAXIMUM STARTING GRADE (GRADEABILITY) Greatest grade on which a vehicle or machine is able to start from a complete stop. Approximately 9/10ths of **Maximum rolling gradeability.**

MAXIMUM SUSTAINED VEHICLE SPEED Highest speed a vehicle can maintain under full load conditions on level ground.

MAXIMUM-TEMPERATURE PERIOD Time interval throughout which the maximum temperature is held constant in an autoclave or steam-curing room.

MAXIMUM WORKING PRESSURE *See* **Pressure.**

MAX' TURN In helicopter logging, a turn of one or several logs with weight approaching the helicopter's maximum lifting capacity.

MAXWELL Obsolete unit of magnetic flux, not permitted for use with the SI system of measurement. *See also the appendix*: **Metric and nonmetric measurement.**

mb *Abbreviation for:* millibar.

MB *Abbreviation for:* machine bolt.

MBF *Abbreviation for:* thousand board feet.

MBh *Abbreviation for:* thousand British thermal units per hour.

mbl *Abbreviation for:* marble; mobile.

MBM *Abbreviation for:* thousand (feet) board measure.

mbr *Abbreviation for:* member.

MBtu *Abbreviation for:* thousand British thermal units.

MC *Abbreviation for:* medium cure; metal clad; moisture content.

MCA *Abbreviation for:* machine accessory.

McDERMID PLATE Part of a pile hammer, seated directly under the ram, that receives the blow of the ram and transmits it to the pile. Normally used only with wood piles. *Also called* **Anvil,** and **Cap block.**

Mcf *Abbreviation for:* thousand cubic feet.

mcflm *Abbreviation for:* microfilm.

McGEE An inexperienced worker, or inexperienced act. Also a **Joe.**

MCI *Abbreviation for:* malleable cast iron.

McLEOD TOOL Short-handled combination hoe, or cutting tool and rake, with or without removable blades, used in building fire lines. *See also* **Barron tool,** and **Pulaski tool.**

MCO *Abbreviation for:* mill culls out.

mdl *Abbreviation for:* model.

mdl. *Abbreviation for:* middle.

MDL *Abbreviation for:* master drawing list.

ME *Abbreviation for:* maintenance equipment; mechanical engineer.

MEADOW Flat to gently rolling plain that occurs on a hillside or along a ridge line. *Also called* a **Terrace.**

MEAN Intermediate between extremes, equidistant to each.

MEAN ANNUAL INCREMENT (MAI) Average annual increase in volume of individual trees or stands up to the specified point in time. The MAI changes with different growth phases in a tree's life, being highest in the middle years and then slowly decreasing with age. The point at which the MAI peaks is commonly used to identify the biological maturity of the tree and its readiness for harvesting.

MEAN DEPTH Cross-sectional area of a stream divided by its surface width.

MEANDERING CHANNEL Bed of a slow-flowing stream in easily erodible material, on flat ground, with many curves and abandoned channels. *See also* **Channel.**

MEANDERING LINE Survey line at the high water mark on navigable lakes and streams; the line at which continuous vegetation ends and sandy or muddy shore begins.

MEAN FILTRATION RATING Measurement of the average size of the pores of a filter medium.

MEAN HIGH TIDE Average daily maximum water surface elevation. *See also* **Mean high water, Mean low tide,** and **Mean low water.**

MEAN HIGH WATER Average high elevation to which the surface of a body of water rises. *See also* **Mean high tide, Mean low tide,** and **Mean low water.**

MEAN LOW TIDE Average daily minimum water surface elevation. *See also* **Mean high tide, Mean high water,** and **Mean low water.**

MEAN LOW WATER Average low elevation to which the surface of a body of water falls. *See also* **Mean high tide, Mean high water,** and **Mean low tide.**

MEAN SEA LEVEL Average 19-year height of the surface of the sea for all stages of the tide.

MEANS OF EGRESS Continuous path of travel provided by a doorway, hallway, corridor, exterior passageway, balcony, lobby, stair, ramp, or other egress facility for the escape of persons from any point in a building, floor area, room, or contained space to a public thoroughfare or other approved open space.

MEAN STACK TEMPERATURE Average temperature of flue gases in a chimney.

MEAN STRESS Average of the maximum and minimum stress in one cycle of fluctuating loading (as in a fatigue test).

meas. *Abbreviation for:* measure; measured.

MEASUREMENT FOR PAYMENT Determination of work done, materials supplied, or services provided to a point in time for authorization of payment under a contract.

MEASURING 1. Determination against an established standard; **2.** Process of dividing the merchantable tree stem into segments of specific length for the purpose of bucking.

MEASURING CHAIN *See* **Chain.**

MEASURING TAPE Literally, any flexible ribbon imprinted or impressed with a regular series of marks by which to measure. In practice, a wide range of such tapes made of fabric, plastic, or metal, of varying lengths and widths, and contained in cases of appropriate size. Likely the most common type consists of a slightly concave steel tape up to 10.97 m (36 ft) or more long, spring coiled inside a metal case that can be clipped to a workers belt. *Also called* a **Tape measure,** and **Tape rule.**

mech. *Abbreviation for:* mechanic; mechanical; mechanism.

MECHANIC Worker skilled with tools.

MECHANICAL ADHESION Bond between surfaces dependent on the adhesive holding the parts together.

MECHANICAL ADVANTAGE Increase in force obtained at the expense of speed or distance.

MECHANICAL ANALYSIS Process of determining particle-size distribution of an aggregate. *See also* **Sieve analysis.**

MECHANICAL ANCHORAGE 1. Any mechanical device capable of developing the strength of the reinforcement without damage to the concrete. **2.** *See also* **Tooth.**

MECHANICAL APPLICATION Application of plaster or mortar using equipment, generally pumping and spraying, as against hand working.

MECHANICAL ATOMIZATION Breaking down of oil into small particles to allow optimum combustion.

MECHANICAL BOND 1. Physical interlock between adjacent surfaces that limits or prevents movement. *See also* **Adhesive,** and **Specific adhesion; 2.** In general concrete construction, the physical interlock between cement paste and aggregate, or between concrete and reinforcement (specifically, the sliding resistance of an embedded bar and not the adhesive resistance); **3.** In plastering, the physical key of a plaster coat to (a) another, (b) to the plaster base by means of plaster keys to the lath, or (c) through interlock with adjacent plaster coats created by means of scratching or cross raking. *Also called* **Mechanical key.** *See also* **Specific adhesion.**

MECHANICAL COLLECTOR Device that separates entrained dust from gas through the application of inertial and gravitational forces.

MECHANICAL CONNECTION Complete assembly of an end-bearing sleeve, a coupler, or a coupling sleeve, and possibly additional intervening material or other components to effect the connection of reinforcing bars. *See also* **Bar-end check, Coupler, Coupling sleeve,** and **End-bearing sleeve.**

MECHANICAL COUPLING LINK Nonwelded, mechanically closed steel link used to attach master links, hooks, etc., to alloy-steel chain.

MECHANICAL DELAY TIME *See* **Machine time.**

MECHANICAL DRAFT Pressure differential produced by machinery, as a fan or blower.

MECHANICAL DRAWING Drawing showing layout of: piping for water, sanitary, gas, and drains; heating, ventilating, and air conditioning ducts and equipment; and other mechanical equipment. *See also* **Mechanical equipment.**

MECHANICAL EFFICIENCY 1. Ratio of an engine's useful horsepower available at the flywheel or power takeoff, to the horsepower developed in the engine cylinders, expressed in percent; **2.** Ratio of energy or work of the output of a machine to the energy of input.

MECHANICAL ENGINEERING Design and construction of engines, machines, and mechanical equipment of all kinds.

MECHANICAL EQUIPMENT In architectural and engineering practice: all equipment under the general heading of plumbing, heating, air conditioning, gas fitting, and electrical work.

MECHANICAL EQUIVALENT OF HEAT Amount of mechanical energy that can be transformed into one unit of heat.

MECHANICAL HARVESTING Tree cutting with mechanized equipment, such as a carrier-mounted shear or feller-buncher, instead of by hand with a power saw.

MECHANICAL KEY *See* **Mechanical bond.**

MECHANICAL PRESSURE-ATOMIZING BURNER Burner in which oil under pressure is permitted to expand through a small orifice, causing the oil to break into a spray of fine droplets.

MECHANICAL PROPERTIES Description of a material's behavior when force is applied.

MECHANICAL SEPARATION Separation of waste into various components using mechanical means, such as cyclones, trommels, and screens.

MECHANICAL SERVICE PIPE Welded steel pipe, available in standard, extra-strong, and double-extra-strong weights in sizes up to 300 mm (12 in.) internal diameter.

MECHANICAL SHIFT TRANSMISSION *See* **Transmission.**

MECHANICAL SPECIFICATIONS Detailed descriptions applicable to the supply, installation, operation, and maintenance of mechanical systems.

MECHANICAL STABILITY *See* **Shear stability.**

MECHANICAL TROWEL *See* **Power trowel.**

MECHANICAL UNIT Plumbing, heating, air conditioning, or electrical system that may be assembled on- or off-site and then installed.

MECHANIC'S LIEN Charge placed against a project for satisfaction of unpaid debts on work performed or materials supplied. *See also* **Lien.** *Also called* **Material lien.**

MECHANISM Articulated system able to deform without an increase in load, used in the special sense that the linkage may include real hinges or plastic hinges, or both.

MECHANISM METHOD Method of plastic analysis in which equilibrium between external forces and internal plastic hinges is calculated on the basis of an assumed mechanism.

MECHANIZED LOGGING Logging in which most or all of the hand labor is replaced by machines.

med. *Abbreviation for:* medical; medium.

MEDIA MIGRATION Material passed into an effluent stream composed of the materials making up a filter medium.

MEDIAN Planted or paved section between two or more lanes in the center of a highway. *See also* **Depressed median.**

MEDIAN BARRIER Double-sided guardrail or similar divider in the median between adjacent lanes of traffic carrying opposing streams of traffic.

MEDIAN OPENING *See* **Intersection elements.**

MEDIAN LANE *See* **Lane.**

MEDICAL LOCK Special chamber in which workers are treated for decompression illness.

MEDIUM 1. Usually a liquid or semiliquid ingredient of a sealant or waterproofing material that controls ease of application, appearance, adhesion, durability, and chemical inertness; **2.** Liquid that binds pigments together and holds them (bonds them) to a surface; **3.** Porous material that performs the actual process of filtration.

MEDIUM ALLOWANCE *See* **Fit classifications.**

MEDIUM-CURING ASPHALT *See* **Asphalt.**

MEDIUM-DENSITY FIBERBOARD Compressed fiberboard of 11.3 to 23 kg/0.02 m³ (25 to 50 lb/ft³) density.

MEDIUM-DENSITY OVERLAY Exterior-type plywood finished with an opaque resin-treated fiber overlay to provide a smooth surface as a paint base.

MEDIUM-DUTY SCAFFOLD Scaffold designed and constructed to carry a working load not exceeding 22.6 kg/0.092m² (50 lb/ft²).

MEDIUM FORCE FIT *See* **Fit classifications.**

MEDIUM GRAVEL Rock fragments ranging in size from 6 to 20 mm (0.25 to 0.75 in.).

MEDIUM-HAZARD INDUSTRIAL OCCUPANCY *See* **Industrial occupancy.**

MEDIUM LOAD RATING *See* **Load rating.**

MEDIUM SAND Rock fragments ranging in size from 0.2 to 0.6 mm (0.008 to 0.025 in.).

MEDIUM SCALING *See* **Scaling.**

MEDIUM SETTING Designation for a standard type of anionic emulsified asphalt.

MEDIUM-STOCKED STAND *See* **Stocking classes.**

MEDIUM-TEMPERATURE WATER HEATING *See* **Heating system.**

MEDULLARY RAY Wood tissue that usually runs continuously from the pitch to the bark, particularly prominent in quarter-cut oak.

M.E.E. *Abbreviation for:* machined end end.

MEETING OF THE MINDS When all parties to a contract agree to its exact terms.

MEETING RAIL Rail of a sashe that meets the matching rail of a second sash when the pair of sashes are closed.

MEETING STILE Door stile in which the lock is set; in a double door, one of the stiles that meet when the doors are closed.

meg *Abbreviation for:* megohm.

MEGA Prefix representing 10⁶. Symbol: M. Used in the SI system of measurement. *See also the appendix:* **Metric and nonmetric measurement.**

MEGAGRAM Unit of mass, equal to one million grams, of the SI system of measurement. Symbol: Mg.

See also the appendix: **Metric and nonmetric measurement.**

MEGAHERTZ Unit of frequency, equal to one million hertz, of the SI system of measurement. Symbol: MHz. *See also the appendix:* **Metric and nonmetric measurement.**

MEGAJOULE Unit of energy, equal to one million joules, of the SI system of measurement. Symbol: MJ. *See also the appendix:* **Metric and nonmetric measurement.**

MEGALITH Large hewn or unhewn stone.

MEGALITHIC MASONRY Masonry constructed of particularly large stones, laid with very close joints, usually without mortar.

MEGANEWTON Unit of force, equal to one million newtons, of the SI system of measurement. Symbol: MN. *See also the appendix:* **Metric and nonmetric measurement.**

MEGASCOPIC Visible to the unaided eye.

MEGAWATT Unit of power, equal to one million watts, of the SI system of measurement. Symbol: MW. *See also the appendix:* **Metric and nonmetric measurement.**

MEGOHM Unit of resistance equal to one million ohms. Symbol: MΩ.

MEGOHMMETER Instrument for measuring resistance, giving a reading in ohms.

MELAMINE Laminate that derives its name from the melamine resin system used to saturate the paper laminate and adhere it to the substrate.

MELAMINE FORMALDEHYDE Thermosetting resin with exterior capabilities, used as a saturating resin for paper laminates. *Also called* **MF.**

MELT 1. Change of state from solid to liquid; **2.** Molten portion of the raw material mass during the burning of cement clinker, firing of lightweight aggregates, or expanding of blast-furnace slags.

MELTING POINT Temperature, for a given pressure, at which the solid and liquid phases of a substance are in equilibrium.

memb. *Abbreviation for:* membrane.

MEMBER 1. Part of an order or of a building; **2.** Definite part of a building (entablature, cornice, column, etc.); **3.** Different parts of a structure (joists, rafters, beams, etc.).

MEMBRANE Layer of usually impervious material sandwiched between two other materials to prevent the transmission of moisture or vapor.

MEMBRANE BARRIER Thin layer of material impermeable to the flow of gas or water.

MEMBRANE CURING Process that involves either a liquid sealing compound (e.g., bituminous and paraffinic emulsions, coal-tar cutbacks, pigmented and non-pigmented resin suspensions, or suspensions of wax and drying oil) or a nonliquid protective coating (e.g. sheet plastics or waterproof paper), either of which functions as a film to restrict evaporation of mixing water from fresh concrete surface.

MEMBRANE FIREPROOFING Lath-and-plaster

membrane constructed, wholly or in part to provide a barrier to fire and intense heat.

MEMBRANE THEORY Theory of design for thin concrete shells, based on the premise that a shell cannot resist bending because it deflects, the only stresses in any section being shear, compression, or tension.

MEMBRANE WATERPROOFING Coating the external faces of the underground portions of basement walls with a waterproof substance.

memo. *Abbreviation for:* memorandum.

MEMORANDUM OF UNDERSTANDING Communication that sets out the basic framework of an agreement, that is, the basis for a more detailed text and possibly a contract, but that is not binding at that stage.

MEMORY Circuits in which a microprocessor stores data and instructions.

MENDING PLATE *See* **Braces and plates.**

MENISCUS Curved edge of a liquid at its plane of contact with a perpendicular surface.

mensur. *Abbreviation for:* mensuration.

MENSURATION Branch of mathematics dealing with length, area, or volume.

mep *Abbreviation for:* mean effective pressure.

mer. *Abbreviation for:* meridian.

MERCANTILE OCCUPANCY Occupancy or use of a building or part of a building for displaying or selling retail goods, wares, or merchandise.

MERCHANTABLE TIMBER Tree or stand that has attained sufficient size, quality, and/or volume to make it suitable for harvesting.

MERCHANTABLE TOP Smallest utilizable top of a tree.

MERCHANTABLE VOLUME Amount of sound wood in a single tree or stand that is suitable for marketing under given economic conditions.

MERCURY VAPOR ILLUMINATION Long-life, high-intensity electric light source.

MERGED LOT *See* **Through lot.**

MERGING Converging of separate streams of traffic into a single stream.

MERGING END *See* **Interchange elements.**

MERIDIAN Line on the surface of the earth having the same longitude at every point (a line running north and south along the astronomical meridian).

MERIDIAN LINE North-south reference line, often appearing on maps. Meridian lines are also etched into the bearing plate on a compass.

MESH 1. Network formed by the crossing and/or interweaving of strands; **2.** Number of openings per given length of a woven fabric. *See also* **Welded-wire fabric.**

MESH NUMBER Method of identifying aggregate sieve sizes according to the number of openings per unit of length in either a screen or sieve in which the openings are 6 mm (1/4 in.) or less.

MESH REINFORCEMENT *See* **Welded-wire fabric reinforcement.**

MESH TAPE *See* **Joint reinforcing mesh.**

MESNAGER HINGE Permanent semiarticulatied or flexible joint in a reinforced concrete arch wherein the angles of rotation at the hinge are very small. *See also* **Hinge joint,** and **Semiflexible joint.**

met. *Abbreviation for:* metal; metallize; metallurgical; metallurgy.

METAKAOLIN Aluminosilicate material, used to increase the durability and life of concrete.

METAL 1. Any of a class of chemical elements characterized by ductility, malleability, luster, and conductivity; **2.** Broken stone used as ballasting for roadbeds.

METAL ARC CUTTING Method of cutting metal with an electric arc.

METAL CARBON ARC CUTTING Metal cutting using an electric arc and a jet of oxygen.

METAL DECK Sheet metal forming a flat roof or the permanent formwork for a reinforced concrete floor or roof.

METAL DEFECTS Flaws developed within the body of sheet glass during the forming process that do not reach the surface.

METAL DETECTOR Device for detecting concealed metal objects by radiating a high-frequency electromagnetic field and detecting the change produced in the field by the presence of ferrous objects.

METAL FORM Sheet metal shape used to provide a forms for floor joist construction. *Also called* **Pan.**

METAL HEAT-CIRCULATING FIREPLACE Prefabricated fireplace unit incorporating hollow walls and vented openings to promote the circulation of hot gases and the generation of heat.

METAL LATH Metal sheet perforated with narrow slits, which are then drawn out to form openings. Used as a key for plaster or other types of screed. *Also called* **Expanded metal.**

METAL LATHER Workman who installs furring and metal lath and places inserts in concrete for support of ceilings. *Also called* **Bar placer, Iron worker,** and **Rebar contractor.**

METALLIC AREA Sum of the cross-sectional areas of all the wires either in a wire rope or in a strand.

METALLIC SHEAVE Metal grooved wheel for guiding wire rope.

METALLOGRAPHY Study of the constitution and structure of metals and alloys through observation.

METALLURGY Science and technology of metal.

METAL PRIMER Paint, used as a protective and first coat on metal. Two basic formulations are:

> **Zinc chromate:** A white or yellow primer, used on metal not expected to be exposed to moisture.

> **Zinc oxide:** A red primer, used on metal exposed to moisture.

METAL QUARRY-TILE RACK Metal grid used to

maintain the same width between quarry tiles during the laying process.

METAL SHINGLE *See* **Shingle.**

METAL STUD Preformed metal section produced to the same nominal dimensions as its lumber equivalent.

METAL TRIM Preformed metal section used to expose, hide, reinforce, or finish.

METAL VALLEY Valley gutter lined with sheet metal.

METAL WALL TIE *See* **Wall tie.**

METAMERISM Change in a color when it is seen under different lighting conditions (e.g., incandescent light, fluorescent light, daylight).

METAMORPHIC Classification of rock that is formed from sedimentary and igneous rocks by intense pressure, high heat, or chemical reaction.

METER 1. Device used for measuring the amount of a material consumed, passed or conveyed; **2.** *See* **Indicator; 3.** One of the seven base units of the SI system of measurement: a unit of length equal to 1 650 763.73 wavelengths in vacuum of the radiation corresponding to the transition between the $2p_{10}$ and $5d_5$ of the krypton-86 atom. Symbol: m. Derived units of area are the square meter, symbol m^2, and the hectare (10 000 m^2), symbol ha; and of volume and capacity the cubic meter, symbol m^3, and the liter, symbol L. Multiply by 39.370 to obtain inches, symbol in.; by 3.280 84 to obtain feet, symbol ft; and by 1.0936 to obtain yards, symbol yd. *See also the appendix*: **Metric and nonmetric measurement.**

METERED FLOW Flow at a controlled rate. *See also* **Flow.**

METERING PIN Valve plunger that regulates the rate of flow of a liquid or gas.

METER PER SECOND A derived unit of speed or velocity with a compound name of the SI system of measurement. Symbol: m/s. Multiply by 196.85 to obtain foot per second, symbol: fps; by 3.2808 to obtain foot per minute, symbol: fpm; by 2.2369 to obtain mile per hour, symbol: mph. *See also the appendix*: **Metric and nonmetric measurement.**

METER PER SECOND SQUARED A derived unit of acceleration with a compound name of the SI system of measurement. Symbol: m/s^2. Multiply by 3.2808 to obtain foot per second squared, symbol: ft/s^2. *See also the appendix*: **Metric and nonmetric measurement.**

METER STOP Valve on a water service between the street main and the water meter.

METER TO THE FOURTH POWER A derived unit of the second moment of area with a compound name of the SI system of measurement. Symbol: m^4. *See also the appendix*: **Metric and nonmetric measurement.**

METES AND BOUNDS (Metes = length; bounds = boundary lines.) Means of describing the location of land by defining boundaries in terms of directions and distances from one or more specific points of reference, such as natural or artificial monuments at corners, or natural or cultural boundary lines.

meth. *Abbreviation for:* methane.

METHANE Odorless, colorless, nonpoisonous flammable and explosive gas, typically resulting from the anaerobic decomposition of organic waste matter; the major component of natural gas.

METHANOGENS Organisms carrying out methogenesis, requiring completely anaerobic conditions for growth.

METHOD OF ACCOUNTING Manner and style in which records of monies and materials about a project are maintained. There are many variations, including:

> **Completed contract:** Accounting method that accounts for a contract when it is completed, that is, all revenue (contract billings) and costs are recognized in the statement of income when the contract is completed.

> **Cost ratio:** Accounting method that uses the ratio of actual contract costs incurred during the reporting period to the total estimated contract costs.

> **Effort-expended:** Accounting method that uses the ratio of some measure of the work input during the reporting period, such as labor hours, labor cost, machine hours, or material quantities, to the units of that measure of work required to complete the contract.

> **Percent-of-completion:** Accounting method that recognizes revenue and costs throughout the life of each contract, based on a periodic measurement of progress.

> **Units-of-work:** Accounting method that uses the ratio of units of work performed to total units of work to be performed under the contract.

METHOD SPECIFICATION Specification that outlines the equipment, materials, and procedures to be used to fulfill a contract.

METRIC MEASUREMENT *See appendix* **Metric and nonmetric measurement.**

METRIC TON Non-SI unit of weight equivalent to 1000 kg, 2,205 lb, or 1.102 short tons. *See also* **Long ton,** and **Short ton.**

metro. *Abbreviation for:* metropolitan.

MEWS Originally, royal stables built on the site where hawks were 'mewed' (put in cages while molting). More recently, rows of stables converted to dwellings. Latterly, housing built to resemble such (often fronting directly to an alley or passage).

MEXICAN PAVER TILE Terra-cotta-like tile, used mainly for floors, and handmade.

mezz. *Abbreviation for:* mezzanine.

MEZZANINE Floor placed between two others (usually between the ground and first floors in a high-ceilinged building) that does not extend to the full perimeter of the building.

mf. *Abbreviation for:* microfilm.

MF *Abbreviation for:* melamine formaldehyde; mill finish.

mfd *Abbreviation for:* manufactured.

mfg *Abbreviation for:* manufacturing.

mfh *Abbreviation for:* maximum fork height.

mfr *Abbreviation for:* manufacturer.

mg *Abbreviation for:* milligram.

MG *Abbreviation for:* mixed grain; motor generator.

mgmt *Abbreviation for:* management.

mgr *Abbreviation for:* manager.

mgt *Abbreviation for:* management.

mh. *Abbreviation for:* manhole.

MHD *Abbreviation for:* manufactured (mobile) home decking.

MHO Metric unit of conductance; reciprocal of the ohm.

MHW *Abbreviation for:* mean high water.

MI *Abbreviation for:* malleable iron; manufacturing instruction.

mi. *Abbreviation for:* mile, mill ($0.001).

mic. *Abbreviation for:* microphone.

MICA Silicate mineral, used in crushed form as a surfacing and backing material on composition roofing.

MICHIGAN TEST Comprehensive field test, principally aimed at determining load-bearing criteria for a driven pile.

MICRO Prefix representing 10^{-6}. Symbol: μ· Used in the SI system of measurement. *See also the appendix*: **Metric and nonmetric measurement.**

MICROAMPERE One millionth of an ampere (0.000 001 ampere).

MICROBALLOON Plastic or glass sphere used, in quantity, to decrease the density of ANFO.

MICROCLIMATE Localized climate of a given area, that may differ from surrounding general climatic conditions, being influenced in natural surroundings by topography, drainage, vegetation, and orientation to the sun, and within urban environments by the mass and disposition of large buildings, airways created by roads and streets, non-porous surfaces such as paved parking lots, etc.

MICROCLIMATOLOGY Study of the effects on climate of the relationship of structures, artificial land forms, and other man-made obstructions to the free flow of air.

MICROCONCRETE Mixture of portland cement, water, and suitably graded sand for simulating concrete in small-scale structural models.

MICROCRACKS Microscopic cracks within concrete.

MICROFARAD One millionth of a farad.

MICROHENRY One millionth of a henry.

MICROMETER 1. Unit of length, equal to one millionth of a meter, of the SI system of measurement. Symbol: μm. *See also the appendix*: **Metric and nonmetric measurement; 2.** Precision measuring device that measures thicknesses and diameters.

MICRON Obsolete metric unit of length, equal to 1 μm, that should not be used with the SI system of measurement. Symbol: μ. *See also the appendix*: **Metric and nonmetric measurement.**

MICRON EFFICIENCY CURVE Curve showing how well a collector traps micrometer-size particles.

MICROPILE *See* **Pile.**

MICROSAND Fine aggregate passing the No. 100 (0.149 mm, 0.0059 in.) sieve, and essentially free of clay and shale.

MICROSCOPIC Discernible only with the aid of a microscope; particles whose diameter is below the threshold of normal vision.

MICROSITE Small area that exhibits localized characteristics different from the surrounding area.

MICROSWITCH Electrical switch that reacts to a very small physical movement of the actuator.

MICROTUNNELING Technique for conduit installation using a steerable, remote-controlled tunnel boring machine by pipe jacking, the excavated material being removed either by mechanical auger or as a slurry.

MICROVOLT One millionth of a volt.

mid. *Abbreviation for:* middle.

MIDDLE SECTION Segments of a telescoping boom located between the base and tip sections. *See also* **Midsection.**

MIDDLE STRIP In flat-slab framing, the slab portion that occupies the middle half of the span between columns. *See also* **Column strip,** and **Edge strip.**

MIDFEATHER Central wythe of a chimney.

MID-ORDINATE Ordinate midway between extremes.

MIDPOINT Average value of two extreme observations.

MIDPOINT HITCH *See* **Intermediate suspension (line).**

MIDPOINT PENDANT Wire rope pendant used to support the center portion of a long lattice boom at lift off.

MIDPOINT SUSPENSION *See* **Intermediate suspension (line).**

midpt *Abbreviation for:* midpoint.

MIDRAIL Horizontal member approximately midway between the top rail and platform of a guardrail system.

MIDSECTION Intermediate, powered telescoping section(s) of a telescoping boom, mounted between the base and fly sections. *See also* **Middle section.**

MIDSPAN Point halfway between the supports of a beam or floor.

mig *Abbreviation for:* metal inert gas.

MIGRATION 1. Contaminant released downstream; **2.** In a rubber compound, the movement of more or less rubber-soluble materials from a point of high concentration to one of low or zero concentration. Migration is applied to the movement of accelerators, antioxidants, antiozonants, sulfur, softeners, and organic colors. It is a form of diffusion.

MIGRATION STAIN Discoloration of a surface by a hose that is adjacent to but not touching the discolored surface.

MIL Nonmetric unit of length, equal to 0.001 in. Symbol: mil. Multiply by 25.4 to obtain micrometers, symbol: μm. *See also the appendix*: **Metric and nonmetric measurement.**

mil. *Abbreviation for:* military.

MILDEW Surface mold, often a green-black loose powdery mass; may occur on both interior and exterior surfaces and is evidence of improper ventilation or condensation.

MILDEW INHIBITED Containing material to prevent or retard the propagation of a fungus growth.

MILDEW RESISTANCE Withstanding the action of mildew and its deteriorating effect.

MILD STEEL *See* **Steel.**

MILE Nonmetric unit of length, equal to 1,760 yards. Symbol: mi. Multiply by 1.609 344 to obtain kilometers, symbol: km; by 1609.344 to obtain meters, symbol: m. *See also the appendix*: **Metric and nonmetric measurement.**

MILE (NAUTICAL) Nonmetric unit of length, equal to 1852 meters. Symbol: n.mi. Multiply by 1.852 to obtain kilometers, symbol: km. *See also the appendix*: **Metric and nonmetric measurement.**

MILE PER HOUR Nonmetric unit of velocity. Symbol: mph. Multiply by 1.609 344 to obtain kilometers per hour, symbol: km/h; by 0.447 04 to obtain meters per second, symbol: m/s. *See also the appendix*: **Metric and nonmetric measurement.**

MILESTONE Key or important intermediate goal in the Network system of project analysis and control.

MILE YARD Measure of payment for excavation representing the movement of 1 yd^3 of material through a horizontal distance of 1 mile.

MILITARY CREST Point on the slope of a hill that will not allow any person or object to be silhouetted against the horizon, but that also allows a full view of the slope below. *See also* **Crest.**

MILKINESS Whitish translucent defect in a varnish film.

MILKING Progressive movement of strands along the axis of a wire rope, resulting from the rope's movement through a restricted passage such as a tight sheave. *Also called* **Ironing.**

MILL 1. Machine for grinding or crushing; 2. Machine with two horizontal rolls revolving in opposite directions, used for mastication or mixing; 3. To cut or machine metal using rotating-tooth cutters; 4. One tenth of a cent, $0.001. A monetary unit used in calculating, but not as coinage.

MILLED Serrated surface pattern machined into a metal surface, usually to facilitate gripping with the hand.

MILLED LEAD Sheet lead rolled from slabs.

MILLED REFUSE Solid waste that has been mechanically reduced in size.

MILL FILE Coarse file used to remove excess metal.

MILL FINISH Surface finish of a bar, rod, or wire

material as received from a mill.

MILLI Prefix representing 10^{-3}. Symbol: m. Used in the SI system of measurement. *See also the appendix*: **Metric and nonmetric measurement.**

MILLIAMPERE Unit equal to 1/1000 of an ampere. Symbol: mA. *See also the appendix*: **Metric and nonmentric measurement.**

MILLIBAR Non-SI unit of pressure, equal to 1000 dynes per cm^2. Symbol: mbar. The millibar is used as a unit of measure of atmospheric pressure; a standard atmosphere being equal to 1013.25 mbar. *See also the appendix*: **Metric and nonmetric measurement.**

MILLIFARAD Unit equal to 1/1000 of a farad. Symbol: mF. *See also the appendix*: **Metric and nonmetric measurement.**

MILLIGRAM Unit of mass, equal to 1/1000 of a gram, of the SI system of measurement. Symbol: mg. *See also the appendix*: **Metric and nonmetric measurement.**

MILLIHENRY Unit equal to 1/1000 of a henry. Symbol: mH. *See also the appendix*: **Metric and nonmetric measurement.**

MILLILITER Unit of volume, equal to 1/1000 of a liter, permitted for use with the SI system of measurement. Symbol: mL. *See also the appendix*: **Metric and nonmetric measurement.**

MILLIMETER Unit of length, equal to 1/1000 of a meter, of the SI system of measurement. Symbol: mm. *See also the appendix*: **Metric and nonmetric measurement.**

MILLIMETER OF MERCURY Obsolete metric unit of pressure, equal to 133.32 pascals, that should not be used with the SI system of measurement. Symbol: torr. *See also the appendix*: **Metric and nonmetric measurement.**

MILLIMICRON Unit equal to 1/1000 of a micron, 1/1 000 000 of a millimeter, or 10 angstroms; a unit of length for measuring light waves. Symbol: mμ. *See also the appendix*: **Metric and nonmetric measurement.**

MILLING *See* **Cold planing/milling.**

MILLING MACHINE *See* **Cold planer/milling machine.**

MILLION GALLONS PER DAY Nonmetric unit of volume. Symbol: mgd. Multiply by 4546.09 to obtain cubic meters per day, symbol: m^3/d; by 0.052 616 8 to obtain cubic meters per second, symbol: m^3/s. *See also the appendix*: **Metric and nonmetric measurement.**

MILLIPASCAL Unit of pressure, equal to 1/1000 of a pascal, of the SI system of measurement. Symbol: mPa. *See also the appendix*: **Metric and nonmetric measurement.**

MILLISECOND DELAY Electric detonation caps that have a built-in delay element, usually 25/1000th of a second apart, consecutively.

MILL MATERIAL Steel mill products ordered expressly for the requirements of a specified project.

MILL MIXED Materials premixed at the point of manufacture, either ready for use or requiring only the addition of water at the job.

MILL RATE Basis for calculation of property taxes levied by a public authority using the assessed value of the property divided by the mill (a theoretical value equal to one tenth of a cent; one thousandth of a dollar) and multiplied by an arbitrary figure established by the authority.

MILL-RUN MORTAR Mortar mixed in a pug mill or similar equipment.

MILL SCALE Partially adherent layers of oxidation products (heavy oxides) developed on metallic surfaces during either hot fabrication or heat treatment.

MILLWORK Building components made of finished wood and manufactured off-site. Includes such items as inside and outside doors, window and door frames, windows, porch work, mantels, panel-work, stairways, moldings, and interior trim.

MILLWRIGHT Worker specialized in the installation and maintenance of heavy machinery.

mil. spec. *Abbreviation for:* military specification.

mil. std *Abbreviation for:* military standard.

min. *Abbreviation for:* mineral; minimum; minor; minority; minute.

MINARET Slender tower, often among others, specifically of a mosque.

MINERAL Naturally occurring, inorganic substance of definite chemical composition and physical properties.

MINERAL ADMIXTURE Additive consisting mostly of inorganic substances or fly ash, used to modify the reaction of cement during the mixing, placing, setting, and curing of concrete and mortar, such as coefficient of expansion or contraction, density, weight, etc.

MINERAL AGGREGATE Aggregate consisting of a mixture of broken stone, crushed slag, crushed or uncrushed gravel, sand, and mineral dust.

MINERAL DEPOSIT Impurity in water that precipitates on metal and other surfaces, such as the interior of pipes and storage vessels.

MINERAL DUST Finely divided mineral product at least 70% of which will pass a No. 200 (0.74 mm, 0.0029 in.) sieve. Pulverized limestone is the most commonly manufactured filler, although other stone dust, hydrated lime, portland cement, fly ash, and certain natural deposits of finely divided mineral matter are also used.

MINERAL-FILLED ASPHALT See **Asphalt.**

MINERAL FILLER Finely divided mineral product at least 65% of which passes the No. 200 (0.074 mm, 0.0029 in.) sieve. *See also* **Silt.**

MINERAL PIGMENT See **Earth pigment.**

MINERAL RIGHTS Right to search for and take minerals, oil, or gas from under the surface of land or sea.

MINERAL SPIRITS Petroleum solvent used as an alternative to turpentine.

MINERAL SURFACE Crushed rock used to coat composition roofing materials.

MINERAL WOOL Insulating material produced by sending steam under pressure through molten slag or rock.

MINIMUM ACCEPTANCE STRENGTH That strength that is 2.5% lower than the catalog or nominal strength. This tolerance is used to offset variables that occur during a sample preparation and actual physical test of a wire rope. *See also* **Breaking strength,** and **Nominal strength.**

MINIMUM BEND RADIUS Minimum radius over which a metal product can be bent to a given angle without fracture.

MINIMUM BURST PRESSURE Lowest pressure at which rupture occurs under prescribed conditions.

MINIMUM DESIGN JOINT WIDTH Realistic estimation of the width at which a joint should be constructed, obtained by applying manufacturing and constructional tolerances to the minimum theoretical joint width.

MINIMUM FIRING RATE Lowest input rate for a burner or process.

MINIMUM INSIDE FORK SPACING Nominal distance between the inside edges of the adjustable forks of an industrial truck at their least separated position.

MINIMUM LOT AREA Smallest building lot allowed in a subdivision.

MINIMUM OUTSIDE FORK SPACING Nominal distance between the outside edges of the adjustable forks of an industrial truck at their least separated position.

MINIMUM PENETRATION See **Penetration.**

MINIMUM STANDARDS BYLAW See **Building code.**

MINIMUM THEORETICAL JOINT WIDTH Initial approximation of the width at which a joint should be constructed, obtained after consideration of the movements at the joint due to thermal, moisture, and other changes, and the movement capability of the sealant. *Also called* **Theoretical joint width.**

MINIMUM TURNING PATH See **Intersection elements.**

MINIMUM TURNING RADIUS See **Intersection elements.**

MINIMUM YARD DEPTH See **Yard requirement.**

MINING Removal of rock or earth having value because of its chemical composition.

MINING SUBSIDENCE Surface subsidence due to underground workings.

MINING WASTE Residues that result from the extraction of raw materials from the earth, or residues left after ore benefication.

MINOR ARCH See **Arch.**

MINOR CHANGE Change of a minor nature to the work under contract, that does not involve an adjustment in the contract sum or time of completion, but that must be confirmed by a field order or other written order signed by an appropriate authority.

MINOR CONTRACT ITEMS See **Major and minor contract items.**

MINOR DIAMETER Least diameter of a screw thread.

MINOR STRUCTURE 1. In engineering construction, any structure not classed as a bridge or a culvert; **2.** In building construction, an outbuilding or other small building not attached to the main structure.

MINUS STATION Stakes or points on the negative, or far side of the beginning point or point of reference. *See also* **Station.**

MINUTE Non-SI unit of time, equal to 60 seconds, permitted for use with the SI system of measurement. Symbol: min. *See also the appendix*: **Metric and nonmetric measurement.**

MINUTE OF ARC Non-SI unit of angle, equal to 60 seconds of arc, permitted for use with the SI system of measurement. Symbol: '. *See also the appendix*: **Metric and nonmetric measurement.**

mip *Abbreviation for:* malleable iron pipe.

mir. *Abbreviation for:* mirror.

MIRROR Polished plate glass, silvered on one side.

MIRROR GLAZING QUALITY Standard for plate glass where a very high quality is required.

misc. *Abbreviation for:* miscellaneous; miscible.

MISCELLANEOUS IRON Steel items such as lintel angles, inserts, plates, form braces, spreaders, and other structural shapes attached to or embedded in masonry or reinforced concrete or attached to lumber or formwork.

MISCIBLE Extent to which liquids or gases can be mixed or blended.

MISFIRE Explosive charge, or part of a charge that, for any reason, has failed to fire as planned.

MISSION TILE *See* **Roof tile.**

MIST Suspension of finely divided liquid droplets.

MIST COAT Very thin sprayed coat. *See also* **Base coat, Full coat,** and **Glaze coat.**

MISUSE OF LAND Use of land that might injuriously affect the interests of a community.

MITER BOX Solid frame with precut slots to guide a back saw blade when cutting a miter.

MITER BRAD Corrugated fastener used to fasten the two components of a miter joint.

MITERED Internal or external angle joint where the meeting faces of the joint bisect the angle.

MITER JOINT Joint of two pieces at an angle that bisects the joining angle.

MIW *Abbreviation for:* malleable iron washer.

MIX 1. Act or process of mixing; **2.** Mixture of materials such as mortar or concrete. *See also* **Compound.**

mix. *Abbreviation for:* mixture.

MIX DESIGN *See* **Proportioning.**

MIXED FACE Digging in dirt and rock in a tunnel in the same heading at the same time.

MIXED-GRAIN LUMBER *See* **Grain.**

MIXED GRANULES Crushed rock granules of various colors, blended and applied to synthetic roof materials.

MIXED-IN-PLACE 1. Materials blended and mixed ready for use at the place of application; **2.** Asphalt course produced by mixing mineral aggregate and cutback or emulsified asphalt at the road site by means of travel plants, motor graders, or special road-mixing equipment.

MIXED SAND Blend of natural sand with crushed stone or crushed gravel sand.

MIXED MUNICIPAL REFUSE *See* **Solid waste.**

MIXER 1. Machine used for blending materials; **2.** Machine for blending the constituents of concrete, grout, mortar, cement paste, plaster, or other mixtures.

MIXER EFFICIENCY Adequacy of a mixer in rendering a homogeneous product within a stated period.

MIXER TRUCK Vehicle carrying a drum, mounted on a tilted axis and capable of being rotated, used to transport prepared concrete between the point of mixing and place of application, and of completing the mix cycle during transportation.

MIXING CHAMBER Chamber, usually placed between the primary and secondary combustion chambers of an incinerator, in which the products of combustion are thoroughly mixed by turbulence created by increased velocities of the gases.

MIXING CYCLE Time taken for a complete cycle in a batch mixer, i.e., the time elapsing between successive repetitions of the same operation (e.g., successive discharges of the mixer).

MIXING FAUCET *See* **Valve.**

MIXING PLANT *See* **Batch plant.**

MIXING SPEED Rotation rate of a mixer drum or the paddles in an open-top, pan, or trough mixer.

MIXING TIME Time taken to thoroughly mix a batch of concrete in a mixer.

MIXING VARNISH Varnish that can be mixed into oil-base paints to produce additional gloss.

MIXING WATER Water in freshly mixed sand-cement grout, mortar, or concrete, exclusive of any moisture previously absorbed by the aggregate (e.g., water considered in the computation of the net water-cement ratio). *See also* **Batched water,** and **Surface moisture.**

MIXTURE 1. Assembled, blended, co-mingled ingredients, typically of mortar, concrete, or the like; **2.** Proportions for the assembly of ingredients for a mix. *See also* **Compound,** and **Proportioning.**

mk *Abbreviation for:* mark.

MK *Abbreviation for:* master key.

mkd *Abbreviation for:* marked.

Mkd *Abbreviation for:* master keyed.

mkg *Abbreviation for:* marking.

mkt *Abbreviation for:* market.

m/l *Abbreviation for:* marker lamp.

ML *Abbreviation for:* master list.

mld *Abbreviation for:* mold; molded.

mldg *Abbreviation for:* molding.

mltn *Abbreviation for:* molten.

mm *Abbreviation for:* millimeter.

mn *Abbreviation for:* main.

MN *Abbreviation for:* magnetic north.

mo. *Abbreviation for:* masonry opening; month.

M.O.A. *Abbreviation for:* machined over all.

MOBILE Capable of being moved from one location to another.

MOBILE BACKSPAR Crawler tractor or hydraulic excavator at the back end with a fairlead or short spar mounted on it, used to hang tailblocks or to simplify and speed up road changes.

MOBILE COMPACTOR *See* **Compactor.**

MOBILE CONVEYOR Self-propelled, wheel-mounted bulk material handling/delivery system.

MOBILE CRANE Crane, mounted on a self-propelled, wheeled, or tracked undercarriage.

MOBILE HARVESTER Self-propelled multifunction machine that may be capable of operating as a swath cutter but also performs chipping and/or forwarding functions in addition to felling.

MOBILE HOIST Platform hoist mounted on a wheeled undercarriage that can be towed.

MOBILE HOME Detached one-family residence:

(a) Designed for long-term occupancy and containing sleeping accommodation, flush toilet, tub or shower bath, and kitchen, with plumbing and electrical connections provided for attachment to external systems.

(b) Able to be transported, after fabrication, on its own wheels, or detachable wheels, or on a flatbed or other trailer.

(c) Arriving at the site where it is to be occupied as a dwelling complete, including major appliances and furniture ready for occupancy excepting minor unpacking, assembly, location of foundation supports and connection to utilities, etc.

MOBILE HOME PARK Development for the temporary or permanent location of mobile homes that provides all services such as roads, street lighting, water supply, sanitary sewers, and electrical service stands.

MOBILE JOB CRANE Mobile crane employed to lift light loads and pick-and-carry tasks around the job- or work-site.

MOBILE PACKER *See* **Refuse truck.**

MOBILE REFERENCE Device that is attached to, and towed by an asphalt paver to provide a grade reference for the automatic screed controls.

MOBILE SHEAR Attachment commonly fitted to the stick of excavators and backhoes and powered from the hydraulic circuit of the base machine, consisting of jaws designed to shear (cut) rebar, pipe, metal tube, and concrete slab.

MOBILE X-RAY X-ray equipment mounted on a permanent base with wheels and/or casters for moving while completely assembled. Typically used to evaluate completed pipe welds.

MOBILE YARDER *See* **Harvesting machines (single function).**

MOBILIZATION TIME Time required to assemble equipment and personnel and have them available for work at a designated place.

mod. *Abbreviation for:* model; moderate; modification; modify; modular; module; modulus.

MODEL 1. Style or design, usually having a designation or identification; **2.** Theoretical abstraction, usually capable of mathematical manipulation, used to evaluate a problem or subject of interest.

MODEL ANALYSIS 1. Scale model of a structure that is subjected to design loads; **2.** Interpretation of a mathematical representation of a set of conditions and their potential or likely results.

MODEL BUILDING CODE *See* **Building code.**

MODEL METHOD System of estimating the cost of a dwelling based on area, perimeter, and general type of construction.

MODEL UNIT Representative home, apartment, office, or other space, used to demonstrate the design, structure and finish, and for promotional purposes.

MODEL WEIGHT Weight of a vehicle with all items of standard equipment, 68 kg (150 lb) per passenger in each designated seating position, and maximum capacity of fuel, oil, and coolant.

MODERATOR Substance that when added to two or more others in a mixture will slow a reaction or process, or change it.

MODERNIZATION Replacement in current design of outmoded aspects of a structure and equipment.

MODIFICATION 1. Written amendment to an agreement, signed by both (all) parties; **2.** Change order signed by all appropriate parties; **3.** Field order for a minor change in the work, issued by the architect or owner's agent.

MODIFIED BITUMEN MEMBRANE Sheet form of single-ply roofing. *See also* **Single-ply roofing.**

MODIFIED CEMENT Portland cement for use when either moderate heat of hydration or moderate sulfate resistance, or both, is desired.

MODIFIED CUBE Portion of a rectangular beam of hardened concrete previously broken in flexure, used in determining the compressive strength of the concrete.

MODIFIED PROCTOR Moisture-density test of more rigid specifications than Proctor. The basic difference is the use of a heavier weight that is dropped from a greater distance in laboratory determinations. *See also* **Proctor needle,** and **Proctor test.**

MODIFIER 1. Additive that changes some or several aspects of a substance without causing it to loose its essential characteristics; **2.** An asphalt modifier is a material incorporated in a paving mixture in addition to normal asphalt, aggregate, and mineral filler to control rutting, cracking, oxidation, or water damage.

MODILLION Small projecting bracket, often placed in series below a classical cornice.

MODULAR COORDINATION Term coined to describe the production of a wide range of building materials designed to a 100 mm (4 in.) module or grid, or to some other logical measurement compatible with construction practice.

MODULAR DESIGN Planning of buildings and building components in accordance with a two- or three-dimensional grid based on a modular dimension or dimensions. *Also called* **Modular system.**

MODULAR DIMENSION Measurement that can be repeated in one, two, or three planes and to which materials and components will conform.

MODULAR HOUSING Dwelling units constructed from components, many of which are interchangeable, prefabricated in a factory and assembled and erected on site.

MODULARIZE Division of a system into separate modules that can easily be removed and replaced, independently of each other.

MODULAR MASONRY UNIT One whose nominal dimensions are based on a 100-mm (4-in.) module. *Also called* **Masonry unit.**

MODULAR RATIO Ratio of the modulus of elasticity of steel to that of concrete.

MODULAR SYSTEM *See* **Modular design.**

MODULATE Regulate or adjust.

MODULATED FAN Control that operates only when an engine reaches a specified temperature.

MODULE Standard. In architecture and construction, a standard unit that can be repeated, i.e., a wall of specific length, height, thickness, and materials that can be produced any number of times.

MODULUS In the physical testing of a substance, the load necessary to produce a stated change as a percentage.

MODULUS OF COMPRESSION Ratio of compressive stress to cubical compression, always positive for all physical substances. *Also called* **Bulk modulus.**

MODULUS OF DEFORMATION 1. Concept of modulus of elasticity expressed as a function of two time variables: strain in loaded concrete as a function of the age at which the load is initially applied and the length of time the load is sustained, for instance; **2.** Ratio of stress for strain for a material that does not deform in accordance with Hooke's law when subjected to applied load. *See also* **Modulus of elasticity.**

MODULUS OF ELASTICITY Mathematical quantity expressing the ratio, within the elastic limit, between a definite range of unit stress and the corresponding unit elongation. *Also called* **Chord modulus.** *See also* **Hooke's law, Modulus of deformation,** and **Proportional limits.**

MODULUS OF INCOMPRESSIBILITY Ratio of pressure in a soil mass to the volume change caused by the pressure.

MODULUS OF RIGIDITY Ratio of unit shearing stress to the corresponding unit shearing strain

MODULUS OF RUPTURE Measure of the ultimate load-carrying capacity of a beam.

MODULUS OF SUBGRADE REACTION *See* **K-value.** *Also called* **Coefficient of subgrade reaction.**

MOE *Abbreviation for:* mobile equipment; modulus of elasticity.

MOHS SCALE Arbitrary quantitative units, ranging from 1 to 10, by means of which the scratch hardness of a mineral is determined, each unit being represented by a mineral that can scratch any other mineral having a lower number.

MOIL POINT Short length of drill steel sharpened to a conical point and used with a jackhammer for breaking or hole-punching.

MOIST Slightly damp but not quite dry to the touch. *See also* **Damp,** and **Wet.**

MOIST-AIR CURING 1. Curing of a material in moist air; **2.** Curing of concrete in an atmosphere of not less than 95% relative humidity at atmospheric pressure and normally at a temperature approximating 22.8°C (73°F).

MOIST CABINET Upright and compartmented case having doors and shelves of moderate dimensions for storing and curing small test specimens of cement paste, mortar, and concrete in an atmosphere of about 22.8°C (73°F) temperature and at least 95% relative humidity. *See also* **Fog curing.**

MOIST SOIL *See* **Soil types.**

MOISTURE ABSORPTION Assimilation of water by a material or substance.

MOISTURE BARRIER *See* **Vapor barrier.**

MOISTURE CONTENT Amount of water, by weight, contained in a material, usually expressed as a percentage of the weight, such as of oven-dry wood.

MOISTURE CONTENT OF AGGREGATE Ratio, expressed as a percentage, of the mass of water in a given granular mass to the dry weight of the mass.

MOISTURE CONTENT OF CONCRETE MASONRY UNIT Amount of water contained in the hardened concrete at the time of sampling and expressed as a percentage of its capacity for total absorption.

MOISTURE DENSITY CURVE Graph plotted from tests to determine at what point of added moisture the maximum density will occur.

MOISTURE EJECTION VALVE *See* **Valve.**

MOISTURE FREE Condition of a material that has been dried in air until there is no further significant change in its mass.

MOISTURE GRADIENT Condition of graduated moisture content between successive thickness zones of wood that may be losing or absorbing moisture. During seasoning the gradations are between the relatively dry surface zones and the wet zones at the center of the piece.

MOISTURE HOLDING CAPACITY Quantity of water held by compacted solid wastes beyond which the application of additional water will cause it to drain rapidly to underlying material.

MOISTURE MOVEMENT Movement of moisture through a porous medium.

MOISTURE PENETRATION Depth to which moisture penetrates soil before the rate of downward movement becomes negligible.

MOISTURE REGAIN Reabsorption following a loss of moisture.

MOISTURE RESISTANT ADHESIVE *See* **Adhesive.**

MOL Molecular weight of a substance, in grams.

MOLD Device containing a cavity into which a fluid or material in suspension can be poured to produce a designed shape. *See also* **Plaster mold.**

mold. *Abbreviation for:* molding.

MOLDBOARD 1. Curved surface of a plow, bulldozer blade, grader blade, or excavation device. Material moved over it by passage of the carrier is given a rotary, spiral or twisting motion; **2.** Scraper that rides on or near the planed/milled surface and confines material; **3.** Front portion or vertical face of a screed.

MOLDED ANGLE Angle between the molded face of a seal lip and the seal axis.

MOLDED INSULATION Quick-setting cellular insulation that can be shaped to the void it has to fill. *Also called* **Sectional insulation.**

MOLDING 1. Profile applied to a projecting member or element; may consist of a single angle or curve, or combined forms of angles and/or curves; **2.** Specific shape applied or struck on a surface; **3.** Wood strip having a curved or projecting surface, used for decorative purpose. Among the standard molding shapes are:

Base: Wide, tapered molding, square at its thickest section; rounded at its thinest. Often with a concave or decorated face. Used at the junction of floor and wall.

Base cap: Square bottomed section with a concave or convex face, used immediately above a base molding for added trim.

Base shoe: Small concave or convex molding, used at the foot of a base molding to mask any slight irregularities.

Carpet strip: Quarter-round strip placed over fitted carpet against the baseboard.

Casing: Wide molding worked to a variety of sections, used to trim around doors and windows.

Casing bead: Molding around openings against which plaster is dressed.

Cavetto: Concave molding with a rabbet at its lower end.

Chair rail: Wide, flat, grooved, or fluted molding applied to walls some 90 mm (3 ft) above the floor.

Congé: Concave molding struck on one corner of a rectangular section, leaving a shoulder on each adjacent face.

Corner: V-shaped channel, used to protect external corners from damage.

Cove: Curved, concave molding, used as a general trim to mask joints and enhance appearance.

Crown: Convex curve on a flat back with 45° beveled edges, used to mask the junction of walls and ceilings.

Cyma-recta: Ogee-type molding consisting, from the top, of a rabbet, a concave molding followed by a convex molding that ends in a second rabbet.

Cyma-reversa: Ogee-type molding consisting, from the top, of a rabbet, a convex molding followed by a concave molding that ends in a second rabbet.

Half-round: Half-circle section, used to mask butt joints and to enhance appearance.

Mullion: Fluted section, typically used as vertical trim between windows.

Ovolo: Wide convex molding, *Also called* a **Quarter-round.**

Quarter-round: Quarter-circle section, used to mask right-angle junctions.

Roll: Molding that is semi-circular, or more than semi-circular in section.

Shoe: Small molding, usually quarter-round, used to cover the joint between the baseboard and floor or floor covering.

Stool: Section with one square and one rounded edge, used to mask right-angle junctions.

Stop: Section with one square and one tapered edge, used in door and window construction to 'stop' or arrest movement.

Threshold: Wide, beveled section, used beneath doors and over the junction of floors between rooms.

MOLD OIL Mineral oil that is applied to the interior surface of a clean mold to facilitate its removal after the object has been cast.

MOLD RELEASE AGENT Lubricant used to coat a surface to prevent another material from adhering to it during a casting process.

MOLE 1. Tunnel-boring machine; **2.** Egg-shaped device pulled behind the tooth of a subsoil plow to open passages for drainage or for the insertion of piped services or cable; **3.** One of the seven base unit of the SI system of measurement: a unit of substance of a system that contains as many elementary entities as there are atoms in 0.012 kg of carbon-12. Symbol: mol. *See also the appendix:* **Metric and nonmetric measurement.**

MOLECULAR WEIGHT Relative average weight of a molecule of a substance, expressed by a number in a scale on which the weight of the oxygen atom was arbitrarily set at 16.

MOLECULE Smallest particle into which a substance can be divided and still retain its original characteristics.

MOLE PER CUBIC METER A derived unit of concentration with a compound name of the SI system of measurement. Symbol: mol/m^3. *See also the appendix:* **Metric and nonmetric measurement.**

MOLE PER KILOGRAM A derived unit of molality with a compound name of the SI system of measurement. Symbol: mol/kg. *See also the appendix:* **Metric and nonmetric measurement.**

MOLING Installing services with the aid of a mole plough.

MOLLE Circle of twisted strands of wire rope used as a temporary line to connect the eye splices of two lines; a ring of wire to replace a cotter key.

MOLLY 1. Single strand from a wire rope rolled into a circle with six wraps, used in most pin shackles in place of a cotter key. Also used as a temporary method of connecting the eyes of two lines; **2.** Threaded holder that is inserted into plaster or concrete to receive a bolt.

MOLYBDENUM *See* **Alloying elements.**

MOMENT Product of a force or loading and the nominal distance to a particular or arbitrary axis, producing in some instances rotational effects (or loading in a member resisting rotation) and in other instances bending effects. *See also* **Bending moment.**

MOMENTARY-CONTACT SWITCH *See* **Switch.**

MOMENTARY-PRESSURE OPERATION *See* **Operation.**

MOMENT CONNECTION Connection between two members that transfers moment and maintains continuity.

MOMENT DISTRIBUTION Method of structural analysis for continuous beams and rigid frames whereby successive converging corrections are made to an assumed set of moments until the desired precision is obtained. *Also called* **Hardy cross method.**

MOMENT SENSOR Sensor, generally located in the structural supporting member, or in the hydraulic system of a crane, whose signal output is proportional to the load moment applied to the base of the crane boom.

MOMENTUM Mass times velocity.

momt *Abbreviation for:* movement.

mon. *Abbreviation for:* monitor; monument.

MONETARY ITEM Money or claim to money, the value of which (in terms of the monetary unit) is fixed by contract or otherwise.

MONIAL Upright framing members: a mullion.

MONITOR 1. Type of gable roof commonly used on industrial buildings that has a raised portion along the ridge with openings for light and/or air; **2.** Nozzle mounted in a swivel that generates extremely high pressures in water jetted through it. *See also* **Giant,** and **Hydraulic monitor.**

MONITORING 1. Capture, organization, and reporting of measures of performance in comparison to standards for project scope, time, and cost; **2.** Process in the project management system.

MONITORING WELL Well developed specifically to permit monitoring of water levels and/or water quality.

MONK BOND *See* **Bond.**

MONKEY STICK *See* **Lead.**

MONKEY WRENCH 1. Pipe wrench; **2.** To repair a machine or piece of equipment.

MONOCOQUE CONSTRUCTION Lightweight type of trailer construction where the sides of the vehicle bear a substantial part of the load in shear that is transmitted to the upper coupler and undercarriage assemblies through side rails, cross members, and end structures.

MONOCOTTURA (SINGLE FIRED) Tile manufactured by a process that allows the simultaneous firing of the clay with the glaze, producing a finished tile with a single firing.

MONOLITH Single, massive block, often as a monument or column; block of concrete in a dam separated from others by a bulkhead form.

MONOLITHIC CONCRETE Concrete cast with no joints other than construction joints. *Also called* **Integrally cast.**

MONOLITHIC CONCRETE CONSTRUCTION *See* **Construction types.**

MONOLITHIC LINING Refractory lining made onsite in large sections without conventional layers and joints; may be formed by pouring or casting, gunniting, ramming, or sintering a fine granular material into place.

MONOLITHIC SLAB *See* **Slab on grade.**

MONOLITHIC SURFACE TREATMENT *See* **Dryshake.**

MONOLITHIC TERRAZZO Application of a 16-mm (0.63-in.) thick terrazzo topping directly to a specially prepared concrete substrate, eliminating an underbed.

MONOLITHIC TOPPING On concrete flatwork, a higher quality, more serviceable topping course placed promptly after the base course has lost all slump and bleed water.

MONOLITHIC UNIT Prefabricated construction unit characterized by a high degree of finish and completeness, resulting in a minimum of site time being necessary for erection.

MONOMER Organic liquid of relatively low molecular weight that creates a solid polymer by reacting with itself or other compounds of low molecular weight, or both.

MONOMOLECULAR Composed of single molecules, specifically films that are one molecule thick.

MONOPITCH ROOF Roof that is pitched in one plane only.

MONORAIL *See* **Crane.**

MONOSTABLE MULTIVIBRATOR Multivibrator that delivers one output pulse for each input pulse.

MONOSTYLE Single column; a shaft or monument.

MONOTRUSS *See* **Truss.**

MONOTUBE Open-top, longitudinally fluted tapered steel tube, driven without a mandrel and filled as a cast-in-place concrete pile.

MONTEE CAISSON Steel caisson with sawtooth cutting edge, that is rotated to cut its way through rock. Loosened material is washed out of the caisson.

MONTMORILLONITE Swelling clay mineral of the smectite group.

MONUMENT Permanent marker of stone or metal set to mark a property or reference line; also used for elevation. *See also* **Bench mark.**

MOONBEAM Curved or horizontal beam attached to a pile driver in a way to permit swinging and holding the leads for driving batter piles at an angle to the machine.

MOONEY SCORCH Measure of the incipient curing characteristics of a rubber compound using the Mooney viscometer.

MOONEY VISCOSITY Measure of the plasticity of a rubber or rubber compound determined in a Mooney shearing-disk viscometer.

MOON WRENCH *See* **Wrench.**

MOORING PILE Usually a treated or greenheart wood pile, driven out from a pier or dock to steady marine vessels from wind and wave action, or to hold a floating dock in position.

MOPBOARD Baseboard.

MOPSTICK Handrail of circular cross section, slightly flattened at the bottom.

MOPPING UP Act of making a fire safe after it has been controlled by extinguishing or removing burning material along or near a control line.

MOR *Abbreviation for:* modulus of rupture.

MORAINE *See* **Glacial till.**

MORATORIUM Action by a jurisdiction to suspend legal enforcement of contractual obligations.

MORE-OR-LESS When used in connection with distance and quantity, such as in a deed, it covers unimportant inaccuracies of measurement.

mort. *Abbreviation for:* mortise.

MORTALITY Number of sound-wood-volume healthy trees that have died from natural causes during a specified period.

MORTALITY OF GROWING STOCK *See* **Quality classes.**

MORTALITY OF SAWTIMBER *See* **Quality classes.**

MORTAR Mixture of sand and/or lime and cement and water that hardens after use. Used between bricks, stones, or blocks in buildings, or as a plaster. *See also* **Cementitious materials, Fat mortar, high-bond mortar,** and **Lean mortar.** The following are the generally accepted mortar types:

Type M: General use, recommended for below grade use. Average compressive strength at 28 days: 17.2 MN/m² (2,500 psi).

Type S: General use, recommended for resistance to high lateral thrust. Average compressive strength at 28 days: 12.4 MN/m² (1,800 psi).

Type N: Exposed masonry above grade, recommended for walls subject to extreme exposure. Average compressive strength at 28 days: 5.2 MN/m² (750 psi).

Type O: Walls of solid load-bearing units where the compressive strength does not exceed 0.69 MN/m² (100 psi). Average compressive strength at 28 days: 2.4 MN/m² (350 psi).

Type K: Bulk fill where no stress is to be applied. Average compressive strength at 28 days: 0.52 MN/m² (75 psi).

MORTARBOARD Platform or tray, about 90 mm (3 ft) square, used for holding freshly mixed mortar. *See also* **Hawk,** and **Hod.**

MORTAR BOX Box in which mortar is mixed and softened by water for use.

MORTAR CUBE Test sample of mortar measuring 50 x 50 x 50 mm (2 x 2 x 2 in.) taken to determine its compressive strength.

MORTAR HOE Hoe used for hand-mixing mortar, consisting of a perforated blade and handle about 1.67 m (66 in.) long.

MORTAR MIXER Power-driven machine that mixes mortar by causing paddles to rotate in a drum.

MORTAR PUMPING MACHINE Device for pumping mortar through hoses to a remote location or to a higher elevation.

MORTAR STRINGING Procedure of spreading enough mortar on the bed joint to ensuring laying several masonry units.

MORTGAGE 1. Pledge of real property to secure a debt; **2.** Conveyance of real property as security for a debt; **3.** An instrument that is evidence of a pledge of conveyance; **4.** Pledge of real property as security for a debt. There are several types, including:

Adjustable-rate: Mortgage that allows the interest rate to be changed at specific intervals over the life of the loan.

Assumable: Mortgage contract that allows the purchaser of the mortgaged equity to undertake the loan obligation with no change in terms.

Balloon: Loan arrangement that calls for a series of regular repayments followed by a final liquidation of the debt by a single payment that is significantly larger.

Blanket: Single loan covering more than one piece of real property.

Budget: Mortgage arranged with repayment through monthly payments for taxes and insurance in addition to principal and interest.

Chattel: Mortgage secured by pledge of personal property.

Closed-end: Mortgage in which the principal amount cannot be increased during its term.

Conventional: Mortgage loan that calls for fixed principal and interest repayments throughout its life.

Convertible adjustable-rate: Mortgage that allows the borrower the option to convert the payments to a fixed-rate schedule at a specific point within the term.

Direct-reduction: Mortgage arranged so both interest and principal are made at each payment such that the level rate of payment will be sufficient for full amortization over the contract period.

First: Mortgage that has priority as a lien over all other mortgages; in the case of foreclosure, the first mortgage to be satisfied.

Fixed payment: Mortgage secured by real property, arranged with period payments of principal and interest that are constant over the term of the loan.

Fixed rate: Mortgage secured by real property that features a constant interest rate for the term of the loan.

Graduated payment: Mortgage requiring lower payments at the beginning of its term than toward the end.

Growing-equity: Loan in which payments increase by a specified amount each year with the extra amounts applied to retirement of the principal.

Interest-only: Mortgage loan requiring regular payments of the interest due until maturity, at which point the entire balance is due.

Junior: Mortgage whose claim against the property can be satisfied only after prior mortgages have been repaid.

Leasehold: Mortgage loan secured by a tenant's interest in real estate.

Level-payment: Mortgage that requires the same payment each payment period for full amortization.

Open: Mortgage that has matured, or that is overdue and is therefore open to foreclosure at any time.

Open-end: Where the borrower may secure additional funds from the lender, usually within a stipulated ceiling.

Package: Mortgage arrangement in which the principal amount loaned also covers the cost of non-structural items such as appliances.

Partially amortized: One that requires some payment towards principal in addition to interest, but that does not fully retire the debt within its term.

Participating: Mortgage that allows the lender to share in the income or proceeds of sale of the property.

Permanent: Mortgage contract spanning a long period (usually considered to be more than 10 years).

Pledged account: Home purchase loan in which a defined sum is deposited by the owner in an account pledged to the lender and used to supplement the initial periodic mortgage payments.

Price-level-adjusted: Loan whose payments are adjusted according to the rate of inflation.

Purchase money: Mortgage given by a buyer to a seller in part payment of the purchase price of real estate.

Reverse annuity: Mortgage that pays a regular amount to the borrower based on his or her equity in the property used as security.

Rollover loan: Mortgage in which the amortization or principal is based on a long term, but

the interest rate is established for a shorter period with the loan being 'rolled over' when a new rate is established.

Second: A subordinated lien, created by a mortgage loan over the amount of a first mortgage.

Self-amortizing: Loan that will retire itself through regular payments of principal and interest.

Senior: *See* **First mortgage,** above.

Shared appreciation: Loan with a fixed rate of interest set below market rates, with the lender entitled to a defined share of any appreciation in value of the secured property over a specific time.

Shared equity: Loan where the lender is entitled to a defined share of the equity on its resale, following satisfaction of any outstanding balance of the loan.

Stable: Mortgage that combines fixed and adjustable rates in the same loan, the rate applied being a blend of a fixed rate and a rate that varies according to an index.

Variable-maturity: Long-term loan under which the interest rate may be adjusted periodically with the level of repayments remaining the same but the length of the loan being adjusted as necessary.

Variable-rate: Long-term mortgage under which the interest rate may be adjusted, within defined limits, each six months (or other defined period) throughout the duration of the loan.

Wraparound: Arrangement in which an existing mortgage is retained and an additional loan, larger than the existing loan, is made with the whole loan bearing a different interest rate than the original loan.

MORTGAGE COMMITMENT Written confirmation by a lender of the terms and conditions of a proposed mortgage loan.

MORTGAGE NOTE Written agreement by a borrower to repay a loan secured by a mortgage that serves as proof of indebtedness.

MORTGAGEE Creditor or lender under a mortgage.

MORTGAGE REDUCTION CERTIFICATE Dated document certifying the status, and balance owing on a mortgage.

MORTGAGOR Debtor or borrower under a mortgage.

MORTISE Notch, hole, or space, cut to receive a mating tenon so as to form a joint.

MORTISE CHISEL *See* **Wood chisel.**

MORTISE GAUGE Scribing tool with two pins.

MORTISE LOCK Lock that fits into a mortise cut into the edge of a door stile.

MORTISING MACHINE Equipment that cuts a mortise using a combination of a cutting bit and chisel-edged bits.

mos. *Abbreviation for:* mosaic.

MOSAIC Surface decoration for walls, floors, ceilings, etc., formed of small pieces of glass, stone, marble, tile, ceramic, or other material different from the background, set in a mastic to form patterns or a design.

mot. *Abbreviation for:* motor; mottle; mottled.

MOTEL Building or buildings providing sleeping accommodation and parking for vehicles, intended for use by motorists as transient lodgers.

MOTIVE FLUID *See* **Fluid power.**

MOTIVE POWER Result of the conversion of energy to motion.

MOTOR Rotating machine that transforms chemical, electrical or explosive energy into mechanical energy.

MOTOR-CIRCUIT SWITCH *See* **Switch.**

MOTOR CONTROL CENTER Assembly of one or more enclosed sections having a common power bus, and principally containing motor control units.

MOTOR-GENERATOR Electrical generator propelled by an electric motor.

MOTOR-GENERATOR SET Machine that converts mechanical energy into electrical energy by electromagnetic induction, typically an AC drive motor coupled directly to a DC generator.

MOTORIZED HAND TRUCK Powered industrial truck that is designed to be controlled by a pedestrian.

MOTOR TRUCK Self-propelled vehicle carrying its load on its own wheels.

MOTTLER Thick brush used for graining.

MOTTLING 1. Film defect having the appearance of blotches; **2.** Irregular appearance in an area or entire surface of a finished particleboard due to heavy application of finishing material, poor drying, or incompatible solvents.

MOUND Small hill, usually round or oval in shape.

MOUND BREAKWATER Rocks heaped into a furrow to serve as a breakwater.

mount. *Abbreviation for:* mounted; mounting.

MOUNTABLE CURB *See* **Rolled curb.**

MOUNTED POINTS AND WHEELS Small bonded abrasive shapes and wheels that are mounted on steel spindles.

MOUNTED TILE Tile assembled into units or sheets by a suitable material to facilitate handling and installation; may be face-mounted, back-mounted, or edge-mounted.

MOUNTING DEVICE One of a wide range of fittings and assemblies used to hold and/or permanently attach components to a frame or larger member.

MOUNTING HEIGHT Distance from the top of a chassis cab frame to the bottom of the body floor.

MOUSE HOLE Socket in the substructure of a rotary drill that holds a single piece of drill pipe ready to be added to the string.

MOUSING To enclose the nose and collar of a hook with several turns of wire to prevent a sling or choker from slipping off the hook when the load is slackened.

MOUSING DEVICE Safety latch, part of a lifting hook, that prevents the load line from slipping off the hook.

MOVABLE BRIDGE Bridge deck that can be moved to allow passage below.

MOVABLE BULKHEAD REFUSE TRUCK *See* **Refuse truck.**

MOVABLE DAM Barrier that may be opened in whole or in part, usually through the movement of vertically hinged gates that provide an adjustable weir.

MOVABLE GRATE *See* **Grate.**

MOVING FORMS Large, prefabricated units of formwork incorporating supports and designed to be moved horizontally on rollers or similar devices with a minimum amount of dismantling between successive uses.

MOVING LOAD Live load.

MOVING WALK SYSTEM Series of moving walks, end-to-end or side-by-side, with no landings between treadways. *Also called* **Electric walk.**

MOVING WALKWAY Passenger transporter on which pedestrians may stand or walk while the passenger-carrying surface moves on rolllers in the direction of its length. There are several types:

> **Belt-pallet-type:** A moving walk having a series of connected and power-driven pallets to which a continuous belt treadway is fastened.

> **Belt-type:** A moving walk having a power-driven continuous-belt treadway.

> **Edge-supported belt-type:** A moving walk with the treadway supported near its edges by a succession of rollers.

> **Pallet-type:** A moving walk having a series of connected and power-driven pallets that together constitute the treadway.

> **Roller-bed-type:** A moving walk with the treadway supported throughout its width by a succession of rollers.

> **Slider-bed -type:** A moving walk with the treadway sliding on a supporting surface.

mp *Abbreviation for:* melting point.

MP *Abbreviation for:* matched pair; milepost; multipurpose.

mph *Abbreviation for:* miles per hour.

MPH Velocity (speed) in miles per hour.

mpt *Abbreviation for:* male pipe thread.

MR *Abbreviation for:* mill run.

mrtr *Abbreviation for:* mortar.

MS *Abbreviation for:* machine screw; master switch; maximum security.

MS CONNECTOR Nonelectric millisecond delay device used with detonating cord for delaying shots from the surface.

MSG *Abbreviation for:* manufacturer's standard gauge.

MSH *Abbreviation for:* Moh's scale hardness.

MSL *Abbreviation for:* mean sea level.

msnry *Abbreviation for:* masonry.

msnte *Abbreviation for:* masonite.

mt *Abbreviation for:* metric ton; mount.

mt. *Abbreviation for:* moountain.

mtd *Abbreviation for:* mounted.

mtg *Abbreviation for:* mounting.

mtge *Abbreviation for:* mortgage.

mthd *Abbreviation for:* method.

MTO *Abbreviation for:* manufactured to order.

mtr *Abbreviation for:* miter.

mtrg *Abbreviation for:* metering.

MUCK 1. Soft mud containing vegetable matter; **2.** General term for all kinds of excavated material; **3.** Finely blasted rock, particularly from tunnels.

MUCK BUCKET Cylindrical tool used for cleaning drilled shafts of muck and water.

MUCKER Machine for loading muck into haulage units.

MUCKING TOOL Any of a range of tools used to excavate, remove, and transport earth from beneath underpinings or from within shafts.

MUCK PILE Pile of excavated broken material or dirt that is to be loaded for removal.

MUD 1. Wet, soft, sticky earth; **2.** In rotary drilling, a mixture of drill cuttings, water, and added material that is pumped through the drill string to clean the hole and cool the cutting bit. *See also* **Joint filler**; **3.** Common name for drywall joint compound.

MUD BALL Ball of clay or silt.

MUD CAP Charge of explosive fired in contact with the surface of a rock, no borehole having been drilled, after being covered with a quantity of mud, wet earth, or similar substance. *Also called* **Adobe shot**, and **Plaster shot.**

MUDDING-IN Technique of stirring soil and water (sometimes also containing commercial driller's mud or bentonite) by an auger to form a slurry as the hole is advanced by auger drilling.

MUD 'EM IN Drag logs when there is not enough lift from the spar tree.

MUD FLAP *See* **Splash guard.**

MUD FLOW Flow of water so heavily charged with earth and debris that the mass is thick and viscous.

MUD JACKING *See* **Slab-jacking.**

MUD LINE River bed at the interface of water and soil.

MUD MAT Thin cover of crushed stone or concrete placed over a muddy area below the planned structure to provide a work platform and keep reinforcing clean. *See also* **Mud slab.**

MUD PAN Hand-held container in which a small quantity of joint compound is held during drywall finishing.

MUD PIT Shallow pit or excavation adjacent to a boring or drilling location, used to contain slurried drilling mud.

MUD PUMP *See* **Pump.**

MUD PUMPING Ejection of water, or water and solid materials such as clay or silt, along transverse or longitudinal cracks in pavement, and along pavement edges, caused by downward slab movement activated by the passage of loads over the pavement after the accumulation of free water on or in the base course, subgrade, or subbase.

MUDSILL Short pieces of horizontal logs placed transversely over the sill log or foundation log to spread loading over a large area to prevent settlement. *Also called* **Crossbearer.**

MUD SLAB 50 to 150 mm (2 to 6 in.) layer of concrete below a structural concrete floor or footing over soft, wet soil. *Also called* a **Mud mat.** *See also* **Base**, and **Subbase.**

MUD WALL 1. Earth diaphragm or impervious cutoff wall in a dam; **2.** Wall above the beam seats of a bridge abutment, designed to support the approach slab and retain the earth behind the abutment.

MUD WAVE Front face of a mud slide.

muf. *Abbreviation for:* muffler.

MUFFLE FURNACE *See* **Furnace.**

MUFFLER Expansion chamber used to muffle the noise of engine combustion, oil, or gas flow noise.

mul. *Abbreviation for:* mullion.

MULCH Layer of vegetable materials, such as straw, leaves, bark, or wood chips, spread on the surface of the soil to discourage weed growth, reduce water loss due to evaporation, and protect roots.

MULE Template used to shape the profile of a curb and gutter.

MULLET Grooved piece of wood, used to check the thickness of the edge of a panel.

MULLION *See* **Molding.**

mult. *Abbreviation for:* multiple.

MULTIAXLE *See* **Axle.**

MULTIAXLE SUSPENSION *See* **Suspension.**

MULTICHORD Cable of two or more wires.

MULTICYCLONE COLLECTOR Dust collector consisting of a number of cyclones operating in parallel.

MULTIDECK ELEVATOR *See* **Elevator.**

multidir. *Abbreviation for:* multidirectional.

MULTIELEMENT PRESTRESSING Prestressing accomplished by stressing an assembly of several individual structural elements as a means of producing one integrated structural member.

MULTIFUEL ENGINE Engine designed to operate on more than one type of fuel without engine modification.

MULTILEG INTERSECTION *See* **Intersection types.**

MULTILEVEL SUSPENDED SCAFFOLD Two-point or multi-point adjustable suspension scaffold with a series of platforms at various levels supported by common stirrups.

MULTIPASS FILTER PERFORMANCE TEST Test designed to obtain consistent and repeatable information on a filter's ability to control specific size particles.

MULTIPERIL INSURANCE Policy providing property, casualty, and inland marine coverage.

MULTIPIECE RIM Wheel rim composed of two (rim base and side ring) or three (rim base, side ring, and lock ring) pieces.

MULTIPLE-CHAMBER INCINERATOR *See* **Incinerator.**

MULTIPLE-DISC SWING BRAKE *See* **Brake.**

MULTIPLE DRIFT Tunnel excavation where two small drifts along each side of the tunnel allow side support to be placed, followed by a top drift that is then widened out slowly to take the roof support.

MULTIPLE ENTRY Entering a forest stand for commercial harvesting more than once in any one continuous rotation.

MULTIPLE-FAMILY DWELLING Dwelling with separate living units for three or more families, with joint services and facilities.

MULTIPLE HEARTH FURNACE *See* **Furnace.**

MULTIPLE HOISTWAY Hoistway for more than one elevator or other vertical lifting device.

MULTIPLE-LAYER ADHESIVE *See* **Adhesive.**

MULTIPLE LINES Single line reeved around two or more sheaves to increase pull.

MULTIPLE LISTING Representation for sale or lease of real property by more than one broker.

MULTIPLE-OPENING DAMPER Type of damper used in a multiple-opening fireplace.

MULTIPLE-POINT SUSPENDED SCAFFOLD Suspension scaffold consisting of a platform(s) suspended by more than two ropes from overhead supports and equipped with means to permit the raising and lowering of the platform to desired work levels. This definition does not include two-point adjustable suspension scaffolds, which are bridged one to another.

MULTIPLE-SHEAVE DRIVE Double set of sheaves controlling the travel of the conveying line of a cableway.

MULTIPLE SURFACE TREATMENT 1. Two or more surface treatments using asphalt and aggregate placed one on another. The aggregate maximum size of each successive treatment is usually one-half that of the previous one, and the total thickness is about the same as the nominal maximum size aggregate particles of the first course; **2.** Multiple surface treatment comprising a series of single treatments that produces a pavement course up to 38 mm (1.5 in.) or more. A multiple surface treatment is a denser wearing and waterproofing course than a single surface treatment, and adds some strength, but is not normally assigned a structural coefficient.

MULTIPLE UNDERREAM Additional underream cut in a bearing soil, at a level above the bottom underream, to force shearing resistance in the soil into a larger peripheral surface.

MULTIPLE USE Practice of forestry that combines two or more objectives.

MULTIPLE-USE FORESTRY Concept of forest management that combines two or more objectives, such as production of wood or wood-derivative products, forage and browse for domestic livestock, proper environmental conditions for wildlife, landscape effects, protection against floods and erosion, recreation, and protection of water supplies.

MULTIPLE-USE MANAGEMENT Management of land resources with the objective of achieving optimum yields of products and services from a given area without impairing the productive capacity of the site.

MULTIPLE-WRAP HOISTING DRUM Drum that accumulates suspension rope in more than one layer on the surface of the drum.

MULTIPLEXING Carrying multiple signals along one pair of conductors.

MULTIPURPOSE BUCKET *See* **Bucket.**

MULTISPAN BRIDGE Bridge containing a pier or piers and more than one span.

MULTISPAN SKYLINE Skyline having one or more intermediate supports.

MULTISTAGE STRESSING Prestressing performed in stages as the construction progresses.

MULTISTEM Operation handling two or more stems at the same time.

MULTIUSE BIT *See* **Bit.**

MULTIWALL BAG Flexible container for transporting any cementitious material, usually consisting of four plies of kraft paper previously treated to ensure resistance to moisture.

MULTIWYTHE CONSTRUCTION Masonry wall made up of at least three thicknesses of masonry units in width.

muni. *Abbreviation for:* muncipal; municipality.

MUNICIPAL COLLECTION *See* **Collection.**

MUNICIPAL ENGINEERING Design and construction of publicly owned services and facilities.

MUNICIPAL SERVICES Services provided to the residents and visitors to a municipality and including water supply, sanitary and storm sewerage, roads and streets, curbs, sidewalks, parks, garbage collection, etc., and may include electrical distribution, hospitals, local airports, harbor facilities, etc.

MUNICIPAL SOLID WASTE *See* **Solid waste.**

MUNICIPAL STOP Valve installed in a building water service line at the water main.

MUNICIPAL TAX Taxes levied by the municipal government, distinct from property taxes, and including general rates, special rates, and local improvement rates.

MUNIMENTS OF TITLE Deeds, contracts, or other documents used to establish ownership.

munt. *Abbreviation for:* muntin.

MUNTIN Vertical or horizontal member dividing the panels within a door, or lights in a window frame, butting into or stopped by the horizontal rails and stiles. *See also* **Door.**

MUSHING *See* **Burring.**

MUSHROOM STAKE Dome-topped steel anvil.

MUSHROOM SYSTEM OF FLAT-SLAB CON-STRUCTION Four-way, reinforced-concrete, girderless floor slab in which the column reinforcing bars are bent down into the slab around the column head in radial directions and additional reinforcing bars are bent into rings laid upon the radials, thus forming a spider web to provide additional reinforcement at the column head and to support the slab steel.

MUSIC WIRE High-tensile steel wire used to plumb critical installations. *See also* **Piano wire.**

MUSKEG Tract of partly forested peatland supporting mosses, shrubby plants, and scattered trees. *Also called* **Peatland.**

MUTT-AND-JEFF Flat hook and round hook combination facing in opposite directions and joined by a link.

MUTTONCLOTH Knit fabric used for removing excess glaze and for imprinting a texture. Each side of the cloth has a different texture. *Also called* **Stockinette.**

momt *Abbreviation for:* movement.

MV *Abbreviation for:* mean variation.

mw *Abbreviation for:* mixed width.

MW *Abbreviation for:* metal wall; moderate weathering.

MWO *Abbreviation for:* modification work order.

MWP *Abbreviation for:* maximum working pressure.

mxd *Abbreviation for:* mixed.

MxM *Abbreviation for:* metal by metal.

mxr *Abbreviation for:* mixer.

N

n *Abbreviation for:* revolutions per minute.

N *Abbreviation for:* negative; north.

n/a *Abbreviation for:* not available.

NA *Abbreviation for:* neutral axis; not applicable.

NADIR Point on the ground vertically below an observer, or survey instrument. *Also called* **Plumb point.**

NAIL Metal spike, pointed at one end with an enlarged flat or rounded head at the other, used to fasten two or more pieces of material together. Nails are available in a wide range of sizes, weights, and styles. Among the more common lengths are:

25 mm (1.0 in.), 2 penny

31 mm (1.25 in.), 3 penny

38 mm (1.5 in.), 4 penny

44 mm (1.75 in.), 5 penny

50 mm (2 in.), 6 penny

57 mm (2.25 in.), 7 penny

64 mm (2.5 in.), 8 penny

69 mm (2.75 in.), 9 penny

75 mm (3 in.), 10 penny

89 mm (3.5 in.), 12 penny

100 mm (4 in.), 20 penny

114 mm (4.5 in.), 30 penny

127 mm (5 in.), 40 penny

140 mm (5.5 in.), 50 penny

150 mm (6 in.), 60 penny

178 mm (7 in.), 70 penny

203 mm (8 in.), 80 penny

(**NOTE:** As applied to nails, penny derives from an English measure based on the pennyweight or 1/20th of an ounce. The term now serves as a measure of nail length (the higher the number the longer the nail) and is abbreviated by the letter **d.**)

Of the many types of nail available, the following are some of the most widely used:

Aluminum: Used for the same purpose as wire nails. Also, a special-purpose nail for roofing, aluminum flashing, etc.

Box: Nail having a uniform shaft that is thinner than that of a common nail, with a flat head. Often galvanized, zinc coated, phosphate plated, or otherwise treated. Typically used for fasten-

ing subflooring and sheathing.

Brad: Thin nail with a small head, used for small parts assembly, panel-molding, etc. Brads vary from 5 to 50 mm (0.2 to 2 in.) in length.

Casing: Nail having a uniform shaft with the head significantly smaller than on common nails, and swelling from the shank; the head is sometimes dimpled to permit countersinking.

Clout: Similar to a roofing nail, but with a longer shank.

Common: Nail having a uniform shaft with a flat head. Typically used for framing.

Cut: Stamped from a strip of rolled iron or steel of the same thickness as the nail and a little wider than its depth.

Cut flooring: Flat nail with a tapered shank, used for nailing into the sides of floorboards and in restoration work.

Double-headed: *See* **Duplex head,** below.

Drywall: Galvanized nail with ridges formed around the shaft for its entire length.

Duplex-head: Common nail with a flat ring formed about the shaft some 6 mm (0.25 in.) below the head. Used for temporary work as the raised head permits easy removal.

Finishing: Nail with a very small, cupped head.

Masonry: Case-hardened and tempered-steel nail with a short, stout shank that may be fluted, knurled, or specially shaped, and that can be driven into concrete and masonry using a heavy hammer or mallet.

Panel pin: Very slender wire nail with a small head that is nearly invisible when punched below the surface.

Ring: Nail with a serrated or roughened shaft.

Ring-shank: Nail with rings formed about the shank for a length above the point.

Roofing: Nail with a short, stout shaft and large flat head, galvanized. Used to fasten roofing materials to the sheathing.

Scaffold: Double-headed nail (one below the other on the shaft) used to fasten materials that must later be disassembled.

Screw-shank: Speciality nail with a shank that has a slow-pitch screw thread along its length.

Siding: Nonstaining nail of a size specified for siding thickness.

Spiral: Nail formed with annular rings about the shank, used to install or repair wood flooring and in rough carpentry.

Tack: Short, very sharp nail with a large flat, domed or decorated head.

Underlayment: Special nail types with greater holding power than ordinary varieties, often incorporating a deformed shank or a shank with a slow-pitch spiral thread.

NAILABLE CONCRETE Concrete, usually made with a suitable lightweight aggregate, with or without the addition of sawdust, into which nails can be driven.

NAILBAR Small steel prybar having one end flat

tapered and the other fishtailed and turned up through 90°, used to remove nails and disassemble formwork.

NAIL CLAW Hexagonal or round steel bar approximately 250 mm (10 in.) long, spoon-shaped at one end. *Also called* a **Cat's paw**, and **Tack claw.**

NAILER Strip of wood or other fitting to facilitate making nailed connections, often attached to steel or concrete members.

NAIL HEAD PULLTHROUGH Load that will pull a panel off the wall over the top of the nail head.

NAIL HOLDING Holding power of a nail determined by the load it can carry when hammered into the material.

NAILING BLOCKS Wood blocks set in masonry to anchor other members with nails or screws.

NAIL POP Condition where nails appear to 'pop' up beyond the surface of a panel (plywood, underlayment, drywall, etc.) that has been fastened in place. Usually caused by shrinkage of the stud or joist away from the nail shank after installation. *Also called* **Back-out.**

NAIL PULLER Pinch bar with a forked end.

NAILSET Tool used to complete the final driving of nails and avoid damage to the surface of the material being nailed. *See also* **Punch.**

NAIL SPOTTER Applicator for applying joint compound to the dimpled areas around the nails fastening drywall panels to their supports.

NAME PLATE RATING Full load continuous rating of an electric generator and its prime mover, or other electrical or mechanical equipment, used under specific conditions as designated by the manufacturer.

NAND DEVICE Control device that has its output in the logical 0 state if, and only if, all the control signals assume the logical 1 state. *See also* **NOR device,** and **NOT device.**

NANO Prefix representing 10^{-9}. Symbol: n. Used in the SI system of measurement. *See also the appendix:* **Metric and nonmetric measurement.**

NAPHTHENIC Petroleum fluid derived from naphthenic crude oil characterized by a high proportion of cyclo-paraffenic structures and, in some cases, by an absence of wax molecules.

NAPPE Unbroken film of water flowing over the crest of a weir or dam.

NAPTHA Volatile solvent derived from petroleum or coal tar. *Also called* **Solvent naptha.**

NARROW AISLE TRUCK Type of industrial truck primarily intended for right-angle stacking in aisles narrower than those normally required by counterbalanced trucks of the same capacity.

NARROW-RINGED TIMBER Wood in which the annual growth rings are more closely spaced than is common for the species; usually denotes wood of greater-than-normal strength.

nat. *Abbreviation for:* natural.

NATIONAL BUILDING CODE *See* **Building code.**

NATIVE ASPHALT *See* **Asphalt.**

NATIVE-ASPHALT-BASE SEALANT Gilsonite ore, a native asphalt, liquefied by the addition of a volatile carrier, used as a surface sealant to reactivate the binder in old pavement and improve the binder in new pavement.

natl *Abbreviation for:* national.

NATURAL ABRASIVE Hard mineral found in nature. *See also* **Abrasive.**

NATURAL AIR DRYING Process of drying cured concrete masonry units without any special equipment (e.g., the drying that occurs in a covered storage area).

NATURAL ASPHALT *See* **Asphalt.**

NATURAL BARRIER Area where the lack of flammable materials obstructs the spread of a forest fire.

NATURAL BED The surface of a stone parallel to the stratification.

NATURAL CEMENT Hydraulic cement produced by calcining an argillaceous limestone at a temperature below the sintering point and then grinding to a fine powder.

NATURAL CONVECTION Transfer of heat by currents set up in fluids, or in air, by the difference in density resulting in differences in temperature.

NATURAL DRAFT *See* **Draft.**

NATURAL FINISH Transparent coating that does not alter the original color or texture of the material it is applied to.

NATURAL FLUORESCENT *See* **Fluorescent lamp.**

NATURAL GAS Any gas found in the earth, as opposed to manufactured gas.

NATURAL GROUND Original ground profile and elevation before any excavation is done.

NATURAL GROWTH ASSET Asset whose value is expected to increase through natural growth, e.g., a stand of trees.

NATURAL HARBOR Sheltered inlet in a coastline.

NATURALLY ASPIRATED ENGINE *See* **Engine.**

NATURALLY DEVELOPED WELL Water well in which the screen is in direct contact with the aquifer materials, no filter pack having been installed.

NATURAL MOISTURE CONTENT *In situ* moisture at the time of measurement.

NATURAL PIGMENT *See* **Earth pigment.**

NATURAL POZZOLAN Either a raw or calcined natural material that has pozzolanic properties (e.g., volcanic ash or pumicite, opaline chert and shales, tuffs, and some diatomaceous earths).

NATURAL REGENERATION Renewal of a tree crop by natural, as opposed to human, means, e.g. seed on-site from adjacent stands, or brought in by winds, birds, or animals.

NATURAL RESIN Paint vehicle obtained by distilling animal or vegetable matter. *See also* **Synthetic resin.**

NATURAL SAND Sand resulting from natural disintegration and abrasion of rock.

NATURAL SEASONING Timber dried by being

stacked and exposed to currents of air while sheltered from direct exposure to the elements.

NATURAL STONE Stone quarried and cut, as against remanufactured from stone residue.

nav. *Abbreviation for:* navigable.

NB *Abbreviation for:* nested bundling; *nota bene* (mark well).

NBC *Abbreviation for:* National Building Code.

NBFU *Abbreviation for:* National Board of Fire Underwriters.

NC *Abbreviation for:* national coarse thread; no connection; noise criteria; normally closed.

N/C *Abbreviation for:* no charge.

NCA *Abbreviation for:* noise criteria alternative.

nch *Abbreviation for:* notch.

NCK *Abbreviation for:* no change keys.

nd *Abbreviation for:* no date.

ND *Abbreviation for:* natural draft.

NDFS *Abbreviation for:* nondwelling floor space.

NDT *Abbreviation for:* nondestructive testing.

NE *Abbreviation for:* northeast.

NEAR FIELD Region near a sound source in which there is direct transmission of sound.

NEAT Without admixture or dilutent.

NEAT CEMENT In masonry, cement and water mixture, uncut by the addition of sand.

NEAT CEMENT GROUT Fluid mixture of hydraulic cement and water, with or without admixture.

NEAT CEMENT PASTE Plastic mixture of hydraulic cement and water, both before and after setting and hardening. *See also* **Cement paste,** and **Paste content.**

NEAT LINE Line defining the limits of a proposed excavation: material removed beyond the neat line is overbreak. *Also called* **Net line.**

NEAT PLASTER Plaster devoid of sand.

NEAT SIZE Dimension of a piece of worked lumber, after cutting and planing.

nec. *Abbreviation for:* necessary.

NEC *Abbreviation for:* National Electrical Code.

NECK That part of a column immediately below the capital and above the astragal.

NECKING Localized and permanent reduction of the cross-sectional area of a test specimen of a material, due to stretching produced by applied tensile load.

NECKING DOWN Localized decrease in the cross-sectional area of a hose resulting from tension.

NEEDLE BEAM 1. Steel or wood beam suspended by ropes or cables to support scaffold planks; **2.** Structural beam used in underpinning a building.

NEEDLE BEARING *See* **Bearing.**

NEEDLE GAUGE Pressure gauge with a needle stem for measuring air or steam pressure in pressure hoses,

by puncturing the wall of the hose with the needle stem.

NEEDLE-NOSED PLIERS *See* **Pliers.**

NEEDLE PILE *See* **Pile.**

NEEDLE SCAFFOLD Scaffolding suspended from needles driven into a wall.

NEEDLE VALVE *See* **Valve.**

NEEDLE WEIR Fixed frame weir supporting heavy vertical timbers in contact. The timbers can be withdrawn vertically to lower the water level.

NEEDLING Inserting a needle into a wall.

NEEDS *See* **Requirements.**

NEF *Abbreviation for:* national extra-fine thread.

neg. *Abbreviation for:* negative.

NEGATIVE ALLOWANCE *See* **Fit classifications.**

NEGATIVE BATTER *See* **Aft batter.**

NEGATIVE BOOM ANGLE Ability of a crane to lower its boom below horizontal. Negative boom angle facilitates wire rope reeving and boom maintenance.

NEGATIVE MOMENT Condition of flexure in which the top fibers of a horizontally placed concrete member, or the external fibers of a vertically placed concrete member, are subjected to tensile stress.

NEGATIVE REINFORCEMENT Steel reinforcement of concrete for negative moment.

NEGATIVE SKIN FRICTION Downward force exerted on the buried portion of a pier, pile, or other structural element by settling soil. *Also called* **Downdrag.** *See also* **Slip layer.**

NEGATIVE-SLUMP CONCRETE Concrete of a consistency such that it not only has zero slump but still has zero slump after adding additional water. *Also called* **Zero-slump concrete,** and **No-slump concrete.**

NEGATIVE TERMINAL Battery terminal toward which a positive electric charge flows in the external circuit, from the positive terminal when the cell discharges.

negl. *Abbreviation for:* negligible.

NEGOTIATION Bargaining to reach an agreement, as in land purchase, etc. *See also* **Acquisition, Conveyance, Dedication, Eminent domain, Expropriation, Option, Remainder,** and **Severality.**

neg. pr. *Abbreviation for:* negative print.

NEIGHBORHOOD Ill-defined and changing area whose residents are conscious, through common intercourse, of a community of like social fabric and institutions.

NEOCLASSIC Revival of Classical architectural styles and forms.

NEON ILLUMINATION Lighting system relying on the discharge of an electrical current through a glass tube containing neon gas.

NEOPRENE Synthetic, rubber-like material able to resist oil, grease, and antifreeze, typically used for sleeve, injection pump, and other engine seals, as bridge bearing pads, etc.

NEOPRENE-RUBBER ADHESIVE *See* **Adhesive.**

NERVE Measure of toughness of recovery from deformation in unvulcanized rubber or rubber compounds.

NESC *Abbreviation for:* National Electrical Safety Code.

NET *Abbreviation for:* not earlier than.

NET ANNUAL GROWTH Increase in a volume of trees during a specified year. Components of net annual growth include the increment of net volume of trees at the beginning of the specified year that survive the year's end, plus the net volume of trees reaching the minimum size class during the year, minus the volume of trees that die during the year, and minus the net volume of trees that become rough or rotten during the year.

NET ANNUAL GROWTH OF GROWING STOCK *See* **Quality classes.**

NET ANNUAL GROWTH OF SAWTIMBER *See* **Quality classes.**

NET AREA Minimum effective load-bearing area of the material composing a structural unit, included in any section perpendicular to the direction of the stress to be resisted. *See also* **Area, Area of building, Floor area, Gross area,** and **Net room area.**

NET CORRECTED FILL Net fill after making allowance for swell or shrink from cut to fill. *See also* **Net fill.**

NET CROSS-SECTIONAL AREA OF MASONRY Gross cross-sectional area of a section of masonry, minus the area of cavities, cells, or cored spaces.

NET CUT 1. Amount of material to be removed from a project, less that required for fills; **2.** In sidehill excavation, the fill required less the cut required at a particular station. *See also* **Cut,** and **Gross cut.**

NET FILL In sidehill excavation, the total measure of fill required at any station, minus the volume of cut obtained at the same station. *See also* **Net corrected fill.**

NET HEATING VALUE Gross heating value minus the latent heat of vaporization of water vapor formed by the combustion of the hydrogen in a fuel.

NET HORSEPOWER *See* **Horsepower, SAE net.**

NET LEASABLE AREA Project floor area available to be rented. This generally excludes common areas and space dedicated to mechanical rooms.

NET LEASE *See* **Lease.**

NET LINE Line defining the proposed or specified limits of an excavation or structure. *Also called* **Neat line.**

NET LISTING Price below which an owner will not sell and at which a broker will not receive a commission. The broker receives the excess above the listing as commission.

NET-NET LEASE *See* **Lease, net.**

NET-NET-NET LEASE *See* **Lease, net.**

NET PENETRATION *See* **Penetration.**

NET POWER Power output of a 'fully equipped' engine.

NET PRESENT VALUE Tree stand's present worth before harvesting, once costs associated with its establishment and tending have been subtracted.

NET ROOM AREA Floor area of a room measured from finished wall to finished wall. *See also* **Area, Area of building, Floor area, Gross areas,** and **Net area.**

NET SCALE Actual amount of merchantable wood contained in a log, as opposed to the gross scale, which includes defect.

NET SETTLEMENT Total downward movement of a structure, pile, or group of piles, minus the rebound after removal of a load: the gross settlement minus rebound. *See also* **Settlement,** and **Gross settlement.**

NET TON *See* **Short ton.**

NET TORQUE *See* **Torque.**

NET VOLUME Volume of the main stem of a tree, excluding the stump and top as well as defective and decayed wood.

NETWORK Tool used for planning and scheduling a project. The Network is a schematic display of the sequential relationships among the activities that a project comprises. Two popular drawing conventions or notations for scheduling are arrow networks and precedence networks.

NETWORK ANALYSIS *See* **Analysis.**

NETWORK DIAGRAM *See* **Diagram.**

NETWORK PLANNING Identification of the sequence, timing, and interrelationships of the activities comprising a project, including a graphical display of this information.

NETWORK PLANNING TECHNIQUE Planning tool to provide information with which to plan, monitor, and control the time to accomplish specified goals.

neut. *Abbreviation for:* neutral.

NEUTRAL 1. Point common to all phases of a polyphase circuit; a conductor connected to that point, or the return conductor in a single-phase circuit; **2.** Gears in a disengaged position.

NEUTRAL AXIS Line in the plane of a structural member subjected to bending where the longitudinal stress is zero.

NEUTRALIZATION NUMBER Measure of the total acidity or basicity of a fluid; this includes organic and inorganic acids or bases, or a combination of them.

NEUTRAL OIL Refined distillate fractions used in the manufacture of most lubricants.

NEUTRALIZING Preparation of surfaces of materials containing lime (concrete, cement mortar, plaster, etc.) so that the lime they contain does not attack such applied materials as paint.

NEUTRAL PLANE Plane within a house envelope that connects points on its perimeter where the interior and exterior air pressures are equal. Above the neutral plane exfiltration can occur; below it infiltration is likely to occur.

NEUTRAL PRESSURE Hydrostatic pressure in the pore water of the soil.

NEUTRAL REFRACTORY Refractory that is resistant to chemical attack by either acidic or basic substances.

NEW CONSTRUCTION An entirely new project at a given location; one that does not involve the alteration, salvage, or rehabilitation of existing works. *Also called* **New work.**

NEWEL 1. Post at the top or bottom of a flight of stairs, supporting the handrail; **2.** Central post around which a winding staircase turns. *Also called* **Rail post.**

NEWEL CAP Cap or terminal to a newel post.

NEWEL DROP Projection of a newel post through a soffit, usually ornamented.

NEWEL JOINT Joint made between a handrail and a newel post.

NEWEL STAND Upright metal mounting that supports an escalator's newel wheels.

NEWEL WHEEL Wheel that carries the handrail around the top and bottom ends of an escalator.

NEWTON One of 17 derived units with special names of the SI system of measurement: a unit of force which, when applied to a mass of 1 kg gives it an acceleration of 1 m/s^2. ($1N=1$ $kg \cdot m/s^2$). Symbol: N. *See also the appendix*: **Metric and nonmetric measurement.**

NEWTONIAN BEHAVIOR Property of a simple liquid in which the shear rate or flow rate is directly proportional to the shear stress or the pressure. This constant proportion is the viscosity of the liquid. *See also* **Non-Newtonian behavior.**

NEWTON METER A derived unit of the moment of force with a compound name of the SI system of measurement. Symbol: N·m. *See also the appendix*: **Metric and non-metric measurement.**

NEWTON PER METER A derived unit of surface tension with a compound name of the SI system of measurement. Symbol: N/m. *See also the appendix*: **Metric and non-metric measurement.**

NEWTONIAN OIL Oil in which the viscosity is independent of the shear rate. Generally true of straight-grde oils containing no polymeric VI improvers.

NEW WORK *See* **New construction.**

NEXT AVAILABLE FLOOR First floor in the direction of travel that an elevator is programmed to serve.

NF *Abbreviation for:* national fine thread; near face.

NFC *Abbreviation for:* national fire code; not favorably considered.

NG *Abbreviation for:* no good.

NIB Small projection, usually at right angles to the principal plane of an object, used to position or attach.

NIBBLER 1. Hydraulically operated equipment attachment comprising a hinged, hardened steel jaw that closes against a fixed anvil, used to 'nibble' into the edge of concrete to reduce the mass to workable rubble; **2.** Shearing tool that removes particles of sheet metal along a line.

NIC *Abbreviation for:* not in contact.

NICHE Recess sunk in the face of a wall, not extend-

ing to the floor, traditionally to hold a carving or statue.

NICKEL *See* **Alloying elements.**

NICKEL-IRON STORAGE BATTERY Storage battery in which the positive active material consists of oxides of nickel, the negative active material consists of a mixture of iron and iron oxides, and the alkaline electrolyte is usually an aqueous solution of potassium hydroxide. *Also called* **Alkaline storage battery.**

NICKER Mason's broad chisel, used to score or groove a stone prior to splitting it.

NIDGED ASHLAR Stone, particularly granite, roughly dressed with a pointed tool.

NIFO Next-in, first-out. *See* **Cost flow.**

NIGHT INSULATION Movable insulation that can be easily placed over glazing to reduce heat loss at night and during cloudy periods, and that may also be used to reduce heat gain during sunny periods. Typically: insulated drapes, shades, panels, or shutters.

NIGHT LATCH Door lock that can be actuated only by a knob from the inside of the room. *See also* **Deadbolt.**

NIGHT VENT Small opening light at the top of a casement window.

NIP 1. Seizing of stone between either the jaws or the rolls of a crusher. *Also called* **Bite; 2.** Clearance between rolls of a mixing mill or calender. *See also* **Angle of nip.**

nip. *Abbreviation for:* nipple.

NIPPER Power-actuated tongs that grab a pile-driving hammer to raise it.

NIPPLE 1. Short length of pipe threaded on both ends to allow for joining pipe elements. *See also* **Close nipple; 2.** Small valve at the high points of a hot-water distribution system through which air can be released; **3.** Cylindrical, pipe-like attachment, one end of which is securely inserted and retained in the end of a hose, serving the same purpose as a hose coupling; **4.** Shaped tube attached to machinery and equipment through which grease may be injected. *Also called* **Grease nipple.**

NIPPLE WIRE *See* **Binding-in wire.**

ni-sil. *Abbreviation for:* nickel-silver.

NISSEN HUT *See* **Quonset hut.**

NITRAMON Insensitive, safe explosive that can be detonated only by a detonator.

NITRIDING Process of heating a ferrous metal in an atmosphere of ammonia gas, causing nitrogen to fix with the molecules of metal, producing an extremely hard surface finish.

NITRILE Copolymer that is often combined with other plastics during the formation of seals.

NITROCELLULOSE 1. Constituent of lacquers, plastic wood, and some glues; **2.** A solid explosive, used with nitroglycerin in gelatine explosives.

NITROGLYCERINE Liquid explosive that is dangerously unstable unless combined with other materials, such as kieselguhr.

NITROUS FUMES Poisonous, reddish fumes produced when nitroglycerine explosives burn instead of detonating.

n/l *Abbreviation for:* no list price.

n/l/a *Abbreviation for:* no longer available.

nld *Abbreviation for:* nailed.

nlg *Abbreviation for:* nailing.

NLGI NUMBER Numerical scale for classifying the consistency range of lubricating greases. The numbers are in order of increasing consistency (hardness), beginning at 000, and progressing through 00, 0, 1, 2, 3, 4, 5, and 6. Greases both softer and harder than these consistency ranges and numbers do not bear an NLGI Number.

NLT *Abbreviation for:* not later than.

NNE *Abbreviation for:* north-northeast.

NNW *Abbreviation for:* north-northwest.

No. *Abbreviation for:* number.

NO *Abbreviation for:* normally open.

NODE In CPM networks, events indicating a point in time when all activities leading up to the event have been completed and when all activities leaving the node may start (i.e., junction points).

NO-FINES CONCRETE Concrete mixture containing little or no fine aggregate. *Also called* **Popcorn concrete.**

No. 5 COMMON *See* **Lumber grades.**

No. 4 COMMON *See* **Lumber grades.**

NOGGING 1. Short horizontal timbers that stiffen the studs of a framed partition; **2.**Brickwork employed as infilling for half-timbered buildings. *See also* **Back filling.**

NO-HUB PIPE Pipe having smooth ends (without bell or spigot).

NOISE 1. Any undesired sound; **2.** Any random disturbance in an electric or electronic circuit that changes or interferes with the signal.

NOISE ABSORPTION Construction techniques (discontinuous construction, etc.) and the use of sound-absorbing materials (cork, stucco, insulation, etc.) so as to reduce the transmission of sound and its reflection.

NOISE ENERGY Cumulative effect when continuous noise in an enclosed space is reflected off of plane surfaces.

NOISE-RATED SYSTEMS Construction designed to reduce sound transmission to a specified level.

NOISE REDUCTION COEFFICIENT Expression of the ability of a material to absorb or attenuate sound waves.

NO LIEN BOND *See* **Bond.**

nom. *Abbreviation for:* nomenclature; nominal.

NOMINAL Dimensional value assigned for the purpose of convenient designation; existing in name only.

NOMINAL CAPACITY *See* **Fuel tank capacity.**

NOMINAL DIAMETER Diameter of a plain round bar of the same weight per linear foot as a deformed bar.

NOMINAL DIMENSION 1. Dimension equal to the actual dimension; **2.** In masonry, plus the thickness of one mortar joint.

NOMINAL FILTRATION RATING Arbitrary value indicated by a filter manufacturer. Due to lack of reproducibility, this rating is deprecated.

NOMINAL LOADS *See* **Loads.**

NOMINALLY HORIZONTAL In plumbing, at an angle less than 45° with the horizontal.

NOMINAL MAXIMUM SIZE OF AGGREGATE Smallest sieve opening through which an entire amount of aggregate is permitted to pass.

NOMINAL MIXTURE Proportions of the constituents of a proposed concrete mixture.

NOMINAL OWNER Person who holds title to an asset, usually a security, on behalf of the beneficial owner.

NOMINAL RATING Arbitrary classification of truck capacity in tons, such as five ton, half ton, one ton, etc.

NOMINAL SIZE Size by which lumber is known and sold, often greater than actual size.

NOMINAL STEEL Reinforcing bars in a concrete member, needed for purposes other than resisting stress.

NOMINAL STRENGTH 1. Published strength calculated by a standard procedure; **2.** Strength of a member or cross section calculated in accordance with provisions and assumptions of the strength design method, before application of any strength reduction factor. *See also* **Breaking strength, Design strength,** and **Minimum acceptance strength**; **3.** Capacity of a structure or component to resist the effects of loads, as determined by computations using specific material strengths and dimensions and formulas derived from accepted principles of structural mechanics or by field tests or laboratory tests of scaled models, allowing for modeling effects and differences between laboratory and field conditions.

NOMINAL VALUE Reference value selected to establish equipment ratings.

NOMINAL VOLTAGE Nominal value assigned to a circuit or system for the purpose of conveniently designating its voltage class (as 120/240, 480Y/277, 600, etc.).

NONAGITATING UNIT Truck-mounted container for transporting central-mixed concrete, not equipped to provide agitation (slow mixing) during delivery.

NONAIR-ENTRAINED CONCRETE Concrete in which neither an air-entraining admixture nor air-entraining cement has been used.

NONASPHALTIC ROAD OIL Nonhardening petroleum distillate, used as dust-laying oil.

NONBEARING WALL *See* **Wall.**

NONBONDED TENDON Prestressing strand cut to length at the mill and coated with a protecting and lubricating compound before being placed in a plastic sheath in which it is free to move when stressed.

NONCIRCULATING HOT WATER SYSTEM Water heating system consisting of storage tank connected to the potable water distribution system, equipped with a means of heating the water (electrical immersion element, gas-fired burner, etc.). Hot water is drawn off the top of the tank with replacement cold water admitted to the bottom of the tank. *See also* **Circulating hot water system,** and **Hot water supply system.**

NONCOHESIVE SOIL Frictional soil: sand, gravel, etc.

NONCOINCIDENT (ELECTRIC) DEMAND *See* **Demand (electric).**

NONCOLLUSION AFFIDAVIT Notarized statement by a bidder that his bid has been prepared without collusion of any kind.

NONCOMBUSTIBLE Any material that will neither ignite nor actively support combustion in air at a temperature of 648°C (1,200°F) when exposed to fire.

NONCOMBUSTIBLE CONSTRUCTION Type of construction in which a degree of fire safety is attained by the use of noncombustible materials for structural members and other building assemblies.

NONCOMBUSTIBLE WASTE *See* **Solid waste.**

NONCOMMERCIAL FOREST LAND *See* Land use classies (forest management).

NONCOMMERCIAL SPECIES Tree species in which small size, poor form, or inferior quality is typical. These species do not normally develop into trees suitable for conventional forest products.

NONCOMPACT SECTION Steel section that can develop yield stress in compression elements before local buckling occurs, but that will not resist inelastic local buckling at strain levels required for a fully plastic stress distribution.

NONCONCORDANT TENDON In statically indeterminate structures, a tendon, the center of gravity of which is not coincident with the pressure line due to prestressing alone.

NONCONDUCTOR Material or substance that does not readily conduct electrical energy.

NONCONFORMING USE Use for which land or a building is lawfully occupied at the time a new regulation becomes effective but that does not comply with the new regulation.

NONCONFORMING WORK Work that does not meet the standards or requirements of regulations or specifications applicable to the site or application.

NONCONSTANT-MESH TRANSMISSION *See* **Transmission.**

NONCRITICAL ACTIVITIES OR WORK ITEMS Activities or work items that have positive float (i.e., within defined limits, can take longer to complete than is planned without affecting the overall project duration).

NONDEPLETABLE ENERGY SOURCE Renewable energy source, such as wind, solar energy, water power, etc.

NONDESTRUCTIVE TESTING Methods of examining the structure of manufactured objects to determine their freedom from flaws and capability of performing to their designed capacities that does not dam-

age them.

NONDIRECTIONAL TREAD Tread design of a tire that is equally effective in either direction of rotation.

NONDISPLACEMENT PILE *See* **Pile.**

NONDRYING Compound that neither dries nor skins on exposure to the atmosphere or the elements.

N1E *Abbreviation for:* nosed one edge.

NONELECTRIC-DELAY BLASTING CAP Blasting cap with an integral delay element in conjunction with, and capable of being detonated by, a detonation impulse or signal from miniaturized detonating cord.

NONEVAPORABLE WATER Water that is chemically combined during cement hydration. *See also* **Evaporable water.**

NONFAMILY HOUSEHOLD One person living alone in a dwelling, or a number of unrelated persons occupying one dwelling.

NONFERROUS Relating to metals not containing iron: aluminum, brass, and copper are examples.

NONFOREST LAND *See* **Land use classes (forest management).**

NONFLAMMABLE Materials that will not burn when exposed to a flame.

NONFLAMMABLE FLUID *See* **Fire resistant fluid.**

NONFOREST LAND Land not primarily intended for growing or supporting a forest.

NONFREE SWING Crane swing system in which the brake applies automatically with the centering of a control lever.

NONHABITABLE ROOM Any room in a dwelling that cannot be lived in, i.e., bathroom, laundry, lobby, stairway, boiler room, or other space for service or maintenance of the structure.

NONHYDRAULIC LIME High-calcium lime.

NONIONIC ASPHALT Emulsified asphalt in which the asphalt globules are neither electronegatively nor electropositively charged.

NONIONIZING RADIATION Amplification of light through ultraviolet and infrared rays by means of instruments such as lasers.

NON-LOAD-BEARING TILE *See* **Tile.**

NON-LOAD-BEARING WALL *See* **Wall.**

NONMANIPULATIVE JOINT Compression joint in pipework requiring no work other than cutting the ends of the pipe square.

NONMETALLIC MINERAL Mineral containing no metal, i.e., asphalt, clay, lime, coal, fossil gums, natural gas, etc.

NONMODULAR BRICK *See* **Brick.**

NONMOVEMENT JOINT Joint designed for minimal or no movement.

NON-NEWTONIAN BEHAVIOR Property possessed by some fluids and many plastic solids, including lubricating greases, of having a variable relationship between shear stress and rate of shear. *See also* **Newtonian behavior.**

NONPOSITIVE DISPLACEMENT PUMP *See* **Pump.**

NONPOWERED AXLE *See* **Axle.**

NONPRESSURE DRAINAGE Drainage system constructed to handle design capacities by gravity flow.

NONPRESSURE-TREATED TIMBER Application of wood preservative by brush, spray, dipping, steeping, etc.

NONPRESTRESSED REINFORCEMENT Reinforcing steel not subjected to either pretensioning or post-tensioning.

NONRECOVERABLE CREEP Residual or non-reversible deformation remaining in hardened concrete after removal of a sustained load.

NONRETURN VALVE *See* **Valve.**

NONRISING-STEM VALVE Valve where the stem does not rise when opening the valve.

NONROTATING CYLINDER *See* **Cylinder.**

NONSAG SEALANT Sealant formulation havng a consistency that will permit application in vertical joints without appreciable sagging or slumping at temperatures between 4°C and 38°C (40°F and 100°F).

NONSELF-SUPPORTING STEEL FRAME Steel framework that requires interaction with other elements to provide the required stability or resistance to wind and seismic forces.

NONSIMULTANEOUS PRESTRESSING Post-tensioning of tendons individually, rather than simultaneously.

NONSKID *See* **Nonslip.**

NONSLIP Surfaces specially treated or prepared to minimize pedestrian or vehicular slipping. *Also called* **Nonskid.**

NONSLIP CONCRETE Floor, pavement, or walkway of concrete, the surface of which has been roughened before final set either by (1) sprinkling fine particles of abrasive material thereon and then trowelling, or by swirling with either a coarse-bristled brush or a trowel, or (2), after final set, by acid etching, mechanically abrading, or grooving.

NONSLIP TILE Tile having greater nonslip characteristics due to an abrasive admixture, abrasive particles in the surface, grooves or patterns in the surface, or because of natural nonskid surface characteristics.

NONSOAP THICKENER Any of several specially treated or synthetic materials, excepting the metallic soaps, that can be either thermally or mechanically dispersed in liquid lubricants to form lubricating grease. *Also called* **Inorganic thickener, Organic thickener,** and **Synthetic thickener.**

NONSPIN DIFFERENTIAL *See* **Differential.**

NONSTAINING CEMENT Masonry cement that contains not more than a stipulated amount of water-soluble alkali, as measured by a stipulated test method.

nonstd *Abbreviation for:* nonstandard.

nonstn. *Abbreviation for:* nonstaining.

NONSTOCKED AREAS *See* **Stand size classes**

NONTELESCOPING MAST Mast in which the support member or members providing the guideways for vertical movement of the fork carriage of an industrial truck do not move vertically with respect to the truck.

NONTILTING MIXER Horizontal rotating drum mixer that charges, mixes, and discharges without tilting.

NONTIMBER RESOURCES Forest resources other than timber, such as recreation, aesthetics, wildlife, fish, forage, range, water, and soils.

NONUSE OF LAND Declaration by a public authority that land is not being used, or is disused, empowering it to levy a tax to enforce development, to clear unmarketable titles, to restrain owners of occupied dwellings from discontinuing their use, etc.

NONUSE, OR DISUSE *See* **Land-use control.**

NONVIBRATING WEIGHT Static weight of a vibratory roller measured at the drum(s) on the ground, minus the vibrating weight.

NONWARP Adhesive system that does not curl, distend, shrink, wrinkle, or cause the laminate to which it is applied to distort.

NONWOVEN MEDIUM Filter medium composed of a mater of fibers.

NOOK Three-sided recess in a room.

No. 1 COMMON *See* **Lumber grades.**

NOP *Abbreviation for:* not otherwise provided.

NOR DEVICE Control device that has its output in the logical 1 state if, and only if, all the control signals assume the logical 0 state. *See also* **NAND device,** and **NOT device.**

norm. *Abbreviation for:* normal.

NORMAL Load applied perpendicular to a member.

NORMAL CAPACITY *See* **Capacity.**

NORMAL CEMENT General-purpose portland cement.

NORMAL CONDITION One during which exposure to compressed air is limited to a single continuous working period followed by a single decompression in any given 24-hour period.

NORMAL CONSISTENCY 1. Degree of wetness exhibited by a freshly mixed concrete, mortar, or neat cement grout when the workability of the mixture is considered acceptable for the purpose at hand; **2.** Physical condition of neat cement paste as determined with the Vicat apparatus.

NORMAL CONSOLIDATION Condition that exists if a soil deposit has never been subjected to an effective stress greater than the existing overburden pressure, and if the deposit is completely consolidated under the existing overburden pressure.

NORMAL DISTRIBUTION Bell-shaped symmetrical curve used in statistical analysis to generalize the relative frequency of occurrence of events.

NORMAL-DUTY AIR CLEANER Engine air cleaner for applications where there is a relatively light concentration of dust.

NORMAL FAULT Steeply inclined rock fracture in

which two intersections of a faulted seam are separated at a width equal to the heave, with one section having moved relatively downward. *See also* **Gravity fault.**

NORMAL HAUL Haul, the cost of which has been included in the cost of excavation.

NORMALIZING Heating steel above the temperature used for annealing, then cooling it in still air at ambient temperature in preparation for further heat treatment.

NORMAL LAPSE RATE Rate of decrease of temperature upward through the atmosphere.

NORMALLY CLOSED CONTACT Electrical switch that completes a circuit when it is not energized or activated.

NORMALLY OPEN CONTACT Electrical switch that does not complete a circuit until it is energized or activated.

NORMAL POSITION Default position of a switch; its position prior to being operated.

NORMAL STRESS Stress component that is perpendicular to the plane on which the force is applied, designated tensile if the force is directed away from the plane and compressive if the force is directed toward the plane. *See also* **Stress.**

NORMAL THREAD ENGAGEMENT Amount of overlap necessary to insure a tight connection between threaded pipes and fittings.

NORMAL WEAR AND TEAR Physical depreciation resulting from age and normal use of a property.

NORMAL-WEIGHT AGGREGATE *See* **Aggregate.**

NORMAL-WEIGHT CONCRETE Concrete having a unit weight of approximately 68 kg/0.28 m³ (150 lb/ft³), made with normal-weight aggregates.

NORMAL-WEIGHT REFRACTORY CONCRETE Refractory concrete having a unit weight (bulk density) greater than 45 kg/0.28 m³ (100 lbs/ft³).

NORMAN BRICK *See* **Brick.**

NORTH BEND Yarding method where the mainline passes through a fall block, then connects to the carriage. This configuration allows side blocking and gives extra block purchase for lift.

NORTH-LIGHT ROOF Sloping roof with the ridge running approximately east-west, the north-facing slope of which is wholly or partially glazed. In the southern hemisphere it is the south-facing slope that is glazed.

nos. *Abbreviation for:* nose.

NOSE CONE Front housing on a rock drill hammer that retains the striker bar and contains the flushing swivel (in hammers with bypass-tube flushing).

NOSING 1. Half-round edge; **2.** Portion of a stair tread that extends beyond the face of the riser.

NOSING LINE Line touching the edges of the nosings on a stair.

NO-SLUMP CONCRETE Freshly mixed concrete exhibiting a slump of less than 6 mm (0.25 in.) *See also* **Negative-slump concrete.**

NOTARY PUBLIC Person qualified and authorized to draw up or attest to contracts, deeds, and mortgages, and before whom affidavits may be sworn and witnessed.

NOTCH 1. Crosswise rabbet at the end of a board; **2.** Groove cut in timber to receive another timber: **3.** Wedge-shaped piece cut out of a stump to prevent a guyline or block strap from lifting off.

NOTCH BRITTLENESS Tendency of a material to break at points where stress is concentrated.

NOTCHED TROWELS Trowels having notches cut in the working edge, used to apply bonding materials. The notches are sized, configured, and spaced according to the type of material to be applied.

NOTCH EFFECT Local increase in stress at the point where a member changes section at a sharp angle.

NOTCHER Steel fabricating machine that strips the flanges from the ends of rolled joists.

NOTCHING 1. Joining two pieces of lumber by cutting away one or both; **2.** Excavating by cutting a series of horizontal steps.

NOTCH PLATE 1. Horizontal metal plate cut to form V- or U-shaped notches and acting as a weir over which a contained body of water is permitted to discharge at a calculated rate; **2.** Small weir used in laboratory models of hydraulic structures.

NOT DEVICE Control device that has its output in the logical 1 state if, and only if, the control signal assumes the logical 0 state. The NOT device is a single NOR device. *See also* **NAND device,** and **NOR device.**

NOTHINGS An expenditure that is ignored for tax purposes.

No. 3 COMMON *See* **Lumber grades.**

NOTICE OF AWARD 1. Formal notification of the awarding of a contract to a successful bidder; **2.** Public notification of the company to which a contract has been awarded, usually as the result of competitive bidding for a publicly funded project.

NOTICE OF CHANGE *See* **Change order.**

NOTICE OF CLAIM Formal notification to a contractor and surety that a claim will be lodged against the protected entity.

NOTICE TO BIDDERS Formal notification to prospective bidders giving the bidding requirements and procedures.

NOTICE TO PROCEED Written notice to a contractor to begin contract work; when applicable, includes the date of beginning of contract time.

NOT SATISFACTORILY RESTOCKED Productive forest land that has been denuded and has failed partially or completely to regenerate naturally or to be artificially regenerated.

notvail. *Abbreviation for:* not available.

No. 2 COMMON *See* **Lumber grades.**

NOVELTY SIDING Siding having a decorated or scrolled lower edge.

noz. *Abbreviation for:* nozzle.

NOZZLE 1. Small mouthpiece or spout of a hose or pipe; **2.** Metal or rubber tip attached to the discharge end of a hose; **3.** Assembly of parts employed to atomize and deliver fuel to an engine.

NOZZLE-AND-HOLDER ASSEMBLY Complete apparatus that injects pressurized fuel into the combustion chamber of an engine.

NOZZLE END End of a hose in which both the inside and outside diameters are reduced.

NOZZLE LINER Replaceable rubber lining fitted into the nozzle tip to prevent abrasion of the interior surface of the nozzle from which a continuous stream of shotcrete is ejected at high velocity.

NOZZLEMAN Operator who manipulates the nozzle and controls placement of shotcrete.

NOZZLE VELOCITY Rate at which a material is ejected from a nozzle.

np. *Abbreviation for:* nameplate.

NP *Abbreviation for:* nameplate; national pipe thread.

NPA *Abbreviation for:* normal pressure angle.

NPS *Abbreviation for:* nominal pipe size.

NPSC *Abbreviation for:* national standard pipe straight coupling.

NPSH *Abbreviation for:* national standard pipe straight-hose coupling.

NPSI *Abbreviation for:* national standard pipe straight-internal dryseal.

NPSL *Abbreviation for:* national standard pipe straight-lock nut.

NPSM *Abbreviation for:* national standard pipe straight mechanical.

NPT *Abbreviation for:* national standard pipe tapered.

NPTF *Abbreviation for:* national standard pipe tapered fuel dryseal.

NPTR *Abbreviation for:* national standard pipe tapered-railroad.

n/r *Abbreviation for:* not replaceable.

nr *Abbreviation for:* near.

NR *Abbreviation for:* natural rubber; noise reduction; not recommended.

NRC *Abbreviation for:* noise reduction coefficient.

NRP *Abbreviation for:* nonremovable pin.

NS *Abbreviation for:* narrow stile; not specified.

N-S *Abbreviation for:* north-south.

NSI *Abbreviation for:* nonstandard item.

NSR *Abbreviation for:* not satisfactorily restocked.

NT. *Abbreviation for:* nontight; normalized and tempered.

NTS *Abbreviation for:* not to scale.

N2E *Abbreviation for:* nosed two edges.

nt wt *Abbreviation for:* net weight.

nucl. *Abbreviation for:* nuclear.

NUCLEAR DENSITY GAUGE Highly accurate device used to measure the density of asphalt pavement at the job site during and after compaction.

NUDGING System for automatic door operation which will, if the door(s) remains open longer than a preset time, continuously sound a warning signal and close the doors at a reduced speed and torque until the obstruction preventing the door(s) from closing is removed.

NU-LINER Liquid of very thin viscosity used in both tubeless and tube-type tires to seal porosity and rim leaks, reduce liner oxidation, and decrease tire operating temperature.

NULL Zero set on an automatic sensing mechanism.

num. *Abbreviation for:* numeral; numerator.

NUMBER ONE Top-grade log, such as number-one peeler, number-one saw log, etc.

NURSE TANKER Water tank truck used to supply a pumper stationed at a forest fire.

NUT Internally threaded component that can be mated with an externally threaded member, such as a bolt, and rotated to produce a high pressure between the nut and bolt bearing surfaces. There are many types, including:

> **Acorn:** Nut with closed, rounded end to conceal mating threads.
>
> **Anchor (fixed):** Cylindrical, nonwrenching nut with integral ears by which it is either riveted or welded to the structure.
>
> **Anchor (floating):** Nonwrenching nut held loosly within a cage or anchor plate welded or riveted to the structure.
>
> **Barrel:** Nut in which the thread is located in the center of, and perpendicular to the axis of, a barrel-shaped piece of material.
>
> **Blind:** One-piece internally threaded hollow rivet.
>
> **Captive:** Nut, usually with a splined grip, that can be pressed into an undersized hole.
>
> **Castle:** Nut, not normally self-locking, but having four or more transverse slots laterally in its tail section designed to anchor a cotter pin or lock wire.
>
> **Jam:** Low height, free-spinning nut.
>
> **Lock:** Nut screwed down tightly on another nut, preventing the first nut from jarring loose.
>
> **Spline:** Self-broaching captive nut used in limited-access areas.
>
> **Wing:** Nut with two opposed wings by which it can be turned by hand.
>
> **Wrenchable:** Six- or twelve-sided nut with, or without a captive washer in its base.

NUT SPLITTER Ring mounted on a shaft containing a threaded rod having a hex drive on its exposed end and a sharpened cutting blade on the other. Placing the ring over a nut and turning the hex head with a wrench causes the blade to split the nut.

N VALUE Number of blows required to drive a 50-mm (2-in.) OD, 35-mm (1.375-in.) ID, 600-mm (24-in.) long, split soil-sampling spoon 300 mm (12 in.)

with a 63.5 kg (140 lb) weight freely falling 750 mm (30 in.). The count is recorded for each of three 150-mm (6-in.) increments. The sum of the second and third increments is taken as the N value in blows per foot. *See also* **Blow count,** and **Driller's stroke.**

NW *Abbreviation for:* northwest; no weathering exposure.

NX CORE Rock core taken with a 60-mm (2.375-in.) diameter 'NX' core barrel.

nyl. *Abbreviation for:* nylon.

NYLON Generic term for all synthetic polyamides.

oa *Abbreviation for:* on account.

OA *Abbreviation for:* overall; oxyacetylene.

O/A *Abbreviation for:* on approval; on or about.

OAKITE Acid used to clean grease and dirt from metal parts.

OAKUM Treated hemp used for caulking joints.

oal.h *Abbreviation for:* overall lowered height.

O&M *Abbreviation for:* operation and maintenance.

OAW *Abbreviation for:* overall width.

OBELISK Tall, tapering shaft of square or rectangular section, often ending in a pyramid.

obj. *Abbreviation for:* object; objective.

OBJECTIVE LENS Lens of a telescope or other optical instrument that is nearest the object; the lens that receives light from the object and forms the first or primary image.

oblgn *Abbreviation for:* obligation.

OBLIGEE Entity or individual to whom a bond is given; in the context of construction, usually the project owner.

OBLIGOR Person who is bound to another; one who makes a bond.

OBLIQUE BUTT JOINT Butt joint at an angle other than 90°.

OBLIQUE GRAIN Diagonal grain.

OBLIQUE OFFSET Distance from a survey line taken at an angle other than 90°.

OBLIQUE PHOTOGRAPH Aerial photograph taken with the camera axis inclined away from a plane vertical to the surface of the ground.

OBLIQUE PROJECTION Pictorial view of an object showing its elevation or plan to scale with parallel lines projected from the corners to indicate the other side.

OBM *Abbreviation for:* ordinance bench mark.

obs. *Abbreviation for:* observation; observed; obsolete.

OBS *Abbreviation for:* open back strike.

obsc. *Abbreviation for:* obscure.

OBSCURING PROCESSES Processes such as sandblasting and acid embossing that obscure vision through glass to varying degrees, increasing the light-diffusing properties of the material.

OBSERVATION ELEVATOR *See* **Elevator.**

OBSERVATION WELL Perforated pipe installed in the ground for monitoring groundwater levels.

OBSERVATORY 1. Room with a lantern on the upper level of a building; **2.** Building specifically for astronomical observations.

OBSERVED POWER Power actually developed by an engine under the atmospheric conditions existing during the test.

OBSIDIAN Natural volcanic glass of relatively low water content.

OBSOLESCENCE 1. Changes in methods or design that make a structure or machinery that predates the changes less valuable or less desirable for the task for which it was acquired (a factor in calculations for depreciation); **2.** In work methods and techniques, development of new methods that make the old ones less efficient, productive, or safe in comparison.

obst. *Abbreviation for:* obstacle.

obstr. *Abbreviation for:* obstruction.

OBSTRUCTION LIGHT Red flashing, or strobe, light that indicates the presence of a fixed object dangerous to aircraft in motion.

obt. *Abbreviation for:* obtain; obtuse.

obv. *Abbreviation for:* obverse.

oc *Abbreviation for:* off center

OC *Abbreviation for:* on center.

OCCUPANCY Use or intended use of a building or part of a building for the shelter or support of persons, animals, or property. *See also* **Major occupancy.**

OCCUPANCY (DWELLING) Stage in the development of a dwelling when it has progressed to the point where it meets all pertinent codes and regulations regarding construction, safety, and health.

OCCUPANCY RATE Relationship of the number of occupants to the number of rooms and the number of dwellings.

OCCUPANCY STANDARD Standard relating to the social composition of a household, i.e., residence of an unrelated roomer, minimum number of bedrooms, minimum size in square meters (square feet), etc.

OCCUPANT Person residing in a dwelling who either owns the property or who is the leaseholder or who pays rent.

OCCUPANT LOAD Number of persons for which a building or part of a building has been designed.

OCCURRENCE Accidental event or condition.

oct. *Abbreviation for:* octagon; octane.

OCTANE NUMBER Percent of isooctane by volume in a mixture of isooctane and normal heptane that has the same antiknock character in a standard variable-compression test engine as the fuel under test. A mixture having 75% octane and 25% heptane is said to have an octane rating of 75.

OCULAR *See* **Eyepiece.**

OCULUS Circular opening occasionally formed at the top of a dome.

OD *Abbreviation for:* outside diameter; outside dimension; overall diameter; overdrive.

ODOMETER Instrument for recording distance covered by a moving vehicle.

odorl. *Abbreviation for:* odorless.

ODOR THRESHOLD Lowest concentration of a vapor that can be detected by smell.

OE *Abbreviation for:* original equipment.

OEM *Abbreviation for:* original equipment manufacturer.

OF *Abbreviation for:* outside face; oxygen-free.

off. *Abbreviation for:* office.

OFF-CENTER *See* **Eccentricity.**

offen. *Abbreviation for:* offensive.

OFF-GAS Gaseous products from chemical decomposition of a material.

OFF GAUGE Not conforming to a specified thickness.

OFF GRADE Bonded abrasive materials that are not of exact grade.

OFFHAND GRINDING Where the work is held in the operator's hand. Also known as **Freehand grinding,** and **Hand grinding.**

OFF HIGHWAY Vehicle operation over private roads of asphalt or maintained crushed rock surface or similar material at variable grades. Not subject to legal weight and dimensional limitations. *See also* **Off road.**

OFF-HIGHWAY SERVICE Operation exclusively upon private roads or job sites, typically involving unprepared surfaces and steep grades. Not subject to federal and/or state or provincial weight and length regulations.

OFF-HIGHWAY TRUCK *See* **Truck.**

OFFICE Space within a building primarily used for administrative or clerical work.

offl *Abbreviation for:* official.

OFF-PEAK ELECTRICAL HEATING *See* **Heating system.**

OFF-PREMISES SIGN Outdoor sign, display, or device advertising a service or product at a location other than on the property where such service or product may be obtained. *See also* **On-premises sign.**

OFF ROAD Slow-speed operations over uneven surfaces. Not subject to legal weight and dimensional limitations. *See also* **Off highway.**

OFFSET 1. Abrupt change in alignment or dimension, either horizontally or vertically. *See also* **Lateral deflection,** and **Longitudinal centerline; 2.** Something that is set off, or has sprung or developed, from something else, i.e., a ledge or recess formed in a wall by reducing its thickness; **3.** In plumbing, a combination of elbows or bends that brings one section of a pipe out of line but parallel with the other section; **4.** Horizontal distance measured at right angles to a survey line.

OFFSET ANGLE Angle between the longitudinal centerline of a jib and the longitudinal centerline of the boom on which it is mounted.

OFFSET BASE *See* **Longitudinal centerline.**

OFFSET BEND 1. Intentional distortion from the normal straightness of a steel reinforcing bar in order to move the center line of a segment of the bar to a position parallel to the original position of the center line; **2.** Mechanical operation commonly applied to vertical bars that reinforce concrete columns.

OFFSET DIGGING Digging with the boom of a ladder ditcher not centered in the machine.

OFFSET FITTING Pipe fitting having two changes of direction, one offsetting the other.

OFFSET FLY Fly section that is capable of being pinned at differing angles.

OFFSET SCREWDRIVER Screwdriver, the blade of which works at right angles to the handle, used in confined places.

OFFSET WOODRUFF KEY Woodruff key that is offset so that the gear it turns on a shaft allows the shaft to be retarded or advanced a set amount.

OFFSET YIELD STRENGTH Stress at which the strain exceeds, by a specified amount, an extension of the initially proportional part of the stress-strain curve, expressed as a percentage of the original gauge length in conjunction with the strength value, or as force per unit area.

OFFSHORE LEAD *See* **Lead.**

OFFSIDE 1. Side of a tree opposite the side where the faller stands when falling or bucking; **2.** Side of the body opposite to that normally used to hold a chain saw.

OFF-SITE COST Construction-related expenditure incurred away from the construction site.

OFF-SITE IMPROVEMENT Work that is necessary for the completion of a subdivision or development but that is not directly on the lots to be sold.

OFF-THE-ROAD TIRE Tire designed primarily for use over unpaved roads, or where no road exists. Built for ruggedness and traction rather than speed.

Og. *Abbreviation for:* ogee.

OG *Abbreviation for:* on grade.

OGEE Profile made up of intersecting concave and convex curves. *See also* **Arch,** and **Spillway.**

OH *Abbreviation for:* oval head; overhead.

O/H *Abbreviation for:* overhaul.

oh.g *Abbreviation for:* overhead guard.

OHM One of 17 derived units with special names of the SI system of measurement: a unit of electric resistance between two points of a conductor when a constant difference of potential of 1 V applied between these two points produces in this conductor a current of 1 A, the conductor not being the source of any electromotive force. Symbol: Ω ($1\Omega = 1$ V/A). *See also the appendix:* **Metric and nonmetric measurement.**

OHMMETER Meter used to measure electrical resistance.

OHMS-PER-VOLT Rating showing the sensitivity and accuracy of an electric meter.

OI *Abbreviation for:* operating instructions.

OIL Fluid lubricant; a liquid petroleum derivative less volatile than gasoline.

OIL-BASE PAINT Paint containing resins and other ingredients derived from various natural and synthetic oils.

OIL-BOUND DISTEMPER Distemper containing a drying oil.

OIL BUFFER Type of buffer, for elevators with speeds of more than 1.02 m/s (200 fpm), that uses a combination of oil and springs to cushion the elevator. It is located in the elevator pit.

OIL-BUFFER STROKE Amount an oil-displacing buffer, plunger, or piston moves under load, excluding the travel of the buffer-plunger accelerating device.

OIL CHECK Device that uses oil as the control medium to dampen a sudden stop.

OIL CLUTCH *See* **Clutch.**

OIL COOLER Heat exchanger used to remove heat from hydraulic fluid or engine oil.

OILED AND EDGE-SEALED Concrete form panels that are oiled and edge-sealed.

OILER Device for injecting oil in air or steam lines. *Also called* a **Lubricator.** *See also* **Line oiler,** and **Sight-feed lubricator.**

OILFIELD BODY *See* **Truck.**

OILFIELD FLOAT *See* **Trailer.**

OILFIELD WINCH *See* **Winch.**

OIL FIRED FURNACE Furnace designed to burn oil.

OIL GALLERY Drilled or cored passages used to pipe oil to various engine parts.

OIL IMMERSED Object or part wholly covered by oil at all times.

OIL IMMERSION HEATER Device used to heat engine lubricating oil.

OILINESS Property of an oil that causes a difference in friction when lubricants of the same viscosity, at the same temperature and pressure on the film, are used with the bearing surfaces.

OILINESS AGENT Material that reduces friction by formation of an adsorbed film.

OIL LINE Pipe that channels oil from a reservoir to a point of work or application, or from a power unit to a jack.

OIL POT Small wheeled, towable tank equipped with an air compressor, used to supply road oil under pressure to hand-held spray nozzles.

OIL PROOF Not affected by exposure to oil.

OIL PUMP 1. Mechanism that pumps oil from an engine's through galleries, from whence it is distributed to lubricate the engine's moving parts; **2.** Oil pump placed on single- or tandem-drive axles to permit extra flow of filtered lubricant in an axle to lubri-

cate the inter-axle differential assembly.

OIL RING Solid ring about a piston, used to distribute lubricating oil evenly throughout the traveled portion of a cylinder wall.

OIL SEPARATOR Device for separating oil from water, or oil from vapor.

OIL SLINGER Metal disk that fits over the end of a camshaft or crankshaft tp prevent oil from overloading a seal.

OIL STAIN Thin oil paint containing very little pigment, used to stain wood.

OILSTONE Natural or manufactured abrasive stone impregnated with oil and used for sharpening keen-edged tools.

OIL SWELL Change in volume of a rubber article resulting from contact with oil.

OIL VARNISH *See* **Varnish.**

OILWELL CEMENT Hydraulic cement suitable for use under high pressure and temperature in sealing water and gas pockets and setting casing during the drilling and repair of wells, often containing retarders to meet the requirements of use.

OITICICA OIL Vehicle used in the manufacture of enamels and varnishes.

OLD GROWTH Timber in or from a mature, naturally established forest.

OLD MAN Jig fastened to provide leverage for backing of a powered drill.

OLEORESINOUS VARNISH Varnish containing vegetable drying oil and natural or synthetic hardening resin.

ON CENTER 1. Term used when taking measurements; **2.** Distance from the center of one structural member to the center of a corresponding member. Same as **Center-to-center.**

ONE Binary value for high, or true. *See also* **Zero.**

1C *See* **Lumber grades.**

ONE-FLOOR-RUN CIRCUIT Special circuit used when an elevator must run from one floor to another wherein the distance is so short that acceleration and deceleration might overlap.

ONE-HUNDRED-YEAR FLOOD Flood that has a 1% chance of occurring in any given year, based on historical records for a given location. *See also* **Fifty-year flood.**

ONE:ONE ROPING Elevator hoist ropes arranged so that one end of each hoist rope passes from the car hitch over the machine sheave to the counterweight hitch.

ONE-PART LINE *See* **Parts of line.**

ONE-PART SEALANT Chemically curing sealant containing a reactive polymer base that cures upon exposure to the air and/or humidity. It requires no mixing.

ONE-PIPE SYSTEM 1. In drainage, two vertical pipes with waste and soil water flowing down the same pipe, and all branches connected to the same antisiphon pipe; **2.** Heating circuit in which flow and return connections to the radiators connect to the same pipe.

ONE-TOUCH DECELERATION/ACCELERA-TION Machine control system in which, by the touch of a button, an operator can reduce engine speed to a preselected level, returning to full power by again touching the button. *See also* **Fine inching control.**

ONE:TWO (1:2) Slope where the elevation rises one unit in two horizontal units.

ONE-WAY *See* **Snow plow.**

ONE-WAY AUTOMATIC ELEVATOR LEVELING *See* **Elevator leveling device.**

ONE-WAY FLOOR AND ROOF SYSTEM One of the two major reinforced concrete floor and roof systems of design and construction. The one-way system consists of a solid slab supported by reinforced beams running in one direction that, in turn, are supported by columns. The supporting beams may or may not be integral with the slab. *See also* **One-way-joist floor.** *Also called* **Pan floor.**

ONE-WAY-JOIST FLOOR A slab, supported by reinforced concrete beams, made up of narrow reinforced concrete beams or joists, joined together at their upper edge by a thin web. The floor is cast on a formwork of steel pans preformed to the specific shape as a series of troughs. The floor and joists are poured together to a form a monolithic whole. A variation involves the use of precast, and often pretensioned concrete units that span between the structural beams and that are capped with a thin, reinforced concrete slab to form the finished floor. *See also* **One-way floor and roof system.**

ONE-WAY RAM *See* **Ram.**

ONE-WAY SLAB Concrete slab with reinforcing bars providing a bearing on two opposite sides only.

ON-HIGHWAY SERVICE Vehicle operation over hard-surfaced or graded roads. Subject to federal and/or state or provincial weight and length regulations.

ON-HIGHWAY TRUCK *See* **Truck.**

ON LINE REPLACEMENT The breaking out of an existing service and the installation of a new conduit in the same place.

ON/OFF CONTROL Control that turns the input on or off, but does no proportioning or throttling, as is the case with a modulating control.

ON/OFF-HIGHWAY TRUCK *See* **Truck.**

ON/OFF VALVE *See* **Valve.**

ON-PREMISES SIGN Outdoor sign, display, or device advertising activities conducted on the property on which it is located or the sale or lease of that property. *See also* **Off-premises sign.**

ON-RUBBER CAPACITY Load that a wheeled crane can lift in various stationary or static configurations without outriggers positioned.

ON-SITE DISPOSAL Includes all means of disposal, but more commonly the volume reduction of refuse on the premises before removal for ultimate disposal.

ON-SITE INCINERATOR *See* **Incinerator.**

ON-THE-FLY Any activity done while in motion on mobile equipment.

o-o *Abbreviation for:* out-to-out.

O-1 WAFERBOARD Panel where the wafers on the face and the back are oriented in the long direction of the panel.

OP *Abbreviation for:* oilproof.

OPACITY Degree of hiding power of a media. *Also called* **Hiding power.**

OPALESCENCE Characteristic of an otherwise transparent or translucent sheet or film to refract light into a limited range of colors.

OPAQUE Impervious to radiant energy, particularly light energy.

OPAQUE GLAZE *See* **Glaze.**

OPEN 1. Not closed, covered, or clogged; **2.** Electrical circuit through which electrical energy will not pass; **3.** Electrical switch in which the contacts are separated; **4.** Braking system in which there is sufficient clearance between the shoe(s) or pad(s) so as not to impede motion.

OPEN ASSEMBLY TIME Interval between application of an adhesive film and completion of an assembly.

OPEN BIDDING/TENDERING System of public advertising of construction and material supply requirements. *Also called* **Open tendering.**

OPEN BUBBLE POINT Differential gas pressure at which gas bubbles are profusely emitted from the entire surface of a wetted filter element under specified test conditions. *Also called* **Boil point.**

OPEN CAISSON Caisson in the form of a tube or drop shaft, open at both top and bottom.

OPEN-CENTER CIRCUIT Hydraulic circuit where the pump continuously circulates fluid through the control valves when they are in a neutral position.

OPEN-CENTER VALVE *See* **Valve.**

OPEN-CIRCUIT CRUSHING Crushing system in which material passes through the crusher without recycling of oversize particles.

OPEN-CIRCUIT GROUTING Grouting system with no provision for recirculation of grout to the pump.

OPEN-CIRCUIT VOLTAGE Voltage existing when no load is attached to the voltage source, such as a generator.

OPEN CORNICE Eaves overhang in which the rafter soffits and roof covering (and sheathing) can be seen.

OPEN CUT Excavation where the working area is kept open, as against cut-and-cover and underground work.

OPEN-CUT TRENCHING Traditional method for conduit installation involving the excavation of a trench, laying the conduit, refilling, and reinstating the surface.

OPEN-CYCLE GAS-TURBINE ENGINE Gas turbine engine in which the working fluid enters the engine from the atmosphere and is discharged to the atmosphere.

OPEN DEFECT Irregularity such as a split, an open joint, or knothole that interrupts the smooth continuity of a veneer.

OPEN-END BLOCK Concrete block with an end web removed for placing the block around steel reinforcement.

OPEN-END MORTGAGE *See* **Mortgage.**

OPEN-END WRENCH *See* **Wrench.**

OPEN FLOOR Floor construction in which the underside of the joists are left exposed.

OPEN GRADED Asphalt mix, often used for lower quality types of construction, in which the aggregate gradation has not been as closely controlled as with dense-graded mixes.

OPEN-GRADED AGGREGATE *See* **Aggregate.**

OPEN-GRAINED WOOD *See* **Grain.**

OPENING LINE Line used to open a grapple, clamshell, or orange-peel bucket.

OPEN LISTING Contract to represent real property for sale made with any number of brokers without liability to compensate any except the one who first secures a qualified buyer.

OPENING OF TENDERS *See* **Bid opening.**

OPEN MORTGAGE *See* **Mortgage.**

OPEN PEDIMENT *See* **Pediment.**

OPEN PIT Surface operation for the mining of metallic ores, coal, clay, aggregates, etc.

OPEN-PIT INCINERATOR *See* **Incinerator.**

OPEN PLAN Interior design of dwellings with a minimum of walls and doorways between living areas.

OPEN ROOF Roof construction in which the beams and any ties of trusses can be seen from below. *Also called* **Cathedral ceiling.**

OPEN SAW One that is stretched too much in the inner area for the speed at which it is running. *Also called* **Dished saw.**

OPEN SANDWICH-TYPE PANEL Sandwich panel with top and bottom edges closed.

OPEN SEAM Seam whose edges do not meet, creating a void.

OPEN-SIDE CARRIAGE Skyline carriage that is open on one side, allowing it to travel over intermediate support jacks.

OPEN SIDES AND ENDS Edges of a platform along which no guardrail system is erected.

OPEN SLATING Slates or tiles laid with a gap between those in the same course.

OPEN SOCKET A wire rope fitting that consists of a basket and two ears with a pin.

OPEN SPACE Portion of a lot not occupied by buildings or structures and available to all occupants of the building.

OPEN SPLICE A retread tire defect caused by failure of the rubber to knit together properly at the tread splice.

OPEN SPOOL VALVE *See* **Valve.**

OPEN STAIRWAY Stairway that is not separated by walls and partitions from other areas in the building, including hallways.

OPEN STEAM CURE Method of vulcanizing in which steam comes in direct contact with the product being cured.

OPEN-STRING STAIR Stairs with the ends of the treads visible from the side, sometimes constructed without risers.

OPEN SYSTEM Construction system having interchangeability of subsystems, components, or building elements.

OPEN TENDERING *See* **Open bidding/tendering.**

OPEN TIME Period during which a bond coat retains its ability to adhere to a tile and bond the tile to a substrate.

OPEN TOP *See* **Open top truck.**

OPEN-TOP MIXER Truck-mounted mixer consisting of a trough or a segment of a cylindrical mixing compartment within which paddles or blades rotate about the horizontal axis of the trough. *Also called* **Pan mixer.**

OPEN-TOP VAN Van that can be loaded from above and that is often covered by a tarp or mesh.

OPEN TRAVERSE Survey traverse that begins at a determined station but does not end at that station.

OPEN TYPE (BRIDGE SOCKET) *See* **Bridge socket.**

OPEN VALLEY Valley in a hipped roof where the shingles or tiles intersecting the slopes leave an open space covered by a metal flashing.

OPEN-WEB STEEL JOIST Lightweight fabrication of steel members, usually consisting of parallel top and bottom chords of steel angles held apart by diagonally placed round steel bars.

OPEN-WELL STAIR Stair that returns through 180° with a significant clearance between the flights, creating an open well.

oper. *Abbreviation for:* operate; operation; operative.

OPERATING BAND Range of pressures above and below the operating pressure within which it is desired to keep a supply output.

OPERATING COSTS Costs that result when a machine is being used, i.e., fuel, repair, wear items, operator wages, etc. *See also* **Variable costs.**

OPERATING DEVICE Push button, lever, or other manual device used to actuate a control.

OPERATING ENGINEER Person certified to operate static machinery.

OPERATING INSTRUCTIONS Details of how something should be operated under specific conditions, a condition of warranty protection.

OPERATING INTERVAL Average time between calling for an elevator and its arrival at that landing.

OPERATING LEASE *See* **Lease.**

OPERATING LOAD Load not exceeding 50% of the full turn static tipping load for wheel loaders, and not exceeding 35% for track loaders.

OPERATING PRESSURE *See* **Pressure.**

OPERATING RADIUS *See* **Load radius.**

OPERATING SECTOR Portion of a horizontal circle about the axis of rotation of mobile equipment providing the limit of a zone where over-the-side, over-the-rear, or over-the-front ratings are applicable.

OPERATING SPEED Optimum specified speed of a machine or engine, usually designated by the manufacturer, but also dependant on any fitted accessories or attachments.

OPERATING SPEED RANGE Speed range of a machine while performing any of the functions for which it is designed, under conditions and in a manner described by its speciifcation and/or operating manual: typically a cold planing/milling machine.

OPERATING STRESS Stress to which a structural unit or component is subjected in service.

OPERATING TEMPERATURE RANGE Ambient temperature range over which a mechanical or measuring system can operate and meet its specified operating performance.

OPERATING TIME *See* **Machine time.**

OPERATING WEIGHT Gross machine weight with full mechanical operating systems, plus a tank full of fuel, plus a half tank of water (if so equipped), plus an 80-kg (175-lb) operator.

OPERATION System by which an elevator responds to passenger requests, comprising:

> **Constant pressure:** A very simple form of elevator operation where the elevator runs only as long as a button is pushed, and can handle only one call at a time.

> **Full automatic:** *See* **Momentary-pressure,** below.

> **Momentary-pressure:** A simple form of elevator operation that accepts only one call at a time, remembers that call, and dispatches the car in the proper direction. *Also called* **Full automatic.**

> **Selective collective:** A form of elevator operation wherein the system accepts and remembers an infinite number of calls and answers them as the car moves in the appropriate direction.

OPERATIONAL CRUISE Estimate, to a specified degree of accuracy, of the volume of timber on an area to be harvested.

OPERATIONAL DELAY Delay caused by interference between components of traffic. *See also* **Delay,** and **Fixed delay.**

OPERATIONAL LOST TIME *See* **Machine time.**

OPERATIONAL STRETCH Stretching of a wire rope caused by compression of its core and embedment of the strands when supporting weights up to its design maximum.

OPERATIONAL TEST *See* **Integrity test.**

OPERATION SELECTOR SWITCH *See* **Switch.**

OPERATIONS RESEARCH Scientific approach to decision making that involves the operations of organizational systems.

OPERATOR 1. One whose work is to operate a machine; **2.** Mechanism that actuates another device.

OPERATOR PLATFORM Platform or area from which a standing person controls the functions of a machine.

OPERATOR PROTECTIVE STRUCTURE Equipment for mobile equipment that protects the operator from falling objects and if the machine rolls over. *See also* **Roll-over protective structure,** and **Falling object protective structure.**

OPERATOR'S CAB Housing that covers the operator's station. *Also called* **Upper cab.**

OPERATOR'S MANUAL Book that accompanies equipment, a machine, or mechanical tools that lays out operating conditions, methods, sequences and limits, gives maintenance schedules and procedures, describes recommended safety measures, and lists guarantee or warranty coverage.

opn *Abbreviation for:* open.

opng *Abbreviation for:* opening.

opns *Abbreviation for:* operations.

opp. *Abbreviation for:* opposite.

opr. *Abbreviation for:* operate.

OPS *Abbreviation for:* operator protective structure.

opt. *Abbreviation for:* optical; option; optional.

OPTICAL COATING Substance applied to glass and other transparent materials used in solar heating systems to increase their transmission of sunlight.

OPTICAL DENSITY Method of expressing degrees of contamination of a fluid by removing contaminants through filtration and measuring change in optical transmission of the filter disk or fluid, or both.

OPTICAL LOSS Loss resulting from solar radiation reflected from the surface of a cover plate.

OPTICAL PROPERTY Amount of transmission of visible ultraviolet light.

OPTIMUM Best obtainable; highest quality.

OPTIMUM AIR SUPPLY Quantity of air that will give the greatest thermal efficiency under actual conditions. With perfect mixing of fuel and air, the optimum air supply is equal to the chemically correct amount of air.

OPTIMUM CURE State of cure when a rubber compound exhibits the best physical properties. Usually expressed in minutes curing time at a specified temperature.

OPTIMUM MOISTURE CONTENT Moisture content in percent by weight of dry rock and earth that results in the least voids and greatest density when the material is compacted.

OPTIMUM REVERBERATION TIME Reverberation time for any given area that will yield the best acoustical conditions for the intended use.

OPTIMUM ROAD SPACING Distance between parallel forest roads that gives the lowest combined cost of

skidding and road construction costs per unit of log volume.

OPTION Right, given for consideration, to purchase or lease a property upon specified terms within a defined period. There are several types, including:

Bargain purchase: Lease provision allowing the lesssee, at his option, to purchase the leased property for a price that is sufficiently lower than the expected fair value of the property, at the date the option becomes exercisable.

Bargain renewal: Lease provision allowing the lesssee, at his option, to renew the lease for a rental that is sufficiently lower than the expected fair rental of the property at the date the option becomes exercisable.

Call: Option giving the right to demand delivery, within a specified time, of a stock or commodity at a specified price and in specified amounts.

Put: Option giving the right to make delivery within a specified time of a particular stock or commodity at a specified price and in specified amounts.

Stock: Right granted by a limited company to purchase specified numbers of shares of the company's capital stock at a stated price within a stated period of time.

See also **Acquisition, Conveyance, Dedication, Eminent domain, Expropriation, Negotiation, Remainder,** and **Severality.**

OPTIONAL FEATURE Any substitution for the standard attachment accessories or components, or any deviation from standard operating methods.

OR *Abbreviation for:* outside radius.

ORANGE-PEEL Unintentionally rough surface.

ORATORY Small, private chapel, usually in a house.

ORBITAL SANDER *See* **Sander.**

ord. *Abbreviation for:* order; ordinary.

ORDER 1. Column with base, shaft, capital, and entablature, ornamented and proportioned according to one of the five accepted orders: **Tuscan, Doric** (Greek and Roman; the Greek Doric lacks a base), **Ionic, Corinthian,** and **Composite,** each of which is superior to the preceding in height, lightness, and decoration of pillar and capital; **2.** Series of concentric steps receding towards the opening of a doorway or window; **3.** An approval, decision, determination, permit, or the exercise of discretion made under the terms of a regulatory code.

ORDER LENGTH Lengths of pile expected for a project, ordered by either the owner, architect, engineer, or contractor from the supplier.

ORDER PICKER TRUCK, HIGH LIFT High-lift truck controllable by the operator stationed on a platform movable with the load-engaging means and intended for (manual) stock selection. The truck may be capable of self-loading and/or tiering.

OR DEVICE Control device that has its output in the logical 0 state if, and only if, the control signals assume the logical 0 state.

ORDINARY HAZARD *See* **Fire hazard.**

ORE Mineral, or association of minerals that, under favorable circumstances, may be worked commercially.

OREGON BLOCK Using a stump instead of a block to change the direction of a line.

OR EQUAL *See* **Approved equal.**

orf. *Abbreviation for:* orifice.

org. *Abbreviation for:* organic; organization.

ORGANIC ADHESIVE *See* **Adhesive.**

ORGANIC BOND Bond made of organic materials, such as synthetic resins, rubber, or shellac.

ORGANIC CONTENT Synonymous with volatile solids, except for small traces of some inorganic materials such as calcium carbonate, that lose weight at temperatures used in determining volatile solids.

ORGANIC ESTER SYNTHETIC FLUID Fluid composed of esters that are compounds of carbon, hydrogen, and oxygen only. It may contain additives.

ORGANIC FELT Combination of organic shredded wood fibers and felted papers that is saturated with asphalt. *See also* **Built-up roof, Glass-fiber felt,** and **Tarred felt.**

ORGANIC FRICTION MATERIAL Nonasbestos substance that is used on clutch disks. Organic friction materials are softer and smoother than most ceramic facings, resulting in smoother, softer clutch engagement.

ORGANIC MASTIC Premixed petroleum- or latex-based tilesetting adhesive.

ORGANIC PIGMENT Pigment usually characterized by brightness and brilliance (though not so permanent as inorganic pigment). Organic pigments are divided into two classes: toners and lakes.

ORGANIC SILT Mineral particles ranging in size from 0.05 to 0.074 mm (0.002 to 0.003 in.) containing appreciable quantities of organic materials.

ORGANIC SOIL Soil containing an appreciable content of decayed vegetable matter.

ORGANIC THICKENER *See* **Nonsoap thickener.**

ORGANIZATION CHART Graphic representation of the functional and administrative relationships, lines of authority, and responsibility within an organization.

ORGANOSOL Suspension of finely divided resin in a volatile organic liquid. The resin dissolves mainly at elevated temperatures. At that point the liquid evaporates and the residue becomes a homogeneous plastic mass when cool. Plasticizer can be dissolved in the volatile liquid.

ORIEL WINDOW Window, or group of windows that projects beyond the wall of a building and is usually carried on brackets or corbels.

orien. *Abbreviation for:* orientation.

ORIENT Location of the principal face of a building by compass points.

ORIENTATION 1. Planning, in relationship to the sun's angle; 2. Direction of wafers in a panel; 3. Solar collector panel position with relation to the points of the compass.

ORIENTED-CORE BARREL Instrument that marks a core sample to show its orientation, simultaneously recording the bearing and slope of the test hole.

ORIENTED STRAND BOARD Waferboard with oriented face and back wafers, made of wafers whose length is at least twice their width. *Also called* **OSB.**

ORIFICE 1. Literally, an opening, but more commonly used to designate a constriction in a passage, circular in shape unless otherwise specified. **2.** Small hole in the tip of a nozzle.

ORIFICE METER Pierced plate placed across a pipeline to create a pressure differential, the amount of which is related to the flow through the pipe.

ORIFICE RESTRICTOR Restrictor, the length of which is relatively small with respect to its cross-sectional area. The orifice may be fixed or variable. Variable types are noncompensated, pressure-compensated, or pressure- and temperature-compensated.

orig. *Abbreviation for:* origin; original.

ORIGINAL EQUIPMENT References a part, component, or assembly that was made by the manufacturer of the equipment itself.

O-RING Pliable seal used around shafts and stems, circular in cross section.

orn. *Abbreviation for:* ornament; ornamental.

ORNAMENT Detail added for the purpose of embellishment.

ORNAMENTATION In masonry, a design or pattern formed by laying stones, bricks, or tiles.

ORSAT Apparatus used to analyze gases volumetrically by measuring the amounts of carbon dioxide and carbon monoxide present.

orthog. *Abbreviation for:* orthogonal.

ORTHOGONAL PERSPECTIVE Type of orthogonal projection in which the rays are assumed to be parallel to each other and perpendicular to the plane of projection.

ORTHOGONAL PROJECTION *See* **Linear perspective.** *Also called* **Cabinet projection.**

ORTHOGRAPHIC PROJECTION Graphic representation in which the projecting lines are perpendicular to the plane of projection.

ORTHOSTYLE Series of columns in a straight row.

ORTHOTROPIC Having three mutually-perpendicular planes of elastic symmetry; a contraction of the terms 'orthogonal anisotropic' as in the phrase 'orthogonal anisotropic plate', a hypothetical plate consisting of beams and a slab acting together with different flexural rigidities in the longitudinal and transverse directions, as in a composite-beam bridge.

OS. *Abbreviation for:* outside.

os&dh *Abbreviation for:* oil suction and discharge hose.

os&y *Abbreviation for:* outside screw and yoke.

OSB *See* **Oriented strand board.**

OSCILLATING CURRENT Electrical current that alternately increases and decreases in magnitude, and that reverses polarity to a definite pattern.

OSCILLATING FIFTH *See* **Fifth wheel.**

OSCILLATING GRATE *See* **Grate.**

OSCILLATION 1. Swinging or flexing about a central point; independent movement through a limited range; **2.** Variation between maximum and minimum values.

OSCILLOSCOPE Electronic testing device that shows a range of measurable conditions in the form of a wave pattern on a CRT. Used to test and evaluate engine performance, etc.

OSKP *Abbreviation for:* outside knob pinned.

OSKR *Abbreviation for:* outside knob rigid.

OSLO POINT Shaft, 75 to 125 mm (3 to 5 in.) in diameter, made of hard steel, set into the lower end of an H-pile by slitting the shaft or cutting the web of the pile for some distance, and then welding the shaft to the H-pile. Used to secure a toe-hold in sloping rock. Can also be fitted to the end plate of a pipe pile.

OSMOSIS Diffusion of a solvent or liquid through a skin (permeable in one direction only) into a more concentrated solution.

OST *Abbreviation for:* outside trim.

O/T *Abbreviation for:* overtime.

otfl. *Abbreviation for:* outflow.

oth. *Abbreviation for:* other.

OTHER PRODUCTIVE TIME *See* **Machine time.**

O to O *Abbreviation for:* out to out.

otr *Abbreviation for:* outer.

OTTAWA SAND Silica sand produced by processing a material obtained by hydraulic mining of massive quartzite situated in open-pit deposits near Ottawa, IL, composed almost entirely of naturally rounded grains of nearly pure quartz, used in mortars for testing of hydraulic cement. *See also* **Standard sand,** and **Graded standard sand.**

O-2 WAFERBOARD Panel where the wafers on the face and the back are oriented in the long direction of the panel, while the wafers in the core are oriented in the cross direction.

OUNCE Nonmetric unit of mass, equal to 1/12 pound. Symbol: oz. Multiply by 28.349 523 to obtain grams, symbol: g. *See also the appendix*: **Metric and nonmetric measurement.**

out. *Abbreviation for:* outer; outlet; outside.

OUTAGE 1. Difference between the rated capacity of a storage tank and what it will contain, allowing for expansion of the contents; **2.** Interruption in the supply of electrical current; **3.** Failure or interruption in use or function.

OUTBAND Stretcher stone visible in a reveal.

OUTBOARD BEARING Bearing and its support at the end of a geared machine drive sheave shaft, opposite from the gear end and closest to the drive sheave. *Also called* **Outboard stand.**

OUTBOARD STAND *See* **Outboard bearing.**

OUTBUILDING Minor structure separate from the principal building.

OUTCROP Underlying bedrock that comes to the surface of the ground and is exposed to view.

OUTDOOR LIVING AREA Outside space immediately adjacent to, and accessible from a dwelling and capable of accommodating a variety of individual outdoor activities for the residents.

OUTER CONNECTION *See* **Interchange elements.**

OUTER HEARTH Part of the hearth that is built out in front of a fireplace.

OUTER MIDSECTION Segment of a four-section hydraulic boom that is attached to the inner midsection and manual sections.

OUTER SEPARATION Portion of an arterial highway between the traveled ways of a roadway for through traffic and a frontage street or road.

OUTER STRING String of a staircase farthest from the wall.

OUTER WIRE In wire rope, a wire in the outer layer.

OUTFALL Discharge end of a drain or sewer.

OUTFALL SEWER Sewer that receives flow from a collector system and conducts it to a point of final discharge.

OUTHAUL CYCLE Moving the butt rigging, carriage, and grapple away from the spar tree.

OUTHOUSE Detached privy.

OUTLET 1.Portion of a pipe, pump, or structure through which water or air exits; 2. Place in a wiring distribution system where electrical energy can be drawn to supply a service.

OUTLET BOX Box in which electric wires are joined to one another and to fixtures.

OUTLET PIPE Pipe that conveys the effluent from a sewage treatment plant to its point of final disposal.

OUTLINE SPECIFICATIONS *See* **Draft specifications.**

OUT OF LEAD When sheaves are out of alignment or lines will not spool properly onto a winch.

OUT-OF-PHASE 1. Condition in which AC voltage waves of two generating systems do not coincide; a condition not suitable for synchronization; 2. Relationship between two sine waves of the same frequency where one leads the other. *Also called* **Phase difference.**

OUT-OF-PLUMB Anything that should be in a perfectly vertical position, but isn't.

OUT-OF-SHIFT REPAIR TIME *See* **Machine time.**

OUT-OF-SHIFT SERVICE TIME *See* **Machine time.**

OUT-OF-TRUE Anything that is not as it should be: warped or twisted lumber, misaligned components, not parallel, lacking uniform thickness, etc.

OUTPUT Total of anything generated or produced.

OUTPUT CLUTCH Friction clutch between a torque converter and its drive train.

OUTPUT RESISTANCE Internal resistance of a circuit or device as measured at the output terminals.

OUTPUT SHAFT Shaft that transmits power from a transmission or clutch.

OUTPUT SHAFT GOVERNOR Regulator on the output shaft of a torque converter that regulates the throttle according to travel speed rather than engine speed.

OUTPUT STAGE Final stage of hydraulic amplification, used in a servovalve.

OUTPUT VOLTAGE Voltage produced by a device creating a current flow.

OUTPUT YOKE Component (synonymous with the end yoke) that serves as a connecting link, transferring torque from a transmission's output shaft through the vehicle's driveline to the rear axle.

OUTRIGGER 1.Formed steel piece attached to, and protruding outward from the longitudinals of an interlaced or flush-type subframe to provide added support to a truck body floor and running boards, particularly trailer-type units; 2. Folding or sliding leg-type device attached to a dump truck or trailer to furnish added stability and support from the ground during dumping operations; 3. Beams projecting out from buildings from which scaffolds and swing stages are hung; 4. Extendable beams provided with a frame or undercarriage, typically carrier-mounted mobile cranes, to increase equipment stability and lift capacity. There are several types, including:

Beam and jack: Hydraulically activated horizontal beams with vertically mounted outrigger jacks enclosed in bolt-on housings located at the end of the outrigger beams, providing out-and-over capabilities as well as leveling abilities.

Bolt-on housing: Vertical outrigger jacks enclosed in a bolt-on housing that simplifies maintenance of the cylinders.

Cantilever: Members projecting beyond the point of suspension that provide a fast and effective outrigger system. A disadvantage is the large amount of set-up space required and the destructive influence on conditioned surfaces due to the binding forces produced during outrigger extension and retraction.

Double box-beam: Outrigger beams mounted in boxes in a side-by-side arrangement to increase the distance of outrigger extension, and to provide a more stable lifting foundation.

Single-box beam: Two outrigger beams housed in one outrigger box, with individual telescopic outrigger cylinders, to provide the machine with an increased outrigger spread for a more stable lifting foundation.

Two-stage control: A means of reducing the possibility of accidental outrigger actuation by employing a two-phase operation. The operator is first required to select the outrigger function he wishes to engage and then to select either the outrigger extension or retraction mode.

OUTRIGGER BEAM Part of the outrigger that extends horizontally and acts as the support for the outrigger jack.

OUTRIGGER JACK Hydraulic cylinder or screw jack on the outer point of the outrigger beam that extends vertically to raise and lower the crane.

OUTRIGGER PAD Float used for supporting a machine on the ground.

OUTRIGGER PIN SYSTEM Hydraulic pin system that facilitates outrigger box removal by means of hydraulic cylinders, used in place of the standard outrigger box mounting pins.

OUTRIGGER REMOVAL SYSTEM Means of reducing the overall weight of a crane (usually for the purpose of transporting) by removing the front and rear outrigger boxes.

OUTRIGGER SCAFFOLD Scaffold consisting of a platform supported by outriggers or thrustouts projecting beyond the wall or face of a building or structure, the inboard ends of which are secured within the perimeter of such a building or structure.

OUTSIDE AIR Air drawn from the atmosphere and not previously circulated through an air handling or air conditioning system.

OUTSIDE CASING Boards forming the outside of a cased frame.

OUTSIDE GLAZING Glass panes installed from outside the frame.

OUTSIDE-IN FLOW Filter element designed for normal flow perpendicular and toward the axis of the filter element.

OUT-TO-OUT Overall dimension.

ov. *Abbreviation for:* oval; over.

OVAL Marble chip that has been tumbled until a smooth oval shape has resulted.

OVEN-DRY Condition resulting from a material having been dried to an essentially constant mass in an oven at a temperature that has been established as a standard.

OVEN-DRY BASIS Measurement of moisture content related to the oven-dry or bone-dry weight of the material.

OVEN-DRY CONCRETE Concrete dried in an oven at a temperature between 105°C and 115°C (221°F and 239°F).

OVEN-DRY TON *See* **Bone-dry ton.**

OVEN-DRY WOOD Wood dried to a constant weight in an oven, or above the temperature of boiling water, usually 101°C to 105°C (213°F to 221°F), generally considered to be free of water that can be readily removed at those temperatures.

OVEN LOSS ON HEATING TEST Method of testing the volatility of asphalt cement.

OVERALL Outside to outside dimension.

OVERALL COEFFICIENT OF HEAT TRANS-FER Time rate of heat flow through a body, per unit area, under steady conditions, for a unit temperature difference between the two sides.

OVERALL DIAMETER 1. Maximum diameter of a complex rounded shape; **2.** Diameter of a buffed tire or the diameter of an unloaded new tire, usually made on an inflated tire using calipers or a diameter-type rule.

OVERALL LOWERED HEIGHT Maximum vertical dimension from the vehicle supporting plane (the ground) to the extreme top point of the mast of an industrial truck with the fork carriage in the fully lowered position and unloaded.

OVERALL RATIO Ratio of the lowest to the highest forward gear in a transmission.

OVERALL TRAVEL SPEED *See* **Speed.**

OVERALL TRAVEL TIME Time of travel, including stops and delays (except those off the traveled way).

OVERALL WIDTH 1. Overall width of a chassis from the widest point of the cab; **2.** Maximum width in cross section of an unloaded tire.

OVER-AND-UNDER TILES *See* **Roof tile.**

OVERBREAK Excessive breakage of rock from an explosion, beyond the desired excavation limit.

OVERBURDEN 1. Surface material, usually top soil, that must be removed and disposed of prior to excavation of a site; **2.** Material lying on top of rock to be excavated or shot; usually refers to dirt and gravel, but can involve another type of rock.

OVERBURDEN PRESSURE Vertical pressure at a point in a soil mass due to the bulk of material above it.

OVERCONSOLIDATION Condition that exists if a soil deposit has been subjected to an effective stress greater than the existing overburden pressure.

OVERCURE Vulcanizing longer than necessary. A condition that can result in deterioration of physical properties.

OVERCURRENT Any electrical current in excess of the rated current of the equipment or the ampacity of a conductor.

OVERDESIGN To require adherence to structural design requirements higher than service demands as a means of compensating for statistical variation or for anticipated deficiencies, or both.

OVERDIG To excavate deeper or further than intended.

OVERDRIVE Gearing of a transmission so that in its highest gear(s), one revolution of the engine produces more than one revolution of the transmission's output shaft.

OVERDRIVING Driving in a manner that causes damage to a pile, often by continuing to pound after penetration of the pile has stopped.

OVERFALL That part of a dam over which water can be discharged; the overpouring water. *Also called* **Overtopping.**

OVERFILL Extra fill, temporarily placed over permanent fill to accelerate consolidation and removed once data indicates consolidation is complete, or nearly so.

OVERFIRE AIR Combustion air blown into an incinerator through nozzles over the flame to ensure complete burnoff of combustion gases.

OVERFIRE AIR FAN *See* **Fan.**

OVERFLOW 1. Flow or spread over or across; **2.** To flow over an edge or brim; **3.** Amount that overflows; **4.** Spew-out of tread compound of a tire at the mold parting line or at the edge of the matrix skirt. Should be trimmed or buffed off the finished product.

OVERFLOW TUBE Vertical tube in the tank of a water closet that prevents overflow.

OVERGLAZE DECORATION *See* **Decoration.**

OVERHAND WORK An entire masonry wall built with staging located on only one side of the wall.

OVERHANG 1. The projection of one part of a structure over another; **2.** Projecting parts of a bank, face, or wall; **3.** Horizontal distance from the center line of the hinge to the rear of the body on a tilting-body truck; **4.** Distance from the center of the rear axle(s) to the rearmost surface of the truck body; **5.** Extension of the top chord of a truss beyond the heel joint.

OVERHANG LAP Method of plane-lapping welded fabric sheets where one sheet is provided with a lap length overhang of main bars without welded intersecting bars.

OVERHAUL 1. Transport of excavation beyond specified limits. *See also* **Average haul, Freehaul, Haul,** and **Station yards**; **2.** Ability of a weight on the end of a hoist line to unwind cable from the drum when the brake is released; **3.** To check thoroughly and make necessary repairs and adjustments.

OVERHAULING LOAD Negative load imposed in a hoisting machine resulting from the overbalance of the load or counterweight.

OVERHAULING WEIGHT Weight added to a load fall to overcome resistance and permit unspooling at the rope drum when no live load is being supported. *Also called* **Headache ball.** *See also* **Cheek weight.**

OVERHEAD 1. Indirect expenses that cannot be charged to individual costs or bid items except by proration; **2.** Upper end of an elevator hoistway.

OVERHEAD DOOR *See* **Door types.**

OVERHEAD GUARD Framework fitted to an industrial truck over the head of the riding operator. *Also called* **Canopy guard.**

OVERHEAD LEAD *See* **Lead.**

OVERHEAD MACHINE Elevator drive machine mounted above the hoistway.

OVERHEAD PROTECTION Rigid screen installed in an elevator hoistway during installation to protect workers from falling objects.

OVERHEAD SHOVEL Bucket loader that loads at one end, swings the bucket overhead, and dumps at the other end; used in closely confined areas such as tunnels.

OVERHEAD STRUCTURE Total of the structural members, machinery, and equipment at the top of an elevator hoistway.

OVERHEAD TRAVELING CRANE Gantry-mounted hoist that can travel to either side of a structure, and be transported the entire length of the rails on which the gantry travels.

OVERHEAD VALVE (ENGINE) Engine having its inlet and outlet valves in the cylinder head.

OVERHEAD WORK 1. Task that requires workers to perform work at a level higher than their shoulders; **2.** Weld on the underside of a joint with the face of the weld in a horizontal position.

OVERHEATING 1. Engine operating condition where coolant temperature exceeds design intent; **2.** Damaging the properties of metal by too much heat.

OVERHUNG SUSPENSION Suspension where the spring is positioned over the axle.

OVERLAID PLYWOOD Plywood panels with factory-applied, resin-treated fiber faces on one or both sides. May also apply to metal and other overlay panels.

OVERLAND FLOW Surface runoff.

OVERLAY 1. Anything laid over another thing; **2.** Layer of concrete or mortar, seldom thinner than 25 mm (1 in.), placed on and usually bonded onto the worn or cracked surface of a concrete slab to either restore or improve the function of the previous surface; **3.** Layer of asphalt mix placed over a deteriorated pavement to restore its strength and riding qualities; **4.** Any factory-applied material bonded to one or both sides of a panel. May include resin treated fiber, fiberglass, plastic, metal, hardboard, etc.

OVERLOAD 1. Operation of equipment in excess of its normal, full-load rating, or of a conductor in excess of its rated ampacity that, when it persists for a sufficient length of time, would cause damage or dangerous overheating; **2.** Mass greater than a structure is designed to carry.

OVERLOAD POWER That load, in excess of rated load, that a generator set is capable of delivering for a specified period of time; the voltage, frequency, and operating temperature may differ from normal rated values.

OVERLOAD RELAY Relay that responds to electric load and operates at a preset value of overload.

OVERLOAD TRIP Protective device that causes power to be reduced or cut when preset limits have been reached.

OVERMANTLE Decoration to the front of a chimney breast, above the mantlepiece.

OVER MATURE Point at which timber has begun to lessen in commercial value, because of size, decay, or other factors.

OVERPRESSURE Pressure generated by a sound wave that produces a variation in atmospheric pressure, measured in psi or decibels.

OVERRIDE PRESSURE *See* **Pressure.**

OVERRUN 1. To run over or beyond certain limits. *See also* **Time overrun.** *Also called* **Construction time overrun;** **2.** Difference between the log scale of a shipment of timber and the volume of actual lumber obtained from it.

OVERRUN BRAKE *See* **Brake.**

OVERRUNNING CLUTCH *See* **Clutch.**

OVERSAIL The projection of an upper story of a building beyond the perimeter of a lower story.

OVERSAILING COURSES Series of stone, brick, or block courses, each one projecting beyond the course below it. *See also* **Corbel.**

OVERSANDED Cement paste, grout, or mortar containing more sand than would be necessary to produce adequate workability and a satisfactory condition for finishing.

OVERSHOOT Amount by which voltage or frequency exceeds the nominal value after a sudden load change on a generator set.

OVERSHOT Condition resulting from more than the necessary amount of explosives being used in a shot, usually characterized by excessive fragmentation, flyrock, and noise.

OVERSITE CONCRETE Layer of unreinforced concrete placed directly on grade below the ground floor of a house, used to seal the soil. (This is not a basement floor).

OVERSIZE In aggregate classification, the larger of two sizes; the smaller being undersize.

OVERSIZE WASTE *See* **Solid waste.**

OVERSLUNG CAR FRAME Elevator car frame with the hoisting-rope fastenings or hoisting-rope sheaves attached to the crosshead or top member of the car frame.

OVERSLUNG SUSPENSION *See* **Suspension.**

OVERSPEED GOVERNOR *See* **Governor.**

OVERSPEED PROTECTION Speed-of-frequency sensing switch that is connected to the emergency engine shutdown circuit and which operates when the engine speed of a generator set exceeds the speed at which the switch is set. The switch is typically set to operate at 15% above the engine full-load operating speed.

OVERSPEED SWITCH *See* **Switch.**

OVERSTOCKED STAND *See* **Stocking classes.**

OVERSTORY That portion of the trees in a forest of more than one story forming the upper or uppermost canopy layer.

OVERSTORY REMOVAL Any silviculture treatment with the desired end result being the removal of the overstory component from the growing stock of a multi-storied stand. Examples are outright harvest, girdling, and simply felling the overstory.

OVERSTRETCHING Stressing of tendons to a value higher than designed for the initial stress to (a) overcome frictional losses, (b) temporarily overstress the steel to reduce steel creep that occurs after anchorage, and (c) counteract loss of prestressing force that is caused by subsequent prestressing of other tendons.

OVERTHROW Fixed panel or arch, often elaborately decorated, above a wrought-iron gate.

OVERTIME Time worked beyond an agreed-to maximum for a given period and qualifying for a premium rate of payment.

OVERTONES All audible frequencies emitted from a sound source at one time, excepting the fundamental tone.

OVERTOPPED *See* **Suppressed.**

OVERTOPPING 1. Flow of water over the top of a dam or embankment. *Also called* **Overfall;** **2.** Vegetation higher than the favored species, as in brush or deciduous species shading and suppressing more desirable coniferous trees.

OVERTRAVEL Distance a moving object may travel beyond its normal extreme of movement.

OVERTURNING Result of an combination of forces tending to overcome equilibrium.

OVERTURNING MOMENT Moment of the load plus the boom weight about the tipping fulcrum.

OVERVIBRATION Excessive use of vibrators during placement of freshly mixed concrete, causing segregation, stratification, and excessive bleeding.

OVERVOLTAGE PROTECTION Protection of a system against excessive voltage fluctuations on the main AC lines.

OVERWINDING Cable or rope wound and attached so that it stretches from the top of a drum to the load. *See also* **Underwinding.**

ovfl. *Abbreviation for:* overflow.

ovhd *Abbreviation for:* overhead.

ovhg *Abbreviation for:* overhang; overhanging.

ovhl *Abbreviation for:* overhaul.

ovld *Abbreviation for:* overlaid; overload.

OVOLO *See* **Molding.**

ovps *Abbreviation for:* overpass.

OVUM The egg in an egg-and-dart molding.

OWNER Corporation, association, partnership, individual or public body or authority that the contractor enters into agreement with and for whom the work is provided.

OWNER-ARCHITECT AGREEMENT Contract for professional services between an architect and client.

OWNER-CONTRACTOR AGREEMENT Contract to complete a described project between an owner and contractor.

OWNER DEFAULT Failure by an owner to pay a contractor under the terms of a construction contract.

OWNER OCCUPANT Tenant of a residence who also owns the property.

OWNER OF RECORD Person or persons stated in the public records as the owner(s) of a properly described property.

OWNER'S AND CONTRACTORS' PROTECTIVE LIABILITY Third-party legal liability insurance protecting owners and contractors from claims arising from the activities of subcontractors.

OWNER'S LIABILITY INSURANCE Protection of an owner against claims arising while construction is underway.

OWNER'S REPRESENTATIVE Person authorized to act for the owner. *Also called* **Field representative,** and **Project representative.**

OWNING COSTS Costs that result from owning a machine, i.e., depreciation, interest cost, taxes, storage, insurance, replacement cost escalation. These costs accrue whether a machine works or not.

OXBOW Abandoned part of a meandering stream.

oxd *Abbreviation for:* oxidized.

oxd. *Abbreviation for:* oxide.

OXIDATION Reaction of oxygen on an organic substance, usually evidenced by a change in the appearance or feel of the surface or by a change in physical properties.

OXIDATION INHIBITOR Additive designed to increase the oxidation stability of a product.

OXIDATION STABILITY Resistance of lubricants to chemical reaction with oxygen.

OXIDIZED ASPHALT *See* **Asphalt.**

OXIDIZED SEWAGE Waterborne wastes that have been aerated to the point where organic matter has become stabilized.

OXIDIZER Supplier of oxygen.

OXIDIZING Combining oxygen with any other substance.

OXIDIZING SUBSTANCE Product or substance that:

(a) Causes or contributes to the combustion of another material by yielding oxygen or other oxidizing substances, whether or not the product or substance is itself combustible, or

(b) Is an organic compound that contains the bivalent O-O structure.

OXYCETYLENE CUTTING/WELDING *See* **Oxygen-acetylene cutting/welding.**

OXYGEN-ACETYLENE CUTTING/WELDING Method of cutting or welding metal using a fuel combination of oxygen and acetylene. *Also called* **Oxycetylene cutting/welding.**

OXYGEN BOMB Chamber capable of holding oxygen at an elevated pressure that can be heated to an elevated temperature. Used for an accelerated aging test.

OXYGEN BOMB AGING Means of accelerating a change in the physical properties of rubber compounds by exposing them to the action of oxygen at an elevated temperature and pressure.

OXYGEN HOSE Reinforced, multilayered, flexible tube, usually of rubber, used to carry high-pressure gases.

oz *Abbreviation for:* ounce.

oz-ft *Abbreviation for:* ounce-foot.

oz-in. *Abbreviation for:* ounce-inch.

OZONE CRACKING Surface cracks, checks, or crazing caused by exposure to an atmosphere containing ozone.

OZONE RESISTANCE Ability to withstand the deteriorating effects of ozone (generally cracking).

P

p *Abbreviation for:* part; per; pipe; port; post; power.

p *Abbreviation for:* mass density.

P *Abbreviation for:* planned; pole; pressure.

PA *Abbreviation for:* parallel arm; profiling accessories; public address (system).

PACE Staircase landing.

PACHE System of color-coding construction drawings to assist the process of quantity take-off and bid estimating.

PACKAGE BUILDER Contractor who provides a range of services in addition to construction toward the completion of a project. This may include all or any of such items as financing, site selection, site provision planning and design, supply of structures and materials, project management, etc. *Also called* **Turnkey contractor.**

PACKAGED AIR CONDITIONER Room air conditioner, usually designed to fit into a window opening.

PACKAGED CONCRETE/MORTAR/GROUT Mixtures of dry ingredients in packages, requiring only the addition of water to produce concrete, mortar, or grout.

PACKAGE MORTGAGE *See* **Mortgage.**

PACKAGE POLICY Policy that includes several different types of insurance protection.

PACKED TOWER Pollution control system device that forces dirty air through a tower packed with crushed rock or wood chips while liquid is sprayed over the packing material. The pollutants in the air stream either dissolve or chemically react with the liquid.

PACKER 1. Device inserted into a hole in which grout is to be injected that acts to prevent return of the grout around the injection pipe, usually an expandable device actuated mechanically, hydraulically, or pneumatically; **2.** *See* **Packer truck.**

PACKERHEAD PROCESS Process for producing concrete pipe that uses a rotating device that forms the interior surface of the pipe as concrete is fed into the form from above. *See also* **Centrifugal process, Dry-cast process, Spun concrete, Tamp process,** and **Wet-cast process.**

PACKER TRUCK Type of solid waste collection vehicle, usually used for residential collection, that compacts refuse into a high density mass for maximum collection efficiency; can incorporate a rear loading or top loading device.

PACKING 1. Material used in the stuffing box of a valve to keep a leakproof seal around the stem; **2.** Pad of resilient material contained between the helmet and the top of a reinforced concrete pile to minimize damage to the head during driving; **3.** Plate inserted between two others to fill a gap; **4.** Small stones used to fill gaps in rubble walls. *Also called* **Flange packing**; **5.** Stuffing of water-resilient material to prevent leaking at a valve stem.

PACKING NUT Nut on the stem of a valve that compresses the packing.

PAD 1. Footing; sometimes a block of concrete to support machinery, or a block of timber to support an outrigger or jack; **2.** Ground contact part of an outrigger or a crawler-type track. *Also called* **Shoe** and **Plate.** *See also* **Helipad,** and **Pile cushion.**

PADDLE LOADER *See* **Loader.**

PADDLE-WHEEL SCRAPER Excavation machine that employs a conveying device to dislodge soil and move it into a storage and transporting bowl.

PADDOCK Fenced enclosure adjoining a stable.

PAD FOUNDATION Isolated foundation to spread a concentrated load, such as from a column.

PADLOCK Self-contained, portable lock having a pivoted arch that is closed to lock.

PADLOCK SHEAVE *See* **Sheave.**

PAD SAW *See* **Saw.**

PADSTONE Block to carry the weight of a truss or beam.

PAI *Abbreviation for:* periodic annual increment.

PAINT Surface coating applied in liquid form containing pigments in a binder.

PAINT BASE Substance to which pigment and thinners are added.

PAINT DRIER Substance added to paint to enhance its drying capabilities.

PAINTED FINISHE Paint coat that is used to enhance a surface.

PAINTER'S PUTTY Filler, similar to glazier's putty.

PAINT REMOVER Liquid solvent that softens a paint or varnish film, causing it to release from the surface to which it was applied.

PAINT SYSTEM 1. Succession of coats and procedures designed to protect a surface and produce a designed finish; **2.** Color scheme prepared for a room or area, including materials selection and specification.

PAINT THINNER Petroleum-based solvent used in place of turpentine in formulas and for thinning and cleaning.

PALING Upright member of a picket fence, fastened to top and bottom rails between posts.

PALISADE Fence of poles driven into the ground.

PALLADIANA *See* **Berliner.**

PALLET 1. Portable platform for moving, storing or handling materials, usually of wood, fabricated with spaces below the load-bearing platform to allow passage of the tines of a forklift or other lifting device; **2.** Soffit formwork of molds in a stressing bed that allows for slight horizontal movement of the concrete following release of tension in the prestressing tendons; **3.** One of a series of rigid platforms that together form an articulated treadway or the support for a continuous treadway.

PALLET BRICK *See* **Brick.**

PALLET TRUCK Self-loading, low-lift industrial truck equipped with wheeled forks of dimensions to go between the top and bottom boards of a double-faced pallet and having wheels capable of lowering into spaces between the bottom boards so as to raise the pallet off the floor for transporting.

PALLET-TYPE MOVING WALKWAY *See* **Moving walkway.**

pam. *Abbreviation for:* pamphlet.

PAN 1. Prefabricated form unit used in concrete-joist floor construction. *See also* **Metal form**; **2.** Container that receives particles passing the finest sieve during mechanical analysis of granular materials; **3.** Occasional name for a carrying scraper. *See also* **Scraper**; **4.** Large, flat, upward-curving metal plate on which log ends or pallets are placed to make skidding easier and prevent digging in and rutting.

pan. *Abbreviation for:* panel; pantry.

PAN BREAKER *See* **Subsoil plow.**

P&A *Abbreviation for:* price and availability.

P&C *Abbreviation for:* purchasing and contracting.

P&L *Abbreviation for:* profit and loss.

P&O *Abbreviation for:* paint and oils.

P&T *Abbreviation for:* posts and timbers.

PAN-AND-ROLL TILE *See* **Tile.**

PANE Glass surface of a light; a window may include a number of panes.

PANEL 1. Large board or sheet handled as a single unit during assembly and erection; **2.** Thin board with its edges inserted in the groove of a surrounding frame of thicker material; **3.** Portion of a flat surface in a different plane; **4.** Section of form sheathing constructed from boards, plywood, metal sheets, etc., that can be erected and stripped as a unit; **5.** Concrete member, usually precast, rectangular in shape and relatively thin with respect to other dimensions. *See also* **Frame**; **6.** Plate or surface for mounting components; **7.** Truss chord segment between two adjacent joints; **8.** *See also* **Frog.**

PANELBOARD Center for controlling a number of electrical circuits by means of fuses or circuit breakers, usually contained in a metal cabinet. Switches are commonly added to control each circuit

PANEL BOX Box in which electric switches and circuit breakers or fuses for branch circuits are located.

PANEL CLIP Specially shaped metal device for supporting panel edges to reduce differential deflection in roof construction.

PANEL CURTAIN WALL *See* **Wall.**

PANEL GRADE May refer to the letter-graded qual-

ity of face and back veneers used in plywood manufacture, or to panels manufactured for specific construction applications, e.g., **Underlayment.**

PANEL GRID Prefabricated panel of radiant heating coils.

PANEL HEATING *See* **Heating system.**

PANELING Wood panels joined in a continuous surface, especially decorative panels for interior wall finish.

PANELIZED CONSTRUCTION Building components fabricated in wall, floor, or roof sections, etc., to be assembled into a completed structure at the building site.

PANEL LENGTH Distance between two adjacent joints along either upper or lower truss chords.

PANEL MARK Mark on a panel product consisting of an end-use mark and a span mark.

PANEL MOLD Mold used for casting plaster panels.

PANEL MOUNTING Panel on which a number of components may be mounted.

PANEL PIN *See* **Nail.**

PANEL POINT Intersection of two or more members of a truss.

PANEL RADIATOR Domestic heating unit placed on, or flush with a flat surface and intended to function essentially as a radiator.

PANEL SAW *See* **Saw.**

PANELS (DOOR) *See* **Door.**

PANEL SPACING Gap left between installed panels in a structure. (Proper spacing helps prevent buckling and warping.) *Also called* **Edge spacing**, and **End spacing.**

PANEL STRIP Strip extending across the length or width of a flat slab for structural design and construction or for architectural purposes.

PANEL WALL *See* **Wall.**

PANEL ZONE Zone in a beam-to-column connection that transmits moments by a shear panel.

PAN FLOOR *See* **One-way floor and roof system.**

PAN FORM Pan-like metal or fiberglass shape used as a form in the construction of a one- or two-way floor or roof slab.

PANHANDLE Heavy concrete beam between footings, used to balance column loads that cannot be centered on their own footings.

PAN HEAD Head of a screw or rivet that has a truncated cone shape.

PANIC BOLT Special form of door-bolt located at waist height on the interior face of an outward-opening door or pair of doors that is released by pressure. Commonly used on exit doors from public places.

PANIER Corbel that breaks the angle between a pilaster and the beam it supports.

PAN MIXER *See* **Open-top mixer.**

PANNING Channeling of water down the sides of a

tunnel behind metal strips, prior to concreting.

PANOPTICON Building designed around a central hub, usually with radiating wings.

PAN STEP Prefabricated concrete step form.

PANTHEON Temple.

pant. *Abbreviation for:* pantograph.

PANTILE *See* **Roof tile.**

PANTOGRAPH Drawing tool for enlarging or reducing.

PANTS Slotted plates fastened to the sides of a pile hammer to engage the heads of template-supported sheet pile to keep the hammer centered upon and in vertical alignment with the pile and to permit the hammer's operation to be free-hanging. *Also called* **Fish tails, Fingers,** and **Skirts.** *See also* **Hairpin.**

PAP Vertical outlet from an eaves gutter.

PAPER AND WIRE Tar paper and wire mesh (or metal lath) used as a backing for the installation of tile.

PAPERBOARD General term describing sheets made of fibrous material 0.3 mm (0.012 in.) or more in thickness.

PAPER FORM Heavy paper mold used for casting concrete columns and other structural shapes.

PAPERHANGER Worker who hangs wallpaper and related materials, including any necessary preparatory work.

PAPER LATCHER Worker who fits rolls of glass-fiber and Kraft paper onto a pipeline wrapping machine.

PAPER ROLLER Blemish to the face of gypsum wallboard, usually caused by sliding one board over another, face-to-face or face-to-back, and characterized by a tightly curled strip coiled against the surface.

par. *Abbreviation for:* paragraph; parallel; parapet.

parab. *Abbreviation for:* parabola; parabolic.

PARABOLA 1. Shape made by cutting a cone parallel to its vertical axis; **2.** Bending moment curve for a uniformly distributed load on a simply supported beam.

PARABOLIC REFLECTOR Reflector that directs reflected light in a narrow beam.

PARAFFINIC Petroleum fluid derived from a paraffinic crude and characterized by a high proportion of linear paraffinic molecules. Paraffinic base oils form the basis of most lubricants due to their viscosity indexes and resistance to oxidation.

PARALLAX 1. Apparent displacement or shift between two things when viewed from two different angles or positions; **2.** Change in the apparent position of an object in relation to the cross hairs when viewed through a telescope, due to maladjustment of the eyepiece focus.

PARALLEL Arrangement of electric blasting caps in which the firing current passes through all of the caps at the same time.

PARALLEL APPLICATION Drywall installation with the long dimension applied in the same direction as the framing members. *Also called* **Vertical applica-**tion.

PARALLEL BATTERIES Where two or more same-voltage batteries have their positive terminals connected on a separate wire and all negative terminals on another, separate wire. This means that four identical-capacity six-volt batteries, for instance, will still produce only six volts but will have four times the energy potential.

PARALLEL CABLEWAY Cableway with movable head and tail towers that maintain the same relative position on parallel or concentric runways.

PARALLEL CIRCUIT Electrical circuit in which components are connected across a power source so that the current divides between them, but the voltage remains the same across each branch. *See also* **Series circuit.**

PARALLEL CLAMP *See* **Clamp.**

PARALLEL CONNECTION Electrical connection in which the input terminal of one element is connected to the input terminal of another element, and the output terminals are similarly connected together, thereby providing two paths for current flow.

PARALLEL DIRECTION *See* **Machine direction.**

PARALLEL FLOW Arrangement within a heat exchanger between two fluids whereby the hottest portion of one flow meets the coldest portion of the other.

PARALLEL GUTTER *See* **Box gutter.**

PARALLELING Procedure used to connect two or more generators to a common load.

PARALLEL LAMINATED Laminate in which all the separate layers of material are oriented in the same direction.

PARALLEL METHOD Indirect method of forest fire suppression in which a fire line is constructed approximately parallel to, and just far enough away from the fire edge to enable men and equipment to work effectively, though the line may be shortened by cutting across unburned fingers.

PARALLELOGRAM BRACE Brace to change pile hammer leads from side to side without the use of a moonbeam.

PARALLELOGRAM-TYPE RIPPER *See* **Ripper.**

PARALLEL OPERATION Two or more generators, or other power sources, of the same phase, voltage, and frequency characteristics supplying the same load.

PARALLEL SERIES Two or more series of electric blasting caps arranged in parallel.

PARALLEL THREAD Screw thread of uniform diameter, commonly used on machine bolts.

PARALLEL-WIRE UNIT Post-tensioning tendon composed of a number of wires or strands that are approximately parallel.

PARAPET 1. Low wall protecting a sudden change in elevation; **2.** Low wall along the top of a dam or at the edge of a bridge deck.

PARAPET GUTTER Gutter formed or placed behind a parapet wall.

PARAPET WALL *See* **Wall.**

PARASITIC LOAD Extra load caused by an engine-driven accessory such as a cooling system fan and battery-charging alternator.

PARBUCKLE 1. Method of locating and setting a choker to overcome a yarding difficulty; **2.** Dumping a log truck by having the load in the bight of a pair of lines, and picking up the lines to force the load sideways and slide it off the truck; **3.** Holt set with a choker to clear a log over a stump.

PARCEL 1. Property under a single ownership; **2.** Lot in a subdivision.

PARCEL PLAT Map of a single parcel of property or portion thereof needed for highway purposes, showing the boundaries, areas, the remainder, improvements, access, ownership, and other pertinent information.

PARE Cut away in thin slices.

PARENT TREE Any tree whose seeds are used to produce progeny for use in genetic experimentation. Usually the parent tree is selected because it displays characteristics either interesting from a research standpoint or desirable in an operational forest management program.

PARGE To roughly cover with mortar, i.e., to parge the inside of a chimney or flue. *See also* **Back plastering.**

PARGETING Ornamental plasterwork to walls or ceilings.

PARING CHISEL *See* **Wood chisel.**

PARING GOUGE *See* **Wood chisel.**

PARK 1. Act of stopping and keeping a motor vehicle immobile; **2.** Public open areas, usually landscaped and often containing a range of facilities, such as recreational areas, ornamental landscaping, etc.

PARKING System by which an elevator(s) receives a signal to always return to a preselected landing after all of its car or landing signals have been answered and canceled.

PARKING BRAKE Secondary chassis brakes, independent in application from service brakes, used for parking the vehicle and for controlling it when the service brake is not functioning.

PARKING DEVICE Electrical or mechanical device that permits the opening or closing of an elevator cab from the landing side of the hoistway door.

PARKING FACILITY Multistory structure designed and used specifically for the temporary accommodation of unattended vehicles.

PARKING LANE *See* **Lane.**

PARKING LOT Grade-level space dedicated to the temporary use of unattended vehicles.

PARKING STAND Adjustable leg, normally stored in a retracted position, used to support the front end of a trailer when unhitched.

PARKING TRACK Section of track provided for rail-mounted traveling cranes.

PARKWAY Arterial highway for noncommercial traffic, with full or partial control of access, and usually located within a park or ribbon of park-like developments.

PARQUET Flooring of narrow strips of thin, approximately 6-mm (0.25-in.), hardwood, laid in patterns and highly polished.

PARQUETRY Inlaid woodwork in different colors, particularly in flooring.

PARQUET STRIP In a hardwood floor, a single board of distinctly different color laid as a band around the perimeter of the area and secret nailed to the other boards.

PARSHALL FLUME Type of Venturi flume, used in hydraulics for measuring flows in open channels.

part. *Abbreviation for:* partial; partially; particle; particulate; partition.

PARTIAL ACCEPTANCE Acceptance from the contractor of a portion of a construction project that has been completed according to requirements, while waiting for final completion of the entire project. *See also* **Acceptance of work, Final acceptance,** and **Interim acceptance.**

PARTIAL CONTROL OF ACCESS Authority to control access exercised to give preference to through traffic to a degree that, in addition to access connections with selected public roads, there may be some crossings at grade and some private driveway connections. *See also* **Control of access,** and **Full control of access.**

PARTIAL CUT Logging area in which only part of the trees are felled and bucked, as opposed to clearcut.

PARTIAL EVICTION *See* **Eviction.**

PARTIAL FLOW FILTER Filter that filters only part of a total system fluid. *Also called* **Bypass oil filter.**

PARTIAL LOSS Loss that neither completely destroys an insured property nor renders it worthless.

PARTIALLY AMORTIZED MORTGAGE *See* **Mortgage.**

PARTIALLY COMPOSITE BEAM Composite beam for which the shear strength of shear connectors governs the flexural strength.

PARTIALLY FIXED End support of a beam or column that cannot develop full fixing moment.

PARTIALLY KNOCKED DOWN Truck disassembled to some degree for shipment. *See also* **Completely knocked down.**

PARTIAL OCCUPANCY Occupation by the owner of a portion of a project prior to final completion.

PARTIAL PAYMENT *See* **Progress payment.**

PARTIAL PENETRATION Situation when the intake of a water well does not penetrate the full thickness of the aquifer.

PARTIAL PRESSURE *See* **Pressure.**

PARTIAL PRESTRESSING Prestressing to a stress level such that, under design loads, tensile stresses exist in the precompressed tensile zone of the prestressed member.

PARTIAL RELEASE 1. Release into a prestressed concrete member of a portion of the total prestress initially held wholly in the prestressed reinforcement;

2. Provision in a mortgage that allows some of the property pledged to be freed from serving as collateral as the remaining principal is reduced through repayment.

partic. *Abbreviation for:* participate; participating; particular.

PARTICIPATING MORTGAGE *See* **Mortgage.**

PARTICLE Small piece; fragment.

PARTICLEBOARD Manufactured structural sheet composed of wood chips bonded into an adhesive.

PARTICLEBOARD UNDERLAYMENT *See* **PBL**

PARTICLE CHARGE TEST Test made to identify cationic emulsions.

PARTICLE COUNT BLANK Allowance for the determinable background contamination.

PARTICLE-SIZE ANALYSIS Mechanical determination of the proportion by weight of different particle sizes in soil or sand.

PARTICLE SIZE DISTRIBUTION Tabular or graphical listing of the number of particles according to particle size ranges.

PARTICLE VELOCITY 1. Velocity at which the earth vibrates; **2.** Velocity at which a water particle or molecule travels through a formation or porous material.

PARTICULATE CONTAMINATION Foreign matter in a fluid taken in the form of solid particles.

PARTICULATE MATTER Airborne matter of a finely divided solid, liquid, or combined material.

PARTING BEAD *See* **Parting strip.**

PARTING LINE Mark on a mold or cast where halves or sections of the mold meet in closing.

PARTING *See* **Joint.**

PARTING SLIP *See* **Parting strip.**

PARTING STRIP Thin strip of wood set into the head and jamb of a window frame for holding the sashes apart. *Also called* **Parting bead,** and **Parting slip.**

PARTING TOOL Turning tool used for cutting narrow recesses, grooves and channels.

PARTITION 1. Wall that subdivides spaces within a building or room; **2.** Division of real property between those who own it with undivided interest.

PARTITION BLOCK Hollow masonry unit made as thin as 50 mm (2 in.), usually scored for plastering on both sides.

PARTITION PLATE Horizontal member serving as a top and bottom cap to studding.

PARTITION SYSTEM Assembly of components that perform as a wall or system of walls, or cubicles.

PARTITION TILE Tile for use in interior partitions.

partn *Abbreviation for:* partition.

PARTS OF LINE Roping configuration that is determined by the amount of reeving performed on the boom nose and accompanying hookblock. Each revolu-

tion of rope reeved between the sheaves on the hookblock and boom nose is equivalent to two parts of the line. As parts of line increase, the machine's lifting ability also is increased. Although lifting performance is increased, each additional part of line will reduce the machine's hookspeed. Therefore, a machine with multiple parts of line will realize lower cycle times. There are several configurations, including:

> **One-part line:** Single strand of rope or cable.
>
> **Two-part line:** Single strand of cable or rope doubled back around a sheave so that two parts of it pull a load together.
>
> **Three-part line:** Single strand of cable or rope doubled back around two sheaves so that three parts of it pull a load together.
>
> **Four-part line:** Single rope or cable reeved around pulleys so that four strands connect the fixed and movable units.

PART-SWING SHOVEL *See* **Shovel.**

PARTY CHIEF Person in charge of a survey party and responsible for survey work done in the field.

PARTY WALL *See* **Wall.**

PASCAL One of 17 derived units with special names of the SI system of measurement: a unit of pressure or stress resulting from a force of 1 N acting uniformly over an area of $1 m^2$. ($1 Pa = 1 N \cdot m^2$). Symbol: Pa.

PASCAL SECOND A derived unit of dynamic viscosity with a compound name of the SI system of measurement. Symbol: Pa·s. *See also the appendix:* **Metric and nonmetric measurement.**

PASCAL'S LAW Principle that states that pressure applied to a confined fluid at rest is transmitted with equal intensity throughout the fluid.

PASS 1. Working trip of an excavating machine; **2.** Layer of shotcrete placed in one movement over the area of operation; **3.** Weld metal created by one progression along the weld; **4.** Single progression of a compactor over the work area (number of passes equals the number of progressions); **5.** One-way working trip or passage of a machine. A round trip in the same path is two passes.

pass. *Abbreviation for:* passage; passenger.

PASSAGE Machined or cored fluid-conducting path that lies within or passes through a component.

PASS BLOCK Lightweight block hung at the top of a spar tree and used to lift the bull block and other gear when rigging the tree.

PASS CHAIN Chain with an open hook at one end and a ring at the other, used to grip wire rope that is to be pulled. *Also called* **Rigging chain.**

PASSENGER CAR EQUIVALENCE Representation of larger vehicles, such as trucks and buses, as equal to a quantity of passenger cars for use in computing the level of service and capacity analyses. The magnitude of the equivalency is dependent upon the vehicle size and weight, vehicle operating characteristics, vehicle speeds, and roadway characteristics such as gradient.

PASSENGER ELEVATOR *See* **Elevator.**

PASSING LANE Widening of a two-lane, opposing

traffic highway to three lanes over a limited distance to provide an additional lane for faster traffic. *Also called* **Climbing lane.**

PASSIVE COMPONENT Device that reacts only to voltage and current changes.

PASSIVE EARTH PRESSURE *See* **Pressure.**

PASSIVE SOLAR ENERGY Collection of solar energy through fixed, south-facing panels, thermal storage devices and a transfer medium; one in which no pumps are used to accomplish the transfer of thermal energy.

PASSKEY *See* **Master key.**

PASTE *See* **Adhesive.**

PASTE CONTENT Proportional volume of cement paste in concrete, mortar, or the like, expressed as volume percent of the entire mixture. *Also called* **Paste volume.** *See also* **Neat cement paste.**

PASTE VOLUME *See* **Paste content.**

PAT Specimen of neat cement paste, about 75 mm (3 in.) in diameter and 13 mm (0.5 in.) thick at the center and tapering to a thin edge, on a flat glass plate, used for indicating setting time.

PAT. *Abbreviation for:* pattern.

pat. *Abbreviation for:* patent.

PATCH *See* **Repairs.**

PATCHING Removal of poor surface areas and making good the surface.

PATCH LOGGING Staggered settings on a claim, for fire prevention and reforestation.

PATCH TEST Any method of evaluating fluid contamination wherein the sample is passed through a standardized laboratory filter, and the change in color, reflectivity, etc. of the laboratory filter is compared with previously established standards.

PATENT GLAZING Any of various methods of dry, puttyless glazing.

PATH OF PRESTRESSING FORCE Locus of points defining the resultant effective prestress force in a concrete member.

PATHOLOGICAL WASTE *See* **Solid waste.**

PATINA Color and texture changes to an exposed metal surface from oxidation and weathering.

PATIO Inner courtyard or paved or decked area, open to the sky.

PATIO BLOCK Precast concrete paving slab available in a variety of shapes, sizes, and finishes. Normally larger than paving stones.

pat. pend. *Abbreviation for:* pattern pending.

patt. *Abbreviation for:* pattern.

PATTERN CRACK *See* **Craze crack,** and **Map cracking.**

PATTERNED LUMBER Lumber that is shaped to a pattern or to a molded form in addition to being dressed, matched, or shiplapped, or any combination.

PATTERN REPEAT Distance between a point of a pattern and the next point where the pattern is identical.

PATTERN SCRIBING Scribing the outline of a complex shape to either an intermediary material prior to cutting the material to be fitted to the shape, or directly to the final material.

PAVEMENT Designed surface and substructure to support vehicles.

PAVEMENT BASE PLATE Metal or plastic plate placed under a concrete pavement joint to prevent ingress of soil or water.

PAVEMENT BIT Pavement engaging tool held by a working tool holder attached to the cutter drum that performs the action of pavement planing/milling.

PAVEMENT CONCRETE Layer of concrete over such areas as roads, sidewalks, canals, playgrounds, or areas used for storage or parking. *See also* **Rigid pavement.**

PAVEMENT CUTER Pavement engaging tool held by a working tool holder attached to the cutter drum that performs the action of pavement planing/milling.

PAVEMENT CUTTING TOOL Pavement engaging tool held by a working tool holder attached to the cutter drum that performs the action of pavement planing/milling.

PAVEMENT LIGHT Precast, transparent or translucent panel let into pavements, especially sidewalks, to light a vault or other space beneath. *Also called* **Vault light.** *See also* **Glass-concrete construction.**

PAVEMENT MARKING Marking set into the surface of, applied upon, or attached to the pavement for the purpose of regulating, warning, or guiding traffic, including the following:

> **Centerline:** Line indicating the division of the roadway between traffic moving in opposite directions.
>
> **Edgeline:** Line that indicates the edge of the traveled way.
>
> **Laneline:** Line separating two lanes of traffic traveling in the same direction.
>
> **Raised pavement marker:** Individual unit marker, reflectorized or nonreflectorized, generally less than 25 mm (1 in.) in height, attached to and extending above the normal pavement surface for the purpose of regulating, warning, or guiding traffic.

PAVEMENT PICK Pavement engaging tool held by a working tool holder attached to the cutter drum that performs the action of pavement planing/milling.

PAVEMENT PROFILING Technique of cold planing/milling a pavement to prepare or modify the riding surface.

PAVEMENT SAW Power-operated equipment used to cut concrete or asphalt pavement, consisting of a diamond-grit or abrasive wheel up to 1.2 m (48 in.) or more in diameter, mounted within a cart equipped to supply cooling water to the cut.

PAVEMENT SEALER Material used to seal areas of asphalt that have become soft and/or spongy due either to dilution of the binder from oil drippings or from evaporation of the binder. Sealants are of three basic types: native-asphalt-base materials, coal-tar bases, and vinyl-base coatings.

PAVEMENT STRUCTURE Combination of subbase, base course, and surface course placed on a subgrade to support the traffic load and distribute it to the roadbed. *See also* **Base course, Subbase, Subgrade, Subgrade treatment,** and **Surface course.**

PAVEMENT STRUCTURE COMBINATION Asphalt pavement constructed on an old portland cement pavement, a portland cement concrete base, or other rigid-type base or on a granular base. *Also called* a **Composite-type pavement.**

PAVEMENT TEXTURIZING Technique of cold planing/milling a pavement to improve skid resistance.

PAVEMENT TOOTH Pavement engaging tool held by a working tool holder attached to the cutter drum that performs the action of pavement planing/milling.

PAVER 1. Unglazed porcelain or natural clay tile formed by the dust-pressed method and similar to ceramic mosaics in composition and physical properties but relatively thicker with 3870 mm² (6 in.²) or more of facial area; **2.** Stone, concrete, brick, or other material shaped or cast to a modular size, sometimes to an interlocking pattern. *See also* **Paving stone.**

PAVILION 1. Building, or part of a building, used for exhibits or entertainment; **2.** Decorative, lightly built shelter or summerhouse.

PAVILION ROOF Sloped roof with equal hipped areas all round.

PAVING 1. Simple road surface constructed directly on natural ground; **2.** Regularly placed stones or bricks forming a floor.

PAVING BREAKER Air hammer in which the cutting steel does not rotate.

PAVING BRICK *See* **Brick.**

PAVING STONE Paving material of stone or concrete, in sizes from approximately brick size to 600 x 600 mm (24 x 24 in.), or larger, constructed with carefully controlled dimensions to permit narrow joints between adjacent blocks. *Also called* **Paver.**

PAVING TRAIN Succession of mobile processing units designed to place and finish asphalt or concrete pavement structures on site.

PAWL 1. Tooth or set of teeth designed to lock against a ratchet; **2.** Pivoting locking lever that will permit movement in only one direction. Movement in the opposite direction can be achieved only by releasing the mechanism.

PAY-AS-CUT Timber purchase based on a dollar amount for a certain amount/volume of wood, such as dollars per MBF. Payment is made only as timber is cut and transported.

PAYBACK Length of time before money invested to create a project is recovered through rents or other earnings derived from the project.

PAY FORMATION Body of rock, earth, or ore, the value of which is enough to justify excavation.

PAY ITEM Unit in a contract that is specified as a bid item and listed as a separate unit for payment.

PAYLOAD 1. Total weight of the commodity being carried on a truck at a given time; **2.** Excavated load for which payment is made.

PAYLOAD AND BODY ALLOWANCE Payload capacity of a truck with allowance for the weight of the truck body.

PAYMENT BOND *See* **Bond.**

PAYMENT REQUEST Request made by a contractor or supplier for payment for work done or materials supplied.

PAYMENT SCHEDULE Sequence and times of partial payments made as part of a total contract sum.

PAYNE'S PROCESS Method of fireproofing lumber by treating it first with an injection of sulfate of iron, then with a solution of sulfate of lime or soda.

PAY OUT To reel line from a drum.

PB *Abbreviation for:* power brake; prebored; pull box; push button.

p-bar *Abbreviation for:* push bar.

PBI Industrial grade particleboard manufactured primarily for use in industrial applications such as furniture and cabinet manufacture.

PBS *Abbreviation for:* protected box strike; push-button switch.

PBU Particleboard underlayment, manufactured for interior use as the smooth surface layer under finish floor coverings in double-layer floor construction.

p-but *Abbreviation for:* push button.

pc *Abbreviation for:* printed circuit.

pc. *Abbreviation for:* piece.

PC *Abbreviation for:* part catalog; portland cement; power circuit; purchasing and contracting.

P/C *Abbreviation for:* poured concrete.

pcb *Abbreviation for:* printed circuit board.

PCB *Abbreviation for:* power circuit breaker.

PCBS *Abbreviation for:* protected cast box strike.

pcc *Abbreviation for:* portland cement concrete.

PCE *Abbreviation for:* pyrometric-cone equivalent.

pcf *Abbreviation for:* pounds per cubic foot.

pch *Abbreviation for:* pitch; punch.

pchg *Abbreviation for:* punching.

pci *Abbreviation for:* pounds per cubic inch.

pc. mk *Abbreviation for:* piece mark.

pcs *Abbreviation for:* pieces.

pct *Abbreviation for:* percent.

pd *Abbreviation for:* paid; period; pressure drop.

PD *Abbreviation for: per diem* (by the day); plastic deformation; preliminary draft.

PDA *Abbreviation for:* proposed development approach.

pdb *Abbreviation for:* power disc brake.

P-DELTA EFFECT Secondary effect of column axial loads and lateral deflection on the moments in members.

pdic *Abbreviation for:* periodic.

pdlk *Abbreviation for:* padlock.

pe *Abbreviation for:* polyethylene.

PE *Abbreviation for:* plain end; polyethylene; porcelain enamel.

PEACH BASKET Template against which the entire head of a tall chimney is built.

PEA GRAVEL Screened gravel, most of the particles of which pass a 10-mm (0.38-in.) sieve and are retained on a No. 4 (4.76-mm, 0.197-in.) sieve.

PEAK 1. Highest point of a gable; 2. Point on a truss where the sloped chords meet.

PEAKED ROOF Roof rising to a point.

PEAKER Top log on a load of logs.

PEAK FLOOD LEVEL Water surface elevation of a design flood.

PEAK HORSEPOWER *See* **Horsepower.**

PEAKING POWER ELECTRIC SET *See* **Electric set.**

PEAK LOAD Maximum instantaneous demand for electrical power that determines the generating capacity required.

PEAK LOAD SHAVING Process by which a utility customer minimizes demand charges either by generating power or by shedding load.

PEAK PARTICLE VELOCITY Maximum particle velocity.

PEAK-TO-PEAK VALUE Total value between the positive and negative peaks on AC sine waves.

PEAK VALUE Maximum voltage or current value measured from zero.

PEAK VOLTAGE Maximum instantaneous value of a voltage.

PEARL MOLDING Molding resembling a string of beads.

PEAT Lightweight mixture of decomposed plant tissue in which parts of the plant are easily recognized.

PEATLAND *See* **Muskeg.**

PEAVEY Long-handled cant hook-type tool used to roll small logs.

peb. *Abbreviation for:* pebble; pebbled.

PEBBLE Sediment particles between 2 mm (0.06 in.) and 65 mm (2.5 in.) in diameter.

PEBBLEDASH *See* **Roughcast.**

PECKER POLE Small log.

ped. *Abbreviation for:* pedestal.

pedes. *Abbreviation for:* pedestrian.

PEDESTAL 1. Upright compression member whose height does not exceed three times its average least dimension; 2. Base supporting a column; 3. Base for a statue.

PEDESTAL PILE *See* **Pile.**

PEDESTRIAN Person traveling on foot.

PEDESTRIAN PATH Footway or track reserved for use by pedestrians or joggers.

PEDIMENT 1. Triangular part crowning the front of a building, formed by running the top member of the entablature along the sides of a gable; 2. Also, a similar feature above doors, windows, etc. There are two principal types:

> **Broken pediment:** Has a gap in the base molding.

> **Open pediment:** One where the sloping sides are returned before reaching the apex.

PEELED Tree trunk from which all bark has been removed before treating and/or driving as a pile.

PEELER 1. Large log without defects, suitable for making plywood; 2. One of a series of blades that channel and direct water expelled by the impeller of a centrifugal pump.

PEELING *See* **Scaling.**

PEEN 1. To distort or bend a steel sheet by pounding with the peen of a hammer; 2. Blunt, wedge- or ball-shaped end of a hammer head, opposite the striking face (except on a double-peen hammer). *Also called* **Pein.**

PEENING 1. Creating a concave or convex shape in sheet metal by beating with a peen hammer; 2. Permanent distortion of wire rope resulting from cold plastic metal deformation of the outer wires. Usually caused by pounding against a sheave or machine member, or by heavy operating pressure between rope and sheave, rope and drum, or rope and adjacent wrap of rope.

PEEP DOOR Small furnace door, usually provided with a shielded glass opening through which combustion may be observed.

PEE-WEE Small-diameter, merchantable log.

PEG 1. Short, pointed wooden stick driven into the ground to mark a line, level, or station; 2. Piece of wood used in place of a nail to hold parts together; 3. Metal pin used to secure glass to a metal window frame.

PEG ADJUSTMENT Process of adjusting a leveling instrument to make the collimation line parallel with the axis of the bubble level, using two pegs the length of one instrument sight apart.

PEGGING OUT *See* **Stakeout.**

PEG POINT *See* **Steady point.**

PEIN *See* **Peen.**

PELICAN HOOK *See* **Finger link.**

PELLET POWDER Hollow cartridges comprised of black powder.

PELLET-TYPE THERMOSTAT Thermostat operated by a heat-sensitive pellet.

PELTIER EFFECT Evolution or dissipation of heat that occurs when an electric current is passed across the junction between dissimilar metals.

PENAL AMOUNT Maximum liability of a bond; often limited to 100% of the contract amount.

PENALTY FOR DELAY Contractural clause that imposes a penalty, usually financial, if a contract is not completed on the date stipulated in the contract documents.

PENCILING Coloring the mortar joints of brickwork to add contrast.

PENCIL ROD Plain metal rod about 6 mm (0.25 in.) in diameter.

pend. *Abbreviation for:* pendant; pending.

PENDANT Supporting rope that, under tension, maintains a constant distance between its points of attachment. *See also* **Guy rope.** *Also called* **Guyline, Stay rope,** and **Standing line.**

PENDANT BOLT Machine bolt used to fasten the hanger sheave assembly to the top of a door.

PENDANT LINE One of thetsraight pull support lines for a boom. Always in pairs, these lines do not go through sheaves, but terminate at a babbitted knob and dee.

PENDANT SPRINKLER Type of automatic sprinkler with its deflector below the orifice.

PENDANT SWITCH *See* **Switch.**

PENDENT Elongated boss projecting down from the intersection of groins, or at the end of hammer beams.

PENDENTIVE ARCH *See* **Arch.**

PENDENTIVE BRACKETING *See* **Cove bracketing.**

PENDENT POST Timber that projects down the inside face of a wall from the plate in hammer-beam trusses, and that receives the hammer braces.

PENDULUM LEAD *See* **Lead.**

PENETRANT 1. Protective treatment that lines masonry pores leaving no film on the surface; **2.** Liquid or gas that, when applied to the surface of a metal, enters cracks to make them visible.

PENETRATING OIL Sealer used to protect the surface of unglazed ceramic tile.

PENETRATION 1. Rate of drilling at 100% efficiency with the drill bit working continuously; **2.** Downward axial movement of a pile per hammer blow as measured at an established point on the pile. There are several variations, including:

> **Gross penetration: (a).** Total downward movement of a pile caused by the hammer blow before rebound; **(b).** Maximum movement during test loading.
>
> **Minimum penetration:** Depth or length of pile specified to theoretically develop the required load bearing or uplift capacity of the soil, the required lateral strength, or a desired bearing strata.
>
> **Net penetration: (a).** Gross movement of a pile, less the rebound from a hammer blow; **(b).** Net measured settlement after unloading from a test.

3. Measure of the hardness of paving asphalts, determined by a test that gives the distance that a standard needle with a weight attached will penetrate a sample at a given temperature in a given length of time; **4.** Extent to which weld metal combines with base metal, as measured from the surface of the base metal; **5.**

Arbitrary measure of consistency (hardness). *See also* **Block penetration, Lubricating grease penetration, Prolonged worked penetration, Unworked penetration,** and **Worked penetration.**

PENETRATION CONSTRUCTION Similar to penetration macadam but employing denser graded mineral aggregate in which void spaces are much smaller.

PENETRATION MACADAM Oldest asphalt construction method. It involves three aggregate applications, progressing from the largest to the smallest sizes, and two binder applications, for pavement thicknesses of 100 mm (4 in.) or more.

PENETRATION NEEDLE *See* **Proctor needle.**

PENETRATION NUMBER Consistency of grease as defined by its resistance to deformation under an applied force.

PENETRATION PROBE Device for obtaining a measure of the resistance of concrete to penetration, customarily determined by the distance a steel pin is driven into the concrete from a special gun by a precisely measured explosive charge.

PENETRATION RATE Length of blast hole produced by a drill in a unit of time.

PENETRATION RESISTANCE 1. Resistance of either mortar or cement paste to penetration by a plunger or needle under standard conditions, usually expressed in megapascals (MPa) or pounds per square inch (psi); **2.** Resistance to penetration from driving expressed in blows per increment of advance of a pile; **3.** N-value in the Standard Penetration Test.

PENETRATION TEST *In situ* soil test to give an indication of its load-bearing capacity and density. *See also* **N-value.**

PENETROMETER 1. Device used to measure the consistency of greases; **2.** Instrument for measuring the resistance to penetration of a soil to a point of defined size and shape. *Also called* **Dutch cone.**

PENITENTIARY Prison.

PENNY 1. *See* **Nail;** **2.** *See* **Anvil.**

PENSTOCK Pipeline or pressure shaft in a dam, leading from the headrace or reservoir to the turbines.

pent. *Abbreviation for:* penetrate; pentagon.

PENTAERYTHRITE TETRANITRATE Substance used in detonating fuse.

PENTASTYLE Having five columns across the main facade.

PENTHOUSE 1. Subsidiary structure attached to a main building, particularly with a sloping roof; **2.** House or apartment built on the roof of a building; **3.** Machine room above the hoistway on traction elevators; **4.** Projecting hood over a door or window.

PENT ROOF Roof, other than a lean-to roof, that has a single sloping surface.

PEPTIZER Compounding ingredient used in small proportions to accelerate by chemical action the softening of rubber under the influence of mechanical action, heat, or both.

per *Abbreviation for:* perimeter.

PERCENTAGE DROOP *See* **Governor droop.**

PERCENTAGE LEASE Commercial lease under which a tenant's rent is based on a percentage of the tenant's gross or net income.

PERCENTAGE-OF-COMPLETION METHOD OF ACCOUNTING *See* **Method of accounting.**

PERCENTAGE OF REINFORCEMENT Ratio of cross-sectional area of reinforcement steel to the effective cross-sectional area of a member, expressed as a percentage. *Also called* **Reinforcement ratio.**

PERCENT AGREEMENT Contract for professional services in which compensation is based on a percentage of the construction cost.

PERCENT AIR Actual amount of air supplied to a combustion process, expressed as a percentage of the amount theoretically required for complete combustion.

PERCENT ARTICULATION 1. Amount of total possible angular movement of an articulated joint; **2.** Measure of the number of discernible speech sounds correctly identified by a group of listeners with average hearing in a defined space.

PERCENT FINES 1. Amount, expressed as a percentage, of material in an aggregate finer than a given sieve, usually the No. 200 (0.074 mm, 0.0029 in.); **2.** Amount of fine aggregate in a concrete mixture expressed as a percent by absolute volume of the total amount of aggregate.

PERCENT GRADE *See* **Expressions of slope.**

PERCENT HUMIDITY *See* **Humidity.**

PERCENT LEASE Rent based on a percent of total revenue realized from from the property, usually with a floor minimum.

PERCENT OF COMPLETION *See* **Method of accounting (percent of completion).**

PERCENT OF SLOPE *See* **Expressions of slope.**

PERCENT POSSIBLE SUNSHINE Amount of radiation available compared to the amount that would be present if there were no cloud cover over the same period.

PERCHED WATER *See* **Groundwater.**

PERCOLATING FILTER Type of aerobic sewage treatment bed.

PERCOLATION Movement of gas or liquid through the void spaces of soil without a definite channel.

PERCOLATION TEST 1. Standard test to measure the rate at which the soil of an area will absorb and disperse moisture, used to determine the suitability of the area to receive the effluent from a domestic waste disposal system. *Also called* **Perk test; 2.** Test carried out on a small sample slab made from a proposed concrete mix to check its suitability for use in a water-retaining structure.

PERCUSSION BORING Means of advancing a bore using air- or cable-driven impact tools or excavating tools that are repeatedly dropped onto the face or base of the bore.

PERCUSSION DRILL *See* **Drill.**

PERCUSSION TOOL Tool that operates by striking rapid blows.

PERENNIALLY-FROZEN SOIL Soil that has been continuously frozen for more than three years.

perf. *Abbreviation for:* perfect; perforate; perforated; performance.

PERFECT COMBUSTION Combination of the chemically correct proportions of fuel and air in combustion so that the fuel and oxygen are both totally consumed.

PERFECT FRAME Frame that is stable under loading from any direction, but that would become unstable if one of its members were removed or if one of its fixed ends became unhinged.

PERFORATED Pierced with holes.

PERFORATED BRICK *See* **Brick**

PERFORATED DRYWALL Plasterboard sheet that has uniformly spaced holes giving it high sound absorption.

PERFORATED WALL *See* **Wall.**

PERFORATOR TEST Test that extracts free formaldehyde from particleboard with tolulene.

PERFORMANCE BOND *See* **Bond.**

PERFORMANCE FACTOR Ratio of the useful output capacity of a system to the input required to obtain it. Input and output units need not be consistent. *See also* **Tractive effort.**

PERFORMANCE INSURANCE *See* **Insurance.**

PERFORMANCE RATED Panel that has been tested to determine if it meets specific loading and deflection conditions from impact, point loads, and uniform loads when the panel spans two or more supports.

PERFORMANCE SPECIFICATION Statement of the desired operation or function of a product or process that does not specify the material from which the product must be constructed.

PERFORMANCE STANDARD Formally established criterion for special activity that (a) outlines the work involved, (b) describes work methods and the composition of efficient crews, and (c) lists the expected accomplishments or productive rate.

PERFORMANCE TEST *See* **Service test.**

PERGOLA Covered walk in a garden formed by a double row of posts or pillars connected longitudinally by a continuous beam that supports joists above, usually supporting climbing plants.

PER HECTARE FACTOR Number used to convert sample plot information to per-hectare information (as in converting a plot volume to be a per-hectare volume).

peri. *Abbreviation for:* perimeter.

PERIL Cause of loss insured against: explosion, collapse, fire, etc.

PERILLA OIL Substitute for linseed oil in paint.

PERIMETER Length of the periphery of an area or object.

PERIMETER GROUTING Injection of grout, usually at relatively low pressure, around the periphery of an area that is subsequently to be grouted at greater pressure.

PERIMETER HEATING *See* **Heating system.**

PERIMETER-INSULATED RAISED FLOOR SYSTEM Crawl space foundation system where insulation is applied only to the inside of the perimeter foundation stem wall, designed to save construction costs and improve energy performance.

PERIMETER JOINT Joint formed by the outer edge of one panel or material and the leading edge of another. *Also called* **Peripheral joint.**

PERIMETER RELIEF Construction method to allow for building movement at the intersections of walls and ceilings through the use of gasketing.

PERIMETER WARM-AIR HEATING SYSTEM System of the combination panel and convection type. Warm air ducts embedded in the concrete slab of a basement-less house, around the perimeter, receive heated air from a furnace and deliver it to the heated space through registers placed in or near the floor. *See also* **Heating system.**

PERIOD 1. Time for one cycle to be completed; 2. Time for one complete vibration or oscillation of a wave, measured in seconds.

PERIOD FOR COMPLETION *See* **Contract time.**

PERIODIC ANNUAL INCREMENT Mean annual growth or increase in volume during a specific period of time.

PERIODIC CUT *See* **Periodic harvest.**

PERIODIC HARVEST Removal of several years' accumulated average allowable cut in one year or other period. *Also called* **Periodic cut.**

PERIPHERAL JOINT *See* **Perimeter joint.**

PERIPHERAL SPEED Speed at which any point or particle on the perimeter of a wheel is traveling when the wheel is revolved, expressed in surface meters per minute (sm/min), or surface feet per minute (sfpm).

PERIPHERY Outside edge of an object.

PERISTALTIC HYDRAULIC PUMP *See* **Pump, hydraulic.**

PERISTALTIC PUMP *See* **Pump.**

PERK TEST *See* **Percolation test.**

PERLITE Volcanic glass that, when expanded by heating, is used as an insulating material and as a lightweight aggregate in concretes, mortars, and plasters.

PERLITE PLASTER Lightweight mixture of perlite aggregate, and gypsum plaster: a product having good sound absorption characteristics.

PERLITIC STRUCTURE Structure produced in homogeneous material by contraction during cooling, consisting of a system of irregular convolute and spheroidal cracks, generally confined to natural glass.

PERM Measure of water vapor movement through a material: one grain per square foot per hour per inch of mercury difference.

perm. *Abbreviation for:* permanent; permeable.

per M *Abbreviation for:* by the thousand.

PERMAFROST Permanently frozen ground.

PERMANENT BLASTING WIRE *See* **Electric blasting circuitry.**

PERMANENT DOLLY *See* **Dolly.**

PERMANENT FORM 1. Any form that remains in place after the concrete has developed its design strength; 2. Form that may or may not become an integral part of the structure.

PERMANENT LOADS *See* **Loads.**

PERMANENT MORTGAGE *See* **Mortgage.**

PERMANENT SCAFFOLD *See* **Suspended scaffold.**

PERMANENT SET Inelastic elongation or shortening.

PERMEABILITY Rate at which moisture vapor or free water will enter and/or pass through a material.

PERMEABLE Soil or material that permits the passage of water or gas.

PERMEABLE COATING *See* **Breathable coating.**

PERMEAMETER Instrument for measuring the coefficient of permeability of a soil sample: a constant-head unit is used for permeable materials like gravel; a falling-head unit for impermeable materials like clay.

PERMEANCE Ratio of the rate of water vapor transmission through a material or assembly to the vapor pressure differential between the surfaces.

PERMISSIBLE Explosives approved for nontoxic fumes and allowed in underground work.

PERMISSIBLE LINE PULL Line pull, less than the available pull, restricted by rope strength, clutch or brake ability, or other equipment limitations.

PERMIT Document issued by a regulatory authority permitting the bearer to take a defined action.

perm. sp. *Abbreviation for:* permanent split.

perp. *Abbreviation for:* perpendicular.

perp. bis. *Abbreviation for:* perpendicular bisector.

PERPEND Vertical joint, particularly in brickwork and other masonry, but also between shingles and roofing tiles and slates.

PERPEND BOND *See* **Bond.**

PERPENDICULAR 1. Exactly vertical; 2. At right angles to a given plane.

PERPEYN Projecting pier, buttress, or pilaster.

PERRON Exterior entrance steps, usually elaborated.

pers. *Abbreviation for:* personnel.

PERSONAL PROPERTY Property other than real property.

PERSONNEL TIME *See* **Machine time.**

persp. *Abbreviation for:* perspective.

PERSPECTIVE Appearance of objects and their surroundings as determined by their orientation, and relative position to each other and to the observer.

PERSPECTIVE DRAWING Depiction in two dimensions of three-dimensional objects and their surroundings.

PERT *Acronym for:* Program Evaluation and Review Technique; Project Evaluation and Review Technique.

PERT SCHEDULE Project Evaluation and Review Technique (PERT) charting of activities and events anticipated in a work process.

PEST Organism capable of causing material damage.

PETCOCK Small faucet or valve.

PETN *See* Pentaerythrite tetranitrate.

petr. *Abbreviation for:* petrified.

petro. *Abbreviation for:* petroleum.

PETROGRAPHY Branch of petrology dealing with the description and systematic classification of rocks aside from their geologic relationships. *See also* **Lithology,** and **Petrology.**

PETROLEUM Oily, liquid solution of hydrocarbons occurring naturally in certain rock strata, and occasionally as a surface discharge. When fractionally distilled, it yields such products as benzene, kerosene, fuel oil, gasoline, naptha, etc. *See also* **Crude oil.**

PETROLEUM ASPHALT Asphalt refined from crude petroleum.

PETROLEUM FLUID Fluid composed of petroleum oil. It may contain additives.

PETROLOGY Science of rocks, treating their origin, structure, composition, etc., from all aspects and in all relations. *See also* **Litholgy,** and **Petrography.**

PEW Fixed bench with back in the auditorium of a church.

pf. *Abbreviation for:* profile.

PF *Abbreviation for:* post forming; power factor.

pfd *Abbreviation for:* preferred.

PF resin *Abbreviation for:* phenol-formaldehyde resin.

pfw *Abbreviation for:* per foot width.

pg *Abbreviation for:* projected grade.

ph. *Abbreviation for:* phase; photostat.

pH Simplified system of measuring acidity or alkali irrespective of the acid or alkali involved; in which neutrality is 7.0; e.g., mineral acid solution is 1.0 to 2.8, acetic acid solution or citric acid solution is 3.0 to 4.0, ammonia is 9.0, and lime water is 12.0.

PH *Abbreviation for:* Phillips-head; powerhead.

PHASE 1. Two items in correct relationship to each other; **2.** Portion of the life of a project. *See also* **Construction phase,** and **Feasibility phase; 3.** Windings of a generator that determine the number of complete voltage and/or current sine waves generated per 360 electrical degrees; **4.** *See* **Traffic signal; 5.** Distinct functional operation during a cycle. Some typical sequential phases are: neutral, rapid advance, feed or pressure stroke, dwell, and rapid return.

PHASE BALANCE Amount of voltage difference between phase voltages under balanced load conditions.

PHASE BALANCE WITH UNBALANCED LOADS Amount of voltage unbalance between phase voltages when one phase is loaded to a specified level and the other two phases are unloaded.

PHASE-CHANGE MATERIAL Material that stores heat by melting and releases heat by solidifying.

PHASED CONSTRUCTION *See* **Accelerated design and construction.**

PHASED DESIGN AND CONSTRUCTION *See* **Accelerated design and construction.**

PHASE DIFFERENCE *See* **Out-of-phase.**

PHASE ROTATION Sequence in which the phases of a generator or network pass through the positive maximum points of their waves. Typically 1-2-3 or 3-2-1.

ph. brz. *Abbreviation for:* phosphor-bronze.

phen. *Abbreviation for:* phenolic.

PHENOL Class of acid organic compounds used in the manufacture of epoxy resins, phenol formaldehyde resins, plasticizers, plastics, and wood preservatives.

PHENOL FORMALDEHYDE Water-resistant thermosetting resin system, commonly used to bond softwood plywoods, oriented strand board (OSB), and exterior particleboard.

PHENOLIC RESIN Class of synthetic, oil-soluble resins (plastics) produced as condensation products of phenol, substituted phenols, and formaldehyde, or some similar aldehyde that may be used in paints.

PHI FACTOR Capacity reduction factor in structural design, a number less than 1.0 (usually 0.65 to 0.90) by which the strength of a structural member or element, in terms of load, moment, shear, or stress, is required to be multiplied in order to determine its design strength capacity. *Also called* **Capacity reduction factor.** *See also* **Design strength,** and **Load factor.**

PHILLEO FACTOR Distance, used as an index of the extent to which hardened cement paste is protected from the effects of freezing, so selected so that only a small portion of the cement paste (usually 10%) lies farther than that distance from the perimeter of the nearest void. *See also* **Protected paste volume.**

PHILLIPS-HEAD SCREW *See* **Screw.**

PHLOEM Layer of tree tissue just inside the bark that conducts food from the leaves to the stem and roots. *See also* **Cambium.**

phoc. *Abbreviation for:* photocopy.

PHOLAD *See* **Marine borer.**

PHON 1. Unit of loudness of sound as perceived by the average human ear; **2.** Loudness level of a sound in phons (numerically equal to decibels of a sound at 1000 cycles).

PHOSPHATE ESTER SYNTHETIC FLUID Fluid composed of phosphate esters. It may contain some additives.

PHOSPHATING Protecting a bare metal surface by application of hot phosphoric acid.

PHOSPHOR-BRONZE Alloy of copper, tin, and phosphorus.

PHOSPHORESCENT PAINT Paint containing phosphors that absorb energy at one wavelength and emit it over a period of time at a wavelength within the visible spectrum. *See also* **Fluorescent paint,** and **Luminous paint.**

photo. *Abbreviation for:* photograph; photographic.

PHOTOCELL FLAME DETECTOR Device that generates or rectifies an electric current while exposed to the light from a flame. Failure of the current or lack of rectification may be used to trip shut a safety shutoff valve.

PHOTOCONDUCTIVE CELL Photoelectric cell whose resistance is proportional to the intensity of light striking it.

PHOTODEGRADABLE Process where ultraviolet radiation degrades the chemical bond or link in the polymer or chemical structure of a plastic.

PHOTODIODE Diode that can switch and regulate electrical current in proportion to the intensity of the light striking the PN junction.

PHOTOELASTICITY Examination by transparent model analysis of the distribution of stresses in an object through the use of polarized light that reveals isochromatic and isoclinic lines. The lines indicate the direction of axis of principal stresses at any point and the magnitude of the difference of principal stresses.

PHOTOELECTRIC CELL Switch that is held in one position by a fixed beam of light and actuated by an interruption to that beam.

PHOTOELECTRIC DOOR OPENER Photoelectric control system used to effect the opening and closing of power-operated doors.

PHOTOELECTRIC LOAD INDICATOR Device comprising a cylinder of optical glass that is placed in a hole in a steel cell body, the load being applied through end plates. When the glass is strained under load, photoelastic interference fringe patterns are visible when the glass is illuminated with polarized light.

PHOTOGRAMMETRY Surveying technique employing a photo-theodolite or other photographic equipment to take a successive series of photographs from a fixed station along a fixed camera axis to record movement.

PHOTOGRAPHING Bas-relief or outline of a reinforcement that appears on the cover of a hose after vulcanization. *Also called* **Profiling.**

PHOTOMETER *See* **Flame photometer.**

PHOTOVOLTAIC ARRAY Group of interconnected photovoltaic modules acting as a composite unit.

PHOTOVOLTAIC CELL Device that converts light into electricity through the excitation of electrons.

phys. *Abbreviation for:* physical.

PHYSICAL DETERIORATION Depreciation of a structure due to the passage of time, or the action of the elements, or the wear and tear to which it has been subjected in use, or any combination.

PHYSICAL INVENTORY Inventory shown to be present by observation and counting and/or sampling. *See also* **Book inventory.**

PHYSICAL LIFE Estimated life expectancy of a structure.

PHYSICAL MAINTENANCE Preservation and up-keep of a property, equipment, machinery, structures, a highway, etc., including all of its elements, in as nearly as practicable its original, as-constructed condition or its subsequently improved condition.

PHYSICAL PROPERTIES Properties or qualities, other than mechanical properties, that have to do with the physics of a material.

PHYSICAL RESTRAINT Physical activity or work item that must be completed before the next activities or items in a sequence can begin (i.e., concrete must be cured before forms can be removed).

PHYSICAL SURVEY 1. On-ground activity that determines or confirms the physical state and/or condition of an area, facility, or structure; **2.** Activity of a sewer system evaluation survey designed to determine specific flow characteristics, groundwater levels, and the physical condition of the system.

PHYSICAL TESTING Examination of a material to determine its physical properties.

pi (π) Number (approximately 3.1416) that, when multiplied by the diameter of a circle, will give its circumference, and that, when multiplied by the square of the circle's radius, will give its area.

pi *Abbreviation for:* point indicated.

PIANO WIRE Thin steel wire used in aligning. *See also* **Music wire.**

PIAZZA 1. Public square or marketplace, usually surrounded by buildings; **2.** Long covered walk with a roof supported by columns.

pic. *Abbreviation for:* picture.

PICK 1. Hand tool with two long, sharp points, used for digging; **2.** Device other than a key, used to open a lock; **3.** Individual filling yarn of a fabric or woven jacket. *See also* **Working tool.**

PICK AND DIP Method of laying brick whereby the bricklayer simultaneously picks up a brick with one hand and, with the other hand, enough mortar on a trowel to lay the brick. Sometimes called the Eastern or New England method.

PICK Pavement engaging tool held by a working tool holder attached to the cutter drum that performs the action of pavement planing/milling. *Also called* **Pavement bit, Pavement cutter, Pavement cutting tool,** and **Pavement tooth.**

PICKET Upright wooden fence member, often decorated, supported by top and bottom rails fastened to posts.

PICKING Surface patterning of rubble stone with a steel point struck at right angles to the surface to produce closely spaced pits.

PICKING TABLE OR BELT Table or belt on which solid waste is manually sorted and certain items are removed.

PICKLING Removal of surface oxides from metals by chemical or electrochemical reaction.

PICKUP POINT Position under a cableway where

concrete buckets are transferred or where the bucket is filled by a transfer car.

PICKUP ROLLER Device on a hoistway door that mates with the clutch on the car door to allow the hoistway doors to be pulled open and closed.

PICKUP TRUCK Small, cargo-carrying vehicles in the 0.25- to 1.0-ton capacity class.

PICO Prefix representing 10⁻¹². Symbol: p. Used in the SI system of measurement. *See also the appendix*: **Metric and nonmetric measurement.**

PICTORIAL DIAGRAM *See* **Diagram.**

PICTURE PLANE Imaginary plane between the eye and the object being drawn in perspective.

PICTURE WINDOW *See* **Window.**

PIECE MARK Identification given to individual components of a structure that have been fabricated off-site.

PIECE RATE Payment for labor where income is related to output.

PIEDMONT Located or formed at the base of a mountain range; an example would be a piedmont terrace.

PIER 1. Structure supporting the spans of a bridge; **2.** Structure built out from the shore, over water, supported on piles; **3.** Square column used to support weight; **4.** Part of a wall between windows, doors, or other openings; **5.** Strengthening mass built out from the face of a wall; **6.** A buttress.

PIER CAP Member at the top of a bridge pier, transverse to the deck, that distributes the concentrated loads from the bridge uniformly over the pier(s).

PIERCING *See* **Probing.**

PIER GLASS Wall mirror hanging between two windows in close proximity, creating the effect of a pier between them.

PIEZOMETER Device for measuring the pressure head of pore water at a specific point within a soil mass.

PIEZOMETER NEST Two or more piezometers installed in a common borehole or in close proximity to measure the vertical hydraulic gradient

PIG 1. Air manifold comprising a number of pipes that distribute compressed air delivered by a single, large duct; **2.** Lead or iron block cast at the smelting furnace. *See also* **Ingot.**

pig. *Abbreviation for:* pigment.

PIGGY-BACK TRUSS *See* **Truss.**

PIGMENT Coloring matter, usually in the form of an insoluble fine powder.

PIGTAIL 1. Y-shaped connection between two or more electrical connectors; **2.** Lamp socket with two attached wires, used for lighting and testing.

PIKE POLE Pole, 3 m to 6 m (10 ft to 20 ft) long, with a sharp twisted point and a hook at one end, used for hauling floating logs.

pil. *Abbreviation for:* pilaster.

PILASTER Shallow pier or rectangular column projecting only slightly from the face of a wall.

PILASTER BLOCK Precast foundation block on which masonry pilasters and columns may be built.

PILASTER FACE Form for the front surface of a pilaster parallel to the wall.

PILASTER SIDE Form for the side surface of a pilaster perpendicular to the wall.

PILE 1. *See* **Harvest functions. 2.** Slender timber, concrete or steel structural element driven, jetted, or otherwise embedded on end in the ground to support a load or to compact the soil. There are many types, including:

Anchor: (a). Pile connected to a structure by one or more ties that furnish lateral support or resist uplift; **(b).** Reaction pile for load testing.

Augered: Concrete pile cast-in-place in an augered hole. *Also called* **Drilled pile,** and **Uncased pile.**

Batter: Pile installed at an angle to the vertical; a **Raker pile** or **Raking pile.** *Also called* **Spur pile.**

Bearing: Pile resting on a solid stratum and supporting a load. *See also* **Friction pile,** below.

Box: Pile fabricated from two deep-arch sheet piles, channels, or other shapes, welded along their contact lines.

Button-bottom: Precast concrete tip of about 430 mm (17 in.) diameter driven with a 355 mm (14 in.) heavy-wall pipe. For a cased pile a corrugated shell is lowered and fastened to the base and the pipe withdrawn. For an uncased pile, concrete is forced out of the bottom and the pipe withdrawn.

Caisson: (a) Cast-in-place pile made by driving a tube, excavating it, and filling the cavity with concrete; **(b)** Augered pile with or without permanent casing. *Also called* **Drilled shaft.**

Cast-in-place: Concrete pile concreted either with or without a casing in its permanent location, as distinguished from a **Precast pile.**

Caudil drive-point: Thin shell pipe driven with a mandrel striking on the point.

Compacted concrete: Cast-in-place pile formed with an enlarged base, where concrete is placed in small batches, each compacted by heavy blows prior to attaining initial set.

Composite: Pile made up of different materials fastened together end-to-end to form a single pile.

Compound batter: Pile driven at an angle in two directions from the principal line of the piles; the angle is normally expressed as a ratio of horizontal to vertical of the piles' centerline along with horizontal orientation angle from the principal line. *Also called* **Skew batter pile.**

Compression: Pile designed to resist an axial force such as would cause it to penetrate further into the ground.

Displacement: Solid or hollow pile with a closed driving end that displaces soil volume as it is driven.

Drilled-in caisson: Open-end pipe driven to rock,

423

cleaned out, and, with a socket, driven into the rock to receive a steel core (H, WF, or bars), then the socket and pipe are filled with concrete. Core, concrete, and pipe contribute to high carrying capacity.

Drilled pier/pile: Large diameter, up to 3 m (10 ft) or more, concrete pier or pile cast in place in a hole previously bored in soil or rock.

End bearing: Pile that transmits forces to the ground mainly by compression on its base or from its base resting on or penetrating into a firm stratum.

Fender: Pile driven in front of a structure to protect it from damage from floating objects or to absorb shock from impact.

Foundation: Relatively long, slender column installed in the ground to provide support from friction along its periphery and/or end bearing on firm material.

Friction: Pile transmitting forces to the ground principally by friction between the surface of the pile and the surrounding ground. *See also* **Bearing pile,** above.

Friction end-bearing: Pile that achieves support from the combination of side friction and tip (end) bearing.

Guard: *See* **Fender,** above.

Guide: Heavy vertical square pile driven close to sheet piles, carrying the walings that first guide, and then support the sheet piles.

H-pile: Rolled steel section with web and flange the same thickness (wide-flange and other steel sections also are used). *Also called* **H-beam.**

Indicator: Piles, often prestressed concrete, driven in advance of major work to determine needed length before ordering.

In situ: Pile formed with or without a casing by forming a hole in the ground and filling it with plain or reinforced concrete, or precast concrete sections that are grouted in place.

Low displacement: *See* **Nondisplacement pile,** below.

Micropile: Small-diameter piles; most often used in underpinning.

Needle: Very small diameter, slender, driven steel tubular or rail section piles, used in underpinning.

Nondisplacement: Piles formed by boring or other methods of excavation. H, open-end pipe, and sheet piles are considered low displacement piles.

Pedestal: Cast-in-place concrete pile constructed so that concrete is forced out into a widened bulb or pedestal shape at the foot of the pipe that forms the pile.

Pilot: Heavy-duty steel pile used to drive through obstructions before driving concrete piles in the same location.

Pipe: Steel cylinder, usually between 250 and 600 mm (10 and 24 in.) in diameter, generally driven with open ends to firm bearing and then excavated and filled with concrete. This pile

may consist of several sections from 1.5 to 12 m (5 to 40 ft) long joined by special fittings, such as cast steel sleeves, and is sometimes used with its lower end closed by a conical steel shoe.

Point-bearing: Pile that relies on end bearing, with no allowance for surface friction.

Precast: Reinforced concrete pile manufactured in a casting plant or at the site but not in its final position. *See also* **Cast-in-place pile,** above.

Precast segmental: Precast pile manufactured in lengths that enable it to be extended, on site, relatively quickly with a mechanical splice.

Pressure: Bored pile formed by placing concrete in the prepared hole under compressed air pressure.

Rail: Pile fabricated from railroad rails, usually three rails with the edges of their bases welded together and the heads out, driven as a unit.

Raker: *See* **Batter pile,** above.

Raking: *See* **Batter pile,** above.

Sand: Column of sand installed by driving a sand-filled pipe with an openable end into soil, then forcing the sand out as the pipe is withdrawn. The resulting sand column serves as a drain or wick to speed consolidation and improve the bearing value of the soil.

Screw: Spiral blade fixed on a shaft and screwed into the ground by a rotating force.

Sheet: Pile in the form of a plank driven in close contact or interlocking with others to provide a tight wall to resist the lateral pressure of water, adjacent earth, or other materials.*See also* **Wakefield pile,** below.

Short-bored: Bored piling about 3 to 5 m (10 to 16 ft) deep, used to support light loads on poor strength soil, usually placed where they will support wall and floor loads.

Skew batter: *See* **Compound batter pile,** above.

Slide: Pile driven into the earth to consolidate the soil and help prevent it from sliding down a slope.

Socketed: *See* **Drilled-in caisson pile,** above.

Spliced: Pile composed of two or more sections secured end-to-end to form a single, composite unit.

Spud: Short, strong member driven vertically or to a batter and then removed and replaced with a pile that is too long to be placed directly in the pile driver leads, or to break through a crust of hard material.

Spur: *See* **Batter pile,** above.

Steel H-section: Rolled-steel section with web and flanges nearly the same thickness. The width and depth are approximately equal. *Also called* **H-pile.**

Steel pipe: Pipe in any wall thickness or diameter; it may be driven closed end or open end. *Also called* **Tube pile.**

Swage: Thin-wall pipe pile with its bottom closed by a slightly tapered precast point. Driving is done by a mandrel on the point, dragging the pile down, with an assist from the shoulder on the exact length mandrel.

Tapered: Pile that decreases in diameter along its length to the tip.

Test: (a). Pile driven to determine driving conditions and probable required lengths; **(b).** Pile on which a loading test is made to determine the carrying capacity of the soil; **(c).** Pile driven as a guide for the efficient design of pile foundations.

Thin shell: Corrugated steel-shell pipe filled with concrete.

Thin-wall shell: Corrugated shell or thin-wall pipe, concrete filled, driven with the aid of a mandrel. *See also* **Swage pile** (above), and **Helical shell.**

Timber: Cut tree, usually debarked and pressure treated with creosote or other preservative, usually driven small-end down. Probably the earliest form of a pile.

Tube: *See* **Steel pipe pile,** above.

Uncased concrete: Column of concrete placed and left in the ground without an encasement. *See also* **Augered pile,** above.

Wakefield: Timber sheet piles consisting of three planks bolted or spiked together, with the middle plank offset so as to form a tongue along one edge and a corresponding groove on the other. *See also* **Sheet pile,** above.

Wing: Bearing pile, usually of concrete, widening in the upper portion to form part of a sheet pile wall.

PILE BENT Two or more piles driven in a row transverse to the long dimension of the structure and fastened together by capping and (sometimes) bracing. *See also* **Bent.**

PILE BRIDGE Bridge deck carried on piles.

PILE BUCKS *See* **Pile crew.**

PILE BULKHEAD Pile structure generally consisting of vertical piles, with brace or anchor piles, wales, and a sheet-pile wall, framed together and capable of resisting earth or water pressure. *See also* **Seawall.**

PILE BUTT 1. Larger end of a tapered pile; **2.** Section of pile cut off at the top.

PILE CAGE Prefabricated rigid steel reinforcement that is inserted vertically into a bored hole prior to concreting.

PILE CAGE SPACER Cement or plastic wheel that is threaded over the main longitudinal reinforcement bar of a pile cage to space the cage away from the perimeter of the bored hole and to ensure correct concrete cover.

PILE CAP 1. Block at the head of one or more piles conducting forces from the superstructure to one or several piles. *Also called* **Girder,** and **Rider cap; 2.** Masonry, timber or concrete footing resting on a group of piles. *See also* **Bent cap; 3.** Metal cap or helmet temporarily fitted over the head of a precast pile to protect it during driving, often incorporating some form of shock-absorbing material. *Also called* **Pile helmet.**

PILE CLAMP 1. One of a pairs of timbers bolted to the sides of piles in a bent to tie the piles together and also to carry loads to the piles, as in the case of pile caps; **2.** Mechanical device for gripping piles, used with vibratory hammers and extractors.

PILE CLUSTER Group of piles, standing free as a dolphin or as mooring piles, or piles in contact forming part of a fender system for mooring or breasting-off vessels for better control.

PILE CORE *See* **Mandrel.**

PILE CREW On-site personnel who will physically install piling on a project, including the crane crew, as applicable. *Also called* **Bridgemen, Dock builders,** and **Pile bucks.**

PILE CUSHION Cushion placed between the drive cap and the top of concrete piles to protect the pile from crushing and spalling. *Also called* **Cushion, Hammer cushion,** and **Pad.**

PILE CUTOFF Length of pile removed above the cutoff elevation.

PILE DRIVER *See* **Pile hammer.**

PILE DRIVE SHOE Metal protection for the foot of a timber pile, to obtain better penetration and to improve the contact area of end-bearing piles.

PILE DRIVING CAP Forged steel or steel casting designed to fit over and around the top or butt end of a pile to prevent damage to the head of the pile while driving. *Also called* **Hood** and **Bonnet.**

PILE DYNAMICS *See* **Case method.**

PILED FOOTING Reinforced concrete footing supported on piles.

PILED RAFT Stiff reinforced concrete foundation raft without a periphery skirt, supported by piles.

PILE ENCASEMENT Protective cover for a steel or timber pile, usually concrete, added at the water or ground line. *Also called* **Encasement.**

PILE EXTRACTOR Device for pulling piles out of the ground, usually a pile driving impact or vibratory hammer fitted with a yoke so as to transmit upward blows to the pile body, or a specially built extractor.

PILE FALL *See* **Pile line.**

PILE FOOT Lower extremity of a pile.

PILE FORMULA Equation from which the static resistance or the allowable load of a pile may be estimated from the driving record of the pile. *See also* **Dynamic formula.**

PILE FOUNDATION Piles and structural members designed and installed primarily to transmit loads directly from a superstructure to the soil or rock stratum some distance below the ground surface that is capable of sustaining the load.

PILE FRAME *See* **Skid rig.**

PILE GATE Device at the base of the pile hammer leads that is closed around the pile to maintain alignment between the end of the pile and the leads.

PILE GROUP Several driven or bored piles placed so that they work together to support a greater load than any one of the piles could carry alone.

PILE HAMMER General term for a machine that drives piling by impact or vibration. Its power source may be mechanical, air, steam, diesel, or hydraulic. *Also called* **Pile driver**, and **Pile rig**. There are different types, including:

Air: Movable ram attached to a piston, operating in a cylinder, that in turn is mounted in a metal frame with grooves that engage the leads of a pile driver, and having a hood or bonnet on the lower end with a cushion block that fits the pile head. In a single-acting hammer, air pressure is used only to raise the moving parts, which then fall by gravity, striking the cushion block. In a double-acting and differential-acting hammer, air pressure is used also to accelerate the downward movement of the ram.

Automatic: Power pile hammer in which the fluid valve is automatically thrown at both ends of the ram stroke to produce a regular cycle.

Caisson: Gravity hammer that is mounted on top of a caisson and that contains a cushioning system and a rudimentary lead. *Also called* **Tapette.**

Convertible diesel: Diesel hammer that can be operated either open-end (single-acting) or closed-end (double-acting).

Diesel: An integrally powered pile hammer activated with diesel fuel oil. A movable ram is raised initially by outside means. When released the ram falls onto an impact block or anvil that itself rests on the pile. The falling ram actuates ports for the admission of fuel and air and compresses the air, raising it to a higher temperature. The oil then vaporizes and ignites from heat and pressure. The instantaneous expansion of gases gives additional drive and raises the ram for the next stroke.

Differential acting: Fluid-powered hammer in which fluid lifts the ram on the up stroke and is exhausted into and combined with additional fluid entering over the piston to take advantage of differential piston areas to accelerate the ram and to act in concert with gravity on the down stroke.

Double-acting: Pile hammer in which fluid lifts the ram on the up stroke and additional fluid, redirected by valving, acts in concert with gravity on the down stroke.

Drop: Heavy weight, usually a metal casting with grooves in the sides to mate with the leads of a pile driver. It is raised in the leads by ropes or cables and allowed to drop on the pile. *Also called* **Gravity hammer.**

Flying: *See* **Free hanging hammer**, below.

Free-hanging: Pile hammer held by a crane line without leads and used to drive a supported pile. *Also called* **Flying hammer, Swinging hammer,** and **Wild hammer.**

Internal: Gravity hammer that acts internally on a plug at the base of a pile casing and that is used to advance the casing by driving at the bottom to a given set or penetration.

Semiautomatic: Power pile hammer for which the valve is manually thrown at one end of the stroke and automatically thrown at the other end.

Single-acting: Pile hammer in which fluid lifts the ram on the up stroke and gravity alone acts on the down stroke.

Sonic driver: Pile hammer designed to produce from 3600 to 9000 cycles/min to match the sonic response of soil for very rapid driving.

Swinging: *See* **Free-hanging hammer**, above.

Vibratory driver/extractor: Pile-driving and -extracting machine that is mechanically connected to a pile or casing and that, through a controlled rate of vibration, drives it into the soil or loosens a driven pile so that it may be pulled. The machine has eccentric weights mounted on shafts that are rotated in different directions and that apply unbalanced forces to a pile shaft. Electric or hydraulic motors rotate the shaft between 700 and 2000 vpm.

Wild: *See* **Free hanging hammer**, above.

PILE HEAD Upper portion of a pile when it is in position for driving, or after it has been driven. *Also called* **Top of pile.**

PILE HELMET *See* **Pile cap.**

PILE HOOP Driving band.

PILE JOINT Means for joining lengths of piles. There are various methods:

Composite: Joining two types of piles.

Full-penetration: Butt weld, or close fitting mechanical device plus welding across flanges and additions.

Pipe: Butt weld, or driving into a tapered circular sleeve.

Precast concrete: Variety of types, including matched male-female ends locked by mechanical means, drilled holes and dowels, or drive sleeve similar to that for pipe.

PILE LINE 1. Rope used to lift a pile and hold it in place while being placed in the leads; **2.** One of the hoisting lines on a crane, assigned to pile handling. *Also called* **Pile fall.**

PILE LOAD TEST Load test under true on-site conditions (same design pile, same driving equipment, etc.) on one or more piles to check the design assumptions made for the project.

PILE LOG *See* **Driving log.**

PILE NOSE Fabrication of angle iron attached to the front of a fender pile or dolphin to protect it from damage by floating debris or ice.

PILE PENETRATION 1. Depth of subsurface elevation reached by the tip of a pile. *Also called* **Tip elevation**; **2.** Embedment of the pile; **3.** The set of the pile.

PILE POINT Metal drive shoe, which may be pointed, fixed to the tip of a pile shaft for easier driving, improved penetration, protection against damage in dense material or boulders, and improved bearing at the tip. *Also called* **Concical point, Drive shoe,** and **Driving point, Pile shoe.** *See also* **Shoe,** and **Shoe plate.**

PILE PULLING TEST Test to determine the pull-rise curve of a friction pile to supplement the load-settlement curve.

PILE REFUSAL *See* **Refusal.**

PILE RIG *See* **Pile hammer.**

PILE RING Metal hoop used to bind the head of a timber pile during driving to prevent splitting and brooming.

PILE SET *See* **Set.**

PILE SHOE *See* **Pile point.**

PILE SPLICER Close-fitting device for quickly aligning an additional length of similar or compatible section of pile to increase its length. *See also* **Pile.**

PILE SPOTTING Preparing the location for a pile and placing it in the correct position for driving.

PILE TAKE-UP Increase in the observed resistance to penetration during driving.

PILE TEMPLATE Prefabricated structure or site-erected frame into which a pile is set and firmly held in position to permit driving with a free-hanging pile hammer. Usually rectangular in form, used to space and position piles so that they will be in the specified position relative to one another. *Also called* **Grid.**

PILE TIP Lower, and usually smaller end of a pile.

PILE WATER JET Water jet provided ahead of the pile toe to assist in driving in compact sand.

PILING Picking up tree-length logs or bolts and depositing them in large piles so that the logs are horizontal and parallel to each other and the ends are approximately in the same vertical plane.

PILING AND BURNING Piling logging slash by hand or machine and subsequently burning the individual piles.

PILLAR Free-standing, vertical support.

PILLOW Cushion-shaped support, usually at the top of a column.

PILLOW BLOCK 1. Rubber block, restrained by a metal bracket, that allows limited motion of an imposed load (such as an engine); **2.** Block that supports the journal of a shaft, spindle, etc.

PILOT BEARING *See* **Bearing.**

PILOT CIRCUIT Control circuit in electrical work.

PILOT CONTROL Servomechanism by which all major hydraulic functions of mobile mechanical equipment can be actuated and controlled through fingertip pressure from the operator's position.

PILOT HOLE Hole drilled as a guide for a wood screw and to relieve the stress attendant with the screw thrusting its way into the stock. The hole is narrower than the screw and just over half its length.

PILOTI Heavy column supporting a structure above an open ground-level area.

PILOT LIGHT 1. Relatively small flame that burns constantly, used to ignite a larger flame immediately when the combustible flow is released; **2.** Low-output electric lamp indicating a closed and energized circuit.

PILOT LINE Line that conducts control fluid.

PILOT NAIL Nail driven to temporarily hold lumber in place while the main fasteners are driven.

PILOT-OPERATED CHECK VALVE *See* **Valve.**

PILOT-OPERATED VALVE *See* **Valve.**

PILOT PILE *See* **Pile.**

PILOT PRESSURE *See* **Pressure.**

PILOT VALVE *See* **Valve.**

PIMPLINE Small diameter line, less than 6 mm (0.25 in.), used in steep terrain to string strawline.

PIN Cylinder of metal or wood, used to hold two or more pieces together by passing through a hole in each of them, as in a pinned joint, or the track of a tracked machine. *See also* **Taper pin.**

PINCERS Gripping tool consisting of jointed handles with opposing jaws at one end.

PINCHBAR Lever, on one end of which a pointed projection serves as a fulcrum.

PINCH DOG *See* **Clamp.**

PINCHERS Pair of poling boards strutted apart to support a trench wall in good soil.

PIN CONNECTION In structural analysis, a member connected to a foundation, another member, or structure, designed so that free rotation is assumed.

PINE OIL Strong solvent made from the oleoresin of pine trees.

PIN HOLE 1. Tiny hole made by or as by a pin; **2.** Holes, usually in pairs, in the body of a saw specifically for driving the saw from the collars.

PINION GEAR *See* **Gear.**

PIN KNOT Blemish in boards consisting of a small knot, 12 mm (0.5 in.) or less in diameter.

PIN LUG COUPLING Hose coupling where the lugs are pin-shaped.

PINNACLE Terminal ornament to a gable, etc.

PINNACLED ROCK Rock extending upward into the line of an excavation but not filling the entire area.

PINTLE 1. Vertical pin, fastened at the bottom, that serves as a centre of rotation; **2.** Tip of any injector needle that protrudes through the hole in an injector nozzle. The pintle may be shaped to assist in dispersing the injected fuel in a desired pattern in the combustion chamber.

PINTLE HOOK C-shaped hook welded or bolted on the rear frame of a truck, crane carrier, or other self-propelled vehicle, frame, or outrigger box of a machine to enable it to tow another vehicle, or to be towed if necessary. Such hooks are not intended for lifting. They consist of a fixed lower part and hinged upper part that, when locked together, form a round opening that can hold a tow ring or receive a loop-shaped fitting at the end of a trailer towbar.

PIN TUMBLERS Pins in a lock cylinder, held in position by springs, that prevent the cylinder plug from rotating until the correct key is inserted.

PIONEERING First working of rough or overgrown areas prior to excavation.

PIONEER PLANTS Plants capable of invading bare sites, e.g., a newly exposed soil surface, and persisting there, i.e., colonizing until supplanted by invader or other succession species.

PIONEER ROAD Semipermanent road built to provide access for workers, vehicles, and machinery.

PIPE Cylindrical tube. Pipe is fabricated from a wide range of materials, including:

ABS: Rigid plastic pipe made of acrylonitrile-butadiene-styrene (ABS)

Alloy: Steel pipe fabricated with one or more elements in addition to carbon that give the metal greater strength and a higher resistance to corrosion than carbon steel pipe.

Asbestos-cement: Pipe fabricated from concrete containing an amount of asbestos fibers.

Black iron: In fact, steel pipe that has not been galvanized.

Brass: Pipe composed of 84% to 86% copper and up to 0.06% lead, plus zinc.

Butt-weld steel pipe: Pipe intended for low-pressure steam, water, gas, or air service.

Cast iron: Cast pipe fabricated from a composite of alloys, primarily of iron, carbon, and silicon.

Concrete: Pipe formed by casting, extruding, or spinning a concrete mix, made in diameters ranging from approximately 75 mm to 2 m (3 in. to 6 ft) or more.

Copper wayer pipe: Solid copper pipe produced in three types:

Type K: Heavy-wall tube identified by a 6.4 mm (0.25 in.) green stripe along the entire length of the pipe, used primarily for underground services, plumbing, heating and cooling systems, and for steam, oil, air, oxygen, and hydraulic lines.

Type L: Medium-wall tube, identified by a blue stripe, used for underground services, interior plumbing, heating and cooling systems, snow-melting systems, and steam, oil, and oxygen lines.

Type M: Light-wall tube, identified by a red stripe, used for interior water distribution, heating and cooling systems, steam lines, interior waste lines, and drainage lines.

CPVC: Semirigid plastic pipe formed of chlorinated polyvinyl chloride; a type of plastic pipe that will carry hot water and some chemicals.

Ductile iron: Cast iron pipe that can sustain not less than 5% elongation before fracturing. Especially refers to cast ductile iron that has less tendency to crack when subjected to high stress.

Electric-fusion-welded: Pipe where coalescence is produced in the longitudinal seam of the preformed tube by manual or automatic electric-arc welding.

Electric-resistance-welded: Individual or continuous lengths of pipe produced from skelp with the longitudinal butt joint coalesced by pressure and an electric-resistance circuit of which the pipe is one element.

Galvanized steel: Steel pipe produced in three grades of wall thickness (standard-weight, extra-strong, and double-extra-strong), galvanized to resist corrosion.

Hubless: Plain-end pipe without built-in facility for mating with adjacent lengths.

Lap weld: Pipe formed by welding along a scarfed longitudinal overlapping seam.

Malleable iron: Cast-iron pipe, heat treated to reduce its brittleness.

Polybutylene (PB): Pipe formed of a light-colored, liquid, straight-chained aliphatic hydrocarbon polymer.

Polyethylene (PE): Thermoplastic high-molecular-weight organic compound used in formulating pipe.

Polypropylene (PP): Pipe formed by the polymerization of high-purity propylene gas in the presence of an organometallic catalyst at relatively low pressures and temperatures. It has exceptional flex life, good surface hardness, scratch and abrasion resistance, and high chemical resistance. It will not stress crack and will take high temperature fluids.

Polyvinyl chloride (PVC): Flexible pipe formed of synthetic resin prepared by the polymerization of vinyl chloride.

Seamless: Pipe or tube formed by piercing a billet of steel and then rolling.

Seamless carbon-steel: Drawn pipe having a high carbon and manganese content giving it greater tensile strength but less ductability.

Spiral: Pipe made by coiling a plate into a helix of uniform diameter and riveting or welding the overlapping edges.

Spiral weld: Strip steel or plate turned into an open pipe with the longitudinal, spiral joint welded or crimped into a tight seam.

Stainless steel: Corrosion-resisting alloy-steel pipe high in nickel and chromium.

PIPE BRACE Brace extending between opposite hangers on a spring-type suspension. *Also called* an **Angle brace.**

PIPE BRANCH *See* **Branch fitting.**

PIPE BUFFER Worker who buffs the ends of pipe to be joined prior to welding.

PIPE BURSTING Technique for conduit installation in which an existing, defective pipeline is first used as a pilot and is then broken up and removed, after which an enlarged excavation is made for a larger pipeline.

PIPE CLAMP *See* **Clamp.**

PIPE COLUMN Metal pipe used as a column to support a load, sometimes filled with concrete. *Also called* **Lally column.**

PIPE COUPLING Short threaded collar, used to connect the threaded ends of two pipes.

PIPE CUTTER Tool for cutting pipes that leaves a smooth end.

PIPE DRILL Drill bit that cuts round holes in masonry.

PIPE DUCT Enclosed interior space reserved for pipework.

PIPE ENAMEL Asphaltic coating given to buried pipes.

PIPE FITTER Worker skilled in the installation of piping systems for water, steam, gas, oil, chemicals, etc.

PIPE FITTING Fitting used to connect lengths of pipe, to support them and permit them to change direction.

PIPE KEY *See* **Barrel key.**

PIPE KOOK Spiked fastener, driven into a stud or masonry joint, with a curved end.

PIPE JACKING Technique of installing a line of pipes through the ground in a previously excavated bore by means of hydraulic jacks from the drive shaft. After pushing a pipe length into the ground a new pipe is positioned and the process repeated. *Also called* **Pipe pushing.**

PIPE JOINT *See* **Pile joint.**

PIPE JUMBO Traveling support in a tunnel for the discharge line from a concrete placer.

PIPE LAYER 1. Tracked equipment fitted on one side with a hinged boom and on the other with a counterweight and winch. The equipment can lift, support, and transport sections of pipe, or assembled pipeline (working in series with other units), and lower the pipe into a trench offset to the side; **2.** Worker skilled in installing all types of pipe in trenches, laying to correct levels, making joints and connections.

PIPE LEAD *See* **Lead.**

PIPE PILE *See* **Pile.**

PIPE PUSHING *See* **Pipe jacking.**

PIPE RAMMING Technique for conduit installation involving a casing driven through the ground by a percussive hammer.

PIPE STOPPER Screw plug for sealing a drain.

PIPE STRAP Metal strap used to support or position a pipe.

PIPE THREAD Screw threads for joining pipe. *See also* **Dry-seal pipe thread,** and **Tapered pipe thread.**

PIPE WRENCH *See* **Wrench.**

PIPING 1. Arrangement of pipes of the same or different diameters as part of a scheme or layout; **2.** Movement of soil particles as a result of unbalanced seepage forces produced by percolating water, leading to the development of boils or erosion channels. *See also* **Blowing.**

PISTON Disk or short cylinder closely fitted in a hollow cylinder. *See also* **Free-running piston,** and **Slave piston.**

PISTON BOSS Part of a piston reinforced to support the piston pin.

PISTON DISPLACEMENT Volume displaced by moving all the pistons of an engine through their full stroke.

PISTON HEAD Closed end of a piston, above the piston ring(s). Sometimes specially shaped to enhance the swirl of combustible gases and to enhance their compression in advance of ignition.

PISTON LAND Area of the exterior piston wall between the piston rings.

PISTON PIN Cylindrical connection that passes through the piston boss and upper end of the connecting rod, used to transmit piston movement to the connecting rod.

PISTON PUMP *See* **Pump.**

PISTON RING Split metal ring seated in a grove around a piston to improve the fit between the piston and the cylinder wall. There are several types, including:

> **Compression ring:** Ring that exters sufficient pressure against the cylinder wall to prevent passage of the compressed fuel mixture, or the products of combustin from escaping past the piston into the crankcase.
>
> **Scraper ring:** Ring that removes material from the surface of a cylinder by a scraping action.
>
> **Wiper ring:** Ring that removes material from the surface of a cylinder by a wiping action. *Also called* **Exclusion device.**

PISTON RING EXPANDER Ring placed behind a piston ring so as to increase its pressure against the cylinder wall.

PISTON RING GAP Designed clearance between the end of a piston ring when mounted on a piston and inserted in the cylinder.

PISTON RING GROOVE Slots formed around the perimeter of a cylinder into which piston rings are fitted.

PISTON ROD Rod that moves, or is moved by the piston to which it is attached.

PISTON SKIRT Portion of a piston that is below the piston ring and that resists the side thrust of the piston.

PISTON SPEED Total distance traveled by a piston in one minute.

PISTON STROKE Length a piston travels (twice its crank radius).

PIT 1. Open excavation that is deep with respect to its area; **2.** Area at the bottom of a hoistway under an elevator car; **3.** Tiny void in the surface of particleboard, between a fine and a flake.

PIT BOARD Piece of sheeting placed, with others, horizontally about the perimeter of a pit to retain the walls.

PITCH 1. Inclined slope of a roof, or the ratio of the total rise to the horizontal length covered by the roof; for example, a 1.8 m (6 ft) rise over an 5.5 m (18 ft) distance gives a 1:3 or 1/3 pitch. *See also* **Roof slope**; **2.** To square a stone. *See also* **Lay**; **3.** Slope of a tooth relative to its direction of movement; **4.** Any of several bituminous or resinous substances consisting of fusible, viscous to solid materials, especially coal tar; **5.** Degree of inclination of a flight of steps or stairs; **6.** Distance from any point on a helix or screw thread to a corresponding point measured parallel to the axis of the shaft; **7.** Dip of a geological stratum or vein; **8.**

Distance between two corresponding points of two adjacent gear teeth; 9. Quality of sound determined by the frequency of sound vibrations; 10. Resin common to certain evergreen trees; 11. Of a continuous series of equally spaced like components or objects, the distance center-to-center between any two. *See also* **Lay** and **Spacing**; 12. Center-to-center spacing between turns of a spiral; 13. Distance between centerlines of two adjacent grooves in a sheave.

PIT CHANNEL Steel channel on the pit floor to anchor one of the guide rails or other pit-mounted devices.

PITCH ARM Adjustable rod that helps determine the digging angle of a blade or bucket. *Also called* **Pitch brace**, and **Pitch rod.**

PITCH BLENDE Brown to black lustrous mineral containing radioactive materials.

PITCH BOARD Template for repetitive cuts, typically on a stair stringer where the shortest side is the riser and the longer side the tread.

PITCH BRACE *See* **Pitch arm.**

PITCH CIRCLE Circle touching the teeth of a gear wheel at points where they mesh with the teeth of another gear wheel.

PITCH DIAMETER Diameter of a sheave or rope drum measured at the centerline of the rope.

PITCHED ROOF Roof slope that is greater than 10° or 15% to the horizontal. *See also* **Double-pitched roof, Gambrel roof,** and **Mansard roof.**

PITCHER FITTING Plumbing fitting, tee, elbow, etc., with a gently curved turn instead of the sharp angles of conventional tees and elbows

PITCHING 1. Elevating piles or sheeting into a vertical position prior to driving; 2. Bricks, concrete blocks, pebbles, etc. on a soil slope to protect against erosion from runoff.

PITCHING CHISEL Heavy-duty chisel for cutting and dressing stone.

PITCHING FERRULE Short length of pipe cast into a reinforced-concrete pile and used to lift and maneuver it.

PITCHING PIECE Horizontal timber supporting the upper ends of the rough strings for a flight of stairs.

PITCH LINE VELOCITY Circumference in feet at the pitch line of a gear multiplied by the rpm of that gear at 1,000 engine rpm. A small PTO driving gear in the transmission gives a low pitch line; a larger gear gives a correspondingly higher pitch line velocity. *ALso called* **PLV.**

PITCH PINE Any of several resinous pines from which pitch or turpentine is obtained.

PITCH POCKET Well-defined opening in lumber, usually between annual growth rings, that contains pitch in either liquid or solid form.

PITCH ROD *See* **Pitch arm.**

PITCH STONE Stone having the arris clearly defined by a line beyond which the rock is cut away by the pitching chisel so as to make approximately true edges.

PITCH STREAK Well-defined accumulation of pitch in a more-or-less regular streak in the wood of certain conifers.

PITCH TRIANGLE A right triangle whose horizontal base is always 300 mm (12 in.) in length (run) and whose hypotenuse is parallel to the roof incline, and whose altitude (rise) is expressed in millimeters or inches.

PITCH TUBES Tubular mass of resin that forms on the surface of bark at bark-beetle entrance holes.

PITCHY Full of pitch; leaking or oozing pitch.

PITH Small, soft core occurring in the structural center of a tree trunk, branch, twig, or log.

PITH KNOT Minor blemish in lumber consisting of a pith hole smaller than 6 mm (0.25 in.).

PITLESS ADAPTER Threaded or welded fitting to a well casing that provides underground connection between the well and the service pipe with ready access to the drop pipe and well shaft.

PITMAN ARM Arm with limited movement about a pivot.

PITOT TUBE Tube having an opening that is inserted into a stream of water or air and to which a gauge is attached indicating the discharge pressure of the stream.

PIT RUN GRAVEL Material as it comes from a gravel pit, without reduction or sizing. *See also* **Gravel.**

PIT TECHNIQUE *See* **Sanitary landfilling.**

PITTED TUBE Surface depression on the inner tube of a hose.

PITTING 1. Development of relatively small cavities in a surface. *See* **Blowing**; 2. In concrete, local disintegration, such as a popout; 3. In steel, localized corrosion evident as minute cavities on the surface.

piv. *Abbreviation for:* pivot.

PIVOT Nonrotating axle about which something turns.

PIVOT BRIDGE *See* **Swing bridge.**

PIVOTED CASEMENT Casement window pivoted horizontally about its central axis.

PIVOTED LUFFING JIB *See* **Articulated jib.**

PIVOTED SASH *See* **Sash.**

PIVOT METHOD Technique of maneuvering a loader/backhoe using the loader bucket as a pivot point.

PIVOT POINT Point where a dump body is hinged to the chassis and about which it rotates or pivots.

PIVOT SHAFT Fixed shaft that acts as a hinge pin.

PIVOT TUBE Hollow hinge-pin.

pk *Abbreviation for:* pack; park; peak; plank.

PK *Abbreviation for:* paneling, knotty.

pkg *Abbreviation for:* packing; parking.

pkg. *Abbreviation for:* package.

pklg *Abbreviation for:* pickling.

pkwy *Abbreviation for:* parkway.

pl *Abbreviation for:* parking lamp.

pl. *Abbreviation for:* place; plan; plane; plural.

PL *Abbreviation for:* plate; pipeliner; planer; private lane; property line.

PLACEABILITY *See* **Workability.**

PLACE BRICK *See* **Brick.**

PLACEMENT 1. Process of placing and consolidating concrete. *Also called* **Pouring; 2.** Quantity of concrete placed and finished during a continuous operation.

PLACER (plasser) Deposit of heavy minerals concentrated mechanically, as by stream flow, wave action, or wind.

PLACER (plasser) DIGGINGS Areas where placer mining has overturned or removed the earth and left a rough, eroded, and scarred surface.

PLACING Deposition, distribution, and consolidation of freshly mixed concrete in the place where it is to harden. *Also called* **Pouring.**

PLACING DRAWING Detailed drawing that gives the size, location, and spacing of reinforcing bars, plus all other information required by an ironworker.

PLACING PLANT Equipment for depositing mixed concrete in position.

PLAIN ASHLAR Smooth-faced stones.

PLAIN BAR Reinforcing bar without surface deformations, or one having deformations that do not conform to the applicable requirements.

PLAIN CHISEL *See* **Cold chisel.**

PLAIN CONCRETE 1. Concrete without reinforcement; **2.** Reinforced concrete that does not conform to the definition of reinforced concrete; **3.** Concrete containing no admixture and prepared without special treatment.

PLAIN CONCRETE WALL Vertical, load-bearing, unreinforced concrete member whose greatest lateral dimension is more than four times its least lateral dimension.

PLAIN END Uncapped, or otherwise unprotected, straight end of a hose.

PLAIN MASONRY 1. Masonry without reinforcement; **2.** Masonry reinforced only for either shrinkage or thermal change.

PLAIN SAWING Converting logs by saw cuts that are parallel to the squared side.

PLAIN TILE Rectangular roofing tile of relatively uniform thickness and only a slight convex camber.

PLAN View from directly above, as of any floor, roof, or foundation of a building, or of a site.

PLAN AND PROFILE Drawing showing both horizontal (plan) and vertical (profile) delineation of a road survey.

PLANE 1. In woodworking, a flat surface where any line joining two points will lie entirely on the surface; **2.** Carpenter's tool used to level, smooth or shape wood. There are two basic categories: bench planes and special-purpose planes. They include:

BENCH PLANES

Block: A small, 125 to 150 mm (5 to 6 in.) long, one-handed plane used primarily for planing end grain and small pieces of wood. The blade is set at a lower angle than other planes.

Fore: From 250 to 450 mm (10 to 18 in.) long, used for making glue joints on long pieces of wood. Lighter in weight than a jointer plane.

Jack: General purpose plane for smoothing and jointing. From 300 to 350 mm (12 to 14 in.) long.

Jointer: Power-driven plane, used for long work and for truing up edges and surfaces. Primarily used for making glue joints on long pieces of wood, this plane is from 500 to 600 mm (20 to 24 in.) long.

Smoothing: Light in weight and from 175 to 250 mm (7 to 10 in.) long, this also is a general purpose plane for extremely fine smoothing work

SPECIAL PURPOSE PLANES

Bullnose: Small, compact metal block plane with a small handle and blade mounted near the front. Used for detailed work and on inside corners.

Combination: Designed to accept a wide variety of specially shaped blades, this plane is used for making cuts and moldings.

Compass: Hand plane, the sole of which can be adjusted to form an arc about the cutting edge of the iron.

Rabbet: Designed to cut rabbet or rectangular groves on the faces and ends of boards.

Router: For removing wood from between two sawn or chiseled edges and for smoothing an otherwise rough dado cut.

PLANE FRAME Structural system, assumed for the purpose of analysis and design to be two-dimensional.

PLANE IRON Cutting blade of a plane.

PLANE OF RUPTURE Plane along which a retained soil is considered to fail when a wall is being designed to retain it.

PLANE OF SATURATION Ground within the water table.

PLAIN OF WEAKNESS Plane along which a body under stress will tend to fracture. May exist by design, by accident, or because of the nature of the structure and its loading.

PLANER 1. Power-driven tool for surfacing timber, metal, stone, and other materials; **2.** Self-propelled machine used to smooth a preheated road surface.

PLANER TYPE Type of surface grinding machine built similar to an open-side planer.

PLANE STOCK Body of a carpenter's plane that holds the cutting and back irons.

PLANE SURVEY *See* **Surveying.**

PLANE TABLING Mapping with a plane table and alidade.

PLANETARY AXLE *See* **Axle.**

PLANETARY GEAR SET *See* **Gear.**

PLANETARY TRANSMISSION *See* **Transmission.**

PLANIMETER Instrument that measures the area of a plane figure when its boundary is traced.

PLANING MILL Plant dedicated to smoothing the rough surface of wood, cutting, fitting, and matching boards.

PLANISH Toughen, smooth, or harden metal by hammering.

PLANISHING HAMMER *See* **Hammer.**

PLANK Long, flat, heavy piece of timber, usually more than 150 mm (6 in.) wide and 25 mm (1 in.) thick, intended to be loaded on its wide face, commonly used as part of a system to support workers standing at a level above grade or above the floor.

PLANK FRAMING Type of construction that employs flat vertical structural members with horizontal beams let into them and having an infilling of planks on edge.

PLANK TRUSS Truss fabricated with planks as the heaviest members.

PLANNING CONSULTANT Qualified independent professional who contracts to study, report on, and plan within defined terms of reference.

PLANNING GRID Reference grid by which to locate structures, elements, services, features, etc.

PLANNING-PROGRAMMING-BUDGET SYSTEM Planning and management process in which resources of an organization are allocated to individual programs, and where relative effectiveness in achieving program goals is compared, in contrast to the administrative budget grouping of resource by type of disbursement. *Also called* **Program budgeting.**

PLANOMETRIC PROJECTION Pictorial view showing an object in plan with oblique parallel lines from the corners to indicate depth and height.

PLANS Contract drawings that show the location, character, and dimensions of the prescribed work, including layouts, profiles, cross sections, and details. *See also* **Standard plans,** and **Working drawings.**

PLANT Generally, power-operated machinery and equipment used in engineering and construction.

PLANTATION Forest stand regenerated artificially either by sowing or planting; a man-made forest.

PLANT BYPRODUCTS Wood products, such as pulpwood chips, obtained incidental to the production of other manufactured products.

PLANTED Fastened to or tongued in, as a molding, separately fabricated and then applied.

PLANTING Artificial regeneration method in which a new stand of trees is established by restocking the area with tree seedlings, by hand or by machine.

PLANTING EASEMENT *See* **Easement.**

PLANTING PLAN Plan indicating the locations, types, and numbers of plants to be installed on a site.

PLANT LIST (SCHEDULE) Chart used with a planting plan to summarize the plant quantities, their botanic names, size or caliper, and the manner of root preparation.

PLANT MIX Mixture, produced in an asphalt mixing facility that consists of mineral aggregate uniformly coated with asphalt cement, emulsified asphalt, or cutback asphalt.

PLANT-MIX SEAL *See* **Hot-mix seal coat.**

PLANT-MIXED SURFACE TREATMENT *See* **Hot-mix seal coat.**

PLANT-MIXED SURFACING Designed combination of mineral aggregate and bituminous material mixed in a central plant.

PLANT RESIDUE Wood material from manufacturing plants not utilized as a product.

PLAQUE Applied or inset decoration.

plas. *Abbreviation for:* plaster.

PLASMA Temporary physical condition of a gas after it has been exposed to and has reacted to an electric arc.

PLASMA ARC CUTTING Cutting metal using an electric arc and fast-flowing ionized gases.

PLASTER Mixture based on lime, hydraulic cement, or gypsum, sometimes with the addition of aggregates, fibers, or other materials that, when mixed, is of a doughy or pasty consistency, which is used to cover a surface, and which hardens after application. *See also* **Stucco.**

PLASTER BASE Working ground for plaster.

PLASTER BEAD Applied edging, usually of metal, to strengthen a plaster angle.

PLASTERBOARD Rigid board made of gypsum plaster, covered on both sides with heavy paper.

PLASTER DAB Pat of plaster applied to brickwork or lathing and used as a fixing for wall ties, etc.

PLASTERER Tradesman who applies and finishes plaster, lathing, drywall, and related work.

PLASTER GROUND Wood strip nailed to walls to serve as a guide.

PLASTER LATH Narrow strips of wood nailed to rafters, joists, or studding as a groundwork for plaster.

PLASTER MOLD Mold or form made from gypsum plaster. *See also* **Form,** and **Mold.**

PLASTER OF PARIS 1. Calcined gypsum (calcium sulfate hemihydrate) without the addition of material to control set, principally used in casting and industrial applications. *Also called* **Bassanite,** and **Hemihydrate; 2.** Gypsum from which three-quarters of the chemically bound water has been driven off by heating and that, when wetted, recombines with water and hardens quickly. *See also* **Hemihydrate.**

PLASTER SHOT *See* **Mud cap.**

PLASTIC 1. Substance that can be permanently formed or deformed under external stress or pressure, usually accelerated by the application of heat. The newly created form retains its shape by cooling, chemical action, or the removal of a solvent through evaporation; **2.** Possessing plasticity; **3.** Coating applied by brush or spray, containing plastic resins.

PLASTIC ANALYSIS Determination of load effects (force, moment, stress, as appropriate) on members and connections, based on the assumption of rigid-plastic behavior, i.e., that equilibrium is satisfied throughout the structure and yield is not exceeded anywhere. Second order effects may need to be considered.

PLASTIC CEMENT Special product manufactured for plaster and stucco application.

PLASTIC CEMENTS *See* **Adhesive.**

PLASTIC CENTROID Centroid of the resistance to load computed for the assumptions that the concrete is stressed uniformly to 85% of its design strength and that the steel is stressed uniformly to its specified yield point.

PLASTIC CLAY Fire clay having sufficient natural plasticity to bond nonplastic material.

PLASTIC CONCRETE Easily molded concrete able to change its form slowly when the mold is removed.

PLASTIC CONSISTENCY Condition of freshly mixed cement paste, mortar, or concrete such that deformation will be sustained continuously in any direction without rupture.

PLASTIC CRACKING Cracking that occurs in the surface of fresh concrete soon after it is placed and while it is still plastic.

PLASTIC CREEP Flow of soil particles due to an imposed load. *See also* **Elastic compression, Soil compression,** and **Soil consolidation.**

PLASTIC DEFORMATION Change in shape that does not disappear when the force causing it is removed.

PLASTIC DESIGN SECTION Cross section of a member that can maintain a full plastic moment through large rotation so that a mechanism can develop; the section suitable for plastic design.

PLASTIC EMULSION *See* **Latex emulsion.**

PLASTIC EQUILIBRIUM State of stress of a soil mass that has been loaded and deformed to such an extent that its ultimate shearing resistance is mobilized at one or more points.

PLASTIC FIRECLAY 1. Fireclay of sufficient natural plasticity as to bond nonplastic material; **2.** Fireclay used as a plasticizing agent in mortar. *Also called* **Bond fireclay.**

PLASTIC FLOW Increase in concrete strain for members subject to constant stress and a decrease in concrete stress for members subject to constant strain. *See* **Creep,** and **Stress relaxation.**

PLASTIC FOAM Expanded-plastic formulation used for insulating.

PLASTIC FRACTURE Tension break in metal caused by drawing out, called necking, that causes a more gradual failure than an abrupt cleavage failure.

PLASTIC GLUES *See* **Adhesive.**

PLASTIC HINGE Region where the ultimate moment capacity in a member may be developed and maintained with corresponding significant inelastic rotation as the main tensile steel elongates beyond yield strain.

PLASTIC INDEX Numerical difference between a soil's liquid limit and its plastic limit. *Also called* **Plasticity index.**

PLASTICITY 1. Complex property of a material involving a combination of qualities of mobility and magnitude of yield value; **2.** Property of freshly mixed cement paste, concrete, or mortar that determines its resistance to deformation or ease of molding; **3.** Quality of being able to be shaped by plastic flow; **4.** Property of soil that allows it to deform beyond the point of recovery without cracking or appreciable volume change; **5.** Measure of the resistance to shear of an unvulcanized elastomer; **6.** Property of vulcanized rubber to retain a shape or form imparted to it by a deforming force.

PLASTICITY INDEX 1. Range in water content through which a soil remains plastic. *See also* **Plastic index**; **2.** Numerical difference between the liquid limit and the plastic limit. *See also* **Atterberg limits.**

PLASTICIZE To produce plasticity or to render plastic.

PLASTICIZER Solvent that does not evaporate and makes a plastic material pliable, softer, easier to mold into shapes; any material capable of combining, but not chemically.

PLASTIC LIMIT Water content at which a soil will just begin to crumble when rolled into a thread approximately 3 mm (0.125 in.) in diameter. *See also* **Atterberg limits.**

PLASTIC LIMIT LOAD Maximum load that is attained when a sufficient number of yield zones have formed to permit a structure to deform plastically without further increase in load. It is the largest load a structure will support, when perfect plasticity is assumed and when such factors as instability, second-order effects, strain hardening, and fracture are neglected.

PLASTIC LOSS *See* **Creep.**

PLASTIC MEMORY Quality of some thermoplastics to return to their original form after reheating.

PLASTIC MODULUS Section modulus of resistance to bending of a completely yielded cross section. It is the combined static moment about the neutral axis of the cross-sectional areas above and below that axis.

PLASTIC MOMENT Resisting moment of a fully yielded cross section.

PLASTIC MORTAR Mortar of plastic consistency.

PLASTIC-RESIN GLUE *See* **Adhesive.**

PLASTIC SHRINKAGE Shrinkage that takes place before cement paste, mortar, grout, or concrete sets.

PLASTIC SOIL *See* **Soil types.**

PLASTIC STRAIN Difference between total strain and elastic strain.

PLASTIC WELDING Fusing plastic materials with heated air.

PLASTIC WOOD Quick-drying paste of nitrocellulose, plasticizers, wood flour and other materials dispersed in a solvent, used for repairs, filling holes, etc.

PLASTIC YIELD Extent of plastic deformation.

PLASTIC ZONE Yielded region of a member.

PLASTIFICATION Process of successive yielding of fibers in the cross section of a member as bending moment is increased.

PLASTIGAGE Thin plastic thread used to check clearances or tolerances.

PLASTOMETER 1. Instrument for measuring the viscosity of raw or unvulcanized rubber; **2.** Instrument for measuring the hardness of vulcanized rubber.

PLAT Determined by surveying, a diagram drawn to scale showing all data pertinent to the boundaries and subdivision of a parcel of land, sufficient for a legal description.

PLAT BAND Flat band or string course projecting less than its breadth.

PLAT BOOK Public record containing maps of land showing its division into streets, blocks, and lots.

PLATE 1. Structural member, the depth of which is substantially less than its length and width; **2.** In formwork for concrete, a flat, horizontal member at either the top or bottom (or both) of studs or posts, a mud sill if on the ground; **3.** Horizontal member anchored to a masonry wall. *See also* **Sill, Soleplate,** and **Top** plate; **4.** Nonstructural protective unit. *Also called* **Kick plate, Pad,** and **Push plate; 5.** Ground contact part of a crawler-type track.

PLATE BAND In classical buildings, a rectangular molding of shallow projection, usually denoting externally the horizontal division between storys.

PLATE BEARING TEST Testing method used to determine required pavement thickness.

PLATE CUT Double cut, one face being horizontal, the other being parallel to the slope of the roof, at the lower end of a rafter where it fits against the plate. *Also called* **Bird's mouoth, Foot cut,** and **Seat cut.**

PLATE FIFTH *See* **Fifth wheel.**

PLATE GIRDER Built-up structural steel beam.

PLATE GLASS Glass flowed onto an iron plate, rolled to required thickness, annealed and polished; used for large windows.

PLATE GLASS INSURANCE Indemnity against breakage of glass, except under defined circumstances (fire, war, etc.).

PLATE LOAD TEST Load test made on a steel plate of given dimensions resting on a surface.

PLATE MOUNTED Disk, segment, or cylinder wheels cemented to a steel back plate having projecting studs or other means for mounting on a machine.

PLATEN Plate, used to attach the jack to the bolster in a hydraulic elevator.

PLATE SCREW Foot screw that permits adjustment of an optical survey instrument.

PLATE VIBRATOR Self-propelled mechanical vibrator with a sole plate, used to compact fill.

platf. *Abbreviation for:* platform.

PLATFORM 1. Raised horizontal surface; **2.** Flooring or stage for performers, speakers, etc.; **3.** Temporary or permanent wood deck used to support machines on soft ground; **4.** Horizontal working surface of a scaffold; **5.** Operator's station on large mobile machinery; **6.** Floor of an elevator, that is placed in the sling and is further supported by brace rods in each corner.

PLATFORM CONSTRUCTION *See* **Frame construction.**

PLATFORM ELEVATOR Elevator in which the platform is directly supported at three or more points by suspension members that keep the platform substantially level.

PLATFORM/FLATBED Load-carrying truck bed with or without removable sides. May be equipped with hydraulic cylinders to tilt and/or slide the platform.

PLATFORM FLOOR EXTENSION *See* **Bumper.**

PLATFORM FRAMING Construction technique in which the floor platforms are framed independently of the walls. Also, upper floors are supported by studs one story in height. *Also called* **Western framing.**

PLATFORM GANTRY Gantry for a portal crane.

PLATFORM HOIST Powered hoist that lifts a platform carrying a load of 90 to 2500 kg (200 to 5500 lb) up to 60 m (200 ft).

PLATFORM TRAILER *See* **Trailer.**

PLATFORM TRUCK *See* **Truck.**

PLATFORM UNIT Truck or trailer deck consisting of individual wood planks, fabricated planks, fabricated decks, or fabricated platforms.

PLAY Clearance for moving parts.

PLAY SPACE Area furnished with play equipment or play-inducing features for children.

PLAY STRUCTURE Structure providing different play opportunities for children, such as climbing, crawling, sliding, and swinging.

plcmt *Abbreviation for:* placement.

plcy *Abbreviation for:* policy.

PLEDGED ACCOUNT MORTGAGE *See* **Mortgage.**

PLENNAR Creating islands of clearcut within a forest. The cutting around the edges of these islands is extended after regeneration has been established.

PLENUM 1. Air compartment maintained under positive pressure and connected to heating or cooling ducts; **2.** Use of compressed air to hold soil from slumping into an excavation; **3.** Air compartment connected to one of more inlets or outlets.

PLENUM METHOD Excavation under air pressure.

PLENUM SYSTEM Ventilation of an interior space with fresh or conditioned air, forced into a plenum chamber and ducted to various areas.

plf *Abbreviation for:* pounds per linear foot.

pl. gl. *Abbreviation for:* plate glass.

plgr *Abbreviation for:* plunger.

pli *Abbreviation for:* pounds per linear inch.

PLIED YARN Yarn made by twisting together in one operation two or more single yarns.

PLIERS Scissor-like gripping tool, usually with cutting blades. There are many types, including:

Diagonal side-cutting: Relatively small pliers with curved handles and short, stout, pointed jaws with cutting edges.

End-cutting nippers: Pliers having wide, curved barrel-shaped jaws that meet at right angles to the handles, used to remove or cut off nail heads that are above the surface.

Fence: Long-handled, heavy, multipurpose pliers approximately 300 mm (12 in.) long with jaws containing grooves of several diameters for gripping wire, a flat hammer-like end, a hooked end, and mating slots with which to shear wire.

Locking: Gripping pliers containing a spring-loaded mechanism that allows adjustment of the width between the toothed, parallel jaws, and that also causes the jaws to clamp onto an object.

Long-nose: Pliers with short, curved handles and long, thin, tapered jaws that sometimes are offset at their tips. *Also called* **Needle-nosed,** and **Thin-nosed.**

Needle-nosed: *See* **Long-nose pliers,** above.

Side-cutting: Heavily-built, square-jawed pliers with toothed jaws. Some models have a concave set of teeth built into the jaws to grip rods and small pipes. The bottom of the jaws are fashioned as guillotine-type cutting edges.

Slip-joint: Pliers with curved, toothed jaws that can be slipped across a fixed pin to make the jaw opening wider or narrower.

Thin-nosed: *See* **Long nose pliers,** above.

Tongue-and-groove: Pliers with parallel, toothed jaws. One leg can be moved up or down a slot milled into the other leg, giving the jaws a wider or narrower minimum and maximum opening.

PLINTH 1. Projecting course(s) at the base of a wall, from foundation level to above ground, on which the damp-proof membrane is placed; **2.** Square block at the base of a column, pedestal, etc.; **3.** Base upon which a statue or monument is mounted.

PLINTH BLOCK Block, thicker and wider than the casing for the interior trim of a door.

PLINTH COURSE Base course, slightly wider than the wall it supports.

plk *Abbreviation for:* plank.

plmb *Abbreviation for:* plumb.

plmb. *Abbreviation for:* plumbing.

pln. *Abbreviation for:* plenum.

PLOT Carefully measured area laid out for experimentation or measurement. *See also* **Building site.**

plot. *Abbreviation for:* plotting.

PLOT PLAN *See* **Site development plan.**

PLOTTAGE Increment in unit value of a plot of land created by assembling smaller ownerships into one.

PLOTTING 1. Conversion of field-taken measurements into a scale drawing; **2.** Laying out of the governing points of a curve or graph.

PLOUGH To cut a groove.

PLOW Attachment, mounted to the rear of such equipment as a trencher, that is pushed vertically into the ground to a predetermined depth under hydraulic pressure, and that is used to draw pipe, cable, and other small-diameter flexible lines into the ground as the prime mover travels forward. A common variation is to provide the plow with a vibratory force, the amplitude of which may often be varied to best suit the ground being worked.

PLOW STEEL Steel that is tougher and stronger than mild steel, used in making hoist rope.

plstc *Abbreviation for:* plastic.

plstcy *Abbreviation for:* plasticity.

Plt. *Abbreviation for:* plate.

pltg *Abbreviation for:* plating.

plth *Abbreviation for:* plinth.

PLUG 1. Cap used for shutting of a tapped opening. *See also* **Magnetic plug; 2.** Soil inside a pile driven open end; **3.** Device that can be inserted in a socket outlet to continue an elecrical circuit through an attached cord. *See also* **Charging plug; 4.** Cylindrical mechanism in a lock cylinder that houses the keyway; **5.** Seedling grown in a small container under carefully controlled (nursery) conditions; *6. See* **Bit; 7.** *See* **Repair.**

PLUG-AND-DOWEL CUTTER *See* **Bit.**

PLUG-AND-FEATHERS Device for splitting drilled boulders consisting of two half-round pieces of hard steel and a gradual-taper wedge.

PLUG-CENTER BIT *See* **Bit.**

PLUG COCK *See* **Valve.**

PLUGGED Filter element that has collected a sufficient quantity of insoluble contminants such that it can no longer pass rated flow without excessive differential pressure.

PLUGGED CORE Plywood inner ply construction of solid C-Plugged veneer pieces.

PLUGGED CROSSBANDS UNDER FACE Designation denoting a sanded plywood panel of special construction making is suitable for use as an underlayment.

PLUGGING Hole drilled in masonry that is filled with a plug of wood or lead to receive a screw, nail, etc.

PLUGGING CHISEL Short, star-shaped steel handbar struck with a mallet to form holes in concrete, masonry, brick, etc.

PLUG RETAINER Device that retains the plug in a cylinder lock.

PLUG TENON Short tenon projecting from the head or foot of a post to stabilize it.

PLUG WELD Weld made in a hole in one piece of metal as it is lapped over another piece of metal.

PLUM Large, random-shaped stone dropped into freshly placed mass concrete to economize on the volume of concrete. *See also* **Cylopean concrete.**

PLUMB 1. Exactly perpendicular; **2.** Vertical.

PLUMB BOB Pointed weight suspended from a line, used for vertical alignment.

PLUMB CUT 1. Vertical cut at the foot of a rafter where it projects beyond the wall plate; **2.** Vertical cut where a rafter meets the ridge.

PLUMBER Worker skilled in all aspects of plumbing.

PLUMBER'S DOPE Soft, nonhardening compound used to seal pipe threads.

PLUMBER'S FURNACE Heating source, used to heat soldering irons, melt lead or solder, etc.

PLUMBER'S RULE Measure with a standard scale on one side and a scale for measuring the length of 45° offsets on the other.

PLUMBER'S SOIL Mixture of glue and lampblack, used in lead work to prevent lead from bonding to lead pipe and fittings in unwanted areas.

PLUMBING Pipes, fixtures, and other apparatus and appurtenances for the supply of water and the removal of waterborne wastes.

PLUMBING FIXTURE Device that receives and discharges water and liquid wastes.

PLUMBING GUY Cable used to pull structural steel into plumb.

PLUMBING INSPECTOR Person authorized to inspect plumbing works for compliance with a regulatory code.

PLUMBING SYSTEM Drainage system, a venting system, and a water system.

PLUMBING TILE (PIPE) Glazed tile, with bell joints, for below-grade drains.

PLUMB LINE Strong cord weighted at one end, used to establish a perpendicular face or line. *Also called* **Bob.**

PLUMB PILE Vertical pile.

PLUMB POINT *See* **Nadir.**

PLUMB RULE Tool (mason's level) used to aid in building surfaces in a vertical plane.

PLUMB SCRATCH Additional scratch coat that has been applied to obtain a uniform setting bed on a plumb vertical plane.

'PLUMB THE ROD' Instruction given by a survey instrumentman to the rodman requiring the rod to be swayed slowly back and forth along the line of sight so as to permit a reading at the moment when the rod is plumb.

PLUNGE Small swimming pool.

PLUNGE CUT Starting a cut in the center of a log using the tip of a chain-saw blade. *Also called* **Boring.**

PLUNGER 1. Hand-operated suction cup, used for clearing plugged drains; **2.** Piston or moving part of the jack of a hydraulic elevator; **3.** Moving screw column of a screw lift; **4.** Small round steel piston contained in the barrel of a riveting gun and chipping hammer that supplies ramming force.

PLUNGER CYLINDER *See* **Cylinder.**

PLV *Abbreviation for:* pitch line velocity.

plwd *Abbreviation for:* plywood.

PLY 1. Number of thicknesses of building paper; **2.** In plywood, the number of thicknesses of wood veneer (as three-ply, five-ply, etc.); **3.** Layer of rubber-coated parallel cords, part of tire construction; **4.** Layer of rubberized fabric; **5.** Layer formed by a single pass through a single deck of a yarn, cord, or wire braiding machine; **6.** Layer formed by a single pass through a single head of a yarn, cord, or wire knitting machine; **7.** Seamless woven jacket consisting of warp and filler yarns and/or copper wire; **8.** Layer consisting of warp and filler yarns and/or wire; **9.** Layer consisting of multiple strands of cord or wire closely spaced; **10.** Layer formed by winding a single strand of cord or wire closely spaced; **11.** Single yarn in a composite yarn; **12.** Layer of unvulcanized rubber.

PLY ADHESION Force required to separate two adjoining reinforcing members of a hose.

PLYFORM Special plywood used to make forms for concrete.

PLYMETAL Plywood faced on one or both sides with sheet metal.

PLYPON Arched space over a door or window.

PLY RATING Strength of a tire in terms of cotton ply strength. Does not necessarily indicate actual number of plies. It is an index of tire strength and load-carrying ability and does not necessarily represent the number of cord plies in the tire.

PLYRON Plywood panel manufactured with a hardboard face for extra-smooth painting and for a tough-wearing surface.

PLY SEPARATION Parting of the rubber compound between adjacent plies.

PLYWOOD Panel of wood comprising three or more layers of veneer joined with glue and usually laid with the grain of adjoining plies at right angles.

PLYWOOD ADHESIVE *See* **Adhesive.**

pm *Abbreviation for:* paint and material.

p.m. *Abbreviation for: post meridian,* aftern noon..

pmf *Abbreviation for:* probable maximum flood.

PMH *Abbreviation for:* productive machine hour; production machine hour.

pmp *Abbreviation for:* perforated metal pipe; probable maximum precipitation.

pmt *Abbreviation for:* payment.

Pn *Abbreviation for:* partition.

P/N *Abbreviation for:* part number.

pndg *Abbreviation for:* pending.

pneu. *Abbreviation for:* pneumatic.

PNEUMATIC Powered or inflated by compressed air.

PNEUMATICALLY-APPLIED MORTAR *See* **Shotcrete.**

PNEUMATIC CAISSON Caisson where the working chamber is kept under a pressure of air approximately equal to the external water pressure.

PNEUMATIC CONVEYOR Tube through which material or objects are conveyed by an air blast.

PNEUMATIC DRILL Drill operated by compressed air.

PNEUMATIC EJECTOR Means of raising the level of a liquid by admitting it through a check valve into the bottom of a chamber and then ejecting it through another check valve into a discharge pipe by admitting compressed air to the chamber above the liquid.

PNEUMATIC FEED Shotcrete delivery equipment in which material is conveyed by a pressurized air stream.

PNEUMATIC FLUID Fluid suitable for use in a pneumatic system. *See also* **Fluid.**

PNEUMATIC MORTAR Mortar applied in layers by successive passes of an air-powered gun.

PNEUMATIC PICK Light, 9 to 14 kg (10 to 30 lb) concrete breaker.

PNEUMATIC RIVETER Percussive, compressed-air tool fitted with a rivet snap.

PNEUMATICS Engineering science pertaining to gaseous pressure and flow.

PNEUMATIC-TIRED ROLLER *See* **Rubber-tired roller.**

PNEUMATIC TOOL Tool activated by compressed air.

PNEUMATIC VALVE *See* **Valve.**

PNEUMATIC WATER SUPPLY SYSTEM Water supply system, usually confined to one dwelling, in which a cistern connected to the main supply is pressurized by air that forces water through the internal distribution system.

pnl *Abbreviation for:* panel.

pnl bd *Abbreviation for:* panel board.

pnlg *Abbreviation for:* paneling.

pns *Abbreviation for:* part number service.

pnt *Abbreviation for:* paint.

pntd *Abbreviation for:* painted.

pntg *Abbreviation for:* painting.

pntr *Abbreviation for:* painter.

PO *Abbreviation for:* planning objective; post office; purchase order.

PO Box *Abbreviation for:* post office box.

poc. *Abbreviation for:* pocket.

POCKET 1. Opening into which something is fitted; **2.** Naturally-occuring void, as in soil, or within the formation of a tree, etc.

POCKET BELT Endless, power-driven belt fitted with pocket-shaped projections that hold dry materials; used to convey bulk materials at steep angles between elevations.

POCKET CHISEL *See* **Wood chisel.**

POCKET DOOR *See* **Door types.**

POCKET ROT Rot in a tree or log that can be located at any point in its length.

POCK MARK Uneven blister-like elevation, depression, or pimpled appearance.

PODIUM 1. Raised platform; **2.** Continuous bench projecting from the walls of a room; **3.** Low wall serving as a pedestal or foundation; **4.** Wall separating the seats from the arena in an amphitheatre.

POGO STICK 1. Generic term for a powered hand tamper; **2.** Stand used to hold the air hose and electrical connections mounted behind the cab on a truck tractor. *Also called* a **Hitchhiker; 3.** *See also* **Lead.**

POINT 1. Earth contacting teeth and bits of excavating machinery; **2.** Sharp end of a saw tooth; **3.** Electrical outlet from which power can be drawn; **4.** Outlet in a gas system; **5.** Terminal to a bored well having a drive point surmounted by a fine-mesh screen; **6.** *See* **Glazing point.**

POINT-BEARING PILE *See* **Pile.**

POINT CHISEL *See* **Cold chisel.**

POINT COUNT Method for determination of the volumetric composition of a solid by observation of the frequency with which areas of each component coincide with a regular system of points in one or more planes intersecting a sample of the solid. *See also* **Linear-traverse method.**

POINT COUNT (MODIFIED) Point count method supplemented by a determination of the frequency with which areas of each component of a solid are intersected by regularly spaced lines in one or more planes intersecting a sample of the solid.

POINTING 1. Troweling mortar into a joint after the masonry units are laid. *Also called* **Masonry joint; 2.** Strong mortar finishing given the exterior of joints in brickwork or blockwork. Can be of several types:

> **Bastard:** Pointing with a projection from the face of the brickwork of the same mortar as the joints.
>
> **Flat:** *See* **Flush pointing,** below.
>
> **Flush:** Where the enriched mortar is struck off flush with the surface. *Also called* **Flat joint.**
>
> **Keyed:** Concave pointing of a mortar joint.
>
> **Raked:** Mortar removed with a tool while it is still wet.
>
> **Rodded:** Convex shape made with a steel rod or special tool.
>
> **Rolling bead:** Mortar joints tooled with a concave striking tool that forms a rolled projecting profile.
>
> **Roughcut:** Joint made by rubbing with a burlap sack. *Also called* **Sacked pointing.**
>
> **Sacked:** *See* **Roughcut pointing,** above.
>
> **Shoved:** Mortar is squeezed out of the joint as the brick is laid.
>
> **Stripped:** Wood strip placed in the joint as the brick is laid.
>
> **Struck:** Where the enriched mortar is struck at a 45° angle undercutting from the horizontal.
>
> **Tuck:** Where the enriched mortar is fully recessed a uniform amount and finished with a vertical face.

V-joint: Made with a V-joint tool.

Weathered: The enriched mortar is struck at a 45° angle laying back from the horizontal.

POINTING MIX Mortar with a consistency of stiff paste.

POINTING TROWEL *See* **Trowel.**

POINT LOAD Load whose area of contact with the resisting body is negligible in comparison with the area of the resisting body.

POINT OF ACCESS All areas used by workers for work-related passage from one area or level to another.

POINT OF BEGINNING In metes and bounds, the first point on the border of the land being described.

POINT OF CONTRAFLEXURE *See* **Point of inflection.**

POINT OF CURVATURE *See* **Geometry of circular curves.**

POINT OF FIXITY *See* **Depth of fixity.**

POINT OF INFLECTION Point on the length of a structural member subjected to flexure where the curvature changes from concave to convex or, conversely, at which the bending moment is zero. *Also called* **Point of contraflexure.**

POINT OF INTERSECTION *See* **Geometry of circular curves.**

POINT OF SALE Specific area where a product (i.e., tree, log) becomes the property of the buyer.

POINT OF TANGENCY *See* **Geometry of circular curves.**

POINT SAMPLING Forest sampling by variable-sized plots, using prisms or other optical devices. *Also called* **Prism sampling.**

POINT-TOOLED FINISH Architectural rough-surface finish to hardened concrete produced by pitting the surface to a regular degree using a hand- or power-operated steel tool.

pois *Abbreviation for:* poisonous.

pois. *Abbreviation for:* poison.

POISE Obsolete unit of dynamic viscosity, not permitted for use with the SI system of measurement. Symbol: P. *See also the appendix:* **Metric and nonmetric measurement.**

POISSON'S RATIO Ratio of transverse (lateral) strain to the corresponding axial (longitudinal) strain resulting from uniformly distributed axial stress below the proportional limit of the material, The value will average about 0.2 for concrete and 0.25 for most metals.

pol. *Abbreviation for:* polish.

POLARISCOPE Instrument for examining the degree of stress in a sample of glass.

POLARITY Polarity of a battery is an electrical condition determining the direction in which current tends to flow on discharge. By common usage, the discharge current is said to flow from the positive electrode through the external circuit.

POLARIZED GLADHAND Air brake connector between a tractor and trailer designed to prevent the crossing or improper connection of supply and control lines. *See also* **Gladhand.**

POLARIZED LIGHT Light in which the vibrations are in one plane only.

POLE 1. Either extremity of an axis of a sphere; **2.** Terminal of an electrical supply; **3.** Young tree at least 100 mm (4 in.) and less than 200 to 300 mm (8 to 12 in.) in dbh (diameter at breast height).

POLE DERRICK Small portable derrick, guyed by ropes, having a hand winch on which a cable is mounted, for raising moderately heavy objects.

POLE PLATE Horizontal beam resting on, and perpendicular to the principal rafters of a roof truss and supporting the feet of the common rafters and the inner edge of a box gutter.

POLE SHORE *See* **Post shore.**

POLE SIZE Trees from 125 to 275 mm (5 to 11 in.) in dbh (diameter breast height).

POLETIMBER Arbitrary term for small sawtimber trees, generally trees 300 to 450 mm (12 to 18 in.) in dbh (diameter breast height). *Also called* **Small sawtimber.**

POLETIMBER STANDS *See* **Stand size classes.**

POLETIMBER TREES *See* **Tree classes.**

POLE TRAILER *See* **Trailer.**

POLICE POWER Right of an elected body to enact and enforce laws.

POLICY RESTRAINT Order in which all or some part of the work on a project is carried out as dictated by someone in authority.

POLING BACK Process of excavating behind existing timber shoring and timbering a new face.

POLING BOARD Flat member in contact with the ground and supporting the face or sides of an excavation.

POLING FRAME Frame in which the walings support the midpoint of the poling boards.

POLISH OR FINAL GRIND Final operation in which fine abrasives are used to hone a surface to its desired smoothness and appearance.

POLISHED PLATE Plate glass.

POLISHING WHEEL Wheel that can be made of several different kinds of materials, and that has been coated with abrasive grain and glue.

poll. *Abbreviation for:* pollution.

POLLUTION Presence in a body of water, soil, or air of substances of such character and in such quantities that the natural quality of the body (water, soil, or air) is degraded so it impairs the body's usefulness or renders it offensive to the senses of sight, taste, or smell.

POLYAMIDE Thermoplastic material, often fiber forming.

POLYBUTYLENE Light-colored, liquid, straight-chained aliphatic hydrocarbon polymer. Nondrying and widely used as a major component in sealing and caulking compounds (made up of many different synthetic rubbers).

POLYBUTYLENE PIPE *See* **Pipe.**

POLYESTER One of a large group of synthetic resins, mainly produced by reaction of dibasic acids with dihydroxy alcohols, commonly prepared for application by mixing with a vinyl-group monomer and free radical catalysts at ambient temperatures and used as binders for resin mortars and concretes, fiber laminates (mainly glass), adhesives, and the like, and as the vehicle in paint. *See also* **Polymer concrete.**

POLYETHYLENE Thermoplastic high-molecular-weight organic compound used in formulating protective coatings or, in sheet form, as a protective cover for concrete surfaces during the curing period, or to provide a temporary enclosure for construction operations.

POLYETHYLENE PIPE *See* **Pipe.**

POLYGLYCOL SYNTHETIC FLUID Nonaqueous fluid composed of polyglycol derivatives. It may contain additives.

POLYGONAL ROOF Roof that in plan forms a figure bounded by more than four straight lines.

POLYMER Rubber or resin consisting of large molecules formed by polymerization.

POLYMER CONCRETE Concrete in which an organic polymer serves as the binder (*See* **Concrete**). Also known as **Resin concrete**, and sometimes erroneously employed to designate hydraulic cement mortars or concretes in which part or all of the mixing water is replaced by an aqueous dispersion of a thermoplastic copolymer.

POLYMER-CEMENT CONCRETE Mixture of water, hydraulic cement, aggregate, and a monomer or polymer (polymerized in place when a monomer is used).

POLYMERIZATION Reaction in which two or more molecules of the same substance combine to form a compound containing the same elements, and in the same proportions, but of higher molecular weight, from which the original substance can be generated, in some cases only with extreme difficulty.

POLYMER-MODIFIED EMULSION Emulsion formed with an asphalt previously modified with a polymer or formed with a latex in the continuous phase during emulsification.

POLYP GRAPPLE Bucket (orange-peel grapple) of a waste handling crane.

POLYPROPYLENE PIPE *See* **Pipe.**

POLYSTYRENE INSULATING BOARD Rigid insulating board produced in two forms: bead board (formed by enclosing expanded beads within forms and bonding them through heat and pressure), and expanded board (produced by extruding styrofoam in the shape of planks). *See also* **Insulating foam board,** and **Polyurethane insulating board.**

POLYSTYRENE RESIN Synthetic resins, varying from colorless to yellow, formed by the polymerization of styrene on heating, with or without catalysts, that may be used in paints for concrete, or for making sculptured molds, or as insulation.

POLYSULFIDE COATING Protective coating system prepared by polymerizing a chlorinated alkyl polyether with an inorganic polysulfide.

POLYSULFIDE POLYMER Long-chain aliphatic polymer containing disulfide linkage combined to make an elastic rubber-type material.

POLYURETHANE Reaction product of an isocyanate with any of a wide variety of other compounds containing an active hydrogen group, used to formulate tough, abrasion-resistant coatings.

POLYURETHANE INSULATING BOARD Rigid insulating board that is stronger and more flexible than polystyrene products. *See also* **Insulating foam board,** and **Polystyrene insulating board.**

POLYVINYL ACETATE Colorless, permanently thermoplastic resin, usually supplied as an emulsion or water-dispersible powder characterized by flexibility, stability toward light, transparency to ultraviolet rays, high dielectric strength, toughness, and hardness.

POLYVINYL CHLORIDE Synthetic resin prepared by the polymerization of vinyl chloride, used in the manufacture of nonmetallic waterstops for concrete; a colorless solid resistant to water and to concentrated acids and alkalis. *Also called* **PVC.**

POLYVINYL-CHLORIDE PIPE *See* **Pipe.**

POLYVINYL-RESIN GLUE *See* **Adhesive.**

POND 1. Small lake. *See also* **Settling pond; 2.** Stretch of water between two canal locks operating in series; **3.** Area where discharge water from a dredge is held long enough for particulate matter to settle out.

PONDING Gathering of water at low or irregular areas of an ostensibly flat surface.

POND VALUE Market price of logs delivered to a wet site, log pond, or tidewater.

PONTOON 1. Float supporting part of a structure; **2.** Support that attaches to an outrigger jack to increase the supporting area.

PONTOON BRIDGE Bridge, usually temporary, over a river or body of water, supported on floating pontoons.

PONY LEAD *See* **Lead.**

POOL 1. Small pond, often artificial and used as an ornamental feature in a garden; **2.** Deep area of a river or stream; **3.** Swimming pool.

POOR ADHESION Condition that develops when a finish or laminate is applied over a filler and sufficient adhesion does not occur.

POORLY STOCKED STAND *See* **Stocking classes.**

pop. *Abbreviation for:* population.

POPCORN Open-graded asphaltic concrete having 20-mm (0.75-in.) aggregate with very little fine material.

POPCORN CONCRETE No-fines concrete containing insufficient cement paste to fill the voids among the coarse aggregate so that the particles are bound only at the points of contact. *See* **No-fines concrete.**

POP-OFF VALVE *See* **Valve.**

POPOUT Breaking away of small portions of a concrete surface due to localized internal pressure that leaves shallow, typically conical depressions. *See also*

Blemish.

POPPET VALVE *See* **Valve.**

POPPING Blowing of plaster. *See also* **Nail pop.**

POP RIVET *See* **Blind rivet.**

POPULATION DISTRIBUTION Population density per stated geographic area.

POPULATION FORECAST Population projection based on historical data about the population of a stated geographical area.

POPULATION GROWTH Percent increase (or decrease) of the population of a stated area over time.

POPULATION STUDY Demographic study of a stated geographic area. *Also called* **Demographic study.**

por. *Abbreviation for:* porous; porosity.

porc. *Abbreviation for:* porcelain.

PORCELAIN Ceramic material made of kaolin, quartz, and feldspar.

PORCH Covered area extending beyond the main dwelling. May be open or closed.

PORE Tiny passage in material, usually microscopic, through which water may be absorbed or discharged.

PORE FILLER Additive consisting of a finely ground powder such as diatomaceous earth, bentonite, fly ash, etc.

PORE PRESSURE *See* **Pressure.**

PORE SIZE DISTRIBUTION Ratio of the number of holes of a given size to the total number of holes per unit area, expressed as a percent and as a function of hole size.

PORE SPACE Naturally-occurring space or void in a body of rock or sediment.

POROSITY 1. Ratio, usually expressed as a percentage, of the volume of voids in a material to the total volume of the material including the voids; **2.** Ability of a material to absorb a fluid; **3.** Ability of an aggregate to absorb a liquid, generally asphalt.

POROUS FILL *See* **Drainage fill.**

POROUS TUBE 1. Physical condition of a hose tube due to the presence of pores; **2.** Hose tube that has low resistance to permeation.

PORT 1. Small drilled hole; **2.** Opening in an engine cylinder head or cylinder through which air, fuel/air mixture, or exhaust gases pass; **3.** Opening in a storage or process vessel; **4.** Harbor attached to a community.

port. *Abbreviation for:* portable.

PORTABLE Something that is not permanently installed and that can be moved from one location to another.

PORTABLE BELT CONVEYOR Material conveyor with a motor-driven belt, mounted on wheels permitting it to be towed.

PORTABLE CONTAINER Reusable container that has a capacity of 30 L (7.92 gal) or less, but excluding a container that is integral with, or permanently attached to, any appliance, equipment, or vehicle.

PORTABLE CRANE Towable crane, that may have a power-driven hoist and power slewing and derricking.

PORTABLE CRANE SCALE Portable load measuring device suitable for use on a crane. *See also* **Dynamometer,** and **Hook scale.**

PORTABLE GENERATING SET Any wheel-, skid-, truck- or railroad car-mounted, but not self-propelled, equipment designed to supply electric current. This consists of an electric geneator and a prime mover mounted on a common frame with all equipment necessary to constitute a complete, self-contained unit. In larger sources, the fuel source may be separate from the unit.

PORTABLE LADDER Ladder that can be readily moved or carried.

PORTABLE TANK Closed container having a liquid capacity more than 230 L (60 gal), and not intended for fixed installation.

PORTABLE WEIGH SCALE Means for weighing either the gross weight or the axle loadings of vehicles, that can be moved from one location to another and set up with minimum installation.

PORTAL 1. Nearly level entrance to a tunnel; **2.** Elevated base for a crane.

PORTAL CRANE Jib crane carried on a four-legged gantry (portal), that itself runs on rails.

PORTAL FRAME Structural wood, steel, or concrete frame comprising two vertical, or near vertical side members connected by a third member, that may be curved, or form the angle necessary for a pitched roof. Used for a multibay structure where cross-bracing is not permitted. *See also* **Three-hinged portal frame.**

PORTAL-IN To begin tunnel excavation from the portal face.

PORT CLIPPING Destructive mechanical fault occurring when the piston ring on a two-stroke diesel engine impacts the upper or lower edge of the intake or exhaust ports on the cylinder liner. The impact of the ring with the liner/port edge will crack the piston ring and destroy the ability of the ring to contain compression pressure or combustion pressure.

PORTE-COCHERE Porch or portico large enough for vehicles to pass under.

PORTICO Roofed space, open or partly enclosed, forming the entrance and feature of the facade of a building, often with attached or detached columns. There are two basic forms:

    Prostyle: when it projects from the face of the building.

    In-antis: When it recedes into the building.

PORTLAND BLAST-FURNACE-SLAG CEMENT Hydraulic cement consisting of an intimately interground mixture of portland-cement clinker and granulated blast-furnace slag or an intimate and uniform blend of portland cement and fine granulated blast-furnace slag in which the amount of the slag constituent is within specified limits.

PORTLAND CEMENT Cement manufactured by combining, burning, and finely grinding a combination of lime, silica, alumina, and iron oxide. It is capable of hardening through a chemical reaction when mixed with water.

PORTLAND-CEMENT CLINKER Partially fused clinker consisting primarily of hydraulic calcium silicates. *See also* **Clinker.**

PORTLAND LIMESTONE Stone quarried on the Isle of Portland, England; the name source for portland cement.

PORTLAND-POZZOLAN CEMENT Hydraulic cement consisting of an intimate and uniform blend of portland cement or portland blast-furnace slag cement and fine pozzolan produced by intergrinding portland-cement clinker and pozzolan, by blending portland cement or portland blast-furnace slag cement and finely divided pozzolan, or a combination of intergrinding and blending, in which the pozzolan constituent is within specified limits.

PORT SPLITTING *See* **Cushion blasting.**

pos. *Abbreviation for:* position; positive.

POSITION INDICATION Function of monitoring the location of an elevator and notifying the controller and the passengers of that location.

POSITION INDICATOR Remote device that indicates the position of a movable device.

POSITIVE BATTER *See* **Fore batter.**

POSITIVE CLUTCH *See* **Clutch.**

POSITIVE CUTOFF Line of tight sheeting, or a barrier of impervious material, extending downward to an essentially impervious lower boundary to intercept completely the path of subsurface seepage.

POSITIVE DISPLACEMENT 1. Displacement of a constant volume of material. **2.** Wet-mix shotcrete delivery equipment in which the material is pushed through the material hose in a solid mass by a piston or auger.

POSITIVE-DISPLACEMENT HYDRAULIC PUMP *See* **Pump.**

POSITIVE DRIVE Driving connection to two or more wheels or shafts that cause them to turn at approximately the same relative speeds under any conditions.

POSITIVE FLOAT Amount of time available to complete noncritical activities or work items without affecting the overall project duration.

POSITIVE MOMENT Condition of flexure in which, for a horizontal, simply-supported member, the deflected shape is normally considered to be concave downward and the top fibers subjected to compression stresses. *Also called* **Flexural moment.**

POSITIVE REINFORCEMENT Reinforcement for positive moment.

POSITIVE TERMINAL Terminal of a battery from which the positive electrical charge flows through the external circuit to the negative terminal when the cell discharges.

POST 1. Upright member in frame construction; a pillar or column; **2.** Piece of wood, metal, etc., usually long and square or cylindrical, set upright as a support; **3.** Vertical formwork member used as a brace. *Also called* **Prop,** and **Shore.**

POST-AND-BEAM CONSTRUCTION *See* **Con-**

struction types.

POST-AND-GIRT CONSTRUCTION *See* **Construction types.**

POST-BUCKLING STRENGTH Load that can be carried by an element, member, or frame after buckling.

POSTCOMPLETION SERVICES Additional contractural services rendered after issuance of the final certificate for payment. Typically consultation on equipment not part of the prime contract; maintenance services; systems selection and/or installation, etc.

POSTCONSUMER RECYCLING Reuse of materials generated from residential and commercial waste, excluding recycling of material from industrial processes that has not reached the consumer, such as glass broken during manufacturing processes.

POSTCONSUMER WASTE Material or product that has served its intended use and has been discarded for disposal after passing through the hands of a final user.

POSTCURE Treatment, often involving heat, applied to a material or assembly following initial chemical cure to further modify certain characteristics.

POSTDISASTER BUILDING Building essential to provide services in the event of a disaster.

POSTERN Secondary door or gate to a dwelling, often adjacent to a larger principal door.

POSTHOLE AUGER Hand- or power-operated earth auger capable of drilling a hole to a depth sufficient to support a post for fencing or signage.

POSTPURGE Short period following operation of a furnace when exhaust fans are operated to remove unwanted products of combustion.

POST SHORE Individual vertical member used to support loads. *Also called* **Pole shore.** *See also* **Adjustable timber single-post shore, Fabricated single-post shore,** and **Timber single-post shore.**

POSTSPLITTING *See* **Cushion blasting.**

POST-TENSIONING Method of prestressing reinforced concrete in which tendons are tensioned after the concrete has hardened.

POT 1. Steel pressure vessel used to transport concrete; at the point of placement the concrete is ejected by compressed air; **2.** Hollow clay tile.

pot. *Abbreviation for:* potential; potentiometer.

POTABLE WATER Water that is safe for human consumption.

POTENTIAL Voltage existing between any two points in a circuit.

POTENTIAL DIFFERENCE Difference of electrical pressure which establishes a flow of electrical current.

POTENTIAL DROP Loss of electrical pressure due to resistance.

POTENTIAL ENERGY *See* **Energy.**

POTENTIAL SWITCH *See* **Switch.**

POTENTIAL VERTICAL RISE Potential ability of a soil, when exposed to capillary or surface moisture, to swell and thereby increase the elevation of its upper surface along with anything resting on it.

POTENTIAL YIELD Estimated maximum sustained-yield cutting level (stated for a period of time such as a year or decade) attainable with intensive forestry; considers productivity of the land, conventional logging technology, standard cultural treatments, and interrelationships with other resource uses and the environment. Excluded in the estimates are the effects of fertilization, genetic improvement, and irrigation.

POTENTIOMETER 1. Resistance unit having a variable or sliding contact that is positioned by the rotation or sliding of a shaft; **2.** Three-terminal rheostat, used to adjust voltage levels.

POTENTIOMETRIC SURFACE Level to which water will rise in tightly cased wells.

POTHOLE Bowl-shaped hole in pavement resulting from localized disintegration under traffic caused by improper construction methods, unsuitable mix formula, or poor drainage.

POT LIFE Time interval after preparation during which a liquid or plastic mixture is usable.

pott. *Abbreviation for:* pottery.

POUND Nonmetric unit of mass, equal to 16 ounces. Symbol: lb. Multiply by 453.592 37 to obtain grams, symbol: g; by 0.453 592 37 to obtain kilograms, symbol: kg. *See also the appendix*: **Metric and nonmetric measurement.**

POUND PER CUBIC FOOT Nonmetric unit of density. Symbol: lb/ft^3. Multiply by 16.018 46 to obtain grams per liter, symbol: g/L, or kilograms per cubic meter, symbol: kg/m^3. *See also the appendix*: **Metric and nonmetric measurement.**

POUND PER CUBIC INCH Nonmetric unit of density. Symbol: lb/in.^{3}. Multiply by 27.679.9 to obtain grams per cubic centimeter, symbol: g/cm^3, or kilograms per liter, symbol: kg/L. *See also the appendix*: **Metric and nonmetric measurement.**

POUND (FORCE) PER SQUARE INCH Nonmetric unit of pressure. Symbol: psi. Multiply by 6.894 757 to obtain kilopascals, symbol: kPa. *See also the appendix*: **Metric and nonmetric measurement.**

POUND (MASS) PER SQUARE INCH Nonmetric unit of force. Symbol: psi. Multiply by 703.0696 to obtain kilograms per square meter, symbol: kg/m^2. *See also the appendix*: **Metric and nonmetric measurement.**

POUR Total volume of mortar or concrete placed at one time in a continuous operation.

POURED FITTING Method of completing an attachment to wire rope by separating the wires, expanding them in a conical socket, and filling it with molten zinc. *See also* **Wedge socket fitting.**

POUR PLANE Undesirable plane markings caused by wet concrete being placed at a single location and being allowed to flow along the formwork to fill the area.

POUR POINT 1. Temperature of an oil at which it will just flow under prescribed conditions; **2.** Lowest temperature at which a liquid will flow.

POURING (OF CONCRETE) *See* **Placement,** and Placing.

POWDER 1. Any dry substance in the form of very fine, dust-like particles, produced by crushing, grinding, etc.; **2.** Any of various solid explosives. *See also* **Black powder.**

POWDERED ASPHALT *See* **Asphalt.**

POWDER FACTOR Units that describe the amount of explosive used relative to the volume of rock; expressed in kg/m^3 or lb/yd^3. These are empirical values based on test blasting and practical experience and vary with the strength of explosive and rock conditions.

POWDERMAN Worker who charges or loads blastholes with explosives but does not initiate the explosion (fire the blast).

POWER Time rate at which work is done, expressed in horsepower or kilowatts.

POWER ARM That part of a lever between the fulcrum and the point where force may be applied.

POWER AVAILABLE Power that a machine can exert at the drawbar, for track machines, or at the wheels, for rubber-tired machines.

POWER BENDER Power-operated (reinforcing) bar-bending machine.

POWER BRAKE *See* **Brake.**

POWER CENTER Air compressor, hydraulic compressor, electric generator, or PTO-equipped gas or diesel engine that acts as a source of power for one or more equipment items or attachments.

POWER CIRCUIT Circuit transmitting electrical energy to a motor or to a heating unit too large to be served by an ordinary circuit.

POWER CONSUMPTION Power used, multiplied by time.

POWER CONTROL UNIT One or more winches mounted on nonhydraulic tractor-driven equipment and used to manipulate the working parts.

POWER CONTROL WINCH *See* **Winch.**

POWER CURVE Graphic illustration of maximum output of horsepower and torque at all operating speeds, established from data obtained by running a sample engine on an engine dynamometer. Curves are established using both a bare operable engine and one with standard accessories. Net power figures (those using standard accessories) are used in vehicle performance calculations.

POWER CUT *See* **Climb cut.**

POWER DIVIDER 1. Assembly that contains an inter-axial differential and equally divides the driving power between two rear axles; **2.** Small auxiliary gearbox or chain-driven device that allows distribution of driveshaft power to several different mechanical devices mounted on the same vehicle.

POWER DIVIDER LOCKOUT Device that cuts out the operation of an interaxle differential to provide maximum traction when road conditions are unfavorable.

POWER DOOR *See* **Door types.**

POWERED AXLE *See* **Axle.**

POWERED INDUSTRIAL TRUCK Mobile, self-propelled truck used to carry, push, pull, lift, stack, or tier material. *Also called* **Industrial lift truck.**

POWERED SCAFFOLD *See* **Suspended scaffold.**

POWER ELEVATOR *See* **Elevator.**

POWER FLOAT Power-operated machine, usually consisting of a fan of blades up to 2 m (6 ft) in diameter, that is propelled over the surface of newly stiffened concrete to produce a dense, flat, polished surface.

POWER GATE *See* **Elevating gate.**

POWER OF ATTORNEY Written statement authorizing another to act as one's legal agent.

POWER OUTLET Enclosed assembly that may include receptacles, circuit breakers, fuseholders, fused switches, buses, and watt-hour meter mounting means, intended to serve as a means for distributing power required to operate mobile or temporarily-installed equipment.

POWER PACK Prime mover, composed of an engine and generator, hydraulic pump, or compressor to provide electricity, hydraulic power, or air to portable construction equipment.

POWER PANEL Electrical cutout box used for power circuits rather than lighting circuits.

POWER PIN BOOM Full-powered telescopic boom sections combined with manually controlled fly sections that are pinned in the retraction or extension positions. The section is power extended and retracted by pinning to the innermost telescopic cylinder and extending or retracting the outer midsection.

POWER PLANT The prime power source, which may be an internal combustion engine or electric motor, and the power takeoff, which may be direct drive, friction clutch, fluid coupling, hydrodynamic torque converter, hydrostatic, or an electric generator type.

POWER RAMMER Mobile, power-operated device consisting of an air-cooled, single-cylinder containing a piston connected to a spring-loaded tamping foot. Mobility is achieved by directing the unit by hand pressure while it is clear of the ground between compressions.

POWER RELAY Relay that responds to power flow in an electric circuit.

POWER REMOVABLE COUNTERWEIGHT Counterweight powered by cables connected to the telescopic cylinders located in the boom, and control levers located near the valve banks of a crane. The device allows quick removal of the counterweight for increased crane roadability. *See also* **Counterweight,** and **Hydraulically-extendible counterweight.**

POWER REQUIRED Amount of power needed to overcome rolling resistance and grade resistance.

POWERSHIFT TRANSMISSION *See* **Transmission.**

POWER SHOVEL Crane equipped with a boom and dipper stick, on the end of which a shovel bucket is mounted, used for moving soil.

POWERSTAT Variable transformer.

POWER SUPPLY Principal source of electric power for any electrical device.

POWER TAKEOFF Place in an engine or transmission to which an external tool or device may be attached and from which it may be powered. There are several types:

> **Crankshaft-driven:** Draws power from the front end of the crankshaft.

> **Flywheel-driven:** Installed in the flywheel housing ahead of the clutch, able to provide full engine torque for powering heavy-duty rear-mounted equipment.

> **Split-shaft:** Self-contained units installed in the driveline between the transmission and rear axle.

> **Transmission-driven:** Attached directly to the transmission; may have reverse-drive capability.

Also called **PTO.**

POWER TRAIN Moving parts that transfer energy from one point to another, accomplished by a mechanical or fluid coupling such as gears, chains, hydraulic pump, drive shafts, torque converters, etc. The power train includes all the components from engine to actuator, whether the actuator is the tire or hydraulic cylinder.

POWER TROWEL Mechanical device consisting of rotating metal or rubber blades, used to compact and smooth plaster finish coats. *Also called* a **Mechanical trowel.** *See also* **Trowel.**

POWER UNIT 1. Combination of pump, pump drive, reservoir, controls and conditioning components that may be required for its application. *See also* **Hydraulic power unit; 2.** Device on hydraulic elevators that supplies the motive force to run the car.

POWER USABLE Amount of power that can be used; a function of power available and the coefficient of traction.

POWER WRENCH *See* **Wrench.**

POZZOLAN Siliceous or siliceous and aluminum material that, in itself, possesses little or no cementitious value but that will, in finely divided form and in the presence of moisture, chemically react with calcium hydroxide at ordinary temperatures to form compounds possessing cementitious properties.

POZZOLANIC Of, or pertaining to, pozzolan.

POZZOLANIC-ACTIVITY INDEX Index that measures pozzolanic activity based on the strength of cementitious mixtures containing hydraulic cement with, and without the pozzolan, or containing the pozzolan with lime.

P-P *Abbreviation for:* push-pull.

PP-AC *Abbreviation for:* air-conditioning power panel.

ppb *Abbreviation for:* parts per billion.

ppd *Abbreviation for:* prepaid.

ppm *Abbreviation for:* parts per million.

pptn *Abbreviation for:* precipitation.

PQ *Abbreviation for:* physical quality.

pr *Abbreviation for:* pair.

pr. *Abbreviation for:* payroll.

PR *Abbreviation for:* pressure recorder; purchase request.

PRACTICAL CAPACITY *See* **Capacity.**

PRACTICAL REFUSAL *See* **Refusal.**

PRACTICAL WORKING LOAD Established by a crane user, with due allowance for operating conditions, including supporting surface and other factors affecting stability, wind, hazardous surrounds, experience of personnel, etc.

prc. *Abbreviation for:* pierce.

PREACTION SYSTEM *See* **Automatic sprinkler system.**

PREALARMS Warning prior to actual actuation of automatic engine safety measures to indicate impending shutdown.

PREAUGERING 1. Boring through obstructions or materials too dense to penetrate with the planned pile type. This is commonly done for displacement piles in stiff clays where heave may occur; **2.** Exploratory drilling below utility lines to locate and minimize damage to unknown facilities.

preb. *Abbreviation for:* prebend.

PREBORING Boring or spudding through obstructions or materials too hard to drive the designated type of pile through.

prec. *Abbreviation for:* precast; precedence; preceding; precision.

PRECAMBER Camber applied to a concrete beam or slab at time of casting.

PRECAST 1. Concrete member that is cast and cured in other than its final position; **2.** Process of placing and finishing precast concrete.

PRECAST CONCRETE Concrete elements that are factory cast and delivered to site ready for use or installation.

PRECAST CONCRETE JOINT *See* **Pile joint.**

PRECAST PILE *See* **Pile.**

PRECAST SEGMENTAL PILE *See* **Pile.**

PRECEDENCE PROGRAMMING Activity-oriented system to more effectively display the logic and interrelationships of work items than is possible using arrow diagramming.

PRECHAMBER Chamber adjacent to the main combustion chamber of an internal combustion engine in which fuel and air begin to burn.

PRECHARGE PRESSURE *See* **Pressure.**

PRECIPITATE Physical separation from a fluid as a result of a chemical or physical change.

PRECIPITATION Total volume, measured over a stated period, of moisture in the form of rain, snow, hail, and sleet.

PRECIPITATION NUMBER The number of milliliters of precipitate formed when 10 ml (0.35 oz) of lubricating oil are mixed with 90 ml (3.15 oz) of a precipitation naptha and centrifuged under prescribed conditions.

PRECISE SETTLEMENT GAUGE Instrument that monitors long-term ground settlement.

PRECISION Relates to the quality of procedures used, fineness of measurement, and the repeatability of a result.

PRECISION WORK Work that is required to be exact in measurement, finish, etc.

PRECOMBUSTION CHAMBER Any of a variety of chambers, not located in the piston, in which fuel injection takes place. In general, the purpose of any precombustion chamber is to increase the speed, turbulence, and completeness of combustion.

PRECOMMERCIAL THINNING Cutting trees from a young stand so that the remaining trees will have more room to grow to marketable size. The cut trees are those having no commercial value and normally none of the felled trees are removed for utilization. The primary intent is to improve growth potential for the trees left after thinning.

PRECOMPRESSED ZONE Area of a flexural member that is compressed by the prestressing tendons.

PRECONSTRUCTION STAGE Work done, in design, scheduling, and on site, prior to the start of new construction.

PRECONSTRUCTION THAWING Use of heat to thaw, to a specific depth, frozen ground on which construction is intended.

PRECOOLER Heat exchanger that reduces the temperature of a working fluid before initial compresion.

prect *Abbreviation for:* precast.

PRECURE 1. Curing of a resin before pressing; **2.** Incomplete sanding of a pressed board. *See also* **Scorch, and Semi-cure.**

PRECURING PERIOD *See* **Presteaming period.**

PREDECESSOR WORK ITEM Work item that directly precedes one or more work items in the logic sequence of a precedence network.

PREDRAINAGE Dewatering of a site by lowering the water table by pumping in the vicinity of a proposed excavation.

PREDRILLING Excavating soil and other material in the space to be occupied by a pile.

PREEXCAVATION 1. Advance excavation of a general site; **2.** Removal by augering of soil that may heave; **3.** Removal of soil by driving and cleaning out an open-end pipe.

pref. *Abbreviation for:* preference.

prefab. *Abbreviation for:* prefabricate; prefabricated.

PREFABRICATED CONSTRUCTION *See* **Construction types.**

PREFABRICATED MOBILE SCAFFOLD Scaffold system that is manufactured and shipped with all necessary components, with the sole purpose of being used as a mobile scaffold.

PREFACED CONCRETE MASONRY UNIT Lightweight concrete block with a thermoplastic-resin finish to the face.

PREFERRED DIMENSIONS Dimensions that have become rationalized within a trade or industry specialization.

PREFERRED NUMBERS Numerical series derived from a geometric progression, frequently used to select and standardize sizes for a series of things. *Also called* **'R'-series**, and **Renard series**.

PREFILL VALVE *See* **Valve**.

prefin. *Abbreviation for:* prefinish; prefinished.

PREFINISHED PANEL Panel having a factory-applied decorative or protective coating.

PREFIRING 1. Raising the temperature of refractory concrete under controlled conditions prior to placing it in service. **2.** *See* **Dieseling**.

PREFLOAT Mortar that has been placed and allowed to harden prior to bonding tile to it with thin-set materials.

PREFORM 1. To cause a material to adopt a shape similar to its final shape in advance of the final manufacturing stage; **2.** A product shaped in anticipation of the function it will have to perform, but not necessarily to its final shape; **3.** To shape the wires of wire rope so they lie in place.

PREFORMED ASPHALT JOINT SEALER Premolded strip of asphalt mixed with fine mineral substances, fibrous materials, cork, sawdust, or similar materials, manufactured in dimensions suitable for construction joints.

PREFORMED CAVITY Void formed in a concrete member, precast or cast *in situ*, to receive pre- or post-tensioning tendons, building services, etc.

PREFORMED FOAM Foam produced in a foam generator prior to its introduction into a mixer with other ingredients to produce cellular concrete. *See also* **Cellular concrete**.

PREFORMED ROOFING AND SIDING Large-size sheets of metal and other materials, formed with interlocking and watertight seams in a range of decorative and structural profiles, used as roofing and wall cladding. *See also* **Sheet-metal roofing**.

PREFORMED SEALANT Flexible, premolded shape of weatherproof and waterproof material, used alone or in combination with other sealants to join materials in a nonstructural application.

PREFORMED STRUCTURAL GASKET Flexible, preformed shape that holds a material in its intended position and that forms a watertight seal.

PREFORMED WIRE ROPE Wire rope in which the strands are permanently formed during fabrication into the helical shape they will assume in the wire rope.

PREFRAMED Panelized building in which wall, floor, or roof sections are framed and sheathed at the factory.

PREGROUTED TILE Surface unit consisting of an assembly of ceramic tiles bonded together at their edges by a material, generally elastomeric, that seals the joints completely.

PREHARVEST SILVICULTURE ASSESSMENT (or SURVEY) Survey carried out on a forest stand prior to logging to collect specific information on the silvicultural conditions such as planting survival, free-growing status, stocking, etc.

PREHARVEST SILVICULTURE PRESCRIPTION Planning system that collects site-specific field data and develops a forest management prescription for cut blocks in advance of logging.

PREHAULING Moving pulpwood from stump site to truck loading site by carrying it off the ground. *Also called* **Forwarding**.

PREHEATED AIR Air heated prior to its use in combustion, frequently done by hot flue gases.

PREHEATER Ancillary device used to warm an item or area prior to initiation of the machine it is attached to; a small oil burner used to heat the air intake of a diesel engine for cold starting.

PREHYDRATION Process of premixing water with pointing mortar in proportions to form a damp ball when squeezed in the hand. After the mortar has set for about one hour, more water is added to bring the mix to the consistency needed for use.

PREIGNITION *See* **Dieseling**.

prelim. *Abbreviation for:* preliminary.

PRELIMINARY DESIGN Design submitted as part of the project brief. The drawings and specifications that comprise the preliminary design are clearly specified as a matter of policy and provide the basis for the design presentation estimate. *Also called* **Rough sketch**.

PRELIMINARY SPECIFICATIONS *See* **Draft specifications**.

PRELIMINARY SURVEY Fact-finding survey of an area or site, carried out in advance of a detailed survey or any subsurface investigation.

PRELOAD 1. To temporarily heap fill to a calculated depth on a site to be developed so as to compress the natural surface materials; **2.** To circle several logs with binders so the entire unit can be hauled as one log.

PRELOADED TRAILER Log trailer left at a landing to be loaded; the trailer is then hauled at a later time.

PRELOG To remove small understory trees, windfalls, or special products such as poles or pilings from a stand ahead of the main logging to prevent breakage.

PRELUBE Auxiliary to the standard lube oil pump that provides lubrication to the engine prior to starting.

prem. *Abbreviation for:* premium.

PREMATURE Explosive charge that detonates before it is intended to.

PREMATURE STIFFENING *See* **False set**, and **Flash set**.

PREMISES Building and its grounds.

PREMISES WIRING SYSTEM That interior and exterior wiring, including power, lighting, control, and signal circuit wiring together with all of its associated hardware, fittings, and wiring devices, both permanently and temporarily installed, that extends from the load end of the service drop, or load end of the service lateral conductors to the outlet(s).

PREMIXED Two or more materials mixed at the point of manufacture and requiring only the addition of a commonly available substance (such as water) to produce a complete product, such as plaster, concrete, etc.

PREMIXING TIP *See* **Hamm tip.**

PREMOLDED ASPHALT PANEL Panel, generally made with a core of asphalt, minerals, and fibers, covered on each side by a layer of asphalt-impregnated felt or fabric, coated on the outside with hot-applied asphalt. These panels are made under pressure and heat to a width of 0.9 to 1.2 m (3 to 4 ft), to a thickness of 3 to 25 mm (0.125 to 1 in.), and to any desired length.

prenotvail. *Abbreviation for:* previously not available.

prep. *Abbreviation for:* preparation; prepare; prepared.

PREPACKAGED CONTAINER Container not intended for reuse.

PREPARED LOG STRINGER Stringer that is barked, trimmed, hewed, or ripped, ready to be placed in position.

PREPAYMENT Payment of any part of a mortgage loan before its due date; an act that may be limited or precluded under the terms of the mortgage, or subject to a penalty.

prepg *Abbreviation for:* preparing.

PREPLACED-AGGREGATE CONCRETE Concrete produced by placing a coarse aggregate in a form and later injecting a portland cement-sand grout, usually with admixtures, to fill the voids.

prepn *Abbreviation for:* preparation.

PRE/POST-TENSIONING Technique of producing prestressed concrete in which some of the tendons are pretensioned and others are post-tensioned.

PREPRIMED Panel with a factory-applied primer or undercoat needing only final finish after installation.

PREPRODUCTION INSPECTION OR TEST Examination of samples from a trial run of a product to determine adherence to a given specification, for approval to produce.

PREPURGE Air blown through the combustion chambers of a furnace prior to ignition to remove unwanted gases.

PREQUALIFICATION Assessment of a potential bidder or supplier prior to an invitation to bid. Prequalification is based on past record, present capability and availability, etc.

pres. *Abbreviation for:* present; president.

PRESCRIBED BURNING Knowledgeable application of fire to a specified land area to accomplish designated land management objectives. *See also* **Area ignition, Backfire,** and **Controlled burning.**

PRESCRIBED MIX One of a standard range of concrete mixes.

PRESELECTIVE TRANSMISSION *See* **Transmission.**

PRESENTATION DRAWING Drawing that presents a project as it will look on completion, usually shown in perspective, surrounded by the buildings and land forms existing at the site, and in a manner that depicts a realistic form to human scale. *Also called* **Display drawing.**

preser. *Abbreviation for:* preservation; preserve.

PRESERVATION Process of maintaining a structure in its present condition and arresting further deterioration. *See also* **Rehabilitation, Repair,** and **Restoration.**

PRESERVATIVE Any substance that, for a reasonable length of time, will prevent the action of wood-destroying fungi, borers, and similar destructive agents.

PRESERVATIVE TREATMENT Treatment of wood that prevents decay from moisture or bacteria.

PRESET *See* **Set point.**

PRESET PERIOD *See* **Presteaming period.**

PRESHEARING *See* **Presplitting.**

PRESHRUNK CONCRETE (MORTAR, GROUT) **1.** Concrete that has been mixed for a short period in a stationary mixer before being transferred to a transit mixer; **2.** Grout, mortar, or concrete that has been mixed one to three hours before placing in order to reduce shrinkage during hardening.

PRESLOTTING *See* **Presplitting.**

PRESPLITTING Stress relief involving a single row of holes, drilled along a neat excavation line, where detonation of explosives in the holes causes shearing of the web of rock between the holes, usually fired in advance of the production holes. *Also caled* **Preshearing,** and **Preslotting.**

PRESS 1. Machine designed to exert uniform pressure over a significant area, compacting materials held within its platens; **2.** Steel fabricating machine for punching holes, shearing, notching, etc.

press. *Abbreviation for:* pressure.

PRESS BRAKE Machine used to cold-form metal sheet or strip.

PRESS DRILL *See* **Drill.**

PRESSED Wire rope eyes secured to the line by crimping, either directly or with a metal sleeve.

PRESSED BRICKS *See* **Brick.**

PRESSED EDGE Edge of a footing along which the greatest soil pressure occurs under conditions of overturning.

PRESSED FACE/BACK One of the surface textures available on OSB panels, having a relatively smooth, low-friction surface.

PRESSING Act of squeezing a panel together.

PRESS RAM Movable part of the cylinder of a hydraulic press.

PRESS-TO-TEST Manual action on a test switch or button that initiates a system integrity test.

PRESSURE Equals force per unit area, usually expressed in kilograms per square meter (kg/m^2), or in pounds per square inch (psi). There are many ways of determining pressure, including:

Absolute: Ordinary outside air pressure (barometric pressure); the sum of atmospheric pressure and gauge pressure.

Atmospheric: Pressure of air at sea level, usually 101.325 Kp (14.7 psia) (1 atmosphere), or 0 psig.

Back: (a). Measured loss in efficiency in an engine exhaust system due to friction or pumping losses; **(b).** Earth pressure exerted on the vertical walls of an abutment by the fill, or the lateral loading of a piling or face log in an abutment; **(c).** In a plumbing system, compressing of trapped air, that resists the flow of waste through the drain, waste, and vent piping.

Burst: Pressure at which rupture occurs.

Charge: Pressure at which replenishing fluid is forced back into a fluid power system.

Critical: Vapor pressure corresponding to the critical state of the substance at which the liquid and vapor have identical properties.

Differential: Difference in pressure between two points in a system.

Gauge: Pressure measured by a gauge and indicating the pressure exceeding atmospheric.

Hydrostatic: 1. Pressure at any point in a liquid at rest; **2.** State of stress in which all the principal stresses are equal (and there is no shear stress), as in a liquid at rest; the product of the unit weight of the liquid and the difference in elevation between the given point and the free water elevation; **3.** Pressure exerted on the underside of a concrete slab when the water table is above its elevation.

Maximum working: Pressure as stated by the machine manufacturer as the maximum at which a hydraulic circuit shall be operated. This pressure may be limited by a relief valve or other means.

Operating: Maximum pressure permitted to be developed in a hydraulic system, as limited by the setting of a main relief valve to the desired pressure.

Override: Difference between the cracking pressure of a valve and the pressure reached when the valve is passing at full flow.

Partial: Portion of total gas pressure of a mixture attributable to one component.

Passive earth: Resistance to deformation due to active earth pressure.

Pilot: 1. Pressure in a pilot circuit; **2.** Hydraulic pressure used to actuate or control hydraulic components.

Pore: Water pressure developed in the voids of a soil mass.

Precharge: Pressure of a compressed gas in an accumulator prior to admission of a liquid.

Proof: Specified pressure that exceeds the manufacturer's recommended working pressure applied to a hose to indicate its reliability at normal working pressure. (Proof pressure is usually twice the working pressure.) *Also called* **Test pressure.**

Rated: Qualified operating pressure recommended for a component or system by the manufacturer.

Rated fatigue: Pressure that a pressure-containing component is represented to sustain 10 million times without failure.

Rated static: Pressure that a component can withstand without failure.

Saturation: For a pure substance, that pressure at which vapor and liquid, or vapor and solid, can coexist in stable equilibrium.

Service: *See* **Working pressure,** below.

Static: Water pressure head available at a specific location.

Suction: Operating pressure measured in the suction line at a compressor inlet.

System: Pressure that overcomes the total resistance in a system. It includes all losses as well as useful work.

Test: *See* **Proof pressure,** above.

Total: Pressure on a horizontal plane of soil from the mass of material above it, plus any superimposed load.

Vapor: 1. Component of atmospheric pressure, caused by the presence of vapor, expressed in inches, centimeters, or millimeters of height of a column of mercury, or in pascals; **2.** Pressure, at a given fluid temperature, in which the liquid and gaseous phases of the fluid are in equilibrium.

Velocity: In moving fluid, the pressure capable of causing an equivalent velocity, if applied to the same moving fluid through an orifice such that all pressure energy expended is converted into kinetic energy.

Working: Maximum pressure to which a component will be subjected, including the momentary surges in pressure that can occur during service. *Also called* **Service pressure.**

PRESSURE-ACTUATED SEAL Sealing device in which sealing action is aided by fluid pressure.

PRESSURE BALANCE Pressure in a system or container equal to that which exists outside.

PRESSURE-BALANCED EXPANSION JOINT *See* **Joint.**

PRESSURE CONTROL VALVE *See* **Valve.**

PRESSURE-DIFFERENTIAL SWITCH *See* **Switch.**

PRESSURE DROP Reduction in pressure between two points in a line or passage due to the energy lost in maintaining flow.

PRESSURE EQUALIZING Permitting high- and low-side pressures to equalize during idle periods through the use of an unloading valve or vapor lock liquid control; or nearly equalizing inlet and discharge pressures on a compressor to reduce starting torque load.

PRESSURE GAUGE *See* **Gauge.**

PRESSURE GROUTING *See* **Foundation grouting.**

PRESSURE HEAD Amount of force or pressure created by a depth of one meter (or one foot, or other dimension) of water.

PRESSURE INJECTED FOOTING *See* **Enlarged base.**

PRESSURE-LIMITING DEVICE Pressure-responsive mechanism designed to arrest operation of the pressure-imposing element at a predetermined level.

PRESSURE LINE 1. Locus of force points within a structure resulting from combined prestressing force and externally applied load; **2.** Line conducting pressurized fluid to a working device or devices.

PRESSURE LOSS Reduction in pressure due to expenditure of pressure energy required to move water or air through a line, including loss from back pressure, elevation, friction, etc.

PRESSURE METER Instrument for *in-situ* testing of the mechanical properties of soil or rock by hydraulically expanding a probe in a bore hole and measuring the volume changes produced by successive increments of pressure.

PRESSURE PILE *See* **Pile.**

PRESSURE PLATE Plate in a clutch that is driven by the flywheel or rotating housing and that can be slid toward the flywheel to squeeze a lined plate or plates against it.

PRESSURE PLATE ASSEMBLY Assembly that applies the force from the diaphragm-spring and which clamps the clutch disks between the center plate and the flywheel. It also houses the adjusting ring and parts of the release bearing assembly.

PRESSURE-PRESERVATIVE TREATED Wood treated with preservative or fire retardants by pressure-injecting treating solutions into wood cells, e.g., creosote, pentachlorophenol, or ammoniacal copper arsenate or chlorinated copper arsenate.

PRESSURE-REDUCING (GAS) VALVE *See* **Valve.**

PRESSURE-REDUCING (WATER) VALVE *See* **Valve.**

PRESSURE-REDUCING VALVE *See* **Valve.**

PRESSURE REGULATOR *See* **Valve.**

PRESSURE-RELIEF VALVE *See* **Valve.**

PRESSURE SEAL Liner spray used in both tubeless and tube-type tires to seal porosity and rim leaks and reduce liner oxidation.

PRESSURE SENSING VALVE *See* **Valve.**

PRESSURE-SENSITIVE ADHESIVE *See* **Adhesive.**

PRESSURE SWITCH *See* **Switch.**

PRESSURE TANK Storage or process vessel capable of holding its contents under a maximum design pressure.

PRESSURE TRANSDUCER Transducer that converts pressure to a proportional electrical signal.

PRESSURE-TYPE AIR COOLER Cooler used with one or more external elements that impose air resistance.

PRESSURE-UNLOADING VALVE *See* **Valve.**

PRESSURE VESSEL Container designed to operate at pressures greater than 100 kPa (14.5 psi) gauge.

PRESSURE WELDING Welding done by pressing the joint parts together while the weld metal is plastic.

PRESTEAMING PERIOD In the manufacture of concrete products, the time between molding of a concrete product and start of the temperature-rise period. *Also called* **Holding period, Precuring period,** and **Preset period.**

prestr. *Abbreviation for:* prestressed.

PRESTRESS 1. To place a hardened concrete member or an assembly of units in a state of compression prior to application of service loads; **2.** Stress developed by prestressing, such as by pretensioning or post-tensioning.

PRESTRESSED CONCRETE Reinforced concrete in which preliminary stresses are placed, using high-strength steel wire in tension, before a load is applied.

PRESTRESSED PILE Precast concrete pile that is either pretensioned or post-tensioned to reduce or eliminate tensile stresses to which piles are subjected during transportation, driving, and in service.

PRESTRESSING Technique of jacking prestressing tendons.

PRESTRESSING STEEL High-strength steel used to prestress concrete, commonly seven-wire strands, single wires, bars, rods, or groups of wires or strands.

PRESTRETCHING Subjecting a wire rope or strand to tension prior to its intended application, for an extent and over a period of time sufficient to remove most of the constructional stretch.

PRESTRIKEOFF An adjustable lower portion added to a paver moldboard to help control the material that passes under the screed. This unit is often contoured (with a radius similar to that of the auger mounted on the rear of the paver) to assist the lateral flow of material across the screed.

PRETENSIONING Method of prestressing reinforced concrete in which the tendons are tensioned before the concrete has hardened.

PRETENSIONING BED (or BENCH) Casting bed on which pretensioned members are manufactured and that resists the pretensioning force prior to release. *See also* **Bench.**

PRETEST Process of testing with hydraulic jacks the bearing capacity of a fabricated component prior to placing the unit in service.

PRETIMED SIGNAL *See* **Traffic signal.**

prev. *Abbreviation for:* prevent; previous.

PREVAILING WIND Direction from which the wind blows most often during a specific season of the year.

PREVENTER In rigging practice, a wire rope used to prevent unwanted movement, or as a safety strap.

PREVENTION Activities aimed at minimizing or reducing occurrences of a circumstance, such as fire, etc.

PREVENTIVE MAINTENANCE Maintenance measures taken in advance to avoid breakdowns.

prf *Abbreviation for:* proof.

pri. *Abbreviation for:* primary; primed; priority.

PRICE ADJUSTMENT Adjustment to a negotiated price.

PRICE BREAKDOWN Breakdown of a lump sum. Typically this might reveal the components for labor, insurance, supplied materials, added value, transportation, etc.

PRICED BILL OF QUANTITIES Listing of all materials required for a project, with prices and costs added.

PRICE-LEVEL-ADJUSTED MORTGAGE *See* **Mortgage.**

PRICE SCHEDULE List of charges showing any variations for specific quantities or special conditions. *Also called* **Schedule of prices.**

PRICE VARIANCE *See* **Variance.**

PRICK *See* **Punch.**

PRICKER MARK Perforation of the cover of a hose performed before or after vulcanization.

PRIMACORD Proprietary name for a detonating fuse.

PRIMARY BLAST Main explosive blast executed to sustain production.

PRIMARY BRAKE Brake used to stop a mobile device or vehicle under normal operating conditions.

PRIMARY BURNER *See* **Burner.**

PRIMARY COMBUSTION AIR That portion of the combustion air introduced with or through the combustible materials.

PRIMARY COMBUSTION CHAMBER Chamber wherein primary ignition and burning occurs.

PRIMARY COMPRESSION FAILURE Failure of reinforced concrete initiated by crushing of the concrete.

PRIMARY CONSOLIDATION Compression of soil under load that occurs while excess pore pressure dissipates with time.

PRIMARY CONVEYOR Conveyor that transports materials to the discharge conveyor.

PRIMARY CRUSHER *See* **Crusher.**

PRIMARY DISTRIBUTION ZONE Zone immediately adjacent to the end face of a prestressed concrete beam in which the force from an end block is dispersed uniformly into the beam.

PRIMARY DRAIN Single sloping connection from the base of a soil or waste stack to its junction with the main building drain or with another branch.

PRIMARY DRIVE CAP Drive cap that requires an adapter for different shapes of piles. It is used only with an adapter.

PRIMARY EXCAVATION Digging in undisturbed soil.

PRIMARY FILTER First stage of a multistage filter system.

PRIMARY FLOW-AND-RETURN PIPE One of the pipes through which water circulates in a hot-water heating system.

PRIMARY GLUING Glue used in the manufacture of plywood and other veneering work, as distinguished from assembly gluing.

PRIMARY LOGGING ROAD Road designed and maintained for a high level of use. Typically an all-weather gravel road that is part of a permanent road system.

PRIMARY MANUFACTURING RESIDUE Waste products that results from the improvement of a raw material; sawdust, chips, slabs, and the like from lumber conversion, for instance.

PRIMARY MARK Occasional deep scratche or mark that feels rough on finish sanded board that are caused by the primary coarse-grit sanding heads, and not totally removed by the finish sanding heads.

PRIMARY MATERIAL Virgin or new material used for manufacturing a basic product.

PRIMARY MEMBER One of the main load-carrying elements of a structural system, such as a column, end wall post, rafter, etc.

PRIMARY RECYCLING Return of secondary material to the same industry from which it came and processing of that secondary material so that it will yield the same or similar product as that to which it was a secondary material.

PRIMARY SPILLWAY *See* **Spillway.**

PRIMARY SURFACING HOPPER Part of a roofing machine where surfacing granules are applied following application of the primary coating.

PRIMARY TANK First in line of two storage tanks in an air brake system.

PRIMARY TENSION FAILURE Failure of reinforced concrete through yielding of the steel.

PRIMARY TRANSPORTATION Movement of a felled tree from the stump to the landing.

PRIMARY WINDING Transformer winding that is connected to a power source.

PRIME 1. Provision of means to start a process; **2.** To supply sufficient fluid (water, gasoline, etc.) to start a pump; **3.** Manual injection of a volatile fuel into an engine to help start the mechanical process; **4.** Placing of a detonator in a cartridge or explosive charge.

PRIME COAT *See* **Bituminous coating.**

PRIME CONTRACT Contract to the party having principal responsibility for his own actions and work, plus that of contractors and suppliers to whom he subcontracts.

PRIME CONTRACTOR Contractor having a contract directly with the owner; the contractor responsible for the project until its completion. *Also called* **Main contractor.**

PRIME LOG Log that is a given size and free from defects.

PRIME MOVER 1. Machine used to pull or push other machines; **2.** Engine or motor, used to drive a machine or machines.

PRIME POWER *See* **Generator set.**

PRIME POWER ELECTRIC SET *See* **Electric set.**

PRIMER 1. Explosive unit containing a suitable firing device that is used for the initiation of an entire explosive charge; **2.** First layer of paint to be applied to a surface to reduce absorbency and to ensure adhesion of subsequent coats. *Also called* **Sealer,** and **Undercoat; 3.** Paint used for such purpose; **4.** Material applied to joint faces to improve the bond (adhesion) of sealants.

PRIMER-SEALER First coat of paint placed on raw unpainted wood, formulated to seal the pores of the wood and to provide a suitable substrate for subseqyent applications of paint..

PRIMING PUMP *See* **Pump.**

prin. *Abbreviation for:* principal.

PRINCESS POST Any subsidiary vertical timber between the queen post and the wall used to stiffen a queen-post truss.

PRINCIPAL 1. Most important element in a complex, design, organization, etc.; **2.** Basic element of a loan; the amount upon which interest is paid; **3.** Person named in a surety bond for whose obligations the surety agrees to be liable.

PRINCIPAL PLANES *See* **Principal stress.**

PRINCIPAL POINT Point on an aerial photograph where the optical axis of the camera intersects the film.

PRINCIPAL RAFTER One of the main rafters in a roof truss that carry the purlins on which common rafters are laid.

PRINCIPAL STRESS Maximum and minimum stresses at any point acting at right angles to the mutually perpendicular planes of zero shearing stress, that are designated as the **Principal planes.**

PRINCIPAL SUPPLIER Manufacturer, fabricator, distributor, or supplier of any material, equipment, or system for a part of the designated work having a purchase order or other agreement with the contractor with a total contract value in excess of ____ thousand dollars ($____,000).

PRINT Reproduction of a drawing or photographic negative.

PRINTED BRAND *See* **Brand.**

PRIORITY SEQUENCE Order in which settings are to be yarded.

PRISM Small masonry assemblage made with masonry units and mortar, primarily used to predict the strength of full-scale masonry members.

PRISMATIC BEAM Beam having both flanges parallel to its longitudinal axis.

PRISMATIC COMPASS Magnetic compass having a prism mounted above its face, permitting a sight to be taken on a distant object while simultaneously observing the magnetic bearing.

PRISM CONSTANT Additive correction made to distances observed through an optical instrument to account for different types of prism.

PRISM GLASS Glass designed to change the direction of light rays a predictable amount by refraction.

PRISMOIDAL FORMULA Means of determining a volume of earth based on the length of an excavation, the two end areas, and the area at midpoint.

PRISM SAMPLING *See* **Point sampling.**

PRIVACY ZONE Area adjacent to a dwelling that is restricted to exclusive use by the residents of the dwelling.

PRIVATE With respect to a room or other space within a building, an area intended solely for the use of an individual tenant or family.

PRIVATE COLLECTION *See* **Collection.**

PRIVATE RESIDENCE Detached dwelling or a separate apartment in a multiple dwelling, occupied by the members of a single family unit.

PRIVATE RESIDENCE ELEVATOR Passenger elevator installed in a private residence, having a load rating less than 317 kg (700 lb), rated speed less than 18 m/min (40 ft/min), net inside platform area less than 1.1 m² (12 ft²), and rise less than 15 m (50 ft).

PRIVATE RESIDENCE INCLINED LIFT Power passenger lift installed on a stairway for raising and lowering individuals from one floor to another.

PRIVATE SEWAGE DISPOSAL SYSTEM Privately owned plant for the treatment and disposal of sewage (such as a septic tank with an absorption field).

prmr *Abbreviation for:* primer.

prob. *Abbreviation for:* problem.

PROBABILISTIC DESIGN Method of design of structures using the principles of statistics (probability) as a basis for evaluation of structural safety.

PROBABLE MAXIMUM FLOOD Flood that may be expected from the most severe combination of meteorologic and hydrologic conditions possible for a region.

PROBABLE MAXIMUM PRECIPITATION Maximum amount and duration of precipitation that can be expected to occur on a drainage basin.

PROBE Device that can be used to investigate an area for one or more specific conditions or states (temperature, humidity, resistance to penetration, etc.).

PROBING Sensing a state or condition using a probe sensitive to the required data. *Also called* **Piercing.** *See also* **Sounding.**

PROBLEM WASTE *See* **Solid waste.**

proc. *Abbreviation for:* procedure; proceed; process; procure.

Proc. *Abbreviation for:* proceedings.

PROCESSABILITY Relative ease with which raw or compounded rubber can be handled in or on rubber processing machinery.

PROCESS ADDITION Material or agent added during a manufacturing process.

PROCESSED SHAKE Sawn shingle, surface-textured on one face to resemble a split shingle.

PROCESS OIL Oil used in the manufacturing process rather than for lubrication.

PROCESSOR *See* **Harvesting machines**

(multifunction).

PROCESS PLANT Industrial occupancy intended for the handling, processing, assembly, and manufacture of substances, materials, products, etc.

PROCESS WASTE Waste material from an industrial process.

PROCTOR (TEST) Method developed for determining the density-moisture relationship in soils. *See also* **Modified Proctor.**

PROCTOR NEEDLE Instrument for measuring the resistance of a soil to penetration. *Also called* **Penetration needle.** *See also* **Modified Proctor.**

PROCUREMENT 1. Process in the project management system that consists of establishing contractual relationships for carrying out program objectives; **2.** Assembly, tendering, and award of contracts or commitment documents.

prod. *Abbreviation for:* product; production.

PRODUCER Independent operator who produces and delivers pulpwood to a dealer or a pulpwood company.

PRODUCER GAS Gas manufactured by burning coal using insufficient oxygen for complete combustion.

PRODUCT DATA Illustration, schedule, performance chart, instruction, brochure, diagram or other information submitted to describe and illustrate any material, equipment, or system.

PRODUCTION Use of raw materials to create producer and consumer goods through various refining, processing, and manufacturing activities; the quantity of product turned out or the amount of work done in a given time.

PRODUCTION PILE Pile that is part of a specified pile foundation, as opposed to a preliminary test pile.

PRODUCTION WELL Well used for the production of water from an aquifer or water-bearing formation.

PRODUCTIVE LAND Forest land that is capable of producing a merchantable stand within a reasonable length of time.

PRODUCTIVE MACHINE TIME *See* **Machine time.**

PRODUCTIVE TIME *See* **Machine time.**

PRODUCT PIPE Permanent pipe used in a conduit installation.

PRODUCTS OF COMBUSTION Gases, vapors and solids that result from the combustion of a material.

PRODUCT STANDARD Standard that states how a product must be manufactured.

prof. *Abbreviation for:* professional.

PROF *Abbreviation for:* profiler.

prof. cut. *Abbreviation for:* profile cutter.

PROFESSIONAL ENGINEER Person registered or licensed to practice as a professional engineer under applicable legislation.

PROFESSIONAL FEE Charge made by a professional for services rendered, according to an established scale, or as negotiated for the specific service(s).

PROFESSIONAL LIABILITY INSURANCE Indemnity against legal liability or such as an architect or engineer for damages arising from an error, omission, or negligence in his professional capacity.

PROFILE 1. Cross section of a molding; **2.** Contour or outline of a building or any part of it; **3.** Intersection of a vertical plane through the centerline with the surface of the ground and the plane or planes of the finished earthwork so as to indicate grades and distances, depth of cut, and height of fill. **4.** Shoulder-to-shoulder area of the tread cross section of a tire.

PROFILE DENSITY Variation of density of a panel from face to core.

PROFILE GRADE Trace of a vertical plane intersecting the top surface of the proposed wearing surface of a road, usually along the longitudinal centerline of the roadbed. *See also* **Grade.**

PROFILE LEVELING Taking of elevations at frequencies sufficient to establish a representative profile of the ground surface along a line that is not necessarily continuous in its direction. *See also* **Differential leveling.**

PROFILER Plastic or metal spacing block positioned at varying centers along a prestressing cable to separate individual strands.

PROFILING Automatically controlled cold milling of pavement to restore the surface to a specified grade and slope; to remove bumps, ruts, and other imperfections, and to leave a textured surface that can be either opened immediately to traffic or overlaid with new pavement materials. *See also* **Photographing.**

PROFILOGRAPH 1. Wheeled instrument for automatically recording the profile of the land over which it travels; **2.** Instrument for measuring the smoothness of a surface by amplification of any variations from the plane or arc of smoothness.

PROFILOMETER Mechanical device used to measure the smoothness of a surface. The California 25-ft profilograph is a widely used type for roadways.

PROFITABILITY STUDY Study of the anticipated revenues and costs of a project.

prog. *Abbreviation for:* program.

PROGRAM 1. Criteria and/or requirements that must be met; **2.** Plan of operations by which to meet an objective; **3.** Instructions in logical sequence that directs a computer to perform a given task.

PROGRAM BUDGETING Short form of the term **Planning, Programming, Budgeting, System.**

PROGRAM EVALUATION AND REVIEW TECHNIQUE Event-oriented system generally used in the research and development field where, at the planning stage, activities and their durations between events are difficult to define, but where completion of these activities by a specific date is essential to the success of the project. Typically, these projects involve massive programs, many large organizations, and extensive operations in many different locations. *Also called* **PERT.** *See also* **Arrow,** and **Critical-path method.**

PROGRAMMABLE Component that can store a repetitive series of instructions.

PROGRAMMING Preparation of instructions for a computer-driven or -controlled device.

PROGRAMMING DEVICE Electronic or mechanical device used to program or insert operational information into a computing device.

PROGRESS CHART Diagrammatic presentation of a project represented by a number of stages and elements with an indication of the stage of completion of each represented on a time line.

PROGRESSIVE BURNING Disposal of slash by burning it as it is piled.

PROGRESSIVE KILN Kiln in which a total charge of lumber is not dried as a single unit, but as several units, such as kiln truckloads, that move progressively through the kiln.

PROGRESSIVE SIGNAL SYSTEM *See* **Traffic signal.**

PROGRESSIVE SYSTEM Well-pointing system, used in trenchwork, that dewaters the ground prior to excavation. As each trench section is completed, the well points are extracted and relocated in a new section.

PROGRESSIVE TAX Tax rate that becomes progressively higher with each successively higher bracket of the tax base. *Also called* **Graduated tax.**

PROGRESSIVE TYPE SPRING *See* **Spring.**

PROGRESSIVE (VARIABLE-RATE) SUSPENSION *See* **Suspension.**

PROGRESS PAYMENT Payment in response to an invoice submitted by a construction contractor, asking for payment for works completed, or in progress, or materials required for the project that are in inventory and on site, to a specific date. The invoice is usually supported by a progress payment certificate. *Also called* **Partial payment.**

PROGRESS REPORT Regular report that describes the progress made to that point in time of a construction project, and that describes any variation from an established construction program. *Also called* **Construction progress report, Construction status report, Project status report,** and **Status report.**

PROGRESS SCHEDULE Diagram, graph, or other presentation showing proposed and actual times of starting and completing described events.

proj. *Abbreviation for:* project; projecting; projection.

PROJECT 1. Entire sequence of activities associated with satisfying a client request through new construction or renovation; **2.** Two or more activities or tasks which, when performed, lead to a common goal, within the concept that a project has a single starting point and a single ending point.

PROJECT ACCOUNTANT Title of an organizational position designating the person responsible for financial accounting and reporting of a project.

PROJECT ANALYSIS Evaluation of a project, at any stage from conceptualization to final completion, according to any criteria established for the purpose.

PROJECT BRIEF Capsulated description of a project defining its purpose, scope, anticipated completion date, cost, etc.

PROJECT BUDGET *See* **Appropriation,** and **Budget.**

PROJECT CLOSEOUT Full completion of a project.

PROJECT COMPLETION Stage at which a project is completed as described and illustrated in the contract documents. *Also called* **Completion of project.**

PROJECT CONTROL A management function that consists of taking actions (or directing or motivating others to take actions) that will yield an outcome in accordance with a predetermined performance standard.

PROJECT COST Any of the cost types (appropriation, commitment, expenditure, or estimate to complete) associated with the total scope of work in a project; that is all cost classes. The project cost comprises: (a) the costs of the feasibility phase, (b) construction costs, and (c) management and overhead costs during the construction phase.

PROJECT DEFINITION Mission statement and overall scope of a project.

PROJECT DRAWING One of the drawings that, along with the project specification, completely describe the construction of the work required or referred to in the contract documents.

PROJECTED BEARING AREA Area equivalent to the length of a bearing multiplied by its diameter. *Also called* **Bearing area,** and **Circumferential bearing area.**

PROJECT ENGINEER Professional engineer responsible for on-site works.

PROJECT EVALUATION AND REVIEW TECHNIQUE A probabilistic, event-oriented control technique, events are identified within the nodes themselves. *Also called* **PERT.**

PROJECT HAND-OVER Acceptance of a building or project by the operator or owner for occupancy or use, even though construction may not be fully complete.

PROJECTING BELT COURSE Plain band course of cut stonework projecting beyond the face of a wall.

PROJECTING SCAFFOLD Working platform, such as a bracket scaffold, that is built out from an upper story, and that does not reach the ground.

PROJECT INSPECTOR Person responsible to the prime or general contractor for inspecting materials supplied to a job and work completed on the job to ensure that they meet the appropriate specifications and local regulations.

PROJECTION 1. Extension of one member beyond another; **2.** Representation of an object on a plane.

PROJECTION WELDING Electric resistance welding, similar to resistance spot welding except that projections are formed at the places to be welded.

PROJECT MANAGER Person given specific duties and assignments on a project.

PROJECT MANAGEMENT Planning, scheduling, measuring physical progress and controlling the entire project: engineering, purchasing, and construction as one coordinated entity.

PROJECT MANUAL Volume assembled by the architect for the work and including, among other things, bidding instructions and requirements, sample forms, the agreement, any special or supplementary conditions, the specifications, contract, and other documents.

PROJECT REPRESENTATIVE *See* **Owner's representative.**

PROJECT SITE *See* **Job site.**

PROJECT SPECIFICATIONS Written documents that describe the requirements for a project in accordance with the various criteria established by the owner.

PROJECT STATUS REPORT *See* **Progress report.**

proj/w *Abbreviation for:* projection weld.

PROLONGED WORKED PENETRATION Penetration of a sample of lubricating grease after it has been worked more than 60 double strokes in a standard grease worker at a temperature of 15° to 30°C (59° to 86°F). After the prescribed number of double strokes, the worker and contents are brought to 25°C (77°F), worked an additional 60 double strokes, and penetrated without a delay. *See also* **Penetration.**

PROMENADE TILE *See* **Quarry tile**

PROMONTORY *See* **Bay,** and **Headland.**

PROMOTERS Chemicals, themselves a feeble catalyst, that greatly increase the activity of a given catalyst.

PROOF LOAD 1. Load (up to 150% of design), applied to a selected portion of a structure, or to a pile, to confirm that it can carry the load specified; **2.** Load at which the links of a chain start to yield.

PROOF LOAD TEST 1. Nondestructive tension test of a sling; **2.** Nondestructive test performed by qualified personnel under defined conditions in a prescribed manner.

PROOF PRESSURE *See* **Pressure.**

PROOF PRESSURE TEST Nondestructive pressure test applied to a hose to determine its reliability at normal working pressures by applying pressures that exceed the manufacturer's rated working pressure.

PROOF ROLLING Use of heavy rubber-tired rollers, usually as a supplement to initial compaction by conventional means, to locate unstable areas, and achieve additional compaction.

PROOF STRESS 1. Stress applied to materials sufficient to produce a specified permanent strain; **2.** Specific stress to which some types of tendons are subjected in the manufacturing process as a means of reducing the deformation of anchorage, reducing the relaxation of steel, or insuring that the tendon is sufficiently strong.

PROOF-TESTING OF ROCK Soundness test of rock at a pier bottom performed by drilling a hole and probing the rock in the walls of the hole by use of a chisel-pointed probe bar.

PROP *See* **Post,** and **Shore.**

prop. *Abbreviation for:* property; proposal; proposed.

PROPAGATION Movement of a detonation wave, either in a column or from hole to hole.

PROPAGATION RATE Speed of flame front or pressure wave progress.

PROPAGATION VELOCITY Rate at which a vibration or seismic wave travels outward from a source.

PROPEL SHAFT Shaft in a revolving shovel that transmits engine power to the walking mechanism.

PROPELLER FAN *See* **Fan.**

PROPELLER SHAFT Drive shaft used to transmit torque from an engine or transmission to the rear or front driving axles.

PROPERTY DAMAGE INSURANCE General liability insurance covering injury to or destruction of property.

PROPERTY LINE Line, established by survey, that sets the legal boundaries of a property; may include several lots.

PROPERTY MANAGEMENT Operation of property as a business.

PROPERTY SURVEY Survey of a parcel of land that sets out its physical dimensions, establishes it in relationship to adjacent lots and principal features, establishes its orientation, and shows the principal features on the property, such as buildings, roads, etc., in relationship to each other and the boundaries of the property.

PROPERTY TAX Annual *ad valorem* tax levied on the owners of taxable real and personal property.

propn *Abbreviation for:* proportion.

propnl *Abbreviation for:* proportional.

PROPORTIONAL CONTROL 1. Mode of control in which there is a continuous linear relationship between the value of the controlled variable and the position of the final control element; **2.** Electronic grade and slope controls designed to generate an electric signal that is proportional to the detected error. The signal displaces a proportional hydraulic valve on the paver that causes the hydraulic cylinder to move the tow point (change screed angle of attack) in proportion to the error or deviation in the grade and/or slope.

PROPORTIONAL LIMIT Greatest stress that a material is capable of developing without any deviation from proportionality of stress to strain. *See also* **Hooke's law,** and **Modulus of elasticity.**

PROPORTIONAL SLACKLINE CARRIERS Wheeled carriers traveling on the track cable of a cableway, supporting the operating ropes, proportionally spaced across the cableway span by differences in travel-speed gearing and actuated by movement of the endless line.

PROPORTIONAL TAX Tax in which the rate remains fixed, regardless of the size of the taxpayer's tax base.

PROPORTIONING Selection of proportions of ingredients to make the most economical use of available materials to produce mortar or concrete of the required properties. *Also called* **Mix design.** *See also* **Mixture.**

PROPOSAL Offer of a bidder to perform stated construction work at the prices quoted.

PROPOSAL CALL *See* **Call for proposals.**

PROPOSAL FORM Prescribed form on which the offer of a bidder is to be submitted.

PROPOSAL GUARANTY Security furnished to assure that the bidder will enter into the contract if his offer is accepted.

PROPOSAL REQUEST Request from the designer to the contractor for a proposal to undertake alterations to the work.

PROPRIETARY CONTROL CENTER Continually supervised station under the control of the owner or others interested in the building or buildings to be protected.

propy *Abbreviation for:* proprietary.

PROPYL ACETATE Medium-fast solvent used with lacquers.

PROSCENIUM The front portion of a stage, in front of the curtain, visible to the audience.

PROSCENIUM ARCH Arch or opening separating the auditorium from the stage.

PROSPECTING Detailed exploration and evaluation of the earth's surface and crust. *See also* **Seismic exploration.**

PROSTYLE *See* **Portico.**

prot. *Abbreviation for:* protect; protected; protection.

PROTECTED Insurance term: in fire insurance, a risk located in an area protected by a fire department; in burglary insurance, a risk located in an area equipped with a burglar alarm.

PROTECTED CORNER Corner of a slab having adequate provision for load transfer, such that at least 20% of the load from one slab corner to the corner of an adjacent slab is transferred by mechanical means or aggregate interlock.

PROTECTED OPENING In an internal, fire-resistant wall, an opening that can be closed by a door or shutter of appropriate fire-rating.

PROTECTED PASTE VOLUME Portion of hardened cement that is protected from the effects of freezing by proximity to an entrained air void. *See also* **Philleo factor.**

PROTECTED POWER CIRCUITS Critical load circuits that are separated from the remainder of the normal load and protected by an emergency power system (standby power system).

PROTECTION BOARD Semirigid sheet material placed on top of a waterproofing membrane to protect it against damage during subsequent construction, and to provide a protective barrier against compressive and shearing forces induced by materials placed above it.

PROTECTION FLANGE *See* **Safety flange.**

PROTECTION HOOD *See* **Hood.**

PROTECTION RELAY Device used to detect defective or dangerous conditions and to initiate suitable switching or give warning.

PROTECTIVE EQUIPMENT 1. Personal safety equipment, including hard hats, safety goggles and shoes, ear plugs, etc.; **2.** Electrical switches and circuits that protect a machine and its operator from faults or overload.

PROTECTIVE GUARD Temporary, or permanent guard to prevent damage to locations susceptible to injury by vehicular traffic, such as external corners of structures, exterior corners in corridors, door jambs, etc.

PROTECTIVE STRIP Strip of nonadhesive material applied to an otherwise adhesive surface as protection during shipping and handling. It is removed immediately prior to use or installation.

PROTECTIVE SYSTEM Method of protecting employees, typically from cave-ins, from materials that could fall or roll from an excavation face into an excavation, or from the collapse of adjacent structures.

PROTECTOR Protective end reinforcement for steel sheet piling.

pro tem *Abbreviation for: pro tempore* (temporarily).

prov. *Abbreviation for:* province; provincial; provision; provisional.

PROVINCIAL BUILDING CODE *See* **Building code.**

PROVINCIAL FOREST INVENTORY Description of the quantity and quality of forest trees, nonwood values, and many of the characteristics of the land base compiled from statistical data for the forest lands of the province.

PROVING RING Device for calibrating load indicators of testing machines consisting of a calibrated elastic ring and a mechanism or device for indicating the magnitude of deformation under load.

prs *Abbreviation for:* pairs; press.

prt *Abbreviation for:* part.

prt. *Abbreviation for:* partial.

PRUNING Selective removal of twigs and branches from woody plants, shrubs, or trees.

PRV *Abbreviation for:* pressure reducing valve.

PRY BAR Nail puller and lever consisting of a thin, flat steel bar approximately 300 mm (12 in.) long with one curved and one bent-up end, both chisel faced and notched.

ps *Abbreviation for:* pieces; power steering.

PS *Abbreviation for:* polystyrene; pressed steel.

P/S *Abbreviation for:* power saw.

PSA *Abbreviation for:* pressure-sensitive adhesive.

PSC *Abbreviation for:* public service commission.

psf *Abbreviation for:* pounds per square foot.

psi *Abbreviation for:* pounds per square inch.

psia *Abbreviation for:* pounds per square inch absolute.

psig *Abbreviation for:* pounds per square inch gauge.

p. sl. *Abbreviation for:* pipe sleeve.

PSR *Abbreviation for:* plow-steel rope.

pst. *Abbreviation for:* paste.

psy *Abbreviation for:* per square yard.

PSYCHROMETER Instrument consisting of two thermometers, one with a dry bulb and the other with a wet bulb, used to determine relative humidity.

PSYCHROMETRIC CHART Graphic representation of the thermodynamic properties of moist air.

PSYCHROMETRY Branch of physics relating to the measurement or determination of atmospheric conditions, particularly regarding moisture mixed with air.

pt *Abbreviation for:* paint; part; payment; pint; point; port.

PT *Abbreviation for:* paneling, thin; pipe thread; pipe trap.

ptn *Abbreviation for:* partition.

PTO *Abbreviation for:* power take off.

P-TRAP Sanitary fitting that provides a water seal where the inlet is vertical and the outlet horizontal (or up to 5° below horizontal). *See also* **S-trap.**

PU *Abbreviation for:* pickup.

pub. *Abbreviation for:* public.

publ. *Abbreviation for:* publication.

PUBLIC With respect to a room or other space within a building, an area intended to be used in common by the occupants of the building, their guests, or tradesmen.

PUBLIC AUTHORITY Any applicable governmental entity having jurisdiction over the work.

PUBLIC BUILDING Any building that the public has the right to enter.

PUBLIC CONTRACT Contract for works or the supply of materials by a public authority in which the terms of the contract and scope of the work are open for public inspection and the awarding of a contract is public knowledge.

PUBLIC CORRIDOR Corridor that provides access to or exit from individually owned or rented rooms, suites of rooms, or dwelling units.

PUBLIC HEARING Open discussion (in this context) about a proposed project, some aspects of which will either affect members of the community or may be at variance with existing regulations or codes.

PUBLIC HOUSING Housing built, owned, and operated by a public authority and occupied by low-income families at a reduced rent, the difference in rent paid and housing costs being made up by government subsidy.

PUBLIC PROTECTION *See* **Land-use control.**

PUBLIC SALE Auction of property with notice to the public.

PUBLIC SEWER Publicly owned and maintained sewerage system to which private sewers may be connected under specific conditions.

PUBLIC SPACE Publicly owned land and facilities that are open to use by the general public, such as a street right-of-way.

PUBLIC WORKS Works owned by the members of a community for their collective benefit and enjoyment, constructed and maintained from the tax revenues.

PUDDLE To compact clay by wetting, so as to render it firm and solid.

PUDDLED WHEEL Grinding wheel made by a process wherein the mixture is of such consistency that it can be poured into molds.

PUDDLE FLANGE Intermediate flange on a pipe that passes through a liquid-retaining concrete wall to ensure a watertight junction.

PUDDLING Technique for achieving an uncontrolled degree of compaction of granular fill by saturating the soil after it is dumped into the excavation.

PUFF BLOWING Blowing chips out of a drill hole by means of exhaust air from the drill.

PUFF PIPE Antisiphon pipe.

PUGGING Coarse kind of mortar laid on boarding between floor joists to prevent the passage of sound. *Also called* **Deafening.**

PUGMILL Mechanical device for mixing materials, usually with paddles attached to rotating shafts. *See also* **Horizontal-shaft mixer.**

pul. *Abbreviation for:* pulley.

PULASKI TOOL Axe-type of tool with an axe head on one side and a mattock blade on the other. *See also* **Barron tool,** and **McLeod tool.**

PULL 1. Amount of advance in a heading by drilling and blasting; **2.** To loosen rock around the bottom of a hole by blasting.

PULL BOX Box placed in a length of conduit for the express purpose of accessing and pulling through electrical cables.

PULLED-DOWN TUBE *See* **Loose tube.**

PULLER Tool used to remove a pulley or gears from the shaft on which they are mounted.

PULLEY Wheel on an axle that carries a cable or belt on part of its surface.

PULLEY STILE Part of a sash window frame that contains the pulleys; the edges of the sashes slide between the stiles.

PULLING 1. Paint that has a tendency to drag while being applied; **2.** Installing and connecting electrical cables in conduit.

PULLING TENSION Longitudinal force exerted on a cable during installation.

PULL-ON CONTAINER Detachable container system in which a large container (around 15 to 23 m^3 (20 to 30 yd^3)) is pulled onto a service vehicle or tilt frame or hoist truck by mechanical or hydraulic means and carried to a disposal site for emptying.

PULL-OUT RESISTANCE *See* **Uplift capacity.**

PULL RIGGING Work done by a rigging slinger.

PULL SHEET List that associates wire identity with the terminal to which it connects.

PULL SHOVEL *See* **Backhoe,** and **Hoe.**

PULL STRAP Strap attached to the upper panel of manually operated, vertical biparting doors.

PULL WIRE Single wire placed in a conduit, one end of which is attached to a group of wires to be pulled through the tube.

PULP Mechanically ground or chemically digested wood used in manufacturing paper and allied products.

PULP HOOK Curved steel hook with a wooden cross handle; used in handling pulpwood.

PULPIT 1. Raised, ornamental platform, as in a church; **2.** Crane operator's location when remote from the equipment, designed to allow the operator a complete but protected view of the work area.

PULP LOG Log that does not meet the one-third merchantability standard for a saw log but that contains a minimum of 50% sound wood fiber by volume.

PULP MILL Mill that converts pulpwood to wood pulp.

PULPWOOD Wood suitable for making paper. *See also* **Fiber.**

PULPWOOD DEALER Middleman who buys pulpwood from the producer and sells it to the pulp mill company or acts as a commission broker for the company in producing pulpwood.

PULPWOOD FORKS *See* **Forks.**

PULSE VELOCITY Velocity at which compressional or other waves are propagated through a medium.

PULTRUSION Process in which glass fiber is pulled through a die in which resin is injected. Heat is applied in the die and a fully structural section is produced continuously.

pulv. *Abbreviation for:* pulverize; pulverized.

PULVERIZATION Crushing or grinding of material into very fine particle size.

PULVERIZING MIXER Rotating mixer equipped with tines, used in soil stabilization for pulverizing the earth through which it passes and mixing it with a previously spread stabilizer.

PUMICE Highly porous and vesicular lava, usually of relatively high silica content, composed largely of glass drawn into approximately parallel or loosely entwined fibers, that themselves contain sealed vesicles. Used as an abrasive in finely powdered form.

PUMICITE Naturally occurring finely divided pumice.

PUMP Mechanism receiving power from an engine or motor to create a flow. There are many types, including:

> **Centrifugal:** Pump that uses centrifugal force developed by the rapid rotation of an impeller to increase flow or pressure of a fluid. *See also* **Self-priming centrifugal pump,** and **Trash-handling centrifugal pump,** both below.

> **Concrete:** Apparatus that forces concrete to the placing position through a pipeline or hose.

> **Crescent:** Half-moon-shaped pump used in hydraulic and lubrication systems.

> **Dewatering:** Pump used to remove ponded water and to counter inflow to a site or excavation. A rugged, heavy-duty unit capable of tolerating a high silt content in the water being pumped. May be self-priming centrifugal, trash-handling centrifugal, electric submersible, or diaphragm type.

> **Diaphragm:** Pump that develops pressure by the reciprocating motion of a diaphragm in a chamber having inlet and outlet check valves.

> **Direct-acting:** Reciprocating pump having two opposing pistons connected to the same crankshaft: one piston is powered by steam, compressed air, etc., the other pumps water at each cycle.

> **Displacement:** Ram-operated or piston pump in which compressed air displaces the water to be pumped.

> **Electric:** Any pump driven by a direct-coupled electric motor.

> **Electric submersible:** Vertical, close-coupled, motor-driven pump designed to be totally immersed in the fluid it is designed to pump.

> **Excess pressure:** Pump required to maintain a higher pressure than municipal pressure in a wet-pipe automatic sprinkler system, located above the alarm valve to prevent false alarms.

> **Feed:** Pump that provides feed water to a boiler.

> **Fuel transfer:** Integrally mounted and driven pump on an engine that supplies fuel to the operating system.

> **Gear:** Hydraulic pump that creates hydraulic flow by moving oil from the inlet to the outlet through the spaces between the teeth of the pump's meshing gears. Gear pumps are considered to be more reliable than other forms of hydraulic pump due to their ability to handle higher levels of contamination.

> **Hydraulic:** Device that converts mechanical force and motion into hydraulic fluid power by means of producing flow. There are several types, including:

>> **Axial-piston:** Pump comprising several pistons in a rotating cylinder block. Each piston makes one stroke per revolution of the block.

>> **Gear:** Pump housing containing one or more sets of meshed gears that, as they rotate, force the oil between them in the direction of rotation.

>> **Peristaltic:** Pump that employs two diametrically-opposed rollers inside a circular chamber subjected to subatmospheric pressure.

>> **Positive-displacement:** Pump that for each cycle or revolution, positively displaces (usually by mechanical means) a specific amount (volume) of fluid.

>> **Vane:** Pump housing an offset, rotating cylinder having a number of sliding radial vanes mounted around its perimeter. The vanes slide in and out to form a seal with the inner face of the housing, creating a chamber that changes in volume as the drum rotates.

> **Injection:** Device that meters fuel and delivers it under pressure to the nozzle and holder assembly of an engine.

> **Jet:** Water pump that develops very high discharge pressure.

Lift: Suction pump that discharges slightly above or below the elevation of the suction inlet.

Mud: Circulating pump supplying fluid to a rotary drill.

Nonpositive displacement: Pump that displaces a variable amount each revolution.

Peristaltic: Pump that transmits force through the regular pulsing of a diaphragm, inducing a wave-like reaction.

Piston: Available in radial and axial designs for both fixed- or variable-displacement systems. All piston pumps operate on the principle that a piston reciprocating in a bore will draw in fluid as it is retracted and expel fluid on the forward stroke. By varying the stroke length of the piston, the pump displacement varies, thus varying pump output.

Priming: Small-capacity pump used to prime another pump.

Ram: Single action reciprocating pump employing a ram in place of a piston.

Reciprocating: Pump utilizing a piston within a cylinder.

Rotary: Pump employing geared wheels to drive fluids between the gear teeth.

Rotary gear: Positive-displacement pump employing closely fitting rotors or gears to force water through the pump chamber.

Sand: Tube, fitted with a check valve at its lower end, lowered into a borehole to extract mud and cuttings.

Self-priming centrifugal: High-volume, high-head pump with the capability of priming itself provided the suction is within a defined height/distance of the water surface. May be powered either by an engine or electric motor; often designed as a portable unit. When engine driven the pump speed can be adjusted to vary pump flow rate.

Shell: Type of pump used to evacuate sand from boreholes.

Single-acting: Reciprocating pump, or compressed air or steam engine, in which only one side of the piston works; every second stroke is a power stroke.

Spout-delivery: Pump designed to deliver water to the level at which it is working.

Submersible: Pump that can operate while partially or wholly submerged.

Sump: Pump, usually electrically operated, to remove water that collects in a sump.

Supply: Pump for transferring fuel from a tank and delivering it into the injection pump of an engine.

Trash-handling centrifugal: Centrifugal pump capable of handling small sticks, stones, sand, and other solid particles, up to 25% by volume, carried in the water being pumped.

Vacuum: Device that uses mechanical force and motion to evacuate gas from a connected chamber to create subatmospheric pressure.

Vane: Pumps that generate volume by employing small vanes that force a fluid through the opening between the oval pump housing and circular vane housing, normally used only with closed, pressurized hydraulic systems due to their lower tolerance to contamination.

Wellpoint: Centrifugal pump that can move a variable mixture of air and water, used to dewater a subterranean area or to temporarily lower the water table over a limited area.

PUMPABILITY Ability of a lubricating grease to flow under pressure through the lines, nozzles, and fittings of grease dispensing systems.

PUMP CAN Five-gallon water can with an attached, hand-operated pump, used to extinguish small fires.

PUMP DISCONNECT Mechanism used to interrupt a pump drive, typically for roading a mobile crane or easier cold-weather starting.

PUMPED CONCRETE Concrete that is transported through hose or pipe by means of a pump.

PUMPED STORAGE RESERVOIR Reservoir filled entirely with water pumped from outside its natural drainage area.

PUMPING 1. Rolling motion in unstable ground, usually created as heavy equipment passes over it; **2.** Rapid raising and lowering of a scraper bowl to force a larger load into it; **3.** Ejection of foundation material, either wet or dry, through joints or cracks along the edges of rigid slabs, due to vertical movements of the slab under traffic.

PUMPING LEVEL Level to which water rises in a well when pumping is in progress.

PUMP-JACK SCAFFOLD Supported scaffold consisting of a platform supported by vertical poles and movable support brackets.

PUMPKIN In helicopter logging, a large (one turn) solid turn.

PUNCH Small, hardened-steel driving tool. There are various types including:

Center: Short steel cylinder having a square head and knurled or milled shaft with a point at the end of a short bevel.

Drift: Similar to a center punch except that the working end consists of a long taper with a flat end that has a larger diameter than that of a pin punch.

Pin: Similar in shape to a drift but having a very small, flat end.

Prick: Similar to a pin, but ending in a point.

Self-centering: A pointed punch spring mounted within a sleeve.

Tracer: Narrow-ended punch used to produce decorative lines on sheet metal.

See also **Nailset**.

PUNCHED WORK Ashlar with rough, diagonal strokes punched across the face.

PUNCHEON 1. Vertical struts that transmit the mass of a bracing to an excavated ground surface inside a cofferdam or excavation; **2.** Log adzed to show one plane surface; **3.** Slabs of wood or small logs sometimes used when building subgrade in wet areas with deep overburden. The subgrade machine places puncheon to enable passage of equipment and to prepare a firm base for the application of blast material.

PUNCHING SHEAR 1. Shear stress calculated by dividing the load on a column by the product of its perimeter and the thickness of its base or cap, or by the product of the perimeter taken at one half the slab thickness away from the column and the thickness of the base or cap; **2.** Failure of a base when a heavily loaded column punches a hole through it.

PUNCH LIST *See* **Inspection list.**

PUNCH-OUT Tightly fitting hole made in gypsum drywall to allow installation of service pipes.

PUNCTURE Rip, tear, or other perforation of an otherwise impervious membrane: of a tire or innertube, for instance.

PUNCTURE SEAL Compound, held under pressure in a container, that can be injected into a punctured tire or innertube to temporarily seal holes up to 5 mm (0.2 in.) in diameter.

PUNKAH Manual ceiling fan.

PUNKINESS Soft or spongy condition.

PUNNING Obsolete term designating a light form of ramming.

pur. *Abbreviation for:* purchase; purlin.

purch. agt *Abbreviation for:* purchasing agent.

PURCHASE 1. Mechanical advantage gained when a rope passes over a sheave. Varies with the amount of sheave contacted; **2.** *See* **Tooth.**

PURCHASE AGREEMENT *See* **Agreement of sale.**

PURCHASE BLOCK Block used in rigging to obtain greater pull on a line.

PURCHASE MONEY MORTGAGE *See* **Mortgage.**

PURCHASE ORDER Commitment document used for the procurement of materials and equipment.

PURCHASE REQUISITION Request for a purchase within the scope of a purchase order or budgetary item.

PURE COAT Thin coat of pure portland cement, used to bond tile to mortar.

PURE GUM Rubber compound containing only those ingredients necessary for vulcanization; particularly applicable to natural rubber.

PURGE 1. Remove or clean out one substance prior to the introduction of another; or one color or type of material before using a second color or material; **2.** Removal of air from a hydraulic or fuel system. *Also called* **Bleed.**

PURLIN Horizontal timber placed parallel with a wall plate and the ridge, some way up the slope of a roof, to support the rafters at one or more points. *See also* **Beam.**

purp. *Abbreviation for:* purpose.

PUSH Foreman or boss.

PUSH BUMPER Device mounted on the front of a machine or vehicle, designed to permit the prime unit to push, or be pushed, by another machine or vehicle without damage to either unit.

PUSHBUTTON Button-operated electric switch.

PUSHBUTTON STATION Decorative device in an elevator car containing one or more hand-operated devices (buttons or switches) by which passengers tell the controller what action is desired.

PUSHER 1. Tractor that pushes a scraper to help it pick up its load; **2.** Common expression for ironworker foreman.

PUSHER TANDEM AXLE *See* **Axle.**

PUSHING When a tree has been undercut and backcut and will not fall, the faller may, as a last resort, fall this tree by 'pushing' another into it.

PUSH PIT Hydraulically powered bulkhead that traverses the length of a storage pit and periodically pushes the stored waste into the hopper of a compactor or onto a conveyor.

PUSH PLATE *See* **Plate.**

PUSH ROD Connecting link in an operating mechanism that transmits pressure.

PUSH ROLLER One of the two rollers on the front of a paving tractor that contact the truck tires during the unloading process. Oscillating push rollers are mounted on a beam that is attached to the paver with a pivot pin that allows minor truck misalignment without adversely affecting the paver.

PUSH SHOVEL Face shovel.

PUSH TRACTOR Crawler-tractor equipped with a push block, used to load push-loaded scrapers, or help load self-propelled scrapers.

PUTLOG 1. Cross-support of a scaffold that holds the scaffold planks or platform; **2.** Crosspiece in a scaffold, one end of which rests in a hole in the wall.

PUTLOG HOLE Small recess in masonry to support scaffolding.

PUT OPTION *See* **Option.**

PUTTY 1. Type of cement made of whiting (calcium carbonate) and boiled linnseed oil, beaten or kneaded to the consistency of dough, used in sealing glass in a sash, filling small holes and crevices in wood, etc.; **2.** Plaster composed of quicklime or hydrated lime and water, with or without plaster of paris or sand.

PUTTY KNIFE Glazier's knife, used to apply putty when glazing.

PVC *Abbreviation for:* polyvinyl chloride.

PVC TUBE Rigid, thin-walled plastic tube that may be used to line a borehole. With a cap fixed on the bottom end, a dry environment is ensures allowing the use of ANFO in wet holes.

pvd *Abbreviation for:* paved.

pvt *Abbreviation for:* pivot.

pvt. *Abbreviation for:* private.

p/w *Abbreviation for:* pressure weld.

PW *Abbreviation for:* pierced wall; plate washer.

PWBS *Abbreviation for:* protected wrought box strike.

pwd. *Abbreviation for:* powder.

pwr *Abbreviation for:* power.

pwr pnl *Abbreviation for:* power panel.

pwt *Abbreviation for:* pennyweight.

PYCNOMETER Vessel for determination of the specific gravity of liquids or solids.

PYLON Slender, towering structure, used to mark, or to support.

PYRAMID 1. Structure with a polygonal base and sloping sides that meet at an apex; **2.** Hip roof that has four sloping surfaces, usually of equal pitch, that meet at a peak.

PYRAMID CUT Method of blasting several rings of holes where the holes of the center ring are inclined toward a common center.

PYRAMID ROOF Pavilion roof.

PYRANOMETER Instrument that measures sunlight intensity.

PYRGEOMETER Instrument that measures infrared radiation.

PYRHELIOMETER Instrument that measures the intensity of direct insolation.

PYROLYSIS Chemical decomposition by heat.

PYROMETER Instrument that measures heat.

PYROMETRIC CONE Small, slender, three-sided oblique pyramid made of ceramic or refractory material for use in determining the time-temperature effect of heating, and in obtaining the pyrometric cone equivalent (PCE) of refractory material.

PYROMETRIC-CONE EQUIVALENT Numbe of that cone whose tip would touch the supporting plane simultaneously with that of a cone of the refr ctory material being investigated. *Also called* **PCE.**

Q *Abbreviation for:* flow rate (gpm).

Q BLOCK Concrete masonry unit made to special specifications.

qch *Abbreviation for:* quench.

qlty *Abbreviation for:* quality.

qlty cotrl *Abbreviation for:* quality control.

qnty *Abbreviation for:* quantity.

qp *Abbreviation for:* quarter panel.

QPL *Abbreviation for:* qualified products list.

qr *Abbreviation for:* quarter; quarter round.

qs *Abbreviation for:* quarter sawn.

QS EMULSIONS Quick-set emulsions designed primarily for use in slurry seals.

qt *Abbreviation for:* quart.

QT *Abbreviation for:* quarry tile; quenched and tempered.

QTB *Abbreviation for:* quarry tile base.

qtd *Abbreviation for:* quartered.

QTF *Abbreviation for:* quarry tile floor.

qtr *Abbreviation for:* quarter.

QTR *Abbreviation for:* quarry tile roof.

qtrs *Abbreviation for:* quarters.

qty *Abbreviation for:* quantity.

qtz *Abbreviation for:* quartz.

QUAD One quadrillion (1×10^{15}) Btu.

quad *Abbreviation for:* quadrangle; quadrant; quadrilateral; quadruplicate.

QUADRANGLE 1. Rectangular courtyard surrounded by building(s); **2.** Square-shaped land area, 38.624 km (24 miles) on each side, used in the rectangular survey method of land description.

QUADRANT 1. Survey instrument consisting of a graduated 90° arc with an index or vernier; **2.** Quarter section of a circle; **3.** Curved guide for a lever; **4.** Curved casement stay.

QUADRANT SENSOR *See* **Quadrant transducer.**

QUADRANT TRANSDUCER Specialized type of azimuth transducer that measures the rotational position of the revolving superstructure of a crane or other equipment with respect to its base (mounting). *Also called* **Quadrant sensor.**

QUADREL Square tile.

QUADRENTAL BEARING *See* **Reduced bearing.**

QUADRIPUNCTATA *See* **Marine borer.**

QUAD-TRAC Two connected crawler tractors, usually controlled by one operator.

QUAKE Elastic compression of soil during pile driving.

qual. *Abbreviation for:* qualify; qualitative.

QUALIFICATION OF BIDDER Means of determining if a prospective bidder is capable of, and competent to complete a contract within the terms specified.

QUALIFICATION TEST Examination of samples from a typical production run to determine adherence to a given specification; performed for approval as a supplier.

QUALIFIED BIDDER Bidder for a contract who has met the requirements for prequalification at the time of invitation to bid.

QUALIFIED PERSON One who, by possession of a recognized degree, certificate, or professional standing, or who, by extensive knowledge, training, and experience, has successfully demonstrated the ability to solve or resolve problems relating to the subject matter, the work, or the project.

QUALIFIED TESTING LABORATORY Properly equipped and staffed testing laboratory that has capabilities for, and that provides, the following services:

(a) Experimental testing for safety of specified items of equipment and materials to determine compliance with appropriate test standards of performance in a specified manner;

(b) Inspecting the run of such items of equipment and materials at factories for product evaluation to assure compliance with test standards;

(c) Service-value determinations through field inspections to monitor the proper use of labels on products, and with authority for recall of the label in the event a hazardous product is installed;

(d) Employing a controlled procedure for identifying the listed and/or labeled equipment or material tested; and

(e) Rendering creditable reports or findings without bias of the test methods employed.

QUALITY ASSURANCE Actions taken by an owner or his representative to provide assurance that what is being done and what is being provided are in accordance with the applicable standards of good practice for the work.

QUALITY CHARACTERISTICS Property that is actually measured to determine conformation of a unit or product with a given requirement.

QUALITY CLASSES Classifications of trees within a forest, including:

Mortality of growing stock: Volume of sound wood in live sawtimber and poletimber trees dying annually from natural causes, that includes fire, insects, disease, animal damage, weather, and suppression.

Mortality of sawtimber: Net board-foot volume of sawtimber trees dying annually from natural causes.

Net annual growth of growing stock: Annual change in volume of sound wood in live sawtimber and poletimber trees, plus total volume of sound wood in live sawtimber and poletimber trees, plus total volume of trees entering these classes through growth, minus volume losses resulting from natural causes.

Net annual growth of sawtimber: Annual change in volume of live sawtimber trees plus total volume of trees reaching sawtimber size, minus volume losses resulting from natural causes.

QUALITY CONFORMANCE INSPECTION OR TEST Examination of samples from a production run to determine adherence to a given specification, for acceptance of that production run.

QUALITY CONTROL Actions taken by a producer or contractor to provide control over what is being done and what is being provided so that the applicable standards of good practice for the work are followed.

QUALITY DISTANCE TABLES *See* **American table of distances.**

QUALITY STANDARD 1. Formally established criterion for a specific activity that (a) describes a deficiency, condition, or schedule that established the need for work, (b) outlines the work involved, (c) tells how to achieve good workmanship, and (d) lists expected end results; **2.** Formally established criterion for a specific activity that (a) outlines the work involved, and (b) lists the number of work units that are usually required to meet the quality standards for various categories.

QUALITY STANDARDS Stipulations of measurable physical properties or characteristics that materials, equipment, or construction items must have as a minimum.

quant. *Abbreviation for:* quantitative.

QUANTITY 1. In estimating, the total occurrence in a project, in terms of numbers, hours, volume, etc., of identical items; **2.** Amount of work to be performed, expressed by measurement, e.g., cubic yards per linear foot, or hour, etc.

QUANTITY DISCOUNT *See* **Discount.**

QUANTITY DISTANCE TABLES *See* **American Table of Distances.**

QUANTITY OVERRUN/UNDERRUN Difference between the original estimated contract quantities and the quantities in the completed work.

QUANTITY SURVEY Detailed analysis and listing of all items of material and equipment necessary to construct a project.

QUANTITY SURVEYOR Person who estimates the types and quantities of materials (including labor content) required for a project, and who measures those materials as they are incorporated into a project. *Also called* **Estimator.**

QUANTITY TAKEOFF Estimation and measurement of the quantities of separate materials and tasks, including labor, required for a project.

QUANTITY VARIANCE *See* **Variance.**

quar. *Abbreviation for:* quarterly.

QUARREL Small rectangular item, typically a pane of glass, particularly when set diagonally. *Also called* **Quarry.**

QUARRY 1. Rock pit; **2.** Open or surface mine for the extraction of aggregate, minerals, or rock; **3.** *See* **Quarrel.**

QUARRY BODY Dump body with sloped sides. *See also* **Body.**

QUARRY FACE 1. Working face of an open-pit rock or aggregate mine; **2.** Ashlar as it comes from the quarry, squared off for the joints only, with split face.

QUARRY SAP Natural moisture in quarried rock.

QUARRY, TECHNIQUE *See* **Sanitary landfilling.**

QUARRY TILE Unglazed, machine-made paving tile not less than 19 mm (0.75 in.) thick. *Also called* **Promenade tile.**

QUARTER 1. Work area of a faller. *Also called* **Strip; 2.** Area that a guyline supports, e.g., back quarter; **3.** To fell trees across a hill at an angle rather than straight down the slope. *See also* **Sidehill.**

QUARTER BEND Pipe fitting passing through an arc of 90°.

QUARTER CROWN Area between the center line of a road and the curb or shoulders running parallel to it.

QUARTER-GIRTH RULE Technique for computing the volume of timber in a log. It is approximately equal in cross-sectional area to a square of the side equal to the quarter girth of the log at the middle of its length.

¼hd *Abbreviation for:* quarter hard.

QUARTERING Method of obtaining a representative sample by dividing a circular pile of a larger sample into four equal parts and discarding opposite quarters successively until the desired size of sample is obtained.

QUARTER PACE Landing interrupting a stair where a turn of 90° is made.

QUARTER PEG Peg set at the quarter-width of a road; in conjunction with center pegs it defines the road surface.

QUARTER-QUARTER SECTION One-sixteenth of a section of land, made by dividing a quarter section into four parts, each containing 16.79 ha (40 acres).

¼rd *Abbreviation for:* quarter round.

QUARTER-ROUND *See* **Molding.**

QUARTER SAWN Lumber sawn along the radius of the annual rings or at an angle less than 45° to the radius.

QUARTER SECTION One-fourth of a section of land, made by dividing a section into four parts, each containing 160 acres (64.7 ha).

QUAY Wharf built parallel to the shore line.

QUEEN BOLT Long bolt serving in a roof truss in place of a queen rod.

QUEEN CLOSER A cut brick having a nominal 50 mm (2 in.) horizontal face dimension.

QUEEN-POST One of a pair of upright posts placed symmetrically on a tie beam or collar, connected at the top with a collar beam, and supporting a purlin.

QUEEN-POST TRUSS Truss framed with two vertical tie posts, as distinguished from the king post, which has but one.

QUENCH 1. Rapid cooling of metal in a heat treating process; **2.** Process of shock-cooling thermoplastic materials from the molten state.

QUENCH AGING Changing the composition of a metal by rapid cooling after heat treating.

QUENCH TROUGH Water-filled trough into which hot residue drops.

quest. *Abbreviation for:* question; questionnaire.

QUESTIONNAIRE Specified forms on which a contractor furnishes required information as to his ability to perform and finance the work being considered.

QUICK CLAY Any of a range of clays that undergo a dramatic reduction in strength upon disturbance.

QUICK COUPLER Equipment subattachment that permits the rapid coupling and uncoupling of virtually any attachment, often without need for the operator to leave the cab or operator's station.

QUICK DISCONNECT Coupling that can quickly join or separate a fluid line without the use of tools or special devices.

QUICK DISCONNECT COUPLER Hydraulic fitting that permits an oil line to be easily connected and disconnected, normally equipped with an automatic valve that shuts off the fluid flow from either half of the connection in case of breakage or at time of disconnect.

QUICK-LEVELING HEAD Ball-and-socket mount for a leveling instrument.

QUICKLIME Calcium oxide (CaO).

QUICK RELEASE VALVE 1. Valve that exhausts air directly to the atmosphere immediately upon release of brake application pressure; **2.** Valve in a trailer service brake line that accelerates the release of pressurized air in the line.

QUICKSAND Semifluid mass of silt and fine sand, thoroughly saturated with water. *See also* **Liquefac-tion.**

QUICK SET Drywall joint treatment that hydrates prior to drying. *Also called* **False set, Fast set,** and **Flash set.** *See also* **Early stiffening.**

QUICK-SET EMULSION Asphalt emulsion (normally used in slurry seal applications) in which the set of the mixture may be controlled by the use of small amounts of additives.

QUICK-SETTING CEMENT Cement obtained by mixing high alumina cement with lime or ordinary portland cement: the setting time will depend on the proportions of the mix.

QUICK TEST Maintained pile load test with time intervals of less than 20 minutes between adding load increments.

QUIESCENT CURRENT Direct current that is present in each servovalve coil when using a differential coil connection, the polarity of the current in the coils being in opposition such that no electrical control power exists.

QUIET ENJOYMENT Right of person owning or legally occupying property to do so without interference of possession.

QUIET TITLE SUIT Court action to establish clear title to real property.

QUILL Drive shaft rotating within a heavier shaft, and turning independently of it.

QUILTING Insulation sealed between two sheets of moisture-resistant paper.

QUIRK Sharp V-shaped incision in a molding, or between moldings.

QUITCLAIM DEED Deed conveying, without warranty, any title, interest, or claim that the grantor may have in the estate conveyed. *See also* **Deed,** and **Warranty deed.**

QUOIN 1. Brick or block at the corner or return of a wall, having the long, and short faces simultaneously exposed. Usually laid so that the long and short faces alternate vertically; **2.** Large squared stone, such as at a buttress, set at an angle of a building.

QUOIN POST Heel post.

QUONSET HUT Semicylindrical building clad in corrugated metal sheeting fastened to steel sections at intervals along its length. *Also called* a **Nisson hut,** after its inventor. (Quonset is the place in Rhode Island where the buildings were first manufactured in North America.)

quot *Abbreviation for:* quotient.

quot. *Abbreviation for:* quotation.

QUOTATION Offer to supply something or do something for a given price.

QUITIENT FACTOR *See* **Form class.**

qw *Abbreviation for:* quarter window.

R

R 1. Unit of thermal resistance: the higher the R value, the higher the insulating ability. It is the reciprocal of U, which is the overall coefficient of heat transmission (R=1/U); **2.** Symbol for pile capacity in tons. 2R means twice design capacity.

r *Abbreviation for:* riser.

R *Abbreviation for:* radius; range; Reynold's number; right; ring.

RA *Abbreviation for:* recording accuracy.

rab. *Abbreviation for:* rabbeted.

RABBET Cut made in the edge of a board, forming an open-sided groove.

RABBET JOINT Rabbet into which another piece fits, or where two rabbets fit together.

RABBET PLANE *See* **Plane.**

RABBLE ARM Radial rotating arm used to transport material within a multiple hearth furnace.

RACE 1. Inner or outer ring of a ball or roller bearing; **2.** Channel to or from a headrace, tailrace, or turbine.

RACEWAY Channel designed expressly for holding wires, cables, or busbars, with additional functions as permitted.

RACK 1. Metal grid used to properly space and align floor tiles; **2.** Trash rack in a waterway; **3.** Metal or wooden support for that portion of a reinforcing bar that extends beyond a construction joint and is not already supported by existing formwork or normal tying to adjacent bars.

RACKBACK Angle the bottom of a machine's bucket's cutting edge will rotate above horizontal at any given point, usually specified at ground level, carry position, and maximum height.

RACK GEAR *See* **Gear.**

RACK SAW *See* **Saw.**

RACKING 1. Lateral stress exerted on an assembly; **2.** Laying the lead or end of a wall with a series of steps so that when the work is resumed, the bond can easily be continued.

RACKING RESISTANCE Ability of a panel to resist forces in the panel's plane tending to distort it from its regular shape.

RACKING TEST Test that determines the strength of a panel when mounted in a stud frame like a wall.

rad. *Abbreviation for:* radial; radian; radiant; radiator; radical; radius.

RADIAL One of the lines converging on a common center.

RADIAL-ARM SAW *See* **Saw.**

RADIAL BRICK Compass brick.

RADIAL CABLEWAY Cableway with a fixed tower and a movable tail tower that travels on a track that is a constant distance from the fixed tower.

RADIAL CARCASS *See* **Tire construction.**

RADIAL CLEARANCE Angular clearance on the sides of a saw tooth or saw blade.

RADIAL CRACKING Cracking of a tire, usually in or near a rib, resulting from underinflation or ozone exposure.

RADIAL ENGINE Engine with its cylinders arranged in a circle around the crankcase.

RADIAL EXPANSION Tendency of communities experiencing rapid growth to expand radially from their urban center.

RADIAL GATE Dam gate with a curved water face and horizontal pivot axis.

RADIAL HIGHWAY *See* **Highway.**

RADIAL PLY Ply or plies of a tire in which the cords run at right angles to the bead.

RADIAL PRESSURE Pressure exerted radially, typically around the entire wall of a cylinder under internal pressure.

RADIAL SECTION Lengthwise section in a plane that passes through the centerline of a tree trunk.

RADIAL STEP Step that is wider at its outside edge: a winder.

RADIAL STRESS Stress normal to the tangent to the boundary of any opening.

RADIAL TIRE Tire in which the cords in the fabric casing are laid at approximately right angles to the centerline of the tread.

RADIAN One of the two supplementary units of the SI system of measurement: a unit of plane angle equal to the measure of a plane angle with its vertex at the center of a circle and subtended by an arc equal in length to the radius. Symbol: rad. *See also the appendix:* **Metric and nonmetric measurement.**

RADIAN PER SECOND A derived unit of angular velocity with a compound name of the SI system of measurement. Symbol: rad/s. *See also the appendix:* **Metric and nonmetric measurement.**

RADIAN PER SECOND SQUARED A derived unit of angular acceleration with a compound name of the SI system of measurement. Symbol: rad/s². *See also the appendix:* **Metric and nonmetric measurement.**

RADIANT ENERGY Energy that travels outward in all directions from its source.

RADIANT HEATING *See* **Heating system.**

RADIATION 1. Mode of heat transfer in which heat energy travels very rapidly in straight lines without heating the intervening space, but which raises the temperature of materials upon which it falls; **2.** Survey technique of plotting surrounding features on a plane table by radiating lines drawn with an alidade and marking the distance of the object to scale.

RADIATION PYROMETER Device that determines temperature by measuring the intensity of radiation from a heat-generating body.

RADIATOR 1. Heat transferring device; **2.** That part of a heating system, exposed or concealed, from which heat is radiated to a room or other space within the building.

RADIATOR COOLING Engine coolant heat that is dissipated to the atmosphere through a radiator.

RADIATOR SHUTTERS Slatted mechanism to reduce or completely shut off air flow to a radiator.

RADIOACTIVE Materials giving off alpha, beta, and gamma particles and shortwave radiations, either due to inherent activity (such as radium) or due to exposure to materials that are radioactive.

RADIOACTIVE PAINT Paint containing radioactive and phosphorescent compounds. Such paints are self-luminous during the active life of the radioactive compound that permanently activates the phosphor, which emits light in the visible spectrum.

RADIO FREQUENCY INTERFERENCE External interference/false signals that can be picked up by unshielded sensors. *Also called* **RFI.**

RADIO OUTLET Electrical outlet with an aerial and ground for the use of a radio connected to it.

RADIO WHISTLE Transmitter-receiver signalling system used by workers for yarding operations.

RADIUS Horizontal distance from the center of rotation of a crane to its hoisting hook. *See also* **Geometry of circular curves.**

RADIUS- AND SAFE-LOAD INDICATOR Pendulum that hangs from a crane jib over a graph on which the crane radius for any angle and the resulting safe load limit are marked.

RADIUS BENT Reinforcing bars bent to a radius larger than that specified for standard hooks; a bar curved to fit into circular walls, as the horizontal bars in a silo.

RADIUS DIFFUSION Horizontal axial distance an air stream travels after leaving an air outlet before the maximum stream velocity is reduced to a specified terminal level.

RADIUS LIMIT ALARM Visual and/or audible indication that one of the limits (minimum or maximum) of radius or load has been reached.

RADIUS OF CURVATURE Distance between the center line of a circular section of road, walkway, or curb and the center line of the corresponding circle.

RADIUS OF GYRATION Value used in calculating the slenderness ratio of a strut.

RADIUS OF INFLUENCE Distance from the center of a wellbore to the point where there is no lowering of the water table when the well is being pumped.

RADIUS OF LOAD Horizontal distance from a projection of a crane's axis of rotation to the center of the

vertical hoist line or tackle with load applied. *Also called* **Working radius.**

RADIUS-OF-LOAD INDICATING SYSTEM Crane-mounted device that measures and displays the radius of load.

RADIUS OF RUPTURE Distance from the center of a blasthole to the limit of radial cracking produced by a charge detonated in that blasthole. *Also called* **Fracture radius.**

RADIUS ROD Component used for keeping the rear axle in correct position when starting and stopping a vehicle or machine.

RADIUS SHOE Trowel with a radius face, used to round the edges and corners of plaster, concrete, or mortar.

RADIUS TEST Test of a flexible material by bending it over a curve of fixed radius.

radn *Abbreviation for:* radiation.

raft. *Abbreviation for:* rafter.

RAFT FOUNDATION Continuous structural concrete slab, usually reinforced, extending over the whole of a structure, laid over soft ground or where heavy loads must be supported to form a foundation. *See also* **Mat foundation.**

RAFTER Roof timber sloping up from the wall plate to the ridge and extending down to the eaves. There are several classes:

> **Angle:** *See* **Hip rafter,** below.
>
> **Common:** Rafter extending from the ridge, past the walls to form the eave.
>
> **Hip:** Rafter that forms the intersection of an external roof angle. *Also called* **Angle rafter.**
>
> **Jack:** Rafter that spans the distance from a wallplate to a hip or from a valley to a ridge.
>
> **Valley:** Rafter that forms the intersection of an internal roof angle.

RAFTER FILLING Material (brick, lumber) used to fill the space between rafters at plate level.

RAFTER PLATE Plate at the top of a masonry or concrete wall supporting a rafter or roof joist.

RAG BOLT *See* **Lewis bolt.**

RAGGING OFF Procedure of spreading damp cheesecloth and pulling it over a tile surface during the tile grouting process in order to clean the tile.

RAGGLE Groove in a mortar joint or special masonry unit to receive roofing or flashing. *See also* **Reglet.**

RAG ROLLING Glazing technique that involves rolling crumpled material into wet glaze.

RAG WORK Rubble masonry using small, thin stones.

RAG-WRAP *See* **Wrapped cure.**

RAIL 1. Transverse member extending from one vertical element to another in the construction of doors, fences, balustrades, staircases, etc.; **2.** Chain, or inner surface of a crawler track; **3.** Steel T-section with machined guiding surfaces.

RAIL BACKING Any structural member that strengthens a guide rail.

RAIL BRACKET Steel plate, angle, strap or beam to which elevator guide rails are attached.

RAIL CLIP Clamp that fastens a guide rail to a guide rail bracket.

RAILING 1. Metal fence or guard; **2.** In cabinetwork, a banding.

RAIL PILE *See* **Pile.**

RAIL POST *See* **Newel.**

RAILROAD CROSSING ANGLE Angle of 90° or less where a railroad and highway intersect.

RAILROAD GRADE CROSSING General area where a highway and railroad cross at the same level, within which are included the railroad, roadway, and roadside facilities for traffic traversing that area.

RAILROAD PILE DRIVER Pile driver mounted on a railroad undercarriage for transportation on the rails and operation from the track.

RAILROAD TIE Parallel crossbeam to which the rails of a railroad are fastened, commonly of creosoted wood, more currently of reinforced concrete. Discarded or replaced wooden ties are commonly used for general site construction and particularly for landscaping.

RAILS (DOOR) *See* **Door.**

RAIL-STEEL REINFORCEMENT Reinforcing bars hot-rolled from standard T-section rails.

RAILWAY GRADE Roadbed upon which ties and rails are laid.

RAINDROP FIGURE Mottling figure in lumber that may alternate with ribbon grain.

RAINFALL SIMULATION Determination of the impact of rainfall and/or runoff on a sewer system.

RAIN GAUGE Instrument that collects falling rain and indicates the total fallen over a given period.

RAINWATER Water resulting from precipitation.

RAINWATER HEAD Box-shaped structure at the top of a down pipe or downspout into which gutters discharge rainwater.

RAINWATER HOPPER Hopper-shaped rainwater head.

RAINWATER LEADER Down pipe from a roof or gutter that carries water from roofs to a drain or to the ground surface.

RAISE Shaft dug upward vertically or at an angle from a tunnel.

RAISE BIT Bit for enlarging a pilot hole ending in a cavity, cavern, or mine working, by drilling from the bottom up.

RAISED FLAT JOINT Mortar joints tooled to form a smooth, raised, flat joint.

RAISED GRAIN Roughened condition of the surface of dressed lumber in which the hard summerwood is raised above the softer springwood, but not torn loose from it.

RAISED HEARTH Masonry hearth built above the level of the finished floor.

RAISED PAVEMENT MARKER *See* **Pavement** marking.

'RAISE FOR RED' Instruction by the instrumentman to the rodman to lift the rod vertically to permit observation of the relevant whole number, a measure necessary only when the observation is of a point so close to the instrument that only a limited portion of the rod can be viewed through the instrument.

RAISER BAR *See* **Support bar.**

RAISE TIME *See* **Hydraulic cycle time.**

RAISING GANG Crew of workers who erect the basic skeleton of a steel-framed building. *See also* **Filling-in gang.**

RAISING HAMMER *See* **Hammer.**

RAISING PLATE Horizontal timber resting on part of a structure and supporting a superstructure.

RAKE 1. Slanting or inclined from the perpendicular; **2.** Angle made by the edge of a cutting tool and a plane perpendicular to the surface being worked; **3.** Dozer blade or attachment consisting of a series of tines. *See also* **Brush rake,** and **Rock rake**: **4.** Board or molding between the edges of the gable end and the soffit to trim the ends of siding; **5.** Hand tool used to remove mortar to a set depth from a joint.

RAKED JOINT Masonry-wall joint that has the mortar raked out to a specified depth while it is only slightly hardened.

RAKED OUT Cutting out of joints, as in concrete or mortar, in preparation for tuckpointing or caulking.

RAKED POINTING *See* **Pointing.**

RAKER Sloping brace for a shore head.

RAKER PILE *See* **Pile.**

RAKING BOND *See* **Bond.**

RAKING CORNICE Cornice or coping on an inclined face, typically on a gable.

RAKING FLASHING Cover flashing, typically between a chimney and sloping roof.

RAKING PILE *See* **Pile.**

RAKING RISER Stair riser that is not vertical, inclining backward at the foot so as to provide more foothold on the tread.

RAKING SHORE Long balk erected as a temporary support to a wall or other structure.

RAM 1. Drive into place or compact; **2.** Attachment comprising a cantilever member, extending horizontally forward from a lift truck carriage, used to handle hollow core, cylindrical, or similar loads such as rugs, coils, pipe, etc.; **3.** Moving weight of a pile-driving hammer; **4.** Reclaimed aggregate material; **5.** One of a pair of timbers, usually 100 x 150 mm (4 x 6 in.), used as the main supports of a scaffold; **6.** Hydraulic cylinder and piston. There are several types:

> **One-way, or single-acting:** Hydraulic cylinder in which fluid can be supplied only to one end so that the piston can be moved only one way by supplied hydraulic power.

> **Two-way, or double-acting:** Hydraulic cylinder in which hydraulic fluid can be supplied to either end, with the piston moving under power

in two directions.

RAM CYLINDER *See* **Cylinder.**

RAMMED-EARTH WALL Technique of building walls by tamping earth of suitable composition into forms.

RAMMER Compacting tool.

RAMMING Form of heavy tamping by means of a blunt tool forcibly applied.

RAMP 1. Slope joining two different levels; **2.** Part of a staircase handrail that rises steeper than normal, usually where there are winders in the staircase. *See also* **Landing**; **3.** Short roadway connecting two or more legs of an intersection or connecting a frontage road and main lane of a highway; **4.** Length of drainpipe laid more steeply than the remainder; **5.** Articulated or removable bridge-type structures used to load or unload lowbed or drop-frame trailers.

RAMP ARCH *See* **Arch.**

RAMPART Earthen embankment surmounted by a parapet.

RAM PLUG Sleeve into which an anchor is pounded.

RAMP METERING Process of facilitating traffic flow on freeways by regulating the amount of traffic entering the freeway using control devices on entrance ramps.

RAM POINT Part of the ram that comes closest to the impact surface of the pile. Sometimes a separate unit of the ram.

RAMP TECHNIQUE *See* **Sanitary landfilling.**

RAM PUMP *See* **Pump.**

R&D *Abbreviation for:* research and development.

RANDOM 1. Without uniformity; **2.** Not laid in courses.

RANDOM ASHLAR *See* **Ashlar.**

RANDOM COURSES Masonry or ashlar courses of varying depth.

RANDOM FILL Fill placed in an embankment that may consist of a variety of materials.

RANDOM RUBBLE Masonry wall of unsquared or roughly squared stones, irregular in size and shape.

RANDOM SAMPLE Small part of a lot that is used to represent the whole, so chosen that each portion of the lot has an equal probability of being selected.

RANDOM SHINGLES Shingles of different widths.

RANDOM WORK Any work where the materials used are not of uniform dimension.

R&R *Abbreviation for:* rules and regulations.

RANGE Cooking appliance equipped with a cooking surface and one or more ovens.

RANGE DIAGRAM Diagram showing the luffing path for all boom and jib lengths for a crane configuration, plus radial lines marking boom angles; a vertical scale indicating height above the ground; and a horizontal scale marked with operating radii. Used to determine lift heights, clearance of the load from the boom, and clearance for lifts over obstructions.

RANGED RUBBLE Masonry of stones rudely dressed to a nearly uniform height. *See also* **Rubble,** and **Scabbled rubble.**

RANGE LINES Lines parallel to a principal meridian, dividing the land into 9.66-km (6-mile) wide strips (called ranges) that are consecutively numbered to the east or west of the principal meridian.

RANGE MASONRY Regular coursed rubble.

RANGE PILE Pile serving as a guide for locating piles or other structures or for marine surveying or dredging.

RANGE POLE Pole marked in alternate 1-ft (0.304-m) high red and white bands.

RANGER *See* **Wale.**

RANGE SHIFT CYLINDER Located in the auxiliary section of a transmission, this component, when directed by air pressure via low and high ports, shifts between high and low range of gears.

RANGE SHIFT LEVER Located on the shift knob, this lever allows the driver to select low or high gear range.

RANGE WORK In masonry, a course of any thickness that, once started, is continued across the entire face (but all courses need not be of the same thickness).

RANGING LINE Cord stretched taut between batter boards to mark an elevation or alignment.

RAP *Abbreviation for:* Reclaimed asphaltic pavement.

RAPID CURING ASPHALT *See* **Asphalt.**

RAPID-HARDENING CEMENT Finely ground portland cement; the finer the grinding the faster the rate of hydration.

RAPID SETTING Designation for a standard type of anionic emulsified asphalt.

RASP-CUT *See* **File.**

RASPER Grinding machine in the form of a large vertical drum containing heavy hinged arms that rotate horizontally over a rasp-and-sieve floor.

RASPING SYSTEM Procedure in which refuse is ground through a screen partly covered with steel pins, that have the effect of a rasp.

RATCHET Set of teeth, vertical on one side and raked on the other, allowing a pawl to move in one direction but holding it in the other.

RATED CAPACITY 1. Maximum weight of a homogeneous cube with a given load center that a truck or other vehicle can safely transport and/or stack to a specified height; **2.** Load that a new wire rope or wire rope sling may handle under given operating conditions and at an assumed design factor.

RATED CAPACITY INDICATOR *See* **Rated load indicator.**

RATED CAPACITY LIMITER *See* **Rated load limiter.**

RATED CURRENT 1. Rated nameplate current of a machine or apparatus; the value of current that it can carry without exceeding the allowable temperature rises; **2.** Specified servovalve input current of either polarity to produce rated flow. Rated current must be specified for a particular coil connection differential, series, or parallel, and does not include null-bias current.

RATED ENERGY Manufacturer's specified energy of a powered pile hammer.

RATED FATIGUE PRESSURE *See* **Pressure.**

RATED FLOW Maximum flow that a power supply is capable of maintaining at a specified operating pressure.

RATED HORSEPOWER *See* **Horsepower.**

RATED LOAD Manufacturer's recommended maximum load. Rated crane loads at specified radii are the lesser of a specified percentage of tipping loads, or of the machine's hydraulic or structural competence, as established by the manufacturer's rating charts, and are maximum loads covered by the manufacturer's warranty. Weight of hook, hook blocks, slings, and other load-handling devices are not considered part of the load to be handled. *Also called* **Lift capacity.**

RATED LOAD INDICATOR System intended to aid an operator in the efficient and safe operation of a crane by continually monitoring the load and warning of an approach to an overload condition. *Also called* **Rated capacity indicator.**

RATED LOAD LIMITER System with function limiter (cutout) ability intended to aid the operator in safe and efficient operation of a crane by continually monitoring the load and warning of an approach to an overload condition. *Also called* **Rated capacity limiter.**

RATED NORMAL DRAWBAR PULL Greatest sustained towing force in pounds at a specified coupler height that a vehicle will develop on a level surface, and within a given duty cycle, without exceeding the allowable continuous temperature rating for the components. *See also* **Drawbar pull,** and **Maximum drawbar pull.**

RATED POWER 1. Power specified by a manufacturer for a given application at a given (rated) speed; **2.** Stated or nameplate net electric output that is obtainable from a generator set when it is functioning at rated conditions.

RATED PRESSURE *See* **Pressure.**

RATED SHEATHING Structural panel product that has been tested to perform in a certain manner, e.g., a 24/16-rated sheathing panel must be used on 600-mm (24-in.) centers for roofs and 400-mm (16-in.) centers for floors. Rated waferboard is sheathing.

RATED SPEED 1. Revolutions per minute at which an engine or apparatus is designed to operate; **2.** Specified operating speed for a pile hammer in blows per minute; **3.** Speed at which an elevator, dumbwaiter, escalator or inclined lift is designed to operate.

RATED STATIC PRESSURE *See* **Pressure.**

RATED TRAILER CAPACITY Total of the payload and vehicle weight.

RATED VOLTAGE Voltage of electrical apparatus at which it is designed to operate.

RATE OF DECAY Rate of diminution of a reverberant sound: the slope of the decay curve plotted against time and intensity.

RATE OF GROWTH Rate at which a tree has accumulated wood, measured radially in the trunk or in the lumber cut from the trunk. The unit of measure is the number of annual growth rings per given measure.

RATE OF PENETRATION Distance penetrated per second during driving of a pile into the ground using a vibratory hammer.

RATE OF SPREAD Relative activity of a forest fire in extending its horizontal dimensions, expressed as the rate of increase of the total perimeter of the fire, as rate of forward spread of the fire front, or as rate of increase in area.

RATEPAYER Person liable to pay an assessment or tax to the municipality in which he resides.

RATE VARIANCE *See* **Variance.**

RAT HOLE 1. Hole, 12 m (40 ft) deep, drilled in advance of an oil-well rig moving onto an oil-well site. *See also* **Doodle; 2.** Socket in the substructure of a rotary drill that supports the kelly and swivel when they are not in use.

RATING PERIOD Period of time during which a forest fire danger rating value is considered valid or representative for administrative purposes.

RATIO Relationship between two similar things in respect to the number of times the first contains the second, either integrally or fractionally.

RATIONAL FORMULA Formula used in drainage computations to determine the amount of runoff within a watershed.

RATIO OF REDUCTION Relationship between the maximum size of rock that a crusher can accommodate and the resulting size of its product.

RATIO REGULATOR Proportional control device that regulates the downstream pressure in a pipeline in which it is located, used to maintain proportional pressures in fuel and air lines in a pressure control system.

RAT-TAIL FILE Slender round file tapering to a point at its end.

RAT-TRAP BOND *See* **Bond.**

RAVELING 1. Cumulative process in which rock separates from finer material on a road surface, caused by the passage of wheeled traffic. **2.** Loosening and falling of materials from a face or bank.

RAVINE Deep, more-or-less-linear depression or hollow worn by running water.

RAW GLAZE *See* **Glaze.**

RAW LAND Unimproved land, i.e., without landscaping, drainage, utilities, structures, etc.

RAW LINSEED OIL Unrefined product obtained from flax seed.

RAW MATERIAL *See* **Material.**

RAW MIX Blend of raw materials, ground to desired fineness, correctly proportioned and blended ready for burning, such as that used in the manufacture of cement clinker.

RAW SEWAGE SLUDGE Solids concentrated by various methods in wastewater treatment plants, usually containing 90% to 95% water.

RAW WATER Water as obtained from its source, particularly a surface source such as a lake, river, etc.; untreated water.

RAZE Level everything above ground level by destruction.

rb *Abbreviation for:* roller bearing.

RB *Abbreviation for:* reverse bevel; Rockwell hardness B-scale.

rblt *Abbreviation for:* rebuilt.

rbm *Abbreviation for:* resisting bending moment.

RBM *Abbreviation for:* reinforced brick masonry; reinforced clay masonry.

R/C *Abbreviation for:* reinforced concrete.

RC *Abbreviation for:* remote control; Rockwell hardness C-scale; rounded corners; rubber cushion.

rcb *Abbreviation for:* reinforced concrete box.

rcp *Abbreviation for:* reinforced concrete pipe.

rcpt *Abbreviation for:* receipt.

rd *Abbreviation for:* road; rod; round.

rd² *Abbreviation for:* square rod.

RD *Abbreviation for:* roof drain; rural district.

RDF-1 *See* **Refuse-derived fuel.**

RDF-2 *See* **Refuse-derived fuel.**

RDF-3 *See* **Refuse-derived fuel.**

RDF-4 *See* **Refuse-derived fuel.**

RDF-5 *See* **Refuse-derived fuel.**

RDF-6 *See* **Refuse-derived fuel.**

RDF-7 *See* **Refuse-derived fuel.**

rdg. *Abbreviation for:* ridge.

rdm *Abbreviation for:* random.

rdwy *Abbreviation for:* roadway.

rdy *Abbreviation for:* ready.

REACH 1. Distance from the axis of rotation of a crane or derrick; **2.** Wood or metal structural member connecting a logging trailer to a truck tractor.

REACH, FULLY RAISED Horizontal distance from the foremost point on a machine (including tires, tracks, or frame) to the cutting edge (including teeth or bolt-on cutting edges) with the hinge pin at maximum height and the bucket at 45° dump angle.

REACH TRUCK Self-loading outrigger-type industrial truck, generally highlift, having load engaging means mounted so that they can be extended forward under control to permit a load to be picked up and deposited in the extended position.

REACTANCE Opposition to the flow of alternating current by induction or capacitance of a component or circuit.

REACTION Resistance to a force or load; countertendency.

REACTION TURBINE Turbine in which the jets or nozzles are on the moving part.

REACTION WOOD Wood with distinct anatomical and physical characteristics, formed typically in parts of leaning or crooked stems and in branches, that tends to restore to the original position of the branch or stem if this has been disturbed. *Also called* **Compression wood** (in conifers), and **Tension wood** (in broadleaved trees).

REACTIVE Substance that is normally unstable and readily undergoes violent change; or reacts violently with water; or generates toxic gases, vapors, or fumes in quantity sufficient to present danger to human health or the environment when mixed with water.

REACTIVE AGGREGATE *See* **Aggregate.**

REACTIVE DIFFERENTIAL COMPENSATION Series differential connection of the various generator parallel current transformer secondaries (and thereby the voltage regulator reactive droop compensation circuits) that act to modify generator excitation so as to minimize its differential reactive current with the end result that reactive load sharing among generators is obtained without voltage droop. Its effect on voltage is similar to that of a parallel isochronous governor's operation effect on speed or frequency.

REACTIVE DROOP COMPENSATION Voltage regulator circuit that acts to affect generator excitation so as to create a droop in generator voltage proportional to the inductive reactive current. This characteristic is used to obtain reactive load sharing among generators operating in parallel. Its effect on voltage is similar to the effect that a droop-type governor has on speed or frequency.

REACTIVE LOAD (VAR) SHARING Process of regulating excitation that causes the reactive load to be shared proportionally between generator sets.

REACTIVE SILICA MATERIAL Several types of materials that react at high temperatures with portland cement or lime during autoclaving.

READILY ACCESSIBLE Capable of being reached quickly for operation, renewal, or inspection without requiring those to whom ready access is required to climb over or remove obstacles. *See also* **Accessible.**

readj. *Abbreviation for:* readjust; readjustment.

READOUT Visual display of data and/or information derived, collected, relayed, or transmitted by a mechanical, electric, or electronic device.

READY-MIXED COMPOUND Premixed drywall joint compound.

READY-MIXED CONCRETE Concrete manufactured for delivery to a purchaser in a plastic and unhardened state.

REAL (ACTIVE) LOAD (WATT) SHARING Process of governing that causes the real load to be shared proportionally between generator sets.

REAL ESTATE Real property; land and what is built upon it.

REAL ESTATE APPRAISER One who makes an evaluation of the worth of real property.

REAL ESTATE BROKER Agent who, on a commission basis, buys and sells real estate for another.

REAL ESTATE DEVELOPER One who owns real estate and develops it to a higher and better use for profit.

REALIGNMENT Alteration to something already established or built: alteration to the line of a highway, for instance.

REAL PROPERTY Land, and generally whatever is built upon it, including all water within its boundaries, minerals and other deposits of value beneath its surface, and the air above it.

REALTOR Licensed and registered real estate broker.

REAM To enlarge a hole by cutting.

REAMER 1. Tool used to enlarge a hole; **2.** Steel tool with a tapered shank used for freeing 'stuck' increment cores from an increment borer; **3.** Tool attached to the drilling bucket or auger to cut or enlarge the bell at the base of a drilled shaft or caisson.

REAMER SHELL Cutter mounted just above a diamond bit, used to assure a full-size hole.

REAR ARCH *See* **Arch.**

REAR AXLE *See* **Axle.**

REAR AXLE RATIO Ratio of the speed of the propeller shaft to the speed of the rear axle shaft.

REAR BOLSTER *See* **Apron.**

REAR DOUBLE-REDUCTION AXLE *See* **Axle.**

REAR-DUMP SCRAPER *See* **Scraper.**

REAR-DUMP TRUCK *See* **Truck.**

REAR-END RADIUS *See* **Tailswing.**

REAR-END SYSTEM Chemical, thermal, or biological system and its supplementary facilities used for the conversion of preprocessed wastes into useful or nonhazardous products.

REAR HANGER Bracket for mounting the rear of a truck or trailer suspension to the truck or trailer frame. Made to accommodate the end of the spring on spring suspensions. Normally of three types: flange-mount, straddle-mount, and under- or side-mount.

REARING Tendency of the front of a tractor to rise when pulling a heavy load.

REAR LOADER, DETACHABLE CONTAINER Detachable container system in which roll-out containers, typically 0.8 to 2.3 m³ (1 to 3 yd³) capacity, are hoisted at the rear of the collection vehicle and mechanically emptied; the container being left with the customer.

REAR-LOADING REFUSE TRUCK *See* **Refuse truck.**

REAR OF A FIRE Portion of the edge of a forest fire opposite the head.

REAR PACKER *See* **Refuse truck.**

REAR STAND Rear-mounted, box-like member that attaches to the main frame rails of a mixer, and which furnishes support to the drum track rollers; a tripod structure off the rear stand or 'cradle' providing a support on which the main chute pivots.

REAR YARD Open space between the back of a developed property and the rear lot line.

REBAR *Abbreviation for:* reinforcing bar.

REBAR CONTRACTOR A contractor or subcontractor who handles and places steel reinforcement and bar supports. *Also called* **Bar placer, Iron worker,** and **Metal lather.**

REBATE Continuous rectangular notch cut on the edge of a surface to receive the edge of another piece.

REBORE To drill or cut an engine cylinder to an oversize.

REBOUND 1. Amount of maximum upward movement of the head of a pile following a blow from the hammer, by reason of the elastic properties of the pile and of the soil into which it is driven; **2.** Upward movement of the head of a pile following removal of a static load; **3.** First upward movement of a hammer after a blow on a pile; **4.** Aggregate and cement, or wet shotcrete that bounces away from the surface against which shotcrete is being projected.

REBOUND HAMMER Apparatus that provides a rapid indication of the mechanical properties of concrete based on the distance of rebound of a spring-driven plunger. *Also called* **Impact hammer.**

REBURN Subsequent burning of an area in which a forest fire has previously burned but which still has left flammable fuel that ignites when burning conditions are more favorable.

REBUTTED-REJOINTED *See* **Shingle.**

rec. *Abbreviation for:* recess; record.

recap. *Abbreviation for:* recapitalize; recapitulation.

recd *Abbreviation for:* received.

RECEIVER Air tank or reservoir of an air compressor.

RECEPTACLE Wall-mounted electrical outlet. *See also* **Charging receptacle.**

RECEPTACLE OUTLET Electrical outlet where one or more receptacles are installed.

RECEPTION SHAFT Excavation into which trenchless technology equipment is driven and recovered following conduit installation. *Also called* **Exit shaft.**

RECEPTOR Metallic or nonmetallic waterproof support for a shower stall.

RECESS Indentation of some inches in the thickness of a wall, such as a niche.

RECESSED WHEELS Grinding wheels made with a depression in one side or both sides to fit special types of flanges or sleeves.

RECHARGE Addition of water to a zone of saturation.

recip. *Abbreviation for:* reciprocal; reciprocate.

RECIPROCAL LEVELING Surveying method that eliminates instrument error between two points by taking readings from two setups, one near each point.

RECIPROCATE To move back and forth or up and down in a repetitive manner.

RECIPROCATING DRILL *See* **Drill.**

RECIPROCATING ENGINE Engine that produces rotary movement through the repetitive motion of a crank-connected piston within a cylinder.

RECIPROCATING GRATE *See* **Grate.**

RECIPROCATING PUMP *See* **Pump.**

RECIPROCATING SAW *See* **Saw.**

recir. *Abbreviation for:* recirculate; recirculating.

recl. *Abbreviation for:* reclaim; reclaimed; reclose.

RECLAIMED AGGREGATE MATERIAL Removed pavement materials containing no reusable binding agent.

RECLAIMED ASPHALT PAVEMENT Material removed from a roadway by cold planing or ripping.

RECLAIMING 1. Removing material from a stockpile; **2.** Reprocessing material previously rejected; **3.** Process of recovering used oils for re-use by filtration, clay or chemical treatment. Not to be confused with re-refining where the used oil is distilled and chemically treated to recover base oils similar to those used in the original manufacture of the lubricant.

RECLAIM TUNNEL Tunnel, often equipped with a belt conveyor, under stockpiled material that facilitates loading out of the material.

RECLAMATION Restoration to a better or more useful state, such as land reclamation by sanitary landfilling, or the obtaining of useful materials from solid waste. *See also* **Road reclamation.**

recm. *Abbreviation for:* recommend; recommendation.

recnd *Abbreviation for:* reconditioned.

recog. *Abbreviation for:* recognize; recognized.

recomp. *Abbreviation for:* recomputation.

recond. *Abbreviation for:* recondition.

reconst *Abbreviation for:* reconstruct; reconstructed.

RECONSTITUTED WOOD Wood that has been first reduced to small fragments and then put back together again by a special manufacturing process into panel products of relatively large sizes and various thicknesses, such as particleboards, medium-density fiberboard, and hardboard.

RECONSTRUCTED STONE Cast stone made from rock dust and chippings.

RECORD DRAWING Drawing that shows something as constructed or fabricated and that records any changes made since first delivery or completion.

RECORDER OF DEEDS *See* **Land Registrar.**

RECORDING Act of entering instruments affecting title to real property into the public record.

RECORDING GAUGE Gauge that continuously plots the value it is reading.

recov. *Abbreviation for:* recover; recovered.

RECOVERABLE RESOURCE Material that still has useful physical or chemical properties after serving its original purpose and that can be reused or recycled for the same or other purposes.

RECOVERED ENERGY Energy utilized that would otherwise be wasted.

RECOVERY 1. Process of retrieving materials or energy resources from waste; **2.** Rescue of a vehicle that has broken down and is unable to move under its own power; **3.** Degree to which a hose returns to its normal dimensions or shape after being distorted.

RECOVERY PEG Survey marker having a known relationship in level, direction, and distance to another marker that enables one or the other to be accurately replaced.

RECOVERY RATE Speed at which a water heater will raise cold water to a desired temperature.

RECOVERY VEHICLE/WRECKER Vehicle used to retrieve and lift/tow other vehicles.

recp. *Abbreviation for:* receptacle.

rec. rm *Abbreviation for:* recreation room.

rect. *Abbreviation for:* rectangle; rectangular; rectified.

RECTANGULAR COFFERDAM Temporary wood or steel sheet piling, braced, single-walled, rectangular-shaped enclosure installed to permit construction of a foundation below ground or below water level.

RECTANGULAR DUCT Four-sided air duct.

RECTIFIER Device that changes alternating current into direct current.

RECUPERATOR Heat exchanger in which energy is transmitted from flowing hot fluid to a flowing cold fluid through a wall whose function is to separate the two fluids.

RECURRENCE PERIOD Interval between occurrences of repeating events.

RECYCLED CONCRETE Hardened concrete that has been processed for reuse, usually as aggregate.

RECYCLED MATERIAL Material that can be utilized in place of a raw virgin material in manufacturing a product, consisting of materials derived from postconsumer waste, industrial scrap, material derived from agricultural wastes, or other items, all of which can be used in the manufacture of new products.

RECYCLING Separating a given waste material (e.g., glass) from a waste stream and processing it so that it may be used again as a raw material for products which may or may not be similar to the original.

RECYCLING AGENT Organic material with chemical and physical characteristics selected to restore aged asphalt to desired specifications. *Also called* **Rejuvenating agent,** and **Softening agent.**

RECYCLING BODY Multicompartment dumping body, typically installed on a low-profile chassis, used primarily for residential pickup of recyclable refuse.

RECYCLING TRAIN Succession of mobile processing units used to recycle old pavement on site.

red *Abbreviation for:* reduced.

red. *Abbreviation for:* reduce; reducing.

REDEMPTION Right of a mortgagor to recover a property lost through foreclosure of a mortgage, tax forfeiture, or other legal process by paying off any outstanding debts.

redesig. *Abbreviation for:* redesignate; redesignated;

redesigned.

REDEVELOPMENT Planning, design, reconstruction and rehabilitation of a structure or structures or an area that is scheduled for upgrading.

REDISTRIBUTION OF MOMENT Process that results in the successive formation of plastic hinges so that less highly stressed portions of a structure may carry increased moments.

RED LABEL *See* **Shake**, and **Shingle**.

RED LEAD Textroxide of lead; when used as a pigment in linseed oil it provides a protection to metal against rust.

RED TOP Subgrade stake indicating finish level.

REDUCED BEARING Bearing of less than 90°. *Also called* **Quadrantal bearing**.

REDUCED LEVEL In surveying, an elevation calculated from stipulated datum.

REDUCED PRESSURE RANGE Adjustment range of a regulator.

REDUCER 1. Pipe fitting with a smaller opening at one end; **2.** Trim unit used to reduce the radius of a bullnose or cove to another radius or to a square; **3.** Paint thinner.

REDUCING AGENT Chemical that lowers the state of oxidation of other chemicals.

REDUCTION Extraction of mineral from its ores.

REDUCTION (SINGLE) Gear set that causes one shaft to turn another at a reduced speed.

REDUCTION COEFFICIENT Factor applied to the theoretical capacity of a pile due to such factors as disturbance of the soil during driving.

REDUCTION (DOUBLE) Two sets of gears in series that both reduce speed and increase power.

REDUCTION FACTOR Structural factor for a given slenderness ratio.

REDUCTION IN AREA Contraction in the area of a tensile test piece.

REDUCTION OF LEVELS Calculation of the differences in level between various points based on survey readings recorded in a field book.

REDUCTION RATIO Difference in size between an aggregate particle entering a crusher and the size of the particle leaving the machine.

REDUCTION-TYPE TRANSMISSION *See* **Transmission.**

REDUNDANT FRAME Structural frame having more members or more stability than is required for it to be a perfect frame.

REED CLIPS Light wire sections fastened to structural steel to hold fireproofing to the faces of structural steel beams and column flanges.

REEDING Decoration consisting of adjacent parallel convex moldings.

REEF To pull hard.

REEFER Truck, semitrailer, or trailer with a heavily insulated body and equipped with a refrigeration unit.

REEL 1. Flanged spool on which wire rope, hose, or strand is wound for storage or shipment; **2.** Revolving rack on which hose is stored; **3.** The winches of a churn drill. *See also* **Casing reel, Dead reel, Live reel, Sand reel,** and **Spudding reel.**

REEL CARRIER Attachment, commonly to a trencher, used to support and discharge the reel carrying the pipe or cable being buried.

REEL STAND Device that supports reels of wire rope or hose of varying sizes.

REEVE To pass a rope through a hole or around a system of sheaves.

REEVING Rope system where the rope travels around drums and sheaves.

REEVING DIAGRAM Diagram showing the path of rope through a system of sheaves.

REEVING DRUM Any winch or drum used for pulling in wire rope. Normally the topping line drum.

REEVING LINE Any line pulled in by a reeving drum.

ref. *Abbreviation for:* refer; reference; refine; refined; refinish.

REFACE Grinding a machined face to original tolerances.

REFERENCE LINE 1. Line marked on something that relates to another object, part, or component; **2.** One of the lines of intersection of the image planes in an orthographic projection.

REFERENCE MARK Distant point used for taking survey bearings and measurements. *See also* **Match mark.**

REFERENCE PEG Survey recovery peg.

REFERENCE STAKE Stake from which measurements and grades are established. *Also called* **RS.**

REFERENCE STUD Hard, chrome-plated stud mounted on a cadmium-plated steel body permanently attached to a structure to act as a datum point to monitor superficial movement.

REFILL TUBE Tube from the ball cock to the overflow tube in a water closet tank.

REFINERY Process plant in which flammable liquids or combustible liquids are produced from crude petroleum.

REFINING Process by which crude petroleum is converted to products.

ref. 1 *Abbreviation for:* reference line.

refl. *Abbreviation for:* reflect; reflection; reflector.

REFLECTANCE Ratio of the light reflected by a surface to the light falling upon it.

REFLECTED PLAN Graphic horizontal section, shown looking up.

REFLECTION COEFFICIENT Fraction of sound returned after a sound wave strikes a surface.

REFLECTION CRACK Crack appearing in a pavement resurface or overlay caused by movement at joints or cracks in the underlying base or surface. *See also* **Crack.**

REFLECTIVE CRACKING Fissures in an asphalt overlay that reflect the crack pattern in the pavement structure underneath.

REFLECTIVE INSULATION Sheet material with one or both surfaces finished, or faced to give low heat absorption and high thermal reflectivity.

REFLECTIVE TAPE Adhesive tape with the exposed side consisting of light-reflecting material.

REFLECTOR 1. Glass or plastic prism that reflects light; **2.** Polished surface to reflect light or heat in a specific direction.

REFLECTOR LAMP Spot or flood lamp with an internally silvered body.

REFLUX VALVE Check valve.

REFORESTATION Natural or artificial restocking (i.e., planting, seeding) of an area with forest trees. *Also called* **Forest regeneration.**

refr. *Abbreviation for:* refractory; refrigeration.

REFRACTIVE INDEX Ratio of the angle of light in the first of two media to its angle in the second.

REFRACTORIES Materials, usually nonmetallic, used to withstand high temperatures.

REFRACTORY AGGREGATE *See* **Aggregate.**

REFRACTORY CONCRETE Hardened hydraulic-cement concrete that has refractory properties and that is suitable for use with temperatures between 315°C and 1315°C (600°F and 2,400°f). *Also called* **Heat-resistant concrete.**

REFRACTORY FURNACE WALL *See* **Wall.**

REFRACTORY INSULATING CONCRETE Refractory concrete having low thermal conductivity.

refrig. *Abbreviation for:* refrigerator.

REFRIGERANT Fluid used for heat transfer in a refrigerating system.

REFRIGERATING PLANT Complete refrigerating system.

REFUSAL Condition reached when a pile or soil sampler being driven by a hammer has zero penetration per blow, as when the point of the pile reaches an impenetrable bottom, or when the effective energy of the hammer blow is no longer sufficient to cause penetration. *Also called* **Pile refusal,** and **Practical refusal.**

REFUSE *See* **Solid waste.**

REFUSE BURNER *See* **Burner.**

REFUSE-DERIVED FUEL Shredded refuse fuel, used principally as a supplement in industrial or utility boilers that have ash handling capabilities. Using a separation system, much of metal, glass, and other inorganics are first removed, the remaining organic fraction is processed to relatively uniform size particles that are classified as:

RDF-1: Waste used as a fuel in as-discarded form with only bulky wastes removed.

RDF-2: Waste processed to coarse particle size with or without ferrous metal separation.

RDF-3: Combustible waste fraction processed to particle sizes, 95% passing 50-mm (2-in.) screening.

RDF-4: Combustible waste fraction processed into powder form, 95% passing 10 mesh screening.

RDF-5: Combustible waste fraction densified (compressed) into the form of pellets, slugs, cubettes, or briquettes.

RDF-6: Combustible waste fraction processed into liquid fuel.

RDF-7: Combustible waste fraction processed into gaseous fuel.

REFUSE HANDLING What is done to prepare refuse for disposal or for processing, which is conversion of wastes into something useful.

REFUSE TRUCK Vehicle specially designed for the collection and transport of refuse. Includes the following types:

Batch loader: Type of enclosed compactor truck equipped with a loading hopper at the rear end and a large mechanized panel that sweeps the solid wastes into the body of the unit.

Front loading: Used for commercial and industrial pick-up; capacities range from 15 to 38 m³ (20 to 50 yd³). Usually installed on a tilt-cab tandem-axle chassis. Front fork arms lift steel containers over the cab to empty the waste into a hopper atop the body; a hydraulic ram and push-plate system compacts the waste rearward inside the body.

Mobile packer: An enclosed vehicle provided with special mechanical devices for loading, compressing and distributing refuse within the body.

Movable bulkhead: Type of side-loading, enclosed compactor truck equipped with a movable bulkhead that pushes the solid wastes from the front loading area to the rear of the vehicle.

Rear packer: Primarily used for residential refuse collection by two- or three-man crews. Capacities range from 7.65 to 24.5 m³ (10 to 32 yd³) of ram-compacted waste; hopper swings up for unloading and waste is pushed out by a movable front wall; may include a container-attachment and winch system for collecting waste from apartment complexes and small businesses in containers of up to 7.65 m³ (10 yd³).

Side loading: Designed for residential and some commercial applications using a one-man crew. Sizes from 5.4 to 30.1 m³ (7 to 40 yd³).

Stationary packer: An adjunct of a refuse collection system that compacts refuse into a pull-on, detachable container at the site of generation.

reg. *Abbreviation for:* region; register; regular; regulate; regulation; regulator.

REGAIN OF MOISTURE Amount of moisture absorbed by a material as a percent of its original weight.

REGELATION Refreezing of moist ice under pressure at a temperature above freezing.

REGENERATION Renewal of a tree crop, either by natural or artificial means; a young tree crop.

REGENERATION DELAY Maximum time allowed for initial restocking of a denuded forest area (as a result of harvesting, fire, etc.) with the minimum number of acceptable trees, measured in growing seasons from the beginning of denudation.

REGENERATION PERFORMANCE ASSESS-MENT Sampling survey carried out to collect field data on the height growth, competition, and stocking of young (5 to 10 years) forest stands.

REGENERATION SURVEY Survey to determine the initial restocking of a site, used to describe the number of trees on a site that have reached acceptable standards.

REGENERATIVE CIRCUIT Circuit in which pressurized fluid discharged from a component is returned to the system to reduce power input requirements. On single end cylinders, the discharge from the rod end is often directed to the bore end to increase rod extension speed. *See also* **Circuit.**

REGENERATIVE-CYCLE GAS-TURBINE EN-GINE Gas-turbine engine employing exhaust heat recovery in the thermodynamic cycle consisting of successive compression, regenerative heating, combustion, expansion, and regenerative cooling (heat transferred to compressor discharge air) of the working fluid.

REGENERATIVE HEATING/COOLING Utilization of heat that must be rejected or absorbed by one part of the cycle, to perform a useful function in another part of the cycle, by heat transfer.

REGENERATOR Heat exchanger in which energy is transmitted from a flowing hot fluid to a flowing cold fluid by alternately passing these fluids through the same mass of material.

REGISTER Device to regulate the discharge of warm air from ducts to rooms.

REGISTER BOX Metal pan at the end of a duct into which, or over which a register is mounted.

REGISTERED LAND Land recorded in a public register.

REGISTRAR OF DEEDS *See* **Land Registrar.**

REGISTRATION 1. Recording of something in a public record; **2.** Mating parts when in their proper position; **3.** Process of determining and certifying competency to practice a self-governing profession.

regl *Abbreviation for:* regional.

REGLET Recess to receive and secure metal flashing. *See also* **Raggle.**

REGLETTE Surveyor's short scale divided into tenths and hundredths of a foot, used with a steel band marked in feet only.

REGOLITH Noncemented, residual and deposited rock fragments, soil and other materials that overlie bedrock in most places. *See also* **Crust.**

REGRESSION Statistical technique used to evaluate relationships among variables.

REGRESSIVE TAX Situation where the ratio of tax liability to net income or net worth declines as the income or worth increases.

REGROOVE Refinishing worn groves on the drive sheave of a traction machine.

REGROOVING (RECUTTING) Cutting of a tread design into the worn down tread of a tire, or cutting a deeper design in an existing tread.

REGULAR LAY *See* **Lay types.**

REGULATED-SET CEMENT Hydraulic cement containing fluorine-substituted calcium aluminate, capable of very rapid setting.

REGULATED VOLTAGE Voltage that is stable over a wide range of conditions.

REGULATION 1. Controlling or directing principle or rule established in writing; **2.** Manipulation of growing stock so that it contains a proper proportion of young, middle-aged, and mature trees in order to obtain continuous production, or sustained yield.

REGULATOR Control instrument,: typically to control voltage and current in a vehicle's electrical system. *See also* **Controller.**

REHABILITATION 1. Process of repairing or modifying a structure to a desired useful condition. *See also* **Preservation, Repair,** and **Restoration; 2.** All aspects of maintaining or upgrading the performance of existing pipeline systems.

REHEATER Accessory to steam or compressed-air engines that reduces consumption of air or steam by superheating the steam, or reheating the air between expansion stages.

reim. *Abbreviation for:* reimburse.

REIMBURSABLE EXPENSES Moneys spent by the contractor on a project that, under the terms of the contract, will be reimbursed by the owner.

reinf. *Abbreviation for:* reinforce; reinforcement.

REINFORCE Strengthen by the incorporation of additional materials or through the application of one material to another.

REINFORCED CONCRETE Concrete that has metal rods, or fibers of synthetic matial, in it to withstand greater stress and strain.

REINFORCED CONCRETE CONSTRUCTION *See* **Construction types.**

REINFORCED EARTH Densified earth to which a reinforcing material may have been added, consolidated to support a calculated load.

REINFORCED MASONRY Masonry units, reinforcing steel, grout, and/or mortar combined together in resisting forces.

REINFORCEMENT Bars, cables, fabric, fibers, mesh, rods, or wires embedded in a material so as to resist imposed forces.

REINFORCEMENT DISPLACEMENT Movement of reinforcing steel from its specified position in the form.

REINFORCEMENT RATIO Ratio of the effective area of reinforcement to the effective area of concrete at any section of a structural member. *See also* **Percentage of reinforcement.**

REINFORCEMENT REPAIR Tire casing repairs that require both hole filling material and reinforcing patches when an injury has extended through more than 25%, but less than 75%, of the tire body.

REINFORCING AGENT Ingredient (not basic to the vulcanization process) used in a rubber compound to increase its resistance to mechanical forces.

REINFORCING BAR Steel rod embedded in concrete to provide resistance to tension stresses.

REINFORCING MESH Grid of welded steel wires used to resist tension stresses in concrete.

REINFORCING STEEL Steel used in concrete construction to give added strength.

REINSURANCE *See* **Insurance.**

rej. *Abbreviation for:* reject.

REJOINTING Pointing.

REJUVENATING AGENT *See* **Recycling agent.**

REKINDLE Reignition due to latent heat, sparks, or embers or due to the presence of smoke or steam.

rel. *Abbreviation for:* relative; release; relief.

RELATED TRADE Different or allied trade whose work is necessary to complete the work of any one trade.

RELATING DEVICE Device that connects two or more elevator door panels and determines the speed and travel of each.

RELATIVE COMPACTION Dry density of a soil sample divided by the maximum dry density of the same sample as determined by a standard compaction test.

RELATIVE DENSITY State of compaction of a granular soil mass relative to the loosest and most dense conditions possible.

RELATIVE HUMIDITY *See* **Humidity.**

RELATIVE PHYSICAL INTENSITY SCALE Instrument showing the relationship between decibels and physical intensity.

RELATIVE SETTLEMENT Differential settlement.

RELATIVE UTILIZATION FACTOR Ratio of the utilization efficiency of a fuel to the utilization efficiency of the base (comparative) fuel.

RELAXATION 1. Phenomenon where some piles decrease their load-carrying capacity after being driven; **2.** Loss of stress in prestressed concrete tendons following initial imposition of the design load.

RELAY 1. Valve or switch that reflects, amplifies, or restores initial strength to a hydraulic, pneumatic, or electrical impulse; **2.** Electrical magnetic switch employing an armature to open or close contacts.

RELAY EMERGENCY VALVE *See* **Valve.**

RELAY VALVE *See* **Valve.**

RELEASE 1. Written surrender of a claim or interest; **2.** Relinquishment of a right or claim for consideration.

RELEASE AGENT Material used to prevent bonding of concrete to a surface. *See also* **Bond breaker,** and **Form oil.**

RELEASE BEARING ASSEMBLY Assembly that transfers the movement of a clutch linkage to engage or disengage the clutch.

RELEASE CLAUSE Blanket mortgage clause giving an owner the right to pay off a portion of the debt, freeing an appropriate portion of the property from debt.

RELEASE FOR CONSTRUCTION Release by the owner permitting the fabricator to commence work under the contract, including ordering material and the preparation of shop drawings.

RELEASE OF LIEN *See* **Lien waver.**

RELEASE OF LIEN BOND *See* **Bond.**

RELEASE OF RETAINED PERCENTAGE BOND *See* **Bond.**

RELICTION Gradual, and relatively permanent, subsidence of water leaving dry land.

RELIEF Difference in elevation between high and low parts of a land surface.

RELIEF HOLE Hole drilled close to others along a line, not loaded with explosives but serving to weaken the rock so that it shears along the line when it is blasted.

RELIEF PLATFORM Platform on the soil side of a retaining wall to transmit heavy superloads directly to the wall and prevent them from becoming imposed on the back of the wall as additional earth pressure.

RELIEF VALVE *See* **Valve.**

RELIEF VENT Branch from a vent stack, connected to a horizontal branch between the first fixture branch and the soil or waste stack, whose primary function is to provide for circulation of air between the vent stack and the soil or waste stack.

RELIEF WELL Borehole drilled at the toe of an earthen dam to relieve pressures created by the dam's weight.

RELIEVING ARCH *See* **Arch.**

RELIEVING PLATFORM Decking on the high side of a retaining wall to transmit loads vertically and prevent them from becoming a surcharge on the wall.

RELOAD Central location where log loads are transferred from one mode of transport to another.

reloc. *Abbreviation for:* relocate.

RELOCATION 1. New alignment, varying from the original; **2.** Movement of something from one location to another.

RELOG To log a setting again because it was not logged clean enough the first time.

rem. *Abbreviation for:* remainder; remit.

REMAINDER Giving up of a right or claim to the person against whom it could have been exercised; **2.** Estate of property that takes effect only after termination of any prior estate, such as a life estate. *See also* **Acquisition, Conveyance, Dedication, Eminent domain, Expropriation, Negotiation, Option,** and **Severality.**

REMAINDERMAN Person who is to receive a property following the death of the life tenant.

REMAINING ECONOMIC LIFE Time remaining from the date of appraisal to the expiry of the economic life of equipment, a machine, or a structure.

REMEDIAL ACTION Work done or steps taken to halt or reverse a development, usually a development that is deleterious, harmful, or dangerous.

REMIXING Working concrete that has begun to stiffen prior to placing in the forms.

Rem. Mul. *Abbreviation for:* removable mullion.

remod *Abbreviation for:* remodeled.

remod. *Abbreviation for:* remodel.

REMODEL Change the plan, form, or style of a structure.

REMOLDABILITY Readiness with which freshly mixed concrete responds to a remolding effort such as jigging or vibration causing it to reshape its mass around reinforcement and to conform to the shape of the form. *See also* **Flow.**

REMOLDING INDEX Ratio of the modulus of elasticity, or deformation of an undisturbed soil to its modulus of elasticity once remolded.

REMOLDING TEST Test to measure remoldability.

REMOTE-CONTROL CIRCUIT Any electric circuit that controls any other circuit through a relay or an equivalent device.

REMOTE RADIATOR Radiator and fan that are not mounted to, or driven by the unit.

remov. *Abbreviation for:* removable; remove.

REMOVABLE ATTACHMENT Attachment that can be mounted on mobile equipment, such as on the forks of an industrial truck or in place of the forks on the carriage, by means of such conventional fasteners as bolts, pins, etc., and that does not require a disassembly of any other portion of the equipment to install or remove.

REMOVABLE BRACKET *See* **Side bracket.**

REMOVABLE FLIGHTING Flighting that is attached to the cutter drum other than by welding.

REMOVABLE GOOSENECK *See* **Gooseneck.**

REMOVABLE KINGPIN Kingpin, part of a fifth-wheel assembly, designed to be readily removed and relocated to other positions or settings in the upper coupler assembly.

REMOVABLE UNDERCARRIAGE *See* **Undercarriage.**

REMOVAL TECHNIQUE Technique where a glaze or other medium that covers a surface is lifted off or wiped out.

remvd *Abbreviation for:* removed.

ren. *Abbreviation for:* renew; renewable; renewed.

RENARD SERIES *See* **Preferred numbers.**

RENDER To apply a coat of mortar by a trowel or float.

RENDER AND SET Two-coat plaster work; rendering covered by a finishing coat.

RENDER, FLOAT, AND SET Three-coat plaster work.

RENDERING 1. Covering an outer wall with plaster, etc.; **2.** Presentation drawing of a proposed project, usually showing it in its likely setting.

RENDERING MATERIAL Mixture of fine aggregate with cement or lime and water, possibly with the addition of other binders, colorants, texturizers, or admixtures, that hardens following application.

RENEWABLE ENERGY SOURCE Source of energy, such as wind power and solar heat, that is inexhaustible or which are derived from organic matter that reproduces itself continually, such as wood or moss.

renov. *Abbreviation for:* renovate.

RENOVATION 1. To make new or like new; **2.** Methods by which a new pipeline is constructed replacing the original fabric on the same line.

RENT Periodic payment made by a tenant to a landlord or his agent for the use of land, buildings, dwellings or other property. *See also* **To rent.**

RENTAL HOUSING Accommodation designed and built specifically to be rented to tenants.

RENTAL VALUE Monetary amount that can be reasonably expected for the right to the specified use of real property.

RENTAL VALUE INSURANCE Policy that pays an owner or occupant of a building for the cost of renting another premises should the insured property become untenable as a result of damage from an insured peril.

RENT SUPPLEMENT Direct subsidy to a tenant based on the difference between the rent stipulated by a landlord and the tenant's ability to pay.

reorg. *Abbreviation for:* reorganize.

rep. *Abbreviation for:* repair; represent; representative.

REPAIR 1. To replace or correct deteriorated, damaged, or faulty materials, components or elements of a structure or machine. *See also* **Preservation, Rehabilitation,** and **Restoration; 2.** Any patch, plug, or shim in a veneer. There are several types, including:

> **Patch:** Sound wood inserts that replace a defect in veneer. They consist of the following types:
>
>> **Boat** - Patches that are oval shaped with sides tapering to points or small rounded ends.
>>
>> **Router** - Patches that have parallel sides and rounded ends.
>>
>> **Sled** - Patches that are rectangular with featured ends.
>
> **Plug:** May be circular or dogbone-shaped, or a synthetic filler of fiber and resin to fill openings and provide a smooth, level, durable surface.
>
> **Shim:** A long, narrow wood or synthetic repair not more than 4.8 mm (0.189 in.) wide.

REPAIRED TIRE Any tire with punctures, cuts or other types of injuries that has been reconditioned to provide additional safe service life.

REPAIR GUM Tire repair material used for filling voids, or covering reinforcing material.

REPAIR PLUG Rubber material that fills the cavity of a tire injury.

REPAIR PATCH Reinforcing material used to

strengthen the area around a tire injury.

REPAIR TIME *See* **Machine time.**

REPEATABILITY Variability among replicate test results obtained on the same material within a single laboratory by one operator, a quantity that will be exceeded in only about 5% of the repetitions by the difference, taken in absolute value, of two randomly selected test results obtained in the same laboratory on a given material.

repel. *Abbreviation for:* repellent.

REPETITION Method of measuring angles requiring accumulation of the angle on the circle of a survey instrument two or more times, then dividing by the number of accumulations.

repl. *Abbreviation for:* replace; replacement.

REPLACEMENT COST Cost necessary to completely replace equipment, machinery, or a structure to the same state as when use of the original facility ceased. *Also called* **Replacement value.**

REPLACEMENT COST INSURANCE Policy that pays the undepreciated replacement value of the damaged property.

REPLACEMENT VALUE *See* **Replacement cost.**

replm. *Abbreviation for:* replacement.

repln. *Abbreviation for:* replenish.

REPO' Repositioning of a log to a new position, where it can be safely handled.

REPOINTING Process of removing dry, loose mortar from the exposed section of a masonry joint and replacing it with fresh cement grout.

REPOUSSÉ Ornamental metalwork raised into relief by hammering from the back.

REPOUSSÉ HAMMER *See* **Hammer.**

REPRESENTATIVE SAMPLE Sample drawn from a fair proportion of a whole.

repro. *Abbreviation for:* reproduce; reproduction.

reproc. *Abbreviation for:* reprocess; reprocessed.

REPROCESSING 1. Changing the character of secondary materials, i.e., minor (such as crushing or shredding) or major (such as biochemical conversion); **2.** Renewal of an existing highway surface by scarifying, remixing with or without additional material, and relaying.

REPRODUCTION COST Cost of construction at current prices of an exact duplicate or replica.

rept *Abbreviation for:* report.

req. *Abbreviation for:* request; require; requirement; requisition.

REQUEST FOR PAYMENT Application for interim or final payment under the terms of a contract.

REQUIRED STRENGTH Strength of a member or cross section required to resist factored loads or related internal moments and forces in such combinations as are stipulated in the applicable code or specification.

REQUIRED TRACTIVE EFFORT Tractive effort needed to propel a vehicle, and any trailer being towed

by it, while overcoming (a) the rolling resistance of all wheels, including all wheel bearing friction, (b) grade resistance, and (c) the inertia of the truck and any trailers being towed by it.

REQUIREMENTS Client's or building user's needs to be satisfied by a proposed building. *Also called* **Needs,** and **User needs.** *See also* **Architectural program, Conceptual program,** and **Functional program.**

REQUISITION Written request for something authorized but not automatically available.

REREDOS Ornamental wall or screen rising behind an altar.

res. *Abbreviation for:* resawn; research; reserve; reservoir; reset; residence; resident; residential; resonator.

resc. *Abbreviation for:* rescind.

RESCREENING PLANT Plant set up and equipped to rescreen already sized material.

RESERVE FOREST LAND *See* **Land use classes (forest management).**

RESERVE FOR SCOPE CHANGES Sum of money set aside for changes in a client's requirements and to be used only at the direction of the client.

RESERVOIR 1. Place where anything (usually a fluid) is collected and stored, especially a natural or artificial impoundment; **2.** Receptacle or part for holding fluid, such as hydraulic fluid. *Also called* **Sump tank.**

RESERVOIR FILTER Filter installed in an oil reservoir in series with a suction or return line. *Also called* **Sump filter.**

RESERVOIR ROUTE *See* **Collection method.**

RESETTING (OF FORMS) Setting of forms separately for each successive lift of a wall to avoid offsets at construction joints.

RESHORING Construction operation in which the original shoring or posting is removed and replaced in such a manner as to avoid deflection of the shored element or damage to partially cured concrete.

resid. *Abbreviation for:* residual; resident; residential.

RESIDENCE Place where a person or family lives; dwelling, apartment, etc.

RESIDENCE TIME Time that a substance or condition remains unchanged from a given state.

RESIDENT ARCHITECT Architect at a job site supervising the work and representing the owner.

RESIDENT ENGINEER Professional engineer, usually in the employ of the owner, whose place of work is at the site of a contract for work.

RESIDENTIAL INCINERATOR *See* **Incinerator.**

RESIDENTIAL OCCUPANCY Occupancy or use of a building or part of a building by persons for whom sleeping accommodation is provided but who are not harbored or detained to receive medical care or treatment or are not involuntarily detained.

RESIDENTIAL SILENCER Exhaust muffler used to produce the silencing level usually associated with or required for residential areas.

RESIDENTIAL WASTE *See* **Solid waste.**

RESIDUAL One of the trees remaining after an intermediate or partial cutting of a tree crop or stand. In general, residuals are by-products of some operation.

RESIDUAL DIRT FACTOR Dirt capacity remaining in a service loaded filter element after use, but before cleaning, measured under the same conditions as the dirt capacity of a new filter element.

RESIDUAL DRAWDOWN Distance that the water table present in a well is below the initial static water level following a period of pumping.

RESIDUAL ERROR Minor surveying error that cannot be eliminated by measurement.

RESIDUAL FACTOR Percentage of a machine's adjusted tax basis left on the books after the machine has been depreciated.

RESIDUAL MOISTURE Moisture content remaining in a sample after it has been milled down to an analysis sample.

RESIDUAL OIL Oil that is too heavy to be evaporated in any normal evaporation and distillation process and so is left over from that process.

RESIDUAL SOIL Soil formed *in situ* by rock decay and left as residue after the leaching out of more soluble products.

RESIDUAL STAND Trees remaining in an area after the cutting operation has been completed.

RESIDUAL STRESS Stress that remains in an unloaded structural steel member after it has been formed into a finished product.

RESIDUAL VALUE Actual or assumed value of a machine after it has been fully depreciated.

RESIDUAL WASTE Those materials (solid or liquid) that still require disposal after the completion of a resource-recovery activity, e.g., slag and liquid effluents, plus the discards from front-end separation systems.

RESIDUE 1. Material that remains after gases, liquids, or solids have been removed; 2. Wood or bark that is left after a manufacturing process.

RESIDUE FROM DISTILLATION Test that determines the relative proportion of asphalt cement and water in an emulsified asphalt.

resil. *Abbreviation for:* resilient.

RESILIENCE 1. Work done per unit volume of a material in producing strain; 2. Ability of a plastic to quickly regain its original shape after having been strained or distorted.

RESILIENT *See* **Texture.**

RESILIENT CHANNEL Metal furring used to reduce sound transmission or impact noise by isolating surfacing membranes from framing.

RESILIENT FLOOR COVERING Vinyl sheet material, usually patterned, glued to a subfloor or underlayment.

RESILIENT MODULUS OF SOIL Testing method used to determine required pavement thickness.

RESIN Natural or synthetic, solid or semisolid organic material of indefinite and often high molecular weight having a tendency to flow under stress, usually has a softening or melting range, and usually fractures conchoidally.

resin. *Abbreviation for:* resinous.

RESIN ADHESIVE *See* **Adhesive.**

RESIN BONDED Lumber glued with a synthetic resin.

RESIN CONCRETE *See* **Polymer concrete.**

RESINOID BOND Bonding material described commercially as synthetic resin.

RESIN SPOT Hard piece of dark or black foreign material in a face layer of a particleboard that is composed of glue and wood dust.

RESISTANCE 1. Power to resist applied force(s); 2. Force opposing another or others; 3. In electricity, the property of a conductor to oppose the passage of a current; 4. In fluid flow, the opposition to flow that makes it inevitable that there will be a pressure drop when a fluid is flowing; 5. Sum total of all the forces that oppose the penetration of a pile under a hammer blow or a static force.

RESISTANCE FACTOR Factor that accounts for unavoidable deviations of the actual strength from the nominal value, and the manner and consequences of failure.

RESISTANCE TO CONTROL Relative difficulty of constructing and holding a forest fire control line, as affected by resistance to line construction and by fire behavior.

RESISTANCE VALUE METHOD Testing method used to determine the required pavement thickness for a particular application.

RESISTANCE WELDING Welding of two pieces held tightly in contact, by electrodes through which a heavy alternating current momentarily flows, causing the pieces to fuse together.

RESISTING BENDING MOMENT Calculation used to compare beams and frames of different section modulus and of different materials. It is the product of the section modulus times the yield strength of the material being studied.

RESISTING BUTT-WELDING Welding two parts butted together. *See also* **Resistance welding.**

RESISTING FORCE Forces in a system that tend to resist failure.

RESISTOR Electrical device that creates a resistance to the flow of electricity.

resnt *Abbreviation for:* resonant.

RESOLUTION 1. Sharpness of the image seen through a telescope; 2. Minimum incremental measurement possible in a given measuring device.

RESONANCE 1. State when the frequency of an applied dynamic load coincides with a natural frequency or load support; 2. Neutral condition in an alternating current circuit when the inductive reactance is just equal to the capacitive reactance.

RESONANT FREQUENCY Actual frequency at which the combination of a vibratory roller drum and the material exhibits the greatest amplitude; that is when the generated frequency coincides with natural frequency of the material being compacted.

RESOURCE RECOVERY General term used to describe the extraction of materials or energy from wastes.

RESOURCES Personnel, equipment, and material available for application to a given situation.

RESOURCINOL FORMALDEHYDE RESIN Type of synthetic resin.

RESOURCINOL RESIN ADHESIVE See **Adhesive.**

resp. *Abbreviation for:* respective; respectively; responsibility.

RESPIRABLE DUST Airborne dust in sizes capable of passing the upper respiratory system to reach the lower lung passages.

RESPIRATOR Filtering mask for individual protection against smoke and fumes.

RESPOND Half-pillar or half-pier attached to a wall to support an arch.

RESPONSE Act of responding to a stimulus or alarm.

RESPONSE-SPECTRA Mathematical method for estimating the response of a structure to different frequencies.

RESPONSE TIME Time required for effective transition.

REST That part of a grinding wheel stand that is used to support the work, dresser, or turning tool when applied to the grinding wheel.

RESTAURANT Building or portion of a building where food is offered for sale for immediate consumption on the premises, but not including premises where drink or prepackaged food requiring no further preparation before consumption is offered for sale.

RESTORATION Process of reestablishing the materials, form, and appearance of a structure to those of a particular era of the structure. See also **Preservation, Rehabilitation,** and **Repair.**

restr. *Abbreviation for:* restaurant; restrict.

RESTRAINED SLAB Reinforced concrete floor slab in which the corners are prevented from lifting and provision is made for torsion.

RESTRAINING DEVICES See **Check chains.**

RESTRAINT 1. Materials or attachments that prevent or impede the movement or passage of other materials or parts; **2.** Restriction of free movement of fresh or hardened concrete following completion of placing in formwork or molds or within an otherwise confined space.

RESTRAINT EQUIPMENT See **Tiedown assemblies.**

RESTRICTED AREA Area accessible only to authorized personnel.

RESTRICTION 1. Defined limit or limitation to a course of action; **2.** Obstacle or other hazard that prevents free movement or operation of mechanical equipment; **3.** Reduced cross-sectional area in a fluid- or vapor-carrying line that produces a pressure drop.

RESTRICTIVE COVENANT Limitation on the use of real property, created by deed, which may 'run with the land,' binding all subsequent purchasers, or which may be 'personal' and binding only between the original seller and buyer.

RESTRICTOR Device that reduces the cross-sectional flow area. See also **Orifice restrictor.**

resup. *Abbreviation for:* resupply.

RESURFACING Placing of one or more new courses on an existing surface.

RESUSCITATOR Approved mechanical device for assisting the respiration of an unconscious person.

ret. *Abbreviation for:* retain; retainer; retard; return.

RETAINAGE Amount withheld from progress payments in accordance with the terms of the construction contract, paid following completion and acceptance of the project.

RETAINED DIRT CAPACITY In fluid filter evaluation, the amount of dirt that is captured by the filter in a test system before the terminal differential pressure is reached.

RETAINER Sum of money paid to secure services.

RETAINING WALL 1. Wall, usually battered, that supports or retains a weight of earth; **2.** Enclosing wall built to resist the lateral pressure of internal loads.

retard. *Abbreviation for:* retardant.

RETARDANT Substance that by chemical or physical action slows or reduces a tendency.

RETARDATION Reduction in the rate of either hardening or setting of a substance, or both.

RETARD CHAMBER Device provided in an automatic sprinkler system to dampen normal fluctuations in municipal water supply to prevent false alarms caused by pressure surges.

RETARDER 1. Fluid-resistance or electrical device that can be used to slow a vehicle **2.** Admixture used to slow the setting process of concrete or mortar; **3.** Compounding ingredient used to reduce the tendency of a rubber compound to vulcanize prematurely.

RETARDING ADMIXTURE See **Admixture.**

RETARD SWITCH See **Switch.**

RETEMPERING Addition of water and remixing of concrete or mortar that has started to stiffen.

RETENSIONING Additional tension applied to a tendon to compensate for losses and to increase the contained stress following imposition of the dead load.

retent. *Abbreviation for:* retention.

RETENTION BOND Substitution of a bond for retainage toward the end of a project.

RETENTION MONEY See **Holdback.**

RETENTION POND Basin in which sudden influxes of surface runoff are held temporarily before being released gradually into a drainage system.

RETENTION TIME Time that a liquid waste takes to pass through a process being performed upon it.

rethd *Abbreviation for:* rethread.

RETICLE Cross hairs or wires in the focus of the

eyepiece of a telescope.

RETICULATED Decoration dressed into the face of masonry that resembles a network of small squares.

RETIRING CAM Retractable cam mounted to an elevator car, used to actuate landing interlocks. It is retracted when the car is in motion, then contacts and unlocks the landing interlocks by moving against the lock roller arm.

retn *Abbreviation for:* retain.

retr. *Abbreviation for:* retraced; retract; retractable.

RETRACT Mechanism by which a dipper shovel bucket is pulled back out of the digging.

RETRACTION Fuel that is allowed to drain from an injection nozzle, following injection, to prevent nozzle dribbling.

RETRACTION LAND Small piston that seals the fuel injection line from the injection pump; part of the delivery valve.

RETRACTION VALUE Amount of retraction travel of the delivery valve.

RETREADING Repairing roads whose surface is breaking up or is misshapen, by scarifying, reshaping, rolling, and surfacing.

RETREAD TIRE A tire built of a used casing and new tread, which extends the casing's usable life.

retro. *Abbreviation for:* retroactive.

RETROFIT Modify, add to, update, etc. a machine, building, equipment that is already complete and working to its original specifications.

RETURN 1. Part receding from the line of front;continuation of a molding, projection, member, etc., in a different direction, usually at a right angle; **2.** Ending of a small splash wall or a wainscot at right angles to the major wall.

RETURN AIR Air returned from heated, conditioned, or refrigerated space.

RETURN AIR DUCT Ducting through which air delivered from a furnace is returned to the unit having warmed an area.

RETURN CORNER BLOCK Concrete masonry unit having finished faces on two adjacent sides.

RETURN JAMB Portion of a door frame behind which a sliding door passes during opening and closing. *Also called* **Side jamb.**

RETURN LINE Line conducting fluid from working devices to a reservoir.

RETURN MAIN Pipes that return a heating or cooling medium from a heat transfer unit to the source of heat or refrigeration.

RETURN NOSING Mitered, overhanging end of a stair tread outside the balusters.

RETURN OFFSET Double offset that allows a pipe to return to its original line having passed over an obstacle. *Also called* **Jumpover.**

RETURN PIPE Pipe by which water is returned to a heating source.

RETURN SHEAVE Pulley, distant from a haulage

drum: a tail rope from the drum passes around the return sheave, enabling the haulage engine to pull away from itself.

RETURN TIME Time it takes to travel from the dump area to the load area.

RETURN WALL Short length of wall perpendicular to the end of a longer wall.

REUSABLE HOSE FITTING *See* **Fitting.**

REUSE Return of a commodity or product into the economic stream for use in exactly the same form and kind of application as before, without any change in its identity.

REUSE CONTROL *See* **Land-use control.**

rev. *Abbreviation for:* revenue; reverse; reversible; review; revise; revision; revolution; revolve.

REVALUATION LEASE Lease that provides for periodic revaluation of the leased property and reestablishment of its rental.

REVEAL Internal side surface of an opening or recess. If cut diagonally, it is called a **Splay.**

REVEAL LINING Finish covering to a reveal.

REVERBERATION Multiple reflection of sound waves in a confined space.

REVERBERATION PERIOD Time required for sound of a specific frequency to become inaudible after the source is silenced.

REVERSE 1. Moving or acting in an opposite direction or manner; **2.** In plastering, a template cut to the reverse shape of a molding to check the accuracy of work.

REVERSE ANNUITY MORTGAGE *See* **Mortgage.**

REVERSE BEND Reeving a wire rope over sheaves and drums so that it bends in opposing directions.

REVERSE BOARD AND BATTEN Siding surface treatment: deep, wide grooves are cut into textured siding surfaces during manufacture to create striking, sharp shadow lines.

REVERSE CIRCULATION Counterflow method of circulating drilling fluid and spoil in a drill hole where drilling fluid is pumped out of the drill stem at the top, circulated through a pit where cuttings are removed, and returned to the annular space around the drill stem, circulation being upward inside the drill stem and downward outside it. *Also called* **Circulation,** and **Direct circulation.**

REVERSE CURVE Ogee, cyma recta or cyma reversa curve; an S-shape.

REVERSED LAYER LAP Lap for nesting sheets of welded fabric where one layer of fabric is inverted so all the transverse bars of the two layers are in one plane.

REVERSED LOADER *See* **Loader.**

REVERSE HEAD Means of stabilizing a borehole by maintaining a head of water in the bore above the level of the surrounding water table.

REVERSE LAY *See* **Lay types.**

REVERSE-PHASE RELAY Device normally

mounted on an elevator control panel that prevents power from being applied to the motor if the building power has reversed phase or has an open phase.

REVERSIBILITY Ability of a grease to return to its normal consistency after temporary exposure to temperature near or above the dropping point of that grease.

REVERSIBLE *See* **Snow plow.**

REVERSIBLE LOCK Lock in which the latchbolt can be rotated, adapting it to doors of either hand.

REVERSIBLE MOTOR Motor whose direction of rotation can be changed by varying the electrical connections or by mechanical means.

REVERSING CLUTCH TRANSMISSION *See* **Clutch,** and **Transmission.**

REVERSING DRUM MIXER Single, nontilting drum with fixed internal blades that rotates on rollers about a horizontal axis for concrete mixing and discharge: dry materials are loaded into the drum and mixed through rotation in one direction while moisture is added. When completely mixed, the contents are discharged by reversing the direction of rotation.

REVERSING SHAFT Shaft whose direction of rotation can be reversed.

REVERSION Softening of vulcanized rubber when it is exposed to an elevated temperature; a deterioration in physical properties.

REVET Face with masonry.

REVETMENT 1. Wall sloped back steeply from its base; **2.** Masonry or steel facing of a sloped bank; **3.** Layer of solid material (rock, aggregate, concrete block, etc.) placed on the bottom or banks of a river or waterway to minimize erosion.

REVIBRATION One or more applications of vibration to fresh concrete after completion of placing and initial consolidation, but preceding initial setting of the concrete.

REVISIONARY INTEREST Interest or right of a person in real property upon termination of the preceding estate.

REVOCATION Recall of a conferred power.

REVOLUTIONS PER MILE Number of tire revolutions in a mile (varies with speed, load, and inflation).

REVOLUTIONS PER MINUTE Non-SI unit of rotational speed. It is equal to the number of complete cycles, typically of an engine, pulley, shaft, etc., completed in one minute; the customary term used to describe engine speed. Symbol: rpm.

REVOLUTIONS PER SECOND Unit of rotational speed of the SI system of measurement. Symbol: r/s. *See also the appendix*: **Metric and nonmetric measurement.**

REVOLVING DOOR *See* **Door types.**

REVOLVING FUND Working capital established to facilitate operations of special functions, that is replenished by reimbursement from other funds.

REVOLVING SCREEN Circular trash rack that is turned mechanically or by the force of the water passing through it.

REVOLVING SHOVEL *See* **Shovel.**

REVOLVING SUPERSTRUCTURE Rotating frame of a crane and components located thereon. *Also called* **Upperstructure.**

revrs. *Abbreviation for:* reverse; reversible.

revs *Abbreviation for:* revolutions.

REYNOLD'S NUMBER Numerical ratio of the dynamic forces of mass flow to the shear stress due to velocity. Flow usually changes from laminar to turbulent between Reynold's numbers 2,000 and 4,000. *See also* **Turbulent flow.**

REZONING Act of changing the designation of land on a zoning map, affecting its permitted use.

rf *Abbreviation for:* roof.

RF *Abbreviation for:* radio frequency; radius front; raised face; right front; rounded front.

RFA *Abbreviation for:* request for alteration.

RF&S *Abbreviation for:* rounded front and strike.

rfg *Abbreviation for:* roofing.

RFI *Abbreviation for:* radio frequency interference; request for information.

RFP *Abbreviation for:* request for proposal.

RFQ *Abbreviation for:* request for quotation.

rftr *Abbreviation for:* rafter.

rgh *Abbreviation for:* rough.

Rh *Abbreviation for:* Rockwell hardness.

RH *Abbreviation for:* relative humidity; right hand; round head.

rhbdr. *Abbreviation for:* rhombohedral.

rheo. *Abbreviation for:* rheostat.

RHEOLOGY Science dealing with the flow of materials. *See also* **Flow.**

RHEOPECTIC GREASE Lubricating grease that has the property of increasing in consistency (hardening appreciably) upon being subjected to shear.

RHEOSTAT Device that regulates the flow of electricity by appling variable resistance to the circuit.

rhomb. *Abbreviation for:* rhombic; rhomboid.

rhr *Abbreviation for:* reheater.

RHR *Abbreviation for:* right-hand reverse.

RHSB *Abbreviation for:* right-hand reverse bevel.

RI *Abbreviation for:* reflective insulation; refractive index.

RIB 1. Projecting band separating the cells of a groined vault. *See also* **Arch,** and **Vault; 2.** One of a number of parallel structural members backing sheathing; **3.** Portion of a T-beam that projects below the slab; **4.** In deformed reinforcing bars, the deformations or the longitudinal parting ridge; **5.** Ridge projecting above grade in the floor of a blasted area; **6.** Section of a tunnel between the spring line and back.

rib. *Abbreviation for:* ribbed; ribbon.

RIBBED METAL LATHING Steel sheets slit and

expanded to form the mesh and ribs of lathing; used primarily as reinforcement for concrete floor slabs.

RIBBED PANEL Panel composed of a thin concrete slab reinforced by a system of ribs in one or two directions, usually orthogonal.

RIBBON 1. Narrow board let into studs to support joists; **2.** Narrow strip of wood or other material used in formwork.

RIBBON COURSE Alternate course of tiles, laid to shorter or longer gauge, alternately showing long and short exposed depth.

RIBBON DEVELOPMENT Urban expansion in the form of single-depth commercial development and/or housing along roads radiating from the central core. *Also called* **Corridor pattern,** and **Strip development.**

RIBBON GRAIN Alternating light and dark stripes in quarter-sawn timber.

RIBBON LOADING Method of batching concrete in which the solid ingredients, and sometimes also the water, enter the mixer simultaneously.

RIBBON SAW *See* **Saw.**

rib. gl. *Abbreviation for:* ribbed glass.

RIB HOLE One of a series of blast holes in tunneling or shaft sinking, drilled at the sides of the tunnel or shaft. Rib holes are fired last, after the relief holes.

RIB VAULT *See* **Vault.**

RICH CONCRETE Concrete of high cement content.

RICH MIXTURE 1. Air-fuel mixture containing too much fuel or too little air for perfect combustion; **2.** Concrete mixture containing a high proportion of cement.

RICHTER SCALE Logarithmic scale from 0 (the smallest) to 10, used to measure earth tremors.

RICK Pile of evenly stacked cordwood, staves, bolts, or other short-length wood.

RIDER CAP *See* **Drive cap,** and **Pile cap.**

RIDER TRUCK Powered industrial truck that is designed to be controlled by a riding operator.

RIDGE Horizontal line formed by the junction of two sloping surfaces of a roof.

RIDGE-BACKED TILE Wall or floor tile manufactured with a series of ridges formed into the back to increase the surface area of the tile contacting the adhesive during application.

RIDGE BEAM Horizontal structural member, usually 38 mm (1.5 in.) thick, supporting the upper ends of rafters.

RIDGE BOARD or RIDGEPIECE Horizontal member, usually 19 mm (0.75 in.) thick, to which the upper end of rafters are fixed.

RIDGE CAPPING Weatherproof covering to a ridge.

RIDGE CUT Cut at the top of a hip rafter.

RIDGED SURFACE FINISH Concrete road surface finish produced by a tamping action of a beam spanning the slab while under construction.

RIDGE POLE 1. Member that can be located longitudinally in the center of an open-top truck body to support a tarpaulin in a tent-like manner; **2.** Horizontal pole of a long tent.

RIDGE ROLL Metal roofing trim consisting of a half-cylinder with a flat extension to each side, used for finishing the ridge of a roof.

RIDGE TERRACE Ridge built along a contour line of a slope to form a containment area for rainwater.

RIDGE TILE Tile shaped to fit over the ridge, covering the tops of the common tiles at the apex of a roof.

RIDGE VENT Sheet metal or plastic vent along the ridge of a roof.

RIDGING *See* **Beading.**

RIFFLE BOX Segmented box through which aggregate can pass; used in aggregate grading.

RIFFLER One of a series of small rasps or files with bent or curved tips, thus able to reach concave surfaces. *See also* **Files.**

RIFLE BAR Cylinder with curved splines.

RIFLE NUT Splined nut that slides on a rifle bar.

RIFLING Spiral thread formed on the wall of a drill hole making it difficult to pull out the bit. *See also* **Spiraling.**

RIFT Split, crack, fissure, etc.

RIG 1. Generic term for a machine or item of mobile equipment; **2.** Attachment or front of a revolving shovel. *See also* **Drilling rig; 3.** To install the blocks and lines used in a cable logging system.

rig. *Abbreviation for:* rigid.

RIGGER Mechanic whose function is to brace, guy, and arrange for hoisting materials.

RIGGING 1. Lines, blocks, chokers, and all gear used in cable logging systems; **2.** Performing rigging jobs.

RIGGING CHAIN *See* **Pass chain.**

RIGGING CREW Workers who set chokers.

RIGGING CUT Partial bucking cuts, usually two log lengths apart, made in a tree lying in such a position that a normal bucking cut cannot be made safely. Done to facilitate yarding or skidding. *Also called* **Weakening cut.**

RIGGING SLINGER Supervisor of the choker setters who directs which logs are to be choked.

RIGGING SWITCH *See* **Switch.** *Also called* **System override switch.**

RIGHT ANGLE 1. Angle of 90 degrees; **2.** Meeting of two pieces of wood of similar dimension at right angles to each other. This joint has no structural strength and requires some mechanical means to ensure a permanent attachment.

RIGHT ANGLE STACK AISLE WIDTH Aisle width dimensionally equal to the sum of the load length (measured lengthwise of a counterbalanced industrial truck), plus load moment constant, plus minimum outside turning radius.

RIGHT BANK Bank bounding flowing water that is on the right when looking downstream.

481

RIGHT LAY *See* **Lay types.**

RIGHT OF ACCESS Right of an abutting land owner for entrance to or exit from a public road. *Also called* **Access right.**

RIGHT OF IMMEDIATE POSSESSION Right to immediately enter upon and use property for highway purposes.

RIGHT OF SURVEY ENTRY Right to enter property temporarily to make surveys and investigations for proposed highway improvements.

RIGHT OF SURVIVORSHIP Right of a surviving joint owner to succeed to the interests of a deceased joint owner.

RIGHT-OF-WAY 1. Route that is lawful to use; **2.** Strip of land acquired or used for utility installation and service; **3.** Land or water rights necessary for construction; **4.** General term denoting land, property, or interest therein, usually in a strip, acquired for or devoted to transportation purposes.

RIGHT-OF-WAY APPRAISAL Expert opinion of the market value of property including damages, if any, as of a specified date, resulting from an analysis of facts.

RIGHT-OF-WAY ESTIMATE Approximation of the market value of property including damages, if any, in advance of an appraisal.

RIGHT-OF-WAY LINE Line marking the limit between land secured for public use and adjacent private property.

RIGHT-OF-WAY MAP Plan of a highway improvement showing its relation to adjacent property, the parcels or portions thereof needed for highway purposes, and other pertinent information.

RIGID 1. Stiff, unyielding; **2.** Vehicle where the engine compartment, cab, and cargo compartment are mounted on a continuous chassis.

RIGID DISK Clutch disk that has no vibration adsorption qualities.

RIGID FRAME *See* **Structural steel framing.**

RIGID INSULATION Dense insulation material that is structurally rigid, commonly available in sheets 1200 x 2400 mm (4 x 8 ft). *Also called* **Board insulation.**

RIGIDITY Resistance to bending or twisting.

RIGID PAVEMENT Pavement that will provide high bending resistance and distribute loads to the foundation over a comparatively large area.*See also* **Flexible pavement,** and **Pavement concrete.**

RIGID PIER ACTION Behavior of a pier that is so inelastic in comparison with its surrounding material that the distortion of the pier under load exerts negligible influence on stress distribution in the surrounding soil or rock.

RIGID SLAB Section of portland cement concrete pavement bounded by joints and edges, designed for continuity of tensile stress.

RIGID TORQUE ARM Member used to retain axle alignment, and in some cases, control axle torque. Normally one adjustable and one rigid torque arm are used per axle so the axle may be aligned.

RIG UP 1. To fit a machine with required rigging; **2.** To string lines and hang blocks once a yarder is in position.

RIG-UP CREW Workers who erect and rig the home spar. *Also called* the **Bull gang.**

rii *Abbreviation for:* rainfall-induced infiltration.

RIM Wheel component that supports a tire.

RIM BASE Unit of a multipiece wheel rim assembly that supports the tire bead on one side and provides a locking mechanism for the side ring or lock ring.

RIM BEAD SEAT Surface of a wheel rim that contacts the tire bead seat.

RIM DIAMETER Diameter of a wheel rim corresponding to the tire bead heel.

RIM FLANGE That part of a wheel rim that supports the tire bead heel and resists lateral internal pressure.

RIM LOCK Lock that is fastened to the surface of a door.

RIMPULL A wheeled machine's power available measured at the tire.

RIM SPACER Device used with spoke wheel assemblies to ensure proper dual tire spacing with demountable rims.

RIM SPEED Speed of a saw blade at the extreme periphery of the rim.

RIM TAPER The slanting of a tire rim bead seat area.

RING *See* **Arch.**

RING-AND-BALL SOFTENING POINT Method of measuring the softening point of asphalt material.

RING BEAM Continuous beam around the perimeter of a domed roof structure that carries the vertical loads to the supporting columns or wall and that balances the lateral thrust by tensile force in the beam.

RING COMPRESSION Principal stress in a confined circular ring subjected to pressure.

RING COURSE Course closest to the extrados in an arch, itself composed of several courses of brick, masonry, etc.

RING DEBARKER *See* **Harvesting machines (single function).**

RINGELMANN CHART Set of five charts that emulate smoke densities in percentages of black, used to assess the opacity of smoke issuing from stacks, etc.

RING GEAR *See* **Gear.**

RING GEAR RUNOUT Measurement that ensures the ring gear is correctly mounted to the main differential housing.

RING NAIL *See* **Nail.**

RING ROAD Highway that wholly or partially encircles a major city, having principal connections with all arterial roads and highways leading to the city.

RING ROT Circular rot in a log; any rot localized mainly in the springwood of the growth rings, giving a concentric pattern of decayed wood in the cross section of a tree or log.

RING SHAKE Separation of the wood between annual growth rings of a tree.

RING-SHANK NAIL *See* **Nail.**

RING SYSTEM Well pointing system around the periphery of the area in which construction is to take place that creates a dry zone.

RING TEST Test that estimates the cracking tendencies of a sample of cement paste or mortar.

RIP 1. To dislodge or fracture soil and rock by pulling through it a ripper tooth or teeth mounted on a tractor; **2.** To cut a board along the length of the grain.

rip. *Abbreviation for:* ripped.

RIPARIAN OWNER One who owns land bounding upon a body of water with right of access to the water and its enjoyment.

RIPARIAN RIGHTS Rights of a land owner to water on or bordering his property, including the right to prevent upstream diversion or misuse.

RIPARIAN ZONE Timber left standing on lake and river banks to give shade and protection.

RIP HAMMER *See* **Hammer.**

RIPPABILITY Characteristic of dense and rocky soils that can be excavated without blasting after ripping with a rock rake or ripper.

RIPPER 1. Towed machine fitted with teeth for loosening hard soil and soft rock; **2.** Log that must be sawn (ripped) in half lengthwise to obtain an acceptable helicopter lift weight; **3.** Narrow strips of gypsum drywall used for soffits, window returns, etc. **4.** Tooth, or set of teeth, mounted on the rear of a tractor or other machine, used to loosen dirt and break rock (*See also* **Back ripper**). There are several designs, including:

Adjustable parallelogram: Design having features of both the hinge-type and parallelogram designs. It can vary the tip angle beyond vertical for enhanced penetration and can be hydraulically adjusted while ripping to provide optimum ripping angle in most materials.

Adjustable radial: Design that combines features of hinge-type rippers with the advantages of shank adjustment capability, providing a greater range of shank angles and more reach.

Hinge type: Where the linkage carrying the beam and shank pivots about a fixed point at the rear of the tractor. The shank is also fixed to the linkage at a certain angle. As the shank enters the ground and penetrates to maximum depth, the tooth angle is constantly changing. The hinge-type ripper uses a beam capable of holding from one to five shanks that can be adjusted to meet depth and tooth angle requirements.

Parallelogram-type: Design that allows the linkage carrying the beam and shank to maintain an essentially constant tip-ground angle regardless of tooth depth.

RIPPER Stud cut diagonally lengthwise and nailed on top of a roof joist on a flat roof to give the roof a slight pitch.

RIPPER SHANK Tooth-shaped part attached to a shank assembly, itself an attachment to track-type tractors, used to rip and tear the surface by being forced into the ground while the tractor is moving forward. There are two basic designs:

Curved: Shank designed to work in less dense material. It produces less ripping resistance while lifting the ripped material, which is further broken before passing the vertical portion of the shank.

Straight: Shank that provides a lifting action in tight, laminated materials, plus the ripping ability required in blocky or slabby material.

Also called **Shank.**

RIPPER TIP Replaceable tip for the ripping shank. There are several types, including:

Short: A tip for use in extreme impact conditions only.

Intermediate: Tip for use in moderate impact and abrasion conditions.

Long: Tip designed for low impact, highly abrasive conditions where breakage is not a problem.

RIPPING 1. Sawing or cutting with the grain (as with a rip saw); **2.** Fragmentation of rock by a crawler-tractor equipped with ripper shanks and points.

RIPPING CHISEL Hexagonal steel bar, approximately 450 mm (18 in.) long, having one end shaped as a wide chisel face containing a notch plus a teardrop-shaped nail slot, and the other offset at 90°, beveled and notched.

RIPPLE Periodic variation of pressure above and below an operating pressure. It is defined as a percentage of the operating pressure in terms of the maximum peak-to-peak value obtained at the point of rating.

RIPPLE AMPLITUDE Amount of voltage variation in the output waveform of a DC power supply.

RIPPLE FINISH Intentional and uniformly wrinkled painting finish.

RIPRAP Rough stones of various sizes, from about 150 kg (330 lb) up to 10 t (9.07 tons), placed irregularly and compactly to prevent scour by water.

RIP/S *Abbreviation for:* ripsaw.

RIPSAW *See* **Saw.**

RISE 1. Amount of height gained by a single step; **2.** Total vertical height of a staircase; **3.** Vertical distance between the plate and the ridge in a roof; **4.** Height of an arch from springing to crown; **5.** Vertical height from the plate to the ridge of a roof; **6.** Vertical height of the camber of a road above the level at the verge.

RISE AND FALL Survey method of reducing staff readings by calculating the rise or fall from each point to the one following it, noting the readings in a column dedicated for that use in the field book.

RISE-AND-FALL TABLE Table of a bench saw that can be raised or lowered relative to the saw arbor.

RISE AND RUN Vertical and horizontal dimensions that set the angle of a sloped roof.

RISER 1. Vertical part of a step; **2.** Vertical conduit containing electrical wires or an electrical distribution between storys; **3.** Vertical water supply pipe extending from a horizontal water supply pipe to a fixture; **4.** Vertical pipe supplying water or steam for heating an upper story; **5.** Steel or wood section between the chassis frame and body underframe of a vehicle to give proper tire clearance and/or required ground-to-floor height. *See also* **Longitudinals**; **6.** Spacer between stacked units to allow entry of fork lift blades; **7.** Vertical raceway in the hoistway for an elevator; **8.** Tube in a water-tube boiler that conducts heated water upwards.

RISE RATE Ratio of pressure rise to time.

RISER PIPE Vertical pipe rising from one story to another.

RISE TEST Determination of the distance a fire hose, under a specified internal pressure, lifts from the surface on which it rests.

RISING AND LATERAL CONDUCTOR One of the electrical cables in a branch circuit.

RISING BUTT HINGE Door hinge, the mating faces of which form a helix so that the door rises as it is opened. *Also called* **Helical hinge.**

RISING CURRENT SEPARATOR Unit housing a flowing current of water to carry off or wash away organic materials.

RISING DAMP Groundwater that travels upward through a masonry wall by means of natural capillary action.

RISING MAIN Electrical, gas, or water supply service that passes through two or more floors.

RISING STEM VALVE Valve whose stem rises when the valve is opened.

RISK The degree of probability of injury, damage, or loss. *See also* **Buyer's risk,** and **Seller's risk.**

riv. *Abbreviation for:* river; rivet; riveted.

RIVE Split, tear apart.

RIVER GRAVEL Gravel having a rounded shape and found at the site of a current or former river bed.

RIVET Steel fastener made of either an annealed, soft, carbon steel or carbon-manganese steel, and consisting of a rounded buttonhead plus a shank long enough to pass through the members to be joined and form a head at the other end.

RIVET CATCHER Worker who catches in a bucket a hot rivet thrown by the rivet heater and passes it to the holder-up.

RIVETED CONSTRUCTION Structural steel members fastened together by riveting following initial positioning and clamping by pinning or bolting.

RIVETER Worker who forms the head on a hot rivet.

RIVETING HAMMER Pneumatic or hydraulic riveting device.

RIVET SNAP Recessed punch.

RIVING KNIFE Blade projecting up from a saw table, in line with the back of the blade, protecting the back of the blade and preventing the sawn lumber from binding the saw.

rj *Abbreviation for:* road junction.

RL *Abbreviation for:* red lead.

R/L *Abbreviation for:* random lengths.

rld *Abbreviation for:* rolled.

rlg *Abbreviation for:* railing.

rly *Abbreviation for:* relay.

rm *Abbreviation for:* ream; room.

R/M *Abbreviation for:* ready mixed.

rms *Abbreviation for:* root mean square.

rnd *Abbreviation for:* round.

rndm *Abbreviation for:* random.

rntl *Abbreviation for:* rental.

ROAD 1. Designated and prepared way, mainly for vehicles. *See also* **Street,** and **Toll road**; **2.** In forestry operations, the haul road; **3.** Area the width of a choker on both sides of the mainline from the yarder to the back end.

ROADBED 1. Finished surface of a highway, between shoulder lines, prior to paving; **2.** Finished surface of the roadway for a railway upon which the ballast rests.

ROADBED MATERIAL Material below the subgrade in cuts and embankments and in embankment foundations extending to such depth as affects the support of the pavement structure.

ROADBUILDER Excavator designed with a heavier attachment to be used in clearing heavily wooded areas.

ROAD CLEARANCE 1. Removal of brush, other vegetation, debris, and obstructions in preparation for layout and development of a road; **2.** Distance between the lowest part of a vehicle between the wheels and the road surface.

ROAD FORM Movable form that establishes the location and thickness of a pavement slab.

ROAD HEATER Mobile machine that heats a road surface preparatory to resurfacing.

ROAD METAL Crushed stone used to surface roads.

ROAD-MIXED SURFACING Designed combination of material components of a flexible pavement mixed on the roadbed or in a traveling plant.

ROAD-MIX TYPE Open-graded aggregate and liquid bituminous material mixed on the roadway by a motor grader or similar tool.

ROAD OIL *See* **Asphalt.**

ROAD PATTERN In forestry, a characteristic arrangement of spur roads in relation to each other.

ROAD PROTRUSION Any lump or protruding rock left in a road surface after it has been graded.

ROAD RECLAMATION Process that pulverizes and mixes existing pavement structure in-place with a certain amount of underlying base material to form a sound base. *See also* **Reclamation.**

ROADSIDE 1. General term denoting the area adjoining the outer edge of the roadway. Extensive areas between the roadways of divided highways may also be considered roadside; **2.** Left or driver's side of a vehicle when viewed from the rear; opposite side from

Curbside.

ROADSIDE CONTROL Public regulation of the roadside to improve highway safety, expedite the free flow of traffic, safeguard present and future highway investment, conserve abutting property values, or preserve the attractiveness of the landscape.

ROADSIDE DEVELOPMENT Those items necessary to the complete highway that provide for the preservation of landscape materials and features; the rehabilitation and protection against erosion of all areas disturbed by construction through seeding, sodding, mulching, and the placing of other ground covers; such suitable planting and other improvements as may increase the effectiveness and enhance the appearance of the highway.

ROADSIDE ZONING Application of zoning for roadside control.

ROAD-SPEED GOVERNOR Device that allows operators to limit vehicle speed to a preset maximum.

ROAD SURFACE Traveled surface.

ROADWAY 1. Entire construction area for a highway; **2.** That part of the right-of-way of a railroad prepared to receive construction of ditches, shoulders, and roadbed.

ROAD WIDENER Specialized machine used to place asphalt mix, aggregate, or stabilized materials adjacent to a roadway structure in order to increase its width.

ROB 1. Take out supporting pillars or walls of pay rock in an underground mine; **2.** *See* **Cannibalize.**

ROBERTSON-HEAD SCREW *See* **Screw.**

ROCK Naturally occurring mineral cohesively bound by chemical bonds and forming the basic structure of the earth's crust.

ROCK ANCHOR High-tensile bar or cable grouted into a hole drilled into rock.

ROCK ASPHALT *See* **Asphalt.**

ROCK ASPHALT PAVEMENT Pavement constructed of rock asphalt, natural or processed, and treated with asphalt or flux as may be required for construction.

ROCK AUGER Auger-type drilling tool fitted with hard-metal teeth to enable it to penetrate soft or weathered rock and hardpan.

ROCK BODY Dump body with hardwood-plank flooring set inside a double-steel floor. *See also* **Body.**

ROCK BOLT Steel rod cemented into a drill hole to be used as an anchor.

ROCK CLEANUP Cleanup of a rock surface on which concrete is to be placed.

ROCK CUT FACE Building stone cut with a pitching chisel to leave the face in a roughly squared shape.

ROCK DRILL *See* **Drill.**

ROCK DRILLING Use of mechanical equipment to produce holes in rock preparatory to the use of explosives. There are three basic techniques:

Drifter: Uses pneumatically- or hydraulically-powered drifters located on a drill guide outside

the hole. The drifter piston discharges energy to the rock through the striker bar, couplings, extension rods and the bit. This energy crushes the rock into small chips (drill cuttings). The drifter rotation motor turns the bit so it encounters fresh rock; compressed air flushes drill cuttings from the hole.

Down-the-hole: The drill is located on the end of the drill string. The hammer piston directly contacts the bit. A hydraulic or air rotary head or kelly-bar drive outside the hole provides rotation. Drilling pipes conduct compressed air to the hammer; the same air flushes cuttings out of the hole.

Rotary: Generally used for large-diameter holes, the rotary drill uses high pull-down pressure on the bit and top drive rotation outside the hole. A hydraulic rotary head or kelly-bar drive provides rotation. Feed pressure and rotation crush and grind the rock; compressed air, mud, or foam carry cuttings out of the hole.

ROCKER End-dump haul truck specifically designed to haul rock from a blast area.

ROCKER ARM 1. Bell crank with the fulcrum at the bottom; **2.** Device to open an engine's inlet and exhaust valves, actuated by cam followers and push rods.

ROCKER BEARING Support of a bridge or truss that is free to rotate but not move horizontally, unless itself carried on rollers.

ROCKER LUG COUPLING Hose coupling in which lugs, used for tightening or loosening, are semicircular in shape and designed to pass over obstructions.

ROCKER SHOVEL High-speed mechanical shovel used in tunneling.

ROCK EXCAVATION *See* **Excavation.**

ROCK FACED Natural face of a rock, or a dressing resembling it.

ROCK-FILL DAM Dam constructed with a central core of loose rock, faced with rolled earth, concrete, or other impervious surfacing.

ROCK FLOUR Rock crushed to a silt.

ROCKING Pushing a resistant object repeatedly backwards and forwards across its original position to gain momentum.

ROCKING FRAME Oscillating frame on which molds are placed during concrete or plaster placement to enhance settlement and compaction.

ROCKING GRATE *See* **Grate.**

ROCK LADDER Steel framework under the discharge of aggregate stockpiling conveyor belts that restricts the free fall of aggregate to less than 1 m (3 ft).

ROCK NECKLACE Large rocks drilled and threaded on a steel cable so that they will better withstand the erosive force of water when placed in a water diversion opening.

ROCK POCKET 1. Porous, mortar-deficient portion of hardened concrete consisting primarily of coarse aggregate and open voids, caused by leakage of mortar from the form, separation during placement, or insufficient consolidation. *See also* **Blemish,** and **Honeycomb; 2.** Portion of a drilled shaft that penetrates into a rock formation beneath less dense overburden.

ROCK POINT Pile point protection specifically designed to develop a toe hold on rock, or to improve penetration into boulder-infested soils.

ROCK RAKE Heavy-duty blade equipped with teeth along its cutting edge, an attachment for a crawler-tractor.

ROCK-RIPPING BUCKET *See* **Bucket.**

ROCK SOCKET That portion of a pile bore that penetrates into a hard formation beneath less competent overburden.

ROCK THROW Distance broken rock will be thrown when a round is detonated.

ROCKWELL HARDNESS Surface hardness test method in which a diamond point is forced into a material to a specified depth. The pressure required to create the depth measurement is the hardness reference.

ROCK WOOL Type of fibrous insulation made from rock and molten slag.

ROCOCO Lavish ornamentation and decoration.

ROD 1. Small diameter, solid, rigid, metal section; **2.** Sharp-edged cutting screed used to trim shotcrete to forms or ground wires. *Also called* **Cutting screed**; **3.** Surveying instrument graduated in feet and tenths of a foot, in meters and/or centimeters and millimeters, used with various optical instruments to determine differences in elevation between two points; **4.** Slender length of wood marked with regular spacings specific to masonry courses, stair spacing, etc.

RODABILITY Susceptibility of fresh concrete or mortar to consolidation by means of a tamping rod.

ROD ASSEMBLY Location where the gland, wipers, O-rings, and packing are found in a hydraulic cylinder rod.

ROD CHANGE TIME Time it takes to add rods after drill penetration stops, or the time to detach rods after reaching the required hole depth.

RODDED POINTING *See* **Pointing.**

RODDING 1. Up-and-down action with a tamping rod to compact concrete and to cause air pockets to be released; **2.** Use of a flexible section of metal rod capped with one of several devices to clear a blocked drain; **3.** Leveling plaster work with a floating rule.

RODDING DOLLY Rig used to check that preformed ducts in prestressed concrete members are clear and of the correct size.

RODDING EYE Access in a pipe fitting for cleaning purposes. *See also* **Access eye,** and **Cleanout.**

ROD JOURNAL Smooth ground surface on a crankshaft to which a connecting rod and bearing are connected.

RODMAN Person who holds the leveling rod and assists the instrument man.

ROD MILL Horizontal, cylindrical, rotating mill charged with steel rods for grinding. *See also* **Ball mill.**

ROD PUSHING Technique for conduit installation

using a steerable piercing head that is pushed through the ground to form a bore.

ROD SAW *See* **Saw.**

ROD SOUNDING Simple method of determining subsurfce soil and rock conditions by driving a metal rod into the ground: the material encountered is assessed by the ease of penetration and sound produced by the blows.

ROD STOCK Round steel rod.

ROLL 1. Compaction wheel of a roller. *See also* **Compression roll, Drum,** and **Guide roll**; **2.** Placing a choker on a log is such a way as to cause the log to roll in a desired manner when the line is tightened; **3.** Wooden batten over which sheet roofing is lapped and folded.

roll. *Abbreviation for:* rolled; roller.

ROLL-AND-FILLET MOLDING Molding comprising a convex face with a square fillet projecting from its center.

ROLL-BACK *See* **Curl.**

ROLL BAR Steel protection over the cab of a tractor or vehicle to prevent injury to the operator.

ROLLCRETE No-slump concrete that can be hauled in dump trucks, spread with a dozer or grader, and compacted with a vibratory roller.

ROLL CRUSHER *See* **Crusher.**

ROLLED A brick laid with an overhanging face.

ROLLED CURB Curb that is tapered to one side to permit free passage of wheeled vehicles between road surface and pavement. *Also called* a **Mountable curb.** *See also* **Curb.**

ROLLED GLASS Rolled flat glass, one surface of which bears a pattern in the form of narrow parallel ribs.

ROLLED STEEL JOIST Structural steel section formed from a hot billet in a rolling mill.

ROLLER 1. Heavy compacting machine, self-propelled or towed, that may be equipped with rubber-tired wheels, rubber-tired wobble wheels, steel rolls, or special types of rolls such as sheepsfoot or grid, etc. Vibratory or oscillating equipment often is added. Rollers for asphalt work are always self-propelled; many types for earth and subbase compacting are towed; **2.** Relatively small-diameter cylinder, or wide-faced sheave, that serves as a support.

ROLLER BEARING Bearing that uses rollers that turn freely as the shaft revolves.

ROLLER-BED-TYPE MOVING WALKWAY *See* **Moving walkway.**

ROLLER BIT *See* **Bit.**

ROLLER CHAIN *See* **Chain.**

ROLLER-COMPACTED CONCRETE 1. Concrete densified by roller compaction; **2.** Concrete that in its unhardened state will support a roller while being compacted.

ROLLER COMPACTION 1. Process of employing heavy mechanical rollers to compact a material or surface; **2.** Process for compacting concrete using a roller, often one equipped for vibration.

ROLLER CRUSHER Machine that crushes material between two opposing steel rollers that rotate slowly on horizontal axes.

ROLLER GATE Hollow cylindrical crest gate used in dam spillways. It operates by being driven up and down geared racks in the side piers or walls. *See also* **Sector gate.**

ROLLER OPERATING ZONE *See* **Rolling zone.**

ROLLER SKID Wheeled support that can be used singly, in pairs, or groups (individually or in linked combinations) to support a load and permit it to be moved, transferring the weight in a specific manner.

ROLLER TAPPET Valve lifter with a roller mounted on the end of the contacting cam.

ROLLER TRAIN Sizes and types of compaction units used to perform breakdown, intermediate and finish rolling of an asphalt mat on a particular project.

ROLL GRINDING MACHINE Machine for grinding cylindrical rolls.

ROLLING **1.** Use of heavy metal or stone rollers on an applied finish (such as a floor covering) to ensure complete and even adhesion to the subbase; **2.** On terrazzo topping, to extract excess matrix.

ROLLING BEAD JOINT *See* **Pointing.**

ROLLING CIRCUMFERENCE Calculated from the revolutions per mile: 63,360 divided by revolutions per mile = rolling circumference in inches.

ROLLING CURRENT Current drawn by an engine starter motor while it is cranking the engine.

ROLLING DOOR Door supported on, or by, wheels running in tracks, and moving in the plane of the wall in which it is installed.

ROLLING LIFT BRIDGE Lift bridge with a shore-end section resting on a flat or rolling surface.

ROLLING RADIUS Tire dimension from the center of the axle to the ground; normally measured with the tire properly inflated and loaded to its rated capacity.

ROLLING RESISTANCE Sum of forces (ground condition, machine weight, internal friction, tire flexing, tire penetration) that prevent a machine from moving on a level surface, expressed in terms of kg/t (lb/ton) of weight of the machine, and in percentage of weight of the machine.

ROLLING TAIL GATE Roller running the full width of a the rear deck of a vehicle that permits the loading and unloading of long loads or, on a float, allows a winch line to run freely.

ROLLING TOWER Composite structure of frames, braces, platforms, guardrail systems, and accessories supported by casters, and intended to be moved by rolling.

ROLLING ZONE Area behind an asphalt paver where compaction is in process. *Also called* **Roller operating zone.**

ROLL MARK Fine parallel blemishe on the surface of rolled glass in the direction of the draw, caused by a difference in speed between rolls and the glass during manufacture.

ROLL MOLDING *See* **Molding.**

ROLL-OFF BODY Truck body used for heavy-duty, high-volume commercial and industrial pickup. There are two basic types:

Hook-arm: Uses hydraulic boom arms with hook ends to grab and pull containers or other body types onto the chassis.

Tilt frame: Employs subframe assemblies that rise and roll rearward for lowering containers by cable and winch.

ROLL-OFF CONTAINER Steel box with wheels used to collect waste at a site, such as a construction site, that can be rolled onto a truck using a winch and then taken to a disposal facility for discharge. The empty container can then be returned or taken to another site.

ROLL-ON/ROLL-OFF *See* **Waste containers.**

ROLL-OVER Accidental violent tipping of mechanical equipment.

ROLLOVER LOAN *See* **Mortgage.**

ROLL-OVER PROTECTION STRUCTURE Structure on mobile equipment to protect the operator in the event that the machine rolls over. *Also called* **ROPS.**

ROLL RATIO Ratio of the surface speeds of two adjacent mill or calender rolls.

ROLL ROOFING Roofing material, saturated with asphalt, composed of fiber, supplied in rolls containing 10.03 m^2 (108 ft^2) in 750- and 914-mm (30- and 36-in.) widths. Generally furnished in weights of 20 to 40 kg (45 to 90 lb) per roll.

ROLL SQUARE 10.03 m^2 (108 ft^2) of roofing.

ROLLWAY **1.** Chute or path along or down which objects can be rolled; **2.** Spillway of a dam carrying the overflow.

ROLOK *See* **Rowlock.**

Rom. *Abbreviation for:* Roman.

ROMAN ARCH *See* **Arch.**

ROMAN BRICK *See* **Brick.**

ROMAN CEMENT Misnomer of a hydraulic cement made by calcining a natural mixture of calcium carbonate and clay, such as argillaceous limestone, to a temperature below that required to sinter the material but high enough to decarbonate the calcium carbonate, followed by grinding. So-named because its brownish color resembles ancient Roman cements produced by the use of lime-pozzolan mixtures.

ROOF Covering over an open space or over a building.

ROOF BOARD Board laid butting others, nailed to the common rafters as a base for some types of finish roofing materials or to support roofing felt under slates or tiles.

ROOF BOW Transverse member in the roof of a truck body.

ROOF BUCK Scaffold frame made of lumber, built to fit the slope of a roof.

ROOF COVERING Exposed, weatherproof exterior skin of a roof.

ROOF DECK Area designed for residents' communal use on the roof of a building or other structure

ROOF DRAIN Drain built into a flat or nearly flat roof to collect water and convey it to a leader or downspout.

ROOF FAIRING Large, highly efficient, parabolic contoured aerodynamic improvement device mounted on the roof of a cab.

ROOF FRAMING Parts of a roof in position: rafters, ridge, plates, valleys, and hips, etc.

ROOF GUARD *See* **Snow guard.**

ROOF HOOK Anchoring means used to attach the suspension system employed in a two-point swing scaffold or a single-point cage or boatswain's chair to a building or structure.

ROOFING Upper layers of a roof providing a weatherproof surface.

ROOFING BRACKET Bracket used in sloped roof construction, having provisions for fastening to the roof or supported by ropes fastened to the roof or supported by ropes fastened over the ridge and secured to some suitable object. *Also called* **Bearer bracket.**

ROOFING FELT Material used as an underlay, and sheathing, and reinforcement in built-up roofing. It is fabricated from a combination of mineral fibers, shredded wood fibers, or glass fibers saturated with asphalt or coal-tar pitch. *See also* **Built-up roof, Glass-fiber felt, Organic felt,** and **Tarred felt.**

ROOFING NAIL *See* **Nail.**

ROOFING PAPER Asphalted building paper used as an underlay.

ROOFING SATURANT Saturating asphalt used in the manufacture of roll roofing and shingles. *See also* **Saturating asphalt.**

ROOFING SQUARE Amount of roofing materials required to finish 9.29 m² (100 sq²).

ROOF/FLOOR INSULATION (SOLID) Low-density concrete used for insulating purposes only and placed over a structural system.

ROOF JACK Sleeve with a flashing used to flash vent pipes piercing a flat or pitched roof surface.

ROOF JOIST Structural member of a flat roof, designed to support roof loads.

ROOF LADDER *See* **Cat ladder.**

ROOFLIGHT Glazed opening in a flat or pitched roof, often consisting of a factory-made unit of wood and/or aluminum framing and one or more formed sheets of transparent plastic. *See also* **Skylight.**

ROOFLIGHT SHEET Corrugated sheet of transparent or translucent plastic, used as a substitute for solid corrugated sheets to form a rooflight.

ROOF OVERHANG Roof extension beyond the exterior walls of a building.

ROOF PITCH *See* **Roof slope.**

ROOF RAIL Member running longitudinally that connects the roof to the side of a truck body.

ROOF SHEATHING Boards or sheet material fastened to roof joists, rafters, or battens that close in a roof preparatory to attachment of weatherproof coverings.

ROOF SLOPE Angle of a sloped roof to the horizontal, expressed as a percentage, or in inches of rise per 300 mm (12 in.) of run. *Also called* **Roof pitch.** *See also* **Pitch.**

ROOF SPAN Clear distance between plates supporting opposing common rafters.

ROOF TERMINAL Open roof end of a ventilation pipe.

ROOF TILE Individual, interlocking or lapping units in a range of materials and designs used to conduct water from sloped roofs. There are many shapes, including:

> **Bonnet:** A semicylindrical roofing tile used on hips and ridges.
>
> **Concrete:** Roofing tiles made from regular and lightweight concrete in a range of profiles and finishes.
>
> **English:** Corrugated roofing tile with interlocking, flush side joints.
>
> **Hip:** *See* **Bonnet tile,** above.
>
> **Mission:** Roofing tile curved to the arc of a circle, slightly tapered lengthwise, laid alternately with the convex and concave side uppermost.
>
> **Over-and-under:** Rounded roofing tiles laid in each course with the convex and concave surfaces facing upwards in alternations. *See also* **Pantile tile,** below.
>
> **Pan-and-roll:** Roofing tile in two shapes: one a flat rectangular section, the other a half-round section for covering the joints.
>
> **Pantile:** Curved roofing tile, somewhat like a prone letter S, uniform in width throughout its length. *Also called* **Over-and-under tile.**
>
> **Spanish:** Rounded-top roofing tile having an interlocking side joint.

ROOF TRUSS Combination of members in tension and compression designed to span an opening and support an imposed load.

ROOM Enclosed space within a story used for other than circulation.

ROOM AIR CONDITIONER Factory-sealed unit designed to mount in a window opening, through a wall, or as a console, in capacities scaled to air condition individual rooms or a suite of connected spaces.

ROOM DRY BULB Dry-bulb temperature of an air conditioned room or space.

ROOM-TEMPERATURE SETTING ADHESIVE *See* **Adhesive.**

ROOM THERMOSTAT Wall-mounted thermometer-controlled instrument for regulating the temperature within a room or house.

ROOSTER *See* **Top tower.**

ROOSTER SHEAVE *See* **Auxiliary boom nose.**

ROOT 1. Section of a dam that merges into the ground; **2.** Bottom surface joining the sides of adjacent threads; **3.** Part where a tenon widens out at the shoulders; **4.**

Inner fold of a filter pleat.

ROOT BOUND Plant with roots that have become so crowded that plant growth is affected.

ROOT COLLAR Portion of a tree where the roots and stem merge, at or near the ground line.

ROOT CRACK Crack in either the weld or the heat-affected metal at the root of a weld.

ROOT DIAMETER *See* **Tread diameter.**

ROOTER Towed machine equipped with teeth, used for loosening hard soil and soft rock.

ROOT HOOK Heavy hook that catches large roots and tears them out of the ground.

ROOT OF JOINT Point at which metals to be joined by welding are closest together.

ROOT OF THE FLANGE Location on the web of the corner radius termination point or the toe of the flange-to-web weld, measured as the *k*-distance from the far side of the flange.

ROOT PENETRATION Depth to which weld metal extends into the root of a welded joint.

ROOT ROT Disease that destroys tree roots, often killing the tree.

ROOTS BLOWER Blower using a pair of rotors, driven by the engine, that forces air into the intake system of an engine.

ROOT WAD Torn-up mass of dirt and rock caught in the root system of an uprooted tree.

ROP *Abbreviation for:* record of production.

ROPE 1. Thick, strong cord made of intertwisted strands of fiber, wire, or synthetic materials; **2.** Refers to wire rope in many instances, particularly in craning and hoisting.

ROPE BAND Flexible steel band used to seize the end of a wire rope.

ROPE CONSTRUCTION Twisting arrangement of wires in each strand, and number of strands twisted together to form a rope. *See also* **Lay.**

ROPE CREEP Relative movement between ropes and the sheaves that drive them.

ROPED HYDRAULIC DRIVE *See* **Driving machine.**

ROPED HYDRAULIC ELEVATOR *See* **Elevator.**

ROPE DIAMETER Diameter across the outer edges of the strands.

ROPE DRUM That part of a drum hoist consisting of a rotating cylinder with side flanges on which a hoisting rope is spooled in or out.

ROPE FALL Block-and-tackle device used to raise and lower a load.

ROPE GRADE Quality of the materials from which wire rope is manufactured.

ROPE MOLDING Molding that simulates twisted cordage.

ROPE RETAINING GUARD Guard installed close to the face of a sheave to prevent the rope from jumping from its grooves.

ROPE-SUSPENDED LEAD *See* **Lead.**

ROPE TENSION INDICATOR Device that determines the tension on hoist ropes.

ROPE TIE 1. Ring, hook, cleat, or knob attached to a body wall frame member of a truck body for use with lashings either inside or outside; **2.** Liner slat or bar attached to a truck body wall frame member to which a lashings may be attached.

ROPE-TO-SHEAVE PITCH DIAMETER RATIO Ratio of the diameter of a rope to the pitch diameter of the sheave over which it runs.

ROPEY Paint that remains as applied and does not flow out, drying with slight ridges.

ROPS *Abbreviation for:* roll-over protection structures.

ROSE 1. Escutcheon behind a doorknob; **2.** Ornamental centerpiece in a ceiling.

ROSE BIT Countersink for wood.

ROSEBUD Special tip for a gas welding torch, used for heating.

ROSETTE Shape of the end of a wire rope that has had the strands turned in preparatory to babbitting when socketing a rope.

ROSSER Machine that peels bark using knives.

ROSTRUM Raised dais or pulpit.

ROT Decay of lumber.

rot. *Abbreviation for:* rotary; rotate; rotating; rotation.

ROT *Abbreviation for:* rule-of-thumb.

ROTARY *See* **Traffic circle.**

ROTARY BORING Method of boring using a rotational means of excavation.

ROTARY BROOM Hydraulically powered attachment for a range of mechanical equipment, used for street cleanup, snow removal, jobsite cleanup, clearing runways, etc. Can be angled left or right, typically up to 30°.

ROTARY CUT VENEER Veneer cut by revolving a log against a knife running the length of the log.

ROTARY DRILL *See* **Drill,** and **Rock drilling.**

ROTARY ENGINE 1. Engine producing direct rotational motion, with reciprocating parts; **2.** Radial engine in which reciprocating pistons rotate around a stationary crankshaft.

ROTARY EXCAVATOR Machine used to bore large circular tunnels.

ROTARY FLOAT Motor-driven revolving disk that smooths, flattens, and compacts the surface of concrete floors and floor toppings.

ROTARY GEAR PUMP *See* **Pump.**

ROTARY INTERSECTION *See* **Intersection types.**

ROTARY JOINT *See* **Rotating joint.**

ROTARY KILN Long steel cylinder with a refractory lining, supported on rollers so that it can rotate about its own axis, and erected with a slight inclination from the horizontal so that prepared raw materials fed into the higher end move to the lower end, where fuel is blown in by air blast.

ROTARY LINE On a rotary drilling rig, the wire rope used for raising and lowering the drill pipe, as well as for controlling its position.

ROTARY OIL BURNER Centrifugally-driven oil-burning heating unit.

ROTARY PLUG VALVE *See* **Valve.**

ROTARY PUMP *See* **Pump.**

ROTARY SCREEN *See* **Screen.**

ROTARY SWITCH *See* **Switch.**

ROTARY TABLE Part of a rotary drill that turns the kelly and drill string.

ROTARY TILLER Machine equipped with rotating tines that loosen and mix soil and vegetation.

ROTATING-BALL FAUCET Single-handle faucet that regulates and mixes the flows of hot and cold water through a single spout.

ROTATING CYLINDER Cylinder in which relative rotation of the cylinder housing and the piston and piston rod, plunger, or ram, is recommended.

ROTATING JOINT Component that transfers fluid, air, or electricity between a stationary and a rotating member. *Also called* **Rotary joint.**

ROTATION AGE Age at which a forest stand is considered mature and ready for harvesting. *See also* **Biological maturity.**

ROTATIONAL FAILURE Foundation failure caused when soil below different parts of the structure are compressed at different rates and cause the settling foundation to rotate about a horizontal axis.

ROTATIONAL FIRING Blasting a block of rock nearest the face with a first explosion followed by other blasts successively further away from the face, in timed sequence, so as to throw their burden toward the space created by the preceding blast. *Also called* **Buffer blasting,** and **Row shooting.**

ROTATION CAPACITY Incremental angular rotation that a given shape can accept prior to local failure.

ROTATION RECORDER Instrument that measures the minute rotation of a bridge support during loading.

ROTATION-RESISTANT ROPE A wire rope consisting of an inner layer of strand laid in one direction covered by a layer of strand laid in the opposite direction. This has the effect of counteracting torque by reducing the tendency of finished rope to rotate.

ROTATOR Power-actuated attachment that provides rotary motion about an axis that is usually perpendicular to the face of the carriage.

ROTOR 1. Component that works by spinning; the rotating part of a motor, dynamo, etc.; **2.** Propeller of a wind machine. It is moved by the wind and, in turn, turns the generator to produce electricity; **3.** Component of the distrubutor of a spark plus-equipped gas engine that provides continuity between the coil and the spark plug in the proper firing order and at the exact moment of spark advance.

ROTOR WASH Downward draft caused by a heli-copter's main rotor blades.

ROTTENSTONE Finely ground limestone used as an abrasive in the process of rubbing down finish coats.

ROTTEN TREES *See* **Tree classes.**

ROTUNDA Round building or room, especially one capped by a dome.

ROUGE Fine-grained reddish powder, mainly ferric oxide, used for polishing metal and stone.

ROUGH *See* **Bulk appearance.**

ROUGH ARCH *See* **Arch.**

ROUGH BORE HOSE Wire-reinforced hose in which a wire is exposed in the bore.

ROUGH BRACKET Support bracket in a concealed position; an under-stair bracket, for instance.

ROUGHCAST External rendering of rough mortar, usually applied in two coats of cement, sand, and water on to which gravel, crushed stone, or pebbles are thrown before the second coat is dry. *Also called* **Pebbledash.**

ROUGH CAST GLASS Flat glass, one surface of which has a texture produced by rolling.

ROUGH COAT Rendering coat of plaster.

ROUGH CUT Lumber that has not been dressed (surfaced) but which has been sawn, edged, and trimmed to at least show saw marks in the wood on the four longitudinal surfaces of each piece for its overall length.

ROUGHCUT POINTING *See* **Pointing.**

ROUGH FLOOR Rough floor boards on which the finished floor or subfloor is laid, separated by a layer of building paper.

ROUGH FRAME Framing of an enclosure in which a finished fitment (door, window) will subsequently be placed.

ROUGH GRADING First stage of excavation, when the grade in the cut and fill is held to about ±0.03 m (±0.1 ft) prior to finish grading.

ROUGH GRIND 1. Initial operation in which coarse abrasives are used to cut projections from a manufactured surface; **2.** To remove the projecting stone chips in hardened terrazzo down to a level surface.

ROUGH HARDWARE All concealed fasteners used in construction: nails, bolts, hangers, etc.

ROUGH HORSE *See* **Stair carriage.**

ROUGH IN First stage of a plumbing or electrical installation, consisting of the installation of primary lines and circuits that are left exposed and not connected to appliances or fittings. *See also* **Finishing,** and **Roughing in.**

ROUGHING IN First rough work associated with any part of a construction project, e.g., the carcass of a staircase. *See also* **Rough in.**

ROUGH LUMBER Lumber that has been sawn to size but not dressed. It will be equal to or larger than the stated dimensions, but may not be true throughout its length.

ROUGHNESS COEFFICIENT Effect on water flow of the degree of roughness of the inner surface of a conduit.

ROUGHOMETER Device for measuring the roughness of a pavement surface.

ROUGH OPENING Unfinished window or door opening.

ROUGH SAND Area of a sanded panel that was not sanded with the finish sanding heads. The surface will appear and feel rough.

ROUGH SAWN Siding treatment imparting a rough, rustic appearance by saw-scoring the surface of a panel during manufacture.

ROUGH SKETCH *See* **Preliminary design.**

ROUGH TERRAIN EQUIPMENT Equipment (cranes, haulers, trucks, etc.) designed to travel and operate on unimproved or unfinished surfaces.

ROUGH WORK Brickwork that will eventually be covered or concealed.

ROUND 1. Explosive blast, including any delay shots; 2. Group or set of blastholes constituting a complete cut in underground headings, tunnels, etc.

ROUNDED STEP Tread having a bullnose edge.

ROUND-HEADED BUTTRESS DAM Concrete dam consisting of parallel buttresses thickened at their inner faces until they touch, the spillway being a curved slab overlapping the exterior or downstream face of the buttresses.

ROUND HOOK Hook with a smooth inner surface that permits it to slide along a chain. *Also called a* **Choker hook.** *See also* **Cable hook, Grab hook, Hook, Pintle hook, Safety Hook,** and **Swivel hook.**

ROUND-NOSE CHISEL *See* **Cold chisel.**

ROUND-TOPPED ROLL Metal roofing joint formed over a wood roll having vertical sides and a rounded top.

ROUND TREE Live tree of any size that does not contain at least one mechantable 3.65-m (12-ft) saw log, now or prospectively, because of roughness or poor form. Only commercial species are considered.

ROUNDWOOD A length of cut tree, generally having a round cross section, such as logs or bolts.

ROUNDWOOD PRODUCTS Logs, bolts, or other sections cut from trees for industrial or consumer use.

ROUT 1. Cut and mold using a router; 2. To deepen and widen a crack to prepare it for patching or sealing.

rout. *Abbreviation for:* routed; routine.

ROUTER Portable, high-speed power tool used to cut and shape wood and other materials. *See also* **Plane.**

ROVING Continuous strands of glass fiber that can be grouped together and wound on a tube.

ROW HOUSING *See* **Attached dwelling.**

ROWLOCK A brick laid on its face edge so that the normal bedding area is visible in the wall face (frequently spelled **Rolok**).

ROWLOCK COURSE Bricks set on edge.

ROW SHOOTING *See* **Rotational firing.**

RP *Abbreviation for:* radius point; reference point; reinforced plastic.

RPC *Abbreviation for:* regional planning commission.

rpm *Abbreviation for:* revolutions per minute.

rpmn *Abbreviation for:* repairman.

RPQ *Abbreviation for:* request for price quotation.

rpt *Abbreviation for:* repeat; report.

rptdly *Abbreviation for:* repeatedly.

rr *Abbreviation for:* right rear.

RR *Abbreviation for:* railroad.

rrl *Abbreviation for:* reroll.

rs *Abbreviation for:* right side.

RS *Abbreviation for:* radio station; reference stake; rough sawn.

R-SERIES *See* **Preferred numbers.**

RSJ *Abbreviation for:* rolled-steel joist.

rsn *Abbreviation for:* reason.

rstprf *Abbreviation for:* rustproof.

rt *Abbreviation for:* right.

RT *Abbreviation for:* raintight; road tar; room temperature; rough turned.

rt ang. *Abbreviation for:* right angle.

rtd *Abbreviation for:* rated.

rtg *Abbreviation for:* rating.

rtr *Abbreviation for:* reinforced thermosetting resins.

RTR *Abbreviation for:* router.

rub. *Abbreviation for:* rubbed; rubber; rubbing.

RUBBED FINISH Finish obtained by using an abrasive to remove surface irregularities.

RUBBED JOINT Joint between two narrow boards that are glued to make one wide board. The edges of both boards are planed smooth on a jointer, coated with glue, and rubbed together to expel air and excess glue.

RUBBER 1. Material that is capable of recovering from large deformations quickly and forcibly and which can be, or already is, modified to a state in which it is essentially insoluble (but can swell) in a boiling solvent; 2. *See* **Brick.**

RUBBER-BASED CEMENT *See* **Adhesive.**

RUBBER BOND Bonding material, the principal constituent of which is natural or synthetic rubber.

RUBBER-EMULSION PAINT Paint incorporating a vehicle consisting of natural or synthetic rubber dispersed in fine droplets in water.

RUBBER FLOOR Sheet flooring material, available in a variety of widths, primarily used in heavy-traffic commercial and industrial applications.

RUBBER GROMMETS Cylindrically shaped rubber pieces that absorb shock and sound, used to isolate the operator's cab, valve banks, wires passing through metal bulkheads, etc. from machine vibration.

RUBBERIZED ASPHALT Asphalt mix containing new or reclaimed rubber, designed to produce a more resilient pavement.

RUBBER SET *See* **False set.**

RUBBER SPACER Cross- and T-shaped objects used to space tile on floors or walls.

RUBBER-TIRED ROLLER Roller equipped with rubber tires, commonly used for compacting trimmed subgrade or aggregate base. *Also called* **Pneumatic-tired roller.**

RUBBER TROWEL Nonporous, synthetic-rubber-faced float mounted on an aluminum back with a wood or plastic handle used to force material deep into tile joints and to remove excess material.

RUBBER WHEEL Grinding wheel made with rubber bond.

RUBBING BRICK Silicon-carbide brick used to smooth and remove irregularities from surfaces of hardened concrete.

RUBBING STONE Carborundum stone used to smooth the rough edges of tile.

RUBBISH *See* **Solid waste.**

RUBBISH CHUTE Pipe, duct, or trough through which waste materials are conveyed by gravity from above to a storage area preparatory to burning, compaction, or removal.

RUBBLE 1. Field stone or rough stone as it comes from the quarry; **2.** Masonry of rough, undressed stones. *See also* **Ranged rubble,** and **Scabbled rubble**; **3.** Demolition wastes; broken pieces of masonry and concrete, asphalt roofing, etc. *See also* **Solid waste.**

RUBBLE ASHLAR Rubble-filled, ashlar-faced wall.

RUBBLE CONCRETE 1. Concrete similar to cylopean concrete except that small stones (such as one person can handle) are used. *See also* **Cylopean concrete; 2.** Concrete made with rubble from demolished structures.

RUBBLE COURSED Masonry composed of roughly shaped stones laid approximately level and well bonded.

RUBBLE MASONRY Walling or facing of rough, unhewn building stones or flints, generally not laid in courses. Coursed rubble is where the same material is laid to courses.

RUB RAIL Member running longitudinally providing a protective surface on the side of a truck body. *See also* **Running board.**

RUB TEST Test that determines how firmly granules are embedded in the surface coating of roofing materials.

RUB TREE Tree used as a fender or pivot to protect the remaining stand during yarding.

RULE 1. Straightedge, plain or marked for measuring; **2.** Straightedge for working plaster to a plane surface.

RULE-OF-THUMB Statement of formula that, while not accurate, is sufficiently precise for an approximation.

RUMBLE STRIP Means on a road surface (uneven paint lines, button markers, etc.) to produce a sensory (audible, tactile) warning to motorists.

RUMBLE SURFACE Rough-textured surface constructed for the purpose of causing the tires of a motor vehicle driven over it to vibrate audibly as a warning to drivers.

RUN 1. Width of a step; **2.** Horizontal distance covered by a flight of stairs from the face of the first or lower riser to the face of the last or upper riser; **3.** That part of a pipe or fitting that continues in the same straight line as the direction of flow; **4.** Horizontal distance between the outer face of a wall and the ridge of the roof; **5.** Cut made in a log by a bucker; **6.** Passing plaster or lime putty through a sieve; **7.** Vertical streak caused by excess paint; **8.** Subdivision of a test, the operational phase.

run. *Abbreviation for:* runner, running.

RUNBY 1. Distance between an elevator car buffer striker plate and the striking surface of the car buffer when the car floor is level with the bottom terminal landing; **2.** Distance between an elevator counterweight buffer striker plate and the striking surface of the counterweight buffer when the car floor is level with the top terminal landing; **3.** Distance an elevator car can run above its top terminal landing before the plunger strikes its mechanical stop.

RUNG Step or horizontal rod of a ladder.

RUN LEVELS To survey an area to determine relative elevations.

RUNNER 1. Horizontal scaffold member that extends from post to post and that supports putlogs or bearers forming a tie between the posts; **2.** Metal or wood members at floor or ceiling that receive framing members; **3.** Rotating part of a turbine; **4.** Channel in which metal studs are anchored; **5.** Lengthwise horizontal bracing or bearing members, or both; **6.** Sheet timber pile ahead of a digging at the edge of an excavation.

RUNNER LINE Single hoist line, usually extending from the jib. *Also called* **Whipline.**

RUNNING 1. Operating a machine or item of equipment; **2.** Forming a plaster molding in place with a horsed mold.

RUNNING BOARD Horizontal extension to a dump body that extends outboard at floor level from the body sides. *Also called* **Rub rail.**

RUNNING BOND *See* **Bond.**

RUNNING GROUND Water-bearing, or very dry sand that will flow down even a very slight slope. *See also* **Running sand.**

RUNNING HEAD Forest fire spreading rapidly with a well-defined head or front.

RUNNING LINE 1. Rope that moves over sheaves or drums; **2.** Moving wire rope in logging operations.

RUNNING LINE TENSIOMETER Device normally consisting of an assembly of deflector and idler sheaves and a load cell that measures the load (tension) in a crane's running load line. Other special sheave configurations, including fixed or movable sheaves, are also used to measure running line load. *Also called* **Line rider, Load line tensiometer,** and **Tensiometer.**

RUNNING OFF Application of finish plaster coat to a molding.

RUNNING SAND Sand below the natural ground water level that flows into excavations. *See also* **Running ground.**

RUNNING SKYLINE System of two or more suspended moving lines, generally referred to as main lines and haulback lines. Will provide lift and travel to the load carrier when tension is properly applied.

RUNNING SPEED *See* **Speed.**

RUNNING TIME Time a vehicle is in motion.

RUNNING TRAP Sanitary fitting that provides a water seal where both the inlet and outlet are horizontal. *See also* **House trap.**

RUNOFF Excess surface water that flows over a site instead of percolating through the soil after precipitation.

RUN-OF-PIT (QUARRY OR MINE) Material as it comes directly from a pit, quarry, or mine before any processing.

RUN-ON SLAB Reinforced concrete road slab acting as a bridge approach slab, one end of which is supported by the bridge abutment, the other on the road subgrade.

RUNOUT Deviation from flatness of a circular saw near the periphery when rotated.

RUN PERIOD Point at which the timer controlling a boiler releases the burner to modulation.

RUN RELAY CIRCUIT Portion of an elevator wiring diagram containing the hoist motor direction relays.

RUNWAY 1. Platform extending from the inside floor level to the outside grade level of a structure; **2.** Platform extending a portion of a stage into the auditorium; **3.** Decking over the area of concrete placement, usually of movable panels and supports, on which buggies containing concrete travel to points of placement.

RUPTURE Tear or split in a filter medium.

RUPTURE MEMBER Safety device that automatically fails at a predetermined pressure.

RURAL FIRE PROTECTION Fire protection and fire-fighting problems that are outside of areas under municipal fire prevention and building regulations and that are usually remote from public water supplies.

rus. *Abbreviation for:* rustic.

Russ. *Abbreviation for:* Russian.

RUSSIAN COUPLING Incomplete bucking cut resulting from an unsafe bucking situation where the faller only partially cuts through the tree.

RUSTICATION 1. Groove in a concrete surface; **2.** Masonry cut in massive blocks, separated from each by deep joints. Various dressings are applied to the exterior faces of the blocks, including:

Cyclopean (or rock-faced): Natural, or carved rough-hewn shapes.

Diamond-pointed: Each exposed face is cut in the form of a low pyramid.

Frosted: Having the margins reduced to a plane parallel to the plane of the wall, and the intermediate area left with an irregular surface.

Smooth or chamfered: Showing a flat face with edges chamfered at an angle of 135°.

Vermiculated: Carved with shallow curling channels like worm tracks.

RUSTICATION STRIP Strip of wood or other material attached to a form surface to produce a groove or rustication in the concrete.

RUSTIC FINISH Type of terrazzo topping in which the matrix is recessed by washing prior to setting so as to expose the chips without destroying the bond between chip and matrix. *Also called* **Washed finish.**

RUSTIC SIDING *See* **Waney-edge siding.**

RUSTIC WOODWORK Garden panels, screens, furniture, etc., made of unpeeled logs and saplings.

RUST INHIBITOR Additive that helps provide protection against rusting by neutralizing harmful acids or by forming a water-resistant layer on the metal surface.

RUT Channel that develops in wheel tracks on an asphalt pavement.

RV *Abbreviation for:* relief valve.

R VALUE 1. Resistance to heat flow: the higher the R number the more efficient the insulation; **2.** Stability of soils and paving materials as determined by a **Stabilometer.** It represents the resistance of a material to plastic deformation, thus reflecting its load-carrying capacity.

r/w *Abbreviation for:* right-of-way.

R/W *Abbreviation for:* random widths.

R/W&L *Abbreviation for:* random widths and lengths.

rwd *Abbreviation for:* rewind.

rwg *Abbreviation for:* rain water goods.

rwy *Abbreviation for:* railway.

RYEGRASS *See* **Grass.**

ryn *Abbreviation for:* rayon.

S

s *Abbreviation for:* second; sink.

S *Abbreviation for:* seamless; south.

S/A *Abbreviation for:* shipped assembled.

S= *Abbreviation for:* slope.

SA *Abbreviation for:* scale accuracy; single acting; stress annealed; supply air.

SABER SAW *See* **Saw.**

SACK JOINT *See* **Pointing.**

SACK RUB Finish for formed concrete surfaces designed to produce an even texture and fill all pits and air holes, produced by dampening the surface and rubbing mortar over it and then, before the surface dries, rubbing a mixture of dry sand and cement over it with either a wad of burlap or a sponge-rubber float to remove surplus mortar and fill voids. *See also* **Bug hole.**

SACRIFICIAL ANODE Anode, connected electrically to a cathode, used for cathodic protection.

sad. *Abbreviation for:* saddle.

SADDLE 1. Flanged channel attached to the top of a square pile or timber as a bearing plate for the pile cap, and to which one or more angles may be attached for retaining the pile cap; **2.** Frame that locks the hammer in the leads of a pile hammer; **3.** Support that the snorkel, used in cable logging, rests on, located approximately 3 m (10 ft) from the snorkel pocket; **4.** The liveline support section of a wire rope clip; **5.** Low point between two domes or knolls along a ridge line; **6.** Pipe fitting; **7.** Short horizontal member on top of a post, in line with a girder whose weight it is designed to help distribute; **8.** Bends in a pipe allowing it to pass over other pipes in the same plane or similar obstacles; **9.** Ridged shape behind a chimney that supports a flashing, between it and a sloped roof.

SADDLEBACK ROOF Roof with a slope on both sides of its ridge and two gables.

SADDLE BAR In stained glass windows, a small iron or other metal bar to which the leadwork is tied.

SADDLE BEAD Glazing bead for mounting glass to the sides of curved glazing bars.

SADDLE BLOCK Boom swivel block in a dipper shovel through which the stick slides when crowded or retracted.

SADDLE FITTING Plumbing fitting used to install a branch from an existing run of pipe.

SADDLE JIB Horizontal live-load-supporting members of a hammerhead-type tower crane having the load falls supported from a trolly that traverses the jib. *Also called* **Load jib.**

SADDLE JOINT Saddle-shaped joint between masonry blocks in a cornice that throws water away from the joint. *Also called* **Water joint.**

sadl. *Abbreviation for:* saddle.

S.A.E. NET HORSEPOWER *See* **Horsepower.**

S.A.E. PORT Straight-thread port used to attach tube and hose fittings. It employs an 'O' ring compressed in a wedge-shaped cavity.

saf. *Abbreviation for:* safety.

SAFE BEARING CAPACITY Load per unit area, inclusive of a factor for safety, that the soil can carry. *Also called* **Admissible load.**

SAFE CARRYING CAPACITY Weight or load that a part or construction will support without failure.

SAFE EDGE 1. Edge or face of a file without teeth; **2.** Mechanical elevator door protective and automatic reopening device used with electric power-door operators. If the car door approaches or meets an obstruction when closing, the safe edge automatically stops the closing action and reopens the door. *Also called* **Safety edge**; **3.** Turned-in edge of sheet metal.

SAFE FRAME HORIZONTAL MEMBER LOAD Load that can safely be directly imposed on a horizontal member of a scaffold system.

SAFE LEG LOAD Load that can safely be directly imposed on the frame leg of a scaffold. *See also* **Allowable load.**

SAFE LIMIT Amount of vibration that a structure can safely withstand.

SAFE LOAD Load less than that which would cause failure, after allowance for the factor of safety.

SAFETY Large clamp that anchors an elevator car to to the building to keep it from falling.

SAFETY ARCH *See* **Arch.**

SAFETY BELT 1. Single belt or harness-type system with means for securing it about the waist or body and for attaching to a lanyard or lifeline; **2.** Flexible means to permit an operator or helper engaged in 'order picking' or building maintenance freedom to perform those tasks while offering restraint from a free drop to the ground in the event of a fall.

SAFETY BULKHEAD Closure at the end of a hydraulic cylinder, above the head, having an orifice for regulating the loss of fluid in the event of cylinder head failure.

SAFETY CAN Approved closed fuel container, maximum 18.5-L (5-gal) capacity, having a flash-arresting screen, spring-closed lid, and spout cover, designed to safely relieve internal pressure when subjected to fire exposure.

SAFETY CHAIN Chain assembly used to connect the towing and towed vehicles as a secondary coupling system.

SAFETY CIRCUIT *See* **Basic safety circuit.**

SAFETY CLUTCH *See* **Clutch.**

SAFETY DEVICE Any device for the protection of operators, equipment, and machines in case of accident.

SAFETY EDGE *See* **Safe edge.**

SAFETY FACTOR Ratio betwen breakage resistance and load.

SAFETY FENCING Temporary wood, metal, or plastic fencing intended to prevent unauthorized entrance or proximity to a hazardous or restricted area for a limited period.

SAFETY FLANGE Any of several special types of flanges designed to hold together the broken parts of a grinding wheel in case of breakage, thus protecting workers. *Also called* **Protection flange.**

SAFETY FLOORING Flooring materials that, through their surface design or type of materials used, or both, incorporate slip-resistance or sanitation features, or both. Other features that may be built into such flooring include underfoot comfort, impact and cut resistance, etc.

SAFETY FUSE Flexible cord containing an internal burning medium by which fire is conveyed at a continuous and uniform rate for the purpose of firing blasting caps.

SAFETY GLASS Two or more sheets of glass laminated with transparent plastic under heat and pressure. If fractured, it produces crystals of glass rather than shards.

SAFETY GUY Line rigged under the bull block to take it to the ground if the holding straps break.

SAFETY HARNESS Harness worn by a user to restrain a fall.

SAFETY HEAD Cylinder of a compressor, held in place by a spring that will not compress under normal operation, but that will be moved by solid or liquid matter or abnormal gas pressure between it and the piston.

SAFETY HOOK Round hook, fitted with a hinged arm that normally holds the opening closed by means of a spring and which allows a line to readily enter the hook, but which must be manipulated to remove it. *See also* **Cable hook, Grab hook, Hook, Pintle hook, Round hook,** and **Swivel hook.**

SAFETY ISLAND Area used for escape in the event a forest fire line is outflanked or in case a spot fire causes fuels outside the control line to render the line unsafe.

SAFETY LINTEL Load-bearing lintel that protects a second, more ornamental unit.

SAFETY OFFICER Officer responsible to the plans chief for the safety and welfare of fire-fighting personnel.

SAFETY-OPERATED SWITCH *See* **Switch.**

SAFETY PLANK Bottom member of a sling for a traction elevator that contains the safety.

SAFETY PLATFORM *See* **Work platform.**

SAFETY RELEASING CARRIER Clamping device, usually attached to an elevator car crosshead, that protects accidental application of the safeties.

SAFETY REST AREA Roadside area with parking facilities separated from the roadway, provided for motorists to stop and rest for short periods.

SAFETY SCREEN Air- and water-tight diaphragm placed across the upper part of a compressed-air tunnel between the face and bulkhead, in order to prevent flooding the crown of the tunnel between the safety screen and the bulkhead, thus providing a safe means of refuge and exit from a flooding or flooded tunnel.

SAFETY SHUTOFF VALVE *See* **Valve.**

SAFETY STEEL SHIELD Strands of flexible steel, sealed in rubber and positioned in a double layer in the tread area of a tire.

SAFETY STRAP Short piece of wire rope secured to a block or other rigging to prevent the block or rigging from falling into the work area due to a connection failure.

SAFETY SWEDE Lever used to tighten binders on loaded logging trucks.

SAFETY VALVE *See* **Valve.**

SAFETY WORK SURFACE Surface intended to reduce the possibility of foot slippage.

SAFE YIELD Sustained pumping rate from an aquifer or water-bearing formation in which the yield does not exceed the aquifer storage and recharge rate over a defined period.

SAG 1. Depression in a horizontal line; **2.** Amount that a bricklayer's line, over a long distance, will fall slightly below the level because of its own weight, no matter how tightly it is stretched; **3.** Applied wall surface that has developed a slide before stiffening or setting; **4.** Flow of an uncured sealant within a joint resulting in loss of the sealant's original shape.

SAG BAR Rigid bar to prevent sagging of a less rigid member.

SAGGING *See* **Slough.**

SAG ROD Tension member used to limit deflection of a girt or purlin.

SAILER Masonry unit laid on end with the normal bed surface showing.

SAILING COURSE String course.

SALAMANDER Type of portable metal stack heater that burns fuel oil and which radiates heat from the chimney stack pipe.

SALES AGREEMENT *See* **Agreement of sale.**

SALES AND USE TAX BOND Bond that guarantees payment of sales and use taxes where required.

SALES-TYPE LEASE *See* **Lease.**

SALIENT Projecting mass.

SALINOMETER Hydrometer calibrated for salt solutions.

SALMON BRICK *See* **Brick.** *Also called* **Chuff brick,** and **Place brick.**

SALON Reception or exhibition room.

SALT-AND-PEPPER BLEND Mineral granules of

two or more colors blended in a mix prior to application during manufacture of composite roofing or shingles.

SALT GLAZE Gloss finish obtained by thermomechanical reaction between silicates of clay and vapors of salt or chemicals.

SALT HAY Coarse grass growing in salt marsh areas, used for rough caulking of breast boarding and around cofferdams.

SALT STABILIZATION *See* **Soil stabilization.**

salv. *Abbreviation for:* salvage.

SALVAGE Quantity of materials, sometimes of mixed composition, no longer useful in its present condition or at its present location, but capable of being recycled, reused, or used in other applications. *See also* **Salvage scrap.**

SALVAGE AND RECLAMATION Refuse disposal process in which discarded material is separated mechanically or by hand into various categories.

SALVAGE LOGGING 1. Forest operations cleanup operation, generally with a small crew and light equipment, that collects merchantable material too small to be handled economically with big equipment; **2.** Salvaging timber damaged by wind, insect, fire, ice, or other natural causes.

SALVAGE SCRAP Materials, products, or equipment beyond repair that have to be sold or disposed of as scrap. *See also* **Salvage.**

samp. *Abbreviation for:* sampling.

SAMPLE 1. Either a group of units, or a portion of material taken respectively from a larger collection of units or a larger quantity of material that serves to provide information that can be used as a basis for action on the larger collection or quantity or on the production process; **2.** Sample of observations.

SAMPLE DIVISION Process of extracting a smaller sample from a sample so that the representative properties of the larger sample are retained.

SAMPLE PREPARATION Process that includes drying, size reduction, division, and mixing of a laboratory sample for the purpose of obtaining an unbiased analysis sample.

SAMPLE REDUCTION Process whereby sample particle size is reduced without change in sample weight.

SAMPLE SPLITTER Device used to reduce the size of an aggregate sample to obtain a smaller sample that is representative of the original.

SAMPLING PLAN 1. Procedure that specifies the number of units of product from a lot that is to be inspected in order to establish acceptability of the lot; **2.** Pre-arranged program stipulating location and procedures for securing samples of a material for testing purposes.

SAMPLING SPOON Split shell, 50 mm (2 in.) OD, 35 mm (1.375 in.) ID, 600 mm (24 in.) long, for taking earth samples.

SAMPLING TRAIN Series of devices into which a known volume of exhaust or stack gas is drawn for collection and analysis.

san. *Abbreviation for:* sanitary.

SanC *Abbreviation for:* sanitary code.

SAND *See* **Soil types.**

SAND ASPHALT *See* **Asphalt.**

SANDBAG 1. Hessian or plastic sack averaging 300 x 700 mm (12 x 28 in.), holding between 7 and 13.5 kg (15 and 30 lb) of sand; fabric bags are closed with a draw string, plastic bags by stapeling. Both types are used for a wide range of tasks, principally to form a low dike or water barrier to exclude floodwater; **2.** Tubular leather container holding the amount of dry sand appropriate to counterbalance scenery in a theater; **3.** Leather bag filled with sand on which the hollowing of sheet metal is done.

SAND BLASTING Use of sand or other sharp-faceted abrasive grit in a compressed air stream to clean (by abrading dirt and scale), mark, or score (typically, to etch a decorative pattern into a glass surface), or to cut (concrete, stone, etc.).

SAND BOLSTER Part of a trailer landing gear that rests on the ground, across and between the two upright portions.

SAND BOX Tight box filled with clean, dry sand on which rests a tight-fitting timber plunger that supports the bottom of posts used in centering. Removal of a plug from a hole near the bottom of the box permits the sand to run out when it is necessary to lower the centering. *Also called* a **Sand jack.**

S&CM *Abbreviation for:* surfaced one or two sides and center matched.

SAND/COARSE AGGREGATE RATIO Ratio of fine to coarse aggregate in a batch, by mass or volume.

SAND CONE Test for determining the compaction level of soil.

SAND DRAIN Vertical sand columns installed to speed drainage and rapid consolidation of marshy land. *See also* **Pile, sand,** and **Sand or earth wick.**

S&E *Abbreviation for:* surfaced one side and edge.

SANDED BITUMEN FELT Bitumen impregnated roofing felt, the surface of which is embedded with sand-like granules.

SANDED GROUT Grout in which fine aggregate is incorporated into the mixture.

SANDED PANEL Interior or exterior plywood panel factory-sanded for applications where smoothness and appearance are important.

SANDED/SIZED Operation of reducing the thickness of a panel.

SAND EQUIVALENT Measure of the relative proportions of detrimental fine dust or clay-like material, or both, in soils or fine aggregate.

SANDER Portable power tool that comes in a range of styles including:

> **Belt:** Sander designed to accept a belt finished with a sanding grit on its exterior surface that rotates on rollers.

> **Disk:** Rotating (approximately 100 mm (4 in.) diameter) disk, either an integral part of the

equipment or an attachment to a hand drill, to which disks of abrasive-coated material can be attached.

Drum: Cylinder to which an abrasive-coated sleeve can be attached, equipped with an integral spindle or able to be attached to a madrel, for attachment to a power device.

Orbital: Power-actuated plate to which a sheet of abrasive-coated material can be attached, and that oscillates in a forward-and-backward plus side-to-side motion.

Vibratory: Power-actuated plate to which a sheet of abrasive-coated material can be attached, and that vibrates horizontally.

SANDER HESITATION Sander head marks that appear across a particleboard panel width. The marks are low, concave indentations within the radius of the sander head, caused when the panel stopped under a sander head. At times there will be primary sanding marks on the other side.

SANDER SKIP Area of a sanded particleboard panel that was not sanded when surrounding areas were sanded. These areas are usually low indentations in the panel.

SAND FILTER Large-scale equipment for the filtration of potable water consisting of layers of increasingly fine aggregate ranging from coarse stone at the bottom to fine quartz grains at the top. Raw water is fed to the bottom of the tank, flows through the graded media, and decants over weirs at the top. When the filter bed becomes sufficiently clogged with trapped material the flow is reversed with the wash water fed to a drain.

SAND FINISH Final coat of plaster, usually lime and sand, floated smooth.

S&G *Abbreviation for:* studs and girts.

SAND-GRAIN METER Hydrographic instrument used to measure the quantity of sand in flowing water.

SAND GROUT Cement grout incorporating fine sand.

S&H *Abbreviation for:* staple and hasp.

SAND HOG Worker who works in compressed air.

SANDING 1. Use of an abrasive, usually an abrasive grit glued to a paper or other flexible base, to smooth a surface; 2. Application of sand to a freshly oiled road surface.

SANDING POLE Abrasive paper holder attached to a long pole by a swivel, used to finish drywall on ceilings and high walls.

SANDING SEALER First coat of paint to a sanded wood surface that seals the surface but which does not hide the grain.

SAND ISLAND Temporary structure formed in a body of water by creating a perimeter of sheet piling into which bottom silts are pumped.

SAND JACK *See* **Sand box.**

SAND LINE In well drilling, the wire rope that operates the bailer that removes water and drill cuttings.

SAND-LIGHTWEIGHT CONCRETE Concrete made with a combination of expanded clay, shale, slag,

or slate or sintered fly ash and natural sand. Its unit weight is generally between 47 and 54 kg/0.28 m³ (105 lb and 120 lb/ft³).

S&M *Abbreviation for:* surfaced and matched.

SAND OR EARTH WICK Ribbon-like woven strip that is inserted into small-diameter holes drilled vertically into water-bearing surface strata. Used to raise the water to the surface by capillary action for subsequent removal.

SANDPAPER Generic term covering a range of abrasive materials supported by various weights of paper and fabrics. Within the construction trades, six principal abrasives are used for woodwork and metalwork: emery, flint, garnet, and iron oxide are naturally occurring minerals; aluminum oxide and silicon carbide are manufactured products. Sandpapers are graded into seven grades according to the size of a the abrasive particles on their surface: extra coarse (12, 16, 20), very coarse (24, 30, 36), coarse (40, 50), medium (60, 80, 100), fine (120, 150, 180), very fine (220, 240, 280), and extra fine (320, 360, 400, 500, 600). Most sandpapers are available with 'open' or 'closed' coats. Open coats cover 60% to 70% of the surface and are used for hand sanding and have spaces between the grains to allow cuttings to drop from the backing; closed coats cover 100% of the surface and are designed for machine sanding.

SAND PILE *See* **Pile.**

SAND PLATE Flat steel plate or strip welded to the legs of bar supports for use on compacted soil.

SAND POCKET Zone in concrete or mortar containing fine aggregate with little or no cement.

SAND PUMP *See* **Pump.**

SAND PUMP DREDGE Suction dredge.

SAND REEL High-speed winch in a churn drill that lifts the bailing cylinder. *See also* **Reel.**

SAND SEAL COAT Single application of asphalt material to an existing pavement with a light covering of fine aggregate.

S&SM *Abbreviation for:* surfaced one or two sides and standard matched.

SANDSTONE Cemented or otherwise compacted sedimentary rock composed predominantly of quartz grains. *See also* **Material density.**

SAND STREAK Streak of exposed fine aggregate in the surface of formed concrete, caused by bleeding. *See also* **Blemish.**

SAND-STRUCK BRICK *See* **Brick.**

SAND THROUGH Condition where the face layer of a particleboard panel has been sanded off. These areas will appear to be darker and larger particles will be exposed.

SAND TRAP Trap in a water line that permits sand and other heavy particles to settle out from the flow.

sandw. *Abbreviation for:* sandwiched.

SANDWICH CONSTRUCTION Prefabricated construction materials and components consisting of two or more materials layered together so as to take best advantage of the special properties of each.

SANDWICH PANEL Prefabricated panel that is a layered composite formed by attaching two thin facings to a thicker core.

SANDY CLAY *See* **Soil types.**

SANDY LOAM *See* **Soil types.**

sanit. *Abbreviation for:* sanitation.

SANITARIUM Building for convalescent invalids.

SANITARY APPLIANCE Fixed appliance, normally supplied with water and connected to a drain.

SANITARY ENGINEERING Professional responsibility for the design and maintenance of facilities for the provision of potable water and the collection and disposal of liquid and solid wastes.

SANITARY LANDFILL Land area where municipal solid wastes are disposed of under regulatory control. *Also called* **Landfill.**

SANITARY LANDFILL COMPACTOR *See* **Compactor.**

SANITARY LANDFILLING An engineered method of disposing of solid waste on land in a manner that protects the environment, by spreading the waste in thin layers, compacting it to the smallest practical volume, and covering it with soil by the end of each working day. There are several methods, including:

> **Area technique:** Where refuse is deposited on the ground level or upon an earlier lift of solid wastes.
>
> **Canyon technique:** An area method in a depression where cover material is obtained within the depression.
>
> **Quarry or pit technique:** An area method in a depression where the cover material generally is obtained from within the depression.
>
> **Ramp technique:** An area method where cover is obtained by excavating in front of the working face.
>
> **Trench technique:** A method where a trench is excavated specifically for placement of solid wastes and the excavated soil used as cover material.
>
> **Wet or low-lying area technique:** A method of operating in swampy ground where precautions are made to avoid water pollution before proceeding with area landfill.

SANITARY SEWAGE Waterborne wastes containing biological matter.

SANITARY SEWER Sewer intended to carry only sanitary and industrial wastewaters from residences, commercial buildings, industrial plants, and institutions.

SANITATION CUTTING Removal of damaged or diseased tree stems to prevent the spread of insects or disease.

SAP 1. Water fluid that circulates beneath the bark of woody plants and trees; **2.** Moisture in freshly quarried stone.

Sap. *Abbreviation for:* sapwood.

SAP *Abbreviation for:* soon as possible.

SAPLING Immature tree less than 100 mm (4 in.) in diameter and 6 m (20 ft) tall. *See aslo* **Tree classes.**

SAPLING-SEEDLING STANDS *See* **Stand size classes.**

SAPONIFICATION Interaction of fats or esters, generally with an alkali, to form a metallic salt that is commonly called soap.

SAP ROT Decay in the outer, living zone of the wood of a tree, below the bark.

SAP STREAK Streak on a finished wood surface that reveals the presence of sapwood.

SAPWOOD Pale-colored living wood near the outside of a log. Considered to be more susceptible to decay than heartwood.

SARCOPHAGUS Stone tomb.

sas *Abbreviation for:* side, angle, side.

SASH Framework that holds the glass in a window. There are several types, including:

> **Awning:** A partially movable sash hinged at the top and opening either inward or outward.
>
> **Double-hung:** A window frame containing a pair of vertical sliding sashes.
>
> **Fixed:** A single sash fastened permanently in a frame so that it cannot be raised, lowered, or swung open.
>
> **Hopper:** A partially movable sash hinged at the bottom and opening inwards.
>
> **Pivoted:** A sash that swings open or shut by revolving on pivots at either side of the sash or at the top or bottom.
>
> **Single-hung:** A window frame containing a pair of vertical sliding sashes in which only one sash is movable, usually the lower, in contrast to a double-hung sash.
>
> **Sliding:** A sash that moves horizontally on a tongue or track.

SASH BALANCE In double-hung windows, a device, usually operated with a spring, designed to counterbalance the window sash without the use of weights, pulleys, and cord.

SASH BRACE Horizontal member secured to the piles of a bent.

SASH CHAIN *See* **Sash cord.**

SASH CORD Cord or chain in a double-hung sash window that supports the counterweight. *Also called* **Sash chain.**

SASH DOOR Door in which the upper portion is an opening glazed panel.

SASH FRAME Outer frame with sill in which the sliding sashes or casements are suspended.

SASH GANG Frame saw in which one or several straight blades are clamped in a reciprocating frame.

SASHLESS WINDOW Window with a wood frame, containing at least two lights of glass with polished or ground edges, or sash with light metal or plastic edges. At least one light of glass slides horizontally or vertically.

SASH PIN Stout, headless nail or pin used to fasten the mortise-and-tenon joints of window sashes and doors.

SASH PULLEY Pulley-wheel installed in the window jamb on each side at the top of a double-hung sash opening to carry the sash cord. *Also called* **Axle pulley.**

SASH WEIGHT Counterweight for a window sash.

SASH WINDOW Window formed with glazed frames running in vertical grooves.

sat. *Abbreviation for:* satellite; satin; saturate.

satd *Abbreviation for:* saturated.

SATELLITE COMMUNITY Relatively self-contained community, dependent upon an adjacent town or city for many services and facilities.

SATELLITE VEHICLE Small refuse collection vehicle that transfers its load into a larger vehicle operating in conjunction with it.

SATIN (FINISH) Any finish with a sheen between flat (matte) and high gloss. *See also* **Low luster.**

SATISFACTION PIECE Document acknowledging payment of a debt secured by a mortgage.

satn *Abbreviation for:* saturation.

SATURANT Asphaltic material used to impregnate, or saturate, felt used in roofing materials.

SATURATE To fill all the voids in a material with fluid; to form the most concentrated solution possible under a given set of physical conditions in the presence of an excess of substance.

SATURATED Soaked to capacity.

SATURATED AIR Air containing all the water vapor it can hold under existing conditions.

SATURATED FELT Felt material impregnated with tar or asphalt.

SATURATED FLOW Flow of water through a porous material under saturated conditions.

SATURATED PAPER Decorative surface paper, generally weighing between 60 and 130 gm/m^2. Such papers are saturated with melamine or polyester resins and partially cured at the point of manufacture. Final curing is done at the time of hot-press lamination.

SATURATED SOIL *See* **Soil types.**

SATURATED STEAM Steam at the boiling point of water at the existing pressure.

SATURATED, SURFACE-DRY Condition of an aggregate particle or other porous solid when the permeable voids are filled with water and with no water on the exposed surfaces.

SATURATED ZONE Subsurface zone in which all voids are completely filled with water.

SATURATING ASPHALT *See* **Asphalt,** and **Roofing saturant.**

SATURATION 1. Condition of coexistence or stable equilibrium of either a vapor and a liquid or a vapor and solid phase of the same substance at the same temperature. *See also* **Degree of saturation; 2.** At-

tribute by which a perceived color is judged to depart from gray of equal lightness toward a pure hue. *See also* **Color, Hue,** and **Lightness.**

SATURATION COEFFICIENT *See* **C/B ratio.**

SATURATION PRESSURE *See* **Pressure.**

SAUCER WHEEL Shallow, saucer-like grinding wheel.

SAW Toothed steel blade. There are many types, including:

Backsaw: Rectangular, fine-toothed (up to 20 points per inch) hand saw reinforced along its upper edge.

Band: Variable-speed, power-driven saw where the blade is in the form of a continuous loop of tempered steel mounted vertically on pulleys. Band saws are rated according to the width of the throat across which the blade passes. Sizes between 300 and 450 mm (12 and 18 in.) are common. At the bottom of the throat, a table can be adjusted to produce an angle up to 45° relative to the blade, and is equipped with one or more chases to accept such attachments as a miter gauge. A wide range of blades is available, variable as to width of blade (which dictates the minimum arc that can be cut) as well as teeth/mm (teeth/in.) and set, to cut wood, metal, or composite materials.

Board: Short, pointed, coarse-toothed saw for cutting drywall.

Bow: 1. Thin, narrow, slightly tempered, extremely coarse-toothed saw blade mounted in a sturdy U-shaped metal frame. Used for cutting logs, etc.; **2.** Hand saw, fitted with a thin, narrow blade, used for making curved cuts.

Chain: Gasoline- or electric-powered motor that drives a toothed chain around a wide bar up to 700 mm (28 in.) long, or longer.

Chop: A circular saw ('Skillsaw') mounted in a hinged frame that can be lowered, causing the saw to cut the piece. The saw can be angled to make miter and compound-miter cuts. On some models, the saw can also be drawn through the work, similar to a radial arm saw.

Circular: Commonly called a 'Skillsaw' (although this is based on a registered trade name). This is a portable power saw with a removable circular blade measuring 100 to 300 mm (4 to 12 in.) in diameter. Interchangeability of blades permits selection of a cutting or abrasive edge best suited to the job at hand. The base plate of the saw is equipped to accept an adjustable fence, and can be rotated through an arc up to 45° relative to the plane of the blade. In addition to abrasive disks for cutting thin metal and ceramic materials, and special-purpose blades for cutting plywood and composite materials, wood cutting blades are designed for rip cutting, crosscutting, and as planer or finishing blades, sometimes fitted with tungsten carbide teeth.

Compass: Long, thin, pointed saw blade fitted into a curved handle. *Also called* **Keyhole saw.**

Coping: Very thin, fine-toothed blade held under

tension in a U-shaped metal frame.

Crosscut: Hand saw with a blade from 500 to 700 mm (20 to 28 in.) long, and from 8 to 10 points/in., designed specifically to cut wood across the grain.

Double-faced: *See* **Double handed.**

Double-handed: Saw, usually with a blase longer than 1 m (3 ft) and with a handle at each end, pulled through the work by two workers, one at each end of the blade. *Also called* **Double faced.**

Dovetail: Small hand saw, similar to but smaller than a backsaw, with finer teeth and a different handle.

Floor: Portable motorized circular saw used to cut grooves in a hardened concrete floor.

Flooring: Short, 8-point crosscut saw with a curved bottom cutting edge.

Fret: Hand saw used for cutting sharp curves in thin wood or metal, consisting of a thin, narrow, replaceable blade held in a U-shaped frame.

Gang: A machine with multiple blades used to saw rough quarry blocks into slabs.

Grout: Saw-toothed carbide-steel blade mounted on a wooden handle, used to remove old grout.

Grub: Hand saw used for cutting soft stone.

Hacksaw: Thin, narrow, fine-toothed (14 to 32 points/in.), tempered-steel blade mounted in a U-shaped frame, used for cutting metal, plastic, etc.

Hand: Nonpower operated saw used with one hand.

Hole: Cup-shaped saw blade with teeth on the perimeter and with a drill bit mounted through the center of the base. When mounted in a drill it will form a pilot hole and then cut a circular hole equal to the external diameter of the cup.

Jigsaw: This power-operated saw consists of a fine-toothed blade about 150 to 200 mm (6 to 8 in.) long gripped vertically by chucks mounted on either extremity of a C-shaped arm that offeres some 200 mm (8 in.) or more clearance between the blade and the curve of the arm. The saw works by the C-arm oscillating up-and-down at high frequency; in variable-speed models, the frequency is dictated by the material being cut and the blade used. About midway up the length of the blade a work table is mounted. The table is sometimes milled to accept such accessories as a miter gauge, and can be rotated against the plane of the blade by up to 45° in one direction. Blades are available for cutting wood, thin metal, and composite materials.

Keyhole: *See* **Compass,** above.

Pad: Small compass saw with a detachable handle that also serves as a holder for the blade when not in use.

Panel: Hand saw with fine teeth, used particularly for cutting thin sheets of wood or other fibrous material.

Rack: Hand saw with widely-spaced teeth.

Radial arm: A circular saw mounted on a movable head on an arm so that it can be moved along the length of the arm, being drawn through the material to be cut. The arm can be raised or lowered above the work surface. The head and arm rotate left or right across the work table so that any angle cut can be made perpendicular to the surface. The head, too, can be rotated independent of the arm, and can be swung through 60° left or right from perpendicular. In this way the blade can be positioned at any angle, in any plane, at any height relative to the work surface. With the head rotated through 90° on the arm and locked in position, work can be passed through the blade, which can also be angled from the vertical. In some models, one or more auxiliary spindles mounted on the opposite face of the motor housing to the blade arbor are driven via a gearbox at different speeds to the blade arbor. This permits slow-speed sanding, for instance, on one spindle, or high-speed routing on the other. *See* **Circular saw** for an indication of the types of blades available for this saw.

Reciprocating: Power saw having a straight blade projecting from its tubular housing. In use, the blade reciprocates backward and forward. Usually variable speed. Can be fitted with a wide range of wood- and metal-cutting blades.

Ribbon: Narrow blade band saw.

Rip: Hand saw with a blade from 500 to 700 mm (20 to 28 in.) long and from 5 to 6 points/in.; saw with coarse, chisel-shaped teeth, used in cutting wood with the grain.

Rod: Steel saw, approximately 3 mm (0.125 in.) in diameter, having tungsten carbide particles embedded in its surface, used to cut circles or irregular curves in tile.

Saber: Portable power saw in which the blade reciprocates vertically through the base, which can be angled left or right through 45°. As well, the blade can be rotated through 360° in the vertical plane, locking at will at the 90° quadrents. Often variable speed, this versatile saw can be equipped with blades to cut wood, metal or composite materials and is particularly useful when cutting free curves or irregular shapes. *Also called* **Scroll saw.**

Scroll: *See* **Saber saw,** above.

Swing: Circular saw mounted on a pivoted frame that allows the saw to be pulled into the work.

Table: A circular saw mounted vertically below a horizontal table or bed in a manner that permits infinite adjustment of the amount to which the blade protrudes above the surface of the table. The blade can also be angled up to 60° from the vertical. The table is milled at least once to allow for such attachments as a miter gauge, and is equipped for a fence that can be locked at any position parallel to the blade, extended to its right or left. For the types and range of blades commonly available for a table saw see the list under **Circular saw.** In addition, such special-purpose devices as dado heads can be fitted to the arbor.

Tenon: Small, fine-toothed backsaw.

Veneer: Small, approximately 100 x 50 mm (4 x 2 in.) flat saw blade with one long face slightly bowed and toothed, and having a handle mounted at an angle from the face of the blade. Used to make cuts flush with a surface.

SAW ARBOR Spindle on which a circular saw is mounted.

SAW BENCH Table in which a circular saw is mounted.

SAW BOSS Supervisory office in a forest fire suppression crew responsible for saw crews using hand and power saws to cut snags or logs on a fire.

SAW CUT 1. Cut made by a saw blade; **2.** Cut in a hardened material using an abrasive blade or disk.

SAWDUST CONCRETE Concrete in which the aggregate consists mainly of sawdust.

SAWED JOINT 1. Joint made by sawing the appropriate shapes; **2.** Joint cut in hardened concrete, generally not to the full depth of the member, by means of special equipment.

SAW GUIDE Supporting device above and/or below the cut to restrain the saw from deviating off line.

SAW GULLET Throat at the bottom of a saw tooth.

SAW GUMMER Grinding wheel used for gumming saws.

SAW GUMMING Saw and other cutting edge sharpening with a grinding wheel.

SAWHORSE Four-legged stand, usually of 50 x 100 mm (2 x 4 in.) lumber with two pairs of legs in the shape of inverted Vs, joined across the top. *Also called* **Horse.**

SAWLOG PORTION That part of the bole of sawtimber trees between the stump and the sawlog top.

SAWLOG Log meeting minimum regional standards of diameter, length, and defect. Logs must be at least 2.4 m (8 ft) long, and a minimum diameter inside bark of 150 mm (6 in.) for softwoods and 200 mm (8 in.) for hardwoods, and a maximum defect as specified by regional standards.

SAW SET Adjustable hand tool used to give the left- and right-hand set to saw teeth.

SAWTIMBER *See* **Tree classes.**

SAWTIMBER STAND *See* **Stand size classes.**

SAWTIMBER TREE Live tree of a commercial species containing at least one 3.65-m (12-ft) sawlog or two non-continuous sawlogs, each at least 2.4 m (8 ft) long, and having a maximum allowable defect of 67% of the gross tree volume. Softwoods must be at least 225 mm (9 in.) dbh (diameter breast height); hardwoods at least 275 mm (11 in.) dbh.

SAW-TOOTHED ROOF Continuous series of pitched roofs, often with the ridges oriented east-west and with the north-facing slopes glazed.

SAWYER Worker in a saw mill who operates the saw and decides how a log will be converted for greatest yield and least waste.

SAYBOLT VISCOSITY SCALE Standard measurement of the flow characteristics of a liquid. A standard container with a calibrated hole size outlet is used for Saybolt measurement. The liquid to be measured is put in the container and heated to a standard temperature. The time, in Saybolt seconds, for 60 cc of liquid to pass through the orifice determines the viscosity index of the liquid. Two scales of reference are used: Universal for light oils and fuels, and Furol for heavier, more viscous oils. SSU and SSF are the identifiers for the Universal or Furol scales, respectively.

SB *Abbreviation for:* sheet board; slab bolster; soot blower; standard bead.

SBR *Abbreviation for:* slab bolster with runners; styrene-butadiene rubber.

sc *Abbreviation for:* soil cock.

sc. *Abbreviation for:* scale; score; screw.

SC *Abbreviation for:* solid core; speech communication; suspended ceiling.

SCAB Short piece of wood or plywood fastened to two abutting timbers splicing them together.

SCABBING Effect when sections of a road surface spall.

SCABBLED RUBBLE Masonry of undressed stones from which only the roughest irregularities have been knocked off. *See also* **Ranged rubble,** and **Rubble.**

SCABBLING Roughening of a surface with a machine tool.

SCAB BLOCK Block hung between the butt rigging and bight of the haulback to give extra lifting capacity.

SCAB STRAP Short piece of line or chain that secures the scab block to the butt rigging.

SCAFFOLD Temporary elevated platform and its supporting structure, or a suspended platform, used to support users or materials, or both.

SCAFFOLD ACCESS Separate, attachable or built-in means of access to and from a scaffold platform.

SCAFFOLD HEIGHT The height of an unfinished wall that requires another raising of the scaffold to continue the building.

SCAFFOLD HITCH Hitch used to hang planks or scaffolds and tied so that the plank cannot roll.

SCAFFOLDING 1. Elevated platform for supporting workers, tools, and materials; **2.** Temporary structure for the support of deck forms, cartways, or workers. *Also called* **Staging.**

SCAFFOLDING LAYOUT Engineering drawing prepared prior to erection showing the proper arrangement of scaffolding equipment.

SCAFFOLD NAIL *See* **Nail.**

SCAFFOLD PLANK Board of sawn lumber or laminated wood.

SCALE 1. Reduction or enlargement of size to which drawings and models are made; **2.** Divided line on a drawing or map indicating the length used for a larger unit of measure; **3.** Oxide formed on the surface of metal during heating; **4.** To measure the weight or volume of a log or load of logs; **5.** Linear measuring tool on which dimensions are represented proportionally smaller than actual size, to a set sequence of reductions: 1:100, 1:250, 1:500, for instance.

SCALE BOX A derrick box made with an open top and one open end.

SCALED ACCELERATION Acceleration multiplied by the square root of the explosive weight of a cylindrical charge; acceleration multiplied by the cube root of the explosive weight of a spherical charge.

SCALED DISTANCE Distance from some point to a blast, divided by the square root of the explosive weight of a cylindrical charge; distance from some point to a blast, divided by the cube root of the explosive weight of a spherical charge. Relates to seismic disturbance.

SCALE DEDUCTION Volume in a log that is deducted for rot, breaks, or insufficient trim.

SCALE DRAWING Drawing made to other than life size.

SCALER 1. Worker who removes loose rock and debris from a slope; **2.** Person qualified and licensed to calculate the volume of usable lumber in a log.

SCALE UP Application of information gathered from a test facility to a full-scale prototype facility or entity.

SCALING 1. Local flaking or peeling away of the surface or near-surface layer of a material. There are varying degrees, including:

Light scaling: Local flaking or peeling away of the near-surface portion of hardened concrete or mortar that does not expose the coarse aggregate.

Medium scaling: Local flaking or peeling away of the near-surface portion of hardened concrete or mortar that involves loss of surface mortar to 5 to 10 mm (0.2 in. to 0.4 in.) in depth and exposure of coarse aggregate.

Peeling: 1. Process in which thin flakes of mortar are broken away from a concrete surface, such as by deterioration or by adherence of surface mortar to forms as they are removed; **2.** Final stage in the failure of a coat of paint due to excessive moisture in the wood behind the paint, or to incompatibility of successive coats. Some types of multipigment paints fail by peeling.

Severe scaling: Local flaking or peeling away of the near-surface portion of hardened concrete or mortar that involves loss of surface mortar to 5 to 10 mm (0.2 to 0.4 in.) in depth with some loss of mortar surrounding aggregate particles to 10 to 20 mm (0.4 to 0.8 in.) in depth.

Spalling: 1. Erosion of a concrete surface characterized by chipping, usually rather large chips, compared with other failure types where the removed particles are very small; **2.** Chipping or crumbling of a pavement at cracks, joints, or edges.

Very severe scaling: Local flaking or peeling away of the near surface portion of hardened concrete or mortar that involves loss of coarse aggregate particles as well as mortar, generally to a depth greater than 20 mm (0.78 in.).

2. Prying loose pieces of rock off the face of a cut or

roof of a tunnel to avoid danger from their falling unexpectedly; **3.** Measuring of lengths and diameters of logs and calculating deductions for defect to determine volume; **4.** Measuring and calculating dimensions and volumes from a scale drawing.

SCALLOP 1. Unevenly wavy edge; **2.** Undesirable remnant of rock remaining at the toe of a bench due to blasting inefficiency.

SCALPER Sieve for removing oversize particles.

SCALPING 1. Removal of residuals and weathered rock from a cut prior to excavation of the hard rock; **2.** Removal of particles larger than a specified size by sieving; **3.** Removing small plants and duff or ashes from around the spot where a tree seedling will be planted.

SCALPING SCREEN *See* **Screen.**

SCALP ROCK Rock that has passed over a screen, but not through it, i.e., rock rejected because of oversize. *Also called* **Waste rock.**

SCANDINAVIAN TECHNIQUE *See* **Lever.**

SCANT Dimensions in sawn lumber slightly under the standard dressed dimensions.

SCANTLE Strip of wood with two nails projecting through it, used to mark the position of holes measured from the back of slates.

SCANTLING Lumber that ranges in size from 50 x 100 mm (2 x 4 in.) to 100 x 100 mm (4 x 4 in.)

SCARF Cut made to the ends of reinforcing bars preparatory to mechanical splicing.

SCARF CONNECTION Connection made by precasting, beveling, halving or notching two pieces to fit together. After overlapping, the pieces are secured by bolts or other means.

SCARF CUT Sloping cut of the upper portion of the bottom chord of a truss at the heel joint.

SCARF JOINT Joint made by notching, grooving, or otherwise cutting the ends of two pieces of lumber and fastening them so that they lap over and join firmly into one continuous piece.

SCARIFICATION 1. Method of seedbed preparation that consists of exposing patches of mineral soil through mechanical action; **2.** Shallow loosening of the soil surface.

SCARIFIER 1. Mobile equipment accessory attachment used to loosen a compacted surface to a shallow depth; **2.** Piece of thin sheet metal with teeth or serrations cut in the edge, used to roughen fresh mortar surfaces to achieve a good bond for tile.

SCARIFY Make scratches or cuts in; break up and loosen a surface.

SCATTERED DEVELOPMENT Unorganized development of land for a variety of purposes.

scav. *Abbreviation for:* scavenge; scavenger.

SCAVENGE 1. To clean out thoroughly; **2.** To remove unwanted liquids or gases through vacuum or pressure; **3.** In a two-stroke diesel, the use of a blower or supercharger to push exhaust gases out of the exhaust port or valve.

SCAVENGER SYSTEM That part of an asphalt mixing plant dryer draft system that pulls dust-laden air from the screen, weigh hopper, and hot elevator.

SCAVENGER VALVE Device that automatically recovers hydraulic oil that leaks from a jack packing head.

SCAVENGING EFFICIENCY Ratio of a new air charge trapped in a cylinder to the total volume of air and exhaust gases in the cylinder at port closing position.

SCE *Abbreviation for:* schedule compliance evaluation.

SCENIC EASEMENT *See* **Easement.**

SCENIC OVERLOOK Roadside area provided for motorists to stop their vehicles beyond the shoulder, primarily for viewing the scenery in safety.

scfm *Abbreviation for:* standard cubic feet per minute.

sch. *Abbreviation for:* schedule; school.

SCHEDULE 1. Table or list on working drawings detailing number, size, and positioning of similar items such as doors, windows, columns, beams, etc.; **2.** Time sequence of activities and events.

SCHEDULED MACHINE HOUR *See* **Machine time.**

SCHEDULED NONOPERATING TIME *See* **Machine time.**

SCHEDULED OPERATING TIME *See* **Machine time.**

SCHEDULE OF PRICES *See* **Price schedule.**

SCHEDULE OF VALUES List compiled by a contractor reflecting the allocation of a contract sum to various parts of the work, used as a basis for reviewing applications for progress payments.

SCHEDULE OF WORK *See* **Construction schedule.**

SCHEDULER Person who organizes the timing of tasks and the flow of materials.

schem. *Abbreviation for:* schematic.

SCHEMATIC Diagram showing the general principles of construction or operation, usually without accuracy of scale or mechanical representation.

SCHEMATIC DESIGN PHASE Interpretation in graphic form, supported by various types of documentation, by an architect or other designer, of a client's physical requirements for a project.

SCHEMATIC DIAGRAM *See* **Diagram.**

SCHEMATIC WIRING DIAGRAM *See* **Diagram.**

SCHEME Preliminary design and description for the resolution or solution of a proposal.

SCHIST Finely layered metamorphic rock that splits easily and in which the grain is coarse enough to permit identification of the principal minerals.

SCHMIDT HAMMER Tool for the nondestructive testing of hardened concrete. The rebound of a steel hammer applied under controlled conditions is measured; the amount of rebound being proportional to the compressive strength of the concrete.

SCHOKLITSCH FORMULA Mathematical means of determining the volume of water lost to seepage through a dam.

SCHOOLMARM Tree whose stem branches into two; the area of such a tree where the two main stems join.

SCHOOL MASTER Tree that has forked out into three or more separate tops.

sci. *Abbreviation for:* scientific; scientist.

SCISSOR TRUSS *See* **Truss.**

scler *Abbreviation for:* scaler.

SCLEROSCOPE Instrument for determining the relative hardness of materials by a drop and rebound method.

SCONCE 1. Protective cover or screen; **2.** Originally a wall-mounted candleholder mounted in a reflector, more recently an electric light fixture patterned on the same principle.

SCOOP Attachment to handle loose or bulk material by a shoveling action. Can be manually or power manipulated and dumped.

SCOOT Two-runner sled, without tongue or shafts, used to haul logs or bolts from the woods.

SCOPE Work content of a project or any component of a project, such as a work package of a cost class. Scope is fully described by naming all activities performed, the end product(s) that result, and the resources consumed.

SCORCH Premature vulcanization of a rubber compound.

SCORE 1. Mark a surface by scratching; **2.** Cut the face of a gypsum board with a knife prior to snapping away the waste.

SCORIA Vesicular volcanic ejecta of larger size, usually of basic composition and characterized by dark color.

SCORING 1. Surface failure characterized by relatively deep grooves, typically caused by foreign particles between adjacent moving parts; **2.** Cutting partway through the thickness of concrete flat work to control shrinkage cracking; **3.** Marking with a sharp tool.

SCORING SIZE Particulate whose dimensions are such that it is capable of entering a working clearance.

SCOTIA Concave molding similar to a section of a parabola.

SCOUR 1. Erosion caused by fast-flowing water containing abrasive particles or solids; **2.** Removal of sand, earth, or silt from the bottom or banks of a river; **3.** Erosion of a concrete surface exposing the aggregate.

SCOURING SLUICE Opening in the foot of a dam, regulated by a gate, through which accumulated silt, etc., may be ejected.

SCOUR PROTECTION Mechanical means (sheet piling, riprap, revetments, etc.) to protect submerged organic silts.

SCOW Large, flat-bottomed, square-ended boat used for hauling bulk material.

SCOW END Raised portion of the floor in the rear of a dump body, particularly on rock bodies, that replaces the conventional tailgate in preventing load spillage.

scp *Abbreviation for:* scrap.

SCR *See* **Brick.**

scr. *Abbreviation for:* screw.

SCRAPER Digging, hauling, and grading machine equipped with a cutting edge, carrying bowl, movable front wall, and a dumping mechanism, used to dig, transport, and spread soil. *Also called* **Pan,** and **Tractor scraper.** There are a number of designs, including:

> **Auger:** Self-loading system incorporating a hydraulically powered auger located in the center of the scraper bowl. As material flows over the scraper's cutting edge, it is lifted by the auger.
>
> **Bottom-dump:** Carrying scraper that ejects its load over the cutting edge.
>
> **Drag: a.** Digging and hauling equipment consisting of a bottomless bucket working on the cable of a mast and anchor, corresponding to headworks and tailworks; **b.** Towed bottomless scraper used for leveling and maintaining haul roads.
>
> **Rear-dump:** Two-wheel scraper that discharges its load at the rear.
>
> **Self-powered:** Tractor and scraper built and operating as a single unit.
>
> **Two-axle:** Scraper mounted on or built into a full trailer.

SCRAPER/EXCAVATOR Excavator equipped with multiple buckets.

SCRAPER RING *See* **Piston ring.**

SCRAP LOADER Machine equipped with a magnet or grapple attachment for handling metal in scrap yards.

SCRATCH Mark left on a surface that has been ground, caused by a dirty coolant or a grinding wheel unsuited for the operation.

SCRATCH COAT Roughened first coat of cement mortar parging, scratched with a stiff brush to form a keyed surface for subsequent coats.

SCRATCH COURSE Layer of asphalt mix used to fill in low spots in a deteriorated pavement prior to resurfacing.

SCRATCHER 1. Any serrated or sharply tined object used to roughen the surface of one coat of material to provide a mechanical key for the next coat; **2.** Tool used to detect rock defects in a core hole.

SCRATCH HARDNESS Hardness of metal determined by measuring the width of a scratch made by a cutting point under known pressure.

SCRATCHING Application of a scratch coat and its combining with a scratcher.

SCRATCH LINE Unfinished preliminary control line established as an emergency measure to check the spread of a forest fire.

SCR BRICK *See* **Brick.**

SCREED 1. Nonstructural layer of cement in which a finishing material (such as flooring tiles) is bedded; **2.** Device or material at the perimeter of a horizontal or vertical surface that acts as a thickness guide for subsequent applications of plaster, cement, or other materials; **3.** Tool for striking off a concrete surface. *Also called* **Strike-off; 4.** That part of an asphalt paver that smooths and compacts the asphalt mix.

SCREED CHAIR Support to fix the depth of slab and to hold guides for leveling off concrete.

SCREED EXTENSION Short section of asphalt paver screed that can be attached to the main screed to obtain a wider paving width. *Also called* **Hard extension.**

SCREED GUIDE Firmly established grade strips or side forms for unformed concrete that will guide the strikeoff for producing the desired plane or shape. *Also called* **Screed rail.**

SCREEDING Operation of leveling a concrete surface. Performed by moving a straightedge across the top of a template or the side forms. *Also called* **Strike-off.**

SCREED PIVOT Point at which the leveling arms of an asphalt paver screed are attached to the tractor unit.

SCREED PLATE Bottom part of an asphalt paver screed that rides on the mix being placed.

SCREED RAIL *See* **Screed guide.**

SCREED WIRE *See* **Ground wire.**

SCREEFING Removing weeds and small plants, together with most of their roots, from the area immediately surrounding a planting hole.

SCREEN 1. Anything that serves to shield, protect, conceal, or shelter in the manner of a curtain; **2.** Production equipment for separating granular material according to size using woven-wire cloth or other similar device with regularly spaced apertures of uniform size. There are many types, including:

> **Deck:** Two or more screens, commonly vibrating, placed one above the other, used to classify the same run of material.
>
> **Rotary:** An inclined, meshed cylinder that rotates on its axis and screens material placed in its upper end.
>
> **Scalping:** Vibrating grizzly.
>
> **Shaking:** Screen that is moved back-and-forth, or in rotary motion, causing material to move along it or through it.
>
> **Vibrating:** An inclined screen that is vibrated to move material along it and/or through it.

SCREEN ANALYSIS *See* **Sieve analysis.**

SCREENED MATERIAL Material that has passed through or over a screen.

SCREENINGS Undersized or oversized rejects from a screening process.

SCREEN TEXTURED One of the surface textures available on OSB panels to provide traction/slip resistance.

SCREEN TILE *See* **Tile.**

SCREW Metal rod with a thread along all or part of its

length and with a means for turning it. There are several configurations, including:

Machine screw: Metal rod of uniform diameter, threaded for all or part of its length, with an enlarged head shaped to receive a tool with which to turn it.

Set screw: Externally threaded fastener, usually headless and usually having an internal hex-wrenching recess. Available in a range of point styles including: cup point, cone point, dog point, flat point, half-dog point, and oval point. Used for retaining shaft driven gears and for control adjusting.

Sheet-metal screw: Sharply pointed screw with a coarse thread capable of drilling a hole in light-gauge metal and then taping a thread that serves to hold it tightly in place. *Also called* **Self-tapping screw.**

Wood screw: Tapered metal rod with a coarse thread along all or part of its length, and with an enlarged head recessed to receive a tool with which to turn it. There are several head patterns, including:

Phillips - Recessed-head screw having an X-shaped indentation that requires a specially shaped driver.

Robertson - Recessed-head screw having a square indentation that requires a special driver.

Slot - Recessed-head screw having a channel formed across the entire width of its head.

Torque - Recessed-head screw with a hexagonal indentation. It is driven by a specially shaped driver or by an Allen key.

SCREW ANCHOR Bushing of malleable material that is driven into a predrilled hole in masonry, concrete, etc., and that expands to lock tightly when a fastener is screwed into it.

SCREW AUGER Auger with a threaded, sharp point at its lower end.

SCREW CLAMP *See* **Clamp.**

SCREW CONVEYOR *See* **Conveyor.**

SCREW DRIVE *See* **Driving machine.**

SCREWDRIVER Tool for driving or withdrawing screws by turning them, the working end of which is shaped to match the type of recess formed in the head of the screw: slot, Phillips, Robertson, etc.

SCREWDRIVER BIT *See* **Bit.**

SCREW EYE Wood screw with a head formed into a closed loop.

SCREW GUN Power operated, hand-held tool, used for driving screws.

SCREWHOLDING CAPACITY 1. Ability of a panel to hold screws so they do not tear out; **2.** Measure of the force required to withdraw a screw directly from the face or edge of the board into which it is mounted.

SCREW HOOK Wide range of hooked shapes having a tapered thread formed on the end of one leg.

SCREW JACK Lifting jack operated by a threaded shaft that is turned by a handle working a cog.

SCREW MACHINE Electric driving machine, the motor of which raises and lowers a vertical screw through a nut, with or without suitable gearing.

SCREW-ON BEAD Metal glazing bead fastened with machine screws, as compared to similar sections that snap into position.

SCREW PILE *See* **Pile.**

SCREW PLUG Expandable flexible ring used to block the end of a drain for test purposes.

SCREW SHANK NAIL *See* **Nail.**

SCREW STAIR *See* **Spiral stair.**

SCREW THREAD Helical groove cut into the surface of a rod or cone.

SCREW-TYPE ELEVATOR Elevator that is raised and lowered by a screw machine.

SCREW WASHER Cupped or countersunk washer in a range of external and internal diameters, used to increase the bearing area of wood screws and nails.

SCRIBE 1. Making and fitting woodwork to an irregular surface; **2.** Mark a line by scratching.

SCRIBE AWL Sharp-pointed hand tool used for marking on wood, metal, etc.

SCRIBER 1. Hard sharp steel point for marking metal; **2.** Two-legged hand tool used to trace an irregular contour by offsetting the profile to another piece.

SCRIBNER RULE Diagram log rule, one of the oldest in existence, that assumes 25-mm (1-in.) boards and a 12-mm (0.5-in.) kerf, makes a liberal allowance for slabs, and disregards taper.

scrn *Abbreviation for:* screen.

SCROLL Ornamentation in the form of a scroll of paper partly rolled.

SCROLL SAW *See* **Saw.**

SCRUB Poor, unmerchantable timber.

SCRUBBED FINISH Architectural exposed aggregate surface to concrete produced by removing the cement skin from green concrete with a stiff brush.

SCRUBBER 1. Device for removing impurities from aggregates, generally employing cascading water with mechanical devices to produce a scrubbing and rinsing action; **2.** Equipment for removing fly ash and other objectionable materials from the products of combustion by means of sprays, wet baffles, wetted packing, etc. Also reduces elevated temperatures.

sctd *Abbreviation for:* scattered.

SCUFFING Surface failure characterized by patches of erosion, not necessarily in a regular or repeated pattern.

sculp. *Abbreviation for:* scuplture; sculpted.

SCULPTURED Multilevel design incised or cast into a surface.

SCULPTURED TILE Tile with a decorative design of high and low areas molded into the finished face.

SCUMBLING Opaque colors painted over a dry undercoat, with surplus color then wiped off, stippled, dabbled, or rubbed with brush, rag, or fingers.

SCUPPER 1. Opening in the base of a wall or parapet that allows surface water to drain away; **2.** Catch basin at the low point of a bridge slab to collect and drain water.

SCUTCH Two-edged bricklayer's tool used for trimming, cutting, and dressing.

SCUTTLE Openable covering, usually spring-loaded and sometimes lockable, used to provide a watertight seal to a wall or roof opening.

SCYTHE STONE Long, narrow grinding stone for sharpening or whetting.

sd *Abbreviation for:* storm drain.

Sd *Abbreviation for:* seasoned.

SD *Abbreviation for:* semidiameter; shank diameter; shop drawing; standard deviation.

sdbl. *Abbreviation for:* sandblast.

sdg *Abbreviation for:* siding.

SDK *Abbreviation for:* sheet deck.

sdl. *Abbreviation for:* saddle.

sdppr *Abbreviation for:* sandpaper.

sdr *Abbreviation for:* sender.

sdwlk *Abbreviation for:* sidewalk.

SE *Abbreviation for:* service equipment; square edge; stress equalized.

SEAL 1. To permanently close off; **2.** Material used to cover or close the joint between materials to prevent passage of dust, moisture, wind, etc.; **3.** Depth of water held in a trap to prevent the passage of air or gass; **4.** Airtight or watertight joint; **5.** Gland, stuffing box, or similar mechanical device between a static and moving part.

SEALABLE EQUIPMENT Equipment enclosed in a case or cabinet that is provided with a means of sealing or locking so that electrically live parts cannot be made accessible without opening the enclosure.

SEALANT An elastomeric material with adhesive qualities that joins components of similar or dissimilar nature to provide an effective barrier against the passage of the elements.

SEALANT RESERVOIR Cavity, indentation, channel, or formed joint into which a sealant is placed.

SEAL BORE Device, the outside diameter of which mates with a bore surface to provide sealing between the two surfaces.

SEAL CAGE *See* **Lantern ring.**

SEAL CASE Rigid member to which a seal lip is attached.

SEAL COAT *See* **Bituminous coating.**

SEAL CONTACT APPROACH *See* **Seal outside lip angle.**

SEALE Name for a type of wire rope strand pattern that has two adjacent layers laid in one operation with any number of uniform sized wires in the outer layer,

and with the same number of uniform but smaller sized wires in the inner layer.

SEALED BEARING Bearing with seals on both sides to retain lubricant within the device and prevent ingress of foreign matter.

SEALED BID Contract for which interested parties submit written bids at the time and place specified.

SEALED HYDRAULIC SYSTEM Leakproof hydraulic system with static oil pressure.

SEALER Finishing material, either clear or pigmented, usually applied directly over uncoated wood and other absorbent surfaces to seal the surface and reduce their absorptive capacity. *See also* **Primer.**

SEA-LEVEL CORRECTION Calculation to correct a measurement to its equivalent at mean sea level.

SEAL HELIX ANGLE Angle between a helical rib and the line of contact in the plane of the surface of the seal outside lip.

SEAL HELIX CONTACT ANGLE Angle between the contact surface of the rib leading edge and the line of contact.

SEALING COMPOUND Liquid that is applied as a coating to a surface to either prevent or decrease the penetration of liquid or gaseous media.

SEALING GROOVE That part of a joint groove in which the joint sealant is placed.

SEAL INSIDE LIP ANGLE Angle between the inside lip surface and the axis of the seal case.

SEAL OUTER CASE Thin-wall metal structure that encases a lip seal assembly and contains the inner case, primary seal ring, spring parts, and secondary seal.

SEAL OUTSIDE LIP ANGLE Angle between the outside lip surface and the axis of the seal case. *Also called* **Seal contact approach.**

SEAM 1. Stratum or bed of mineral; **2.** Stratification plane in a sedimentary rock deposit; **3.** Line formed by the joining of the edges of a material to form a single ply or layer.

SEAMING STRIP Strip of material laid over a seam to act as a binder.

SEAMLESS CARBON-STEEL PIPE *See* **Pipe.**

SEAMLESS PIPE *See* **Pipe.**

SEAM SEALING Method of joining vinyl sheets.

SEAM WELDING Resistance welding of seams and joints.

SE&S *Abbreviation for:* square-edge and sound.

seas. *Abbreviation for:* seasonal.

SEASONAL EFFICIENCY Ratio between solar energy collected and used to that striking a solar collector, measured over a heating season.

SEASONALLY Four times a year.

SEASONED (WOOD) Wood that has been dried to a certain moisture content to improve its serviceability.

SEASONING 1. Removing moisture from green wood to improve its serviceability. *Also called* **Air dried,** and **Kiln dried**; **2.** Drying of quarried stone.

SEAT Machined surface upon which a mating part rests.

SEAT ANGLE Small steel angle joined to one member to support the end of another.

SEAT CUT *See* **Plate cut.**

SEATING Prepared surface supporting a heavy load.

SEAT OF SETTLEMENT Soil thickness below a loaded foundation within which 75% of settlement occurs.

SEAWALL Work constructed along a shore line consisting of loose mounds or heaps of rubble, or masonry walls supplemented with treated timber, steel, or reinforced concrete sheet piling driven into the beach and strengthened by wales and guide and brace piles. Intended as a barrier to prevent the encroachment of the sea upon land by direct wave action. *See also* **Dike,** and **Pile bulkhead.**

sec. *Abbreviation for:* secant; second; secondary; secure.

sec^{-1} *Abbreviation for:* inverse secant.

SECOND 1. One of the seven base units of the SI system of measurement: a unit of time equal to the duration of 9 192 631 770 periods of the radiation corresponding to the transition between two hyperfine levels of the ground state of the caesium-133 atom. Symbol: s. Derived units are the hertz (symbol: Hz), a unit of frequency representing one period (cycle) per second; speed or velocity, measured in meters per second; and acceleration, measured in meters per second per second. *See also the appendix:* **Metric and nonmetric measurement; 2.** One of the divisions of a minute of an angle or a minute of time into 60 equal parts; **3.** Second-quality material.

SECONDARY BEAM Minor beam supported by other beams.

SECONDARY BLASTING Explosives used to break up large masses of rock resulting from a primary blast.

SECONDARY BRAKE Brake that is intended to stop the descent of suspended scaffold under emergency conditions only.

SECONDARY BRANCH Branch off the primary branch of a drain.

SECONDARY BURNER *See* **Burner.**

SECONDARY COMBUSTION AIR That portion of combustion air introduced to already burning material to assure completeness of combustion, particularly of the resulting gases.

SECONDARY COMBUSTION CHAMBER Chamber where unburned combustible gases and particulate from the primary chamber are burned to completion.

SECONDARY CRUSHER *See* **Crusher.**

SECONDARY DRILLING Drilling of oversize fragments following an initial blast.

SECONDARY FILTER Second stage of a multistage filter system.

SECONDARY LEVEL Enclosed space below the machine room in certain elevator installations, containing the deflector and secondary sheaves, when used.

SECONDARY LINE Forest fire line constructed at a distance from the fire perimeter concurrently with or after a line constructed on or near to the perimeter of the fire.

SECONDARY MANUFACTURING RESIDUE Sawdust, planer shavings, and the like, created by converting lumber, plywood, and veneer into manufactured products.

SECONDARY MATERIAL Material that is used in place of a primary, virgin, or raw material in manufacturing a product; materials that might go to waste if not collected and processed for reuse.

SECONDARY MEMBER Any structural member that carries a load to the primary members: purlins, girts, struts, diagonal braces, wind bents, flanges, sag members, etc.

SECONDARY MOMENT In statically indeterminate structures, the additional moments caused by deformation of the structure due to the applied forces.

SECONDARY PROCESS Where components separated from solid waste may be further processed to allow reuse in their original form or use in an entirely different form.

SECONDARY RECYCLING Use of a secondary material in an industrial application other than that in which the material originated.

SECONDARY REINFORCEMENT Reinforcement provided in a concrete member at right angles to the main reinforcement.

SECONDARY SHEAVE *See* **Sheave.**

SECONDARY TRANSPORT Movement of wood from the landing or transfer point, includes movement by truck, rail, or water.

SECONDARY TREATMENT In wastewater processing, a biological treatment process that follows physical screening and settlement, and which precedes chemical, or tertiary treatment.

SECONDARY USE Use of a material in an application other than that in which it originated; however, the material is not changed significantly by processing and retains its identity.

SECONDARY WINDING Transformer winding normally connected to the load circuit.

SECOND-CLASS LEVER *See* **Lever.**

SECOND FLOOR Story located immediately above the first story. *See also* **First floor.**

SECOND-FOOT In hydraulics, a unit of flow; one cusec (ft^3/sec).

SECOND GRADE CERAMIC TILE Ceramic tile with appearance defects not affecting wearing or sanitary qualities.

SECOND GRADE REFERENCE Second sensor that can be used when both sides of a screed are to be controlled by grade, e.g., stringline on one side and an existing mat on the other.

SECOND GROWTH Timber that has grown after the removal, by natural or human means, of all or a large part of the previous stand.

SECOND LOADER Worker who assists a loader/

operator in loading a log truck.

SECOND MOMENT OF AREA Moment of inertia in a section.

SECOND MORTGAGE *See* **Mortgage.**

SECOND OF ARC Non-SI unit of angle permitted for use with the SI system of measurement. Symbol: ". *See also the appendix*: **Metric and nonmetric measurement.**

SECOND-ORDER ANALYSIS Analysis based on second-order deformation, in which equilibrium conditions are formulated on the deformed structure.

SECRET DOVETAIL Dovetail joint on a miter where the wood on the exposed face is undisturbed.

SECRET FIXING Various methods of attachment whereby the means are not visible in the finished work.

SECRET GUTTER Enclosed gutter built into the eaves of a roof and not visible from below.

SECRET NAILING Nailing through the tongue of tongue-and-groove flooring or matched sheathing.

SECRET SCREWING Method for fabricating a wide board out of two or more narrow ones where the screws joining the pieces are not visible in the finished work.

sect. *Abbreviation for:* section; sector.

SECTION 1. View obtained at an intersecting plane surface; **2.** Shape produced in a continuous process to a definite cross section that is small in relation to its length; **3.** Area equal to 259 ha (640 a or 1 mile^2); **4.** Portion of an area being worked; **5.** Sub-subdivision of a specification dealing with the work of a single trade.

SECTIONAL INSULATION *See* **Molded insulation.**

SECTIONALLY SUPPORTED FURNACE WALL *See* **Wall.**

SECTION HALF *See* **Half section.**

SECTION LINES Thin, evenly-spaced slant lines on a drawing, used to indicate cut surfaces in section view.

SECTION MODULUS Term pertaining to the cross section of a flexural member.

SECTION REPAIR Casing repairs made when a tire injury has extended through 75% or more of the plies or completely through the casing in the tread or sidewall areas.

SECTION WIDTH Distance between outside surfaces of sidewalls of an inflated tire.

SECTOR BOSS Officer responsible for two or more suppression crews on a specific sector of a forest fire.

SECTOR GATE Roller-type gate where the roller is a sector of a circle. *See also* **Roller gate.**

secy *Abbreviation for:* secretary.

sed. *Abbreviation for:* sediment.

SEDIMENT Regolith that has been transported and deposited by water, air, or ice; predecessor of sedimentary rocks.

SEDIMENTARY ROCK Classification of rock formed by sedimentation in water of fine material deposited by the wind.

sedmt *Abbreviation for:* sedimentary; sediment.

SEED 1. To broadcast or sow seed over an area on which grass or other crop is to be grown; **2.** Sprinkle aggregate on the surface of weak concrete producing an exposed aggregate finish.

SEEDBED In natural regeneration, the soil of the forest floor on which seed falls; in nursery practice, a prepared area over which seed is sown.

SEED BLOCK Generally, uncut blocks of trees that are left between and around small clearcut blocks to provide seeds for natural regeneration.

SEEDLING 1. Young tree grown from seed, from the time of germination until it reaches sapling size. *See also* **Tree classes**; **2.** In nursery practice, a young tree that has not been transplanted.

SEEDLING AND SAPLING STAND Where 10% of a stand consists of growing-stock trees and saplings, and/or where seedlings constitute more than half this stocking.

SEED ORCHARD Area of specially planted trees that have been selected for their superior characteristics (i.e., growth, volume, branching, pest resistance, etc.) to breed genetically improved seed.

SEED TREE Tree that produces seeds; usually a superior tree left standing at the time of cutting to produce *Seed*s for reforestation.

SEEPAGE Water percolating through a soil deposit or soil structure.

SEEPAGE COLLAR Projecting collar around a pipe, tunnel, or conduit under an embankment dam that lengthens the seepage path along the exterior of the conduit.

SEEPAGE FORCE *See* **Capillary pressure.**

seg. *Abbreviation for:* segment; segmental.

SEGMENT 1. Part of a circle, smaller than a semicircle; **2.** Bonded abrasive section, any of various shapes, to be assembled with others to form a continuous or intermittent grinding surface.

SEGMENTAL ARCH *See* **Arch.**

SEGMENTAL MEMBER Structural member made up of individual elements prestressed together to act as a monolithic unit under service loads.

SEGMENTAL SLUICE GATE Radial hydraulic gate.

SEGMENTED SAW Round saw composed of pie-shaped sections.

SEGREGATION Tendency of coarse aggregate (stone) to separate from the mortar (cement paste and sand) as concrete is placed.

SEISMIC EXPLORATION Underground exploration conducted by analyzing the results of vibrations caused by explosions set off in drill holes. *See also* **Prospecting.**

SEISMIC LOAD Earthquake force on a structure or component part of a structure.

SEISMIC-REFRACTION SURVEY Measure of the time required for a shock wave to travel a specific distance from an explosion point to the ground surface.

SEISMIC STUDIES Analysis and appraisal of sub-terranean strata by seismic means by which shock-wave velocities at different depths are recorded.

SEISMIC VELOCITY Velocity at which a seismic wave travels outward from its source.

SEISMIC WAVE Wave that travels through the earth.

SEISMOGRAPH Instrument that measures and records earthborne vibrations.

SEISMOGRAPH TRACE Line on a seismographic record showing the frequency or vibration of a seismic wave.

seismol. *Abbreviation for:* seismology.

SEISMOMETER Instrument to detect linear (vertical, horizontal) or rotational displacement, velocity, or acceleration.

SEIZE 1. Grab and fail to move due to expansion; **2.** To make a secure binding at the end of a wire rope or strand using seizing wire or seizing strand. *See also* **Annealed wire.**

SEIZIN Possession of land by somebody claiming ownership.

SEIZING WIRE *See* **Annealed wire.**

sel. *Abbreviation for:* select.

SELECT 1. Exterior unsanded plywood of uniform surface with minor open splits; **2.** Brick accepted as one of the best after culling; **3.** Lumber of a superior grade; for certain species, particularly hardwoods, it refers to a specific grade.

SELECTED BIDDERS *See* **Invited bidders.**

SELECTED MATERIAL Suitable native material obtained from roadway cuts or borrow areas, or other similar material, used for subbase, roadbed material, shoulder surfacing, slope cover, or other specific purposes.

SELECT GRADE High-quality lumber. This grade is recommended for all uses where fine appearance is essential. Widely used for high-quality interior trim and cabinet work with natural, stain, or enamel finishes.

SELECTION CUTTING Even-aged silviculture system in which trees are harvested individually or in small groups continuously at relatively short intervals.

SELECTION SYSTEM Uneven-aged silvicultural system in which single or small groups of trees are periodically selected to be removed from a large area so that age and size classes of the reproduction are mixed.

SELECTION THINNING Removal of dominants that have exceeded the diameter limit prescribed, in favor of smaller trees with good growth form and condition. This will promote conversion to a selection forest.

SELECTIVE BIDDING System of restricted bidding for a contract whereby those being invited to bid are selected according to some criteria (prequalification, previous experience, etc.). *Also called* **Closed bidding, Invitational bidding,** and **Limited tendering.**

SELECTIVE COATING Coating having high solar radiation absorbtivity and low emissivity or infrared emittance.

SELECTIVE COLLECTIVE OPERATION *See* **Operation.**

SELECTIVE CUT Type of timber harvesting that removes only certain species above a certain size or value.

SELECTIVE DIGGING Separating two or more types of soil while excavating them.

SELECTIVE LOGGING Logging an area taking out only specific types of trees, leaving the rest standing.

SELECTIVE SURFACE Coating applied to solar flat-plate collectors that absorbs most of the incoming solar energy.

SELECT MERCHANTABLE *See* **Lumber grades.**

SELECTOR Electrical device, driven by an elevator, that simulates elevator movements.

SELECT TIGHT FACE Exterior plywood with permissible surface openings.

SELF-AMORTIZING MORTGAGE *See* **Mortgage.**

SELF-CENTERING FORMWORK *See* **Telescopic centering.**

SELF-CENTERING PUNCH *See* **Punch.**

SELF-CLEANING TREAD Tire tread pattern that stays open when running in dirt and slush.

SELF-CLOSING FIRE DOOR *See* **Fire door.**

SELF-CONTAINED AIR-CONDITIONING UNIT Independent air conditioner having the means for ventilation, air circulation, air cleaning, air warming, and air cooling.

SELF-CONTAINED BREATHING APPARATUS Device enabling an individual to have air or oxygen independent of the atmosphere in which he is working.

SELF-DESICCATION Removal of free water by chemical reaction so as to leave insufficient water to cover the solid surfaces and to cause a decrease in the relative humidity of the system.

SELF-DRILLING SCREW *See* **Sheet-metal screw.**

SELF-ENERGIZING BRAKE *See* **Brake.**

SELF-EXTENDING SCREED Type of asphalt paver screed that can be extended and retracted hydraulically for fast changes of paving width.

SELF-FURRING Metal lath or welded wire fabric formed in the manufacturing process to include means by which the material is held away from the supporting surface, thus creating a space for keying of the insulating concrete, plaster, or stucco.

SELF-FURRING NAIL Nail with a flat head and a washer or a spacer on the shank, for fastening reinforcing wire mesh and spacing it from the nailing member.

SELF-HEALING Tendency of a hairline crack in concrete or cement mortar to heal due to hydration of the exposed particles of undecomposed cement.

SELF-HELP HOUSING Housing construction in which the potential owners/occupiers complete all or some of the work.

SELF INSURANCE *See* **Insurance.**

SELF-LEVELING (SCREED) Action of a floating

asphalt paver screed that permits it to reduce humps and fill in low spots while paving.

SELF-LEVELING SEALANT Sealant formulation having a consistency that will permit it to achieve a smooth level surface when applied in a horizontal joint at temperatures between 4°C and 38°C (40°F and 100°F).

SELF LOADER Logging truck with a loading device, generally a knuckleboom loader, mounted behind the cab.

SELF-LOADING Capability of a powered industrial truck to pick up, carry, and deposit its load without the aid of external handling means.

SELF-POWERED SCAFFOLDING Powered scaffold having the raising and lowering mechanism located on the working stage.

SELF-POWERED SCRAPER *See* **Scraper.**

SELF-PRIMING CENTRIFUGAL PUMP *See* **Pump.**

SELF-SIPHONING Condition where a partial vacuum in an unventilated or inadequately ventilated plumbing system causes the seal to be drawn from a trap(s), allowing the passage of air or gases through the fitting. *Also called* **Siphonage.**

SELF-SPACING TILE Tile with lugs, spacers, or protuberances on the sides that automatically space the tile for grout joints.

SELF-STRESSING CEMENT *See* **Expansive cement.**

SELF-STRESSING CONCRETE Expansive mix that, if properly restrained, induces persistent compressive stresses in the finished product. *Also called* **Chemically prestressed concrete.**

SELF-SUPPORTING SCAFFOLD Load-bearing scaffold.

SELF-SUPPORTING STEEL FRAME Steel framework that provides the required stability and resistance to gravity loads and design wind and seismic forces without interaction with other elements of the structure.

SELF-SUPPORTING WALL *See* **Wall.**

SELF-TAPPING SCREW *See* **Sheet-metal screw.**

SELF-TEST *See* **Integrity test.**

SELF-TONE Two or more shades of the same color.

SELF-VULCANIZATION Vulcanization activated by chemical agents without the application of heat.

SELLER'S MARKET Condition with an industry or geographic area where the demand for a product or service exceeds the supply. *See also* **Buyer's market.**

SELLER'S RISK Probability of having acceptable material or construction rejected as a result of using a particular acceptance plan. *See also* **Buyer's risk.**

sels. *Abbreviation for:* selsyn.

selsyn. *Abbreviation for:* self-synchronous.

SELSYN GENERATOR Electric device used to transmit electric power to a selsyn motor moving synchronously with the movements of the generator.

SELSYN MOTOR Electric motor driven synchronously with a selsyn generator.

SELVAGE 1. Finished edge of woven-wire screen cloth or fabric produced in the weaving process of the finer meshes; **2.** Edge of a sheet of roll roofing left free of granules.

SELVAGE-EDGE ROOFING Roll roofing with a 50-mm (2-in.) uncoated surface along one edge to allow for overlap.

SEMIANNUALLY Twice per year.

SEMIAUTOMATIC BATCHER *See* **Batcher.**

SEMIAUTOMATIC PILE HAMMER *See* **Pile hammer.**

SEMICIRCULAR ARCH *See* **Arch.**

SEMICONDUCTOR 1. Material that is neither a good insulator nor a good conductor; **2.** Solid-state device.

SEMICURE Preliminary but incomplete cure applied to a tube or hose in the process of manufacture to cause the tube or hose to acquire a degree of stiffness or to maintain some desired shape.

SEMIDETACHED DWELLING One of a pair of residences joined into one structure by a common sidewall or party wall. *See also* **Attached dwelling.**

SEMIDIESEL ENGINE Engine that employs fuel injection, but which also relies on spark ignition.

SEMIDOME Half dome.

SEMIELIPTICAL SPRING *See* **Spring.**

SEMIFIXED LEAD *See* **Lead.**

SEMIFLEXIBLE JOINT Concrete connection in which the reinforcement is arranged to permit some rotation of the joint. *See also* **Hinge joint,** and **Mesnager hinge.**

SEMIFLOATING AXLE *See* **Axle.**

SEMIGELATIN Type of explosive that partially resembles a gelatin, but is more economical.

SEMIGLOSS Finish with a sheen between eggshell and high-gloss.

SEMIGROUSER Crawler track shoe with one or more shallow cleats.

SEMIHOUSED STAIR Stair having a wall on one side only.

SEMIHOUSED STRINGER Board cut and fastened to the face of a solid stringer and concealing the ends of the risers and treads.

SEMIMAT GLAZE *See* **Glaze.**

SEMIRIGID FRAMING *See* **Structural steel framing.**

SEMISOLID ASPHALT *See* **Asphalt.**

SEMISOLID MANDREL *See* **Mandrel.**

SEMITRAILER Towed vehicle whose front rests on the towing unit. *See also* **Trailer.**

SENIOR CITIZEN HOUSING Housing with design features and special facilities intended for use by persons sixty years of age and over. *Also called* **Elderly**

housing.

SENIOR MORTGAGE *See* **Mortgage.**

SENSIBLE HEAT *See* **Heat.**

SENSIBLE HORIZON Visible horizon.

SENSIBILITY RECIPROCAL Change in applied load required to change the position or equilibrium of a weighbeam or indicate a definite amount.

SENSING UNIT Device that senses the condition, state, or position of something.

SENSITIVE SWITCH *See* **Switch.**

SENSITIVENESS Responsiveness of an instrument to the signal or condition being recorded or monitored.

SENSITIVITY Responsiveness to change.

SENSITIVITY ANALYSIS *See* **Analysis.**

SENSITIZER Ingredient used in explosive compounds to promote greater ease in initiation or propagation of the reactions.

SENSOR Device designed to respond to a physical stimulus (as temperature, illumination, motion, etc.) and transmit a resulting signal for interpretation, measurement, or for operating a control.

sep. *Abbreviation for:* separate.

SEPARATE-APPLICATION ADHESIVE *See* **Adhesive.**

SEPARATE CONTRACT Contractural agreement independent of the main construction contract but pertaining to the same project.

SEPARATE CONTRACTOR Person or entity who has contracted with the owner to perform other construction, work or operations for the project that is not included within the main work.

SEPARATELY DERIVED SYSTEM Premises wiring system whose power is derived from generator, transformer, or converter windings and that has no direct electrical connection, including a solidly connected grounded circuit conductor, to supply conductors originating in another system.

SEPARATE SYSTEM Drainage system in which rainwater and sewage are carried in separate pipes.

SEPARATION 1. Pulling apart, such as ply separation (from each other) or tread separation (from plies); **2.** Tendency, as concrete is caused to pass from the unconfined ends of chutes or conveyor belts or similar arrangements, for coarse aggregate to separate from the concrete and accumulate at one side; **3.** Tendency for solids to separate from water by gravitational settlement.

SEPARATION SPACE Open space provided around dwelling units to ensure access, privacy, and exposure to sun.

SEPARATOR 1. Sections of steel pipe forming spacers between I-beams. When bolted together the composite members serve as a structural unit; **2.** Device whose primary function is to isolate contaminants by physical properties other than size. There are several types, including:

   Absorbent: Separator that retains certain soluble and insoluble contaminants by molecular adhesion.

   Ballistic: Device that drops mixed materials having different physical characteristics onto a high-speed rotary impeller; they are hurled off at different velocities and land in separate bins.

   Centrifugal: Separator that removes nonmiscible fluid and solid contaminants that have a different specific gravity than the fluid being purified by accelerating the fluid in a circular path and using the radial acceleration component to isolate these contaminants.

   Coalescing: Separator that divides a mixture or emulsion of two nonmiscible fluids, using the interfacial tension between the two liquids and the difference in wetting of the liquids on a particular porous medium.

   Electrostatic: Separator that removes contaminants from dielectric fluids by applying an electrical charge to the contaminant that is then attracted to a collection device of different electrical charge.

   Interial: a. Device that relies on ballistic or gravity separation of materials having different physical characteristics; **b.** Device that removes particles from a gaseous stream by imparting a centrifugal motion with fixed mechanical parts.

   Magnetic: Separator that uses a magnetic field to attract and hold ferromagnetic particles.

   Vacuum: Separator that utilizes subatmospheric pressure to remove certain gases and liquids from another liquid because of their difference in vapor pressure.

SEPARATOR WOOD Material that has been separated from the whole tree during the chipping process and is unacceptable for pulp and paper manufacture. Usually used as energy wood.

sepd *Abbreviation for:* separated.

sepg *Abbreviation for:* separating.

sepn *Abbreviation for:* separation.

SEPTIC BED Area adjacent to a septic tank in which an underground network of perforated pipes distributes the effluent from the tank into the surrounding soil.

SEPTIC TANK Tank, embedded in the earth, into which sewage is allowed to drain, divided into compartments and of an internal configuration that promotes anaerobic decomposition of organic wastes suspended in water, and discharging to a tile bed or drainage field; sludge from settled solids is retained for sufficient time to secure satisfactory decomposition of organic solids by bacterial action.

SEPTIC TREATMENT SYSTEM Small-scale sanitary treatment system consisting of a septic tank and tile bed in which sewage flows are subjected to bacterial treatment in an anaerobic environment. The resulting sludge settles to the bottom of the tank, the liquid effluent decants over a weir to the tile bed, and a foamy by-product of the bacterial action produces a blanket within the tank that prevents oxygen from approaching the contents.

seq. *Abbreviation for:* sequence.

seql *Abbreviation for:* sequential.

SEQUENCE Order in which operations take place or are scheduled.

SEQUENCE CIRCUIT Circuit that establishes the order in which two or more functions of a circuit occur. *See also* **Circuit.**

SEQUENCE-STRESSING LOSS In post-tensioning, the elastic loss in a stressed tendon resulting from the shortening of the member when additional tendons are stressed.

SEQUENCE TEST Series of engine tests used to evaluate engine oil performance.

SEQUENCE VALVE *See* **Valve.**

SEQUENTIAL FIRING Method of firing the explosives within a series of holes in a sequence to reduce the burden and provide many independent blasts.

ser. *Abbreviation for:* serial; series.

SERIAL STAGE Series of changes occurring in the ecological succession of a plant community, e.g., pioneer stage or climax stage.

SERIES 1. Group or number of similar things arranged in a row; **2.** Number of things produced by identical means of identical materials; **3.** Electrical circuit in which the parts are connected end-to-end, positive to negative poles; **4.** Arrangement of blasting caps in which the firing current passes through them in succession.

SERIES CIRCUIT Electrical circuit in which the various components are connected end-to-end across a power source. *See also* **Parallel circuit.**

SERIES FIELD Field coils connected in series with the armature of an electric motor.

SERIES MOTOR Electric motor with the armature circuit in series with the field windings.

SERIES-PARALLEL Two or more parallel circuits of blasting caps, each arranged in series.

SERIES-PARALLEL CIRCUIT Electrical circuit having a combination of series and parallel connections.

SERPENTINE Curving alternately either side of a centerline.

SERPENTINE COIL Numerous flow paths in each grid of a circulator pump that lower fluid friction.

SERPENTINE WALL Masonry wall (usually brick) built on an alternating curve from a centerline. Usually 100 mm (4 in.) thick, the configuration of the wall helps it to resist movement.

serr. *Abbreviation for:* serrate; serrated.

SERRATED Line of teeth cut into an edge.

serv. *Abbreviation for:* service.

servd *Abbreviation for:* serviced.

SERVE To cover the surface of a wire rope or strand with a fiber cord or wire wrapping.

SERVICE Water or sewer pipes, electrical or gas lines that connect public services to a building.

SERVICEABILITY LIMIT STATE Limiting condition affecting the ability of a structure to preserve its appearance, maintainability, durability, or the comfort of its occupants or function of machinery under normal usage.

SERVICE AREA Geographic area over which service will be provided under specific conditions. This includes permanent and fixed service, such as water supply or electrical energy as well as mobile service, such as that offered on site to owners of machinery and equipment.

SERVICE BODY Truck cargo bed incorporating enclosed compartments in which to carry tools, parts, or crew members in addition to a payload other than bulk materials.

SERVICE BOX *See* **Curb box.**

SERVICE BRAKE *See* **Brake.**

SERVICE CHIEF Officer in a forest fire suppression organization responsible for procuring, maintaining and distributing men, equipment, supplies and facilities at the times and places specified by the suppression plan.

SERVICE CONDUCTOR Electrical supply conductor that extends from the street main or from a transformer to the service equipment of the premises supplied.

SERVICE DEAD LOAD Dead weight supported by a member.

SERVICED LOT Surveyed lot provided with domestic water, sanitary sewers, storm drainage and proximity to electrical distribution lines, plus a curb crossing where roads and curbs have been developed. *Also called* **Fully-serviced lot.**

SERVICE DROP Overhead service conductors from the last pole or other aerial support to and including the splices, if any, connecting to the service-entrance conductors at the building or other structure.

SERVICE ELL *See* **Street elbow.** *Also called* **Service L.**

SERVICE ENTRANCE CONDUCTOR (OVERHEAD SYSTEM) Service conductor between the terminals of the service equipment and a point, usually outside the building, clear of building walls, where joined by a tap or splice to the service drop.

SERVICE ENTRANCE CONDUCTOR (UNDERGROUND SYSTEM) Service conductor between the terminals of the service equipment and the point of connection to the service lateral.

SERVICE ENTRANCE SWITCH *See* **Switch.**

SERVICE EQUIPMENT Necessary equipment, usually consisting of a circuit breaker or switch and fuses, and their accessories, located near the point of entrance of supply conductors to a building or other structure, or an otherwise defined area, and intended to constitute the main control and means of cutoff of the supply.

SERVICE HOIST Lifting device designed to raise and support a machine or vehicle free of the ground so that the entire underbody may be inspected and serviced. May be mechanically, hydraulically, or pneumatically actuated.

SERVICE L *See* **Street Elbow.**

SERVICE LIFE 1. Period for which a product is

guaranteed, warranted, or can be reasonably expected to perform in the manner for which it was designed; **2.** In fluid filtration, the length of time that a filter will survive in an actual system before the terminal differential pressure is reached.

SERVICE LIVE LOAD Live load specified by the general building code or bridge specifications, or the actual non-permanent load applied in service.

SERVICE LOAD Load expected to be supported by the structure under normal usage; often taken as the nominal load.

SERVICE PANEL Electric box containing the disconnect switch, and from which the circuits are taken via circuit breakers.

SERVICE PIPE Water or gas supply pipe connecting the public supply main to the building.

SERVICE PRESSURE See **Pressure.**

SERVICE RACEWAY Electrical raceway that encloses the service-entrance conductors.

SERVICE RISER Vertical pipe through which electrical service conductors are connected to the meter box.

SERVICE ROAD Minor road parallel to a main road, used primarily by local traffic.

SERVICE ROOM Room or space provided in a building to accommodate building service equipment such as air-conditioning or heating appliances, electrical services, pumps, compressors, and incinerators.

SERVICES Systems comprising equipment, pipes, cables, ducts, fittings, appliances, and appurtenances that supply drainage, water, steam, electricity, gas, and communication services to a building or structure.

SERVICE SPACE Space provided in a building to facilitate or conceal the installation of building service facilities such as chutes, ducts, pipes, shafts, or wires.

SERVICE STAIR Secondary stair.

SERVICE SYSTEMS Heating, ventilating, air conditioning, water supply, electrical supply, and gas supply to a building.

SERVICE T See **Street T.**

SERVICE TEST Test in which a product is used under actual service conditions. *Also called* **Performance test.**

SERVICE TIME See **Machine time.**

SERVICE WATER HEATER Device for heating water for plumbing services.

SERVICE WATER HEATING Heating of water for domestic and commercial purposes other than comfort heating.

servo. *Abbreviation for:* servomechanism.

SERVOCIRCUIT Closed-loop circuit that is controlled by some type of feedback, i.e., the output of the system is sensed or measured and is compared with the input. The actual output and the input control the circuit. The system output may be position, velocity, force, pressure, level, flow rate, temperature, etc. *See also* **Circuit.**

SERVOCONTROL Control actuated by a feedback system that compares the output with a reference signal and makes corrections to reduce the difference.

SERVOPISTON Piston that moves another part on command from a servovalve.

SERVOVALVE See **Valve.**

SERVOVALVE OVERLAP Lap condition that results in a decreased slope of the normal flow curve in the null region.

SERVOVALVE UNDERLAP Lap condition that results in an increased slope of the normal flow curve in the null region.

SERVOVALVE NULL Condition where the servovalve supplies zero control flow at zero load pressure drop.

SE Sdg *Abbreviation for:* square-edge siding.

SET 1. Change from a plastic to a hard state. *See also* **Cure; 2.** Chisel used for cutting brick, *Also called* **Bolster; 3.** Permanent or semipermanent deformation; **4.** Net penetration of a pile struck by a pile-driving hammer. *Also called* **Pile set; 5.** Final penetration of a pile or final set, expressed in millimeters or inches per blow; **6.** Point at which the breaking process has advanced so that an asphalt mix will no longer track when blotted with white paper. The mix may be too tender for traffic at this point; **7.** To place a choker around a log. *Also called* **Add-to-set; 8.** Cutting crew; may consist of one faller who falls and bucks timber or one faller and one bucker working as a team. Sometimes used to describe right-of-way fallers (The term was more common in the 'hand' falling era. Two fallers and two buckers formed a four-man set or gang.); **9.** Amount of strain remaining after complete release of a load producing a deformation; **10.** To drive a fastener below the surface of the material into which it is being driven; **11.** Direction of the flow of water; **12.** Permanent bend in metal; **13.** Correct angle for the purpose, to left or right, of saw teeth; **14.** Wide, beveled chisel for cutting masonry; **15.** Tunnel-supporting system consisting of a roof beam or arch, and two posts; **16.** Hammer-like wooden head on a handle that is struck with a hammer to shape metal.

set. *Abbreviation for:* setting; settling.

SET ACCELERATOR Admixture used to produce a quick-setting grout or mortar.

SETBACK 1. Horizontal distance between the faces of the exterior wall of one story and the exterior wall next above it, where a lower story extends beyond the higher story; **2.** Horizontal distance between the wall of a building and the adjacent street line; **3.** Occurs when a tree being felled sets back opposite to the intended direction of fall.

SET-BACK AXLE See **Axle.**

SETBACK BUTTRESS See **Buttress.**

SETBACK LINE Line outside the right of way, established by public authority or private restriction, on the highway side of which the erection of buildings or other permanent improvements is controlled.

SET-CONTROL ADDITIVE 1. Small amounts of materials (mineral fillers or chemical) that, when added to a slurry/micro-surfacing mixture, speed or retard its setting characteristic; **2.** Material, composed essentially of calcium sulfate in any hydration state, interground with clinker during the manufacture of cement to modify the setting time of the cement.

SET-FORWARD AXLE *See* **Axle.**

setg *Abbreviation for:* setting.

SET-IN The amount that the lower edge of a brick on the face tier is back from the line of the top edge of the brick directly below it.

SET OF DRAWINGS Two or more drawings that complement each other, usually depicting various aspects of a product or project.

SET OFF Line on a plan where a wall is reduced in thickness.

SETOUT/SETBACK COLLECTION *See* **Collection.**

SET POINT Any of the variable functions, such as load, radius-of-load, or boom angle, that may be set or entered into an operational device for the purpose of alerting the operator to a specific condition. *Also called* **Preset.**

SET/RESET RELAY Relay capable of holding itself residually until the magnetic field holding it is electrically neutralized.

SET SCREW *See* **Screw.**

SETT Type of grease (cold sett greases) that changes from a fluid to a semifluid or plastic state after component combination and often after packaging. *Also called* **Cold sett.**

SETTING 1. Area logged by one yarder; **2.** Temporary location of a cable yarding system, portable mill, or other machine used for logging; **3.** Area yarded to one landing; **4.** Placement of lights or panels in sashes or frames; **5.** All boards held in position by one frame or, in the case of tucking or piling frames, by two adjacent frames.

SETTING BED 1. Layer of mortar on which a tile is set; **2.** The final coat of mortar on a wall or ceiling.

SETTING DERRICK Mobile derrick used to assist in setting stone blocks. Mounted on wheels, it can be moved from one location to another.

SETTING OUT Positioning on the site the pegs, batter boards, profiles, and other markers from which centerlines and levels of the proposed construction can be taken.

SETTING SHRINKAGE Reduction of volume of a hydraulic mixture due to settling of solids and chemical combination of water.

SETTING TEMPERATURE Temperature needed to set a liquid resin to a solid state.

SETTING THE SAFETY Act of triggering the safety to stop an elevator from falling.

SETTING TIME *See* **Final setting time,** and **Initial setting time.**

SETTING-TYPE JOINT COMPOUND Drywall joint compound that hardens by chemical reaction prior to drying.

SETTING-UP 1. Change in a material from fluid to a firm, hard, or fixed state; **2.** Gelling of paint during storage; **3.** Procedures necessary in establishing an organizational base on a new job site.

SETTLEMENT 1. Downward movement of a structure or part of a structure, a pile or group of piles, during or following construction. *See also* **Gross settlement,** and **Net settlement; 2.** Downward vertical movement experienced by structures or a soil surface as the underlying supporting earth compresses; **3.** Sinking of solid particles in grout, mortar, or fresh concrete after placement and before initial set; **4.** Tendency of asphalt globules to settle during storage of an emulsified asphalt.

SETTLEMENT JOINT Joint that permits adjacent concrete members to settle or deflect relative to each other.

SETTLEMENT SHRINKAGE Reduction in volume of concrete prior to the final set of cementitious mixtures, caused by settling of the solids.

SETTLING Lowering in initial elevation of structures or sections of pavement due to their mass, the loads imposed upon them, or shrinkage or displacement of the support.

SETTLING CHAMBER Any chamber designed to reduce the velocity of a material (fluid, gas, products of combustion, etc.) to promote the settlement of particulate matter.

SETTLING POND Natural or artificial depression where water is held long enough to allow the precipitation of particulates.

SETTLING VELOCITY Terminal rate of fall of a particle through a fluid, or a given dust from a dust-laden gas, as induced by gravity or other external force.

SET-UP 1. Exact positioning of mobile equipment so that it can complete the task(s) required of it; **2.** Premature vulcanization of a rubber compound during processing or storage. *See also* **Freeze; 3.** Concrete that has become firm, but not solid; **4.** Positioning and adjustment to level of a survey instrument.

sev. *Abbreviation for:* severe.

SEVERALTY Sole ownership of real property. *See also* **Acquisition, Conveyance, Dedication, Eminent domain, Expropriation, Negotiation, Option,** and **Remainder.**

SEVERE-DUTY BUCKET *See* **Bucket.**

SEVERE-DUTY DUMP BODY *See* **Truck.**

SEVERE SCALING *See* **Scaling.**

sew. *Abbreviation for:* sewer; sewage.

SEWAGE Liquid waste that contains animal, mineral, or vegetable matter in suspension or solution, conveyed in a sewer.

SEWAGE DISPOSAL Process of collecting, conveying, and treating waterborne organic wastes originating from domestic, commercial, and industrial sources, but not necessarily including industrial wastes, or stormwater runoff.

SEWAGE DISTRIBUTOR Mechanism that applies sewage, or sewage effluent to the top of a filter.

SEWAGE GAS By-product of bacterial action on organic material, predominantly methane.

SEWAGE SLUDGE Semiliquid substance consisting

of settled sewage solids combined with varying amounts of water and dissolved materials.

SEWAGE TREATMENT PLANT Processing facility that receives untreated waterborne wastes from a community, and that discharges treated effluent into a receiving body of water and treated sludge cake to a disposal area.

SEWAGE TREATMENT RESIDUE Coarse screenings, grit, and dewatered or air-dried sludge solids from sewage treatment plants, or pumpings of cesspool or septic tank sludges, which require disposal as putrescible wastes.

SEWER Pipe or other construction, and its associated manholes, vents, and other appurtenances, forming part of a sewerage system.

SEWERAGE System for the collection and conveyance of surface water and wastewater from an area.

SEWER BRICK *See* **Brick.**

SEWER SYSTEM EVALUATION SURVEY Systematic examination of tributary sewer systems or their subsections to determine the location, flow rate, and cost for correction for each definable element of a demonstrated total infiltration and inflow problem.

SEWER TILE *See* **Tile.**

SEWER TRAP *See* **Trap.**

SEXAGESIMAL SYSTEM System of division by sixties: the circle is 360 degrees, the degree is 60 minutes, and the minute is 60 seconds.

SEXTANT Hand-held instrument for measuring vertical or horizontal angles.

sf *Abbreviation for:* surface foot; semifinished.

SF *Abbreviation for:* safety factor.

sfgd *Abbreviation for:* safeguard.

SFO *Abbreviation for:* service fuel oil.

S4S *Abbreviation for:* surfaced four sides.

S4S&CS *Abbreviation for:* surfaced four sides and caulking seam.

sfpm *Abbreviation for:* surface feet per minute.

sft *Abbreviation for:* shaft.

sftwd *Abbreviation for:* softwood.

sg. *Abbreviation for:* subgrade.

SG *Abbreviation for:* spike grid; structural glass.

sgd *Abbreviation for:* signed.

SGD *Abbreviation for:* sliding glass door.

sgl. *Abbreviation for:* single.

sgn *Abbreviation for:* sign.

SGRAFFITO Decoration on plaster of incised patterns, the top coat being cut through to show a differently colored coat beneath.

sh. *Abbreviation for:* share; sheet; shingle; shower.

SH *Abbreviation for:* single-hung; skin hard.

SHACK *See* **Site office.**

SHACKLE 1. U- or anchor-shaped fitting with pin; **2.**

Curved portion of a padlock that passes through a hasp; **3.** One of the threaded rods (in elevator construction) to which the hoist cables are socketed and that bolts to the hitch plate and counterweight.

SHACKLE SPRING *See* **Spring.**

SHADE Gradation of a color; any hue plus black.

SHADED POLE Pole of an AC motor produced by cutting away a section of a field pole and inserting one turn of a conductor.

SHADE TOLERANT Capacity of a tree or plant species to develop and grow in the shade of, and in competition with, other trees or plants.

SHADING COEFFICIENT Ratio of solar heating gain through a glazing system, corrected for external and internal shading, to the solar gain through an unshaded light.

SHADING COIL Continuous loop of electrically conductive material in the frame of an AC relay to retard the magnetic flux and keep the relay from chattering.

SHADOWING Appearance of a poorly-completed drywall joint through the surface decoration.

SHAFT 1. Trunk of a column between the base and the capital; **2.** Vertical passage connecting the surface to the underground workings of a mine; **3.** Chimney stack; **4.** Round bar that rotates or provides an axis of revolution; **5.** Elevator well; **6.** That part of a rotor that carries other rotating members and that is supported by bearings; **7.** *See* **Spillway.**

SHAFT CARRIAGE Carriage of an industrial truck on which forks or attachments are mounted by a shaft that passes through the vertical members of the carriage.

SHAFT HORSEPOWER *See* **Horsepower.**

SHAFT RESISTANCE Static positive soil resistance along the shaft of a pile. *Also called* **Side resistance.**

SHAFT SEAL Rubbing seal or stuffing box to prevent leakage between a shaft and its bearing, and the ingress of foreign material to the bearing.

SHAFTWALL Assembly of metal sections holding drywall boards that form vertical shafts.

SHAKE 1. Imperfections in wood characterized by a separation along the grain, the greater part of which occurs between the rings of annual growth, caused by high winds or adverse conditions, or during felling; **2.** Thick, split shingle used for roofing and siding, may be straight split, tapersplit, or handsplit and resawn. *See also* **Shingle.** The following descriptions apply to the various grades of shake product:

No. 1 Handsplit and Resawn: These shakes have split faces and sawn backs. Cedar logs are first cut into desired lengths. Blanks or boards of proper thickness are split and then run diagonally through a bandsaw to produce two tapered shakes from each blank.

No. 1 Tapersplit: Produced largely by hand, using a sharp-bladed steel froe and a mallet. The natural shingle-like taper is achieved by reversing the block, end-for-end, with each split.

No. 1 Straightsplit: Produced in the same manner as tapersplit shakes except that by splitting from the same end of the block the shakes acquire the same thickness throughout.

No. 1 Blue Label: A product having both faces sawn and not as precisely manufactured as shingles. Also, some variations in butt thickness are permitted, which result in a more rustic appearance, hence the termed 'shakes.'

No. 2 Red Label: Permitted for roofs (at reduced exposure) and wall coverings.

No 3 Black Label: Permitted for wall coverings or roofs of accessory buildings.

SHAKING SCREEN *See* **Screen.**

SHALE *See* **Soil types.**

SHALE PIT Pit into which coarse material screened out of rotary drill mud is dumped.

SHALE SHAKER Screen in the mud circulating system of a rotary drill.

SHALLOW FOOTING Footing founded at less depth below grade than its width.

SHALLOW FOUNDATION Foundation unit that derives its support from soil or rock located close to the lowest part of the building that it supports.

SHALLOW MANHOLE Inspection chamber not containing branch connections.

SHALLOW WELL Well less than 6 m (20 ft) deep.

SHANK 1. Permanent mounting to which bucket teeth are attached; **2.** That part of a tool between the implement and the handle or hold; **3.** Bar or standard that connects a rooter tooth with the frame; **4.** That part of drill steel that fits into the drill; **5.** That portion of a coupling that is inserted into the bore of a hose; **6.** Part of a bit key between the bow and the bit; **7.** *See* **Ripper shank.**

SHANK ADAPTER *See* **Striker bar.**

SHANK PROTECTOR Replaceable protector pinned to the leading edge of a ripper shank, designed to protect the shank from excessive wear.

SHAPED SHINGLE *See* **Shingle.**

SHAPED WORK Curved carpentry work.

SHAPE FACTOR Ratio of the plastic moment to the yield moment, or the ratio of the plastic modulus to the section modulus for a cross section.

SHAPE Flanged product produced in a rolling mill.

SHARED APPRECIATION MORTGAGE *See* **Mortgage.**

SHARED EQUITY MORTGAGE *See* **Mortgage.**

SHARP-CRESTED WEIR Measuring weir topped with a thin metal plate.

SHARPENING STONE Natural or manufactured abrasive stone, usually of oblong shape, used for sharpening or whetting tools.

SHARP SAND *See* **Soil types.**

SHAY SWIVEL Fitting used to attach the slack-pulling line to the main haul line on a skyline system.

sh. bl. *Abbreviation for:* shotblast.

shd *Abbreviation for:* shaded.

shd. *Abbreviation for:* shade.

Sh. D *Abbreviation for:* shipped dry.

SHEAR 1. Internal force tangential to the plane on which it acts; **2.** To cut off as by two equal and opposed forces; **3.** Large, bench-mounted snips used to cut sheet metal; **4.** Hydraulically operated, scissor-like device for crosscutting the stem of a tree.

SHEAR CONNECTOR Bolt or rod welded to the top of a beam.

SHEAR DIAPHRAGM Membrane-like member capable of resisting deformation by in-plane shear forces.

SHEARED PLATE Structural steel plate that has been rolled and sheared or gas cut on all edges.

SHEAR FACTOR Weight of explosives required to produce 1 m² (1 ft²) of presplit surface area.

SHEAR FORCE *See* **Force.**

SHEAR FORCE DIAGRAM Diagram that shows the amount of the shearing force at any position along a beam or other structural member.

SHEAR FRICTION Friction between the embedment and the concrete that transmits shear loads. The relative displacement in the plane and the shear load is considered to be resisted by shear-friction anchors located perpendicular to the plane of the shear load.

SHEARHEAD Assembled unit in the top of the columns of flat slab or flat plate construction to transmit loads from the slab to the column.

SHEARING 1. Slipping or sliding of one part of a substance relative to an adjacent part. In a solid, such action requires cutting or breaking of the crystal structure; in a fluid or plastic, shearing does not necessarily destroy the continuous nature of the substance; **2.** In Christmas tree culture, to prune the branches to make dense foliage and give the tree a conical shape.

SHEARING STRESS Stress resulting from forces that tend to cause two contiguous components to move toward each other in a direction parallel to their plane of contact.

SHEAR JOINT Joint in which one opposing face may move parallel to the other.

SHEAR LINE Space between the shell and the plug of a lock cylinder.

SHEAR LUG Plate, welded stud, bolt or other steel shape that is embedded in concrete and located transverse to the direction of the shear force and which transmits shear loads introduced into the concrete by local bearing at the shear lug-concrete interface.

SHEAR MODULUS Ratio of the shear stress to the resulting shear strain. May be either static or dynamic.

SHEAR PIN *See* **Antirotation device.**

SHEAR PLATE Metal plate, used primarily when connecting wood to nonwood in heavy timber construction.

SHEAR RATE Rate of slip within a substance engaging in flow. The average or mean shear rate in a pipe or tube is the average velocity divided by the radius of the tube. It, therefore, has the dimensions of the reciprocal of the time and is usually expressed in the unit of reciprocal seconds (sec^{-1}).

SHEAR REINFORCEMENT Reinforcement designed to resist shear or diagonal tension stresses.

SHEAR SHREDDER Size reduction machine that cuts material between two large blades or between a blade and a stationary edge.

SHEAR SLIDE Landslide where a mass of earth separates and moves down a slope.

SHEAR SLUMP Slump test in which the top half of the released molded concrete shears off and slides down an inclined plane.

SHEAR STABILITY Ability of a lubricating grease to resist changes in consistency (hardness) during mechanical working. *Also called* **Mechanical stability.**

SHEAR STRAIN Angular displacement of a member due to a force across it.

SHEAR STRENGTH Maximum shearing force a flexural member can support at a specific location, as controlled by the combined effects of shear forces and bending moment.

SHEAR STRESS Stress component acting tangentially to a plane.

SHEAR TEST There are several, including:

Box shear test: Measure of the shear strength of soil, determined using a box split into two parts, on one half of which pressure and a measured shearing force is applied.

Triaxial compression test: Test in which a specimen is subjected to a confining hydrostatic pressure and then loaded axially to failure.

Vane shear test: Technique where a four-bladed vane is forced into undisturbed soil ahead of a sampling tube and rotated so the soil shears. Peak and remolded shearing resistance is recorded.

SHEAR WALL *See* **Wall.**

SHEAR WAVE Seismic wave whose motion is at right angles to the direction of travel, generated by a rock's resistance to shear or change in shape.

SHEAR ZONE Portion of a rock mass traversed by closely spaced surfaces along which shearing has occurred; rock that may be crushed and brecciated.

SHEATH Enclosure in which post-tensioning tendons are encased to prevent bonding during concrete placement.

SHEATH COUPLER Metal coupler for joining and sealing lengths of prestressing tendon.

SHEATHING 1. Structural sheet covering over studs or rafters of a wood-frame structure; **2.** Material forming the contact face of forms. *See also* **Form lining; 3.** Exterior unsanded plywood, the face of which may contain limited-size knots, knotholes, and other minor defects.

SHEATHING PAPER Tar paper sheeting applied between the sheathing and outer covering of a wall to prevent wind and water infiltration.

SHEAVE Grooved pulley over which a rope or cable runs. *See also* **Padlock sheave,** and **Traveling sheave.** There are many types, including:

Bucket: Pulley attached to a shovel bucket, through

which the hoist or drag cable is reeved. *Also called* a **Padlock sheave.**

Compounding sheave: Pulley located on an elevator car, and on the counterweight, under which the hoist cables run to double the capacity and reduce the speed of an elevator.

Deflector sheave: Pulley, aligned with the drive sheave, that provides a path for the cables to drop straight to an elevator counterweight.

Drive sheave: Grooved wheel of a traction-type hoisting machine over which the hoist ropes pass.

Padlock sheave: Bucket sheave on a dipper or hoe shovel.

Secondary sheave: **a.** Sheave used to enable double wrapping of an elevator hoisting ropes, used to increase traction or deflect the ropes; **b.** Pulley on a gearless machine that serves two purposes: (a) to allow each cable a second pass over the drive sheave, and (b) to deflect the cable for a straight drop to the counterweight.

Traveling sheave: Sheave block that slides in a track; or is mounted on an elevator car and the counterweight when multiple roping is used.

SHEAVE BLOCK Pulley and case provided with a means to secure it.

SHEAVE FRICTION Friction associated with the rotation of a sheave.

SHEAVE HEAD ASSEMBLY *See* **Head block.**

SHEAVE PITCH DIAMETER Diameter between the center of the rope from one side of a sheave or the other.

SHE BOLT Type of form tie and spreader bolt in which the end fastenings are threaded into the end of the bolt, thus eliminating cones and reducing the size of holes left in the concrete surface.

SHED Roof with only one set of rafters, falling from a high to a lower wall.

SHEEN Gloss seen at a glancing angle on an otherwise matt surface.

SHEEP'S FOOT Projection from the drum of a compactor that leaves an imprint resembling that of a sheep's foot. *Also called* **Foot.**

SHEEP'S FOOT ROLLER Compacting roller with feet expanded at their outer tips, used in compacting soil.

SHEEPSHANK Method of shortening cord or fiber rope without cutting it. The knots will hold under a steady pull but will release if the load is slackened.

SHEET ASPHALT Well-graded sand combined with asphalt cement and filler. Its use ordinarily is confined to a surface course, usually laid on an intermediate or leveling course.

SHEET ASPHALT SURFACING Designed mixture of well-graded sand, mineral filler, and asphaltic cement processed in a central plant, laid and compacted while hot.

SHEET DRAINAGE Where water flowing on the surface stays in a thin layer, uniformly covering the entire area.

SHEETED PIT Excavation where all the sides have been sheeted.

SHEET EROSION Lowering of land elevation by the almost uniform removal of soil particles by flowing water.

SHEET GLASS Generic term for flat glass.

SHEET GROOVE Notch or reglet along the outside edge of a foundation to receive and support wall panels.

SHEETING 1. Members of a shoring system that retain the earth in position and which in turn are supported by other members of a shoring system; **2.** Tongue-and-groove board, 22 mm (0.875 in.) thick, used in shoring and bracing.

SHEETING DRIVER Air hammer attachment that encloses plank ends, enabling them to be driven without splintering.

SHEETING JACK Push-type turnbuckle, used to set trench bracing.

SHEET LATH Metal lath formed by punching geometrical perforations in sheet steel.

SHEET METAL Thin-gauge metal, usually galvanized. Typically used in the manufacture of pipe, ductwork and fittings.

SHEET-METAL ROOFING Metal sheets preformed in a range of profiles that incorporate an interlocking, waterproof seam, used for covering pitched roofs. *See also* **Corrugated roofing**, and **Preformed roofing and siding.**

SHEET-METAL SCREW *See* **Screw.** *Also called* **Self-drilling screw**, and **Self-tapping screw.**

SHEET PILE *See* **Pile.**

SHEET STEEL SHAPES Cold-formed sheet or strip steel shaped as a structural member for the purpose of carrying live and dead loads.

SHELBY TUBE Thin-walled sampler used in cohesive soils.

SHELF 1. Ledge or setback; **2.** Horizontal board or slab supported on a wall.

SHELF ANGLE Structural angle with holes or slots in one leg for bolting to a structure to support brickwork, stone, terra cotta, etc.

SHELF LIFE Length of time that a perishable product may remain in stock before serious deterioration takes place.

SHELF RETAINING WALL Retaining wall having a relieving platform built into its upper section.

SHELL 1. Thin curved reinforced concrete slab; **2.** Outer framework of a block.

SHELLAC Transparent coating made by dissolving lac, a resinous secretion of the lac bug, in alcohol.

SHELLAC BOND Bonding material, the principal constituent of which is shellac.

SHELL AND TUBE Heat exchanger comprising a nest of tubes through which the primary fluid is pumped, contained within a shell through which the secondary fluid is pumped.

SHELL CONSTRUCTION Construction using thin curved slabs.

SHELLED GRAIN Grain that has separated from the wood beneath, usually at the transition zone between spring wood and summer wood.

SHELLING UP Act of placing a corrugated shell on the core or mandrel before it is set for driving into the ground. *Also called* **Doodle hole.**

SHELL PUMP *See* **Pump.**

SHELL ROOF Thin, reinforced concrete curved roof used to span large areas.

SHELTER Basic accommodation.

SHELTERWOOD Any harvest cutting of a more-or-less regular and mature crop designed to establish a new crop under the protection of the old.

shgl. *Abbreviation for:* shingle.

SHIELD 1. Structure that is able to withstand the forces imposed on it by a cave-in and thereby protect workers within a structure. *Also called* **Trench box**; **2.** Eye and face protector incorporating a special lens that enables a person to look directly at an electric arc. **3.** *See also* **Drive cap.**

SHIELDED ARC Form of electric welding in which a heavy flux-coated electrode is used.

SHIELDED CABLE Electrical or signal cables enclosed in a woven metal sheath that is designed to prevent noise from entering the line(s).

SHIELDED CONDUCTOR Conductor or cable enclosed by an integral metallic covering designed to prevent interference.

SHIELDING CONCRETE Concrete, employed as a biological shield to attenuate or absorb nuclear radiation, usually characterized by high specific gravity or high hydrogen (water) content or boron content, having specific radiation attenuation effects. *See also* **Biological shielding**, and **Boron-loaded concrete.**

SHIFT Work period.

SHIFT BAR HOUSING Transmission component that houses the shift rails, shift yokes, detent balls and springs, interlock balls and pin, and neutral shaft, available in standard and forward-position configurations.

SHIFT FORK Y-shaped component of a transmission that is located between the gears on the mainshaft and that, when actuated, causes the gears to engage or disengage via the sliding clutches. Shift forks typically are located between low and reverse, first and second, and third and fourth gears.

SHIFT RAIL Rail that guides a shift fork in a transmission using a series of grooves, tension balls, and springs to hold the shift fork in gear. The grooves in the forks allow them to interlock the rails, and the transmission cannot be accidentally shifted into two gears at once.

SHIFT TOWER Main interface between the driver and a transmission, consisting of a gearshift lever, pivot pin, spring, boot, and housing.

SHIM Thin strip or wedge of metal, wood, or other material used under or between members for fine adjustment to plum, or level. *See also* **Repair.** *Also called* **Insert.**

SHIM PACK Pack of thin washers used in conjunction with a spring.

SHIN Replaceable edge of a moldboard.

SHINGLE 1. Roof or wall covering of asphalt, asbestos, slate, or wooden wedge-shaped slat or board of varying widths.

> **Asphalt shingle:** Roofing and siding shingle manufactured on an organic felt or inorganic glass-mat base, the wearing surface being coated with mineral granules, the bottom surface with sand, talc or mica.
>
> **Fiberglass shingle:** Roof and wall shingle fabricated with fiberglass as the base material.
>
> **Metal shingle:** Roof and wall shingle fabricated of aluminum in interlocking shapes.
>
> **Shaped shingle:** Cedar shingle produced in a range of special shapes to produce decorative effects on roofs and walls.
>
> **Strip shingle:** Roof and wall shingle fabricated in strips to resemble three individual units.

2. Thin, rectangular pieces of wood, sawn along the grain and tapering in thickness, used like tiles for roofing. *See also* **Shake.** The following descriptions apply to various grades of shingle product:

> **No. 1 Blue Label:** The premium grade of shingle for roofs and sidewall. They are 100% heartwood, 100% clear, and 100% edge grain.
>
> **No. 2 Red Label:** A proper grade for some applications. Not less than 254 mm (10 in.) clear on 406-mm (16-in.) shingles, 279 mm (11 in.) clear on 457-mm (18-in.) shingles and 406 mm (16 in.) clear on 609-mm (24-in.) shingles. Flat-grain and limited sapwood permitted. Reduced weather exposures recommended.
>
> **No. 3 Black Label:** A utility grade for economy applications and secondary buildings. Not less than 152 mm (6 in.) clear on 406-mm (16-in.) and 457 mm (18 in.) shingles, 254 mm (10 in.) clear on 609-mm (24-in.) shingles.
>
> **No. 4 Undercoursing:** A utility grade for undercoursing on double-coursed sidewall applications or for interior accent walls.
>
> **No. 1 or No. 2 Rebutted-Rejointed:** Same specification as for No. 1 and No. 2 grades but machine trimmed for exactly parallel edges with butts sawn at precise right angles. For sidewall applications where tightly fitting joints are desired.

SHINGLE FLASHING *See* **Step flashing.**

SHINGLE SIDING Variety of shingle types, often ornamental in outline, used as an exterior wall cladding.

SHINGLE STYLE Term used to describe the Domestic Revival of the 1870s and 1880s, influenced mainly by Norman Shaw, but replacing his tile-hanging on the exterior of residences with shingle-hanging.

SHINGLE TILE Flat clay tiles, used chiefly for roofing.

SHINGLING HATCHET Roofers hatchet with a notch in the blade for pulling nails.

Ship. *Abbreviation for:* shiplap.

SHIP AND GALLEY TILE Quarry tile having an indented pattern on its face to produce an antislip effect.

SHIPLAP 1. *See* **Lumber; 2.** Offset lamination of two layers of drywall.

SHIPPER SHAFT Hinge on which the stick of a dipper shovel pivots when the bucket is hoisted.

SHIPPING DRY Shipping lumber having a moisture content (oven-dry basis) of 14% to 20%. This results in reduced shipping weight and less susceptibility to decay; used in the international lumber trade.

SHIPPING LIST Detailed list itemizing individual items in goods shipped or consigned.

SHIPPING WEIGHT 1. Gross weight, including all packaging, of an article to be shipped; **2.** Dry weight of a complete truck or machine with all standard equipment including grease and oil but without any fuel or coolant; **3.** Gross machine weight with full mechanical operating systems.

SHIP SCAFFOLD *See* **Float scaffold.**

SHIPWORM *See* **Marine borer.**

SHIVE Wood sliver piercing roofing felt when laid.

SHIVERING Splintering that occurs in fired glaze or other ceramic coatings due to critical compressive stress.

shk *Abbreviation for:* shank.

shl *Abbreviation for:* shell; shellac.

shld *Abbreviation for:* shield; shoulder.

shlv. *Abbreviation for:* shelves.

shlvg *Abbreviation for:* shelving.

SHOCK ABSORBER Vibration-damping device; on vehicles, used with chassis springs to lessen road bounce.

SHOCK HAZARD Circuit that is accessible simultaneously with a ground having a potential of more than 42.4 V peak and where the current through a 1,500-ohm load is more than 5 milliamperes.

SHOCK LOAD 1. Stress created by a sudden force; **2.** Impact of material such as aggregate or concrete as it is released or dumped during placement.

SHOCK LOSS Condition where, in converting fluid velocity to static pressure, the fluid is stopped or slowed down too rapidly, some of the energy of fluid velocity being converted to useless heat instead of the desired static pressure.

SHOCK PLY *See* **Breaker strip/ply.**

SHOCK WAVE Pressure wave front that moves at a sonic velocity.

SHODDY WORK Careless and unworkmanlike.

SHOE 1. Ground plate forming the link of a track, or bolted to a track link; **2.** Support for a dozer blade or other digging edge to prevent it cutting down; **3.** Cleanup device following the buckets of a ditching machine. *See also* **Pad,** and **Tile shoe; 4.** Short length of pipe at an almost horizontal angle at the base of a downpipe to direct flow away from a wall; **5.** Metal socket enclosing the end of a rafter or other load-bearing timber; **6.** Fitting that holds the lower end of a glazing bar to a roof member. *See also* **Pile point.**

SHOE MOLD *See* **Molding.**

SHOE PLATE Reinforcing plate added to the flanges, and possibly the web, of rolled steel H-section when it is used as a pile so as to reinforce the point to improve penetration into dense materials. *See also* **Pile point.**

SHOE TILE Box towed behind a ditcher from which tile can be laid in the ditch bottom.

SHOO-FLY Temporary overhead electric line.

S-HOOK Metal fastener for ductwork connections. *See also* **S-lock.**

SHOOK OUT Procedure of running steel sheet piles up and down in the interlocks to insure they are free sliding before driving.

SHOOT 1. Detonation of a blast; **2.** Truing the edge of a board with a plane.

SHOOTER *See* **Shot firer.**

SHOOTING Placing of shotcrete.

SHOOTING BOARD Board framed to hold another board steady while it is being planed.

SHOOTING POINTS Establishing level marks with the aid of a builder's level.

SHOOTING ROCK Material or rock that requires blasting.

SHOOT WIRE Wire running across the width of sieve cloth, as woven. *Also called* **Fil, Filler, Weft,** and **Woof.**

SHOP 1. Space for retail sales; **2.** Building equipped for fabrication, manufacturing and repair.

SHOP CUTTING PANEL Plywood panel rejected as not conforming to grade requirements. Normally such a panel's defects may be eliminated by cutting the panel into smaller pieces for applications not governed by building codes.

SHOP DRAWING Drawing showing construction details, furnished by the contractor, subcontractor, or supplier. *Also called* **Working drawings.**

SHOP PANEL Plywood or wood-composite panel in which 1.95 m² (21 ft²) is suitable for industrial uses: it may have broken corners or edges.

SHOPPING CENTER Location predominantly dedicated to retail activity.

SHOP PRIMED Fabricated part that has received a coat of protective paint prior to shipment.

SHOP WELD Weld done during the fabrication or manufacture of an assembly or part, prior to shipping to site. *See also* **Field weld.**

SHOPWORK Work completed in a shop, as against on site.

SHORE 1. Lumber used as a prop, support or brace; **2.** Temporary support for formwork and fresh concrete, or for recently built structures, that have not developed full design strength. *Also called* **Post, Prop, Tom,** and **Strut.**

SHORE-A HARDNESS Reading of a material's hardness on a durometer, the scale of which is 0-100, used on elastomers as polyacrylic esters and natural rubber. Consists of a pinpoint depression into the material, the material being at least 100 mils thick. A Shore-A reading of 80 equals a Shore-D reading of 30.

SHORE-D HARDNESS Reading of a material's hardness on a durometer, the scale of which is 0-100, used on rigid and semirigid materials such as polystyrene.

SHORE HEAD Wood or metal horizontal member placed on and fastened to vertical shoring members. *See also* **Raker.**

SHORE PROTECTION Various types of construction aimed at protecting the shore at its waterline.

SHORING 1. Materials used to support a vertical, or near-vertical face cut in earth as part of an excavation. *See also* **Bracing,** and **Horizontal shoring; 2.** Props or posts of timber or other material in compression used for the temporary support of formwork, or unsafe structures; **3.** Support for structures, especially underpinning; **4.** Process of erecting shores.

SHORING LAYOUT Drawing prepared prior to erection showing the arrangement of equipment for shoring.

SHORT *See* **Short-circuit.**

SHORT-BORED PILING *See* **Pile.**

SHORT BREAK LINE Heavy freehand line on a drawing that indicates a short break in the objects depicted.

SHORT CAP Cap in a pile abutment or pier that is placed in the direction of travel over two or three piles to hold them together and to accommodate a long cap.

SHORT CHAIN Method of attaching a tow sling to a towed vehicle so that the tow chains support the entire load.

SHORT CIRCUIT Accidental connection of two sides of an electrical circuit through which nearly all the current will flow. *Also called* a **Short.**

SHORT CIRCUIT CURRENT Current magnitude at the output terminals of a generator set when the terminals are short circuited.

SHORT CIRCUITED AND DUMPED Condition of a storage battery that has been discharged, short circuited, and with the electrolyte drained. Filling with electrolyte and an initial charge are required before it is ready for use.

SHORT COLUMN 1. Column whose load capacity is limited by strength rather than buckling; **2.** Column that is customarily so stocky and sufficiently restrained that at least 95% of the cross-sectional strength can be developed.

SHORT CONVENTIONAL CAB *See* **Cab.**

SHORTENING Decrease in length.

SHORT FIBER *See* **Texture.**

SHORT-FOURTH TRANSMISSION *See* **Transmission, close ratio.**

SHORT LENGTH Lumber measuring less than 2.4 m (8 ft) in length.

SHORT LIST List developed from a longer list through a process of exclusion for cause or reason.

SHORT-LOG TREES *See* **Tree classes.**

SHORT OIL Varnish having a low ratio of oil to resin.

SHORT PERIOD DELAY Electric blasting cap that explodes 1/50th to ½ second after being initiated. *See also* **Delay.**

SHEET RIPPER TIP *See* **Ripper tip.**

SHORT-ROTATION ENERGY PLANTATION Plantings established and managed under short-rotation intensive culture practices.

SHORT-TERM EXPOSURE Period of time, less than or equal to 24 hours, that an excavation is open.

SHORT-TERM LEASE Generally a lease for a term of less than 10 years.

SHORT TON Measure equal to 907 kg (2,000 lb). *Also called* **Net ton.**

SHORT WALL Wall where the ratio of effective height to thickness does not exceed 12.

SHORTWOOD 1. Pulpwood less than 3.048 m (10 ft) in length; **2.** Trees or stemwood portions of trees delivered in product lengths of less than 4.57 m (15 ft) and normally considered only for pulpwood.

SHOT *See* **Blast.**

SHOT BARREL *See* **Calyx.**

SHOT BLASTING Steel shot projected under compressed air onto a steel surface to remove scale and other debris prior to painting or metal coating.

SHOTCRETE Mortar or concrete pneumatically projected at high velocity onto a surface. Also known as **Air-blown mortar, Gunned concrete, Pneumatically-applied mortar,** and **Sprayed mortar.**

SHOTCRETING Application of concrete with a pneumatic gun.

SHOT FIRER Person who initiates an explosion (fires a blast). *Also called* **Blaster,** and **Shooter.**

SHOTGUN 1. Two-drum, live skyline yarding system used in uphill logging, in which the carriage moves down the skyline by gravity, is lowered to attach logs, and is then raised and pulled to the landing by the main line; **2.** Hot stick used as a ground while working on live exposed electrical conductors.

SHOT PEENING Method of increasing the strength and service life of a metal component, accomplished by increasing the density and strength of the crystalline structure on the surface of the part by impacting the surface with steel shot.

SHOT PIN Fastener projected into a dense material, such as concrete, by an explosive charge.

SHOT ROCK Rock that has been blasted.

SHOT SAWED A finish obtained by using steel shot in the gang sawing process of structural stone to produce random markings for a rough surface texture.

SHOULDER 1. Outer edges of the tread of a tire; **2.** Portion of a roadway between the finished traffic lanes and the top of the foreslope of a ditch or embankment;

3. Unintentional offset in a formed concrete surface, usually caused by bulging or movement of formwork; **4.** Area between the tapered edge and face of a drywall panel; **5.** Side of a horizontal pipe at the level of its centerline; **6.** Surface of a tenon that abuts the wood beside the mortise.

SHOULDERED ARCHITRAVE Architrave around a door that is wider at the top.

SHOULDER NIPPLE Pipe nipple with a space of approximately 19 mm (0.75 in.) between threads at the middle.

SHOULDER PAD Leather, canvas, or felt pad threaded through the suspenders on one shoulder to protect the body from contact with a saw being carried.

SHOVED JOINT *See* **Pointing.**

SHOVEL Digging and loading machine or tool. There are several configurations, including:

> **Dipper:** Revolving shovel with a push-type bucket rigidly fastened to a stick that slides on a pivot in the boom.
>
> **Hoe:** Revolving shovel with a pull-type bucket rigidly attached to a stick hinged on the end of a live boom.
>
> **Hydraulic:** Revolving shovel where all except the propel systems are hydraulically actuated.
>
> **Part-swing:** Excavating shovel in which the upper works can rotate horizontally only through part of a circle.
>
> **Revolving:** Mechanical shovel in which the superstructure, or deck, revolves independently of the lower works about a center-pin.
>
> **Track:** Excavating shovel mounted on a tracked undercarriage.

SHOVEL DOZER Tractor fitted with a front-mounted bucket that can be used for pushing, digging, and truck loading.

SHOVING Formation of a bulge in pavement due to movement of one or more of the top layers; shoving can occur during roller compaction or under traffic.

SHOW Any unit of operation in the woods associated with timber harvesting; logging operations.

SHOWER BATH Floor receptacle, plumbing, and cubicle for bathing by overhead water spray.

SHOWER RECEPTOR Floor and side walls of a shower up to and including the curb of the shower.

SHOW RAFTER False rafter below a cornice.

SHOWROOM Enclosed space for the display of goods.

shp *Abbreviation for:* shaft horsepower.

shp. *Abbreviation for:* shape.

shpmt *Abbreviation for:* shipment.

shr *Abbreviation for:* shear.

SHREDDER Size reduction machine that tears or grinds materials to a smaller and more uniform particle size.

SHRINE Place or area held sacred.

SHRINKAGE 1. Decrease in wood dimensions due to loss of water in the wood cell walls. Shrinkage across the grain of wood occurs when the moisture content falls below 30%, the fiber saturation point. Below the fiber saturation point, shrinkage is generally proportional to moisture content, down to a moisture content of 0%. Shrinkage is expressed as a percentage of the green wood dimensions; **2.** *See* **Deformation**; **3.** Reduction in volume of concrete due to normal drying, or due to thermal movement; **4.** Reduction in volume of a soil caused by evaporation or dispersion of moisture; **5.** Reduction in height of recently placed fill due to consolidation and moisture loss; **6.** Undesirable slight concave depression in drywall joint treatment or over the fasteners.

SHRINKAGE-COMPENSATING CEMENT *See* **Expansive cement.**

SHRINKAGE-COMPENSATION Characteristic of grout, mortar, or concrete made using an expansive cement in which the volume increases after setting. If properly elastically restrained, shrinkage-compensation induces compressive stresses that are intended to approximately offset the tendency of drying shrinkage to induce tensile stresses.

SHRINKAGE CRACK Crack due to restraint of shrinkage.

SHRINKAGE CRACKING 1. Cracking of a concrete structure or member due to failure in tension caused by external or internal restraints as reduction in moisture content develops, or as carbonation occurs, or both; **2.** Interconnected cracks in a pavement surface that form large rectangles, generally caused by volume changes in either the paving material or the subgrade.

SHRINKAGE LIMIT Maximum water content at which a reduction in water content will not cause a decrease in volume of the soil mass.

SHRINKAGE LOSS Reduction of stress in prestressing steel resulting from shrinkage of concrete.

SHRINKAGE RATIO Ratio between a given volume change, expressed as a percentage of the dry volume, and the corresponding change in water content above the shrinkage limit; expressed as a percentage of the weight of oven-dried soil.

SHRINKAGE REINFORCEMENT Reinforcement designed to resist shrinkage stresses in concrete.

SHRINKAGE TEST Determination of the maximum water content below which a reduction will not cause a decrease in volume of a soil mass.

SHRINK FACTOR Calculation obtained by dividing the density required per compacted cubic meter (yard) of a material by the density per bank cubic meter (yard) of the material.

SHRINK FIT Fit that allows an outside member, when heated to a practical temperature, to assemble easily with an inside member, and which, when cooled, forms a particularly tight connection.

SHRINK-MIXED CONCRETE Ready-mixed concrete mixed partially in a stationary mixer and then mixed in a truck mixer.

SHROUD Housing located around a fan to direct the flow of air.

sht *Abbreviation for:* sheet; short.

shth *Abbreviation for:* sheath.

shthg *Abbreviation for:* sheathing.

SHUNT 1. Piece of metal connecting two ends of leg wires, deliberately shorting any portion of an electrical blasting circuit, to prevent stray electrical currents from causing the accidental detonation of a blasting cap; **2.** Resistive element placed across the terminals of an ammeter to bypass a portion of the current.

SHUNT CIRCUIT Parallel circuit connection.

SHUTDOWN 1. Work stoppage for any reason; **2.** To shut off a machine.

SHUTOFF NOZZLE Common type of fire hose nozzle permitting flow of the stream to be controlled at the nozzle rather than only at the source of supply.

SHUTOFF VALVE *See* **Valve.**

SHUTTER BAR Pivoted bar that fastens a shutter in position over a window.

SHUTTERING *See* **Formwork.**

SHUTTERS Louver-like doors; used to restrict air and/or light from entering a window, or to regulate the passage of air through a radiator, etc. *See also* **Blinds.**

SHUTTERSTAT Temperature-operated device that controls the opening and closing of radiator shutters.

SHUTTING STILE *See* **Door.**

SHUTTLE 1. Repeated movement back and forth over a predetermined route; **2.** Back and forth motion of a machine that continues to face in one direction.

SHUTTLE HAULING Use of preloaded trailers to reduce truck turnaround time.

SHUTTLE VALVE *See* **Valve.**

shv. *Abbreviation for:* shave; sheave.

shwr *Abbreviation for:* shower.

SI *Abbreviation for:* Système International. (*See appendix* **Metric and nonmentric measurement.**)

SIAMESE CONNECTION Hose fitting for combining the flow from two or more lines of hose into a single stream; one male coupling to two female couplings.

SIC *Abbreviation for:* standard industrial classification.

sid. *Abbreviation for:* siding.

SIDE Men and equipment needed to yard and load any one logging unit of an operation.

SIDE BATTER Batter (slope angle) left or right from the bottom direction.

SIDE BIND One of five basic conditions affecting the lay of a tree. It generally occurs when both ends of the tree are supported by a solid object, with a stump, tree, or other object in between. As a result, the bole of the tree is in a bind and will spring sideways when the run is completed. *See also* **Bottom bind, Drop, End pressure,** and **Top bind.**

SIDEBOARD 1. Removable board used with a swinging or pivoting side bracket to increase the effective width of the loading deck of a lowbed or drop deck trailer; **2.** Loose, plank-type extension, used to increase the height of a dump body side, which increases payload capacity.

SIDE-BORING BACKCUT Intentional variation to the standard backcut to prevent loss of control of a tree and/or the possibility of it barber-chairing. The nose of the bar is pushed into the tree behind the face and 50 mm (2 in.) above the horizontal cut.

SIDE BRACKET 1. Cantilevered arm unit, supported by scaffolding; **2.** Extension to the side of a lowbed or drop-deck trailer that increases the vehicle's loading deck width. There are two types:

Removable: Extensions designed to be installed in sockets or clips provided at intervals along the sides of a trailer.

Swinging: Pivoted or hinged extensions installed at intervals along the sides of a trailer.

SIDECASTING Piling spoil alongside the excavation from which it is taken.

SIDE CHANNEL 1. *See* **Spillway**; **2.** One of the part of a pile hammer fitted to the sides of the hammer that engages the rails of a set of leads to hold the hammer in the leads. *Also called* **Angle-iron guides, Gibs, Grooves, Jaws, Side guides,** and **Ways.**

SIDE CLEARANCE Distance that a tooth projects sideways beyond the body of a saw.

SIDE CONSTRUCTION TILE Tile intended for placement with the axes of its cells horizontal.

SIDE CUTTER Portion of the working tools on the cutter drum that cuts the perpendicular joint.

SIDE-CUTTING PLIERS *See* **Pliers.**

SIDE DITCH Open drain alongside a road, designed to receive water from it and the adjacent ground.

SIDE-DUMP BUCKET *See* **Bucket.**

SIDE-DUMP DUMP BODY *See* **Truck.**

SIDE-ENTRANCE MANHOLE Deep manhole having a horizontal access shaft to an inspection chamber.

SIDE FRAME Supporting structure of a track mechanism. Side frames are attached to the crawler carbody and may be extendable and/or removable.

SIDE GEAR Gear that meshes with the differential cross and splines on the axle shafts to transmit power to the wheels.

SIDE GUIDE *See* **Side channel.**

SIDE GUTTER Small section of gutter on a roof slope at the intersection of a dormer, chimney, etc.

SIDEHILL 1. Slope that crosses the line of work; **2.** To fell a tree at an angle to the hill instead of straight down the slope. *See also* **Quarter.**

SIDEHILL CUT Excavation in a hill involving one cut slope and, usually, one fill slope.

SIDE JAMB *See* **Return jamb.**

SIDE LAP 1. Lap by which single-lap tiles cover each other at the side; **2.** Amount by which the vertical joint in one course is covered by the slate, tile, or shingle above.

SIDE LEAN Sideways lean of a standing tree. This is the least pronounced lean of a tree. *See also* **Lean.**

SIDE LIGHT 1. One or a pair of windows flanking a door; **2.** Artificial light in an interior wall or partition.

SIDE LOAD *See* **Lateral deflection.**

SIDE LOADER Self-loading industrial lift truck, generally high-lift, having load engaging means mounted in such a manner that they can be extended laterally under control to permit a load to be picked up and deposited in the extended position and transported in the retracted position.

SIDE-LOADER DETACHABLE CONTAINER System similar to a rear loader except that it is loaded at the side of the collection vehicle.

SIDE LOADING Crane loading applied at any angle to the vertical plane of the boom.

SIDE-LOADING ATTACHMENT Attachment used to pick up loads and deposit them at right angles to the longitudinal axis of a lift truck. Primarily intended for narrow aisle operations.

SIDE-LOADING PACKER *See* **Refuse truck.**

SIDE-NOTCH Additional side saw cuts made to prevent a 'barber chair' or to facilitate sawing large trees into logs.

SIDE-NOTCHING BACKCUT Intentional variation to the standard backcut to prevent loss of control and/or the possibility of it barber-chairing by reducing the amount of holding wood remaining to be cut by cutting each side prior to the final backcut.

SIDE OUTLET Fitting having an outlet or opening in the side.

SIDE POST 1. Normal practice of mounting rails on opposite hoistway walls midways, front-to-back; **2.** Princess post.

SIDE RAIL Upper or lower extension of a truck body's side that runs longitudinally front to back.

SIDE RESISTANCE *See* **Shaft resistance.**

SIDE RING Removable component of a multipiece wheel rim assembly that provides lateral support for one tire bead.

SIDE *See* **Face.**

SIDESCAR Tendency of some felled trees to tear out fibers of wood at the edges of the notch.

SIDE SEAL Longitudinal seam of a filter medium in a filter element.

SIDESHOT Survey readings taken to each side of a traverse to establish the relative direction of something that intersects the line of traverse.

SIDE STAKE Stake on the line of the outer edge of a proposed pavement.

SIDESWAY Lateral movement of a structure under the action of lateral loads, unsymmetrical vertical loads, or unsymmetrical properties of the structure, or by wind.

SIDESWAY BUCKLING Buckling mode of a multistory structural steel frame precipitated by the relative lateral displacement of joints, leading to failure by sidesway of the frame.

SIDE VALVE (ENGINE) Engine having its inlet and outlet valves housed in the engine block.

SIDE VENT Vent connected to a drain at an angle of 45° or less.

SIDEWALK That portion of the roadway primarily constructed for the use of pedestrians.

SIDEWALK ELEVATOR *See* **Elevator.**

SIDEWALL 1. Walls at each end of a culvert; 2. *See* **Tire construction.**

SIDEWALL GROOVING Circular or spiral grooves cut in the walls of a pier hole in rock or soil to improve the sidewall support of the pier.

SIDEWALL SHEAR Frictional resistance to axial movement of a pier or pile, developed between the peripheral surface of the pier and the surrounding soil.

SIDEWALL SPRINKLER Automatic sprinkler with a deflector so arranged that the sprinkler may be installed on the walls near the ceiling and spray water out into the room.

SIDEWINDER 1. Tree that is pushed over to the side; a limb or sapling that is bent under a tree that has been felled; 2. Type of boom boat.

SIDING Finish covering of the outside of a frame building. *Also called* **Weatherboard.**

SIDING NAIL *See* **Nail.**

SIEMENS One of 17 derived units with special names of the SI system of measurement: a unit of electrical conductance between two points of a conductor when a constant current of 1 A in the conductor produces a potential difference of 1 V, and when the conductor itself is not the source of any electromotive force. (1 $S=1$ $A/V=1$ Ω^{-1}). *See also the appendix:* **Metric and nonmetric measurement.**

SIEVE Metallic plate or sheet, a woven wire cloth, or other similar device with regularly spaced apertures of uniform size, mounted in a suitable frame or holder for use in separating granular material according to size. The following are the opening sizes of US standard sieves:

COARSE SERIES

	mm	in.
4 in.	101.6	4.0
3-1/2 in.	88.9	3.5
3 in.	76.2	3.0
2-1/2 in.	63.5	2.5
2 in.	50.8	2.0
1-3/4 in.	44.4	1.75
1-1/2 in.	38.1	1.5
1-1/4 in.	31.7	1.25
1 in.	25.4	1.0
7/8 in.	22.2	0.875
3/4 in.	19.1	0.75
5/8 in.	15.9	0.675
1/2 in.	12.7	0.5
3/8 in.	9.52	0.375
1/4 in.	6.35	0.25

FINE SERIES

	mm	in.
No. 4	4.76	0.187
No. 5	4.00	0.157
No. 6	3.36	0.132
No. 8	2.238	0.0937
No. 10	2.00	0.0787
No. 16	1.19	0.0469
No. 20	0.84	0.0331
No. 30	0.59	0.0232
No. 40	0.42	0.0165
No. 80	0.177	0.0070
No. 100	0.149	0.0059
No. 200	0.074	0.0029
No. 500	0.029	0.0011

SIEVE ANALYSIS Particle size distribution, usually expressed as the weight percentage retained upon each of a series of standard sieves of decreasing size and the percentage passed by the sieve of the finest size. *Also called* **Screen analysis.**

SIEVE CORRECTION Correction of a sieve analysis to adjust for deviation of sieve performance from that of standard calibrated sieves.

SIEVE FRACTION That portion of a sample that passes through a standard sieve of specified size and is retained by some finer sieve of specified size.

SIEVE NUMBER Number used to designate the size of a sieve, usually the approximate number of openings per linear inch, applied to sieves with openings smaller than 6.3 mm (0.25 in.).

SIEVE SHAKER Mechanically vibrated table on to which a bank of standard sieves is clamped.

SIEVE SIZE Nominal size of openings between cross wires of a testing sieve.

SIEVE TEST Test used to determine quantitatively the percent of asphalt present in a mix in the form of relatively large globules.

SIFTINGS Fine materials that fall from a fuel bed through its supporting grate openings during combustion, comprised of fuel, ash, etc.

sig. *Abbreviation for:* signal.

SIGHT DISTANCE Minimum distance required for passing vehicles to *See* each other on a highway; distance measured from a point 100 mm (4 in.) above the pavement to a point 1.3 m (4.5 ft) above the pavement.

SIGHT-FEED LUBRICATOR Lubricator containing a view of the oil flow, which is installed in the fluid line to supply oil to lubricate a pile hammer. *See also* **Line oiler, Lubricator,** and **Oiler.**

SIGHT GLASS Glass tube inserted into piping to show the liquid level in pipes, tanks, bearings, etc.

SIGHTING Technique of aligning the handlebars and/or gunning mark on a chain saw with the desired falling direction. Since the gunning mark and handlebars are at a 90° angle to the bar, the exact position of the undercut in relation to the desired falling location is constant. *Also called* **Gunning.**

SIGHT-LINE EASEMENT *See* **Easement.**

SIGHT RULE *See* **Alidade.**

SIGHT SIZE Actual size of an opening that admits light.

SIGHT TRIANGLE Area of a corner lot on which, to preserve unobstructed vision for motorists, neither shrubs, trees, nor structures may be placed or erected.

SIGN 1. Visible means of communicating; 2. Warnings of hazard, temporarily or permanently fixed or placed at the location where the hazard exists.

SIGNAGE Any device intended to convey information as to location, direction, action, etc. *See also* **Audible signage.**

SIGNAL 1. Audible, hand, or verbal sign used to direct actions; **2.** Moving sign (warning of possible or existing hazard).

SIGNAL CHANGE INTERVAL *See* **Traffic signal.**

SIGNAL DEVICE One that provides a signal light in an elevator car and that is illuminated when the car approaches a landing at which a landing signal registering device has been actuated.

SIGNAL HEAD *See* **Traffic signal.**

SIGNAL INDICATION *See* **Traffic signal.**

SIGNALING CIRCUIT Any electric circuit that energizes signaling equipment.

SIGNAL REGISTERING DEVICE Button or other device, located at an elevator landing, that when actuated by a waiting passenger, causes a stop signal to be registered in the car.

SIGNAL SYSTEM One consisting of buttons or other devices located at elevator landings that, when activated by a waiting passenger, illuminates a flash signal or operates an annunciator in the car indicating floors at which stops are to be made.

SIGNAL TRANSFER DEVICE Device by which a signal registered in an elevator car is automatically transferred to the next car following, in case the first car passes a floor for which a signal has been registered without making a stop.

SIGNAL TRANSFER SWITCH *See* **Switch.**

SIGNIFICANT LOSS Any loss that introduces a bias in the final result.

SIGN-OFF Formal acceptance of a planned or proposed course of action, constituting authority for the planner or proposer to proceed with the action.

sil. *Abbreviation for:* silence; silver.

SILANE Generally refers to alkyltrialkoxysilanes, a monomeric organosilicon compound that forms a chemical bond with siliceous minerals providing water repellent protection to masonry substrates.

silct. *Abbreviation for:* silicate.

SILENCER Device for reducing gas flow noise; noise is decreased by tuned resonant control of gas expansion.

SILENT CHAIN *See* **Chain.**

silic. *Abbreviation for:* silicate; siliceous.

SILICA Silicon oxide (SiO_2); Mineral contained in the clay used for brickmaking.

SILICA FLOUR Very fine-divided silica.

SILICA FUME Very fine noncrystalline silica produced in electric arc furnaces as a by-product of the production of elemental silicon or alloys containing silicon.

SILICA GEL Dehumidifying substance that readily absorbs moisture at ambient temperature, and gives up absorbed moisture when heated.

SILICATE Salt of a silicic acid.

SILICATE BOND Type of bond matured by baking, in which silicate of soda is an important bonding constituent.

SILICATE ESTER SYNTHETIC FLUID Fluid composed of organic silicates. It may contain additives.

SILICEOUS AGGREGATE CONCRETE Concrete made with normal-weight aggregates having constituents composed mainly of silica or silicates.

SILICON *See* **Alloying elements.**

SILICONATE Organic modified alkali silicates, generally applied in aqueous solution to harden and/or protect masonry substrates.

SILICON CARBIDE An artificial product, granules of which may be embedded in concrete surfaces to increase resistance to wear or as a means of reducing skidding or slipping on stair treads or pavements; used as an abrasive in saws and drills for cutting concrete and masonry.

SILICONE Resin, characterized by water-repellent properties, in which the main polymer chain consists of alternating silicon and oxygen atoms, with carbon-containing side groups.

SILICONE RESIN Vehicle used in the manufacture of heat-resistant paint.

SILICONE SYNTHETIC FLUID Fluid composed of silicone. It may contain additives.

SILICOSIS Lung disease, principally caused from inhaling rock dust.

SILL 1. Heavy horizontal timber. *See also* **Plate**; **2.** Lowest member of a frame structure; **3.** Line of masonry supporting a wall; **4.** Lower horizontal part of a window frame or door opening; **5.** Horizontal overflow line of a dam, spillway, weir, etc.; **6.** Submerged structure across a river to control upstream water levels; **7.** *See also* **Hoist frame side rail.**

SILL ANCHOR Bolt anchored in a concrete or masonry foundation to which the plates of framed construction are fastened.

SILL BEAD Deep bead.

SILL BLOCK Solid masonry unit, used for the sills of openings.

SILL COCK *See* **Bib.**

SILL COURSE *See* **Belt course.**

SILL HIGH The level upon which the window frame rests.

SILL LOG Log placed transversely under the stringer ends to form a simple abutment; the bottom log of a post or frame bent.

SILL PLATE Structural member anchored to the top of a foundation wall, upon which the floor joists rest.

SILL SEALER Impermeable barrier between a foundation wall and mudsill.

SILL STICK Shortened version of a story pole, usually about 1.2 m (4 ft) in length, used for gauging the height of corners to sill height.

SILO Large vessel, whose cross section is considerably smaller than its height, for the storage of loose or granular material.

SILO BLOCK Curved masonry unit.

SILO LOADOUT System that weighs materials as they are discharged from a silo, stopping flow when a predetermined weight has been reached.

SILT 1. Contaminated particles 5 μm and less in size; 2. See **Soil types.**

SILT BOX Metal box at the bottom of a road gully for collecting sand and other debris, that can be removed and emptied.

SILT DISPLACEMENT Tunneling technique in nearly fluid silts in which, as the shield is driven forward, silt is forced into the completed tunnel through openings, loaded, and removed.

SILTING Filling with soil or mud conveyed by water.

SILT LOAM See **Soil types.**

SILT TEST See **Decantation test.**

SILT TRAP Settling basin that prevents waterborne soil or mud from entering a drainage system or waterway.

silv. *Abbreviation for:* silver, silvery.

SILVER SOLDER Alloy of copper and zinc plus a small amount of silver that has a lower melting point than hard solder.

SILVICS Study of the life history and general characteristics of forest trees and stands.

SILVICULTURAL SYSTEM Process of tending, harvesting, and replacing forest trees, that results in the production of forests with distinct compositions. Systems are classified according to the method of harvest cutting used for stand reproduction.

SILVICULTURE Art and science of growing and tending a forest. *See also* **Incremental silviculture.**

SILVICULTURE SURVEY Sampling procedure to determine silvicultural conditions such as planting survival, free-growing status, stocking, etc., leading to management decisions.

sim. *Abbreviation for:* similar; simulate; simulated.

simp. *Abbreviation for:* simplified.

SIMPLE BEAM Beam without restraint or continuity at its supports. *Also called* **Simply supported beam.**

SIMPLE CURVE Arc joining two straights without a transition curve.

SIMPLE FRAMEWORK Perfect frame.

SIMPLE FRAMING See **Structural steel framing.**

SIMPLE SPAN Superstructure between abutments.

SIMPLY SUPPORTED BEAM See **Simple beam.**

simul. *Abbreviation for:* simultaneous.

SIMULATED SERVICE TEST See **Bench test.**

sin *Abbreviation for:* sine.

sin^{-1} *Abbreviation for:* inverse sine.

SINGLE-ACTING CYLINDER See **Cylinder.**

SINGLE-ACTING HAMMER See **Pile hammer.**

SINGLE-ACTING PUMP See **Pump.**

SINGLE-ACTING RAM See **Ram.**

SINGLE-ACTION SHEAR Mechanized cutting tool that uses one hydraulic cylinder to push a cutting blade through a tree while a fixed anvil provides support for the blade on the tree's opposite side.

SINGLE-AXLE SUSPENSION See **Suspension.**

SINGLE-AXLE WEIGHT Weight transmitted to the road by all wheels whose centers fall within two parallel transverse lines, 1016 mm (40 in.) apart, extending across the full width of the vehicle.

SINGLE-BLADE GATE Vertical sliding, counterweighted device used to provide entrance protection on freight elevators and consisting of one panel, usually made of expanded metal. *See also* **Double-blade gate.**

SINGLE-BEAM OUTRIGGER See **Outrigger.**

SINGLE-CHAMBER INCINERATOR See **Incinerator.**

SINGLE-CLEAT LADDER Ladder consisting of a pair of side rails, connected together by cleats, rungs, or steps.

SINGLE-COATING TECHNIQUE Method of producing composition roofing in which the surface coating is applied in one layer.

SINGLE CONTRACT Construction contract let to a single prime contractor who is responsible for all the work.

SINGLE CORNER BLOCK Concrete masonry unit that has one flat end, used at the end or corner of a wall.

SINGLE CURVATURE Deformed shape of a member having one smooth continuous arc, as opposed to a double curvature that contains a reversal.

SINGLE-CUT FILE See **File.**

SINGLE-DRIVE AXLE See **Axle.**

SINGLE EXPANSION JOINT See **Joint.**

SINGLE-FACE SLEDGEHAMMER See **Hammer.**

SINGLE-FAMILY HOUSE Dwelling designed for occupancy by one family.

SINGLE FLOOR Single-layer structural wood panel flooring system combining subflooring and underlayment.

SINGLE FOOTING Footing that carries a single column.

SINGLE-FUNCTION MACHINE Machine, fixed or mobile, designed to perform one type of function.

SINGLE GLAZING Light consisting of a single thickness of glass.

SINGLE-HUNG SASH See **Sash.**

SINGLEJACK 1. One faller in a quarter who falls and bucks; 2. 1.8-kg (4-lb) sledge hammer.

SINGLE-LAP TILE Tile that overlaps only the next course below it.

SINGLE LAYER One layer of drywall applied as the finish to walls and ceilings.

SINGLE-LAYER HOISTING DRUM Drum that accumulates a suspension rope in a single layer on the surface of the drum.

SINGLE-LINE DIAGRAM *See* Diagram.

SINGLE LOAD *See* **Collection method.**

SINGLE-LOCK WELT Flexible roofing seam.

SINGLE-PASS SOIL STABILIZER Machine equipped with several wheels or paddles that rotate rapidly in contact with the soil, pulverizing it to a measured depth, and mixing it with any materials spread ahead of the machine, such as binders, stabilizers, conditioners, etc.

SINGLE-PIECE CONVEYOR One-piece conveyor that transports and deposits material away from the machine.

SINGLE-PITCH ROOF Lean-to roof.

SINGLE-PLY ROOFING Group of roofing materials that offer a finished roof covering with one sheet or layer. *See also* **Modified bitumen membrane, Thermoplastic single-ply membrane,** and **Thermoset single-ply membrane.**

SINGLE-PLY ROOF SYSTEM ASSEMBLIES Combination of materials that forms a complete roof system over the structural supports.

SINGLE-POINT ADJUSTABLE SUSPENSION SCAFFOLD Suspension scaffolding consisting of a platform suspended by one rope from an overhead support and equipped with means to permit the raising and lowering of the platform to desired work levels.

SINGLE-POLE SCAFFOLD Supported scaffold consisting of platforms resting on putlogs or bearers, the outside ends of which are supported on runners secured to a single row of posts or uprights, and the inner ends of which are supported on or in a wall.

SINGLE-POLE SWITCH *See* **Switch.**

SINGLE-RAIL LADDER Portable ladder with rungs, cleats, or steps mounted on a single rail instead of the normal two rails.

SINGLE-REDUCTION AXLE *See* **Axle.**

SINGLE-ROD CYLINDER *See* **Cylinder.**

SINGLE-SHAFT TURBINE ENGINE Gas turbine engine in which the compressor and turbine are mechanically coupled to the same shaft and mechanically connected to the power output shaft either directly or through gearing.

SINGLE-SIZED AGGREGATE *See* **Aggregate.**

SINGLE SLING Lifting sling with a hook at one end and a ring at the other.

SINGLE SPAN Building or structural member without intermediate support.

SINGLE-SPAN SKYLINE Skyline without intermediate support spars.

SINGLE SPEED DOOR *See* **Door types.**

SINGLE SPREAD Type of adhesive that need be applied to only one of the two surfaces forming a joint.

SINGLE-STAGE COMPRESSOR Machine that compresses air to its full capability in one cylinder.

SINGLE-STAGE CURING Autoclave curing process in which precast concrete products are put on metal pallets for autoclaving and remain there until stacked for delivery or yard storage.

SINGLE-STEM PROCESSING Operation handling one tree at a time.

SINGLE SURFACE TREATMENT Single application of asphalt to any kind of road surface followed immediately by a single layer of aggregate of as uniform size as practicable. The thickness of the treatment is about the same as the nominal maximum size aggregate particles. A single surface treatment is used as a wearing and waterproofing course. *Also called* **Inverted penetration.**

SINGLE THROW Switching device that has only open and closed circuit positions.

SINGLE WRAP Roping arrangement on a traction elevator machine where one end of the hoist rope fastens to the car, continues over the drive sheave only once, with the other end fastened to the counterweight.

SINGLE-WYTHE WALL *See* **Wall.**

SINGULATE 1. To separate logs or lumber; **2.** To make single.

SINK 1. Kitchen or laundry plumbing fixture, usually rectangular in plan with relatively vertical sides and that can contain a reasonable volume of water; **2.** Collapsed blister or bubble leaving a depression in a product.

SINKAGE Recess in an otherwise flat surface.

SINK BIB Tap designed for a kitchen sink.

SINKER Log too heavy to float in water.

SINKING Recess cut into a surface, typically to receive butt hinges.

SINKING-CUT Set of blastholes (round) drilled, loaded, and timed to be lifted vertically, due to the fact that no open face is available.

SINKING FUND Pool of cash and investments reserved for the redemption of debt or capital stock.

SINKING PUMP Pump designed to operate in a well or shaft under construction to keep it dry.

SINTER Ceramic material or mixture fired to less than complete fusion, resulting in a coherent mass.

SINTERED Metallic or nonmetallic filter medium processed to cause diffusion bonds at all contacting points.

SINTERED MEDIUM Metallic or nonmetallic filter medium processed to cause diffusion bonds at all contacting points.

SINTERING Process of making metal parts. Powdered metal is placed in a mold under extreme heat and pressure, causing fusion of the metal powder.

SINTERING GRATE Grate on which material is sintered.

SINUOUS FLOW Turbulence.

SIPE Any of the small, often hook- or bracket-shaped grooves in the tread of a tire that provide extra traction and skid prevention.

SIPHON 1. Tube or pipe through which a fluid flows

over a high point by gravity; **2.** *See* **Siphon spillway,** and **Spillway.**

SIPHONAGE *See* **Self-siphoning.**

SIPHON BREAK Small groove to arrest capillary action between adjacent surfaces.

SIPHON SPILLWAY Spillway constructed as a siphon. Water levels must rise to the crest of the siphon, which then primes itself and flows until the water level falls below its inlet, that is, below the crest. *See also* **Spillway.**

SIPHON TRAP *See* **S-trap.**

S-IRON S-shaped iron strap fastened to the end of a rod, used to give surface reinforcement to masonry walls, or to secure interior framing to a masonry wall.

SISTER BLOCK One of two sheaves or pulleys arranged in tandem.

sit. *Abbreviation for:* situated; situation.

SIT BACK Tree that settles back on the stump, closing the kerf of the backcut, generally a result of improper determination of the tree's lean and/or of wind.

SITE 1. Metes and bounds of an area designated for a project; **2.** Location where construction takes place; the area to be occupied by the project and all adjacent and related areas to be used by the contractor during performance of the work including easements, rights-of-way, buildings not scheduled for demolition, and storage, staging, production, and disposal areas. *Also called* **Job site.**

SITE ASSEMBLY Putting together on site components fabricated elsewhere.

SITE CLASS *See* **Land classification (forest management).**

SITE CLEANUP Removal of construction debris, refuse, and other unwanted materials from a site following completion of the works and prior to handover.

SITE CLEARING Stripping a site of unwanted vegetation, rubble, debris, and structures ready for further development. *Also called* **Clearing the site.**

SITE CONDITIONS Overall description of the site for a project: ground cover, relative elevations, surrounding environment, presence of surface water and the normal water table, soil borings, known history, as well as available access and egress, etc.

SITE DESIGN Planning of an area of land in anticipation of development.

SITE DEVELOPMENT Stage of construction that sees the land on which a project is to be built prepared in anticipation of the construction phase.

SITE DEVELOPMENT PLAN Detailed plan illustrating the proposed arrangement of a site, including site layout, grading, hard materials, and planting. *Also called* **Site plan,** and **Plot plan.**

SITE DRAINAGE Removal of surface water from a site by natural runoff or through a storm sewer system.

SITE EXPLORATION *See* **Site investigation.**

SITE FENCE Temporary fence around the perimeter of a construction site, not necessarily on its boundary.

SITE FURNITURE All accessories that are provided on a site, such as benches, refuse containers, and light standards.

SITE INDEX *See* **Land classification (forest management).**

SITE INSPECTION On-site, physical review of the progress of work made in the completion of a construction project. *Also called* **Construction inspection.**

SITE INVESTIGATION Detailed survey and physical determination of the characteristics of a site, on and below grade, according to terms of reference or a specification particular to the site. *Also called* **Site exploration.**

SITE LOCATION Physical description or drawing of the whereabouts of a site.

SITE OFFICE Location of person responsible for the site and works. *Also called* **Shack,** and **Site shack.**

SITE ORGANIZATION Disposition on site of entities, trades, personnel, materials, etc., plus establishment of a line of authority and responsibility.

SITE PLAN *See* **Site development plan.**

SITE PREPARATION Disturbance of an area's topsoil and ground vegetation to create conditions suitable for construction, development, forest regeneration, etc.

SITE REHABILITATION Conversion of the existing unsatisfactory cover on highly productive forest sites to a cover of commercially valuable species.

SITE SHACK *See* **Site office.**

SITE SIGN Sign visible from outside the site fence stating, as a minimum, the name and location of a firm responsible for the project. In some cases, it may be required that details of the owner, architect (and other professionals), and general contractor, together with information about the applicable building, development, or planning permits also be displayed. The sign may also list the names of some or all of the entities contributing to the development, including principal suppliers, etc.

SITE SPECIFIC Data such as surface or subsurface soil and water testing taken on the property of interest.

SITE SUPERVISION Responsibility, on site, for day-to-day management of operations and supervision of construction. *Also called* **Field supervision.**

SITE UTILIZATION 1. Use to which a site is put, or intended to be put; **2.** Indication of the proportion of a usable forest site occupied by healthy, vigorous forest crop trees at any point in time.

siv. *Abbreviation for:* sieve.

SIWASH 1. Unintentional bight in a line caused by stumps or other objects, preventing the line from running straight; **2.** A line not running in a straight line by being bent around a tree, stump, or rock.

SIX-BY-SIX Truck with six powered wheels, two in front and four in back.

SIX-CHANNEL ABS Antilock braking system (ABS) that uses six sensors and six brake control valves to individually monitor and control all six wheel ends of a 6x2 or 6x4 truck or tractor. *See also* **Four-channel ABS.**

SIX-PANEL DOOR *See* **Door types.**

SIX WHEELER Truck with one steering axle and two rear axles, one or both of which may be driving axles.

SIZE 1. Magnitude of a defined unit, may be linear, square, or cube; **2.** Thin paste applied to porous surfaces to seal them and reduce their absorptive capacity.

SIZE ANALYSIS Grading curve.

SIZE CONSISTENCY The particle size distribution of a product necessary to be consistent with the standard method of sieve analysis.

SIZED SLATES Slates of regular size, not random.

SIZE FACTOR Sum of a tire's section width at the rim and its overall diameter.

SIZE REDUCTION Conversion of a material into smaller pieces through mechanical means.

SIZE STICK Slater's scantle.

SIZING 1. Calculating the size or capacity of a component required to perform a specific task; **2.** Estimating the demand to be placed on a system; **3.** Spreading diluted glue or size on a surface before gluing it.

sk Abbreviation for: sack; sink; sketch; skew.

SK Abbreviation for: spanner key.

skd Abbreviation for: skilled.

SKELETON 1. Network of survey lines obtained by triangulation; **2.** Completed structural framework for a building, prior to cladding.

SKELETON CONSTRUCTION Construction system in which all loads and stresses are transferred to the foundations by a rigidly connected framework of structural elements.

SKELETON CORE Hidden internal frame of a hollow-core door.

SKELETON KEY Warded lock cut especially thin to bypass the wards in several warded locks, making it common to them all.

SKELETON STEP Stair consisting of a tread with no riser.

SKETCH Preliminary, freehand presentation drawing.

SKETCH PLATE Structural steel plate marked out with all dimensions and necessary cuts.

SKEW ANGLE Complement of the acute angle between two centerlines that cross.

SKEWBACK Sloping surface against which the end of an arch rests; a thrust block. *See also* **Chamfer strip.**

SKEW BATTER *See* **Pile, compound batter.**

SKEW CHISEL *See* **Wood chisel.**

SKEWED On an oblique course or direction.

SKEWED BRIDGE Bridge with a superstructure forming an angle other than 90° with the direction of the stream channel or the substructure.

SKEW FLASHING Flashing between a gable coping and the roof below.

SKEW NAILING Driving nails on a slant or obliquely.

SKI ANGLE Angle at which an asphalt paver screed floats on the mix being placed.

SKID 1. *See* **Harvest functions;** **2.** Rollers placed under equipment or structures to allow moving them; **3.** Wooden platform on which goods are placed for shipment; **4.** Fixed quantity of a given class or type of goods.

SKIDDER *See* **Harvesting machines (single function).**

SKIDDING Process of sliding/dragging logs from the stump to a landing, usually applied to ground-based operations.

SKIDDING LINE Main haulage line from a carriage to which chokers are attached.

SKIDDING PAN Plate of heavy steel, round in front, placed under the front end of logs being skidded to prevent them from digging into the ground.

SKIDDING TONG Tong used in skidding to grasp a log.

SKID LOG Short, small log, a series of which are laid transversely at intervals under big logs or stringers. They support logs being stored above ground, and support the end of stringers being prepared.

SKID NUMBER Coefficient of skid resistance (locked tire) times 100.

SKID POLE Log or pole, commonly used in pairs, on which logs are rolled.

SKID RESISTANCE Measure of the frictional characteristics of a surface.

SKID RIG 1. Pile driver rig consisting of a fluid power supply, hoisting apparatus, pile hammer, and leads, all mounted on a common sled for movement. *Also called* **Pile frame;** **2.** Soil boring rig on skids for movement by pushing or dragging.

SKID ROAD Path or trail a skidder or crawler tractor uses to move logs on.

SKID-STEER LOADER Small wheel loader that employs skid steering. The equipment may be fitted with a range of attachments.

SKID TRAIL Rough-formed, temporary forest trail suitable for use by horses or equipment such as dozers or skidders in bringing trees or logs from the actual place of felling to a landing.

SKIM *See* **Skim coat.**

SKIM COAT 1. Thin coat of drywall joint treatment applied over an entire surface; **2.** Layer of rubber material laid on a fabric but not forced into the weave. Normally laid on a frictioned fabric. *Also called* **Skim.**

SKIMMED FABRIC Fabric coated with rubber on a calender. The skim coat may or may not be applied over a friction coat.

SKIMMING 1. Removing surface irregularities, typically from soil; **2.** Diverting surface water by shallow overflow to avoid disturbing sediment.

SKIN Tire used on mobile equipment.

SKIN ENCLOSURE Weatherproof enclosure that is minimal in size, and that usually follows the contours of the equipment or structure being protected.

SKIN FRICTION Resistance to shear between the concrete of a shaft and the soil or rock in contact with it.

SKINNING 1. Removing covering or insulation from electrical cables; **2.** Formation of a hard layer on the surface of paint or varnish by oxidation of the drying oil.

SKINTLED BRICKWORK Brickwork showing irregular variations of projections on the exterior.

SKIP 1. Nondigging bucket or tray that lifts material; **2.** Area of lumber missed by a surfacing machine; **3.** Area unintentionally left unpainted.

SKIP HOIST Material handling system comprising an open-top bucket that can be hoisted along tracks mounted at any angle (including vertical) to a designated height, at which point the bucket pivots to discharge its load.

SKIP LOADER Rubber-tire or tracked equipment having a bucket on the front for picking up material, carrying it for short distances, and loading it onto other equipment.

SKIP TROWEL Technique of plaster texturing giving a rough 'Spanish stucco' effect.

SKIRT 1. Border or molded piece under a window; **2.** Vertical strip placed to the side of a conveyor belt, or to the side of a cutting edge to prevent spillage or to increase capacity.

SKIRTING *See* **Baseboard.**

SKIRTING BLOCK Architrave block.

SKIRTS *See* **Pants.**

SKIVE 1. Cut made on an angle to the surface of a sheet of rubber to produce a tapered or feathered cut; **2.** Removal of a short length of cover to permit the attachment of a fitting directly over the hose reinforcement; **3.** Dig in thin layers.

SKIVE EDGE Drywall paper joint tape, the edges of which have been lightly sanded to improve adhesion and reduce waviness.

SKIVING 1. To dig in thin layers; **2.** Removal of damaged material prior to repair of a tire.

skt *Abbreviation for:* skirt.

SKULCH Logs of little value for lumber.

sky. *Abbreviation for:* skylight.

SKY BOUND Tree that fails to fall after being faced and backcut, generally the result of not correctly estimating the lean.

SKY HOOK Used jokingly, but actually a small bent steel pole for mounting on steel roof beams, used to hoist light material.

SKYLIGHT Glazed opening in a roof or ceiling for admitting daylight. *Also called* **Roof light.**

SKYLINE 1. Type of cable logging in which the mainline is stationary as a carriage moves along it carrying logs from the felling site to the landing; **2.** Line on a yarder that supplies lift blocks, rigging, carriage, and logs.

SKYLINE CARRIAGE Wheeled device that rides back and forth on the skyline for yarding and loading.

SKYLINE CRANE Yarding system capable of moving logs laterally to a skyline as well as transporting logs either up or down a skyline to a landing.

SKYLINE CRANE CARRIAGE Skyline carriage that incorporates provisions for pulling slack in the skidding line.

SKYLINE ROAD Area bounded by the length and lateral yarding width of any given skyline setting.

SKYLINE SLOPE Slant or inclination of the skyline chord, generally expressed as a percent.

SKYSCRAPER Multistory building of considerable height. The term originated in the U.S. in the 1880s with the all-masonry Pulitzer Building that reached 26 storys or 309 feet. The opportunity to build higher was opened with the introduction in 1883-1885, on the Home Insurance Building in Chicago, of metal framing that included iron lintels and girders. For many years, New York's Empire State Building (1930-1932; 1,250 ft) was the tallest building in the world. More recently 'tallest self-supporting structure' and 'tallest man-made structure' has replaced the term 'skyscraper.'

sl. *Abbreviation for:* slate.

Sl. *Abbreviation for:* sleeve.

SL *Abbreviation for:* sea level; shank length; snow load; spacer lug.

S/L *Abbreviation for:* shiplap.

SLAB 1. Flat area of plain or reinforced concrete; **2.** Lateral split in a log being bucked; **3.** Thick plate of material: stone, glass, etc.

SLABBING *See* **Cushion blasting.**

SLAB BOLSTER 1. Continuous wire bar support used to support bars in the bottom of slabs; **2.** Top wire, corrugated at 25-mm (1-in.) centers to hold bars in position. *See also* **Bar support, Bat,** and **Chair.**

SLAB CONSTRUCTION Form of construction without excavation, with a concrete slab as the floor usually supporting the superstructure.

SLAB JACKING Process of either raising concrete pavement slabs or filling voids under them, or both, by injecting a material under pressure. *Also called* **Mud jacking.**

SLAB ON GRADE Nonsuspended, ground-supported concrete slab, reinforced or plain. There are several types:

> **Edge-supported:** The slab rests on a perimeter foundation wall.

> **Floating slab:** The slab terminates at the inside face of the perimeter wall and is considered as floating independently of the foundation wall.

> **Monolithic slab:** The slab and foundation wall are formed as one integral mass.

SLAB SCHEDULE Table on the reinforcement placing drawings giving the quantity and mark of the slabs, and the number of pieces, size, length, and bending details of the reinforcement in each slab.

SLAB SPACER Bar support and spacer for slab reinforcement, similar to a slab bolster but without corrugations in the top wire.

SLACK ADJUSTER Lever that transfers motion from the brake chamber to the brake cam shaft. Manually or automatically actuated, the device ensures correct clearance between brake shoes and brake drum.

SLACKLINE Skyline yarding system where the skyline can be tensioned at the operator's discretion.

SLACKLINE CABLEWAY Cable excavator having a track cable that is loosened to lower the bucket, and tightened to raise it.

SLACKLINE SYSTEM 1. Live skyline system employing a carriage, main line, and haulback line. Both main and haulback lines attach directly to the carriage. The skyline is lowered by slackening it to permit the chokers to be attached to the carriage. Lateral movement is provided by side blocking; **2.** Four-drum standing skyline yarding system in which either the slack pulling line pulls the main line through the carriage or a carriage containing a skidding line is used. The haulback line returns the carriage and holds it in place during lateral yarding.

SLACK-PULLING LINE Line used to pull the main line through a logging carriage.

SLACK-ROPE SWITCH *See* **Switch.**

SLAG Nonmetallic by-product of a blast furnace operation that is lighter in weight and more absorbent than gravel and stone, commonly used as an aggregate for road construction. *See also* **Fouling,** and **Material density.**

SLAG BLOCK *See* **Concrete masonry unit.**

SLAG CEMENT Hydraulic cement consisting mostly of an intimate and uniform blend of granulated blast-furnace slag and hydrated lime in which the slag constituent is more than a specified minimum percentage.

SLAG CONCRETE Concrete in which blast-furnace slag is used as an aggregate.

SLAG STRIP Strip of wood nailed to the edge of a graveled roof to give the edge a finish and prevent gravel from rolling off the roof.

SLAG WOOL Insulation made by blowing steam through fluid slag.

SLAKING Process of combining quicklime with water.

SLAMMING STILE Upright strip at the edge of a door opening against which a door closes.

SL&C *Abbreviation for:* shipper's load and count.

S/LAP *Abbreviation for:* shiplap.

SLASH *See* **Harvest functions.**

SLASH BURNING Prescribed burning of slash and debris left by logging operations. *See also* **Broadcast burning.**

SLASHER *See* **Harvesting machines (single function).**

SLASHER-BUNCHER *See* **Harvesting machines (multifunction).**

SLASH GRAIN Exposed grain of flat-sawn timber.

SLASHING 1. Cutting felled and limbed trees into lengths; **2.** *See also* **Harvest functions.**

SLAT Thin strip, usually of wood, used in a louver or blind or for lattice work.

SLAT BUCKET *See* **Bucket.**

SLATE Fine-grained metamorphic rock possessing a well-developed slaty cleavage. Used for paving or roofing.

SLATE CLAD Composition roofing surfaced with slate granules.

SLATE CRAMP Piece of roofing slate shaped like an hourglass.

SLATE HANGING Roofing slates used as a wall cladding.

SLAVE CYLINDER *See* **Cylinder.**

SLAVE PISTON Small piston having a fixed connection with a larger one. *See also* **Free-running piston,** and **Piston.**

SLAVE UNIT Machine that is controlled by or through a similar machine.

SLAVE VALVE *See* **Valve.**

sld *Abbreviation for:* sealed; slide; soldered.

SLEDGEHAMMER *See* **Hammer.**

SLED PATCH *See* **Repair.**

SLEEPER 1. Heavy timber beam that supports the subfloor and rests directly on a concrete slab; **2.** Cross log or timber in a corduroy road supporting the stringers (longitudinal supports); **3.** Large horizontal beam on which a joist rests.

SLEEPER BEAM Beam that does not transfer load but that serves to provide additional support at the end of a long slab or grade.

SLEEPER CLIP Metal strip used to anchor a sleeper to concrete.

SLEEPER WALL Underground wall either supporting sleepers, or between two walls, piers, or a pier and a wall, to prevent them from shifting.

SLEEVE 1. Tube that encloses a bar, dowel, anchor bolt, etc.; **2.** Metal adapter used to splice pipe by driving the two pipes into, or onto the sleeve. An inside sleeve reduces the ID at the splice while maintaining the pipe OD; an outside sleeve increases the OD while maintaining the pipe ID.

SLEEVE BEARING Bearing in the shape of a stationary tube around a shaft that allows the shaft to revolve in the tube.

SLEEVE PIECE Short, thin-walled brass or copper tube used in plumbing to solder pipes of different metals.

SLENDER Pile, post or upright structure that has a slenderness ratio (height to width or diameter) over 10.

SLENDER BEAM Beam that, if loaded to failure without lateral bracing of the compression flange, would fail by buckling rather than in flexure.

SLENDER COLUMN Column whose load capacity is reduced by the increased eccentricity caused by secondary deflection moments. *See also* **Long column.**

SLENDERNESS RATIO Effective unsupported length of a uniform column divided by the least radius of gyration of the cross-sectional area.

SLENDER SECTION Cross section of a structural steel member that will experience local buckling in the elastic range.

SLENDER WALL Wall where the ratio of effective height to thickness exceeds 12.

SLEW To rotate a boom or superstructure about its axis. *Also called* **Swing.**

SLICKENSIDES Soil mass surfaces that have been smoothed and striated by shear movements.

SLICKER 1. Metal tool used to strike a flat joint; **2.** Plasterer's tool consisting of a thin board about 1.2 m (4 ft) long by 150 to 200 mm (6 to 8 in.) wide, beveled on both sides.

SLICK HOLE Hole column loaded with explosive, without springing.

SLICK LINE End section of a pipeline used in placing concrete by pump that is immersed in the placed concrete and moved as the work progresses.

SLICK SHEET Thin steel plate laid on a tunnel floor before a blast to make hand mucking easier.

slid. *Abbreviation for:* sliding.

SLIDE 1. Small landslide; **2.** Fresh tile wall that has buckled or sagged; **3.** Rigging device fitted on the end of a choker that slides on the bight of the mainline.

SLIDEBACK CYLINDER Hydraulic cylinder, usually a long stroke, mounted horizontally at the front of a truck body, used to slide the body forward or rearward.

SLIDE COUPLING Slip joint.

SLIDE HAMMER *See* **Hammer.**

SLIDE PILE *See* **Pile.**

SLIDER-BED-TYPE MOVING WALKWAY *See* **Moving walkway.**

SLIDING DAMPER *See* **Damper.**

SLIDING DOOR *See* **Door types.**

SLIDING FIFTH WHEEL *See* **Fifth wheel.**

SLIDING FORM *See* **Slipform.**

SLIDING GEAR TRANSMISSION *See* **Transmission.**

SLIDING JOINT 1. Similar to a fillet in shape except that it is a moving joint in the corner of which backer rods are placed before a sealant is applied. **2.** *See also* **Fifth wheel.**

SLIDING-PANEL WEIR Frame weir containing panels that slide between grooved uprights.

SLIDING RESISTANCE Shear strength between unconnected, dissimilar materials.

SLIDING SASH *See* **Sash.**

SLIDING TANDEM Two-axle assembly capable of being moved forward or backward on a trailer body to obtain optimum load distribution.

SLIDING-T BEVEL Measuring tool similar to a try-square except that it has a slotted blade that can be clamped at any angle relative to the handle. *See also* **Bevel.**

SLIDING UNDERCARRIAGE *See* **Undercarriage.**

SLIDING WINDOW *See* **Window.**

SLIGHT NEGATIVE ALLOWANCE *See* **Fit classifications.**

SLIME Soft viscous deposit or coating.

SLING 1. Lifting hold consisting of two or more strands of chain or cable, or a suitably reinforced fabric length; **2.** An assembly that connects the load to the lifting device; **3.** Flexible strap attached to lifting cables used to lift heavy objects; **4.** Basic structural frame that consists of two stiles, a crosshead, and a bolster or safety plank that supports the platform and cab of an elevator.

SLING BLOCK Frame containing two sheaves mounted on parallel axles so that they will line up when pulled from opposite directions. *See also* **Block.**

SLINGING HOLE Small hole cut in a pile for attachment of a shackle for handing.

SLIP 1. Movement occurring between the adjacent faces of materials or objects in intimate proximity to each other; **2.** Movement occurring between steel reinforcement and concrete in stressed reinforced concrete, indicating anchorage breakdown; **3.** Amount or degree of mechanical inefficiency in a mechanism; **4.** Ramp leading from the shore to below water level; **5.** Fluid grout; **6.** Long thin strip of material, usually wood; **7.** Physical property of an adhesive to permit movement of one or both of the adherends after application and prior to initial set.

SLIP CLUTCH *See* **Clutch.**

SLIP COUPLING Pipe coupling without a stop, permitting it to slip over a pipe.

SLIP-CRITICAL JOINT Bolt joint in which the slip resistance of the connection is required.

SLIP FACTOR Coefficient of friction between friction-grip fasteners and the members they are gripping.

SLIP FEATHER Tongue in a joint such as a feather joint.

SLIPFORM Form that is moved progressively, horizontally or vertically, as concrete is placed, producing a continuous section. *Also called* a **Sliding form; 2.** Loose vertical panel in a wall mold, used temporarily to separate two types of concrete during simultaneous placing operations; a colored or special-aggregate facing mix from a backing mix, for instance. The panel is removed as soon as the form is full.

SLIPFORM PAVER Machine that lays concrete road slabs on prepared subgrade, pulling the forms with it.

SLIPFORM PAVING TRAIN Assembly of equipment and machinery that continuously produces road or airfield pavement slabs. A typical train consists of: (a) a side feeder into which mixed concrete is placed; (b) a spreader to distribute the concrete over the width of the pavement; (c) a slip-form paver; (d) longitudinal groover; (e) transverse joint groover; (f) texturing equipment; (g) curing compound sprayer.

SLIP GRAB Pear-shaped link, attached by a swivel to

a chain. The chain runs freely through this link when the large end is down but catches and holds when the small end is down.

SLIP HITCH Means of freeing or moving a log too tight to pass a choker under by placing the choker on part of the log butt.

SLIP HOOK Rounded hook that permits a chain to run freely through it.

SLIP-IN BEARING Bearing that can be placed in position without additional fitting; a precision-fit bearing.

SLIP JOINT 1. Connection that permits axial movement between adjoining parts; **2.** Splined connection loose enough to allow its two parts to slide on each other; **3.** Masonry joint between old and new work consisting of a channel cut in the old wall to receive the brick of the new construction.

SLIP-JOINT PLIERS *See* **Pliers.**

SLIP LAYER Viscous liquid (normally a bituminous coating) applied to a pile surface. It is expected to shear under the downward force of setting soil and minimize additional load from downdrag. *See also* **Negative skin friction.**

SLIPLINING Insertion of new pipe, usually of polyethylene (PE), within an existing defective pipe. Developments provide for the new PE line to be temporarily reduced in diameter before insertion, and subsequently enlarged to provide a tight fit into the original pipeline.

SLIP MORTISE Open or chase mortise.

SLIP NUT Nut used on P traps and similar connections where a gasket is compressed around the joint to form a watertight seal.

SLIP-ON FLANGE Flange that slips on to the end of a plain pipe and is soldered or welded in place.

SLIP-ON TANKER Complete pumping unit including tank, pump, and plumbing that can be readily loaded or removed from the bed of a pickup truck or other suitable vehicle.

SLIPPAGE CRACK Crack in a pavement surface that curves in the direction of wheel thrust.

SLIPPER BRACKET Frame-mounted leaf-spring hanger or bracket in which the spring end is not pinned or shackled and can automatically compensate for changes in spring length due to load variations.

SLIP RAMP Angular connection between an expressway and a parallel frontage road.

SLIP RING Electrical contact that rotates within a survey instrument, used to transmit information from one section to another.

SLIP SILL Door or window sill that does not project into the wall beyond the jambs.

SLIP SPLINE Driveshaft coupling device that permits changes in shaft length due to articulation of the rear axle.

SLIP SURFACE Plane of failure in an earth bank.

SLIP TONGUE *See* **False tongue.**

S-LOCK Metal fastener for ductwork connections.

See also **S-hook.**

SLOOP Two-runner sled used to haul logs or bolts out of the woods. Similar to a scoot except that the sloop is equipped with a tongue.

SLOPE 1. Deviation from the horizontal or vertical; **2.** Incline of a bank, roof, or other inclined plane, expressed as a ratio of horizontal distance to vertical rise.

SLOPE ANGLE 1. Angle up or down from horizontal; **2.** Natural angle of repose of fill material, measured from a horizontal plane.

SLOPE CLASS Code that designates the most common slope encountered in the primary forest fire problem area on a protection unit.

SLOPE CORRECTION TABLE Table with conversions from slope distance to horizontal distance.

SLOPED FOOTING Footing having sloping top or side faces.

SLOPE DISTANCE *See* **Expressions of slope.**

SLOPE EASEMENT *See* **Easement.**

SLOPE RATIO *See* **Expressions of slope.**

SLOPE SENSOR Portion of an automation system that controls cross slope.

SLOPE STAKE Surveyor's stake set at the point where the finished slope of a cut or fill intersects the surface of the ground.

SLOPING Digging of excavation walls in such a way, and at such an angle, that the risk of bank collapse or cave-in is eliminated.

SLOPING FACE CUT Second of the two cuts required to face or undercut a tree. It is angled sufficiently to allow a wide-mouthed face opening. The sloping cut's location is above the horizontal cut when using the conventional face and below the horizontal cut when making a Humboldt face.

SLOPS *See* **solid waste.**

SLOP SINK Large, rectangular deep sink for service or janitorial use.

SLOT Narrow slit or groove; groove machined into a shaft.

slot. *Abbreviation for:* slotted.

SLOT CUT Tile that has been cut to fit around pipes or switch boxes.

SLOT-HEAD SCREW *See* **Screw.**

SLOUGH (slew) 1. Secondary river channel through which flow is usually sluggish; **2.** Shallow depression created to store water.

SLOUGH (sluff) 1. Sliding of overlying material such as overburden upon rock. **2.** Subsidence of shotcrete, plaster, or the like due, generally, to excessive water in the mixture. *Also called* **Sagging; 3.** Release of contaminant from the upstream side of a filter element to the upstream side of the filter enclosure.

SLOW CURING ASPHALT Asphalt containing little or no volatile portions, or a blend of asphalt cement and residual oil.

SLOW-DOWN SWITCH *See* **Switch.**

SLOW POWDER Black powder; a slow acting dynamite.

SLOW SET EMULSION Asphalt emulsion that demonstrates very stable properties; must be stable to dilution in addition to having a high resistance to chemical breakdown.

SLOW TEST Maintained pile load test with time intervals greater than 20 minutes.

slp. *Abbreviation for:* slope.

slpg. *Abbreviation for:* slippage.

slr *Abbreviation for:* sealer.

SLUDGE 1. Waste from a wet grinding process; **2.** Particulate contaminant or a mixture of particulate and liquid contaminant separated from a fluid in an unconsolidated state. *See also* **Solid waste; 3.** Undesirable, insoluble material that accumulates in static zones in an engine as a result of overloading the oil with soot and other insoluble combustion products.

SLUDGE PUMP Pump designed to handle highly viscous materials.

SLUDGER Tool for cleaning material out of holes drilled in rock prior to inserting explosives.

SLUDGE SAMPLE Mud sample from a rotary drill, or sand from a churn drill, used to yield data about the formation being drilled.

SLUGGER Tooth on a roll-type rock crusher.

SLUGGING Pulsating and intermittent flow of shotcrete material due to improper use of delivery equipment and materials.

SLUICE Steep, narrow waterway.

SLUICING Moving granular materials and soil by flowing water, often under pressure.

SLUM Dwelling(s) that lack the minimum services and conveniences set out in the National Building Code and that are also deteriorated, hazardous, and unsanitary.

SLUM CLEARANCE Demolition of buildings declared unfit for habitation and unsuitable to be renovated.

SLUMP 1. Test to evaluate viscosity; **2.** Downward slipping of a coherent body of earthen material along a surface of rupture. **3.** Measure of consistency or fluidity of concrete equal to the measured subsidence of a truncated cone of concrete released immediately after molding in a standard slump cone.

SLUMPABILITY *See* **Feedability.**

SLUMP CONE *See* Slump mold.

SLUMP LOSS Amount by which the slump of freshly-mixed concrete changes during a period of time after an initial slump test was taken, made on a sample or samples.

SLUMP MOLD Standard mold in the form of a truncated cone with a base diameter of 200 mm (8 in.), a top diameter of 100 mm (4 in.), and a depth of 300 mm (12 in.), used to fabricate a concrete specimen for a slump test. *Also called* **Slump cone.**

SLUMP TEST Procedure for measuring a slump.

SLURRY 1. Mixture of water and any finelydivided insoluble material, such as portland cement, slag, or clay in suspension; **2.** Explosive consisting of ammonium nitrate, water, thickeners, and a high-energy sensitizer such as TNT.

SLURRY COAT Application of a thick, often glutinous coat of material. *Also called* **Slush coat.**

SLURRY EXPLOSIVE *See* **Water gel.**

SLURRY SEAL *See* **Bituminous coating.**

SLURRY TRENCH Narrow trench with vertical, unshored walls, in which caving or sloughing of the earth walls is prevented by the hydrostatic pressure of the slurry or mud with which the trench is filled.

SLURRY TUNNEL-BORING MACHINE Type of microtunneling machine in which the soil is turned to slurry and is used to counterbalance water pressure to stabilize the face, before being pumped to the surface.

SLURRY WALL Wall constructed from the ground surface to an impermeable layer; consisting of slurry, usually soil, bentonite, and/or cement. *Also called* **Diaphragm wall.**

SLUSH 1. Broken or crushed ice mixed with water; **2.** Mixture having a high water content.

SLUSH COAT Pure coat of a very soft consistency. *Also called* **Slurry coat.**

SLUSHED JOINT Vertical joint in brickwork filled, after units are laid, by 'throwing' mortar in with the edge of a trowel.

SLUSHER Mobile drag scraper with a metal slide to elevate the bucket to dump height.

SLUSHER TRAIN Muck train composed of a locomotive and articulated muck cars.

SLUSH GROUTING Distribution of grout, with or without fine aggregate, as required over a rock or concrete surface that is subsequently to be covered with concrete, usually by brooming it into place to fill surface voids and fissures.

SLUSHING Hydraulic filling.

SLUSH PUMP Mud pump of a rotary drill.

slv. *Abbreviation for:* sleeve.

sly *Abbreviation for:* slowly.

sm. *Abbreviation for:* small.

s.m. *Abbreviation for:* surface measure.

Sm *Abbreviation for:* standard matched.

SM *Abbreviation for:* standard matched; super multiplate.

SMALL SAWTIMBER *See* **Poletimber.**

SMALL-SCALE FORESTRY Nonindustrial forestry operations.

SMALLTREE Live tree 25 to 125 mm (1 to 5 in.) in DBH (diameter breast height).

SMALLWOOD General term describing small-diameter trees (such as might be removed through precommercial thinning) that are typically unsuitable for commercial roundwood products.

SMH *Abbreviation for:* scheduled machine hour.

smk. *Abbreviation for:* smoke.

smkls *Abbreviation for:* smokeless.

s. mld *Abbreviation for:* struck mold.

smls *Abbreviation for:* seamless.

smog *Abbreviation for:* smoke and fog.

SMOKE Suspended particles and aerosols of the products of combustion, mostly incomplete combustion, less than 0.1 micron in size.

SMOKE ALARM Electrical device that sounds an alarm when sensing the products of combustion; a combined smoke detector and audible alarm device.

SMOKE CHAMBER Space in a fireplace immediately above the throat, and narrowed by corbeling to the size of the flue lining above, where the smoke gathers before passing into the flue.

SMOKE CONTROL ZONE Compartment within a floor area that is separated from the remainder of the floor area in such a way as to be smoke tight for a predicted period of time.

SMOKE DETECTOR Device for sensing the presence of visible or invisible particles produced by combustion, and automatically initiating a signal indicating this condition.

SMOKE JUMPER Trained and equipped fire fighter who travels to forest fire sites by aircraft and parachute.

SMOKE PIPE Pipe conveying the products of combustion from a firebox or hearth to a chimney flue or to the outside atmosphere.

SMOKE SHELF Ledge or shelf, laid level with the bottom of the damper and back of the firebox, that serves to deflect the air currents that come down a chimney back up the opposite side of the flue.

SMOKE TEST 1. Test of a pipe system to determine its tightness by filling it with smoke and sealing all inlets and outlets; **2.** Test conducted following completion of a fireplace, consisting of building a fire at the bottom of the flue and then tightly covering its top.

SMOOTH *See* **Rustication.** *See also* **Bulk appearance.**

SMOOTH ASHLAR Squared, smooth-faced stone.

SMOOTH-BORE HOSE Wire-reinforced hose in which the wire is not exposed on the inner surface of the tube.

SMOOTH COVER Cover having an even and uninterrupted surface; a commercial finish.

SMOOTHER BAR Drag towed behind a leveling machine to break up lumps.

SMOOTH-FACED DRUM Drum with a plain, ungrooved surface.

SMOOTHING IRON Heated iron tool, used to smooth asphalt and seal joints.

SMOOTHING PLANE *See* **Plane.**

SMS *Abbreviation for:* sheet metal screws.

smth *Abbreviation for:* smooth.

SNAG Any dead or dying tree more than 3 m (10 ft) tall.

SNAG BOAT Boat equipped with a hoist and grapple for clearing obstacles from the path of a dredge.

SNAGGING Grinding the gates, fins and sprues from castings.

SNAKE Spring-steel wire or tape used to travel through pipes and conduit when pulling wires or to clear an obstruction. *See also* **Fish.**

SNAKE EYES In helicopter logging, an eleven-thousand pound turn.

SNAKE HOLE Hole drilled or bored under a rock or tree stump for the placement of explosives.

SNAKING 1. Towing a load on a long cable; **2.** Inserting a hoist or tow line under an object without moving it.

SNAP Die that fits in the end of a riveting gun, shaped to form the head of a rivet.

SNAP HEADER Half length of brick, sometimes used in brick facing placed end on. *Also called* **False header.**

SNAPHEAD RIVET Rivet with a rounded head.

SNAPLINE *See* **Chalk line.**

SNAP TIE Proprietary concrete walltie, the end of which can be twisted or snapped off after the forms have been removed.

SNAP TOP Broken-off top of a tree resulting from wind and/or rot.

SNARE Choker used in forest operations.

SNATCH BLOCK Block that opens on one or both cheek plates, permitting the block to be reeved without having to use a free rope end, used to change the direction of pull. *See also* **Block.** *Also called* **Gate block.**

snd *Abbreviation for:* sound.

s.n.d. *Abbreviation for:* sap no defect.

sndprf *Abbreviation for:* soundproof.

SNECK Small stone, used to fill in between larger stones in rubblework masonry.

snflk. *Abbreviation for:* snowflake.

SNIPE 1. To bevel the leading edge of a skid log so it will not hang up; **2.** Allowance for falling and bucking cuts; **3.** Extra length added to a regular log length.

SNIPS Scissor-action shears for cutting sheet metal.

SNORKEL Wooden or steel boom extension mounted on a loader to increase the distance that logs can be reached for loading.

SNORKEL CAP Steel cap fitted on the furthest end of a snorkel to which the snorkel block and guyline are fastened.

SNORKEL POCKET Support and holdfast for the machine at the end of the snorkel.

SNOW DENSITY Moisture content of snow.

SNOW FENCE Open mesh wood, metal, or plastic fencing designed to obstruct the free passage of blowing snow and to cause it to form a drift.

SNOW GUARD Rigid mesh barrier set vertically at the lower end of a sloped roof to prevent snow from sliding off. *Also called* **Roof guard.**

SNOW LOAD Component of a live load attributable to the weight of snow anticipated to accumulate on a surface, particularly a roof, bridge deck, etc.

SNOW PLOW Shaped metal shield that, when propelled by a vehicle, will move snow ahead or to one side. There are several types, including the following:

Leveling wing: Auxiliary plow that folds flat against the side of the vehicle when not in use; when extended, it adds to the width of the primary plow to clear both sides of a road in one pass.

One-way: Having a truncated cone shape designed to discharge in one direction only.

Reversible: Designed to discharge snow to the right or the left; can be used for high- or slow-speed plowing.

Underbody scraper: Dual-purpose device used to remove modest snow accumulations in the winter, and to grade and maintain dirt and gravel roads in the summer.

V-blade: Will discharge snow to the right and left simultaneously.

SNUB 1. To wrap rope several times around an object to provide means of controlling the load; **2.** To lower anything; **3.** To assist one machine down a hill by holding it back with another, the two being connected by a line.

SNUBBER CHAINS *See* **Check chains.**

SNUBBING LINE Line used for lowering a load.

SNUB GABLE Roof gable that is truncated at the top by a hip.

SNUG *See* **Fit classifications.**

snw *Abbreviation for:* snow.

snwfl *Abbreviation for:* snowfall.

SO *Abbreviation for:* special order; south.

SOAKER Strip of flexible metal cut to interlock with slates or tiles to make a watertight joint at hip or valley. *Also called* **Lead soaker.**

SOAKING PERIOD In high-pressure and low-pressure steam curing, the time during which the live steam supply to the kiln or autoclave is shut off and the concrete products are exposed to the residual heat and moisture.

SOAP 1. A brick or tile of normal face dimensions, having a nominal thickness of 50 mm (2 in.); **2.** Interaction of fats or esters, generally with an alkali, to form a metallic salt. *See also* **Complex soap, Saponification,** and **Thickener.**

SOAPING TILE Method of applying a soapy film to newly tiled walls to protect them from paint and plaster during construction.

soc. *Abbreviation for:* society; sociology; socket.

SOCKET 1. Enlarged end, shaped to receive an unenlarged end, as in a pipe joint; **2.** Type of wire rope shackle that has a tapered cavity to receive the rosette and babbitt, used to terminate a wire rope; **3.** Device designed to support an electric light bulb and mechanically connect it electrically to a circuit.

SOCKETED PIER or PILE *See* **Pile, drilled-in caisson.**

SOCKETING Process of preparing shackles consisting of putting wire rope through a shackle, forming a rosette, and pulling it into the shackle, then pouring babbitt to secure the wire rope to the shackle basket.

SOCKET OUTLET Electrical fixture into which a plug may be inserted to obtain a supply of electrical energy.

SOCKET WRENCH *See* **Wrench.**

SOD Matting of grass and soil, that is cut just below the roots and then used on a new site to provide quick grass cover.

SODIUM-SILICATE ADHESIVE *See* **Adhesive.**

SODIUM-VAPOR LAMP High-intensity, electric light source of distinctive yellow color, primarily used for street lighting.

SOFFIT Underside of a structural part.

SOFFIT BOARD Purpose-made wood, metal, or gypsum board product intended for use on exterior overhangs.

SOFFIT FORM Form supporting the bottom of a concrete beam.

SOFFIT SPACER Metal piece used to hold a beam bottom a fixed distance from structural steel.

SOFFIT VENT 1. Opening in the underside of a roof overhang to permit the passage of air into the roof space; **2.** Small, round metal louvered vents.

SOFT-BURNED Clay products that have been fired at low temperature ranges, producing relatively high absorption and low compressive strengths.

SOFT END Hose end in which the rigid reinforcement of the body, usually wire, is omitted.

SOFTENER 1. Additive that prevents a material such as an adhesive from becoming too brittle on final set or cure; **2.** Pipe, wood, or rubber cut to fit the flange of heavy steel beams to prevent damage to a sling.

SOFTENING AGENT *See* **Recycling agent.**

SOFTENING POINT TEST Test used as the basic measurement of consistency for grading tars and certain asphaltic cements.

SOFT-FACED HAMMER *See* **Hammer.**

SOFT FALL System that allows a vehicle's conventional brake system to function normally should an ABS malfunction.

SOFT GROUND Ground unable to bear even a moderate imposed load due to such factors as a high water content, lack of cohesion between soil particles, etc.

SOFT HAMMER *See* **Hammer.**

SOFT JAWS Covers of a soft or malleable material placed over vise jaws to prevent damage to materials held in a vise.

SOFT JOINT Joints between panels of prefabricated masonry in curtain wall construction. Joints are either pointed with mortar or filled with an elastic sealant.

SOFT-MUD BRICK *See* **Brick.**

SOFT PARTICLE Aggregate particle possessing less than an established degree of hardness or strength as determined by a specific testing procedure.

SOFT ROT Rot occurring in outer wood layers under very wet conditions.

softwd *Abbreviation for:* softwood.

SOFTWOOD Wood of any tree with true cones; a conifer.

SOH *Abbreviation for:* stocks on hand.

SOHC *Abbreviation for:* single overhead camshaft.

SOIL *See* **Soil types.**

SOIL ADHESION Sticking of soil to foreign materials such as soil implements, tracks, or wheels.

SOIL ANALYSIS Investigation of earth and the preparation of a report giving conclusions and recommendations, a fundamental stage prior to the design of foundations for most commercial and industrial structures. The analysis covers, at least, soil density, moisture content, load-bearing capacity, shear strength, plasticity index, organic content, and grain size. *See also* **Soil types.**

SOIL ANCHOR High-strength steel tendon installed in the earth with anchoring provision in the soil that will resist movement.

SOIL BORING Small-diameter hole drilled into the soil for the purpose of obtaining earth samples and exploring the subsurface conditions. *Also called* **Test boring.**

SOIL BORING LOG Complete record of what was found from drilling one soil exploration hole.

SOIL BRANCH Sewer branch leading to a soil pipe.

SOIL-CEMENT *See* **Soil stabilization.**

SOIL-CEMENT CYLINDER TEST Steel mold, 71 mm (2.75 in.) ID by 229 mm (9 in.) high, that contains an upper and lower piston to produce a soil-cement sample 142 mm (5.6 in.) high for testing.

SOIL CLASSIFICATION SYSTEM Method of categorizing soil and rock deposits in a hierarchy of Stable Rock, Type A, Type B, and Type C, in decreasing order of stability. The categories are determined based on an analysis of the properties and performance characteristics of the deposits and the environmental conditions of exposure. Such a classification includes:

Stable rock: Natural solid mineral matter that can be excavated with vertical sides and remains intact when exposed.

Submerged soil: Soil that is underwater or is free seeping.

Type A: Cohesive soils with an unconfined compressive strength of 144 kPa (1.5 tons/ ft²) or greater.

Type B: Cohesive soil with an unconfined compressive strength greater than 48 kPa (0.5 tons ft²) but less than 144 kPa (1.5 tons ft²); granular cohesionless soils including angular gravel, silt, silt loam, sandy loam, and in some cases silty clay loam and sandy clay loam; previously disturbed soils (except those classed as Type C); soil that meets the unconfined compressive strength or cementation requirements for Type A, but is fissured or subject to vibration; dry rock that is not stable; or material that is part of a layered system where the layers dip into an excavation on a slope less steep than 4:1.

Type C: Cohesive soil with an unconfined compressive strength of 48 kPa (0.5 tons ft²), or less; granular soils, gravel, sand and loamy sand; submerged soil or soil from which water is freely seeping; submerged rock that is not stable; or material in a sloped, layered system where the layers dip into an excavation on a slope of 4:1 or steeper.

SOIL COHESION The mutual attraction exerted on soil particles by molecular forces and moisture films.

SOIL COMPACTION Compression of soil as a result of heavy equipment traffic.

SOIL COMPACTOR *See* **Loader.**

SOIL COMPRESSION *See* **Plastic creep.**

SOIL CONSOLIDATION 1. Techniques and materials that assist in the removal of moisture from a soil mass, or that prevent the entry of moisture to the mass; **2.** Moving together of soil particles as water and air are squeezed out due to an imposed load. *See also* **Elastic compression,** and **Plastic creep.**

SOIL COVER Covering of lightweight plastic film, roll roofing, or similar material used over the soil to minimize moisture permeation.

SOIL DAMPING Dynamic soil resistance force that is a direct function of pile velocity.

SOIL DAMPING PARAMETER Often designated by 'J' with a unit of measure of s/m (s/ft). Soil damping equals J times pile velocity times static soil resistance.

SOIL EROSION Detachment and movement of soil from the land surface by wind or water.

SOIL FAILURE Alteration or destruction of the soil structure by mechanical forces such as in shearing, compression, or tearing.

SOIL FILL *See* **Backfill,** and **Fill.**

SOIL HORIZON Any layer of soil that may be distinguished from adjacent layers because it differs in physical, chemical, or biological characteristics, usually designated as A, B, and C horizons.

SOIL INVESTIGATION Study of the earth in the area of a foundation consisting of sampling, classification, preparation of logs of borings, and a report setting forth conclusions and recommendations; a basic practice preparatory to the design of foundations as required under most building codes.

SOIL MECHANICS Investigation of the composition of soils, their classification, consolidation, strength, etc.

SOIL-OR-WASTE PIPE Pipe in a sanitary drainage system.

SOIL-OR-WASTE STACK Vertical soil-or-waste pipe that passes through one or more storie and that includes any offset that is part of the stack.

SOIL PIPE Vertical drainage stack into which branch

lines drain plumbing fixtures.

SOIL PLASTICITY Property that allows soil to be deformed or molded in a moist condition without cracking or falling apart.

SOIL PLUG Material that rises in an open-end pile as it is driven.

SOIL PRESSURE *See* **Contact pressure.**

SOIL PROFILE Vertical section showing the succession of soils at a location.

SOIL REPORT Report based on samples, tests, and analysis done to determine the various material composition and structural capabilities of soils in a given area. *See also* **Soil types.**

SOIL SAMPLE Representative specimen.

SOIL SAMPLER Equipment used to extract soil samples from borings or test pits made in subsurface investigations.

SOILS ENGINEER *See* **Geotechnical engineer.**

SOIL SERIES Basic unit of soil classification, consisting of soils that are alike in all major profile characteristics save texture of the surface layer, and that have similar horizons. *See also* **Soil types.**

SOIL STABILIZATION Chemical or mechanical treatment designed to either increase or maintain the stability of a mass of soil or to otherwise improve its engineering properties. There are several techniques, including:

  Cement-modified soil: Soil mixed with a relatively small amount of portland cement; less than would be necessary to produce a hardened compound. *Also called* **Cement stabilization.**

  Fly-ash stabilization: Technique of mixing controlled quantities of fly ash with soil during the process of preparation, compaction of base courses and subbases for road construction, embankments and other earth works, to limit or control the rate of moisture loss.

  Salt stabilization: Use of sodium chloride to lower the rate of evaporation of water during the compaction of soil.

  Soil-cement: Mixture of soil and measured amounts of portland cement and water, compacted to a high density.

SOIL STACK General term for the vertical main of a system of soil, waste, or vent piping.

SOIL TEST Sampling of an area to determine the characteristics of its soils and to map their location; usually accomplished by borings and subsequent laboratory analysis. *See also* **Soil types.**

SOIL TEST BORING Subsurface sample-taking program under controlled conditions using equipment that obtains samples of predictable size and shape to predetermined depths, maintaining the sample size and shape for transport to a laboratory for analysis.

SOIL TYPES Classification of soils into identifiable groups or types. A typical grouping is:

  Cemented: Soil in which the particles are held together by a chemical agent, such as calcium carbonate, such that a hand-size sample cannot be crushed into powder or individual soil parti-

cles by finger pressure.

Clay: Natural cohesive soils having plastic properties and composed of very fine particles that are firmly coherent, compact, and hard when dry, but usually stiff, viscid and ductile when moist. *See also* **Material density.**

Clay loam: Mixture of sand, clay, and silt, having a large percentage of clay.

Cohesive: Clay (fine-grained soil) or soil with a high clay content, that has cohesive strength. Cohesive soil does not crumble, can be excavated with vertical sideslopes, and is plastic when moist. Cohesive soil is hard to break up when dry, and exhibits significant cohesion when submerged. It includes clayey silt, sandy clay, silty clay, clay, and organic clay.

Compact coarse sand: Soil consisting of coarse particles, 0.063 mm (0.0024 in.) or less in diameter, that have been compacted by the weight of overburden or weather. The grains are generally spherical or angular in shape, depending on the extent of weathering and/or decomposition.

Compact fine sand: Sand predominantly retained on the No. 200 (0.074 mm, 0.0029 in.) sieve that, when confined, whether wet or dry, will bear heavy loads. Fine sand has a lower bearing value than coarse sand, since the fine particles can be rearranged and tend to squeeze together, However, water will make fine sand flow more readily than coarse sand.

Dry: Soil that does not exhibit visible signs of moisture content.

Fissured: Soil material that has a tendency to break along definite planes of fissure with little resistance, or material that exhibits open cracks, such as tension cracks, in an exposed surface.

Granular: Gravel, sand, or silt (coarse-grained soil) with little or no clay content. Granular soil has no cohesive strength. Some moist granular soils exhibit apparent cohesion. Granular soil cannot be molded when moist and crumbles easily when dry.

Hardpan: 1. Dense, heterogeneous mass of clay, sand and gravel of glacial drift origin; **2.** Hard layer of consolidated or cemented earth underlying surface soil.

Layered system: Two or more distinctly different soil or rock types arranged in layers. Micaceous seams or weakened planes in rock or shale are considered layered.

Loam: *See* **Topsoil,** below.

Moist soil: Condition in which a soil looks and feels damp. Moist cohesive soil can easily be shaped into a ball and rolled into small-diameter threads before crumbling. Moist granular soil that contains some cohesive material will exhibit signs of cohesion between particles.

Plastic: A property of a soil that allows it to be deformed or molded without cracking or appreciable volume change. A test is to roll it into 3-mm (0.125-in.) diameter strings without it crumbling.

Sand: 1. Small grain of mineral, largely quartz, that is the result of disintegration of rock; **2.** Granular material passing the 9.6-mm (0.375-in.) sieve and almost entirely passing the No. 4 (4.76-mm, 0.187-in.) sieve and predominantly retained on the No. 200 (0.074-mm, 0.0029-in.) sieve. *Also called* **Fine aggregate.** *See also* **Material density.**

Sandy clay: Soil type characterized by sand containing enough clay to act as a binder, thus exhibiting the properties of a sandy soil but lacking the movement of loose sand, and without the slipping or shearing qualities of clays.

Sandy loam: Soil type containing enough clay and silt to render it cohesive.

Saturated: Soil in which the voids are filled with water. Saturation does not require flow. Saturation, or near saturation is necessary for the proper use of instruments such as a pocket penetrometer or sheer vane.

Shale: Laminated and fissile sedimentary rock, the constituent particles of which are principally in clay and silt sizes. *See also* **Material density.**

Sharp sand: Coarse sand consisting of particles of angular shape.

Silt: Granular material resulting from the disintegration of rock, with grains passing a No. 200 (0.074-mm, 0.0029-in.) sieve. *See also* **Mineral filler.**

Silt loam: Sandy soil having a moderate amount of clay and sand, in which over 50% of the sand is composed of extremely fine particles.

Soil: Generic term for unconsolidated natural surface material above bedrock.

Topsoil: 1. Uppermost layer of soil; **2.** Adequately drained soil containing humus and capable of supporting good plant growth. *Also called* **Loam.** *See also* **Material density.**

SOL Suspension of particles of colloidal dimensions in a liquid. These systems possess the gross properties of a liquid.

sol. *Abbreviation for:* solar; solenoid; solid; solution.

SOLAR ACCESS 1. Exposure to, and collection of sunlight; **2.** Right to maintain exposure to sunlight already enjoyed.

SOLAR CELL Device that converts sunlight into electrical energy, or that captures and transmits the thermal component of sunlight.

SOLAR COLLECTOR Device that transforms solar radiation to usable heat. There are various types, including:

Concentrating: Curved collector, that may either increase the available surface area of the collecting surface, or reflect the collected sunlight and focus it on a solar cell.

Flat-plate: Fixed flat-shape collector, oriented to receive maximum exposure to sunlight.

Tracking: Moving collector that orients itself toward the sun.

SOLAR CONSTANT Average solar radiation reaching the earth's atmosphere per minute.

SOLAR ENERGY Heat that is derived from sunlight.

SOLAR FRACTION Percentage of a building's heat energy requirement provided through a solar system.

SOLAR FURNACE Heat exchanger by which the solar energy focused through a concentrating collector is transferred to another medium: rocks, water, etc.

SOLAR GAIN Percentage of a building's heating or cooling load contributed by solar radiation striking the structure or entering it.

SOLAR INSOLATION Total available solar radiation, composed of direct, diffuse, and reflected radiation.

SOLARIUM Glassed-in porch or room.

SOLAR LIGHT Floodlight powered by batteries that are recharged during daylight hours by solar energy via a photovoltaic array.

SOLAR NOON Moment of the day that divides daylight hours exactly in half; half of the time between sunrise and sunset.

SOLAR ORIENTATION Positioning and design of a building to take advantage of maximum exposure to the winter sun.

SOLAR PUMP Mechanical device driven by energy obtained through a solar collector.

SOLAR RADIATION Sun's energy as received on the earth's surface.

SOLAR REFLECTING SURFACE Exterior finish to a roof or wall intended to reduce the effects of solar heating.

SOLAR RIGHTS Right to continue to enjoy the ability to receive direct sunlight.

SOLAR SCREEN Perforated wall used as a sunshade.

SOLAR SCREEN TILE Tile manufactured for masonry screen construction.

SOLAR STILL Desalination plant driven by solar heat.

SOLAR STORAGE Medium used to hold heat obtained through a solar collector.

SOLAR SYSTEM Assembly of equipment designed to collect solar radiation and convert it to another form of usable energy.

SOLDER Metal alloy that can be melted at relatively low heat and used for joining or mending metal surfaces. The most commonly used type of solder is an alloy of tin and lead. It is available in a wide range of proportions, including:

Composition	Melting point
20% tin 80% lead	277°C (532°F)
40% tin 60% lead	230°C (446°F)
50% tin 50% lead	205°C (401°F)
60% tin 40% lead	187°C (369°F)
70% tin 30% lead	185°C (365°F)
90% tin 10% lead	215°C (419°F)

SOLDER JOINT Connection made by fusing with solder. *See* **Solder.**

SOLDER-LINK-TYPE SPRINKLER *See* **Auto-**

matic sprinkler.

SOLDIER 1. Vertical wale used to strengthen or align formwork or excavations; **2.** Stretcher set on end with its face showing on the wall surface.

SOLDIER ARCH *See* **Arch.**

SOLDIER BEAM Steel section driven into the ground to support horizontal sheeting used to support the face of an excavation.

SOLDIER COURSE Masonry units with the normally horizontal face placed vertically.

SOLDIER PILE Strong rolled steel H or WF section driven at intervals of a few feet to hold horizontal lagging, which is placed as excavation proceeds.

SOLENOID Iron core about which is wound a coil of wire, designed to move linkage when an electrical current is passed through the coil; an electromagnet that operates another device.

SOLENOID CONTROLLED, PILOT OPERATED Valve that is operated by a solenoid-operated pilot valve.

SOLENOID OPERATED Valve that is positioned by one or more solenoids.

SOLEPLATE Horizontal member placed on the subfloor upon which the wall and partition studs rest. *Also called* **Ground plate.**

SOLID ASPHALT *See* **Asphalt.**

SOLID BEARING *See* **Bearing.**

SOLID BRIDGING Bridging between floor joists consisting of short lengths of joist fastened at right angles in each space at midspan. *Also called* **Block bridging.** *See also* **Strut.**

SOLID CORE DOOR *See* **Door types.**

SOLID DRILLING Use of a diamond drilling bit to grind the whole face, without preserving a core.

SOLID FLOOR Floor comprising a concrete slab without voids.

SOLID LOADING Completely filling a drill hole with explosive, except for a stemming space at the top.

SOLID MANDREL *See* **Mandrel.**

SOLID MASONRY UNIT *See* **Masonry unit.**

SOLID MASONRY WALL *See* **Wall.**

SOLID MOLDING Molding formed on one or both edges of a board; not applied.

SOLID PARTITION Partition wall formed of solid materials: brick, solid block, filled block, etc.

SOLID ROCK Rock that can only be moved or processed after being blasted.

SOLIDS Dry ingredients remaining after evaporation of all volatile solvent or water. Not a fluid and not flowable.

SOLIDS CONTENT Percentage by weight or by volume of nonvolatile components in a solution.

SOLID SOIL Soil having a constant density and internal resistance that is little affected by temperature changes, moisture variations, or vibration. *See also* **Soils types, plastic.**

SOLID STATE Electrical control and power circuits that have no moving parts and that rely on any combination of optical, electrical, or magnetic phenomenon within a solid.

SOLID STATE DEVICE Element that can control current flow without moving parts. The electrical function is carried out by semiconductors, resistors, capacitors, etc.

SOLID STOP Stop rebated into a solid frame.

SOLID STREAM Water used for fire fighting; discharged by an open round orifice at sufficient pressure to provide a stream having impact and range.

SOLID-SURFACE FASTENER *See* **Fastener.**

SOLID TIRE Tire such as 'cushion solid tire' or 'standard solid tire' made of resilient solid material.

SOLID TOOTH Saw tooth that is formed from and is an integral part of a saw blade.

SOLID-UNIT MASONRY Masonry consisting wholly of solid masonry units laid in mortar.

SOLID VINYL TILE Flooring tile with a high vinyl content consisting of vinyl resins, plasticizer, filler, and color pigments. *Also called* **Luxury vinyl tile.**

SOLID WASTE General term for discarded materials destined for disposal, but not discharged to a sewer or to the atmosphere. Solid wastes can be composed of a single material or a heterogeneous mix of various materials, including semisolids. It can be grouped under the following categories:

Ash: Residue from burning of combustible materials; may include extraneous noncombustibles, unburned carbon, as well as mineral matter inherent in the combustible material.

Bulky waste: Large discarded materials: appliances, furniture, junked automobile parts, diseased trees, large branches, stumps, etc.

Combustible waste: The organic content of solid waste, including paper, cardboard, cartons, wood, boxes, excelsior, plastic, textiles, bedding, leather, rubber, paints, yard trimmings, leaves, and household waste, all of which will burn.

Commercial waste: From businesses, office buildings, apartment houses, stores, markets, theaters, and hospitals and institutional facilities.

Domestic waste: Putrescible and nonputrescible waste originating from a residential unit, and consisting of paper, cans, bottles, food wastes, and may include yard and garden wastes.

Food waste: Animal and vegetable discards from handling, storage, sale, preparation, cooking, and serving of foods.

Garbage: *See* **Food waste,** above.

Hazardous waste: Any waste material, or combination thereof, which poses a substantial present or potential hazard to human health or living organisms because such wastes are nondegradable or persistent in nature or because they can be biologically magnified, or because they can be lethal, or because they may otherwise cause or tend to cause detrimental cumulative effects.

Household solid waste: *See* **Domestic refuse,** above.

Industrial waste: Discarded waste materials from industrial processes and/or manufacturing operations.

Infectious waste: Waste materials from a medical facility, hospital, or laboratory which may contain pathogens or other disease infected wastes which could be transmitted to another human during the collection, transportation, or disposal cycle.

Mixed municipal refuse: *See* **Municipal solid waste,** below.

Municipal solid waste: Domestic refuse and some commercial waste.

Noncombustible waste: Inorganic content of solid waste, including glass, metal, tin cans, foils, dirt, gravel, brick, ceramics, crockery, and ashes.

Oversize waste: *See* **Bulky waste,** above.

Pathological waste: *See* **Infectious waste,** above.

Problem waste: A general term used to describe bulky wastes, dead animals, abandoned vehicles, construction and demolition waste, hazardous and infectious waste, and any other waste that requires special considerations in the collection, handling/transportation, or disposal cycle.

Refuse: Term sometimes used in place of **Solid waste.** Used to define general community waste, which includes kitchen/food wastes and rubbish.

Residential waste: Discarded materials originating from residences.

Rubbish: Nonputrescible materials collected from residences, commercial establishments, and institutions.

Rubble: Rough stones of irregular shape and size, broken from large masses either naturally or artificially, as by weathering action or by demolition of buildings, pavements, roads, etc.

Slops: *See* **Swill,** below.

Sludge: A semiliquid sediment resulting from the accumulation of settleable organic/inorganic solids deposited from wastewaters or other fluids in tanks or basins.

Special wastes: Wastes that require special care and consideration in the storage, collection, handling/transportation, and disposal cycle by reason of their pathological, explosive, or toxic/hazardous nature.

Street refuse: Material collected by manual and mechanical sweeping of streets and sidewalks; litter from public receptacles and dirt removed from catch basins.

Swill: Semiliquid waste material consisting of food waste and free liquids.

Trash: Larger (nonputrescible) residential solid wastes unsuitable for routine pickup by refuse collection.

Unconventional waste: Hazardous wastes by rea-

son of their pathological, explosive, radioactive, or toxic nature.

White goods: Discarded appliances such as stoves and washing machines.

Yard waste: Plant clippings, prunings, grass clippings and leaves, and other discarded material from yards and gardens.

SOLID WASTE DERIVED FUEL Fuel derived from solid waste that can be used as a primary or supplementary fuel in conjunction with, or in place of, fossil fuels.

SOLID WASTE DISPOSAL Disposal of all solid wastes through landfilling, incineration, composting, chemical treatment, and any other method that prepares solid wastes for final disposition.

SOLID WASTE MANAGEMENT A planned program for effectively controlling the generation, storage, collection, transportation, processing, and reuse, conversion or disposal of solid wastes in a safe, sanitary, aesthetically acceptable, environmentally sound, and economic manner.

soln *Abbreviation for:* solution.

SOLUBILITY Amount of one material that will dissolve in another.

SOLUTION Liquid consisting of at least two substances, one of which is a liquid solvent in which the other or others, which may be liquid or solid, are dissolved. *See also* **Eutectic solution.**

SOLUTION TREATMENT Immersion of a metal part in a liquid bath for the purpose of improving some physical characteristic of the metal. The metal is heated to the point where a chemical in the bath is able to enter the metal, then cooled in such a way that the chemical does not precipitate out of the metal during the cool-down process.

solv. *Abbreviation for:* solvent.

SOLVENT Liquid in which another substance may be dissolved.

SOLVENT-ACTIVATED ADHESIVE *See* **Adhesive.**

SOLVENT ADHESIVE *See* **Adhesive.**

SOLVENT ALCOHOL *See* **Denatured alcohol.**

SOLVENT NAPTHA *See* **Naptha.**

SOLVENT REFINING Process in the refining of lubricants in which a selective solvent such as phenol or N-methyl pyrrolidone (NMP) is used to remove undesirable components.

SOLVENT-RELEASE SEALANT Sealant that does not change chemically while curing to a flexible or semirigid state.

S1E *Abbreviation for:* surfaced one edge.

S1S *Abbreviation for:* surfaced one side.

S1S1E *Abbreviation for:* surfaced one side and one edge.

S1S2E *Abbreviation for:* surfaced one side and two edges.

SONIC DRIVER *See* **Pile hammer.**

SONOTUBE Trade name for a cylindrical form of treated cardboard, used for forming round columns of concrete.

SOOT Black substance, consisting mostly of very small particles of carbon, that appears in smoke resulting from incomplete combustion.

SOOT DOOR Metal door at the base of a chimney through which soot and ash can be removed.

SORBENT Material capable of extracting one or more substances from an atmosphere or liquid.

SORTING 1. Breaking up of bundles of reinforcing bars after unloading, so that all items are readily accessible; **2.** Separation of forest products, usually at the landing.

SORTING FORK *See* **Forks.**

SOUND ABSORPTION COEFFICIENT Fraction of sound energy absorbed by a material. *Also called* **Acoustic absorptivity.**

SOUND ATTENUATING INSULATION Unfaced mineral or glass-fiber material used to impede the transmission of sound waves.

SOUND ATTENUATION Reduction of objectionable noise to acceptable limits. *See also* **Attenuate.**

SOUND BOARD Reflective surface designed and placed to direct or deflect sound toward a specific point.

SOUNDING 1. Determination of the measurement between the surface of a body of water and the bottom immediately below; **2.** Method of examining soil to 6 to 10 m (20 to 30 ft) depth by driving or hydraulically pushing a cone, steel rod, or small-diameter pipe (gas pipe) into the ground with a hammer or maul. With experience, the movement of the rod or pipe under each hammer blow can give an indication of the approximate types of soil materials that are being penetrated. *Also called* **Probing.**

SOUNDING WELL Vertical conduit in the mass of coarse aggregate for preplaced aggregate concrete, provided with continuous or closely spaced openings to permit the entrance of grout.

SOUND ISOLATION Various means to prevent transmission of noise and vibration.

SOUND KNOT Solid, tight, undecayed knot at least as sound as the wood around it.

SOUND LEVEL Value of the sound level pressure, in psi or decibels.

SOUND LOCK Acoustically treated vestibule intended to prevent or diminish the transmission of sounds from one area to another; similar to an air lock.

SOUNDNESS 1. Freedom from flaws; **2.** Solid, free from cracks, flaws, fissures, or variations from an accepted standard.

SOUNDPROOFING Construction techniques and materials used exclusively to minimize or eliminate the transmission or reflection of sound waves.

SOUND REDUCTION FACTOR Value, usually in decibels, of the measure of sound transmission intensity reduction of a material. *Also called* **Acoustical reduction factor.**

SOUND TRANSMISSION CLASS Numerical rating of the ability of an assembly of materials to resist the transmission of airborne sound.

SOUND WAVE Pressure disturbance in air proceeding at approximately 341 m/sec (1120 ft/sec).

SOUND WOOD Wood that is free from defect.

SOURCE 1. To investigate the location from which something may be obtained; **2.** Point or place from which something originates or emanates.

SOURCE REDUCTION Design, manufacture, acquisition, and reuse of materials including products and packaging, so as to minimize the quantity and/or toxicity of waste produced.

SOURCE SEPARATION Sorting at point of generation of specific discarded materials into specific containers for separate collection.

SOUR GAS Fuel that contains a relatively large proportion of sulfur or sulfur compounds.

SOUTHERN STANDARD BUILDING CODE *See* **Building code.**

SOUTHING Distance measured southwards from an east-west axis.

SOV *Abbreviation for:* shutoff valve.

SOYBEAN OIL Oil used in varnish and enamels, and when modified, in alkyd exterior paint.

sp. *Abbreviation for:* single pitch; soil pipe; spare; special; specific; speed; spool; sump pit.

SP *Abbreviation for:* self-propelled; shear plate; single pole; soil pipe; speciality grade; special quality; splashproof; standpipe.

SPACE 1. Contained volume or area; **2.** Horizontal distance from one point of a saw tooth to the next; **3.** Dimension of the clear opening between adjacent parallel wires or bars of a screen.

SPACED LOADING Loading of explosive so that cartridges or groups of cartridges are separated by open spacers that allow the concussion from one charge to reach the next in line.

SPACE-FRAME Three-dimensional framework for enclosing spaces, in which all members are interconnected and act as a single entity, resisting loads applied in any direction.

SPACE HEATING Method of heating individual rooms or living units by equipment located entirely within these rooms or living units; such equipment consists of a single unit without ducts or piping.

SPACER 1. Device that maintains reinforcement in its proper position; **2.** Device for keeping wall forms apart at a given distance before and during concreting; **3.** T- and Y-shaped units, used during installation to separate tiles on walls and floors.

SPACER BAND Metal band separating two demountable rims on spoke wheels.

SPACE STRUCTURE Three-dimensional, self-supporting structure without interior supports.

SPACING 1. Arrangement of spaces; the distance between things; **2.** Distance between boreholes or blasting charges in a row; **3.** Act of removing trees from a stand to decrease the stand density, distribute the crop trees more evenly over the growing site, and create more growing room. *See also* **Intertree distance, Juvenile spacing,** and **Thinning**; 4. Space between adjacent turns of helically wound wire. (Differs from **Pitch** in that the diameter or width of wire is not included.); 5. Distance between the keyhole and spindle hole of a door lock.

SPACING CONTROL Act of creating, within the limits of the existing stand, a uniform distribution of trees that provides optimum growing space for each tree by eliminating overcrowding.

SPACING LUG Built-in projection on the back edge of a wall or floor tile that ensures constant spacing, and with a consistent width grout trim.

SPACING MIX Dry or damped mixture of one part portland cement and one part extra-fine sand, used as a filler in the joints of mounted ceramic mosaic tiles to keep them evenly spaced during installation.

SPAD Surveyor's nail.

SPADE One or more hydraulically operated, ground-penetrating feet designed primarily to resist rearward movement of a vehicle or machine.

SPADE BIT *See* **Bit.**

SPADING Consolidation of mortar or concrete by repeated insertion and withdrawal of a flat, spade-like tool.

SPALL Small fragment removed from the face of a masonry unit by a flow or by action of the elements.

SPALLING *See* **Scaling.**

SPAN 1. Distance between the support reactions of members carrying transverse loads. *See also* **Arch**; 2. Horizontal distance between skyline supports.

SPAN-DEPTH RATIO Numerical ratio of total span to member depth.

SPANDREL 1. Space between the outer curve of an arch and a rectangular frame or mold enclosing it, or between shoulders of adjoining arches and the molding above; 2. Part of a wall between the windowsill of a window and the head of a window immediately below it.

SPANDREL BEAM Beam in the perimeter of a building spanning between columns and usually supporting floors or roof. *See also* **Beam.**

SPANDREL WALL *See* **Wall.**

SPANISH TILE *See* **Roof tile.**

SPAN RATING Load-carrying capacity of a panel.

SPARK ARRESTER Device designed to help prevent the emission of hazardous flames and hazardous sparks.

SPARK CHASER Worker designated to watch for fires during high fire hazard conditions. *Also called* **Fire watch.**

SPARK ENCLOSING EQUIPMENT Auxiliary units that completely enclose the electrical equipment on a vehicle to the emission of electric sparks. It includes the enclosure of the generator, electric motors, contactors, and protection of certain other electrical equipment.

SPARROW PECK Pattern imposed on a surface: to a wet mixture, by pitting with a stiff brush; to a solid surface, by striking with a light pick.

SPAR TREE Tree or mast on which rigging is hung for one of the many cable hauling systems.

spat. *Abbreviation for:* spatter.

SPATTERDASH Rich mixture of portland cement and coarse sand thrown onto a background by a trowel, scoop, or other appliance so as to form a thin, coarse-textured continuous coating.

SPATULA Iron rod with a flattened end, used to place flux and spelter in brazing or for hard soldering.

spcr *Abbreviation for:* spacer.

spd *Abbreviation for:* speed.

spdl *Abbreviation for:* spindle.

spdr *Abbreviation for:* spider.

spdt *Abbreviation for:* single-pole, double-throw.

spec. *Abbreviation for:* specialist; specification; specify.

SPECIAL ASSESSMENT Extraordinary property tax levied to cover the costs of public works in the immediate area.

SPECIAL BENDING All bending of reinforcing bars to special tolerances, all radius bending in more than one plane, all multiple-plane bending containing one or more radius bends, and all bending for precast units. *See also* **Heavy bending,** and **Light bending.**

SPECIAL BENEFIT Advantage accruing from a given highway improvement to a specific property and not to others generally. *See also* **Benefit,** and **General benefit.**

SPECIAL CONDITION Description within a construction contract peculiar to work, materials, procedures, prices, etc., not adequately covered by the general conditions.

SPECIAL DECOMPRESSION CHAMBER Chamber to provide greater comfort of workers when the total decompression time exceeds 75 minutes.

SPECIAL FOREST PRODUCTS Poles, posts, pilings, shakes, shingle bolts, Christmas trees, building logs, mining timbers, props and caps, cribbing, firewood and fuel, hop poles, orchard props, car stakes, round stakes, sticks and pickets, split stakes, pickets, palings, shake bolts, blocks and blanks, and shingle blocks.

SPECIAL HAZARDS INSURANCE Property insurance coverage of such additional hazards as collapse, physical loss, sprinkler leakage, water damage, etc., or of materials, supplies, and equipment at other locations and/or in transit.

SPECIAL LIEN Lien binding a specific piece of property. *Also called* **Specific lien.**

SPECIAL LOAD RATING *See* **Load rating.**

SPECIAL MILEAGE TIRE Tire with an extra layer of rubber between the cords and the tread for the purpose of recutting and regrooving.

SPECIAL-ORDER, NATURAL-FINISH VENEER *See* **Veneer grades.**

SPECIAL PROVISION Addition or revision to the standard and supplemental specifications applicable to an individual project.

SPECIAL-PURPOSE FABRICATED-STAGE, SCAFFOLD, PLANK, or PLATFORM Platform unit that represents either a modification or a combination of design or construction features in one of the general-purpose types of stages or planks otherwise defined, in order to adapt the plank or platform to special or specific use.

SPECIAL PURPOSE OUTLET Electrical outlet used for purposes other than ordinary lighting and power, usually fused separately.

SPECIAL STEEP ASPHALT Roofing asphalt having a softening point of 88°C (190°F).

SPECIAL WARRANTY DEED Deed in which the grantor agrees to protect the grantee against title defects or claims that arose during the period that the grantor held title to the property.

SPECIAL WASTES *See* **Solid waste.**

SPECIAL WEARING SURFACE Surface overlay applied to a panel or back to increase resistance to surface abrasion.

SPECIES Group of similar individuals having a number of correlated characteristics and sharing a common gene pool. The species is the basic unit of taxonomy on which the binominal system has been established. The scientific name of a plant or animal gives the genus first and then the species.

SPECIFIC ADHESION Chemical bond between glued or cemented surfaces. *See also* **Adhesive,** and **Mechanical bond.**

SPECIFICATION Compilation of provisions and requirements for the performance of prescribed work, material, or product.

SPECIFICATION OF WORKS Written document describing all aspects of the construction to be carried out, the materials to be used, and the manner of their finishing. *Also called* **Architectural specification,** and **Book of specifications.**

SPECIFICATION WRITER One who specializes in the preparation of specifications.

SPECIFIC CONDUCTIVITY Quantity of electricity transferred across a unit area per unit potential gradient per unit of time. In the SI system, K=amp/cm^2 divided by V/cm.

SPECIFIC CAPACITY Rate of discharge of water in a well divided by the drawdown of the water level.

SPECIFIC DISCHARGE Rate of discharge of goundwater per unit cross-sectional area measured at right angles to the direction of flow.

SPECIFIC FUEL CONSUMPTION Amount of fuel consumed to produce a unit of work, usually expressed in pounds per horsepower or kilowatt hours, or grams per kilowatt hour.

SPECIFIC GRAVITY Comparison of the density of one material with a reference material under specific test conditions. The reference material for specific gravity is the maximum density of water, that is, water at 4°C (39.2°F). The device used for this, typically a comparison of fuel, or coolant, or battery electrolyte with water, is a hydrometer. *Also called* **Apparent specific gravity.** *See also* **Baumé scale, Bulk density, Bulk specific gravity, Density (dry),** and **Liquid specific gravity.**

SPECIFIC HEAT *See* **Heat.**

SPECIFIC HEAT REJECTION Heat rejection of an engine expressed essentially in BTU's/minute/brake horsepower.

SPECIFIC IDENTIFICATION *See* **Cost flow.**

SPECIFIC HUMIDITY *See* **Humidity.**

SPECIFIC LIEN *See* **Special lien.**

SPECIFIC PERFORMANCE Court order requiring a defendant to carry out the terms of a contract or agreement.

SPECIFIC RETENTION Ratio of the volume, or weight of water that a soil will retain against the force of gravity, having once been saturated, to its dry volume or weight.

SPECIFIC STORAGE Volume of water released from, or taken into storage, per unit volume of the porous medium.

SPECIFIC SURFACE Surface area of particles contained in a unit weight or volume of a material. *See also* **Blaine fineness.**

SPECIFIC WEIGHT Relative weight of a fluid or solid compared to that of water.

SPECIFIC YIELD Ratio of the volume of rock or soil to the volume of water that, following saturation, will be yielded by gravity.

SPECIFIED DIMENSION Dimension for a manufactured or supplied part that must not be varied.

SPECIMEN Piece or portion of a sample used to make a test.

SPECK Dark dot on a tile, less than 4 mm (0.16 in.) in diameter, and noticeable at a distance of more than three feet.

SPECKLED GLAZE *See* **Glaze.**

SPECTRAL ANALYSIS Method of analyzing the vibration frequencies present in a vibration record.

SPECTRUM Band of colors formed when radiant energy is broken up; by being passed through a prism or other means.

SPECULATIVE BUILDER One who develops and constructs building projects for subsequent sale or lease.

SPEED Rate of vehicular movement, generally expressed in kilometers per hour or miles per hour, including the following classifications:

Average overall travel speed: For all traffic or component thereof, the summation of distance divided by the summation of overall travel times.

Average running speed: For all traffic or component thereof, the summation of distances divided by the summation of running times.

Average spot speed: Arithmetic mean of the speeds of all traffic or component thereof, at a specified point.

Design speed: Speed determined for design and correlation of the physical features of a highway that influence vehicle operation. It is the maximum safe speed that can be maintained over a specified section of highway when conditions are so favorable that the design features of the highway govern.

Overall travel speed: Speed over a specified section of highway, being the distance divided by overall travel time.

Running speed: Speed over a specified section of highway, being the distance divided by the running time.

SPEEDABILITY Speed a vehicle will attain based on engine power, gross weight, power train efficiency, air resistance, grade resistance, and road type.

SPEED-CHANGE LANE *See* **Lane.**

SPEED CHANGER Device that adjusts a speed governing system so as to change engine speed.

SPEED DROP 1. Progressive lowering of the rpm of an engine as the load applied to it increases; **2.** Percentage of engine rpm lost when the load on it goes from zero to full load.

SPEED INDICATOR Measure of the travel speed of hauling equipment. When laden, it equals the engine brake horsepower divided by the gross vehicle weight in thousands of pounds; when unladen it equals the engine brake horsepower divided by the tare weight in thousands of pounds.

SPEED OF SOUND Speed at which sound waves travel through still air at sea level: 341 m/sec (1,120 ft/sec) or 1227 km/hr (763 mi/hr).

SPEED REDUCER Gearing device used to reduce the output speed of a prime mover.

SPELTER Form of brass used in brazing.

SPENDING VARIANCE *See* **Variance.**

spg *Abbreviation for:* spring.

sp. gr. *Abbreviation for:* specific gravity.

sph. *Abbreviation for:* sphere.

Sp. Hd. *Abbreviation for:* spanner head.

spher. *Abbreviation for:* spherical.

SPHERE Round body, the surface of which is equidistant at all points to its center.

sp. ht *Abbreviation for:* specific heat.

SPIDER 1. Fitting mounted to a pivot at the top of a derrick mast, providing attachment points for guy ropes. *Also called* **Mast cap; 2.** Temporary support to maintain a penstock in a circular shape while it is being transported and erected. *See also* **Brace.**

SPIDER GEAR Differential gear that rotates on its shaft in a rotating case.

SPIDER MARK 1. Cleavage or weak spot caused by the failure of a compound to reunite after passing a spoke of the spider of an extrusion machine; **2.** Grain produced at a point of joining of stock after passing the spoke of the spider of an extrusion machine.

SPIGOT Plain end of a bell and spigot pipe joint.

SPIKE 1. Large nail. Spikes have a larger diameter than nails and are used for fastening large pieces of timber; **2.** Railroad-type spike used to secure wire rope when splicing; **3.** Very short, abnormal, and unwanted high-amplitude distortion of voltage or frequency.

SPIKE BAR Sturdy, steel pry bar used to extract spikes from a log or stump.

SPIKED GRID Timber fastener with spikes on both sides, used between bolted timbers to prevent movement between the two timbers.

SPIKE IT Verbal order to stop action.

SPIKE KNOT Knot sawed lengthwise.

SPIKE TOP Dead top of a tree with few if any branches remaining.

SPILE Small wooden peg. *See also* **Forepole.**

SPILING Supporting wood or steel members that support the roof of a tunnel at the face.

SPILL-THROUGH ABUTMENT Two or more columns carrying a beam that supports a bridge with the fill extending on a natural slope from behind the bridge beam, through the openings between the columns.

SPILLWAY Overflow channel or chute. When part of a dam, it may be of several types:

Fuse plug: Auxiliary or emergency spillway comprising a low embankment or saddle that is overtopped only during exceptionally high water.

Ogee: Overflow whose channel, in longitudinal section, is in the shape of an S or ogee curve.

Primary: Principal spillway over which water flows during flood flows.

Shaft: Vertical or inclined shaft passing through, under, or around a dam and into which floodwater or surplus water is spilled.

Side channel: Spillway whose crest is approximately parallel to the downstream channel.

Siphon: Spillway having one or more siphons at crest level.

SPINDLE *See* **Arbor.**

SPINE WALL Internal load-bearing wall parallel to the main axis of a building.

SPIN LOCK Mechanical outrigger locking system designed to provide an increased measure of outrigger reliability.

SPINNING 1. Essential factor of the process of producing spun concrete; **2.** Process of shaping sheet metal using a lathe.

SPINNING LINE Line wrapped around a drill pipe so that a pull will rotate the pipe, causing the threaded end to fasten or unfasten from another length of pipe.

SPINNING PROCESS *See* **Centrifugal process.**

SPIN-ON FILTER Disposable filter that mates to a permanent base and that is attached by turning onto a threaded base stud.

SPIRAL 1. Continuous coil; **2.** Method of applying hose reinforcement in which there is no interlacing between individual strands of the reinforcement.

SPIRAL BEVEL GEAR *See* **Gear.**

SPIRAL BLADE MIXER Concrete, mortar, and other wet or dry materials mixer that incorporates a cross current mixing action, constantly forcing the materials into themselves to produce a homogeneous mix of all constituents.

SPIRAL CLEANER Device for removing dirt from a conveyor belt.

SPIRAL COLUMN Reinforced concrete column in which the vertical bars are enclosed within a spiral.

SPIRAL FERRULE Steel knob used for chokers and lines. Can be attached quickly using wedges to secure it on the line.

SPIRAL-GRAINED WOOD *See* **Grain.**

SPIRAL GROOVE Continuous helical groove that follows a path on and around a drum face.

SPIRALING Drill hole twisting into a spiral around its intended dead center. *See also* **Rifling.**

SPIRAL LAY *See* **Lay types.**

SPIRAL LINING Technique in which a ribbed PVC strip lining is spirally wound by machine into an existing pipeline.

SPIRALLY REINFORCED COLUMN Column in which the vertical bars are enveloped by spiral reinforcement, i.e., closely spaced continuous hooping.

SPIRAL NAIL *See* **Nail.**

SPIRAL PIPE *See* **Pipe.**

SPIRAL REINFORCEMENT *See* **Helical reinforcement.**

SPIRAL SPACER Usually made of a channel or angle, punched to form a hook, and bent over the coiled spiral reinforcing to maintain it at a definite pitch.

SPIRAL STAIR Series of steps attached to a vertical pole and progressing upward in a winding fashion within a cylindrical space. *Also called* a **Circular stair**, **Helical stair**, and **Screw stair.**

SPIRAL-WELD PIPE *See* **Pipe.**

SPIRE Tall pyramidal, polygonal, or conical structure terminating in a point.

SPIRIT LEVEL Transparent tube containing fluid and an air bubble.

SPIRIT STAIN Dye dissolved in alcohol with a binder, usually shellac or other resin.

SPIRIT VARNISH *See* **Varnish.**

spk *Abbreviation for:* spark.

spkprf *Abbreviation for:* sparkproof.

spkr *Abbreviation for:* speaker.

spl *Abbreviation for:* special; spline.

SPL *Abbreviation for:* sound pressure level.

SPLASH BLOCK Small masonry block laid with the top close to the surface of the ground to receive roof drainage and divert it away from a building.

SPLASHBOARD 1. Weather board projecting from the exterior bottom of a door that throws water clear of the threshold; **2.** Board placed on edge against a wall, or beside a scaffold, to protect a finished surface during construction.

SPLASH GUARD Deflecting shield at the rear of wheels. *Also called* a **Mud flap.**

SPLASH LAP Part of a flexible-metal drip or roll that extends over the flat surface of an adjacent sheet.

SPLASH WALL Wall of a tile drainboard or bathtub.

SPLAY 1. Slope or bevel, particularly at the sides of a window or door; **2.** To cut at an angle. *See also* **Reveal.**

SPLAY ANGLE Two surfaces meeting at an angle greater than 90°.

SPLAY BRICK *See* **Brick.**

SPLAYED GROUND Ground, that also acts as a screed for plaster, having a beveled or rebated edge that acts as a key.

SPLAYED HEADING JOINT Joint between the ends of floorboards that are cut at overlapping 45° angles.

splc. *Abbreviation for:* splice.

SPLICE 1. Connection of one reinforcing bar to another by lapping, welding, mechanical couplers, or other means; **2.** Connection of welded-wire fabric by lapping; **3.** Connection of piles by mechanical couplers; **4.** Connection between two structural steel elements joined at their ends to form a single, longer element; **5.** Line where two ends of a precured tire tread are joined; **6.** Joint or junction of rubber sheet made by lapping or butting, straight or on a bias, and held together through vulcanization or mechanical means; **7.** *See* **Connection.**

SPLICE BOX Enclosed box in a raceway permitting the connection of electrical conductors.

SPLICED PILE *See* **Pile.**

SPLICE MARK Defect on the face of a drywall panel where one roll of paper is joined to another.

SPLICE POINT Location where the chord member of a truss is spliced to form one continuous member.

SPLICER One who joins two pieces of cable together by intertwining the wire strands.

SPLICING 1. Making a loop or eye in the end of a wire rope by tucking the ends of the strands back into the main body of the rope; **2.** Formation of loops or eyes in a rope by means of mechanical attachments pressed onto the rope; **3.** Joining of two rope ends so as to form a long or short splice in two pieces of rope.

SPLINE 1. Set of parallel grooves running lengthwise of a shaft; **2.** Thin strip of wood that is fitted into matching grooves cut in the joining faces of a joint.

SPLINED SHAFT Shaft with equally spaced grooves machined into the whole or part of its length.

SPLINE NUT *See* **Nut.**

SPLIT 1. Lengthwise separation of wood due to the tearing of the wood cells; **2.** Half height masonry unit.

SPLIT A QUARTER Start the first road on a setting halfway between fairlead and square lead.

SPLIT-BATCH CHARGING Method of charging a mixer in which the solid ingredients do not all enter the mixer together.

SPLIT BLOCK Concrete masonry units with one or more faces having a rough surface from being split during manufacture.

SPLIT BUSHING Bushing made in two pieces. *See also* **Bushing.**

SPLIT CABLE GRIP Woven wire basket device used for gripping wires, cables, etc.

SPLIT-COEFFICIENT Road condition where one side of a lane has high friction, the other low friction (i.e., the outside of the lane is ice-covered, the inside of the lane is dry).

SPLIT COURSE Course of bricks or masonry units cut to a depth less than regular courses.

SPLIT FACE Rough face created by shearing cast or natural stone blocks.

SPLIT-FACE BLOCK Concrete masonry unit with one or more faces purposely fractured to provide architectural effects in masonry wall construction.

SPLIT HEATING *See* **Heating system.**

SPLIT LAGGING Drum lagging constructed in two sections to allow changing it without removing the drum. *See also* **Lagging.**

SPLIT LEVEL HOUSE Residence in which individual floors are on more than one level, or where individual floors are less than a story apart.

SPLIT-RING WASHER *See* **Washer.**

SPLIT-SHAFT TRANSMISSION *See* **Power take-off.**

SPLIT SHAKE Wood shingle split rather than sawn.

SPLIT SPOON SAMPLER Type of drill core used in soil exploration.

SPLIT SPROCKET Two-piece sprocket that can be assembled on a shaft without removing the shaft bearings.

SPLITTER 1. Shearing device consisting of a disk or knife blade mounted behind a circular saw to prevent boards from falling onto the saw, or those with a spring from binding or pinching the saw; **2.** Hand tool used to shear stone, wood, or other materials.

SPLITTING TENSILE STRENGTH Tensile strength of a material determined by a splitting tensile test.

SPLITTING TENSILE TEST Indirect tensile test consisting of the application of compressive loads to a cylindrical specimen along a diametrical plane, which causes failure by splitting along that plane. *Also called* **Diametrical compression test.**

SPM *Abbreviation for:* superpower multiplate.

spndr. *Abbreviation for:* spandrel.

SPOIL 1. Material removed from an excavation; **2.** Rock-earth wasted because it has no use. *See also* **Cuttings.**

SPOIL BANK Bank or pile of spoil, or of wasted material.

SPOKESHAVE Short-bottomed plane with wing-like handles whose main purpose is smoothing curved surfaces. The blade is adjusted similar to a plane; the device operates like a drawknife except that it is pushed away from the user.

SPOKE WHEEL Wheel casting with 3, 5, or 6 spokes that provides for mounting and support of one or two demountable rims.

spont. *Abbreviation for:* spontaneous.

SPOOL 1. Drum to hold cable; **2.** To wind-in a winch cable; **3.** Cylindrically shaped part of a hydraulic valve that moves to direct flow through a particular section of the hydraulic system; **4.** Spacer between timbers, usually of cast iron or of steel.

SPOOL CABLE HANGER Wheel-like insulating element support having a circumferential groove for the hanging and support of traveling cables.

SPOOLING IRON Tool used for spooling line onto a winch so that hands are not placed directly onto the line.

SPOON Pipe-shaped tool split in half for part of its length, used to obtain soil samples when driven into the ground.

SPOON BLOW Blow of a 63.5-kg (140-lb) hammer falling 760 mm (30 in.) onto a 50-mm (2-in.) diameter OD by 35-mm (1.375-in.) ID split spoon sampler. *See also* **N Value.**

SPOT To indicate or designate an exact place or location.

SPOT BURNING Modified form of broadcast slash burning in which only the greater accumulations are fired and the fire is confined to those spots.

SPOT FIRE Fire set outside of the perimeter of the main fire by flying sparks or embers.

SPOT LEVEL Elevation of a point above or below datum.

SPOT LOG Log or marker placed to indicate the spot where trucks should stop to be loaded (by an excavator, for instance).

SPOT REPAIR Replacement of rubber in a tire injury that penetrates less than 25% of the body plies.

SPOTTER 1. Worker who directs truck drivers into the proper loading or dumping position; **2.** Horizontal connection in a pile driver between the machinery deck and the lead. *Also called* **Brace.**

SPOTTING 1. Positioning trucks for loading or dumping; **2.** Placement of fire-fighting equipment for effective operation and attack on a forest fire; **3.** To cover drywall fastener heads and fill the dimple with joint compound.

SPOT ZONING Zoning of a lot or parcel of land in a manner that is distinctly different to that of adjacent lots or land parcels.

SPOUT Water outlet of a faucet.

SPOUT-DELIVERY PUMP *See* **Pump.**

sp. pl. *Abbreviation for:* space plate; splash plate.

spr. *Abbreviation for:* spray; spring; sprinkler.

SPR *Abbreviation for:* simplified practice recommendation.

SPRAY Fluid applied through specially designed orifices in the form of finely divided particles.

SPRAY APPLICATION Application of surface texturing material by use of mechanical equipment.

SPRAY BAR Pipe fitted with jets along its length, used horizontally behind and below a pressure tank to apply materials such as road oil and bitumen to a road surface.

SPRAY CHAMBER Chamber equipped with water sprays that cool and clean combustion products passing through the chamber.

SPRAY DRYING Method of evaporating liquid from a solution by spraying it into a heated gas.

SPRAYED MINERAL FIBER Blend of mineral fibers and inorganic binders to which water is added during the spraying operation.

SPRAYED MORTAR *See* **Shotcrete.**

SPRAY GUN Tool designed to convert a body of liquid into a fine mist for application to a surface, such as paint, powder, cement mortar, etc., usually but not necessarily powered by compressed air.

SPRAY LANCE Hand-held pipe with a spray nozzle on one end, connected by flexible hose to a pressurized tank containing road oil, bitumen, etc.

SPRAY LIME Hydrated lime of such fineness that at least 95% of the particles pass a No. 325 sieve.

SPRAY-ON LINING Technique for applying a lining or coating by a rotating head that is winched through a pipeline.

SPRAY PAINTING Applying paint and other surfacing materials as a mist of fine droplets.

SPRAY TEXTURE Result of materials mechanically applied to walls and ceilings to produce different effects.

sprdr *Abbreviation for:* spreader.

SPREAD 1. Range of mobile, mechanical equipment necessary or assembled to complete a job of work or contract; **2.** Horizontal width or diameter of the head of a tree or shrub; **3.** Quantity of adhesive per unit area applied to an adherent, usually expressed in pounds of adhesive per thousand square feet of area; single spread refers to application of adhesive to only one adherent, double spread refers to application of adhesive to both adherents; **4.** Thin coat of material in solvent form applied on a fabric surface by means of knife, bar, or doctor blade.

SPREADABILITY Capacity of a mixture to be applied in a layer of intended thickness and/or texture.

SPREADER 1. Beam used to support long or limber members that might tip, slide or bend during lifting, eliminating the necessity for low sling angles where long members are involved; **2.** Piece of lumber, usually about 25 x 50 mm (1 x 2 in.), cut to the thickness of a wall and inserted into a form to hold it temporarily against the tension of form ties; **3.** Device consisting of reciprocating paddles, a revolving screw, or other mechanism for distributing asphalt or concrete to a required uniform thickness in a paving slab. *See also* **Asphalt paving machine,** and **Paving train machine**; **4.** Multiarm device that spreads a tire at the bead area; **5.** Short piece of line or chain between the butt rigging of a grapple used to separate the haulback eye from the

carriage. *See also* **Floating harness**; **6.** Devices used to apply salt, chemicals and abrasive materials to icy surfaces. They include the following types:

Hopper/conveyor: In models for various sizes of truck, these devices include a conveyor at the bottom that moves material to the rear spinner assembly for dispersal over the road surface; spread patterns and flow rates are precisely adjustable.

Tailgate: This type mounts directly beneath the tailgate of a standard dump body and is fed by gravity with the body in a raised position.

SPREADER BAR *See* **Brace,** and **Floating harness.**

SPREADER BEAM Beam suspended by a crane hook, fitted with hooks or chains along its length by which to lift long loads suspended at more than a single point.

SPREADER BOX Device attached to an asphalt-mix haul truck that spreads an uncompacted layer of asphalt on the surface to be paved.

SPREADER BRACKET U-shaped elevator counterweight bracket fastened to two counterweight rails but not to the building.

SPREADER CHAIN 1. Chain used to limit a dump truck's tailgate opening to, in turn, control the flow of material during spreading operations; **2.** Chain connecting the side walls of a trailer; used at intervals along its length, to prevent the walls being pushed out by cargo.

SPREADER HOPPER Device that spreads granule surfacing or backing material on the hot bitumen-coated sheet of roofing material during manufacture.

SPREAD FABRIC Fabric, the surface of which is coated with a rubber solution and dried.

SPREAD FOOTING Footing of larger lateral dimensions than the column or wall it supports, used to distribute loads to the subgrade.

SPREADING Troweling mortar onto masonry units to form a bed in which to lay additional units.

SPREADING RATE Area of surface covered by a given volume of material, such as paint.

SPREAD RECORDER Instrument used to measure the outward spread of a loaded abutment during bridge testing.

SPREAD-TANDEM AXLE *See* **Axle.**

SPREAD-TANDEM SUSPENSION *See* **Suspension.**

SPRIG Small, headless wire nail.

SPRING 1. Flexible or elastic members that support the weight of a vehicle or machine; **2.** Locked-in tension in a log or cant released during sawing, causing the piece to deviate from a straight line; **3.** Elastic body or device that recovers its original shape when released after being distorted. There are various forms, including:

Air: Flexible pneumatic cushion incorporating a heavy spring enclosed within a flexible, sealed compartment containing air chambers and regulating valves that is pressurized by a vehicle's on-board air system.

Auxiliary: Spring that does not come into operation until a predetermined load is applied to the part being supported. On a vehicle they are designed to provide riding comfort whether the truck is empty or under partial load. *Also called* **Helper spring.**

Auxiliary rear: Spring that provides increased stability for high center of gravity and/or shifting loads.

Compression: Spring that, when not in tension, forms an open spiral. When tensioned by pushing the ends toward the center in the plane of the helix, the spiral works to resume its original shape.

Extension: Spring that, when not in tension, forms a compact, closed spiral. When tensioned by pulling the ends away from the center in the plane of the helix, the extended spiral works to resume its original shape.

Progressive (variable rate): Spring that automatically adjusts to load or road conditions.

Semieliptical: Spring basically consisting of one main leaf with eyes at each end for connection to spring shackles and brackets and a number of shorter leaves of uniformly decreased length shaped in the form of an arc.

Shackle: Spring that ejects the shackle from an unlocked padlock case.

Torsion: Length of spring steel, one end of which is anchored, the other end being attached to a part that is free to rotate about the axis of torsion spring, that works to return to its original orientation.

SPRINGBACK Tendency of a pressed particleboard panel to return to its original uncompressed state.

SPRING-BEAM SUSPENSION *See* **Suspension.**

SPRINGBOARD Board with a bolted steel 'nose.' placed in a notch cut into a tree to be falled and used as a platform to allow the faller to work above a large butt, or for use while falling timber on very steep terrain.

SPRING BUFFER One type of buffer, for elevators with speeds less than 200 fpm, that cushions the elevator. It is located in the elevator pit.

SPRING CAPACITY 1. Capacity at pad: the total weight that a spring can support in its maximum position; **2.** Capacity at ground: the total weight that a spring can support in its maximum position, plus a portion of the weight not supported by the spring.

SPRING CHAIR *See* **Axle seat.**

SPRING CLAMP *See* **Clamp.**

SPRING DEFLECTION Depression of a trailer suspension when a spring is placed under the load.

SPRING DEFLECTION RATE Weight necessary to deflect a spring a stated amount.

SPRINGER Stone from which an arch springs.

SPRING FENDER Fender built so that impact on it is taken up in part by steel springs, plates, or rubber, or by elastic bending of the construction.

SPRING GUARD Helically wound wire applied internally or externally to reinforce the end of a hose.

SPRING HINGE Hinge with a built-in spring that will cause the hinge to close, or open.

SPRINGING *See* **Chambering.**

SPRINGING COURSE Masonry course from which an arch springs.

SPRINGING LINE The level at which an arch springs from its supports. *See also* **Arch.**

SPRING LAY *See* **Lay types.**

SPRING LINE Meeting of the roof arch and sides of a tunnel.

SPRING LOADED Held in contact or engagement by springs.

SPRING POLE Tree trunk or branch that is held in a bent position such that it may release suddenly and cause damage or injury.

SPRING RATING AT GROUND Total weight a spring is designed to carry, measured at the ground.

SPRING RATING AT PAD Manufacturer's rating of a spring, measured at the spring mounting pad on the axle.

SPRING SEAT Seat on which a spring is anchored.

SPRING SET Alternately bending saw teeth to make the kerf wider than the blade.

SPRING SPACER Riser block, often used on top of the spring seat to obtain increased mounting height.

SPRING STEEL High-carbon steel that, in strip form, resists pressure to change shape.

SPRING WASHER *See* **Washer.**

SPRINGWOOD Portion of the annual growth ring that is formed during the early part of the season's growth, usually less dense and mechanically weaker than summerwood.

SPRINKLERED Building or part of a building equipped with a system of automatic fire extinguishing sprinkler heads.

SPRINKLER LEAKAGE INSURANCE Indemnity against property loss or damage caused by the discharge or leakage from an automatic sprinkler system caused by other than fire.

SPRINKLER SYSTEM 1. Fire extinguishing system consisting of pipes mounted in or under the ceilings of rooms and equipped with nozzles. The pipes are connected to a water main (but not necessarily charged with water) and the nozzles will eject a spray of water when the system is activated by heat or some other criteria; **2.** System, usually underground, by which lawns and beds can be irrigated.

sprkt *Abbreviation for:* sprocket.

SPROCKET 1. *See* **Gear; 2.** Short length of timber attached to the face of a rafter a little above the eaves to give the lowest part of a roof a flatter pitch.

SPROCKETED EAVES Eaves that project beyond the wall of a building at a flatter angle than the main slope of the roof.

SPRUNG FOUNDATION Foundation incorporating springs, designed to support machinery producing dynamic loading or having excessive vibration.

SPRUNG SAW Twisted or bent saw.

SPRUNG WEIGHT Total weight support by a spring or springs. *See also* **Nonvibrating weight,** and **Unsprung weight.**

SPS *Abbreviation for:* standard pipe sizes.

spst *Abbreviation for:* single-pole single-throw.

SPT *Abbreviation for:* standard penetration test.

SPUD 1. Movable vertical pipe, H-section or pile, placed through a frame on a floating pile driver or dredge and driven into the bottom silt to hold the vessel in position; **2.** Tool with a narrow-shaped, curved blade used in removing bark by hand. *Also called* **Barking iron; 4.** Steel tube, pointed at the bottom and fitted with lifting tackle at the top; spuds are the means by which a dredge is held in position while operating.

SPUD CLIP Device bolted to a pole hammer to engage it with and permit it to slide along a spud lead. *Also called* a **Gib.**

SPUDDING Act of opening a hole through dense material by dropping or driving a spud.

SPUD DRILL *See* **Drill.**

SPUDDING DRUM In a churn drill, a winch that controls the drilling line. *Also called* **Spudding reel.**

SPUDDING REEL Winch that controls the drilling line of a churn drill. *Also called* a **Spudding drum.** *See also* **Reel.**

SPUD KEEPER Framework on the back of a dredge that holds spuds or legs dropped down to anchor the vessel while dredging.

SPUD LEAD *See* **Lead.**

SPUD VIBRATOR Vibrator having a vibrating casing or a vibrating head, used to consolidate freshly placed concrete.

SPUD WELL Pair of guide collars for a spud on a dredge.

SPUD WRENCH Open-end connecting wrench with a tapered handle used to 'spud' or spear holes.

SPUN CONCRETE Concrete compacted by centrifugal action. *Also called* **Centrifugally cast concrete.** *See also* **Dry-cast process, Packerhead process, Tamp process,** and **Wet-cast process.**

SPUR Rock ridge projecting from a side wall, sometimes the result of inadequate blasting.

SPUR DIKE Riprap or rock fill shaped as a quarter of an ellipse, built in front of or behind abutments or piers to prevent scouring. Its length is more than 1.5 times its width. *See also* **Groin.**

SPUR GEAR *See* **Gear.**

SPUR PILE *See* **Pile, batter.**

SPUR ROAD Branch of a main or secondary forest road.

SPUR VALEY Short branch valley.

sp. vol. *Abbreviation for:* specific volume.

sq. *Abbreviation for:* square.

sq. ch. *Abbreviation for:* square chain.

sq cm *Abbreviation for:* square centimeter.

sq. E. *Abbreviation for:* square edge.

sq. edg. sd *Abbreviation for:* squarer edge and sound.

sq ft *Abbreviation for:* square foot.

sq. in. *Abbreviation for:* square inch.

sq km *Abbreviation for:* square kilometer.

sq. li. *Abbreviation for:* square link.

sq m *Abbreviation for:* square meter.

sq mi *Abbreviation for:* square mile.

SQUARE Non-SI unit of measurement, such as 100 ft^2 (10 x 10 ft) usually applied to roofing materials (that may also be used as sidewall covering).

SQUARE CENTIMETER Derived unit of area with a compound name of the SI system of measurement. Symbol: cm^2. *See also the appendix:* **Metric and nonmetric measurement.**

SQUARE EDGE Factory-formed gypsum board edge that has no taper.

SQUARE FOOT Nonmetric unit of area. Symbol: ft^2. Multiply by 0.092 903 04 to obtain square meters, symbol: m^2. *See also the appendix:* **Metric and nonmetric measurement.**

SQUARE INCH Nonmetric unit of area. Symbol: in.^{2}. Multiply by 6.4516 to obtain square centimeters, symbol: cm^2; by 645.16 to obtain square millimeters, symbol: mm^2. *See also the appendix:* **Metric and nonmetric measurement.**

SQUARE KILOMETER Derived unit of area with a compound name of the SI system of measurement. Symbol: km^2. *See also the appendix:* **Metric and nonmetric measurement.**

SQUARE LEAD At right angles to a yarding machine.

SQUARE METER A derived unit area with a compound name of the SI system of measurement. Symbol: m^2. Multiply by 0.0001 to obtain hectares, symbol: ha; by 10.7639 to obtain square feet, symbol: ft^2; by 1.195 99 to obtain square yards, symbol: yd^2. *See also the appendix:* **Metric and nonmetric measurement.**

SQUARE METER PER SECOND A derived unit of kinematic viscosity with a compound name of the SI system of measurement. Symbol: m^2/s. *See also the appendix:* **Metric and nonmetric measurement.**

SQUARE MILE Nonmetric unit of area. Symbol: mi^2. Multiply by 258.9988 to obtain hectares, symbol: ha; by 3.589 9988 to obtain square kilometers, symbol: km^2. *See also the appendix:* **Metric and nonmetric measurement.**

SQUARENESS At right angles at the corners, having equal diagonals from corner to corner.

SQUARE YARD Nonmetric unit of area. Symbol: yd^2. Multiply by 0.836 127 4 to obtain square meters, symbol: m^2. *See also the appendix:* **Metric and nonmetric measurement.**

SQUAW HITCH Holt set with a choker on the end of a log when the choker cannot be placed completely around the log.

SQUEEZE FILM LUBRICATION State of lubrication in which surfaces thickly coated or flooded with lubricant move toward each other at sufficient speed to develop fluid pressure sufficient to support a load of short duration. Because of viscosity (or apparent viscosity), the lubricant cannot immediately flow away from the area of contact. This action occurs, for example, between gear teeth and between wrist pins and their bushings.

SQUEEZEOUT Bead of excess material, such as adhesive or caulk, forced out from a joint upon application of sufficient pressure to ensure a complete bond.

SQUEEZE RIVETER Single-stroke, air-driven mechanism that closes rivets through a toggle mechanism.

SQUEEZING GROUND Soil formation, usually clay, silt, or organic material, that tends to bulge or squeeze into a hole during drilling, or afterwards if the hole is left uncased.

SQUIB Detonator comprising a firing device and a chemical that will ignite black powder.

SQUINT *See* **Two-inch-piece.**

SQUINT BRICK Brick molded, cut, or shaped to a special shape.

SQUIRREL-CAGE MOTOR Alternating current motor in which the rotor has a number of parallel bars on its perimeter, joined to end rings.

sqrs *Abbreviation for:* squares.

sq. rt *Abbreviation for:* square root.

sq/up *Abbreviation for:* square up.

sq yd *Abbreviation for:* square yard.

sr *Abbreviation for:* senior.

SR *Abbreviation for:* sedimentation rate; shower receptor; split ring; stateroom; stress relieved.

SRD *Abbreviation for:* standard reference data.

SRO *Abbreviation for:* single room occupancy.

SRT *Abbreviation for:* steel-reinforced tread.

ss *Abbreviation for:* sewer service; single strength; slope stake.

s/s *Abbreviation for:* shear to size; silica sand.

SS *Abbreviation for:* single strength; slab spacer; slop sink; stainless steel; summary sheet.

SSA-B *Abbreviation for:* single strength A and B quality.

SSE *Abbreviation for:* sound, square-edged; south-southeast.

sses *Abbreviation for:* sewer system evaluation study.

SSF *Abbreviation for:* Saybolt Furol scale.

ssk *Abbreviation for:* soil stack.

ssol. *Abbreviation for:* semisolid.

sss *Abbreviation for:* side, side, side.

sst *Abbreviation for:* standard seam tin.

SSU *Abbreviation for:* Saybolt Universal scale.

SSW *Abbreviation for:* sheet-steel wainscot; south-southwest.

St *Abbreviation for:* strut.

st *Abbreviation for:* single throw.

st. *Abbreviation for:* stair; start; steam; steel; stone; street.

S/T *Abbreviation for:* short ton.

sta. *Abbreviation for:* single-tape-armored; station; stationary.

stab. *Abbreviation for:* stabilize; stable.

STAB Action of lining up and catching the threads of the loose piece of a drill string.

STABBING POINTS Fabricated steel points fastened to the very bottom of swinging box leads to facilitate rigidly positioning the bottom of the leads by stabbing the points into the ground.

STABILITY 1. Resistance of a structure to tilting, sliding, overturning, collapsing, etc.; **2.** Ability of a mobile crane to resist tipping (does not normally apply to a stationary mounting); **3.** Industrial truck's resistance to overturning; the recognized method of measurement is the 'tilting platform Test'; **4.** Resistance of material in a cut or fill to movement downslope due to inherent characteristics or to the weight of a superimposed load; **5.** Ability of a governor to maintain desired engine speed without fluctuations.

STABILITY-LIMIT LOAD Maximum (theoretical) load a structure can support when second-order instability effects are included.

STABILIZATION 1. Preparation of road base materials to increase their bearing strength and resistance to moisture, generally by compaction or mixing with asphalt, cement, salt, or chemicals; **2.** Process by which wastes are rendered relatively inert, uniform, biologically inactive, nuisance free, or harmless.

STABILIZE To make soil firm and prevent it from moving.

STABILIZED Condition of equilibrium.

STABILIZER 1. One of a set of two, or four arms, manually or hydraulically actuated, that can be extended laterally from the substructure of mobile equipment to provide a widened and level operating base; **2.** Device used to provide 'no-hop' during vehicle braking by positioning the radius rod below the axle to resist brake wind-up; **3.** Substance that makes either a solution or suspension more stable, usually by keeping particles from precipitating.

STABILIZING JACK Device that provides temporary support to the front or rear of a parked vehicle.

STABILIZING MOMENT *See* **Machine resisting moment.**

STABILOMETER Triaxial testing device used in the Hveem method of asphalt mix design in which vertical loads are applied to compacted specimens and the resulting lateral pressures measured. *See also* **R value.**

STABLE DOOR Door cut in half horizontally with each leaf separately hinged.

STABLE MORTGAGE *See* **Mortgage.**

STABLE ROCK *See* **Soil classification system.**

STACK 1. Any structure that contains a flue or flues for the discharge of gases; **2.** Vertical run of drain-waste-and-vent piping; **3.** Accurate, vertical line of rails located in a hoistway; **4.** Rainwater downpipe.

STACK BOND *See* **Bond.**

STACKED ASHLAR *See* **Ashlar.**

STACK EFFECT Rise of air through a tall space caused by temperature and/or pressure difference.

STACKER 1. Mobile elevating belt; **2.** Mobile machine for unloading and stacking or decking logs.

STACK GAS Conglomerate of gaseous, solid and liquid particles generated by a source and contained within a ventilating stack.

STACK HEIGHT Vertical distance between the bottom of a heating unit and the top of the outlet opening.

STACKING TILE Method of installation whereby glazed tiles are placed on a wall so that they are in direct contact with the adjacent tiles. The width of the joints is not maintained by the use of string or other means. The tiles may be set either straight or broken joint.

STACKING TUBE Slender, free-standing tubular structure used to store granular materials.

STACK PARTITION Partition wall that contains the stack or soil waste pipe, usually constructed of 50 x 100 mm (2 x 6 in.) studs or larger.

STACK SAMPLING Collection of representative samples of gaseous and particulate matter that flows through a duct or stack.

STACK VENT Extension of the waste pipe above the highest horizontal connecting drain, through the roof to atmosphere.

STADIA Device used in calculating distance through measurement of angles; specifically, by calculating the distance by proportion to the space on a vertical rod seen between upper and lower instrument cross hairs in a surveyor's level or transit.

STADIA ROD Leveling staff marked for use with stadia.

STADIUM Amphitheater designed and used for sports and athletics.

STAFF Slender metal or wood rod marked with graduations, usually about 3 m (12 ft) long and telescoping or folding, used in survey work.

STAFF GAUGE Graduated rod or plate, or marked masonry of a pier, etc., by which the level of water can be read.

STAGE 1. Stages in the life of a project, separated by milestone events that are sign-offs or approvals by authorities. The seven stages in the life-cycle of a construction project are:

    (a) Investigations.

    (b) Preliminary analysis.

    (c) Conceptual design.

    (d) Project brief.

    (f) Control and drawings.

    (g^1) Construction.

    (g^2) Construction close-out.

    (h) Operations.

Of these, stages **(c)** and **(d)** are the development process or feasibility phase, and stages **(e)**, **(g^1)**, and **(g^2)** are the construction process or construction phase; **2.** Suspended structure that supports the working load of a single, two-point or multiple-point scaffold; **3.** Hydraulic amplifier used in a servovalve: may be single stage, two stage, three stage, etc.; **4.** Space designed primarily for theatrical performances with provision for quick-change scenery and overhead lighting, including environmental control for a wide range of lighting and sound effects and that is traditionally, but not necessarily, separated from the audience by a proscenium wall and curtain opening.

STAGE CONSTRUCTION Building a roadway in successive treatments, applications, or layers suited to increasing traffic requirements and available funds.

STAGE CUT Two-phase cutting in which the forest understory is cut and removed first to reduce breakage when the overstory is felled.

STAGE GROUTING Sequential grouting of a hole in separate steps or stages in lieu of grouting the entire length at once.

STAGGED PANTS Work pants with the cuff cut off to prevent limbs from catching the cuffs and tripping the worker.

STAGGER Arrange in parallel rows but with the joints in one row opposite solid faces in the next.

STAGGERED COURSES Roofing shingles laid in courses with the butts not in a horizontal line.

STAGGERED JOINT *See* **Broken joint.**

STAGGERED SETTING Clear-cut settings separated by uncut timber.

STAGGERED SPLICES Splices in reinforcing bars not made at the same point.

STAGGERED STUD PARTITION Partition constructed using two rows of studding at staggered centers, one row supporting the sheeting for one side, the other for the reverse face of the wall, the two sides separated by sound-deadening insulation.

STAGING *See* **Scaffolding.**

STAIN 1. Discoloration by foreign matter. *See also* **Migration stain**; **2.** Penetrating or nonpenetrating medium used for staining woods and for making graining media.

STAINED GLASS Glass in which color is infused during the molten stage, used to form decorative panels and windows.

STAINLESS STEEL *See* **Steel.**

STAINLESS STEEL PIPE *See* **Pipe.**

STAIR Set of steps leading from one level to another.

STAIR CARRIAGE Supporting member for stair treads. *Also called* **Rough horse.**

STAIRCASE Flight of steps leading from one level to another, particularly from one story to another.

STAIRCASE FORMULA Rule governing the relationship between the treads and risers of a staircase that states that the aggregate of one step dimension plus that of two risers shall not be less than 550 mm (21 in.) or more than 700 mm (27 in.), and that the sum of one step dimension multiplied by one riser dimension shall be approximately 63.

STAIRHEAD Top of a stair.

STAIR LANDING Platform between flights of stairs.

STAIR PLATFORM Extended step or landing breaking a continuous run of stairs.

STAIR RAIL Vertical barrier erected along the unprotected sides and edges of a stairway.

STAIR RISE Vertical distance from a floor level to the top of a landing or to a floor above.

STAIR RUN Horizontal distance between the face of the first rise and the face of the platform or stair opening above.

STAIR SLAB Reinforced concrete stair flight.

STAIR TREAD Upper horizontal boards of a flight of steps.

STAIR WELL Area extending vertically through a building in which a staircase is housed.

STAKE 1. Sharpened or pointed wood or metal length used to mark a position; **2.** Metal or wood posts by means of which sides are attached to truck platforms; when used alone, stakes are a means of retaining loads on flat-deck platforms; **3.** Shaped block used as an anvil in sheet metal work.

STAKE BODY See **Truck**.

STAKE NOTE Record of the field location of survey stakes.

STAKEOUT Locating on the ground, previously determined survey points such as boundary corners and construction stakes. *Also called* **Construction stakeout,** and **Pegging out.**

STAKE POCKET Cutout in the side rail of a vehicle body or trailer into which wood or metal stakes may be inserted to provide a vertical restraint to a load being carried.

STAKING PLAN Drawing that shows the locations of stakes to be set in the ground as a preliminary to construction.

STALK 1. Vertical section of a reinforced-concrete retaining wall; **2.** Central leg of a T-section.

STAMP Identification/code mark embedded into a log end that indicates the setting it came from. *See also* **Brand.**

STAMP HAMMER Hammer used to place a stamp. *Also called* **Branding axe.**

stan. *Abbreviation for:* stanchion.

STANCHION Vertical supporting member.

STAND Community of trees sufficiently uniform in species, age, arrangement or condition to be distinguishable as a group from the forest or other growth in the area.

STAND AGE *See* **Land classification (forest management).**

STAND-ALONE SYSTEM Collection of devices and mechanisms that will complete the designed function without assistance from an external source.

STANDARD 1. Document, or an object for physical comparison, for defining product characteristics, products, or processes, prepared by a consensus of a properly constituted group of those substantially affected by, and having the qualifications to prepare the standard for use; **2.** Free-standing vertical post, usually metal, that supports something else; **3.** U-shaped mounting on the upper plate of a theodolite that carries the telescope trunnions.

STANDARD AIR Air at a temperature of 20°C (68°F), a pressure of 1.03 kg-cm² (14.70 psi) absolute, and a relative humidity of 36%. (In gas industries, the temperature of standard air is usually given as 21.1°C (70°F).) *See also* **Compressed air,** and **Free air.**

STANDARD ATMOSPHERE Non-SI unit of pressure, equal to 101.325 kPa, permitted for use with the SI system of measurement for a limited time. Symbol: atm. *See also the appendix:* **Metric and nonmetric measurement.**

STANDARD BEND Right angle bend at the end of a reinforcing bar to act as an anchorage.

STANDARD BRICK *See* **Brick.**

STANDARD CONDITION 1. Atmospheric pressure at sea level of 760 mm (29.92 in.) of mercury; **2.** Reference for compressible fluids of 15.6°C (60°F) at 1.0 atmospheres (total pressure) at dry-gas conditions.

STANDARD COOL-WHITE FLUORESCENT *See* **Fluorescent lamp.**

STANDARD COUPLING Fire hose coupling having NS (American National Standard) fire hose threads.

STANDARD CURING Exposure of test specimens to specified conditions of moisture and temperature.

STANDARD DENSITY Typical density of OSB when manufactured to meet industry standards.

STANDARD DETAIL Design drawing detail common to many jobs.

STANDARD DEVIATION Constant allowance due to mechanical misadjustment or other permanent condition. *See also* **Coefficient of variation,** and **Deviation.**

STANDARD DIMENSION Manufacturer's designated dimension.

STANDARD DRYING DAY Day that produces the same net drying as experienced during a 24-hour period under laboratory conditions where the dry-bulb temperature is maintained at 32°C (90°F) and the relative humidity at 20%.

STANDARD DUTY DUMP BODY *See* **Truck.**

STANDARD HOOK Hook at the end of a reinforcing bar made in accordance with a standard.

STANDARDIZE Reduce to or compare to a standard.

STANDARD MATCHED Tongue-and-groove lumber with the tongue and groove offset rather than centered as in center-matched lumber.

STANDARD PENETRATION RESISTANCE Number of blows of a 63.5-kg (140-lb) hammer falling

750 mm (30 in.), required to advance a 50-mm (2-in.) OD, split-barrel sampler 300 mm (12 in.) through a soil mass.

STANDARD PENETRATION TEST Number of blows required to drive a 50-mm (2-in.) O.D., 35 mm (1-3/8 in.) I.D., 600-mm (24-in.) long, split, soil-sampling spoon 300 mm (1 ft) with a 63.5-kg (140-lb) weight falling freely 750 mm (30 in.). The count is recorded for each of three 150-mm (6-in.) increments. The sum of the second and third increments is taken as the N value in blows/mm (blows/ft).

STANDARD PLAN Drawing approved for repetitive use, showing details to be used, where appropriate. *See also* **Plans,** and **Working drawings.**

STANDARD PRESSURE Working pressure of 861.8 kP (125 psi); the minimum pressure for which plumbing fixtures are designed.

STANDARD PROVISION Contract and construction clause that is common to many types of project and that is used unless the situation requires some special consideration or treatment.

STANDARD RAILING Vertical barrier erected along exposed edges of floor and wall openings, ramps, platforms, or runways.

STANDARD RATING Rating based on tests performed under standard rating conditions.

STANDARD RIM A rim that has been calibrated and found to meet the precise measurements specified by the Tire and Rim Association, or the European Tire & Rim Association.

STANDARD SAND Ottawa sand accurately graded to pass a No. 20 (0.84-mm, 0.0331-in.) sieve and be retained on a No. 30 (0.59-mm, 0.0232-in.) sieve. *See also* **Graded standard sand,** and **Ottawa sand.**

STANDARD SEAL COAT Single application of asphalt material followed immediately by a single layer of aggregate of a uniform size.

STANDARD SOLID TIRE Solid tire having a low, flat profile in cross section.

STANDARD SPECIFICATIONS Book of specifications approved for general application and repetitive use.

STANDARD WARM-WHITE FLUORESCENT *See* **Fluorescent lamp.**

STANDARD WHITE FLUORESCENT *See* **Fluorescent lamp.**

STANDBY CREW Group of workers organized, trained and placed for quick suppression of forest fires.

STANDBY POWER *See* **Generator set.**

STANDBY POWER ELECTRIC SET *See* **Electric set.**

STANDBY POWER SUPPLY Power supply that is selected to furnish electric energy when the normal power supply is not available.

STANDBY SERVICE Generating equipment exclusively utilized in the event of failure of the utility-supplied service.

STAND CONDITION General health of a stand of trees reflected by its development relative to the site potential. A good stand condition refers to a fully stocked stand that is producing fiber at a high rate based on specific site conditions such as moisture, soil quality, and other biological variables.

STAND DENSITY Relative measure of the amount of stocking on a forest area, often described in terms of stems per hectare. *See also* **Density.**

STANDEE Special reinforcing bar bent to a U-shape with 90° bent legs extending in opposite directions at right angles to the U-bend, used as a high chair resting upon a lower mat of bars and supporting an upper mat.

STAND IMPROVEMENT Any silviculture treatment that increases the growth, quality, or value of trees in a stand.

STANDING LINE Fixed-length line that supports a load without being spooled on or off a drum. *Also called* **Guyline, Guy rope, Pendant,** and **Stay rope.**

STANDING PIER Bridge pier with spans on either side (as distinct from an abutment pier).

STANDING RIGGING Fixed lines or rigging used to support, hold, or secure something.

STANDING SEAM Flexible-metal roofing seam, generally between eaves and ridge.

STANDING TIMBER Timber still on the stump.

STANDING WATER LEVEL Level at which ground water stands in a hole or pit left open for a prolonged period.

STANDPIPE 1. Vertical water storage pipe used to establish uniform pressure in a water distribution system; **2.** Pipe or tank connected to a closed conduit and extending to or above the hydraulic grade line.

STAND SIZE CLASSES Designations for various stand sizes, including:

> **Nonstocked areas:** Commercial forest land on which the stocking of growing-stock trees is less than 16.7%.

> **Poletimber stands:** Stands at least 16.7% stocked with growing-stock trees, with half or more of this stocking in sawtimber and/or poletimber trees. Stocking of trees exceeds that of sawtimber stands.

> **Sapling-seedling stands:** Stands at least 16.7% stocked with growing-stock trees, with saplings and/or seedlings comprising more than half of this stocking.

> **Sawtimber stands:** Stands at least 16.7% stocked with growing-stock trees, with half or more of this stocking in sawtimber or poletimber trees. Sawtimber stocking at least equals poletimber stocking.

STAND TABLE Table showing the number of trees by species and diameter classes, generally per unit area of a stand. Such data may be presented in the form of a frequency distribution of diameter classes.

STAND TENDING Variety of forest management activities carried out at different stages in the life of a stand. Treatments may include: juvenile spacing, brushing, commercial thinning, fertilization, conifer release, site rehabilitation, mistletoe control, seed tree control, and pruning.

STAND-UP TIME Time an unsupported excavation can be maintained in a tunnel or drill pier.

STAPLE 1. Fastener consisting of a double-pointed U-shaped piece of metal; **2.** Textile fiber of relatively short length that when spun and twisted forms a yarn; **3.** Length of such a textile fiber.

STAPLER or STAPLE GUN Spring-, air-, or electric-powered device that ejects staples with considerable force. Will accept a wide range of staple sizes and shapes.

STAR DRILL Hand-held drill consisting of a hexagonal or octagonal rod shaped at its cutting end into a four-cornered star. The drill is tapped with a heavy hammer and rotated through a part circle after each blow.

STARLING Pile driven into the river bed around a bridge pier.

STARTABILITY Vehicle's grade-pulling capability from a dead stop at full GVW. *Also called* **Starting gradability.**

STARTER 1. Electrical motor or internal combustion engine used to start a larger motor or engine; **2.** Short steel used to start a drill hole.

STARTER BAR Reinforcing bar left projecting from concrete in order to locate and provide continuity for further reinforcement.

STARTER PIECE Masonry unit or glazed tile cut specially for use at a corner to establish the bond.

STARTER STRIP First layer of roofing material placed around the edge of the roof.

STARTING GRADABILITY *See* **Startability.**

STARTING NEWEL Post at the bottom of a stair supporting the balustrade.

STARTING STEP First step of a stair.

STARTING SYSTEM Group of components used to initially rotate a prime mover at sufficient speed to get it started.

STARTING THROTTLE LOCK Device on a chainsaw that keeps the throttle partially open during starting.

START OF CONSTRUCTION Moment when work on construction begins, following such stages as site preparation, site clearing, pegging out, etc.

START-UP Initiation of a mechanical device or system.

START-UP COST Aggregation of the costs, excluding acquisition costs, incurred to bring a new facility into production.

START-UP TIME Period of time needed to reach a steady-state condition within the operating band, starting from a long-term off condition.

STARVED 1. Condition where an internal combustion engine does not perform according to specification due to an insufficiency of air, or of fuel. *Also called* **Hungry; 2.** Surface that is too absorptive for the amount or kind of paint applied to it resulting in a paint film that is thin and patchy.

STARVED GLUELINES Condition caused by an insufficient adhesive spread to adhere two materials together, as in a laminated panel construction.

stat. *Abbreviation for:* statistic; statistical; statuary.

STATE BUILDING CODES *See* **Building code.**

STATE FOREST INVENTORY Description of the quantity and quality of forest tres, nonwood value, and many of the characteristics of the land base compiled from statistical data for the forest lands of the state.

STATE-OF-THE-ART TECHNIQUE One that incorporates the most recently developed thinking on the subject and/or processes and materials.

STATICALLY DETERMINATE FRAME Frame in which the reactions and bending moments can be determined by the laws of statics.

STATIC BALANCE 1. Masses, forces, or bodies at rest, or in equilibrium. *See also* **Balance,** and **Dynamic balance; 2.** Wheel balance on a nonrotating tire.

STATIC BASE Tower-crane base mounting where the crane mast is set into a foundation.

STATIC BONDING Use of a grounded conductive material to eliminate static electrical charges.

STATIC CONDUCTIVE Having the capability of furnishing a path for a flow of static electricity. *Also called* **Antistatic.**

STATIC CONE PENETRATION TEST Test in which a standard cone is pushed into soil to a depth of 76 mm (3 in.) with the pressure required to reach that penetration being noted.

STATIC DISCHARGE HEAD *See* **Head.**

STATIC ELECTRICITY Negative or positive electrical charge accumulated by an object that can be transmitted to another object, or to a person, or discharged to ground through a conductor.

STATIC HEAD *See* **Head.**

STATIC JOINT A nonmovement or nonworking joint.

STATIC LINE Horizontal wire rope attached to a scaffold stirrup (or structural member) to which a lanyard is attached. *Also called* **Trolley line.**

STATIC LOAD Weight of a single stationary body or the combined weights of all stationary bodies in a structure (such as the load of a stationary vehicle on a roadway), or, during construction, the combined weight of all materials and appurtenances used to erect the structure.

STATIC MOMENT First moment of area about the axis of a section: the sum of the products obtained by multiplying each element of area by its distance (x) from YY.

STATIC PILE SYSTEM Windrow composting method in which air ducts are generally installed under or in the base of compost piles so air can be blown or drawn into the pile.

STATIC PRESSURE *See* **Pressure.**

STATIC ROCK STRENGTH Amount of stress that a rock mass can withstand from a stationary load without failing.

STATICS Study of forces and bodies at rest.

STATIC SEAL Sealing device used between parts that have no relative motion between them. *See also* **Dynamic seal.**

STATIC TIPPING LOAD Minimum weight that tips a machine. On track loaders this means the rollers are clear of the track; on wheel loaders, the rear wheels are off the ground.

STATIC WATER LEVEL Level at which water stands in a well when no water is being pumped from or being added to the aquifer.

STATIC WATER SUPPLY Supply of water at rest that does not supply a pressure head for fire fighting, but that may be employed as a suction source for fire pumps.

STATIC WEIGHT That portion of the operating weight (mass) of a vibratory roller exerted on the ground at the drum(s).

STATIC WIRE Wire incorporated in a hose or other product to conduct static electricity.

STATION 1. Any of a series of stakes or points indicating the distance from a beginning point or point of reference. *See also* **Minus station**; **2.** Distance of 100 ft (30.48 m) measured along a centerline and designated by a stake bearing its sequential number.

STATIONARY COMPACTOR *See* **Compactor.**

STATIONARY EMISSION SOURCE Stationary facility that releases combustion gases or vapors to the environment.

STATIONARY FIFTH *See* **Fifth wheel.**

STATIONARY HOPPER Container used to receive and temporarily store freshly mixed concrete.

STATIONARY PACKER *See* **Refuse truck.**

STATION BATTERY Power supply utilized for control of switchgear.

STATIONING Variation of the coordinate method of survey, used for long, narrow projects such as pipelines and highways. A line, which may be straight or contain angles or curves, is established from a zero point. Stations are established at regular intervals along the line with points on the line designated as such and points off the line designated to left or right perpendicular to it.

STATION-POINT PROJECTION *See* **Linear perspective.**

STATION YARD Unit of quantity multiplied by distance, corresponding to 1 yd³ moved horizontally through a distance of 100 ft. *See also* **Haul, Overhaul,** and **Free haul.**

STATISTICAL SAMPLING Selection of sample units from a range and the measurement and/or recording of information on these units, to obtain estimates of population characteristics, for instance.

STATISTICALLY SIGNIFICANT Values of test statistics that lie outside of predetermined limits of test precision and so are taken to indicate a difference between populations.

STATISTICS Numerical facts and data.

statnry *Abbreviation for:* stationary.

STATOR 1. Set of fixed vanes in a torque converter that change the direction of flow of fluid entering the pump or at the next stage turbine; **2.** Fixed or stationary plate of a colloid mill; emulsions are formed when two immersible liquids are introduced into a small clearance cavity between the stator and a high-speed rotor creating high shear forces; **3.** Stationary portion of an alternator that provides the magnetic field.

STATUS REPORT *See* **Progress report.**

STATUTE Law established by an act of a legislature.

STATUTE OF FRAUDS Law providing that certain types of contract must be in writing for them to be legally enforceable.

STAUNCHING BEAD Vertical gap left between successive bays in a concrete dam and not concreted until most shrinkage has occurred in adjoining bays.

STAUNCHING ROD Flexible rod of compressible material between the crest gate and dam structure to form a watertight joint.

STAVE 1. Vertical slat in a column form; **2.** Slat used in the manufacture of a wooden storage tank or wooden conduit.

STAY Support: flexible when in tension; rigid when in compression.

STAY BAR Temporary brace holding something in place.

STAYLATHING Horizontal timbers placed on opposite sides of a row of piles and drawn together by bolts, used to pull pile heads into line.

STAY PILE Driven or cast pile acting as an anchor.

STAY ROPE *See* **Guyline, Guy rope, Pendant,** and **Standing line.**

stch *Abbreviation for:* stitch.

std *Abbreviation for:* standard.

stdby *Abbreviation for:* standby.

stdg *Abbreviation for:* standing.

Std M *Abbreviation for:* standard matched; standard mast.

stdn *Abbreviation for:* standardization.

Stds Engr *Abbreviation for:* standards engineer.

std. tran. *Abbreviation for:* standard transmission.

stdy *Abbreviation for:* steady.

stdzn *Abbreviation for:* standardization.

STEADY FLOW Stream line flow.

STEADY POINT Pointed bar in a slide clamp, used to brace a machine being worked on. *Also called* **Peg point.**

STEADYREST Support for pieces being ground on a cylindrical grinding machine.

STEADY STATE Operating conditions under constant load.

STEADY-STATE FREQUENCY Governed frequency occurring when an engine generator is operating with a steady-state electrical load.

STEADY-STATE PRESSURE REGULATION
Band indicating maximum and minimum pressure on a single curve with maximum deviation indicated in percent of operating pressure, all as a function of flow.

STEADY-STATE SPEED Mean governed speed occurring when an engine generator is operating with a steady-state electrical load.

STEADY-STATE VELOCITY Chemically compounded rate of detonation of an explosive, governed by diameter, degree of confinement, temperature, etc.

STEADY-STATE VOLTAGE Value of voltage output when a generator set is operating at a steady-state load.

STEAM Invisible water vapor at temperatures of 100°C (212°F) or more; also condensing white vapor.

STEAM ATOMIZATION Steam used to break oil into small particles preparatory to combustion.

STEAM BOX Enclosure for steam-curing concrete products.

STEAM CLEANING Use of steam under pressure to remove grease and oil.

STEAM CURING Curing of concrete, mortar, grout, or neat-cement paste in water vapor at atmospheric or higher pressures and at temperatures between 40°C and 215°C (100°F and 420°F).

STEAM DRUM That part of a water-wall furnace where water is evaporated into saturated steam.

STEAM FITTER Tradesman certified to work on steam or hot water heating systems.

STEAM HEATING *See* **Heating system.**

STEAM LEAD *See* **Lead.**

STEAM TRAP Pipe arrangement allowing the passage of condensate, or air and condensate, and preventing the passage of steam.

STEARATE Salt or ester of stearic acid that functions as a water repellent.

STEEL 1. Malleable alloy of iron and carbon (up to 2%) with substantial quantities of manganese. There are many types, including:

Alloy steel: Steel compounded with other metals to increase strength, wearing capability, or rust resistance, or other qualities.

Carbon steel: Hardened steel not alloyed with other metals.

Galvalume sheet: Steel roofing and siding sheet with an aluminum-zinc coating.

High-strength low-alloy steel: Specially formulated steel that can develop up to 40% more strength than structural carbon steel, producing thinner sections offering less weight per unit of length.

Mild steel: Similar in structure to wrought iron, but capable of being reheated and rolled into shapes, causing it to develop a finer grain structure with less slag.

Stainless steel: Iron, to which various quantities of chromium and other elements have been added to improve corrosion resistance, strength, toughness, ease of fabrication, and weldability, among others.

Structural steel: Steel beams, girders, and columns used for building purposes. May be hot-rolled, cold-rolled, cold-formed, or built-up shapes.

Weathering steel: Type of high-strength, low-alloy steel that can be used in normal environments (not marine) and outdoor exposures without protective paint covering. This steel develops a tight adherent rust at a decreasing rate with respect to time.

2. Hollow or solid steel bar connecting the hammer of an air hammer with the cutting tool.

STEEL CENTRALIZER Guide to hold the starting steel of a drill in proper alignment.

STEEL CHANGE Difference in length between two of the successive steels used in drilling a hole.

STEEL ERECTOR Worker who erects steel, including those who climb steel members and complete attachments as components are lowered to them by crane.

STEEL FIBER Filament produced from steel and used as continuous or chopped fiber reinforcement in cement paste and concrete.

STEEL FRAME CONSTRUCTION *See* **Construction types.**

STEEL H-SECTION PILE *See* **Pile.**

STEEL JOIST Prefabricated, structural secondary member fabricated from hot- or cold-rolled steel sections.

STEEL PIPE PILE *See* **Pile.**

STEEL PULLER Hinged clamp on the bottom of a hand drill.

STEEL-REINFORCED TREAD Tire tread with an undertread and sidewall protective layer of thick tread rubber containing many short lengths of brass-coated, hardened steel filaments. *Also called* **SRT.**

STEEL RING *See* **Tubbing.**

STEEL SHEET Cold-formed sheet or strip, used as a structural member to carry live and dead loads in lightweight concrete floor and roof construction.

STEEL SHEET PILING Intelocking profiled rolled-steel sections driven vertically into the ground prior to excavation.

STEEL SPRING SUSPENSION *See* **Suspension.**

STEEL SQUARE Layout tool consisting of a large arm, 50 mm (2 in.) wide and 600 mm (24 in.) long, called the body or blade, and a smaller arm, 38 mm (1.5 in.) wide and 400 mm (16 in.) long, set at right angles to the blade and called the tongue. The point where the outside edges of the blade and the tongue join is called the heel. The surface with the manufacturer's name is called the face; the opposite surface is called the back. The surfaces are scribed with a range of markings from which calculations can be made.

STEEL TROWEL Flat steel hand tool used to create a dense finish on concrete.

STEEL WOOL Loosely woven steel threads, used to abrade and polish a surface. Available in a range of grades between super-fine (#0000) through extra-fine

(#000), very fine (#00), fine (#0), medium (#1), medium coarse (#2), coarse (#3), and extra coarse (#4).

STEELWORKER Tradesman skilled in working with structural steel and steel reinforcing.

STEEP ASPHALT Roofing asphalt having a softening point of approximately 104°C (220°F).

STEEPLE Tower and spire or lantern of a church.

STEEPLEJACK Worker who builds or repairs steeples and similar structures.

STEERING Mechanism by which a moving vehicle or machine is guided. There are several methods, including:

> **Ackerman:** Steering system used on motor graders and off-highway trucks. In an Ackerman system, the wheels can be turned relative to the axle either mechanically, mechanically with a hydraulic boost, or fully hydraulically.

> **Articulated:** Steering system used on wheel loaders, wheel dozers, compactors, motor graders, skidders, and scrapers.

> **Track-type:** There are three types of steering used on track-type machines: Conventional clutch brake, a combination clutch and gear system, and hydrostatic. In the conventional system, steering clutches are located between the bevel pinion and final drive and are normally in the engaged position. When the clutch is disengaged, power is interrupted. A brake can also be applied for a pivot turn. The gear steer system uses a gear and clutch set on each side of the bevel gear along with conventional steering clutches and brakes. This system provides power turns in one of two fixed ratios. The hydrostatic system uses hydraulic pumps and motors to power the tracks. Turns are made under power by slowing the speed of one track versus the other. Counterrotation also is possible for spot turns.

STEERING AXLE *See* **Axle.**

STEERING BRAKE *See* **Brake.**

STEERING CLUTCH Clutch used to disconnect power from one side or the other of a tractor.

STEERING KNOB Device mounted near the rim of the hand steering wheel, designed to rotate with the driver's hand to aid in steering the vehicle.

STEERING WHEEL LOCK Independent device used to secure the steering wheel of a towed vehicle.

STEM 1. Main body of a tree from which branches grow; **2.** Shaft of a faucet to which the handle is attached; **3.** Vertical part of a reinforced concrete retaining wall.

STEM BAR Bar used in the wall section of a cantilevered retaining wall or in the web of a box.

STEMMING 1. Inert material, such as drill cuttings, used in the collar portion or elsewhere in a blasthole so as to confine the gaseous products formed by an explosion; **2.** Length of blasthole left uncharged; **3.** Prevention of soft or loose soil from escaping through joints or openings in timbering.

STEM WALL That part of a wall extending from the foundation to the ground floor.

STEMWOOD Wood from the main part of a tree, not from the branches, stump, or root.

sten. *Abbreviation for:* stencil.

steno. *Abbreviation for:* stenographer.

STEP Difference in height between panels when butted together due to thickness variations in panels nominally of the same thickness.

STEP BRACKET Bracket ornamenting the end of a step in an open-string stair.

STEP-DOWN TRANSFORMER Transformer having a higher voltage on the primary windings than at its secondary windings.

STEP DUTCHMAN Intentional alteration of the standard falling technique to solve problems in maintaining a lead by sawing off the lean side holding wood and placement of a step (rock, wood block) into the face to force the tree to pivot to the desired direction.

STEP EXCAVATION Excavation at the abutments of a dam to provide level foundation surfaces forming steps up the abutment.

STEP FLASHING Rectangular or square pieces of flashing used at the junction of shingled roofs and walls. *Also called* **Shingle flashing.**

STEP JOINT *See* **Broken joint.**

STEPLADDER Free-standing ladder, usually supported by a frame that hinges out from the top to form an inverted V.

STEPPED COLUMN Column with changes in cross section occurring abruptly at certain points within its length.

STEPPED FLASHING Flashing between a vertical and sloped face that is tucked into successive courses of the vertical face.

STEPPED FOOTING Change in level of a footing to compensate for a rise or drop in ground level.

STEPPED SKIRTING Asphalt skirting placed at the sloped intersection with a roof.

STEPSTOOL Self-supporting, foldable, portable ladder, nonadjustable in length, 800 mm (32 in.) or less in overall size.

STEP-TAPERED PILE Cast-in-place concrete pile formed by installing successively larger sections of corrugated shell or thin-wall pipe; ordinarily installed with a mandrel that bears on a connector between lengths. A cuneiform pile.

STEP-UP TRANSFORMER Transformer having a higher voltage at its secondary winding than at its primary winding.

ster. *Abbreviation for:* sterilizer.

STERADIAN One of the two supplementary units of the SI system of measurement: a unit of solid angle equal to the measure of a solid angle with its vertex at the center of a sphere and enclosing an area of the spherical surface equal to that of a square with sides equal in length to the radius. Symbol: sr. *See also the appendix:* **Metric and nonmetric measurement.**

STERE Obsolete metric unit of volume, equal to 1 m³, that should not be used with the SI system of measurement. Symbol: st. *See also the appendix:* **Metric and nonmetric measurement.**

STEREOMETRIC MAP Map depicting valleys and elevations in a way that can be easily visualized.

STEREOSCOPIC PAIR Two aerial photographs with sufficient overlap and consequent duplication of detail to make possible stereoscopic (three-dimensional) examination of an object, or an area, common to both.

STEREOSCOPY Viewing of objects in three dimensions.

stf *Abbreviation for:* staff.

stg *Abbreviation for:* starting.

stg. *Abbreviation for:* stage; storage.

stgr *Abbreviation for:* stringer.

STICK 1. In a shovel or backhoe, the arm connecting the boom to the dipper, bucket or other attachment. *Also called* **Crowd**; **2.** Piece of short pulpwood; **3.** Waxed paper cartridge of explosive measuring approximately 38 x 200 mm (1.5 x 8 in.); **4.** *See* Sticker.

STICK CYLINDER *See* **Curl,** and **Crowd force.**

STICKER Slip of (approximately) 12 x 38 mm (0.5 x 1.5 in.) lumber used to space pieces or lifts of lumber and sheet goods to allow air to circulate. *Also called* **Kiln stick** and **Stick.**

STICKY CEMENT Finished cement that develops low- or zero-flowability during or after storage in silos, or after transportation in bulk containers, hopper-bottom cars, etc.; may be caused by (a) interlocking of particles, (b) mechanical compaction, or (c) electrostatic attraction between particles. *See also* **Warehouse set.**

stiff. *Abbreviation for:* stiffener.

STIFFENER Member, usually an angle or plate, attached to a plate or web of a beam or girder to distribute load, to transfer shear, or to prevent buckling of the member to which it is attached.

STIFFENER LIP Short extension at an angle to the flange of cold-rolled structural steel sections, added to increase its strength.

STIFFENING GIRDER Structural steel section designed into a suspension bridge to help distribute loads more uniformly and reduce local deflection under concentrated loads.

STIFF LEG 1. Type of derrick; **2.** Loader with a boom that does not swing.

STIFF-MUD BRICK *See* **Brick.**

STIFFNESS Resistance to deformation of a member or structure measured by the ratio of the applied force to the corresponding displacement.

STIFFNESS FACTOR Measure of the stiffness of a structural member.

STILE 1. Vertical framing member of a paneled door or paneling; **2.** In elevator construction, one of the vertical members of the sling, one on each side.

STILE END Junction between glazing bars of roof lights, and the roof, including the flashing.

STILLING POOL Artificial deepening of a river bed

at the foot of a dam spillway that reduces hydraulic velocity and minimizes scour.

STILLSON WRENCH Pipe wrench.

STILTS Adjustable extensions attached to the legs and feet of a drywall finisher to permit him to reach higher-than-normal areas.

STILTED ARCH *See* **Arch.**

STINGER 1. Welding rod holder or clamp; **2.** Metal nail-like attachment at the end of a logger's measuring tape, used to secure one end of the tape and permit the faller to proceed down the tree to accurately determine the desired length of the log to be cut. *See also* **Brace,** and **Forks.**

STIPPLE Rough-textured surface coating.

STIPULATED PRICE CONTRACT Contract for work to be done that does not allow for variation, for whatever reason, from the negotiated price. *Also called* **Fixed price contract,** and **Lump sum agreement.** *See also* **Lump sum contract.**

STIPULATION Requirement or term within a written contract.

stir. *Abbreviation for:* stirrup.

STIRRUP 1. Drop support attached to a wall or hung from a girder to carry a load such as a beam end or the end of a joist. *Also called* **Bridle iron,** and **Stirrup iron**; **2.** Support used to hold reinforcing bars. *See also* **Batter brace, Hook, Hooked bar,** and **Tie**; **3.** Metal strap hanger on a staging.

STIRRUP IRON *See* **Bridle iron.**

stk *Abbreviation for:* stack; stock; strike.

stl *Abbreviation for:* steel.

stm *Abbreviation for:* storm.

STMS *Abbreviation for:* strike to template with machine screws.

stn *Abbreviation for:* stain; stainless; station; stone.

stn. I *Abbreviation for:* stainless iron.

STOCK 1. Handle of a peavey or cant hook; **2.** Toolholder for an external-thread cutting die; **3.** Uncured rubber compound of definite composition from which a given article is manufactured.

STOCKINETTE *See* **Muttoncloth.**

STOCKING Degree of utilization of land by trees. Measured in terms of basal area and/or the number of trees in a stand compared to the basal area and/or number of trees required to fully utilize the growth potential of the land. A stocking percent of 100 indicates full utilization of the site and is equivalent to 7.43 m^2 (80 ft^2) of basal per 0.4 ha (1.0 a) in trees 125 mm (5 in.) in dbh and larger. A stocking percent of 100 in a stand of trees less than 125 mm (5 in.) would indicate that the present number of trees is sufficient to produce 7.43 m^2 (80 ft^2) of basal area per 0.4 ha (1.0 a) when the trees reach 125 mm (5 in.) dbh.

STOCKING CLASS Numeric code representing a range of stems per hectare, sometimes estimated by crown closure on aerial photographs, e.g. stocking class 1 is mature with 76 stems/ha of ±27.5 cm (11 in.) dbh; class 2 is mature with <76 stems/ha; class 3 is immature.

STOCKING CLASSES Land containing varying densities of trees, including the following:

Fully stocked stand: Stand in which the stocking of trees is from 100% to 133%.

Medium-stocked stand: Stand in which the stocking of trees is from 60% to 100%.

Nonstocked area: *See* **Stand size classes.**

Overstocked stand: Stand in which the stocking of trees is 133% or more.

Poorly stocked stand: Stand in which the stocking of trees is from 16.7% to 60%.

STOCK OPTION *See* **Option.**

STOCKPILE 1. Material excavated and piled for future use. **2.** Accumulation of spoil or other materials on the job site.

STOCK ROTATION The use of stock on a 'first in, first out' basis so that the oldest stock is used first and inventories are kept fresh.

STOICHIOMETRIC RATIO The air/fuel ratio that produces the maximum amount of fuel burn with the minimum amount of unburned exhaust by-products, that is 14.7 pounds of air for each pound of fuel.

STOKE Obsolete metric unit of kinematic viscosity that should not be used with the SI system of measurement. Symbol: St. *See also the appendix:* **Metric and nonmetric measurement.**

STOKER Mechanism for feeding solid fuel to a furnace.

STOKER TIMER Timing device that causes a stoker to operate periodically, regardless of the setting of a thermostat, to ensure adequacy of combustion.

STOMATA Pores in plant leaves that control the respiration of a plant.

STONE 1. Natural rock formation of igneous, sedimentary and/or metamorphic origin, either in its original or altered form; **2.** Natural or artificial abrasive used to abrade or sharpen; **3.** Natural rock having been separated from its original mass and readied for use. *See also* **Material density.**

STONE BOAT Flat steel sled with an up-curved front.

STONE COLUMN Coarse crushed stone or gravel injected into poor-bearing soil by jetting and vibro-displacement to improve carrying capacity. *See also* **Vibroflotation.**

STONE DRAIN Rubble drain; trench filled with stones and aggregate so as to drain groundwater or surface water.

STONE DUST Residue from stone crushing; used for the finished surfacing of secondary walkways, and as a levelling layer immediately below paving stones.

STONE MACADAM Roadway built by placing granular material over a dirt or soil base. *Also called* **Gravel macadam.**

STONE MASON Tradesman skilled in working with stone.

STONE SAND Fine aggregate (usually less than 6 mm (0.25 in.) in diameter) resulting from the mechanical crushing and processing of rock.

STONE SEAL Crushed rock, slag, or gravel chips applied at a rate of 8 to 18 kg/0.76 m² (18 to 40 lb/yd²) over a heavy bituminous material applied at approximately 1.1 L/0.76 m² (0.3 gal/yd²).

STONE SETTER Mason who specializes in setting large, precut stone.

STONE SETTER'S ADJUSTABLE MULTIPLE-POINT SUSPENSION SCAFFOLD Two-point or multiple-point adjustable suspension scaffold designed and used for stone setters' operations.

STONE TONGS Two-armed hinged device for moving and lifting large blocks of stone.

STONEWARE Channels, drains, and other sanitary fittings of salt-glazed fired clay.

STOOL *See* **Molding.**

STOOP Low platform, with or without steps, outside the entrance of a house.

STOP 1. Device intended to prevent movement beyond a certain point; **2.** Projections against which moldings finish. *See also* **Molding.**

STOP-AND-WASTE COCK Valve used to stop the flow of water in a pipe and permit the water downstream of the valve to be drained by allowing air into the line.

STOP BEAD Molding added to a door or window frame against which the door closes or sash slides.

STOP CHAINS *See* **Check chains.**

STOP-CHAMFER Ornamental termination to a chamfer created by bringing the edge of the pared-off angle back to a right-angle.

STOPCOCK Small valve.

STOPE Underground excavation made in a series of steps or benches.

STOP END Temporary shutter in formwork, used when concreting has to be briefly suspended.

STOPER Hand-sized air drill mounted on a column or other support.

STOP LOG Horizontal member (not necessarily lumber) fitted into vertical slides attached to walls or piers across a water channel. Adding additional stop logs raises the upstream water level. *Also called* **Flashboard.**

STOP MOLDING *See* **Stop.**

STOP MORTISE Blind mortise.

STOP PAYMENT Order by the drawer of a check directing the bank on which it is drawn not to pay the amount shown.

STOPPER 1. Plate fastened between the flanges of an H or steel-pipe pile at right angles to the vertical axis, used to increase friction area. *See also* **Lagged pile**; **2.** Filler, used to fill cracks or small holes when painting or glazing.

STOPPING Action of final motion of an elevator from leveling speed to an accurate floor level.

STOPPING DISTANCE Distance traveled by a vehicle from the point of application of force to the brake control to the point at which the vehicle reaches a full stop.

STOPPING KNIFE Glazier's putty knife, similar to a chisel knife with one rounded edge and one splayed edge, meeting at a point.

STOP SWITCH *See* **Switch.**

STOP VALVE *See* **Valve.**

STOP WORK ORDER Written instruction by the owner's representative, or by a regulatory authority, to cease work on a project for specified cause, e.g., failure to meet the terms of the contract documents, failure to comply with building regulations or codes, labor disputes, etc.

stor. *Abbreviation for:* storage.

STORAGE BIN Bin, usually cylindrical, insulated, and heated, used for long-term storage of asphalt paving mixes, sometimes at a site away from the plant.

STORAGE CAPACITY Volume that a vessel or container is designed to hold.

STORAGE COEFFICIENT Volume of water released from an aquifer per unit area per unit decline of head.

STORAGE PIT Pit in which solid waste is held prior to processing.

STORAGE TANK Closed container that has a capacity of more than 250 L (66 gal), designed to be installed in a fixed location.

STORM CELLAR Reinforced underground room for protection against violent windstorms.

STORM DOOR Additional door, usually partially or fully glazed, mounted outside an exterior door.

STORM DRAIN Drain that conveys rain or groundwater.

STORM OVERFLOW Weir that permits discharge from a combined sewer of flows in excess of design capacity.

STORM SEWER Sewer that conveys water collected by storm drains.

STORMWATER Excess rain water that is not absorbed into the ground.

STORMWATER TANK Sedimentation tank through which storm water flows prior to discharge to a receiving body of water, to permit settlement of solids.

STORM WINDOW Additional window mounted on the exterior of an existing window.

STORY (or STOREY) Space between any two floors (i.e., between the top of any floor and the top of the floor next above it) or the floor and roof (i.e., the top of the floor and the underside of the ceiling above it) of a building.

STORY DRIFT Difference in horizontal deflection at the top and bottom of a story.

STORY HIGH The height for the floor joists.

STORY POLE Piece of wood or bar marked with the story height and vertical distances for spacing horizontal reinforcing bars.

STORY ROD Pole cut to the proposed clear height between finished floor and ceiling.

STOVE Appliance intended for cooking and space heating.

STOVE BOLT Bolt threaded for the full length of its shaft.

stp *Abbreviation for:* stamp; standpipe; strap.

STP *Abbreviation for:* standard temperature and pressure.

stpl. *Abbreviation for:* staple; stipple.

str. *Abbreviation for:* straight; strain; strength; stringer; strip; structural; strut.

STR *Abbreviation for:* strike.

STRADDLE POLE Sloping scaffolding pole placed along a roof line at the ridge and on the slope in a straddle scaffold.

STRADDLE TRAILER *See* **Trailer.**

STRADDLE TRUCK Self-loading, outrigger-type industrial truck, generally high-lift, for picking up and handling loads between its outrigger arms.

STRAIGHT COURSE Shingles laid with butts in line.

STRAIGHT DYNAMITE *See* **Dynamite.**

STRAIGHTEDGE Rigid, straight piece of wood or metal; used to strike off or screed a concrete surface to proper grade or to check the planeness of a finished surface or grade.

STRAIGHT-EDGE CHISEL *See* **Wood chisel.**

STRAIGHT END Hose end with an inside diameter the same as that of the main body of the hose.

STRAIGHT FLIGHT Uninterrupted stair without winders.

STRAIGHT-GRAINED WOOD *See* **Grain.**

STRAIGHT JOINT Usual style of laying floor and wall tile where all the joints are in alignment.

STRAIGHT-JOINT TILE Single-lap roofing tile laid so that the edges in successive courses run in a line from eaves to ridge.

STRAIGHT-LINE DEPRECIATION Specific amount set aside annually from income over the economic life of a project to pay for the cost of improvements, without regard for the interest it earns.

STRAIGHT-LINE THEORY Assumption in reinforced-concrete analysis stating that strains and stresses in a member subject to flexure vary in proportion to the distance from a neutral axis.

STRAIGHT-LINE WIRING DIAGRAM Wiring diagram according to circuits, not according to physical layout.

STRAIGHT RIPPER SHANK *See* **Ripper shank.**

STRAIGHT-SHAFT PIER Pier, cased or uncased, poured in a drilled hole without underream or bell.

STRAIGHTSPLIT SHAKE *See* **Shake.**

STRAIGHT TEE Pipe fitting consisting of a tee with all openings of the same size.

STRAIGHT TRUCK Truck having the body and engine mounted on the same chassis.

STRAIGHT WARP In the curing process, a wrap of lightweight fabric in which the warp threads of the fabric are parallel to the axis of the hose.

STRAIGHT WHEEL Grinding wheel of any dimension that has straight sides, a straight face, and a straight or tapered arbor hole, and is not recessed, grooved, dovetailed, beveled, or otherwise changed.

STRAIN Measured deformation of a material or body as the result of an applied load.

STRAIN AGING Phenomenon seen in some materials (steel, iron, light alloys, etc.) whereby they increase in strength and hardness over time.

STRAIN ENERGY Energy stored in an elastic body under load.

STRAINER Filtering device for the removal of coarse solids from a fluid.

STRAIN GAUGE Instrument for measuring relative motion (compression, elongation, shear, etc.) between two points in a mechanism or in a structural member.

STRAIN HARDENING Phenomenon wherein ductile steel, after undergoing considerable deformation at or just above yield point, exhibits the capacity to resist substantially higher loading than that which caused initial yielding.

STRAIN-HARDENING STRAIN For structural steels that have a flat (plastic) region in the stress-strain relationship, the value of the strain at the onset of strain hardening.

STRAIN WIRE Steel center supporting wire in a traveling cable.

STRAKE Metal cleat or lug attached to a pneumatic-tired wheel to improve traction.

STRAND 1. Plurality of round or shaped wires helically laid about an axis; **2.** Prestressing tendon composed of a number of wires twisted about a center wire or core; **3.** Specialized knife-cut wood flake of controlled thickness and of a length along the grain direction of at least twice, and usually many times, its width.

STRANDED CONDUCTOR Conductor consisting of a group or groups of wires where the individual wires are twisted or braided together.

STRAND GRIP Mechanical attachment used to anchor strands.

STRAND-LAID ROPE Wire rope made with six to eight strands wrapped around a fiber, wire strand, or independent wire-rope core.

STRAND-LAID SLING Wire rope sling made endless from one length of rope with the ends joined by one or more metallic fittings.

STRAND WRAPPING Application of a high-tensile strand, wound under tension by machine, around circular concrete or shotcrete walls, domes, or other tension-resisting structural components.

S-TRAP Sanitary fitting that provides a water seal where the inlet and outlet are vertical, offset from each other. *See also* **P-trap.** *Also called* **Siphon trap.**

STRAP 1. Short piece of wire rope with an eye in each end used to hold blocks; **2.** One of the bands connecting the cover to the clutch pressure plate, transmitting engine torque through the rotation of the flywheel and clutch cover to the disk.

STRAP HINGE Hinge having one very long leaf and one regular leaf; the leaves are attached to the face of the door or gate and its support, rather than to the edges.

STRAPPED ELBOW Drop elbow: a plumbing fitting.

STRAPPING 1. High-tensile steel banding used to band timber piles to prevent splitting while driving; **2.** Grounds used as base for lath and plaster work; **3.** 25 x 50 mm (1 x 2 in.) lumber fastened perpendicular to rafters as support for preformed metal-sheet roofing; **4.** High-tensile steel banding used to secure materials (typically to a skid) during shipping and storage.

STRATA TITLE *See* **Condominium.**

STRATIFICATION 1. Separation of a material into layers; **2.** Separation of over-wet or over-vibrated concrete into horizontal layers with increasingly lighter material toward the top; **3.** Occurrence in aggregate stockpiles of layers of differing grading or composition.

STRATUM 1. Homogeneous soil layer in a stratified soil deposit; **2.** Subdivision of a forest area to be inventoried based on a group of trees with the same or similar species composition, age, and height class.

STRAW DRUM Small drum on a yarder that handles the strawline.

STRAWLINE Small-diameter wire rope used in rigging up or moving larger lines or blocks.

strchr *Abbreviation for:* stretcher.

strd *Abbreviation for:* strand.

STREAK Line-like mark that appears the length of a particleboard panel, parallel to the feed direction through the sander. These streaks are narrow, and slightly higher than the sanded surface, and are caused by metal in previous boards stripping grit off the sanding belt.

STREAM Body of running water in a narrow, clearly defined natural watercourse or channel. Includes rivers, creeks, and brooks.

STREAM GRADIENT General slope, or rate of vertical drop per unit of length of a flowing stream.

STREAMLINE Path of water or air that is flowing without turbulence; the device causing such flow.

STREAMSIDE MANAGEMENT ZONE *See* **Buffer strip.**

STREAM SIZE Width or depth of the channel, and volume of water flowing.

STREET 1. Highway, road, boulevard, square, or other improved thoroughfare, 9.1 m (30 ft) or more in width, that has been dedicated or deeded for public use, and is accessible to fire department vehicles and equipment; **2.** General term denoting a public way for purposes of vehicular travel, including the entire area within the right of way, including the following classifications:

> **Cul-de-sac:** Local street open at one end only and with special provision for turning round.

Dead-end: Local street open at one end only without special provision for turning round.

Frontage: Local street or road auxiliary to and located on the side of an arterial highway for service to abutting property and adjacent areas and for control of access.

Local: Street or road primarily for access to residence, business, or other abutting property.

Major: Arterial highway with intersections at grade and direct access to abutting property, and on which geometric design and traffic control measures are used to expedite the safe movement of through traffic.

Through: Highway on which vehicular traffic is given preferential right of way and at the entrance to which vehicular traffic from intersecting highways is required by law to yield right of way to vehicles on such throughways in obedience to either a stop sign or a yield sign, when such signs are erected.

STREET ELBOW Elbow pipe fitting with one male end and one female end. *Also called* **Service L** and **Street L.**

STREET FURNITURE Fittings and fixtures installed in streets, such as lamp posts, fire hydrants, street signs, and similar municipal structures, at or above grade level.

STREET L *See* **Street elbow.**

STREET LINE Dividing line between a lot and the street bounding it.

STREET REFUSE *See* **Solid waste.**

STREET SWEEPER Specialized vehicle equipped to sweep debris from paved streets and collect it into a self-contained vessel prior to discharge. The vehicle also has a sprinkler system to lay dust while sweeping. *Also called* a **Sweeper.**

STREET T (or Tee) Tee fitting with one internal and one external threaded opening, plus an outlet opening with an internal thread. *Also called* **Service T.**

STRENGTH 1. Ability of a member to sustain stress without failure; **2.** In a specific mode of test, the maximum stress sustained by a member loaded to failure; **3.** Explosive energy of an explosive relative to a like amount of nitroglycerine dynamite.

STRENGTH-DESIGN METHOD Design method that requires service loads to be increased by specified load factors and computed nominal strengths to be reduced by the specified phi factors. *See also* **Limit design.**

STRENGTH LIMIT STATE Limiting condition affecting the safety of a structure, in which the ultimate load-carrying capacity is reached.

STRESS 1. Load that is applied to a material object; **2.** Force of resistance within any solid body against alteration of form.

Stress. Anal. *Abbreviation for:* stress analyst.

STRESS ANALYSIS Calculation of stresses in a loaded structure.

STRESS CONCENTRATION Localized stress considerably higher than average (even in uniformly loaded cross sections of uniform thickness) due to abrupt changes in geometry or localized loading.

STRESS-CORROSION CRACKING Cracking process that requires the simultaneous action of a corrosive substance and of sustained tensile stress.

STRESS CRACKING Long hairline cracks in hardened concrete due to premature loading, moisture loss, or thermal contraction.

STRESS DIAGRAM Diagram showing the direction and amount of each stress in a structure.

STRESSED-SKIN PANEL 1. Engineered structural panel assembly for roof deck or floor applications, built of plywood sheets glued to framing members; **2.** Engineered building panel composed of two sheets of waferboard mounted on either side of a lumber frame; such panels may be left empty or filled with foam.

STRESSED-SKIN STRUCTURE Structure clad with thin elements that combine with the structural frame to resist static and imposed forces.

Stress Engr *Abbreviation for:* stress engineer.

STRESSING END In prestressed concrete, the end of the tendon at which the load is applied when the tendons are stressed from one end only.

STRESS-NUMBER CURVE Curve developed in fatigue testing showing the range of stresses in a material, plotted against the number of cycles to failure.

STRESS RELAXATION Time-dependent decrease in stress in a material held at constant strain.

STRESS RELIEVING 1. Even heating of a structure to a temperature below the critical temperature, followed by a slow, even cooling; **2.** *See* **Presplitting.**

STRESS-STRAIN Relationship of force and deformation of a unit area of a body during compression, extension. or shear.

STRESS-STRAIN CURVE Curve plotting test results on a metal sample in which strains are plotted against stresses.

STRETCH 1. An increase in dimension; an elongation; **2.** End load applied to a fire hose during vulcanization to reduce hose elongation.

STRETCHER The long face of a brick.

STRETCHER BOND *See* Bond.

STRIA Ridge separating adjacent flutes in a column shaft.

STRIATION Narrow, parallel grooves cut into the face of a panel.

STRIKE 1. Course or bearing of the outcrop of an inclined bed or geologic structure on a level surface; **2.** Process of removing formwork; **3.** To cut off mortar at the face of a masonry joint with a trowel; **4.** Withdrawal of labor by a workforce.

STRIKE FAULT Rock fault whose strike line is approximately parallel to the stratum that it cuts.

STRIKE-OFF 1. Action of removing concrete in excess of that which is required to fill the form evenly, *Also called* **Screeding**; **2.** Name applied to the straightedge used to level concrete with the form edges. *See also* **Screed.**

STRIKE PLATE *See* **Striker plate.**

STRIKER BAR Device that fits into the hammer and transmits impact energy and rotation to the drill steel. *Also called* **Shank adapter.**

STRIKER PLATE Part of a door lock set that is set into the jamb and that engages the lock bolt or latch. *Also called* **Keeper,** and **Strike plate.** *See also* **Anvil.**

STRIKE THROUGH 1. In coated frictioned fabric, a penetration of rubber compound through the fabric; **2.** In woven fire hose, the penetration of the rubber backing through the jacket.

STRIKING Release or lowering of centering or other temporary support.

STRIKING HAMMER *See* **Hammer.**

STRIKING JOINTS Process of removing excess grout from tile joints by wiping with a sponge or cloth or scraping with a curved instrument.

STRIKING WEDGE Folding wedge.

STRING Tools suspended on the drilling cable of a churn drill.

STRING BOARD Board or built-up facing that covers the ends of steps in a staircase.

STRING COURSE Continuous projecting horizontal band set in the surface of a wall, often molded or of patterned brick. Also *See* **Belt course.**

STRINGER 1. Long, heavy, horizontal timber or reinforced concrete beam resting on vertical supports. *See also* **Beam.** *Also called* **Ledger; 2.** Inclined member supporting the treads and risers of a stair. *See also* **Longitudinal.**

STRINGING MORTAR Method whereby a mason picks up mortar for a large number of units and spreads it before laying the units.

STRING LEVEL Spirit level fitted with prongs permitting it to be hung from a string line.

STRING LINE 1. Line strung tightly between supports to indicate both direction and elevation. *See also* **Grade line; 2.** To pull a line by hand in preparation for yarding.

STRING LOADING Filling a drill hole with cartridges smaller in diameter than the hole, without slitting them or tamping them.

STRINGY *See* **Texture.**

STRIP 1. To remove formwork or a mold. *Also called* **Demolding; 2.** Long, thin piece of wood, metal, or other material; **3.** Remove overburden or thin layers of pay material; **4.** To remove lines, blocks, and other rigging from a spar; **5.** Area allotted to each faller; **6.** Designated area of trees established by natural boundaries (roads, streams), or ribbons within which fallers are assigned. *Also called* **Quarter; 7.** Band of reinforcing bars in flat slab or flat plate construction. The column strip is a quarter-panel wide each side of the column centerline and runs either way of the building, from column to column. The middle strip is a half-panel wide, filling in between two column strips, and runs parallel to the column strips to fill in the center part of a panel.

STRIP BURNING Setting fire to a narrow strip of fuel adjacent to a forest fire control line, then burning successively wider adjacent strips inside as the preceding strip burns out.

STRIP DEVELOPMENT *See* **Ribbon development.**

STRIP FIRING Setting fire to more than one strip of fuel and providing for the strips to burn together; frequently done in backfiring against a wind where inner strips are fired first to create drafts to pull flame and sparks away from a forest fire control line. *See also* **Line firing.**

STRIP FLOORING Parquet strip.

STRIP FOOTING Continuous footing under a wall, or supporting two or more columns in a row.

STRIP FOUNDATION Long narrow foundation that does not return at either end.

STRIP LAYOUT Best method of falling the trees of an area in relation to themselves and the terrain.

STRIP MINING Open-pit method of mining material by removal of the overburden.

STRIPPED POINTING *See* **Pointing.**

STRIPPER Liquid compound formulated to remove coatings by either chemical or solvent action, or both.

STRIPPER CUT Longitudinal cut in the cover of lead-press hose caused by an improperly set stripper knife.

STRIPPING 1. Removal of undesirable material from a surface to be used for another purpose or finished in another manner or with a different material; **2.** Separation of the asphalt film from aggregate in the presence of water, a common cause of roadway failure; **3.** Removal of forms following the hardening of poured concrete.

STRIPPING SHOVEL Shovel with an especially long boom and stick, enabling it to reach further and pile higher.

STRIP RUNNER In helicopter logging, someone who sets chokers in conjunction with the hooker.

STRIP SEALANT Cellular or elastic compression seal that must be cut, then heated with an approved hot iron to form corners.

STRIP SHINGLE *See* **Shingle.**

STRIP TEST In fabric testing, a tensile strength test made on a strip of fabric raveled down to a specified number of threads or width of fabric, all of which are firmly held in grips wider than the test piece.

strngr *Abbreviation for:* stringer.

STROBE *See* **Stroboscope.**

STROBOSCOPE An electronic light source of extremely short duration that can be triggered to illuminate at precise intervals a moving system of parts. Often used to 'freeze' the action of moving parts to examine them at some specific point in their motion. *Also called* **Strobe.**

STROKE 1. Movement of a piston from top dead center to bottom dead center, in inches or metric measurement; **2.** Full mechanical movement distance of the plunger in an injector, as opposed to effective stroke, which indicates the duration of fuel delivery to the injector.

STRONG AXIS Major 'principal' axis of a cross section.

STRONGBACK Frame attached to the back of a form or precast member to stiffen or reinforce the form or member during concrete placing operations or handling.

STROP Leather strap used to hone a sharpened steel-tool edge.

STROPER Air drill mounted on a support.

strtn *Abbreviation for:* straighten.

struc. *Abbreviation for:* structural; structure.

STRUCK CAPACITY Capacity of a container, measured as if it were full of water; rated capacity. *See also* **Bucket capacity,** and **Bucket rating.**

STRUCK JOINT *See* **Pointing.**

STRUCK MEASURE Capacity of a bucket, dump body, hauler, or other load-carrying device under full load and in a level position with the load struck level with the sides or gunwales. *Also called* **Line-of-plate,** and **Water level.**

Struct. Desgnr *Abbreviation for:* structural designer.

Struct. Engr *Abbreviation for:* structural engineer.

STRUCTURAL Those aspects of a building that carry their own weight plus all imposed weights and forces placed by other parts of the structure.

STRUCTURAL ADHESIVE Bonding agent used for transferring required loads between adherents exposed to service environments typical for the structure involved.

STRUCTURAL ALTERATIONS Any change or revision to the supporting frame of a structure.

STRUCTURAL ANALYSIS Analysis and evaluation of the structural components of a building or structure.

STRUCTURAL BOND Means by which otherwise independent components are interlocked or tied together, causing them to act as a single structural unit.

STRUCTURAL CLAY UNIT Fired clay, load-bearing building unit.

STRUCTURAL COMPETENCE Ability of a machine, equipment, or material to withstand the loads it was designed and manufactured to resist.

STRUCTURAL CONCRETE 1. Concrete used to carry a structural load or to form an integral part of a structure; **2.** Concrete of a quality specified for structural use.

STRUCTURAL COVERING Construction above the purlins of a roof truss, such as rafters and sheathing, designed to support the weathering surface.

STRUCTURAL DEFECT Crack or lamination in the body of a tile that detracts from its aesthetic appearance and/or the structural soundness of the tile installation.

STRUCTURAL DESIGN Design of the structural elements of a building or structure.

STRUCTURAL DESIGN DOCUMENT One of the documents prepared by the designer (plans, design details, and job specifications).

STRUCTURAL DRAWING One of the drawings that show framing plans, sections, details, and elevations required to construct a building. For reinforced concrete structures they include the sizes and general arrangement of all reinforcement from which the bar fabricator prepares placing drawings. *Also called* **Engineering drawing.**

STRUCTURAL ENGINEER Professional who specializes in the design and analysis of structures.

STRUCTURAL EXCAVATION Excavation and removal of material necessary for the construction of specific works (foundations, pipelines, substructures, etc.), distinct from mass excavation.

STRUCTURAL FILL Material that is placed and compacted in layers under carefully controlled conditions to achieve a uniform and dense soil mass that is capable of supporting structural loading. *See also* **Backfill,** and **Fill.**

STRUCTURAL FRAME Load-bearing skeleton of a structure that resists all imposed and applied forces and loads.

STRUCTURAL GLASS Glass blocks used to form walls and capable of carrying limited loads.

STRUCTURAL INTEGRITY Ability of a member, component, or machine to withstand the stresses imposed by the applied load.

STRUCTURAL LIGHTWEIGHT CONCRETE Structural concrete made with lightweight aggregates, having an air-dry unit weight of not more than 52 kg/ 0.028 m³ (115 lb/ ft³) and a 28-day compressive strength of more than 17 238 kPa (2,500 psi).

STRUCTURAL LUMBER Lumber that is intended for use where working stresses are required.

STRUCTURAL MEMBER Component of a structure designed to resist a force.

STRUCTURAL 1 Unsanded grade of plywood used where shear and cross-panel strength properties are of maximum importance, such as panelized roofs and diaphragms. All plies in such panels are special improved grades and marked PS 1 and are limited to Group 1 species. Other panels, marked Structural 1 Rates, qualify through special performance testing.

STRUCTURAL RAMP Ramp built of steel, concrete, or wood, usually made for vehicle access.

STRUCTURAL SANDWICH CONSTRUCTION Laminar construction comprising a combination of alternating dissimilar simple or composite materials assembled and intimately fixed in relation to each other so as to use the properties of each to attain specific structural and thermal advantages for the whole assembly.

STRUCTURAL SLAB Suspended, self-supporting, reinforced concrete floor or roof slab.

STRUCTURAL SPECIFICATIONS Detailed descriptions of the materials to be used, methods of assemblage and installation, finishes, etc., of the structural systems of a building or of a structure.

STRUCTURAL STEEL *See* **Steel.**

STRUCTURAL STEEL FRAMING System comprising all vertical load-bearing structural steel members, all primary horizontal load-supporting structural members rigidly connected to vertical members, and all other primary members essential to the stability of the structural frame. There are three designations:

Type 1 — Rigid Frame: A continuous frame that assumes that beam-to-column connections have sufficient rigidity to hold virtually unchanged the original angles between intersecting members.

Type 2 — Simple Framing: An unrestrained free-ended system that assumes for gravity loading that the ends of the beams and girders are connected for shear only and are free to rotate.

Type 3 — Semirigid Framing: A partially restrained system that assumes that the connections of beams and girders possess a dependable and known moment capacity intermediate in degree between the rigidity of Type 1 and the flexibility of Type 2.

STRUCTURAL SYSTEM Assemblage of load-carrying components that are joined together to provide regular interaction or interdependence.

STRUCTURAL TIMBER Piece of wood of relatively large size, the strength of which is the controlling element in its selection and use.

STRUCTURAL WOOD FASTENER Any of a range of fabricated metal shapes designed to hold a wide range of common sizes of finished lumber in position to form a structural frame, fastened to individual members by nailing.

STRUCTURE 1. Arrangement of all the parts to a designed whole; **2.** Load-bearing part of a building.

STRUCTURE SECTION Includes all road materials placed from subgrade level to finished road surface.

STRUT 1. Member in compression, typically normal to a purlin and rising from the tie beams. *See also* **Brace**; **2.** Any piece of lumber fixed between two other pieces to keep them apart. *See also* **Shore**, and **Solid bridging**; **3.** Short column; **4.** Structural member designed to resist compression.

STS *Abbreviation for:* self-tapping screw.

St. Sz. *Abbreviation for:* stock size.

STUB 1. Extension to another line; **2.** A very short spur road; **3.** That portion of a grinding wheel left after having been worn down to the discarding diameter for a particular operation or machine.

STUB COLUMN Short, structural steel compression-test specimen, long enough for use in measuring the stress-strain relationship for the complete cross section, but short enough to avoid buckling as a column in the elastic and plastic ranges.

STUB LENGTH Short end of pipe that extends beyond the floor level.

STUB TENON Tenon that fits into a blind mortise.

STUB WALL Low wall, usually 100 to 200 mm (4 to 8 in.) high, placed monolithically with a concrete floor or other members to provide control and attachment of wall forms.

STUCCO Any of a wide range of mortar and plaster mixes used to cover both exterior and interior walls. Can be finished in a variety of fashions: by brushing and scratching, through the addition of colored aggregates, by casting or throwing different kinds of stone and chips onto the wet surface, and by the addition of colorants to the mix. *Also called* **Cement plaster.**

STUD 1. Upright timber in frame construction, most often nominal 50 x 100 mm, 50 x 150 mm, or 50 x 200 mm (2 x 4 in., 2 x 6 in., or 2 x 8 in.). *See also* **C stud**; **2.** Member of appropriate size and spacing to support sheathing and concrete forms; **3.** Headed steel device used to anchor steel plates and shapes to concrete members; **4.** Headless bolt firmly anchored at one end.

STUDDING Dimension lumber from which studs are cut.

STUDENT HOUSING Housing, usually on-campus, designed specifically to accommodate students and their specific needs.

STUD FINDER Magnetic device that locates the positioning of metal studs through sheathing, or that signals the position of metal fasteners used to attach paneling to a stud wall, thus indicating the position of wooden studs.

STUD GUN Explosive-operated hand tool used to drive male- or female-threaded studs into hardened concrete, masonry, steel studs, etc.

STUDIO Traditionally, an artist's workroom with windows facing north and a higher-than-normal ceiling. Currently, a term loosely applied to a large room intended for special activities.

STUDIO APARTMENT Small, self-contained living unit, usually consisting of a single room with alcoves, intended for a single occupant.

STUD SHEAR Hand tool used for cutting metal studs.

STUD-TYPE CHAIN *See* **Chain.**

STUD WALL Wall made up of studs fastened at regular spacings (center-to-center) between a top and bottom plate.

STUD WELDING Use of a special type of resistance welding gun to fasten studs to a structural steel frame following erection.

STUDY Development of written and graphic material intended to explore in greater detail a preliminary idea or concept.

STUFFING BOX Space around a shaft filled with pliable packing to prevent fluids or gases from leaking along it.

STUMPAGE 1. Value of timber as it stands uncut in the woods; **2.** Price that must be paid to a provincial government for timber harvested from Crown land.

STUMPAGE APPRAISAL Process by which the stumpage to be charged for harvesting on any given area is estimated.

STUMPER Narrow, heavy, dozer attachment used to push out tree stumps.

STUMP JUMPER Heavy plate underneath a skidder that protects the back housing from contact with high stumps.

STUMP PULL Slivers of wood remaining attached to the stump after a tree is felled; the slivers are considered as having been pulled from the butt of the log.

STUMP RIG To hang blocks on stumps at the back

end.

STUMP SCALING Measurement of stump diameters to estimate the volume of removed trees using special or adjusted volume tables, frequently in the case of timber trespass or cutting of unauthorized tress.

STUMP SHOT Difference of 50 mm (2 in.) or more in height between the horizontal cut of the face or undercut and the backcut, establishing an antikick step that will prevent a tree from jumping back over the stump toward the faller.

STUMPWOOD Stumps harvested after conventional logging or separated from stemwood after complete harvesting.

STUMPWOOD CHIP Chip manufactured from stumpwood.

STUNT END Vertical formwork across a wall, slab or trench to form a construction joint at the end of a pour.

STUNTHEAD Thin board or plate with a lower serrated edge temporarily fitted over dowel bars of a joint during construction of a concrete slab. After the concrete has set the board is carefully removed vertically.

S2E *Abbreviation for:* surfaced two edges.

S2S *Abbreviation for:* surfaced two sides.

S2S&CM *Abbreviation for:* surfaced two sides and center matched.

S2S&M *Abbreviation for:* surfaced two sides and standard or center matched.

S2S&SL *Abbreviation for:* surfaced two sides and shiplapped.

S2S&SM *Abbreviation for:* surfaced two sides and standard matched.

S2S1E *Abbreviation for:* surfaced two sides and one edge.

stwy *Abbreviation for:* stairway.

sty *Abbreviation for:* story.

sty ht *Abbreviation for:* story height.

STYLE (OF ARCHITECTURE) Design incorporating characteristics typical of a period, approach, school, or authority.

STYRENE-BUTADYLENE Manufactured resin used in masonry paint and interior paints.

STYROFOAM Expanded polystyrene in plank form, used as insulation.

s/u *Abbreviation for:* set up.

S/UA *Abbreviation for:* shipped unassembled.

SUB Threaded thread protector used with drill pipe. *Also called* **Joint protector.**

sub. *Abbreviation for:* submerge; submit; substitute; substitution; suburb.

SUBASSEMBLAGE Truncated portion of a structural frame.

SUBBASE Layer in a pavement system between the subgrade and base course, or between the subgrade and a finished pavement. *See also* **Base, Mud slab,** and **Pavement structure.**

SUBBASEMENT Story or storys immediately below a basement.

SUBBIDDER Contractor or supplier who submits a price to a prime contractor.

subcontr. *Abbreviation for:* subcontractor

SUBCONTRACT Agreement that is subsidiary to a prime contract, usually between a prime contractor and another contractor or a supplier.

SUBCONTRACT BOND *See* **Bond.**

SUBCONTRACTOR Contractor who contracts to a principal or general contractor to complete specific parts of a structure or be responsible for specific trades, such as plumbing or electrical distribution. *See also* **Contractor.**

SUBDIVIDE (LAND) To divide a plot of land into building lots.

SUBDIVISION Parcel of land divided into blocks, lots, or plots and developed as an entity, often for dwellings, but including such communal facilities as churches, schools, etc.

SUBDIVISION BOND *See* **Bond.**

SUBDIVISION BYLAW Local bylaw that regulates the manner in which land may be subdivided for development and/or building purposes.

SUBDRAIN Drain built beneath a sewer to intercept groundwater.

SUBDRILL To drill blastholes beyond the planned grade lines or below floor level to ensure breakage to the required elevation.

subfl. *Abbreviation for:* subfloor.

SUBFLOOR 1. Any course or layer immediately below the finished floor; 2. Boards or plywood laid on joists over which a finished floor is to be laid.

SUBFLOORING Sheathing panels applied directly over floor joists that will receive an additional underlayment layer. *See also* **Flooring.**

SUBFLUORESCENCE Accumulation of soluble salts under or just beneath the masonry surface, formed as moisture evaporates.

SUBFRAME SUSPENSION *See* **Suspension.**

SUBGRADE 1. Soil prepared and compacted to support aggregate, paved surface, a concrete slab, or other construction. *Also called* **Basement soil,** and **Foundation soil.** *See also* **Improved subgrade,** and **Pavement structure;** 2. Top surface of a roadbed upon which the pavement structure and shoulders, including curbs, are constructed.

SUBGRADE DRILLING That part of blasthole drilling in which the depth of the hole is extended past the planned surface of the underlying bench.

SUBGRADE MACHINE Machine employed on the pioneering phase of forest road construction; establishes the shape of the subgrade.

SUBGRADE MODULUS *See* **Coefficient of subgrade reaction.**

SUBGRADE RESTRAINT Frictional resistance between the subgrade and an overlying pavement slab.

SUBGRADE TREATMENT Modification of road-bed material by stabilization. *See also* **Pavement structure.**

subj. *Abbreviation for:* subject.

SUBLET To lease or rent all or part of an already leased or rented property to another person.

SUBMERGED SOIL *See* **Soil classification system.**

SUBMERSIBLE PUMP *See* **Pump.**

SUBMITTAL Shop drawing, schematic, fabrication drawing, diagram, layout, descriptive literature, illustration, schedule, product, performance and test data, template, test, sample, or other material or data related to the materials, equipment, systems, or methods proposed for performance of the work that is required by the contract documents, and prepared or furnished by the contractor to the architect or owner's representative to illustrate some part of the work.

subord. *Abbreviation for:* subordinate.

SUBORDINATION CLAUSE Mortgage or lease clause making the rights of the holder secondary, or subordinate, to subsequent encumbrances.

SUBPAN Horizontal surface installed either between or above crossmembers prior to insulation and installation of flooring in truck bodies.

SUBPLATE Auxiliary ported plate for mounting components. *Also called* a **Back plate.**

SUBPOST CAR FRAME Elevator car frame all of whose members are located below the platform.

SUBPURLIN 1. Light structural section used as a secondary member; **2.** In lightweight concrete roof construction, used to support the formboards over which the lightweight concrete is placed.

SUBRAIL Molding on the upper face of a stair string that receives and carries the lower end of the balusters.

SUBROGATION Substitution of legal rights and responsibilities from one person to another.

SUBSAVER Protector for the thread protector of the kelly of a rotary drill.

SUBSCRIBING WITNESS Person writing their name as witness to the execution of an instrument.

subsec. *Abbreviation for:* subsection.

subseq. *Abbreviation for:* subsequent.

subsid. *Abbreviation for:* subsidiary.

SUBSIDENCE Reaction of soil to an imposed load. *See also* **Elastic compression, Plastic creep,** and **Soil consolidation.**

SUBSIDIZED HOUSING Dwellings, the construction and/or maintenance costs of which is in part borne by other than the tenant; housing for low-income groups.

SUBSIEVE FRACTION Particles, all of which will pass through a No. 325 sieve.

SUBSILL Additional sill on the external face of a window that causes rainwater to be shed further from the face of the wall below.

SUBSISTENCE Allowance in addition to salary or wages that compensates the cost of living away from home.

SUBSOIL Weathered portion of the earth's crust between the topsoil and the unweathered material below.

SUBSOIL DRAIN Drain that receives only groundwater and conveys it to a storm drain.

SUBSOIL DRAINAGE PIPE Perforated pipe that is installed underground to intercept and convey groundwater.

SUBSOIL EXPLORATION Determination of the disposition and characteristics of the materials below the surface of the ground to a specified depth, to a defined degree of detail, and for an express purpose.

SUBSOIL PLOW One-tooth ripper. *Also called* **Pan breaker.**

SUBSTANTIAL COMPLETION Date the works or any designated part thereof that the owner has agreed to accept separately, is sufficiently complete in accordance with the contract documents for the owner to occupy or utilize it for the purpose for which it is intended, and is so certified by the designing authority. *See also* **Substantial performance.**

SUBSTANTIAL PERFORMANCE Is achieved when:

(a) The work or a substantial part of it is ready for use or is being used for the purpose intended, and

(b) The work to be done under the prime contract is capable of completion or correction at a cost of not more than:

   (i) 3% of the first $250,000 of the contract price

   (ii) 2% of the next $250,000 of the contract price, and

   (iii) 1% of the balance of the contract price and is so certified by the Certificate of Substantial Performance.

SUBSTATION Building or room containing electrical equipment that receives high voltage from a transmission system, transforms it to lower voltage, and connects it to a local distribution system.

SUBSTITUTION Material or process accepted as being equal to that specified in the contract documents.

substr. *Abbreviation for:* substrate; substructure.

SUBSTRATE Material that provides a supporting surface for other materials.

SUBSTRUCTURE Part of a structure generally below the level of the adjoining ground; all that part of a structure below the bearings of simple and continuous spans, skewbacks, or arches and the tops of footings or rigid frames, including backwalls, wingwalls, and wing protection railings.

SUBSURFACE INVESTIGATION Investigative and analytical techniques using samples obtained from boring, aimed at revealing the nature and characteristics of the subsurface materials likely to influence the design of a structure.

SUBSYSTEM Physically integrated series of parts that function as a unit whose product or output is employed by a larger system.

SUBTENSE BAR Bar of known length, used in surveying to calculate distances by measuring the angle that it subtends at the instrument.

SUBTERRANEAN TERMITE *Reticuliterme*: an insect that lives in soil but that feeds on wood, constructing tubes to make earth-to-wood contact.

subtr. *Abbreviation for:* subtraction.

SUBURB Ill-defined area of a city, usually residential but containing its own retail and commercial (and sometimes industrial) development, often identified by a name, and that may be further subdivided into communities.

SUBURBAN EXPANSION Tendency of cities to expand at their outer fringes through the development of bedroom communities. *Also called* **Suburban sprawl.**

SUBURBAN SPRAWL *See* **Suburban expansion.**

SUBWAY 1. Underground public transit system; **2.** Passageway for pedestrian and/or vehicular traffic that passes under another structure.

suc. *Abbreviation for:* succeeding.

SUCCESSION Replacement of one plant community by another in progressive development toward climax vegetation.

SUCCESSOR WORK ITEM Work item that directly follows one or more work items in the logic sequence of a precedence network.

SUCK Shape of the bottom of a cutting edge or tooth that tends to pull it into the ground as it is moved forward.

SUCKER Large dead or live branche on the stem of a tree, or a lone branch below the crown.

SUCTION 1. Inlet, pull side of a pump; **2.** Effect of atmospheric pressure that causes objects to resist being lifted from or pulled out of soft soil or mud; **3.** Atmospheric pressure against a partial vacuum; **4.** Partial vacuum on the downwind face of a structure; **5.** Adhesion of a brick to mortar or of wet plaster to a porous surface.

SUCTION BOOSTER Type of jet siphon device used to bring water to a pumper from greater distances and to higher elevations than is possible with suction (depending on atmospheric pressure).

SUCTION-CUTTER DREDGE Suction dredge having a rotating cutter at the working end of its suction pipe.

SUCTION FAN Fan positioned in a cooling system so that air passes through the radiator before entering the fan.

SUCTION HAT Strainer in the bottom of an oil reservoir tank, used to prevent foreign matter from entering an oil line.

SUCTION HEAD *See* **Suction lift.**

SUCTION HOSE Hose reinforced against collapse due to atmospheric pressure and used for drafting water into fire pumps where a partial vacuum is created in the pump, causing atmospheric pressure to push water through the hose upward into the pump.

SUCTION LINE 1. Tubular connection between a reservoir or tank and the inlet of a hydraulic pump; **2.** Pump intake line in which the fluid is below atmospheric pressure.

SUCTION LIFT Distance from the surface of the water supply to the center of the pump impeller. *Also called* **Suction head.**

SUCTION PIPE Pipe connecting the suction inlet of a pump with its source of supply, which is commonly at a lower elevation than that of the pump.

SUCTION PRESSURE *See* **Pressure.**

SUCTION-TYPE MIXER *See* **Aspirator mixer.**

SUCTION VALVE Check valve on a suction pipe.

SUDDEN DRAWDOWN Rapid drop in water level behind a dam or alongside a quay or earth embankment that may result in an unstable condition.

suit. *Abbreviation for:* suitable.

SUITABLE That which fits, and has the qualities or qualifications to meet a given purpose, occasion, condition, function, or circumstance.

SUITE Single room or series of rooms of complementary use, operated under a single tenancy, and including dwelling units, individual guest rooms in motels, hotels, boarding houses, rooming houses, and dormitories as well as individual stores and individual or complementary rooms for business and personal service occupancy.

SULFATE ATTACK Either a chemical or physical reaction between sulfates, usually in soil or groundwater, and concrete and mortar.

SULPHATED ASH Ash content determination in which the oil is burned and treated with sulphuric acid.

SULFATE RESISTANCE Ability of concrete or mortar to withstand sulfate attack.

SULFATE-RESISTANT CEMENT Portland cement low in tricalcium aluminate to reduce the susceptibility of concrete to attack by dissolved sulfates in water or soils.

SULFOALUMINATE CEMENT *See* **Expansive cement.**

SULKY Logging arch equipped with wheels instead of crawler tracks and towed behind a skidding machine.

sum. *Abbreviation for:* summary.

SUMMER Large horizontal timber supporting a superstructure.

SUMMER SWITCH *See* **Switch.**

SUMMERWOOD Portion of the annual growth ring of a tree that develops largely during the latter part of the season's growth, but not necessarily in the summer. It is less porous and is usually harder and heavier than springwood. *Also called* **Latewood.**

SUMMIT Highest point of any area or grade.

SUMP 1. Pan attached to the lower part of an engine from which lubricating oil is pumped for circulation of the engine's moving parts; **2.** Tank or pit that receives and holds the discharge from a drainage pipe.

SUMP FILTER *See* **Reservoir filter.**

SUMP PUMP *See* **Pump.**

SUMP TANK *See* **Reservoir.**

SUN CHECKING Surface cracks, or crazing caused in rubber by direct exposure to direct or indirect sunlight.

SUNDECK Exterior flat area attached to or part of a building, used for sunbathing.

SUN GEAR *See* **Gear.**

SUNK DRAFT Margin of a building stone dressed below the rest of the face.

SUNK FACE Face of a building stone dressed below the level of the margin.

SUN TRACKING Mechanism that causes a solar collector to follow the position of the sun through its orbit.

sup. *Abbreviation for:* supplementary; supply; support.

SUPER Continuous slope in one direction on a road.

SUPERCHARGED GAS TURBINE ENGINE Gas turbine engine containing two mechanically independent rotors, each containing a driving turbine; one compressor operating with an air inlet at atmospheric pressure, that supercharges the second compressor inlet to a higher pressure. Useful power may be taken from either the rotors, or from a free power turbine.

SUPERCHARGER Gear-driven air blower that pressurizes air entering the intake manifold of an engine; any device designed to force more air into an engine cylinder than would be taken in as the result of barometric air pressure.

SUPER-DUTY FIRECLAY BRICK *See* **Brick.**

SUPERELEVATION Rise of the outside curve above the level of the inside curve of a finished road surface to accommodate the centrifugal force present in vehicles traveling round the curve.

SUPERELEVATION RATE Rate of rise in cross section of the finished surface of a roadway on a curve, measured from the lowest or inside edge to the highest or outside edge.

SUPERFICIAL MEASURE Face measure of lumber.

SUPERIMPOSED LOAD Load, other than its own weight, that is resisted by a structural member or system. *Also called* **Superload.**

SUPERINTENDENT Contractor's authorized representative in responsible charge of the work. *See also* **Contractor.**

SUPERIOR DIAMETER Diameter of the upper end of a column. *See also* **Diameter,** and **Inferior diameter.**

SUPERLOAD *See* **Superimposed load.**

SUPERMARKET Large retail food store.

SUPER SNORKEL Steel boom extension of two or more sections, up to 36 m (120 ft), mounted on a loader to increase the distance that logs can be reached for yarding purposes.

superstr. *Abbreviation for:* superstructure.

SUPERSTRUCTURE 1. All of that part of a structure above grade, supported on a foundation. *See also* **Turntable,** and **Upperstructure; 2.** That part of a bridge above the bridge seats, spring line of the arches, or bottom of the caps.

SUPERSULFATED CEMENT Hydraulic cement made by intimately intergrinding a mixture of granulated blast-furnace slag, calcium sulfate, and a small amount of lime, cement, or cement clinker, so named because the equivalent content of sulfate exceeds that for portland blast-furnace slag cement.

SUPERVISION Organization and direction of work by contractor's personnel.

SUPERVISORY INDICATOR Visual or audible signaling indicator that advises of a condition.

SUPERVISORY SIGNAL Signal indicating the need for action in connection with the supervision of sprinkler and other extinguishing systems or equipment, or with maintenance features of other protection systems.

SUPERVISORY SWITCH *See* **Switch.**

supg *Abbreviation for:* supporting.

suplr *Abbreviation for:* supplier.

suppl. *Abbreviation for:* supplement.

SUPPLEMENTAL AGREEMENT Negotiated agreement constituting a modification of the originally executed contract and covering the performance of work beyond its general scope.

SUPPLEMENTAL SPECIFICATIONS Approved conditions and revisions to standard specifications.

SUPPLEMENTARY CONDITIONS Section of contract documents that supplements (and may also modify) provisions of the general conditions.

SUPPLEMENTARY PLATFORM Any platform that can be mounted on a high-lift fork truck or other elevating device, and that does not require disassembly of any portion of the lifting system to install or remove, but not intended to elevate personnel.

SUPPLEMENT OF AN ANGLE Angle equal to the difference between the given angle and 180°.

SUPPLIER One who is contracted to supply materials, goods, or services within a general contract.

SUPPLY BOND *See* **Bond.**

SUPPLY DUCT Duct for conveying air from a heating, ventilating, or air-conditioning appliance to a space to be heated, ventilated, or air-conditioned.

SUPPLY MAIN 1. Primary piping system bringing water, gas, etc. into a facility; **2.** Conduit through which a heating or cooling medium flows from the source of heat or refrigeration to the runouts and risers.

SUPPLY PUMP *See* **Pump.**

SUPPLY TANK 1. Oil reservoir used in a hydraulic system; 2. Air tank or reservoir charged directly by the compressor in an air brake system. *Also called* the **Wet tank.**

SUPPORT BAR Reinforcing bar that rests upon an individual high chair or bar chair to support top slab bars in slabs or joists, respectively. They are usually #4 bars and may replace a like number of temperature bars in slabs when properly lap spliced; also used longitudinally in beams to provide support for the tops of stirrups. *Also called* **Raiser bar.** *See also* **Continu-**

ous **high chair,** and **Individual high chair.**

SUPPORT CLEARANCE Distance between the rearmost interference point of a truck tractor or dolly and the trailer landing gear or its support structure during turning maneuvers.

SUPPORTED-FILM ADHESIVE See **Adhesive.**

SUPPORTED FRAME Frame that depends upon adjacent braced or unbraced frames for resistance to lateral load or frame instability.

SUPPORT LINE Cable reeved from a second hoist drum used to hold clamshell buckets suspended during dumping and lowering operations. See also **Cross support.**

SUPPORT ROLLER Roller that supports the slack upper part of crawler track.

SUPPORT SYSTEM Structure such as underpinning, bracing, or shoring, that provides support to an adjacent structure, underground installation, or the sides of an excavation.

SUPPRESSANT Agent used to extinguish the flaming and glowing phases of combustion by direct application to the burning fuel.

SUPPRESSED Trees with crowns entirely below the general level of the crown cover, receiving little or no direct light from above or from the sides. Also called **Overtopped.**

SUPPRESSED WEIR Measuring weir notch whose sides are flush with the channel, eliminating or suppressing end contractions of the overflowing water.

SUPPRESSION Work of extinguishing or confining a fire, beginning with its discovery.

SUPPRESSION CREW Two or more workers stationed at a strategic location, either regularly or in an emergency, for initial action on forest fires.

supsd Abbreviation for: superseded.

supsd. Abbreviation for: supersede.

supt Abbreviation for: superintendent.

supv. Abbreviation for: supervise.

supvr Abbreviation for: supervisor.

sur. Abbreviation for: surface.

SURCHARGE 1. Fill, temporarily placed on a site soon to be developed, in sufficient quantities to impose a load calculated to compact the natural soils to a calculated degree; **2.** Static or live elevated load above the top of a retaining wall; **3.** Charge above the customary cost.

SURCHARGED WALL Retaining wall carrying a surcharge, such as an embankment.

SURETY 1. Security for payment or for the performance of some act; **2.** Corporation, partnership, or individual, other than the contractor, executing a bond furnished by the contractor.

SURETY BOND See **Bond.**

SURFACE To make smooth or regular.

SURFACE ACTIVE Having the ability to modify surface energy and to facilitate wetting, penetrating, emulsifying, dispersing, solubilizing, foaming, froth-

ing, etc., of other substances.

SURFACE-ACTIVE AGENT Substance that affects markedly the interfacial or surface tension of solutions even when present in very low concentrations. See also **Surfactant.**

SURFACE AREA Area of the surface of each particle in a given amount of aggregate, used to calculate the optimum asphalt content in the Hveem method of mix design.

SURFACE BONDING OF MASONRY Bonding of dry-laid masonry by parging with a thin layer of fiber-reinforced mortar.

SURFACE BURNING CHARACTERISTICS Rating of construction materials providing indices for flame spread and smoke development. Also called **Flame spread classification.**

SURFACE COMPACTION Increasing the dry density of surface soil by applying a dynamic load.

SURFACE COURSE Top, or riding surface of a pavement structure. See also **Pavement structure.**

SURFACE CRACKING Creation of discontinuities in the cover material of a sanitary landfill as a result of settlement and decomposition of solid wastes and/or a change in the moisture content of the cover material.

SURFACE DEPRESSION Indentation on the face of a drywall panel.

SURFACED LUMBER Lumber that is dressed on at least one face by passing it through a planer. There are several standard designations:

> **S1E:** Surfaced on one edge.
>
> **S1S:** Surfaced on one side.
>
> **S2E:** Surfaced on two edges.
>
> **S2S:** Surfaced on two sides.
>
> **S4S:** Surfaced on four sides.

SURFACE DRAIN Surface channel that primarily removes surface water.

SURFACE DRESSING Pavement wearing surface consisting of a layer of chips or gravel on a thin layer of bitumen.

SURFACE DRYING Material that dries faster on its surface than in the body of the film.

SURFACE FILTRATION Filtration which primarily retains contaminant on the influent face.

SURFACE FINISH 1. Smoothness of a finished face; **2.** Smoothness of filing or grind, particularly on a saw tooth.

SURFACE FIRE Forest fire that burns litter, other loose debris, and small vegetation.

SURFACE GRINDING Grinding a plane surface.

SURFACE GRINDING MACHINE Machine for grinding plane surfaces.

SURFACE GROUTING Grouting the rock below the surface of a dam foundation.

SURFACE MEASURE Face measure.

SURFACE MOISTURE Free water retained on surfaces of aggregate particles and considered to be part of the mixing water in the concrete, as distinguished from absorbed moisture. *See also* **Free moisture**, and **Mixing water.**

SURFACE PLANER Power-operated equipment consisting of flail-type cutters mounted within a metal cart, employing high-speed steel and tungsten carbide insert flails to mill misaligned concrete and asphalt slabs, remove thermoplastic and paint markings, clean surfaces, etc.

SURFACE PROTECTIVE TREATMENT Application of a chemical solution to the surface of hardened concrete, to seal the pores and reduce attack from aggressive agents.

SURFACE RECYCLING Process in which an asphalt pavement surface is heated in place, scarified, remixed, relaid, and compacted.

SURFACE RETARDER Retarder applied to the concrete surface of a form or to the surface of newly placed concrete, to delay setting of the cement, to facilitate construction joint cleanup, or to facilitate production of exposed-aggregate finish.

SURFACE RIGHTS Right to use and modify the surface area of· real property.

SURFACE TENSION Property, due to molecular forces, that exists in the surface film of all liquids and tends to prevent the liquid from spreading.

SURFACE TEXTURE Visible aspect of a surface that appears to be other than perfectly flat or perfectly even.

SURFACE TREATMENT 1. Application of various materials to provide a wearing surface over stone and gravel base courses; **2.** Treatment to panels to make them less slippery when used as roof sheathing.

SURFACE VIBRATION Portable horizontal platform on which a vibrating element is mounted, used to assist in consolidating concrete in a form or mold. *See also* **External vibration**, **Internal vibration**, and **Vibration.**

SURFACE VOID Cavity visible on the surface of a solid.

SURFACE WATER Water lying on or flowing over the surface of the ground.

SURFACE-WATER DRAIN Pipe to convey runoff and rainwater.

SURFACE WAVE Seismic wave that travels over the surface of the earth or rock rather than through the rock mass.

SURFACING Upper layers of a pavement. *See also* **Plant-mixed surfacing**, **Road-mixed surfacing**, and **Sheet asphalt surfacing.**

SURFACTANT Any substance that alters the energy relationship at interfaces; organic compounds displaying surface activity such as detergents, wetting agents, dispersing agents and emulsifiers. *Also called* **Emulsifier**, and **Surface-active agent.**

SURGE Sudden and temporary increase in a state or condition.

SURGE BIN 1. Compartment for the temporary storage of granular materials that allows converting a variable rate of supply to an even rate of discharge of the same average amount; **2.** Bin, usually cylindrical, used to store hot asphalt mixes at a plant site for faster truck loadout and continuous, more efficient plant operation.

SURGE PIPE Open-top standpipe, used to release surge pressure.

SURGE TANK Separate, vented tank in a cooling system provided to perform one or more of the following functions: (a) filling, (b) coolant reservoir, (c) deaeration, (d) retention of coolant expelled from a radiator by expansion and/or after boil, and (e) visible fluid level indication. *Also called* **Coolant recovery tank.**

SURRENDER Cancelation of a lease by mutual consent of the involved parties.

SURROUND Enclosure or enframement.

surv. *Abbreviation for:* survey.

SURVEY 1. Boundary and/or topographic mapping; **2.** Measurements of an existing structure; **3.** Determination of criteria; **4.** Analysis of function and use; **5.** Program of investigation and reporting.

SURVEYING Gathering of all important measurements so as to establish the relative position of defined factors, features, or objects. There are several divisions, including:

> **Construction:** Measurement taken while construction is in progress to control altitude, horizontal positions, and dimensions.

> **Geodetic:** Surveying that takes into account the shape and size of the earth.

> **Land:** Process of determining the lengths and directions of boundary lines and the areas included within such boundaries, or to establish such lines on the ground.

> **Plane:** Branch of surveying that considers the earth's surface a flat plane. The technique is suitable for the measurement and plotting of small areas. However, accuracy will decrease over distance.

> **Topographical:** Survey to designate the relief of the earth's surface and the position of natural and man-made features and objects.

SURVEYING SYSTEM Group of measuring, calculating, and recording instruments and computers that automate the task of measuring and plotting.

SURVEYOR Individual qualified and usually licensed to undertake land survey and certify his findings.

SURVEYOR'S CHAIN *See* **Chain.**

SURVEYOR'S LEVEL *See* **Level.**

SURVEY PARTY Group under the direction of a party chief, engaged in physical survey in the field.

SURVIVAL ASSESSMENT Survey that estimates survival, that is, the percentage of trees living after a period of growth (often 2 to 5 years) following planting.

susp. *Abbreviation for:* suspended; suspension.

susp. clg *Abbreviation for:* suspended ceiling.

SUSPENDED CEILING False or lowered ceiling. *Also called* **Drop ceiling.**

SUSPENDED FLOOR Floor that spans between supports.

SUSPENDED FORMWORK Floor forms carried on the floor supports rather than propped up from below.

SUSPENDED SCAFFOLD Elevated work platforms of the following types:

Manual: Manually operated suspension scaffold suspended by rope from an overhead supporting system so arranged and operated as to permit raising or lowering to desired working positions.

Permanent: Permanently dedicated powered equipment for building maintenance of a specific structure.

Powered: Power-operated suspension scaffold suspended by wire rope from an overhead supporting system so arranged and operated as to permit raising and lowering to desired working positions.

Swing: *See* **Two-point suspended scaffold,** below.

Temporary: Suspended scaffold used to service structures on a temporary basis for construction, alteration, demolition, and maintenance.

Two-point: Suspension scaffold consisting of a platform supported by hangers (stirrups) suspended by two ropes from overhead supports and equipped with means to permit the raising and lowering of the platform to desired work levels. *Also called* **Swing scaffold.**

SUSPENDED SPAN Short, middle, freely supported span of a cantilever bridge.

SUSPENDER Vertical hanger of a suspension bridge.

SUSPENSION 1. Solid particles distributed through a liquid; **2.** Emulsion in which one liquid carries another in the form of small, separate drops; **3.** Articulated mechanical assembly used to position and secure axle and wheel assemblies to a vehicle's or trailer's frame or subframe. There are several types, including:

Air-ride: Suspension employing flexible pneumatic cushions, or air springs, pressurized by the vehicle's on-board air system and regulated by height control valves or variable-pressure regulators.

Multiaxle: Suspension consisting of more than three axles with equalizing means for transferring weight between the axles.

Overslung: Suspension employing flat plate or tapered springs where the springs are attached above the top of the axle bar.

Progressive (variable-rate): Leaf-type spring with a variable deflection rate obtained by varying the effective length of the spring, accomplished using a cam-type mounting bracket.

Single axle: Suspension usually employing two springs and designed to accommodate one axle.

Spread-tandem: A two-axle assembly in which the axles are spaced to allow maximum axle loads under existing regulations; the distance generally being more than 1.4 m (55 in.).

Spring-beam: Two-axle suspension that employs

a spring beneath each walking beam to enhance ride.

Steel spring: Any of several types of suspension employing either flat plate or tapered-steel springs.

Subframe: Structural assembly to which suspension and axle assemblies are attached.

Tandem-axle: Suspension consisting of two axles with an equalizing means for transferring weight between axles.

Three-point: Two-axle suspension equipped with elliptical or plate-type springs where the springs on one axle are caused to interact with the springs on the other by pivotal rockers or equalizer beams.

Triaxle: Suspension consisting of three axles with an equalizing means for transferring weight between axles.

Two-point: Tandem two-axle suspension consisting of two springs deployed between the axles and pivoted halfway between the axles on a trunnion shaft.

Underhung: Suspension where the spring positions under the axle.

Underslung: Suspension in which the springs or air bags are attached below the bottom of the axle bar.

Walking beam: Underslung suspension comprising parallel beams pivoted on a common transverse axle or trunnion between the axles, and attached at the ends to the axle and wheel assemblies.

SUSPENSION BRIDGE Bridge deck supported by suspenders hung from steel cables suspended from towers.

SUSPENSION CABLE Wire rope holding a load; the wire rope carrying a suspension bridge.

SUSPENSION-CABLE ANCHOR Means by which a suspension cable carrying a bridge deck is secured to an appropriate foundation on the land side of a bridge tower.

SUSPENSION MOUNTING HEIGHT Vertical distance from, and perpendicular to, the horizontal centerline of the axle(s) in a suspension to the horizontal line representing the lower surface of the suspension subframe.

SUSPENSION ROPE EQUALIZER Device to automatically equalize the tension in elevator car or counterweight hoisting wire ropes.

SUSPENSION SCAFFOLD *See* **Suspended scaffold.**

SUSPENSION SEAT Type of equipment operator's seat designed to isolate the occupant from machine vibration and impact.

SUSPENSION SETTING Distance from the rear extreme of a trailer or dolly to the vertical centerline of the axle, if single axle, or to the vertical centerline of the axle arrangement when two or more axles are employed in the suspension.

SUSPENSION SYSTEM 1. Light metal framing system that enables a finished ceiling to be constructed a predetermined distance below the structural framing; **2.** Vehicle or machine springs, shock absorbers, and their attachments. *See also* **Suspension.**

SUSTAINED MODULUS OF ELASTICITY Elastic and inelastic net effects of stress/strain.

SUSTAINED YIELD Method of forest management that calls for an appropriate balance between net growth and amount harvested.

s.v. *Abbreviation for: sub verbo* or *sub voce* (under the word).

SV *Abbreviation for:* safety valve.

svc. *Abbreviation for:* service.

sw *Abbreviation for:* saltwater; switch.

s/w *Abbreviation for:* shear to width; spot weld.

Sw. *Abbreviation for:* swivel.

SW *Abbreviation for:* severe weathering; single width; sound wormy; southwest.

SWAGE 1. Shaping or marking metal by hammering; **2.** Method of shaping a saw tooth to provide side clearance on both sides of each tooth. *Also called* **Swedge.**

SWAGE BLOCK Metal block containing a range of holes and hollows, used as a form when shaping sheet metal.

SWAGED FITTING Fitting into which wire rope can be inserted and then permanently attached by cold pressing (swaging) the shank that encloses the rope.

SWAGED LINE Wire rope that has been reduced one size in diameter by a pounding action to replace the same nominal-sized rope, but have a greater strength.

SWAGE PILE *See* **Pile.**

SWAGE SETTING Set given the teeth of circular saws used for ripping.

SWALE Shallow dip made to allow the passage of surface water.

SWAMP Area saturated with water throughout much of the year, but with the surface of the soil usually not deeply submerged. Usually characterized by tree or shrub vegetation.

SWAMP BOAT Shallow-draft, boat-shaped vehicle powered by an aircraft-type, propeller-equipped engine mounted at the rear, and steered by vanes mounted to deflect the air flow produced by the propeller, used to navigate swamp and marsh.

SWAMP BUGGY Skidder equipped with high flotation tires.

SWAMPER 1. Person employed to help a machine operator; driller's helper; **2.** Axe man who cuts and clears away brush, limbs, small trees, and down timber. May also use saws.

SWAMPING AXE Double-bitted axe; one edge is usually kept sharp for wood cutting while the other is used as a utility blade.

SWAMP OUT To clean out brush and other material around the base of a tree to be felled or logs to be bucked. *Also called* **Brush out.**

SWAN'S NECK 1. Pipe shape forming an S-bend,

particularly between an eaves gutter and downpipe; **2.** Ramp and knee joint of a handrail.

SWAPLINE Procedure where the position of the mainline and haulback are exchanged.

SWARF Filings and other waste removed from metal during the process of being worked.

SWATH CUTTER *See* **Harvesting machines (single function).**

SWAY Sideways lean.

SWAYBRACE Diagonal brace used to resist wind or other lateral forces. *See also* **Cross bracing,** and **X-brace.**

swbd *Abbreviation for:* switchboard.

swd *Abbreviation for:* sawed.

SWEAT To unite two closely fitting pieces of metal by enlarging the outer one by heating.

SWEATING 1. Formation of water droplets on the outside of a pipe or tank containing still or flowing water at a significantly lower temperature than the surrounding atmosphere, due to condensation; **2.** Exuding of droplets of liquid, usually plasticizer, on the surface of a plastic sheet.

SWEAT JOINT Capillary pipe joint.

SWEAT SOLDERING Soldering in which the parts to be joined are first tinned (coated with a thin layer of solder) and then joined while exposed to heat sufficient to melt the solder.

SWEDES Method of setting grades at a center point by sighting across the tops of three laths. Two laths are placed at a known elevation and the third is adjusted until it is in line.

SWEDGE *See* **Swage.**

SWEDISH BREAK Slide or movement of a clayey soil embankment showing a cylindrical or spherical fracture surface.

SWEEP 1. Deviation from a straight line lengthwise; **2.** Gradual bend in a standing tree or in a log, pole, or piling; **3.** Method for removing numerous small branches in which the saw is constantly moving or 'sweeping.'

SWEEPER 1. Log or tree swinging sideways from the direction being yarded or skidded; **2.** *See* **Street sweeper.**

SWEEP T Plumbing fitting in which the T branch leaves via a curve.

SWELL 1. Percent a material expands when disturbed or removed from its bank state; **2.** Thickness increase in a panel that can occur from excessive moisture pickup or wetting.

SWELL BUTT Tree larger than normal at ground level.

SWELL FACTOR Ratio of the volume of material in its solid state to that when broken.

SWELLING Increase in either length or volume. *See also* **Elongation,** and **Expansion.** *Also called* **Volume swell.**

SWELLING SOIL Material subject to volume increase caused by wetting, oxidation, buildup of crystals, or relaxation after load removal.

SWEPT VALLEY A valley in which each course is carried round horizontally by means of specially cut wedge-shaped tiles, slates, or shingles.

S.W.G. *Abbreviation for:* British Standard wire gauge.

SWG *Abbreviation for:* standard wire gauge.

SWIFT Reel or turntable on which prestressing tendons are placed to facilitate handling and placing.

SWILL *See* **Solid waste.**

SWIMMING POOL Artificial enclosure designed for recreational swimming.

SWING 1. Lateral side-to-side movement of a pivoted member of an item of mechanical equipment. *Also called* **Slew**; **2.** Hoe function that permits side movement. **3.** To rotate a revolving shovel on its base; **4.** In cable or churn drills, to operate a string of drilling tools; **5.** To haul or yard logs from one landing to another.

SWING-AND-DUMP Ability of a feed beam to rotate on different planes (swing means horizontal plane, dump means vertical plane).

SWING ANGLE 1. Distance, measured in degrees, that a shovel must swing horizontally between digging and dumping points; **2.** Angle formed between the longitudinal centerlines of a tractor and trailer during a tight turn.

SWING AXIS *See* **Axis of rotation.**

SWING AXLE *See* **Axle.**

SWINGBEARING Two-part ring containing balls or rollers, the internal or external face of which features gear teeth to enable rotation, capable of sustained radial and axial loads of the revolving superstructure of a machine (crane, dragline, etc.).

SWING BRAKE *See* **Brake.**

SWING BRIDGE Bridge having part of its deck capable of swinging through a horizontal or vertical arc. *Also called* **Pivot bridge.**

SWING CLEARANCE 1. Distance that an item of equipment must maintain from an obstruction so as to allow for normal swing or rotation of its superstructure, turret, or attachment; **2.** Distance between the back of the cab, or other point of possible interference, and the corner of its trailer during a sharp turn. *Also called* the **Jackknife clearance.**

SWING CREW *See* **Collection method.**

SWING DOOR *See* **Door types.**

SWING DUTCHMAN Special falling technique to minimize breakage and maintain a lead, caused by severing the holding wood on the lean side.

SWINGER 1. Pointed bar used to move runners in trench shoring; **2.** Log that, when being skidded or yarded, hangs up against something and swings in an arc overhead. *Also called* **Upender.**

SWING EXPANSION JOINT *See* **Joint.**

SWING FRAME Undercarriage frame of a crawler tractor.

SWING FRAME GRINDER Grinding machine suspended by a chain at the center point so that it may be turned and swung in any direction.

SWING GEAR External or internal gear portion of the swing bearing with which a swing pinion meshes to provide swing motion. If the gear is attached to the chassis, the pinion is attached to the swing box in the superstructure.

SWINGING To haul or transfer logs from one landing to another.

SWINGING BRACKET *See* **Side bracket.** *See also* **Removable bracket.**

SWINGING HAMMER *See* **Pile hammer.**

SWINGING LEAD *See* **Lead.**

SWING-JIB CRANE Stationary or track-mounted crane having a counterweighted boom capable of pivoting through 360°.

SWING LOADER *See* **Loader.**

SWING MOTOR Hydraulic device that uses a planetary gear system to rotate the upper on a carrier.

SWING MECHANISM Machinery involved in providing dual directional rotation of a revolving superstructure.

SWING PARK BRAKE *See* **Brake.**

SWING RADIUS Distance between the kingpin and the corner of a semitrailer body.

SWING-RETURN ANGLE Angle for swing-return through which a revolving superstructure must pass in a load-swing-dump-return cycle.

SWING ROLLER One of several tapered wheels in a revolving shovel that roll on a circular turntable and support the upper works.

SWING SAW *See* **Saw.**

SWING SCAFFOLD *See* **Suspended scaffold.**

SWING TREE Spar to which logs are yarded for temporary storage, then swung to another landing.

SWING YARDER Any yarder that swings on a turntable, as opposed to a stationary-spar yarder.

SWIRL Direction induced in the flow of a gas; rotation of the mass of air as it enters the cylinder of an engine.

SWIRL FINISH 1. Nonskid texture imparted to a concrete surface during final trowelling by keeping the trowel flat and using a rotary motion; **2.** Technique of applying gypsum texturing material in decorative circular patterns.

SWITCH Device to open and close an electrical circuit, including the following:

   Access: 1. Keyed switch by which a circuit may be activated. **2.** Keyed switch by which an elevator car may be operated from a landing with the car and hoistway doors open at the landing to provide access to the car top or hoistway pit.

   Alarm: 1. Switch that, when activated by a condition or signal, causes an alarm to sound or show; **2.** Automatic sprinkler system switch used to open or close (generally close) an electric circuit to sound an electric alarm.

Broken step-chain: Switch mounted in the lower end of an escalator designed to stop the travel mechanism if the step chain breaks.

Buffer: Switch to prevent normal operation of an elevator car in the event the buffer fails to return.

Bypass: Specific device or combination of devices designed to bypass a regulator or an automatic transfer switch.

Compensating-rope sheave: Device that automatically cuts electrical power to a driving-machine motor and brake when the compensating sheave approaches the upper limit of its travel.

Direction: Contactor that determines the direction of rotation of an electric motor when power is applied.

Direction limit: 1. Mechanical switches that are activated when a moving part reaches the design limit of travel, cutting off power to the part; **2.** Switch that prevents an elevator from further travel in one direction only.

Directional start: Key-operated switch that selects the direction of travel of an escalator or conveyor.

Disconnecting: Mechanical switching device used for isolating a circuit or equipment from a source of electric power. *Also called* an **Isolating switch.**

Door: Electrical switch that is automatically operated by the opening and closing of a door.

Double-pole: Switch that opens or closes two isolated circuits simultaneously.

Double-throw: Switch that connects one circuit to either of two other isolated circuits.

Emergency stop: 1. Switch that, when activated, interrupts the circuit of, or to a motor or engine; **2.** Hand-operated switch in the car push-button station that, when thrown into the off position, stops the elevator and prohibits running.

Float: Electric switch that is responsive to liquid level.

Flow: Electric switch operated by a liquid flow.

Four-way: *See* **Three-way switch,** below.

Fused: Switch containing a fuse.

General-use: Switch intended for use in general distribution and branch circuits.

General-use snap: Form of general-use switch so constructed that it can be installed in flush device boxes or on outlet box covers, or otherwise used in conjunction with wiring systems.

Governor: 1. Mechanically operated switch mounted on a governor that removes power from a motor when the object the motor is driving or powering overspeeds; **2.** Mechanically operated switch mounted on a governor that actuates a control circuit to reduce speed.

Hoistway access: Electrical switch, located at a landing, that permits operation of an elevator car with the hoistway door at that landing and the car door or gate open, in order to permit access to the top of the car or to the pit.

Ignition: Keyed switch, part of the ignition circuit of an engine, that commonly also serves to control other functions, such as initiation of the starter motor, lights, auxiliary equipment, etc. *Also called* **Ignition key.**

Isolating switch: Switch intended for isolating an electric circuit from the source of power.

Key: Electric switch operated by a removable key.

Key operated: Switch that incorporates a keyed lock as part of its mechanical operation.

Knife: Electrical switch consisting of a thin blade that makes contact between two flat surfaces.

Knife-blade: Electrical switch where the moving contact is sandwiched between the arms of a fixed contact.

Limit: Electric switch that restricts the mechanical travel of an electrically operated or electrically controlled device.

Linked: Two or more switches physically joined so that they open and close simultaneously.

Master: Switch controlling two or more electrical circuits.

Momentary-contact: Electrical switch whose pole will always return to its default position when operating pressure is removed.

Motor-circuit: Switch rated in horsepower, capable of interrupting the maximum operating overload current of a motor of the same horsepower rating as the switch at the rated voltage.

Operation selector: Multiposition switch that can be set to the selected mode of operation.

Overspeed: Switch that is actuated to remove power from an electric motor when the device the motor is powering exceeds a preset speed.

Pendant: Electrical toggle switch, usually fitted to a lighting fixture, operated by a hanging cord.

Potential: Contactor whose coil is in series with a safety circuit: the contacts are in the power line so that if the safety circuit is not closed, the power supply to the principal device cannot be completed.

Pressure: Electric switch operated by a fluid or physical pressure.

Pressure-differential: Electric switch operated by a difference in pressure.

Retard: Switch used to regulate the speed at which an elevator car strikes the buffer.

Rigging: Switch that can be used to override any or all of the function limiters (cutouts) that have been activated on a crane during rigging/set-up.

Rotary: Electrical switch where a rotating central lever causes alternating make and break of contacts.

Safety-operated: Mechanically operated switch that removes power from an engine or motor when a safety mechanism is actuated.

Sensitive: Switch actuated by a spring mechanism that operates independently of the amount of force applied to it.

Service entrance: Switched main panel through

which service conductors are brought into a building and fed to a fused distribution box.

Signal transfer: Manually operated switch, located in an elevator car, by means of which the operator can transfer a signal to the next car approaching in the same direction, when he desires to pass a floor at which a signal has been registered in the car.

Single-pole: Switch that makes or breaks one side of an electric circuit.

Slack-rope: Device that automatically causes the electric power to be removed from a driving-machine motor when the hoisting ropes of a winding drum become slack.

Slow-down: Elevator limit switch, located at the terminal landing to slow down the car.

Stop: Switch that turns off the engine's ignition system and stops the engine, typically on a chainsaw.

Summer: Switch on the electrical circuit of a forced-air furnace permitting manual operation of the fan alone; used to circulate air without the furnace operating.

Supervisory: Monitoring switch activated by the closing of a valve, power failure, abnormal air pressure drop, etc.

Terminal motion: Switch used to stop an elevator car from running by the terminal floors under power if normal circuits fail.

Terminal slow-down: Mechanical switch that triggers an electrical contact to decelerate an elevator car.

Three-way: Switch used to make or break a circuit at more than one location.

Toggle: Electrical switch in which contacts are made and opened through the action of a toggle.

Transfer: Switch designed so that it will disconnect the load from one power source and reconnect it to another source.

Tumbler: Lever-actuated switch that causes a connected part to make, or break an electrical contact.

Water flow: Switch, activated by either pressure drop or water flow within an automatic sprinkler system that causes a local or remote alarm to indicate system operation.

Zone: Assembly of electrical contacts used to indicate to the door controller that an elevator is in the proper zone to permit door operation.

SWITCHBACK Hairpin turn of about 180° in a road, usually on a steep grade.

SWITCHBOARD Large single panel, frame, or assembly of panels that have switches, buses, instruments, overcurrent, and other protective devices mounted on the surface or the back or both.

SWITCHGEAR Devices, including switches and circuit breakers, protecting and/or controlling a power circuit.

SWITCHING DEVICE Device designed to close and/or open one or more electric circuits.

SWITCHING SPEED Time required to turn a device on or off.

SWIVEL Rotary connection joining power circuits (electric, air, hydraulic) between a carrier and its superstructure while allowing a continuous 360° rotation of the superstructure.

SWIVEL HEAD Mechanism in a diamond drill that rotates the kelly and drill string.

SWIVEL HOOK Hook having a swivel connection to its base or eye. *See also* **Cable hook, Grab hook, Hook, Pintle hook, Round hook,** and **Safety hook.**

SWIVEL JOINT Joint that permits variable operational positioning of lines.

SWIVEL LEAD *See* **Lead.**

swp *Abbreviation for:* sweep.

SWP *Abbreviation for:* safe working pressure; solvent-welded plastic pipe.

SWSI *Abbreviation for:* single-width single-inlet.

swt *Abbreviation for:* single-wrap traction; sweat.

swvl *Abbreviation for:* swivel.

SYLVESTER Jack-like tool used to withdraw timbers and steel posts from the ground, apply tension, or move heavy objects.

sym. *Abbreviation for:* symbol; symmetrical.

SYMBOL Graphic design, mark, character, letter, or figure, singularly or in combination, on a drawing that represents a material or thing or procedure.

SYMMETRICAL Having identical forms balanced about a central axis.

symp. *Abbreviation for:* symposium.

syn. *Abbreviation for:* synchronize; synchronous; synthetic.

SYNCHROMESH TRANSMISSION *See* **Transmission.**

SYNCHRONISM State where connected alternating current systems, machines, or a combination, operate at the same frequency and where the phase-angle displacements between voltages in them are constant.

SYNCHRONIZE Two or more events or operations occurring in the proper sequence in relation to each other.

SYNCHRONIZED-GEAR TRANSMISSION *See* **Transmission.**

SYNCHRONIZER Device that equalizes the speeds of a sliding clutch and the mating gear, preventing gear clashing in a nonsynchronized transmission.

SYNCHRONIZING (ROTATING MACHINERY) Process whereby a synchronous machine, with its voltage, frequency, and phase suitably adjusted, is parallel with another synchronous machine or system.

SYNCHRONOUS GENERATOR Synchronous alternating-current machine that transforms mechanical power into electric power.

SYNCHRONOUS INVERTER Device in an electrical system that converts the direct current generated by a wind machine or photovoltaic cells into alternating current, drawing its frequency signal from the public utility grid.

SYNCHROSCOPE Instrument that provides a visual indication of the proper time for closing the switch when synchronizing generators that are connected in parallel to the load.

SYNERESIS 1. Contraction of a gel, usually evidenced by the separation from the gel of small amounts of liquid; **2.** Loss of liquid lubricant from lubricating grease due to shrinkage or rearrangement of the structure.

SYNERGY Combined effect of two or more forces, substances, agencies, etc., that is greater than the sum of the effect achieved individually by each.

SYNGAS Synthetic gas resulting from pyrolysis of organic material produced by incomplete combustion of organic matter. The combustible components are primarily carbon monoxide and hydrogen (usually about 300 Btu/scf, but less than 900 Btu/scf).

SYNTHETIC GREASE Grease composition in which the liquid lubricant is other than mineral oil.

SYNTHETIC LUBRICANT Lubricant in which the base fluid is derived by synthesizing specific molecules from components, typically polyalphaolefins and esters.

SYNTHETIC RESIN Manufactured resin used as a vehicle in paint manufacture. *See also* **Natural resin.**

SYNTHETIC RESIN ADHESIVE *See* **Adhesive.**

SYNTHETIC THICKENER *See* **Nonsoap thickener.**

sys. *Abbreviation for:* system.

SYSTEM Prefabricated assemblies, components and parts, assembled into a complete operating unit.

SYSTEM ACCURACY Accuracy of an 'installed device' (system) that includes the errors associated with its being applied or installed on a machine. *See also* **Accuracy.**

SYSTEM AIR FLOW RESTRICTION Static pressure differential that occurs at a given air flow from air entrance through an air exit in a system, generally measured in millimeters (inches) of water.

SYSTEMATIC ERROR Survey error that is always positive or always negative, as distinguished from compensating errors. *See also* **Compensating error.**

SYSTEM DISPLAY MONITOR Cathode ray tube (CRT) mounted in proximity to and visible from the operator's position that displays information relating to machine operating state, conditions, selection, etc.

SYSTEME INTERNATIONAL Metric system of measurement. *See the appendix* **Metric and nonmentric measurement.**

SYSTEM OVERRIDE SWITCH *See* **Rigging switch.**

SYSTEM PRESSURE *See* **Pressure.**

SYSTEMS BUILDING *See* **Industrialized building.**

SYSTEM SCAFFOLD Supported scaffold consisting of individual fabricated posts with fixed connection points for runners, bearers, and diagonals, equipped with locking devices to connect to the posts at predetermined levels.

SYSTEMS ENGINEERING Design and specification of the means to achieve an objective part of a whole: heating/ventilating/cooling of a building, for instance, production of a given quantity per hour of concrete, warehousing of a type of goods, etc.

sz. *Abbreviation for:* size.

szg *Abbreviation for:* sizing.

T

t *Abbreviation for:* time (in seconds); tonne.

T *Abbreviation for:* tap; tee; temperature; township; truss.

tab. *Abbreviation for:* tablet; tabulate.

TABBY Section of cord fabric with closely woven pick yarns, enabling the woven cord to be cut without the individual cords in the rest of the roll becoming displaced.

TABER ABRADER Instrument used to test the abrasion resistance of a material.

TABLE 1. The working surface of a table saw through which the power-driven saw blade protrudes, and that is grooved to receive attachments such as a miter gauge and fence; **2.** That part of a grinding machine that directly or indirectly supports the work being ground.

TABLE SAW *See* **Saw.**

TABLE TRAVERSE Reciprocating movement of the table of a grinding machine.

TABULATED DATA Tables and charts approved by a registered professional engineer and used to design and construct a protective system.

TACHEOMETER Surveying instrument that rapidly determines distance, direction, and elevation differences from one observation.

TACHOGRAPH Instrument that replaces the speedometer and tachometer, giving a visible speed readout and a permanent record of vehicle mph or rpm in relation to time for a vehicle trip.

TACHOMETER Device for sensing and displaying rotational speed.

TACK 1. Tackiness, a stage beyond fluid and prior to set; **2.** Stickiness of an adhesive measurable as the force required to separate an adherent from it by viscous or plastic flow of the adhesive; **3.** *See* **Nail**; **4.** Act of hitting sheet piles down with a hairpin so they will have some penetration into the ground to stabilize the wall.

tackbd *Abbreviation for:* tackboard.

TACK CLAW *See* **Nail claw.**

TACK CLOTH Cloth (usually cheesecloth) impregnated with resins, used to remove dust and particles from a surface prior to coating it with a paint or finish coat. *Also called* **Tack rag.**

TACK COAT *See* **Bituminous coating.**

TACK HAMMER *See* **Hammer.**

TACKLE Assembly of ropes and sheaves designed for pulling or lifting.

TACK PULLER Hand tool consisting of a short metal shaft with a flaired, curved, and notched end.

TACK RAG *See* **Tack cloth.**

TACK RIVET Non-load-bearing rivet.

TACK WELD Small weld used to temporarily hold components together.

TACKY Sticky, gummy.

TAG 1. Joining of two or more chokers end-to-end for extended reach; **2.** System or method of identifying circuits, systems, or equipment for the purpose of alerting persons that the circuit, system, or equipment is being worked on; **3.** Temporary sign, usually attached to a piece of equipment or part of a structure to warn of existing or immediate hazard; **4.** Folded copper strip used as a wedge to hold copper sheet into a masonry joint.

TAG AXLE *See* **Axle.**

TAGLINE 1. Small wire rope used to prevent rotation of a load; **2.** Rope tied or hooked to a load being hoisted to guide it while in the air; **3.** Line from a clamshell bucket to a crane boom that holds the bucket from rotating.

TAIL 1. Rear section of a shovel deck; the anchor end of a cable excavator; **2.** Built-in end of a stone step.

TAIL ANCHOR Anchor for a track cable, or the turn point for the backhaul line of a cable excavator.

TAIL BAY 1. End span of a roof deck spanning several bays; **2.** Section of a canal downstream of the tailgate of a lock.

TAIL BEAM Relatively short beam or joist supported in a wall on one end and by a header at the other. *Also called* **Tail piece.**

TAILBLOCK 1. Boom for an idler sprocket assembly of a ladder ditcher; **2.** Block used to guide the haulback line at the back end of a yarding area.

TAIL CUT Vertical cut made to the end of a rafter at the eaves overhang.

TAILGATE 1. Hinged rear wall of a truck body or dump body; **2.** Sliding or tilting rear wall of a scraper bowl.

TAILGATE LIFT *See* **Elevating gate.**

TAILGATE SPREADER *See* **Spreader.**

TAIL HEATER Heater used to regulate the temperature of heavy oil for optimum combustion.

TAILHOLD 1. Point of anchor of the skyline; **2.** Stump or tree used to secure line blocks. *Also called* **Tailholt.**

TAILHOLT *See* **Tailhold.**

TAILING Part of a stone or brick projecting from a wall.

TAILING CRANE In a multimachine operation, the crane controlling the base end of the object.

TAILING IN Securing a cantilevered member by

weighting it down at the wall end.

TAILING IRON Steel section built into a wall to secure the end of a cantilever projecting below it.

TAILINGS Waste material separated from pay material during processing or screening.

TAIL JOIST Any joist with one end fitted against a header joist.

TAIL PIECE *See* **Tail beam.**

TAIL PLATE Rearmost part of a towing or recovery vehicle body.

TAIL ROPE Hoisting rope that passes around a return sheave.

TAILSHAFT Output shaft of a torque converter.

TAIL SPAR *See* **Backspar.**

TAILSWING Distance from the centerline of rotation of the upper frame of a crane or machine to the extreme rear swing arc of the counterweight. *Also called* **Rear-end radius.**

TAIL TOWER Tower of a cableway system that does not contain the operating machinery.

TAIL TREE *See* **Backspar.**

TAIL TRIMMER Trimmer adjacent to a wall, into which the ends of joists are fastened.

TAIL WATER Water immediately downstream from a structure, having been spilled from it or lost from a system.

TAKEOFF List of materials prepared from working drawings and specifications giving the type of material, number, weight, volume, size, etc.

TAKE-OFF LINE Pipeline leading from the high pressure side of a main circulating oil loop to the branch circuits.

TAKE-UP Mechanism for adjusting chain or belt tension.

TAKING *See* **Acquisition.**

TAKING UP Process of removing slack from a line; spooling in a line.

TALBOT PROCESS Protective coating of sand and bitumen inside cast iron pipe.

TALC Very soft mineral with a greasy or soapy feel.

TALK IN THE HOOK Directions to a helicopter, by radio communication, to the load hook-up area; usually by use of 'clock-coordinate system (e.g., 'I'm at your two o'clock low').

TALL BOY Down-draft preventer attached to the top of a chimney, consisting of a galvanized steel, straight-sided arch some 1.5 m (5 ft) high.

TALL OIL By-product of paper made from pine trees that, when distilled and blended with other oleoresinous vehicles or alkyds, is used as a vehicle or the liquid portion of paint.

TALLY Brass label attached to every tenth link of a 100-ft-long survey chain, having a notch cut in it for every 10 ft from the end of the chain.

TALLY COUNTER Any device for recording numbers mechanically.

TALUS Slope of loose rock and gravel formed by disintegration of a rock face.

TAMP To compact soil or any other material by applying repeated vertical blows, either manually or with a mechanical device.

TAMPER 1. Hand tool used to pack and consolidate soil. Sizes range from simple hand operated devices to self-contained, powered-operated units; **2.** Implement used to consolidate concrete or mortar in molds or forms. *See also* **Jitterbug; 3.** Hand-operated device for consolidating floor topping or other unformed concrete by impact from the dropped device in preparation for strikeoff and finishing.

TAMPING 1. Operation of consolidating freshly placed concrete by repeated blows or penetration with a tamper; **2.** Process of compressing the stemming or explosive in a blasthole.

TAMPING FEET Specially shaped and arranged projections on a compactor drum, used to achieve compaction by a series of blows.

TAMPING ROD Straight steel rod of circular cross section, having one or both ends rounded to a hemispherical tip.

TAMPING ROLLER Steel drum fitted with projecting feet, used singly or in a group in a machine, self propelled, or to be towed.

TAMPING SCREED Type of asphalt paver screed that utilizes a tamper bar mounted in front of the screed plate to achieve compaction of the mix.

TAMP PROCESS Process for producing concrete products such as pipe that uses direct mechanical action to consolidate the concrete by the action of tampers that rise automatically as the form is rotated and filled with concrete from above. *See also* **Centrifugal process, Dry-cast process, Packerhead process, Spun concrete,** and **Wet-cast process.**

tan. *Abbreviation for:* tangent.

tan⁻¹ *Abbreviation for:* inverse tangent.

T&B *Abbreviation for:* turned and bored.

t&c *Abbreviation for:* threads and couplings.

T&E *Abbreviation for:* test and evaluate.

TANDEM 1. Two connected items where one follows the other; **2.** Double-axle drive unit. *See also* **Bogie.**

TANDEM-AXLE SUSPENSION *See* **Suspension.**

TANDEM-AXLE WEIGHT Weight transmitted to the ground by two or more consecutive axles whose centerlines may be included between two transverse planes spaced more than 1.02 m (40 in.) and less than 2.44 m (96 in.) apart, extending across the full width of the vehicle.

TANDEM-CENTER VALVE *See* **Valve.**

TANDEM CYLINDER *See* **Cylinder.**

TANDEM-DRIVE AXLE *See* **Axle.**

TANDEM-DRIVE UNIT Three-axle vehicle, with two of the axles driving axles. *See also* **Bogie.**

TANDEM GEARED Geared machine in which there

are two gears driven by two worms on the same shaft.

TANDEM ROLLER Two or more rollers of approximately similar diameter, mounted behind each other on a common track.

TANDEM TRAILER *See* **Trailer.**

T&G *Abbreviation for:* tongue and groove.

T&P *Abbreviation for:* treated and primed.

TANG Sharp or pointed end of a metal tool that is driven into a handle.

TANGENT 1. Straight line from one point to another that passes over the edge of a curve; **2.** Knob on a survey instrument for making fine adjustments in the position of the alidade. **3.** *See also* **Geometry of circular curves.**

TANGENT DISTANCE Distance from an intersection point to a tangent point.

TANGENTIAL SHRINKAGE Loss of dimension in timber parallel to the growth rings.

TANGENTIAL STRESS Stress parallel to the tangent to the boundary of any opening.

TANGENT MODULUS At any given stress level, the slope of the stress-strain curve of a material in the inelastic range as determined by the compression test of a small specimen under controlled conditions.

TANGENT POINT Point at which a curve changes its curvature or become a straight line.

TANGENT SCREW Fine adjustment screw with minimal backlash.

TANK Container for the storage of fluid or gas. *See also* **Air-oil tank,** and **Vacuum tank.**

TANK BODY Fully enclosed truck, tractor, or trailer body designed to transport fluid.

TANKER Vehicle fitted with a storage tank capable of storing, and often of distributing, liquids.

TANKER BOSS Person in a forest fire suppression organization responsible for supervising usually three to five tanker units to get efficient and productive use of water in either direct attack or mop-up work.

TANKING Waterproof coating or layer laid below a basement floor and up the walls to above grade.

TANK SPRAYER Pressure tank on wheels fitted with a means of distributing its contents in the form of small droplets.

TANK VEHICLE Any vehicle, other than a railroad tank car, or a boat, with a cargo tank having a capacity of more than 450 L (118 gal), mounted or built as an integral part of the vehicle.

TAP 1. Tool for forming internal or female threads; **2.** Connection made at some intermediate point in a winding, coil, or resistor; **3.** Building connection to a gas or water service main; **4.** A faucet.

TAPE 1. Continuous ribbon of steel, cloth, metal, or some other material, marked on one or both sides with graduations of the metric or nonmetric systems of measurement, used to measure distances or lengths; **2.** Drywall joint reinforcing paper product; **3.** Continuous ribbon of material.

TAPE CORRECTION Correction applied to all survey lengths measured by an invariable tape.

TAPE CREASER Device that folds drywall joint tape longitudinally for use on inside corners.

TAPE MEASURE *See* **Measuring tape.**

TAPER 1. Gradual and uniform decrease in the size of a rectangular shape, hole, or cylinder; **2.** Factory edge applied to the long dimension of gypsum board used for drywall that produces a shallow V-shaped depression when two long edges are butted together and into which joint tape and joint compound are placed to form a permanent joint; **3.** Worker who finishes drywall; **4.** *See also* **Butt cut.**

TAPERED-FLANGE BEAM Rolled-steel joist section having its flanges tapered from their root with the web.

TAPERED PARAPET GUTTER Flexible-metal box gutter behind a parapet, narrower at its lower end to allow for the roof slope.

TAPERED PILE *See* **Pile.**

TAPERED PIPE THREAD Pipe threads in which the pitch diameter follows a helical cone to provide interference in tightening. *See also* **Dry-seal pipe thread,** and **Pipe thread.**

TAPERED REAMER Tool for deburring and cleaning the inside ends of pipes.

TAPERED SHAKE Roofing shake split from the bolt, that is turned end-for-end for each successive split. *See also* **Shakes, hand-split and resawn,** and **Straight split.**

TAPERED TIP *See* **Boom tip section.**

TAPERED WASHER Beveled washer.

TAPERED WHEEL Grinding wheel shaped similar to a straight wheel but having a taper from the hub of the wheel to the face and thus being thicker at the hub than at the face.

TAPER FILE Fine-toothed file narrower at one end than at the other, typically used for sharpening saws.

TAPER-LEAF SPRING Leaf-type spring in which the center of the spring leaf is thicker than the ends: gives lighter weight, softer ride, and permits fewer leaves for a given application.

TAPER PIN Straight-sided pin that is smaller at one end than at the other. *See also* **Pin.**

TAPER-ROLLER BEARING Bearing in which the rollers have a uniform taper.

TAPE RULE *See* **Measuring tape.**

TAPERSPLIT *See* **Shake.**

TAPETTE *See* **Caisson hammer.**

TAPING 1. Measuring in a straight line between two points using a steel, plastic or fabric tape reinforced with steel mesh; **2.** Masking of joints between sheets of drywall by means of tape that is smoothed over with mud or joint cement.

TAPING COMPOUND Drywall joint compound especially formulated to embed joint tape.

TAPING STRIP Strip of roofing felt placed over

joints between precast roofing slabs prior to bonding and sealing.

TAPING TOOL Any of the range of tools designed to assist in the application of joint and finishing materials and compounds.

TAPPED COIL Coil with a center tap fastened to the coil windings, giving a single coil the effectiveness of two coils, used to set and rest a single relay.

TAPPED T Cast iron T with at least one branch tapped to receive a threaded pipe or fitting.

TAPPET Projection in a machine that moves, or is moved by another part.

TAPPET CLEARANCE Clearance between the rocker arm (tappet) and the valve stem.

TAPPING TILE Inspection technique whereby a small metallic object is tapped against an installed tile to determine by sound whether the tile is completely bonded to its backing.

TAPROOT Principal root of certain species of trees that grows downward for a considerable depth from the base of the tree.

TAR Bituminous material, liquid or semisolid, that has adhesive and waterproofing properties; a product of distillation, extracted from coal instead of petroleum. There are two types: coal tar, and water-gas tar, which is usually combined with coal tar as a flux.

tar. *Abbreviation for:* tarred.

TAR-AND-GRAVEL ROOFING Roof covering composed of felt sheets mopped with hot tar or pitch and covered with gravel or sand.

TAR CEMENT Heavy-grade tar prepared for direct use in pavement construction and maintenance.

TAR CONCRETE Bituminous concrete.

TARE Allowance made for the weight of a container.

TARE WEIGHT Working weight of a vehicle or machine, less its self weight. *Also called* **Chassis weight,** and **Curb weight.**

TARGET 1. Portable marker, often mounted on a pole or tripod, used for sighting on, or to reflect electronic signals in surveying; **2.** Wood frame located at the bottom of a hoistway to hold plumb lines steady.

TARGET BOARD Wood fabrication with a point marked on it, fixed to the ground or a solid, horizontal surface, used in pairs or multiple numbers to align structures or mark the bounds of excavations. The marks on them are used as sight points.

TARGET FOREST Type of forest, in terms of species mixture, size, stocking, and harvest age, considered best for a particular site in order to economically produce fiber in the qualities and quantities desired on a perpetual basis.

TARGET ROD Leveling rod.

TARGET STRING Line tightly stretched between two target boards to align structures.

TARMACADAM Pavement surfacing consisting of low-fines aggregate coated with tar or a tar-bitumen mixture.

tarp *Abbreviation for:* tarpaulin.

TARPAULIN Waterproof, weatherproof flexible sheet, used to temporarily protect exposed materials and work.

TARP BASKET Open framework rack or compartment across the front of a trailer, used to store a tarpaulin.

TARP HOOK Fittings attached at intervals around the edge of a flatbed truck or trailer deck and used as anchor points for securing a tarpaulin.

TARRED FELT Organic felt saturated with coal-tar pitch, used in built-up roofing where coal-tar pitch is also used between the layers of roofing felt. *See also* **Built-up roofing, Buillt-up roofing, Glass-fiber felts, Organic felts,** and **Roofing felts.**

TASK Item of work assigned or required.

TASK-ORIENTED LIGHTING Artificial lighting fixtures designed for and positioned to illuminate one or more task locations.

TAX Compulsory contribution levied upon residents, property, or businesses.

TAXABLE HORSEPOWER *See* **Horsepower.**

TAX SALE Sale of property, usually at public auction, for nonpayment of property taxes.

TAYWOOD SHEET-PILE DRIVER/EXTRACTOR Heavy device with hydraulic rams attached to several individual sheet piles; rams are operated separately to push or pull a sheet while others furnish reaction.

tb. *Abbreviation for:* turnbuckle.

TB *Abbreviation for:* technical bulletin; through bolt.

TB&S *Abbreviation for:* top, bottom, and sides.

T-BAR 1. T-shaped bar used in place of steel pins to support a string line over trenches and other excavations: **2.** Preformed, light metal section that is the principal element in a suspended ceiling system, and that supports ceiling panels.

tbc *Abbreviation for:* top back of curb.

T-BEAM Beam composed of a stem and a flange in the form of a tee. *See also* **APA glued floor system,** and **Tee beam.**

T-BEAM FOOTING Footing that combines an inverted T-beam with a column footing.

TBGN *Abbreviation for:* through bolt and grommet nut.

tbl. *Abbreviation for:* table.

TBM *Abbreviation for:* tunnel boring machine.

TBN *Abbreviation for:* through bolt and nut.

tbrs *Abbreviation for:* timbers.

tc *Abbreviation for:* top of curb.

TC *Abbreviation for:* terra cotta; thermal conductivity; tin-clad; toilet case.

td *Abbreviation for:* tinned.

tdc *Abbreviation for:* top dead center.

tdm *Abbreviation for:* tandem.

TE *Abbreviation for:* table of equipment; trailing edge.

TEAR RESISTANCE Property of a rubber tube or cover of a hose to resist tearing forces.

TEAR STRIP Stout paper ribbon under bundling tape to facilitate opening of bundles of drywall.

TEC *Abbreviation for:* total estimated cost.

tech. *Abbreviation for:* technical.

techn *Abbreviation for:* technician.

TECHNICAL LIFE LENGTH Time from when a machine goes into operation until it is no longer used in any operation; machine productive time, expressed in hours.

technol *Abbreviation for:* technological.

Tech. Wrtr *Abbreviation for:* technical writer.

TECTONIC STRESS Stress caused by deformation of the earth's crust; may occur near the surface and may greatly exceed the stress in the rock due to gravity.

TEE Three-way pipe fitting shaped like the letter T.

TEE BEAM Rolled-steel, reinforced concrete, pretensioned concrete or post-tensioned concrete section shaped like the letter T. *See also* **T-beam.**

TEEPEE Unintentional lodging of two or more trees in another standing tree, generally the result of poor falling technique.

TEEPEE BURNER *See* **Burner.**

TEE SQUARE *See* **T-square.**

TEETH 1. Hardened steel cutting edges attached to an excavating device such as a bucket; 2. Tension filaments that appear between two adjoining plies of rubber as they are pulled apart.

TEJ *Abbreviation for:* transverse expansion joint.

tel. *Abbreviation for:* telephone.

TELEGRAPHING Show-through on a smooth overlaid plywood panel surface of underlying grain or defects.

TELEMETRY Science concerned with measurement, transmitting the results to a distant station, and interpreting, indicating, or recording the transmitted data.

teles. *Abbreviation for:* telescoping.

TELESCOPE 1. To slide one piece inside another; 2. Magnifying optical instrument, part of several types of surveying instruments.

TELESCOPE HYDRAULIC ELEVATOR Direct-plunger hydraulic elevator having a set of coaxial plungers.

TELESCOPE LEAD *See* **Lead.**

TELESCOPIC CENTERING Pressed-steel, pan-type, interlocking forms, adjustable for length by sliding one over the other. *Also called* **Self-centering formwork.**

TELESCOPIC FORM 1. Full-circle or arch form that, when stripped and collapsed, will pass through other similar forms erected in place in a tunnel; 2. Slab formwork made up of nesting steel channels that can be telescoped to increased or decrease the total length (span).

TELESCOPIC HOIST *See* **Truck hoist.**

TELESCOPIC STICK Special excavator and backhoe stick arrangement consisting of hydraulically extendable and retractable outer, middle, and inner tubes offering additional digging depth and dump height.

TELESCOPING BOOM Boom, with sections that extend or retract by use of a hydraulic telescoping cylinder or other mechanical means and that are contained within a nontelescoping base section.

TELESCOPING CYLINDER *See* **Cylinder.**

TELESCOPING FLY Extension to the fly section of a crane that is stored through its center. Mounted on rollers, the extension is erected by telescoping (pulling) it out of the center of the fly and then pinning it into position.

TELESCOPING HOIST *See* **Truck-mounted crane.**

TELESCOPING MAST Multiple mast wherein one member is stationary and the other(s) movable vertically with respect to the stationary member and supporting the fork carriage of a lift truck in its vertical movement. This mast permits maximum lifts substantially greater than the overall lowered height.

TELLTALE 1. Any device designed to indicate movement; 2. Marking of formwork or of a point on the longitudinal surface of a pile under load to indicate movement from a given reference.

TELLUROMETER Electronic measuring device that measures the time taken by microwaves on a round trip between the transmitter and a target receiver.

TELPHER Electrically powered hoist suspended from a wheeled cart rolling on a single overhead rail.

tem. *Abbreviation for:* temper; template.

TEM *Abbreviation for:* total energy management.

temp. *Abbreviation for:* temperature; temporary.

TEMPER 1. To mix mortar in the proper condition for use; 2. Heat treatment of a material to develop required qualities; 3. Part of heat treating in which hardened steel or hardened cast iron is heated to a temperature below its melting point for purposes of decreasing the hardness and increasing the toughness.

TEMPERATURE Degree of hotness or coldness. *See also* **Celsius,** and **Fahrenheit.**

TEMPERATURE COEFFICIENT Expected change in value per degree of temperature difference from a specified temperature.

TEMPERATURE CRACKING Cracking caused by a temperature drop in members subject to external restraint, or to a temperature differential in members subject to internal restraints, due to tensile failure.

TEMPERATURE GRADIENT Measured changes in temperature over time.

TEMPERATURE REINFORCEMENT Steel reinforcement in concrete to carry stresses resulting from temperature changes. *Also called* **Temperature rod.**

TEMPERATURE RISE Increase in temperature caused by absorption of heat or by its internal generation.

TEMPERATURE-RISE PERIOD Time interval during which the temperature of a concrete product rises at a controlled rate to the desired maximum in autoclave or atmospheric-pressure steam curing.

TEMPERATURE ROD *See* **Temperature reinforcement.**

TEMPERATURE STRESS Stress in a structure or member due to changes or differentials in temperature.

TEMPERED 1. Thoroughly mixed mortar or cement; **2.** Case-hardened metals.

TEMPERED GLASS Glass that has been rapidly cooled under rigorous control from near the softening point to increase its mechanical and thermal endurance.

TEMPERING AIR Ambient air added for cooling by dilution.

TEMPILSTICK Crayon-like material that melts at a given temperature.

TEMPLATE 1. Any form of pattern or shaped guide that replicates a design or an original; **2.** Thin plate or board frame used as a guide in positioning or spacing form parts, reinforcement, or anchors; **3.** Short piece placed in a wall under a beam to distribute pressure; **4.** Frame at the top of an elevator hoistway to locate the lines, checking size, plumb, and square of the hoistway.

TEMPORARY CASING 1. Casing left in place until concrete has been placed; **2.** Casing placed as protection for workers.

TEMPORARY POWER Electrical service provided for lighting and construction needs before a permanent supply is in place.

TEMPORARY SCAFFOLD *See* **Scaffold.**

TEMPORARY STRESS Stress that may be produced in a precast concrete member or in a component of a precast concrete member during fabrication or erection, or in cast-in-place concrete structures due to construction or test loadings.

TEMPORARY STRUCTURE General term for anything that is built or constructed (usually to carry construction loads) that will eventually be removed before or after completion of construction and that does not become part of the permanent structural system.

ten. *Abbreviation for:* tenon.

TENANCY AT WILL Tenancy of land and/or structures at the will of the owner.

TENANCY BY THE ENTIRETY Estate existing solely between husband and wife, each enjoying equal rights and with right of survivorship.

TENANCY IN COMMON Ownership of real property by two or more persons, each having undivided interest, without the right of survivorship.

TENANT Person who rents or leases accommodation from another.

TENANT AT SUFFERANCE One who comes into legal possession of land and who keeps it without title being registered.

TENDER 1. Offer or bid to complete work or supply materials; **2.** Laborer who tends masons. A general name covering hod and pack carriers and wheelbarrow handlers.

TENDER BOND *See* **Bond, bid/tender.**

TENDER CALL *See* **Call for tenders.**

TENDER DOCUMENTS *See* **Bid documents.**

TENDERER *See* **Bidder.**

TENDERERS' LIST *See* **Bidders' list.**

TENDER FORM *See* **Bid form.**

TENDER MIX Soft asphalt mix that does not readily support the weight of an asphalt paver screed.

TENDER PACKAGE *See* **Bid package.**

TENDON Steel element such as wire, cable, bar, rod, or strand, or a bundle of such elements, primarily used in tension to impart compressive stress to concrete.

TENDON PROFILE Path or projection of the prestressing tendon.

TENEMENT Building containing low-rent apartments; such an apartment.

TEN-HOUR TIME LAG FUELS Dead forest floor fuels consisting of roundwood measuring 6 to 25 mm (0.25 to 1 in.) in diameter, and very roughly the layer of litter extending from just below the surface to approximately 18 mm (0.75 in.) below the surface.

TENON 1. Projecting part cut on the end of a member, shaped so as to fit into a matching mortise; **2.** Head of a timber pile after being cut or shouldered to accommodate a splicer or other device.

TENON SAW *See* **Saw.**

tens. *Abbreviation for:* tensile; tension.

TENSILE BOLT High-strength bolt, made of high-tensile steel.

TENSILE FORCE *See* **Force.**

TENSILE ROCK STRENGTH Amount of tensile stress that a rock can withstand without failing.

TENSILE STRENGTH Strength of a material, as measured by attempting to pull apart a specific amount of the material.

TENSILE STRESS Maximum unit stress that a material is capable of resisting under axial loading, based on the cross-sectional area of the specimen before loading.

TENSILE TEST Test in which a standard piece of material is pulled in a testing rig until it breaks.

TENSIOMETER *See* **Running line tensiometer.**

TENSION Pressure on an object caused by an expansive force.

TENSION CAPACITY *See* **Uplift capacity.**

TENSION FIELD ACTION Behavior of a plate girder panel under shear force in which diagonal tensile stresses develop in the web and compressive forces develop in the transverse stiffeners.

TENSIONING Method of stretching a saw body in the inner area of either a circular saw or band saw to compensate for heating that expands the circular saw periphery or band saw edges.

TENSION PILE Pile installation designed to resist uplift.

TENSION REINFORCEMENT Reinforcement designed to carry tensile stresses such as those in the bottom of a simple beam.

TENSION SET Condition of wood in which a group of fibers, owing to restraint imposed by adjoining fibers or by an external mechanical agency, are fixed or set in a condition of tension as a result of a restraint on normal shrinkage during a drop in moisture content.

TENSION SHEAVE Sheave either anchored via a spring mounting, or suspended from a cable and equipped with a weight, the purpose of which is to maintain tension on a line.

TENSION SLEEVE Screw shackle.

TENSION STRENGTH Strength of a panel when pulled in the long direction.

TENSION WOOD *See* **Reaction wood.**

TENURE Holding, particularly as to manner or term (i.e., period of time) of a property.

ter. *Abbreviation for:* terrace; terazzo; tertiary.

TERA Prefix representing 10^{12}. Symbol: T. Used in the SI system of measurement. *See also the appendix:* **Metric and nonmetric measurement.**

TEREDO *See* **Marine borer.**

term. *Abbreviation for:* terminal; terminate.

TERMINAL 1. End feature or place; **2.** End of a rope or chain provided with an eye or attaching device; **3.** Device attached to a conductor to facilitate a connection; **4.** Station or depot at the end of a public transportation system; **5.** One of the parts of a storage battery to which the external circuit is connected.

TERMINAL CONNECTOR 1. Eyelet attached to a wire in an electrical circuit, used to complete a connection. **2.** Electrical conductor for carrying current from a storage battery to the external circuit. *See also* **Connector.**

TERMINAL EXPENSE One of the expenses incurred during the termination of a contract.

TERMINAL LANDING Top or bottom landing of an elevator or dumbwaiter.

TERMINAL MOTION SWITCH *See* **Switch.**

TERMINAL RAIL Section of door guide rail to guide the lower half of an elevator door at the bottom terminal landing and the upper half of the door at the top terminal landing.

TERMINAL SLOW-DOWN SWITCH *See* **Switch.**

TERMINAL STOPPING DEVICE Device to slow down and stop an elevator or dumbwaiter car automatically at or near a terminal landing, independently of the functioning of the operating device.

TERMINAL VELOCITY Maximum velocity a body can attain falling freely.

TERMITE BOND *See* **Bond.**

TERMITE INSURANCE Insurance protection against termite infestation or damage.

TERMITE SHIELD Barrier, usually of noncorrodible metal, or of flexible plastic sheeting, placed in or on a foundation wall or other masonry mass, or around pipes, to prevent passage of termites from the exterior

to the interior of a structure.

termt. *Abbreviation for:* termite.

TERNE PLATE Sheet steel roofing coated with an alloy of 80% lead and 20% tin.

terr. *Abbreviation for:* territory.

TERRACE 1. Unroofed, paved area immediately adjacent to a house, often with a railing or balustrade on its external perimeter; **2.** Flat, or relatively flat bench, sometimes fronted by a berm or ridge, developed along a ground contour. *Also called* **Meadow.**

TERRA COTTA Hard, brown-red, usually unglazed earthenware frequently used for ornamental facings.

TERRAIN Ground, considered for its fitness or desirability for some purpose.

TERRAZZO Flooring finish made in-situ of marble chips mixed with cement mortar, then ground and polished when set.

TERRAZZO CONCRETE Marble-aggregate concrete, cast-in-place or precast and ground smooth for decorative surfacing purposes.

TERTIARY TREATMENT Phase of wastewater treatment involving chemical reaction.

TESLA One of 17 derived units with special names of the SI system of measurement: a unit of magnetic induction equal to 1 Wb of magnetic flux per m^2. (1 $T = 1$ Wb/m^2). Symbol: T. *See also the appendix:* **Metric and nonmetric measurement.**

TESSELATED Cement floor or wall surfacing in which tesserae are embedded.

TESSERA Small cubes of marble, stone, or glass used in mosaic work.

TEST Trial, examination, observation, or evaluation used as a means of measuring either a physical or chemical characteristic of a material, structural element, or structure.

TEST ANCHOR Ground anchor constructed at the beginning of a project on which a pull-out test is made.

TESTATE To die leaving a valid will.

TEST BORING *See* **Soil boring.**

TEST CERTIFICATE Document certifying that the product it refers to meets the requirements of an established standard or performance criteria.

TEST CORE Core, usually 150 mm (6 in.) in diameter by 300 mm (12 in.) long, or 100 mm (4 in.) in diameter by 200 mm (8 in.) long, cut from hardened concrete using a diamond-tipped corer, used for compressive tests, determination of concrete cover to steel reinforcement, composition of aggregate, degree of compaction, etc.

TESTED DESIGN Design that has been load tested to demonstrate it can support the required load, including an appropriate safety factor.

TESTING MACHINE Device for applying test conditions and accurately measuring results.

TEST LOAD Load, the value of which is known and that is used in testing or calibrating a load indicating system.

TEST PILE *See* **Pile.**

TEST PIT Method of examining soil by excavating a pit to permit direct examination of the materials in place and its degree of compaction, as well as to permit test loads directly on what will be the foundation soils.

TEST POINT One of the specific points in a circuit at which operations may be checked.

TEST PRESSURE *See* **Pressure.**

TESTWEIGHT One of several metal blocks of known weight (usually 22.7 kg (50 lb)), used to load an elevator under test in simulation of its operating capacity.

TEST WELL Well installed to assess aquifer conditions.

tet. *Abbreviation for:* tetrachloride.

TETHERED-ELECTRIC TRUCK An electric truck in which the power source is remote from the vehicle and connected by a flexible cable.

TETHER LINE Line used to restrain a balloon in flight, such as the line from a logging balloon to the butt rigging.

tetr. *Abbreviation for:* tetragonal.

tetrah. *Abbreviation for:* tetrahedron.

TETRAPOD Four-legged equiangular block of reinforced concrete weighing up to 20 t, used in the construction of breakwaters.

TETRASTYLE Classical design of a building having four columns across its facade.

TEX Yarn size system defined as the weight in grams of 1000 m of yarn.

tex. *Abbreviation for:* textile.

TEXAS QUICK-LOAD PILE TEST Similar to Constant-rate-of-penetration test.

text. *Abbreviation for:* texture; textured.

TEXTILE 1. General term applied to that which is or may be woven, as a woven cloth or yarn; **2.** Fibrous material suitable for being spun and woven into cloth or yarn.

TEXTURE 1. Pattern or configuration apparent in an exposed surface. *See also* **Appearance; 2.** In concrete and mortar, roughness, streaking, striation, or departure from flatness; **3.** Property of lubricating grease that is observed when a small separate portion of it is pressed together and then slowly drawn apart, described as follows:

Brittle: Has a tendency to rupture or crumble when compressed.

Buttery: Separates in short peaks with no visible fibers.

Long fibers: Shows tendency to stretch or string out into a single bundle of fibers.

Resilient: Capable of withstanding moderate compression without permanent deformation or rupture.

Short fiber: Shows short break-off with evidence of fibers.

Stringy: Shows tendency to stretch or string out

into long fine threads, but with no visible evidence of fiber structure.

TEXTURED BRICK *See* **Brick.**

TEXTURED FINISH Intentional rough finish to a surface, sometimes involving the application of additional materials for color or texture, and often requiring manipulation to form a distinctive pattern.

TEXTURED PLYWOOD Plywood panels with a variety of machined surface textures. Available in **Exterior type** with fully waterproof glueline for siding and other outdoor uses, and **Interior type** for interior wall paneling.

TEXTURE PAINT Paint that can be manipulated after application to produce a surface texture.

TEXTURING Process of producing a special texture on either unhardened or hardened concrete.

TEXTURING MACHINE Equipment, part of a mobile paving train, that produces a textured surface on freshly placed concrete.

tf *Abbreviation for:* tar felt.

tg. *Abbreviation for:* tangent.

TG&B *Abbreviation for:* tongued, grooved and beaded.

TG&D *Abbreviation for:* tongued, grooved and dressed.

TG. B *Abbreviation for:* toggle bolt.

tgl *Abbreviation for:* toggle.

th. *Abbreviation for:* theory.

TH *Abbreviation for:* true heading.

THATCH Roof covering of reeds or straw.

thd *Abbreviation for:* thread.

T-HEAD 1. In precast framing, a segment of girder crossing the top of an interior column; **2.** Top of a shore formed with a braced horizontal member projecting on two sides forming a T-shaped assembly.

THEATER Place of public assembly intended for the production and viewing of the performing arts or the screening and viewing of motion pictures, and consisting of an auditorium with permanently fixed seats intended solely for a viewing audience.

theo. *Abbreviation for:* theoretical; theodolite.

THEODOLITE Surveying instrument comprising a telescope mounted on a graduated circle, and equipped with levels and a reading device.

THEORETICAL AIR Calculated amount of air required to supply oxygen for complete combustion of a given quantity of a specific combustible material.

THEORETICAL JOINT WIDTH *See* **Minimum theoretical joint width.**

THERM Nonmetric unit of heat, equal to 100,000 British thermal units. Symbol: th. Multiply by 105.506 to obtain megajoules, symbol: MJ. *See also the appendix:* **Metric and nonmetric measurement.**

therm. *Abbreviation for:* thermal; thermometer.

THERMAL BARRIER Strip of thermally nonconducting material placed in locations appropriate to prevent the passage of heat or cold from one face to another.

THERMAL COLUMN Column of smoke and gases given off by forest fires, moving upward because heated gases expand and become lighter and rise, while cooler air, bringing additional oxygen, is drawn in toward the base of a fire.

THERMAL CONDUCTANCE Property (of a particular body or assembly) measured by the ratio of steady-state heat flux in common between two definite surfaces (time-rate of heat flow per unit area of one surface, that must be identified to the difference between the average temperatures of the two surfaces).

THERMAL CONDUCTIVITY Property (of a homogeneous body) measured by the ratio of the steady-state heat flux (time-rate of heat flow per unit area) to the temperature.

THERMAL CONDUCTOR Substance capable of transmitting heat.

THERMAL CONTRACTION Contraction caused by a decrease in temperature.

THERMAL CONVERSION Conversion of organic waste into energy through combustion.

THERMAL DEGRADATION Deleterious change in the chemical structure of a material due to excess heat. *See also* **Thermal shock.**

THERMAL DIFFUSIVITY Thermal conductivity divided by the product of specific heat and unit weight; an index of the facility with which a material undergoes temperature change.

THERMAL EFFICIENCY Efficiency of an engine in converting heat energy from combustion of fuel into mechanical work.

THERMAL ENERGY Latent energy of a temperature differential.

THERMAL EXPANSION Expansion caused by an increase in temperature.

THERMAL FRACTURE Compression crack caused by expansion of peripheral components.

THERMAL INSULATION Materials and construction used to reduce heat loss or gain.

THERMAL LAG 1. Delay between the application of heat and the time when the object or space being heated reaches the intended temperature; **2.** In an indirect-gain solar system, the time delay for heat to move from the outer collecting surface to the inner radiating surface.

THERMAL MASS Heat storage capacity of a material or collection of materials.

THERMAL MOVEMENT Change of dimension of a material resulting from change of temperatures.

THERMAL OVERLOAD Overload relay operated by bimetallic strips, surrounded by heaters carrying the motor current.

THERMAL RADIATION Emission of radiant energy waves in forms transmitting or producing heat.

THERMAL RESISTANCE 1. Reciprocal of thermal conductance expressed by the symbol R; **2.** Insulating property of a material.

THERMAL SHOCK Effect on a material caused by exposure to a sudden or rapid change in temperature, beyond that considered normal.

THERMAL SHOCK RESISTANCE Ability of a material to withstand sudden heating or cooling, or both, without cracking or spalling.

THERMAL STORAGE, PHASE CHANGE Heat storage system based on materials, such as eutectic salts, that change from solid to liquid as they absorb heat and revert from liquid to solid as they lose it.

THERMAL STORAGE ROCK BED An insulated container of small-size pebbles that retain solar heat for later use.

THERMAL STRESS CRACKING Crazing and cracking of some materials resulting from overexposure to elevated temperatures.

THERMAL TRANSMISSION VALUE Resistance factor to the conductance of heat.

THERMAL TURBULANCE Atmospheric disturbance, sometimes rather violent, above and downwind of a major forest fire.

THERMISTOR Resistor whose value can change when heat is applied.

thermo. *Abbreviation for:* thermopane; thermostat.

THERMOCHEMICAL PRESSURE Pressure that theoretically should be created when an explosive is detonated, calculated from thermochemical properties of the explosives.

THERMOCOUPLE Electronic temperature measuring device consisting of two dissimilar electrical conductors so joined as to produce a thermal electromotive force when exposed to temperatures. The electromotive force generated can be calibrated to read in temperature units.

THERMODYNAMICS Study of heat energy and its conversion to other forms of energy.

THERMOGRAPHY Conversion of heat emissions to a visible picture.

THERMOMETER Instrument capable of determining and registering temperature.

THERMOPLASTIC Becoming soft when heated and hard when cooled.

THERMOPLASTIC ADHESIVE *See* **Adhesive.**

THERMOPLASTIC SEALANT Material that becomes more plastic and less elastic with a rise in temperature.

THERMOPLASTIC SINGLE-PLY MEMBRANE Single-ply roofing material that can be shaped and formed following application of heat. *See also* **Single-ply roofing.**

THERMO-REGULATING VALVE *See* **Valve.**

THERMOSET SINGLE-PLY MEMBRANE Single-ply roofing material available in sheet and liquid form. *See also* **Single-ply roofing.**

THERMOSETTING Becoming rigid by chemical reaction and not re-meltable.

THERMOSETTING ADHESIVE *See* **Adhesive.**

THERMOSIPHONING AIR PANEL Passive solar space heater that heats air in a collector located below a living area so that warmed air flows up by convection to the living space without assistance of a fan, and when cooled, flows back down to the collector, forming a convective loop.

THERMOSIPHONING WATER HEATER System in which water heated by a solar collector flows up to a storage tank by convection without assistance from a pump, then back down to the collector, forming a convective loop.

THERMOSTAT Automatic device for controlling the supply of heat.

THERMOSIPHON Phenomenon where warm water rises and cold water sinks, causing a vertical circulation within a body of water.

THI *Abbreviation for:* temperature-humidity index.

THICK-AND-THIN Thickness variation within a panel or between two panels.

THICKBUTT Square-butt shingle having a thicker butt, or exposed section, than its unexposed or covered section.

THICKENER Solid particles that are relatively uniformly dispersed to form the structure of lubricating grease in which the liquid is held by surface tension and other physical forces. (The solid particles may be fibers, as is the case with various metallic soaps, or plates or spheres, as is the case with some of the nonsoap thickeners.)

THICKENING AGENT Substance added to a concrete mix to overcome segregation and, by increasing the viscosity of the water, increase resistance to bleeding.

THICKENING TIME TEST Test to determine the time it is possible to pump cement slurry.

THICK-LIFT ASPHALT CONSTRUCTION Construction practice in which the asphalt course is placed in one or more lifts of 100 mm (4 in.) or more compacted thickness.

THICKNESS Dimension perpendicular to the face of a wall, floor, or other assembly.

THICKNESS CONTROL Control, usually located at the ends of the main asphalt paver screed, by which the screed operator can raise or lower the angle of attack of the screed plate to increase or decrease mat thickness.

THICKNESS SWELL Increase in thickness when a panel is exposed to moisture. Waferboard/OSB is allowed to swell 20% after 24 hours of soaking if 12 mm (0.5 in.) or greater thickness, and 25% if less than 12 mm (0.5 in.) thickness.

THICKNESS VARIATION Difference in thickness within a panel.

THIMBLE 1. Round, terra-cotta or fireclay insert that fits into a chimney to take in the furnace pipe and form a fireproof connection into the masonry wall; **2.** Grooved metal fitting to protect the eye or fastening loop of a wire rope.

THIN COVER 1. Cover, the thickness of which is less than specified; **2.** Wire braid hydraulic hose specifically made with a thin cover to eliminate the need for buffing when attaching couplings.

THINNER Volatile liquid used to regulate the consistency of a finishing material.

THINNING Process of removing excess and poorer quality trees from a stand for the purpose of improving the growth and value of the remaining trees.

THIN-NOSED PLIERS *See* **Pliers.**

THIN OVERLAY *See* **Hot-mix seal coat.**

THINSET No-fines cement-based adhesive that must be mixed with a liquid (water, latex, or acrylic) before use and that produces a joint approximately 3 mm (0.125 in.) thick, used principally with tiles.

THIN SHELL Thin, poured-slab roof.

THIN-SHELL PILE *See* **Pile.**

THIN-SHELL PRECAST Precast concrete characterized by thin slabs and web sections. *See also* **Shell construction.**

THIN TUBE Lining, the thickness of which is less than specified.

THIN WALL SHELL PILE *See* **Pile.**

THIRD-CLASS LEVER *See* **Lever.**

THIRD DRUM Third hoist drum, in addition to two main hoist drums, often used in pile driving.

THIRD FACING CUT Special falling technique of making an additional facing cut to promote a proper face, occasioned by such features as root protrusions, rot, cat faces, etc.

THIRTIETH HIGHEST HOURLY VOLUME *See* **Volume.**

THIXOTROPY Property of some gels to become fluid when stirred and of returning to a jelly-like state at rest.

thk *Abbreviation for:* thick.

thkns *Abbreviation for:* thickness.

thm. *Abbreviation for:* thimble.

T-HOOK Attachment device used for towing.

thou. *Abbreviation for:* thousand.

THOUSAND BOARD FEET Unit of measurement equal to 1,000 ft (304.8 m) of wood having a thickness of 1 in. (25 mm).

THPFB *Abbreviation for:* treated hard-pressed fiberboard.

THREAD 1. Spiral or helical ridge of a screw, bolt, nut, etc.; **2.** To reeve a line through blocks or carriage. *See also* **Lace.**

THREADBAR High-tensile alloy-steel bar with a coarse-pitch thread cold-rolled along its length.

THREADED ANCHORAGE Anchorage device that is provided with threads to facilitate attaching a jacking device and to effect anchorage.

THREADER Device or tool used to cut threads on the end of a pipe.

THREADING *See* **Lacing.**

3C *See* **Lumber grades.**

3/C *Abbreviation for:* three conductor.

THREE-CORE BLOCK Concrete masonry unit having three cores.

3-D *Abbreviation for:* three dimensional.

THREE-HINGED ARCH Arch having hinges at crown and abutments.

THREE-HINGED PORTAL FRAME Portal frame in which identical halves are hinged at the apex with the two vertical legs free to move relative to the floor in pockets. *See also* **Portal frame.**

THREE-LEG INTERSECTION *See* **Intersection types.**

THREE-PART LINE *See* **Parts of line.**

THREE-PART SEALANT Chemically cured sealant supplied in three parts, one containing the reactive polymer base, one its curing agent, and one the color agent. Usually supplied in separate containers.

3-ph. *Abbreviation for:* three-phase.

THREE-POINT SUSPENSION *See* **Suspension.**

THREE-POSITION VALVE *See* **Valve.**

THREE-QUARTER BAT A brick with one end cut off, usually measuring 150 mm (6 in.) long.

THREE-QUARTER HEADER Header brick of length equal to three-quarters of the wall thickness.

THREE-QUARTER S-TRAP Sanitary trap that provides a water seal and that has a vertical inlet and an outlet that is 45° to the horizontal.

THREE-SPEED AXLE *See* **Axle.**

THREE-SPEED TANDEM AXLE -*See* **Axle.**

3/W *Abbreviation for:* three way.

THREE-WAY CLAMP *See* **Clamp.**

THREE-WAY LEAD *See* **Lead.**

THREE-WAY SWITCH *See* **Switch.**

THREE-WAY VALVE *See* **Valve.**

thres. *Abbreviation for:* threshold.

THRESHOLD Horizontal member over which a door opens. *See also* **Molding.**

THRESHOLD LIMIT VALUE Values of airborne toxic materials that are to be used as guides in control of health hazards and which present time-weighed concentrations to which all workers may be exposed eight hours per day over extended periods without adverse effects.

THRIBLE Three sections of drill pipe handled as a unit.

THROAT Opening at the top of a fireplace through which the smoke passes to the smoke chamber and chimney.

THROATING Projecting molding undercut to produce a drip.

THROTTLE Mechanism that actuates an engine carburetor, feeding more fuel/air mixture to the engine.

THROTTLE TRIGGER Control on a chainsaw that adjusts the speed of the engine.

THROTTLE TRIGGER INTERLOCK Lever that must be depressed before the throttle trigger of a chainsaw can be activated.

THROTTLE VALVE Quick-acting valve mounted in the fluid line, used to control the flow of fluid to a steam/air pile hammer.

THROUGH BOND In masonry, a bond that extends across from face to back of a wall.

THROUGH CAR Elevator car having entrances on opposite walls.

THROUGH CUT Cut with up-slopes on both sides.

THROUGH HIGHWAY *See* **Highway.**

THROUGH LINTEL Lintel that extends the full thickness of a wall.

THROUGH LOT Building lot other than a corner lot having frontage on two public highways or streets. *Also called* **Merged lot,** and **Double-frontage lot.**

THROUGHPUT CAPACITY Maximum volume that a process or device can handle over a given period; the amount of waste a process can handle within a given time period, for instance.

THROUGH SHAKE Separation of wood between annual growth rings that extends between two faces of the timber.

THROUGH STONE Bonding stone that appears on both faces of a wall.

THROUGH STREET *See* **Street.**

THROUGH TRAVELED WAY Portion of the roadway for the movement of vehicles, exclusive of shoulders and auxiliary lanes. *See also* **Traveled way.**

THROW 1. Longest straight distance moved in the stroke of a reciprocating or rotary part; **2.** Distribution of blast fragments.

THROW-AND-HEAVE Displacement of rock resulting from detonation of an explosive and the resulting expansion of gases.

THROWOUT BEARING *See* **Bearing.**

thru *Abbreviation for:* through.

THRUST Force tending to push outward.

THRUST ARM Cable-controlled bar able to slide under power in two directions.

THRUST BEARING *See* **Bearing.**

THRUST BLOCK Concrete poured behind a bend or angle of a pipe to support the pipe against the thrust of fluids being transported through the pipe. *Also called* **Kicker block.**

THRUST BORER Equipment that drills an underground hole, primarily to insert pipes or cables.

THRUST RING Mild-steel plate cast into the end of a prestressed concrete member to form an even bearing for the anchor plate.

THRUST WASHER Hardened steel washer that holds a rotating part from sideways movement in its bearing.

THUMB LATCH Door catch operated by thumb pressure on a lever.

THUMBSCREW Adjustment screw that can be easily turned with thumb and finger pressure.

THUMBTACK Tack with a large head and sharp point.

TIDAL MARSH Low flat marshlands traversed by interlaced channels and tidal sloughs and subject to tidal inundation; normally, the only vegetation present is salt-tolerant bushes and grasses.

TIE 1. Loop of reinforcing bars encircling the longitudinal steel in a column. *See also* **Bent bar, Hook, Hooked bar,** and **Stirrup**; **2.** Tensile unit adapted to holding concrete forms secure against the lateral pressure of unhardened concrete; **3.** Any item that connects masonry to masonry or other materials; **4.** Tension/compression member used to securely attach a scaffold to a structure; **5.** Two or more scaled distances from known points that define a third point; **6.** Temporary bottom truss chord brace that is omitted if a ceiling is attached directly to the bottom chords and provides adequate lateral support.

TIEBACK 1. Rod fastened to a deadman; **2.** Rigid foundation; **3.** Rock or soil anchor to prevent lateral movement of formwork, sheet pile walls, retaining walls, bulkheads, etc. *See also* **Pile, anchor**; **4.** To use a twister.

TIE BAR Bar at right angles to and tied to reinforcement to keep it in place, designed to hold abutting edges together without transfer of load.

TIE BEAM Horizontal, transverse beam in a roof, connecting the rafters at the height of the wall plate.

TIED COLUMN Column laterally reinforced with ties.

TIED JOINT Joint where movement between adjacent slabs or parts is restricted by dowels bonded with the concrete.

TIED OUT Process of determining the fixed location of existing objects (manholes, meter boxes, etc.) in a street so that they may be uncovered and raised following paving.

TIEDOWN ASSEMBLIES Any of a wide range of mechanical arrangements used to secure a load or to hold a piece of equipment or materials in place, sometimes permanently but more commonly only for a limited duration or until a permanent measure is installed. *Also called* **Cargo control,** and **Restraint equipment.**

TIER 1. Any of a series of rows or ranks arranged one above or behind another. *See also* **Lifts**; **2.** One of the 100 mm (4 in.), or one-brick layers in the thickness of a wall. *See also* **Wythe.**

TIERING Process of placing one load upon another with each successive load resting directly on the one below.

TIE ROD 1. Connecting rod between the steering arms of the front axle; **2.** Rod, threaded for a short distance at both ends (may be left- and right-hand threads); **3.** Tension members between sets that maintain spacing and which pull the sets against the struts; **4.** Axial external hydraulic cylinder rod that traverses the length of the cylinder. It is prestressed at assembly to hold the ends of the cylinder against the tubing. The rod extensions can be a mounting device.

TIE WIRE Galvanized wire, generally #16, #15, or #14 gauge, used in construction work.

Tig *Abbreviation for:* tungsten invert gas.

TIGHT 1. Rock-earth formations without natural weaknesses, that may require hard ripping or light blasting before excavation; **2.** Blasts or blastholes around which rock cannot break away freely.

TIGHT BRAID 1. Unevenness in a braid reinforcement caused by one or more ends of the reinforcement being applied at a greater tension than the remaining ends; **2.** Localized necking down of braided reinforcement caused by a stop on the braiding operation.

TIGHT FIT *See* **Fit classifications.**

TIGHT KNOT Knot held firmly in the wood around it.

TIGHTLINE To obtain maximum lift on yarding lines by holding one line back and pulling on another.

TIGHTLINING Method of high-lead cable yarding in which the haulback line supports the butt rigging and makes it possible to lift the butt rigging and its load over obstacles.

TIGHT SHEATHING Diagonal matching boards nailed to studs or rafters.

TIGHT SIDE Face of a sheet of sliced veneer that was not in contact with the blade while being cut.

TILE 1. Ceramic surfacing unit, usually relatively thin in relation to facial area, made from clay or a mixture of clay and other ceramic materials, called the body of the tile, having either a glazed or unglazed face and fired above red heat in the course of manufacture. **2.** *See* **Roof tile**; **3.** Clay tile and building units, manufactured for a wide range of purposes and in a variety of styles, including:

> **Facing:** Tile used in exposed masonry construction.
>
> **Land:** Short sections of porous pipe having open butt joints, used for subsoil drainage.
>
> **Load-bearing:** Tile designed for masonry construction carrying superimposed loads.
>
> **Non-load-bearing:** Tile used in masonry construction carrying no superimposed loads
>
> **Screen:** Perforated tile used in screen walls.
>
> **Sewer:** Glazed waterproof clay pipe with bell joints.

TILE BED *See* **Leach field.**

TILEHANGING Wall covering of overlapping rows of roofing tiles on a timber frame.

TILE SHAPE Any of various shapes of structural tile that adapt to special needs in the construction of structural clay tile walls.

TILE SHOE Box towed behind a ditching machine through which land tiles can be laid on a ditch bottom. *See also* **Shoe.**

TILE SPACER Tool used during installation of wall or floor tiles that ensures a constant-sized grout joint.

TILL Dense glacially deposited soil formations consisting of a heterogeneous mixture of fine-grained and coarse-grained material, often including significant quantities of boulders and cobbles.

TILT 1. Longitudinal angular displacement; **2.** Of the mast structure, forks, or carriage of an industrial truck, any variation from vertical or horizontal.

TILT-BED TRAILER *See* **Trailer.**

TILT BLADE Dozer blade that can be tilted in respect to a vertical position.

TILT CAB *See* **Cab.**

TILT CYLINDER Cylinders used to change the attitude of a structure or body.

TILT FRAME *See* **Roll-off body.**

TILT GATE Crest gate on the spillway of a dam. It is opened by water pressure when the upstream water level exceeds a defined level, closing automatically when the water is below that level.

TILTH Tilled or cultivated land.

TILTING DOZER Tractor- or wheel-dozer whose blade can be pivoted on a horizontal center pin to cut low on one side or the other.

TILTING MIXER Revolving-drum mixer that discharges by tilting the drum about a fixed or movable horizontal axis at right angles to the drum axis.

TILT SENSOR Device that senses how out of level a surveying instrument is and corrects the vertical observations being displayed.

TILT-SLAB WALL UNIT Reinforced concrete wall slab containing all door and widow openings, cast in a horizontal position on a deck and then tilted by lifting into a vertical position.

TILT-UP CONSTRUCTION Technique of casting walls and floors on a central deck, then tilting and raising the various components into their permanent position as they reach a designed yield strength.

TIMBER 1. Standing trees of commercial size; **2.** Felled trees or logs suitable for sawing; **3.** Lumber with a cross-section greater than 100 x 150 mm (4 x 6 in.). *See also* **Lumber.**

TIMBER APPRAISAL Economic appraisal of the monetary value of a timber stand.

TIMBERBIND Tension or 'spring' in a log or cant resulting from growth stresses.

TIMBER CRUISING Collection of field data on forests, commonly by the measurement and recording of information in sample plots. Includes the measurement and estimation of volumes of standing trees.

TIMBER FALLER-BUCKER *See* **Faller.**

TIMBER FRAMING Load-carrying frame of appropriately sized timbers, used in frame construction.

TIMBER HITCH Type of knot used for twisters.

TIMBERING Support of the ground in excavations, whether of wood, steel, concrete, or other materials.

TIMBERING SET Tunnel support consisting of a roof beam or arch and two posts.

TIMBER MARK Hammer indentation made on cut timber for identification purposes.

TIMBER PILE *See* **Pile.**

TIMBER PRODUCTS OUTPUT Timber products cut from roundwood and by-products of wood manufacturing plants.

TIMBER REMOVAL Plant byproduct (such as pulpwood chips, obtained incidental to the production of other manufactured products) or plant residue (wood material from manufacturing plants not utilized as a product).

TIMBER REMOVAL FROM GROWING STOCK Volume of sound wood in live sawtimber, forest products (including roundwood products and logging residues), and/or other removals.

TIMBER REMOVAL FROM SAWTIMBER Net board-foot volume of live sawtimber trees removed annually for forest products (including roundwood products and logging residues) and other removals, such as growing-stock trees removed by cultural operations (timber stand improvement work, land clearing, and changes in land use).

TIMBER SINGLE-POST SHORE Timber used as a structural member for shoring support. *See also* **Adjustable timber single-post shore, Fabricated single-post shore,** and **Post shore.**

TIMBER STAND IMPROVEMENT Intermediate thinning of a forest stand, prior to its reaching mature rotation age, generally for the purpose of improving growing conditions or controlling stand composition.

TIMBER VOLUME *See* **Volume of growing stock,** and **Volume of sawtimber.**

TIME CONSTANT Time required to change from one condition to another, usually the time to complete 63.2% of the total rise or decay.

TIME CYCLE 1. Time required for one complete sequence of events; **2.** *See* **Traffic signal.**

TIME-DEPENDENT DEFORMATION *See* **Deformation.**

TIME FOR COMPLETION *See* **Contract time.**

TIME FRAME Time necessary to do something, or the time that something is due to happen, put in perspective of the time of related circumstances.

TIME LAG 1. Time interval between two closely related events or phenomena; **2.** Time necessary for a fuel particle to lose approximately 63% of the difference between its initial moisture content and its equilibrium moisture content.

TIME-LAG FUSE Fuse designed to momentarily absorb large amounts of surge currents without failing, but to fail if such currents persist.

TIME OF COMPLETION Date specified in the contract documents for substantial completion of the works.

TIME OF HAUL In production of ready-mixed concrete, the period from first contact between mixing water and cement until completion of discharge of the freshly mixed concrete.

TIME OVERRUN Time taken to complete a phase or a task beyond that which had been scheduled. *Also called* **Construction time overrun,** and **Overrun.**

TIME SCHEDULE Sequence of points in time at which circumstances will/must occur.

TIMING DEVICE Device responsive to engine speed and/or load to control the timed relationship between fuel injection cycle and engine cycle.

TIN White lustrous metal, typically used to coat sheet steel. *See also* **Alloying elements.**

tinct. *Abbreviation for:* tincture.

TINE Sharp projecting point, typically positioned on equipment to break the surface of the ground, or to rake a surface.

TINGLE Small lead, copper, or zinc strip used to secure panes of glass or roofing slates when making a repair.

TINKER'S DAM In plumbing, a small dam made to enclose a work area that is to be flooded with solder.

TINNING Thin solder coating applied to metals to be soldered.

TINPLATE Bright sheet steel coated on both sides with a thin film of tin.

TIN SNIPS Hand shears designed to cut sheet metal. There are many configurations, including:

> **Aviation:** Shears fitted with a spring-action, self-opening movement. Available in straight, left- and right-hand models. Used for cutting both straight and curved lines.
>
> **Duckbill:** Used to cut straight lines.
>
> **Hawk's bill:** Capable of cutting small-diameter circles without distortion.
>
> **Universal:** Used to cut straight as well as curved lines.

TINT Color made by mixing a small amount of colored pigment in a white base.

T-INTERSECTION *See* **Intersection types.**

TIP 1. Leading end of something; the end having a lesser dimension or shape; **2.** End of a welding torch where the gas burns, producing the high-temperature flame; in resistance welding, the ends of the electrodes; **3.** *See* Pile tip; **4.** Nozzle end for changing the size of the orifice of a hose stream.

TIP ELEVATION *See* **Pile penetration.**

TIP GRADE Toe line of piles.

TIPPING BAY Opening, typically 4 to 5 m (12 to 15 ft) wide, that allows vehicles carrying wastes to discharge their load into a storage pit or transfer station hopper.

TIPPING CONDITION When balance is reached between the overturning moment of the load and the stabilizing moment of the crane.

TIPPING FEE Charge to unload waste materials at a transfer station, processing plant, landfill, or other disposal site.

TIPPING FLOOR Unloading area for vehicles that are delivering waste materials to a transfer station, incinerator, or other waste processing plant.

TIPPING FULCRUM Horizontal line about which a crane or derrick will rotate should it overturn.

TIPPING LOAD Load for a particular operating radius that brings a lifting device to the point of incipient overturning.

TIPPING WAGON Small wagon pivoted for side or end dumping, running on narrow-gauge track.

TIP RESISTANCE *See* **End resistance.**

TIP SECTION Outer-most live segment of a telescopic boom. It is attached to the middle section and contains the head machinery of the attachment.

TIR *Abbreviation for:* technical information release.

TIRE Air-filled flexible casing attached to the rim of a wheel.

TIRE CARRIER Rack or other device attached to a vehicle, used to carry one or more spare tires.

TIRE CHAINS Chain arrangement that can be fasted around the perimeter or road-contacting surface of a mounted tire to provide increased traction on slippery or loose material.

TIRE CLEARANCE Space between tires and the nearest part of the body or underconstruction.

TIRE CONSTRUCTION The two principal tire construction types used on most construction and construction-related equipment are the bias ply and the radial ply. The principal features of each type are:

BIAS PLY

> **Bead:** Each of the tire beads consist of bundles of steel wire (three in larger tires) that are forced laterally by tire inflation pressure and wedge the tire firmly on the rim's tapered bead seat. The nylon plies tie into the bead bundles and the forces inherent in the tire are transmitted from the rim through the bead bundles into the nylon.
>
> **Body ply:** One of several layers of rubber-cushioned nylon cord comprise the carcass of the tire. Alternating plies of cord cross the tread centerline at an angle (bias). The term 'ply rating' is an index of tire strength and is not the actual number of plies in the tire.
>
> **Breaker or tread ply:** These, if used, are confined to the tread area of the tire and are intended to improve carcass strength and provide additional protection to the body plies. Some 'work' tires employ steel breakers or belts to give further protection to the carcass.
>
> **Inner liner:** This is the sealing medium that retains the air in the tire and, combined with the O-ring seal and rim base, eliminates the need for inner tubes and flaps. *See also* **Liner.**
>
> **Sidewall:** Comprises several protective layers of rubber covering the body plies in the sidewall area.
>
> **Tread:** The wearing part of the tire that contacts the ground. It must transmit the vehicle or machine weight to the ground as well as provide traction and flotation.
>
> **Tubes and/or flaps:** There are a few applications where tire life may be improved by the use of tubes and flaps.
>
> **Undertread:** Protective cushion of rubber lying between the tread and the body ply.

RADIAL PLY

> **Bead:** A single bead bundle of steel cables or steel strip (spiraled like clock spring) comprise the bead at each rim interface.
>
> **Belt:** One of several layers or plies of steel cable

form the belts that underlie the tread area around the tire circumference. The cable in each belt crosses the tread centerline at an angle with the angle being reversed from the preceding belt.

Radial carcass: This consists of a single layer or ply of steel cables laid archwise (on the radian), bead to bead.

Sidewall: Comprises several protective layers of rubber covering the body plies in the sidewall area.

Tread: The wearing part of the tire that contacts the ground, transmitting machine weight and providing traction and flotation.

Undertread: Protective cushion of rubber lying between the tread and the steel belts.

TIRE FLEXING Tire squashing under load.

TIRE LOAD CAPACITY Maximum recommended load that may be carried by a vehicle's tires.

TIRE LOADED RADIUS Distance from the center of a wheel to the road with the tire loaded to rated capacity. Static radius applies when the vehicle is at rest; rolling radius for a vehicle when in motion. The latter figure is usually slightly greater than the static radius and is the figure used in determining the tire revolutions per mile.

TIRE PAINT A black paint, compatible to tire bodies, used to enhance appearance.

TIRE PENETRATION Depth a tire sinks into the ground.

TIRE REVOLUTIONS PER MILE Number of times a tire revolves while traveling one mile.

TIRE SIZE Specified by approximate tread width, diameter of wheel (rim), and tire ply rating. For example: 9.00 x 20 10 ply is a 200-mm (9-in.) wide tire of 10 ply construction on a 500-mm (20-in.) diameter wheel.

TIRE SIZE MARKING One of a series of designations that appear on the side of a tire to indicate basic dimensions.

TIRE TYPES Off-the-road tires are classified into three categories:

Load and carry: Tires for wheel loaders in transporting as well as digging.

Transport: Tires designed for earth-moving machines in transporting material.

Work: Tires normally fitted to tractive-type earth-moving machines such as wheel-type tractors and loaders.

TIRFOR Type of cable puller for swing stages.

TITANIUM *See* **Alloying elements.**

TITANIUM OXIDE White pigment, considered to have the greatest hiding power of all white pigments.

TITLE Documents indicating legal ownership.

TITLE INSURANCE Insurance protecting owners or lenders against loss of interest due to legal defects in title.

TITLE PANEL Panel at the lower right of a drawing giving such standard information as (a) title of draw-ing, (b) contract/job name or number, (c) client, (d) scale, (e) date drawing approved, (f) drawing number and revision identification, (g) detailer, (h) checker.

TITLE SEARCH Review and report on the records of a title.

TITLE SHEET First sheet of a set of drawings, or of a legal document such as a contract, establishing principal data such as the name and location of the project, the owner, name of the designer, etc.

tk *Abbreviation for:* tank.

TK *Abbreviation for:* turn knob.

tk. off *Abbreviation for:* take off.

tk/w *Abbreviation for:* tack weld.

tl. *Abbreviation for:* truckload.

TL *Abbreviation for:* throw latch; total load; transmission loss.

t-lamp *Abbreviation for:* tail lamp.

tll *Abbreviation for:* technical life length.

tlr *Abbreviation for:* trailer.

tly *Abbreviation for:* tally.

TM *Abbreviation for:* technical manual; trademark; training manual; treated millwork.

tmd *Abbreviation for:* timed.

TMS *Abbreviation for:* to template with machine screws.

tn *Abbreviation for:* town.

TN *Abbreviation for:* true north.

tnl *Abbreviation for:* tunnel.

t no c *Abbreviation for:* threads no couplings.

TNT *Abbreviation for:* trinitrotoluene.

tntv. *Abbreviation for:* tentative.

TO *Abbreviation for:* takeoff.

T/O *Abbreviation for:* table of organization.

TOE 1. Part of a rafter that does not project over the plate; **2.** Bottom of a sloped surface where the angle changes from vertical, or inclined toward vertical, to horizontal or inclined toward horizontal; **3.** Burden, or distance, between the bottom of a borehole to the vertical free face of a bench in an excavation; **4.** End of a blast hole where explosive is placed; **5.** Bottom of the working face at a sanitary landfill.

TOE BOARD Vertical barrier at floor level erected along exposed edges of a floor opening, wall opening, platform, runway, or ramp to prevent material from falling.

TOE GUARD 1. Guard used to protect a passenger from catching his foot under a hoistway or elevator car sill projection; **2.** Plate fastened to a hoistway edge at a landing sill, or to the edge of an elevator car platform beneath, and in line with the entrance.

TOE-IN Alignment of wheels so that they are closer together at the back than at the front.

TOE JOINT Joint between a horizontal member and another at a vertical angle, typically a rafter on a plate.

TOE LEVEL Depth to which piles are to be driven.

TOE NAILING Driving a nail at a slant to the initial surface so that it can penetrate a second member.

TOE OF DAM Junction of the downstream face of a dam with the ground surface.

TOE OF THE FILLET Termination point of a fillet weld or rolled section fillet.

TOE STEEL Lateral reinforcement for a distance from the toe of a precast concrete pile, intended to prevent damage during driving.

TOGGLE 1. Rod or pin that can be inserted through the links of a chain or a loop in wire rope or rope to cause, or prevent, rotation; **2.** Lever or pin that rocks from side to side, usually to engage and disengage something at the end of each stroke;**3.** *See* **Switch, toggle. 4.** *See* **Wall anchor.**

TOGGLE SWITCH *See* **Switch.**

toil. *Abbreviation for:* toilet.

TOILET SEAL Wax or putty ring used to seal the joint between a water closet and the pipe on which it sits.

tol. *Abbreviation for:* tolerance.

TOLERANCE 1. Permitted variation from a given dimension, quality, or quantity; **2.** Range of variation permitted in maintaining a specified dimension; **3.** Permitted variation from location or alignment. *See also* **Float.**

TOLL Charge made for the use of a facility.

TOLL BRIDGE Bridge open to traffic only upon payment of a direct toll or fee.

TOLL ROAD (OR TUNNEL) Highway or tunnel open to traffic only upon payment of a direct toll or fee.

TOLUENE By-product of coke production or extracted from coal tar, used as a solvent. *Also called* **Toluol.**

TOLUOL *See* **Toluene.**

TOM *See* **Shore.**

TOMMY MOORE Small block with a wide throat, usually used with the strawline as a lead block.

TON (LONG, 2,240 lb) Nonmetric unit of mass. Symbol: ton. Multiply by 1016.046 908 8 to obtain kilograms, symbol: kg; by 1.016 046 908 8 to obtain tonnes, symbol: t. *See also the appendix:* **Metric and nonmetric measurement.**

TON (SHORT, 2,000 lb) Nonmetric unit of mass. Symbol: t(s). Multiply by 907.187 74 to obtain kilograms, symbol: kg; by 0.907 184 74 to obtain tonnes, symbol: t. *See also the appendix:* **Metric and nonmetric measurement.**

TONGS Pair of curved arms hinged to each other, scissor fashion, so that a pull on a chain or ring conecting their short ends will cause the long ends to close and grip an object.

TONG TESTER Clamp-on ammeter used to measure alternating current.

TONGUE 1. Drawbar of a towed vehicle; **2.** Projecting bead on the edge of a board cut to fit into the groove on another piece.

TONGUE-AND-GROOVE 1. Joint in which a protruding bead on the edge of one side fits into a groove in the edge of the other side; **2.** Boards cut or planed to produce a bead on one long edge and a matching groove on the other. *See also* **Keyway.**

TONGUE-AND-GROOVE EDGE Machined panel edge that has the ability to hold firmly with the edge of an adjacent panel.

TONGUE-AND-GROOVE PLIERS *See* **Pliers.**

TON-MILE Movement of a ton of freight or cargo a distance of one mile.

TONNAGE 1. Total weight expressed in tons; **2.** Charge per unit of weight on cargo.

TONNE Non-SI unit of mass, equal to 1000 kg (1 Mg), permitted for use with the SI system of measurement. Symbol: t. Multiply by 2,205 to obtain pounds, symbol: lb; by 0.984 206 5 to obtain long tons (2,240 lb), symbol: ton; and by 1.102 311 to obtain short tons (2,000 lb), symbol: t(s). *See also the appendix:* **Metric and nonmetric measurement.**

TOOL Implement, usually hand-held.

TOOLCARRIER Wheel loader-type equipment fitted with a quick-change attachment and modified as necessary to enable it to carry a wide range of attachments, and to change working attachments quickly and easily, usually without need for the operator to leave the cab. Typically the attachments include a wide range of buckets, forks, and blades, plus brooms, asphalt cutters, hooks, and augers, etc.

TOOLING 1. Working a surface with a tool, usually to produce a pattern or decoration; **2.** Compressing and shaping the face of a mortar or sealant joint with a special tool other than a trowel.

TOOLING UP Assembling and preparing an appropriate selection of tools and equipment to complete a task.

TOOTH 1. Projection, as on a gear wheel or a saw blade; **2.** Texture given to a surface by abrading it so that a subsequent coat will bond to it. *Also called* **Purchase,** and **Mechanical anchorage.** *See also* **Working tool.**

TOOTH BASE Inner portion or base of a two-piece tooth for a digging bucket.

TOOTH BRAKE *See* **Brake.**

TOOTH CHISEL *See* **Cold chisel.**

TOOTHER *See* **Toothing.**

TOOTH-GRAB METHOD Technique in which the operator of an excavator or backhoe may use the bucket teeth to hook or grab objects such as small trees, pipes, reinforcing bars, etc.

TOOTHING Constructing the temporary end of a wall with the end stretcher of every alternate course projecting; each projecting units is a **Toother.**

TOOTH PRESSURE Force each tooth of a saw exerts while cutting.

TOOTH ROTATION Free rotation of the working tool in the working tool holder.

TOOTH WHEEL Wheel containing 100 teeth that, when turning, sends magnetic impulses to an inductive speed sensor, which provides wheel rotation information to the electronic control unit (ECU) of an antilock braking system.

TOP *See* **Harvest functions.**

TOP BEVEL Angle filed across the top of a saw tooth, usually staggered on alternate teeth.

TOP BIND One of five basic conditions affecting the lay of a tree. In top bind, downward pressure is caused by the weight of the suspended tree that is supported on both ends but not in the middle. The top of the tree is compressed while the bottom is under tension. *See also* **Bottom bind, Drop, End pressure,** and **Side bind.**

TOP BLOCK *See* **Head block.**

TOP CAP Retread that covers the crown, or top, of a tire. *Also called* **Top retreading.**

TOP CHORD *See* **Chord.**

TOP CLEARANCE Angle of clearance on the top of a saw tooth.

TOP-COAT SEALER Substance that seals and protects the surface of unglazed tile.

TOP COURSE TILES Roofing tiles sized to position at the ridge and maintain the gauge of the tiles further down the slope.

TOP FORM Form required on the upper or outer surface of a sloping concrete slab or thin shell. *Also called* **Backform.**

TOP HEAD *See* **Head block.**

TOP HEADING 1. Upper section of a tunnel; **2.** Tunnel excavation method where the entire top half of the work is completed before the bottom section is started.

TOP HEADING AND BENCH Method of tunneling where the top heading is carried about 1.5-times the length of one round ahead of the lower heading, or bench.

TOP-HUNG WINDOW Casement window hinged at the top.

TOP LOPPING To cut limbs from downed tree tops so that no limbs are more than a specified length along the tree stem.

TOPO *Abbreviation for:* topographic; topographical.

TOP-OF-CAR INSPECTION STATION Controls on the top of an elevator car used by an elevator mechanic to operate the car at inspection speed.

TOP OF PILE *See* **Pile head.**

TOPOGRAPHICAL SURVEY *See* **Surveying.**

TOPOGRAPHIC INTERPOLATION Determination on a scale drawing of the position of contour lines through interpolation of the data obtained at the intersections of a regularly spaced grid.

TOPOGRAPHIC MAP Map showing ground features and their relative elevation.

TOPOGRAPHY Configuration of the surface of a site; its relief, landforms, and slopes.

TOPPED OUT Moment when the highest piece on a building or structure is fixed in place.

TOP PIN Pin that retains the shackle in a padlock case when unlocked.

TOPPING 1. Layer of concrete or mortar placed to form a floor surface on a concrete base; **2.** Structural, cast-in-place surface for a precast floor and roof system; **3.** Mixture of marble chips and matrix that produces a terrazzo surface; **4.** Fine material forming a dressing or surface layer for a road; **5.** Any material applied to another; **6.** Layer of gradable gravel or other material in the top of the ballast, or sometimes just over the subgrade; **7.** *See also* **Luffing; 8.** Cutting off the top of a felled tree at a predetermined, minimum diameter.

TOPPING COMPOUND Drywall joint compound formulated for the final finishing coat.

TOPPING LINE Support line that is used to hold, raise, and lower a boom. *Also called* **Gantry boom.**

TOPPING OUT 1. Process of laying the last section of masonry to a wall to the finished height of the building; **2.** Erecting, or fastening, the highest member of a roof or completing the roof.

TOP PLATE 1. Top horizontal member of a frame wall supporting ceiling joists, rafters, or other members; **2.** *See* **Anvil.**

TOP RAIL Uppermost horizontal rail of a guardrail system.

TOP RETREADING *See* **Top cap.**

TOP SECTION Uppermost section of a lattice boom or jib that also contains the head machinery of the attachment.

TOP SHADOW LINE Band of dark-colored granules applied to a thickbutt shingle so as to accentuate the shadow cast by the shingles in the course above.

TOPSOIL *See* **Loam,** and **Soil types.**

TOP TOWER Tower mounted above the jibs of some tower cranes, providing means for attachment of the pendants. *Also called* **Towerhead.**

TOP TRAVEL SPEED 1. Maximum horizontal velocity at which a vehicle will operate (subject to a tolerance of ±10%); **2.** Top travel speed of an electric truck, that is usually greater when the truck is empty than when loaded, therefore the word 'empty' or 'loaded' must be added, i.e., 'Top travel speed loaded', or 'Top travel speed empty.'

TOP U-BOLT HANGER Plate that is located on top of a suspension spring and held in place when the U-bolts are tightened. Used to clamp the spring and axle together.

tor. *Abbreviation for:* torsion.

TORCH Tool from which issues the gases used to burn or produce heat in welding.

TORCHING Filling in with a lime mortar the uneven spaces between the undersides of tiles or slates on an unboarded or unfelted roof.

TO RENT To grant possession and enjoyment of property. *See also* **Rent.**

TORICELLI'S THEOREM Statement that the liquid velocity at an outlet discharging into the free atmosphere is proportional to the square root of the head.

TORQUE Turning or twisting force measured in terms of pounds-per-foot (ft/lbs), pounds-per-inch (in./lbs), kilograms-per-centimeter (kg/cm), etc. There are several measures, including:

Engine: Measure of the amount of twist an engine can produce at its crankshaft, as determined on a dynamometer.

Gross: Maximum torque developed by an engine after allowing for the power absorbed by the engine's accessory units, such as the fan, water pump, generator, and exhaust system.

Net: Torque available at the flywheel of an engine after the power required by the engine accessories (fan, water pump, generator, etc.) has been provided.

TORQUE CONVERTER Hydraulic coupling that turns engine horsepower into torque or twisting force. There are three principal variations:

Torque divider: Torque converter with a planetary gear set. By dividing the power flow (the majority goes through the torque converter but some goes through a gear set), the torque divider provides the benefits of both a torque converter and direct drive, i.e., higher efficiency, torque multiplication, and better acceleration.

Twin turbine: Two turbines, one inside the other, work together or separately depending on ground speed. In each gear range both turbines work together to handle the higher initial work loads. As ground speed increases, the transmission up-shifts automatically, allowing the second turbine to work independently.

Variable capacity: Torque converter with two impellers. There is a rotating inner impeller and an outer impeller that can gradually be engaged. The outer impeller is an extension of the inner and is controlled by the operator.

TORQUE CONVERTER TRANSMISSION *See* **Transmission.**

TORQUE DIVIDER *See* **Torque converter.**

TORQUE-HEAD SCREW *See* **Screw.**

TORQUE MOTOR Type of electromechanical transducer having rotary motion, used in the input stages of servovalves.

TORQUE MULTIPLICATION Multiplication of engine torque by use of transmission and rear-axle gears.

TORQUE PIN *See* **Antirotation device.**

TORQUE-PROPORTIONING DIFFERENTIAL *See* **Differential.**

TORQUE RISE Increase in engine torque as the engine is put under increasingly heavy load from high idle to maximum lug.

TORQUE ROD Bar that absorbs twisting strains.

TORQUE-TENSION RELATIONSHIP Wrench torque required to produce specified pre-tension in high-strength bolts.

TORQUE VISCOMETER Apparatus for measuring the consistency of slurries in which the energy required to rotate a device suspended in a rotating cup is proportional to viscosity.

TORQUE WRENCH *See* **Wrench.**

TORSION FORCE *See* **Force.**

TORSIONAL Extremely rapid, back-and-forth oscillation or torsional vibration within a system.

TORSIONAL ANALYSIS Evaluation, either by calculation or test, to determine the level of crankshaft stress caused by torsional vibration and to compare the stress with the manufacturer's design limit.

TORSIONAL STRESS Shear stress on a transverse cross-section resulting from a twisting action.

TORSIONAL VIBRATION 1. Vibration in an operating mechanical system that is excited by the elasticity or springiness of the various parts in the system; **2.** Vibration at the flywheel created by the engine cylinder pressure pulses during fuel burning.

TORSION SPRING *See* **Spring.**

TORTUOSITY Ratio of the average effective flow path length to minimum theoretical flow path length (thickness) of a filter medium.

tot. *Abbreviation for:* total.

TOTAL COST BIDDING Method of establishing the purchase price for movable equipment whereby the buyer is guaranteed that maintenance shall not exceed a set maximum amount during a fixed period of time (typically, five years) and that the equipment will be repurchased by the seller at a set minimum price at the end of the period agreed upon.

TOTAL DYNAMIC HEAD Same as total head and is the dynamic suction head plus the dynamic discharge head.

TOTAL ENERGY 1. *See* **Energy**; **2.** Process whereby independent users generate on-site power and utilize exhaust heat and jacketed water heat in addition to electricity generated.

TOTAL FLOAT Amount of time that an activity can be lengthened without delaying the project completion, assuming that all other activities are done in their normal time.

TOTAL HEAD *See* **Head.**

TOTAL LOAD Sum of all loads imposed or supported.

TOTAL LOSS 1. Destruction beyond repair or salvage; **2.** Destruction sufficient to require maximum settlement of an insurance claim.

TOTAL MOISTURE Total of surface and internal moisture present in an aggregate, generally expressed as a percentage of the aggregate weight.

TOTAL OPERATING WEIGHT Weight of an item of mechanical equipment, including all fitted attachments and optional equipment plus the load it is designed to transport.

TOTAL PERFORMANCE Condition achieved when all work, including all deficiencies listed at the time of the substantial performance inspection, is completed and accepted

TOTAL PRESSURE *See* **Pressure.**

TOTAL RISE (OF A ROOF) Vertical distance between the plate line and ridge.

TOTAL RUN Horizontal distance that a rafter spans.

TOTAL STATISTICAL COUNT Raw count multiplied by a counting calibration factor.

TOTAL STRESS Sum of the net stress across points of soil particles at a given point in a soil mass, plus the pore water pressure at that point.

TOTAL TIME Total elapsed time for the period under consideration: total time for a period of one week is 168 hours (7 days multiplied by 24 hours per day).

TOTAL TREE Tree with crown, main stem, and taproot. Does not include the lateral roots.

TOTAL WATER DEMAND Volume of water required by a complete plumbing system, i.e., volume required to fill all the piping beyond the water meter and to operate all of the attached fixtures.

TOTE BARREL *See* **Waste container.**

TOTE BOX Small- to medium-sized rectangular box used for the collection of solid waste recyclables.

TOUCH-SANDED PANEL Structural wood panel 'sized' to uniform thickness by light surface sanding during manufacture.

TOUGHNESS Property of matter that resists fracture by impact or shock.

TOW ARM One of two rigid attachments, one on each side of a towed machine, typically attaching a tractor to a screed.

TOW BAR Device for positioning a towed vehicle behind a recovery vehicle.

TOW CHAIN Chain assembly used as a primary coupling between towing and towed vehicles, distinct and separate from a safety chain.

TOWER 1. Structure that is relatively high for its length and width; **2.** Composite structure of frames, braces and accessories; **3.** Tower crane mast, comprised of a number of individual frames pinned together and supported by a permanent or removable base; **4.** Steel mast used instead of a spar tree at the landing for cable yarding.

TOWER ATTACHMENT Crane attachment, usually adaptable to a basic crane, consisting of a vertical tower with a working boom and/or jib affixed to the upper part of the tower.

TOWER HEAD *See* **Top tower.**

TOWER LOADER *See* **Loader.**

TOWING CAPACITY Total weight (trailer or trailers plus their cargo) a vehicle is designed to tow.

TOWING WINCH *See* **Winch.**

TOWNSHIP 1. Area of 36 square miles (792 ha) between range and township lines, divided into 36 sections; **2.** Division of a county having certain powers of government; **3.** Land-survey area on which later subdivisions may be based.

TOWNSHIP BUILDING CODE *See* **Building code.**

TOWNSITE 1. Site of a town; **2.** Area of land being developed or available to be developed as a town.

TOWPATH Path along the bank of a canal or river, used by vehicles for towing vessels.

TOW POINT Point on the side of a tractor where the tow arms are attached. The location is usually at or near the midpoint of the tractor. In paving, if automatic grade controls are to be used, the tow point can be displaced vertically by hydraulic cylinders as dictated by the control device.

TOW SLING Device used for lifting and towing vehicles with the load supported on rubberized belts and chains.

TOW VEHICLE Vehicle used to lift/tow other vehicles or machines.

tp *Abbreviation for:* tail pipe; top.

tp. *Abbreviation for:* type.

TP *Abbreviation for:* total pressure; treated and primed.

TPH *Abbreviation for:* toilet-paper holder.

T-PLATE *See* **Brace and plate.**

tpr *Abbreviation for:* taper.

tr. *Abbreviation for:* trace; transitive; transpose; tread; trim.

TR *Abbreviation for:* tie rod; toothed ring.

trac. *Abbreviation for:* tractor.

TRACE Line on a vibration record.

TRACE AMPLITUDE Amplitude of a seismic wave on any of the traces of the vibration record.

TRACER *See* **Punch.**

TRACERY Ornamental work of interlacing or branching lines, as in a Gothic window.

TRACK 1. Assembled crawler tread shoes and connecting pins around idler rollers and drive sprockets; that part of crawler that contacts the ground; **2.** Metal channels used in suspended ceilings.

TRACK CABLE Main cable of a cableway on which the cableway carriage travels.

TRACK CHAIN Sprocket-driven chain that passes across the track rollers of a crawler-equipped machine and to which the grouser pads are connected.

TRACK DRILL *See* **Drill.**

TRACK FRAME Side frame in a crawler mounting to which the track roller and idler are attached.

TRACKING 1. Grooves or wear lines formed on a road surface by traffic; **2.** Movement by a solar collector to follow the sun in its orbit.

TRACKING COLLECTOR *See* **Solar collector.**

TRACK LOADER Mechanical loader on a tracked undercarriage.

TRACK PIN Hinge pin connecting two sections or shoes of a crawler track. *See also* **Master pin.**

TRACK ROLLER Small wheels under the track frame of a crawler-equipped machine that rest on the track.

TRACK ROLLER FRAME Solid or box-section frame that supports the track components, connected at the front to the mainframe by an equalizer spring, equalizer bar, or a hardbar. The rear of the frame is connected to the mainframe by live axles through the final drives or pivot shafts independent of the final drives.

TRACK SHOE Metal plate attached to the track that with others helps distribute the weight of the machine to the ground.

TRACK SHOVEL *See* **Shovel.**

TRACK-TYPE STEERING *See* **Steering.**

TRACK WHEEL One of a set of small, flanged steel wheels resting on a crawler track and supporting a track frame.

TRACTION 1. Total driving force of a vehicle on a given surface; 2. Amount of friction between the tires or tracks and the surface traveled on; 3. Friction developed between the drive sheave and hoist ropes; 4. One of two methods by which an elevator is moved, whereby the elevator is 'pulled' up by cables. *See also* **Hydraulic.**

TRACTION DRIVE *See* **Driving machine.**

TRACTION DRUM Drum that does not accumulate a suspension rope but is designed to climb the rope by application of a friction force between the rope and the drum.

TRACTION MACHINE Electric machine in which the friction between the hoist ropes and machine sheave is used to move an elevator car.

TRACTION RIB Metal or wood strip fastened at intervals to a ramped surface to provide improved traction for equipment moving up, or down the ramp.

TRACTION SHAFT Horizontal shaft in a crawler lower that transfers power from the gear train in the upper to the track mechanism of the carrier.

TRACTION STEEL Wire ropes, used mainly for hoist ropes.

TRACTIVE EFFICIENCY Measure of the proportion of a vehicle's weight resting on its tracks or wheels that can be converted into motion.

TRACTIVE EFFORT Motive force, exerted at the circumference of the driving wheels at the point of contact with the ground.

TRACTIVE FACTOR Tractive effort per given measure of gross vehicle weight: a means of measuring the performance potential of a truck, tractor, or machine. *See also* **Performance factor.**

TRACTIVE RESISTANCE Frictional resistance to motion.

TRACTOR 1. Motorized vehicle equipped with wheels or tracks, used for towing and/or operating fixed equipment or attachments; 2. Truck with a fifth-wheel mounted on the rear frame that supports and pulls a single- or tandem-axle trailer called a semitrailer; 3. Vehicle designed for pulling loads greater than the weight actually applied to the vehicle; 4. Engine-powered (gas or diesel) main paving unit that can be equipped with either track drive assemblies or pneumatic drive wheels and solid-rubber front steering wheels. The tractor provides the power to store and meter material to the screed and to pull the screed to spread, level, smooth, and compact the material being placed.

TRACTOR BREAKAWAY VALVE Coupling between a tractor and the trailer it is pulling that provides an air supply to the trailer emergency system for normal operating conditions. In case of trailer brake system failure, the breakaway valve automatically seals off the flow of air pressure from the tractor braking system and activates the trailer emergency brake. In conjunction with the breakaway valve, a dash-mounted control is used to charge the trailer brake system reservoir for normal operation. In the event of loss of air pressure in the normal braking system, this manual control can be used to seal off the tractor brake system.

TRACTOR-DRAWN Vehicle equipped to be attached to the fifth-wheel of an independently powered vehicle.

TRACTOR-DRIVEN GENERATOR Electric generator so constructed that its rotor is driven by a PTO.

TRACTOR LOADER *See* **Loader.**

TRACTOR-PULLED SCRAPER Scraper towed by a tractor, used for loading, hauling, and discharging material.

TRACTOR-SCRAPER *See* **Scraper.**

TRACTOR-SHOVEL Wheel- or track-mounted shovel.

TRADE 1. Specialized occupation recognized by certification, or training, or membership of a trade/craft organization, or any combination; 2. Members of an organization having common skill levels.

TRADE DISCOUNT *See* **Discount.**

TRADE OFF In an economic analysis or assessment of design alternatives, certain products or certain input resources that are so interrelated that increasing one reduces the others. Between such products or resources there is a 'trade-off' cost or revenue as one is substituted for another. In preparing a building design, for example, there is a relationship between heating capacity and the amount of thermal insulation that can be translated into a set of 'trade-off' costs that vary over the range of feasible heating capacity values for a specific temperature range.

TRADE PRACTICE Way in which something is done within a particular trade or craft.

TRADE UNION Organization representing members having same or similar skills and/or qualifications, commonly to promote such interests as remuneration, benefits, hours of work, etc.

TRADITIONAL MASONRY Masonry in which design is based on empirical rules that control minimum thickness, lateral support requirements, and height without a structural analysis.

TRAFFIC-ACTUATED SIGNAL *See* **Traffic signal.**

TRAFFIC CIRCLE Intersection of several roads where traffic flow is regulated through a one-way circle about a large hub. *Also called* a **Rotary.**

TRAFFIC CONE Cone-shaped device fabricated of easily seen and pliable material, used to define traffic routes and boundaries, usually on a temporary basis, commonly while maintenance and repair work is undertaken.

TRAFFIC CONTROL DEVICE Sign, signal, marking, or other device placed on or adjacent to a street or highway by authority of a public body or official having jurisdiction to regulate, warn, or guide traffic.

TRAFFIC CONTROL SIGNAL Any device, whether manually, electrically, or mechanically operated, by which traffic is alternately directed to stop or permitted to proceed.

TRAFFIC LANE *See* **Lane.**

TRAFFIC MARKING Line, pattern, word, color, or other device, except signs, set into the surface of, applied upon, or attached to the pavement or curbing or to objects within or adjacent to the roadway, officially placed for the purpose of regulating, warning, or guiding traffic.

TRAFFIC OPERATION PLAN Program of action designed to improve the use of a highway, street, or road network, through the application of the principles of traffic engineering.

TRAFFIC SERVICES Operation of highway facilities and incidental services to provide safe, convenient, and economical highway transportation.

TRAFFIC SIGN Device mounted on a fixed or portable support displaying a specific message by means of words or symbols, officially erected for the purpose of regulating, warning, or guiding traffic.

TRAFFIC SIGNAL Power-operated traffic control device by which traffic is regulated, warned, or alternately directed to take specific actions. There are many types, including:

Cycle time: Elapsed time for a complete cycle of events indicated by illuminated or movable traffic direction signals.

Detector: Device that senses the presence of a vehicle either passing a given location, or approaching a location.

Emergency signal: Any type of traffic control device installed solely for the purpose of assigning right-of-way to emergency vehicles at locations where standard traffic control devices are unwarranted.

Green time: That period of any phase assigning right-of-way to the movement(s) of vehicular traffic.

Interval: Discrete portion of traffic signal cycling during which signal indications do not change.

Phase: Those right-of-way and clearance intervals in a cycle assigned to any independent movement(s) of vehicular traffic or pedestrians.

Pretimed signal: Traffic control device that assigns right-of-way to vehicles at locations where standard traffic control devices are unwarranted.

Progressive signal system: Series of traffic control signals timed and coordinated in such a way as to provide optimum movement of traffic through the system.

Signal change interval: That portion of any phase warning the vehicular traffic or pedestrians of the impending termination of the right-of-way.

Signal head: Assembly containing one or more signal lenses that control a vehicular traffic or a pedestrian movement.

Signal indication: Illumination of a traffic signal

lens or of a combination of lenses at the same time.

Time cycle: Elapsed time between events or phases of traffic regulation signals.

Traffic-actuated signal: Traffic control signal whose right-of-way interval selection and interval times are varied by the demands of vehicular traffic for those intervals or movements.

TRAFFIC SURFACE Surface exposed to traffic, either pedestrian or vehicular. *Also called* **Finish wearing surface.**

T-RAIL Steel bar with a T-like cross section.

TRAILER Towed carrier that rests wholly or partially on its own set of two or more wheels, connected by a drawbar to the pintle hook of the towing vehicle or, in the case of a semitrailer, via a fifth-wheel, through which part of the gross trailer weight is transferred to the towing vehicle. There are many configurations, including:

B-train: Type of doubles semitrailer having a specially designed suspension-subframe fifth-wheel combination that obviates the need for a converter dolly when towing a full trailer.

Converted semitrailer: Semitrailer and converter dolly combination.

Converter dolly: Trailer chassis equipped with one or more axles, the lower half of a fifth wheel, and a drawbar.

Convertible: Trailer that can be used either as a flat-bed or open top by means of removable side panels.

Double bottom: *See* **Tandem trailer,** below.

Double drop-frame: Trailer having minimum clearance between the road and the underside of its main floor, excepting a raised section for the rear wheel housings and a raised forward section. *Also called* **Drop-center trailer,** and **Drop-frame trailer.**

Doubles: Trailer combination comprising a truck tractor, semitrailer, and a full trailer in series.

Drop-center: *See* **Double drop-frame trailer.**

Drop-frame: *See* **Double drop-frame trailer.**

Dump: Steel and aluminum dump trailers, available in many styles and capacities, including:

All-conventional: Units that use front-mounted telescopic hoists and anywhere from two to seven axles, depending on payload density.

Hopper-type: Dump trailer that discharges at the bottom for controlled windrow spreading.

Typical 'Michigan train': Doubles combination employing a 7.3-m (24-ft) triaxle lead trailer and a 5.8-m (19-ft) five-axle pup.

Extendable flatbed: Trailer whose length may be readily increased or decreased within prescribed limits and with accompanying variations in load carrying capacity.

Flatbed: *See* **Platform trailer,** below.

Full: Trailing load-carrying vehicle entirely supported by its own suspension systems.

Lowbed: Platform trailer with minimum ground clearance beneath the loading deck and designed to transport heavy, sometimes concentrated loads with a high center of gravity.

Lowbed, level-deck: Lowbed trailer with a deck level from the back of the gooseneck to the rear of the trailer. *Also called* **Lowboy trailer.**

Lowboy: *See* **Lowbed, level-deck trailer,** above.

Oilfield float: Flatbed trailer equipped with a pick-up loop at the front of the winch line and provision for lowering the trailer nose to the ground for loading and unloading, plus a rolling tailpipe across the rear of the deck.

Platform: Flatbed trailers used to haul a wide range of materials and equipment. Construction is mainly of three types:

    Center-frame - These versions have closely spaced main rails positioned near the longitudinal centerline of the platform. This design is required for heavy, concentrated loads positioned down the middle of the flatbed.

    Drop-center lowbed and removable-gooseneck - These versions are used to haul high, concentrated loads such as heavy construction equipment.

    Wide-frame - These versions have main rails positioned near the outer edge of the platform, and are designed for uniformly distributed loads, such as timber.

Pole: Trailer consisting of an upper coupler assembly connected by a tube or pole to a running gear assembly.

Semitrailer: Load-carrying vehicle having one or more axles and constructed so that its front end is supported on the fifth wheel of a truck tractor.

Straddle: Trailer fitted with lifting devices for hoisting and securing a cargo over which it is reversed.

Tandem: Combination consisting of a tractor pulling a semitrailer with a full trailer behind. *Also called* **Double bottom trailer,** and **Twin trailer.**

Tilt-bed: Trailer having one or more axles centered on a platform-type body that may be manipulated to form a loading or unloading ramp.

Transfer: Vehicle used to transport large quantities of waste over a long distance.

Twin: *See* **Tandem trailer,** above.

Van: Trailer having a totally-enclosed cargo space.

TRAILER AXLE *See* **Axle.**

TRAILER CONVERTER DOLLY Trailer chassis equipped with one or more axles, a lower half of a fifth wheel, and a drawbar.

TRAILER KINGPIN Short, heavy pin with a locking flange on its lower extremity, mounted near the front of the underside of a semitrailer that, when positioned in the fifth wheel of a tractor, provides a flexible connection between the two vehicles.

TRAILER LENGTH Distance from the front of the body to the bumper.

TRAILER PUMP Water pumping unit, usually mounted on a two-wheel trailer.

TRAILER SUPPORT Retractable ground support located near the front end of a semitrailer.

TRAILING BOOM KIT Option that allows the operator to swing the boom over the rear and couple it with a dolly for increased roadability. This kit includes valves that allow the boom lift cylinders to float in accordance with road conditions, and connections for the trailer's lights and braking system.

TRAILING CABLE Armored electrical cables between a power source and mobile equipment such as a dragline, excavator, crane, etc.

TRAILING TANDEM AXLE *See* **Axle.**

TRAIN *See* **Equipment train.**

TRAINING WALL Structure constructed along a river consisting of loose mounds or heaps of rubble, with or without a surmounting wall, timber, close timber piling, wood sheet piling, steel sheet piling or reinforced concrete to direct the flow of the river into a more favorable, fixed channel.

TRAMMEL 1. Device for adjusting or aligning part of a machine; **2.** Drafting instrument for drawing ellipses; **3.** Drafting instrument used to draw arcs or circles greater than the radius of conventional compasses.

TRAMMING TIME Time taken from when the number one rod is stored on the drill guide after retracting rods to when the drilling machine is in position to position the boom and drill guide.

TRAMP IRON Scrap metal entering a crusher.

TRAMWAY *See* **Aerial tramway.**

trans. *Abbreviation for:* transaction; transfer; transformer; transit; transmission; transport; transportation.

transc. *Abbreviation for:* transcribe.

TRANSDUCER Device that can change energy from one form to another, i.e., electrical energy into mechanical energy. *See also* **Load sensor.**

transf. *Abbreviation for:* transformer.

TRANSFER Act of transferring the stress in prestressing tendons from the jacks or pretensioning bed to the concrete member.

TRANSFER BOND In pretensioning, the bond stress resulting from the transfer of stress from the tendon to the concrete.

TRANSFER CAR Railroad car, which may be self-propelled or pulled/pushed by a locomotive, divided into compartments of the same approximate size as the concrete buckets used on a cableway; used to transfer concrete from a mix plant to the bucket dock of a cableway.

TRANSFER CASE Device located behind the transmission to divide the driving power between more than one axle. It is usually used with front-driving axles on 4x4 or 6x6 chassis.

TRANSFER LENGTH Length from the end of the member where the tendon stress is zero to the point along the tendon where the prestress is fully effective. *Also called* **Transmission length.**

TRANSFER MEDIUM Substance capable of conveying heat from a solar collector to a storage unit, or from a storage unit to living areas.

TRANSFER POINT Turning point.

TRANSFERRING Lifting an entire load of logs from one mode of transportation and placing the logs on another carrier.

TRANSFER STATION Supplemental transportation system, an adjunct to route collection vehicles to reduce haul costs or add flexibility to a waste collection and disposal operation. Typically, route vehicles empty into a large hopper from which large semitrailers, railroad gondolas, or barges are filled. There may be some compaction of refuse. Transfer stations may be fixed or mobile, since the larger compacting collection vehicles can serve this function.

TRANSFER STRENGTH Concrete strength required before stress is transferred from the stressing mechanism to the concrete.

TRANSFER SWITCH *See* **Switch.**

TRANSFER TAX Tax charged by some jurisdictions on the property belonging to an estate.

TRANSFER TRACK Railroad track running between a concrete mixing plant and a cableway.

TRANSFER TRAILER *See* **Trailer.**

TRANSFORMED SECTION Hypothetical section of one material arranged so as to have the same elastic properties as a section of two materials.

TRANSFORMER Device for changing the voltage characteristics of an AC current supply.

TRANSIENT PROTECTION Protection of a system from transient noise originating on the main AC lines.

TRANSIENT RECOVERY TIME Period of time required for an abrupt change in the power supply output pressure to dampen out to within the operating band.

TRANSIENT RESPONSE Response of current and induced voltage to an immediate change in applied voltage.

TRANSISTOR Semiconductor that conducts current in one direction only, as long as a control current is present at its control lead.

TRANSIT 1. Surveying instrument that can measure angles both vertically and horizontally; **2.** Movement or transport of mobile equipment.

TRANSITION 1. Stages between one defined style to another; **2.** Fitting that converts one shape or size to another.

TRANSITION BELT Short belt carrying material from a loading point to a main conveyor belt.

TRANSITION CURVE Curve connecting a straight line and a circle.

TRANSITION FOREST *See* **Interim forest.**

TRANSITION SLAB Concrete slab constructed between rigid and flexible pavement.

TRANSIT-MIXED CONCRETE Concrete, the mixing of which is wholly or principally accomplished in a truck mixer.

TRANSIT SYSTEM Organization providing any type of interurban or rural intercommunity multiple-occupancy-vehicle passenger service, including fixed-route, variable-route, and unscheduled service, provided for the use of the general public.

transl. *Abbreviation for:* translation.

translu. *Abbreviation for:* translucent.

TRANSLUCENT Letting light pass through but diffusing it so that objects on the other side cannot be distinguished.

TRANSLUCENT CONCRETE Combination of glass and concrete used together in precast and prestressed panels.

transm. *Abbreviation for:* transmission.

TRANSMISSION Arrangement of gears that permits changes in the speed-power ratio and/or direction of rotation. There are many types, including:

> **Automatic:** Separate transmission, often used by logging trucks or other extra-heavy-duty applications to create greater ratios or finer steps. This unit, requiring a second shift tower or lever in the cab, meets demands for greater ratio reduction from the transmission.

> **Close-ratio:** Five-speed transmission where the fourth gear is used to 'split' the axle step between low and high, the result being sustained high engine speed and power availability in the upper speed range. *Also called* **Short fourth transmission.**

> **Clutch-shifted:** Constant mesh transmission in which power is directed through gear trains by engagement of a friction clutch.

> **Compound:** Gear set in which power can be directed through two sets of reduction gears in succession.

> **Constant mesh:** Transmission that operates with all gears of the twin countershaft (main and counter) meshing with the main gear. The unit uses a sliding clutch, riding on the mainshaft, that is splined to the mainshaft. Operators must double clutch (as with a sliding-gear unit) trying to synchronize the relative speeds of the two parts.

> **Countershaft:** Constant mesh-type transmission with hydraulic clutch packs for smoother shifting.

> **Direct drive:** Gearing of a transmission so that in its highest gear, one revolution of the engine produces one revolution of the transmission's output shaft. The top-gear or final-drive ratio of a direct drive transmission would be 1:1.

> **Flywheel clutch:** Friction clutch in an engine flywheel, oil-type or dry-type.

> **Gathered-ratio:** Designates a transmission gear combination that provides relatively close gear steps for the upper speed ranges, and progressively wider steps between successively lower gears. Generally intended for use without supplemental gearing (two-speed axle or auxiliary transmission); can be teamed with engines having a wide operating range: controlled power or high torque-rise engines.

Hydrostatic: System that transfers power by hydraulic oil pressure, the transmission and drive motor taking the place of power train components between the flywheel and final drives.

Inter-axle differential: Gear device that equally divides transmission power between two axles and compensates for unequal tire diameters. *Also called* **Power divider.**

Main: Assembly consisting of an input shaft, floating mainshaft assembly and main drive gears, two countershaft assemblies, and reverse idler gears.

Mechanical shift: Transmission requiring manual actuation of the friction clutch, reversing gears, and speed gears, when used.

Nonconstant mesh: *See* **Sliding gear transmission,** below.

Planetary: Transmission that uses a center or sun gear and three planet gears inside a large ring gear to transfer power.

Power divider: *See* **Interaxial differential transmission,** above.

Power shift: Transmission in which the actual drive shifting of the reversing gear is accomplished by power application triggered by a pilot device such as a fingertip lever control or push button. Usually it is used in conjunction with a torque converter, and hydraulically applied clutches are used to accomplish the gear shifts. Speed changing gears may also be power shifted.

Preselective: Arrangement by which a gear lever can be moved but the resulting speed shift will not take place until the clutch or throttle is manipulated.

Reduction-type: Transmission whose output shaft always turns more slowly than the input shaft.

Reversing clutch: Forward-and-reverse transmission that is shifted by a pair of friction clutches.

Short forth: *See* **Close-ratio transmission,** above.

Sliding gear: Transmission in which the gears are splined to the mainshaft. Two shafts are used in the transmission. The gear will not mesh with another until that actual gear is used. Mainshaft gears are not in mesh with their mate on the countershaft, except for the gearset being used. Mainshaft gears are moved along the mainshaft to engage into their countershaft mate. Sliding clutches are not used. Operators must double clutch with such a unit, trying to synchronize the relative speeds of the two parts. *Also called* **Nonconstant mesh transmission.**

Synchromesh: Application of a synchronizer to a sliding clutch, making the unit a speed-equalizing device between the sliding clutch and gear of a transmission. There is no relative speed difference between the sliding clutch and the selected gear.

Synchronized-gear: Manual transmission featuring synchronizing devices that bring gear speeds into agreement before engagement, for clash-free shifting without double-clutching.

Torque converter: Hydrodynamic drive that transmits power with the ability to change torque (torque ratio being a function of speed ratio).

TRANSMISSION DRIVE *See* **Power takeoff.**

TRANSMISSION GEAR RATIO Ratio of input shaft speed to the speed of the output shaft.

TRANSMISSION LENGTH *See* **Transfer length.**

TRANSMISSIVITY Rate at which water is transmitted through an aquifer.

TRANSOM 1. Opening over a door or window; **2.** Panel or panels that close a hoistway enclosure above an entrance.

TRANSOM BAR Horizontal member across a door or window opening, or as part of a wooden panel.

transp. *Abbreviation for:* transparent; transport; transportation.

TRANSPARENT Transmitting light rays so that objects on the other side can be seen distinctly.

TRANSPARENT COATING Liquid formulation (such as varnish, shellac, or lacquer) that when dry forms a transparent film.

TRANSPIRATION Process by which water absorbed by plants is evaporated into the atmosphere.

TRANSPORT *See* **Tire types.**

TRANSPORT AIR Air employed in pneumatic conveying systems to move materials by entrainment in the moving air stream.

TRANSPORTATION Movement of traffic from one place to another.

TRANSPORTATION EXPENSE Total expense of all labor, materials, fees, and rents required for operating a transportation system.

TRANSPORTATION PLAN Program of action to provide effectively for present and future demands for movement of people and goods. This program must necessarily include consideration of the various modes of travel.

transv. *Abbreviation for:* transverse.

TRANSVERSE Direction at right angles to another.

TRANSVERSE CRACK Crack that develops at right angles to the long direction of the member or pavement.

TRANSVERSE JOINT Joint normal to the longitudinal dimension of a structural element, assembly of elements, slab, or structure.

TRANSVERSE PRESTRESS Prestress that is applied at right angles to the longitudinal axis of a member or slab.

TRANSVERSE REINFORCEMENT Reinforcement at right angles to the longitudinal reinforcement.

TRANSVERSE TRACE Trace on a vibration record that records motion at right angles to the line joining the vibration source and the seismograph.

TRAP 1. Dark-colored, fine-grained, and dense igneous rocks composed of ferromagnesian materials, basic feldspars, and little or no quartz; **2.** U- P- or S-shaped pipe filled with water and located beneath plumbing fixtures to form a seal against the passage of foul odors or gases. *See also* **P-trap, Running trap, S-trap, Sewer trap,** and **Three-quarter S-trap.**

trap. *Abbreviation for:* trapezoid; trapezoidal.

TRAP DOOR Removable or hinged covering for an opening in a floor, ceiling, or roof.

TRAPEZOIDAL BOOM Boom configuration offering a specific strength-to-weight ratio by providing a larger geometric cross section than conventional rectangular booms, a wide boom base, and strategic placement of high-strength steel and wear pads.

TRAP LOCK BOOM Trapezoidal boom that uses only one telescope cylinder to extend and retract all boom sections. As a boom section is extended to its full length, an electronic locking system will engage pins to lock the boom sections together. This boom extension method leaves only one section adjustable from the operator's cab during lifting operations.

TRAPPED AIR Air, trapped during rubber cure that usually causes a loose ply or cover, a surface mark, depression, or void.

TRAPROCK 1. Any of various fine-grained, dense, dark-colored igneous rocks, typically basalt or diabase; **2.** Sized or graded rock used as an underlay between two layers of asphalt.

TRAP SEAL Vertical distance between a crown weir and the dip of a trap.

TRASH *See* **Solid waste.**

TRASH-HANDLING CENTRIFUGAL PUMP *See* **Pump.**

TRASH RACK Grid or screen across a stream or channel, designed to catch floating debris.

trav. *Abbreviation for:* traverse.

TRAVEL 1. Movement of wheeled or tracked equipment under its own power; **2.** In elevator work, the distance from the top to the bottom floor.

TRAVEL ALARM Audible and/or visual warning actuated by the propel motion, in any direction, of mobile mechanical equipment that normally performs the majority of its work in a static mode: typically, an excavator.

TRAVEL BASE Base mounting for a traveling tower crane.

TRAVELED WAY Portion of the roadway reserved for the movement of vehicles, exclusive of shoulders. *See also* **Through traveled way.**

TRAVELER Inverted U-shaped structure, usually mounted on wheels or tracks, that permits it to be moved from one location to another to facilitate the construction of an arch, bridge, or building.

TRAVELING BLOCK Frame for a sheave or set of sheaves that slides in a track. *See also* **Block.**

TRAVELING CABLE Cable consisting of a number of electrical conductors, providing electrical connection between an elevator or dumbwaiter car and a fixed outlet in the hoistway.

TRAVELING FORM Formwork that travels along the path of the structure to be formed.

TRAVELING GRATE *See* **Grate.**

TRAVELING NUT Nut that moves along a screw as the screw is rotated.

TRAVELING SCREEN Rotating trash screen.

TRAVELING SHEAVE *See* **Sheave.**

TRAVELING STRINGLINE Articulated beam with multifoot pads and a string on top that rides on the surface to be improved and controls the screed elevation (mat depth), via a sensor, by averaging the highs and lows on the existing surface.

TRAVEL LINE *See* **Conveying line.**

TRAVEL MOTOR Hydraulic motor dedicated to operating the track or wheel mechanisms used to propel mechanical equipment.

TRAVEL PLANT Type of asphalt mixer that operates on the surface to be paved, picking up windrowed aggregate, mixing it with asphalt, and discharging the mix in a windrow for spreading and compaction.

TRAVEL SPEED RANGE Speed range available for roading a machine.

TRAVEL SWING LOCK Mechanical lock that engages directly with the upper over either the front or rear of the carrier only.

TRAVEL TIME Allowance paid for time spent traveling to and from a workplace.

TRAVERSE 1. Pass over or across; **2.** Series of connected lines of known length, related to each other by known angles; **3.** Surveying technique in which lengths and directions of lines between positions on the earth are derived from field measurements. *See also* **Closed traverse,** and **Open traverse.**

TRAVERSE CLOSURE Line that exactly closes a traverse.

TRAVERTINE Dense to irregularly porous, commonly stratified or banded calcium carbonate, either aragonite or calcite, formed by deposition from hot spring waters.

TRAY Container for one or more storage battery cells.

trd *Abbreviation for:* tread.

trdwy *Abbreviation for:* treadway.

TREAD 1. *See* **Tire construction; 2.** Distance between the left and right tire centerlines of each axle of a vehicle (centerline for dual tires is midway between the tire centers); **3.** Horizontal element of a step; **4.** High-friction lagging on a belt pulley.

TREAD DEPTH Distance, measured near the center of the tread, from the base to the top of the tread of a tire.

TREAD DESIGN The pattern of a tire tread.

TREAD DIAMETER 1. Diameter of a sheave or grooved rope drum measured at the base of the groove; **2.** Diameter of a smooth barrel on a rope drum; **3.** *See also* **Root diameter.**

TREAD GUM Rubber compound used primarily to build up the tread when making a repair to a tire.

TREADLE Foot pedal hinged to a mount at one end.

TREAD NOISE Noise or vibration caused by ropes traveling over a sheave.

TREAD PLY *See* **Tire construction.**

TREAD RADIUS Measure of tread surface curvature from shoulder to shoulder of a tire.

TREAD RIB Tread pattern section that encircles a tire.

TREAD RUBBER Uncured rubber material that replaces the worn off tread portion of a tire.

TREAD SEPARATION Pulling away of the tread from a tire casing.

TREAD TEARING Tearing away of a portion of the tread design from a tire casing.

TREAD WIDTH Horizontal distance from back to front of a step, including nosing.

treas. *Abbreviation for:* treasurer.

TREATMENT Material impregnating an abrasive product aiming to improve its grinding action, often by reducing the tendency for loading when in use.

TREE 1. Woody plant that usually grows to at least 6 m (20 ft) in height at maturity, typically having a single trunk with no branches within 1 m (3 ft) of the ground; **2.** Group of wires tied together in a cluster.

TREE CLASSES Classifications by which trees can be grouped for identification as to potential commercial value, including:

All live trees: Growing stock, rough, and rotten trees 25 mm (1 in.) in dbh (diameter breast height) and larger.

Growing-stock trees: Sum (by number or volume) of all the trees in a forest or in a specified part of the forest.

Poletimber trees: Live, vigorous, and well-formed trees of commercial species at least 125 mm (5 in.) in dbh (diameter breast height), but smaller than sawtimber size.

Rotten trees: Live trees of any size that do not contain a merchantable 3.65-m (12-ft) saw log, now or prospectively, because of rot (more than 50% of the cull volume of the tree is or will become rotten). Only commercial species are considered.

Saplings: Live, vigorous, and well-formed trees of commercial species, usually 25 to 125 mm (1 to 5 in.) in dbh (diameter breast height).

Sawtimber trees: Trees suitable for production of saw logs.

Seedlings: Live trees of commercial species with diameters less than 12 mm (1 in.) that are expected to survive (not diseased and not heavily damaged by logging, browsing, or fire). Only softwood seedlings over 150 mm (6 in.) tall and hardwood seedlings over 300 mm (1 ft) tall are counted.

Short-log trees: Sawtimber-sized trees of commercial species that contain at least one merchantable 2.4- to 3.35-m (8- to 11-ft) saw log (but not a 3.65 m (12-ft) saw log).

TREE FALLER *See* **Faller.**

TREE FARM 1. Parcel of land on which trees are planted, cultured, managed, and harvested as a crop; **2.** Privately owned, managed forest area that has been certified as a tree farm by the American Forest Institute.

TREE FARMING Application of silvicultural practices for the perpetual production of commercial timber crops. Includes all activities from stand establishment through delivery of commercial timber (logs) to a log yard at the initial commercial product processing facility.

TREE FARM LICENCE Form of tenure agreement that allows the long-term practice of sound forest management and harvesting on Crown land or on a combination of Crown and private land by private interests under the supervision of federal and provincial agencies.

TREE JACK Device used to support a skyline or maintain the skyline in an elevated position at a backspar.

TREE LENGTH Entire tree, excluding the unmerchantable top and limbs.

TREE-LENGTH LOGGING Felling and transporting the trimmed bole in one piece, whenever possible, for cross-cutting at a landing or mill.

TREELINE Edge of the falling face; usually refers to the upper edge but can also be the sidelines.

TREE PAINT Spray paint that will stick to wet bark and wood for the purpose of identification.

TREE PLATE J-shaped plate spiked to a spar tree to prevent cutting of the wood by wire rope at the point where guylines and straps are hung.

TREE PULLING Felling method used to overcome the natural lean of timber with the objective of reducing breakage and increasing volume and grade recovery. It may be used for pulling timber uphill, against its natural lean, or simply to maintain the lead when heavy leaners are encountered on less-steep ground.

TREE SHEARER Mechanical device used in felling that cuts or shears the standing tree off at the stump.

TREE SHOE Device used to support a skyline on a backspar.

TREE SNIPPER Crawler machine, fitted with a hydraulic shear or chain, used for felling timber.

TREFOIL Three-lobed ornament resembling foliage.

TRELLIS *See* **Lattice.**

TREMIE Pipe or tube through which concrete is deposited under water, having at its upper end a hopper for filling and a bail for moving the assemblage.

TREMIE CONCRETE Subaqueous concrete placed by means of a tremie.

TREMIE SEAL 1. Depth to which the discharge end of the tremie pipe is kept embedded in the fresh concrete that is being placed; **2.** Layer of tremie concrete placed in a cofferdam for the purpose of preventing the intrusion of water when the cofferdam is dewatered.

TRENCH Long narrow ditch. *Also called* **Ditch.**

TRENCH BOX *See* **Shield.**

TRENCH BRACE Horizontal member of a trench shoring system whose ends bear against the uprights or stringers.

TRENCHER Vehicle or attachment that cuts a ditch. Depending on capacity, may consist of excavating cups attached to a chain rotating about a bar, or buckets or skips mounted to the ladder of an excavator. *Also called* **Bucket ladder excavator.**

TRENCH FORM Vertical sides and semicircular bottom of a trench shaped to provide full, firm and uniform support for the lower 210° of a cast-in-place concrete pipe.

TRENCHING BUCKET *See* **Bucket.**

TRENCH JACK Screw or hydraulic jack used as cross-bracing in a trench shoring system.

TRENCHLESS TECHNOLOGY Techniques for conduit installation, replacement, or renovation that minimize excavation from the surface.

TRENCH SHIELD Trench shoring system that can be moved horizontally as work progresses.

TRENCH SHORING Wood, steel, or aluminum shapes used to support and restrain the walls of a trench. May consist of prefabricated shapes making up a demountable and movable shoring system.

TRENCH SOIL CLASSIFICATION SYSTEM Series of classifications that define soil classes, used to determine the most suitable system of shoring necessary to support and restrain the walls of an excavated trench. The system comprises the following classifications:

Type A: Stiff cohesive soil, 11.33 kg/0.092 m² (25 lb/ft²) per 30 mm (1 ft) of depth. Clay, silty clay, sandy clay, clay loam with an unconfined compressive strength of 1.36 t/0.092 m² (1.5 tons/ft²) or greater. Not Type A if fissured, subject to vibration, previously disturbed, or part of a sloped layered system where the layers dip into the excavation of a slope of four horizontal to one vertical or greater.

Type B: Medium cohesive to granular soil, 20.4 kg/0.092 m² (45 lb/ft²) per 30 mm (1 ft) of depth. Clay with an unconfined compressive strength between 0.45 and 1.36 t/0.092 m² (0.5 and 1.5 tons/ft²). Cohesionless gravel, silt, silt loam, or sandy loam. Previously disturbed soils may be Type B unless they would be classified as Type C. Soil that meets the requirements of type A but that is subject to vibration or is fissured may be Type B. Dry rock that is not stable or soil that is part of a sloped, layered system where the layers dip into the excavation on a slope less steep than four horizontal to one vertical are Type B if the material would otherwise be classified as Type B.

Type C: Soft, cohesive to saturated soil, 27 kg/0.092 m² (60 lb/ft²) per 30 mm (1 ft) of depth. Clay with an unconfined compressive strength less than 0.36 t/0.092 m² (0.5 tons/ft²), saturated sand, clay, or fractured rock that is not stable. Soil in a sloped, layered system where the layers dip into the excavation on a slope of four horizontal to one vertical or steeper may be Type C. Saturated soils or soils from which water is freely seeping, but is not standing in the trench.

Conditions more severe that Type C would require dewatering or sealing on four sides of the excavation and pumping the trench.

TRENCH TECHNIQUE *See* **Sanitary landfilling.**

TREPAN Large, heavy tool dropped on a line down a bore hole to advance an excavation in rock under water.

TRESTLE 1. Bridge or viaduct having a number of closely spaced supports between the abutments; **2.** Temporary work bridge to provide access to a work site and support equipment; frequently used in rivers for access to pier construction; **3.** Timber, reinforced concrete, or steel structure that is used to support a pile rig during driving.

TRIAL BATCH Batch of material prepared to establish or check proportions and constituents.

TRIAL PIT Pit or shaft dug to inspect and determine *in situ* the type of soil below ground level.

TRIANGULAR LEAD *See* **Lead.**

TRIANGULATION Technique that establishes the measurement to a point of a triangle whose other two selected sides have been computed.

TRIAXIAL COMPRESSION TEST *See* **Shear test.**

TRIAXIAL TEST Test in which a specimen is subjected simultaneously to lateral and axial loads.

TRIAXLE *See* **Axle.**

TRIAXLE SUSPENSION *See* **Suspension.**

trib. *Abbreviation for:* tributary.

trib. a. *Abbreviation for:* tributary area.

TRIBOLOGY Science of lubrication, friction and wear.

TRICALCIUM SILICATE Compound having the composition $3CaO \cdot SiO_2$, an impure form of which (alite) is a main constituent of portland cement. *See also* **Alite.**

TRICHLOROETHYLENE Liquid chemical compound used as a cleaning agent.

TRICKLE CHARGER Minimal DC charging device to maintain storage batteries at a continual fixed rate.

TRICKLE DRAIN Pond overflow pipe set vertically with its open top at the same elevation as the water surface.

TRIG Bricks laid in the middle of a wall between the two main leads to overcome sag in the line, and also to keep the center plumb.

trig. *Abbreviation for:* trigonometry.

TRIGGER To start an action in another circuit that then functions for a period under its own control.

TRIG (TRIGONOMETRICAL) STATION Base survey station used in large-scale triangulation.

TRILATERATION Surveying process that measures the length of a triangle's sides and then calculates the angles from the lengths.

TRIM 1. Finish materials in a building, such as moldings, applied around openings like doors and windows, or at the floor and ceiling of rooms; **2.** Stone used as sills, copings, enframements, etc., with the facing of another material.

trim. *Abbreviation for:* trimmer.

TRIM ALLOWANCE Extra length, usually 200 mm (8 in.) of a bucked log to allow for trimming waste in the sawmill.

TRIM BLASTING *See* **Cushion blasting.**

TRIM HOLE Unloaded drill hole, a number of which are closely spaced along a line to limit breakage from a blast.

TRIMMER 1. Beam or joist to which a header is nailed in framing around an opening; **2.** Tile unit in any of various shapes, consisting of such items as bases, caps, corners, moldings, angles, etc., necessary or desirable to make a complete installation and to achieve sanitary purposes as well as architectural design.

TRIMMER ARCH *See* **Arch.**

TRIMMER SCREW Adjusting screw that provides a limited amount of adjustment.

TRIMMING ANGLE Angle between the trimmed face of a seal and the seal axis.

TRIMMING JOIST Timber, or beam, that supports a header.

TRIM TILE Ceramic tile glazed on one or more edges and produced in a range of sizes and shapes to border and finish the main field of an installation.

TRINITROTOLUENE Explosive to which all military explosives are compared. Commonly known as **TNT.**

TRIP 1. Release catch; **2.** Block in the leads of a drop-hammer pile driver used to release the hammer and regulate the height of its fall; also used to drop the ram to start a diesel hammer. *Also called* **Trip block,** and **Tripping device.**

trip. *Abbreviation for:* triplicate; standard triple mast.

TRIP BLOCK *See* **Trip.**

TRIP COIL Solenoid-operated circuit breaker.

TRIPLE Tractor pulling three trailers: semitrailer plus two full trailers.

TRIPLE COURSE Three rows of shingles laid together at an eaves as a finish line.

TRIPLE DRUM Three-drum yarder.

TRIPLE GLAZING Three panes of glass separated from each other by spacers, and hermetically sealed as a unit.

TRIPLE GRIP Sheet metal anchor bracket, perforated for nails, used to tie framing members together.

TRIPLE POINT State point at which three phases of a given substance (e.g. solid, liquid, or gaseous) exist in equilibrium.

TRIPLE SEAL-TAB Strip roofing shingle having three tabs, each being one-half a modified hexagon.

TRIPLE SLIDING HITCH Knot used to tie off a lanyard to a lifeline.

TRIPLE-TIPPER BODY Dump truck body hinged so that it can be raised by hydraulic lift cylinders to the rear, or to either side, with hinged gates on all three vertical faces.

TRIPOD Three-legged support.

TRIPPER Double pulley that turns a short section of conveyor belt upside down so as to dump the load into a side chute.

TRIPPING DEVICE *See* **Trip.**

TRIPPING TIME Time required before excessive motor current forces an overload relay to open its control contact and remove electric power from the motor.

trk *Abbreviation for:* track; truck; trunk.

trn *Abbreviation for:* train.

trnbkl. *Abbreviation for:* turnbuckle.

trne *Abbreviation for:* trainee.

trnsm *Abbreviation for:* transom.

TROFFER Trough-like enclosure for artificial lighting fixtures.

TROLLEY Traveling block used in a skyline.

TROLLEY LINE *See* **Static line.**

TROMBE WALL Masonry or concrete wall behind large floor-to-ceiling glass or other glazing material; its purpose is to absorb and store solar heat to be used later.

TROMMEL Revolving cylindrical screen.

trop. *Abbreviation for:* tropical.

TROUBLE LIGHT Incandescent bulb housed in a protective metal cage and attached to a long electrical cord fitted with a plug.

TROUBLESHOOTING Systematic and methodical approach to determine the cause, extent, and remedy for problems to equipment, structures, systems, etc.

TROUGH 1. Long, narrow container; **2.** Maximum amplitude of a wave in the downward direction below the zero line.

TROUGH GUTTER Sheet metal form of box gutter.

TROUGHING 1. Repeated dozer pushes in one track forming ridges of spilled material that hold dirt on the blade; **2.** Rolled steel U-shaped section, used in bridge decks, welded with the U alternating up and down.

TROWEL Any of a number of shapes used to carry, place and finish mastic materials such as mortar and concrete. They include:

> **Finishing trowel:** A flat, broad-bladed, rectangular steel trowel, used in the final stages of finishing operations to impart a relatively smooth surface to concrete slabs or other concrete surfaces.
>
> **Mason's trowel:** Any of a group of similar-shaped, pointed trowels used by masons to carry and place mortar when laying bricks and blocks.
>
> **Pointing trowel:** Similar to, but smaller than a mason's trowel, used to point the joint between bricks and blocks.
>
> **Power trowel:** Power-driven system of blades mounted on a horizontal spider that is floated over newly set-up concrete to consolidate the surface and produce a smooth, polished finish.

TROWEL FINISH Smooth or textured finish of an unformed concrete surface obtained by trowelling.

TROWELING Smoothing and compacting the unformed surface of fresh concrete by strokes of a trowel.

TROWELING MACHINE Motor-driven device that operates orbiting steel trowels on radial arms from a vertical shaft.

trt *Abbreviation for:* treat.

trtmt *Abbreviation for:* treatment.

TRUCK 1. Vehicle designed for carrying an entire load; **2.** Motorized or towed vehicle for hauling. There are several types, including:

Bottom-dump: Trailer or semitrailer that discharges bulk material by opening doors in the floor of the body.

Closed top van: Van with a sealed top that must be rear loaded.

Dump body: Storage and discharging component of a dump truck. Following are some common types:

Heavy-duty and severe-duty: Designed for damaging payloads and bucket loading. They feature thicker steel floors and sides, extra bracing in sides and tailgate, thicker and stronger corner posts, and boxed side rails to withstand shovel impact.

Hopper/conveyor spreader: Can be used for patching jobs, windrowing material, feeding a paver, and for broadcast-spreading of abrasives or chemical compounds for ice control, dust control, or seal coating.

Light-duty: 2.4 to 3 m (8 to 10 ft) long, 0.76 to 2.2 m³ (1 to 3 yd³) capacity, used for light hauling and clean-up. Drop sides are available; a direct-lift hoist is common.

Side-dump: Having a body with a fixed rear wall and with sides that can hinge open and means for the body to be raised on hydraulic rams to either the left or right for discharge of a load such as soil, sand, gravel, etc.

Standard-duty: Designed to handle routine dirt/aggregate hauling, debris removal, and municipal and highway maintenance applications.

Two-way dump-spread: Can discharge at the rear as a normal dump body, or can be raised at the rear for front discharge through an integral spreader system.

Flatbed: Truck fitted with a bed that is completely flat. The bed may contain stake pockets, and can be equipped with a range of accessories at its rear edge, such as a roller to assist in the loading of pipe or similar objects.

Highway: Truck designed to haul a load that, combined with the net weight of the vehicle, does not exceed legal highway limits.

Off-highway: Truck designed for operations that would not be permitted on facilities with posted weight limits.

Oilfield body: Heavily constructed platform-type truck body equipped with a rear-end roller or bullnose adapted for winch loading.

On-highway: Vehicle designed for operations on paved or improved surfaces only.

On/off-highway: Vehicle designed for operation mostly on hard-surfaced or graded roads with some work over unprepared surfaces.

Open top: Vehicle body without a permanent top assembly.

Platform: Truck with a flat, open body.

Rear-dump: Truck or semitrailer with a box body that can be raised at the front so the load will slide out the rear.

Stake body: Flat platform-type vehicle body with removable stakes around its perimeter.

Utility/Service: Specialized truck body customarily consisting of a narrow platform bed between lockable 'saddlebag' storage compartments (for tools, service parts, special equipment) that open from the exterior side for easy access. Standard bodies are manufactured in a wide range of sizes and configurations for general purpose, electrical, and gas/water plumbing applications. Many options, including aerial devices for overhead work, are available.

TRUCK BED LINER *See* **Bed liner.**

TRUCK CARRIER Specially built truck for mounting a drilling rig or for carrying a crane.

TRUCK CRANE Crane having a rotating superstructure and independent power plant, mounted on a separately powered truck chassis.

TRUCK HITCH 1. Device for positioning and supporting one end of a towed vehicle behind a recovery vehicle; **2.** Oscillating push roller with hydraulically controlled extension arms, one on either end of the push roller beam. In operation, the arms are extended to receive the truck and are then closed to hold the truck captive during the unloading cycle.

TRUCK HOIST Mechanical device for raising truck bodies, including the following types:

Arm hoist: Employs single or dual cylinders and twin mechanical arms to equalize the lifting force on both sides of the dump body, regardless of uneven load distribution or irregular surfaces.

Direct-lift: Single hydraulic cylinders with a piston attached directly to the body's understructure, typically used for light-duty applications.

Double-acting: 1. Hoist capable of providing lift as well as retractive effort or force (power-down) to a dump body by hydraulic or mechanical power; **2.** Power-up and power-down, needed for bodies with heavy equipment mounted on the tailgate.

Telescopic: Used for the heaviest loads. Single cylinder front-lift installations are the most common; twin underbody installations are available.

TRUCKING SILL Heavy metal sill that acts as a bridge between a loading door opening and the hinged dock ramp.

TRUCK MIXER Concrete mixer suitable for mounting on a truck chassis and capable of mixing concrete in transit.

TRUCK-MOUNTED CRANE Hydraulic crane mounted between the cab and bed, or on the bed, that can be used to load and unload the vehicle or to perform ancillary tasks. They are primarily of two types:

  Articulating: Characterized by a joint allowing vertical movement between the first and second boom sections. The first boom section moves horizontally on a rotating base and can also be moved vertically through an arc. The second boom section can telescope to provide additional reach. When not in use these cranes fold into a particularly compact shape. *Also called* **Knuckleboom.**

  Telescoping: Cranes having from two to four boom sections that extend and retract under hydraulic power. May be sequential or proportional: sequential extension causes each boom section to extend only when the preceding, larger boom section in which it is housed has itself been fully extended. Proportional extension causes all boom sections to extend simultaneously at varying rates. The entire crane can be rotated on its base, and the entire unit can often be raised vertically in addition to the boom being swung vertically through an arc.

TRUCK TRACTOR Truck designed primarily for pulling a trailer and not so constructed as to carry a load other than a part of the weight of the towed vehicle.

TRUCK TRAILER Cargo-carrying vehicle designed to be pulled by a truck or tractor.

TRUCK UNDERBODY BLADE Grading/plowing blade that can be mounted under a truck body and that can be stored in a raised position and lowered to ride on the road surface.

TRUE BEARING Horizontal angle between a survey line and true north.

TRUE MERIDIAN Geographic north-to-south plane.

TRUE SECTION Cross-sectional drawing to the same horizontal and vertical scales.

TRUING Grinding wheel that has its cutting face finished so as to run true so that it will produce perfectly round (or flat) and smooth work; or to alter the cutting face for grinding special contours.

trun *Abbreviation for:* trunnion.

TRUNK, TRUNKLINE 1. Main pipe from which building drains or water supply branch; **2.** Detonation cord line on the surface, to which the downline is tied prior to firing; **3.** Large diameter ventilation pipe.

TRUNNION 1. Oscillating bar that allows changes in angle between a unit fastened to its center, and another attached to both ends; **2.** A heavy horizontal hinge.

TRUNNION AXIS Horizontal axis of rotation of a theodolite's telescope.

TRUNNION AXLE *See* **Axle.**

TRUNNION-MOUNTED CYLINDERS Telescopic boom cylinders mounted on a trunnion system so that, as the boom deflects under load, the cylinder can shift to prevent any direct load from being placed on the cylinder.

TRUSS Number of elements, usually triangular, framed together to bridge a space, to be self-supporting, and designed to carry loads. There are many configurations, including:

  Attic: Truss framed to create a usable area within the roof space.

  Bobtail: *See* **Cutoff,** below.

  Common: A simple triangular shape spanning between walls. Normal span range is 3 to 20 m (10 to 60 ft) with roof slopes from 2:12 to 12:12.

  Cutoff: A truncated common truss, used as a chimney split truss or where building plans dictate that a symmetrical truss will not fit. *Also called* **Bobtail truss.**

  Flat: Truss designed to provide a flat roof (usually built with a slight top chord slope to create positive roof drainage).

  Gable-end: Filler truss designed to fill in a roof space at a T-roof junction; this is not true trusses and are not designed to span. They are not triangulated and must be supported along the entire length of the bottom chord. *Also called* **Valley jack.**

  Hip: Specially configured truss to accommodate a roof hip.

  Jack: Short truss used to frame the end of a hip roof system. Jacks usually are supported by a hip girder truss; in turn, jacks provide lateral restraint for the hip girder top and bottom chords.

  Monotruss: Used for single-slope roofs.

  Piggy-back: Combination hip truss and valley jack, used when the overall height of the truss exceeds about 3 m (10 ft), making it difficult to transport by road.

  Scissors: Common truss having bottom chords that angle up from the supports to meet in the center, creating a vaulted ceiling effect.

  Valley jack: *See* **Gable-end truss,** above.

TRUSS BAR Bar bent up to act as both top and bottom reinforcing.

TRUSSED Braced by an arrangement of members into a rigid unit.

TRUSSED ARCH *See* **Arch.**

TRUSSED BEAM Beam stiffened by a truss rod.

TRUSSED ROOF Roof construction supported by trusses.

TRUSS LEAD *See* **Lead.**

TRUSTEE Person given legal responsibility to hold property for another.

TRY SQUARE L-shaped measuring tool most commonly used to check the square of an object, line or position. The tool consists of a thin but wide blade mounted at right angles at the top of a thicker handle.

The blade is usually marked with dimensions; the handle is often cut back to 45° at its upper junction with the blade permitting the blade to be used to scribe a 45° mitre.

ts *Abbreviation for:* tensile strength.

TS *Abbreviation for:* tamping shield; tentative standard; time switch; topsoil and seed.

T-SHORE Shore with a T-head.

tsi *Abbreviation for:* timber stand improvement; tons per square inch.

tsp *Abbreviation for:* township.

tspec. *Abbreviation for:* test specification.

T-SQUARE T-shaped wood, plastic, or metal guide used in drafting, and as a guide in marking panels for cutting. *Also called* **Tee square.**

ttn *Abbreviation for:* tighten.

TUB Base of a walking dragline.

tub. *Abbreviation for:* tubing.

TUBBING Metal lining segments used to line tunnels or shafts. *Also called* **Steel ring.**

TUBE 1. Hollow cylinder or pipe, generally long in proportion to its diameter; **2.** Conductor whose size is its outside diameter (available in varied wall thicknesses); **3.** Innermost continuous all-rubber or plastic element of a hose. *See also* **Lining; 4.** *See also* **Tire construction.**

TUBE-AND-COUPLING SCAFFOLDING Scaffolding consisting of individual pieces of tubing or pipes, erected with special coupling devices to join uprights, braces, bearers, and runners to form an integral load-carrying structure.

TUBE-AND-COUPLING SHORING Load-carrying assembly of tubing or pipe that serves as posts, braces, and ties, a base supporting the posts, and special couplers that connect the uprights and join various members.

TUBES AND FLAPS *See* **Tire construction.**

TUBEAXIAL FAN *See* **Fan.**

TUBE FLOAT Machine used to level and finish the surface of concrete pavement.

TUBE PILE *See* **Pile, steel pipe.**

TUB GRINDER Grinding machine used to process forestry, agricultural, and other organic waste, including yard debris, pallets, etc.

TUBING 1. Lightweight pipe made of materials such as copper, brass, rubber, or plastic; **2.** Component of a hydraulic system by which hydraulic fluid is transmitted.

TUBING MACHINE *See* **Extruder.**

tubl *Abbreviation for:* tubular.

TUBULAR JIB Multiple-section lattice extensions, fabricated from tubular sections, supported by pendants and attached to the main boom head.

TUBULAR SCAFFOLDING Scaffolding fabricated of steel or alloy tube.

TUBULAR WALL ANCHOR *See* **Wall anchor.**

TUBULAR WELDED-FRAME SCAFFOLD *See* **Fabricated tubular frame scaffold.**

TUBULAR WELL Well in which a circular section screen or perforated casing is in direct contact with the water-bearing formation.

TUCK To pass one strand under another in splicing.

TUCK-IN *See* **Curl.**

TUCK POINTING *See* **Pointing.**

TUDOR ARCH *See* **Arch.**

TUMBLER 1. Part that moves a gear into place; **2.** One of the parts of a cylinder lock that hold the locking bar in place until turned by a key.

TUMBLER SWITCH *See* **Switch.**

TUMBLING Operation for deburring, breaking sharp edges, finishing, and polishing in which abrasive, water and the work are 'tumbled' in a rotating barrel, or by other means.

TUMBLING IN Brickwork courses sloped to meet horizontal courses at a gable wall.

tun. *Abbreviation for:* tunnel.

TUNG OIL Oil pressed from the fruit of the tung tree and used in quick-drying, oil-based paint.

TUNGSTEN *See* **Alloying elements.**

TUNGSTEN CARBIDE Type of steel used to form cutting edges where extreme wear and abrasion resistance are required. *Also called* **'Carbide.**

TUNGSTEN-FILAMENT LAMP Incandescent electric light source employing a tungsten wire filament in a glass bulb filled with a mixture of argon and nitrogen.

TUNNEL Long enclosed passage or way not within a building.

TUNNEL-BORING MACHINE Device used to excavate a bore by mechanical means for the purpose of conduit installation.

TUNNEL LINING Structural system of concrete, steel, or other materials to provide support for a tunnel for exterior loads, to reduce water seepage, or to increase flow capacity.

TUNNEL TEST Standard test for flame spread.

TUNNEL VAULT *See* **Vault.**

tur. *Abbreviation for:* turret.

turb. *Abbreviation for:* turbidity.

TURBIDIMETER Device for measuring particle-size distribution of a finely divided material by taking successive measurements of the turbidity of a suspension in a fluid.

TURBINE That component of an engine that produces torque from expansion of the working fluid. Consists usually of a turbine nozzle and a turbine wheel that, together, constitute a turbine stage. A multistage turbine comprises more than one turbine stage.

TURBINE ENGINE Engine in which the working medium is a gaseous fluid throughout the cycle with the principal mechanical parts being driven by turbines.

TURBINE NOZZLE Arrangement of stationary blades for directing the flow of gas into a turbine wheel.

TURBINE WHEEL Rotary component of the turbine stage that consists of a series of blades or buckets through which fluid flows. May be of the axial, radial, or mixed flow type.

TURBOBLOWER Type of centrifugal blower in which air leaving the blade tips passes through a narrow slot into a large-volume chamber, thus efficiently converting velocity energy to static pressure.

TURBOCHARGED ENGINE *See* **Engine.**

TURBOCHARGER A supercharger, driven by an impeller and shaft that are spun by the movement and expansion of hot exhaust gases, forcing air under pressure into an engine intake. *See also* **Centrifugal blower.**

TURBOCHARGER IMPELLER Drive wheel inside a turbocharger housing against which the exhaust gases push.

TURBOCHARGER TURBINE The driven, supercharging fan on the intake air side of a turbocharger.

TURBODRILL *See* **Drill.**

TURBULENCE State of being highly agitated.

TURBULENT FLOW Flow situation in which the fluid or gas particles move in a random manner. *See also* **Flow,** and **Reynold's number.**

TURF Sod.

TURN 1. One or more logs that are yarded to the landing at one time; 2. The choked logs that make up the load for a skidder or tracked vehicle to skid to the landing.

TURN ANGLE Angle measured with a survey instrument.

TURNAROUND TIME Time it takes for a truck or tractor to be loaded and unloaded.

TURNBUCKLE Line-tightening device with a rod at either end, one with a right-hand and the other with a left-end thread.

TURN BUTTON Simple door latch, frequently used on cupboards.

TURNED WORK Woodwork done on a lathe.

TURNING CENTER Center of the circle described by a vehicle moving in a turn.

TURNING DIAMETER Distance two times the turning center. *See also* **Curb-to-curb turning diameter,** and **Wall-to-wall turning diameter.**

TURNING GOUGE Woodworking tool for use on a lathe.

TURNING MOVEMENT *See* **Intersection elements.**

TURNING PATH *See* **Intersection elements.**

TURNING POINT Temporary bench mark; fixed point or object, often temporary in character, used in leveling where the rod is held first for a foresight and then for a backsight.

TURNING RADIUS One-half the diameter of the circle made by the center line of the outermost tire as a mobile machine maneuvers through its tightest turning mode.

TURNING ROADWAY *See* **Intersection elements.**

TURNING ROADWAY TERMINAL *See* **Intersection elements.**

TURNING TRACK WIDTH *See* **Intersection elements.**

TURNKEY Contract that requires the contractor/supplier to be responsible for all aspects of the job, often including such things as financing, design, site acquisition, fabrication, start-up, etc.

TURNKEY CONTRACTOR *See* **Package builder.**

TURN-OF-NUT Procedure whereby the specified pretension in high-strength bolts is controlled (the state when a nut is tightened half-a-turn past snug; the point at which an impact gun commences impact; or the tightness achieved by applying full effort on a spud wrench).

TURNOUT Space adjacent to a road in which vehicles may park or pull into to allow others to pass.

TURNSTILE One-way rotating barrier, used to control and/or count the passage of people passing through.

TURNTABLE Base that supports a part and allows it to rotate or swing. *See also* **Superstructure.**

TURNTABLE BEARING Large bearing that attaches the upper to the carrier, allowing the upper to rotate on the carrier.

turp. *Abbreviation for:* turpentine.

TURPENTINE Light-colored volatile oil distilled from the oleoresin obtained from coniferous trees, used as a thinner in paints and as a solvent in varnishes.

TURRET Small and slender tower.

TURRET STEP Triangular steps, laid on top of each other to form a spiral stair.

TUSCAN Designation of a classical (Roman) order of architecture characterized by unfluted columns with a ring-like capital and by a frieze like the Doric. *See also* **Order.**

TUYERE Air opening or port in a furnace wall or grate system.

TV *Abbreviation for:* television.

tvl *Abbreviation for:* travel.

tw. *Abbreviation for:* twin; twisted.

TWIN CABLE *See* **Duplex cable.**

TWIN PUG Pugmill with two contrarotating horizontal shafts.

TWIN TRAILER *See* **Trailer.**

TWIN TURBINE *See* **Torque converter.**

TWIN-TWISTED BAR REINFORCEMENT Two bars of the same nominal diameter twisted together.

TWIST 1. Turns about the axis, per unit of length, of a fiber, roving, yarn, cord, etc. Twisting is usually expressed as turns per inch; 2. Turn about the axis of a hose subjected to internal pressure; 3. Spiral warp in lumber.

TWIST BIT *See* **Bit.**

TWIST DRILL *See* **Drill.**

TWISTER Line that supports a tailhold stump or tree that does not appear to be strong enough. This is done by connecting the tailhold to another stump or tree opposite by wrapping the two with a line. This line is then tightened by placing a piece of large-diameter limb between the wrappings and twisting them together.

TWIST GIMLET Gimlet having a helical groove cut into its shank by which wood cuttings are removed.

TWISTING BAR Tool to be attached to the kelly for screwing down casing through caving or squeezing soil, and sometimes for pulling casing.

TWITCH To skid logs of tree length on the ground without an antifriction device.

twl *Abbreviation for:* towel.

TWO-AXLE SCRAPER *See* **Scraper.**

TWO-BLOCK On cranes, contact between the load block and boom tip. *See also* **Antitwo-block system.**

TWO-BLOCK WARNING SYSTEM System of electromechanical devices used to warn crane operators of an impending two-block condition.

TWO-BOLT LOCK Door lock operated by a key, combined with a knob-operated latch.

2C *See* **Lumber grades.**

2/C *Abbreviation for:* two conductor.

TWO-COAT WORK Finish applied in two layers.

TWO-CORE BLOCK Concrete masonry unit having two cores, or cells.

TWO-CYCLE An engine where a power stroke occurs during every crankshaft revolution, rather than every two crankshaft revolutions.

2-D *Abbreviation for:* two-dimensional.

TWO-INCH-PIECE Closer about one-quarter of a brick in length, used to start the bond from the corner. *Also called* **Squint.**

TWO-LEG SLING Two chains or ropes made into a sling hanging from one link, having a hook attached at the other end.

TWO:ONE ROPING Arrangement of elevator hoist ropes in which one end of each hoist rope passes from a dead-end hitch in the overhead, down and under a car sheave(s), up over the drive sheave, down around a counterweight sheave(s), and up to another dead-end hitch in the overhead. The car travels at one-half the rope speed. *See also* **Compound roping.**

TWO- OR THREE-POSITION VALVE *See* **Valve.**

TWO-PART LINE *See* **Parts of line.**

TWO-PART SEALANT Chemically cured sealant supplied in two parts, one containing the reactive polymer base and the other the curing agent. It may be supplied in separate containers or in the same container, but separated by an inert layer.

2 ph *Abbreviation for:* two phase.

TWO-PINNED ARCH Arch-shaped or rectangular rigid frame hinged at both supports.

TWO-PIPE SYSTEM Heating system in which one pipe is used for the supply of a heating medium to the heating unit and another for the return of the heating medium to the source of heat supply. The essential feature of a two-pipe system is that each heating unit receives a direct supply of the heating medium, one that has not served a preceding heating unit.

TWO-PLATE CLUTCH Mechanical friction clutch employing two driven disks instead of one, to achieve torque capacity that would otherwise demand single-plate clutch diameters too large for practical installation in a truck.

TWO-POINT SUSPENDED SCAFFOLD *See* **Suspended scaffold.**

TWO-POINT SUSPENSION *See* **Suspension.**

TWO-SHAFT, FREE-POWER TURBINE ENGINE Gas turbine engine in which the compressor and its driving turbine are mounted on one shaft and the output turbine is mounted on a separate shaft supplying useful power.

TWO-SPEED AXLE *See* **Axle.**

TWO-SPEED DIFFERENTIAL *See* **Differential.**

TWO-SPEED DOOR *See* **Door types.**

TWO-STAGE CONTROL, OUTRIGGER *See* **Outrigger.**

TWO-STAGE CURING Process in which concrete products are cured in low-pressure steam, stacked, and then autoclaved.

TWO-STAGE ELEMENT Filter element assembly composed of two filter media in series.

TWO-STAGE FILTER Filter element assembly composed of two filter elements or media in series.

TWO-STORIES STAND Forest stand in which two height classes of considerable difference occur: the overstory and understory. Does not apply to a forest in the process of reproduction, in which the appearance of two stories is due to a seed tree or shelterwood cut before final cut.

TWO-STROKE CYCLE ENGINE Engine that requires one complete revolution of the crankshaft to complete the cycle of events necessary to generate power.

TWO-, THREE-, OR FOUR-WAY VALVE *See* **Valve.**

2/W *Abbreviation for:* two way.

TWO-WAY AUTOMATIC MAINTAINING ELEVATOR LEVELING *See* **Elevator car leveling device.**

TWO-WAY AUTOMATIC NONMAINTAINING ELEVATOR LEVELING *See* **Elevator car leveling device.**

TWO-WAY DUMP-SPREAD BODY *See* **Truck.**

TWO-WAY FLOOR AND ROOF SYSTEM One of the two major reinforced concrete floor and roof systems of design and construction. The two-way system consists of a solid slab supported by reinforced beams running in two directions that, in turn, are supported by columns. The supporting beams may, or may not be integral with the slab.

TWO-WAY RAM *See* **Ram.**

TWO-WAY RAMP *See* **Interchange elements.**

TWO-WAY REINFORCED FOOTING Footing having reinforcement in two directions, generally perpendicular to each other.

TWO-WAY REINFORCEMENT Reinforcement arranged in bands of bars at right angles to each other.

TWO-WAY SYSTEM System of reinforcement consisting of bars, rods or wires placed at right-angles to each other in a slab and intended to resist stresses due to bending of the concrete slab in two directions.

TWO-WAY VALVE *See* **Valve.**

twp *Abbreviation for:* township.

twr *Abbreviation for:* tower.

typ. *Abbreviation for:* typical.

TYPE A SOIL *See* **Soil classification system,** and **Trench soil classification system.**

TYPE B SOIL *See* **Trench soil classification system,** and **Trench soil classificaiton system.**

TYPE C SOIL *See* **Trench soil classification system,** and **Trench soil classification system**

TYPE K COPPER TUBING *See* **Pipe.**

TYPE K EXPANSIVE CEMENT *See* **Expansive cement.**

TYPE K MORTAR *See* **Mortar.**

TYPE L COPPER TUBING *See* **Pipe.**

TYPE M COPPER TUBING *See* **Pipe.**

TYPE M EXPANSIVE CEMENT *See* **Expansive cement.**

TYPE M MORTAR *See* **Mortar.**

TYPE N MORTAR *See* **Mortar.**

TYPE O MORTAR *See* **Mortar.**

TYPE S EXPANSIVE CEMENT *See* **Expansive cement.**

TYPE S MORTAR *See* **Mortar.**

TYPICAL DRAWING End-section view of a road construction, usually showing half of the road if both sides are the same.

TYPICAL 'MICHIGAN TRAIN' DUMP TRAILER *See* **Trailer.**

U Overall coefficient of heat transmission; the reciprocal of R (R=1/U).

u *Abbreviation for:* unit.

UBC *Abbreviation for:* Uniform Building Code.

U-BOLT Member used to clamp the top U-bolt plate, spring, axle, and bottom U-bolt plate together in a suspension system. Inverted U-bolts cross the springs when in place; conventional (nuts up) units wrap around the axle.

UDC *Abbreviation for:* universal decimal classification.

U-GAUGE *See* **Gauge.**

U-GROOVE Shape of the groove machined into the drive sheave of a double-wrap traction machine, approximately equal to the diameter of the hoist rope to be used.

UL *Abbreviation for:* Underwriters Laboratories, Inc.

U-LEAD *See* **Lead.**

ult. *Abbreviation for:* ultimate.

ULTIMATE BEARING PRESSURE Pressure at which shear failure occurs in soil.

ULTIMATE BEARING VALUE (OF A PILE) Maximum load that a single pile will support.

ULTIMATE BEARING VALUE OF A PILE FOUNDATION Maximum load, computed as a total load or as a load per pile, that the foundation will support without objectionable progressive movement. *See also* **Safe load.**

ULTIMATE CARBON DIOXIDE Percentage of carbon dioxide that appears in the dry flue gases when a fuel is burned with its chemically correct air-fuel ratio.

ULTIMATE COMPRESSIVE STRENGTH Maximum compressive strength a material can stand under a gradual and evenly applied load.

ULTIMATE DESIGN RESISTING MOMENT Moment at which a reinforced concrete section reaches its usable flexural strength, commonly accepted for under-reinforced concrete flexural members to be the bending moment at which the concrete compressive strain equals 0.003.

ULTIMATE LOAD Maximum load that may be placed on a structure, pile, pier, or structural element before its failure.

ULTIMATE MOMENT 1. Bending moment at which a section reaches its ultimate usable strength; **2.** In concrete, the moment at which the tensile reinforcement reaches its specified yield strength. *See also* **Flexural strength.**

ULTIMATE SHEAR STRENGTH Loading of a section that results in the member failing in shear.

ULTIMATE STRENGTH 1. Maximum resistance to a load or combination of loadings a member or structure is capable of developing before failure; **2.** Tensile strength measurement at which a sample of material separates.

ULTIMATE STRENGTH DESIGN Structural design based on the ultimate load.

ULTRASONIC Frequencies above those detectable by the human ear; generally, above 20,000 Hz.

ULTRASONIC TESTING Nondestructive testing method that transmits high-frequency sound waves through materials.

ULTRASONIC WELDING Metal fusion process involving high-frequency sound waves.

ULTRAVIOLET Zone of invisible radiations beyond the violet end of the spectrum. Their short wavelengths have enough energy to initiate chemical reactions and degrade some materials and colors.

un. *Abbreviation for:* union.

unaff. *Abbreviation for:* unaffected.

unatt. *Abbreviation for:* unattached.

unauthd *Abbreviation for:* unauthorized.

unavail. *Abbreviation for:* unavailable.

UNBALANCED BID Bid in which the total is appropriate for the work to be done, but where some of the unit prices are either abnormally high or low. *See also* **Balanced bid,** and **Bid.**

UNBALANCED CONSTRUCTION When individual components or layers of a laminate do not respond equally to changes in moisture, thus causing warp.

UNBALANCED HEATING SYSTEM System delivering an unequal distribution of heat to various areas or rooms.

UNBONDED MEMBER Prestressed concrete member post-tensioned with tendons that are not bonded to the concrete between end anchorages after stressing.

UNBONDED POST-TENSIONING Post-tensioning in which the tendons are not grouted after stressing.

UNBONDED TENDON Tendon that is permanently prevented from bonding to the concrete after stressing.

UNBRACED FRAME Frame in which the resistance to lateral load is provided by the bending resistance of frame members and their connections.

UNBRACED LENGTH Distance between braced points of a structural steel member, measured between the centers of gravity of the bracing members.

UNBRACED LENGTH OF COLUMN Distance between lateral supports.

UNBUFFED END Untrimmed, factory-cut end of a drywall panel.

UNBUTTONING Breaking off the heads of rivets

holding structural steelwork during demolition.

UNCASED PILE *See* **Pile.**

UNCHANNELIZED INTERSECTION *See* **Intersection types.**

UNCHARGED AND DRY Condition of a storage battery when it is assembled with formed and dried plates. Filling with electrolyte and an initial charge are required before it is ready for use.

unclas. *Abbreviation for:* unclassified.

UNCLASSIFIED EXCAVATION *See* **Excavation.**

UNCOMMITTED *See* **Cost types.**

uncond. *Abbreviation for:* unconditionally.

UNCONFINED COMPRESSIVE STRENGTH Load per unit area at which a soil will fail in compression.

UNCONFINED WATER *See* **Groundwater.**

UNCONTROLLED FILL Fill not placed under supervision and not compacted.

UNCONVENTIONAL WASTE *See* **Solid waste.**

uncorr. *Abbreviation for:* uncorrected.

UNCOURSED RUBBLE Random or snecked rubble.

und. *Abbreviation for:* under.

undef. *Abbreviation for:* undefined.

UNDERBEAD CRACK Crack in the base metal, near the weld and beneath the surface

UNDERBED Base mortar, usually horizontal, into which strips are embedded and on which terrazzo topping is applied.

UNDERBODY HOIST Generally, any conventional-type hoist with all the parts positioned beneath the floor of a dump body.

UNDERBODY SCRAPERS *See* **Snow plow.**

UNDERBRUSH Brush under a stand of timber.

UNDERCANT Holt set with a choker to free a log by rolling it from underneath.

UNDERCARRIAGE 1. *See* **Bogey; 2.** Structural subframe of a vehicle or trailer, complete with suspension system and axle-wheel assemblies. There are several types, including:

> **Removable undercarriage:** Undercarriage capable of being detached from a vehicle or trailer structure.

> **Sliding undercarriage:** Trailer undercarriage having provision for readily moving it to new locations, forward and backward, to effect weight distribution and/or wheelbase.

UNDERCOAT Coating applied prior to a finishing coat. It may be the first of two or the second of three coats. *See also* **Primer.**

UNDERCONSOLIDATION Condition of a soil deposit that is not fully consolidated under the existing overburden pressure and excess hydrostatic pore pressures existing within the material.

UNDERCOURSING *See* **Shingle.**

UNDERCURE Incomplete vulcanization or curing.

UNDERCUT 1. First cut made, when falling or bucking, that forms a notch. *See also* **Conventional cut; 2.** Notch resulting from the melting and removal of base metal at the edge of a weld.

UNDERCUT BANK Bank that is undermined by the flow of a stream.

UNDERCUT BEDDING Plane along which failure and sliding may occur because the slope or free face is greater than that of the bedding. *Also called* **Undercut joint.**

UNDERCUT JOINT *See* **Undercut bedding.**

UNDERCUT LINE Forest fire line in the form of a trench to catch rolling embers below a fire line or a slope. *See also* **Gutter trench.**

UNDERCUT U-GROOVE Groove formed in a sheave in the shape of a modified V-groove that has its lower section cut in the shape of a U.

UNDERDRAIN Perforated pipe drain laid in a wedge of stone to intercept groundwater.

UNDERDRIVE Lowest ratio in an auxiliary transmission or multispeed transmission.

UNDEREAVES COURSE Course of eaves tiles in the eaves course.

UNDERFIRE AIR Any forced or induced air that is supplied beneath a grate for burning fuel.

UNDERFLOOR HEATING Incorporation of electric heating cables or hot water pipes into the body of a concrete floor slab.

UNDERFLOW 1. Oversize material from a classifier; **2.** Water movement in the soil, under ice or under a structure.

UNDERFOOT CONDITIONS The surface material a machine is working in.

UNDERGLAZE DECORATION *See* **Decoration.**

UNDERHUNG LEAD *See* **Lead.**

UNDERHUNG SUSPENSION *See* **Suspension.**

UNDERLAYMENT Material placed under finish floor coverings to provide a smooth, even surface.

UNDERLAYMENT NAIL *See* **Nail.**

UNDERLAY MINERAL Sand or granules applied to a sheet of roofing felt prior to application of a layer of asphalt.

UNDERLIFT Device used for towing vehicles by lifting one end of the towed vehicle from under the axle or structural member.

UNDERMINE To erode the soil under a structure or at the foot of a bank. *See also* **Scour.**

UNDERPASS Road, walkway, railroad, etc., crossing another at a lower level.

UNDERPINNING Introduction of support to an existing structure.

UNDERREAM Enlargement of the lower end of an augered or drilled pile that increases its bearing area. *Also called* **Bell.**

UNDERREAM BUCKET *See* **Belling bucket.**

UNDERSANDED Concrete containing an insufficient proportion of fine aggregate to produce optimum properties in the fresh mixture, especially workability and finishing characteristics.

UNDERSHOT Condition resulting from not enough explosive, or a pattern too large for the amount of explosive used, characterized by poor fragmentation and lack of movement.

UNDERSIZE Particles of aggregate passing a designated sieve.

UNDERSLUNG CAR FRAME Elevator car frame to which the hoisting-rope fastenings or sheaves are attached at or below the platform.

UNDERSLUNG SUSPENSION *See* **Suspension.**

UNDERSTORY Portion of the trees or other vegetation in a forest stand below the main canopy level.

UNDERTREAD *See* **Tire construction.**

UNDERWATER DRIVING Driving a pile underwater by operating the hammer under water. Provision must be made for exhaust to the atmosphere for air- or steam-powered hammers.

UNDERWINDING Cable or rope wound and attached so that it stretches from the bottom of a drum to the load. *See also* **Overwinding.**

undfl. *Abbreviation for:* underfloor.

undflg *Abbreviation for:* underflooring.

undgrd *Abbreviation for:* underground.

UNDISTURBED PENETRATION Penetration at 25°C (77°F) of a sample of grease in its container with no disturbance.

UNDISTURBED SAMPLE Sample of cohesive soil taken from a test hole in a condition so little changed that it can be used for laboratory tests.

undnth *Abbreviation for:* underneath.

undp. *Abbreviation for:* underpass.

UNDRESSED LUMBER Rough-cut lumber prior to being planed.

undw. *Abbreviation for:* underwater.

UNEARNED INCREMENT Increase in the value of real property not due to any effort by the owner.

UNEVEN-AGED Forest stands with a wide range of ages and sizes.

UNEVEN-AGED MANAGEMENT Silvicultural system in which individual trees originate at different times and result in a forest with trees of all ages and sizes. Harvest cuts are on an individual-tree selection basis.

unexpl. *Abbreviation for:* unexplainable.

unfav. *Abbreviation for:* unfavorable.

unfin. *Abbreviation for:* unfinished.

UNFINISHED BOLT Low-carbon steel bolt and nut with rough, unfinished shank, used to hold structural steel components in place while field rivets are being driven.

UNFRAMED DOOR Batten, or ledged and braced door.

ungl. *Abbreviation for:* unglazed.

UNGLAZED TILE Hard, dense tile of homogeneous composition throughout, deriving color and texture from the materials of which the body is made.

UNIAXIAL COMPRESSION Compression in only one direction; no forces act in other directions.

UNIAXIAL COMPRESSIVE STRENGTH Strength of a rock sample under uniform compressive stress on one axis only.

unif. *Abbreviation for:* uniform.

UNIFIED SCREW THREAD STANDARD Standard for screw threads used in Canada, the U.K., and the U.S.

UNIFORM BUILDING CODE *See* **Building code.**

UNIFORM LOADS *See* **Loads.**

UNIFORM SAND Sand with most particles of uniform size.

UNILATERAL TOLERANCE Tolerance specified in one direction only.

UNIMPROVED PROPERTY Land that has not been cleared or developed in any way.

UNINTERRUPTED POWER SUPPLY Power supply that provides a continuous source of electric power without any voltage or frequency disturbances when switching from utility power to standby power source.

UNION Type of fitting used to join lengths of pipe for easy opening of a pipe line.

UNION BEND Bend or other fitting with a union at one end.

UNION WAGE BOND *See* **Bond.**

UNIT Specific magnitude of a quantity, set apart by specific definition, that serves as a basis of comparison or measurement for other quantities of the same nature.

UNIT COST Cost of producing a unit of work, measured by time, volume, or pieces completed.

UNIT HEATER Self-contained, factory-assembled heating unit including a heating element, fan, and motor with a directional outlet.

UNIT MASONRY Structural element consisting of masonry units bonded by mortar, grout, or both.

UNIT OF BOND Shortest length of a brick or masonry course that repeats itself.

UNIT PRICE Price established for a given quantity of material or labor.

UNIT PRICE CONTRACT Contract based on the unit price established for materials and labor, irrespective of the quantity.

UNIT SALE Sale of timber based on an agreed 'price per unit of material for timber volume' with payment based on scaled volume.

UNITS-OF-WORK METHOD OF ACCOUNTING *See* **Method of accounting.**

UNIT-SUSPENDED FURNACE WALL *See* **Wall.**

UNIT VENT Vent pipe that serves two or more traps.

UNIT WATER CONTENT 1. Quantity of water per unit volume of freshly mixed concrete, often expressed as pounds or gallons per cubic yard; **2.** Quantity of water on which a water-cement ratio is based, not including water absorbed by the aggregate.

UNIT WEIGHT Weight per unit volume of a material, expressed as grams/cm³, kg/m³, or lb/ft³.

univ. *Abbreviation for:* universal; university.

UNIVERSAL EXPANSION JOINT *See* **Joint.**

UNIVERSAL GRINDING MACHINE Machine on which cylindrical, internal, and surface grinding can be done, usually used for tool-room work.

UNIVERSAL JOINT Flexible or articulated coupling providing transmission of power through an angle.

UNIVERSAL-MILL PLATE Steel plate in which the longitudinal edges have been formed by a rolling process during manufacture.

UNIVERSAL MOTOR Electric motor, usually of less than 1 hp capacity, that can run on either AC or DC.

UNIVERSAL PROCESSOR Demolition attachment for concrete cracking and rebar shearing. Commonly fitted to the stick of excavators and backhoes and powered from the hydraulic circuit of the base machine. The equipment consists of jaws that can be fitted with different types of blades for crushing and cutting.

UNIVERSAL SNIPS *See* **Tin snips.**

UNIVERSAL TIED EXPANSION JOINT *See* **Joint.**

unk. *Abbreviation for:* unknown.

unlim. *Abbreviation for:* unlimited.

unliq. *Abbreviation for:* unliquidated.

UNLOADED DEFLECTION Vertical distance between the chord and the unloaded skyline, measured at midspan.

UNLOADED VEHICLE WEIGHT Weight of a vehicle with maximum capacity of all fluids necessary for its operation, but without cargo or occupants.

UNLOADER *See* **Load pusher.**

UNLOADING Release of contaminant that was initially captured by a filter medium.

UNLOADING BULKHEAD Steel plate that ejects waste out the rear doors of an enclosed transfer trailer. The movable wall is propelled by a telescoping, hydraulically powered cylinder that traverses the length of the trailer.

UNLOADING VALVE *See* **Valve.**

unltd *Abbreviation for:* unlighted.

UNMANAGED FOREST LAND Forest land that is not subject to management under a forest management plan.

UNMERCHANTABLE WOOD Material that is unsuitable for conversion to industrial wood products due to size, form, or quality. May include rough, rotten, and dead trees, the top limbs and cull sections from harvested trees, or small and noncommercial trees.

unmkd *Abbreviation for:* unmarked.

unpd *Abbreviation for:* unpaid.

unpntd *Abbreviation for:* unpainted.

UNPROTECTED OPENING Doorway, window or opening other than one equipped with a closure having the required fire-protection rating, or any part of a wall forming part of the exposing building face that has a fire-resistance rating less than that required for an exposing building face.

uns. *Abbreviation for:* unsafe; unsanded; unsymmetrical.

UNSAFE CONDITION Any condition that could cause undue hazard to life, limb or health of any person authorized or expected to be on or about the premises.

UNSANDED PANEL Interior or exterior sheathing-grade plywood panel designed for utility applications and left unsanded for greater stiffness, strength, and economy.

UNSATURATED COMPOUND Compound having more than one bond between two adjacent atoms, usually carbon atoms, capable of adding other atoms at that point to reduce it to a single bond.

unserv. *Abbreviation for:* unserviceable.

unsnd *Abbreviation for:* unsound.

UNSOUND 1. Not firmly made, placed, or fixed; **2.** Subject to deterioration or disintegration during service exposure.

UNSOUND KNOT Knot in lumber that is not as solid as the wood surrounding it, or that is loose.

UNSPRUNG WEIGHT Vehicle weight of all components not supported by the springs, including axle, brakes, tires and wheels. *See also* **Sprung weight,** and **Vibrating mass.**

unstab. *Abbreviation for:* unstable.

UNSTABLE Structure that is apt to fail as a whole, usually by sliding or overturning.

UNSTABLE FRAME Frame containing too few members or of too little stability to be a perfect frame.

UNSTABLE LIQUID Liquid that is chemically reactive to the extent that it will vigorously react or decompose at or near normal temperature and pressure conditions and that is chemically unstable when subject to impact.

UNSTABLE SOIL Earth material, other than running, that because of its nature or the influence of related conditions, cannot be depended upon to remain in place without extra support, such as would be furnished by a system of shoring.

unstdy *Abbreviation for:* unsteady.

UNSUPPORTED-FILM ADHESIVE *See* **Adhesive.**

unsvc. *Abbreviation for:* unserviceable.

UNSYMMETRICAL FOOTING Footing in which the resultant load does not coincide with the center of gravity of the area of the footing.

UNTREATED JOINT Drywall joint without joint compound, tape, or battens.

UNTRIMMED FLOOR Floor laid only on common joists.

UNWORKED PENETRATION Penetration at 25°C (77°F) of a sample of lubricating grease that has received only minimum disturbance in transferring to a grease worker cup or dimensionally equivalent container. *See also* **Penetration.**

up. *Abbreviation for:* upper.

UP *Abbreviation for:* used parts.

UP ANGLE Ceramic trim tile with one rounded or curved corner, used to finish off an inside angle.

UPDATING Revision of a network to reflect progress to date. The date of the update becomes the new start date for the project and completed items are eliminated.

UP-CUTTING *See* **Down-cutting/up-cutting.**

UPEND 1. Swapping of wire rope ends on a yarder or loader to distribute wear; **2.** In yarding, to cause a log to flip end-for-end, either intentionally or inadvertently, by coming in contact with a stationary object.

UPEND A LOG Intentionally turn a log end-for-end, usually for loading onto trucks.

UPENDER *See* **Swinger.**

UPFEED DISTRIBUTION Water system that requires pumps to develop pressure.

UPHOLSTERER'S HAMMER *See* **Hammer.**

UPLIFT Upward force exerted on a pier, pile, or other buried structural element by soil or rock expanded through frost action or by hydraulic pressure.

UPLIFT CAPACITY Ability of a driven pile to resist uplift and overturning forces resulting from wind and hydrostatic pressure. *Also called* **Pull-out resistance,** and **Tension capacity.**

UPPER Portion of a crane located above the turntable bearing. *See also* **Upperstructure.**

UPPER BOUND LOAD Load, computed on the basis of an assumed mechanism, that will always be at best equal to or greater than the true ultimate load.

UPPER CAB *See* **Operator's cab.**

UPPERMOST AQUIFER Geologic formation nearest the natural ground surface that is an aquifer, as well as any lower aquifers that are hydraulically interconnected with this aquifer within the area under consideration.

UPPER REVOLVING FRAME Main structure of the upper section of a crane that serves as a mount for other components of the upper.

UPPER SPREADER *See* **Floating harness.**

UPPER STEM PORTION Sawtimber tree bole extending from above the merchantable top to a minimum 100 mm (4 in.) top diameter outside bark or to the point where the central stem breaks into limbs.

UPPERSTRUCTURE Rotating superstructure of a mobile crane or other mobile equipment, less any front-end attachments. *Also called* **Revolving superstructure, Superstructure,** and **Upper.**

UPRIGHT Vertical member of a trench shoring system placed in contact with the earth and usually positioned so that it and other individual members do not contact each other.

UPRIGHT SPRINKLER Type of sprinkler installed with its deflector above the orifice.

UPSET To enlarge the end of a bar by shortening it.

UPSET PRICE *See* **Guaranteed maximum cost.**

UPSHOT BURNER Burner on the floor of the firing chamber of a gas-fired boiler that shoots upward.

UPSTAND Portion of a flashing or roof covering that turns up beside a wall without being tucked into it and that is usually covered with a stepped flashing.

UPSTAND BEAM Concrete beam that extends above the slab or structure it is supporting.

UPTHRUST Device on the hanger assembly that limits the vertical motion of an elevator panel.

ur. *Abbreviation for:* urinal.

urb. *Abbreviation for:* urban.

URBAN AREA Distribution of community functions (residential, commercial, retail, industrial, public amenity).

URBAN GROWTH Expansion and development of housing, commercial, and light industrial facilities within and beyond an established community. *Also called* **City growth.**

URBAN PLANNER Professional qualified in the orderly and appropriate growth of existing and new communities. *Also called* **City planner,** and **Community planner.**

URBAN PLANNING Planning function that considers all the elements comprising a well-balanced urban area, and their various interrelationships. *Also called* **Community planning.**

URBAN PROPERTY Property over which a duly constituted local government has defined authority, including the provision of such services as potable water, sewerage, etc.

UREA FORMALDEHYDE Interior thermosetting resin system commonly used in the manufacture of particleboard.

UREA FORMALDEHYDE ADHESIVE *See* **Adhesive.**

UREA RESIN ADHESIVE *See* **Adhesive.**

URETHANE Generic term for ethylene carbonate. The material is used extensively for a base polymer in elastic sealants and waterproofing. Gives strength and flexibility that is sensitive to moisture.

urg. *Abbreviation for:* urgent.

URINAL Plumbing fixture used for urinating.

urnl *Abbreviation for:* urinal.

u.s. *Abbreviation for: ubi supra* (in the place above mentioned).

USABLE LIFE 1. Pot life of a mixture prone to setting or hardening; **2.** Useful life of a product as estimated at time of manufacture.

USABLES Secondary materials recovered from discards or waste streams that are salable in their existing form.

USC *Abbreviation for:* under separate cover.

USEFUL LIFE Period over which a building is expected to remain viable.

USER 1. Intended owner of a project or occupant of a building constructed or renovated as the result of a contract for work; **2.** Individual having access to or control over equipment.

USER NEEDS *See* **Requirements.**

USS *Abbreviation for:* ultimate shear strength; United States standard.

USURY Rate of interest charged on a loan greater than that permitted by law.

U-TIE Heavy wire wall tie in the shape of a U.

util. *Abbreviation for:* utilities; utility; utilization.

UTILITY Service provided by a public agency, can include electrical energy, potable water, sewerage, gas, telephone, or cable-vision.

UTILITY-CONNECTED SYSTEM Natural energy system (wind power, photovoltaic cells) that is connected to the utility grid, enabling the system to draw power from a public system and, if permitted, to pass excess electricity into the public system.

UTILITY CRANE Mobile crane designed to perform a range of tasks.

UTILITY GRADE *See* **Lumber grades.**

UTILITY GRID System of electrical generating facilities and transmission lines developed and maintained by a utility company.

UTILITY PALLET *See* **Forks.**

UTILITY ROOM Space in a dwelling where the heating plant, laundry equipment, water heater, and other utilities are grouped.

UTILITY/SERVICE TRUCK *See* **Truck.**

UTILIZATION EQUIPMENT Equipment that utilizes electric energy for mechanical, chemical, heating, lighting, or similar useful purposes.

UTILIZATION STANDARD Utilization limit (stump height and top diameter inside bark) that defines the trees considered commercially salable, and therefore the dimensions of all trees that must be cut and removed from Crown land harvesting operations. *See also* **Close utilization.**

UTILIZATION SYSTEM System that provides electric power and light for employee workplaces, and includes the premises wiring system and utilization equipment.

UTS *Abbreviation for:* ultimate tensile strength.

UV *Abbreviation for:* ultraviolet.

U-VALUE 1. Overall coefficient of heat transmission; **2.** Standard measure of the rate at which heat will flow through a unit area of a material of known thickness.

UV STABILIZER Chemical compound that, when mixed with a thermoplastic, selectively absorbs ultraviolet light rays.

U/W *Abbreviation for:* underwriter.

UYP *Abbreviation for:* upper yield point.

v *Abbreviation for: vide* (See); vent; ventilator.

V *Abbreviation for:* volt.

VA *Abbreviation for:* volt-amperes.

vac *Abbreviation for:* volts, alternating current.

vac. *Abbreviation for:* vacant; vacuum.

VACANCY Housing unit that is not occupied.

VACANCY RATE Ratio between the number of vacant housing units and the total number of developed housing units in a building, complex, area, city, etc.

VACATION *See* **Abandonment.**

VACUUM Air pressure below atmospheric pressure.

VACUUM ASSIST BRAKE *See* **Brake.**

VACUUM BREAKER Device that prevents development of a vacuum in a pipe containing fluids and thus prevents backflow.

VACUUM CLEANING PLANT *See* **Central vacuum system.**

VACUUM CONCRETE Concrete from which excess water and entrapped air are extracted by a vacuum process before hardening occurs.

VACUUM GAUGE *See* **Gauge.**

VACUUM HEATING SYSTEM *See* **Heating system.**

VACUUM LIFTING Use of the negative pressure of a partial vacuum generated in suction cups to lift heavy objects, particularly large sheets of glass and glazing systems.

VACUUM LOADER Specialized vacuum truck body used for street sweeping, sewer cleaning, leaf pickup, and other municipal and industrial applications.

VACUUM MAT Stiff flat metal screen covered by a filtering fabric, the back of which can be kept under a partial vacuum, used in making vacuum concrete.

VACUUM PROCESS Process used to cool concrete aggregates by causing surface water to evaporate in a partial vacuum.

VACUUM PUMP *See* **Pump.**

VACUUM SATURATION Process for increasing the amount of filling of the pores in a porous material with a fluid by subjecting the porous material to reduced pressure in the presence of the fluid.

VACUUM SEPARATOR *See* **Separator.**

VACUUM TANK Tank for storing gas at less than atmospheric pressure.

val. *Abbreviation for:* valley; value.

VALANCE Decorative board across a window head, fastened to the wall, used to conceal the top of window curtains.

VALLEY Interior angle formed by the slopes of two intersecting roof surfaces.

VALLEY BOARD Board fixed on and parallel to a valley rafter.

VALLEY CRIPPLE JACK When the ridges of the two roofs are on different levels, a rafter framed from the supporting valley rafter to the valley of the addition.

VALLEY DRAINAGE Where water collects in a natural or man-made depression, typically along a gutter. *See also* **Crowned drainage.**

VALLEY FLASHING Sheets or strips of metal worked in with shingles or other roofing materials along the valley of a roof so as to form a gutter to conduct rainwater from the roof.

VALLEY GUTTER Channel used to carry water from the junction of multigabled roofs.

VALLEY JACK Roofing member that extends from the valley rafter to the ridge of the roof. *See also* **Truss.**

VALLEY RAFTER *See* **Rafter.**

VALLEY SHINGLE Roofing shingle cut to fit against a valley gutter.

VALLEY TILE Large, concave tile used to form a valley without metal flashing.

VALUATION Estimation of worth.

VALUE 1. Quantity or amount for which a symbol stands; **2.** Amount of light and dark in a hue.

VALUE ANALYSIS *See* **Value Engineering.**

VALUE ENGINEERING Analysis of materials, processes, and products in which functions are related to cost and from which a selection may be made for the purpose of achieving the required function at the lowest overall cost consistent with the requirements for performance, reliability, and maintainability. *Also called* **Value Analysis.**

VALVE 1. Device that directs the flow and direction of hydraulic fluid to a specific actuator; **2.** Mechanism that directs the flow rate and flow direction of a fluid contained within a system. The following are some of the many types of valve:

> **Angle:** Similar to a globe valve but with pipe connections at right angles to each other.
>
> **Angle stop:** Type of shutoff valve, often mounted beneath sinks, lavatories, and toilets.
>
> **Antisiphon:** Type of check valve that prevents siphoning of potentially contaminated water back into a potable supply system.
>
> **Backwater:** Flap valve that prevents the reverse flow of water or sewage in a pipe or channel.
>
> **Ball:** Valve in which the rate of flow is regulated by a drilled ball that rotates against a flexible seat.

Bypass: Device which allows flow to divert around a filter when differential pressure becomes too great. *Also called* **Relief valve.**

Check: Valve that permits flow of fluid in only one direction. *Also called* **Nonreturn valve.**

Closed center: Valve in which all ports are closed in the center position.

Compression: Faucet or valve designed to stop the flow of water by the action of a washer closing against a seat.

Control: Device that controls the flow of oil within a hydraulic system.

Counterbalance: Valve that regulates fluid flows by maintaining resistance in one direction, but allows free flow in the other. *Also called* **Holding valve.**

Dead weight relief: Valve in which the unrelieved weight of the plug is the force that tends to keep the valve closed. The valve opens when the pressure increases sufficiently to lift the plug against the force of gravity.

Detector check: Swing check valve, typically used in an automatic sprinkler system, that has its clapper weighted to divert small flows away from the main line through a by-pass, where it will be measured by a meter.

Differential pressure: Valve operated by opposing pressures of different values.

Directional control: Valve whose primary function is to direct flow through selected passages.

Discharge: Control valve for reducing and increasing flow in a pipe, but not stopping the flow.

Diversion: Valve that permits flow to be directed into any one of several pipes.

Diverter: Hydraulic valve that permits a change in the direction of flow of a fluid.

Dry: *See* **Dry pipe valve,** below.

Dry-pipe: Valve used on an automatic sprinkler system to control water supply to a dry-pipe system and, under defined conditions, to cause an alarm to sound. *Also called* **Dry valve.**

Faucet: Valve that will control the flow of water.

Flap: Check valve with a hinged disk that opens when the flow is normal and closes by gravity or by the flow when the flow tends to reverse.

Flapper action: Valve design in which output control pressure is regulated by a pivoted flapper in relation to one or two orifices.

Float: Valve actuated by a floating ball on a lever; a ball cock.

Flow control: Valve whose primary function is to control flow rate.

Flow control (deceleration): Flow control valve that gradually reduces flow rate to provide deceleration.

Flow divider: Valve that divides a flow into two streams.

Flush: 1. Valve in the bottom of a water closet tank that controls the flow of water into the toilet bowl; **2.** Pressure-controlled valve that controls the flow, and duration of flow, of water into the bowl of a water closet or to a urinal.

Foot: Check valve fitted to the inlet of a pump suction hose.

Footbrake: Driver-controlled valve that controls the air pressure delivered to or released from the service brake chambers.

Fourway: Valve that can be set to direct flow to all or any of four distinct settings.

Gas cock: Shut-off valve on a gas line.

Gate: Valve that regulates flow within a pipe. *Also called* **On-off valve.**

Globe: Valve that allows for throttling the flow of water in a pipe.

Holding: Device placed in a hydraulic circuit to prevent loss of fluid from a load-bearing cylinder. This device also has thermal relief properties that prevent temperature increases from causing cylinder damage.

Holding (integral): Device attached directly to a cylinder to protect against fluid loss from the cylinder.

Hydraulic control: Mechanical device to divert or control the flow of fluid in a hydraulic circuit.

Hydraulic relief: Mechanical device used to limit the pressure in a hydraulic circuit.

Indicator: Valve that indicates by a sign, an open or shut position.

Integral holding: *See* **Holding valve (integral),** above.

Key: Valve operated by a removable key.

Lowering: Valve that allows a supported load to travel downward at a rated speed.

Mixing faucet: Faucet for hot and cold water supply that discharges through a common spout.

Moisture ejection: Valve that automatically expels moisture accumulation from the compressed air tanks of air-actuated brake systems.

Needle: Valve with an externally adjustable tapered closure that regulates the flow passage.

Nonreturn: *See* **Check valve,** above.

On/off: *See* **Gate valve,** above.

Open-center: Valve in which all ports are interconnected in the center position.

Open spool: Valve that permits the flow of hydraulic oil in an actuator when that particular function's control lever is placed in neutral; commonly employed in circuits using hydraulic motors in order to provide the components with enough flow to allow them to glide to a stop once the valve is returned to neutral.

Pilot: Valve applied to operate another valve or control.

Pilot-operated: Valve in which operating parts are actuated by pilot pressure.

Pilot-operated check: Used in conjunction with outrigger jacks, this device is mounted on the side of the vertical jack to prevent the loss of hydraulic fluid from the jack cylinder. Due to the shock loading this cylinder must endure, even slight cylinder retraction cannot be tolerated. Therefore, this cylinder is not equipped with a holding valve that could allow fluid to escape; **2.** One way-flow valve whose operating parts are actuated by pilot pressure.

Plug cock: Valve where the fluid passes through a hole in a tapered plug, and that is closed by turning the plug through 90°.

Pneumatic: Valve for controlling gas flow or pressure.

Pop-off: Safety valve that opens automatically when pressure or some other factor exceeds a predetermined value.

Poppet: Mushroom-shaped valve that rests on a circular seat and that is opened by pressure on the stem (usually by the cam of a rotating shaft) and closed by the action of a confined spring mounted round the stem.

Prefill: Valve that permits full flow from a tank to a 'working' cylinder during the advance portion of a cycle, which permits the operating pressure to be applied to the cylinder during the working portion of the cycle, and permits free flow from the cylinder to the tank during the return portion of the cycle.

Pressure control: Valve whose primary function is to control pressure. There are several types:

Counterbalance: Pressure control valve that maintains back pressure to prevent a load from falling.

Decompression: Pressure control valve that controls the rate at which the contained energy of compressed fluid is released.

Load-dividing: Pressure control valve used to proportion pressure between two pumps in series.

Pressure-reducing: Pressure control valve whose primary function is to limit outlet pressure.

Pressure-reducing (gas): Valve used to reduce gas line pressure to usable limits of a gas carburetor.

Pressure-reducing (water): Valve used to reduce water pressure, typically between the main and an engine cooling system.

Pressure-relief: Pressure control valve whose primary function is to limit system pressure.

Pressure-unloading: Pressure control valve whose primary function is to permit a pump or compressor to operate at minimum load.

Pressure regulator: Valve used to automatically reduce and maintain pressure.

Pressure sensing: Device similar to an electrical pressure switch in which a signal to be sensed enters a control point and actuates a mechanism that, at the proper pressure level, causes one or more flow passages to change condition. Removal of the signal allows the pressure sensing valve to reset.

Relay: Valve that is actuated upon receipt of a signal or because of an activity of another valve or device.

Relay emergency: Combination relay valve with provision for the automatic application of trailer brakes in the event pressure is lost in the trailer supply air line.

Relief: Valve whose primary function is to limit system pressure by opening at a preset pressure and/or temperature. *See also* **Bypass valve.**

Rotary plug: Type of valve in which a ported sleeve or plug is rotated past an opening in the valve body.

Safety: Combination temperature and pressure relief valve required on hot water tanks.

Safety shutoff: Valve that automatically and completely shuts off a process or machine upon detection of an out-of-limit condition.

Sequence: Valve whose primary function is to direct flow in a predetermined sequence.

Servo-: Valve that modulates output as a function of an input command.

Shutoff: **1.** Valve that operates fully open or fully closed; **2.** Valve installed in a water line whenever a cutoff is required.

Shuttle: Connective valve that selects one of two or more circuits because of flow or pressure changes between the circuits.

Slave: Valve that helps protect gears and components in a transmission's auxiliary section by permitting range shifts to occur only when the transmission's main gearbox is in neutral. Air pressure from a regulator signals the slave valve into operation.

Stop: Valve used to turn on or close a supply.

Tandem-center: Valve in which advance and return ports are closed in the center position and the pump and tank ports are open.

Thermo-regulating: Heat-actuated valve that limits the amount of municipal or raw cooling water into a system to conserve water and regulate cooling.

Two- or three-way: Valve having two or three positions to give various selections of flow.

Two-, three-, or four-way: Directional control valve having 2, 3 or 4 ports for direction of flow.

Unloading: Valve that bypasses flow to a tank when a set pressure is maintained on its pilot port.

VALVE ACTUATOR Valve part(s) through which force is applied to move or position flow-directing elements.

VALVE BAG Paper bag, either glued or sewn, made of four or five plies of kraft paper and incorporating a plastic membrane, completely closed except for a self-sealing paper or plastic-film valve through which the contents, usually a powder or granulated compound, are poured.

VALVE BANK Enclosed grouping of valves fed by its own hydraulic pump.

VALVE BODY Main part of a valve into which the stem and other parts are installed.

VALVE BRIDGE Part that allows two valves to be operated by a single rocker arm. *Also called* **Crosshead**

VALVE CLEARANCE Air gap between the end of a valve stem and the valve lifter or rocker arm.

VALVE HOLE Round hole in a single-piece wheel rim to accommodate a valve stem.

VALVE INTERRUPTER Hydraulic control lockout system that restricts the flow of hydraulic oil to a machine's actuators, therefore prohibiting machine movements that may endanger the machine or its operator. *See also* **Control lever lockout.**

VALVE MOUNTING Mounting characteristics of a valve.

VALVE OVERLAP Time in the cylinder cycle operation of an engine when both intake and exhaust valves are open.

VALVE ROTATOR Device installed on intake and exhaust valves permitting them to rotate during engine operation.

VALVE SEAT Stationary portion of a valve that, when in contact with the movable portion, stops flow.

VALVE SLOT Opening in a multipiece wheel rim to accommodate the tire tube valve stem.

VALVE TRAIN Components making up a valve and its operating mechanism.

van. *Abbreviation for:* vanity.

VANADIUM *See* **Alloying elements.**

VANDALISM AND MALICIOUS MISCHIEF INSURANCE Indemnity against loss or damage to property or equipment caused by vandalism or willful damage or destruction.

VANDOSE ZONE Subsurface zone between the surface and the water table that contains water under pressure less than that of the atmosphere.

VANE Any of several flat or curved pieces set around an axle, moved by the passage of air or water.

VANEAXIAL FAN *See* **Fan.**

VANE HYDRAULIC PUMP *See* **Pump, hydraulic.**

VANE PUMP *See* **Pump.**

VANE RATIO Depth of a vane to the shortest opening between two adjacent vanes.

VANE SHEAR TEST *See* **Shear test.**

VANE-TYPE HYDRAULIC PUMP *See* **Pump.**

VANG LINE Side line reeved to a derrick boom, used to swing the boom.

VANISHING POINT Point to which parallel lines converge in a perspective drawing.

VANITY Counter or cabinet to support a basin or sink in a bathroom or lavatory.

VAN TRAILER *See* **Trailer.**

vap. *Abbreviation for:* vapor.

VAPOR Gaseous phase of a substance normally in a liquid or solid state.

VAPOR BARRIER 1. Membrane placed under concrete floor slabs that are placed on grade, intended to retard transmission of water vapor. *See also* **Dampproofing,** and **Waterproofing; 2.** Material used to retard the movement of water vapor into walls and ceilings to prevent condensation. *Also called* **Moisture barrier.**

VAPOR BLASTING Cleaning method for metal parts where a solvent or other chemical is blown, under significant air pressure, against the object being cleaned.

VAPOR DENSITY Weight of a vapor or gas compared to the weight of an equal volume of air.

VAPOR DIFFUSION Transfer of water in a partially dry solid from regions of high concentration to those of low.

VAPOR HEATING *See* **Heating system.**

VAPOR LOCK Vapor trapped in a service line that prevents the normal flow of a fluid; typically, the boiling of gasoline in a fuel line that prevents liquid fuel from reaching the carburetor.

VAPOR PLUME Stack effluent consisting of flue gas made visible by condensed water droplets or mist.

VAPOR PRESSURE *See* **Pressure.**

VAPOR RETARDER Material that retards the flow of water vapor into walls.

VAPOR SEAL Moisture resistant material used in a laminated material to prevent the passage of moisture through the finished product.

vap. prf *Abbreviation for:* vaporproof.

var. *Abbreviation for:* variable; variance; variant; variation; various; varying.

VARIABLE AREA PLOT SAMPLING METHOD Method of timber cruising commonly used for industrial timber cruising in which the sampling area (plot size) varies with tree diameter.

VARIABLE CAN RATE Charge made for solid waste services based on the volume of waste generated measured by the number of containers set out for collection.

VARIABLE CAPACITY *See* **Torque converter.**

VARIABLE COST Operational cost that results from running a machine, calculated on an hourly basis; includes the cost of labor and items such as fuel, oil, wire rope, and other replacement parts. *Also called* **Operating cost.**

VARIABLE DISPLACEMENT System in which hydraulic volume generated from the pump(s) can be varied in accordance with the number of actuators being used at any one time, and the hydraulic system's need for hydraulic fluid.

VARIABLE DROOP Percent of regulation of an engine from high-idle, no-load speed to full-load speed.

VARIABLE INCREMENT SCREED Paving screed that can be added to in combinations of different length extensions.

VARIABLE-MATURITY MORTGAGE *See* **Mortgage.**

VARIABLE-RATE MORTGAGE *See* **Mortgage.**

VARIABLE RESISTOR Device capable of increasing or decreasing resistance to the passage of electrical current, such as a potentiometer or rheostat.

VARIABLE-SIZE CREW *See* **Collection method.**

VARIABLE TIME Those times (i.e., haul, return) that vary with distance and speed.

VARIABLE-WIDTH STRIKE-OFF Asphalt paver screed attachment that hydraulically extends and retracts to adjust paving width.

VARIANCE 1. Square of the standard deviation in survey readings; average of the squares of the deviations of all observations; **2.** Permission granted by a zoning authority for a specified difference to established zoning requirements; **3.** Difference between budgeted and expected performance and actual performance. There are several measures, including:

    Budget: Difference between the total overhead incurred and the total budgeted overhead, based on standard volume for work done.

    Efficiency: 1. Quantity variance for labor; **2.** Variable overhead rate, times the difference between actual volume and standard volume.

    Price: Difference between the standard cost of direct materials and/or labor and the actual cost, resulting from changes in input prices.

    Quantity: Difference between the standard cost of direct materials and/or labor and the actual cost, resulting from changes in the input quantities.

    Rate: Price variance for labor.

    Spending: Equivalent of price variance for overhead costs.

    Volume: Difference between applied fixed overhead and budgeted fixed overhead.

See also **Adjustment.**

varig. *Abbreviation for:* variegated.

varn. *Abbreviation for:* varnish.

VARNISH 1. Preparation of resinous substances dissolved in oil (oil varnishes) or in a quick-evaporating liquid like alcohol (spirit varnish) used to seal and finish wood and, depending on the mix of ingredients, to impart a continuous transparent or translucent coating. Varnish is also mixed with pigments to make enamels. *Also called* **Oil varnish,** and **Spirit varnish; 2.** Materials generated by hydraulic fluid due to oxidation, thermal instability, or other reactions. These materials are insoluble in the hydraulic fluid and generally are found as brownish deposits in the work surfaces; **3.** Non-wipeable deposit on engine parts that can intefere with engine operation.

VARVED SILT or CLAY Fine-grained, glacial lake deposit with alternating layers of silt or fine-grained sand and clay, formed by variations in sedimentation from winter to summer during the year.

VAT *Abbreviation for:* vinyl asbestos tile.

VAULT 1. Enclosure above or below ground that workers may enter, used for installing, operating, or maintaining equipment and/or electric cable; **2.** Arched ceiling or roof. The different configurations can be identified as:

    Annular: Vault formed by a series of semicircular rings placed adjacent to each other.

    Barrel: The simplest form of vault, consisting of a continuous semicircular arch. *Also called* **Tunnel vault.**

    Cloister: A dome rising on a square or polygonal base, the curved surfaces separated by groins.

    Cross: Produced by the right-angle intersection of two identical barrel vaults.

    Double: Vault formed by a double wall.

    Fan: Concave-sided semi-cones, meeting or nearly meeting at the apex.

    Rib: Framework of diagonal arched ribs carrying cells that cover the spaces between.

    Tunnel: *See* **Barrel vault,** above.

VAULT LIGHT *See* **Pavement light.**

VB *Abbreviation for:* valve box; vapor barrier.

V-BEAM SHEETING Rolled light-metal sheet having one or more corrugations formed by three adjacent flat surfaces running longitudinally.

V-BLADE *See* **Snow plow.**

vc *Abbreviation for:* vertical curve.

VCG *Abbreviation for:* vertical center of gravity.

vcp *Abbreviation for:* vitrified clay pipe.

V-CUT Short drill holes inclined to the center of a tunnel face, drilled so that when the first shots are detonated, a wedge of rock is removed allowing relief for the remaining rock when the delay exploders initiate the charges in the other drill holes.

vdc *Abbreviation for:* volts, direct current.

veg. *Abbreviation for:* vegetation.

vehic. *Abbreviation for:* vehicle; vehicular.

VEHICLE 1. Device on wheels or runners for conveying persons or objects; **2.** Any means of carrying, conveying, or communicating; **3.** Liquid carrier or binder of solids.

VEHICLE CURB WEIGHT Actual, or the manufacturer's estimated weight of a vehicle in operational status with all standard equipment and weight of fuel, lubricants, and coolant.

VEHICLE HANDLING Response or controllability of a vehicle under normal driving conditions.

VEHICLE IDENTIFICATION NUMBER (VIN) Seventeen-character alphanumeric inscription assigned to a vehicle by the manufacturer for identification purposes.

VEHICLE MAINTENANCE EXPENSE Total expense of all labor, materials, services, and equipment used to repair and service vehicles and machines.

VEHICLE MAXIMUM LOAD ON THE TIRE Load on any individual tire that is determined by the distribution to each axle of the total vehicle weight, divided by the number of wheels on the axle.

VEIN Layer, seam, or narrow irregular body of material different from surrounding formations.

vel. *Abbreviation for:* velocity.

VELOCITY 1. Rate of change of distance with time; **2.** Rate or speed at which a detonation wave travels through an explosive.

VELOCITY HEAD *See* **Head**

VELOCITY PRESSURE *See* **Pressure.**

ven. *Abbreviation for:* veneer; Venetian.

VENDEE'S LIEN Lien against property under contract of sale to secure the deposit paid by a purchaser.

VENEER 1. Thin sheet of wood; **2.** To cover with a material having an attractive or superior surface; **3.** Single wythe of masonry for facing purposes, not structurally bonded.

VENEER BOLT *See* **Bolt.**

VENEERED CONSTRUCTION Construction system in which thin layers of a facing material are applied to a structural system.

VENEER GRADE Standard grade designation of softwood veneer used in plywood panel manufacture. The six grades are:

Special order natural-finish veneer: Select all-heartwood or all sapwood. Free of open defects. Allows some repairs.

A: Smooth and paintable. Neatly made repairs permissible. Also used for natural finish in less demanding applications.

B: Solid-surface veneer. Router or sled repairs and tight knots permitted.

C plugged: Improved C veneer with splits limited to 3 mm (0.125 in.) in width and knotholes and borer holes limited to 6 x12 mm (0.25 x 0.5 in.).

C: Knotholes to 25 mm (1 in.). Occasional knotholes 12 mm (0.5 in.) or larger permitted provided the total width of all knots and knotholes within a specified section does not exceed certain limits. Limited splits permitted. Minimum veneer grade permitted in exterior-type plywood..

D: Permits knots and knotholes to 75 mm (3 in.) in width and 12 mm (0.5 in.) or larger under certain specified limits. Limited splits permitted.

VENEER SAW *See* **Saw.**

VENEER TIE Wall tie used to attach a veneered facing to its backing.

VENEER WALL *See* **Wall.**

VENETIAN Type of terrazzo topping that incorporates large chips of stone.

VENETIAN BLIND Screen consisting of horizontal or vertical slats that can be pivoted about their long axis to cause more, or less obstruction to light or the passage of air, and that can be raised, or drawn aside to partially or fully vacate the opening in which they are mounted.

VENETIAN DOOR Door with side lights.

V-ENGINE Engine configuration where the cylinders are arranged to form a V about the crankshaft.

venr *Abbreviation for:* veneer.

VENT 1. Device placed to permit the passage of air

from one space to another; **2.** Shallow channel or hole cut into a mold to allow air to escape as it is being displaced by a liquid.

vent. *Abbreviation for:* ventilate; ventilation; ventilator.

VENT CONNECTOR Part of a venting system that conducts the flue gases or vent gases for the flue collar of a gas appliance to the chimney or gas vent, and that may include a draft control.

VENTED FORM Form so constructed as to retain the solid constituents of concrete and permit the escape of water and air.

VENTED MANIFOLD Manifold that is open to the atmosphere and that returns fluid to a reservoir. *See also* **Manifold.**

VENTILATED Provided with a means to permit circulation of air sufficient to remove an excess of heat, fumes, or vapors.

VENTING SYSTEM Assembly of pipes and fittings that connects a drainage system with outside air to assure circulation of air and the protection of trap seals within the system.

VENT PIPE 1. Pipe or flue connecting any interior space to the outside atmosphere for the purpose of ventilation; **2.** Pipe connecting a plumbing fixture or its drain to the vent stack; **3.** Small-diameter pipe used in concrete construction to permit escape of air in a structure being concreted or grouted.

VENT STACK Vertical pipe connecting all the individual vent pipes to carry off foul air and gases from a building, and in particular from the drainage system with which a building is equipped.

VENTURE CAPITAL Imprecise term often used to describe equity investment in projects where there is limited record of past success in similar ventures by the promoters.

VENTURI Pressure jet that draws in and mixes two or more gases or fluids.

VENTURI FLUME Tube or channel having a constricted section to smooth pressure surges in materials being conveyed or to increase pressure or flow through that section.

VENTURI METER Meter for measuring the flow within a closed pipe having a throat followed by an expansion to normal diameter.

VERANDA Open gallery or balcony with a roof.

VERDIGRIS Green basic acetate of copper that forms when copper is exposed to air and that protects the surface from further deterioration.

VERGE The extreme edge of anything, in particular of a roof. *Also called* **Margin.**

VERGE BOARD The vertical board under the verge of a gable, sometimes molded to a decorative design, including piercings, or painted. *Also called* a **Barge board.**

VERGE FILLET Batten fastened on a gable wall to the ends of the roof battens, over which the roofing shingles hang.

VERIENDEEL BOOM Open-side boom design that

provides a lightweight boom with high strength characteristics. The boom is fabricated by cutting large circular holes in the boom's side plate and then reinforcing these holes with strips of T-1 steel welded around the circumference of the circle.

verif. *Abbreviation for:* verification; verify.

VERIFICATION Statements sworn before a qualified authority as to the correctness of the contents of an instrument.

verm. *Abbreviation for:* vermiculite.

VERMICULATED *See* **Rustication.**

VERMICULITE 1. Group name for certain platy minerals, hydrous silicates of aluminum, magnesium, and iron, characterized by marked exfoliation on heating; **2.** Constituent of clays.

VERMICULITE CONCRETE Concrete in which the aggregate consists of exfoliated vermiculite.

vern. *Abbreviation for:* vernier.

VERNIER 1. Device that permits fine measurement or control; **2.** Brake adjustment on a spudding drill that permits a line to pay out automatically as the hole deepens.

VERNIER CALIPER Caliper with a sliding rule for extremely fine measurement.

vers. *Abbreviation for:* versed sine.

vert. *Abbreviation for:* vertical.

VERTICAL 1. Perpendicular to the earth's gravity field; **2.** Helipad built in standing timber that requires the helicopter to approach the pad vertically to land.

VERTICAL ANGLE Angle between two intersecting lines.

VERTICAL APPLICATION Drywall panels applied with the long dimension parallel to the framing members. *Also called* **Parallel application.**

VERTICAL BAR Any reinforcing bar in a vertical or upright position.

VERTICAL BRACING SYSTEM System of shear walls, braced frames, or both, extending throughout one or more floors of a building.

VERTICAL BROKEN JOINT Style of laying tile with each vertical row of tile offset for half its length.

VERTICAL CENTER OF GRAVITY *See* **Center of gravity.**

VERTICAL CIRCLE Ring graduated in degrees and fractions, attached to a survey telescope to show the angle from horizontal of an observation.

VERTICAL CURVE Curvature in a horizontal line to a higher or lower elevation.

VERTICAL DEFLECTION Bending of a boom in the vertical plane caused by a load suspended directly below the boom nose.

VERTICAL DRAIN Column of sand or other porous material used to vent water squeezed out of waterlogged or saturated soil.

VERTICAL EARTH-BORING MACHINE *See* **Continuous flight auger.**

VERTICAL-GRAINED WOOD *See* **Grain.**

VERTICAL HITCH Method of supporting a load by a single leg of a sling.

VERTICAL HOIST *See* **Front-mount hoist.**

VERTICAL JOINT Joint whose central axis lies primarily in a vertical plane.

VERTICAL LINE Plumb line or a line perpendicular to the horizon.

VERTICAL MAST The mast of a fork truck with a tilting mast is considered vertical at any fork height when adjusted so that the intersection of load-carrying surfaces and forward faces or forks or their equivalent is the same horizontal distance from the load axle as it is when at the height of the mast pivot.

VERTICAL PHOTOGRAPH Aerial survey photograph taken of the ground from the air with the camera pointing vertically down.

VERTICAL PICKUP Device to control sag in the wire rope in catenary scaffolds.

VERTICAL SERVICE SPACE Shaft oriented essentially vertically that is provided in a building to facilitate the installation of building services.

VERTICAL-SHAFT MIXER Mixing compartment, cylindrical or annular, stationary or rotating about a vertical axis, having an essentially level floor and containing one or more vertical rotating shafts to which blades or paddles are attached.

VERTICAL SHINGLING Shingles hung on a wall.

VERTICAL SLIP FORM Form that is jacked vertically during the placement of concrete.

VERTICAL TRACE Line on a vibration record that shows the up and down motion of the earth's vibration.

VERTICAL-TRAVEL LEAD *See* **Lead.**

VERY SEVERE SCALING *See* **Scaling.**

ves. *Abbreviation for:* vessel.

vest. *Abbreviation for:* vestibule.

VESTIBULE Small entrance room behind the principal entrance to a building, or to a suite of rooms.

VF *Abbreviation for:* visual field.

VG *Abbreviation for:* vertical grain; very good.

V-GROOVE Traction sheave groove shaped as a straight-sided V.

V-GUTTER Valley gutter.

vhf *Abbreviation for:* very high frequency.

Vhn *Abbreviation for:* Vickers harness number.

VI *Abbreviation for:* viscosity index.

VIADUCT Long series of arches carrying a road or railway.

VIAL Part of a builder's level or survey instrument that contains liquid, used to indicate level or the angle of a surface.

vib. *Abbreviation for:* vibrate; vibration; vibrator.

VIBRATED CONCRETE Concrete consolidated by

vibration during and after placing.

VIBRATING MASS Mass of all the intentionally vibrated parts of a vibratory roller at each drum.

VIBRATING PLATE COMPACTOR Heavy steel plate attached to a vibrating mechanism.

VIBRATING ROLLER Towed or self-propelled compacting roller that is mechanically vibrated.

VIBRATING SCREEN *See* **Screen.**

VIBRATING TABLE Strong, flat, steel table that is mechanically vibrated, used for compacting concrete in molds temporarily attached to the table.

VIBRATION 1. Rapid rhythmic motion back and forth across a position of equilibrium of the particles of a fluid or elastic solid when its equilibrium has been disturbed; **2.** Energetic agitation of freshly mixed concrete during placement by mechanical devices, either pneumatic, sonic or electric, that create vibratory impulses of moderately high frequency to assist in consolidating the concrete in the form or mold. *See also* **External vibration, Internal vibration,** and **Surface vibration.**

VIBRATION CRACK Tension crack characteristic of material failure due to vibration.

VIBRATION DAMPER Any device, mounted to a rotating shaft, that operates to reduce torsional and harmonic vibrations.

VIBRATION LIMIT Age at which fresh concrete has hardened sufficiently to prevent its becoming mobile when subjected to vibration.

VIBRATION PARAMETER Any physical quantity used to describe vibration, commonly: displacement, velocity acceleration, or frequency.

VIBRATION SENSITIVITY Size of an electrical signal generated by a transducer for each unit of vibration; usually expressed in millivolts per inch per second.

VIBRATOR 1. Oscillating machine used to agitate fresh concrete so as to eliminate gross voids, including entrapped air but not entrained air, and to produce intimate contact with form surfaces and embedded materials; **2.** Shaft mounted to a screed frame having off-center or eccentric weights that cause the screed to vibrate to enhance compaction. The shaft speed is adjustable, and in some cases, additional weights can be added to the vibrator shafts to increase the compactive effort.

VIBRATOR COMPACTOR Roller equipped with a mechanism that generates vibrations of determined amplitude and frequency. The vibrations increase the compactive effort and are tuned to the most effective rate for the material being compacted.

VIBRATORY DRIVER *See* **Pile hammer.**

VIBRATORY-PLATE COMPACTOR Manually directed, power-operated compactor incorporating a sole plate that can be vibrated to a predetermined rate of oscillation and which, by varying the direction of rotation of the eccentric shafts, can be made to travel in a forward or backward direction.

VIBRATORY PLOW Blade that is towed behind a tractor, and to which a vibratory force is imparted, permitting it to slice a trench through even densely packed material. The blade is frequently shaped to permit it to bury small-diameter cable and pipe as it progresses.

VIBRATORY ROLLER Self-powered or towed machine containing equipment that vibrates at a set or variable frequency, the pulses of vibration being transmitted to the road wheels or rollers to increase compaction.

VIBRATORY SANDER *See* **Sander.**

VIBRATORY SCREED Type of asphalt paver screed that utilizes vibrators to achieve compaction of the mix.

VIBROFLOTATION Vibration of the natural granular material overlying a site while, at the same time, injecting sand and water into the crevices.

vic *Abbreviation for:* victaulic.

vic. *Abbreviation for:* vicinity.

VICAT APPARATUS Penetration device used in the testing of hydraulic cements and similar materials.

VICAT NEEDLE Weighted needle for determining the setting time of hydraulic cements.

VICKERS HARDNESS Method and scale for defining the comparative surface hardness of a metal. A diamond point with a 136° point angle is pressed into the surface of the metal at a controlled pressure, and the size of the resulting dent provides a scale measurement.

VICTAULIC COUPLING Rubber-gasketed pipe-coupling fitting that permits limited movement after fixing.

VICTAULIC PIPE Cast iron water pipe with ends formed to accept victaulic couplings.

VIF *Abbreviation for:* verify in field.

VI IMPROVER Polymeric additive designed to increase the viscosity index of an oil, used to provide a wide operating temperature range in, for example, multigrade engine and gear oils.

vil. *Abbreviation for:* village.

VILLA Rural or suburban residence, particularly one that is pretentious.

VIN *Abbreviation for:* vehicle identification number.

vin. *Abbreviation for:* vinyl.

VINYL Thermoplastic material formed through additional polymerization.

VINYL-BASE SEALANT Highly durable emulsion used as a protective pavement sealer. Available in a range of colors, they form a tough, weather-resistant coating that resists oxidation and frost damage.

VINYL COMPOSITION TILE Flooring tile composed of vinyl resins, plasticizers, fillers and color pigments. *See also* **Luxury vinyl tile,** and **Solid vinyl tile.**

VINYL FILM Film made of polyvinyl chloride used for decorative surfacing. It may be either clear or solid color. If it is clear, it is printed on the reverse side to protect the print; if it is of a solid color, the printing is on the top.

VINYL RESIN Manufactured resin that is the basis of latex paint.

VINYL SHEET GOODS Flooring material produced in rotovinyls and inlaid vinyls, with and without cushions; can be applied fully adhering, loose laying, or modified loose laying.

VINYL SHEET GOODS WITH FILLED VINYL WEAR SURFACE Flooring material with felt backing or calendered solid vinyl backing.

VINYL SHEET GOODS WITH VINYL WEARING SURFACE Flooring material having a wearing surface made of unfilled, or very lightly filled clear polyvinyl chloride. Any decorative design is printed under the wear surface.

VINYL TRIM Range of extruded sections used to conceal joints between materials at wall-floor, wall-ceiling and other junctions, and to finish such areas as external and internal corners, etc.

VIOLATION Anything contrary to law regarding the use of property.

VIRGIN GROWTH Timber that has not been cut over within known history.

VIRGIN TIMBER Timber from an original forest that has not been previously disturbed or influenced by human activity. *See also* **First growth.**

VIRTUAL SLOPE Hydraulic gradient, showing the loss from friction in pressure per unit length.

vis. *Abbreviation for:* visibility; visible; visual.

visc. *Abbreviation for:* viscosity.

VISCOSIMETER Instrument for determining the viscosity of liquids and slurries.

VISCOSITY *Also called* **Body.** Measure of the internal friction or the resistance of a fluid to flow. There are several measures, including:

   Absolute: Product of a fluid's kinematic viscosity times its density; a measure of a fluid's tendency to resist flow without regard to its density.

   Kinematic: Flow property of liquid.

VISCOSITY INDEX Measure of the viscosity-temperature characteristics of a fluid as referred to that of two arbitrary reference fluids; a number derived from viscosity measurements at 40°C and 100°C that indicates the extent to which the viscosity decreases with increasing temperature.

VISCOUS DAMPER Vibration damper filled with a fluid.

VISE Device having adjustable jaws, used for holding an object being worked on.

VISE CLAMP *See* **Clamp.**

VISIBLE AREA Ground, or vegetation, that can be directly seen from a given lookout point under favorable atmospheric conditions.

VISIBLE AREA MAP Map showing the specific territory in which either the ground surface or the vegetation growing there is directly visible, to practical distances, from a lookout point.

VISIBILITY DISTANCE Maximum distance at which a smoke column of specified size and density can be seen and recognized as a smoke by the unaided eye.

VISTA View or outlook.

VISUAL ALARM Device that produces a visual indication (light, flag, etc.) of a specific operational condition or mode of operation.

vit. *Abbreviation for:* vitreous; vitrified.

vit. ch. *Abbreviation for:* vitreous china.

VITREOUS Material that resembles glass, frequently used to surface sanitary fittings.

VITREOUS TILE Nearly waterproof tile with water absorption of more than 0.5%, but not more than 3.0%.

VITRIFICATION Condition resulting when kiln temperatures are sufficient to fuse grains and close the pores of a clay product, making the mass impervious.

VITRIFIED BOND Bonding material of which the chief constituent is clay.

VITRIFIED CLAY PIPE Fired clay pipe, generally used for sewers.

viz *Abbreviation for:* videlicet (namely).

vj *Abbreviation for:* V-joint.

V-JOINT POINTING *See* **Pointing.**

vlf *Abbreviation for:* very low frequency.

vlr *Abbreviation for:* very long range.

vlv. *Abbreviation for:* valve.

VOID 1. Absence of material, or an area devoid of materials where not intended; **2.** Volume in the wood structure that is not occupied by wood tissue; **3.** To have no force or effect.

VOIDABLE Something that can be made void, but is not so unless such action is taken.

VOID-CEMENT RATIO Volumetric ratio of air plus net mixing water to cement in a concrete or mortar mixture.

VOIDED SLAB Precast reinforced concrete or prestressed concrete floor slab constructed with the largest practicable longitudinal voids to reduce the mass of the whole.

VOID FORMER Temporary formwork designed to produce internal voids in a box beam.

VOID RATIO 1. Ratio of the void space volume to the volume of soil solids; **2.** Spaces in a pavement structure not occupied by solid material but filled instead with air, water, or other gaseous or liquid material.

vol. *Abbreviation for:* volume.

volat. *Abbreviation for:* volatile.

VOLATILE Readily vaporized organic material that, when mixed with oxygen, is easily ignited.

VOLATILE FLAMMABLE LIQUID Flammable liquid having a flash point below 38°C (100°F).

VOLATILE MATERIAL 1. Material that is subject to release as a gas or vapor; **2.** Liquid that evaporates readily.

VOLATILE MATTER Material weight lost from a

dry powdered fuel sample that is heated at 950°C (1742°F) for seven minutes in a closed crucible.

VOLATILE SOLID Sum of the volatile matter and fixed carbon of a sample, as determined by allowing a dried sample to burn to ash in a heated and ventilated furnace.

VOLATILE THINNER Liquid used to thin finishes without changing the relative volumes of pigments, and that evaporates readily.

VOLCANIC TUFF Rock composed of fine, loose volcanic material resulting from rock disintegration.

VOLT One of 17 derived units with special names of the SI system of measurement: a unit of potential difference between two points of a conducting wire carrying a constant current of 1 A, when the power dissipated between these points is equal to 1 W. (1 V=1 W/A). Symbol: V. *See also the appendix*: **Metric and nonmetric measurement.**

VOLTAGE Measure of electric pressure between any two wires of an electric circuit. *See also* **Nominal voltage.**

VOLTAGE AMPLIFIER Amplifier that increases a signal voltage.

VOLTAGE DIP Reduction in voltage resulting from a sudden application of load, usually expressed in percentage.

VOLTAGE DRIFT Gradual deviation of the mean regulated voltage above or below the desired voltage under constant operating conditions.

VOLTAGE DROP Difference in voltage measured across a current-carrying device.

VOLTAGE GAIN Ratio of input voltage to output voltage.

VOLTAGE MULTIPLIER Circuit that produces high-voltage DC from low-voltage AC current.

VOLTAGE OPERATING BAND Span of voltage through which a generator can be adjusted and operated.

VOLTAGE REGULATION Difference between the steady-state no load and steady-state full load output, expressed as a percentage of the full-load voltage.

VOLTAGE REGULATOR Device that automatically controls the voltage output of a generator at its specific value.

VOLTAGE-TO-GROUND For grounded circuits, the voltage between the given conductor and that point or conductor of the circuit that is grounded; for ungrounded circuits, the greatest voltage between the given conductor and any other conductor of the circuit.

VOLT/AMPERE Product of volts and amperes flowing in a circuit: in resistive circuits, the product of volts and amperes equals watts.

VOLTMETER Meter used to measure voltage in an electrical system.

VOLTOHMMETER Multipurpose electrical testing instrument.

VOLT PER METER A derived unit of electric field strength with a compound name of the SI system of measurement. Symbol: V/m. *See also the appendix*:

Metric and nonmetric measurement.

VOLUME 1. Amount of space occupied in three dimensions. *See also* **Hydraulic volume; 2.** Amplitude of a sound in the audible range; **3.** Number of vehicles passing a given point during a specified period of time. There are several means to calculate this, including:

> **Average daily traffic:** Average 24-hour volume, being the total volume during the stated period divided by the number of days in that period. Unless otherwise stated, the period is a year.

> **Design:** Volume determined for use in design, representing traffic expected to use a highway. Unless otherwise stated, it is an hourly volume.

> **Design hourly:** Estimated value based on a projection system or traffic model, usually given as a total number of vehicles with an allowance for the number of trucks in the traffic stream.

> **Thirtieth highest hourly:** Hourly traffic volume that is exceeded by 29 hourly volumes during a designated year. (Corresponding definitions apply to any other ordinal highest hourly volume, as tenth, twentieth, thirtieth, etc.)

VOLUME BATCHING Measuring the constituents of mortar or concrete by volume.

VOLUME CHANGE *See* **Deformation.**

VOLUME DISCOUNT *See* **Discount.**

VOLUME METHOD OF COST ESTIMATION Estimating technique that multiplies the gross building volume by a predetermined cost per unit of volume.

VOLUME OF BOREHOLES Space created when holes are drilled in rock.

VOLUME OF GROWING STOCK Volume of sound wood in the bole of sawtimber and poletimber from a stump to 100 mm (4 in.) minimum top diameter outside bark or to the point where the central stem breaks into limbs. *See also* **Growing-stock trees.**

VOLUME OF SAWTIMBER Net volume of the saw log portion of live sawtimber in board feet. *See also* **Growing-stock trees.**

VOLUME REDUCTION To decrease the volume of solid wastes by incineration (90% to 98% reduction) or compaction (50% to 80% reduction).

VOLUME SWELL *See* **Swelling.**

VOLUME TABLE Table showing the estimated average tree or stand volume based on given tree measurements, usually diameter and height.

VOLUMETRIC DISPLACEMENT Volume for one revolution or stroke.

VOLUMETRIC EFFICIENCY Ability of an engine to breathe or fill its cylinders with air during operation; a comparison between the actual air volume a cylinder can hold and the air volume with which it is filled during engine operation.

VOLUMETRIC PROPORTIONING Batching of dry materials by time and volume.

VOLUME VARIANCE *See* **Variance.**

VOLUME YIELD Volume of finished or usable product as compared to the volume of raw or prime components: Volume of mixed concrete compared to the volume of cement powder used, for instance.

VOLUTE Spiral scroll.

VOLUTE WITH EASEMENT Spiral terminal to a handrail that replaces a newel post.

V1S *Abbreviation for:* vee one side.

VON MISES YIELD CRITERION Theory that states that inelastic action at any point in a body under any combination of stresses begins only when the strain energy of distortion per unit volume absorbed at the point is equal to the strain energy of distortion absorbed per unit volume at any point in a simple tensile bar stressed to the elastic limit under a state of uniaxial stress.

VOUSSOIR One of the wedge-shaped masonry units that form an arch ring. *See also* **Arch**.

v-p *Abbreviation for:* vice president.

VP *Abbreviation for:* vent pipe.

V-PLOW Forest plow with a V-shaped blade, used to prepare strips for hand planting by removing surface debris and competing vegetation.

Vpn *Abbreviation for:* Vickers pyramid number.

vps *Abbreviation for:* vibrations per second.

VPT *Abbreviation for:* vinyl plastic tile.

VR *Abbreviation for:* vulcanized rubber.

v-reg. *Abbreviation for:* voltage regulator.

vs *Abbreviation for: versus* (contrasted with).

v.s. *Abbreviation for: vide supra* (see above).

VS *Abbreviation for:* variable speed; vent stack; vertical sliding.

vsby *Abbreviation for:* visibility.

VT *Abbreviation for:* vaportight.

V2S *Abbreviation for:* vee two sides.

VULCANIZATION Chemical reaction that takes place under an appropriate temperature and pressure. *See also* **Air cure,** and **Hot air cure.**

w *Abbreviation for:* water; watt; weight; wide; width.

w/ *Abbreviation for:* with.

W *Abbreviation for:* west.

WADDING Soft material (paper or cloth, plastic, etc.) placed over explosives in a hole.

WAF *Abbreviation for:* wiring around frame.

WAFER Wood flake produced by a waferizer with a minimum length of 31 mm (1.25 in.), a controlled width, and a controlled thickness. The product is essentially flat with the grain running in the direction of its length.

WAFERBOARD Type of particleboard composed of wafers cut from roundwood bolts (mostly poplar) of uniform length and thickness resembling small pieces of veneer. The wafers are bonded together with resin binder under heat and pressure.

WAFERIZER Machine that converts whole trees into wafers by cutting the tree in the long direction (as compared to a chipper, which cuts the tree in the cross direction).

WAFFLE-PLATE CONSTRUCTION The use of preformed steel domes or pans to form a reinforced concrete slab containing an integral two-way joist system.

WAGNER FINENESS Fineness of portland cement, expressed as total surface area in square centimeters per gram, determined by the Wagner turbidimeter apparatus and procedure.

WAGON Trailer equipped with a dump body.

WAGON DRILL Wheeled frame holding a pneumatic drill and mechanism for feeding it into rock and retracting it.

wains. *Abbreviation for:* wainscot.

WAINSCOT 1. The lower part of an interior wall when finished in a manner different to the upper part; **2.** Interior walls of an elevator cab extending from the platform to the underside of the car top.

WAIST Minimum depth of a concrete slab, usually of a stair tread.

WAIST LINE Portion of a haulback between the two haulback blocks.

WAITING REPAIR TIME *See* **Machine time.**

WAIVER Intentional relinquishing of a right, claim, or privilege.

WAKEFIELD PILE *See* **Pile.**

wal *Abbreviation for:* wider, all length.

WALE 1. Long formwork member (usually double), used to gather loads from several studs (or similar members) to allow wider spacing of the restraining ties; **2.** When used with prefabricated panel forms, this member is used to maintain alignment. *Also called* **Ranger,** and **Waler.**

WALER Horizontal brace, used to hold timbers in place against the sides of an excavation. *Also called* **Wale,** and **Whaler.**

WALKIE-TALKIE Two-way radio hand set.

WALKING BEAM Rigid member that is supported at the ends on supports, that may move vertically, and whose center is hinged to the load it carries.

WALKING BEAM SUSPENSION *See* **Suspension.**

WALKING DRAGLINE Dragline shovel, self-propelled by means of side-mounted shoes or pontoons actuated by overhead cams, the motion being akin to walking.

WALKING EDGER Hand-powered edging machine used on pavement and large slabs.

WALKING OFF Tendency for a rotating bit to deflect laterally when encountering a sloping surface, boulder, cobble, etc.

WALK-UP Tenement, usually of not more than four stories, without elevators.

WALKWAY 1. Permanent gangway with handrails; **2.** Covered or roofed pedestrian thoroughfare used to connect two or more buildings in which the least horizontal dimension of the thoroughfare is less than 9.1 m (30 ft).

WALL Continuous, vertical member of a structure whose horizontal dimension, measured at right angles to the thickness, exceeds three times its thickness and that may contain openings for doors and windows. *See also* **Apron.** Can be constructed of a wide range of materials and can be load-bearing, non-load-bearing, or a partition. Wall types include:

> **Apron:** That part of a panel wall between window sill and wall support.
>
> **Area: 1.** Masonry surrounding or partly surrounding an area; **2.** The retaining wall around basement windows below grade.
>
> **Bearing:** Designed to support imposed vertical loadings in addition to its own weight.
>
> **Cavity:** Masonry wall comprised of two skins or wythes enclosing an air space, and in which the inner and outer wythes are tied together with metal ties.
>
> **Common:** Vertical separation completely dividing a portion of a structure from the remainder of the structure and creating, in effect, a building that, from its roof to its lowest level, is separate and complete unto itself for the purpose for which it is designed, intended, or used. Such a wall is owned by one party, but jointly used by two parties.
>
> **Composite:** Multiple-wythe wall in which at least one of the wythes is dissimilar to the other

wythe or wythes with respect to type or grade of masonry units or mortar.

Curtain : Exterior, non-load-bearing wall, attached to a structural framework.

Demising: Tenant separation wall. *Also called* **Party wall.**

Dividing: Wall joining two dwellings, owned by one owner but subject to an easement or right of the adjoining owner to have it maintained. *Also called* **Party wall.**

Division: *See* **Firewall,** below.

Dwarf: Wall or partition that does not extend to the ceiling.

Enclosure: Exterior nonbearing wall in skeleton frame construction. It is anchored to columns, piers, or floors, but not necessarily built between columns or piers nor wholly supported at each story.

Exterior: Any outside wall or vertical enclosure of a building other than a party wall. *Also called* **External wall.**

Faced: A combination wall in which the masonry backing and the facing are so bonded or connected as to exert a common reaction under load.

Fender: Small wall carrying a ground floor fireplace hearth slab.

Fire division: Any wall that subdivides a building so as to resist the spread of fire. It is not necessarily continuous through all stories to and above the roof.

Fire division party: Fire resistant wall separating adjacent properties.

Fire partition: Wall designed to prevent or restrict the spread of fire or provide an area of refuge in case of fire.

Firewall: Wall constructed to exhibit predictable fire resistance, of materials of known fire rating, extending continuously from the foundation through the roof..

Flank: Side wall; return wall from the front.

Foundation: A wall below, or partly below grade, or below first floor beams or joists, providing support to a superstructure.

Furnace: There are several types, including:

Air-cooled: Refractory wall that has a lane directly behind it through which cool air can flow.

Battery: Double or common wall between two combustion chambers; both faces are exposed to heat.

Bridge: Partial partition between combustion chambers over which pass the products of combustion.

Core: Center courses of brick in a battery wall that are not exposed directly to furnace heat.

Curtain: A hanging or arched refractory construction or baffle that deflects combustion gases downward.

Insulated: Furnace wall behind which insula-

tion material is installed.

Refractory: Wall made of heat-resistant ceramic material.

Sectionally supported: Furnace or boiler wall consisting of special refractory blocks or shapes that are mounted on and supported at intervals of height by metallic hangers.

Unit-suspended: Furnace wall or panel that is hung from a steel structure.

Water-cooled: Wall having water tubes for extracting or absorbing heat, affording cooling.

Hollow: Wall built of masonry units arranged to provide air space within the wall. The separate facing and backing are bonded together with masonry units.

Knee: Partitions of varying length, used to support roof rafters when the span is so great that additional support is required to stiffen them.

Load-bearing: Wall that supports any vertical load in addition to its own weight.

Masonry: Constructed of masonry units.

Masonry panel: Exterior non-load-bearing wall whose outer surface may form the exterior building face or may be used in back of a panel curtain wall as a backup wall.

Nonbearing: Wall that supports no other weight than its own. *Also called* **Non-load-bearing wall.**

Non-load-bearing: *See* **Nonbearing wall,** above.

Panel: Exterior, non-load-bearing wall wholly supported at each story.

Panel curtain: Exterior non-load-bearing wall made of panels either directly attached to the building structure with adjustable attachments, or mounted on a supporting subframe that is itself attached to the structure by adjustable attachments. The exterior surface of the panels forms the face of the building; the interior surface may, or may not form the interior finish.

Parapet: That part of any wall entirely above the roof line.

Party: Wall used for joint service by adjoining buildings. *See also* **Demising wall.**

Perforated: Wall that contains a considerable number of relatively small openings. *Also called* **Pierced wall,** and **Screen wall.**

Self-supporting: Non-load-bearing wall.

Shear: Wall that resists horizontal forces applied in the plane of the wall.

Single-wythe: Wall containing only one masonry unit in wall thickness.

Solid masonry: Wall built of solid masonry units, laid contiguously, with joints between units completely filled with mortar or grout.

Spandrel: That part of a curtain wall above the top of a window in one story and below the sill of the window in the story above.

Veneer: Wall with an exterior face of a material that is not bonded to the body of the wall and that does not exert a common reaction under load.

WALL ANCHOR Device that can be built into, or inserted into a solid wall as a receptacle for a fastener, used to attach something to the wall. *See also* **Anchor.** There are many types, including:

Collapsible anchors: Machine screw built into a deformed sleeve that expands as the screw is tightened.

Expansion shield: Thick sleeve of malleable metal (usually lead) fabricated as a single item or two-piece item that is inserted into a predrilled hole in dense material and that expands to grip through its external face as a fastener is inserted. Single-piece types are used for nails and wood screws; two-piece types with lag bolts and machine screws.

Toggle: Machine screw and a nut that has collapsible or spring-mounted wings. The screw is passed through the item to be mounted prior to the nut being threaded on. The nut, with wings collapsed is then passed through a predrilled hole in thin panels (such as drywall). As the screw is tightened the wings expand against the inner face of the panel.

Tubular: Cylindrical or cone-shaped, plastic or fiber predrilled shape in a diameter and length to match a standard wood screw. It is inserted into a predrilled hole in a dense materials, and expands to grip through its external face as a fastener is screwed in.

WALL BEARING STRUCTURE One with slabs (i.e., the floors or roofs) supported on walls, generally of masonry, rather than a framing system.

WALL BRACKET Light fixture attached to a wall.

WALL CABINET Cupboard suspended from a wall.

WALL COVERING Nonstructural, usually decorative finish to a wall,

WALL FORM Retainer or mold so erected as to give the necessary shape, support and finish to a concrete wall.

WALL FRICTION Friction between the back of a retaining wall and the retained material.

WALL HANGER Metal stirrup built into a wall to carry the ends of wood joists.

WALL OPENING Opening at least 760 mm (30 in.) high and 460 mm (18 in.) wide, in any wall or partition, through which persons may fall, such as a yard-arm doorway or chute opening.

WALL PLATE Timber supported by a wall to which a wooden superstructure (roof, wall) is attached. *Also called* **Head plate.**

WALL PLUG Two- or three-pronged accessory necessary to complete an electrical connection to a socket outlet.

WALL SHEATHING Panels or boards applied directly to the exterior stud walls before applying the finished exterior siding or brick veneer.

WALL SPREADER Accessory, usually fabricated from reinforcing bar to a 'Z' or 'U' shape, used to separate and hold apart two faces or curtains of reinforcement in a wall.

WALL STACK Rectangular duct used in forced-air heating systems, built vertically into a wall thickness to connect the system to upper floors.

WALL STRING Structural component of a stair frame that fastens against a wall.

WALL THICKNESS 1. Thickness of the metal forming a pipe or tube; **2.** Thickness of the metal of a pile.

WALL TIE 1. Bonder or metal piece that connects wythes of masonry to each other or to other materials; **2.** In cavity wall construction, a rigid, corrosion-resistant metal tie that bonds two wythes of the cavity wall. It is usually 'Z' shaped and is designed to cause any moisture to drip from it rather than be conducted from one wythe to the other; **3.** In a veneer wall, a strip or piece of metal used to tie a facing veneer to the backing.

WALL TILE Glazed tile with a body that is suitable for interior use and that is usually nonvitreous, and is not required or expected to withstand excessive impact or abrasion.

WALL-TO-WALL TURNING DIAMETER Smallest circle that will enclose the outermost projecting points of a vehicle or machine while it is executing its sharpest practicable turn.

WALL UNIT Preassembled, modular panel containing all structural wall elements, doors, windows, etc., truck-delivered to site ready to be moved into final position. May be one or more stories high.

WALL VENT Vent to the cavity of a cavity wall or to solid or veneer masonry walls.

WANE Defect in lumber in which either bark is present or wood is lacking on an edge or corner.

WANY-EDGE SIDING Siding with an uneven exposed edge showing a strip of bark or the unfinished exterior surface of the wood immediately below the bark. *Also called* **Rustic siding.**

WARD 1. Baffle in a lock to prevent overrotation of the key; **2.** Area within a hospital.

WAREHOUSE Structure in which goods are kept.

WAREHOUSE RECEIPT Documentary evidence issued by the operator of a public warehouse that goods are being held in storage on behalf of their owner.

WAREHOUSE SET 1. Partial hydration of cement stored for a time and exposed to atmospheric moisture; **2.** Mechanical compaction of cement during storage. *Also called* **Sticky cement.**

WARM AIR HEATING SYSTEM *See* **Heating system.**

WARNING DEVICE Device that warns an operator when a predetermined operating parameter has been reached.

WARNING LIGHT Illuminated indicator that warns of a situation or condition.

WARP 1. Any deviation from a true or plane surface; includes crooking, bowing, cupping, and twisting, or any combination; **2.** Yarn running lengthwise in a woven fabric.

WARPING JOINT *See* **Joint.**

WARPING TORSION That portion of the total resistance to torsion that is provided by resistance to warping of the cross section.

warr. *Abbreviation for:* warranty.

WARRANTY Guarantee or assurance that work or a product shall be or will perform as represented. There are two forms:

Construction: Guarantee that the entity has been constructed in accordance with applicable codes and regulations, with the materials specified, in the manner described in the contract documents, and that installed equipment will operate or perform as described by the manufacturer or supplier.

Expressed: One that is defined and particular to a product.

Implied: One that is not written but exists under law.

WARRANTY DEED Deed containing a covenant by the grantor to the grantee to warrant and defend the title of the estate conveyed. *See also* **Deed,** and **Quitclaim deed.**

WASH Passing water over and through earth, gravel, and other material to cause separation and segregation.

WASHBASIN Sanitary fixture designed to facilitate personal washing.

WASHBOARDING Ripples formed transversely across the width of an asphalt paved, or unpaved road surface.

WASH BORING 1. Method of advancing a boring by means of rotary drilling utilizing water or a bentonite slurry to stabilize the sides of the opening; **2.** Method of examining soil, usually in soft soil or clay, by driving a pipe into the ground and then inserting a small pipe inside of it, through which water is forced to wash out soil particles in water suspension for examination. *Also called* **Jetting,** and **Water jet.** *See also* **Air lift.**

WASHED FINISH *See* **Rustic finish.**

WASHER Ring-shaped device used to increase the bearing area beneath bolt heads and nuts. There are many types, including:

External-tooth: A lock washer that grips from the deformation under pressure of the large number of small teeth projecting out and at an angle from the perimeter.

Flat: Flat, circular shape, one outer rim of which may be rounded or beveled.

Internal-tooth: A lock washer that grips from the deformation under pressure of the small teeth projecting in and at an angle from the interior perimeter.

Split-ring: A partial spiral that, when under tension, creates pressure on a nut to prevent it from loosening.

Spring: Metal washer deformed to a helical curve that, when compressed under a nut, exerts continuous pressure to prevent the nut from unscrewing.

WASH FILTER Filter element in which a larger unfiltered portion of the fluid flowing parallel to a filter element axis is utilized to continuously clean the influent surface which filters the lesser flow.

WASHING DOWN Cleaning the surface of a brick wall with a mild solution of muriatic acid after it is completed and pointed.

WASHITA OILSTONE Natural stone capable of producing smooth, long-lasting cutting edges on tools.

WASHOUT RESISTANCE *See* **Water resistance.**

WASH PRIMER *See* **Etching primer.**

WASHROOM Toilet facility; also a laundry room.

WASH WATER Water carried on a truck mixer in a special tank for flushing the interior of the mixer after discharge of the concrete. *Also called* **Flush water.**

WASTE 1. Material of no value. *See also* **Residual; 2.** Drain line in plumbing; **3.** Overflow line in piping.

WASTE CONTAINER There are several types, including:

Carrying: Receptacle of 9.2- to 13.2-L (35- to 50-gal) capacity, usually constructed of plastic or aluminum, that is carried by a collector in a backyard carryout service. *Also called* **Tote barrel.**

Disposable: Plastic or paper sacks designed for storing solid waste.

Lift-and-carry: A large container that can be lifted onto a service vehicle and transported to a disposal site for emptying. *Also called* **Detachable container,** and **Drop-off box.**

Roll-on/roll-off: Large container of 15- to 30-mm^3 (20- to 40-yd^3) capacity that can be pulled onto a service vehicle mechanically and carried to a disposal site for emptying.

WASTE DISPOSAL UNIT Electrically driven grinder designed to fit under the waste outlet of a kitchen sink and connected to the drain. *Also called* a **Garburator.**

WASTE EXCHANGE Use by one company of an industrial waste generated by another firm.

WASTE PIPE Pipe for carrying off waste flows.

WASTE PROCESSING Operation such as shredding, compaction, composting, or incineration, in which the physical or chemical properties of wastes are changed.

WASTE REDUCTION Practice of producing smaller quantities of disposable waste.

WASTE ROCK *See* **Scalp rock.**

WASTE STACK Vertical pipe that collects waste flows from appliances and their individual waste pipes.

WASTEWATER Water containing biological, organic and/or chemical contaminants, primarily as a result of human activities.

WASTEWAY A spillway.

WASTING ASSET Natural resources that are depleted through the process of consumption or removal.

WATER ABSORBENT *See* **Water resistance.**

WATER ABSORPTION Act of a material, particularly one that has been manufactured, picking up water when wet or exposed to rain.

WATER ABSORPTION CHARACTERISTIC *See* **Water resistance.**

WATER BALANCE Method of determining the gross loss or gain of water based on precipitation, evaporation, transpiration, and runoff.

WATER BAR 1. Metal or plastic bar set in the joint between a wood sill and masonry, or wood sill and sash of a window, to prevent penetration of water. *Also called* **Weather bar; 2.** Shallow trench cut into the surface of a forest road or created by an embankment (e.g., log and soil) to collect and channel water off the surface to avoid erosion. *See also* **Cross-ditch.**

WATER-BASED COATING Water repellents and latex paints containing water-soluble binders.

WATER-BEARING GROUND Permeable ground below the standing-water level.

WATER BLAST System of cutting or abrading a surface by a stream of water ejected from a nozzle at high velocity.

WATERBOUND Material that uses water as the binding agent, such as waterbound macadam.

WATER-CEMENT RATIO Ratio of the amount of water, exclusive only of that absorbed by the aggregates, to the amount of cement in a concrete, mortar, grout, or cement paste.

WATER CLOSET A toilet.

WATER CONDITIONER Device that changes the condition of water passed through it: by removing dissolved minerals, etc.

WATER CONTENT Ratio of the quantity (by weight) of water in a given volume of soil mass to the weight of the soil solids, typically expressed as a percentage.

WATER-COOLED FURNACE WALL *See* **Wall.**

WATER CORROSION RESISTANCE *See* **Water resistance.**

WATER DEMAND Volume of water required by an individual plumbing fixture, or group of fixtures. *See also* **Total water demand.**

WATERFALL FURNACE *See* **Furnace.**

WATER FLOW ALARM INDICATOR Visual and/or audible signaling indicator in an automatic sprinkler system that is activated by a water flow alarm device, located within a premises or at a remote location.

WATER FLOW SWITCH *See* **Switch.**

WATER GAIN *See* **Bleeding.**

WATER GAUGE *See* **Gauge.**

WATER GEL Any of a wide variety of materials used for blasting. They all contain substantial proportions of water and high proportions of ammonium nitrate, some of which is in solution in the water. *Also called* **Slurry explosives.** Two broad classes of water gels are:

(1) Those that are sensitized by a material classed as an explosive, such as TNT or smokeless powder.

(2) Those that contain no ingredient classified as an explosive; these are sensitized with metals such as aluminum or with other fuels.

WATER-GLYCOL FLUID Fluid whose major constituents are water and one or more glycols or polyglycols.

WATER HAMMER Condition occasioned by the sudden stopping of water flow in a pipe resulting in a pressure wave that impacts upon closed valves and pipe walling at extreme direction changes.

WATER-IN-OIL EMULSION Dispersion of water in a continuous phase of oil. *See also* **Emulsion.**

WATER JACKET Voids cast into a cylinder block and cylinder head through which coolant is pumped to maintain an even temperature of the whole.

WATER JET *See* **Jetting,** and **Wash boring.**

WATER JOINT *See* **Saddle joint.**

WATER LEVEL Device for transferring a known level to another place consisting of a length (usually about 15 m (50 ft)) of plastic pipe or hose, 10 to 13 mm (0.375 to 0.5 in.) in diameter, filled with water from which all air has been removed, fitted at both ends with a graduated transparent sight glass. *See also* **Struck measure.**

WATER MAIN Large water supply pipe to which branches are connected. *Also called* **Main water line.**

WATER MARK Water surface elevation of the design peak flow. *See also* **High water mark,** and **Low water mark.**

WATER METER Device for recording the quantity of water flowing through a pipe.

WATER MISCIBLE Water soluble; easily mixed with water.

WATER MOTOR ALARM GONG Alarm gong, mounted on an outside wall adjacent to the sprinkler valve of an automatic sprinkler system, activated by a flow of water from the riser or sprinkler valve through a paddle wheel and clapper.

WATER MUFFLER Muffler incorporating a water trap to prevent the emission of sparks or flames from an exhaust system.

WATER OF CAPILLARITY *See* **Held water.**

WATER OF HYDRATION Chemically combined water forming a hydrate that can be expelled without essentially altering the composition of the substance.

WATER PLATE Metal plate that is bolted to the cylinder head of an engine to plug all coolant outlets so that the head can be pressure checked.

WATER PREHEAT TANK Water tank used to store water that is heated by alternative means, such as passive solar heat, before it is fed into a domestic hot water tank.

WATERPROOF Impervious to water in either liquid or vapor state.

WATERPROOF AND BOILPROOF RESIN *See* **Adhesive.**

WATERPROOF CEMENT Cement mixture that, when set, is watertight, usually as a result of the addition of a waterproofing admixture.

WATERPROOF COATING Surface treatment that excludes liquid water but is permeable to water vapor.

WATERPROOFING Sealing or coating with a material or substance that will prevent the passage of water. *See also* **Dampproofing**, and **Vapor barier**.

WATERPROOFING COMPOUND Material used to impart water repellency to a structure or a constructional unit.

WATERPROOF MEMBRANE Sheet materials applied to a roof or wall surface to prevent the penetration of water, often in several layers or 'plies.'

WATERPROOF PAPER Paper treated with a moisture-resisting compound.

WATER PUTTY Powder that, when combined with water, forms a filler suitable for filling cracks and holes prior to painting.

WATER RATE Amount charged per given volume or per time period for the supply of water. *Also called* **Water tax**.

WATER-REDUCING ADMIXTURE *See* **Admixture**, and **Fluidifier**.

WATER REPELLENT Property of a surface that resists wetting but that permits the passage of water when hydrostatic pressure occurs.

WATER-REPELLENT CEMENT Hydraulic cement having a water-repellent agent added during manufacture.

WATER-REPELLENT PAPER Special paper, treated to minimize wetting of the surface, used as a face in the manufacture of drywall panels.

WATER-REPELLENT PRESERVATIVE Liquid formulated to penetrate into the pores of wood, impregnating the surface fibers causing them to repel moisture.

WATER RESISTANCE Ability of a lubricating grease to withstand the addition of water to the lubricant system without adverse effects. Generally considered to be made up of following components:

　Washout resistance: Ability of a lubricating grease to resist being removed from a bearing when operated fully or partially submerged in water.

　Water absorption characteristic: Characteristic of a lubricating grease when water is added to the lubricating system, causing it to become water soluble or water absorbent.

　Water soluble: The lubricating grease absorbs the water, and then de-gels to a semifluid consistency.

　Water absorbent: The lubricating grease absorbs relatively large quantities of water with small or no change in consistency, and without leaving free water as a separate phase.

　Water corrosion resistance: Ability of a lubricating grease to prevent corrosion of surfaces when water is present in the lubricating system

　Water resistant: The lubricating grease does not absorb more than small amounts of water, does not change appreciably in consistency, and leaves the added water as a second phase in the system.

　Water spray resistance: Ability of a grease to resist displacement from a surface by the impact of a water spray.

WATER RESISTANT 1. Having the ability to withstand the passage, or deteriorating effect of water; **2.** *See* **Water resistance**.

WATER-RESISTANT BACKING BOARD Drywall panels designed for use in areas such as bathrooms and in showers as a base for applied ceramic tile.

WATER-RESISTANT CORE Drywall panel manufactured with a core containing special additives to reduce water absorption.

WATER RETENTIVITY That property of mortar that prevents the rapid loss of water to masonry units of high suction. It prevents bleeding or water gain when fresh mortar is in contact with relatively impervious units.

WATER RING Device in the nozzle body of dry-mix shotcrete equipment through which water is added to the materials.

WATER SEAL Depth of water trapped in a sanitary fitting due to the shaping of the pipe in order to prevent the passage of vapor or gases from the sewer through the fixture.

WATER SEASONING Tempering lumber by soaking it in water for two weeks and then air drying it.

WATER SERVICE Provision to the lot line or water meter by a water authority of the supply of potable water and/or water for fire protection.

WATER SERVICE PIPE Pipe that is part of a water system and that conveys water from a public water main or a private water source to the inner side of the wall or floor through which the system enters the building.

WATERSHED Area that drains into a stream or other water passage.

WATER SHEEN Lustrous surface resulting from the evaporation of bleed water from the surface of concrete.

WATER SOFTENER Attachment to a water distribution system that receives water from the potable supply, passes it through a bed composed of a calcium- and magnesium-absorbing medium (usually zeolite), and sometimes through an additional sand-and-gravel filter, before passing the softened and filtered water to the appliances and faucets.

WATER SOLUBLE *See* **Water resistance**.

WATERSPOUT Pipe or other orifice through which water is spouted or conveyed.

WATER SPRAY RESISTANCE *See* **Water resistance**.

WATERSTOP Thin sheet of metal, rubber, plastic, or other material inserted across a joint to obstruct the seepage of water through the joint.

WATER-STRUCK BRICK *See* **Brick**.

WATER SUPPLY MAP Map showing the location of supplies of water readily available for pumps, tanks, trucks, camp use, etc.

WATER SUPPLY SYSTEM Array of piping, valves, and fittings from the source of water to its point of use.

WATER TABLE 1. Level in the ground above which water will not naturally rise (not a constant measure, influenced by many natural phenomena); **2.** Projection of masonry on the outside of a wall, slightly above ground level. Often a damp course is placed at the level of the water table to prevent upward penetration of groundwater.

WATER TAX *See* **Water rate.**

WATER TEST Pressure test applied to pipes and drains.

WATERTIGHT Impermeable to water except when under hydrostatic pressure sufficient to produce structural discontinuity by rupture.

WATER TOWER Tank mounted above ground level for the storage of water.

WATER TUBE Tube in a boiler having water and steam on the inside and heat applied to the outside.

WATER TURBINE Wheel turned by the force of flowing water.

WATER VAPOR Water in a vaporous form and diffused in the atmosphere.

WATER VAPOR TRANSMISSION Rate of water vapor flow, under defined conditions, through a unit area of a material.

WATER VELOCITY Speed of a stream. Usually refers to the average water speed through a stream's cross section.

WATER VOID Void along the underside of an aggregate particle or reinforcing steel, initially filled with bleed water that formed during the bleeding period.

WATER WALL 1. Side of a boiler furnace consisting of water-carrying tubes that absorb radiant heat and thereby prevent excessively high furnace temperatures; **2.** Wall incorporating water-filled vessels designed to absorb and store solar heat.

WATERWALL FURNACE *See* **Furnace.**

WATERWHEEL Wheel made to rotate by falling water; a wheel designed to lift water to a higher level.

WATERWORKS System by which a public water supply is conditioned to an acceptable level of quality and pumped into the distribution system.

WATT One of 17 derived units with special names of the SI system of measurement: a unit of power available when energy of 1 J is expended in 1 s. (1 W=1 J/s). Symbol: W. *See also the appendix:* **Metric and nonmetric measurement.**

WATT HOUR Non-SI unit of energy. Symbol: W/hr. Multiply by 3.600 to obtain kilojoules, symbol: kJ. *See also the appendix:* **Metric and nonmetric measurement.**

WATT-HOUR METER Indicating instrument that displays the kilowatt-hour output continuously for record purposes.

WATTMETER Instrument that measures the real power of the circuit in which it is connected.

WATT PER METER KELVIN A derived unit of heat conductivity with a compound name of the SI system of measurement. Symbol: W/(m·K). *See also the appendix:* **Metric and nonmetric measurement.**

WATT PER SQUARE METER A derived unit of heat flux density with a compound name of the SI system of measurement. Symbol: W/m². *See also the appendix:* **Metric and nonmetric measurement.**

WATT PER STERADIAN A derived unit of radiant intensity with a compound name of the SI system of measurement. Symbol: W/sr. *See also the appendix:* **Metric and nonmetric measurement.**

WAVE Motion characterized by a regular progression of equally spaced peaks separated by valleys of uniform depth and form.

WAVE EQUATION Mathematical equation that describes the mechanics of force transmission along an elastic rod (pile) that has been subjected to a mass having a specific initial velocity, from which the energy transmission and stress at any point along a pile being driven can be computed.

WAVEFORM Plot showing the magnitude and direction of a current or voltage at every instant of time.

WAVELENGTH Distance between two successive crests or troughs of a wave.

WAVE PARAMETER Any of the mathematical quantities that are used to describe wave motion, commonly amplitude, period, frequency, wavelength, etc.

WAVE PRESSURE Pressure on breakwaters and other marine structures caused by the wave action of the adjacent water body.

WAVE SPEED Speed of a stress wave in various materials.

WAVY EDGE Warp or bow to the edge of a sheet material.

WAVY-GRAINED LUMBER *See* **Grain.**

WAVY TUBE Tube or lining with an inner surface having surface ripples formed by the pattern of the reinforcement.

WAX Ingredient of waferboard/OSB panels that is mixed with the wafers prior to forming for the purpose of reducing the rate of water absorption.

WAYS *See* **Side channels.**

wb *Abbreviation for:* wheel base.

WB *Abbreviation for:* water ballast; welded base; wet bulb.

WBS *Abbreviation for:* wrought box strike.

WC *Abbreviation for:* water closet.

wcr *Abbreviation for:* water cooler.

wd *Abbreviation for:* wind; wood; wound.

WD *Abbreviation for:* wind direction.

wdn *Abbreviation for:* wooden.

wd pnl *Abbreviation for:* wood panel.

wdr *Abbreviation for:* wider.

wdw *Abbreviation for:* window.

wdwk *Abbreviation for:* woodwork.

wea. *Abbreviation for:* weather.

WEAK AXIS Minor principal axis of a cross section.

WEAKENING CUT *See* **Rigging cut.**

WEAR Removal of materials from surfaces in relative motion. Three types of wear are:

  Abrasive wear: Removal of materials from surfaces in relative motion by a cutting or abrasive action of a hard particle (usually a contaminant).

  Adhesive wear: Removal of materials from surfaces in relative motion as a result of surface contact. Galling and scuffing are extreme cases.

  Corrosive wear: Removal of materials by chemical action.

WEAR AND TEAR Physical deterioration of a capital asset through use or exposure to the elements.

WEAR CONTROL FILTER Filter capable of emoving all particles contributing to component wear from a system operating fluid.

WEARING COURSE Topping or surface treatment to increase the resistance of a concrete pavement or slab to abrasion.

WEAR PAD Pad of any of various shapes and sizes, made of a number of materials that are softer than steel, yet that provide smooth, even, nonabrasive wear characteristics.

WEAR PLATE Metal accessory used to protect and extend the life of a dump truck floor used to haul chunked concrete and other damaging materials. *Also called* **Wood cushion.**

WEATHER 1. Changes in state, texture, color or appearance due to exposure to the elements; **2.** Distance, measured up the slope, that a shingle or tile overlaps the next below it.

WEATHER BAR *See* **Water bar.**

WEATHERBOARD A wedge-shaped board used to clad the exterior of a structure, applied in horizontal strips so that the thin, upper edge is covered by the thick lower edge of the board immediately above. *Also called* **Siding.**

WEATHER CHECK Groove on the underside of a projecting member to form a drip to prevent rain from running down the wall or entering the joint.

WEATHER CHECKING Condition that appears as fine cracks in the sidewall rubber of a tire.

WEATHERED POINTING *See* **Pointing.**

WEATHER FILLET Mortar struck to an angle so as to shed water.

WEATHERING 1. Any slight angle applied to an exterior horizontal surface so as to shed water; **2.** Changes in color, texture, strength, chemical composition, or other property of a natural or artificial material due to the action of the weather; **3.** Any process of decay brought about by the effect of weather conditions. *See also* **Deterioration,** and **Disintegration.**

WEATHERING STEEL *See* **Steel.**

WEATHER MOLDING Any molding designed, and positioned to cast water clear of a surface.

WEATHER OMETER Apparatus in which specimen materials can be subjected to artificial and accelerated weathering tests that simulate natural weathering.

WEATHERPROOF Equipment so constructed or protected that exposure to the weather will not interfere with its operation.

WEATHER RESISTANCE Ability of a material to withstand the effects of wind, rain, sun, etc.

WEATHER SHINGLING Vertically hung shingles.

WEATHER SIDE Area of a wheel rim not covered by the tire.

WEATHERSTRIP Specially shaped strip of wood, metal, plastic, felt, or other material, or a combination of such materials that, when fastened to the exterior jamb and head of a doorway, in close proximity to the door, prevents infiltration of air and moisture.

WEATHERVANE To allow a rotating member to swing with the wind when out of service so as to expose a minimal area to the wind.

WEATHER WORKING DAY Average weather conditions over a given period of a day that either constitute or preclude working conditions.

WEAVE Wobble or flutter of a circular saw blade whose rim area is too long for the speed at which it operates. *Also called* **Wobble.**

WEAVING Crossing of traffic streams moving in the same general direction accomplished by merging and diverging.

WEAVING SECTION Length of one-way roadway at one end of which two one-way roadways emerge and at the other end of which they separate.

WEB 1. Cross-wall connecting the face shells of a hollow concrete masonry unit; **2.** Wide part of a structural steel I-beam connecting the flanges; **3.** Diagonal support member in a truss; **4.** Thin plate or panel.

WEB BUCKLING Buckling of a web plate.

WEB CLAMP *See* **Clamp.**

WEB CRIPPLING Local failure of a web plate in the immediate vicinity of a concentrated load or reaction.

WEBER One of 17 derived units with special names of the SI system of measurement: a unit of magnetic flux which, linking a circuit of one turn, produces in it an electromotive force of 1 V as it is reduced to zero at a uniform rate of 1 s. (1 Wb=1 V·s). Symbol: Wb. *See also the appendix:* **Metric and nonmetric measurement.**

WEB MEMBER Secondary member of a truss contained between chords.

WEB REINFORCEMENT Reinforcement placed in a concrete member to resist shear and diagonal tension.

WEB Member that joins the top and bottom chords of a truss to form triangular patterns that give truss action. These members are subjected only to axial compression or tension forces (no bending).

WEB STIFFENER Plate welded between the flanges of structural shapes to reinforce against concentrated loading, as for testing.

WEDGE 1. Piece of wood, metal, or plastic tapering to a thin edge; **2.** Tapered plastic or metal hand tool that is driven into the kerf to keep a tree from setting back on the backcut when felling. It is sometimes used to lever a tree over when felling; **3.** Wedge-shaped block of rock whose boundaries are joint or fault surfaces.

WEDGE ANCHORAGE Device for providing the means of anchoring a tendon by wedging.

WEDGE BRAKE *See* **Brake.**

WEDGE CUT Technique of setting the cut holes in tunneling or shaft sinking so that they slope inwards, forming the faces of a wedge.

WEDGE SOCKET FITTING Wire rope fittings wherein the rope end is secured by a wedge. *See also* **Poured fitting.**

WEDGE-WIRE SCREEN Screen formed of parallel wedge-shaped wires having their thickest faces uppermost.

WEDGING PLATE Steel plate on top of an underpinning or under a footing against which wedges are driven to pick up the load of the structure being underpinned.

WEEP HOLE Opening at the bottom of retaining walls or in mortar joints of facing material immediately above the level of the dampproof flashing to let moisture escape.

WEFT *See* **Filling, thread.** *See also* **Shoot wire.**

WEFTLESS CORD FABRIC Cord fabric either without filling yarns or with a few small filling yarns, widely spaced.

WEIGH BATCHING *See* **Weight batching.**

WEIGH HOPPER Asphalt batch plant component usually located under the hot bins in which the aggregates and asphalt are weighed prior to discharge into the mixing pugmill.

WEIGH SCALE Inground device for weighing vehicles, often up to 21 m (70 ft) long.

WEIGH STATION Facility for weighing vehicles and recording the results.

WEIGHT Vertical downward force exerted by a mass. *See also* **Sprung weight,** and **Unsprung weight.**

WEIGHT BATCHING Measuring the constituent material for a mixture by individual weight. *Also called* **Weigh batching.**

WEIGHT DISTRIBUTION 1. Manner in which an applied load is distributed over the area supporting it; **2.** Total operating weight of an item of mechanical equipment distributed into the area on which it sits; **3.** Portions of the total weight of a vehicle that will be supported by each axle.

WEIGHTED AVERAGE Average obtained after each value has received an appropriate weight.

WEIGHTING 1. Applying a factor to observations; **2.** Adding a dead load to something to meet an objective.

WEIGHT-TO-POWER RATIO Measure of machine performance and acceleration found by dividing operating weight by flywheel power.

WEIR Structure across a waterway, used for measuring flow, stream diversion, or for catching fish or sediment.

WEIR HEAD Depth of water measured from the bottom of the notch in a weir plate to the upstream water surface.

WEIR PLATE Metal plate set vertically into the top face of a weir, often containing one or more V-shaped notches either to regulate the overflow or for measurement purposes.

WEISBACH TRIANGLE Survey technique used to transfer readings vertically down a shaft.

WELCH PLUG *See* **Expansion plug.**

WELD 1. Joint between metal parts formed by bonding of molten metal; **2.** Joining thermoplastic pieces by one of several heat-softening processes, e.g., torch, friction.

WELD BEAD Deposit of a row of filler metal from a single welding pass.

WELD CRACK Crack in weld metal.

WELDED-BUTT SPLICE Joint made by welding the butted ends of two pieces.

WELDED REINFORCEMENT Reinforcement joined together by welding.

WELDED SPLICE Means of joining two reinforcing bars by electric arc welding. Rebar may be lapped, butted, or joined with splice plates or angles.

WELDED-WIRE FABRIC Series of longitudinal and transverse wires arranged substantially at right angles to each other and welded together at all points of intersection. *See also* **Mesh.**

WELDED-WIRE FABRIC REINFORCEMENT Welded-wire fabric in either sheets or rolls, used to reinforce concrete. *Also called* **Mesh reinforcement.**

WELDER/GENERATOR Engine-powered, dual-purpose electric generator that produces low-voltage, high-amperage current for electrical welding, or high-voltage, low-amperage current for lighting and for powering small tools, pumps, etc.

WELDING ROD Wire that is melted into the weld metal.

WELDING SEQUENCE Order in which the component parts of a structure are welded.

WELDMENT Assembly of component parts fastened together by welding.

WELD METAL Fused portion of base metal or fused portion of base metal and filler material.

WELL 1. An open, unoccupied area bounded on all sides by walls passing vertically through at least one story; **2.** Hole dug or drilled into the ground and penetrating the water table, from which water may be drawn or pumped. *See also* **Artesian well,** and **Dry well; 3.** Slot in the front of a hydraulic dredge hull in which the digging ladder pivots; **4.** Protective wall around an isolated object; **5.** Vertical shaft containing stairs.

WELL DRAIN Well to drain surface and groundwater into a lower strata. *Also called* **Absorbing drain.**

WELL DRILL *See* **Drill.**

WELL-GRADED AGGREGATE *See* **Aggregate.**

WELL LOSS Head loss associated with friction and turbulence as water enters the wellbore.

WELL OPENING Opening in a floor to permit passage from one story to another.

WELLPOINT Pipe, 38 to 65 mm (1.5 to 2.5 in.) in diameter, fitted with a driving point and a fine mesh screen, used to remove underground water.

WELLPOINT DRILL Centrifugal pump that can accommodate large quantities of air, used to remove underground water from the vicinity of a planned or completed excavation.

WELLPOINT PUMP *See* **Pump.**

WELLPOINT SYSTEM Machinery and equipment consisting principally of a pump, header pipe, lateral pipes, riser pipe, valves, and a number of wellpoints, laid out in a systematic manner to remove underground water.

WELL SCREEN Filtering device used to keep sediment from entering a water well.

WELT Seam in flexible metal roofing.

WESTERN FRAMING *See* **Platform framing.**

WESTING Coordinate measured westward from an origin.

WET Covered with visible free moisture. *See also* **Damp,** and **Moist.**

WET ANALYSIS Mechanical analysis of soil samples smaller than 0.06 mm (0.0024 in.) made by mixing the sample in a measured volume of water and checking its density at intervals using a hydrometer.

WET-BULB TEMPERATURE Temperature of the wet-bulb thermometer in a relative humidity measurement. It is compared with the dry-bulb temperature to determine relative humidity (RH).

WET-CAST PROCESS Process for producing items such as pipe that uses concrete having a measurable slump, generally placed from above, and consolidated by vibration. *See also* **Centrifugal process, Dry-cast process, Packerhead process, Spun concrete,** and **Tamp process.**

WET CLUTCH *See* **Clutch.**

WET COLLECTOR Asphalt plant dust collection system that uses water and a high-pressure venturi to capture dust particles.

WET DIGESTION Solid waste stabilization process in which solid organic wastes are placed in an open digestion pond to decompose anaerobically. The carbonaceous matter is converted into carbon dioxide and methane; the soluble and suspended fraction is converted aerobically by algae in a bio-oxidation pond.

WET DOCK Ship dock in which water can be retained at tide level.

WET-FILM GAUGE Gauge for measuring the thickness of a wet film.

WET-HEAD HYDRANT Fire hydrant with the valve in the top.

WET HEAT Heating system employing steam or hot water.

WETLAND 1. Transitional area between dry land and aquatic areas, having a high water table or shallow water; **2.** Land with one of the following attributes: (a) periodically supports hydrophytes, (b) substrate is predominately undrained hydric soil, (c) substrate is nonsoil and saturated or covered with water during part of the growing season each year.

WET LINE KIT System used in conjunction with an enclosed transfer trailer to power its unloading bulkhead. The bulkhead's hydraulic pump is driven by a power-take-off unit on the tractor's transmission.

WET LINER *See* **Cylinder liner.**

WET LOCATION Electrical installations underground or in concrete slabs or masonry in direct contact with the earth, and locations subject to saturation with water or other liquids, such as those exposed to weather, and unprotected.

WET MILLING Hydro-mechanical size reduction of solid wastes that have been wetted to soften the binders of the fibers in the paper and cardboard content.

WET MIX Overly moist concrete.

WET-MIX SHOTCRETE Shotcrete in which the ingredients, including water, are mixed before introduction into the delivery hose.

WETNESS 1. The percentage of water in steam; **2.** The presence of a water film on heating surface interiors.

WET OR LOW-LYING AREA TECHNIQUES *See* **Sanitary landfilling.**

WET PIPE ALARM VALVE Swing check valve installed in the main riser of a wet-pipe sprinkler system that causes an alarm to sound when water flows through an opened sprinkler or a break in the sprinkler piping.

WET-PIPE SPRINKLER SYSTEM *See* **Fire sprinkler system.**

WET PROCESS In the manufacture of cement, the process in which the raw materials are ground, blended, mixed, and pumped while mixed with water. *See also* **Dry process.**

WET RETURN That part of a return main in a steam heating system that is filled with condensed water. *See also* **Dry return.**

WET ROT Fungus decay of wood promoted by warm, moist, unventilated conditions.

WET SAND To smooth a finished surface with water and a product or material (such as wet/dry abrasive paper or emery cloth, or a type of natural or synthetic sponge).

WET SCREEN Aggregate screen equipped with water spray bars to cleanse the material during screening.

WET SCREENING Screening to remove from fresh concrete all aggregate particles larger than a certain size.

WET SCRUBBER Emission control device used to neutralize acid gases. An alkaline mixture is sprayed into the combustion gases, condensing and reacting with them.

WET SIEVING Use of water during sieving of granular material on standard sieves.

WET SLEEVE Cylinder wearing surface that contacts the engine coolant when inserted into an engine block.

WET SOIL Soil that contains significantly more moisture than moist soil, but in such a range of values that cohesive material will slump or begin to flow when vibrated. Granular material that would exhibit cohesive properties when moist will lose those properties when wet.

WET SYSTEM *See* **Automatic sprinkler system.**

WET TANK *See* **Supply tank.**

WETTEST STABLE CONSISTENCY Condition of maximum water content at which cement grout and mortar will adhere to a vertical surface without sloughing.

WETTING Thorough impregnation of a material by a liquid.

WETTING ACTION Reduction in the tendency of a solid to repel a liquid flowing over its surface.

WETTING AGENT Substance capable of lowering the surface tension of liquids, facilitating the wetting of solid surfaces and permitting the penetration of liquids into capillaries.

WET VENT Soil or waste pipe serving also as a vent.

WET WELL Sump of a pumping station.

wf *Abbreviation for:* wood frame.

WF *Abbreviation for:* wide flange; wind force.

wg *Abbreviation for:* wire gauge; wing.

wg. *Abbreviation for:* wedge.

WG *Abbreviation for:* wire gauge; with the grain.

WH *Abbreviation for:* water heater.

WHALER *See* **Waler.**

WHARF Generic term for a landing place or platform built at the edge of or out into water for the berthing of vessels.

WHEEL ARM Device that attaches to a lift bar for engaging the tires of a towed vehicle.

WHEELBASE Horizontal dimension from the centerline of the front axle to the centerline of the rear axle on a single-rear-axle vehicle; measured from the center ine of the front axle to the centerline midway between the axles on a tandem-rear-axle vehicle.

WHEEL BEARING Assembly of hardened-steel rollers and races designed to minimize friction at the spindle and wheel-hub interface.

WHEEL CHOCK Wedge-shaped stop applied at ground level in front of or behind a wheel at rest to prevent further motion.

WHEEL DITCHER Machine that digs trenches by rotation of a wheel fitted with toothed buckets. *See also* **Ditcher,** and **Ladder ditcher.**

WHEEL HOP Bouncing motion of an axle and wheel assembly associated with application of service brakes, usually on a lightly loaded vehicle. *Also called* **Brake hop.**

WHEEL HORSEPOWER *See* **Horsepower.**

WHEELHOUSING Housing over the wheels of a vehicle, replacing the floor area where necessary.

WHEEL LIFT Device used for towing vehicles by lifting one end of the towed vehicle from under the tires.

WHEEL LOAD Portion of the gross weight of a loaded vehicle transferred to the supporting structure under a given wheel of the vehicle.

WHEEL SCRAPER Bowl scraper.

WHEEL SLEEVE Form of flange used on precision grinding machines where the wheel hole is larger than the machine arbor. *See also* **Flange.**

WHEEL SPEED Speed at which a grinding wheel is revolving, measured either in revolutions per unit of time or in surface feet per minute.

WHEEL STRAP Used to tie down a wheel when using a wheel-lift or dolly tow equipment.

WHEEL TRAVERSE Rate of movement of a grinding wheel across the work.

WHEELWELL Housings in a vehicle body floor to allow clearance over tires.

WHET Make sharp by rubbing or grinding.

WHETSTONE Natural or artificial fine-grain stone used to sharpen the cutting edge of steel tools.

WHIPLINE A secondary or auxiliary hoist line. *See also* **Runner line.**

WHIPPING Wrapping to prevent unraveling of the strands at the end of a fiber rope.

WHISPERING GALLERY Interior space having the acoustic property of reflecting sound waves in one or more directions with little loss of volume.

WHISTLE Radio transmitting signaling device, used to communicate line movement or requirements to a machine operator in the yarding of logs.

WHITE CAST IRON Cast iron in which most of the carbon is combined chemically with the iron.

WHITE CEMENT Portland cement that hydrates to a white paste, made from raw materials of low iron content, the clinker for which is fired by a reducing flame.

WHITE GOODS *See* **Solid waste.**

WHITE LATEX ADHESIVE *See* **Adhesive.**

WHITEPRINT Reproduction of a drawing in which the background is white and the drawing is represented by lines of a single color. *See also* **Blueprint.**

WHITEWASH Creamy solution of slaked lime in water, applied as a paint. *Also called* **Limewash.**

WHITING Finely ground calcium carbonate used as an extender in paints, as a thickener in glazes, and to make linseed oil glazing putty less sticky and more easily handled.

WHITNEY STRESS DIAGRAM Diagram of stress distribution in a reinforced concrete beam according to the ultimate load theory.

whl *Abbreviation for:* wheel.

WHMIS Workplace Hazardous Materials Information

System, a system which requires evaluation of the portential harmful effects of materials used in the workplace and includes requirements for special labelling, Material Safety Data Sheets (MSDS) and employee training in handling hazardous materials.

WHMIS CLASSIFICATION Classification of products that are controlled under WHMIS according to the nature of the hazard involved in their handling (e.g., flammability, toxicity).

WHND *Abbreviation for:* wormholes not considered as defects.

WHOLE-BRICK WALL Wall having a thickness equal to the length of one brick.

WHOLE-CIRCLE BEARING Survey bearing defined by its horizontal angle from true north.

WHOLE TREE All components of a tree, except the stump. *Also called* a **Full tree.**

WHOLE-TREE CHIP Chip made from a whole tree.

WHOLE-TREE CHIP FIRE Fire occurring in a whole-tree chip pile, usually as a result of heating and spontaneous combustion.

WHP *Abbreviation for:* water horsepower.

whp *Abbreviation for:* weep hole.

w/hr *Abbreviation for:* watt-hour.

whse *Abbreviation for:* warehouse.

wi. *Abbreviation for:* within.

WI *Abbreviation for:* wrought iron.

WICKET Small door or grilled opening in a larger door.

WICKING 1. Capillary action that takes place through a fibrous conductor; **2.** Either the capillary action of air escaping from a tire casing or the porous material used during curing that allows air to escape rather than build up within the casing body of the tire.

WIDE FLANGE Rolled-steel H-shape section where the flange width and flanges are thicker than the web.

WIDE FRAME *See* **Forks.**

WIDE-FRAME PLATFORM TRAILER *See* **Trailer.**

WIDENING Taking an additional strip of timber off the right-of-way or quarter after the road is in. *Also called* **Daylighting.**

WIDE-STRIP FOUNDATION Reinforced concrete strip footing over 750 mm (2 ft 6 in.) wide.

WIDE TRACK Option available for some tracked equipment that gives the equipment greater stability and a lower center of gravity. Usually selected when working steep slopes.

WIDOWMAKER Loose limb or broken top, or anything loose in a tree that may fall on a worker.

WILDFIRE Unplanned fire requiring suppression action. *See also* **Forest fire.**

WILD HAMMER *See* **Pile hammer.**

WILDLING Seedling naturally reproduced outside of a nursery, used in forest planting.

WIM *Abbreviation for:* white Italian marble.

WINCH Mechanism composed of a drum and gears, used for hoisting, towing, or pulling. There are several types, including:

Capstan: Nonwinding winch comprising a revolving spool that exerts pull by the friction of one or more turns of fiber rope about its shaft. *Also called* **Cat head.**

Donkey: Two-drum powered winch used for hoisting, the drums of which are controlled separately by clutches and brakes. *Also called* **Yarder.**

Double drum: Winch consisting of two drums controlled separately: one for the dragline and the other for the haulback line. Sometimes mounted on and powered by a tractor.

Drum: Hoisting mechanism incorporating one or more rope drums.

Drum machine: Geared machine in which the hoisting ropes are fastened to and wind on a drum.

Hoist: *See* **Drum winch,** above.

Hoisting engine: *See* **Drum machine winch,** above.

Oilfield: Powerful, low-speed winch on a crawler tractor.

Power control: High-speed, tractor-mounted winch with one to three drums.

Towing: Heavy-duty winch mounted on the rear of a crawler tractor. *Also called* **Logging winch.**

WINCH DRUM Rotating cylindrical spool with side flanges, used to wrap the winch rope during raising and lowering of the load with the winch. *See also* **Lagging.**

WINCH ROPE Wire rope used to reeve the winch and the attachments for lifting loads.

WIND 1. Twisting warp in lumber; **2.** Turn or bend in an otherwise flat or straight element; **3.** Wrap or twist around a spool.

wind. *Abbreviation for:* window.

WINDBOX Chamber below a furnace grate or surrounding a burner, through which air is supplied under pressure to burn the fuel.

WINDBOX PRESSURE Static pressure of the air in the windbox of a burner or stoker.

WIND BRACE Strut used as a brace to strengthen a frame against wind loads.

WIND CAP Metal or stone cap mounted above the terminal of a chimney, used to increase the draft and to prevent downdraft.

WIND ENERGY Useful energy derived from wind currents.

WINDER Stair tread with one end wider than the other, used in winding staircases or in quarter-turns.

WINDFALL Tree felled by wind. *See also* **Wind throw.**

WINDING DRUM Drum that accumulates a suspension rope in more than one layer on the surface of the drum.

WINDING-DRUM MACHINE *See* **Driving machine.**

WINDING STAIRS Spiral staircase.

WINDLASS Hand- or power-driven hoisting device where the lifting force is generated from the friction between the tensioned lifting rope and a rotating cylinder around which it makes at least two passes.

WINDLEAN Lean of a tree caused by constant prevailing wind.

WIND LOADING Total force exerted by the wind on a structure or a part of a structure or on a building component. Usually used in the form of a test to determine how much wind a certain type of design can withstand in order to meet certain criteria.

WINDOW Opening in a wall to allow for the passage of light and air, usually fitted with movable or opening glazed frames. There are many basic shapes, including:

> **Bay:** Composed of three or more individual windows abutting each other at an angle that, cumulatively, allows them to close an opening less than the sum of their widths by projecting out from the face of the building.
>
> **Bow:** Window composed of three or more windows mounted on a curved or semicircular frame that projects from the face of the building.
>
> **Casement:** Window hinged at the side and usually opening outwards.
>
> **Combination storm and screen:** Pair of vertically opening windows faced with additional sashes that incorporate a screened opening.
>
> **Double-hung:** Two vertically sliding sashes that bypass each other in a single frame, each counterbalanced by springs or weights.
>
> **Picture:** Large, stationary, nonventilating window placed so as to obtain the maximum view.
>
> **Sliding:** Window with two or more sashes that slide past each other within the frame, horizontally or vertically.

WINDOW APRON Plain or molded finish covering the rough edge of plastering below the stool of a window.

WINDOW BAR Glazing bar.

WINDOW BOX Box attached to the outside of a windowsill, usually containing plants.

WINDOW FRAME Parts of a window surrounding the casements, in which they are fixed, hinge, or slide.

WINDOW HEAD The upper portion of a window frame.

WINDOW-JACK SCAFFOLD Platform supported by a bracket or jack that projects through a window opening.

WINDOW JAMB The sides of a window opening.

WINDOWPANE Glass pane in a window.

WINDOW PIPE Dredge discharge pipe with one or more openings in the bottom.

WINDOW SASH Frame in which the window lights are set.

WINDOW SCHEDULE Table giving details of each type of window required for a project.

WINDOW SEAT Seat built into a window recess, or in front of a window.

WINDOWSILL The bottom portion of a window frame.

WINDOW STILE Upright member of a window frame, guiding the sash.

WINDOW STOOL 1. Nosing directly above the window apron; **2.** Horizontal member of the window finish that forms a stool for the side casings and conceals the window sill.

WINDOW STOP Narrow wooden strip that holds the sash in position in the window frame.

WINDOW TRIM Interior finish of a window opening.

WINDROW 1. Ridge of loose material thrown up by a machine; **2.** Continuous pile of material placed on a grade or subgrade for later pickup or spreading; **3.** Composting material stacked in a triangular prism shape.

WINDROW ELEVATOR Device that travels ahead of an asphalt paver and that is generally attached to it, used to pick up windrowed mix and feed it into the paver hopper to achieve more continuous paver operation.

WINDROWING Concentration of slash, branchwood, and debris into rows to clear the ground for regeneration. Windrows are often burned.

WIND SHAKE Crack in a tree caused by high winds.

WINDSHIELD Transparent weather shield in front of the driver's position on a vehicle.

WIND SOCK Bright colored fabric or plastic tube in the form of a truncated cone, fastened by its wider end at the top of a mast, used to indicate wind direction at landing strips, airfields, and helipads.

WIND SPEED Wind velocity, measured at 6 m (20 ft) above the ground or the average height of vegetation cover, and averaged over at least a 10-minute period.

WINDTHROW Stand of trees blown down by wind. *Also called* **Blowdown.**

WINDTHROWN Uprooted by the wind.

WIND TUNNEL Test structure in which wind action can be simulated, used to determine the reaction of structures and components to wind force.

WIND WASH Tendency of wind blowing over an exposed window to increase heat transfer.

WING 1. Section of a building extending from the main part; **2.** Projection on an air-drill bit; **3.** Short length or strip of steel welded near the toe of a steel H-pile to increase its bearing capacity.

WING DAM Breakwater.

WING LOG One of the logs making up the wing walls of a crib.

WING NUT *See* **Nut.**

WING PILE *See* **Pile.**

WING STOCK Tread rubber that is tapered to a feathered edge on each side in order that it may be applied to the shoulder of the tire. Used only on a full retread.

WING TREAD Tire tread with special shoulder extensions that wrap down onto the shoulder of the casing. These extensions provide extra strength along the bond line and give the finished retread a new tire appearance.

WING WALL Short retaining wall at the end of a bridge or culvert to retain the earth; wall that guides a stream into a bridge opening or culvert barrel.

WINTER DESIGN TEMPERATURE Lowest ambient temperature anticipated for a location and for which adequate insulation should be provided.

WIP *Abbreviation for:* work in progress.

WIPE Action of one surface or contact point across the face of another.

WIPED JOINT Finishing action in completing a liquid solder pipe joint.

WIPER RING *See* **Piston ring.** *Also called* **Exclusion device.**

WIPER/SCRAPER Device placed in the head of a cylinder for the purpose of excluding foreign matter from the inside of the cylinder.

WIRE Flexible metal strand.

WIRE AXE Tool used to cut wire rope or rope strands.

WIRE BRAID *See* **Braid.**

WIRE BRUSH Wood or plastic handle with rows of stiff wire bristles embedded in the face. Used to abrade and scrape dirt and debris from a surface.

WIRE CLOTH Screen formed from wire or rod woven and crimped or welded into a mesh.

WIRE-CUT BRICK *See* **Brick.**

WIRED EDGE Wire enclosed in a rolled sheet metal edge.

WIRED GLASS Wire mesh embedded in sheet glass as reinforcement.

WIRE EDGE Burr on the ground edge of a sharpened steel tool, usually removed by honing.

WIRE GAUGE System of defining wire diameter by number: the larger the number the finer the wire.

WIRE LATH Galvanized wire netting used to support plaster.

WIRE LOOP In braided hose, a loop in the wire reinforcement caused by uneven tensions during bobbin winding or braiding.

WIRE REINFORCED Hose containing wires to give added strength, increased dimensional stability, or crush resistance. *See also* **Reinforcement.**

WIRE ROPE A plurality of wire strands helically laid about an axis. *Also called* **Line.**

WIRE ROPE CLIP Reusable wire rope fastener.

WIRE STRAND CORE Wire strand used as the axial member of a wire rope.

WIRE THROW-OUT 1. In braided hose, a broken end or ends in the wire reinforcement protruding from the surface of the braid; **2.** Displaced coil in rough bore hose.

WIRE TIE Short length of 1.6-mm (0.06-in.) diameter soft iron wire used to fasten together the intersection of reinforcing bars and links.

WIREWAY Sheet metal trough with a hinged or removable cover, used for housing and protecting electrical wires and cables.

WIRE WOUND Having a single wire or plurality of wires spiraled in one or more layers as a protective or reinforcing member.

WIRE WOVEN Woven with the wire reinforcement applied helically by means of a circular loom.

WIRE WRAPPING Application of high-tensile wire, wound under tension by machine, around circular concrete or shotcrete walls, domes, or other tension-resisting structural components.

WIRING DIAGRAM Diagram that includes all the devices in an electrical system and showing their functional relationship to each other. Such a diagram gives the necessary information for physically tracing circuits when troubleshooting is necessary.

WIRING HARNESS Preassembled group of wires arranged to facilitate interconnection of electrical circuits.

WIRING SCHEDULE List that associates wire identity with a terminal.

WITCHES' BROOM Abnormal tufted growth of small branches on a tree or shrub caused by fungi or viruses.

WITHE *See* **Wythe.**

WITHOUT RECOURSE Words used in endorsing a negotiable instrument to signify that the endorser will not be liable to a future holder in the event of nonpayment.

WITNESS TREE Tree used by surveyors to mark the location of a survey corner; the tree is located near the survey corner and is inscribed with survey data. *Also called* **Bearing tree.**

wk *Abbreviation for:* week; work.

wkd *Abbreviation for:* worked.

wkg *Abbreviation for:* working.

wkmnshp *Abbreviation for:* workmanship.

wkn *Abbreviation for:* weaken.

wkr *Abbreviation for:* wrecker.

wkt *Abbreviation for:* wicket.

WL *Abbreviation for:* warning light; white lead; wind load.

wlbd *Abbreviation for:* wallboard.

wld *Abbreviation for:* weld.

wlkwy *Abbreviation for:* walkway.

wm *Abbreviation for:* wattmeter.

WM *Abbreviation for:* wire mesh.

wnd *Abbreviation for:* wind.

WNW *Abbreviation for:* west-northwest.

w/o *Abbreviation for:* without; wheel opening.

WOBBLE *See* **Weave.**

WOBBLE COEFFICIENT Coefficient used in determining the friction loss occurring in post-tensioning, which is assumed to account for the secondary curvature of the tendons.

WOBBLE FRICTION In prestressed concrete, the friction caused by the unintended deviation of the prestressing sheath or duct from its specified profile.

WOBBLE SAW Circular saw blade set eccentric to its arbor so that relatively narrow teeth produce a wide rabbet.

WOBBLE-WHEEL ROLLER Pneumatic-tired roller with wheels freely suspended on springs to follow irregularities in a road surface.

WOLF TREE Vigorous tree that has merchantable value but occupies more space than its value warrants, usually broad-crowned, dominant, and very limby. Often a remnant from a previous stand.

WOOD Logs, timber, trees.

WOOD BLOCK One of a number of wood blocks laid with the grain vertical, used for paving and industrial flooring.

WOOD CONVERSION Transformation of natural timber into any kind of commercial product. Includes all activities from commercial timber (log) delivery to the log yard at the initial commercial processing facility to the final product form offered for commercial sale as a consumer product.

WOOD CHISEL Steel cutting tool made from tempered steel. The blade is mounted in a wooden or plastic handle. The free end of the blade is tapered to a sharp, wide edge that is then honed at an angle of 30° to the horizontal. There are several types:

Bevel edge: Where the edges on the upper (non-working) face of the blade are chamfered at a 45° angle.

Butt: A short-bladed, heavy-duty, general-purpose chisel.

Firmer: A square-sided, medium-duty chisel.

Framing: Chisel with an extra-long blade, up to 50 mm (2 in.) wide.

Gouge: Shaped-edge chisel for wood turning, or for hand carving.

Mortise: Chisel with an extra-thick blade.

Paring: Hand chisel with a long blade, commonly employed by patternmakers for slicing or paring cuts.

Paring gouge: Wood chisel with a cutting edge beveled on the inside, concave face.

Pocket: Chisel with a wide, thin blade, honed on both sides.

Skew: Woodworking tool with a straight, inclined cutting edge.

Straight edge: Chisel with a rectangular blade.

WOOD CUSHION *See* **Wear plate.**

WOOD FLOAT Tool consisting of a wooden plate, approximately 225 x 100 mm (9 x 4 in.) with a handle on one flat face, used for compacting and smoothing floor and deck mortar.

WOOD FLOORING Wood, prefinished in some form, installed as the final wearing surface on a finished floor. May consist of strips or planks of wood that are laid one at a time in a continuous row and that usually incorporate some design of edge joint to lock together adjacent strips. Another form is parquet flooring, which comprises small pieces of hardwood grouped together to form a pattern and that is most commonly available in tile form from 150 mm (6 in.) square to 600 mm (24 in.) square and more. Parquet tiles are fastened to a subfloor with a mastic adhesive.

WOOD FLOUR Fine sawdust, used as an extender for glues, in explosive manufacture, as the basis of plastic wood, etc.

WOOD FOUNDATION Residential and light frame foundation system that uses pressure-preservative-treated plywood panels and wood framing in place of poured concrete footing and masonry walls.

WOOD FRAME CONSTRUCTION *See* **Construction types.**

WOODLOT LICENCE Area of forest land containing not more than 400 ha (1000 acres) of Crown land that may be managed in conjunction with some private land.

WOOD PRESERVATIVE Chemical product used to prevent or halt decay in exterior wood, applied by pressure treatment, soaking, or brush.

WOOD PULP Fiber from wood which, with varying degrees of purification, is used for the production of paper, paper board, and chemical products.

WOOD ROLL Round-topped section of wood fixed to roof boarding, around which metal roofing sheets are lapped.

WOODRUFF KEY Half-moon-shaped key.

WOODS Logging area beyond a camp or shop, or the area beyond the landing.

WOOD SCREW *See* **Screw.**

WOOD TURNING Working wood on a lathe.

WOODWORK Finished goods fashioned of wood.

WOOF *See* **Shoot wire.**

WORK 1. The entire project (or separate parts) that is required to be completed under the Contact Documents. Work is the result of performing services, furnishing labor, and supplying and incorporating materials and equipment into the construction, as required by the contract documents; 2. Unit or piece being handled or worked on; 3. Product of force and the distance through which it moves; 4. *See* **Tire types.**

WORKABILITY That property of fresh concrete or mortar that determines the ease with which it can be mixed, placed, and finished. *Also called* **Placeability.**

WORK ARM Part of a lever between the fulcrum and the working end.

WORK BASKET Integral single-point suspended scaffold welded or bolted together with provision for mounting a hoisting machine, and including a platform, toeboard, and guardrails to accommodate users in a standing position. *Also called* **Work cage.**

WORK CAGE *See* **Work basket.**

WORK-COST INDICATOR Measure of the efficiency of a hauling machine, equal to the gross vehicle weight plus tare weight divided by payload weight.

WORK DAY Portion of a 24-hour day during which work is contracted for, and/or paid for. *See also* **Day,** and **Calendar day.**

WORKED PENETRATION Penetration of a sample of lubricating grease that has been brought to 25°C (77°F), subjected to 60 double strokes in a standard grease worker, and penetrated without delay. *See also* **Penetration.**

WORKING Subjection of lubricating grease to any form of agitation or shearing action beyond simple transfer to any test apparatus.

WORKING CHAMBER Space or compartment under air pressure in which work is done.

WORKING CYCLE Complete set of repetitive and productive operations of a machine.

WORKING DAY Calendar day during which normal construction operations could proceed for a major part of a shift; normally excludes Saturdays, Sundays, and holidays. *See also* **Calendar day.**

WORKING DRAWING Drawing that shows sufficient detailed information, including sizes and shapes, from which to properly build the object shown and described. *Also called* **Plans, Shop drawing,** and **Standard plans.**

WORKING EDGE Face edge of lumber.

WORKING FACE 1. Face side of lumber; 2. That portion of the compacted solid wastes at a sanitary landfill that will have more refuse placed upon it or is being compacted prior to placement of cover material.

WORKING HORSEPOWER *See* **Horsepower.**

WORKING LIFE *See* **Application life.**

WORKING LIMIT Maximum breaking strength divided by the factor of safety for cable or chain.

WORKING LOAD Forces normally imposed on a member in service. *See also* **Actual load.**

WORKING PRESSURE *See* **Pressure.**

WORKING RADIUS *See* **Radius of load.**

WORKING SHAFT Shaft sunk to excavate a tunnel or sewer and filled in on completion of the work.

WORKING STRESS Maximum permissible design stress using working-stress design methods.

WORKING-STRESS DESIGN Method of proportioning either structures or members for prescribed service loads at stresses well below the ultimate, and assuming linear distribution of flexural stresses and strains.

WORKING TOOL Pavement engaging tool held by a working tool holder attached to the cutter drum that performs the action of pavement planing/milling. *Also called* **Bit, Cutter, Cutting tool, Pick,** and **Tooth.**

WORKING TOOL HOLDER Device that retains and supports the working tool.

WORKING TO ONE'S HAND Working to the left or right of a center line: a right-handed mason is more efficient working from the left of a wall toward the center, a left-handed person from the right to the center.

WORKING WEIGHT Weight of a machine with full radiator, half-full fuel tank, and attachments installed.

WORK ITEM Portion of the project that can be clearly identified and isolated.

WORKMANSHIP Quality of work done.

WORK MODE SELECTION Machine operating system that permits the operator to choose the operating mode best suited to the task at hand. With many variations available, a typical system controls such functions as engine speed and hydraulic fluid flow, offering the following settings: Heavy-duty (involving 100% of capacity); Standard (at approximately 80% to 85% maximum capacity); and Fine Control (regulated to 60% to 65% of capacity), used when control rather than speed is more important.

WORK ORDER Written instruction authorizing work to be done and/or costs to be incurred.

WORK PACKAGE Generic term for units of work as a subdivision of the total scope of a project.

WORK PLATFORM Platform intended to provide safe working conditions and designed to be mounted on an elevating device to provide an area for one or more persons riding on and working from it. *Also called* **Safety platform.**

WORK SCHEDULE *See* **Construction schedule.**

WORK SPEED In cylindrical, centerless, and internal grinding, the rate at which the work revolves, measured in either revolutions per minute or surface meters (feet) per minute; in surface grinding, the rate of table traverse measured in meters (feet) per minute.

WORK TRIANGLE Sum of the sides of a triangle connecting the refrigerator, range, and sink in a residential kitchen, that should be between 4.5 and 6.4 m (15 and 21 ft).

WORM Cylindrical gear with spiral threads cut into its surface.

WORM CONVEYOR Spiral conveyor system.

WORM GEAR *See* **Driving machine.**

WORM-GEAR MACHINE Direct-drive machine in which energy from the motor is transmitted to the driving sheave or drum through worm gearing.

WORM WHEEL Modified spur gear that has curved teeth and that meshes with a worm.

WOUND MEDIUM Filter medium comprised of layers of helical wraps of a continuous strand or filament in a predetermined pattern.

WOVEN FABRIC Flat structure composed of two series of interlacing yarns or filaments, one parallel to the axis of the fabric and the other transverse.

WOVEN JACKET *See* **Jacket.**

WOVEN MEDIUM Filter medium made from strands of fiber, thread, or wire interlaced into a cloth on a loom.

WOVEN-WIRE FABRIC Prefabricated steel reinforcement composed of cold-drawn steel wires mechanically twisted together to form hexagonally shaped openings.

wp *Abbreviation for:* working pressure.

w/p *Abbreviation for:* water pump.

WP *Abbreviation for:* weatherproof.

WPC *Abbreviation for:* white portland cement.

wpr. *Abbreviation for:* waterproofing; weatherproofing.

wr. *Abbreviation for:* wire; wrench.

WR *Abbreviation for:* wash room.

WRACKING FORCE Force that tends to distort a rectangular shape into a parallelogram.

WRAP One circumferential turn of wire rope around a hoist drum. *See also* **Cross wrap.**

WRAPAROUND Ability of a coating to cover all areas of the substrate to which it is applied, including edges.

WRAPAROUND MORTGAGE *See* **Mortgage.**

WRAPPED CURE Vulcanizing process using a tensioned wrapper (usually of fabric) to apply external pressure. *Also called* **Rag-wrap.** *See also* **Cross wrap,** and **Herringbone wrap.**

WRAPPER *See* **Binder.**

WRAPPER MARK Impression left on the surface of a hose by a material used during vulcanization. Usually shows characteristics of a woven pattern and wrapped edge marks. *See also* **Wrapped cure.**

WRAPPING Reinforcing bars or mesh surrounding a structural steel column or beam to reinforce concrete or plaster fireproofing.

WREATH Section of a stair rail curved both horizontally and vertically, used to join the ascending run of the handrail to the side of a newel post.

WRECKER Truck equipped with winch or winches and boom(s), used for recovering and towing vehicles.

WRECKING Dismantling, demolishing, or razing a building or structure.

WRECKING BALL *See* **Headache ball.**

WRECKING BAR Steel bar, one end of which is slightly bent with a chisel-shaped tip, the other U-shaped with a claw tip for pulling or prying, used for forcing apart and separating. *See also* **Pry bar.**

WRECKING STRIP Small piece or panel fitted into a formwork assembly in such a way that it can be easily removed ahead of the main panels or forms, making it easier to strip those major form components. *See also* **Crush plate.**

WRENCH Hand tool for holding and turning nuts, bolts, pipe, etc. There are many variations, including:

    Adjustable: Tool having parallel jaws that can be opened or closed with a worm gear. Range in total length from 100 to 600 mm (4 to 24 in.) to fit nuts up to 50 mm (2 in.) and greater.

    Allen: L-shaped hexagonal metal bars in a range of diameters, used to turn screws and bolts having hexagonal recesses in their heads.

    Box: Narrow handle having ring-shaped ends with six or twelve interior 'points' or facets.

    Clip: Adjustable wrench. *Also called* **Crescent wrench.**

    Combination: Narrow handle having a box wrench at one end and an open-end wrench at the other, usually to fit the same-size nut.

    Crow's foot: Tool consisting of the head of an open-end wrench having a square recess in a stubby extension, into which a socket wrench can be fitted.

    Moon: Type of box wrench having a handle formed as a hoop.

    Open-end: Narrow handle having open, fixed jaws at both ends.

    Pipe: Heavy-duty, long-handled wrench with toothed, parallel jaws that can be adjusted to a wider or narrower opening by means of a worm gear. One jaw is fixed; the other is free to rock through a limited range so that it will form an ever tighter grip on a pipe to which it is fitted, but will release when contrarotated. Available in sizes from 250 to 1200 mm (10 to 48 in.).

    Power: Wrench actuated by electric, hydraulic, or pneumatic force.

    Torque: Wrench with a gauge that measures the torque (twisting force) put on a bolt.

    Socket: Long handle with a square peg at right angles at one end, often as part of a ratchet mechanism. A wide range of sockets may be attached to the peg and rotated by the handle. The sockets have interior openings with either six or twelve 'points' or facets.

WRENCHABLE NUT *See* **Nut.**

WRENCHING Transmission of torque to a fastener by a tool.

WRINGING FIT *See* **Fit classifications.**

WRINKLE *See* **Cockle.**

WRINKLED PLY *See* **Buckled ply.**

WRIST ACTION 1. Articulating action resembling that of a human wrist; **2.** Hinging of a backhoe bucket so that its digging and dumping positions can be regulated by hydraulic cylinders.

WRITE DOWN Downward adjustment of the recorded amount of an asset or liability for reasons other than the occurrence of a transaction.

WRITE OFF Elimination of the recorded amount of an asset or liability for reasons other than the occurrence of a transaction.

WROUGHT Hammered to a shape.

WROUGHT IRON Iron made up of iron silicate fibers entrained in a ferrite matrix. *See also* **Iron,** and **Malleable cast iron.**

WROUGHT TIMBER Lumber planed on one or more surfaces.

wrp *Abbreviation for:* wrap.

wrpg *Abbreviation for:* wrapping.

wrpg. *Abbreviation for:* warpage.

wrt *Abbreviation for:* wrought.

ws *Abbreviation for:* weatherstripping.

w/s *Abbreviation for:* windshield.

WS *Abbreviation for:* water supply.

wshbl. *Abbreviation for:* washable.

wshr *Abbreviation for:* washer.

wsp *Abbreviation for:* welded steel pipe.

w/srp *Abbreviation for:* weatherstrip.

WSW *Abbreviation for:* west-southwest.

wt *Abbreviation for:* weight.

WT *Abbreviation for:* wall tie; water table; water tank; watertight; water tower.

wtc *Abbreviation for:* whole tree chips.

wth *Abbreviation for:* width.

wtr *Abbreviation for:* water.

wtrprf *Abbreviation for:* waterproof.

wv *Abbreviation for:* wall vent; water valve.

wvd *Abbreviation for:* waived.

WVT *Abbreviation for:* water vapor transmission.

ww *Abbreviation for:* white wash.

ww. *Abbreviation for:* wireway.

WW *Abbreviation for:* wire wound.

wwf *Abbreviation for:* welded wire fabric.

wwm *Abbreviation for:* welded wire mesh.

wxd *Abbreviation for:* waxed.

wxg *Abbreviation for:* waxing.

WYE Three-ended plumbing fitting, usually consisting of two pipes joined to a third at 45° angles.

WYE-DELTA Method of starting an AC, three-phase electric motor.

WYTHE 1. Each continuous vertical section of masonry one unit in thickness; 2. Thickness of masonry separating flues in a chimney, *Also called* **Tier,** and **Withe.**

xarm *Abbreviation for:* crossarm.

xband *Abbreviation for:* crossband.

xbar *Abbreviation for:* crossbar.

X-BRACE Paired set of crossing sway braces. *See also* **Cross bracing,** and **Swaybrace.**

xconn. *Abbreviation for:* cross connection.

X-CRACK Tension crack characteristic of material failure due to vibration.

XC/S *Abbreviation for:* crosscut saw.

xcut *Abbreviation for:* crosscut.

x hvy *Abbreviation for:* extra heavy.

XL *Abbreviation for:* extra large; wood screw and lead expansion shield.

xmbr *Abbreviation for:* crossmember.

xmsn *Abbreviation for:* transmission.

x-ovr *Abbreviation for:* crossover.

X-RAY FLUORESCENCE Characteristic secondary radiation emitted by an element as a result of excitation by X-rays, used to yield chemical analysis of a sample.

xref. *Abbreviation for:* cross-reference.

XS *Abbreviation for:* extra small; machine screw and expansion sleeve.

xsect. *Abbreviation for:* cross-section.

x-shaft *Abbreviation for:* cross shaft.

x str. *Abbreviation for:* extra-strong.

XX str. *Abbreviation for:* double-extra-strong.

XYLEM Principal strengthening and water-conducting tissue of stems, leaves, and roots; the botanical name for wood.

XYLENE Distillation of petroleum or coal tar, used as a solvent. *Also called* **Xylol.**

XYLOL *See* **Xylene.**

Y *Abbreviation for:* Y-branch.

YARD 1. Nonmetric unit of length, equal to three feet. Symbol: yd. Multiply by 0.9144 to obtain meters, symbol: m. *See also the appendix:* **Metric and nonmetric measurement**; **2.** Area for the storage and maintenance of equipment, storage of materials, site office, etc.; **3.** Place where logs are accumulated. *See also* **Harvest functions**; **4.** Open space on a building lot, usually to the rear of the building.

YARDAGE Volume of material filled or excavated, measured in cubic yards.

YARDER Machine that pulls in logs from a stump to a landing. *Also called* **Donkey winch**. *See also* **Winch**.

YARDER WOOD Wood brought into a yard in the form of tree lengths, logs, or bolts, to be cut into shorter lengths.

YARDING Moving trees or logs from felling site to roadside with cable, winch, or helicopter.

YARDING CRANE Crane-type machine engineered and rigged to yard logs.

YARDING OF UNMERCHANTABLE MATERIAL Yarding of cull, rotten, small, or otherwise unsaleable wood material to a designated area for disposal as written into a timber contract.

YARDING ROAD Path followed by a turn of logs yarded by a cable method.

YARDING TOWER Steel tower used on a steel-spar skidder; lightweight tower built on a tractor.

YARD LUMBER Term for lumber of all sizes and patterns intended for general building purposes, having no design property requirements.

YARD MULE Small tractor used to move semitrailers around a terminal yard.

YARD REQUIREMENT Space mandated under by-law or regulation as necessary for the development of a building lot. *Also called* **Minimum yard depth**.

YARD RUBBISH Prunings, grass clippings, weeds, leaves, and general yard and garden wastes.

YARD WASTE *See* **Solid waste.**

YARN Generic term for continuing strands of textile fibers or filaments in a form suitable for knitting, weaving, or otherwise intertwining to form a textile fabric.

YARNING IRON Tool used to pack oakum into bell and spigot joints before they are leaded.

YARN NUMBER Number of hanks in a pound, usually cotton.

Y-BRANCH Plumbing fitting joining a single pipe to two others at 45° angles.

yd *Abbreviation for:* yard.

y-d *Abbreviation for:* yard drain.

ydg. *Abbreviation for:* yardage.

yds *Abbreviation for:* yards.

YEAR RING One of the clearly defined rings in a cross section of a tree trunk showing the amount of annual growth of the tree; each ring represents one year of growth.

YELLOWING Development of a yellow color or cast in white or clear coatings as a consequence of aging.

YIELD 1. Amount yielded or produced; **2.** Volume of freshly mixed concrete produced from a known quantity of ingredients; **3.** Total weight of ingredients divided by the unit weight of freshly mixed concrete; **4.** Number of units produced per bag of cement or per batch of concrete; **5.** Estimate in forest mensuration of the amount of wood that may be harvested from a particular type of forest stand by species, site, stocking, and management regime at various ages; **6.** Of a lubricating grease, the amount of grease of a given consistency that may be made with a definite amount of thickening agent.

YIELD MOMENT In a member subjected to bending, the moment at which an outer fiber first attains the yield stress.

YIELD PLATEAU Portion of the stress-strain curve for uniaxial tension or compression in which the stress remains essentially constant during a period of substantially increasing strain.

YIELD POINT 1. That point during increasing stress when the proportion of stress to strain becomes substantially less than it has been at smaller values of stress; **2.** Minimum force required to produce flow of a plastic material.

YIELD STRENGTH 1. Tensile strength measurement at which any deformation of a sample of material becomes permanent; **2.** Measured strength at which the elastic limit of a material is exceeded, and it cannot return to its previous size or form.

YIELD STRESS Yield point, yield strength or yield-stress level as defined.

YIELD-STRESS LEVEL Average stress during yielding in the plastic range.

Y-INTERSECTION *See* **Intersection types.**

YOKE 1. Tie or clamping device around column forms or over the top of wall or footing forms to keep them from spreading because of the lateral pressure of fresh concrete; **2.** Part of a structural assembly for slipforming that keeps the forms from spreading and that transfers form loads to the jacks. *See also* **Head jamb**; **3.** Heavy U-shaped part of a block by which the pulley is attached.

YOUNG'S MODULUS OF ELASTICITY Stress required to produce unit linear strain. *See also* **Modulus of elasticity**.

YO-YO LINE Short piece of line that runs through the carriage to open a grapple.

YP *Abbreviation for:* yield point.

yr *Abbreviation for:* year.

YS *Abbreviation for:* yield strength.

yt *Abbreviation for:* yoke top.

YUM *Abbreviation for:* yarding of unmerchantable material.

Z

z *Abbreviation for:* zone.

ZAX Slater's edged and pointed hand tool, used for cutting slates and punching nail holes.

Z-BAR Z-shaped wall tie.

ZEITFUCHS CROSS-ARM VISCOMETER Asphalt viscosity testing device, used to measure the viscosity of paving-grade asphalts.

ZENITH Point on the celestial sphere directly above a station on the earth.

ZENITH ANGLE Type of angle measurement in a vertical plane that uses upward vertical as zero.

ZERK FITTING Fitting on mechanical equipment to which a grease gun may be attached for the injection of grease into operating hinge joints.

ZERO Binary value (0) for false or low. *See also* **One.**

ZERO ALLOWANCE *See* **Fit classifications.**

ZERO-BASE BUDGETING Management system in which all programs are reevaluated and must be justified each time a new budget is drafted.

ZERO FLOAT Describes a work item of activity on the critical path (i.e., a delay in its completion would cause a corresponding delay in the overall project completion date).

ZERO GOVERNOR *See* **Atmospheric (ratio) regulator.**

ZERO LOT LINE Form of cluster housing in which individual units are sited on separate lots, but that may have a common party wall.

ZERO-SLUMP CONCRETE *See* **Negative-slump concrete.**

ZERO TERM Condition in multiple circuits where all inputs are off and all switches are open.

Z-FLASHING Z-shaped piece of galvanized steel, aluminum, or plastic installed at the horizontal joints of plywood siding to prevent water from entering the wall cavity.

ZINC Bluish-white metal used mainly in alloys for coating steel.

ZINC CHLORIDE Chemical used as a flux for soldering.

ZINC CHROMATE Yellow, rust-preventing pigment. *See also* **Metal primer.**

ZINC DIALKYLDITHIOPHOSPHATE Type of antiwear additive, used in engine oil formulation.

ZINCKED FASTENINGS Method of finishing wire rope attachments by pouring molten zinc into the socket and allowing it to solidify.

ZINC OXIDE White pigment, used to prevent mold or mildew on a paint film. *See also* **Metal primer.**

ZIRCONIUM *See* **Alloying elements.**

ZONE 1. Area defined by a local authority for specific use and subject to defined conditions and/or restrictions. *Also called* **District; 2.** Space or group of spaces within a building, combined for common control of heating and cooling.

ZONE CONTROL 1. Heating and cooling control system employing two or more thermostats; **2.** Control of air flow into individual zones of the undergrate of the stoker, or the plenums of a burner system.

ZONE OF AERATION Area above a water table where the interstices are not completely filled with water.

ZONE OF CAPILLARITY Area above a water table where some or all of the interstices are filled with water that is held by capillarity.

ZONE OF SATURATION Ground below the water table.

ZONE SWITCH *See* **Switch.**

ZONING Division of an area into districts and the public regulation of the character and intensity of use of land and improvements thereon. *See also* **Roadside zoning.**

ZONING BYLAW *See* **Zoning ordinance.**

ZONING ORDINANCE Regulation of property by local government. *Also called* **Zoning bylaw.**

ZONING PERMIT Permit issued by a regulatory agency authorizing land to be used or developed for a specific use.

METRIC AND NONMETRIC MEASUREMENTS:

The SI, U.S. Customary, and Imperial Systems

America still going metric

Once, the United States almost went metric: in 1975, Congress passed the *Metric Conversion Act* "...to coordinate and plan the increasing use of the metric system in the United States." However, it never happened; the U.S. Metric Board that was established to do the job found significant disinterest in the subject by the general public and, seven years later, in 1982, was 'disestablished.'

In reality, America's involvement with metric measurement is both long and considerable, dating back to the early 1800s when the U.S. Coast and Geodetic Survey used the meter and kilogram. In 1866, Congress authorized the use of the metric system and supplied each state with a set of standard metric weights and measures. In 1875, the United States was one of the original seventeen signatory nations to the *Treaty of the Meter*, and in 1893, the metric measurement standards established by the International Bureau of Weights and Measures, established by the treaty, were adopted as the fundamental standards for length and mass in the U.S.

A century later, not much has happened. According to the U.S. Department of Commerce (which is charged with encouraging the use of metric), 'The United States is the only industrialized country in the world not officially using the metric system.' While this is true of virtually all products used domestically, there is a growing trend among those companies doing business overseas — including many of the country's largest — to wholly convert to metric measurements both for the home market and for exports. Also, most NASA projects are now designed and built to metric specifications; the General Services Administration is establishing metric specifications for products it buys for government — even the Congressional Record and Federal Register will be produced to metric sizes by the Government Printing Office.

There is no law that requires U.S. companies to convert to metric measurements. Increasing global trade, NAFTA, and the other economic trading blocks, and the increasing isolation of American industry in a metric world may at last hasten its move to the International System of Units (SI), the name adopted under the *Treaty of the Meter* in 1960 for the metric system.

Is Canada metric?

Canadians buy their gasoline and milk by the litre (note the spelling), and drive on roads that are measured in kilometres at speeds registered in km/hr. In a bilingual country where all consumer products sold in other than their province of manufacture must, by law, be labeled in English and French, so too must all products be measured in metric terms (although most also carry nonmetric equivalents, which the majority of purchasers continue to prefer). At one time, about a decade ago, merchants in several 'test' communities where full conversion to metric was staged were prosecuted for displaying price tickets in other than metric. But those days have gone, and metric conversion in Canada has gone far enough to make a dent in the old 'Imperial' system, but not far enough to encourage its population to make the full conversion — in their heads or in the marketplace.

The Canadian construction industry was scheduled to go metric on 'M-Day,' January 1, 1978. After the expenditure of millions of dollars in committees, special research, seminars, tons upon tons (or, rather, tonnes upon tonnes) of literature aimed at all those involved in building and the supply of construction materials and equipment, it didn't happen. The country is left with a hodgepodge of often conflicting requirements and standards. Some public authorities insist on all quotations, contracts, and working drawings being prepared in SI — the International System of Units, as the metric system is known. However, only a diminutive number of products are yet produced to 'hard' metric dimensions, most documents are cluttered with a confusion of 'soft' converted numbers, interspersed by the dimensions of the relatively few truly metric products yet available.

(A word of explanation. 'Hard' conversion means that a product traditionally manufactured to a rational imperial set of dimensions would be 'resized' to a set of rational metric dimensions. Plywood, for instance, which is produced in 4 x 8 ft (1219.2 x 2438.4 mm) sheets, in thicknesses like 1/2 in. (12.7 mm) and 3/4 in. (19.05 mm), among others, would in future be manufactured in 1200 x 2400 mm sheets in equivalent thicknesses of 12.5 mm, 18.5 mm, etc.

'Soft' conversion means that the imperial dimensions of the product remain the same. However, those imperial dimensions are converted to metric. Most plywood sheets still typically measure 1219.2 x 2438.4 mm.)

Since Canada's heady days of the 'metric police', some change toward metrification has taken place. Children in school are mostly taught measurement using the metric system, with acknowledgment for the historic (and still used) nonmetric equivalents. A (very) slowly growing range of products is made to fully metric measurements, and there is complete conversion by public agencies to metric. Canada will become a fully metric country — likely later rather than sooner and, as always, subject to the pleasure of U.S. practice.

The metric, or SI system of measurement

The meter was introduced as a unit of linear measurement in 1791, being defined as 10^{-7} of the earth's quadrant. This was changed in 1799 when the meter was defined by the length of a standard meter bar. In 1960 it was again redefined, by the Eleventh General Conference of Weights and Measures, in terms of the wavelength of the radiation produced by a specified quantum transition of the Krypton-86 isotope. Also in 1960, the metric system was renamed at the General Conference of Weights and Measures as *Le Système International d'Unités* or the International System of Units, abbreviated SI.

SI has been described as 'a coherent set of units,' so that when SI units are used throughout the solution of an equation, the answer is automatically equal to the value of the quantity.

Well, it's not quite that simple. The above description works fine so long as only the base, supplementary, and derived units are used, but not their multiples. And what's this about 'supplementary' units? Isn't SI supposed to be a logical and **complete** measuring system. Nonmetric is being thrown out because of such supplementary units as 'peck' 'bushel' and 'skip.' Now we have base units, supplementary units, derived units, derived units with special names, derived units expressed by means of special names, derived units formed by using supplementary units, and units in use temporarily with SI!

Let's put all this in order.

SI is constructed from seven base units for independent quantities, plus two supplementary derived dimensionless units:

SI BASE UNITS

Quantity	Name	Symbol
length	meter	m

Quantity	Name	Symbol
mass	kilogram	kg
time	second	s
electric current	ampere	A
thermodynamic temperature	kelvin	K
amount of substance	mole	mol
luminous intensity	candela	cd

SUPPLEMENTARY UNITS

Quantity	Name	Symbol
plane angle	radian	rad
solid angle	steradian	st

Units for all other quantities are derived from these nine units and are expressed as products and quotients without numerical factors:

EXAMPLES OF DERIVED UNITS EXPRESSED IN TERMS OF BASE UNITS

Quantity	Name	Symbol
area	square meter	m^2
volume	cubic meter	m^3
speed, velocity	meter per second	m/s
acceleration	meter per second squared	m/s^2
wave number	reciprocal meter	m^{-1}
density, mass density	kilogram per cubic meter	kg/m^3
specific volume	cubic meter per kilogram	m^3/kg
current density	ampere per square meter	A/m^2
magnetic field strength	ampere per meter	A/m
concentration (of amount of substance)	mole per cubic meter	mol/m^3
luminance	candela per square meter	cd/m^2

Certain derived units have been given special names, and may themselves be used to express other derived units:

SI DERIVED UNITS WITH SPECIAL NAMES

Quantity	Name	Symbol	Expression in terms of other units
frequency	hertz	Hz	s^{-1}
force	newton	N	$m \cdot kg/s^2$
pressure, stress	pascal	Pa	N/m^2
energy, work, quantity of heat	joule	J	$N \cdot m$
power, radiant flux	watt	W	J/s
electric charge, quantity of electricity	coulomb	C	$s \cdot A$
electric potential, potential differencc, electromotive force	volt	V	W/A
capacitance	farad	F	C/V
electric resistance	ohm	W	V/A
electric conductance	siemen	S	A/V

Quantity	Name	Symbol	Expression in terms of other units
magnetic flux	weber	Wb	V•s
magnetic flux density	tesia	T	Wb/m²
inductance	henry	H	Wb/A
Celsius temperature	degree Celsius	°C	K
luminous flux	lumen	lm	cd•st
illuminance	lux	lx	lm/m²
activity (of a radionuclide)	becquerel	Bq	s⁻¹
absorbed dose, specific energy imparted, kerma, absorbed dose index	gray	Gy	J/kg
dose equivalent, dose equivalent index	sievert	Sv	J/kg

SOME SI-DERIVED UNITS EXPRESSED USING SPECIAL NAMES

Quantity	Name	Symbol
absorbed dose rate	gray per second	Gy/s
dynamic viscosity	pascal second	Pa•s
electric charge density	coulomb per cubic meter	C/m³
electric field strength	volt per meter	V/m
electric flux density	coulomb per square meter	C/m²
energy density	joule per cubic meter	J/m³
exposure (x and y rays)	coulomb per kilogram	C/kg
heat capacity, entropyy	joules per kelvin	J/K
heat flux density, irradiance	watt per square meter	W/m²
kinematic viscosity	square meter per second	m²/s
molar energy	joule per mole	J/mol
molar energy, molar heat capacity	joule per mole kelvin	J/(mol•K)
moment of force	newton meter	N•m
permeability	henry per meter	H/m
permitivity	farad per meter	F/m
radiant intensity	watt per steradian	W/st
specific energy	joule per kilogram	J/kg
specific heat capacity, specific entropy	joules per kilogram kelvin	J(kg•K)
surface tension	newton per meter	N/m
thermal conductivity	watt per meter kelvin	W(m•K)

Supplementary units may be used in the expression of derived units:

EXAMPLES OF SI-DERIVED UNITS FORMED BY USING SUPPLEMENTARY UNITS

Quantity	SI unit	
	Name	Symbol
angular velocity	radian per second	rad/s
angular accelration	radian per second squared	rad/s²
radiant intensity	watt per steradian	W/st
radiance	watt per square meter steradian	W/(m²•st)

In fields where their usage is already well established, the use of the following units is accepted, but subject to future review:

UNITS PERMITTED FOR USE WITH SI

Quantity	Name	Symbol
area	hectare	ha
energy	electronvolt	eV
length	parsec (astronomical unit)	pc
linear density	tex	tex
mass	tonne (metric ton)	t
mass of an atom	unit of unified atomic mass	u
plane angle	degree	°
	minute	'
	second	"
	revolution	r
time	minute	min
	hour	h
	day	d
	year	a
volume	liter	L

UNITS IN USE TEMPORARILY WITH SI

Name	Symbol	Definition
barn	b	$1 \text{ b} = 100 \text{ fm}^2$
knot	kn	1 nautical mile per hour
		$1 \text{ kn} = (1,852/3,600) \text{ m/s}$
millibar	mbar	$1 \text{ mbar} = 100 \text{ Pa}$
nautical mile	M	1 nautical mile = 1,852 m

UNITS NOT ACCEPTABLE FOR USE WITH SI

Quantity	Name	Symbol
absorbed dose of ionizing radiation	rad	rad
acceleration	gal	Gal
activity (radioactive)	curie	Ci
amount of substance	equivalent	eq
area	are	a
conductance	mho	mho
dose equivalent	rem	rem
dynamic viscosity	poise	P
energy	calorie (IT)	cal
	erg	erg
force	kilogram-force	kgf
	kilopond	kp
	dyne	dyn
illuminance	phot	ph
illuminance	stilb	sb
kinematic viscosity	stoke	St
length	angström	Å

Quantity	Name	Symbol
	micron	μ
	fermi	fm
	X unit	-
magnetic field strength	oersted	Oe
magnetic flux	maxwell	Mx
magnetic flux density	gauss	Gs, G
magnetic induction	gamma	γ
mass	metric carat	-
pressure	torr	Torr
	millimeter of mercury	mm Hg
	bar	bar
	standard atmosphere	atm
radiation exposure	röntgen	R
volume	stere	st
	lambda	λ

There are 16 prefixes used in SI, ranging from 10^{18} to 10^{-18}. In practical use, in construction-related activities, the following six prefixes are generally used:

PREFIXES

factor	prefix	symbol
10^6	mega	M
10^3	kilo	k
10^{-1}	deci	d
10^{-2}	centi	c
10^{-3}	milli	m

How to write SI

There are some rules. For instance, each unit of measurement has an individual name: mass is expressed in kilograms; force in newtons (both are expressed in pounds in the nonmetric system of measurement). Likewise, each quantity has only one unit: power, be it electrical, thermal, or mechanical, is expressed only in watts.

Symbols are written in lower case (m for meter, g for gram) except for those derived from proper names (C for Celsius, N for Newton), and for liter (L). When written out in full, the names of all units, excepting Celsius are written in lower case (newton, pascal, liter, etc.). The plural form for symbols is identical to the singular (5 m, not 5 ms); a full space is left between a numeral and the character denoting its symbol (5 m not 5m) except where the numeral is followed by characters other than letters (18°C not 18 °C). Prefixes are used with SI names or symbols to signify multiples or submultiples. The prefixes most commonly used in the construction/contracting industry are: 10^6 mega (M), 10^3 kilo (k), 10^{-2} centi (c), and 10^{-3} milli (m).

When a triad separator is required to facilitate the reading of long numbers, the separator is a space (eg 12 345). However, a space is not necessary with a four-digit group (eg 1234) except when required for consistency, eg, when it is in a column with other numbers having five or more digits. If a numerical value is less than one written in decimal form, a zero precedes the decimal marker (eg, 0.10, not .10).

This dictionary has been written using metric terminology (in American-English), with nonmetric equivalents in parentheses. The following conversion table includes terms commonly used in the fields covered by the dictionary (bold face are exact).

SOME CONVERSION FACTORS

On the following pages, factors are given that permit conversion of nonmetric values to metric values, and metric values to nonmetric.

MULTIPLY THIS	UNIT OF	BY THIS	TO OBTAIN	SYMBOL
acre	area	0.404 685 6	hectare	ha
acre	area	4 046.856	square meter	m²
acre foot	volume	1 233.482	cubic meter	m³
ampre hour	electric charge	3.6	kilocoulomb	kC
ampere turn	electric charge	1.0	ampere	A
ångström	length	0.1	nanometer	nm
arpent (French measure)	area	0.341 889 4	hectare	ha
arpent (French measure)	length	58.471 31	meter	m
astronomical unit	length	149.597 870	gigameter	Gm
atmosphere, standard (760 torr)	pressure	101.325	kilopascal	kPa
atmosphere, technical (1 kgf/cm²)	pressure	98.066 5	kilopascal	kPa
bar	pressure	100	kilopascal	kPa
barrel (oil, 42 U.S. gal)	volume	0.158 987 3	cubic meter	m³
barrel (36 U.K. gal)	volume	0.163 659 2	cubic meter	m³
barrel (U.S. dry, 7056 in.³)	volume	0.115 627 1	cubic meter	m³
biot	electric charge	10	ampere	A
board foot	volume	2.359 737	cubic decimeter	dm³
British thermal unit (Btu) (Int. table)	energy	1.055 06	kilojoule	kJ
Btu foot per (square foot hour °F)	heat flow	1.730 735	watt per (meter kelvin)	W/(m·K)
Btu inch per (square foot hour °F)	heat flow	0.144 227 9	watt per (meter kelvin)	W/(m·K)
Btu inch per (square foot second °F)	heat flow	519.220 4	watt per (meter kelvin)	W/(m·K)
Btu (59°F, 15°C)	energy	1.054 80	kilojoule	kJ
Btu (60.5°F)	energy	1.054 615	kilojoule	kJ
Btu (390°F)	energy	1.059 67	kilojoule	kJ
Btu (mean)	energy	1.055 87	kilojoule	kJ
Btu per cubic foot	heat conductivity	37.258 95	kilojoule per cu m	kJ/m³
Btu per (cubic foot °F)	heat conductivity	67.066 11	kilojoule per (cu m kelvin)	kJ/(m³·K)
Btu per gallon	heat	232.08	kilojoule/cubic meter Celsius	kJ/(m³°C)
Btu per gallon (U.S.)	heat	278.717	kilojoule/cubic meter Celsius	kJ/(m³°C)
Btu per hour	heat flow	0.293 071 1	watt	W
Btu per hour (thermochemical)	heat flow	0.292 875 1	watt	W
Btu per minute (thermochemical)	heat flow	17.572 50	watt	W
Btu per pound	heat capacity	2.326	kilojoule per kilogram	kJ/kg
Btu per (pound °F)	heat capacity	4.186 8	kilojoule per (kilogram kelvin)	kJ/(kg·K)
Btu per second (thermochemical)	heat flow	1.054 350	kilowatt	kW
Btu per (square foot hour °F)	heat flow	5.678 263	watt per (square meter kelvin)	W/(m²·K)
Btu per (square foot hour)	heat flow	3.154 60	watt per square meter	W/m²
Btu per ton (2000 lb)	heat capacity	1.163	kilojoules per tonne	kJ/t
Btu (thermochemical)	energy	1.054 35	kilojoule	kJ
Btu (thermochemical) per (square foot hour)	power/unit area	3.152 481	watt per square meter	W/m²
Btu (thermochemical) per (sq ft minute)	power/unit area	189.148 8	watt per square meter	W/m²
Btu (thermochemical) per (sq ft second)	power/unit area	11.348 93	kilowatt per square meter	kW/m²
Btu (thermochemical) per (sq in. second)	power/unit area	1 634 246	megawatt per square meter	MW/m²
calorie (international)	energy	4.186 8	joule	J
calorie (thermochemical)	energy	4.184	joule	J
candela per square foot	luminance	10.763 91	candela per square meter	cd/m²
candela per square inch	luminance	1.550 003	kilocandela per square meter	kcd/m²
carat	mass	200	milligram	mg

MULTIPLY THIS	UNIT OF	BY THIS	TO OBTAIN	SYMBOL
centimeter	length	0.393 70	inch	in.
centistoke	velocity	**1.0**	square millimeter per second	mm²/s
cental (100 lb)	mass	45.359 237	kilogram	kg
chain (66 ft)	length	**20.116 8**	meter	m
circular mil	area	506.707 5	square micrometer	μm²
cord (stacked wood 128 ft³)	volume	3.624 6	cubic meter	m³
cubic centimeter	volume	0.061 02	cubic inch	in.³
cubic centimeter	volume	0.035 195	ounce (fluid)	fl oz
cubic centimeter	volume	1.0	milliliter	mL
cubic centimeter	volume	**0.001**	liter	L
cubic foot	volume	28.316 85	cubic decimeter	dm³
cubic foot	volume	0.028 316 85	cubic meter	m³
cubic foot	volume	28.316 85	liter	L
cubic foot per hour	volume rate of flow	28.316 85	liter per hour	L/h
cubic foot per hour	volume rate of flow	0.007 865 79	liter per second	L/s
cubic foot per minute	volume rate of flow	0.000 471 947 4	cubic meter per second	m³/s
cubic foot per minute	volume rate of flow	0.471 947 4	liter per second	L/s
cubic foot per second	volume rate of flow	0.028 316 85	cubic meter per second	m³/s
cubic foot per second	volume rate of flow	28.316 85	liter per second	L/s
cubic inch	volume	**16.387 064**	cubic centimeter	cm³
cubic inch	volume	**0.016 387 064**	liter	L
cubic inch per minute	volume rate of flow	0.273 117 7	cubic centimeter per second	cm³/s
cubic inch per second	volume rate of flow	16.387 06	cubic centimeter per second	cm³/s
cubic meter	volume	0.275 9	cord	cd
cubic meter	volume	1.308	cubic yard	yd³
cubic meter	volume	35.314 7	cubic foot	ft³
cubic meter	volume	219.97	gallon (imperial)	gal(imp)
cubic meter	volume	264.17	gallon (U.S.)	gal(U.S.)
cubic millimeter	volume	**0.001**	cubic centimeter	cm³
cubic millimeter	volume	61.023 x 10⁶	cubic inch	in.³
cubic millimeter	volume	1.0 x 10⁻⁹	cubic meter	m³
cubic yard	volume	0.764 555	cubic meter	m³
cubic yard per minute	volume rate of flow	12.742 58	cubic decimeter per second	dm³/s
cunit (100 ft³ solid wood)	volume	2.831 685	cubic meter	m³
darcy	permeability	0.986 923 3	square micrometer	μm²
day (mean solar)	time	**86 400**	kilosecond	ks
day (sidereal)	time	86.164 09	kilosecond	ks
decibel	sound power level	**1.0**	decibel	dB
decibel	sound pressure level	**1.0**	decibel	dB
degree (angle)	plane angle	0.017 453	radian	rad
degree (Celsius)	temperature	(°C x 9/5) + 32	degree Fahrenheit	°F
degree Celsius	interval	1.0	kelvin	K
degree Fahrenheit	temperature	(°F - 32) x 5/9	degree Celsius	°C
degree Fahrenheit	interval	5/9	kelvin	K
degree Rankine	interval	5/9	kelvin	K
dyne	force	**10**	micronewton	μN
electronvolt	energy	0.160 217 7	attojoule	aJ
ell (45 in.)	length	1.143	meter	m
erg	energy	**0.1**	microjule	μJ
fathom	length	1.828 8	meter	m
foot	length	**0.304 8**	meter	m
foot	length	304.8	millimeter	mm
foot candle	illuminence	10.763 91	lux	lx
foot lambert	luminance	3.426 259	candela per square meter	cd/m²
foot of water (39.2°F, 4°C)	pressure	2.988 98	kilopascal	kPa
foot per hour	velocity	0.084 666 7	millimeter per second	mm/s

MULTIPLY THIS	UNIT OF	BY THIS	TO OBTAIN	SYMBOL
foot per hour	velocity	**304.8**	millimeter per hour	mm/h
foot per minute	velocity	**5.08**	millimeter per second	mm/s
foot per minute	velocity	304.8	millimeter per minute	mm/m
foot per second	velocity	**0.304 8**	meter per second	m/s
foot per second squared	acceleration	**0.304 8**	meter per second squared	m/s²
foot poundal	energy	42.140 11	millijoule	mJ
foot pound-force	energy	1.355 818	joule	J
foot pound-force per hour	energy	0.376 616 1	milliwatt	mW
foot pound-force per minute	energy	22.596 97	milliwatt	mW
foot pound-force per second	energy	1.355 818	watt	W
foot (U.S. survey)	length	0.304 800 6	meter	m
furlong	length	0.201 168	kilometer	km
gal	acceleration	1.0	centimeter per second squared	cm/s²
gallon (imperial)	volume	**0.004 546 09**	cubic meter	m³
gallon (imperial)	volume	**4.546 09**	liter	L
gallon (imperial) per day	volume rate of flow	**0.004 546 09**	cubic meter per day	m³/d
gallon (imperial) per day	volume rate of flow	0.000 052 616 8	liter per second	L/s
gallon (imperial) per minute	volume rate of flow	75.768 17	cubic centimeter per second	cm³/s
gallon (imperial) per minute	volume rate of flow	0.075 768	liter per second	L/s
gallon (U.S.)	volume	0.003 785 412	cubic meter	m³
gallon (U.S.)	volume	3.785 412	liter	L
gallon (U.S.) per minute	volume rate of flow	63.090 20	cubic centimeter per second	cm³/s
gallon (U.S.) per minute	volume rate of flow	0.063 090 2	liter per second	L/s
gamma	mass	1.0	microgram	µg
grain	mass	64.798 91	milligrag	mg
gram	mass	0.035 273 96	ounce	oz
gram	mass	0.002 204 62	pound	lb
gram per cubic centimeter	density	1.0	kilogram per liter	kg/L
gram per cubic centimeter	density	62.428 7	pound per cubic foot	lb/ft³
gram per cubic centimeter	density	10.022 41	pound per gallon (imp)	lb/gal
gram per cubic meter	density	0.436 995 7	grain per cubic foot	gr/ft³
gram per cubic meter	density	1.0	milligram per liter	mg/L
gram per cubic meter	density	0.000 062 428	pound per cubic foot	lb/ft³
gram per liter	density	0.062 428 7	pound per cubic foot	lb/ft³
gram per liter	density	0.010 022 41	pound per gallon (imp)	lb/gal
grain per cubic foot	mass density	2.288 352	gram per cubic meter	g/m³
grain per gallon (imperial)	mass density	14.253 77	gram per cubic meter	g/m³
grain per gallon (U.S.)	mass density	17.118 06	gram per cubic meter	g/m³
hectare	area	2.471 054	acre	ac
hectare	area	**10 000**	square meter	m²
horsepower (boiler)	power	9.809 50	kilowatt	kW
horsepower (electric)	power	**746**	watt	W
horsepower (water)	power	746.043	watt	W
horsepower (550 ftúlbf/s)	power	745.699 9	watt	W
horsepower hour	energy	2.684 52	megajoule	MJ
hour (mean solar)	time	3.6	kilosecond	ks
hour (sidereal)	time	3.590 17	kilosecond	ks
hundredweight (100 lb)	mass	**45.359 237**	kilogram	kg
hundredweight (112 lb)	mass	50.802 345	kilogram	kg
inch	length	**2.54**	centimeter	cm
inch	length	0.025 4	meter	m
inch	length	25.4	millimeter	mm
inch³	section modulus	16.387 064	cubic centimeter	cm³
inch of mercury (conventional 0°C)	pressure	3.386 39	kilopascal	kPa
inch of mercury (60°F)	pressure	3.376 85	kiloPascal	kPa
inch of mercury (68°F, 20°C)	pressure	3.374 11	kiloPascal	kPa
inch of mercury (0°C)	pressure	**25.4**	millimeter of mercury	mm

MULTIPLY THIS	UNIT OF	BY THIS	TO OBTAIN	SYMBOL
inch of water (conventional)	pressure	249.082 89	pascal	Pa
inch of water (39.2°F, 4°C)	pressure	249.082	pascal	Pa
inch of water (60°F)	pressure	248.843	pascal	Pa
inch of water (68°F, 20°C)	pressure	248.641	pascal	Pa
inch per minute	velocity	25.4	millimeter per minute	mm/min
inch per second	velocity	25.4	millimeter per second	mm/s
inch per second squared	acceleration	25.4	millimeter per second squared	mm/s²
inch pound-force	work	0.112 985	newton meter	N•m
joule	heat	0.000 947 8	Btu (international)	Btu
joule	work	0.737 562	foot pound-force	ft/lb
joule	work	0.3725 x 10⁶	horsepower hour	hp•hr
joule	power	0.2778 x 10⁶	kilowatt hour	kW•h
joule	work	1.0	newton meter	N•m
joule per liter	heat	0.026 839	Btu per cubic foot	Btu/ft³
joule per liter	heat	1.0	kilojoule per cubic meter	kJ/m³
kilogram	mass	35.273 96	ounce	oz
kilogram	mass	2.204 62	pound	lb
kilogram	mass	0.000 984 21	ton (long)	t(l)
kilogram	mass	0.001 102 3	ton (short)	t(s)
kilogram calorie (international)	heat	3.968 3	Btu (international)	Btu
kilogram calorie (international)	heat	4 186.8	joule	J
kilogram-force	force	9.806 65	newton	N
kilogram per cubic centimeter	mass	32.127 292	pound per cubic inch	lb/in.³
kilogram per cubic meter	mass	1.0	gram per liter	g/L
kilogram per cubic meter	mass	1.685 556	pound per cubic yard	lb/yd³
kilogram per cubic meter	mass	0.010 022	pound per gallon (imp)	lb/gal(I)
kilogram per kilometer	mass	1.0	gram per meter	g/m
kilogram per kilometer	mass	0.6719 x 10³	pound per foot	lb/ft
kilogram per kilometer	mass	3.458	pound per mile	lb/mi
kilogram per liter	mass	1.0	gram per milliliter	g/mL
kilogram per meter	mass	0.671 97	pound per foot	lb/ft
kilogram per meter	mass	2.015 91	pound per yard	lb/yd
kilogram per square centimeter	mass	2 048.16	pound per square foot	lb/ft²
kilogram per square centimeter	mass	14.223	pound per square inch	lb/in.²
kilogram per square meter	mass	0.204 816	pound per square foot	lb/ft²
kilojoule per cubic meter	heat	0.026 839	Btu per cubic ft	Btu/ft³
kilojoule per cubic meter	heat	0.004 309	Btu per gallon (imperial)	Btu/gal
kilojoule per cubic meter	heat	1.0	joule per liter	J/L
kilojoule per kilogram	heat	0.429 923	Btu per pound	Btu/lb
kiloliter	volume	35.315	cubic foot	ft³
kiloliter	volume	219.969	gallon (imperial)	gal
kiloliter	volume	264.172	gallon (U.S.)	gal
kilometer	length	0.621 371	mile	mi
kilometer per hour	velocity	0.539 96	knot	kn
kilometer per hour	velocity	0.277 778	meter per second	m/s
kilometer per hour	velocity	0.621 371	mile per hour	mph
kilonewton	force	0.112 40	ton-force (short ton)	
kilopascal	pressure	0.295 3	inch of mercury (0°C)	
kilopascal	pressure	4.014 74	inch of water (4°C)	
kilopascal	pressure	1 000	newton per square meter	N/m²
kilopascal	pressure	20.885 43	pound-force per square foot	psf
kilopascal	pressure	0.145 037 7	pound-force per square inch	psi
kilopond	force	9.806 65	newton	N
kilowatt	heat	0.947 81	Btu (international) per second	Btu/sec
kilowatt	power	1.340 48	horsepower (electric)	hp(e)
kilowatt hour	heat	3 412	Btu (international)	Btu
kilowatt hour	power	1.340 5	horsepower hour	hp/hr
kilowatt hour	energy	3.6	megajoule	MJ

MULTIPLY THIS	UNIT OF	BY THIS	TO OBTAIN	SYMBOL
kip (thousand pound-force)	force	4.448 222	kilonewton	kN
knot (International)	velocity	1.852	kilometer per hour	km/h
knot (international)	velocity	0.514 444 4	meter per second	m/s
knot (U.K.)	velocity	1.853 184	kilometer per hour	km/h
ksi (kip per square inch)	force per area	6.894 757	megapascal	MPa
lambert	uminance	3 183.099	candela per square meter	cd/m²
langley per minute	power/unit area	0.697 8	kilowatt/square meter	kW/m²
league (International nautical)	length	5.556	kilometer	km
league (U.K. nautical)	length	5.559 552	kilometer	km
league (U.S.)	length	4.828 032	kilometer	km
legal subdivision (40 acres)	area	0.161 874 2	square kilometer	km²
link (1/100 chain)	length	0.201 168	meter	m
liter	volume	1.0	cubic decimeter	dm³
liter	volume	0.035 315	cubic foot	ft³
liter	volume	61.023 744	cubic inch	in.³
liter	volume	0.001	cubic meter	m³
liter	volume	0.219 969	gallon (imperial)	(I)gal
iter	volume	0.264 172	gallon (U.S.)	(U.S.)gal
liter	volume	35.195 1	ounce (fluid)	fl oz
liter	volume	0.879 877	quart (imperial)	qt
liter per second	velocity	2.118 88	cubic foot per minute	ft³/min
liter per second	velocity	13.198 2	gallon (imperial) per minute	gpm
liter per second	velocity	15.850 3	gallon (U.S.) per minute	gpm
lumen per square foot	illuminance	10.763 91	lux	lx
meter	length	39.370	inch	in.
meter	length	3.280 84	foot	ft
meter	length	1.093 6	yard	yd
meter of water (4°C)	volume	3.280 84	foot of water	
meter of water (4°C)	pressure	9.806 378	kilopascal	kPa
meter of water (4°C)	pressure	1.422 29	pound per square inch	psi
meter per minute	velocity	3.280 8	foot per minute	fpm
meter per minute	velocity	0.054 68	foot per second	fps
meter per minute	velocity	0.037 28	mile per hour	mph
meter per second	velocity	2.236 9	mile per hour	mph
meter per second	velocity	196.85	foot per minute	fpm
meter per second	velocity	3.280 8	foot per second	fps
meter per second squared	velocity	3.280 8	foot per second squared	fps²
metric carat	mass	200	milligram	mg
microinch	length	25.4	nanometer	nm
micrometer	length	0.039 370	mil	mil
micron	length	1.0	micrometer	μm
mil (0.001 in.)	length	25.4	micrometer	μm
mile (international nautical)	length	1.852	kilometer	km
mile	length	1.609 344	kilometer	km
mile	length	1 609.344	meter	m
mile (U.K. nautical)	length	1.853 184	kilometer	km
mile per hour	velocity	1.609 344	kilometer per hour	km/h
mile per hour	velocity	0.447 04	meter per second	m/s
mile per minute	velocity	26.822 4	meter per second	m/s
mile per second squared	acceleration	1.609 344	kilometer per second squared	km/s²
millibar	pressure	0.1	kilopascal	kPa
milligram	mass	0.015 432	grain	gr
millgram	mass	35.274 x 10⁶	ounce	oz
milligram	mass	2.204 62 x 10⁶	pound	lb
milliliter	mass	0.061 02	cubic inch	in.³
milliliter	mass	0.035 195	ounce (fluid)	fl oz
millimeter	length	0.039 37	inch	in.
millimeter	length	39.37	mil	mil
millimeter	length	1000	micrometer	μm
millimeter mercury (0°C)	pressure	133.322 4	pascal	Pa

MULTIPLY THIS	UNIT OF	BY THIS	TO OBTAIN	SYMBOL
millimeter water (4°F)	pressure	9.806 378	pascal	Pa
million Btu per ton	heat	1.163	gigajoule per tonne	GJ/t
million gallons (imperial) per day	volume rate of flow	52.616 78	cubic decimeter per second	dm³/s
million gallons (U.S.) per day	volume rate of flow	**4 546.09**	cubic meter per day	m³/d
million gallons (U.S.) per day	volume rate of flow	0.052 616 8	cubic meter per second	m³/s
minute (mean solar)	time	60	second	s
minute (sidereal)	time	59.836 17	second	s
newton	force	0.224 808 9	pound-force	lb
newton meter	force	0.737 562	foot pound-force	ft/lb
ounce	mass	28.349 523	gram	g
ounce (apothecary or troy)	mass	**31.103 476 8**	gram	g
ounce (fluid)	mass	28 413 062	milliliter	mL
ounce inch squared	moment of inertia of mass	0.182 899 8	kilogram square centimeter	kg•cm²
ounce-force	force	0.278 013 9	newton	N
ounce-force inch	torque	7.061 552	millinewton meter	mN•m
ounce-force per square inch	pressure	0.430 922 3	kilopascal	kPa
ounce (mass) per square foot	area density	305.151 7	gram per square meter	g/m²
ounce (mass) per square yard	area density	33.905 75	gram per square meter	g/m²
ounce per cubic foot	mass density	1.001 154	kilogram per cubic meter	kg/m³
ounce per gallon (imperial)	mass density	6.236 023	kilogram per cubic meter	kg/m³
ounce per gallon (U.S.)	mass density	7.489 152	kilogram per cubic meter	kg/m³
ounce per inch	linear density	1.116 123	kilogram per meter	kg/m
parsec	length	30.856 78	petameter	Pm
perch (French measure)	area	34.188 94	square meter	m²
perch (French measure)	length	5.847 130 8	meter	m
perm (23°C)	permeability	57.452 5	nanogram/(pascal second square meter)	ng/(Pa•s•m²)
perm inch (23°C)	permeability	1.459 29	nanogram per (pascal second meter)	ng/(Pa•s•m)
Petrograd standard (165 ft³ sawn timber)	volume	4.672 280	cubic meter	m³
phot	illuminance	10	kilolux	klx
pennyweight	mass	1.555 174	gram	g
pica (printer's)	length	4.217 518	millimeter	mm
pint (imperial)	volume	0.568 261 2	cubic decimeter	dm³
pint (imperial)	mass	0.568 261	liter	L
pint (U.S.)	volume	0.473 176 5	cubic decimeter	dm³
pint (U.S.)	mass	0.473 176	liter	L
point (Didot)	length	0.375 972 9	millimeter	mm
point (pica)	length	0.351 459 8	millimeter	mm
poise	dynamic viscosity	0.1	pascal second	Pa•s
pound	mass	**453.592 37**	gram	g
pound	mass	**0.453 592 37**	kilogram	kg
poundal	force	0.138 255 0	newton	N
poundal per square foot	force per area	1.488 164	pascal	Pa
poundal second per square foot	dynamic viscosity	1.488 164	pascal second	Pa•s
pound (apothecary or troy)	mass	**373.241 721 6**	gram	g
pound foot per second	momentum	0.138 255	kilogram meter per second	kg•m/s
pound foot squared	moment of inertia of mass	42.140 11	gram square meter	g•m²
pound foot squared per second	angular momentum	42.140 11	gram square meter per second	g•m²/s
pound force	force	4.448 222	newton	N
pound-force foot	torque	1.355 818	newton meter	N•m
pound-force inch	torque	0.112 985	newton meter	N•m
pound-force per square foot	force per area	47.888 26	pascal	Pa
pound-force per square foot	pressure	0.047 880 26	kilopascal	kPa
pound-force per square inch	pressure	0.703 1	meter of water (4°C)	m
pound-force per square inch (psi)	force per area	6.894 757	kilopascal	kPa
pound-force second per square foot	dynamic viscosity	47.880 26	pascal second	Pa•s
pound inch squared	moment of inertia of mass	2.926 397	kilogram square	

664

MULTIPLY THIS	UNIT OF	BY THIS	TO OBTAIN	SYMBOL
			centimeter	kg•cm²
pound (mass) per square foot	area density	4.882 428	kilgram per square meter	kg/m²
pound (mass) per square inch	area density	703.069 6	kilgram per square meter	kg/m¹
pound per cubic foot	mass	16.018 46	gram per liter	g/L
pound per cubic foot	mass density	16.018 46	kilogram per cubic meter	kg/m³
pound per cubic inch	mass density	27.679 90	megagram per cubic meter	Mg/cm³
pound per cubic inch	mass	27.679 90	kilogram per liter	kg/L
pound per cubic yard	mass	0.593 276	kilogram per cubic meter	kg/m³
pound per foot	linear density	1.488 164	kilogram per meter	kg/m
pound per (foot second)	dynamic viscosity	1.488 164	pascal second	Pa•s
pound per gallon (imperial)	mass density	99.776 37	kilogram per cubic meter	kg/m³
pound per gallon (U.S.)	mass density	119.826 4	kilogram per cubic meter	kg/m³
pound per hour	force	0.125 997 9	gram per second	g/s
pound per hour	force	**0.453 592 37**	kilogram per hour	kg/h
pound per inch	linear density	17.857 97	kilogram per meter	kg/m
pound per mile	force	0.281 849	kilogram per kilometer	kg/km
pound per minute	force	7.559 87	gram per second	g/s
pound per minute	force	0.007 559 87	kilogram per second	kg/s
pound per second	force	**0.453 592 37**	killgram per second	kg/s
pound per ton (short)	mass	**0.50**	kilogram per tonne	kg/t
pound per yard	linear density	0.496 055	kilogram per meter	kg/m
pound (2000) per acre	area density	0.224 170 2	kilogram per square meter	kg/m²
pounds (2000) per square mile	area density	350.266	milligram per square meter	mg/m²
quart (imperial)	volume	1.136 522	cubic decimeter	dm³
quart (imperial)	volume	1.136 522	liter	L
quart (U.S.)	volume	0.946 352 9	cubic decimeter	dm³
quart (U.S.)	volume	0.946 353	liter	L
quarter (28 lb U.K.)	mass	12.700 58	kilogram	kg
radian	plane angle	180/π	degreee (angle)	°
radian per second	velocity	30/ π	revolution per minute	rpm
radian per second	velocity	1/(2 π)	revolution per second	rps
revolution per minute	velocity	π/30	radian per second	rad/s
revolution per second	velocity	2π	radian per second	rad/s
second (sidereal)	time	0.997 269 6	second	s
section (1 mi², 640 acres)	area	2.589 988	square kilometers	km²
square centimeter	area	0.155	square inch	in.² or sq in.
square centimeter	area	**0.0001**	square meter	m²
square centimeter	area	**100**	square millimeter	mm²
square foot	area	**929.030 4**	square centimeter	cm²
square foot (French measure)	area	1 055.214	square centimeter	cm²
square foot hour °F per Btu	heat	0.176 110 1	square meter kelvin per watt	m²•K/W
square foot per second	kinematic viscosity	92 903.04	millimeters squared per second	mm²/s
square inch	area	**6.4516**	square centimeter	cm²
square inch	area	**645.16**	square millimeter	mm²
square inch per second	kinematic viscosity	645.16	millimeter squared per second	mm²/s
square kilometer	area	247.1	acre	ac
square kilometer	area	**100**	hectare	ha
square kilometer	area	0.386 10	square mile	mi²
square meter	area	**0.000 1**	hectare	ha
square meter	area	**10 000**	square centimeter	cm²
square meter	area	**10.763 9**	square foot	ft² or sq ft
square meter	area	**1.195 99**	square yard	yd² or sq yd
square mile	area	**258.998 8**	hectare	ha
square mile	area	**2.589 988**	square kilometer	km²
square millimeter	area	**0.01**	square centimeter	cm²
square millimeter	area	0.001 550	square inch	in.² or sq in.

MULTIPLY THIS	UNIT OF	BY THIS	TO OBTAIN	SYMBOL
square yard	area	0.836 127 4	square meter	m²
stilb	luminance	1.0	candela per square meter	cd/m²
stoke	kinematic viscosity	100	millimeter squared per second	mm²/s
tex	linear density	1.0	milligram per meter	mg/m
therm	energy	105.506	megajoule	MJ
ton-force (long) per square inch	force per area	13.789 514	megapascal	mPa
ton-force (short ton)	force	8.896 443	kilonewton	kN
ton-force (short) per square foot	force per area	95.760 514	kilopascal	kPa
ton-force (short) per square inch	force per area	15.444 3	megapascal	mPa
ton (long 2240 lb)	mass	1 016.046 908 8	kilogram	kg
ton (long 2240 lb)	mass	1.016 046 908 8	tonne	t
ton (long) per cubic yard	mass density	1.328 939	megagram per cubic meter	Mg/m³
tonne kilometer	force	0.684 944	ton per mile (short)	t(s)/mi
tonne per cubic meter	mass	1.0	kilogram per liter	kg/L
tonne per cubic meter	mass	1685.555	pound per cubic yard	lb/yd³
tonne per cubic meter	mass	0.842 777	ton (short) per cubic yard	t(s)/yd³
tonne per hour	force	0.277 778	kilogram per second	kg/s
tonne (2205 lb)	mass	1 000	kilogram	kg
tonne (2205 lb)	mass	0.984 206 5	long ton (2240 lb)	ton
tonne (2205 lb)	mass	1.102 311	short ton	t(s)
ton mile (long ton)	force	1.635 169	tonne kilometer	t•km
ton mile (short ton)	force	1.459 97	tonne kilometer	t•km
ton (nuclear equivalent of TNT)	energy	4.2	gigajoule	GJ
ton (register)	volume	2.831 685	cubic meter	m³
ton (short 2000 lb)	mass	907.187 74	kilogram	kg
ton (short 2000 lb)	mass	0.907 184 74	tonne	t
ton (short) per cubic yard	mass density	1.186 553	megagram per cubic meter	Mg/m³
ton (short 2000 lb)	mass	907.187 74	kilogram	kg
ton (short 2000 lb)	mass	0.907 184 74	tonne	t
ton (short) per cubic yard	mass density	1.186 553	megagram per cubic meter	Mg/m³
ton (short) per hour	force	0.251 995 8	kilogram per second	kg/s
ton (short) per hour	force	0.907 184 74	tonne per hour	t/h
torr	pressure	133.322 4	pascal	Pa
township (36 section)	area	93.239 57	square kilometer	km²
watt hour	energy	3.600	kilojoule	kJ
watt hour	heat	3.412	Btu (internatonal)	Btu
watt per square foot	heat flow	10.763 91	watt per square meter	W/m²
watt second	energy	1.0	joule	J
yard	length	0.914 4	meter	m
yard per second squared	acceleration	0.914 4	meter per second squared	m/s²